Lines and Electromagnetic Fields
for Engineers

THE OXFORD SERIES IN ELECTRICAL AND COMPUTER ENGINEERING

M. E. Van Valkenburg, Senior Consulting Editor
Adel S. Sedra, Series Editor, Electrical Engineering
Michael R. Lightner, Series Editor, Computer Engineering

Allen and Holberg, *CMOS Analog Circuit Design*
Bobrow, *Elementary Linear Circuit Analysis, 2nd ed.*
Bobrow, *Fundamentals of Electrical Engineering, 2nd ed.*
Campbell, *The Science and Engineering of Microelectronic Fabrication*
Chen, *Linear System Theory and Design*
Chen, *System and Signal Analysis, 2nd ed.*
Comer, *Digital Logic and State Machine Design, 3rd ed.*
Comer, *Microprocessor Based System Design*
Cooper and McGillem, *Probabilistic Methods of Signal and System Analysis, 2nd ed.*
Ghausi, *Electronic Devices and Circuits: Discrete and Integrated*
Houts, *Signal Analysis in Linear System*
Jones, *Introduction to Optical Fiber Communication Systems*
Kennedy, *Operational Amplifier Circuits: Theory and Application*
Kuo, *Digital Control Systems, 3rd ed.*
Leventhal, *Microcomputer Experimentation with the IBM PC*
Leventhal, *Microcomputer Experimentation with the Intel SDK-86*
Leventhal, *Microcomputer Experimentation with the Motorola MC6800 ECB*
McGillem and Cooper, *Continuous and Discrete Signal and System Analysis, 3rd ed.*
Miner, *Lines and Electromagnetic Fields for Engineers*
Navon, *Semiconductor Microdevices and Materials*
Papoulis, *Circuits and Systems: A Modern Approach*
Ramshaw and Van Heeswijk, *Energy Conversion*
Schwarz, *Electromagnetics for Engineers*
Schwarz and Oldham, *Electrical Engineering: An Introduction, 2nd ed.*
Sedra and Smith, *Microelectronic Circuits, 3rd ed.*
Stefani, Savant, and Hostetter, *Design of Feedback Control Systems, 3rd ed.*
Van Valkenburg, *Analog Filter Design*
Vranesic and Zaky, *Microcomputer Structures*
Warner and Grung, *Semiconductor Device Electronics*
Yariv, *Optical Electronics, 5th ed.*

Lines and Electromagnetic Fields for Engineers

GAYLE F. MINER

New York Oxford
OXFORD UNIVERSITY PRESS
1996

OXFORD UNIVERSITY PRESS

Oxford New York
Athens Auckland Bangkok Bombay
Calcutta Cape Town Dar es Salaam Delhi
Florence Hong Kong Istanbul Karachi
Kuala Lumpur Madras Madrid Melbourne
Mexico City Nairobi Paris Singapore
Taipei Tokyo Toronto

and associated companies in
Berlin Ibadan

Published by Oxford University Press, Inc.,
198 Madison Avenue, New York, New York 10016

Library of Congress Cataloging-in-Publication Data
Miner, Gayle F.
Lines and electromagnetic fields for engineers / Gayle F. Miner.
p. cm.
Includes bibliographical references and index.
ISBN 0-19-510409-9
1. Electric lines. 2. Electromagnetic fields. I. Title.
TK3221.M48 1996
621.3—dc20 95-655

9 8 7 6 5 4 3 2 1

Printed in the United States of America
on acid-free paper

Preface

Inasmuch as there are many excellent field theory texts available, some explanation for yet another is appropriate. Three observations prompted the undertaking of this writing: first, the continued pressure of new material into the undergraduate program, which requires that something be removed; second, the position of the fields course in the electrical engineering curriculum; and third, my personal experiences teaching field courses.

With respect to the first point, I made the decision to reduce the time allotted to the solution of static field problems. The introduction of special functions normally met in static boundary value problems can be accomplished just as effectively in conjunction with time-varying boundary value problems. Enough ideas about basic static fields are introduced so that the physical significance of the mathematics can be presented without the encumbrance of the time (or frequency) variable. Thus the curl, divergence, and gradient are easily related to basic field structures. The approach of emphasizing time-varying fields can, of course, present the danger of the student's making erroneous extrapolations to static fields, but this also happens if static fields are extensively covered first (e.g., magnetic fields in conductors or potential). If care is taken to point out the limitations of each derivation there will be no difference in the possibility of misapplication.

The second observation concerns the position of the fields course in the curriculum. The fields course usually closely follows an introductory circuits course. Because of this, it is pedagogically ideal to start the fields course by taking advantage of this circuits background. Transmission line theory is, on that basis, the best place to begin. Thus waves can be developed in a more familiar context, and propagating electromagnetic waves can then be related to those results. Using transmission lines first introduces waves phenomena without the vector complexities. One initial loss with this approach is that transverse electromagnetic (TEM) mode ideas cannot be introduced conveniently; however, they are easily given when TEM mode solutions are presented in waveguide theory. This ''lines first'' approach also avoids students getting the notion that field theory must be understood before any transmission line material can be studied.

Third, as I have taught transmission lines and field theory over several years, I observed that the students seemed to have the same questions each time. They first wanted to know how the concepts applied in practice. Then, similar questions on certain key concepts kept arising. If one flips the pages of texts and looks at the figures, very little of a practical nature is seen. In this text, a conscientious attempt is made to draw examples from physically oriented configurations whenever possible. This avoids a course that appears to be a mere mathematics and symbol manipulation exercise. Many of the approaches taken to the various topics have been devised as a result of student questions and of observing student responses to different presentations. However, it is expected that the student will not hesitate to expend the intellectual effort required to meet with forthrightness the challenges afforded by the mathematics of field theory. For these reasons, the text is written for the student, with many physical discussions and a minimum of ''it can be shown'' introductions. Because of this, the text approach is frequently informal and conversational, which, it is hoped, will put the reader at ease in spite of the detail given. As an additional aid for the student, essentially every section has several exercises. These exercises consist of both simple questions and number problems that use the section material and problems that apply the material to new applications. These have the advantage that the student knows what concepts specifically apply. At the end of each chapter there are general exercises for which the student must determine which approach and concept must be used.

This text is addressed to two groups of students. First, there are those who will terminate their study of lines and fields with only one or two semesters work and who need to obtain a good physical understanding of the principles and applications for more immediate use. The second group consists of those who will do more advanced work for whom a reasonably rigorous treatment is desired. To satisfy two such diverse groups is admittedly an optimistic goal, but I have pursued that end by giving engineering applications and physical interpretations as far as possible parallel with the mathematical development without sacrificing rigor.

Except for starting with transmission lines, the text is organized along the historical line. This was done so that the student might better appreciate the thinking and synthesis methods of the pioneers in the fields area—Faraday, Maxwell, and Coulomb to mention only three. Little of the history is presented, however unfortunate this may be. This is because many students do not really appreciate such pleasant, meaningful diversions in a technical class and reduce such discussions to ''What do I have to remember for the exam?'' (Maybe we need to require some additional reading along this line.) At appropriate points in the text, references are given to some of the original experiments to show the genius of those workers who labored without 500-MHz scopes and computers. It is also my opinion that to postulate Maxwell's equations is to hide the real foundations of the understanding of the methods of creativity, especially Maxwell's contribution and thought. Although most undergraduate physics courses do explain Maxwell's equations, my experience has been that the benefits of some repetition of such derivations outweighs whatever time is gained by postulating them. Additionally, physical insights are strengthened and the role of mathematics put into proper perspective.

An introductory Chapter 0 is intended to help the student understand why something beyond circuit theory is needed. Some suggestions on how to approach this area of study are given along with a discussion of how the text is organized to help the student in that study.

In Chapters 1, 2, and 3 the theory and applications of transmission lines are given.

This includes Smith chart applications for transmission lines and its use as an aid in designing lumped element matching networks (e.g., for transistor input and output matching). Wave phenomena receive significant discussion since they are used in Chapters 8 and 9. These latter chapters can, however, be studied without these analogies.

A review of the basic theorems, identities, and coordinate systems of use in field theory appears in Chapter 4. This can be used as a review or simply referred to at appropriate points in the remaining chapters. There are some who prefer to introduce vector mathematics as it is needed rather than separate the mathematics from the applications. This can be done with Chapter 4 since each operation has its own section.

Chapters 5 and 6 introduce static electric and magnetic fields respectively. Many applications are given to assist the student in obtaining physical insights into the field structures and properties. The properties of materials are also discussed.

The development of Maxwell's equations and the concepts of power flow and wave impedance are given in Chapter 7. The presentations of the Poynting vector and wave impedance follow the suggestions of H. T. Booker in this thought-provoking article on the teaching of electromagnetic theory. (See the reference in that chapter. I have added some details in the developments.)

Chapter 8 is a discussion of the propagation of plane waves and polarization concepts. Reflection, refraction, and transmission are thoroughly studied. The presentation follows standard lines, and it is here that the propagation concepts learned for transmission lines are of great benefit since they transfer directly in terms of propagation, attenuation constant, and wave phenomena.

In Chapter 9 a study of guided wave theory is presented, including dielectric guides and an introduction to optical fibers. The concepts of mode and mode structure and their effects on signal propagation are given. A final section on resonant cavities is included because it gives an example of a shorted transmission line calculation.

Systeme Internationale (SI) units are emphasized throughout the discussions and exercises. An occasional exception to this appears in tables that have been taken from manufacturers' literature and specification materials.

The text can be used in a number of ways.

1. A one-semester, course of three or four credit hours would consist of selected sections from the first eight chapters. The chapters generally begin with the basis of the topic and proceed to more detailed discussions. Thus the instructor could determine the depth of each topic.
2. Those who prefer to begin from a physics background could use the first four or five sections of Chapters 5, 6, and 7. This would be followed by selected sections of Chapters 1–3 and then Chapter 8.
3. For a two-semester course, most of the material can be learned.

As mentioned before, students who have had a fairly complete course in vector analysis could bypass most of Chapter 4 without difficulty. Such students could use this chapter as a review or simply refer to it as the occasion may require in case the terminology and symbolism used herein differ from that used in their math course.

The nature of field theory, with its mathematics, symbolism, and physical visualizations, can provide students with confidence in their abilities and bring them to a point where they will not be hesitant to work with material of any complexity. We must, however, heed the Proverbalist who wrote:

> Wisdom is the principle thing; therefore, get wisdom;
> and with all thy getting, get understanding. (Proverbs 4:7)

Acknowledgments are given to my faculty colleagues and students who have miraculously survived the preliminary notes and duplicated versions of the material. Their helpful comments and suggestions have contributed significantly to whatever merit the text may possess. Their patience and cooperation have been deeply appreciated. Hopefully they were not permanently scarred by working with a challenging subject from handwritten manuscript.

One frequently takes colleagues for granted. The daily academic routine often disguises their sterling qualities. Such a colleague was found in the person of Gene A. Ware, who put me on the right track when it came time to "get it into the computer." Sincere thanks are given him. Reviewers unknown also made many helpful suggestions.

And then: to my wife, Evelyn, who sat patiently at the keyboard and generated the final LaTeX draft. I'm sure she never suspected what she was getting into, especially when it came to reading proof pages.

Authors all realize the countless hours of a publisher's staff. I particularly extend thanks to those at Oxford University Press for the effort expended to set difficult material. Their serenity and willingness to make changes in the face of deadlines is much appreciated. Those with whom I worked directly—Bill Zobrist, Carole Schwager, and Mark Naccarelli—provided a most professional and cordial experience.

Provo, Utah
September 1995

G.F.M.

Contents

Appendices

Lines and Electromagnetic Fields for Engineers

0 Preliminaries: A Preface for the Student

0-1 Purpose, Scope, and Mode of the Text

The purpose of this text is to introduce you to the theory and applications of time- and space-dependent electrical quantities of importance in engineering electronics and communications. Throughout the text there are examples and exercises from such diverse areas of study as computers, solid-state theory, integrated circuits, and fiber optics.

This text has been written with the reader in mind. When I felt that detail would be beneficial and develop maturity in handling mathematics, all the steps, including the "obvious" ones, are given. Likewise, when a more heuristic development is appropriate, such is the direction of the discussion. Showing some heuristic approaches has the advantage of indicating that one may often simply make a guess when solving a problem and not worry about precise mathematics.

In the text when each new idea, definition, or technique is introduced, an example is presented and some exercises using the same concepts are given. From this point of view the text is nearly of the programmed instruction type, although the responses are more involved than most programmed texts require. In fact, this represents an intermediate presentation between that mode and the usual encyclopedic exposition.

This text is not intended to be a reference text for engineering practice. Although there are several useful curves and tables, the exercises often contain much of the more applied results. However, if the exercises are carefully worked, a good set of reference material will be developed as you proceed through the text.

0-2 Course Objectives

At the end of each chapter is a list of the main ideas of that chapter. They are given in essentially the same order as the text material and should provide a good review for checking on the understanding of terminology and techniques. These are not to be in any way interpreted as behavioral objectives since they are not written in the ''tight'' form required of those objectives.

0-3 How to Succeed in Transmission Lines and Field Theory

The material of field theory is of broad application in electronics, power, and computers. As such, it is exciting to pursue. Here we bring all of our physics, mathematics, and circuit theory into one focal point and apply them to real situations.

Notwithstanding the fact that other courses are drawn upon heavily, the material in reality forms a basic course since there are many new ideas not found in other courses. Because of this we meet new symbols and new terms. It would thus be expected that if one wants to become conversant in this area, one would memorize the name, spelling, and physical units of each new symbol introduced. Otherwise, field theory tends to be a symbol manipulation exercise only unless one has a clear understanding of what the symbols represent. I have observed that once a person feels comfortable with the symbolism and terminology, there is much less likelihood of becoming lost in the trees when problems are stated in those terms.

Since the eye does not see the electric and magnetic fields involved, one must attach as much intuitive physical meaning as possible to the mathematical symbols. The ability to form mental pictures and to draw them in a multidimensional system is a very important skill you should develop. For example, after you have completed the usual courses in circuits and electronics, the symbols V and I somehow become the voltage and current, when in fact they are only symbols for electrical quantities we cannot see. Our comfort with these quantities arises from our having formed meaningful visual impressions of them.

0-4 Why Field Theory?

In most undergraduate laboratories, if you took a circuit that had been designed and turned it upside down (assuming things were tied down), the circuit performance remained unchanged. Also, the length of the cable that connected the generator to the circuit did not make much difference. If one of the leads was bent a little, nothing happened. All you usually did was consider the time function (or phase) of the various signals and their amplitudes. You assumed that whatever signal was at the generator output also appeared at the input to the circuit. Wire lengths used to build the circuits had little effect, if any.

Let's look at a few specific cases and show how the ideas of elementary circuit theory may not be sufficient to describe some apparent anomalies. We take examples from several areas.

Case 1

A straight wire contains inductance. The plots of the inductances of 10-cm lengths of #10, #20, and #30 wire sizes (102, 32, and 10 mils diameter, respectively, where 1 mil = 0.001

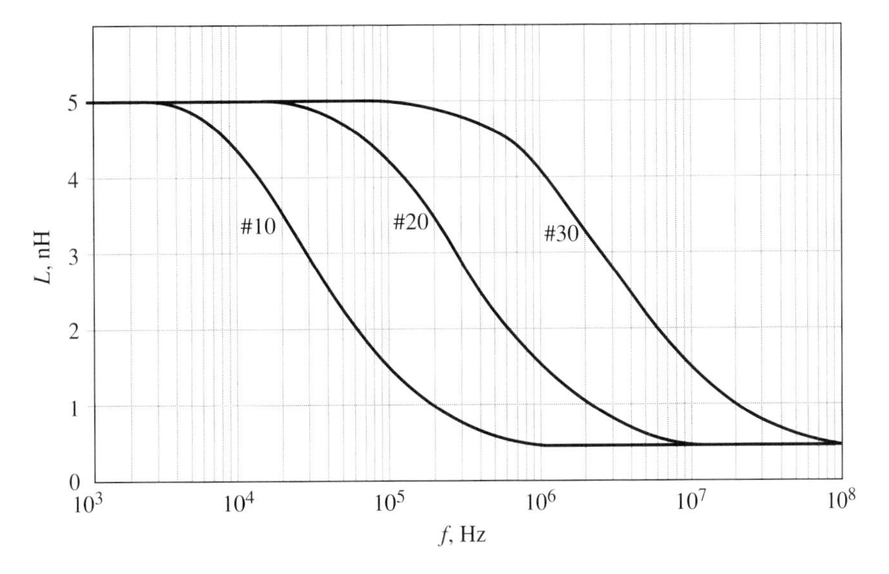

Figure 0-1. Inductance of 10-cm lengths of round copper wires of various wire sizes as a function of frequency.

in.) are given in Figure 0-1. Since there are no turns, where does the inductance come from? Apparently there is more involved than just the total current flow. How do we calculate the inductance? (The resistance, incidentally, is also a frequency-dependent parameter.)

A general definition of inductance and examples of calculating expressions for inductance are given in Section 6-14.

Case 2

Take a 10-cm length of one of the wires described in Case 1 and use it to build a circuit. The physical placement of the wire affects pulses applied to it in ways not completely attributed to the inductance and capacitance to ground of the wire. The input and output pulses have leading edges that do not occur at the same time, and the time difference depends upon wire placement. This happens even when the pulse experiences very little distortion. Apparently, then, space will have to be introduced as a variable in addition to time and amplitude.

This is a typical example of delays and loss effects caused by transmission lines. The theory of transmission lines and their properties are discussed in Chapters 1, 2, and 3. The problem of waveform distortion is discussed in Chapters 3 and 9.

Case 3

Take your transistor radio and place it in different orientations. It plays loud in some positions and soft in others. Thus we are again led to consider spatial coordinates in our theory. Basic circuit theory would say response does not depend upon circuit placement in space.

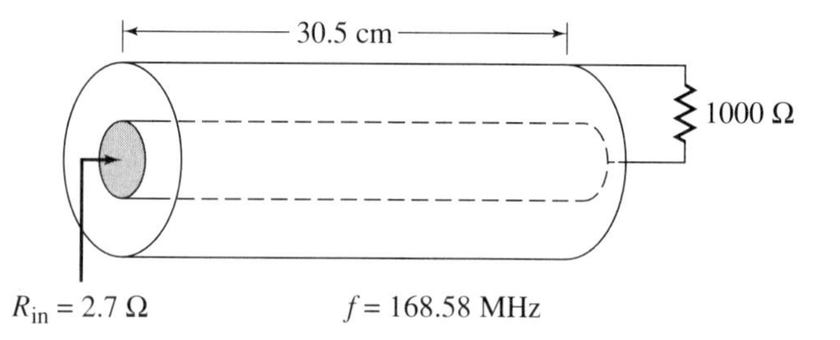

Figure 0-2. Impedance change due to a length of coaxial cable.

This effect is due to the direction of the electric field intensity in a propagating wave, which is called *wave polarization*, and is discussed in Chapter 8.

Case 4

Suppose you design a circuit that requires a 1000-Ω resistance. However, circuit constraints are such that the resistor cannot be physically placed at the desired location. So you run a cable from the resistor to the contact point. Suppose you use 30.5 cm of RG58A/U cable, and the circuit is designed to operate at 168.58 MHz. Upon measuring the cable at the input, a resistance of only 2.7 Ω is obtained. Thus there is nearly a short circuit where originally 1000 Ω was desired. This is shown in Figure 0-2.

Since the circuit doesn't work, you try to juggle other components (somehow) to obtain proper operation. (It turns out that all that would needed to have been done was place a 2.7-Ω resistor in place of the 1000-Ω one.) However, suppose that after you patched up the design, a friend observes a way to run the cable that would require only 14 cm of cable. So the circuit is rebuilt, and lo the system again fails! Why? Because now the impedance at the cable input is $7.8 - j80$ Ω, which means that the source sees essentially a *capacitance*! Again, we see that physical dimenisons will need to enter into the picture. In Chapter 1 you will learn how to handle these situations.

Case 5

For an example from the computer area, we point out that simply connecting chips together can lead to problems. One cannot be content with hardware only, neglecting waves and field effects. Suppose it is necessary to feed two signals simultaneously in the arithmetic logic unit (ALU) and that the signals originate from two different units and are to be coincident at a NAND gate. The pulse widths are specified and equal and one pulse originates from a decoder and another from memory as shown diagrammatically in Figure 0-3. The pulses may leave the two units simultaneously, but they are not coincident at the

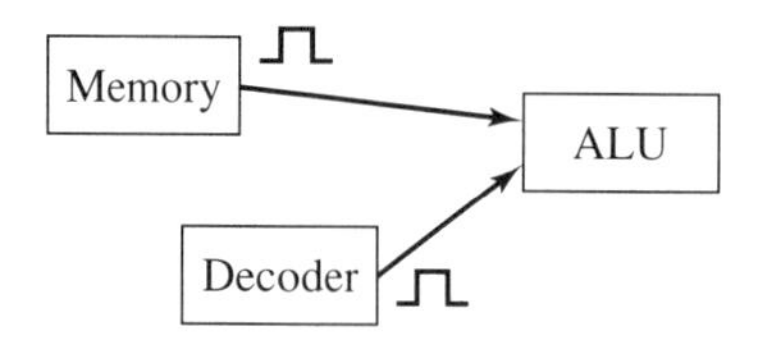

Figure 0-3. Pulses originating at two different units and sent to an arithmetic logic unit.

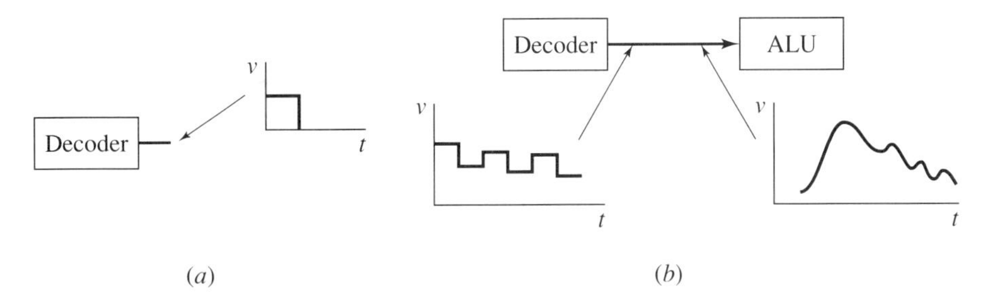

Figure 0-4. Effect of lines connecting units in a digital system. (*a*) Pulse out of unloaded decoder. (*b*) Effect of lines connecting units in a digital system.

ALU. Also, the pulses are distorted. "Simple solution," one counters. "Simply make the two lines the same length or enter a delay in the shorter line." This is partly true, but it solves only one aspect of the problem.

Figure 0-4 shows the true state of affairs. The pulse out of the decoder unit disconnected looks just fine. (See Figure 0-4*a*.) However, when a transmission system is used there appear to be several pulse components and, at the ALU input, a rather "smeared" version of the decoder signal. Again, space and coordinates become significant in our formulations. Our mathematics will thus be a little more difficult because we must include spatial variables in addition to the amplitude and time factors.

The effects described in this last example become noticeable when pulse rates get into the frequency range 30 MHz and above.

Sources of this effect are discussed in Chapters 1, 2, and 3.

Case 6

Our final example concerns the transmission of energy when the frequencies extend from 3 GHz to that of visible light and beyond. At these frequencies any electronic circuit becomes dimensionally large compared with the wavelength of the signal (e.g., wavelength = 3 mm at 100 GHz). In the microwave region (3 to 300 GHz), the signals are often transmitted through hollow tubes, and basic circuit theory alone will not allow you to design and predict the performance of such structures. If we go to the lightwave band of frequencies we find a whole new area of fiber optic communications. Many of the properties of the fiber can be described by solving the same equations that arise in the study of transmission lines and propagation of radio waves. This alone is an indication of the wide range of application of the principles you will learn in this text. Even the operation of a laser diode, which is often used to drive an optical fiber, is characterized in part by the principles of field theory.

Wave propagation and confinement of waves are discussed extensively in Chapters 8 and 9.

You will begin your studies by considering systems that require time and only one spatial coordinate to describe the performance of the system. This will familiarize you with the transmission line, solutions for which answer some of the problem cases described

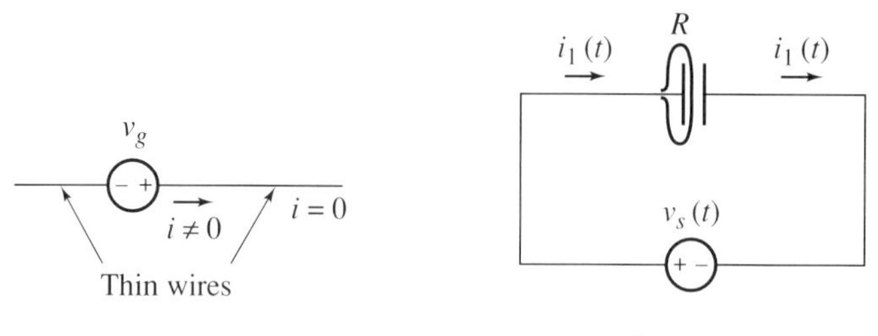

Figure 0-5. **Figure 0-6.**

above. The solutions will also give background for the subsequent study of electromagnetic wave propagation and transmission.

Exercises

0-4.0a Explain why circuit theory is insufficient to describe the operation of the system shown in Figure 0-5. Consider both time and space problems as well as performance.

0-4.0b Give two examples from your own experience that indicate a need for field theory or at least something beyond basic circuit theory.

0-4.0c Figure 0-6 shows a parallel plate capacitor. An imaginary surface R, having balloonlike shape, is around one of the plates. If we think of R as a circuit theory node, explain how Kirchhoff's current law is satisfied. Does it suffice to define current as a flow of charge?

0-4.0d Describe a gravity field and suggest two ways of representing this ''invisible'' quantity on a paper for physical visualization. Remember we have both magnitude and direction, which may vary with position and time.

0-4.0e Suppose you had a problem posed to you and that the solution was not immediately obvious. Discuss a few things you could do to generate some ideas that might assist you in the solution.

0-4.0f From a text on solid-state theory, describe three problems or topics that require a knowledge of field theory to solve. Be sure to state the problem clearly and specifically—don't be too general. (Look for some of the field quantities you learned about in your physics course.)

0-4.0g A digital system is operating at 200 MHz, with the pulse width being one-third of the period. The velocity of propagation on the lines of the system is typically 1×10^8 m/s. For the configuration shown in Figure 0-3, what is the shortest difference in length between the two lines such that two pulses sent out simultaneously from the Memory and Decoder are not overlapping (coincident) at all of the ALU? Now suppose there is a TTL logic gate in the longer of the two lines between the ALU and the unit that originated the pulse. Using a typical gate delay of 3 ns, what would be the shortest difference in the line lengths allowed for which the pulses at the ALU are completely out of coincidence?

1 Transmission Lines: Parameters, Performance Characteristics, and Applications to Distributed Systems

1-1 Introduction

In this chapter we study systems that have only one spatial variable. This, along with time, will provide a complete description of multiconductor systems (usually two conductors as in coaxial and parallel wire lines). The important features of such systems have physical interpretations that can be extended to more complex situations very easily. The notions of wave propagation and wavelength are fundamental to all types of electromagnetic structures, such as antennas and resonant cavities as well as radio wave propagation in space.

Our initial studies will be applied to systems that are physically ''smooth,'' such as coaxial cables or two-wire lines (called *distributed parameter systems*). However, the tools that will be developed to analyze and design such systems will also have applications in the analysis and design of lumped parameter RLC networks used at frequencies in excess of 250 MHz.

1-2 Modeling Distributed Systems

A distributed electrical system has properties of combinations of Rs, Ls, Cs, etc., but the elements do not appear explicitly as two terminal element units. This is in direct contrast to work in circuit theory and electronics where actual elements (called *lumped elements*) are placed in the circuit.

As our first example, we use the coaxial cable shown in Figure 1-1a. Here we have two cylindrical conductors separated by a dielectric (usually polystyrene or similar material). Although there is no lumped capacitance as we usually expect to see it, the two conductors form capacitance plates, and it is obvious that the length of the cable will determine the *total* capacitance value. Thus we may describe the cable by a capacitance

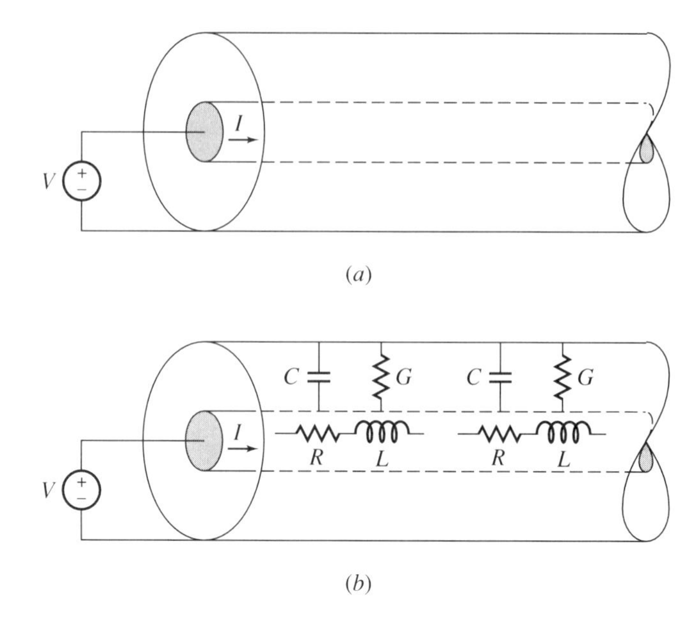

Figure 1-1. Distributed parameter representation of a coaxial cable. (*a*) Coaxial cable physical configuration. (*b*) Coaxial cable distributed parameter definitions.

parameter C measured in capacitance *per unit length* of cable in farad/meter. Using the per unit length measure makes it possible to have a parameter that will be independent of the total cable length. If we consider the current flow in the center conductor, we obtain two effects. The first is a resistive voltage drop due to wire resistance, which we may represent by a series resistance. Since the longer the cable, the greater the total resistance, we use the parameter R in ohms *per unit length* of cable, ohm/meter. The outer conductor can also be described in a similar way. Here we assume that R includes both the inner and outer conductor resistance contributions per meter.

The second effect of current flow in the center conductor is to produce a magnetic flux in and around the conductor, the flux direction being given by the usual right-hand rule. This gives rise to an inductance, which we denote by the parameter L and which is measured in henry/meter.

The final line element shown in Figure 1-1*a* arises from the current flow between the inner and outer conductors, even for an applied DC voltage. This is charge flow and is thus an effective resistance component caused by the imperfect (lossy) dielectric. Since this resistance is in parallel with the capacitance, we think of it as conductance as we do in circuit theory for parallel elements, and give it the symbol parameter G in units of siemens per meter.

With the four parameters discussed, the cable may be represented as an electrical network shown in Figure 1-1*b*, superimposed on the coaxial configuration.

A second example of a distributed system is the parallel-wire or twin-lead transmission

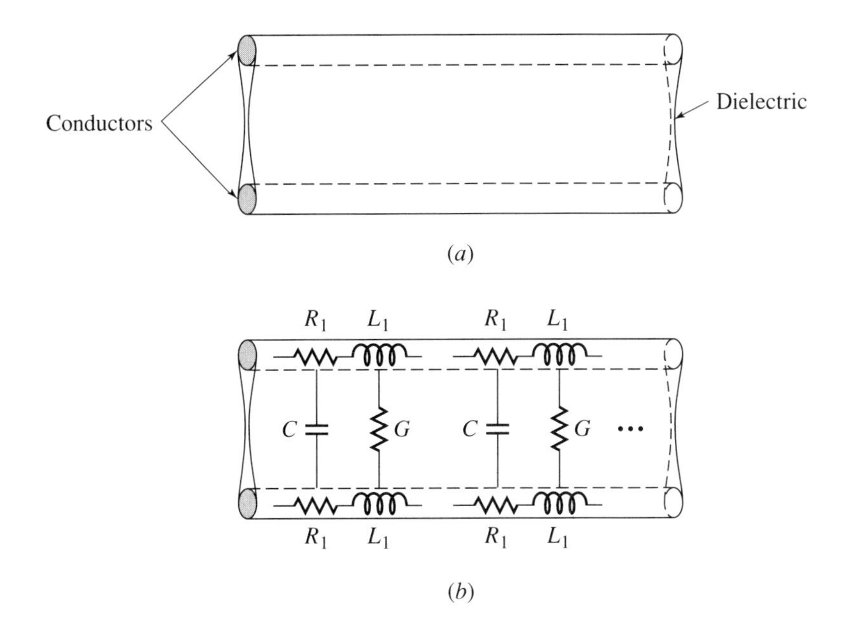

Figure 1-2. Twin-lead transmission line and its distributed parameter representation.

line shown in Figure 1-2. We again have two conductors separated by a rigid dielectric (lossy). An analysis of current flow and voltage drops would produce the symmetrical representation of Figure 1-2*b*. The R, L, C, and G parameters are again on a per meter basis. This transmission line is the common TV antenna lead-in line. As a final example, we consider a slightly more complex transmission system consisting of two parallel conductors shielded between two metal ground planes. The cross section of the system is shown in Figure 1-3*a*. In the usual application of this configuration, the two metal planes are electrically connected by bolts that pass through the dielectric. The bolts are usually placed at a great distance from the two internal conductors compared to the distances between the conductors. Since this configuration has more than two conductors, we must consider the R, L, G, and C terms of all combinations of conductors. The usual parameter representation of the system is shown in Figure 1-3*b*. The ground planes could also have resistance and inductance components drawn on them; however, we consider those effects to be lumped in with the center conductor R and L.

Exercises

1-2.0a The distributed parameter representation of the coaxial cable shown in Figure 1-1*b* was arrived at by supposing that the outer conductor R_0 and L_0 contributions were included in the R and L values. If we represent the total resistance value as the sum $R = R_0 + R_i$, where R_i is the inner conductor resistance per meter, comment on the relative sizes of R_0 and R_i. An RG58B/U cable has a center conductor diameter of 0.813 mm and a braided outer conductor of equivalent thickness 0.89 mm, and the dielectric has a diameter of 2.95

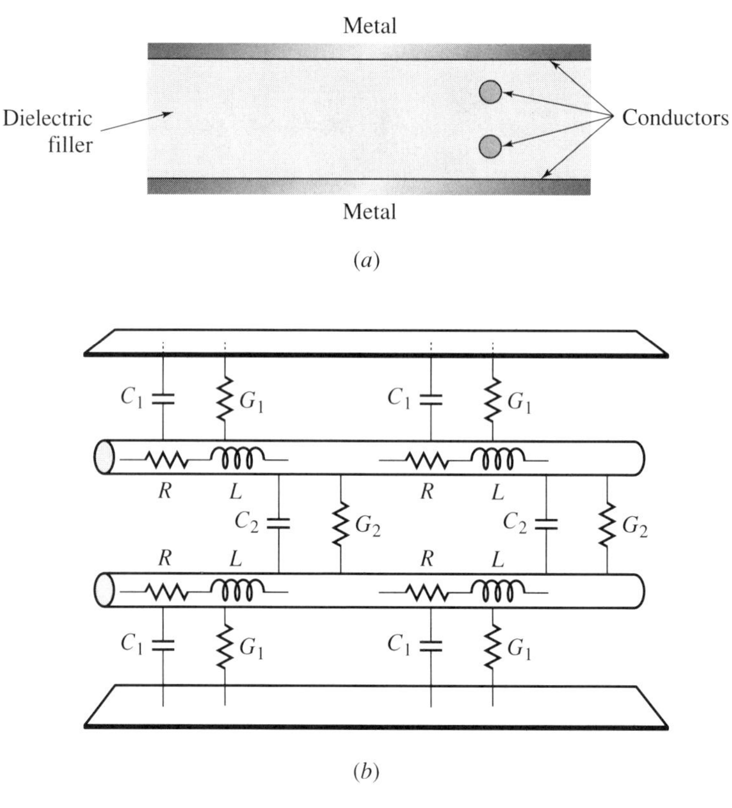

Figure 1-3. Shielded two-wire balanced pair and distributed parameter model. (*a*) Shielded balance pair systems. (*b*) Distributed parameter configuration.

Figure 1-4. (*a*) Single-wire, ground return metal planes. (*b*) Microstrip. (*c*) Balanced parallel lines near a metal plane. (*d*) Shielded pair.

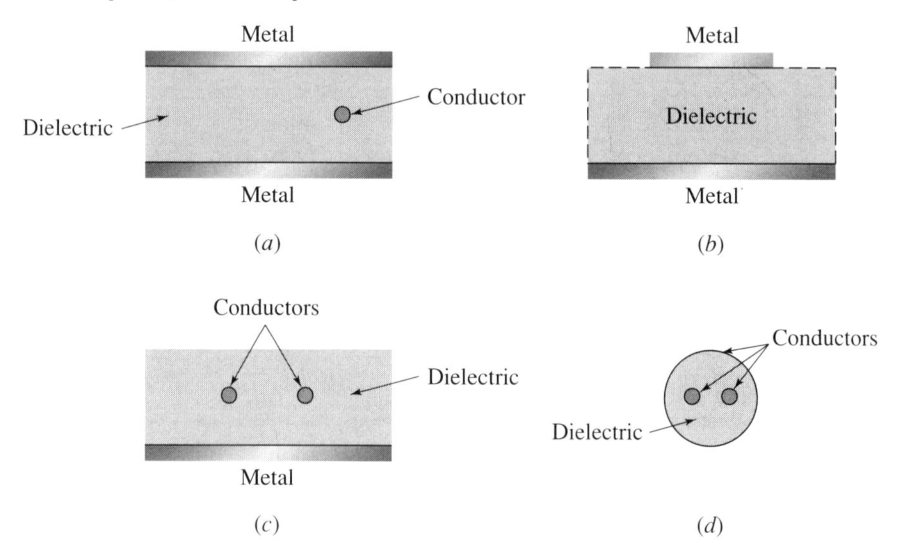

mm. Calculate DC resistance values *per meter* for R_0 and R_i using formulas from physics and compare. Conductors are copper with a conductivity of 5.8×10^7 siemen/meter.

1-2.0b For the distributed systems shown in Figure 1-4, develop possible equivalent circuit component models using R, L, G, and C parameters drawn superimposed on the physical structure (like Figure 1-3*b*). Only cross sections are shown.

Since there are no single capacitance, resistance, inductance, or conductance elements in the line, we cannot analyze the entire cable at once, so we consider only an *incremental* length Δz meters long and represent the cable as a cascaded ladder chain of these sections. Since R, L, G, and C are on a per meter basis, we must multiply by the length Δz to obtain the total element values in ohms, henrys, etc., for each incremental section. This is how we introduce the spatial dimension (in this case z coordinate) into the representation. By

Figure 1-5. Electrical network models of transmission line systems. (*a*) Coaxial cable. (*b*) Twin-lead (parallel wire) lines. (*c*) Shielded wire pair.

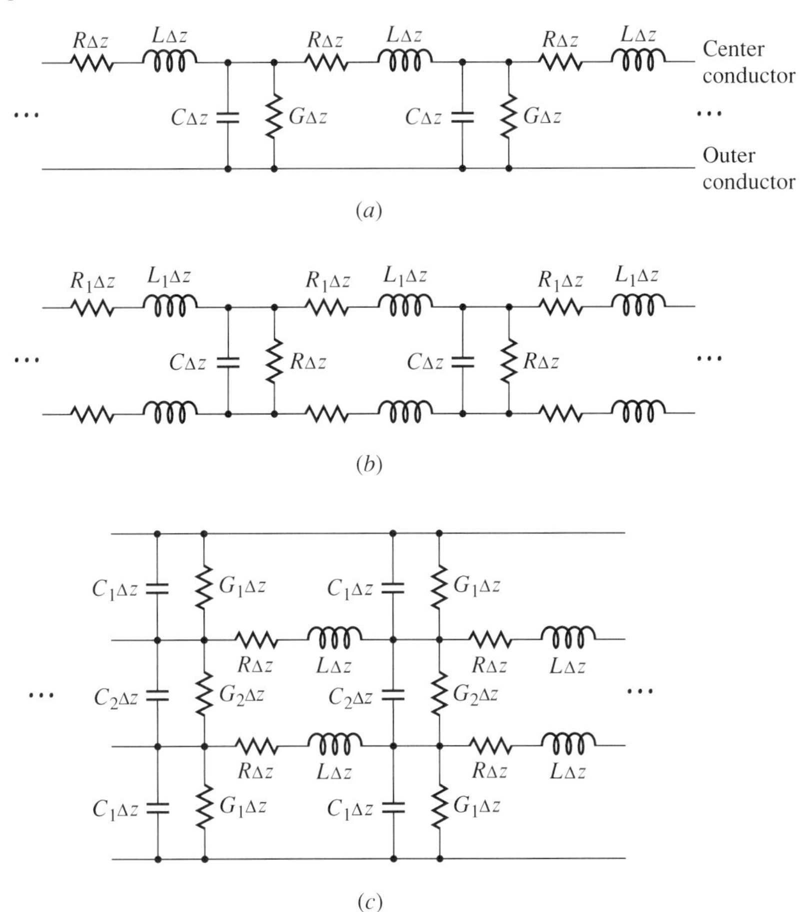

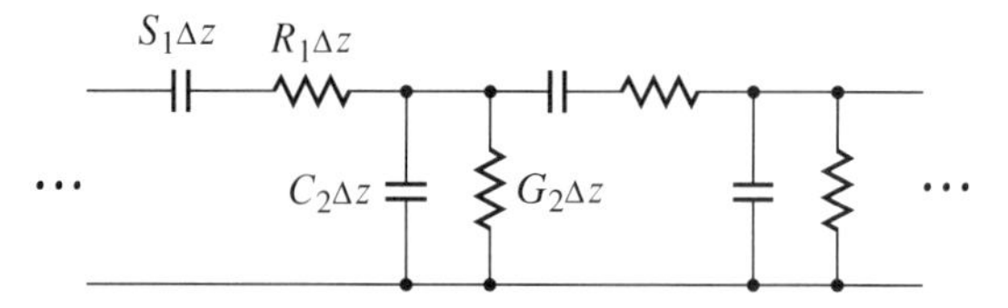

S_1 = reciprocal capacitance, daraf/meter

$$\left(i_S = \frac{1}{S}\frac{dv_s}{dt}\right)$$

Figure 1-6.

these means, we have been able to develop a model for our transmission lines. Using this cascaded section idea for each of the examples of systems given earlier, we would have the three electrical models shown in Figure 1-5.

Exercises

1-2.0c For the distributed parameter systems of Exercise 1-2.0b, draw the single (i.e., one) incremental section electrical network models.

1-2.0d Design a physical system that has the incremental sections equivalent model shown in Figure 1-6. Explain any approximations you use.

1-3 Analysis of Incremental Transmission Line Sections

Figure 1-7 shows one of the incremental sections of the coaxial cable of Figure 1-1. Our goal now is to obtain the differential equations that describe how current and voltage vary along the line. The currents and voltages are now functions of time and coordinate z. Also, we use the lowercase notation for current and voltage to indicate that the time variation may be any general functional form. Physically, the expression for output voltage $v(z + \Delta z, t)$ is obtained by the following reasoning:

1. Output voltage equals the input voltage plus (algebraically) the change that occurs between input and output. As an equation,

$$v(z + \Delta z, t) = v(z, t) + \Delta v \tag{1-1}$$

2. The way that voltage changes with distance is represented by

$$\frac{\partial v}{\partial z} \tag{1-2}$$

where the partial derivative must now be used since v is a function of both z and t. Thus the voltage change from z to $z + \Delta z$ is

$$\Delta v = \frac{\partial v}{\partial z}\Delta z \tag{1-3}$$

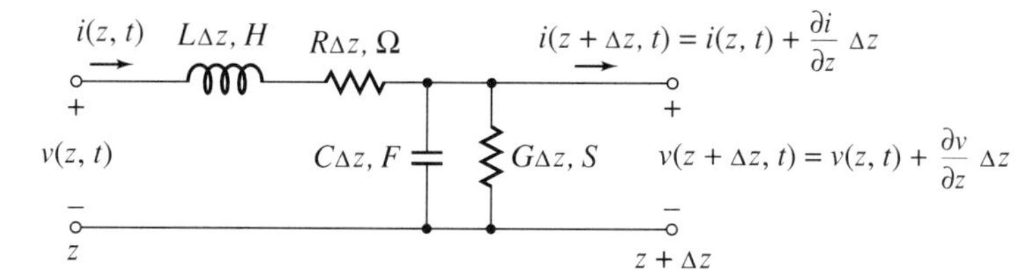

Figure 1-7. Current and voltage conventions on a single incremental section of a transmission line.

3. Substituting this last result into Eq. 1-1 gives

$$v(z + \Delta z, t) = v(z, t) + \frac{\partial v}{\partial z} \Delta z \tag{1-4}$$

The expression for the output current is obtained by similar reasoning using Figure 1-7.

Using Kirchhoff's voltage law (KVL) on the left loop yields

$$v(z, t) = (L \Delta z) \frac{\partial i}{\partial t} + (R \Delta z) i(z, t) + v(z + \Delta z, t)$$

Substituting Eq. 1-4 into this last result and canceling terms where possible yields

$$0 = L \frac{\partial i}{\partial t} \Delta z + Ri(z, t) \Delta z + \frac{\partial v}{\partial z} \Delta z$$

Dividing by Δz and rearranging:

$$\frac{\partial v}{\partial z} = -Ri - L \frac{\partial i}{\partial t} \tag{1-5}$$

Qualitatively, the equation indicates that $\partial v / \partial z$ is negative, which means voltage decreases with distance. This seems physically correct due to the voltage drops across the series R and L. Next we apply Kirchhoff's current law (KCL) to the node at the top of the conductance and capacitance, and cancel terms where possible:

$$i(z, t) = (C \Delta z) \frac{\partial \left[v(z, t) + \frac{\partial v}{\partial z} \Delta z \right]}{\partial t} + (G \Delta z) \left[v(z, t) + \frac{\partial v}{\partial z} \Delta z \right] + \left[i(z, t) + \frac{\partial i}{\partial z} \Delta z \right]$$

or

$$0 = C \frac{\partial \left[v(z, t) + \dfrac{\partial v}{\partial z} \Delta z \right]}{\partial t} \Delta z + G \left[v(z, t) + \frac{\partial v}{\partial z} \Delta z \right] \Delta z + \frac{\partial i}{\partial z} \Delta z$$

Dividing by Δz and rearranging gives

$$\frac{\partial i}{\partial z} = -C \frac{\partial \left[v(z, t) + \dfrac{\partial v}{\partial z} \Delta z \right]}{\partial t} - G \left[v(z, t) + \frac{\partial v}{\partial z} \Delta z \right]$$

Now taking the limit as $\Delta z \rightarrow 0$ (which makes $\partial v/\partial z \, \Delta z \rightarrow 0$) we obtain

$$\frac{\partial i}{\partial z} = -C \frac{\partial v}{\partial t} - Gv \tag{1-6}$$

This equation has the *qualitative* interpretation similar to that for the voltage equation; namely, $\partial i/\partial z$ is negative, which means that current decreases with distance. This is as one would expect, since part of the input current is shunted through the C and G.

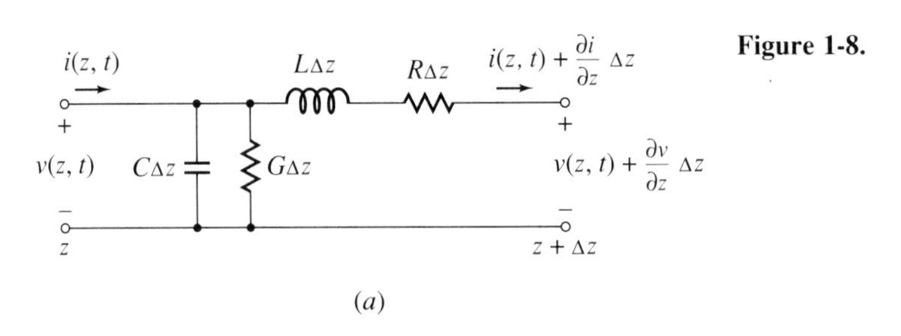

Figure 1-8.

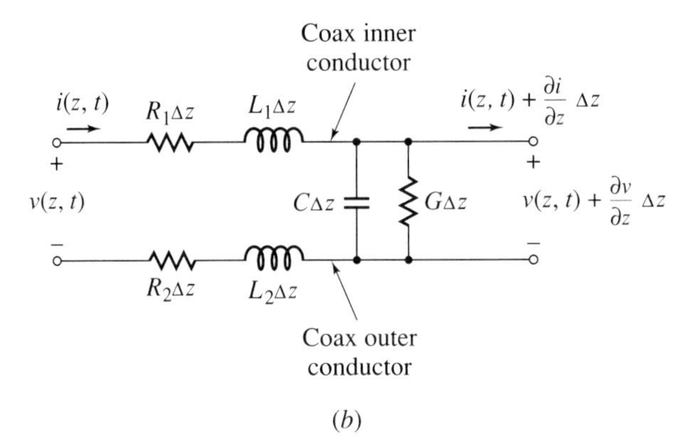

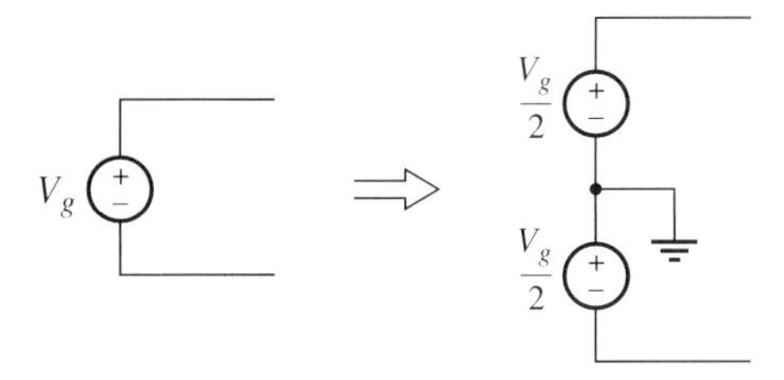

Figure 1-9.

Exercises

1-3.0a Analyze the sections drawn in Figure 1-8 and show that results identical in form to Eqs. 1-5 and 1-6 are obtained. How do you define R and L for the second circuit?

1-3.0b Analyze an increment of the system shown in Exercise 1-2.0d.

1-3.0c Analyze an incremental section of the two-wire shielded pair system of Figure 1-5*c*. Assume that the two shielding planes (see Figure 1-3) are connected electrically and that the voltage source is applied between the two lines. As a hint, notice that the systems shown in Figure 1-9 are equivalent as far as electrical analysis is concerned for a balanced system. Repeat using the single source and show that the same results are obtained.

1-4 Differential Equations for Voltage and for Current: The Wave Equations

Since the results of Exercise 1-3.0a yield equations of the same form as Eqs. 1-5 and 1-6 for any two-conductor system, we use Eqs. 1-5 and 1-6 for the remainder of our work.

The differential equation for voltage is obtained by first differentiating Eq. 1-5 with respect to z and Eq. 1-6 with respect to t to obtain

$$\frac{\partial^2 v}{\partial z^2} = -R\,\frac{\partial i}{\partial z} - L\,\frac{\partial^2 i}{\partial z \partial t}$$

and

$$\frac{\partial^2 i}{\partial t \partial z} = -C\,\frac{\partial^2 v}{\partial t^2} - G\,\frac{\partial v}{\partial t}$$

Interchanging the order of taking partial derivatives on the left side of the second equation and substituting it into the first yields

$$\frac{\partial^2 v}{\partial z^2} = -R\,\frac{\partial i}{\partial z} - L\left(-C\,\frac{\partial^2 v}{\partial t^2} - G\frac{\partial v}{\partial t}\right)$$

Next, using Eq. 1-6, we substitute for $\partial i/\partial z$ to obtain

$$\frac{\partial^2 v}{\partial z^2} = -R\left(-C\,\frac{\partial v}{\partial t} - Gv\right) - L\left(-C\,\frac{\partial^2 v}{\partial t^2} - G\,\frac{\partial v}{\partial t}\right)$$

Collecting terms, we finally have

$$\frac{\partial^2 v}{\partial z^2} = LC \frac{\partial^2 v}{\partial t^2} + (RC + LG) \frac{\partial v}{\partial t} + RGv \tag{1-7}$$

This is the voltage wave equation. The descriptive term *wave* will be more evident later when we examine the physical properties of the solutions of this equation, $v(z, t)$.

The corresponding differential equation for current is developed by first differentiating Eq. 1-5 with respect to t and Eq. 1-6 with respect to z to yield the set

$$\frac{\partial^2 v}{\partial t \partial z} = -R \frac{\partial i}{\partial t} - L \frac{\partial^2 i}{\partial t^2}$$

$$\frac{\partial^2 i}{\partial z^2} = -C \frac{\partial^2 v}{\partial z \partial t} - G \frac{\partial v}{\partial z}$$

Next interchange the order of differentiation on the left term of the first equation and substitute the result into the second equation:

$$\frac{\partial^2 i}{\partial z^2} = -C\left(-R \frac{\partial i}{\partial t} - L \frac{\partial^2 i}{\partial t^2}\right) - G \frac{\partial v}{\partial z}$$

Now using Eq. 1-5 to eliminate $\partial v/\partial z$, there results

$$\frac{\partial^2 i}{\partial z^2} = -C\left(-R \frac{\partial i}{\partial t} - L \frac{\partial^2 i}{\partial t^2}\right) - G\left(-Ri - L \frac{\partial i}{\partial t}\right)$$

Collecting terms on the derivatives yields

$$\frac{\partial^2 i}{\partial z^2} = LC \frac{\partial^2 i}{\partial t^2} + (RC + LG) \frac{\partial i}{\partial t} + RGi \tag{1-8}$$

This is the wave equation for current and is of identical form to that for voltage. One of our main tasks will be to obtain appropriate solutions for v and i. Figure 1-10 is a photograph of transmission line systems to which the wave equation can be applied.

Exercises

1-4.0a The model of a neuron (a body nerve fiber) may be described by a transmission line model as shown in Figure 1-11. Applying KCL and KVL to *one* of the incremental sections, derive the two fundamental partial differential equations, each of which contains both current and voltage, as they depend upon time and distance. Using these results, derive the partial differential equation that contains voltage only with partial derivatives with respect to time and space (a wave equation–like form). Keep these solutions handy since they are used in later exercises.

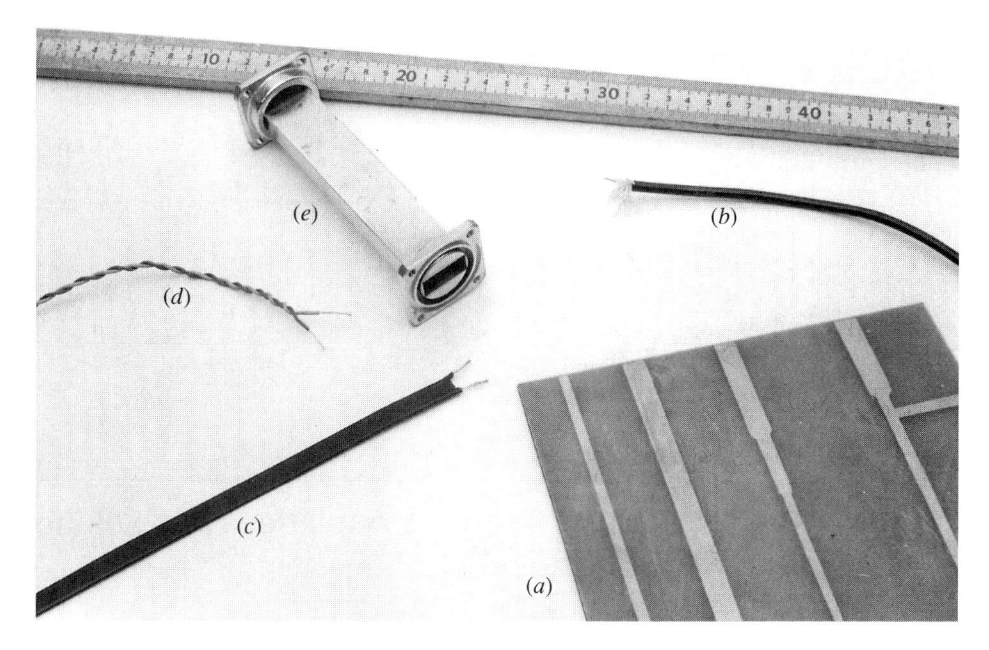

Figure 1-10. Sample transmission lines. (*a*) Microstrip (printed circuit board). (*b*) Coaxial line. (*c*) Twin lead (parallel line). (*d*) Twisted pair. (*e*) Microwave waveguide (Chapter 9).

1-4.0b Using the results of Exercise 1-3.0b, obtain the partial differential (wave) equations for i and for v for the system of Exercise 1-2.0d.

1-4.0c The derivations of the preceding sections have tacitly assumed that the space variable z increases from left to right, with a generator being at the left end of the line ($z = 0$) and some load at the right end ($z = L$). Suppose we wish to measure distances using the load as reference ($l = 0$) and distance l increasing back to the left toward some generator. To do this, we would use the model shown in Figure 1-12. Using this model, derive the wave equations for $v(l, t)$ and $i(l, t)$. Give qualitative physical reasons why your results are plausible. Next, let $z = -l$ and use your knowledge of calculus (chain rule, etc.) to convert

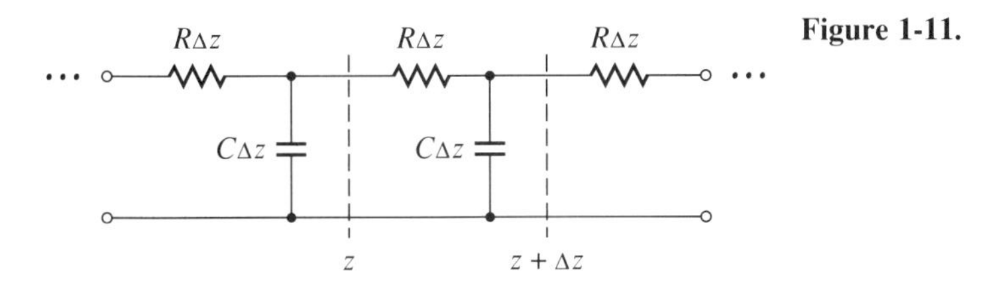

Figure 1-11.

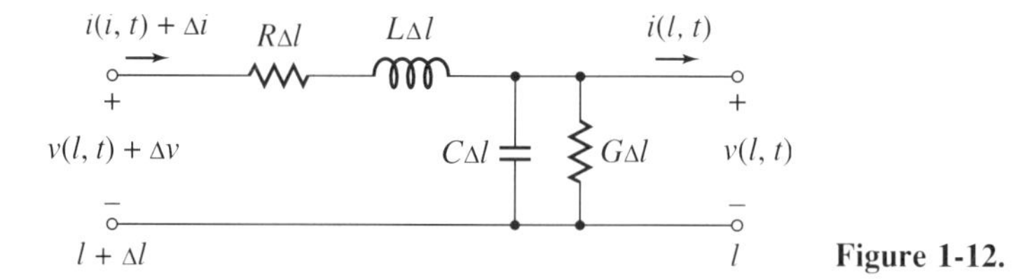

Figure 1-12.

the text wave equation for $v(z, t)$ to a wave equation for $v(l, t)$ and compare with your derived solution.

1-4.0d Analyze an incremental section of the proposed transmission line model shown in Figure 1-13 and obtain the wave equation for $v(z, t)$. Compare your results with Eq. 1-7 and comment on the comparison of the model.

1-4.0e Discuss the conditions under which the order of taking partial derivatives can be interchanged.

1-5 Sinusoidal Steady-State Formulations and Solutions of the Transmission Line Equations

The results of the last section are not easily applied to specific driving sources on a transmission line. It is necessary to apply initial and boundary conditions for each new problem, making the general process quite cumbersome. Some of the most important applications of the line wave equations are obtained from sinusoidal steady-state responses. This formulation also yields a good deal of physical interpretation and insight into line performance. Also, the sinusoidal formulation removes the time variable from the various differential equations so that one can concentrate on the spatial (z) variation. More important is the result that this formulation converts the partial differential equations to the more tractable ordinary differential equations. This yields configurations that allow us to use circuit theory directly in many problems.

It might be suspected that limiting ourselves to the sinusoidal steady state has severely restricted our possible applications. However, if one realizes that other time waveforms can be expressed as a sum of sinusoids (amplitude and phase) via either the Fourier series or Fourier transform and that we have assumed a linear system (R, L, G, and C constants), then we effectively take each frequency component through the transmission line and add up the results at the load (or any location z for that matter).

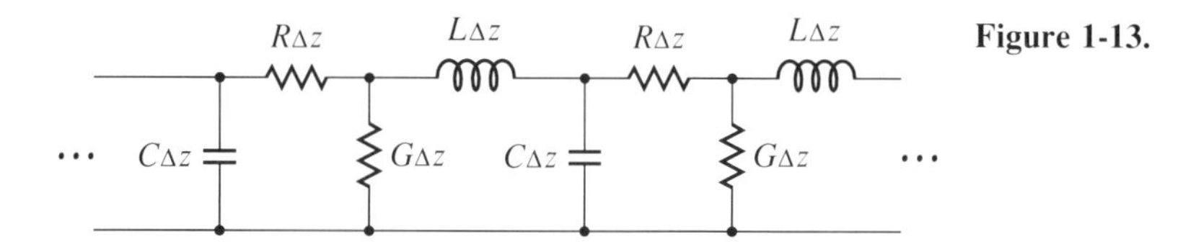

Figure 1-13.

1-5.1 Conversion to the Sinusoidal Steady State

The process for obtaining the sinusoidal representation is the same as that used in circuit theory. A review of the applicable theorems is given in Appendix A. Basically we assume a time dependence of the form $k(t) = K\cos(\omega t + \phi)$ or, equivalently, $\text{Re}\{\overline{K}e^{j\omega t}\}$, where $\overline{K}$ is a complex number $Ke^{j\phi}$ (often written in shorthand form as $K\angle\phi$). The symbol Re{ } is called the *real operator*, which simply takes the real part of the complex number within the braces. The equivalency of the two time expressions is shown by using the usual rules of exponents and Euler's theorem for imaginary exponents. We then proceed as follows:

$$\begin{aligned}\text{Re}\{\overline{K}e^{j\omega t}\} &= \text{Re}\{Ke^{j\phi}e^{j\omega t}\} = \text{Re}\{Ke^{j(\omega t+\phi)}\}\\ &= \text{Re}\{K[\cos(\omega t + \phi) + j\sin(\omega t + \phi)]\} = K\cos(\omega t + \phi) = k(t)\end{aligned}$$

as asserted. $\overline{K}$ is called the *phasor representation* of the sinusoidal time function $k(t)$.

Since partial derivatives with respect to time are important, we consider the first derivative to obtain (see Exercise 1-5.1a)

$$\frac{\partial[K\cos(\omega t + \phi)]}{\partial t} = \frac{\partial[\text{Re}\{\overline{K}e^{j\omega t}\}]}{\partial t} = \text{Re}\left\{\frac{\partial[\overline{K}e^{j\omega t}]}{\partial t}\right\} = \text{Re}\{j\omega\overline{K}e^{j\omega t}\} \tag{1-9}$$

From this we conclude that the first derivative with respect to time multiples by $j\omega$ within the real operator.

We can similarly show that

$$\frac{\partial^2\,\text{Re}\{\overline{K}e^{j\omega t}\}}{\partial t^2} = \text{Re}\{(j\omega)^2\overline{K}e^{j\omega t}\} = \text{Re}\{-\omega^2\overline{K}e^{j\omega t}\} \tag{1-10}$$

and conclude that the second derivative with respect to time multiplies by $-\omega^2$ within the real operator.

For our transmission lines, the current and voltage also vary with z, so for the sinusoidal steady state we write expressions making use of the phasor representation as

$$v(z, t) = \text{Re}\{\overline{V}(z)e^{j\omega t}\} \tag{1-11}$$

and

$$i(z, t) = \text{Re}\{\overline{I}(z)e^{j\omega t}\} \tag{1-12}$$

where

$$\overline{V}(z) \equiv V(z)e^{j\phi_V(z)} \tag{1-13}$$

and

$$\overline{I}(z) \equiv I(z)e^{j\phi_I(z)} \tag{1-14}$$

Using the properties of the real operator shown earlier, the equations for $v(z, t)$ and $i(z, t)$ above have the specific time forms

$$v(z, t) = V(z)\cos[\omega t + \phi_V(z)] \tag{1-15}$$

and

$$i(z, t) = I(z)\cos[\omega t + \phi_I(z)] \tag{1-16}$$

It is important to note that we use *peak* values in our work here and not RMS as in circuit theory. This is common practice in field theory.

We begin by converting Eq. 1-5 to the sinusoidal steady state using Eqs. 1-11 and 1-12 as a first step. Thus

$$\frac{\partial[\mathrm{Re}\{\overline{V}(z)e^{j\omega t}\}]}{\partial z} = -R\,\mathrm{Re}\{\overline{I}(z)e^{j\omega t}\} - L\,\frac{\partial[\mathrm{Re}\{\overline{I}(z)e^{j\omega t}\}]}{\partial t}$$

Interchanging the partial derivatives and real operators (see Exercise 1-5.1a),

$$\mathrm{Re}\left\{\frac{\partial[\overline{V}(z)e^{j\omega t}]}{\partial z}\right\} = -R\,\mathrm{Re}\{\overline{I}(z)e^{j\omega t}\} - L\,\mathrm{Re}\left\{\frac{\partial[\overline{I}(z)e^{j\omega t}]}{\partial t}\right\}$$

Now on the left-hand side, since $e^{j\omega t}$ does not involve z, it may be removed from the derivative. Since $\overline{I}(z)$ is not a function of time, it may be treated as a complex constant as far as the partial by t is concerned in the derivative term on the right-hand side. Thus the last equality in Eq. 1-9 applies to that term. Then

$$\mathrm{Re}\left\{e^{j\omega t}\,\frac{\partial[\overline{V}(z)]}{\partial z}\right\} = -R\,\mathrm{Re}\{\overline{I}(z)e^{j\omega t}\} - L\,\mathrm{Re}\{j\omega\overline{I}(z)e^{j\omega t}\}$$

Dropping the Re operator (this can be proven to be a mathematically rigorous step as in circuit theory; see Appendix A) and noting that since $\overline{V}(z)$ is a function of z only so the partial derivative can be replaced by the total derivative, the preceding equation becomes

$$e^{j\omega d}\,\frac{d\overline{V}(z)}{dz} = -R\overline{I}(z)e^{j\omega t} - j\omega L\overline{I}(z)e^{j\omega t}$$

Dividing through by $e^{j\omega t}$,

$$\frac{d\overline{V}(z)}{dz} = -(R + j\omega L)\overline{I}(z) \equiv -Z\overline{I}(z) \tag{1-17}$$

The combination

$$Z \equiv R + j\omega L \text{ ohm/meter} \tag{1-18}$$

is called the *series impedance* per meter of line.

Using exactly analogous steps, Eq. 1-6 becomes

$$\frac{d\overline{I}(z)}{dz} = -(G + j\omega C)\overline{V}(z) \equiv -Y\overline{V}(z) \tag{1-19}$$

where

$$Y \equiv G + j\omega C \text{ siemen/meter} \tag{1-20}$$

This is called the *shunt admittance* per meter of line.

The wave equations for $\bar{V}(z)$ and $\bar{I}(z)$ are now easily obtained by using these last results directly. For example, if we solve Eq. 1-17 for $\bar{I}(z)$ and substitute the result into Eq. 1-19, we obtain

$$\frac{d\left(-\frac{1}{Z}\frac{d\bar{V}(z)}{dz}\right)}{dz} = -Y\bar{V}(z)$$

or, since Z is a constant depending only upon line parameters and frequency,

$$\frac{d^2\bar{V}(z)}{dz^2} = YZ\bar{V}(z) \tag{1-21}$$

Since Y and Z are constants, we define

$$\gamma^2 \equiv YZ = (G + j\omega C)(R + j\omega L) = (RG - \omega^2 LC) + j\omega(RC + LG) \tag{1-22}$$

where γ^2 has the units of 1/square meter. The wave equation becomes

$$\frac{d^2\bar{V}(z)}{dz^2} = \gamma^2\bar{V}(z) \tag{1-23}$$

The constant γ is called the *transmission line propagation constant* for reasons that will become evident shortly. A check of the units of γ would show, as indicated above, that it is dimensionally 1/meter.

Using Eqs. 1-17 and 1-19 again and following the procedure above would yield the wave equation for $\bar{I}(z)$ as

$$\frac{d^2\bar{I}(z)}{dz^2} = YZ\bar{I}(z) = \gamma^2\bar{I}(z) \tag{1-24}$$

In summary, then, we would solve a transmission line problem as follows:

1. Obtain solutions for $\bar{V}(z)$ and $\bar{I}(z)$ that satisfy Eqs. 1-23 and 1-24.
2. Obtain $v(z, t)$ and $i(z, t)$ using Eqs. 1-11 and 1-12 for the given frequency, $\omega = 2\pi f$.

Exercises

1-5.1a In converting the transmission line equations to the sinusoidal steady state, we made use of the theorem

$$\frac{d\,\mathrm{Re}\{e^{j\omega t}\}}{dt} = \mathrm{Re}\left\{\frac{de^{j\omega t}}{dt}\right\}$$

Prove this theorem. (Hint: Euler's theorem helps nicely.)

1-5.1b As a result of applying the theorem of Exercise 1-5.1a, suppose a particular problem gave the result

$$f(t) = \text{Re}\{j\omega e^{j\omega t}\}$$

Evaluate the right-hand member of this equation to obtain an explicit functional form.

1-5.1c Carry out the steps that result in Eqs. 1-19 and 1-20.

1-5.1d Carry out the steps that yield the wave equation for current, Eq. 1-24.

1-5.1e The sinusoidal steady-state forms of the current and voltage wave equations can also be derived by applying the real operator theorems (Eqs. 1-9 and 1-10) given in this section directly on the general time-wave equations. Carry out the steps that reduce Eqs. 1-7 and 1-8 to their steady-state representations, Eqs. 1-23 and 1-24. (Hint: First express $v(z, t)$ in phasor form with a real operator and substitute into Eq. 1-7.)

1-5.1f Express the equations of Exercise 1-4.0a in sinusoidal steady-state form, and identify a propagation constant. Keep the answer handy for use in a later exercise.

1-5.1g Make a unit (dimensional) analysis of each term in the right-most equality of Eq. 1-22 and show that γ is dimensionally 1/meter.

1-5.2 Solutions of the Sinusoidal Steady-State Equations

As suggested in the preceding paragraph, our first step is to solve for $\overline{V}(z)$ and $\overline{I}(z)$. We begin by solving for $\overline{V}(z)$ using Eq. 1-23; namely,

$$\frac{d^2\overline{V}(z)}{dz^2} = \gamma^2\overline{V}(z)$$

From the theory of ordinary differential equations, we have the characteristic equation corresponding to this second-order differential equation

$$m^2 = \gamma^2 \qquad \text{or} \qquad m = \pm\gamma$$

Thus the general solution is

$$\overline{V}(z) = V^+e^{-\gamma z} + V^-e^{\gamma z} \equiv \overline{V}^+(z) + \overline{V}^-(z) \tag{1-25}$$

where V^+ and V^- are constants, generally complex, that are determined by applying very simple boundary conditions discussed later. Equation 1-25 shows that the total voltage $\overline{V}(z)$ can be written as the sum of two functions of z, the physical interpretations of which are given later.

In the preceding section, the unit of γ was indicated to be 1/meter. The preceding solution shows that this is as it should be since the complete exponent γz must be dimensionless. However, since γ appears in the exponent of e, the Naperian logarithm base, γ is often expressed as neper/meter, or nep/m, where neper is a dimensionless quantity much like the radian in trigonometry.

The companion solution for current in the current wave equation would be identical in form to voltage since the differential equations are identical. Thus we can write directly

$$\bar{I}(z) = I^+e^{-\gamma z} + I^-e^{\gamma z} \equiv \bar{I}^+(z) + \bar{I}^-(z) \tag{1-26}$$

where I^+ and I^- are two *complex* constants. We must note, however, that the two solutions $\bar{I}(z)$ and $\bar{V}(z)$ are not independent since they are related by the basic Eqs. 1-17 and 1-19. Substituting our solution for $\bar{I}(z)$, for example, into Eq. 1-19, we would obtain

$$-\gamma I^+e^{-\gamma z} + \gamma I^-e^{\gamma z} = -Y\bar{V}(z)$$

from which

$$\bar{V}(z) = \frac{\gamma}{Y}I^+e^{-\gamma z} - \frac{\gamma}{Y}I^-e^{\gamma z} \tag{1-27}$$

Comparing this result with the previous solution for $\bar{V}(z)$, Eq. 1-25, we find that

$$\left.\begin{aligned} V^+ &= \frac{\gamma}{Y}I^+ \\ V^- &= -\frac{\gamma}{Y}I^- \end{aligned}\right\} \tag{1-28}$$

The second step in our solution process is to convert the phasors to real-time expressions for current and voltage using Eqs. 1-11 and 1-12. For the voltage, we would represent the complex constants as

$$V^+ = |V^+|e^{j\phi^+} \qquad \text{and} \qquad V^- = |V^-|e^{j\phi^-}$$

and use Eq. 1-11 as follows:

$$\begin{aligned} v(z, t) &= \text{Re}\{\bar{V}(z)e^{j\omega t}\} = \text{Re}\{(V^+e^{-\gamma z} + V^-e^{\gamma z})e^{j\omega t}\} \\ &= \text{Re}\{|V+|e^{j\phi^+} - \gamma z e^{j\omega t} + |V-|e^{j\phi^-}\gamma z e^{j\omega t}\} \end{aligned}$$

Since γ is a complex number $\left(\text{equal to } \sqrt{YZ}\right)$, we express it as a rectangular form

$$\gamma \equiv \alpha + j\beta \text{ nep/m} \tag{1-29}$$

Using this in the equation for $v(z, t)$, we obtain

$$\begin{aligned} v(z, t) &= \text{Re}\{|V^+|e^{j\phi^+}e^{-(\alpha+j\beta)\mathbf{z}}e^{j\omega t} + |V^-|e^{j\phi^-}e^{(\alpha+j\beta)z}e^{j\omega t}\} \\ &= \text{Re}\{|V^+|e^{-\alpha z}e^{j(\omega t+\phi^+-\beta z)} + |V^-|e^{\alpha z}e^{j(\omega t+\phi^-+\beta z)}\} \end{aligned}$$

$$\begin{aligned} v(z, t) = {} & |V^+|e^{-\alpha z}\cos(\omega t + \phi^+ - \beta z) \\ & + |V^-|e^{\alpha z}\cos(\omega t + \phi^- + \beta z) \end{aligned} \tag{1-30}$$

Note that physically αz is an attenuation exponent and βz is a phase shift.

Exercises

1-5.2a Solve the differential equation for the neuron of Exercise 1-5.1f (voltage wave equation) and identify all constants. Using this solution and the differential relation between $\overline{V}$ and $\overline{I}$, obtain the equation for current $\overline{I}$. Identify the propagation constant γ, and give its expression. Keep the results handy for use in later exercises.

1-5.2b Using the current phasor solution, Eq. 1-26, obtain an equation for $i(z, t)$ analogous to Eq. 1-30. Remember that γ is complex and is defined in rectangular form as $\alpha + j\beta$.

1-5.2c It was shown in Eq. 1-28 that the constants in the solutions for $\overline{V}(z)$ and $\overline{I}(z)$ are related by $V^+/I^+ = \gamma/Y$ and $V^-/I^- = -\gamma/Y$. Show that the same results are obtained if we evaluate $\overline{V}^+(z)/\overline{I}^+(z)$ and $\overline{V}^-(z)/\overline{I}^-(z)$.

1-5.3 Attenuation Constant and Phase Constant and Their Interpretations

The real constants α and β play important roles in transmission system performance. It is important to have a physical picture of their effects on the performance with respect to current and voltage. Since α is a real number, the exponential terms $\pm\alpha z$ in the solution for voltage, Eq. 1-30, represent *amplitude* changes in the z direction. To relate α to this description, it is called the *attenuation constant* and given the unit neper/meter, the same as γ. The constant α is thus a measure of the attenuation rate (amplitude change) per meter.

One may wonder about the physical admissibility of a term $+\alpha z$ in the solution, since it represents an increase in amplitude with coordinate z. This can be explained by observing that, in practice, a voltage source, or any other effect that produces a voltage, can appear at any point on the line. If one were then to inquire as to the voltage value to the left of the excitation (*decreasing* values of z) the term αz would in fact decrease away from the excitation. This is shown schematically in Figure 1-14. Thus the general solution of the wave equation can legitimately retain this term, since it represents an amplitude decrease to the left and away from an excitation point, or origin, of the voltage.

To obtain the interpretation of the constant β, we examine the exponential factors in Eq. 1-25, and write them in the form

$$e^{\pm\gamma z} = e^{\pm(\alpha+j\beta)z} = e^{\pm\alpha z}e^{\pm j\beta z} \tag{1-31}$$

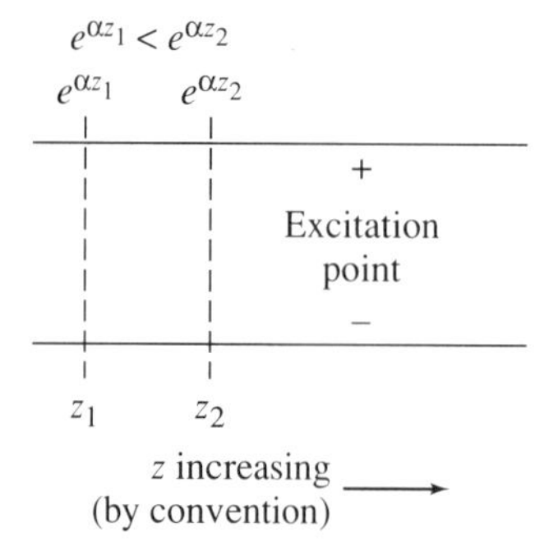

Figure 1-14. Physical admissibility of $e^{+\alpha z}$ in the wave equation solution. ($z_1 < z_2$).

Since β is real, we then observe that (using Euler's theorem)

$$|e^{\pm j\beta z}| = |\cos \beta z + j \sin \beta z| = \sqrt{\cos^2 \beta z + \sin^2 \beta z} \equiv 1$$

and

$$\text{Arg}(e^{\pm j\beta z}) = \text{Arg}(\cos \beta z \pm j \sin \beta z) = \tan^{-1}\left(\frac{\pm \sin \beta z}{\cos \beta z}\right) = \pm \beta z \qquad (1\text{-}32)$$

From these results, we see that β is a measure of the phase change per meter of line. For this reason, β is often called the *phase constant* since it does not affect amplitude but only signal phase at a given point. The units given to β are radian/meter, which may at first seem inconsistent with the fact that γ is given in neper/meter (as is α). This is resolved if we recall that both neper and radian are dimensionless quantities.

As a final note on β, we point out that it is also often called the *propagation constant*, the same as γ. This looseness of terminology arises from the fact that in many engineering systems, the attenuation is often very small so that $\alpha \simeq 0$. This would then make

$$\gamma = j\beta \qquad \text{(lossless system)} \qquad (1\text{-}33)$$

A physical description of the effect of β on the wave equation solution is given in Figure 1-15. If we imagine a two-channel scope connected at two different locations on a transmission line, the equations for, say, the left term in Eq. 1-30 would be

$$\left.\begin{aligned} &\text{Scope Channel A: } |V^+|e^{-\alpha z_1} \cos(\omega t + \phi^+ - \beta z_1) \\ &\text{Scope Channel B: } |V^+|e^{-\alpha z_2} \cos(\omega t + \phi^+ - \beta z_2) \end{aligned}\right\} \qquad (1\text{-}34)$$

where we suppose $z_1 < z_2$. The scope traces would then appear as shown in the figure, the phase difference being $(\beta z_2 - \beta z_1)$ between the two channels and z_1 being the reference

Figure 1-15. Effect of the phase constant β on line voltage time waveform.

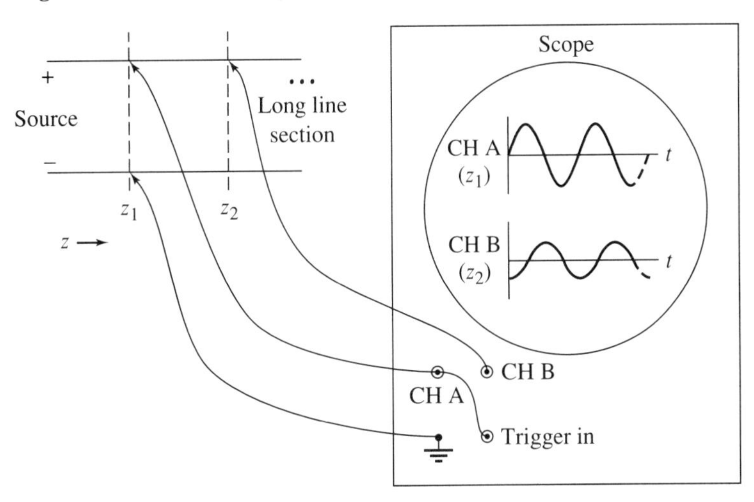

location waveform. Note that the plot variable is now *time*, not coordinate as in Figure 1-14. The effect of the attenuation α is also portrayed in the figure by an amplitude decrease of the waveform at z_2, the source being at the left in this example.

Exercises

1-5.3a From the solutions of the neuron model in Exercises 1-4.0a and 1-5.1f, take the equation for γ and obtain explicit expressions for α and β. Explain the physical significance of these parameters in terms of an electrical signal traveling along a nerve. Multiple sclerosis causes scarring and hardening of the neural fibers. Discuss the effects of these scars on the parameters α and β and also the physical effects of these changes on a person who has this disease. Keep these results handy for use in later exercises.

1-5.3b For γ that is valid for the general uniform line (Eq. 1-22), obtain explicit expressions for α and β in terms of R, L, G, C, and ω.

1-5.3c Consider the system of Figure 1-15 with a source connected at the left end. Explain why the phase of Channel B would be correct for the connections shown for Channels A and B.

1-5.3d Suppose you took a laboratory voltmeter and monitored locations z_1 and z_2 on the line as shown in Figure 1-15. Derive expressions for the voltmeter readings in terms of the parameters in Eq. 1-34. Discuss how you could use two readings to determine α. Derive the equation for α. Is it possible to obtain β from these readings?

1-5.3e For the solutions $\overline{V}(z)$ and $\overline{I}(z)$ for the neuron model (Exercise 1-5.2a), obtain the real-time sinusoidal forms $v(z, t)$ and $i(z, t)$ both showing R and C explicitly and V^+, V^-. Give physical interpretations of the results.

To develop some physical insights into the mathematics, let's examine each of the terms for $v(z, t)$ in Eq. 1-30. Consider first the term with V^+ in it. In Figure 1-16 are plots of that term, which we denote by $v^+(z, t)$, for two different times—specifically t_1 and $t_2 > t_1$. The plots in the figure show two effects. First, at $t = t_2$ the line waveform has shifted to the right; that is, in the positive z direction. This was the reason for using the plus sign superscript on the complex constant V^+. We then identify this term as the positive traveling component of the total solution $v(z, t)$. Frequently this component is referred to as the *incident component*. This latter terminology, however, must be used with care since this traveling wave is an incident quantity only if one were standing at point b. If one were standing at point a, the same component would appear to be leaving as one looked to the

Figure 1-16. Plots of the $V^+(z, t)$ portion of the wave equation solution for two values of time.

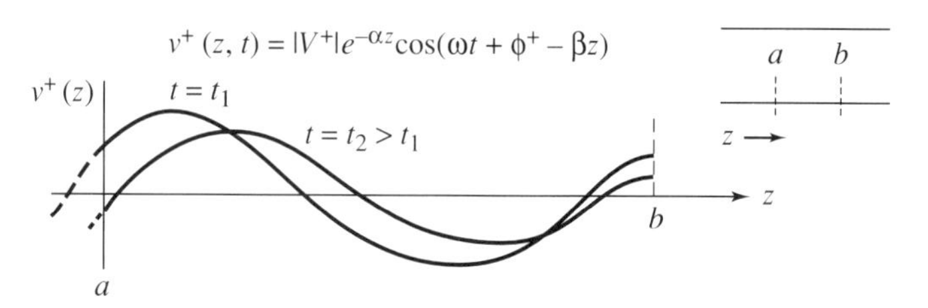

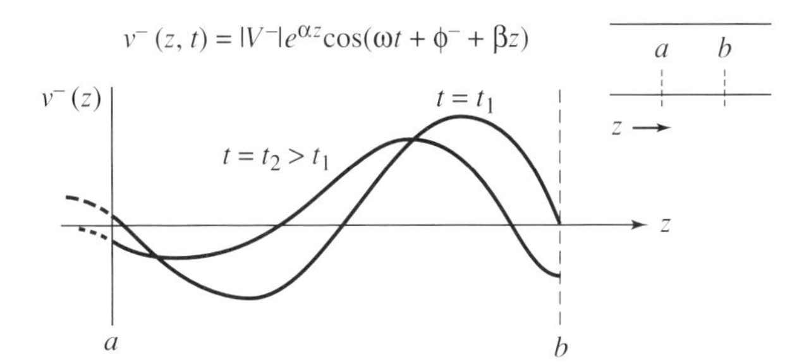

Figure 1-17. Plots of the $V^-(z, t)$ portion of the wave equation solution for two values of time.

right down the line. This becomes clear if one notices that the peaks of the line waveform recede from point a as t increases.

The second effect is one due to attenuation. At time t_2, the amplitudes of the corresponding peaks have decreased as one would expect since the waveform should attenuate as it moves to the right.

As you might suspect as a result of the preceding discussion, the second term containing V^- will turn out to be a wave traveling in the negative z direction. We'll call this component $v^-(z, t)$. Since the attenuation factor in that term—αz—appears to increase in value, it is instructive to examine just what $v^-(z, t)$ is doing physically on the line.

In Figure 1-17 are shown plots of $v^-(z, t)$ for two different values of time, t_1 and $t_2 > t_1$. As expected, the amplitudes are larger as z increases. However, an examination of the figure shows that for t_2, the line waveform has moved to the "left," or negative z direction. Thus the amplitude of that component actually decreases in the direction it moves or propagates as one would expect. We therefore identify the V^- constant as the amplitude factor of the negative traveling component of the total solution. This component is often referred to as a *reflected component*, but again, care must be used with this nomenclature. This is the reflected component only as far as location b is concerned. If one stood at point a, this wave would appear to be an incident wave since the peaks would be approaching.

Exercises

1-5.3f A unit step function is defined by the equation

$$U(x) = \begin{cases} 0, & x < 0 \\ 1, & x \geq 0 \end{cases}$$

Plot the functions $U(6 - t)$ and $U(t - 6)$. Now for the function $v(z, t) = U(t - \frac{1}{3}z)$, plot $v(z, t)$ on the z axis for the two cases $t = 1$ and $t = 2$, and describe the results. If z is in meters and t in seconds, what is the velocity of the function discontinuity in magnitude and direction? (Velocity $= \Delta z/\Delta t$.) Repeat for the function $v(z, t) = U(t + \frac{1}{3}z)$ for the same two values of t.

1-5.3g Plot the function $v(z, t) = 3[U(t - \frac{1}{5}(z - 1)) - U(t - \frac{1}{5}z)]$ for the two times $t = 2$ and $t = 4$. Give a physical description of the results.

1-5.3h For the function given in Exercise 1-5.3g, suppose there is an attenuation multiplying factor $e^{-0.01z}$ put onto $v(z, t)$. Plot the new resulting function, and discuss whether or not the result is physically correct. Use the same two time values.

1-6 Transmission Line Characteristic Impedance and Propagation Constant: Some Cases

The solution for $\overline{V}(z)$ as obtained from $\overline{I}(z)$ in Section 1-5 (Eq. 1-27) resulted in a ratio γ/Y. Since both γ and Y are functions only of the line parameters R, L, G, and C and the operating frequency ω, we use Eq. 1-22 to obtain

$$\frac{\gamma}{Y} = \frac{\sqrt{YZ}}{Y} = \left(\frac{Z}{Y}\right)^{1/2} = \left(\frac{R + j\omega L}{G + j\omega C}\right)^{1/2}$$

The unit of $\sqrt{Z/Y}$ is the ohm, and since it is a function of transmission line parameters, we give it the name *characteristic impedance* and denote it by Z_0. Thus

$$Z_0 \equiv \left(\frac{Z}{Y}\right)^{1/2} = \left(\frac{R + j\omega L}{G + j\omega C}\right)^{1/2} \text{ ohms} \tag{1-35}$$

With this definition, the voltage phasor solution $\overline{V}(z)$ becomes

$$\overline{V}(z) = Z_0 I^+ e^{-\gamma z} - Z_0 I^- e^{\gamma z} \tag{1-36}$$

where

$$I^+ = \frac{V^+}{Z_0} \quad \text{and} \quad I^- = \frac{-V^-}{Z_0} \tag{1-37}$$

It is important to remember that Z_0 is not defined as the ratio of *total* line voltage to *total* line current, but as the ratio of corresponding *components* of $\overline{V}(z)$ and $\overline{I}(z)$ as given by Eq. 1-37 above. Thus it is not what one would always measure with an impedance meter on an arbitrary length of the cable. Z_0 is simply a ''characteristic'' value of a particular transmission line.

The parameter Z_0 is the parameter referred to when we say we want a 50-Ω line or a 300-Ω line. In such cases, however, it is usual to compute and specify Z_0 on the basis of a lossless line where $R = G = 0$ so that the result is a real number

$$Z_0 \approx \sqrt{\frac{L}{C}}\ \Omega \tag{1-38}$$

Exercises

1-6.0a For the general transmission line, the characteristic impedance is given by Eq. 1-35. Express Z_0 in the form $R_0 + jX_0$.

1-6.0b Using the results of Exercise 1-5.1f, determine the expression for Z_0 of the neuron model and express it in the form $R_0 + jX_0$.

1-6.0c For a lossless transmission line ($R = G = 0$) show that $Z_0 = \sqrt{L/C}$ and $\beta = \omega C Z_0$.

1-6.0d A "distortionless" transmission line is defined by the relation

$$\frac{R}{L} = \frac{G}{C}$$

Show that for such a line $Z_0 = \sqrt{L/C}$ and also that $\gamma = \alpha + j\beta = \sqrt{RG} + j\omega\sqrt{LC}$.

1-6.0e The RG63B/U cable has a nominal characteristic impedance of 125 Ω and a capacitance of 32.8 pF/m. Compute the line inductance and the phase shift per meter β for a distortionless line defined in Exercise 1-6.0d. Use an operating frequency of 500 MHz.

There are four important cases for which Z_0 and γ have special expressions that are of significant use in practice. These will next be developed.

1-6.1 Lossless Lines

By definition this is a transmission line that has $R = G = 0$. Although such a line does not exist practically, real lines have such small losses that this assumption is often justified, and results based on this assumption are nearly exact for many applications in some frequency ranges.

For the propagation constant, we obtain, from Eq. 1-22,

$$\gamma^2 = -\omega^2 LC$$

or

$$\gamma = j\omega\sqrt{LC} \qquad \text{lossless line} \tag{1-39}$$

Since we have defined γ as the complex quantity $\alpha + j\beta$, we then conclude that

$$\left.\begin{aligned} \alpha &= 0 \\ \beta &= \omega\sqrt{LC} \end{aligned}\right\} \qquad \text{lossless line} \tag{1-40}$$

The characteristic impedance Eq. 1-35 reduces to

$$Z_0 = \sqrt{\frac{L}{C}} \qquad \text{lossless line} \tag{1-41}$$

which is a real number. This was discussed in the preceding section.

The equations for current and voltage then become

$$\begin{aligned} \overline{V}(z) &= V^+e^{-j\beta z} + V^-e^{j\beta z} \\ \overline{I}(z) &= \frac{V^+}{Z_0}e^{-j\beta z} - \frac{V^-}{Z_0}e^{j\beta z} \qquad \text{lossless line} \\ &\text{where } \beta = \omega\sqrt{LC} \text{ and } Z_0 = \sqrt{\frac{L}{C}} \end{aligned} \tag{1-42}$$

It is interesting to observe that the propagation constant γ is now simply a phase shift per meter. The wave equations for $\overline{V}(z)$ and $\overline{I}(z)$ become the rather simple equations

$$\frac{d^2\overline{V}}{dz^2} = -\beta^2\overline{V} \qquad \text{lossless line} \qquad \text{(1-43)}$$
$$\frac{d^2\overline{I}}{dz^2} = -\beta^2\overline{V}$$

Exercises

1-6.1a An RG58C/U cable has a characteristic impedance of 50 Ω. The capacitance of a 0.2-m length of the cable is measured and found to be 20.2 pF. What is the cable inductance per meter, L, for the lossless case? What is the phase shift of the 0.2-m section at a frequency of 1 GHz? How much time delay does the phase shift correspond to? [2π radians $\sim T$ seconds (period).]

1-6.1b Show that the factor $e^{\pm j\beta z}$ is a phase factor having amplitude of unity for any value of z. (Hint: Try Euler's Theorem for complex exponentials.)

1-6.1c Suppose we have a lossless line for which $V^+ = -V^-$. Make a plot of $|\overline{V}(z)/V^+|$ as a function of βz for βz over the interval $(0,2\pi)$. Also develop an equation for $|\overline{V}(z)/V^+|$ that contains only trigonometric functions but no exponentials. This is called a *voltage standing wave pattern* since it is independent of time.

1-6.1d Repeat Exercise 1-6.1c for $I^+ = I^-$ and $|\overline{I}(z)/I^+|$.

1-6.2 Low-Loss (High-Frequency) Lines

This case can be described in two ways that are equivalent in performance. Low-loss lines have $R \ll \omega L$ and $G \ll \omega C$. This, of course, is the same as high-frequency operation since such frequencies produce the same inequalities. As a technical point, one should realize that the parameters R and G also depend on frequency, but usually increase as $\sqrt{\omega}$. A rough rule of thumb on the "much less than" signs would be when R and G are about one-tenth of the inductive reactance and capacitive susceptance, respectively. Multiplying the two inequalities together yields an equivalent definition of low-loss lines:

$$RG \ll \omega^2 LC \qquad \text{(1-44)}$$

We cannot set R and G to zero, or we would simply revert to the lossless case, but there are some approximations that greatly simplify the results.

Examining first the characteristic impedance, we have

$$Z_0 = \left(\frac{R + j\omega L}{G + j\omega C}\right)^{1/2} = \left(\frac{j\omega L}{j\omega C}\right)^{1/2} \left(\frac{\dfrac{R}{j\omega L} + 1}{\dfrac{G}{j\omega C} + 1}\right)^{1/2} = \left(\frac{L}{C}\right)^{1/2} \left(\frac{1 - j\dfrac{R}{\omega L}}{1 - j\dfrac{G}{\omega C}}\right)^{1/2} \qquad \text{(1-45)}$$

At this point, we could simply note that the definition for a low loss line makes $R/\omega L \ll 1$ and $G/\omega C \ll 1$ and neglect the imaginary terms in the radicand to obtain

$$Z_0 = \sqrt{\frac{L}{C}} \qquad \text{low-loss or high-frequency line} \tag{1-46}$$

This, of course, is what we would have obtained by simply setting R and G to zero, but this can be justified on another basis. The two complex numbers in the radicand of the rightmost expression in Eq. 1-45 will each have, using the 1⁄10 "rule," an angle less than $-6°$ [Arg$(1 - j0.1)$]. Then the difference of the angles required by the division of the complex numbers will thus be much less than $-6°$. Even if the difference was as much as $-5°$, the real part of the radicand would be in error less than 0.5% ($\cos 5° = 0.996$) and the square root in error less than 0.2%.

Another approach that also allows us to introduce some ideas of approximation theory yields the same result. From algebra we have the approximation

$$(1 \pm x)^n \approx 1 \pm nx \tag{1-47}$$

for n any constant (not necessarily an integer) and $x \ll 1$. For example, for $x = 0{,}1$, $1/(1 - x) = 1.1\overline{1}$, whereas the approximation (for $n = -1$) yields $1/(1 - x) \approx 1 + 0.1 = 1.1$ for an error less than 1%! The square root approximation (for $n = -\frac{1}{2}$) gives an error in the evaluation of $1/\sqrt{1 - 0.1}$ of just under 0.4%.

Using the approximation on Eq. 1-45 results in the following:

$$Z_0 = \left(\frac{L}{C}\right)^{1/2} \left(\frac{1 - j\dfrac{R}{\omega L}}{1 - j\dfrac{G}{\omega C}}\right)^{1/2} \approx \left(\frac{L}{C}\right)^{1/2} \frac{\left(1 - j\dfrac{R}{\omega L}\right)^{1/2}}{\left(1 - j\dfrac{G}{\omega C}\right)^{1/2}}$$

$$\approx \left(\frac{L}{C}\right)^{1/2} \left(1 - j\frac{R}{2\omega L}\right)\left(1 + j\frac{G}{2\omega C}\right)$$

$$\approx \left(\frac{L}{C}\right)^{1/2} \left[1 + \frac{RG}{4\omega^2 LC} + j\left(\frac{G}{2\omega C} - \frac{R}{2\omega L}\right)\right]$$

$$= \left(\frac{L}{C}\right)^{1/2} \left[1 + \frac{RG}{4\omega^2 LC} + j\frac{1}{2}\left(\frac{G}{\omega C} - \frac{R}{\omega L}\right)\right]$$

Now $RG/(4\omega^2 LC)$ is second order in effect (See Eq. 1-44), so it may be neglected. The imaginary part is ½ of the difference of two small quantities and may be dropped. The result is the same as the earlier expression for Z_0.

We next examine the propagation constant γ for this case beginning with Eq. 1-22 and using the preceding approximations. In the real part of the radicand of

$$\gamma = \sqrt{(RG - \omega^2 LC) + j\omega(RC + LG)}$$

we observe that $RG \ll \omega^2 LC$ in second order, so we neglect RG. We then factor out $-\omega^2 LC$ to obtain

$$\gamma \approx \sqrt{-\omega^2 LC}\left[1 - j\left(\frac{RC + LG}{\omega LC}\right)\right]^{1/2} = j\omega\sqrt{LC}\left[1 - j\left(\frac{R}{\omega L} + \frac{G}{\omega C}\right)\right]^{1/2}$$

Next we use the approximation developed above:

$$\gamma \approx j\omega\sqrt{LC}\left[1 - j\frac{1}{2}\left(\frac{R}{\omega L} + \frac{G}{\omega C}\right)\right] = \frac{1}{2}\left(R\sqrt{\frac{C}{L}} + G\sqrt{\frac{L}{C}}\right) + j\omega\sqrt{LC}$$

Since $Z_0 = \sqrt{L/C}$ and $Y_0 = 1/Z_0 = \sqrt{C/L}$, this becomes

$$\gamma = \alpha + j\beta \approx \tfrac{1}{2}(RY_0 + GZ_0) + j\omega\sqrt{LC}$$

from which

$$\begin{aligned} \alpha &\approx \tfrac{1}{2}(RY_0 + GZ_0) \\ \beta &\approx \omega\sqrt{LC} \end{aligned} \qquad \text{low-loss or high-frequency line} \tag{1-48}$$

With these results, the expressions for voltage and current become

$$\begin{aligned} \bar{V}(z) &= V^+ e^{-1/2(RY_0+GZ_0)z} e^{-j\omega\sqrt{LC}z} + V^- e^{1/2(RY_0+GZ_0)z} e^{j\omega\sqrt{LC}z} \\ \bar{I}(z) &= \frac{V^+}{Z_0} e^{-1/2(RY_0+GZ_0)z} e^{-j\omega\sqrt{LC}z} \\ &\quad - \frac{V^-}{Z_0} e^{1/2(RY_0+GZ_0)z} e^{j\omega\sqrt{LC}z} \qquad \text{low loss or high frequency lines} \end{aligned} \tag{1-49}$$

where $Z_0 = \dfrac{1}{Y_0} \approx \sqrt{\dfrac{L}{C}}, \quad \beta \approx \omega\sqrt{LC},$

and $\alpha = \dfrac{1}{2}(RY_0 + GZ_0)$

If one compares the results for the lossless case with Eq. 1-49, one observes that the values for Z_0 and β are the same, and the only significant difference is the inclusion of an attenuation factor $e^{\pm\alpha z}$ into the lossless solutions. This is a common feature of low-loss systems, which occur in many practical engineering solutions. We simply solve the problem assuming a lossless solution and then put in an attenuation factor to account for small losses.

Example 1-1

To get an idea of the extent of the approximations made in low-loss cases, let's consider a typical coax, which we will idealize to have solid outer and inner conductors. This

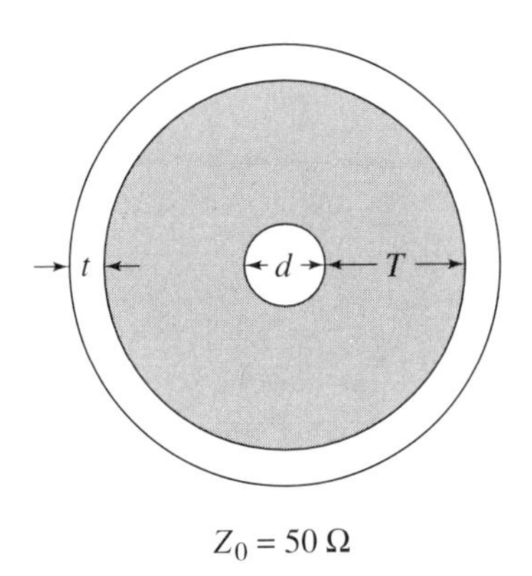

RG58C/U

Inner conductor: Copper (tinned)
$d = 0.9$ mm
Resistivity $= \rho_c = 1.7 \times 10^{-8}\ \Omega \cdot \text{m}$

Dielectric: Polyethylene
$T = 1.02$ mm
Resistivity $= \rho_d = 10^{14}\ \Omega \cdot \text{m}$

Outer conductor: Copper (tinned)
$t = 1$ mm
Resistivity $= \rho_c = 1.7 \times 10^{-8}\ \Omega \cdot \text{m}$

Figure 1-18. Dimensions and material properties of RG58C/U coaxial cable. (Idealized to solid outer conductor.)

idealization will allow us to use our formulas from physics to compute the resistance parameters easily. Figure 1-18 gives the dimensions of an RG58C/U cable and its materials.

From the manufacturer's data we are given that the capacitance is 101 pF/m. Using the given value for Z_0, we compute

$$L = CZ_0^2 = 252.5 \text{ nH/m}$$

The series resistance is the sum of the inner and outer conductor effects for a 1-m length of line. The basic resistance formula from physics is $r = \rho(l/A)$, where ρ is the material resistivity, l the conductor length in the direction of current flow, and A the cross-sectional area normal to current. We then have

$$R = \frac{r}{l} = \frac{\rho}{A} \text{ ohm/m}$$

For our coax then,

$$R = R_{\text{inner}} + R_{\text{outer}} = \frac{\rho}{A_{\text{inner}}} + \frac{\rho}{A_{\text{outer}}} = \rho\left(\frac{1}{A_{\text{inner}}} + \frac{1}{A_{\text{outer}}}\right)$$

The area of the outer conductor can be obtained by subtracting the area corresponding to the dielectric diameter from the area corresponding to the total cable diameter.

$$R = 1.7 \times 10^{-8}\left[\frac{1}{\pi(0.00045)^2} + \frac{1}{\pi((0.00247)^2 - (0.00147)^2)}\right] = 0.028\ \Omega/\text{m}$$

For the dielectric, we will for simplicity assume the parallel plate equivalent and use the mean dielectric radius to determine the area through which current flows from inner to outer conductor. Thus

$$G = \frac{A}{\rho T} \cdot \frac{1}{l} = \frac{2\pi \times 0.00096 \times 1}{10^{14} \times 0.00102 \times 1} = 5.91 \times 10^{-14} \text{ S/m}$$

Now as a check, we determine ω so that $R \ll \omega L$ and $G \ll \omega C$

$$0.028 \ll \omega \times 253 \times 10^{-9} \rightarrow \omega \gg 1.11 \times 10^5 \text{ rad/s}$$

Using 10 times as the "much greater than" criterion, the frequencies for which the high-frequency approximations hold for the series impedance term are $\omega \geq 1.11 \times 10^6$ or $f \geq$ 176 kHz. Similarly, for the shunt admittance term, the frequencies are determined from

$$5.91 \times 10^{-14} \ll \omega \times 101 \times 10^{-12}$$

Again using the criterion of 10 times being satisfied, this last result gives $f \geq 0.001$ Hz, a small value indeed. Thus both inequalities are satisfied for the value $f \geq 176$ kHz. It should be pointed out, however, that R and G are functions of frequency, so for higher frequencies, new R and G values must be computed. These ideas are covered in a later chapter. For typical attenuation variations, see Appendix B, "Nominal Loss Characteristics."

Exercises

1-6.2a For the example above, compare the exact value of Z_0 (in magnitude) with the low-loss (high-frequency) approximation for Z_0 at a frequency of 200 MHz. What is the percentage magnitude error. Also compute the exact and approximate values of α and β and compare.

1-6.2b Using the values computed in the example above, determine the attenuation in decibels that a voltage would be reduced in 30 m length of the cable. Assume that there is only a $+z$ traveling wave (i.e., $V^- = 0$). (Hint: See Eq. 1-48.)

$$\text{Attenuation} = -20 \log_{10}\left|\frac{V_0}{V_{in}}\right|$$

1-6.2c For the low-loss and lossless cases, show that $\beta = \omega C Z_0$.

1-6.3 Mid-Frequency Lines

In this case no approximations can be made since the reactive components are of the same order as the resistive or loss components. We must, therefore, use the entire expressions for Z_0 and γ. For frequencies in this range, ($R \approx j\omega L$ and $G \approx j\omega C$) Z_0 and γ will be complex, and the angles of the complex values will vary considerably. Applying the techniques of complex algebra, we can express Z_0 and γ in rectangular forms as follows:

$$\begin{aligned} Z_0 &= R_0 + jX_0 \\ &= \left(\frac{R^2 + (\omega L)^2}{G^2 + (\omega C)^2}\right)^{1/4} \cos\left[\frac{1}{2}\tan^{-1}\left(\frac{\omega LG - \omega RC}{RG + \omega^2 LC}\right)\right] \\ &\quad + j\left(\frac{R^2 + (\omega L)^2}{G^2 + (\omega C)^2}\right)^{1/4} \sin\left[\frac{1}{2}\tan^{-1}\left(\frac{\omega LG - \omega RC}{RG + \omega^2 LC}\right)\right] \end{aligned} \quad (1\text{-}50)$$

mid-frequency lines

$$\gamma = \alpha + j\beta$$
$$= \left(\frac{(RG - \omega^2 LC) + ((RG - \omega^2 LC)^2 + \omega^2 (RC + LG)^2)^{1/2}}{2}\right)^{1/2}$$
$$+ j\left(\frac{-(RG - \omega^2 LC) + ((RG - \omega^2 LC)^2 + \omega^2 (RC + LG)^2)^{1/2}}{2}\right)^{1/2} \quad \text{mid-frequency lines} \qquad (1\text{-}51)$$

In the last result, the signs of the quantities are taken so that α and β are positive. Some of the properties of these results are left as answers to exercises.

Exercises

1-6.3a The rather vague term *mid-frequency* needs to be examined. For the example parameters given for low-loss lines (Section 1-6.2), determine the frequency ranges for which $0.5R \leq \omega L \leq 5R$ and $0.5G \leq \omega C \leq 5G$. Compare these two frequency ranges and determine whether there is an overlap in the two regions. In most practical cases the dielectric losses, and hence G, are negligible. Determine general expressions for Z_0 (in rectangular form), α, β, and γ for the case $R \approx j\omega L$ and $G \ll j\omega C$. Evaluate each of these terms for the parameters of the example for low-loss lines for the frequency at which $R = \omega L$.

1-6.3b Carry out the details of the complex algebra that give Eqs. 1-50 and 1-51. (Hints: For Z_0, convert the radicand to polar form and use the rules of complex algebra. You will then have to use a trigonometric identity for the difference of two inverse tangents. For γ, note that γ^2 in Eq. 1-22 is of the form $a + jb$. Since $\gamma = \alpha + j\beta$, square this and equate to $a + jb$ and solve for α and β.) What do these expressions for α and β reduce to for $R = G = 0$?

1-6.3c For the parameters given for low-loss lines (Section 1-6.2), plot $\angle Z_0$, $|Z_0|$, α, and β as functions of frequency in the frequency range 100 Hz to 1 MHz.

1-6.3d Several logic gates are connected to a transmission line that consists of a narrow copper line foil on a dielectric sheet (see Figure 1-19). There is solid copper ground foil on the bottom side. The input shunt capacitance of each of the gates between signal and ground pins is 4 pF. Using the parallel plate formula for capacitance from physics, neglecting fringing, what is the capacitance per meter between the narrow strip and the ground plane on the transmission line (microstrip line) with gates removed? Now suppose that the characteristic impedance of the microstrip is 250 Ω without any connections to it. Next, suppose that all of the capacitance of the gates is assumed to be uniformly distributed along the line. What is the new effective value of the transmission line capacitance per meter? What is the apparent value of characteristic impedance of the line now? Assume lossless line.

1-6.4 Low-Frequency and DC Lines

The frequency range described by this case is approximately the high audio range and below. For this case the R and G terms dominate the $j\omega L$ and $j\omega C$ terms; specifically, $R \gg \omega L$ and $G \gg \omega C$. Multiplying these two expressions yields a second (higher) order inequality $RG \gg \omega^2 LC$.

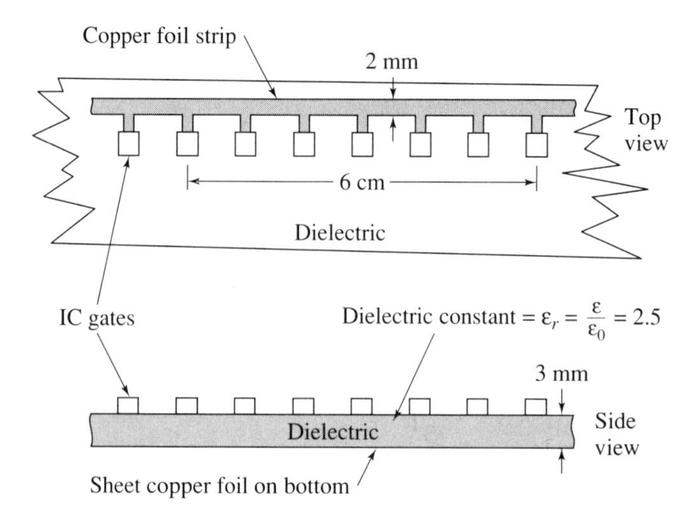

Figure 1-19.

For the characteristic impedance, the general expression reduces to

$$Z_0 \approx \sqrt{\frac{R}{G}} \qquad \text{low-frequency and DC lines} \tag{1-52}$$

Using the results of the idealized coax of the example in Section 1-6.2, we would have

$$Z_0 = \left(\frac{0.028}{5.91 \times 10^{-14}}\right)^{1/2} = 688 \text{ k}\Omega$$

This value is significantly higher than the nominal (high frequency) value of 50 Ω specified by the manufacturer.

For the propagation constant γ, the approximations initially yield

$$\gamma \approx \sqrt{RG + j\omega(RC + LG)} = \sqrt{RG}\left(1 + j\frac{\omega RC + \omega LG}{RG}\right)^{1/2}$$

Now the inequality $R \gg \omega L$, when multiplied by G, gives $RG \gg \omega LG$. Similarly, $G \gg j\omega C$ also means $RG \gg j\omega RC$. These results mean that the j term in the radicand is small compared with 1. Using the binomial power approximation developed earlier results in

$$\begin{aligned}\gamma = \alpha + j\beta &\approx \sqrt{RG}\left(1 + j\frac{\omega RC + \omega LG}{2RG}\right) \\ &= \sqrt{RG} + j\omega\left(\frac{RC + LG}{2\sqrt{RG}}\right) \qquad \text{low-frequency and DC lines}\end{aligned} \tag{1-53}$$

from which expressions for α and β are easily obtained by equating terms. They are

$$\alpha = \sqrt{RG} \qquad \beta = \omega \frac{RC + LG}{2\sqrt{RG}} \qquad \text{low-frequency and DC lines} \tag{1-54}$$

It should be pointed out that the four cases presented are not the only ones of practical interest. For example, maybe only one of the inequalities in Section 1-6.4 is valid, so that while $R \gg \omega L$, $G \sim j\omega C$ so one could not neglect ωC over some range. See Exercise 1-6.4f.

Exercises

1-6.4a For the parameters in the example in Section 1-6.2, determine the frequency range for which the conditions of Section 1-6.4 are valid, ($R \gg \omega L$ and $G \gg \omega C$). Which loss factor is the dominant one for limiting the upper bound on frequency? (Use a factor of 10 as the criterion for "much greater than.")

1-6.4b For most dielectrics, the conductance G is so small that the low-frequency approximation consists of frequencies for which $R \gg \omega L$ but $G \ll \omega C$. For this case, derive general expressions for Z_0, α, and β. Determine this frequency range for the parameters given in the example in Section 1-6.2, and plot $|Z_0|$, α, and β over this frequency range. How long must such a cable be to give 180° phase shift at the midpoint frequency of the range? How long must a length of the same cable be to give 3-dB attenuation? What do these last two results tell you about low-frequency operation?

1-6.4c Prepare a table that gives values of $|Z_0|$, α, and β for the extremes of the four regions of operation. Are the value variations such that you can give a qualitative way in which each parameter varies with frequency (i.e., increases or decreases)? This is to be done for the data in the example in Section 1-6.2.

1-6.4d The RG211A/U cable is a popular high-voltage cable. For the cable dimensions, see Appendix B. (BC means a bare copper conductor.) The dielectric is Teflon with $\epsilon_r = 2.1$ and $\rho = 10^{19}\ \Omega \cdot \text{m}$. Take the outer conductor to have thickness 1.2 mm. $\rho_{cu} = 1.7 \times 10^{-8}\ \Omega \cdot \text{m}$. Determine the four regions of operation, and prepare a table as outlined in Exercise 1-6.4c. Can G be neglected in any case?

1-6.4e An RG58B/U coaxial cable has the following properties: $Z_0 = 53.5\ \Omega$, $C = 93.5$ pF/m, and an inner conductor resistance of 3.28 Ω/100 m. The polyethylene dielectric loss is negligible. Develop and run a program that will plot the reflected (negative traveling) voltage term of Eq. 1-30 as a function of z for three fixed values of time. Use $f = 1$ GHz. From your plot, explain the direction of propagation. For purposes of this problem, recall that Z_0 given in a spec sheet is the lossless one. Let $V^- = 10\angle 0°$ and use z over a range that includes positive and negative values. To exaggerate the effects of α, suppose next that the resistance is increased to 100 and then 1000 times the actual value. Now obtain the plots.

1-6.4f Sketch $|Z_0|$ and $\angle Z_0$ as functions of ω for the case $R \gg j\omega L$. Compare the results with the limiting given in Section 1-6.4.

Although the derivations of the line parameters in terms of R, L, G, and C are correct for the various cases as defined, the numerical evaluations are actually only first order, since the evaluations assumed that R and G were constants. As a matter of fact, these two parameters are frequency dependent, and increase in value with increasing frequency. Practically speaking, this frequency dependence results in actual values of α that are much larger than predicted in the examples and exercises of this section. The large increase in attenuation of higher frequencies is clearly evident from the table of nominal loss characteristics given in Appendix B. The table also shows the wide range of attenuations available, or rather present, among the many cable types. In a later chapter, the evaluations of these frequency-dependent parameters will be investigated.

As an example of the importance of the attenuation α, consider the low-loss performance of the popular RG58A coaxial cable. One would be tempted to conclude that by increasing the frequency sufficiently, the inequalities $R \ll \omega L$ and $G \ll \omega C$ would be much better satisfied and that the line would tend toward lossless performance. The table in Appendix B, however, gives an attenuation of 1.34 dB/m (41 dB/100 ft) at 3 GHz, hardly a negligible value for longer lengths.

At this point we need to develop a relationship between the attenuation constant α (neper/meter) and a common engineering measure of attenuation, the decibel/meter, which we will denote by $\alpha_{\mathrm{dB/m}}$. For a 1-m length of line, the ratio of the output to input voltage would be

$$\left|\frac{\overline{V}_{\mathrm{out}}}{\overline{V}_{\mathrm{in}}}\right| = e^{-\alpha 1}$$

Thus

$$\alpha_{\mathrm{dB/m}} = -20 \log_{10}\left|\frac{\overline{V}_{\mathrm{out}}}{\overline{V}_{\mathrm{in}}}\right| = -20 \log_{10} e^{-\alpha} = 20\alpha \log_{10} e$$

$$\alpha_{\mathrm{dB/m}} = 8.686\alpha \tag{1-55}$$

1-7 Terminated Lossless Transmission Lines in the Sinusoidal Steady State

In most practical engineering systems the losses are very small, so a very good approximation is to assume a lossless system from the beginning. Also, as was shown in Section 1-6.2, low-loss systems can be characterized by simply multiplying the lossless result by the attenuation factor $e^{\pm \alpha z}$. Additionally, a first exposure to transmissin lines that are terminated (finite length) can be adequately understood on the basis of a lossless solution, and many of the concepts apply directly to the lossy cases.

1-7.1 Voltage, Current, Reflection Coefficient, and Standing Wave Patterns

In the preceding sections, we have discussed transmission lines without regard to any sources or loads placed on the lines. We need those if we want to accomplish anything

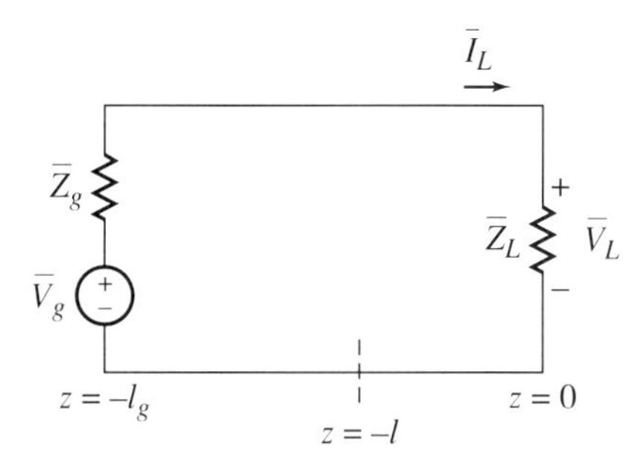

Figure 1-20. Conventions used to model loaded (terminated) lines in the sinusoidal steady state.

useful. For our first investigation, we will limit the results to the sinusoidal steady state. Other waveforms will be discussed in Chapter 3, including transient signals. The configuration and conventions in most common use are shown in Figure 1-20. The first important thing to notice is that we have taken the load location as the coordinate $z = 0$ reference. Since other points will be to the left, they will be at location $-l$ from the load (l positive). In Figure 1-20, an equivalent generator, with its associated Thevenin impedance, is used to represent the source at location $z = -l$.

At the load where $z = 0$, the sinusoidal current and voltage solutions (Eqs. 1-25 and 1-26) become

$$\left.\begin{aligned} \bar{V}_L &= \bar{V}(0) = V^+ + V^- \\ \bar{I}_L &= \bar{I}(0) = I^+ + I^- \end{aligned}\right\} \tag{1-56}$$

Using these and Ohm's law for phasors gives

$$\bar{Z}_L = \frac{\bar{V}_L}{\bar{I}_L} = \frac{V^+ + V^-}{I^+ + I^-}$$

Eliminating I^+ and I^- using Eq. 1-37, we obtain

$$\bar{Z}_L = \frac{V^+ + V^-}{\dfrac{V^+}{Z_0} - \dfrac{V^-}{Z_0}} = \frac{V^+ + V^-}{V^+ - V^-} Z_0 = \frac{1 + \dfrac{V^-}{V^+}}{1 - \dfrac{V^-}{V^+}} Z_0 \tag{1-57}$$

We next define a voltage reflection coefficient at any point on the line as the ratio of the voltage reflected at that point to the voltage incident at that point. From the load point of view, the incident voltage is the positive traveling component $\bar{V}^+$ and the reflected is $\bar{V}^-$ since these two components are moving toward the load (incident) and away from the load (reflected), respectively. Thus Eq. 1-25 gives, where $\gamma \to j\beta$ for a lossless line,

$$\Gamma(z) = \frac{\bar{V}^-(z)}{\bar{V}^+(z)} = \frac{V^- e^{j\beta z}}{V^+ e^{-j\beta z}} = \frac{V^-}{V^+} e^{j2\beta z} \tag{1-58}$$

At the load where $z = 0$, we obtain the load reflection coefficient

$$\Gamma(0) = \frac{V^-}{V^+} \equiv \Gamma_L \tag{1-59}$$

Note that, in general, Γ_L will be complex, since V^- and V^+ are complex. We can express Γ_L in polar form (magnitude and angle) and write (after substituting Eq. 1-59 into Eq. 1-57 and solving for Γ_L

$$\Gamma_L = \rho_L e^{j\theta_L} = \left|\frac{\overline{Z}_L - Z_0}{\overline{Z}_L + Z_0}\right| \angle[\text{Arg}(\overline{Z}_L - Z_0) - \text{Arg}(\overline{Z}_L + Z_0)] \tag{1-60}$$

With Γ_L defined, the expressions for $\Gamma(z)$ and $\overline{Z}_L$ become

$$\Gamma(z) = \Gamma_L e^{j2\beta z} \tag{1-61}$$

$$\overline{Z}_L = \frac{1 + \Gamma_L}{1 - \Gamma_L} Z_0 \qquad \text{or} \qquad \Gamma_L = \frac{\overline{Z}_L - Z_0}{\overline{Z}_L + Z_0} \tag{1-62}$$

At the location $z = -l$, we then obtain

$$\Gamma(z = -l) = \Gamma_L e^{-j2\beta l} \tag{1-63}$$

From the second expression for Γ_L in Eq. 1-62, we observe that:

1. $|\Gamma_L| \leq 1$, for the usual case where $\overline{Z}_L$ and $\overline{Z}_0$ have positive real parts (passive elements) and where Z_0 is pure real. (See General Exercise GE1-35 at the end of the chapter.)
2. If we know $\overline{Z}_L$ and $\overline{Z}_0$, we can easily compute Γ_L, and this is the ratio of the complex amplitudes, V^-/V^+.

When we study the propagation and reflection of electromagnetic waves in Chapter 8 we will obtain results identical to those obtained here.

Exercises

1-7.1a Show that for a general load $\overline{Z}_L = R_L + jX_L$ ($R_L \geq 0$), the magnitude of Γ_L is less than or equal to 1 for a lossless line. Under what conditions (values) on $\overline{Z}_L$ does ρ_L assume its smallest and largest values?

1-7.1b Suppose $\overline{Z}_L$ is purely resistive. For a lossless line, plot Γ_L versus R_L/Z_0 and mark the regions corresponding to $R_L < Z_0$ and $R_L > Z_0$.

1-7.1c For $R_L = Z_0$, show that $V^- = 0$. For this case, write the equation for the total line voltage, $v(z, t)$. Assume a lossless line.

1-7.1d A transmission line has $Z_0 = 300\ \Omega$. Calculate and make a plot of Γ_L for a lossless line for load resistances of value 0 to 1500 Ω.

1-7.1e A transmission line has a characteristic impedance of 75 Ω (lossless). The load is variable and is purely reactive and covers the range $-j300\ \Omega$ to $j300\ \Omega$. Plot ρ_L and θ_L as functions of the load reactance.

1-7.1f The expression for Γ_L can be written in the form

$$\Gamma_L = \frac{\dfrac{\overline{Z}_L}{Z_0} - 1}{\dfrac{\overline{Z}_L}{Z_0} + 1} \equiv \frac{\tilde{Z}_L - 1}{\tilde{Z}_L + 1}$$

where $\tilde{Z}_L$ is called the *normalized load value*. Since Γ_L and $\tilde{Z}_L$ are in general complex, they can be expressed in complex rectangular form as $\Gamma_L = U + jV$ and $\tilde{Z}_L = \tilde{R}_L + j\tilde{X}_L$. Making these complex substitutions for Γ_L and $\tilde{Z}_L$, obtain U as a function of $\tilde{R}_L$ and $\tilde{X}_L$, and also V as a function of $\tilde{R}_L$ and $\tilde{X}_L$.

The next general result to be developed is the voltage at any point on a terminated lossless line. We begin with Eq. 1-25 and substitute other definitions as they arise ($\gamma = j\beta$ again):

$$\overline{V}(z) = V^+e^{-j\beta z} + V^-e^{j\beta z} = V^+e^{-j\beta z}\left(1 + \frac{V^-}{V^+}e^{j2\beta z}\right)$$

$$\overline{V}(z) = V^+e^{-j\beta z}[1 + \Gamma(z)] = V^+e^{-j\beta z}(1 + \Gamma_L e^{j2\beta z}) \tag{1-64}$$

At the load, $z = 0$, this becomes

$$\overline{V}(0) = \overline{V}_L = V^+(1 + \Gamma_L) \tag{1-65}$$

This result is nice, since it gives us an easy way to find V^+. We know the load in many cases, so we can compute Γ_L and then measure the load voltage. V^+ is then easily computed. (Don't forget all quantities have angles since they are complex.) V^- is then computed using Eq. 1-59.

At a location $z = -l$, the voltage is

$$\overline{V}(-l) = V^+e^{j\beta l}[1 + \Gamma(-l)] = V^+e^{j\beta l}(1 + \Gamma_L e^{-j2\beta l}) \tag{1-66}$$

If we now substitute the polar form for Γ_L, from Eq. 1-60, the preceding becomes

$$\overline{V}(-l) = V^+e^{j\beta l}(1 + \rho_L e^{-j(2\beta l - \theta_L)}) \tag{1-67}$$

The magnitude of this voltage expression is what one can easily measure with a voltmeter, so let's evaluate it:

$$\begin{aligned}|\overline{V}(-l)| &= |V^+e^{j\beta l}||1 + \rho_L e^{-j(2\beta l - \theta_L)}| \\ &= |V^+||e^{j\beta l}||1 + \rho_L e^{-j(2\beta l - \theta_L)}|\end{aligned}$$

Using Euler's theorem on the rightmost complex exponential and recalling that $|e^{j\beta l}| = 1$,

$$|\overline{V}(-l)| = |V^+||1 + \rho_L \cos(2\beta l - \theta_L) - j\rho_L \sin(2\beta l - \theta_L)|$$

Thus upon evaluating the magnitude, we obtain

$$|\overline{V}(-l)| = |V^+|[(1 + \rho_L^2) + 2\rho_L \cos(2\beta l - \theta_L)]^{1/2} \tag{1-68}$$

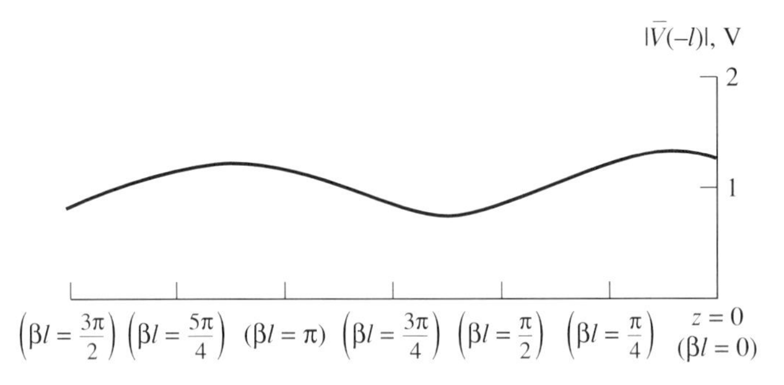

Figure 1-21. Standing wave pattern corresponding to $\Gamma_L = 0.3$ at 45° and $|V^+| = 1$ V.

Note that this result is independent of time and so the amplitude variation is stationary on the line. Because of this stationarity, this equation is called the *voltage standing wave equation* or *standing wave pattern*. In Figure 1-21 this is plotted for $V^+ = 1$ V and $\Gamma_L = 0.3e^{j45^\circ}$.

Example 1-2

A lossless transmission line has $Z_0 = 50\ \Omega$ and is terminated in an impedance of $\overline{Z}_L = 100 + j50\ \Omega$. The voltage is measured at the load and found to be $|\overline{V}_L| = 35$ V. Find the values of the incident and reflected voltage amplitudes at the load (V^+ and V^-), the voltage at the transmission line terminals at $l = 5$ m, the equation for the voltage at any point on the line, and the power dissipated in the load. The line has a capacitance of 101 pF/m, and the operating frequency is 500 MHz.

Solution:

$$\Gamma_L = \frac{100 + j50 - 50}{100 + j50 + 50} = 0.447\angle 26.57^\circ$$

We recall from circuit theory that we can arbitrarily select the phase angle on one current or voltage. Let's pick the load voltage to have $\angle 0^\circ$, so $\overline{V}_L = 35\angle 0^\circ$. Then, using Eq. 1-65,

$$35\angle 0^\circ = V^+(1 + 0.447\angle 26.57^\circ) = 0.414\angle 8.13^\circ V^+$$

$$V^+ = 24.8\angle - 8.13^\circ \text{ V}$$

From the definition of Γ_L,

$$V^- = \Gamma_L V^+ = (0.447\angle 26.57^\circ)(24.8\angle - 8.13^\circ) = 11.1\angle 18.4^\circ \text{ V}$$

Since the line is assumed lossless, $Z_0 = \sqrt{L/C}$ from which

$$L = CZ_0^2 = 101 \times 10^{-12} \times 50^2 = 2.525 \times 10^{-7} \text{ H/m}$$

Then

$$\beta = \omega\sqrt{LC} = 2\pi \times 10^8 \times \sqrt{2.525 \times 10^{-7} \times 101 \times 10^{-12}}$$
$$= 15.865 \text{ rad/m}$$

Equation 1-67 gives the total line voltage at any point $-l$ as

$$\overline{V}(-l) = 24.8e^{-j8.13^\circ}e^{j15.865l}[1 + 0.447e^{-j(31.730l-26.57^\circ)}]$$

$$\overline{V}(-l) = 24.8e^{j(15.865l-8.13^\circ)}[1 + 0.447e^{-j(31.730l-26.57^\circ)}]$$

Note that the exponents have mixed units—part in radians and part in degrees. One must use care in evaluating these terms. Usually if radians are used, working to thousandths is adequate (0.001 rad = 0.06°).

At the source end at l = 5 m,

$$\begin{aligned}\overline{V}(-5\text{ m}) &= 24.8e^{j(79.325-8.13^\circ)}(1 + 0.477e^{-j(158.65-26.57^\circ)}) \\ &= 24.8e^{j(4544.98^\circ-8.13^\circ)}(1 + 0.477e^{-j(9089.97^\circ-26.57^\circ)}) \\ &= 24.8e^{j4536.85^\circ}(1 + 0.447e^{-j63.40^\circ})\end{aligned}$$

Since the complex exponentials are periodic, we can remove all multiples of 360°. (What this means physically is that the phase difference between the input and output signals computed using this equation may be off by a multiple of 360°. If it is necessary to determine the total absolute phase difference, one must compute the velocity of the wave and determine the total time delay. Dividing the total delay by the period of the sinusoid will give the total number of cycles delay. This will be discussed later.) Continuing with the calculation:

$$\overline{V}(-5\text{ m}) = 24.8e^{-j143.5^\circ}(1 + 0.447e^{-j63.4^\circ}) = 31.4\angle -161.6^\circ\text{ V}$$

The power dissipated in the load is computed using the techniques of circuit theory for peak current and voltage values:

$$P = \frac{1}{2}|\overline{I}_L|^2 R_L = \frac{1}{2}\left|\frac{\overline{V}_L}{\overline{Z}_L}\right|^2 R_L = \frac{1}{2}\frac{|\overline{V}_L|^2}{|\overline{Z}_L|^2}R_L = \frac{1}{2}\frac{35^2}{100^2+50^2} \times 100 = 4.9\text{ W}$$

Exercises

1-7.1g Obtain an equation for $\overline{I}(z)$ in terms of V^+, V^-, and Z_0 by substituting Eq. 1-37 into Eq. 1-26. Now, following steps as used on $\overline{V}(z)$, derive the equations for $\overline{I}(z)$ that are of forms similar to Eqs. 1-64–1-68. These results will, of course, also involve Z_0 as well as Γ_L, V^+, α, β, γ, etc. Remember, $\gamma \rightarrow j\beta$ for lossless lines.

1-7.1h For the example above, determine $\overline{I}_L$ (using $\overline{V}_L = 35\angle 0^\circ$), I^+, and I^-. Now compute a current reflection coefficient Γ_L^i and state its relationship to Γ_L.

1-7.1i Using the results of Exercise 1-7.1g, compute the current into the line at the −5-m (source) end of the example above. Using this result and $\overline{V}(-5\text{ m})$ computed in the example, determine the input power. Compare it with the results of the example and explain physically. If a generator having a Thévenin impedance of 50 Ω is present at $l = -5$ m, what is the generator voltage, $\overline{V}_g$?

1-7.1j Without assuming a lossless line (i.e., use $\gamma = \alpha + j\beta$), derive equations for $|\overline{V}(-l)|$ and $|\overline{I}(-l)|$.

1-7.1k Using the values computed in the preceding example obtain the expression for $|\overline{V}(-l)|$ and plot it for $0 \leq l \leq 0.5$ m.

1213

1-7.1l For the coaxial cable RG58B/U (now assumed lossless) described in Exercise 1.64e, let the load $\overline{Z}_L$ have values that make $\Gamma_L = -1, -\frac{1}{2}, 0, \frac{1}{2}, 1$. What are the $\overline{Z}_L$ values? Now plot $|\overline{V}(-l)|$ for each Γ_L at a frequency of 1 GHz for the range $0 \leq \beta l \leq 4\pi$. (A little programming may be in order here.) Assume the incident voltage is 20 V in magnitude.

1-7.1m Plot the voltage standing wave pattern, $|\overline{V}(-l)|$, for the preceding example. Use l from 0 to 0.8 m.

1-7.1n Two theorems were used in developing the expression for $|\overline{V}(-l)|$; viz., $|e^{j\theta}| = 1$ and $|\overline{AB}| = |\overline{A}||\overline{B}|$, where $\overline{A}$ and $\overline{B}$ are complex numbers. Prove these theorems.

1-7.2 Phase Velocity, Line Wavelength, and Properties of the Standing Wave Pattern

In earlier sections the propagating natures of the positive and negative components of the current and voltage were discussed. We now investigate some of the characteristics of these waves and then develop some properties of the standing wave pattern in terms of those characteristics.

For a starting point we examine the voltage on a transmission line and some of its characteristics. Much of what will be done for the line voltage will also apply to the line current. We first determine the velocity of propagation of the $v^+(z, t)$ (positive traveling) component of the total line voltage. Imagine that you are standing on the crest of the positive traveling component as shown in Figure 1-22 and want to know how fast you need to run to stay at that point. Since you are standing at the same point on the sine wave—that is, a constant phase point—the velocity is called the *phase velocity* and is denoted by V_p. The equation for $v^+(z, t)$ is (from the appropriate part of Eq. 1-30 with $\alpha = 0$ for the lossless line).

$$v^+(z, t) = |V^+| \cos(\omega t + \phi^+ - \beta z) \tag{1-69}$$

Now if one stays at the peak, the argument of the cosine function must be a constant. Thus we set

$$\omega t = \phi^+ - \beta z = K$$

Taking the *time* derivative, remembering that ϕ^+ is a constant,

$$\omega - \beta \frac{dz}{dt} = 0$$

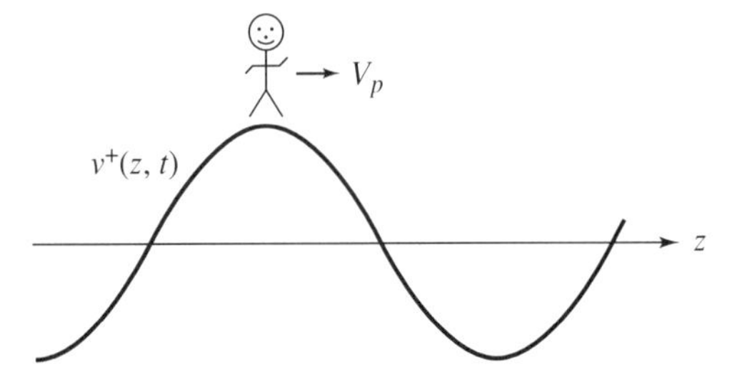

Figure 1-22. Physical description of phase velocity.

from which the phase velocity follows as

$$V_p = \frac{dz}{dt} = \frac{\omega}{\beta} \text{ m/s} \tag{1-70}$$

The next property of the voltage we want to find is the distance we must move along the line to cover one wavelength, which we'll call λ_l, the line wavelength. By definition, then, the line wavelength will be the distance z required to produce 2π phase shift (one complete variation). Thus

$$\beta z|_{z=\lambda_l} = 2\pi \qquad \text{or} \qquad \beta\lambda_l = 2\pi$$

which gives the result

$$\lambda_l = \frac{2\pi}{\beta} \tag{1-71}$$

If we measure the wavelength on the line, we could compute β using

$$\beta = \frac{2\pi}{\lambda_l} \tag{1-72}$$

We still have, for lossless or low-loss lines, the companion formulas

$$\beta = \omega\sqrt{LC} = \omega C Z_0 \tag{1-73}$$

Using these expressions for β, we may develop other forms for V_p using Eq. 1-70:

$$\left.\begin{aligned} V_p &= \frac{\omega}{\omega\sqrt{LC}} = \frac{1}{\sqrt{LC}} && \text{lossless or low-loss} \\ V_p &= \frac{\omega}{2\pi/\lambda_l} = f\lambda_l && \text{general} \\ V_p &= \frac{\omega}{\omega C Z_0} = \frac{1}{C Z_0} && \text{lossless or low-loss} \end{aligned}\right\} \tag{1-74}$$

Example 1-3

A transmission line has a characteristic impedance of 75 Ω and a capacitance of 50 pF/m. The distance along the line for which the phase shift is 360° is measured to be 0.1 m. For a lossless line, compute β, V_P and the operating frequency.

$$\beta = \frac{2\pi}{\lambda_l} = \frac{2\pi}{0.1} = 62.8 \text{ rad/m}$$

Since

$$\beta = \omega C Z_0 = \omega \times 50 \times 10^{-12} \times 75$$

$$\omega = \frac{62.8}{3.75 \times 10^{-9}} = 1.67 \times 10^{10} \text{ rad/s}$$

or

$$f = 2.66 \text{ GHz}$$

Finally,

$$V_P = f\lambda_l = 2.66 \times 10^9 \times 0.1 = 2.66 \times 10^8 \text{ m/s}$$

Note that the velocity approaches the speed of light in vacuum. To determine the consequence of this high velocity, consider a laboratory situation in which a 100-kHz signal generator is applied to a 0.3-m length of the cable. The signal at the input at a time t_1 will then reach the output in a time of only $0.3/2.66 \times 10^8 = 1.13 \times 10^{-9}$ seconds later. However, for a 100-kHz signal, this time represents only about 1/10,000 of the signal period, or about 0.04°, so the input has hardly changed. Thus, the input and output are essentially in phase, which means that for low frequencies, cable delay can be neglected, i.e., line length is not significant. (See also Exercise 1-7.2d.)

Exercises

1-7.2a Following the procedure given in the preceding example, derive the expression for the phase velocity of the negative traveling component of line voltage, $v^-(z, t)$. Explain your result.

1-7.2b Suppose you had a system that allowed you to monitor the voltage on a coaxial line (called a *slotted line*) and an oscilloscope that allowed you to display the output waveform of a generator applied to the system. Show how you would measure or determine the line wavelength, λ_l. Assume single-channel presentation.

1-7.2c RG213/U cable has a characteristic impedance of 50 Ω and a capacitance of 101 pF/m. An 8-m length of this cable is used to connect a generator from a test station to an antenna platform. Determine the highest frequency that can be used such that the phase error between the generator and antenna signals is 4° (about 0.01 cycle). Assume a lossless cable. If the antenna has an impedance of 125 Ω at that highest frequency, what is the impedance seen by the generator? $\overline{Z}(-l) = [\overline{V}(-l)]/[\overline{I}(-l)]$. What is the percentage error in impedance magnitude? Repeat for an antenna impedance of 300 Ω. Comment on the difference in the two results.

1-7.2d The example above showed only that the line length was negligible as far as phase is concerned. Suppose that the generator was a tuned circuit at 100 kHz (maybe in the collector of a transistor), and the tuned circuit consisted of a 100-μH inductor in parallel with a 0.0253-μF capacitor. Is the cable capacitance significant? What is the frequency shift?

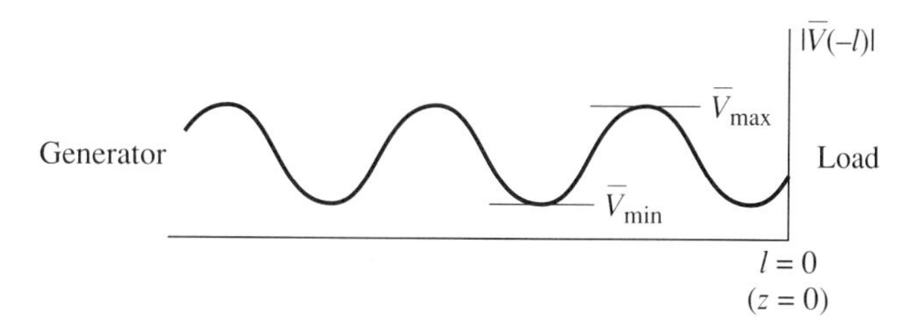

Figure 1-23. Magnitude of total voltage on a lossless line.

Let's return for a moment to the expression for the magnitude of the total line voltage, $|\overline{V}(-l)|$, which we call the standing wave pattern. For our lossless case this is

$$|\overline{V}(-l)| = |V^+|[(1 + \rho_L^2) + 2\rho_L \cos(2\beta l - \theta_L)]^{1/2} \tag{1-75}$$

A sketch of a standing wave pattern is given in Figure 1-23.

At this point, it is important that we obtain a clear physical understanding of the difference between these figures and those drawn earlier when propagating, or traveling, waves were discussed. Note that for the pattern just plotted, the result is *independent* of time and hence does *not* move along the line as time increases. Because of this stationarity, they are called *standing wave patterns*. Physically, these represent the peak amplitude of the *sum* of the positive and negative traveling waves of the sine wave time variations at *each* point along the line. If one were to put a two-channel oscilloscope on the line and monitor two different locations l_1 and l_2, sine waves of different peak amplitudes would be observed. This is shown diagrammatically in Figure 1-24. Since the oscilloscope has a *time* base horizontally, one would see a sine wave at each location. The oscilloscope would also show the phase difference as well as the amplitude difference. Referring to the earlier

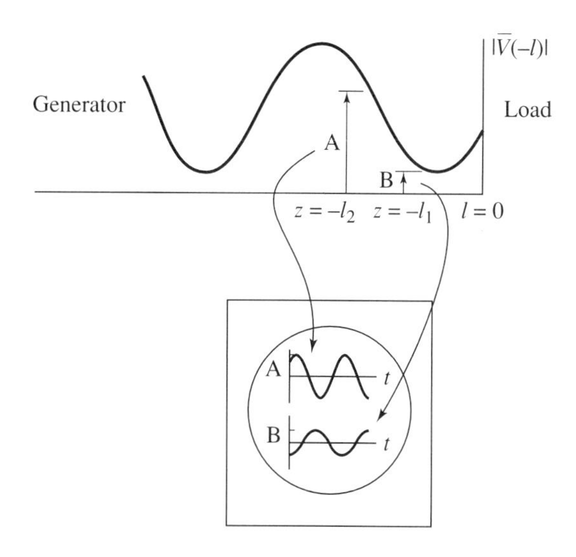

Figure 1-24. Waveforms observed with an oscilloscope at two different line locations.

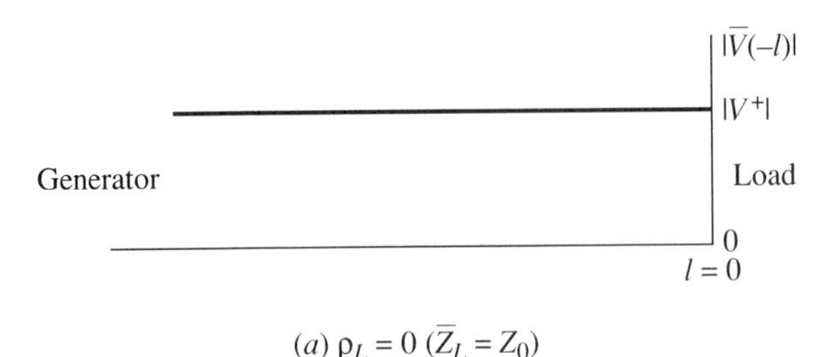

(a) $\rho_L = 0$ ($\bar{Z}_L = Z_0$)

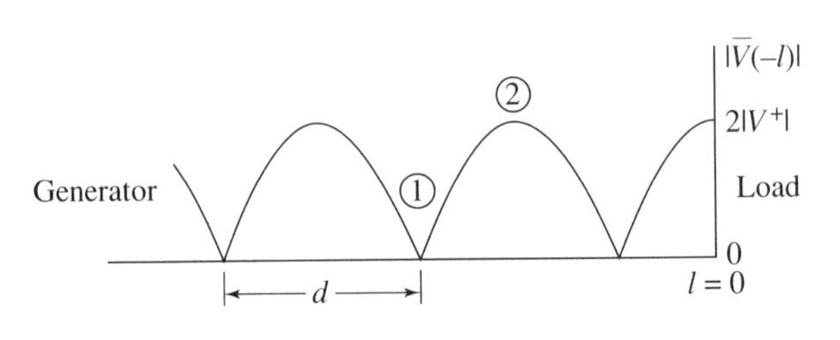

(b) $\rho_L = 1$ ($\bar{Z}_L = \infty$)

Figure 1-25. Standing wave patterns corresponding to limiting values of ρ_L on a lossless transmission line.

figures, observe that the patterns shown were only *one* of the components of the line voltage, namely, either $v^+(z, t)$ or $v^-(z, t)$.

If we take the two extreme values of ρ_L and plot the standing wave patterns, we can obtain the range of variation of the maximum and minimum values. First assume that $\bar{Z}_L = Z_0$. This would give $\rho_L = 0$ and the equation for the standing wave reduces to

$$|\bar{V}(-l)| = |V^+|$$

Since the cosine function is removed, there are no peaks and valleys in the standing wave pattern, and the line is said to be a matched, or "flat," line. The expression *terminated line* is also used occasionally. This is illustrated in Figure 1-25*a*. For the other extreme, we let $\bar{Z}_L = \infty$ (or 0) for which $\rho_L = 1$ and $\phi_L = 0°$ (or 180°). The standing wave equation becomes (after the application of a trig identity for $1 + \cos x$ in the lossless case

$$|\bar{V}(-l)| = |V^+|\left\{2 + 2\cos\left[2\beta l - \begin{pmatrix}0°\\180°\end{pmatrix}\right]\right\}^{1/2} = 2|V^+|\left|\cos\left[\beta l - \begin{pmatrix}0°\\90°\end{pmatrix}\right]\right|$$

The case of an open circuit load ($\phi_L = 0°$) is shown in Figure 1-25*b*.

To describe the standing wave pattern, we need some parameters that can be measured in the laboratory. Perhaps the most obvious of these would be the distances between the maximum and minimum values of the standing wave pattern. Referring to the general Eq. 1-75 for the standing wave, one can see that the peak values will occur where the cosine is +1 and the minimum values where the cosine is −1. Thus the peak values will occur when

$$2\beta l_{\max} - \theta_L = 2n\pi, \qquad n = 0, 1, \ldots$$

from which

$$l_{\max} = \frac{2n\pi + \theta_L}{2\beta} \tag{1-76}$$

The minimum values occur when

$$2\beta l_{\min} - \theta_L = (2n + 1)\pi, \qquad n = 0, 1, \ldots$$

from which

$$l_{\min} = \frac{(2n + 1)\pi + \theta_L}{2\beta} \tag{1-77}$$

The distance between successive minima (see Figure 1-25) is then obtained by using $n = 0$ and $n = 1$ in the expression for $l_{\min}$:

$$d = l_{\min_1} - l_{\min_0} = \frac{3\pi + \theta_L}{2\beta} - \frac{\pi + \theta_L}{2\beta} = \frac{\pi}{\beta} \tag{1-78}$$

If we use Eq. 1-71 to eliminate β, this becomes

$$d = \frac{\lambda_l}{2} \tag{1-79}$$

Thus by simply measuring the distance between adjacent minima (called *nulls* in engineering parlance), we can easily obtain the operating line wavelength.

A similar analysis on the distance between adjacent maxima yields the same result (see Exercise 1-7.2e). In practice, however, it is most common to use the null distance because of the sharper (more sensitive) indication compared with the maxima, as can be seen in Figure 1-25*b*, Points ① and ②.

Exercises

1-7.2e Prove that the distance between adjacent maxima on the standing wave pattern is one-half of the line wavelength. Now show that the distance between adjacent maxima and minima is one-quarter of the line wavelength.

1-7.2f For a lossy line (see Exercise 1-7.1j), plot the standing wave pattern as a function of βl for the case $\alpha = \beta$ and $\rho_L = 0$. Repeat for $\rho_L = 1$, $\theta_L = 0°$.

1-7.2g For the cable of Exercise 1-6.4e (assumed lossless now) and for $\overline{Z}_L = 100\ \Omega$, find the locations of the maximum and of the minimum voltage that are closest to the load. Assume $f = 500$ MHz. Also find λ_l and d.

1-7.2h Plot Eq. 1-75 for a lossless line having $Z_0 = 50\ \Omega$ and $C = 99$ pF/m. Use $|V^+| = 1$ V, and use the five cases $\Gamma_L = \pm 1, \pm 0.5, 0$ for $f = 1$ GHz.

1-7.2i Derive the equation for the magnitude of the total current on a lossless line, which will be similar to the voltage standing wave, Eq. 1-75, in terms of ρ_L and θ_L. Repeat Exercise

1-7.2h for this current standing wave pattern. Plot both $|\overline{V}(-l)|$ and $|\overline{I}(-l)|$ on the same axes and compare. Are the results physically reasonable?

1-7.2j An analysis of a transmission line gave the following result for $\overline{V}(z)$:

$$\overline{V}(z) = 6e^{-jz} - 2e^{jz}$$

Determine the following:

(a) The reflection coefficients $\Gamma(-l)$ and Γ_L
(b) The propagation constant γ
(c) The equation of the positive traveling wave in real time form
(d) The equation of the standing wave
(e) The distance between nulls on the standing wave pattern

The distance between nulls does not give us all the information we need to completely describe the standing wave pattern. It tells us nothing about the amplitudes of the maximum and minimum values indicated in Figure 1-23. We now define the *voltage standing wave ratio*, S, as the magnitude of the ratio of the maximum voltage to the adjacent minimum voltage. For a lossless line, of course, all the maximum values are the same and all the minimum values are the same, so the qualification adjacent is not necessary. As an equation, then,

$$S \equiv \left|\frac{\overline{V}_{\max}}{\overline{V}_{\min}}\right| = \frac{|\overline{V}_{\max}|}{|\overline{V}_{\min}|} \tag{1-80}$$

Sometimes the abbreviations VSWR and SWR are used for S.

From the definition and Figure 1-25, it is easy to determine the range of values of the voltage standing wave ratio. To obtain $\rho_L = 0$ requires $\overline{Z}_L = Z_0$ (see Eq. 1-60). This is the so-called matched line condition, for which Figure 1-25*a* applies. Note that for this case, there are no undulations in the standing wave pattern. For a lossless line, in fact, the maximum and minimum values are seen to be equal. Thus the minimum value of S is 1 for a matched, lossless line.

Figure 1-25*b* shows the other extreme, $\rho_L = 1$, where the load is an open. For this case, we see that for the lossless line the minimum voltage value is 0 so that $S = \infty$. This would also be the value for a shorted line. Summarizing:

$$1 \le S \le \infty \tag{1-81}$$

To develop a working expression for S in terms of system parameters we use the values of l, which give the maximum and minimum values of line voltage magnitude $|\overline{V}(-l)|$. As shown earlier, these values of l give cosine values of $+1$ and -1, so that using Eq. 1-75 for these values yields

$$S = \frac{|\overline{V}_{\max}|}{|\overline{V}_{\min}|} = \frac{|\overline{V}(-l_{\max})|}{|\overline{V}(-l_{\min})|} = \frac{V^+|[(1 + \rho_L^2) + 2\rho_L]^{1/2}}{V^+|[(1 + \rho_L^2) - 2\rho_L]^{1/2}}$$

Recognizing the terms in brackets as perfect squares, we finally have

$$S = \frac{1 + \rho_L}{1 - \rho_L} \tag{1-82}$$

We can now easily compute S since ρ_L is the magnitude portion (always positive) of Eq. 1-60.

If one chooses to measure the maximum and minimum voltages on the line and compute S, then ρ_L can be calculated by casting the preceding equation as

$$\rho_L = \frac{S - 1}{S + 1} \tag{1-83}$$

Exercises

1-7.2k Assuming a lossless line, determine the values of the voltage standing wave ratio S for the cases $\bar{Z}_L = 0, Z_0/2, Z_0, 2Z_0$, and ∞. Sketch the corresponding voltage standing wave patterns using βl as the independent variable for the plots, $0 \leq \beta l \leq 3\pi$.

1-7.2l Suppose that the standing wave ratio at the load on a transmission line is 4. To what load reflection coefficient magnitude does this correspond?

1-7.2m Measurements made on a transmission line are given in Figure 1-26. The voltage measurements are for adjacent maximum and minimum values. Determine the line wavelength, λ_l, β, S, and $|\Gamma_L|$. If the transmission line is a coaxial cable RG58B/U, determine the operating frequency and the phase velocity of the positive traveling component of the voltage. For this lossless line, what is the magnitude of the load impedance? (For cable properties see Exercise 1.6.1a.)

1-7.3 Current and Power Computations

The properties of the current on the transmission line are similar to those of voltage and the derivations of those properties use similar methods. As the starting point, we use the sinusoidal steady-state current Eqs. 1-26 and 1-28 to write

$$\bar{I}(z) = I^+ e^{-j\beta z} + I^- e^{j\beta z} = \frac{V^+}{Z_0} e^{-j\beta z} - \frac{V^-}{Z_0} e^{j\beta z}$$

A current reflection coefficient is defined as

$$\Gamma^i(z) = \frac{I^-(z)}{I^+(z)} = \frac{-\dfrac{V^-}{Z_0} e^{j\beta z}}{\dfrac{V^+}{Z_0} e^{-j\beta z}} = -\frac{V^-}{V^+} e^{j2\beta z}$$

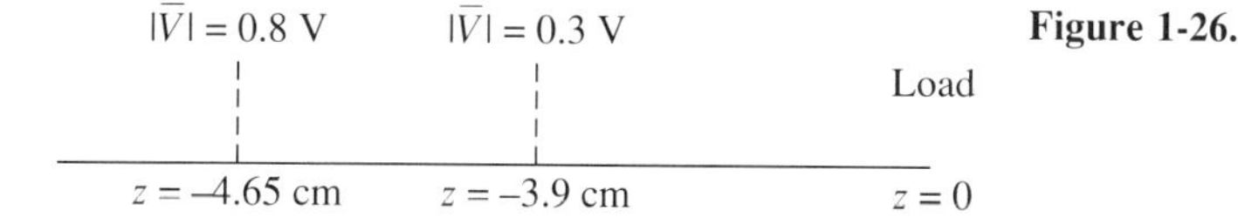

Figure 1-26.

Comparing this with Eq. 1-58 for the voltage reflection coefficient, we obtain

$$\Gamma^i(z) = -\Gamma(z) \tag{1-84}$$

Thus all the facts developed for the voltage reflection coefficient apply with a simple sign change. If we again take the reference for $z = 0$ at the load and use $z = -l$ to move to the left, we have, from Eq. 1-63,

$$\Gamma^i(-l) = -\Gamma(-l) = -\Gamma_L e^{-j2\beta l} \tag{1-85}$$

where Γ_L and β are identical to those defined earlier.

Exercises

1-7.3a For a lossless line suppose $\bar{Z}_L$ is purely resistive. Plot Γ_L^i versus R_L/Z_0 and mark regions corresponding to $R_L < Z_0$ and $R_L > Z_0$. Compare the results with Exercise 1-7.1b.

1-7.3b A transmission line has a characteristic impedance of 75 Ω (lossless). The load is variable and is purely reactive and covers the range $-j300\ \Omega$ to $+j300\ \Omega$. Plot the magnitude and angle of Γ_L^i as functions of load reactance and compare the results with Exercise 1-7.1e.

It is also necessary to be able to compute the current at any point on a transmission line since many solid-state devices can be modeled as current sources and are often used as line drivers. The results we derive here are in terms of the voltage reflection coefficient since that parameter is most often used in practice. Using the preceding current equation, we proceed as follows:

$$\bar{I}(z = -l) = \frac{V^+}{Z_0} e^{j\beta l} - \frac{V^-}{Z_0} e^{-j\beta l} = \frac{V^+}{Z_0} e^{j\beta l}\left(1 - \frac{V^-}{V^+} e^{-j2\beta l}\right)$$

Thus

$$\bar{I}(-l) = \frac{V^+}{Z_0} e^{j\beta l}\,(1 - \Gamma_L e^{-j2\beta l}) = \frac{V^+}{Z_0} e^{j\beta l}\,[1 - \Gamma(-l)] \tag{1-86}$$

At the load, this reduces to

$$\bar{I}(0) = \frac{V^+}{Z_0}(1 - \Gamma_L) \tag{1-87}$$

If we next substitute the polar form for Γ_L, the current at location $z = -l$ becomes

$$\begin{aligned}\bar{I}(-l) &= \frac{V^+}{Z_0} e^{j\beta l}(1 - \rho_L e^{j\phi_L} e^{-j2\beta l}) \\ &= \frac{V^+}{Z_0} e^{j\beta l}(1 + \rho_L e^{j180^\circ} e^{+j\phi_L} e^{-j2\beta l}) \\ \bar{I}(-l) &= \frac{V^+}{Z_0} e^{j\beta l}(1 + \rho_L e^{-j(2\beta l - \phi_L + 180^\circ)})\end{aligned} \tag{1-88}$$

Next we take the magnitude of this current to obtain the current standing wave pattern. Following the same steps used to obtain the voltage standing wave pattern, Eq. 1-68, we have (assuming Z_0 real for this case)

$$|\bar{I}(-l)| = \frac{|V^+|}{Z_0}\,[(1 + \rho_L^2) + 2\rho_L \cos(2\beta l - \theta_L + 180°)]^{1/2}$$

Using the trigonometric result $\cos(A + \pi) = -\cos A$, this reduces to

$$|\bar{I}(-l)| = \frac{|V^+|}{Z_0}\,[(1 + \rho_L^2) - 2\rho_L \cos(2\beta l - \theta_L)]^{1/2} \tag{1-89}$$

Comparing this result with Eq. 1-68 for the voltage standing wave pattern, it is seen that the only difference is a minus sign in front of the cosine term. Physically, this means that at a voltage maximum (where the cosine term is $+1$), the current will be a minimum. This means that the current minima are displayed $\lambda/4$ from the voltage minima.

Example 1-4

For a lossless transmission line with an open circuit load, plot the current and voltage standing wave patterns. The generator has an output voltage of 20 V, the line characteristic impedance is 50 Ω, and the line is 3 wavelengths long.

$$\Gamma_L = \frac{\bar{Z}_L - Z_0}{\bar{Z}_L + Z_0} = 1\angle 0° \qquad \text{for} \quad \bar{Z}_L = \infty \cdot \rho_L = 1, \qquad \theta_L = 0°$$

At the generator,

$$\beta l = \frac{2\pi}{\lambda} \times 3\lambda = 6\pi \text{ rad}$$

Equation 1-75 for $\alpha = 0$ then gives a generator voltage of

$$20 = |V^+|[(1 + 1) + 2\cos(12\pi)]^{1/2} = 2|V^+|$$

from which we obtain

$$|V^+| = 10 \text{ V}$$

At a general location, Eq. 1-75 yields (also for $\alpha = 0$)

$$|\bar{V}(-l)| = 10\sqrt{2 + 2\cos 2\beta l} = 20|\cos(\beta l)| \text{ V}$$

Similarly, Eq. 1-89 gives the current standing wave pattern

$$|\bar{I}(-l)| = \frac{20}{50}|\sin(2\beta l)| = 0.4|\sin(2\beta l)| \text{ A}$$

These results are plotted in Figure 1-27. It is interesting to note that the generator delivers zero current, so it is looking into an open circuit. Describe what would happen if the line length were decreased $\lambda/4$. Also, the current is zero at the ''load'' of $\bar{Z}_L = \infty$.

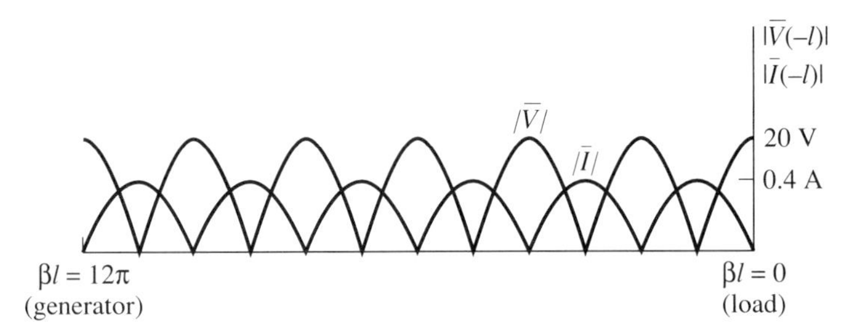

Figure 1-27. Current and voltage standing wave patterns on an open-circuited ($\bar{Z}_L = \infty$) line.

Exercises

1-7.3c A lossless transmission line has a characteristic impedance of 50 Ω. A generator that has an output of 5 V (sinusoidal) is applied to a resistively loaded section of this line, and the line has a length of $2\lambda_l$. For two different loads $\bar{Z}_L = 25$ Ω and $\bar{Z}_L = 75$ Ω, plot the current and voltage standing wave patterns and determine the impedance values (for both loads) at the location of voltage maxima and voltage minima.

1-7.3d A lossless transmission line has a characteristic impedance of 50 Ω. A generator that has an output of 5 V (sinusoidal) is applied to a section of the line $2\lambda_l$ in length, and the line is terminated with loads $\bar{Z}_L = -j25$, $j25$, $j75$, and $-j75$ Ω, successively. Plot the current and voltage standing wave patterns for each case. Write a short discussion of the positions of the nulls by comparing the voltage standing wave patterns with each other, then the current standing wave patterns with each other, and then the voltage patterns with the current patterns. Does it matter whether or not the reactance magnitude is greater or less than Z_0?

The final quantity we investigate in this section is that of power on the transmission lines. It will be found convenient to describe three components of power; namely, an incident power, P_i, which is the power approaching a given point on the line, a reflected power P_r, which represents the part of the incident power reflected back along the line, and a transmitted or output power, P_t, which is that power which actually manifests itself as a heating of the resistive component of a load. One might view the transmitted power as the useful component of power, since it would be the power leaving an antenna at the end of a line, for example.

Since the transmitted power is important, we compute it as the main goal. As one would suspect on the basis of prior work in circuit theory, the result would be obtained on physical principles as

$$P_t = P_i - P_r$$

This, in fact, will be the ultimate result, but what we need are expressions for each of these in terms of line voltages, parameters, and loads that can be measured. The process is rather lengthy and involves several theorems from complex number theory. Some of the steps are given next, and one should justify each step in the derivation and make special note of any restrictions imposed on the final results.

From circuit theory, we have the result

$$P_t = \tfrac{1}{2}\text{Re}\{\overline{V}(-l)\overline{I}^*(-l)\}$$

where * represents the complex conjugate and the $\frac{1}{2}$ factor is used since it is common practice in field theory to deal with peak values rather than RMS as in circuit theory. Next use the lossless line Eqs. 1-66 and 1-86 in the preceding expression for P_t, and then use the rules of complex algebra:

$$\begin{aligned} P_t &= \frac{1}{2}\,\text{Re}\left\{V^+ e^{j\beta l}(1 + \Gamma_L e^{-j2\beta l})\left[\frac{V^+}{Z_0}\, e^{j\beta l}(1 - \Gamma_L e^{-j2\beta l})\right]^*\right\} \\ &= \frac{1}{2}\,\text{Re}\left\{V^+ e^{j\beta l}(1 + \Gamma_L e^{-j2\beta l})\,\frac{(V^+)^*}{Z_0^*}\, e^{-j\beta l}(1 - \Gamma_L^* e^{j2\beta l})\right\} \\ &= \frac{1}{2}\,\text{Re}\left\{\frac{|V^+|^2}{Z_0^*}\,(1 + j2\,\text{Im}\{\Gamma_L e^{-j2\beta l}\} - |\Gamma_L|^2)\right\} \end{aligned}$$

where Im is the imaginary part of the quantity within the braces without the j; that is, $\text{Im}\{a + jb\} = b$.

Since Z_0 is real for a lossless line,

$$P_t = \frac{|V^+|^2}{2Z_0}\,[1 + \text{Re}\{j2\,\text{Im}\{\Gamma_L e^{-j2\beta l}\}\} - \text{Re}\{|\Gamma_L|^2\}]$$

$$P_t = \frac{|V^+|^2}{2Z_0}\,(1 - |\Gamma_L|^2) \qquad \text{lossless line} \tag{1-90}$$

If we write this last result as

$$P_t = \frac{|V^+|^2}{2Z_0} - \frac{|V^+|^2}{2Z_0}\,|\Gamma_L|^2$$

we can identify the first term as the *incident* power and the second term as the *reflected* power, so that

$$\left.\begin{aligned} P_i &= \frac{|V^+|^2}{2Z_0} \\ P_r &= \frac{|V^+|^2}{2Z_0}\,|\Gamma_L|^2 = P_i|\Gamma_L|^2 \end{aligned}\right\} \tag{1-91}$$

From the preceding, we can conclude that the power reflection coefficient is the square of the voltage reflection coefficient. Equation 1-90 can then be written in an alternate form

$$P_t = P_i(1 - |\Gamma_L|^2) \tag{1-92}$$

Example 1-5

In Figure 1-28 is shown a transmission line system which has a resistive termination. The voltage across the resistance is measured as 18 V (peak). For a lossless line, determine the following: load current; transmitted, incident, and reflected powers at the load; incident and reflected current and voltage values; transmitted, incident, and reflected powers at the generator; generator current and voltage; impedance seen by generator.

As in circuit theory, we can select one current or voltage as reference. Here let's select the load voltage as reference and set $\overline{V}_L = 18\angle 0°$ V. Then from Ohm's law,

$$\overline{I}_L = \frac{\overline{V}_L}{\overline{Z}_L} = \frac{18\angle 0°}{200} = 90\angle 0° \text{ mA}$$

By definition, the transmitted power is the actual power dissipation in the load. Thus

$$P_t = P_L = \tfrac{1}{2}|\overline{I}_L|^2 R_L = \tfrac{1}{2} \times 8.1 \times 10^{-3} \times 200 = 0.81 \text{ W (at load)}$$

At the load, $l = 0$, $\Gamma(0) = \Gamma_L = (200 - 300)/(200 + 300) = -0.2$. Using Eq. 1-92,

$$0.81 = P_i(1 - 0.2^2) \qquad \text{or} \quad P_i = 0.8438 \text{ W (at load)}$$

and then

$$P_r = P_i - P_t = 0.8438 - 0.81 = 0.0338 \text{ W (at load)}$$

At the load, $\overline{V}_L = V^+ + V^- = V^+(1 + \Gamma_L)$. Then

$$V^+ = \frac{18\angle 0°}{1 - 0.2} = 22.5\angle 0° \text{ V}$$

$$V^- = \Gamma_L V^+ = (-0.2)(22.5\angle 0°) = 4.5\angle 180° \text{ V}$$

$$I^+ = \frac{V^+}{Z_0} = \frac{22.5\angle 0°}{300} = 75\angle 0° \text{ mA}$$

$$I^- = -\frac{V^-}{Z_0} = -\frac{4.5\angle 180°}{300} = 15\angle 0° \text{ mA}$$

(Incidentally, $\Gamma_L^i = 15\angle 0°/75\angle 0° = 0.2 = -\Gamma_L$, as predicted by Eq. 1-84.

Now the actual power getting to the load must all come from the generator, since the line is lossless. Thus

$$P_t = 0.81 \text{ W (at generator)}$$

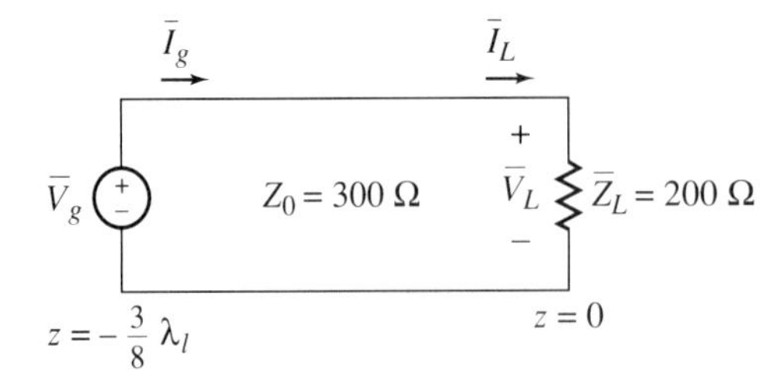

Figure 1-28. Transmission system for power computations.

Using Eq. 1-63 for $z = -\frac{3}{8}\lambda_l$ gives

$$\Gamma_g = \Gamma(-\tfrac{3}{8}\lambda_l) = \Gamma_L e^{-j2\times\frac{2\pi}{\lambda_l}\times\frac{3}{8}\lambda_l} = -0.2e^{-j\frac{3\pi}{2}}$$
$$= 0.2e^{j180^\circ}e^{-j270^\circ} = 0.2\angle -90^\circ$$

$P_i = 0.81/(1 - 0.2^2) = 0.8438$ W (at generator) which is the same as at the load, as expected.

$$P_r = 0.8438 - 0.81 = 0.0338 \text{ W (at generator)}$$

From Eq. 1-67, with $\alpha = 0$, we obtain

$$\overline{V}_g = \overline{V}(-\tfrac{3}{8}\lambda_l) = 22.5e^{+j135^\circ}(1 + 0.2e^{-j(270^\circ-180^\circ)})$$
$$= 22.5e^{j135^\circ}(1 - j0.2) = 22.94\angle 123.7^\circ \text{ V (generator)}$$

Using Eq. 1-88, the generator current is

$$\overline{I}_g = \overline{I}\left(-\frac{3}{8}\lambda_l\right) = \frac{22.5e^{j135^\circ}}{300}(1 + 0.2\angle -270^\circ) = 76.48\angle 146.3^\circ \text{ mA}$$

The impedance seen by the generator is then

$$\overline{Z}\left(-\frac{3}{8}\lambda_l\right) = \frac{22.94\angle 123.7^\circ}{76.48\angle 146.3^\circ}\text{ k}\Omega = 300\angle -22.6^\circ\ \Omega$$

which is quite different from the load of 200 Ω. This will be discussed in the next section.

Two other important definitions of terms are of practical use and will be used later in the text. The first of these is called the *return loss* and is defined as

$$\text{Return loss} \equiv \frac{P_r}{P_i} = |\Gamma(-l)|^2 = \rho^2(-l) \tag{1-93}$$

In practice, this is usually expressed in dB as

$$\text{Return loss, dB} = -10\log_{10}\rho^2(-l) = -20\log_{10}\rho(-l) \tag{1-94}$$

The minus sign is introduced for the convenience of having the return loss be a positive number, since the magnitude of the reflection coefficient is always less than 1. This parameter is a measure of the power returned to the generator. The larger the dB value, the smaller the power being returned to the generator. An ideal value is ∞.

The second quantity is called the *mismatch* or *reflected loss* and is defined as

$$\text{Reflected loss} \equiv \frac{P_t}{P_i} = 1 - |\Gamma(-l)|^2 = 1 - \rho^2(-l) \tag{1-95}$$

This parameter is also commonly expressed in dB as

$$\text{Reflected loss, dB} = -10\log_{10}[1 - \rho^2(-l)] \tag{1-96}$$

Physically, this represents the reduction in transmitted power due to reflection at a mismatch point on a line. The smaller the dB value, the more nearly all incident power is transmitted to the load or down the line. An ideal value is 0.

These two parameters are very convenient because they tell us in a direct way the amounts of power that are being transferred forward and reflected on a line that is not terminated in its characteristic impedance. Even for a mismatched load that results in a standing wave ratio as high as 3, only about 1.5 dB of mismatch loss results.

Exercises

1-7.3e A transmission line with $Z_0 = 50\ \Omega$ is terminated in an impedance $25 + j25\ \Omega$. Find the reflection coefficient, standing wave ratio, and fraction of the incident power delivered to the load. Assume a lossless line and determine the quantities at the load only. Also compute the return loss and mismatch loss in dB.

1-7.3f A transmission line having a characteristic impedance of 75 Ω is terminated in a 50-Ω resistive load. For a measured output power of 25 W, determine the incident power, the reflected power, and the transmitted power.

1-7.3g For the transmission line system shown in Figure 1-29, the incident power is known to be 5 W. Determine the values of the load voltage standing wave ratio, the transmitted power, the reflected power, and the values of the return loss and reflected loss in dB. What is the ratio of the transmitted power to the incident power?

1-7.3h For the system of Exercise 1-7.3e, what is the required incident power to obtain an output power of 100 W?

1-7.3i An RG58C/U cable has an attenuation of 0.656 dB/m at 1 GHz. An 8-m length of this cable is used to connect a generator to an antenna on a test range. The cable has $Z_0 = 50$ Ω and a capacitance of 101 pF/m. The antenna has an input impedance of 75 Ω resistance at the operating frequency. If the generator is set for 5-mW output, how much power is transmitted to the antenna and how much power is dissipated in the cable? Use the low-loss approximation for Z_0.

1-7.3j Complete the details of the steps that yield Eq. 1-90. Begin with the first step for P_t given in the text.

1-7.3k Using the general solutions that include losses, namely, Eqs. 1-25, 1-26, 1-29, and 1-37, show that (for Z_0 complex and given as $R_0 + jX_0$)

$$P_t = \frac{|V^+|^2 e^{2\alpha l}}{2|Z_0|^2}\left[R_0 - R_0|\Gamma_L|^2 e^{-4\alpha l} - 2X_0 \operatorname{Im}\{\Gamma_L e^{-2\gamma l}\}\right]$$

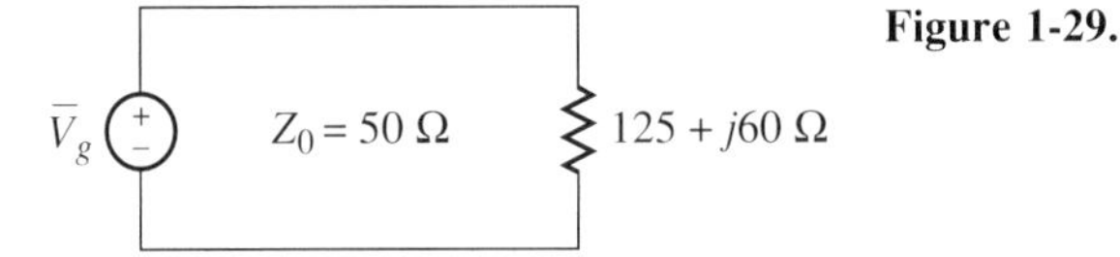

Figure 1-29.

Next show that for low-loss lines that have $Z_0 \approx R_0$

$$P_t = \frac{|V^+|^2}{2Z_0} e^{2\alpha l}[1 - |\Gamma(-l)|^2]$$

Finally, show that for a lossless line this reduced to Eq. 1-90.

1-7.4 Impedance Along Finite Length Lines and Some Special Cases

In this section, we will examine some of the impedance properties of loaded transmission lines. Specifically, we derive the impedance at the terminals of a line looking toward a load. The convention adopted for this impedance property is shown in Figure 1-30. This is the impedance one would measure with a bridge or impedance meter. At $z = -l$ the impedance is defined in the same way as for lumped-element one-port systems. Using Eqs. 1-64 and 1-86 for the lossless case $\gamma = j\beta$, we write

$$\overline{Z}(z = -l) = \frac{\overline{V}(z = -l)}{\overline{I}(z = -l)} = \frac{V^+ e^{j\beta l}[1 + \Gamma(-l)]}{\dfrac{V^+}{Z_0} e^{j\beta l}[1 - \Gamma(-l)]}$$

$$\overline{Z}(-l) = Z_0 \frac{1 + \Gamma(-l)}{1 - \Gamma(-l)} \qquad (1\text{-}97)$$

From this result we see that all information about the impedance at any point can be obtained from the value (complex) of the reflection coefficient, and vice versa.

If we wish to see the effect of the load impedance more explicitly we use Eqs. 1-61 and 1-62 in Eq. 1-97 to eliminate $\Gamma(-l)$ and obtain

$$\overline{Z}(-l) = Z_0 \frac{1 + \dfrac{\overline{Z}_L - Z_0}{\overline{Z}_L + Z_0} e^{-j2\beta l}}{1 - \dfrac{\overline{Z}_L - Z_0}{\overline{Z}_L + Z_0} e^{-j2\beta l}}$$

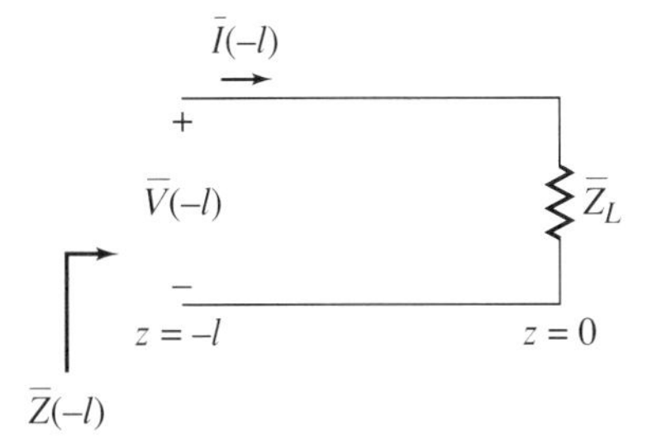

Figure 1-30. Convention for describing the impedance property of a transmission line.

After some algebra and an application of the identity

$$j \tan \beta l = \frac{1 - e^{-j2\beta l}}{1 + e^{-j2\beta l}}$$

we have the form

$$\bar{Z}(-l) = Z_0 \frac{\bar{Z}_l + jZ_0 \tan \beta l}{j\bar{Z}_L \tan \beta l + Z_0} \tag{1-98}$$

Example 1-6

In the transmission system shown in Figure 1-31, the applied voltage is 25 V peak. Find the input power, input reactive power, and input apparent power. Also calculate the power transmitted to the load, the incident power, and the reflected power. Compare these values and comment on them. Assume a lossless RG58/U cable with $Z_0 = 50\ \Omega$ and $C = 93$ pF/m. The operating frequency is 1 GHz.

To find the input power quantities, we must first obtain the input impedance seen by the generator:

$$\beta = \omega C Z_0 = 2\pi \times 1 \times 10^9 \times 93 \times 10^{-12} \times 50 = 29.217 \text{ rad/m}$$

Thus

$$\beta l = (29.217)(2) = 58.434 \text{ rad}$$

Using Eq. 1-98:

$$\bar{Z}(-2) = 50 \frac{100 + j50 \tan 58.434}{j100 \tan 58.434 + 50} = 50 \frac{100 - j3.7}{50 - j7.4} = 29.4\angle 23.9^\circ\ \Omega$$

$$\bar{I}_g = \frac{\bar{V}_g}{\bar{Z}(-2)} = \frac{25\angle 0^\circ}{29.4\angle 23.9^\circ} = 0.85\angle -23.9^\circ \text{ A}$$

Next, using the techniques of circuit theory (peak values again)

$$\bar{S} = \text{apparent power} = \tfrac{1}{2}\bar{V}_g \bar{I}_g^* = \tfrac{1}{2} \times 25\angle 0^\circ \times 0.85\angle 23.9^\circ = 10.6\angle 23.9^\circ \text{ VA}$$

$$P = \text{input power} = \text{Re}\{\bar{S}\} = 9.7 \text{ W} = P_t$$

$$Q = \text{input reactive power} = \text{Im}\{\bar{S}\} = 4.3 \text{ VAR}$$

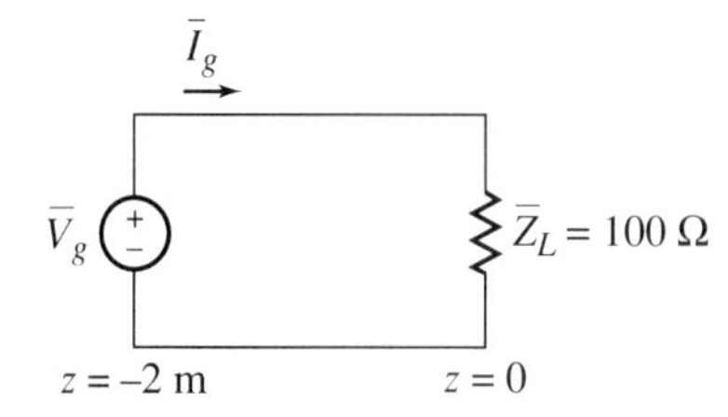

Figure 1-31. Configuration for power and impedance calculations.

The real power must get to the load, since it is a lossless line. At the load, we then have

$$P_t^L = 9.7 \text{ W}$$

$$P_i^L = \frac{P_t^L}{1 - \left|\frac{100 - 50}{100 + 50}\right|^2} = \frac{9.7}{1 - \frac{1}{9}} = 10.9 \text{ W}$$

$$P_r^L = 10.9 - 9.7 = 1.2 \text{ W}$$

These results may seem a little inconsistent overall, and one is tempted to ask how the power incident to the load can be more than the power put out by the generator. We must remember that the physical quantity of importance is the actual heating power in the load, which is 9.7 W, the same as the generator output for a lossless line. The incident power is simply an identification of the power generated by the positive traveling component of line voltage. We could also have computed the transmitted power using the line voltage equation and reflection coefficient as follows:

$$\overline{V}(-2) = \overline{V}_g = 25\angle 0° = V^+ e^{j\beta l} + V^- e^{-j\beta l}$$

$$\Gamma_L = \frac{100 - 50}{100 + 50} = \frac{1}{3} = \frac{V^-}{V^+}$$

$$\overline{V}(0) = \overline{V}_L = V^+ + V^- = V^+ + \tfrac{1}{3}V^+ = \tfrac{4}{3}V^+ \rightarrow V^+ = \tfrac{3}{4}\overline{V}_L$$

Thus

$$\begin{aligned} 25\angle 0° &= \tfrac{3}{4}\overline{V}_L e^{j58.434} + \tfrac{1}{4}\overline{V}_L e^{-j58.434} \\ &= [0.75(\cos 58.434 + j \sin 58.434) + 0.25(\cos 58.434 \\ &\quad - j \sin 58.434)]\overline{V}_L \qquad \text{(remember angles are in radians at this point)} \\ &= 0.567\angle 123°\overline{V}_L \end{aligned}$$

$$\overline{V}_L = \frac{25\angle 0°}{0.567\angle 123°} = 44.1\angle - 123° \text{ V}$$

$$P_t^L = \frac{(44.1)^2}{2 \times 100} = 9.7 \text{ W, as before}$$

Exercises

1-7.4a For the RG58B/U coaxial cable, plot $\overline{Z}(-l)$ for $\overline{Z}_L = 1000\ \Omega$, assuming a lossless cable. Let $f = 1$ GHz and $0 \le l \le 0.3$ m. $Z_0 = 53.5\ \Omega$, $C = 93.5$ pF/m.

1-7.4b Carry out the preceding example for the case where the load resistance is 50 Ω. Comment on your solution as compared with the results of the example.

1-7.4c Assuming a lossless line, plot $|\overline{Z}(-l)|$ for the two cases $\overline{Z}_L = 2Z_0$ and $\overline{Z}_L = \frac{1}{2}Z_0$ for $0 \le \beta l \le 3\pi$. Comment on the values at the load compared with the plot variations. What important feature do you see on the plots for values at locations separated at locations separated a distance $\beta l = \pi$?

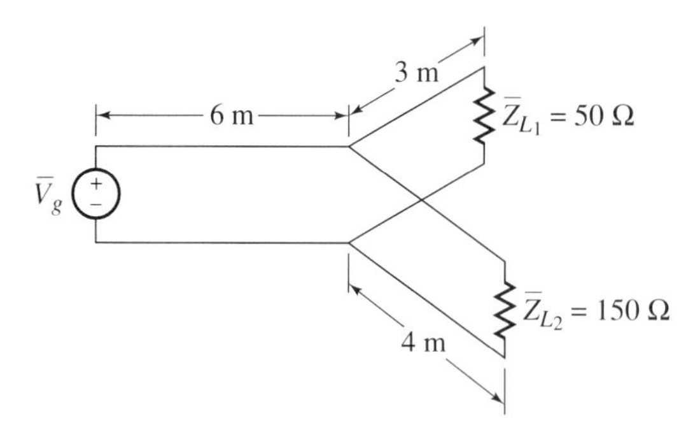

Figure 1-32.

1-7.4d An RG179B/U cable has a characteristic impedance of 75 Ω and a capacitance of 64 pF/m. The input impedance of a loaded 2-m length of cable at 800 MHz is $65 - j80\ \Omega$. Assuming a lossless cable, determine the load impedance.

1-7.4e A 5-m section of RG58C/U cable, which has $Z_0 = 50\ \Omega$ and a capacitance of 101 pF/m, is loaded with a resistance of 100 Ω. Compute the current in the load resistor and the current at the input of the line when a 250 MHz voltage source of amplitude 200 V is placed at the input. The source has an internal resistance of 50 Ω. Assume a lossless line.

1-7.4f Since both $\bar{Z}(-l)$ and $\Gamma(-l)$ are in general complex, we could write the equation for $\bar{Z}(-l)$ in normalized form as

$$\frac{\bar{Z}(-l)}{Z_0} = \tilde{Z}(-l) = \frac{1 + \Gamma(-l)}{1 + \Gamma(-l)}$$

and then set $\tilde{Z}(-l) = \tilde{R}(-l) + j\tilde{X}(-l) \equiv \tilde{R} + j\tilde{X}$ and $\Gamma(-l) = U(-l) + jV(-l) \equiv U + jV$. Make these substitutions and obtain $\tilde{R}$ and $\tilde{X}$ as functions of U and V. (These results form the basis for developing a very important tool for transmission line analysis called the *Smith chart.*)

1-7.4g Prove that if the electrical length of a lossless line is increased by $2n\pi$, n integer (i.e., $\beta l \rightarrow \beta l + 2n\pi$), the value of $\bar{Z}(-l)$ is unchanged. Suppose a transmission line that has a characteristic impedance Z_0 is terminated in a resistance of $2Z_0$. What is the impedance at the line input a distance of 5 wavelengths to the left of the load? (Hint: One wavelength corresponds to how many radians?)

1-7.4h Prove that if the electrical length of a lossless line is increased by $(2n + 1)(\pi/2)$, where n is an integer [i.e., $\beta l \rightarrow \beta l + (2n + 1)(\pi/2)$], the impedance at the input is the reciprocal of the value at the input of the original line times Z_0^2. This is called a *quarter-wave transformer*.

1-7.4i A generator is used to feed two different loads that are physically at different locations. The basic configuration, shown in Figure 1-32, uses RG34A/U cable that has a characteristic impedance of 75 Ω and a capacitance of 67.6 pF/m. The generator is operating at 500 MHz and at a voltage level of 10 V peak. What is the impedance seen by the generator, the generator output power, the magnitude of voltage on each load, and the power received by each load? (Neglect losses.) (Hint: The two load sections present effectively two impedances that are in parallel at the junction and whose parallel combination is the load on the right end of the 6-m section.)

There are special cases of interest and practical use which will now be described. We examine the lossless line results.

Case 1. $\beta l = n\pi$ or $l = \dfrac{n\pi}{\beta} = \dfrac{n\pi}{(2\pi/\lambda_l)} = n\dfrac{\lambda_l}{2}, n = 1, 2, \ldots.$

This means that the line is an integral number of half wavelengths long. Using this βl in the impedance equation gives

$$\overline{Z}\left(-\frac{n\lambda_l}{2}\right) = \frac{\overline{Z}_L + jZ_0 \tan(n\pi)}{j\overline{Z}_L \tan(n\pi) + Z_0} = \overline{Z}_L \tag{1-99}$$

The line thus looks like a 1:1 ideal transformer.

Exercise

1-7.4j On the plot of the solution to Exercise 1-7.4a, determine and mark the distance between points of the same impedance value (as a function of λ_l). (Answer: $\frac{1}{2}\lambda_l$)

Case 2. $\beta l = (2n + 1)\dfrac{\pi}{2}$ or $l = \dfrac{(2n + 1)}{\beta}\dfrac{\pi}{2} = (2n + 1)\dfrac{\lambda_l}{4}, n = 1, 2, \ldots.$

Thus the line is an odd multiple of quarter wavelengths long.

This value makes $\tan \beta l = \pm\infty$ so that in the limit the impedance becomes

$$\overline{Z}\left[-(2n + 1)\frac{\lambda_l}{4}\right] = \frac{Z_0^2}{\overline{Z}_L} \tag{1-100}$$

This reciprocates the load impedance with respect to Z_0^2, a quarter wave transformer.

Example 1-7

Find the input impedance of the RG58B/U transmission line (assumed lossless) shown in Figure 1-33) at a frequency for which the line length is $3\lambda_l/4$. Since the transmission line is an odd multiple of a quarter wavelength long, we use Eq. 1-99 directly to obtain

$$\overline{Z}_{\text{in}} = \overline{Z}\left(3\frac{\lambda_l}{4}\right) = \frac{(53.5)^2}{100} = 28.6\ \Omega$$

It is of interest to note:

1. Resistances transform to resistances.
2. This is *frequency sensitive* since if the frequency were changed, the transmission line length would be other than $3\lambda_l/4$.

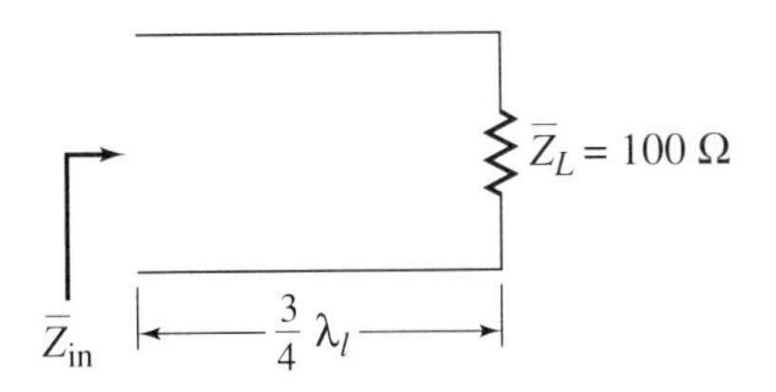

Figure 1-33. A quarter-wave line transformer.

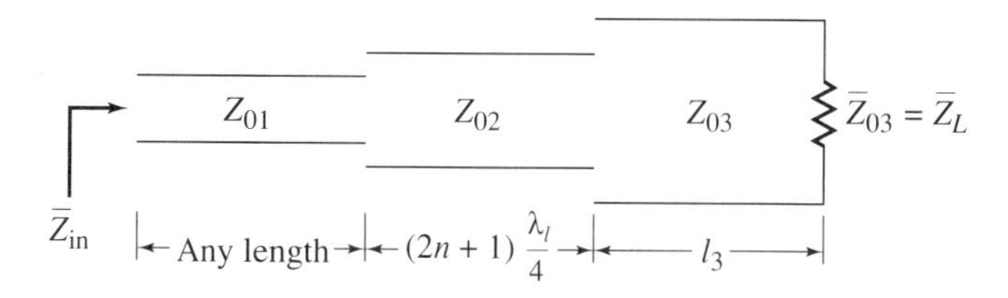

Figure 1-34.

Exercises

1-7.4k A transmission line system has previously been designed and installed. The system has a characteristic impedance Z_{03} (assumed real) and is terminated in that same value of impedance (see Figure 1-34). An engineer finds it necessary to connect into this system with a different cable type having a characteristic impedance Z_{01}, also assumed real. He would like to have his cable see a load impedance Z_{01}. What is the required characteristic impedance Z_{02} (in terms of Z_{01} and Z_{03}) of a quarter-wave transmission line section he could design to connect the two systems together? (Answer: $Z_{02} = \sqrt{Z_{01}Z_{03}}$)

1-7.4l To demonstrate the frequency sensitivity of the quarter-wave matching section described in Exercise 1-7.4k, we consider in this exercise a particular case. Suppose that $Z_{03} = 75$ Ω and that we want to connect an RG58B/U as Z_{01}. What are the required lengths (for $n = 0, 1, 2$) and Z_{02} to provide this match at $f = 1$ GHz? Assume all cables are lossless and $C = 100$ pF/m for Z_{02}. Now plot the input impedance at Z_{02} as a function of frequency for the cases $n = 0$, 1, and 2. (Computer use is suggested; remember that, in general, $\overline{Z}(z = -l)$ may be complex.) Next suppose that a 10% tolerance in impedance magnitude is acceptable as the load on Z_{01}. What are the frequency bandwidths for the three cases?

Case 3. Load Is a Short Circuit, $\overline{Z}_L = 0$

The general equation becomes

$$\overline{Z}(-l) = (j \tan \beta l)Z_0 \tag{1-101}$$

The impedance is *reactive* regardless of the length l.

A graphical description of this case is shown in Figure 1-35. This result shows us that we can obtain a capacitive reactance or an inductive reactance of any value. Again, one must remember that both the value of reactance *and* the corresponding equivalent circuit element value, L_{eq} or, C_{eq}, vary with frequency since

$$\beta l = \frac{2\pi l}{\lambda_l}$$

Thus if one designs a line at a given frequency for the desired reactive value, a frequency shift will alter not only the reactance value but the element value. This can be demonstrated by supposing that a line has been designed for a positive (inductive) reactance. The equivalent inductor value would be determined as follows:

$$jZ_0 \tan \beta l = j\omega L_{eq}$$
$$L_{eq} = \frac{Z_0 \tan \beta l}{\omega} = \frac{Z_0 \tan(\omega C Z_0)}{\omega} \tag{1-102}$$

Case 4. Load Is an Open Circuit, $\overline{Z}_L = \infty$

For this case, the impedance expression reduces to

$$\overline{Z}(-l) = (-j \cot \beta l)Z_0 \tag{1-103}$$

As with Case 3, we can use an open-circuited line to obtain any value of reactance by proper adjustment of the length. The discussion of the effects of frequency variation given in Case 3 also apply here. Figure 1-36 is a plot of the reactance characteristic for the open-circuited section.

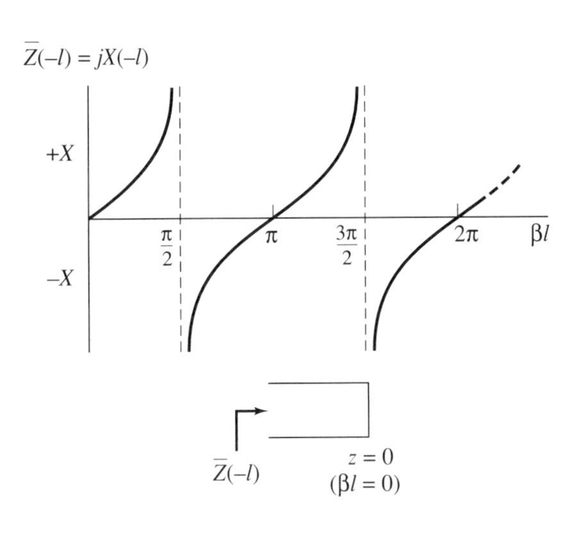

Figure 1-35. Impedance characteristic of a shorted transmission line.

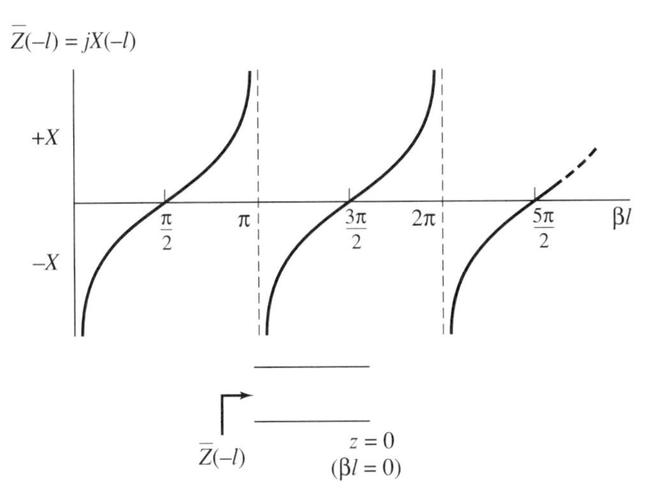

Figure 1-36. Impedance characteristic of an open-circuit transmission line.

One may wonder what practical use can be made of open or shorted lines. We will see in the next section that these elements can be used to match systems to obtain maximum power transfer conditions at specified frequency values.

Example 1-8

For the system shown in Figure 1-37, determine the equivalent series and parallel circuits at the cable input. The operating frequency is 650 MHz, and the cable is RG224/U (assumed lossless here). The cable properties of RG224/U are $Z_0 = 50\ \Omega$ and $C = 101$ pF/m.

$$\beta = \omega\sqrt{LC} = \omega C Z_0 = 2\pi \times 6.5 \times 10^8 \times 101 \times 10^{-12} \times 50$$
$$= 20.624 \text{ rad/m}$$
$$\beta l = (20.624)(0.5) = 10.312 \text{ rad}$$

Using the impedance Eq. 1-98, we compute

$$\overline{Z}_{\text{In}} = \overline{Z}(-0.5) = 50\,\frac{100 + j\tan(10.312)}{j100\tan(10.312) + 50}$$
$$= 44.3\angle -36.3^\circ = 35.7 - j26.2\ \Omega$$

For elements in series, this gives

$$R_s = 35.7\ \Omega$$

$$jX_s = -j26.2 \qquad \text{or} \qquad C_s = \frac{1}{\omega \times 26.2} = 9.34 \text{ pF}$$

To obtain the parallel circuit, we first compute $\overline{Y}_{\text{in}}$

$$\overline{Y}_{\text{in}} = \frac{1}{\overline{Z}_{\text{in}}} = \frac{1}{44.3\angle -36.3^\circ} = .02257\angle +36.3^\circ \text{ S}$$
$$= 0.0182 + j0.0134 \text{ S}$$

For elements in parallel, this gives

$$G_p = 0.0182 \text{ S} \qquad \text{or} \qquad 54.9\ \Omega$$

$$jB_p = j0.0134 \qquad \text{or} \qquad C_p = \frac{0.0134}{\omega} = 3.28 \text{ pF}$$

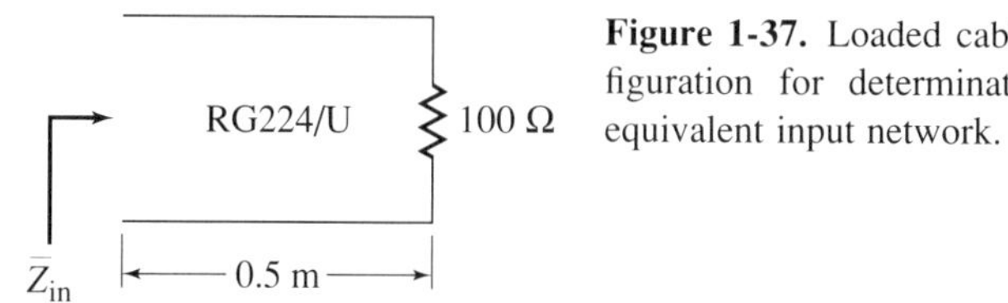

Figure 1-37. Loaded cable configuration for determination of equivalent input network.

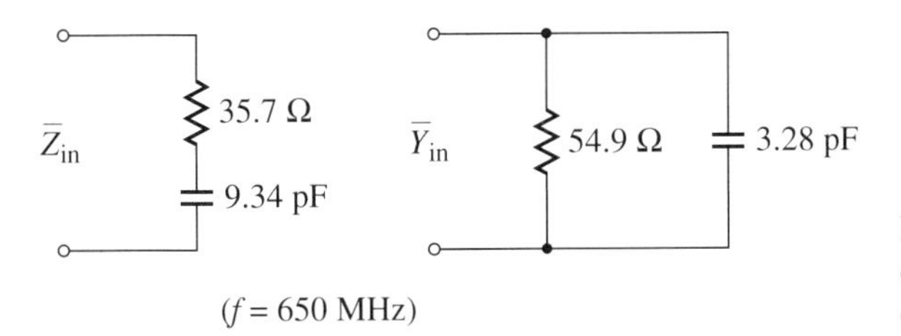

Figure 1-38. Equivalent circuits of lumped elements at the input of the line system of Figure 1-37.

These are summarized in Figure 1-38. It is important to remember that equivalent circuits such as these are valid only at the frequency at which the calculations are made. Both the R and C values vary with frequency and line length. It is a good idea to label any circuits with the appropriate frequency.

Exercises

1-7.4m For the preceding example, suppose that the input frequency shifts $\pm 10\%$. Compute the equivalent series and parallel circuits at the two extremes and determine the error that would occur if one used the same circuit as that at 650 MHz over the entire band represented by the frequency limits. Define the percentage error as

$$\% \text{ error} = \frac{\text{value used} - \text{true value}}{\text{true value}} \times 100\%$$

where the values are *magnitudes* of the impedances and admittances.

1-7.4n Plot the standing wave patterns of current and voltage on a lossless, short-circuited transmission line as a function of βl. Use $V^+ = 20$ V and $Z_0 = 75\ \Omega$. Directly under these plots, plot the input impedance, $\overline{Z}(-l)$, also using βl as the plot variable. What are the current and voltage values at the standing wave nulls? Are these what one would expect physically? Explain.

1-7.4o Plot the curves of $\overline{Y}(-l)$ for lossless open-circuit and short-circuit lines. Use βl as the independent variable.

1-7.4p For an inductor the reactance versus frequency (ω) curve is a straight line through the origin. Figure 1-35, which applies to a shorted-line input impedance, has a portion of the curve passing through the origin and the slope is fairly constant for a short range of βl. Determine the value of βl (which will be in radians) over which the equivalent inductance may be considered constant. Do this by finding the value of βl for which the slope is within 10% of the slope at the origin. For what frequency range is this approximation valid? (Hints: $\beta = 2\pi/\lambda_l$ and $V_p = f\lambda_l$. Your answer will be a function of l.)

1-7.4q For the equivalent circuit example above, determine the new component values for f = 300 MHz.

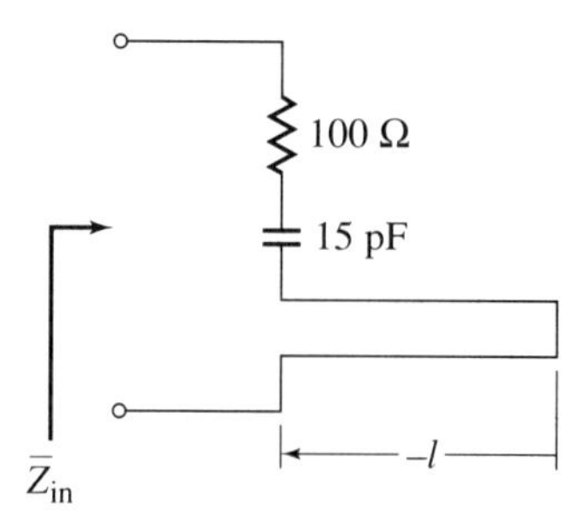

Figure 1-39.

1-7.4r Determine the length of the shorted cable RG58B/U that will make the input impedance of the network shown in Figure 1-39 purely resistive. The frequency is 450 MHz. Do this for the assumption of a lossless cable and for the low-loss case with attenuation (β unchanged). $\alpha_{dB} = 0.33$ dB/m.

1-7.4s Do Exercise 1-7.4r for the case where the shorted line is placed in parallel with the RC combination.

An interesting result can be obtained from the expression for the impedance at any location. When one first encounters this impedance variation with line length one wonders why this effect was not accounted for in the introductory circuits laboratory. In such a lab when a cable was connected from the signal generator to a load or circuit, it was simply assumed that the generator terminals saw the load value, the cable being merely a convenient interconnecting element. From Eq. 1-98 we observe that if the tangent function is small (near zero) the impedance at any point along the line becomes

$$\bar{Z}(-l) = \bar{Z}_L$$

which is independent of cable length! The condition for neglecting the tangent function is that $\beta l \to 0$. This can be given physical significance easily by using expressions for β as follows:

1. For a lossless line

$$\beta l = \omega\sqrt{LC}l$$

Thus for "short" line lengths and low frequencies the product may be brought near zero. For a typical line $\sqrt{LC}$ is about 10^{-8}. If we use frequencies in the 100-kHz range and 10-m lengths of line the value of βl is about 0.01 for which $\tan \beta l = 0.01$. (This is a phase shift of about 0.6° from source to cable output.)

2. For a lossless line

$$\beta l = \frac{2\pi}{\lambda_l} l = 2\pi\left(\frac{l}{\lambda_l}\right)$$

Thus for line lengths that are very small compared with the line wavelengths, the product may be brought near zero. For such cases we can approximate the tangent function by small βl, thereby simplifying the previous equations. Additionally, for $\beta l \approx 0.01$ we can completely neglect the tangent function.

1-8 Improving Transmission System Performance: The Theory of Line Matching

In Section 1-7 it was shown that if a transmission line has a termination other than the characteristic impedance, there are several significant effects. Some of those are:

1. Standing wave patterns of current and voltage that have widely fluctuating current and voltage amplitude values, especially if the load value is several times larger (or smaller) in magnitude than Z_0.
2. The source sees a load at its terminals that is often significantly different from the desired or actual load. Additionally, the apparent load value varies with both frequency and length of transmission line.
3. The power consists of incident, reflected, and transmitted components. It is desirable to have the line such that only a transmitted component is present so that all power a source is capable of generating for the desired load is actually used by the load.

These are, of course, not independent effects but different ways of describing what is called a *mismatched line*. The process or technique for improving line performance is called *line matching*. Since there are practical cases where the load is fixed in value and no cable that has a characteristic impedance equal to the load value is available, it is important to be able to design a system that eliminates (or at least significantly reduces) the effects of load-line mismatch. The three equations that describe the three effects listed above are, respectively (for a lossless line),

$$|\overline{V}(-l)| = |V^+[1 + \rho_L^2 + 2\rho_L \cos(2\beta l - \theta_L)]^{1/2}$$

$$\overline{Z}(-l) = Z_0 \frac{1 + \Gamma(-l)}{1 - \Gamma(-l)}$$

and

$$P_t = P_i[1 - |\Gamma(-l)|^2]$$

In all three cases, it is seen that, as expected, the goal is to make the system appear to have $\Gamma(-l) = 0$ ($\rho_L = 0$). Equivalently, since

$$S = \frac{1 + \rho_L}{1 - \rho_L} \qquad \text{with} \quad \rho_L = \left|\frac{\overline{Z}_L - Z_0}{\overline{Z}_L + Z_0}\right|$$

we would like to have $S = 1$. We can obtain $\rho_L = 0$ if $\overline{Z}_L = Z_0$, but, as pointed out, the load may be specified and not have the value Z_0. We compromise by obtaining $\Gamma(-l) = 0$ for as much of the system as possible.

Since it will be necessary to add elements to the line, we avoid using those elements that absorb power (energy). Obviously, our components must then be purely reactive elements (or susceptive since $B = 1/X$). Such elements can be realized using shorted or open-ended lines as discussed in Cases 3 and 4 in the preceding section. For the practical matter of eliminating radiation out of the end of the line (power loss), it is usual to use shorted-line sections for matching where possible. The reactance values of such shorted sections (often referred to as shorted *stubs*) can be computed from

$$\overline{Z}(-l) = jZ_0 \tan \beta l$$

or, for susceptance,

$$\overline{Y}(-l) = -jY_0 \cot \beta l \qquad \left(Y_0 = \frac{1}{Z_0}\right)$$

We begin by developing the mathematical theory of matching and consider the example of putting a reactive element in parallel to provide the match. The series case is left as an exercise. The configuration and definition of terms used are given in Figure 1-40. The problem is to begin with a *given* load $\overline{Z}_L$ on a *given* length of line l_T, and to design a parallel (shunt) shorted-stub configuration that will make the system present an impedance of value Z_0 at the generator at $z = -l_T$. This means that $\Gamma(-l_T) = 0$.

To obtain the proper matched condition, we must find the location l_1 where a stub is to be placed and the length l_s of the stub, which will make the total admittance at $-l_1$ equal to Y_0. This means that the parallel combination of the stub admittance and the line admittance at $-l_1$, will be Y_0. Thus as far as the remainder of the line to the left of l_1 is concerned, the load on the line appears to be $Z_0(1/Y_0)$ and therefore $\Gamma(-l_T) = 0$ as desired. For simplicity here, we assume that the characteristic impedance of the line used to make the stub is the same as that of the main line. (See Exercise 1-8.0a.) Mathematically, the matched condition means

$$Y_0 = \overline{Y}_s(-l_s) + \overline{Y}(-l_1) \tag{1-104}$$

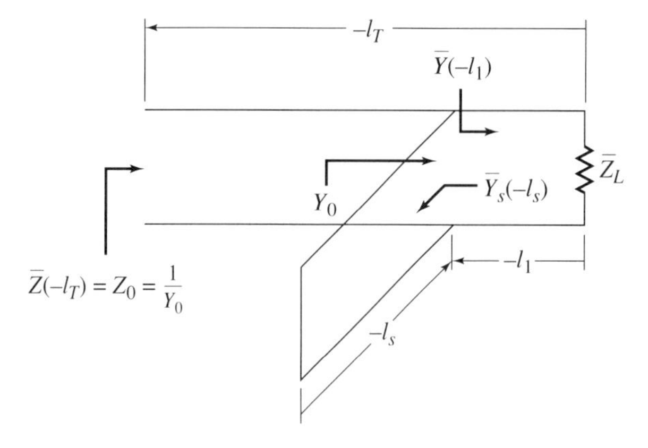

Figure 1-40. System definitions for shunt stub matching.

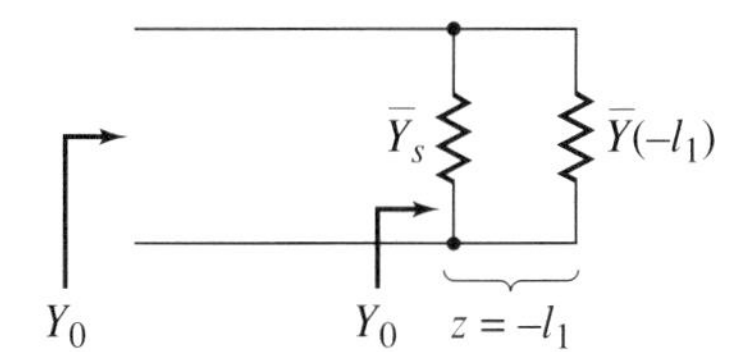

Figure 1-41. Equivalent circuit for the matched line condition.

and the equivalent circuit is shown in Figure 1-41. We begin by normalizing the preceding equation by dividing by Y_0 and using the tilde (˜) over the quantities to distinguish them from the actual values. The matched condition may then be expressed as

$$1 = \frac{\overline{Y}_s(-l_s)}{Y_0} + \frac{\overline{Y}(-l_1)}{Y_0} = \tilde{Y}_s(-l_s) + \tilde{Y}(-l_1) \tag{1-105}$$

Since a shorted stub can only add susceptance, we write the normalized result as

$$1 = j\tilde{B}_s(-l_s) + \tilde{Y}(-l_1)$$

To obtain the expression for the line admittance, we reciprocate Eq. 1-98 to obtain

$$\overline{Y}(-l) = Y_0 \frac{\overline{Y}_L + jY_0 \tan \beta l}{j\overline{Y}_L \tan \beta l + Y_0} \tag{1-106}$$

or, normalized,

$$\tilde{Y}(-l_1) = \frac{\overline{Y}(-l_1)}{Y_0} = \frac{\overline{Y}_L + jY_0 \tan \beta l_1}{j\overline{Y}_L \tan \beta l_1 + Y_0} = \frac{\tilde{Y}_L + j \tan \beta l_1}{j\tilde{Y}_L \tan \beta l_1 + 1} \tag{1-107}$$

(Incidentally, note that the equation for $\overline{Y}(-l)$ has exactly the same form as $\overline{Z}(-l)$. This will prove useful later.) The equation for a matched condition then becomes

$$1 = j\tilde{B}_s(-l_s) + \frac{\tilde{Y}_L + j \tan \beta l_1}{j\tilde{Y}_L \tan \beta l_1 + 1}$$

The right term is an awkward complex quantity since, in general, $\tilde{Y}_L$ is also itself complex. For ease of handling, we simply write the last term in rectangular form and use

$$1 = j\tilde{B}_s(-l_s) + \tilde{G}(-l_1) + j\tilde{B}(-l_1) = \tilde{G}(-l_1) + j[\tilde{B}_s(-l_s) + \tilde{B}(-l_1)]$$

This last result can be satisfied by two constraints:

$$\tilde{G}(-l_1) = 1 \tag{1-108}$$

and

$$\tilde{B}_s(-l_s) = -\tilde{B}(-l_1) \tag{1-109}$$

The steps used to obtain a matched line would then be as follows:

1. Obtain the expression for the real part $[\tilde{G}(-l_1)]$ of $\tilde{Y}(-l_1)$ from Eq. 1-107 and equate that to 1. (Usually one would rationalize the denominator and break $\tilde{Y}(-l_1)$ into real and imaginary components.) Solve the results for l_1 (Eq. 1-108).
2. Using the value of l_1 from Step 1, evaluate the imaginary component of $\tilde{Y}(-l_1)$ and equate the *negative* of that value to $\tilde{B}_s(-l_s)$. Solve the result for l_s. Incidentally, note that since a shorted stub has $\tilde{Y}(-l) = -j \cot \beta l$, we have $\tilde{B}_s(-l_s) = -\cot \beta l_s$.

As a final note, we mention that in engineering practice it is common to specify lengths in terms of fractions of a wavelength. Since the wavelength on a line can be measured very easily (distance between nulls is $\lambda_l/2$), the physical length in meters is readily obtained once the fraction of a wavelength is known. To obtain this length measure, we set

$$\beta l = \frac{2\pi}{\lambda_l} l = 2\pi\left(\frac{l}{\lambda_l}\right) = 2\pi l_\lambda \tag{1-110}$$

where l_λ means fraction of a wavelength. Thus when we see a length given as $l_\lambda = 0.34\lambda$, it means that the physical length corresponding to that value is $l = (0.34)(\lambda_l)$. Note here that the symbol λ with no subscript is simply an indicator of the measure, since l_λ is, of course, dimensionless. This is much like using ''m'' or ''mm'' to indicate the type of distance measure.

Example 1-9

A transmission line has a characteristic impedance of 75 Ω and is terminated in a load that has a value of 85 + j20 Ω. Using the configuration shown in Figure 1-42, determine the location and length of a shorted shunt stub that will provide a match at the generator. Assume a lossless line.

$$\overline{Y}_L = \frac{1}{85 + j20} = 0.011452\angle - 13.24° = 0.011148 - j0.002623 \text{ S}$$

$$Y_0 = \tfrac{1}{75} = 0.0133\overline{3} \text{ S}$$

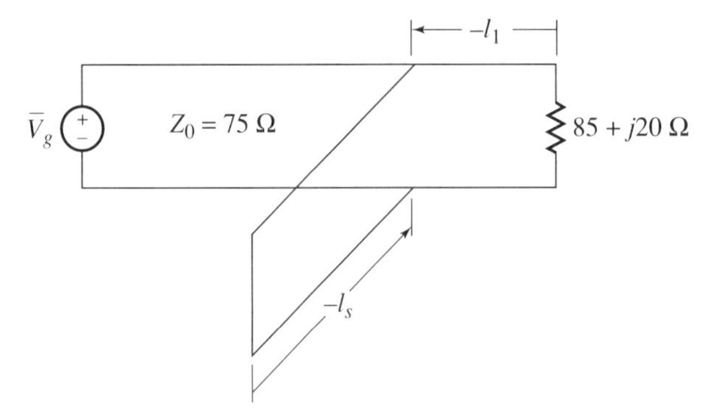

Figure 1-42. Example of a shunt matching configuration.

Using Eq. 1-107 with Eq. 1-110 substituted to obtain the wavelength measure of distances:

$$Y(-l_1) = \frac{\dfrac{0.011148 - j0.002623}{0.01333} + j\tan(2\pi l_{1_\lambda})}{j\left[\dfrac{0.011148 - j0.002623}{0.01333}\tan(2\pi l_{1_\lambda})\right]} + 1$$

$$= \frac{0.8361 + j[\tan(2\pi l_{1_\lambda}) - 0.19672]}{[1 + 0.19672\tan(2\pi l_{1_\lambda})] + j0.8361\tan(2\pi l_{1_\lambda})}$$

Multiplying numerator and denominator by the conjugate of the denominator (i.e., rationalizing the denominator) gives

$$\tilde{Y}(-l_1) = \frac{0.8361 + 0.8361\tan^2(2\pi l_{1_\lambda}) + j[-0.19672 + 0.26224\tan(2\pi l_{1_\lambda}) + 0.19672\tan^2(2\pi l_{1_\lambda})]}{1 + 0.39344\tan(2\pi l_{1_\lambda}) + 0.7378\tan^2(2\pi l_{1_\lambda})}$$

Step 1 in the design process requires that the real part of $\tilde{Y}(-l_1)$ be 1, so we set

$$\frac{0.8361 + 0.8361\tan^2(2\pi l_{1_\lambda})}{1 + 0.39344\tan(2\pi l_{1_\lambda}) + 0.7378\tan^2(2\pi l_{1_\lambda})} = 1$$

Multiplying through by the denominator and collecting coefficients on the tangents yields

$$\tan^2(2\pi l_{1_\lambda}) - 4.002\tan(2\pi l_{1_\lambda}) - 1.6673 = 0$$

Solving this as a quadratic in $\tan(2\pi l_{1_\lambda})$ gives

$$\tan(2\pi l_{1_\lambda}) = \frac{4.002 \pm \sqrt{16.016 + 6.6692}}{2} = 4.382,\ -0.3804$$

Thus

$$\tan(2\pi l_{1_\lambda}) = 4.382 \rightarrow l_{1_\lambda} = 0.2143\lambda$$

or

$$\tan(2\pi l_{1_\lambda}) = -0.3804 \rightarrow l_{1_\lambda} = -0.0578\lambda$$

Mow the stub is to be located at $z = -l_1$. For the two values determined above, this means we have either $z = -0.2143\lambda$ or $z = -(-0.0578\lambda) = 0.0578\lambda$. The positive value of z means physically that the location is to the *right* of the load, which by convention is at $z = 0$. This is not permissible since the stub would be off the line, so that solution is rejected. (See comments after this example.) Thus

$$l_{1_\lambda} = 0.2143\lambda$$

Next, according to Step 2, we evaluate the susceptance component of line admittance at the stub location l_{1_λ}. From the imaginary part of $\tilde{Y}(-l_1)$, we obtain

$$\tilde{B}(-l_1) = \frac{-0.19672 + (0.26224)(4.382) + (0.19672)(4.382)^2}{1 + (0.39344)(4.382) + (0.7378)(4.382)^2} = 0.28$$

The stub susceptance must be the *negative* of this. For the shorted stub, the susceptance is $-\cot(2\pi l_{s\lambda})$, so we require

$$-\cot(2\pi l_{s_\lambda}) = -0.28$$

from which $l_{s_\lambda} = 0.2065\lambda$.

If we have information from which we can obtain the line wavelength λ_l, we then obtain the actual lengths as

$$l_1 = 0.2143\lambda_l$$

$$l_s = 0.2065\lambda_l$$

In the preceding example, we rather hastily rejected $l_{1_\lambda} = -0.0578\lambda$ on the basis that it is physically to the right of the load. However, we showed in an earlier discussion that the impedance is periodic every $\lambda_l/2$ (or π radians) along the line. So if we add 0.5λ to the rejected value, we would have $l_{1_\lambda} = 0.4422\lambda$. If we use this value in the real part of the admittance $\tilde{Y}(-l_1)$, we would have

$$\tilde{G}(-l_1) = \frac{0.8361 + 0.8361\tan^2(2\pi \times 0.4422)}{1 + 0.39344\tan(2\pi \times 0.4422) + 0.7378\tan^2(2\pi \times 0.4422)} = 1$$

which satisfies the condition for a matching location. The corresponding value of the stub length is obtained, as before, by setting

$$\begin{aligned} -\cot(2\pi l_{s_\lambda}) \\ = -\frac{-0.19672 + 0.26224\tan(2\pi \times 0.4422) + 0.19672\tan^2(2\pi \times 0.4422)}{1 + 0.39344\tan(2\pi \times 0.4422) + 0.7378\tan^2(2\pi \times 0.4422)} \\ = 0.28 \end{aligned}$$

From this we get

$$l_{s\lambda} = -0.2065\lambda$$

Again, since this is negative, the location would be beyond (outside) the stub short. We thus add 0.5λ to obtain the physically permissible value

$$l_{s\lambda} = 0.2934\lambda$$

Since we have obtained two solutions, one may wonder which to use. On a theoretical basis, either is acceptable. From a practical standpoint there are several factors to consider. One of these is whether the location l_1 is accessible. One value of l_1 might be inside a unit where the stub cannot be connected. Another factor to consider is the range of frequencies for which the system might be used. The exact match exists only at one frequency, but some acceptable bandwidth performance may be required. Since the lengths (l/λ_l) depend upon frequency (λ_l), the system may perform better for one set of lengths than the other. Generally speaking, the shortest lengths of line and placement are preferred, but even this choice may involve a trade-off since a short length stub may often require a longer distance location l_1. In such circumstances the shortest length is usually given to the element having the largest standing wave ratio (in this case, the stub). A following exercise investigates this point.

Exercises

1-8.0a Show how the shunt stub matching procedure given in this section would be modified if the characteristic impedance of the line used to construct the stub were different from that of the main transmission line. $Z_{0_S} \neq Z_0$. Pay particular attention to Step 2 of the design process.

1-8.0b In some matching situations, it may not be convenient or possible to short the end of the shunt matching stub. Develop the equations in Step 2 of the shunt stub design process that use an open-end-shunt-stub matching element.

1-8.0c Reciprocate the equation for $\overline{Z}(-l)$ and derive the expression for $\overline{Y}(-l)$ given in Eq. 1-106.

1-8.0d In some engineering situations, it may be more convenient to use a series matching stub as shown in Figure 1-43. In this case, since series impedances add, we work with $\overline{Z}$ rather than $\overline{Y}$ as presented in this section. We define the normalized impedance as $\tilde{Z}(-l) = \overline{Z}(-l)/Z_0$. The goal is to obtain $\tilde{Z}(-l_1) = 1 + j0$ just to the *left* of the point where the stub is connected. (Assume that there is zero length between the two dotted planes; i.e., that the width of the stub line is negligible.) Derive expressions analogous to Eqs. 1-108 and 1-109 that yield the lengths l_{1_λ} and l_{s_λ}. Also write explicit steps corresponding to Steps 1 and 2 of the text.

1-8.0e A transmission line (assumed lossless) that has a characteristic impedance of 300 Ω is terminated in a load $\overline{Z}_L = 450$ Ω. Find the location and required length of the shorted shunt stub that will provide a match. There will be two solutions; which one would you select? Obtain the values of the standing wave ratios on all sections of the configuration. Lengths will be in fractions of a wavelength.

1-8.0f Repeat Exercise 1-8.0e for a series shorted stub match.

1-8.0g Prove that adding a line of length $n\lambda_l/2$, $n = 1, 2, \ldots,$ to either l_{1_λ} or l_{s_λ} does not disturb the match condition.

1-8.0h To demonstrate the frequency sensitivity of the system as a function of stub length and stub location, consider an RG58B/U coax (assumed lossless) terminated in $\overline{Z}_L = 100$ Ω. Find the best lengths and l_1 and l_s for a shorted-shunt stub-matching system at $f = 1$ GHz.

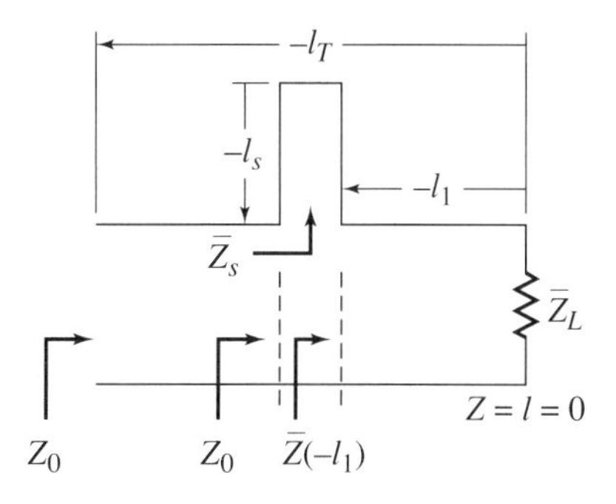

Figure 1-43.

For this system, plot S to the left of the stub (i.e., on the ''matched'' section of line) as a function of frequency for the range 0.5 GHz to 1.5 GHz. Now add $\lambda_l/2$ length to l_1 (at 1 GHz) and plot S vs. f. Repeat by adding $\lambda_l/2$ to l_s only (at 1 GHz). A computer solution might be good here. Although it may make this exercise a little long, it is nevertheless instructive to repeat the entire exercise for the other lengths l_1 and l_s obtained at 1 GHz.

As a practical matter, it is rather bothersome to have to solve such complex equations for each system to be matched. A special chart aid was developed by Phillip H. Smith during the years 1939–1944 for assisting in these calculations. The Smith chart has proven to be of wider use than just for matching, and some of those applications will be considered as well. This is the subject of Chapter 2.

1-9 Printed Circuit and Hybrid Circuit Transmission Lines with Applications

In the preceding sections the main examples were drawn from the more familiar physical systems, namely, coax, parallel wire lines. These lines are from large-scale systems and have many applications there, as have been discussed. As a matter of fact, in later chapters we will learn that the basic ideas of this chapter are applicable to systems that have guided electromagnetic waves such as exist in waveguides (hollow ''tubes'' or ''pipes''). We will also learn there that the voltage and current variables used in multiconductor transmission systems in many cases bear simple relationships to the electric and magnetic fields in and around the conductors.

Some of the exercises in earlier sections were directed at printed circuit boards with chips or other devices attached to the lines. In those exercises capacitance calculations neglected the effects of the fringing capacitance; that is, capacitance outside of the area common to the two conductors. These capacitance components are shown in Figure 1-44. As the physical size is reduced, the width W becomes small and the fringing capacitance is a significant part of the total capacitance. Such narrow lines are in significant use in high frequency and microwave hybrid and integrated circuits (MIC). Submillimeter line widths are common. As the speeds (data rates) of digital systems increase to hundreds of megahertz and into the gigahertz range, the transmission line nature of the connecting lines on either printed circuit boards or integrated circuits must be considered. The two basic configurations in use are shown in Figure 1-45 in cross sections and are called *stripline*

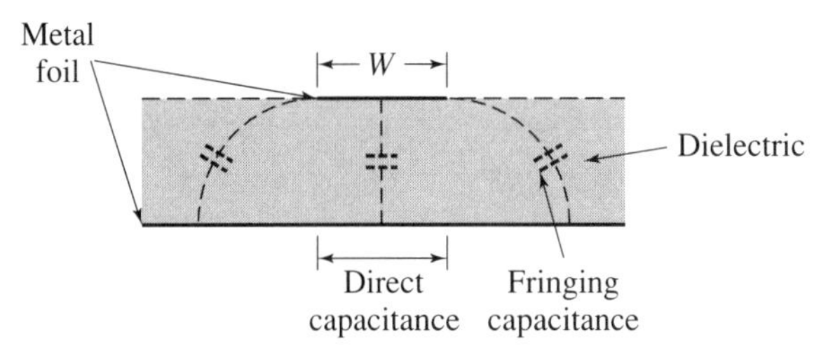

Figure 1-44. Printed circuit trace cross section illustrating direct and fringing components of capacitance.

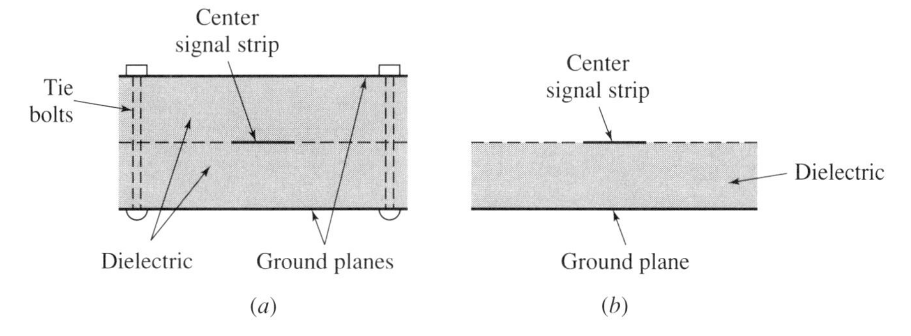

Figure 1-45. Basic printed circuit and hybrid circuit transmission lines. (*a*) Stripline cross section. (*b*) Microstrip cross section.

and *microstrip*. The stripline system is usually made in a sandwich configuration, and the two sections are bolted together to form a shielded center conductor line. The bolts are usually metal so that the outer ground planes are electrically connected. Also, the bolts are usually placed at such distances from the center conductor that they don't affect the performance of the center conductor. Thus the center conductor is essentially embedded in an infinite dielectric horizontally. Analysis of stripline is usually based on this assumption.

A comment is appropriate here concerning the ground planes shown in Figure 1-45. In most of what follows the planes are assumed to be solid. This need not be the case, however. In many practical cases one side of a board consists of many ground lines. If these lines are "reasonably" close together, say, about 1/16 of a wavelength, the set of lines appears to form a solid sheet electrically and the results given here apply very well.

The literature on stripline and microstrip is extensive, and entire books have been written on each of these types of lines. All that can be done at this point is to give some of the results and suggest some of the applications. We treat here initially the case of lossless lines.

Earlier in the chapter results were developed for the characteristic impedance of two-conductor systems and the corresponding phase velocity. These are (for lossless or low-loss systems)

$$Z_0 = \sqrt{\frac{L}{C}} \qquad \text{and} \quad V_P = \frac{1}{\sqrt{LC}}$$

Eliminating the inductance L we obtain

$$Z_0 = \frac{1}{V_P C}$$

As we shall see later, for the propagation of current and voltage on two-conductor systems of most common use, the propagation velocity is also given by

$$V_P = \frac{1}{\sqrt{\mu\epsilon}}$$

where μ and ϵ are the permeability and permittivity, respectively, of the material between the two conductors. Expressing μ and ϵ in terms of relative values the phase velocity becomes

$$V_P = \frac{1}{\sqrt{\mu_0 \mu_r \epsilon_0 \epsilon_r}} = \frac{1}{\sqrt{\mu_0 \epsilon_0}\sqrt{\mu_r \epsilon_r}} = \frac{c_0}{\sqrt{\mu_r \epsilon_r}} \tag{1-111}$$

where $\mu_0 = 4\pi \times 10^{-7}$ H/m and $\epsilon_0 = 1/36\pi \times 10^{-9}$ F/m and c_0 is the velocity of light in vacuum. Putting this result into the expression for Z_0 gives the most common form

$$Z_0 = \frac{\sqrt{\mu_r \epsilon_r}}{c_0 C} \tag{1-112}$$

In most cases the materials used in the construction of stripline and microstrip are dielectrics with $\mu_r = 1$ and values of ϵ_r in the range of 2 to 4. For these cases it is most common to use the expression

$$Z_0 = \frac{\sqrt{\epsilon_r}}{c_0 C} \tag{1-113}$$

From this last result we can see that the line capacitance per meter is the important parameter. Much research has been done to determine expressions for line capacitance that include both the direct and fringe components. Here we will merely state some of the results.

1-9.1 Stripline Parameters

The basic configuration for the results to be given is shown in Figure 1-46. In the figure, C_{PP} represents the direct or plate-to-plate capacitance, and C_f the fringe capacitance of one quadrant. The total capacitance is then

$$C = C_{PP} + 4C_f \tag{1-114}$$

The results most frequently used are those of Cohn [10]. His analysis resulted in

$$C_{PP} = 35.4 \frac{\frac{W}{b}\epsilon_r}{\left(1 - \frac{t}{b}\right)} \text{ pF/m} \tag{1-115}$$

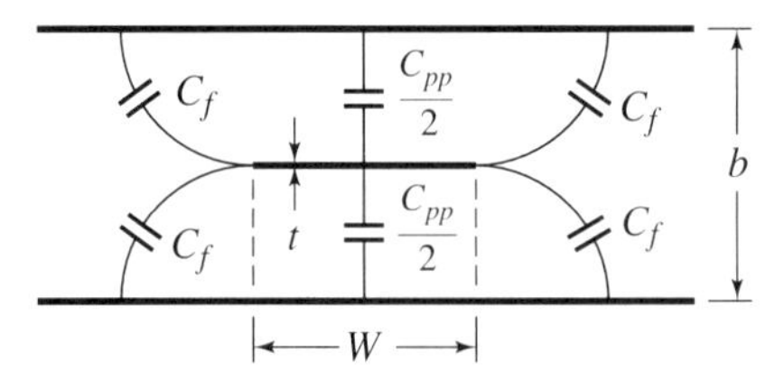

Figure 1.46. Stripline structure, dimensions, and component capacitances.

and

$$C_f = \frac{8.85}{\pi}\,\epsilon_r \left\{ \frac{2}{1-\dfrac{t}{b}} \ln\left(\frac{1}{1-\dfrac{t}{b}} + 1\right) - \left(\frac{1}{1-\dfrac{t}{b}} - 1\right) \ln\left(\frac{1}{\left(1-\dfrac{t}{b}\right)^2} - 1\right) \right\} \text{pF/m} \qquad \text{for } \frac{W}{b-t} \geq 0.35 \tag{1-116}$$

The error on C_f is about 1.2% for $W/(b - t) = 0.35$, and in most practical cases $b >> t$ so that the criterion becomes $W/b \geq 0.35$.

Normalized plots of these equations are in Figures 1-47 and 1-48 (see [27]). The reason that the formula begins to break down for $W/(b - t) < 0.35$ is that the formula is derived on the basis that the center conductor is a semi-infinite sheet as shown in Figure 1-49*a*. However, for a finite-width center conductor, where W is small, there is an additional contribution from the fringing fields at the opposite edge of the sheet (Figure 1-49*b*).

For $W/(b - t) < 0.35$ it is possible to develop an approximate expression for *total C* directly, and it is given by

$$C = \frac{55.5\epsilon_r}{\ln\left(\dfrac{4b}{\pi d_0}\right)} \text{ pF/m} \tag{1-117}$$

Figure 1.47. Plate-to-plate capacitance for stripline.

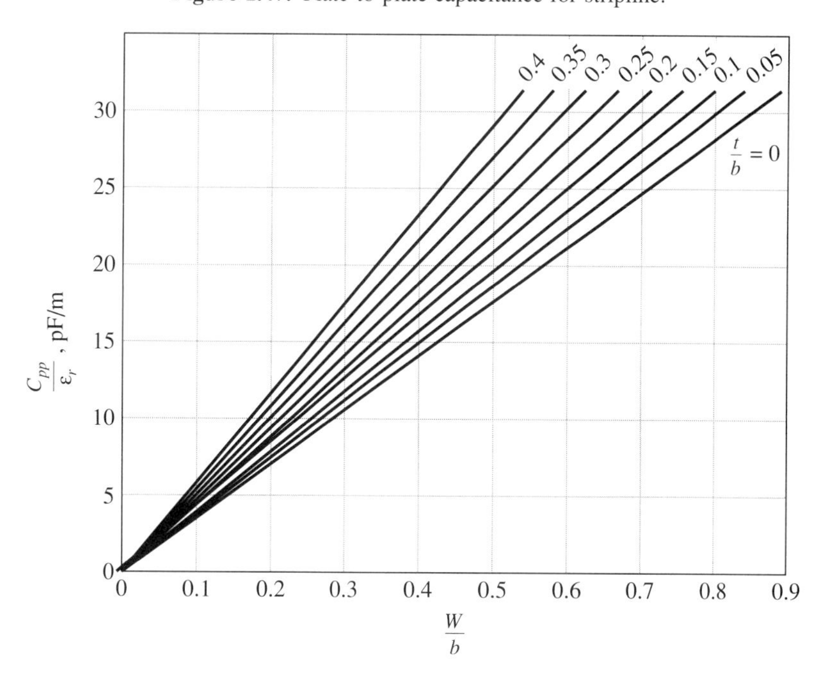

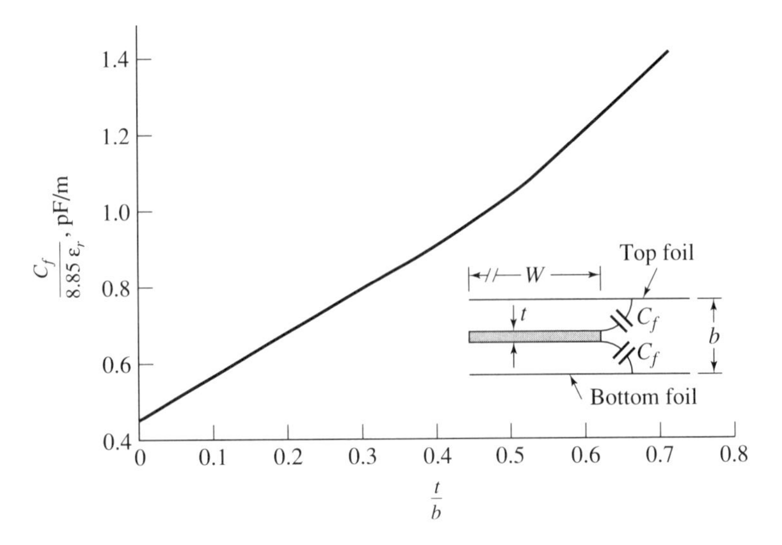

Figure 1.48. Fringing capacitance for stripline.

where d_0 is the diameter of a circular cross section that has the same electrical effect as that of the rectangular center conductor. A plot of the values of d_0/W for various t/W ratios is given in Figure 1-50. The total capacitance is within 1% of the true value for d_0 as large as $b/2$, and this C is plotted in normalized form in Figure 1-51.

With these capacitance formulas available it is an easy matter to determine equations for Z_0 of stripline using Eq. 1-113. The results are

$$Z_0 = \frac{94.15}{\sqrt{\epsilon_r}\left(\dfrac{W/b}{1 - t/b} + \dfrac{C_f}{8.85\epsilon_r}\right)} \quad \text{for} \quad \frac{W}{b - t} \geq 0.35 \tag{1-118}$$

or

$$Z_0 = \frac{60}{\sqrt{\epsilon_r}} \ln\left(\frac{4b}{\pi d_0}\right) \quad \text{for} \quad \frac{W}{b - t} < 0.35$$

Figure 1-49. Configuration for obtaining C_f for the assumption of (*a*) a wide center conductor; and (*b*) interaction effects for narrow strip.

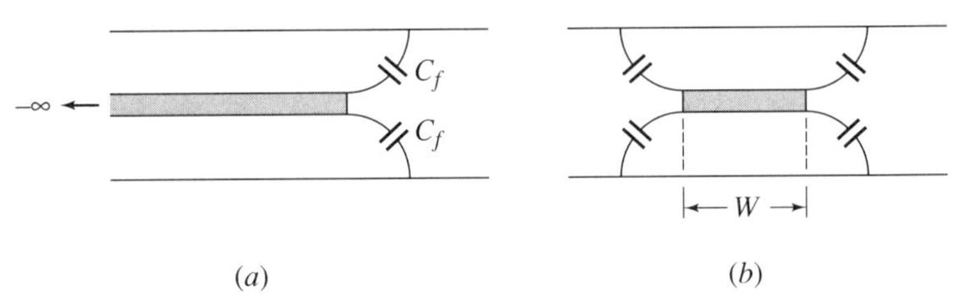

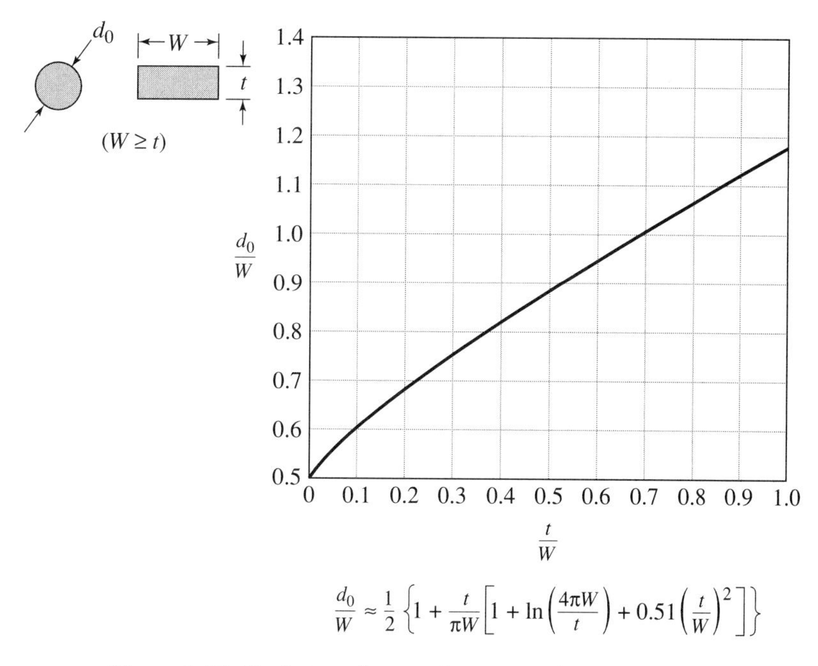

$$\frac{d_0}{W} \approx \frac{1}{2}\left\{1 + \frac{t}{\pi W}\left[1 + \ln\left(\frac{4\pi W}{t}\right) + 0.51\left(\frac{t}{W}\right)^2\right]\right\}$$

Figure 1-50. Equivalent diameter for rectangular cross section.

Figure 1-51. Total stripline capacitance for narrow center strip.

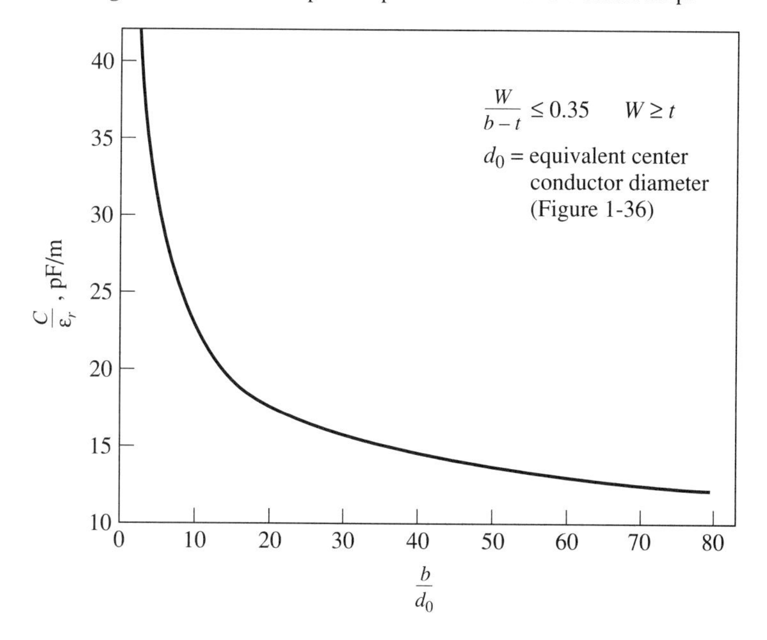

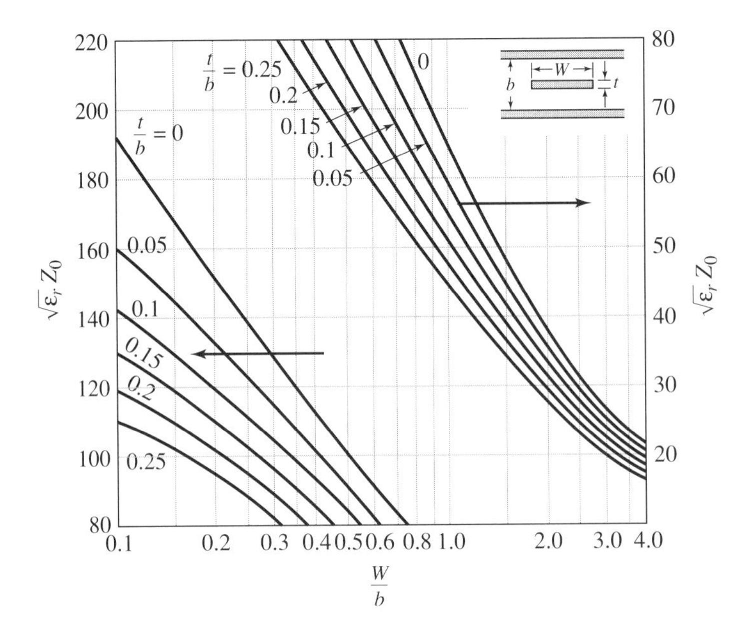

Figure 1-52. Characteristic impedance of stripline.

These are plotted (as $\sqrt{\epsilon_r}\, Z_0$) in Figure 1-52. These curves were obtained (see [8]) using the expression for Z_0 consistent with the value of $W/(b - t)$. These may be used to design a stripline system to obtain a given Z_0.

A companion plot to Figure 1-52 is one in which $Z_0\sqrt{\epsilon_r}$ is the parameter and t/b and W/b are the plot variables. This is given in Figure 1-53. This is a particularly helpful curve

Figure 1-53. Stripline dimensions as a function of characteristic impedance.

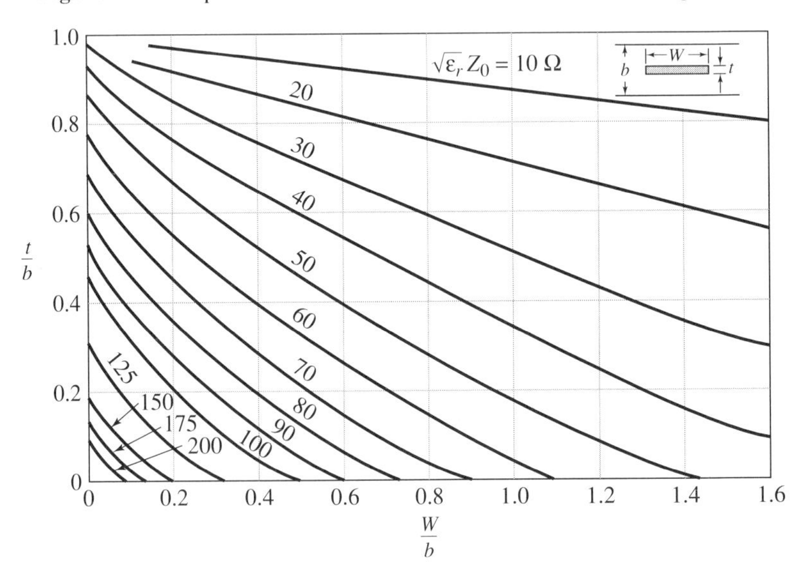

when a desired Z_0 is given and a particular copper-plated material of known dielectric thickness b is available.

The charts given in the figures can be used in many ways. Here we give one example of an approach that might be used to design a stripline unit.

Example 1-10

Suppose that a copper-clad dielectric sheet that has a dielectric thickness of 1.5 mm with $\epsilon_r = 3$ is available. The copper foil has a thickness of 0.2 mm. Design a stripline that has a characteristic impedance of 75 Ω.

Since the total structure consists of *two* sheets put together to cover the center conductor

$$b = 2 \times 1.5 = 3 \text{ mm}$$

Thus

$$\frac{t}{b} = \frac{0.2 \text{ mm}}{3 \text{ mm}} = 0.0\overline{66} \approx 0.07$$

Also

$$\sqrt{\epsilon_r} \times Z_0 = \sqrt{3} \times 75 = 130$$

Now, using Figure 1-52, we locate 130 on the left edge and move to the right horizontally until the $t/b = 0.07$ curve (at least where we imagine it to be) is intercepted. Then read down from that intercept to obtain the value on the bottom horizontal axis of the chart

$$\frac{W}{b} = 0.175$$

Then $W = (0.175)(3)$ mm $= 0.525$ mm. These results yield the design shown in Figure 1-54.

Exercises

1-9.1a Using the parallel plate formula from physics, $C = \epsilon A/d$, where A = plate area and d = plate separation, derive the formula for C_{PP} given in Eq. 1-115.

1-9.1b Repeat the stripline design example of this section for a characteristic impedance of 30 Ω.

1-9.1c Repeat the stripline design example of this section for a characteristic impedance of 75 Ω using Figure 1-53 only.

1-9.1d There is available a copper-clad printed circuit board material on which the copper layer is 0.034 mm thick. The dielectric is phenol-formaldehyde ($\epsilon_r = 4$). Using this material design a 50-Ω stripline that will operate at 1 GHz. The dielectric thickness of the board material is 1.6 mm. Show how you could launch the current and voltage from a coax to the stripline.

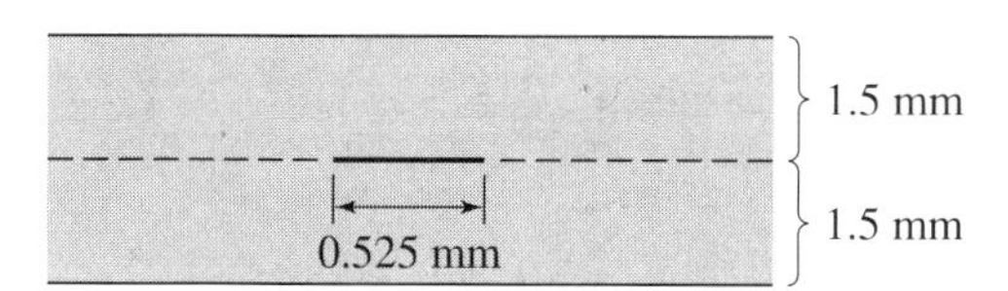

Figure 1-54. Example of stripline design.

1-9.1e For the transmission line designed in Exercise 1-9.1d, design a line section for which the input impedance is $-j10\ \Omega$ at 3 GHz.

1-9.1f There are two possibilities for approximating a rectangular cross section by a circular equivalent. One is to define the equivalent diameter d_0 so that the circle has the same cross-sectional area as the rectangular strip, and the second is to define the equivalent diameter d_0 so that the circle has the same perimeter as the rectangular strip. Plot both of these approximations on Figure 1-50, and compare with the exact plot. Can you make either approximation better by adding and/or multiplying any empirical constants in the equivalent diameter equations?

1-9.2 Microstrip Parameters

The basic configuration for the microstrip is shown in Figure 1-55. One of the most difficult problems associated with this structure arises from the fact that the small strip is not immersed in a single dielectric. On one side there is the board dielectric, and on the top is another—usually air.

The technique others have developed to handle this problem uses the concept of effective relative dielectric constant, ϵ_{eff}. This value represents some intermediate value between ϵ_r and air that can be used to compute microstrip parameters as though the strip were completely surrounded. One obvious advantage of the microstrip structure is the open line, which makes it very easy to connect components. On the other hand, this configuration does not possess the shielded signal line advantage of the stripline.

Aside from the difficulty of calculating the value of ϵ_{eff}, there is another important effect. Since the value of ϵ_{eff} depends upon W (and hence Z_0) the phase velocity is not a constant for a given dielectric ϵ_r. This is shown by considering the expression for V_P in Eq. 1-111. There we must replace ϵ_r by ϵ_{eff} to obtain (with $\mu_r = 1$)

$$V_P = \frac{c_0}{\sqrt{\epsilon_{\text{eff}}}} \tag{1-119}$$

Thus as the characteristic impedance is changed from design to design, the phase velocity will also change! For pulses, as an example, the delays due to finite propagation time (l/V_P) will need to be recomputed for a new design even with the same material. The corresponding wavelength would then be

$$\lambda_l = \frac{V_P}{f} = \frac{c_0}{\sqrt{\epsilon_{\text{eff}}}\, f} \tag{1-120}$$

To get an idea of the range of the values for ϵ_{eff} consider the cases of very wide W and then narrow W. For a wide strip, essentially all of the electric field will be in the

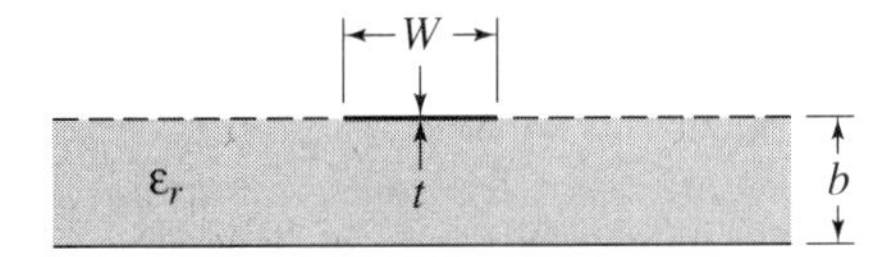

Figure 1-55. Microstrip configuration.

dielectric between the metal planes, similar to the regular parallel-plate capacitor in physics. Thus

$$\epsilon_{\text{eff}_{\max}} = \epsilon_r$$

On the other extreme, for narrow W the electric field lines will be divided equally between the air ($\epsilon_{0_r} = 1$) and the dielectric so that

$$\epsilon_{\text{eff}_{\min}} = \tfrac{1}{2}(\epsilon_r + 1)$$

which gives the range

$$\tfrac{1}{2}(\epsilon_r + 1) \leq \epsilon_{\text{eff}} \leq \epsilon_r \tag{1-121}$$

As with the stripline system, several different equations have been developed for use in microstrip design. For example, [48, 62, 25] and others have developed formulae yielding accuracies on the order of 1%. Here we select

$$\left.\begin{aligned} Z_0 &= \frac{60}{\sqrt{\epsilon_{\text{eff}}}} \ln\left(8\frac{b}{W} + \frac{W}{4b}\right) \\ \text{where}\quad \epsilon_{\text{eff}} &= \frac{\epsilon_r + 1}{2} + \frac{\epsilon_r - 1}{2}\left[\left(1 + 12\frac{b}{W}\right)^{-1/2} \right. \\ &\quad \left. + 0.04\left(1 - \frac{W}{b}\right)^2\right] \qquad \text{for } \frac{W}{b} \leq 1 \end{aligned}\right\} \tag{1-22}$$

or

$$\left.\begin{aligned} Z_0 &= \frac{120\pi}{\sqrt{\epsilon_{\text{eff}}}\left[\frac{W}{b} + 1.393 + 0.667\ln\left(\frac{W}{b} + 1.444\right)\right]} \\ \text{where}\quad \epsilon_{\text{eff}} &= \frac{\epsilon_r + 1}{2} + \frac{\epsilon_r - 1}{2}\left(1 + 12\frac{b}{W}\right)^{-1/2} \qquad \text{for } \frac{W}{b} \geq 1 \end{aligned}\right\} \tag{1-123}$$

For a given Z_0 and ϵ_r, we could then solve for the required ratio W/b to complete the design.

To obtain ϵ_{eff} using measurements made on a given line, which is necessary if we are going to be able to compute the line wavelength (for computing stub lengths for example) and the phase velocity (to compute propagation delay), we proceed as follows.

Assume that the entire system is air so $\epsilon_r = 1$. Then using the Z_0 equation appropriate to the designed W/b compute the characteristic impedance using $\epsilon_r = 1$. Call this value Z_0^{air}. Now, since the system dimensions are unchanged, the dielectric changes only the value of the line capacitance so that

$$C_{\text{line}} = \epsilon_{\text{eff}} C_{\text{air}}$$

(Remember from physics that the capacitance of a system varies directly with the value of dielectric constant between the conductors, such as for parallel planes where $C = \epsilon A/d$, where A = area of plates and d = plate separation.)

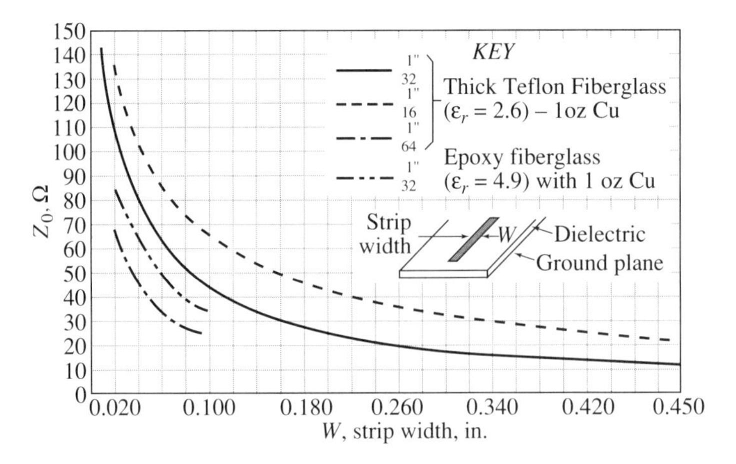

Figure 1-56. Measured microstrip line impedance.

Next, using Eq. 1-113 we would have

$$Z_0 = \frac{\sqrt{\epsilon_{\text{eff}}}}{c_0 C_{\text{line}}} = \frac{\sqrt{\epsilon_{\text{eff}}}}{c_0 \epsilon_{\text{eff}} C_{\text{air}}} = \frac{1}{\sqrt{\epsilon_{\text{eff}}}} \frac{1}{c_0 C_{\text{air}}} = \frac{1}{\sqrt{\epsilon_{\text{eff}}}} Z_0^{\text{air}}$$

or

$$\epsilon_{\text{eff}} = \left(\frac{Z_0^{\text{air}}}{Z_0} \right)^2 \tag{1-124}$$

Thus if we measure Z_0 for the given microstrip system we can compute ϵ_{eff}. From this result, we can then compute the phase velocity and line wavelength from Eqs. 1-119 and 1-120.

Some experimental values of microstrip line impedances are plotted in Figure 1-56. It is also of interest to note that for microstrip, the thickness of the foil is not involved in the determination of Z_0 in the first approximation. The 1-oz copper specification means 1 oz of copper is used to cover 1 sq yd of material, which translates to a thickness of about 0.356 mm.

Example 1-11

Design a microstrip line that has $Z_0 = 50\ \Omega$ using a copper covered phenolformaldehyde (paper) for which the dielectric thickness is 0.8 mm and the copper foil is 0.15 mm thick. Compute the pulse delay of 0.1 m length of line. (Notice it would be nice if Eqs. 1-121 and 1-122 were solved for the ratio *b*/*W*. This would be difficult to do directly.)

From a table we obtain $\epsilon_r = 4$. Using Eq. 1-122 as a guess that it will be the appropriate one:

$$\epsilon_{\text{eff}} = \frac{4+1}{2} + \frac{4-1}{2} \left[\left(1 + 12 \frac{b}{W} \right)^{-1/2} + 0.04 \left(1 - \frac{W}{b} \right)^2 \right]$$

$$= 2.5 + 1.5 \left[\left(1 + 12 \frac{b}{W} \right)^{-1/2} + 0.04 \left(1 - \frac{W}{b} \right)^2 \right]$$

Then

$$50 = \frac{60 \ln\left(\frac{8b}{W} + \frac{W}{4b}\right)}{\left(2.5 + 1.5\left[\left(1 + 12\frac{b}{W}\right)^{-1/2} + 0.04\left(1 - \frac{W}{b}\right)^2\right]\right)^{1/2}}$$

After some algebra and root finding by trial and error we obtain the value $W/b = 2.05$. However, since $W/b > 1$, then the formula used is not the correct one, since it was valid for $W/b < 1$. Thus we try using Eq. 1-123:

$$\epsilon_{\text{eff}} = \frac{4 + 1}{2} + \frac{4 - 1}{2}\left(1 + 12\frac{b}{W}\right)^{-1/2} = 2.5 + 1.5\left(1 + 12\frac{b}{W}\right)^{-1/2}$$

Then

$$50 = \frac{120\pi}{\left(2.5 + 1.5\left(1 + 12\frac{b}{W}\right)^{-1/2}\right)^{1/2}\left[\frac{W}{b} + 1.393 + 0.667 \ln\left(\frac{W}{b} + 1.444\right)\right]}$$

Again using trial and error, we have $W/b = 2.068$. (Note this is close to the earlier value. This means that the transition between the two equations for Z_0 is very smooth.) Thus

$$W = (2.068)(0.8) = 1.65 \text{ mm}$$

To check the theory given earlier, let us compute the value of ϵ_{eff} using Eq. 1-123 directly and then by using the result in Eq. 1-123 for Z_0. Using the expression for ϵ_{eff}

$$\epsilon_{\text{eff}} = 2.5 + 1.5\left(1 + 12 \times \frac{1}{2.068}\right)^{-1/2} = 3.075$$

Now to compute Z_0 for air dielectric, we would first need to use Eq. 1-123 which yields $\epsilon_{\text{eff}} = 1$, as expected, using $\epsilon_r = 1$. Then again using Z_0 from Eq. 1-123,

$$Z_0^{\text{air}} = \frac{120\pi}{[2.068 + 1.393 + 0.667 \ln(2.068 + 1.444)]} = 87.7 \ \Omega$$

Using Eq. 1-124

$$\epsilon_{\text{eff}} = \left(\frac{87.7}{50}\right)^2 = 3.076$$

Notice also that this ϵ_{eff} satisfies the limits Eq. 1-121 which, for $\epsilon_r = 4$, is

$$2.5 \le \epsilon_{\text{eff}} \le 4$$

The preceding example would have been much easier if we could have solved the equations to obtain the ratio W/b as a function of Z_0 and ϵ_r. Of course, since the equations are transcendental they cannot be easily inverted. Results have been obtained, however, by Hammerstad [25] as follows:

For $W/b \leq 2$:

$$\left.\begin{aligned} \frac{W}{b} &= \frac{8e^A}{e^{2A} - 2} \\ \text{where} \qquad A &= \frac{Z_0}{60}\sqrt{\frac{\epsilon_r + 1}{2}} + \frac{\epsilon_r - 1}{\epsilon_r + 1}\left(0.23 + \frac{0.11}{\epsilon_r}\right) \end{aligned}\right\} \tag{1-125}$$

For $W/b \geq 2$:

$$\left.\begin{aligned} \frac{W}{b} &= \frac{2}{\pi}\left[B - 1 - \ln(2B - 1) + \frac{\epsilon_r - 1}{2\epsilon_r}\left\{\ln(B - 1) + 0.39 - \frac{0.61}{\epsilon_r}\right\}\right] \\ \text{where} \\ B &= \frac{377\pi}{2\sqrt{\epsilon_r}\, Z_0} \end{aligned}\right\} \tag{1-126}$$

For the final computation, which will cap the preceding example, we use Eq. 1-111 to compute the phase velocity:

$$V_P = \frac{c_0}{\sqrt{\epsilon_{\text{eff}}}} = 1.71 \times 10^8 \text{ m/s}$$

The time delay for the 0.1 m length would be

$$T = \frac{l}{V_P} = \frac{0.1}{1.71 \times 10^8} = 0.584 \text{ ns}$$

When more accurate results are desired, the thickness of the strip can be taken into account by computing an effective width, W_e, which is used to replace W in the preceding expressions for Z_0 and ϵ_{eff}. The expressions developed by Schneider [48] and Wheeler [62] are:

$$W_e = W + \frac{t}{\pi}\left[1 + \ln\left(\frac{2b}{t}\right)\right] \qquad \text{for} \quad \frac{W}{b} \geq \frac{1}{2\pi} \tag{1-127}$$

or

$$W_e = W + \frac{t}{\pi}\left[1 + \ln\left(\frac{4\pi W}{t}\right)\right] \qquad \text{for} \quad \frac{W}{b} \leq \frac{1}{2\pi} \tag{1-128}$$

These formulas also are both subject to the further restrictions

$$t \leq b \qquad \text{and} \qquad t < \frac{W}{2}$$

which are in practice usually satisfied.

Exercises

1-9.2a Design a microstrip system that is to be built on an aluminum oxide dielectric 0.8 mm thick with a metal thickness of 0.15 mm. $Z_0 = 50\ \Omega$. Compare this result with that of the example above and comment on the effect of the dielectric. How are the dimensions of the aluminum oxide system for practicality?

1-9.2b For the microstrip example, compute the wavelength on the line for $f = 3$ GHz. (Answer: 5.76 cm) Repeat for Exercise 1-9.2a. For each system, compute the length of line required to obtain an impedance of $+j100\ \Omega$ (line open circuit).

1-9.2c A typical IC chip is about 1 cm in length. Suppose 10 such chips tie to a common line in a microstrip configuration. The dielectric is epoxy-glass ($\epsilon_r = 4.9$) and is 1 mm thick. The line to which the chips are connected is 20 cm long and 2 mm wide. What is the approximate value of capacitance added to the pin of each chip, assuming the chips uniformly distributed along the line? The copper foil has thickness 0.04 mm. Estimate the total line capacitance two ways: (1) Using the direct or plate-to-plate capacitance between the signal line strip and the ground plane (using parallel plane formula); and (2) computing Z_0 and ϵ_{eff} and then determining the line capacitance per meter using Eq. 1-113.
Comment on the effect of this capacitance on logic chip delay time, and compare the added capacitance to that already present on a typical logic chip, say, TTL.

1-9.2d From physical considerations one might be led to conclude that the capacitance per meter for a stripline might be about twice that for microstrip. To show that this is fairly accurate, select the 1/32-in. epoxy fiberglass board with a strip width of 0.05 in. from Figure 1-56. Use Eq. 1-124 and prior equations to compute the line C. Now compute the value of C for a stripline made from the same material using the same center strip width. How do they compare?

1-9.2e Discuss the possibility of solving Eq. 1-123 for W/b in closed form. Using ϵ_r as a parameter having values 1, 2, 3, 4, 6, 8, 10, and 15, successively, plot Z_0 as a function of W/b for W/b having the range $1 \leq W/b \leq 20$. (The maximum Z_0 will be on the order of $100\ \Omega$ for $\epsilon_r = 1$ and $W/b = 1$.)

Summary, Objectives, and General Exercises

Summary

In this chapter we have learned many introductory techniques that are useful in their own right for solving transmission-line and transmission-line–related problems. More important, however, is that we have seen how time and space variables can be handled in the same problem. We have learned how to recognize solutions for spatially varying propagating signals and the differential equations that describe their behavior.

For current and voltage sinusoidal steady-state solutions, we learned how to express attenuation and phase as exponentials that have a space variable present. We examined incident and reflected voltages on lines having load terminations different from the characteristic impedance, and spent some time investigating the sinusoidal steady-state properties of lossless line systems.

Although the currents and voltages are functions of time and one coordinate variable, they are not vector quantities since the equations developed did not depend upon orientation in space or a particular coordinate system. The coordinate z is a spatial variable used exclusively to locate points within the transmission line, not to ascribe spatial direction to electrical parameters. It is true that the current flows in the z direction, but the value does not change with respect to the transmission line if the line is moved in space. Angular quantities used to express currents and voltages are phase (time) relationships, not spatial angles. Thus although we used complex numbers, we were dealing with scalars (phasors) in the sinusoidal steady state, not vectors. In a later chapter we investigate, in detail, the properties of scalars and vectors.

In addition to the general theory of transmission lines, some specific applications were given along with an introduction to the general theory of matching transmission systems. A short discussion of the transmission line properties of printed circuit and hybrid circuit lines extended the applications to systems other than coaxial- or parallel-wire lines. In the next chapter, some specific calculation aids and matching techniques are presented. Also, the propagation of pulses and transients on lines is covered in Chapter 3.

Some of the more important concepts of this chapter are presented in the following section as objectives to be learned. Many of these ideas will be used in the next two chapters and in the study of propagating electromagnetic waves.

Objectives

1. Know the definition of a distributed system.
2. Be able to identify the R, L, G, and C parameters in a distributed system, and to draw the equivalent circuit for the system.
3. Be able to obtain the current and voltage differential equations for a distributed system, and be able to solve those differential equations to obtain a wave equation.
4. Be able to give the names and units of all symbols appearing in the solution of the transmission line equations, and to give their physical interpretations.
5. Be able to convert differential equations to their sinusoidal steady-state form, and to use the operator $\mathrm{Re}\{\cdot\}$ to convert phasor representations to explicit time forms.
6. Be able to identify the propagation constant in a sinusoidal steady-state formulation, and to explain its physical significance.
7. Be able to explain the differences in performance of lossless and low-loss transmission lines, and to know the similarities.
8. Be able to give the physical interpretations of V^+, V^-, I^+, and I^-.
9. Know and be able to state the definition of *voltage reflection coefficient* and its range of values. Be able to calculate Γ_L and $\Gamma(z)$ for given $\bar{Z}_L$, Z_0, line parameters and frequency of operation.
10. Know and be able to calculate the distance between a voltage maximum and a voltage minimum on a transmission line.
11. Be able to describe the difference between a standing wave and a traveling wave.
12. Know the definition of *phase velocity* and how to calculate it from a given β. Be able to give a physical interpretation of the phase velocity.
13. Know the definition of *voltage standing wave ratio*, and be able to give its range of values and its relationship to the reflection coefficient. Be able to compute the value of the standing wave ratio for a given line and load and from the reflection coefficient.
14. Be able to calculate the incident, transmitted, and reflected powers at a load on a transmission line.
15. Know the definitions of return loss and mismatch (reflected) loss expressed in dB.
16. Be able to calculate the impedance at any point on a transmission line using the impedance formula $Z(-l)$ and the properties of the transmission line (Z_0, C, etc.).
17. Know and be able to state the effects of placing transmission lines of lengths $n\lambda/4$, $n = 1, 2, 3$ in a line not terminated in Z_0 as far as the input impedance seen by a generator is concerned.

18. Be able to describe the input impedance of a cable that has an open circuit at the load end and a short circuit at the load end, and the corresponding standing wave patterns for current and voltage.
19. Be able to state reasons for matching a transmission system, and be able to state and apply the basic theory of line matching.
20. Know the definition of *characteristic impedance* and its expression for a general line and for a lossless line.
21. Be able to describe and compare the two waveforms that appear on an oscilloscope when simultaneously monitoring the total line voltages at two different points on a transmission line.
22. Given the cable type, load, operating frequency, and voltage source phasor, be able to identify the parameters required to compute the current and voltage phasors at any point on a line.
23. Be able to obtain the element values for a lumped element equivalent circuit corresponding to the input impedance of a loaded transmission line.
24. Be able to use the equations for the design of stripline and microstrip transmission lines.

General Exercises

Note: All of the exercises below can be worked using the equations and concepts developed in this chapter. However, in the next chapter a graphical technique is developed which simplifies the solutions to these exercises.

GE1-1 Determine the phase velocity of a traveling wave in the RG58B/U coax of Exercise 1-6.4e, assuming a lossless condition. How long would it take a traveling wave to move 0.3 m along this line? If a narrow rectangular pulse were applied, what is the time skew (delay) between the input and output pulses? Comment on the consequences of this delay on a 5-ns pulse.

GE1-2 ECL in a computer system is being clocked at a rate of 150 MHz. The chips are mounted on PC boards having lines and dimensions shown in Figure 1-57. The outputs of two NOR gates are sent to a NAND gate. The line characteristic impedance is 350 Ω. Estimate the line capacitance parameter using the parallel plate formula from physics (neglecting fringing), and determine the largest distance D that is allowable to have the NAND gate inputs overlap at least 80%. Discuss qualitatively the effect of reducing the line width to 1 mm. Note that signal sources are placed between the thin strip and bottom copper ground plane.

GE1-3 The line parameter expressions for phase velocity given in Eq. 1-74 assumed lossless or low-loss cables. Using the general expression for β for a lossy line obtain an equation for the phase velocity in terms of line parameters. Using the computer obtain a plot of the phase velocity V_P vs. frequency f for the RG58B/U cable for the conditions given in Exercise 1-6.4e. The frequency dependence of phase velocity is called *dispersion*. Over what frequency range is dispersion negligible?

GE1-4 Derive the expression for the phase velocity of the neuron model given in Exercise 1-4.0a. (See Exercise 1-5.1f, also.) Discuss the effect of neural scarring on V_P.

GE1-5 Derive the expression for phase velocity on a lossless line in terms of Z_0 and C.

Figure 1-57.

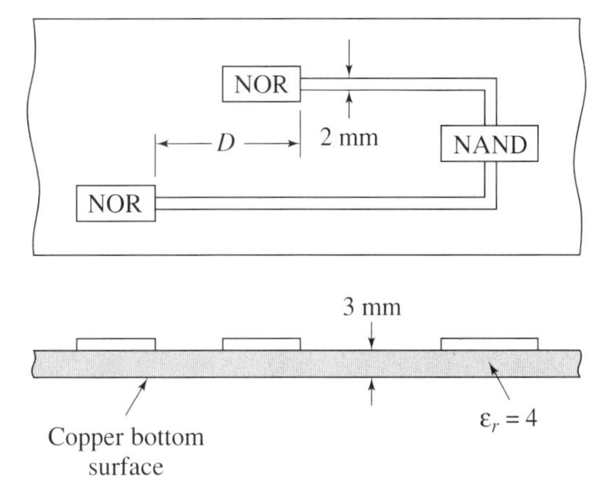

GE1-6 Suppose a lossless transmission line system is set up so the voltage standing wave pattern can be monitored. The line is initially shorted and the position of one of the nulls is recorded. The short is removed and replaced with a comlex load that has a capacitive component. Determine the possible locations of the null if the capacitive reactance varies from $-j0$ to $-j\infty$. Compare the locations with the location of the short circuit null. Repeat for a complex load which has an inductive component varying from $j0$ to $j\infty$. Does it matter whether the real part of the load is less than or greater than Z_0? From your results formulate a rule for determining the nature of an unknown load by observing the null shift from the short circuited line null.

GE1-7 In Section 1-9 it was shown that the phase velocity on a transmission line of two conductors can usually be computed by knowing only the properties μ and ϵ of the material between the conductors. (The conditions for which this is true will be covered in a later chapter.) The expression given was $V_P = c_0/\sqrt{\mu_r \epsilon_r}$ where c_0 is the velocity of light in vacuum and μ_r and ϵ_r are the relative permeability and dielectric constants of the material as you studied in physics. Suppose a coaxial system has air dielectric, and it is 0.3 m long. Compute the electrical length (i.e., the length of the line in wavelengths) and the corresponding values of β for frequencies of 10 MHz, 100 MHz, and 1000 MHz. For each frequency compute the phase shift of a single positive traveling sine wave along the line section. From your results discuss the conditions under which the effect of propagation may be neglected. Repeat for a coax filled with polystyrene, $\epsilon_r = 2.55$. $\mu_r = 1$.

GE1-8 On a transmission line with $Z_0 = 50\ \Omega$ the voltage at a distance $l = 0.4\lambda$ from the load is $40 + j\ 20$ V. The corresponding current is $+2$ mA. Determine the load impedance. Assume a lossless line.

GE1-9 A transmission system is used to couple the output of a tuned circuit to a remote load. The cable is RG58B/U and the system is to be operted at resonance. What is the operating frequency and Q of the network before the cable is connected? Next determine the resonant frequency and Q of the system with the cable connected. The results of this exercise show why we often speak of ''unloaded Q'' and ''loaded Q.'' For the short length of cable involved here we may assume a lossless line. Also determine the operating bandwidths for the two cases. The system is shown in Figure 1-58.

GE1-10 A gallium arsenide field effect transistor (GaAsFET) is used as an oscillator in a system that is built in the microstrip configuration shown in Figure 1-59. The characteristic impedance of the line is 50 Ω, and the dielectric constant of the dielectric is 20. The output of the oscillator is used both in a local mixer FET, which has an input capacitance of 4 pF (also a very high resistance component which is neglected here), and to feed another part of the unit, which is matched to the line. What is the effective impedance seen by the oscillator? What would you expect the effect of the transmission system on the oscillator to be? $f = 3$ GHz, $C = 200$ pF/m. In the configuration shown, signals are applied be-

Figure 1-58.

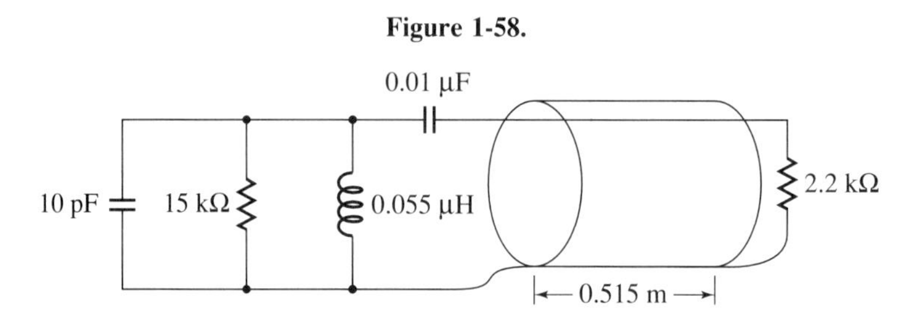

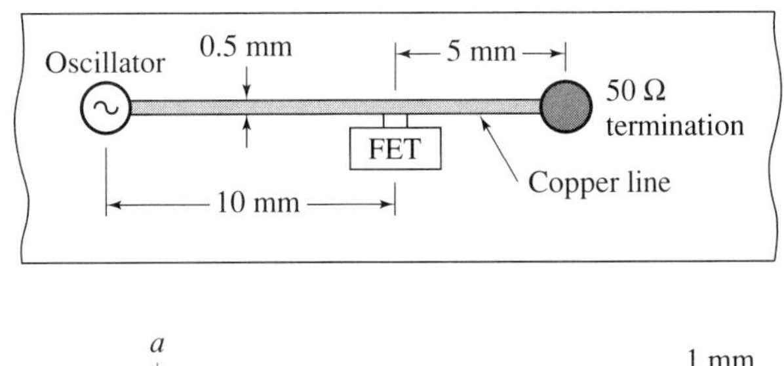

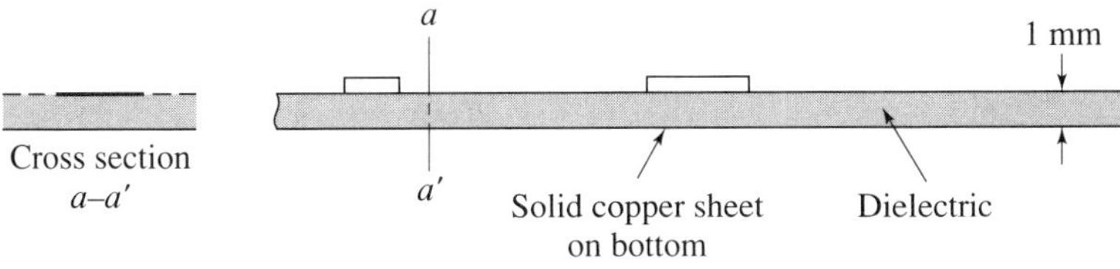

Figure 1-59.

tween the narrow copper trace and the bottom ground plane. This, of course, means that the FET input capacitance appears to be connected between these two conductors, effectively in parallel with the transmission line.

GE1-11 In the circuit of Figure 1-60, what is the smallest value of l_λ that will make the resistive part of $\overline{Z}(l_\lambda)$ equal to 150 Ω at plane P_1? Find the required value of $j\overline{X}$ added in series that will make the impedance 150 Ω real at plane P_2. Assume here that there is essentially zero physical displacement between the planes P_1 and P_2, with only a small reactance placed in the line. Find the lengths of open stub and shorted stub that could be used to obtain the $j\overline{X}$. (Same Z_0 for stubs.) Give lengths in fractions of λ_l.

GE1-12 Prove that the magnitude of the reflection coefficient is always 1 for a lossless transmission line that has a purely reactive load of any value.

Figure 1-60.

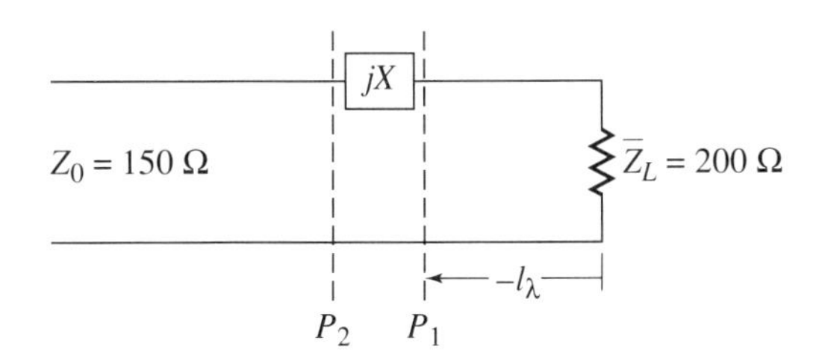

GE1-13 Show that resistive loads of value R_{L_1} and R_{L_2} will produce the same standing wave ratio if $R_{L_1}R_{L_2} = Z_0^2$ (lossless line). Also, show that $S = R_{L_2}/Z_0$ if $R_{L_2} > R_{L_1}$.

GE1-14 The transmission line shown in Figure 1-61 is terminated in a resistance whose value is less than Z_0. Sketch the voltage and current standing wave patterns. Label the distance in fractions of λ. Now suppose that a capacitance is placed in series with R_L. Now sketch the sanding wave patterns for current and voltage and describe what has happened to the nulls in the voltage standing wave pattern compared with those for the resistive load. (I.e., which way have the nulls shifted?) Repeat for the case of an inductor placed in series with the load.

GE1-15 The transmission line shown in Figure 1-62 is terminated in a resistance whose value is greater than Z_0. Sketch the voltage and current standing wave patterns. Label the distance in fractions of λ. Now suppose that a capacitance is placed in series with R_L. Sketch the resulting

Figure 1-61.

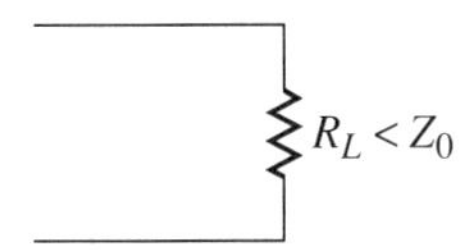

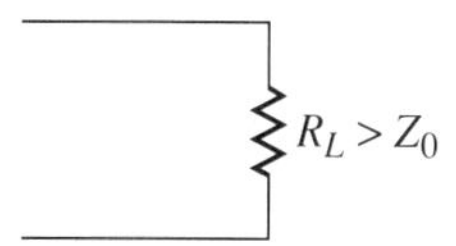

Figure 1-62.

standing wave patterns for current and voltage, and describe what has happened to the nulls in the voltage standing wave pattern compared with those for the resistive load (i.e., which way have the nulls shifted?) Repeat for the case of an inductor placed in series with the load. Compare the results of this exercise with those of GE1.14 and see if you can draw any conclusions.

GE1-16 A transmission line is initially terminated in a short circuit. Sketch the standing wave pattern for voltage. If the short is now replaced by a resistor having value less than Z_0 what is the location of a standing wave null with respect to that for the short circuit? Next, the line is terminated with a resistor having value greater than Z_0. Now what is the location of the null with respect to the short circuit null? If a load having an inductive component is placed as the load where is the standing wave null with respect to the short circuit null? Indicate on a diagram the total range of null shift for inductive values from 0 to ∞. If a capacitive load is placed on the line at the load end, which way is the standing wave null shifted with respect to the short circuit null? On a single diagram summarize the results of this exercise by showing the direction of shift of the various loads with respect to the short circuit nulls.

GE1-17 A uniform coaxial cable has a characteristic impedance of 75 Ω, and is terminated in a load value of 225 + j225 Ω.

(a) What is the normalized impedance 0.25λ toward the generator?

(b) What is the voltage standing wave ratio on the line?

(c) How far from the load (in terms of wavelengths) is the first maximum on the standing wave pattern?

GE1-18 A coaxial line has an air dielectric and as such has a velocity of propagation of signals equal to the speed of light in vacuum ($V_P = c_0$). (Cables having dielectric other than air have phase velocities less than c_0 by a factor of $\sqrt{\epsilon_r}$ where ϵ_r is the relative dielectric constant of the cable dielectric.) A 50-Ω coaxial air filled line is loaded with 125 Ω resistance in series with 50 Ω of inductive reactance. The frequency is 3 GHz.

(a) What is the line wavelength λ_l?

(b) What is the normalized load impedance?

(c) What is the standing wave ratio on the line?

(d) What is the impedance 2.5 cm toward the generator? At 12.5 cm toward the generator?

GE1-19 Figure 1-63 shows one popular way to set up the scale of distances on a slotted line system. In the microwave and UHF regions of frequency the scale is in centimeters. Since the properties of a transmission line (impedance, standing wave pattern, etc.) are periodic along the line and repeat every half wavelength, it is not necessary, or even possible in practice, to place the slotted line directly at the load. Thus, measurements are often made at points remote from load, but this is not important for a lossless line.

For example, if a short circuit were placed at the load, any null that would be observed in the region of the slotted line is just the same as if it were at the load end where the short were placed. Thus, all measurements can be made relative to some reference on the scale of the slotted line section. One particular system is measured and found to yield a standing wave ratio of 1.9 and a minimum is found at 9.0 cm. When a short is placed at the load end the minimums are located at 7.0 and 12.0 cm.

(a) What is the value of λ_l?

(b) How much did the null shift (in fractions of a wavelength, 0.25λ or less)? In which direction?

(c) What is the impedance of the load at the reference plane (i.e., at 7.0 cm)?

GE1-20 A slotted line measurement using a system described in Figure 1-63 results in an S of

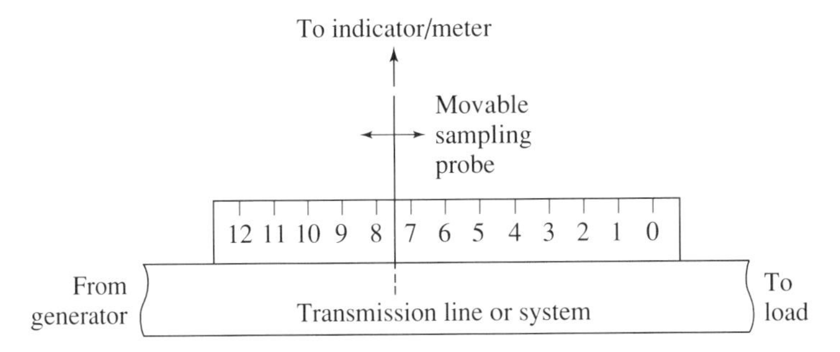

Figure 1-63.

5.1. The *shorted* minimums occur at 9.2 cm and 12.4 cm. The *loaded* minimum occurs at 11.6 cm. What is the normalized load impedance?

GE1-21 A particular transmission line, when terminated in a normalized impedance $1.25 - j0.42$ Ω, has a normalized input impedance of 0.6472 Ω. If the same transmission line were terminated in a normalized admittance $1.25 - j0.42$ S, what would be the value of its normalized input admittance? (Answer: 0.6472 S)

GE1-22 Determine the range of reflection coefficient phase angles that can be produced by any normalized impedance of the form $\tilde{Z} = 5.0 \pm j\tilde{X}$, where $0 < \tilde{X} < \infty$. (Answer: $-11.5° < \phi < 11.5°$.)

GE1-23 Show that the normalized input *impedance* of any section of transmission line terminated in a short circuit is equal numerically to the normalized input *admittance* of the same line section terminated in an open circuit.

GE1-24 A lossless transmission line 1.25 wavelengths long having a characteristic impedance of 50 Ω is terminated by the input terminals of a second lossless transmission line 1.25 wavelengths long having a characteristic impedance of 75 Ω. The second line is terminated in a pure resistance of 100 Ω. Determine the input impedance of the first line. (Answer: 44.4 Ω)

GE1-25 A normalized load $\tilde{Z}_L = 1.5$ is placed on a lossless transmission line. Does a voltage maximum or a voltage minimum occur at the load? Find the location and length of a single shorted shunt stub that will match the load at a frequency f_0, where the corresponding wavelength is λ_0. With the stub length and location and the load value fixed, suppose that the wavelength is increased by 10%. What is the standing wave ratio on the line? If at the wavelength λ_0 the stub is moved $\lambda/2$ toward the generator a match still exists, but now what is the standing wave ratio on the line when the wavelength is increased 10%?

GE1-26 What value of normalized admittance of the form $A - jA$ will produce a reflection coefficient of phase angle 45°? Repeat for $A + jA$ normalized admittance.

GE1-27 A negligible-loss transmission line has a characteristic impedance of 50 Ω. When three different terminal load impedances are separately connected to the line, the resulting standing wave patterns on the line are described respectively by the following data:

(a) $S = 2.35$, $l_{\min}\lambda = 0.395\lambda$
(b) $S = 2.35$, $l_{\min}\lambda = 0.271\lambda$
(c) $S = 2.35$, $l_{\min}\lambda = 0.062\lambda$

What will happen in each case to the voltage standing wave ratios when a 50-Ω resistance is connected in parallel with the terminal load impedance?

GE1-28 A transmission line is initially terminated in a short circuit and the positions of the voltage nulls determined. The short circuit is then removed and replaced by a load that has a capacitive component. Which way did the minima shift with respect to the short circuit nulls, toward the generator or toward the load?

GE1-29 For the general lossy cable ($\gamma = \alpha + j\beta$), start with Eqs. 1-25, 1-26, and 1-27 and show that the impedance at a general line location $z = -l$ is given by

$$\overline{Z}(-l) = Z_0 \frac{\overline{Z}_L + Z_0 \tanh(\gamma l)}{\overline{Z}_L \tanh(\gamma l) + Z_0}$$

GE1-30 For the general lossy line, what is the limiting value of $\overline{Z}(-l)$ for very large l? Explain the physical significance of this result.

GE1-31 In Exercise 1-7.4i the losses were neglected. The RG34A/U cable actually has an attenuation of 0.125 dB/m. Determine the power delivered by the generator and the powers that reach each load. What percentage of the input power is absorbed by the cable? What do you conclude about the assumption of an ideal line? Is your answer to this last question dependent upon cable lengths?

GE1-32 To demonstrate the effects of attenuation and to investigate the validity of assuming the lossless condition for practical engineering systems that still have significant loss, carry through the calculations of Example 1-6 of Section 1.7.4 where the RG58/U cable has an attenuation of 0.574 dB/m at 1 GHz.

GE1-33 Repeat Exercise 1-7.4d under the assumption of a low loss cable with $\alpha = 0.009$ nep/m.

GE1-34 If we define a transmission coefficient as $\tau_L = \overline{V}_L/V^+$, show that

$$\tau_L = 1 + \Gamma_L = \frac{2\overline{Z}_L}{\overline{Z}_L + Z_0}$$

GE1-35 Consider the most general case where both the load $\overline{Z}_L$ and the characteristic impedance are complex, but the real parts of both are positive. Define $Z_0 = R_0 + jX_0$ and $\overline{Z}_L = R_L + jX_L$ with a propagation constant that is complex as $\gamma = \alpha + j\beta$. The usual definitions still apply; namely, $Z_0 = \sqrt{Z/Y}$ and $\gamma = \sqrt{ZY}$, where Z and Y of the line also have positive real parts. Show that if we define

$$\Gamma(-l) = \Gamma_L e^{-2\gamma l} = \rho(-l) e^{j\theta(-l)}$$

it is possible to have $\rho(-l) > 1$. Also find the maximum value of $\rho(-l)$. (Answer: $\rho_{\max}(-l) = 1 + \sqrt{2}$)

2 Graphical Solutions of Transmission Line and Transmission Line–like Systems: The Smith Chart

2-1 Introduction

In Chapter 1 we found it was necessary to solve transcendental (nonlinear) equations to obtain solutions to transmission line problems. Even a simple matching problem involved considerable complex number equation manipulations. In the early 1940s Phillip H. Smith [50] developed a graphical tool that greatly simplifies the solution of the matching problem discussed in Chapter 1.

The Smith chart* and some similar charts developed by others have proven to be among the most useful tools in engineering practice. In this chapter we develop the Smith chart and then in subsequent sections demonstrate some typical applications. We will, in fact, see that the graphical aid is applicable to the design of simple lumped element systems as well as to transmission lines and other distributed systems. The techniques are also applicable to the propagation and reflection of plane electromagnetic waves, but this will be delayed until field theory has been developed.

2-2 Theory of the Smith Chart

In the discussion of the theory of transmission line matching presented in Section 1-8, we learned that the reflection coefficient plays a key role in describing the system performance. Thus we begin with Eqs. 1-25, 1-29, 1-58, and 1-60 and develop the general reflection coefficient as follows, again using $\beta = 2\pi/\lambda_l$ so we can express lengths in terms of fractions of the line wavelength:

*"SMITH" is a registered trademark of the Analog Instruments Co., P.O. Box 950, New Providence, NJ 07974. All Smith charts appearing in figures have been reproduced with the courtesy of Analog Instruments Co., P.O. Box 950, New Providence, NJ 07974.

$$\begin{aligned}\Gamma(z = -l) &= \frac{\overline{V}^-(-l)}{\overline{V}^+(-l)} = \frac{V^- e^{-\gamma l}}{V^+ e^{\gamma l}} = \frac{V^-}{V^+} e^{-2\gamma l} = \Gamma_L e^{-2(\alpha + j\beta)l} \\ &= \rho_L e^{j\theta_L} e^{-2\alpha l} e^{-j2\frac{2\pi}{\lambda_l}l} = \rho_L e^{-2\alpha l} e^{j\left(\theta_L - 4\pi\frac{l}{\lambda_l}\right)}\end{aligned} \tag{2-1}$$

$$= \rho(-l) e^{j\theta(-l)} \tag{2-2}$$

where

$$\rho(-l) = \rho_L e^{-2\alpha l} \tag{2-3}$$

and

$$\theta(-l) = \theta_L - 4\pi\frac{l}{\lambda_l} \tag{2-4}$$

This is a very convenient form since we showed in Chapter 1 that ρ_L has maximum value of 1 (for positive resistance values). Thus a polar plot of $\Gamma(-l)$ can contain all impedance values within a circle of unit radius.

Now at the load $l = 0$, so $\Gamma(0) = \Gamma_L = \rho_L e^{j\theta_L}$. This locates the load reflection coefficient within the unit circle and is indicated as point a in Figure 2-1. Note that we measure angles in the usual way, positive being counterclockwise from the positive right axis and negative being clockwise. For convenience, θ_L is assumed to be a positive angle in the figure. At some point $-l$ from the load (toward the generator), two effects determine the polar coordinates of $\Gamma(-l)$ as compared with $\Gamma(0)$. The first is that the length of the radius has decreased due to the presence of the attenuation factor $e^{-2\alpha l}$, and the phase angle has changed by an amount $-4\pi(l/\lambda_l)$. The new point, denoted by b in the Figure 2-1, is clockwise from point a due to the negative angle change value. The actual angle value would be

$$\theta(-l) = \angle\Gamma(-l) = \theta_L - 4\pi\frac{l}{\lambda_l}$$

Figure 2-1. Plot of $\Gamma(-l)$ in polar form.

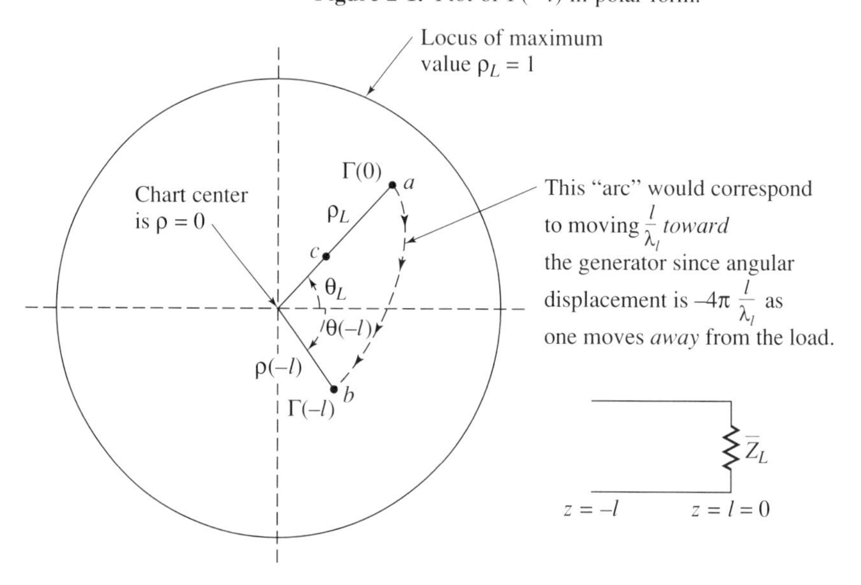

An important fact concerning angular displacement around the plot is obtained by supposing that in particular $l = \lambda_l/2$. The total change in angle from point a would then be

$$\Delta\theta = -4\pi \times \frac{\lambda_l/2}{\lambda_l} = -2\pi$$

Thus a *half* wavelength of distance along the line rotates the reflection coefficient *completely* around the chart (the length, of course, would be reduced due to the attenuation). This would correspond to point c in Figure 2-1.

In summary, for the general lossy line, we have the following plot characteristics:

1. $|\Gamma(-l)| = \rho(-l)$ decreases in value since there is an attenuation factor.
2. Moving toward the generator (away from the load) corresponds to moving clockwise around the chart.
3. Moving $\lambda_l/2$ meters along the line is equivalent to one full rotation around the plot.

A case of special interest is the lossless line. This case is appropriate for most engineering applications since many times relatively short lengths of cables having small attenuation are used. Setting $\alpha = 0$ gives the following result for the reflection coefficient:

$$\Gamma(-l) = \rho_L e^{j\left(\theta_L - 4\pi\frac{l}{\lambda_l}\right)} = \rho_L e^{j\theta(-l)} \tag{2-5}$$

Thus we observe that as we move toward the generator, l numerically larger, the radius (magnitude) remains constant and the angle decreases (moves clockwise). This is shown on the polar chart in Figure 2-2.

Figure 2-2. Movement of $\Gamma(-l)$ in the reflection coefficient plane for lossless line.

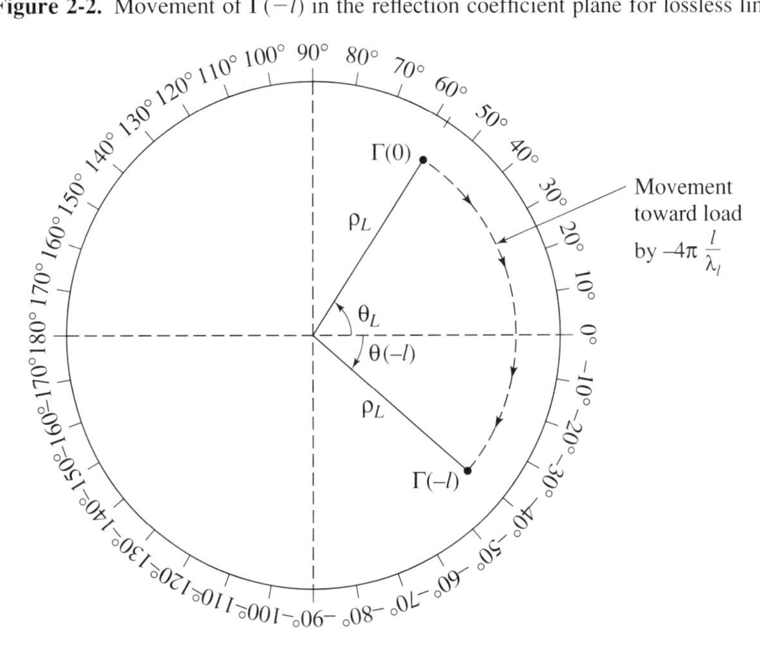

Since all points $\Gamma(-l)$ are located inside the unit circle, it would be permissible to place scales around the outer rim for convenience of locating angles. One would not need to have a protractor to locate a particular reflection coefficient. These angle scales are shown in the same figure. Note that, as expected, they are measured in the conventional way (i.e., from the positive horizontal). For example, to obtain the angle of the load reflection coefficient, $\Gamma(0)$, shown in Figure 2-2, we simply draw a line from the chart center through the point $\Gamma(0)$ to the outside rim. In this case, one would obtain a θ_L of about 57°.

To this point, all we have is the reflection coefficient on our chart. Although $\Gamma(-l)$ is an important parameter, information about the line impedance would be useful. From Eq. 1-97, we can obtain the expression for the normalized impedance at any line location as

$$\tilde{Z}(-l) = \frac{\bar{Z}(-l)}{Z_0} = \frac{1 + \Gamma(-l)}{1 - \Gamma(-l)} \tag{2-6}$$

This result tells us that since all points $\Gamma(-l)$ are inside the unit circle on the reflection coefficient chart, all corresponding impedance values are inside the same unit circle. (As before, we concern ourselves only with positive resistance components of impedance.)

To obtain a convenient form for the impedance, we observe that since both $\tilde{Z}(-l)$ and $\Gamma(-l)$ are complex they may be written in rectangular component form as

$$\tilde{Z}(-l) = \tilde{R}(-l) + j\tilde{X}(-l) \equiv \tilde{R} + j\tilde{X}$$

$$\Gamma(-l) = U(-l) + jV(-l) \equiv U + jV$$

In these results we have dropped the argument list $(-l)$ for ease of writing. We must remember that all components are functions of distance from the load. Putting these two expressions into Eq. 2-6 we obtain, after rationalizing the denominator on the right-hand side:

$$\begin{aligned} \tilde{R} + j\tilde{X} &= \frac{1 + U + jV}{1 - U - jV} \\ &= \frac{(1 + U)(1 - U) - V^2}{(1 - U)^2 + V^2} + j\frac{2V}{(1 - U)^2 + V^2} \end{aligned}$$

Equating the real and imaginary parts, we have

$$\tilde{R} = \frac{(1 + U)(1 - U) - V^2}{(1 - U)^2 + V^2}$$

$$\tilde{X} = \frac{2V}{(1 - U)^2 + V^2}$$

After some algebra these may be written, respectively,

$$\left.\begin{aligned} \left(U - \frac{\tilde{R}}{\tilde{R} + 1}\right)^2 + V^2 &= \frac{1}{(\tilde{R} + 1)^2} \\ (U - 1)^2 + \left(V - \frac{1}{\tilde{X}}\right)^2 &= \frac{1}{\tilde{X}^2} \end{aligned}\right\} \tag{2-7}$$

Now the previous figures we obtained were for $\Gamma(-l)$, so that in these last two results, U and V are the variables. In the reflection coefficient plane, U and V represent the real (horizontal) and imaginary (vertical) axes. Thus to plot the equations, we pick values of $\tilde{R}$ and $\tilde{X}$ and treat them as the parameters. With this point of view we see that both equations are equations of circles in the $U - V$ plane. For the $\tilde{R}$ equation the centers of the circles are at

$$\left(\frac{\tilde{R}}{\tilde{R}+1}, 0\right)$$

with radii

$$\frac{1}{\tilde{R}+1}$$

Similarly, for the $\tilde{X}$ equation the centers of the circles are at

$$\left(1, \frac{1}{\tilde{X}}\right)$$

with radii

$$\frac{1}{|\tilde{X}|}$$

Note that for this latter case, we have really *two* sets of circles, since the reactive component can be positive or negative (inductive or capacitive). Figure 2-3 is a plot of some circles for selected values of $\tilde{R}$ and $\tilde{X}$. You should take a minute to examine this set of circles and remember what happens as $\tilde{R}$ varies from 0 to ∞ and as $\tilde{X}$ varies from 0 to $\pm\infty$. Remember that the outside circle for $\tilde{R} = 0$ corresponds to $\rho_L = 1$.

The only portions of the figure of interest here are for passive linear transmission lines whose characteristics lie inside the $\tilde{R} = 0$ ($\rho_L = 1$) circle. The result is the classical Smith chart using rectangular impedance coordinates ($\tilde{Z}(-l)$) in place of reflection coefficient values. It is important to remember that each point on the chart has a corresponding impedance value and reflection coefficient. For example, point a in Figure 2-3 has $\tilde{Z} = 2 + j1$. The angle of the reflection coefficient at that point is about 28°. Can you verify these values? The magnitude of $\Gamma(-l)$ is about 0.45, being 45% of the distance from the origin to the chart rim.

To illustrate how to obtain the value of the reflection coefficient, Figure 2-4 shows a point $\tilde{Z}(-l)$, located originally at the intersection of the corresponding $\tilde{R}$ and $\tilde{X}$ circles. The $\tilde{R}$ and $\tilde{X}$ circles have been removed for figure clarity. We see that we are back to our original reflection coefficient chart. The angle of $\Gamma(-l)$ is measured with a protractor (or the outer scale as explained before) and the magnitude of $\Gamma(-l)$ computed as the ratio A/B; that is, the ratio of the distance from the center to the location $\tilde{Z}(-l)$ to the total chart radius. Any convenient measuring unit may be used since a ratio is taken.

To avoid having to make such trivial measurements, most charts have scales around the rim (the reflection coefficient angle has already been explained) and along the bottom

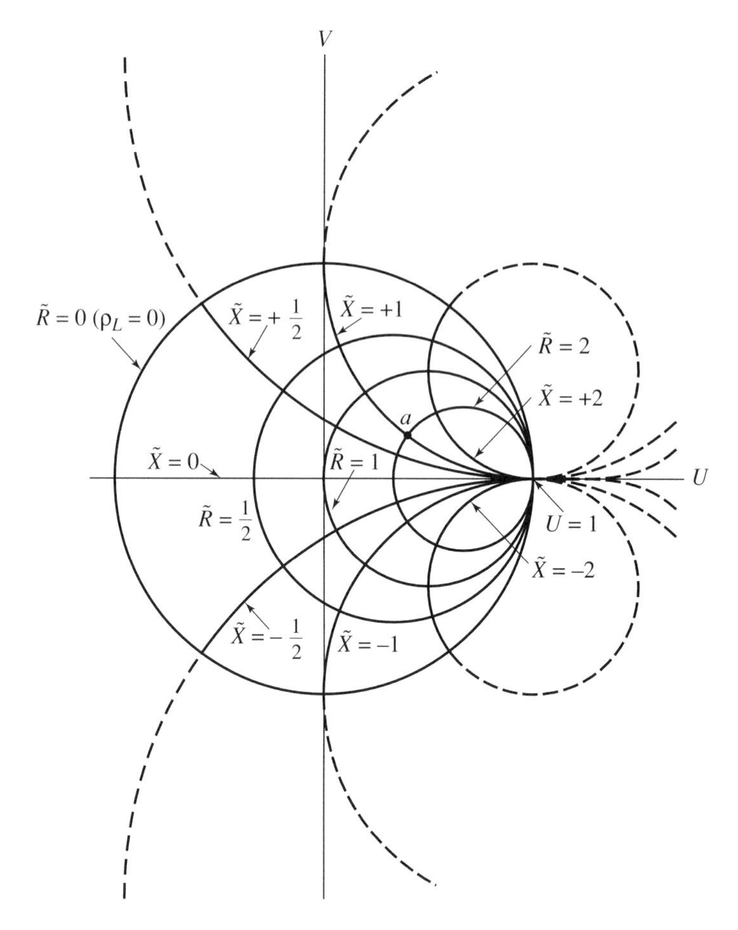

Figure 2-3. Plot of selected values of $\tilde{R}$ and $\tilde{X}$ circles in the reflection coefficient (Γ) plane.

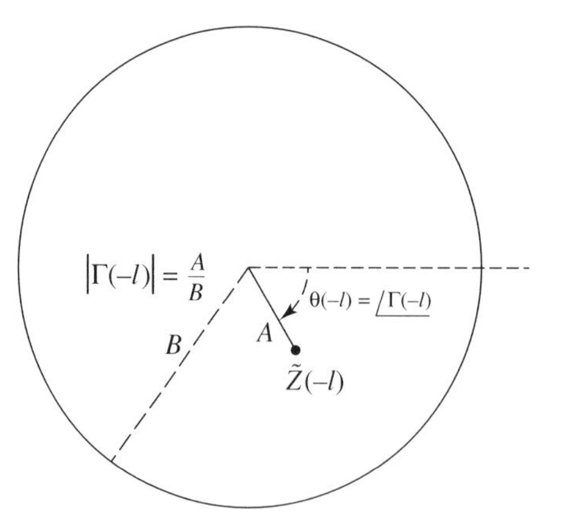

Figure 2-4. Determination of the reflection coefficient corresponding to an impedance at a general point on a line.

for simple direct reading of values. These scales will be explained more fully in the next section.

Exercises

2-2.0a Show that the $\tilde{R}$ = constant and $\tilde{X}$ = constant contours in the reflection coefficient plane are circles (i.e., verify Eq. 2-7).

2-2.0b What is the numerical (complex) value of the reflection coefficient at $\tilde{Z}(-l)$ shown in Figure 2-4? What are the corresponding values of $\tilde{Z}(-l)$ and $\overline{Z}(-l)$ if the transmission line characteristic impedance is 50 Ω? (Answer: $0.4\angle{-65°}$; $1.4\angle{-40.8°}$; $66.8\angle{-40.8°}$)

2-2.0c By writing Eq. 2-6 in admittance form by reciprocating it, show that we may use the Smith chart as an admittance-reflection coefficient chart by simply interpreting $\tilde{R}$ as $\tilde{G}$ and $\tilde{X}$ as $\tilde{B}$ (remembering signs change), reading coordinates in siemens directly. For example, the $\tilde{G} = \frac{1}{2}$ circle coincides with $\tilde{R} = \frac{1}{2}$. We would, of course, now normalize with respect to Y_0. To prove the equivalency, let $\tilde{Y}(-l) = \tilde{G} + j\tilde{B}$ and $\Gamma(-l) = U + jV$, and derive the equations for the circles. How would you obtain the value of reflection coefficient from an admittance value? (Check the angle closely.)

2-2.0d Suppose you are using the Smith chart with impedance values but would like to convert a given point on the chart into admittance. Prove that one can convert to admittance by simply inverting through the chart by 180° and then reading values on the chart directly as admittance. (Hint: π radians on the Smith chart corresponds to what angle change in transmission line movement?)

2-2.0e Point *a* on the Smith chart outline of Figure 2-3 corresponds to the impedance $\tilde{Z}(-l)$ looking to the right at location $z = -l$ on the transmission line. Suppose a resistance were placed in series with the line at that point. Show on the chart the direction one would move from point *a* to obtain the new value of impedance, and give the reason for your answer.

2-2.0f Prove that if we have $|\Gamma| > 1$ we may plot $1/\Gamma^*$ and use the Smith chart by interpreting all $\tilde{R}$ values as negative.

2-3 Fundamental Line Conditions: Impedance, Admittance, Voltage Standing Wave Ratio, Voltage Reflection Coefficient, and Attenuation

In this section, we will show how to use the Smith chart to perform analysis and design without having to carry out all the complex algebra of the equations. You will see that having been freed from the algebra, you can concentrate on what is happening in the system being investigated. As a word of caution, do not let the Smith chart become so mechanical that you lose physical insight into what is actually taking place.

The basic line structure that will be used in this and subsequent sections is shown in Figure 2-5. Note particularly that when a line impedance is being investigated, it is the impedance looking in the positive z direction. Also, $\overline{Z}(-l)$ is the ratio of total line phasor voltage to total line phasor current.

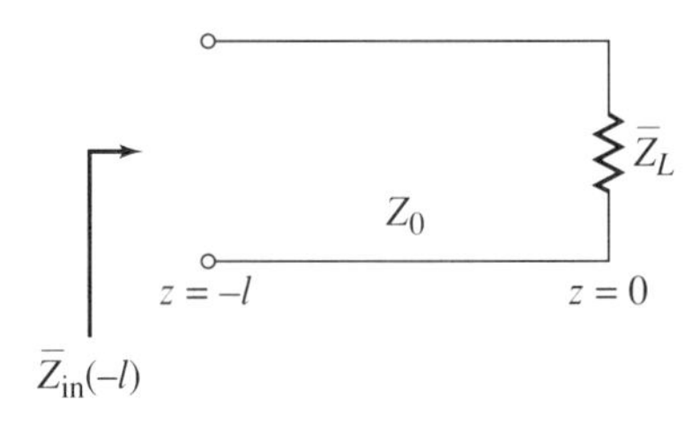

Figure 2-5. Transmission line parameter definitions and conventions.

2-3.1 Plotting $\bar{Z}_L$ on the Smith Chart

First let's learn how to enter the chart by plotting an impedance. Suppose, specifically, we are at the load and want to enter $\bar{Z}_L$. The general steps are as follows:

1. Calculate $\tilde{Z}_L = \bar{Z}_L/Z_0 = \tilde{R}_L + j\tilde{X}_L$.
2. Plot these coordinates on the chart, at the intersection of the appropriate constant resistance and constant reactance circles. The values of the constant resistance circles are marked along the horizontal chart diameter. The values of constant reactance circles are marked along the inside chart rim.

Example 2-1

Let $\bar{Z}_L = 60 + j43$ and $Z_0 = 50\ \Omega$.

$$\tilde{Z}_L = \frac{\bar{Z}_L}{Z_0} = \frac{60 + j43}{50} = 1.2 + j0.86$$

This is point ① in Figure 2-6 at the intersection.

Exercises

2-3.1a Plot the values $\bar{Z}_L = 20 - j35\ \Omega$ and $\bar{Z}_L = 400 + j500\ \Omega$ on a Smith chart using a 75-Ω transmission line. Comment on the accuracies of the plots of these two values.

2-3.1b Using the results of Exercise 2-2.0c, plot the load admittance $\bar{Y}_L = 0.0038 - j0.0089$ S for a 75-Ω line. (Answer: $0.285 - j0.6675$)

2-3.1c Suppose a 50-Ω transmission line is terminated in an impedance $50 + j50\ \Omega$. Plot this value on a Smith chart. Now, using the result of Exercise 2-2.0d, obtain the point on the Smith chart which represents $\tilde{Y}_L$. Convert the Smith chart reading directly to $\bar{Y}_L$. (Answer: $0.5 - j0.5$; $0.01 - j0.01$ S)

2-3.2 Determining the Load Reflection Coefficient, Γ_L, the Return Loss, and Reflected Power Loss

Since the load reflection coefficient and reflected and incident powers are the next important items, we give a step-by-step procedure for obtaining these at a given load. At this point we introduce some special scales at the bottom of the chart, which have been added to reduce the computational effort. Since there are several printed versions of the chart

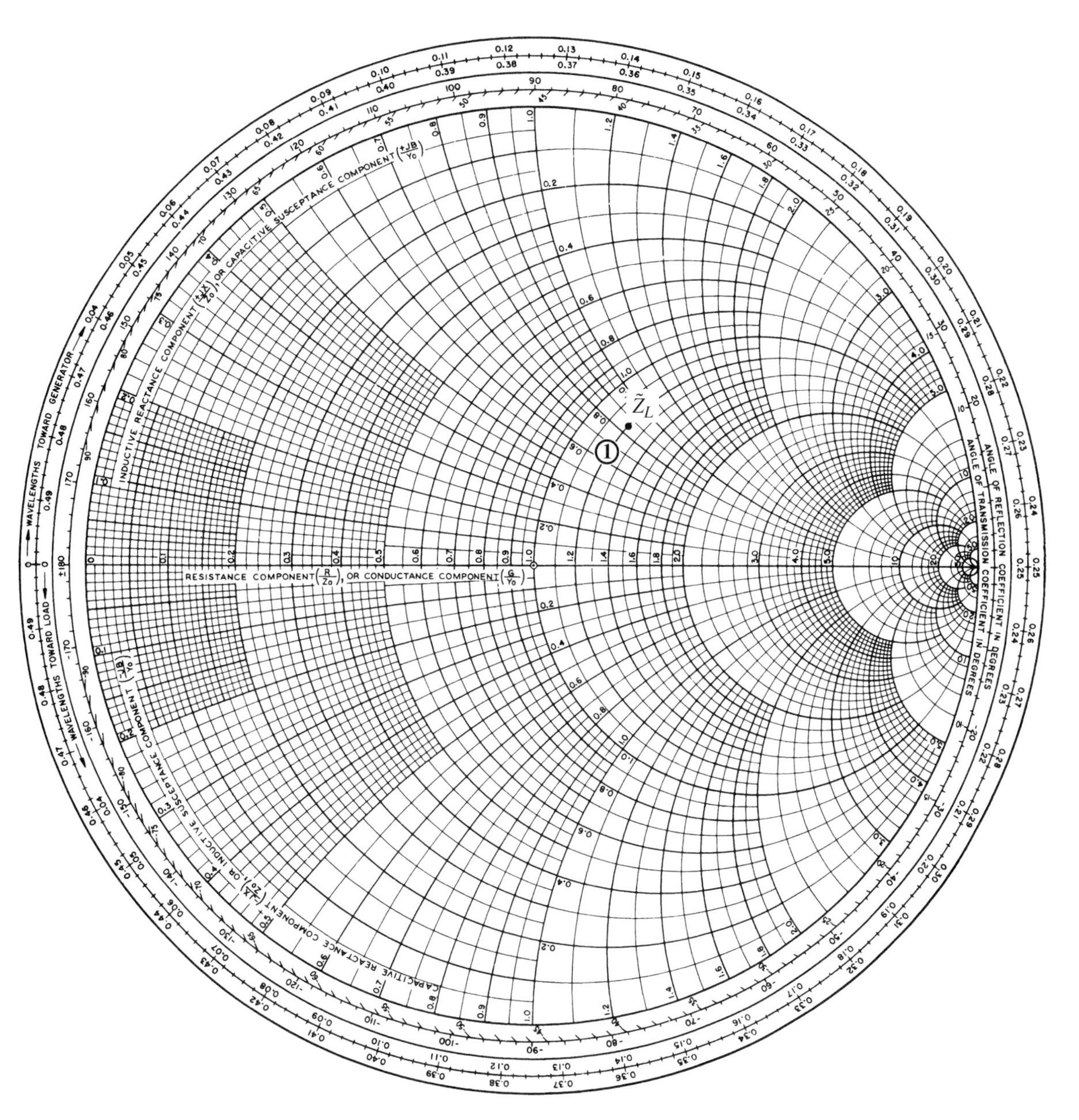

Figure 2-6. Plotting a load impedance.

these lower scales may be in different locations, but the process is the same by simply using the correct scale. We proceed as follows:

1. Plot $\bar{Z}_L$.
2. Draw a line from the chart center ($\tilde{R} = 1$) through the point $\tilde{Z}_L$ to the outside of the chart.
3. With a divider or compass, transfer the distance from the origin to the point $\tilde{Z}_L$ down to the scale labeled ''refl. coef., E or I'' and read $|\Gamma_L| = \rho_L$. The distance is laid out by placing one divider pointer at ''Center'' and the other point on the scale.
4. To obtain $\angle\Gamma_L = \theta_L$, read the degrees at the point on the chart rim where the line intersects the rim identified by the label ''angle of reflection coefficient in degrees.''
5. The ratio of the reflected power to the incident power, P_r/P_i, is read similarly from the contiguous scale labeled ''refl. coef., P.'' Note that this number is simply the square of the voltage reflection coefficient. (See Eq. 1-91.)
6. The return loss in decibels is obtained by reading the number right next to the power reflection coefficient value from the scale labeled ''RET'N LOSS, dB.'' This scale is the value of the defining relationship

$$\text{return loss, dB} = -10 \log_{10} \rho_L^2 = -10 \log_{10}(P_r/P_i)$$

7. The reflected or mismatch loss in decibels is the value read directly from the scale labeled ''REFL. LOSS, dB'' by transferring the distance of step 3 to that scale and is the value of the defining expression

$$\text{reflected loss, dB} = -10 \log_{10}(1 - \rho_L^2) = -10 \log_{10}(P_t/P_i)$$

At this point, a word of help is offered to make the Smith chart examples easier to follow. Of necessity, a text has to show the entire solution at once on the Smith chart. What one should do is either take a separate chart and construct the solution as it is read or else mentally blank out the arcs and lines on the figures and imagine adding them one at a time as the solution is carried out.

Example 2-2

Let $\bar{Z}_L = 60 + j43\ \Omega$ and $Z_0 = 50\ \Omega$. The power being absorbed by the load is measured as 80 mW. Determine the load reflection coefficient and the incident and reflected powers.

1. Plot $\tilde{Z}_L$. This is point ① on the Smith chart in Figure 2-7.
2. Draw a line from origin through $\tilde{Z}_L$ to chart rim, line labeled ②.
3. Transfer the distance from chart center to pont ① down to the ''refl. coeff., E or I'' scale, line ③. Read $|\Gamma_L| = \rho_L = 0.37$, point ④.
4. Where line ② intersects degree scale near chart rim, read $\theta_L = +57°$, point ⑤. Then: $\Gamma_L = V^-/V^+ = 0.37\angle 57°$ (a complex number).
5. The ratio P_r/P_i is read at point ⑥ on the line labeled ''REFL. COEFF, P'' as 0.136, approximately. The values on this scale are simply the squares of the values on the voltage reflection coefficient scale.
6. The return loss is also read from point ⑥ on the adjacent scale, and is about 8.7 dB. Since the transmitted power is the same as the power actually absorbed by the load,

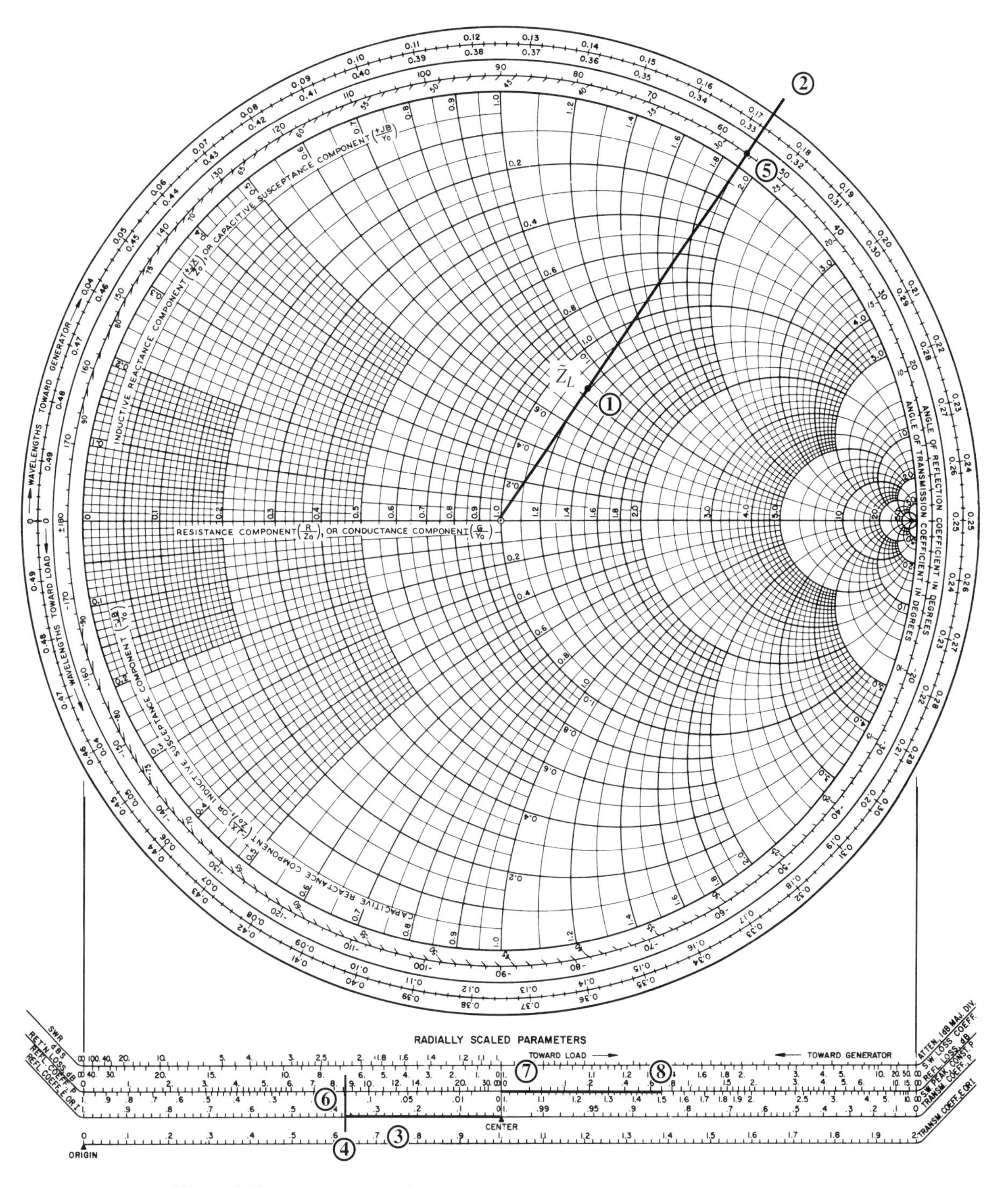

Figure 2-7. Determination of load reflection coefficient, return loss, and reflected loss.

$P_t = 80$ mW. Using the 8.7 dB value and the defining equation for return loss, we calculate the value of P_r as follows using $P_i = P_t + P_r$:

$$\text{return loss, dB} = -10 \log_{10}\left[\frac{P_r}{P_r + P_t}\right]$$

$$P_r = \frac{80 \times 10^{[8.7/(-10)]}}{1 - 10^{[8.7/(-10)]}} = 12.5 \text{ mW}$$

Note this is the same value we would have obtained using the ratio of step 5:

$$\frac{P_r}{P_i} = \frac{P_r}{P_t + P_r} = 0.136$$

$$P_r = \frac{0.136 P_t}{1 - 0.136} = \frac{(0.136)(80)}{0.862} \text{ mW} = 12.6 \text{ mW}$$

The small error is due to scale-reading accuracy.

7. Transfer the distance from the origin to $\tilde{Z}_L$ down to the "REFL. LOSS, dB" scale as line ⑦. Then, from point ⑧, reflected loss is 0.65 dB. From this, we can calculate the incident power from

$$\text{reflected loss, dB} = -10 \log_{10}\left(\frac{P_t}{P_i}\right)$$

$$P_i = \frac{80}{10^{[0.65/(-10)]}} = 92.9 \text{ mW}$$

(Note the order of determining the reflected loss or return loss is not important and which is obtained first is dictated by the particular problem at hand. Actually, in this particular case, it would have been more convenient to obtain P_i first.)

Exercises

2-3.2a Determine the load reflection coefficient of a load $\overline{Z}_L = 200 + j300\ \Omega$ that terminates a 300-Ω line. The admittance of this load is $\overline{Y}_L = 0.00154 - j0.0023\ \Omega$. Plot $\overline{Y}_L$ and determine the Γ_L following the same procedure as for $\overline{Z}_L$. Since the $\overline{Y}_L$ and $\overline{Z}_L$ represent the same load, why are the Γ_L values different? What is the angular difference between the two values? Develop a procedure for determining the load reflection coefficient for a given load admittance $\overline{Y}_L$ using the Smith chart.

2-3.2b Using the Smith chart determine the load impedance and load admittance corresponding to a reflection coefficient $\Gamma_L = 0.8\angle -50°$. Use a characteristic impedance of 50 Ω for the line.

2-3.2c A transmission line that has a characteristic impedance of 75 Ω is terminated with an impedance of value $75 + j75\ \Omega$. The power input from a generator is 2 W. Determine the reflection coefficient of the load, the reflected and return losses, and the incident and reflected powers at the load.

2-3.2d An antenna is fed by a transmission line and is to deliver 10 kW in the radiation field. The reflected power is to be below 10 W. Discuss the allowable reflection coefficient and the permissible load on a 75-Ω transmission line. How close do you think this could be approached in practice?

2-3.2e A generator has internal impedance of 300 Ω and is connected to a transmission line that has a characteristic impedance of 300 Ω. The load on the line is such that the return loss is 12 dB, and the power at the load is to be 100 mW. What is the required power that the generator must be capable of producing at its terminals? Determine the load reflection coefficient and the incident power.

2-3.3 Using the Smith Chart to Determine the Load Admittance $\overline{Y}_L$

Although the procedure for determining $\overline{Y}_L$ using the Smith chart has been used in some of the exercises, the steps are given here formally for completeness.

1. Plot $\tilde{Z}_L$.
2. Using a compass with the origin of the chart as center and distance to $\tilde{Z}_L$ as radius, move around the chart 180° (either direction). Note that in this case, we are not moving along a transmission line but simply inverting a complex number. See Exercise 2-2.0d.
3. Read $\tilde{Y}_L$. Remember that chart values are read as siemens now. As a matter of fact, it is often helpful to mark points on the chart carefully so clear distinction between impedance and admittance is made. If you return to the chart at a future date, it would be marked so a minimum retrace of steps would be required.
4. Calculate $\overline{Y}_L = \tilde{Y}_L/Z_0 = \tilde{Y}_L Y_0$.
(Note that this is valid for any transmission line lossy or lossless, since the rotation is not a movement along a transmission line but merely a reciprocation of a complex number.)
5. The procedure is the same for converting $\overline{Y}_L$ to $\overline{Z}_L$.

Example 2-3

Let $\overline{Z}_L = 60 + j43\ \Omega$ and $Z_0 = 50\ \Omega$. Determine $\overline{Y}_L$.

1. Plot $\tilde{Z}_L$. This is Point ① on the chart in Figure 2-8.
2. Rotate 180° along Arc ②.
3. At Point ③, read $\tilde{Y}_L = 0.56 - j0.375$.
4. $\overline{Y}_L = (0.56 - j0.375)/50 = 0.0112 - j0.0075$ S.

2-3.4 Determining the Input Impedance (or Admittance) at a General Location on a Line

Since the attenuation affects the impedance value (see the equation in General Exercise GE1-29, $\gamma = \alpha + j\beta$), two cases need to be covered: lossless and lossy. Note that for the low-loss case, Z_0 will still be a real number having the same value as for the lossless case. If, however, the frequency is very low so that R and G are not negligible, the complex value for Z_0 (Y_0) must be computed to be used as the normalizing value referred to in the following procedures.

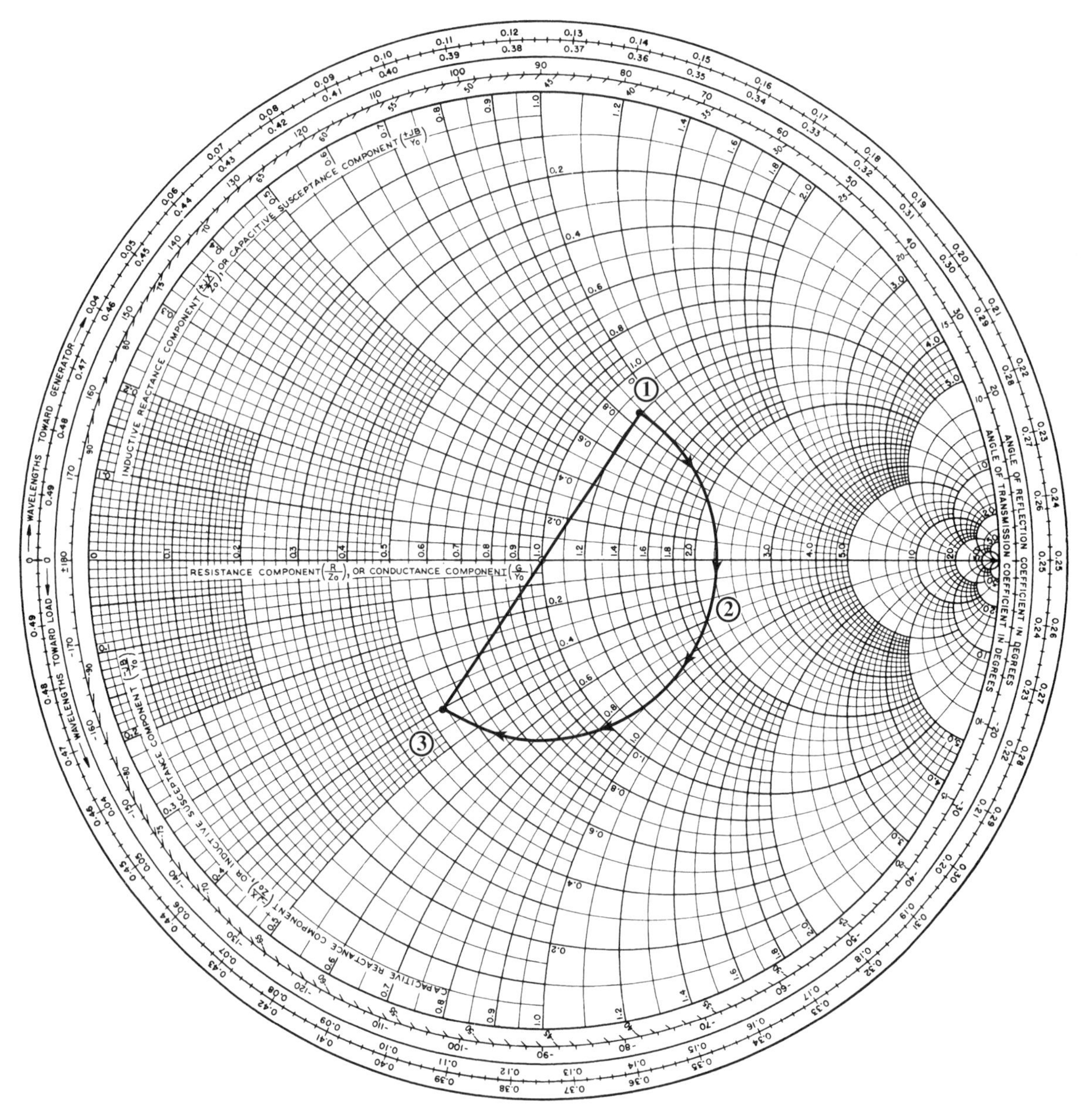

Figure 2-8. Determining $\overline{Y}_L$ from a given $\overline{Z}_L$.

Lossless Line

1. Plot $\tilde{Z}_L$ (or $\tilde{Y}_L$), and draw a line from center through $\tilde{Z}_L$ (or $\tilde{Y}_L$) to chart rim.
2. Calculate the fraction of a wavelength that l is (i.e., $l_\lambda = l/\lambda_l$ where λ_l is the wavelength on the line: $\lambda_l = V_p/f = (\omega/\beta)/f = 2\pi/\beta$).
3. Using a compass, draw an arc in the direction toward the generator (*cw*, see direction arrow on left chart rim) by an amount that increases the chart rim reading by λ_l. Use distance from chart origin to $\tilde{Z}_L$ (or $\tilde{Y}_L$) as radius.
4. Read value of $\tilde{Z}_{in}$ (or $\tilde{Y}_{in}$); $\bar{Z}_{in} = \tilde{Z}_{in} Z_0$. ($\bar{Y}_{in} = \tilde{Y}_{in} Y_0$)

Example 2-4

Let $\bar{Z}_L = 60 + j43\ \Omega$ and $Z_0 = 50\ \Omega$. Find the input impedance with $l = 0.32$ m and a line wavelength of $\lambda_l = 0.854$ m. (See the next Smith chart, Figure 2-9.)

1. Plot $\tilde{Z}_L$ as point ①, and draw line to chart rim, ②.
2. $l_\lambda = 0.32/0.854 = 0.375\lambda$.
3. Rotate a line toward the generator from line ② a distance 0.375λ to the reading $(0.1736 + 0.375)\lambda = 0.548\lambda$. Since chart stops at 0.5000λ, we need to go to an additional $(0.548 - 0.500)\lambda$ or 0.048λ, which is at line ③. The arc drawn from point ① to line ③ gives the solution at point ④.
4. $\tilde{Z}_{in}(-0.32) = 0.5 + j0.25$.

$$\bar{Z}_{in}(-0.32) = Z_{in}(-0.32)(50) = 25 + j12.5\ \Omega$$

Note how much easier this is than having to use the formula for line impedance, Eq. 1-98. Although we lose some accuracy at some points, the use of the Smith chart suffices for many applications.

Exercises

2-3.4a Given a load $\bar{Y}_L = 0.06 - j0.09$ on a 0.45-m length of RG58B/U cable (assumed lossless) at a frequency of 500 MHz, what is the input admittance of the system? What is the input impedance? (See Exercise 1-6.4e.) (Answer: $0.0027 - j0.0137$S; $13.6 + j70.5\ \Omega$)

2-3.4b Prove that $\lambda_l = 1/fCZ_0 = c_0/f\sqrt{\epsilon_r}$ where c_0 is the velocity of light in vacuum and ϵ_r is the relative dielectric constant of the dielectric between the conductors.

2-3.4c Work Exercise 1-7.1e using the Smith chart where possible.

2-3.4d Work Exercise 1-7.4i using the Smith chart where possible.

2-3.4e A transmission line has a characteristic impedance of 300 Ω and is terminated in a load impedance of value $450 - j200\ \Omega$. The line capacitance is 20 pF/m. At a frequency of 100 MHz, determine component values of a series circuit and of a parallel circuit which represent the input impedance of a 10-m length of this loaded line.

2-3.4f A length of RG34A/U cable (see properties in Exercise 1-7.4i) is connected to an antenna that has an impedance of $150 - j80\ \Omega$ at 250 MHz. The cable originally came from a roll marked with approximate ''length: 20 m.'' At the generator end, the impedance is measured and found to be $37.5 + j37.5\ \Omega$. What is the exact distance from the generator to the antenna assuming a lossless line? (Answer: 20.06 m)

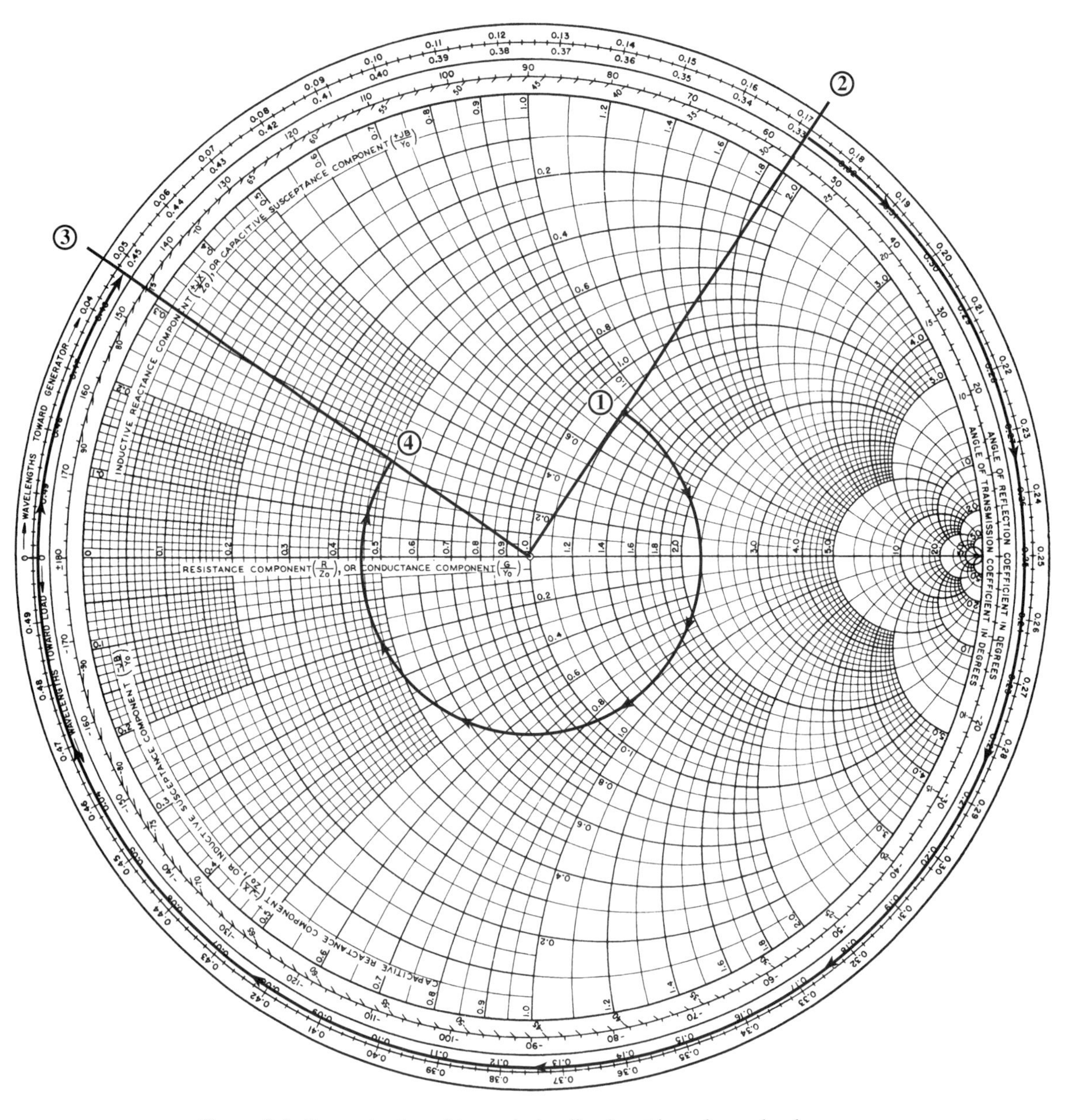

Figure 2-9. Determination of transmission line input impedance: lossless case.

Lossy Line

We have already shown that the reflection coefficient for a lossy line involves a magnitude change as the distance from the load increases. The result is

$$\Gamma(-l) = (\rho_L e^{-2\alpha l})e^{j(\theta_L - 2\beta l)}$$

We see from this equation that it is not possible to rotate at constant radius since now the magnitude of $\Gamma(-l)$, that is,

$$\rho(-l) = |\Gamma(-l)| = \rho_L e^{-2\alpha l} \tag{2-8}$$

decreases as we move toward the generator. We then need to move inward on a spiral.

Values of α at 1 GHz typically run on the order of 0.005 to 0.1 nep/m (0.04 to 0.9 dB/m). For very long cable runs such as those required by remote monitoring stations, however, the total attenuation could reach a significant level in a very few meters. For these cases, it is important to be able to account for the attenuation.

It is customary to specify line attenuation in decibels per meter ($\alpha_{\text{dB/m}}$) rather than in neper per meter, which is the unit of α used in the exponent of the attenuation term $e^{-2\alpha l}$.

The relationship derived in Chapter 1 is Eq. 1-55:

$$\alpha_{\text{dB/m}} = \frac{\alpha_{\text{dB}}}{l} = 8.686\alpha \tag{2-9}$$

Thus if we go to a table of cable attenuations in decibels per meter, we find α using this equation and then calculate $|\Gamma(-l)|$ using Eq. 2-8. However, the Smith chart's scale allows us to avoid making the arithmetic computation. Since several steps are involved, small Smith charts are drawn in Figure 2-10, as appropriate, for ease in following the process.

1. Plot $\tilde{Z}_L$ (or $\tilde{Y}_L$). Draw a line chart center through $\tilde{Z}_L$ (or $\tilde{Y}_L$) to the chart rim.
2. Determine $\alpha_{\text{dB/m}}$ from table or handbook equations, and then calculate $\alpha_{\text{dB}} = \alpha_{\text{dB/m}} \cdot l$ = total one way attenuation.
3. Calculate the length of the line in wavelengths using $l_\lambda = l/\lambda_l$.
4. Rotate the *line* drawn in part (*a*) toward the generator by an amount l_λ.
5. Now with a compass and distance from chart center to $\tilde{Z}_L$ (or $\tilde{Y}_L$) as radius, draw an arc that intersects the right-hand horizontal Smith chart axis (left-hand axis if the 1-dB scale is on the lower left side of the chart). Note that we may rotate either direction since here we are not moving along the line but merely locating a reference point on the horizontal chart axis.
6. Drop a vertical line from the intersection down to the scale labeled ATTEN., 1 dB MAJ. DIV at the lower right.
7. Now move along this scale (toward the generator) by an amount α_{dB}, and then vertically upward to the Smith chart right-hand axis again. The distance between marks is 1 dB per major division. Note that they are not uniformly spaced. Some charts may have subdivisions between the 1-dB marks. Examine your chart carefully.
8. With a compass, draw an arc using the distance from chart center to the final point on

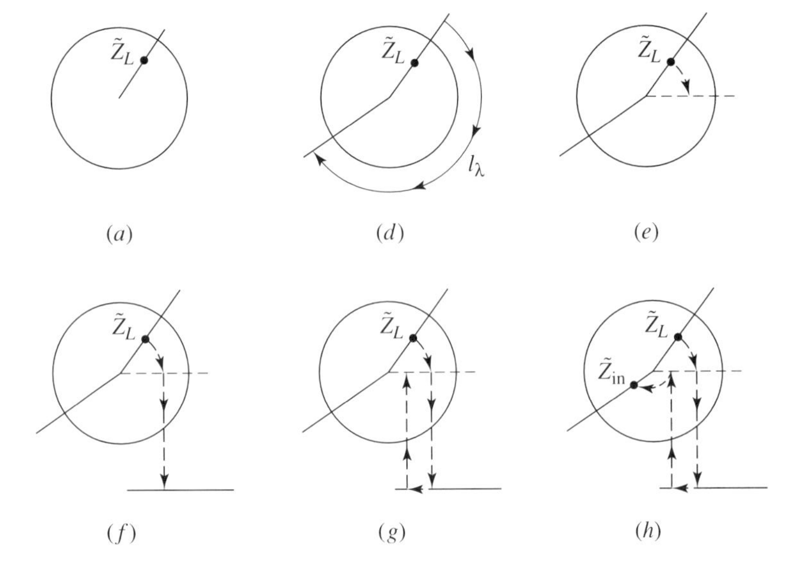

Figure 2-10. Steps for determining the input impedance on a lossy line (letters refer to steps in the procedure).

the Smith chart right axis as radius, to the rotated line drawn in step 4 (it may rotate either direction). This intersection is $\tilde{Z}_{in}$, from which

$$\overline{Z}_{in}(-l) = \tilde{Z}_{in} \cdot Z_0$$

9. We can determine $\Gamma(-l)$ at the generator using the point $\tilde{Z}_{in}$ as described in the procedure for $\Gamma(-l)$ given earlier.

Example 2-5

Let $\overline{Z}_L = 60 + j43\ \Omega$ and $Z_0 = 50\ \Omega$ and $\alpha_{dB/m} = 0.1$ dB/m with $l = 4.7$ m and $\lambda_l =$ 9 cm.

1. Plot $\tilde{Z}_L = \overline{Z}_L/50 = 1.2 + j0.86$ point ①. Chart in Figure 2-11. Draw line ②.
2. $\alpha_{dB} = (0.1)(4.7) = 0.47$ dB.
3. $l_\lambda = 4.7/0.09 = 52.222\lambda$.
4. Rotate line ② by 52.222λ (net 0.222λ) toward the generator to line ③. The final location of line ③ is at $(0.1735 + 0.222)\lambda = 0.3955\lambda$.
5. Rotate $\tilde{Z}_L$ point ① to right chart axis point ④.
6. Drop vertical line down to the ATTEN., 1 dB MAJ. DIV. scale (labeled ⑨) to point ⑤.
7. Move toward the generator (to left) by 0.47 dB to point ⑥, and then draw a line vertical to chart horizontal axis at point ⑦.
8. Rotate point ⑦ at constant radius to point ⑧. This yields $\tilde{Z}_{in}(-l) = 0.7 - j0.56$.

$$\overline{Z}_{in}(-l) = \tilde{Z}_{in}(l) \cdot 50 = 89 - j25\ \Omega$$

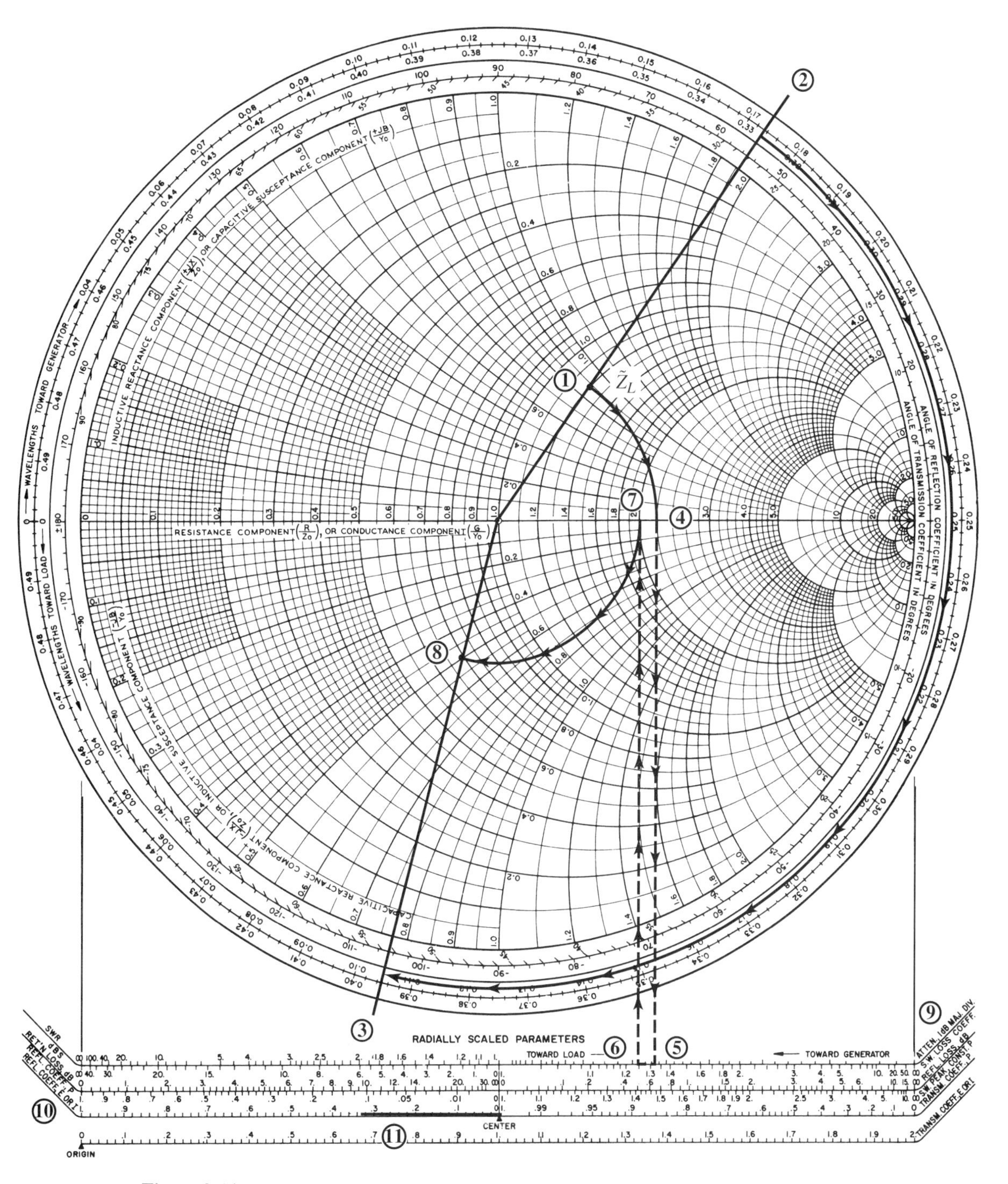

Figure 2-11. Determination of input impedance and reflection coefficient for a lossy line.

9. The reflection coefficient seen by the generator is obtained by taking the distance from the chart center to point (8) and laying it along the scale labeled REFL. COEFF., E or I, called scale (10) on the chart (this is length labeled (11)). We obtain the angle from where line (3) intersects the chart rim angle scale. Thus

$$\Gamma(l) = 0.33\angle -103^\circ$$

Exercises

2-3.4g Work Exercise 1-7.4e using the Smith chart.

2-3.4h Work Exercise 1-7.4c using the Smith chart.

2-3.4i An RG211A/U coaxial cable was used to connect a surface seismometer at ground zero for an underground nuclear test to a permanent recorder located 850 m away. (The reason a radio transmitter was not used was due to the use of rf-sensitive detonators.) The information was transmitted in analog format by FM modulation of a 10-MHz, 10-V sine wave. The cable has a nominal loss of 0.006 dB/m. If the cable was terminated in 200 Ω at the receiving end, estimate the unmodulated signal amplitude at the receiver. What is the generator reflection coefficient? Compute the corresponding voltage standing wave ratios. Comment on these latter two values. Z_0 for the cable is 50 Ω, with $C = 96$ pF/m.

2-3.5 Standing Wave Ratio

Our next step is to examine methods for using the Smith chart to determine the standing wave ratio (SWR). One method involves a simple scale at the bottom of the chart. The second makes use of an important result involving the value of the impedance at the location of a voltage maximum or at a voltage minimum. One obvious way of obtaining the voltage standing wave ratio is to determine the reflection coefficient from the Smith chart as already explained and then compute S using the results derived earlier:

$$S(-l) = \begin{cases} \dfrac{1 + \rho_L e^{-2\alpha l}}{1 - \rho_L e^{-2\alpha l}} & \text{(lossy line)} \\[2ex] \dfrac{1 + \rho_L}{1 - \rho_L} & \text{(lossless line)} \end{cases}$$

The Smith chart, once again, makes it possible to avoid having to make this computation. We first look at the values of the input impedances at voltage maxima and minima.

$\overline{Z}(-l_{\text{max}})$ and Its Relationship to $\Gamma(-l_{\text{max}})$ and S

From Eq. 1-76, the location l_{max} of a voltage maximum is

$$l_{\text{max}} = \frac{2n\pi + \theta_L}{2\beta}$$

Using Eq. 1-63 (lossless line),

$$\begin{aligned}\Gamma(-l_{\text{max}}) &= \Gamma_L e^{-2j\beta l_{\text{max}}} = \rho_L e^{j\theta_L} e^{-j2\beta\frac{(2n\pi+\theta_L)}{2\beta}} \\ &= \rho_L e^{j\theta_L} e^{-j2n\pi} e^{-j\theta_L} = \rho_L e^{-j2n\pi} = \rho_L = |\Gamma_L|\end{aligned}$$

Since $\rho_L = |\Gamma_L|$, then $\Gamma(-l_{max})$ is positive and real (and, as always, ≤ 1). Now from Eq. 1-97 and since $\Gamma(-l_{max})$ is real and positive, the impedance given by

$$\bar{Z}(-l_{max}) = Z_0 \frac{1 + \Gamma(-l_{max})}{1 - \Gamma(-l_{max})} = Z_0 \frac{1 + \rho_L}{1 - \rho_L}$$

is thus positive and real (recall $|\Gamma| \leq 1$).

Normalizing this impedance we have that

$$\tilde{Z}(-l_{max}) = \frac{\bar{Z}(-l_{max})}{Z_0} = \frac{1 + \rho_L}{1 - \rho_L}$$

is positive, real, and >1.

Then using Eq. 1-82, we also have for this case

$$S = \frac{1 + \rho_L}{1 - \rho_L}$$

or, comparing with $\tilde{Z}(-l_{max})$ above,

$$S = \tilde{Z}(-l_{max}) \qquad \text{positive, real, } > 1$$

Thus the standing wave ratio corresponding to an impedance $\tilde{Z}(-l)$ on a line is the value of $\tilde{Z}(-l_{max})$, which occurs at a real positive value greater than 1. On the Smith chart, this is obtained by rotating the load to the real *right* horizontal chart axis. This result is perhaps not too surprising in some ways since if the voltage is a maximum one would expect a large impedance. The remarkable result is that the impedance is also purely real there and equal to the standing wave ratio.

$\bar{Z}(-l_{min})$ and Its Relationship to $\Gamma(-l_{min})$ and S

Analysis similar to the preceding would show that

$$\tilde{Z}(-l_{min}) = \frac{1}{S} \qquad \text{positive, real, } < 1$$

Thus the standing wave ratio corresponding to an impedance $\tilde{Z}(z)$ on a line is the reciprocal of the value of $\tilde{Z}(-l_{min})$ that occurs at a real positive value less than 1 which is on the *left* horizontal Smith chart axis. $\Gamma(-l_{min})$ will be a negative real number. The proof of this result is left as an exercise since it follows steps identical to those in the preceding case.

We can now give one method for determining the standing wave ratio on a lossless line. Even for a line having small losses, the procedure is very accurate and is used in practice.

Determination of Voltage Standing Wave Ratio, S, for a Given Load Impedance, $\overline{X}_L$

1. Plot $\tilde{Z}_L$ (See Figure 2-12.) (This could also be the impedance at any location in general.)
2. Using as the radius the distance from the center of the Smith chart to $\tilde{Z}_L$, draw an arc that intercepts the right real axis of the chart. This intersection is the value of S as discussed earlier. Or draw a circle, of radius from center of Smith chart to $\tilde{Z}_L$, that intercepts the left real axis of the chart. This intersection is $1/S$.

Example 2-6

Let $\overline{Z}_L = 60 + j43\ \Omega$ and $Z_0 = 50\ \Omega$. Find S.

1. Plot $\tilde{Z}_L = \overline{Z}_L/Z_0 = 1.2 + j0.86$ as point ① on the Smith chart in Figure 2-13.
2. Rotate $\tilde{Z}_L$ to right hand horizontal chart axis; point ②. From point ② read $S = 2.18$. Or rotate $\tilde{Z}_L$ to the left-hand horizontal chart axis (point ②′). From point ②′, read $1/S = 0.46$:

$$S = \frac{1}{0.46} = 2.18$$

Finally, we explain the use of the SWR scale on the Smith chart. It consists of two contiguous scales labeled SWR and dBS. The use of these scales is described next, although it is a simple extension of that covered in the previous cases. Other forms of the Smith chart may have slightly different names for these two scales, but their functions are identical.

To obtain the standing wave ratio we proceed as follows (see Figure 2-14):

1. Plot the load impedance, $\tilde{Z}_L$.
2. Using a compass and the distance from the chart center to $\tilde{Z}_L$ as radius, draw an arc from $\tilde{Z}_L$ to intersect the left-hand real axis of the chart.
3. Draw a line vertically downward from the intersection on the real axis to the SWR scale. Read the value of S directly from the SWR scale. (It is interesting to note that this scale has values that are merely reciprocals of the values on the left real axis as expected from the preceding theoretical discussion. For this reason, this set of scales does not have a great deal of use since the right real axis of the chart gives the value directly.)

There is a second scale in the SWR set labeled dBS. This scale is handy when measurements are physically being made on a transmission system. dB readings are usually obtained using a meter that has a dB scale. One usually takes two readings on such a

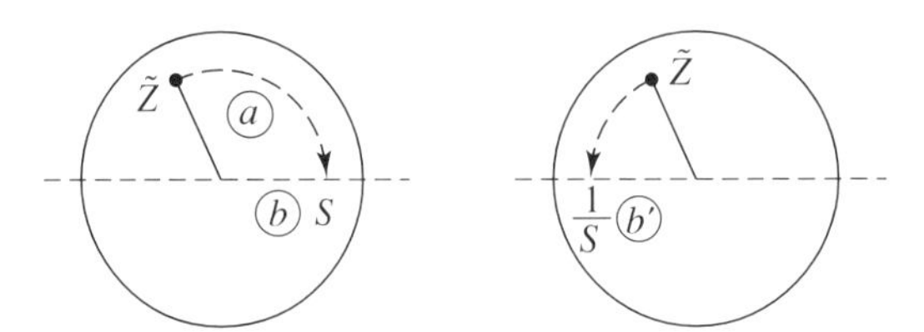

Figure 2-12. Determiniation of the standing wave ratio corresponding to an impedance $\overline{Z}(-l)$.

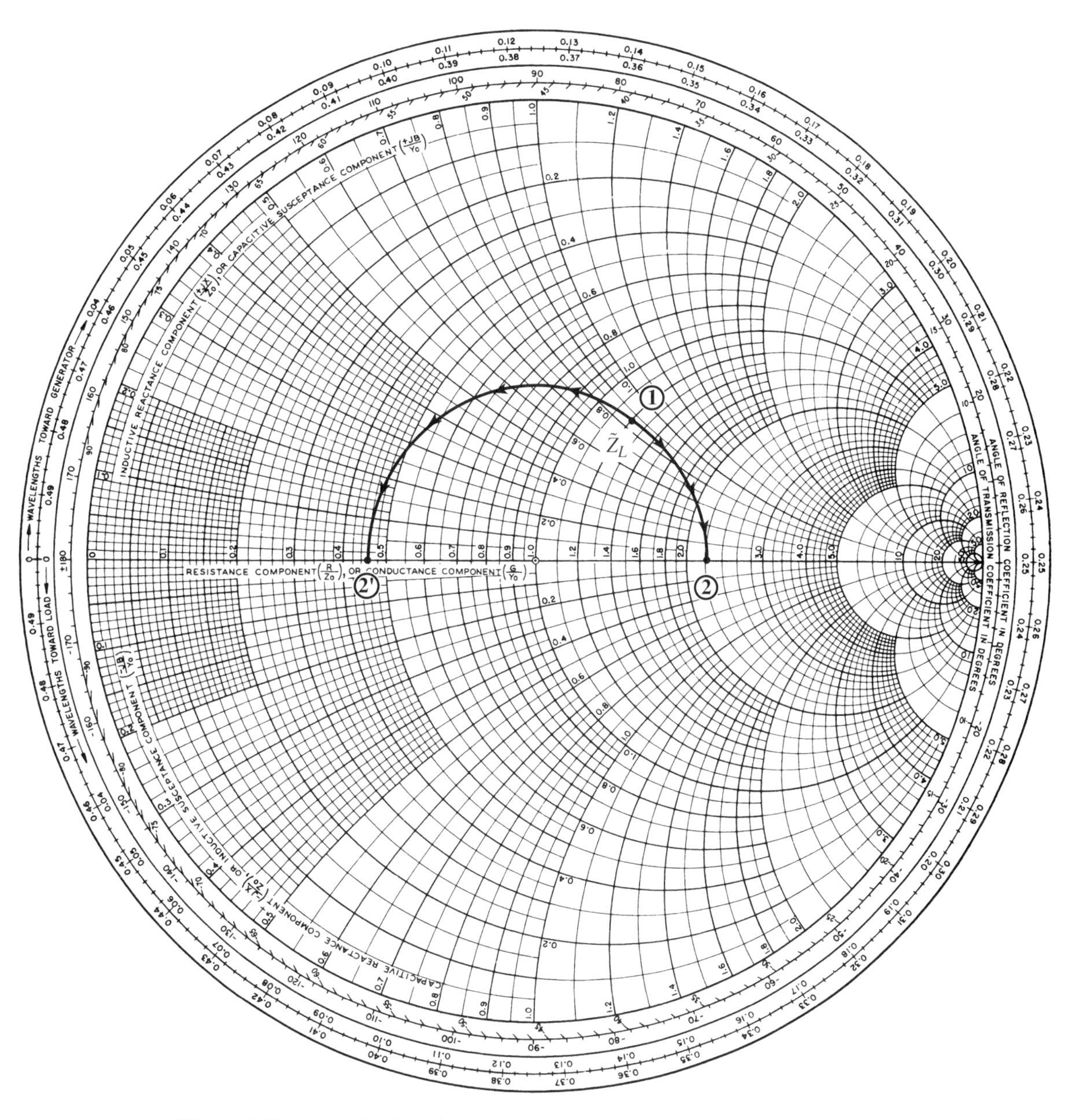

Figure 2-13. Determination of the standing wave ratio for a given load impedance.

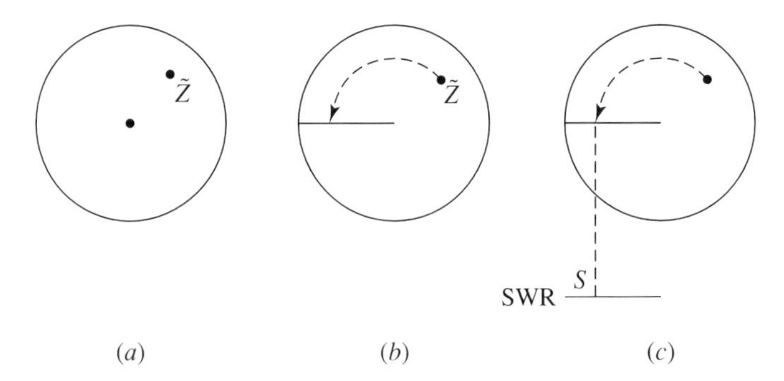

Figure 2-14. Using the SWR scale on the Smith chart.

meter, those being a dB reading at a maximum reading location and one at a minimum reading location. The equivalent dB difference between the maximum and minimum is simply the difference between the two readings obtained by measurement. By definition the standing wave ratio in dB is

$$S_{\text{dB}} = 20 \log_{10}\left|\frac{V_{\max}}{V_{\min}}\right| = 20 \log_{10} S = |\text{dB reading max} - \text{dB reading min}|$$

The dBS scale is the solution of this equation for S when S_{dB} is measured as described earlier. Thus all one needs to do is determine S_{dB} and then go to the dBS scale and obtain S by taking the corresponding value adjacent to it on the SWR scale.

Exercises

2-3.5a Verify that $\tilde{Z}(-l_{\min}) = 1/S$ and is positive, real, and less than 1, by carrying out the derivation.

2-3.5b Suppose a lossless transmission line terminated in $\overline{Z}_L = 60 + j45\ \Omega$ has a length of 0.15λ and characteristic impedance of 75 Ω. Determine the input standing wave ratio and the reflection coefficient (complex). Repeat for the case where the loss on that line is 0.5 dB. (Answer: 2.0, $0.34\angle -18°$, 1.85, $0.29\angle -18°$)

2-3.5c A 75-Ω coaxial cable is found to have a voltage standing wave ratio of 2.5. A voltage maximum occurs at a distance $\lambda/8$ from the load. Find the load impedance $\overline{Z}_L$, the value of impedance at a voltage minimum and the impedance at a voltage maximum. Sketch the standing wave patterns for current and voltage on this line and mark the locations of $\overline{Z}_{\min}$ and $\overline{Z}_{\max}$. Are these results physically reasonable? Explain. (Answer: $50.6 + j54.8\ \Omega$, 30 Ω, 187.5 Ω)

2-3.5d A generator that has an output impedance of 50 Ω is connected to a cable having a characteristic impedance of 50 Ω. The cable is 6 m in length and is found to have a voltage standing wave ratio of 2 when the frequency is 500 MHz. The cable capacitance is 98.4 pF/m. If the load on the cable is purely resistive and if the generator open-circuit voltage is V_0, what is the generator terminal voltage when connected to the cable? Assume the cable is lossless. A voltage minimum occurs at $z = -\lambda/4$.

2-3.5e A 75-Ω line is known to be terminated in a resistance. Measurements with a voltmeter yield a maximum of 28 dB and a minimum of 23.5 dB on the line. What are possible values of the load resistance? Voltage reference is 1 V.

The results obtained in the preceding discussion of the standing wave ratio contain ideas which give a good deal of physical insight into transmission line performance. Often a simple knowledge of the location of a minimum or maximum voltage on a line will reveal considerable information concerning the load.

It was shown earlier that at the location of a voltage minimum the normalized impedance was real and less than 1 (i.e., the resistance is less than Z_0 of the line). If one examines the Smith chart, one observes that not only is the resistive value real and less than 1, but it is also the smallest value that can exist anywhere along the line. To see this, consider the point labeled ① on the Smith chart of Figure 2-15. As one moves away from this point in either direction along the line at constant radius (assuming a lossless line), one sees that the value of the real part increases in addition to picking up a reactive component.

From this observation, we have the following conclusions:

1. At a voltage minimum, the impedance is real so that the current $\overline{I}$ and voltage $\overline{V}$ are in phase at that point.
2. At a voltage minimum, the impedance is the smallest value that it has anywhere along the line.
3. Since we define the impedance at any location along the line as

$$\overline{Z}(-l) = \frac{\overline{V}(-l)}{\overline{I}(-l)}$$

the fact that $\overline{Z}$ is minimum and $\overline{V}$ is minimum also implies that $\overline{I}$ has its maximum there.

Following an analysis similar to the preceding for the case of a voltage maximum location (see point ② on the Smith chart), we would reach the following conclusions:

1. At a voltage maximum, the impedance is real so that the current $\overline{I}$ and voltage $\overline{V}$ are in phase at that point.
2. At a voltage maximum, the impedance is the largest value that it has anywhere along the line. (Incidentally, the value is also the reciprocal of the minimum value resistance.)
3. At a voltage maximum, the current assumes its minimum value.

2-3.6 Final Examples

This section concludes with a couple of examples that show how much easier it is to use the Smith Chart to solve some systems than it is to use the impedance equations.

Example 2-7

A slotted line is a device used to probe the voltage field in a transmission line system. An operational schematic is shown in Figure 2-16. In Figure 2-17, pictures of some actual systems that are used as slotted lines for making transmission line measurements are shown. The coaxial line is an air-filled transmission line with a probe that is inserted into

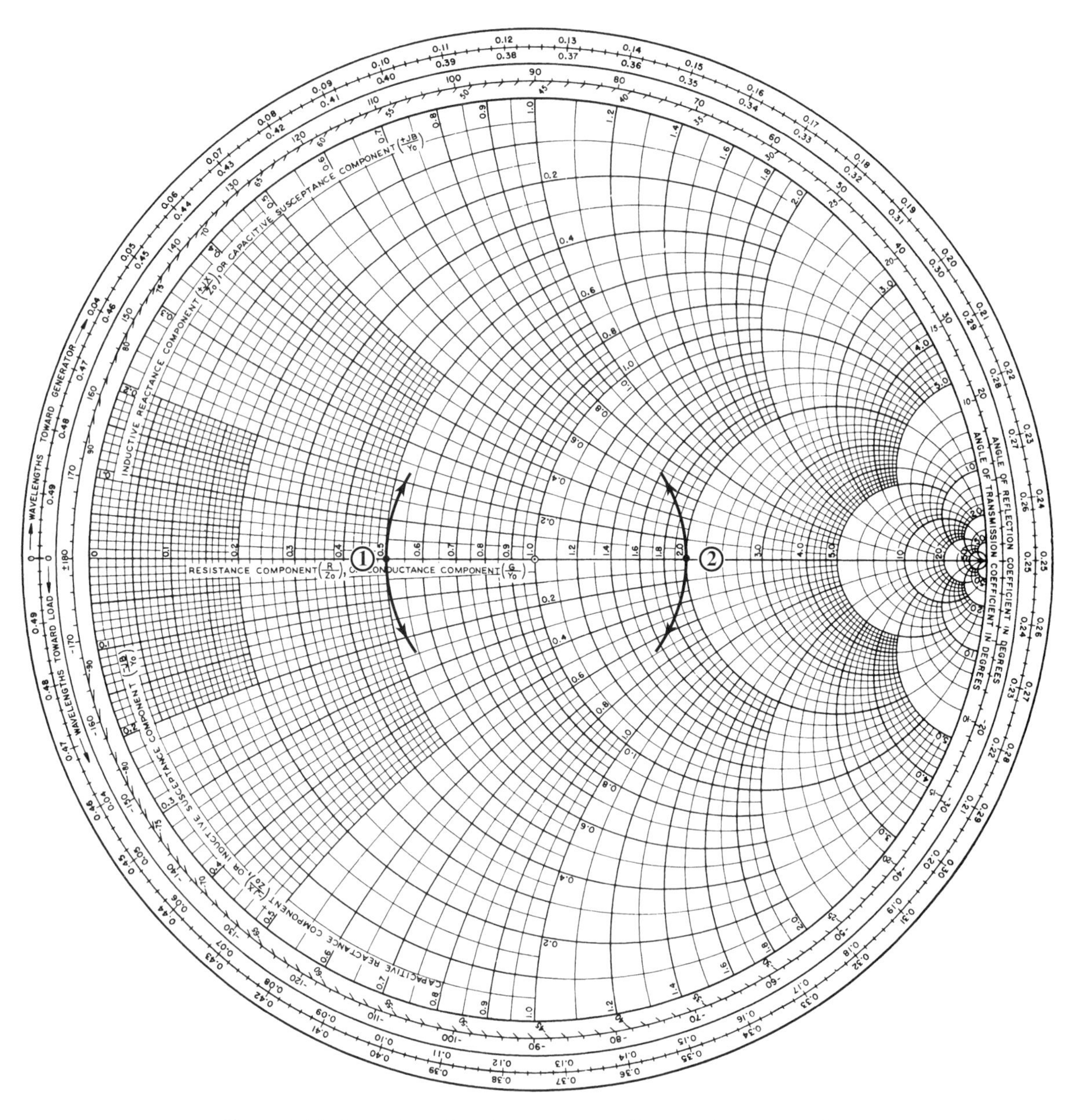

Figure 2-15. Properties of line impedance at voltage maxima and minima.

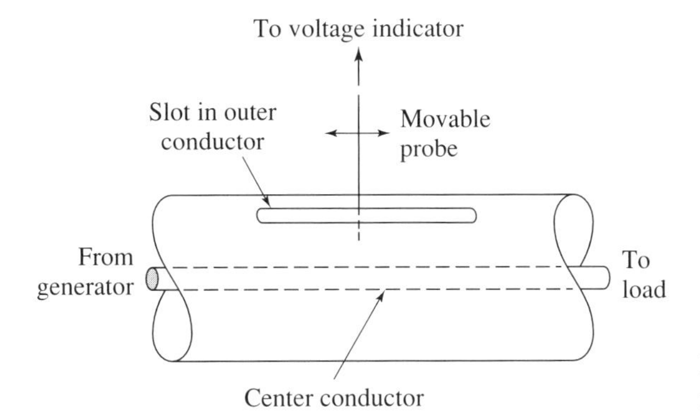

Figure 2-16. Coaxial slotted line functional diagram.

the outer conductor slot to sample the amplitude of the voltage. With such a probe mounted on a moveable carriage, it is possible to determine the locations of and relative values of maximum and minimum voltages. The ratio of the maximum and minimum values is, of course, the voltage standing wave ratio.

A slotted line measurement on an air-filled transmission system yields a standing wave ratio of 1.6. When a short circuit is put in place of the load, the voltage null in the standing

Figure 2-17. Slotted line systems: (*a*) Slab line; (*b*) Coaxial slotted line with stub tuner; (*c*) Waveguide slotted line with stub tuner (Chapter 9).

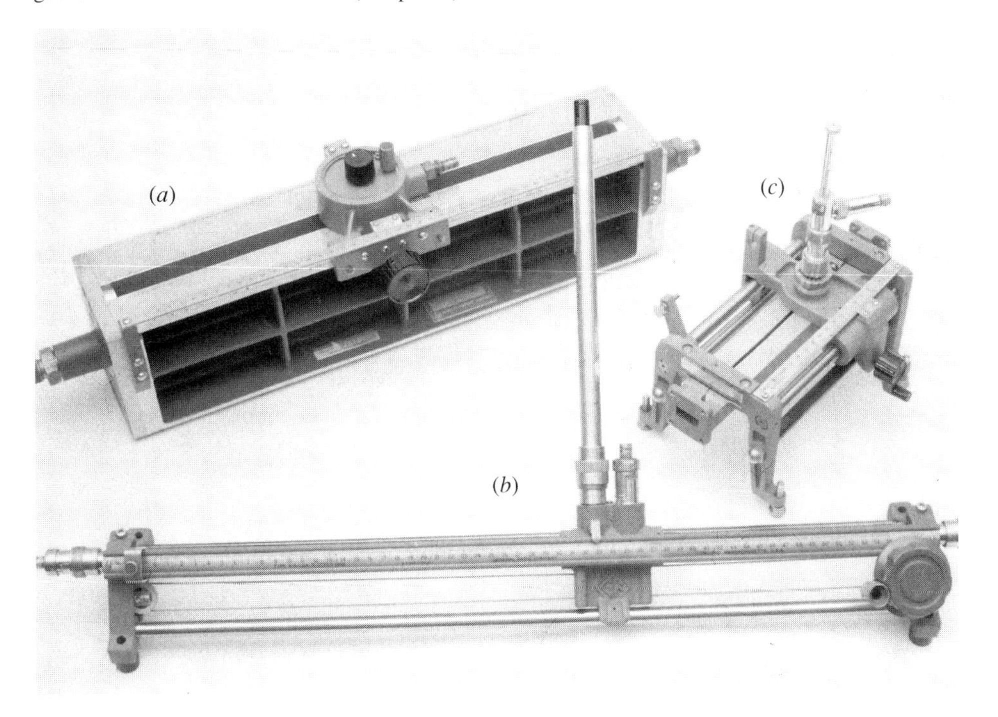

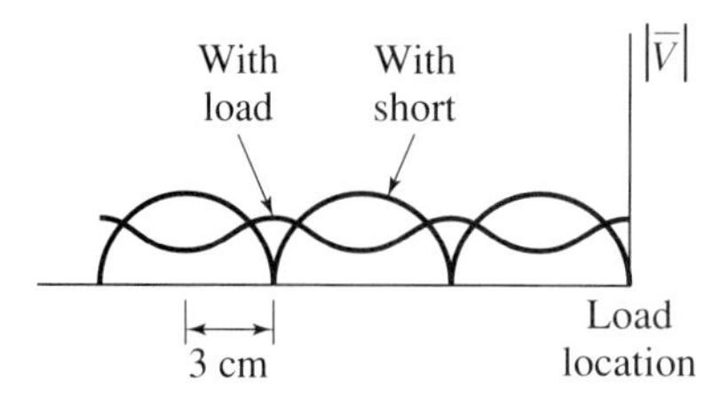

Figure 2-18. Standing wave patterns for Example 2-7.

wave pattern shifts 3 cm toward the load. The source frequency is 1 GHz. Determine the following:

1. Line wavelength
2. Normalized load impedance
3. Suppose next that if the frequency were changed S remained the same. What would be the nature of the load?
4. For item 3, what would be the two possible load values (normalized)?

The standing wave patterns described in the statement are shown in Figure 2-18.

1. In Chapter 9, it will be shown that for the typical *two*-conductor transmission line, the phase velocity can be computed from $V_p = 1/\sqrt{\mu\epsilon}$ as well as $1/\sqrt{LC}$, where μ is the permeability and ϵ the permittivity of the material between the conductors. For air (vacuum), $\mu \equiv \mu_0 = 4\pi \times 10^{-7}$ and $\epsilon \equiv \epsilon_0 = 1/36\pi \times 10^{-9}$, which gives $V_p = 3 \times 10^8$ m/s. Then

$$\lambda_l = \frac{V_p}{f} = \frac{3 \times 10^8}{1 \times 10^9} = 0.3 \text{ m}$$

2. Now when a short circuit is placed at the load, there will be a null at the load and every $\lambda_l/2$ toward the generator. Thus for a lossless line, the impedance at the locations of any one of the short circuit nulls will be the same as at the load end regardless of the load, since those locations are all physically a multiple of a half wavelength from the load, and the impedance on the line repeats every half wavelength. (See Eq. 1-99.)
   The next step is to draw a circle corresponding to a standing wave ratio of 1.6. This is easily done by using the distance from the chart center to the point 1.6 on the real right axis of the Smith chart as shown in Figure 2-19. We assume a lossless line, so $\rho(-l)$ is constant ρ_L as explained in an earlier section. Note that another way of looking at this circle is that we are plotting a circle of all values of $\Gamma(-l)$ for which

$$\rho_L = \frac{S - 1}{S + 1} = 0.231$$

   and use the length of 0.231 from the REFL. COEFF., E or I scale at the chart bottom as the radius to draw the circle. The load value must lie on this circle. Now with the load in place, we know that at a voltage minimum the impedance will be a minimum. On the Smith chart, this will be at the left real axis at value $1/S$, or in this case 0.625 at point ① on the chart. We also know that the load impedance appears at any one of

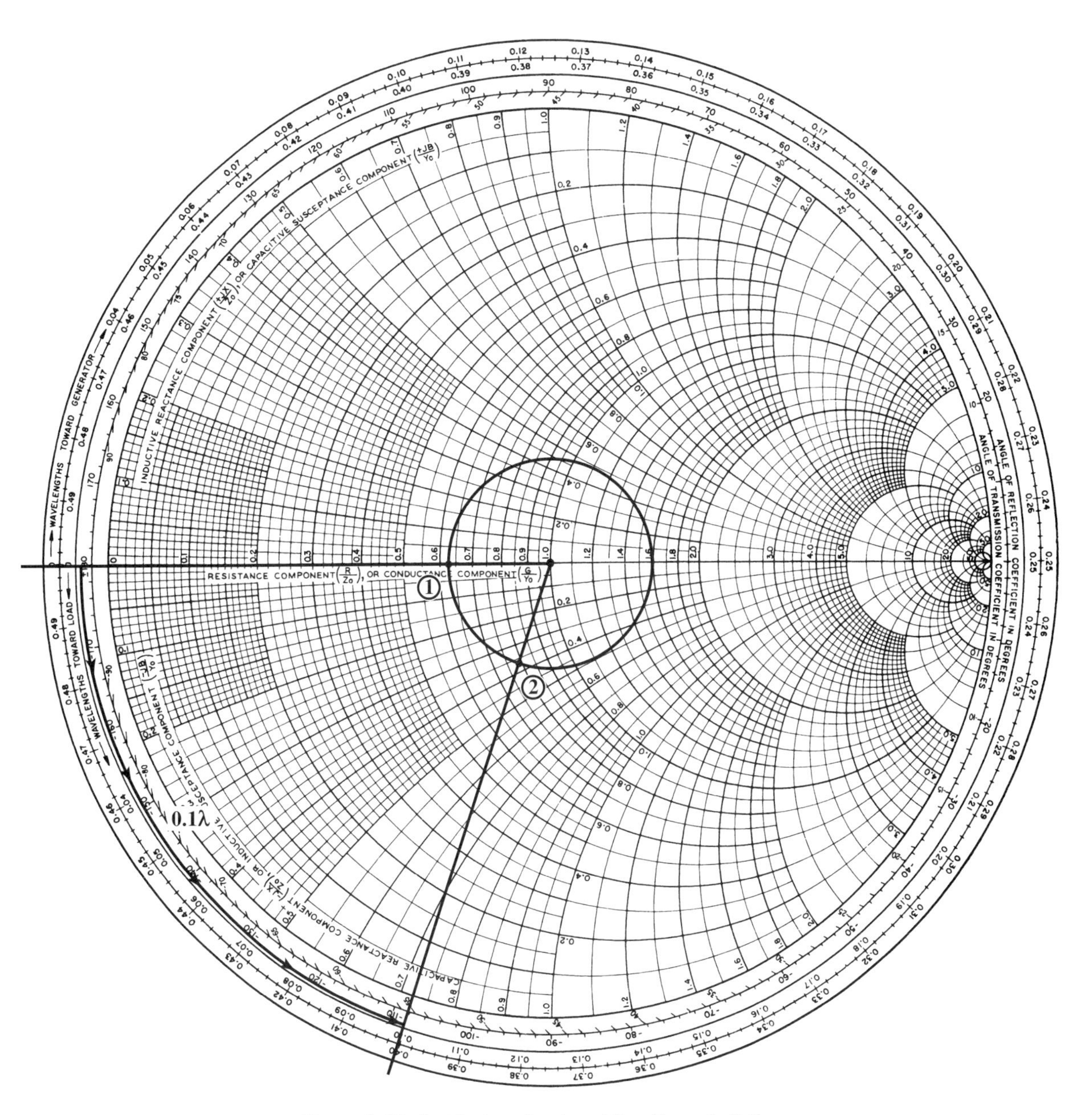

Figure 2-19. Smith chart for slotted line Example 2-7.

the shorted load null locations. Thus, we rotate point ① a distance $3/30\lambda$ or 0.1λ toward the load to point ②. The normalized load impedance is then

$$\tilde{Z}_L = 0.79 - j0.36$$

3. We observe that if S remains constant as frequency changes, then the load must be a constant (i.e., a resistance).
4. Since the load is resistive and the standing wave ratio value is the value of the corresponding pure resistance on the Smith chart axis, we have the two possibilities

$$\tilde{Z}_L = 0.625\ \Omega \qquad \text{or} \quad 1.6\ \Omega$$

Example 2-8

A transmission line of characteristic impedance 50 Ω (lossless) is terminated in a load $\bar{Z}_L = -j25\ \Omega$. Determine the distance from the load where the nearest voltage maximum occurs, and the value of the input impedance at that point.

$$\tilde{Z}_L = -j\tfrac{25}{50} = 0 - j0.5$$

This is plotted on the Smith chart of Figure 2-20 as point ①. The voltage maximum occurs at the impedance maximum, which is located at a point on the right-hand real axis on the Smith chart. Thus we rotate point ① toward the generator along the chart rim as shown to point ②:

$$\text{Distance} = (0.25 + 0.074)\lambda = 0.324\lambda$$

Since the center of the $\tilde{R} = 0$ circle is the chart center, the rotation from ① to ② stays on the $\tilde{R} = 0$ circle. Therefore, at point ②, we have

$$\tilde{Z}_2 = 0 + j\infty \qquad \text{from which} \quad \bar{Z}_2 = 0 + j\infty.$$

Exercises

Most of the latter exercises in Chapter 1 can be solved using Smith chart techniques. In particular, the general exercises at the end of Chapter 1 (denoted by GE) are good applications. The following are recommended:

2-3.6*a*	See Exercise 1-7.3*a*	2-3.6*k*	See Exercise GE1-17
2-3.6*b*	See Exercise 1-7.3*e*	2-3.6*l*	See Exercise GE1-18
2-3.6*c*	See Exercise 1-7.3*g*	2-3.6*m*	See Exercise GE1-19
2-3.6*d*	See Exercise 1-7.3*i*	2-3.6*n*	See Exercise GE1-20
2-3.6*e*	See Exercise 1-7.4*d*	2-3.6*o*	See Exercise GE1-21
2-3.6*f*	See Exercise 1-7.4*e*	2-3.7*p*	See Exercise GE1-22
2-3.6*g*	See Exercise 1-7.4*i*	2-3.6*q*	See Exercise GE1-24
2-3.6*h*	See Exercise 1-7.4*r*	2-3.6*r*	See Exercise GE1-26
2-3.6*i*	See Exercise GE1-9	2-3.6*s*	See Exercise GE1-27
2-3.7*j*	See Exercise GE1-11	2-3.6*t*	See Exercise GE1-33

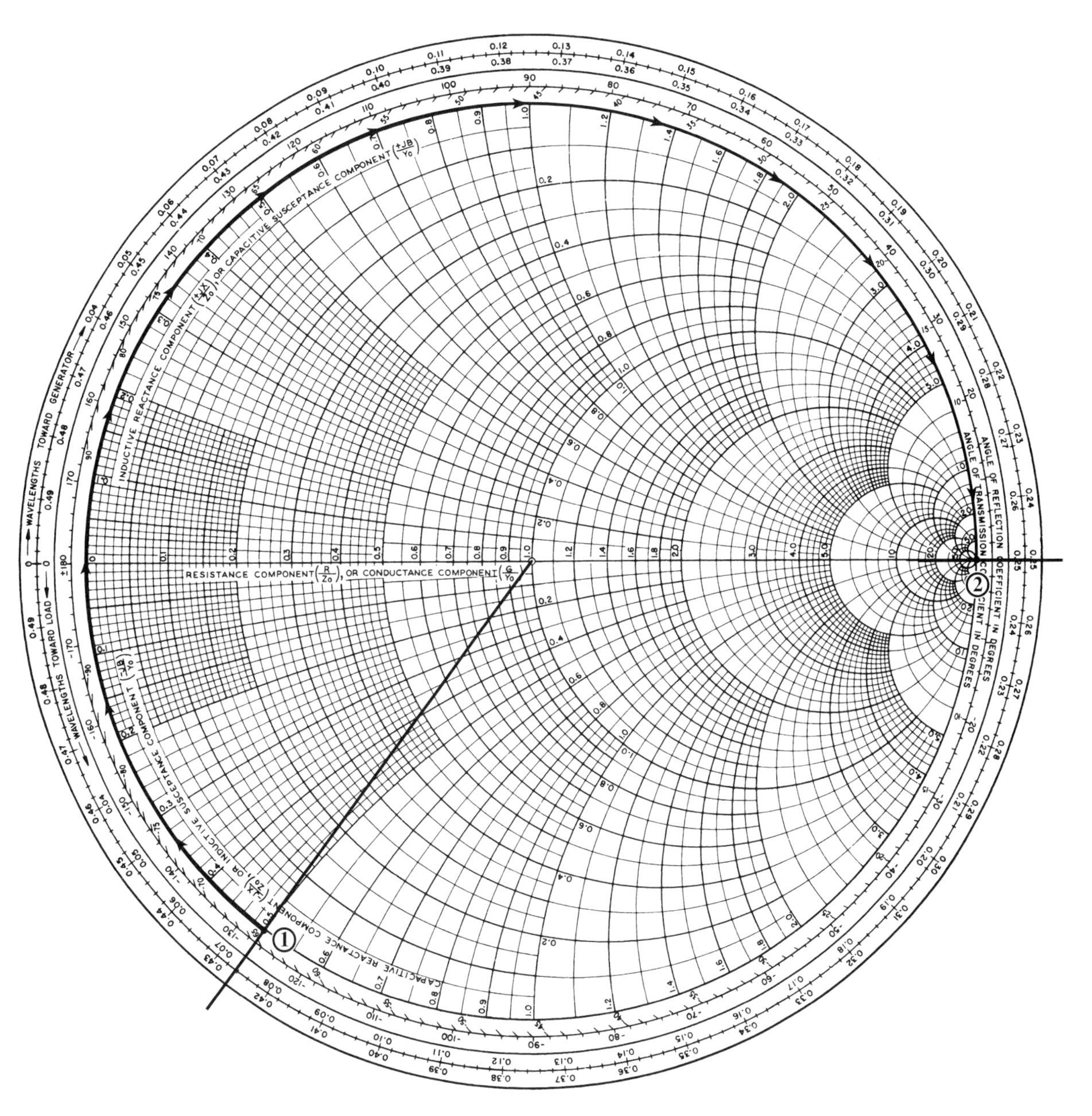

Figure 2-20. Smith chart solution for reactive load of Example 2-8.

2-4 Transmission Line Matching: Single and Double Stub

We have now developed enough Smith chart tools to allow us to learn some matching techniques. These tools include being able to transfer an impedance along a line and to calculate distances in terms of wavelengths.

In Chapter 1 the theory of matching was developed, but to apply that theory it was necessary to solve nonlinear and transcendental equations. We can now use the Smith chart to solve these equations for us and avoid the cumbersome equations.

The examples for matching presented in this section also give additional familiarity with the Smith chart as a computational tool. We begin the discussion with the example of a single shunt stub.

2-4.1 Single Shunt Stub Matching

This is the Smith chart solution of the mathematical theory presented in Section 1.8. (One might profit from a quick review of the ideas.) The basic configuration is shown in Figure 2-21.

Review particularly statements 1 and 2 below along with Eqs. 1-108 and 1-109, which give the values of l_1 and l_s for a matched line:

1. $\tilde{G}(-l_1) = 1$ gives the location $-l_1$ at which the stub is placed. The admittance at that location is

$$\tilde{Y}(-l_1) = \tilde{G}(-l_1) + j\tilde{B}_1(-l_1)$$

2. $\tilde{B}_s(-l_s) = -\tilde{B}_1(-l_1)$ gives the stub length l_s.

Procedure: Lossless Line

Steps given graphically in Figure 2-22.

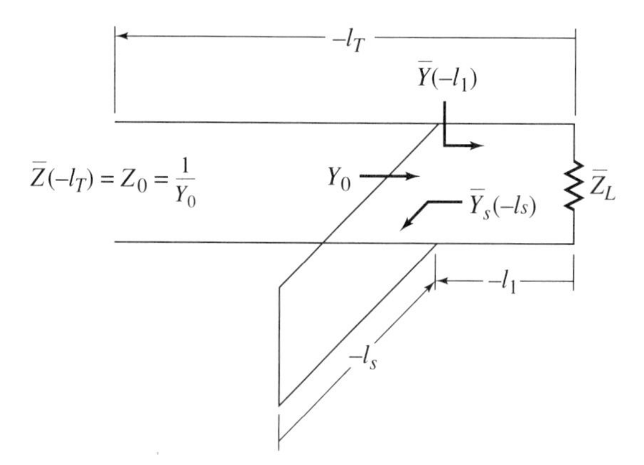

Figure 2-21. Configuration for shunt shorted stub matching.

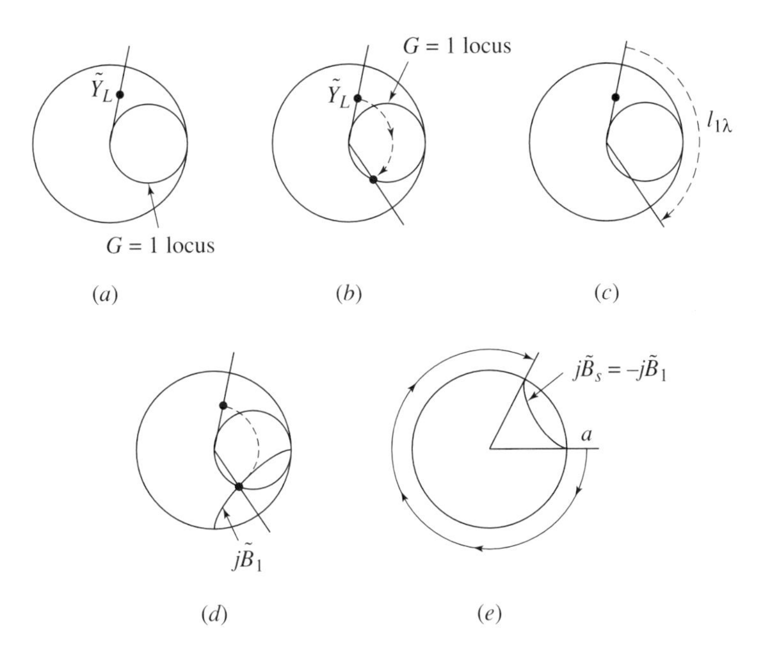

Figure 2-22. Steps in single shunt stub matching.

1. Plot $\tilde{Y}_L$ and draw a line from the origin through $\tilde{Y}_L$ to the outer rim of the chart.
2. Using the distance from the origin to point $\tilde{Y}_L$ as radius, draw an arc toward the generator until it intersects the circle for $\tilde{G} = 1$. Draw a line from the chart center through this intersection to the outer rim of the chart. (Note: There are *two* possibilities. In practice, the distances requiring shortest lengths are preferred. This reduces frequency sensitivity of the match as discussed earlier. For clarity, we show a large movement on diagram (*b*).)
3. The wavelength difference between the two lines is $l_{1\lambda}$. See diagram (*c*). Then $l_1 = l_{1\lambda} \cdot \lambda_l$, where λ_l is the wavelength at the particular frequency of operation.
4. At the point of intersection on the $\tilde{G} = 1$ circle, read the value of susceptance at that point. The susceptance required of the stub is the negative of this: $\tilde{B}_s(-l_s) = -\tilde{B}_1(-l_1)$. See (*d*) in figure.
5. To obtain stub length l_s for the shorted stub, begin at the outer rim of chart where $\tilde{G} = \infty$ (point *a*) and move toward the generator around rim of chart until the value $\tilde{B}_s$ is reached (point *b*). Remember that a short circuit is $\tilde{Y} = \infty$. The wavelength displacement of the two lines through the initial and final points is $l_{s\lambda}$. This could be done on a separate Smith chart to avoid confusion. Then $l_s = l_{s\lambda} \cdot \lambda_l$.

Example 2-9

Let $\overline{Z}_L = 60 + j43\ \Omega$ and $Z_0 = 50\ \Omega$. Design a single shorted stub matching system. See Figure 2-23. (Don't forget to mentally blank out the chart and add the lines and arcs as the solution proceeds.)

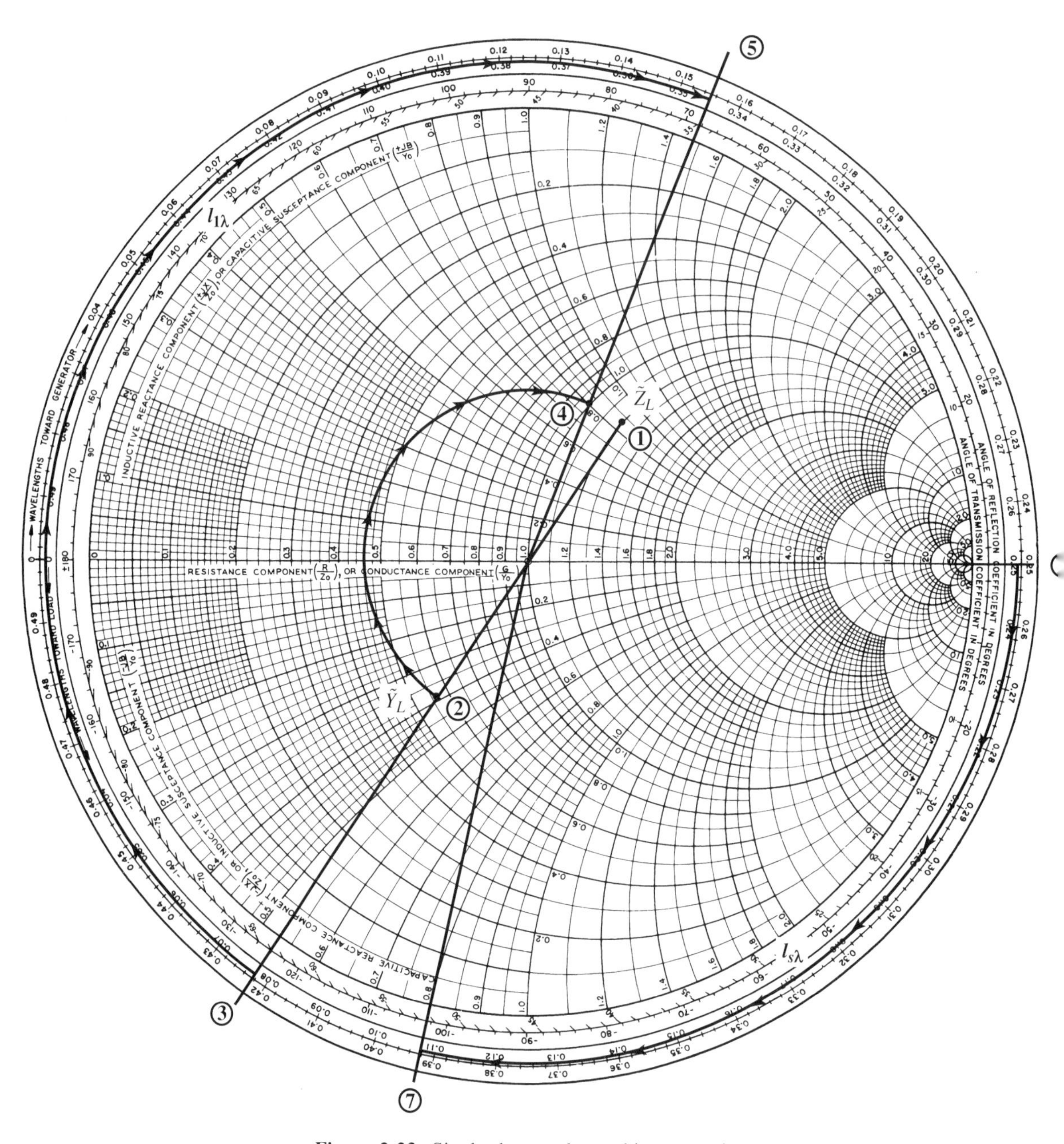

Figure 2-23. Single shunt stub matching example.

1. $\tilde{Z}_L = (60 + j43)/50 = 1.2 + j0.86$. Plot this at point ① on the Figure 2-23 chart. Reciprocate to obtain $\tilde{Y}_L$, ②, and then draw a line through $\tilde{Y}_L$ to outer chart rim, line ③.
2. Rotate $\tilde{Y}_L$ toward the generator to the $\tilde{G} = 1$ circle, point ④. Draw a line through this point to the chart rim, line ⑤.
3. From chart lines ③ and ⑤, we obtain

$$l_{1\lambda} = (0.077 + 0.154)\lambda$$

$$l_{1\lambda} = 0.231\lambda$$

   (If λ_l were given or could be computed from known line-operating conditions, $l_1 = l_{1\lambda} \cdot \lambda_l$)
4. At point ④, $\tilde{Y}_1(-l_1) = 1 + j0.8$. Thus

$$\tilde{B}_1(-l_1) = 0.8$$

   from which

$$\tilde{B}_s(-l_s) = -0.8$$

5. Draw line ⑥ through $\tilde{G} = \infty$ and rotate the line toward the generator to pass through $-j$ on the *rim* at line ⑦. Then

$$l_{s\lambda} = (0.395 - 0.25)\lambda = 0.145\lambda$$

$$l_s = l_s\lambda \cdot \lambda_l$$

Exercises

2-4.1a For the general single shunt stub procedure given in the preceding section, tell what would change if the matching stub were open rather than shorted. Next complete the example given above using an open stub match (shunt). Compare the stub lengths.

2-4.1b A lossless transmission line of characteristic impedance 50 Ω is terminated in an admittance $\overline{Y}_L = 0.008 - j0.012$ S.

(a) Determine the shortest distances from $\overline{Y}_L$ at which the normalized *admittance* is $1 \pm j\tilde{B}$.
(b) Determine the values of $\tilde{B}$ at the locations found in (a).
(c) Find the lengths of *shunt* shorted stubs that will provide single stub matching for each case. (Answer: 0.266λ, 0.422λ, $j1.35$ S, $-j1.35$ S, 0.102λ, 0.3984λ)

2-4.1c A lossless transmission line of characteristic impedance 50 Ω is terminated in an admittance $\overline{Y}_L = 0.008 - j0.012$ S.

(a) Determine the shortest distances from $\overline{Y}_L$ at which the normalized impedance is $1 \pm j\tilde{X}$.
(b) Determine the values of $\tilde{X}$ at the locations found in (a).
(c) Find the lengths of *series* shorted stubs that will provide single stub matching for each case.

(d) Which of the two systems is more broad band (i.e., is more independent of frequency)? (Hint: Let the wavelength increase by 10%. Then analyze the system designs to compute S on the originally matched line section Z_0. The lengths $l_{1\lambda}$ and $l_{s\lambda}$ decrease by 10%.) (Answer: 0.0166λ, 0.1718λ, $j1.35\ \Omega$, $-j1.35\ \Omega$, 0.352λ, 0.1484λ, the case where the stub length is 0.1484λ)

2-4.1d Both design procedures presented in the text assumed lossless systems. Develop a design procedure for a lossy line system (Z_0 still real for the low-loss case) for a single shunt stub configuration. Note that since the line is lossy and that the total attenuation which must be known to determine the spiral decrease is a function of an unknown length, a trial-and-error procedure is called for. Try modifying an initial lossless first guess. Note that the stub input impedance will not be pure imaginary any more and so will add some conductance component on the line. This means that l_1 will not quite be on the $\tilde{G} = 1$ circle.

2-4.1e Develop a design procedure for a series stub match (shorted stub):

(a) Draw the circuit configuration.
(b) Give a step-by-step procedure for determining stub location length. (Hint: Use impedances.)

2-4.1f Using the procedure of Exercise 2-4.1*e*, provide a series shorted-stub match for $\overline{Z}_L = 60 + j43\ \Omega$ and $Z_0 = 50\ \Omega$. Give answers in fractions of λ. Repeat for an open circuit series stub. Compare stub lengths to the example result.

2-4.2 Double Shunt Stub Line Matching

From the discussions of shunt matching by a single stub, we observe that the stub locations were at fixed points from the load. It may be that the particular location is not accessible in the actual system or that it may be more convenient to have the matching section located at another point. By using two stubs we can most often (not always) put the matching system at an arbitrary point $-l$ from the load.

Principles of matching with two shunt stubs should be understood before proceeding. The following basic principles refer to the configuration shown in Figure 2-24.

1. A shorted stub can add only susceptance (See Eq. 1-101).

$$j\tilde{B}_s(l) = -j \cot \beta l$$

2. Just *before* we add stub 2, the admittance at bb' must be of the form $\tilde{Y}_{bb'} = 1 \mp j\tilde{B}_{bb'}$, so that the addition of stub 2 in amount $\pm j\tilde{B}_{bb'}$ will produce $\tilde{Y}_{bb'} = 1$, which is a matched condition.
3. The value $\tilde{Y}_{bb'} = 1 \pm j\tilde{B}_{bb'}$, prior to adding the stub 2 must have resulted from rotating the admittance at aa' (with stub 1 in place) toward the generator by a distance

$$d_\lambda = \frac{d}{\lambda_l}$$

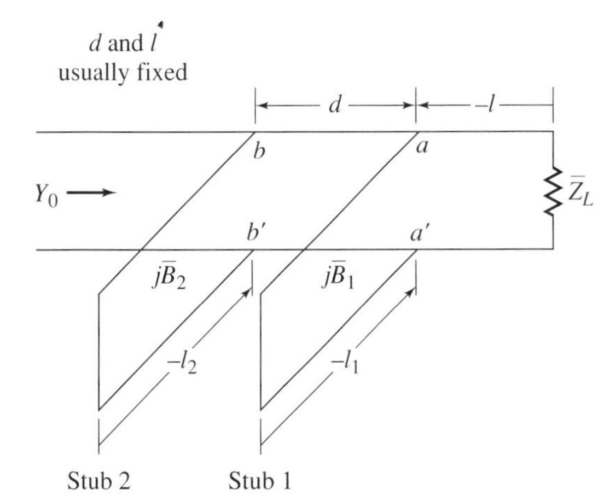

Figure 2-24. Basic configuration for double shunt stub matching.

Thus to find the value of $Y_{aa'}$, with stub 1 in place, we would rotate $Y_{bb'} = 1 \pm j\tilde{B}_{bb'}$, toward the *load* an amount d_λ. Note that all such points can be found by rotating every point on the $\tilde{G} = 1$ circle toward the *load* by d_λ, as shown in Figure 2-25. The dotted circle is thus a locus of all admittances at $_{aa'}$ with stub 1 connected that, when rotated toward the generator to bb', will produce the required admittance $1 \pm jB_{bb'}$ without stub 2.

4. The value of stub 1 added to $\tilde{Y}_{aa'}(l_\lambda$ from the load) must be such as to make the total admittance at aa' lie on the rotated $\tilde{G} = 1$ circle.

Procedure: Lossless Line

See the graphical steps in Figure 2-26.

1. For the given system, calculate $d_\lambda = d/\lambda_l$.
2. Rotate the diameter of the $\tilde{G} = 1$ circle toward the load an amount d_λ and draw in this rotated circle. (Radius of that rotated circle is the distance from chart center to the point 3.0 on real axis.)

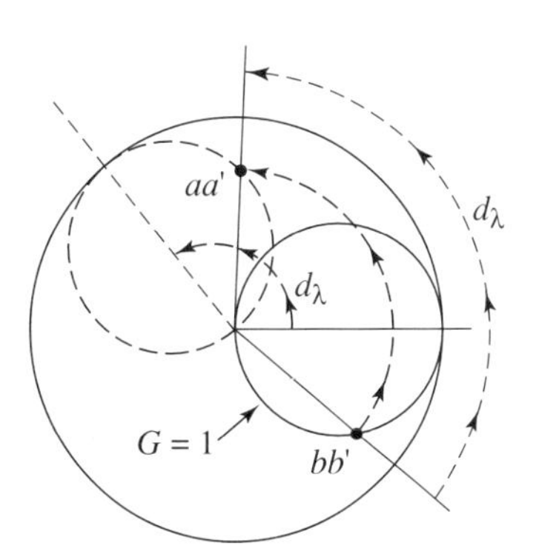

Figure 2-25. Rotation of the $G = 1$ circle to set up final match condition for adding stub 2 at bb'.

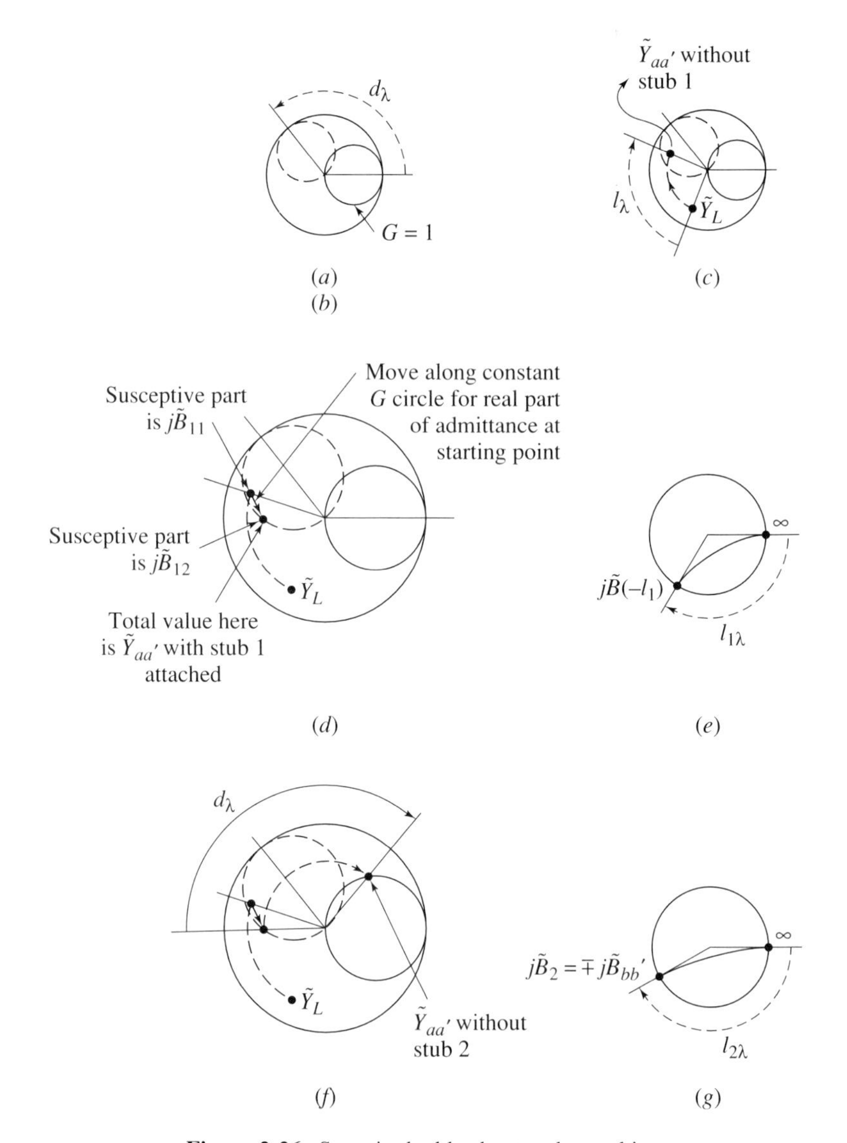

Figure 2-26. Steps in double shunt stub matching.

3. Plot $\tilde{Y}_L$ and rotate toward the generator at *constant radius* an amount

$$l_\lambda = \frac{l}{\lambda_l}$$

4. Now add the appropriate stub 1 susceptance $\pm j\tilde{B}_1(-Il_1)$ so that the value $\tilde{Y}_{aa'}$ without stub 1 is moved to lie on the rotated $\tilde{G} = 1$ circle. There are two possibilities: one by adding capacitive susceptance (+) and another by inductive susceptance (−). Also,

since we are adding susceptance only, we must move along the constant conductance value of the point determined previously in item c. Note this movement on the Smith chart is not a movement physically along the transmission line, but the placing of a susceptance at the point *aa'*. Calculate $j\tilde{B}_1(-l_1)$ by reading the values of susceptance passing through the two points:

$$j\tilde{B}_1(-l_1) = j(\tilde{B}_{12} - \tilde{B}_{11})$$

5. To obtain the stub length for shorted stub 1, rotate a line passing from the chart center through $\tilde{G} = \infty$ toward the generator to the chart rim point of value $j\tilde{B}_1(-l_1)$. (In the case shown, $j\tilde{B}_1(-l_1)$ would be a negative susceptance.) The displacement of these lines is $l_{1\lambda}$. This is the same process we used in the single stub length determination. Thus

$$l_1 = l_{1\lambda} \cdot \lambda_l$$

   It is important to note that stub 1 is not used to cancel any susceptance but only to move to a new location on the Smith chart, specifically to the rotated $\tilde{G} = 1$ circle.
6. Now rotate the point $\tilde{Y}_{aa'}$ *total* toward the generator a distance d_λ, calculated in part *a*, along the *constant radius* since we are now moving along the transmission line. However, by prior construction, this will end up on the $\tilde{G} = 1$ circle. Be certain to go to the proper point on the $\tilde{G} = 1$ circle since there must be a total displacement of d_λ! Read the value at that point as $\tilde{Y}_{bb'} = 1 \pm j\tilde{B}_{bb'}$.
   Required susceptance of stub 2 is $j\tilde{B}_2(-l_2) = -(\pm j\tilde{B}_{bb'}) = \mp j\tilde{B}_{bb'}$, since we want to cancel the remaining susceptance now.
7. To obtain the length of stub 2 for the shorted stub, rotate a line passing from the chart center through $G = \infty$ toward the generator on the chart rim point of value $\mp j\tilde{B}_{bb'}$. The displacement of these lines is $l_{2\lambda}$. Then

$$l_2 = l_{2\lambda} \cdot \lambda_l$$

Example 2-10

Double Shunt Stub Matching

See Figures 2-27 and 2-28 for charts.

Let $\bar{Z}_L = 60 + j43\ \Omega$ and $Z_0 = 50\ \Omega$. Suppose that l_λ is fixed at 0.06λ and that for our double stub device $d_\lambda = 0.1875\lambda = \frac{3}{16}\lambda$. Assume shorted stubs.

1. d_λ given as 0.1875λ.
2. $\tilde{G} = 1$ circle rotated toward load by d_λ, line ① to line ② (Chart A).
3. Plot $\tilde{Z}_L = (60 + j43)/50 = 1.2 + j0.86$ and convert to $\tilde{Y}_L$ by moving across the Smith chart 180°, to point ③. Rotate this point *at constant radius* toward the generator by an amount $l_\lambda = 0.06$ to point ④.
4. Normally, we would move down to the rotated $\tilde{G} = 1$ circle for shorter stub, but for diagram clarity, we add positive susceptance, which moves us up to point ⑤ along a constant conductance circle of value $\tilde{G} = 0.47$.

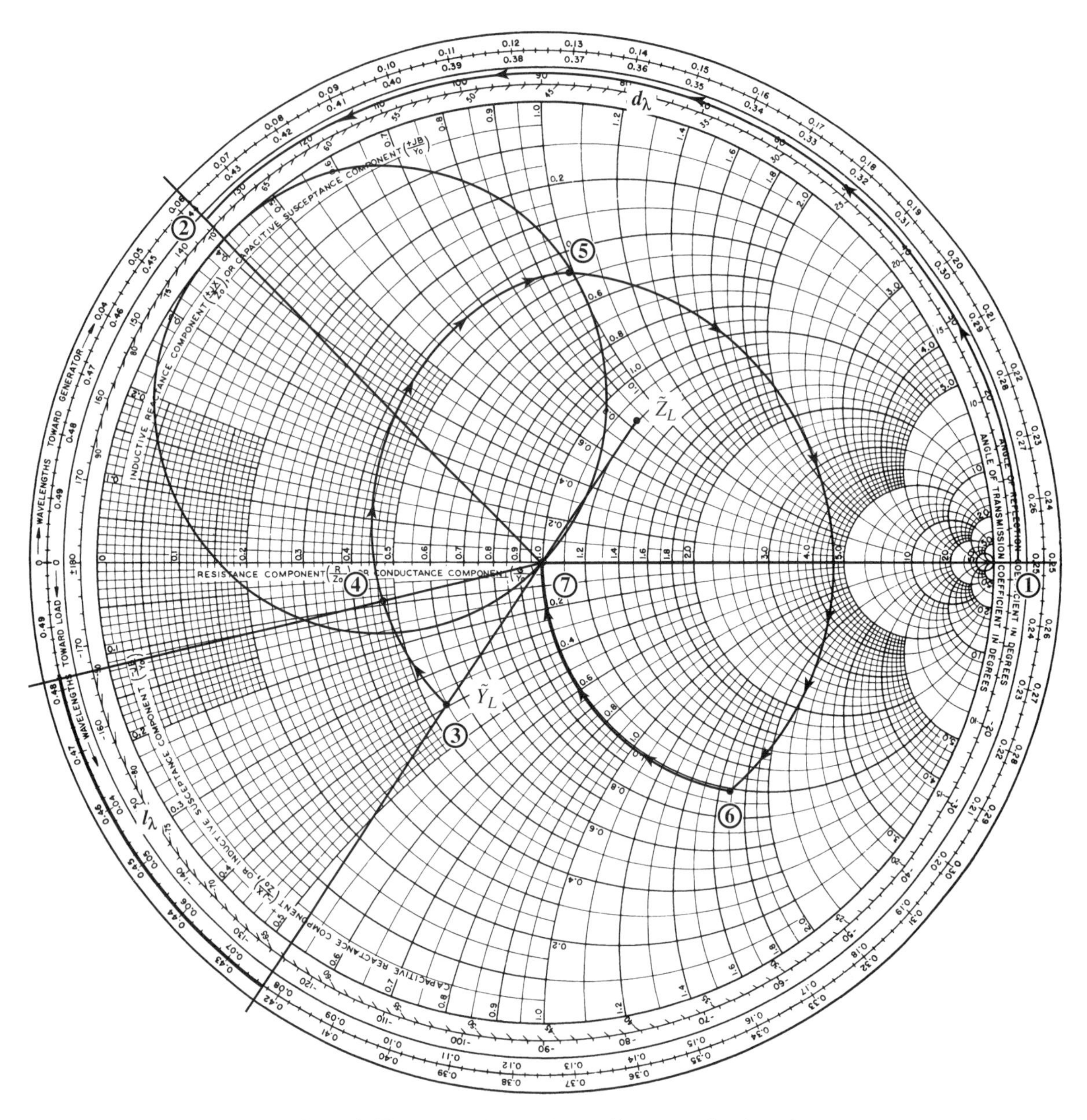

Figure 2-27. Double shunt stub matching example: chart A.

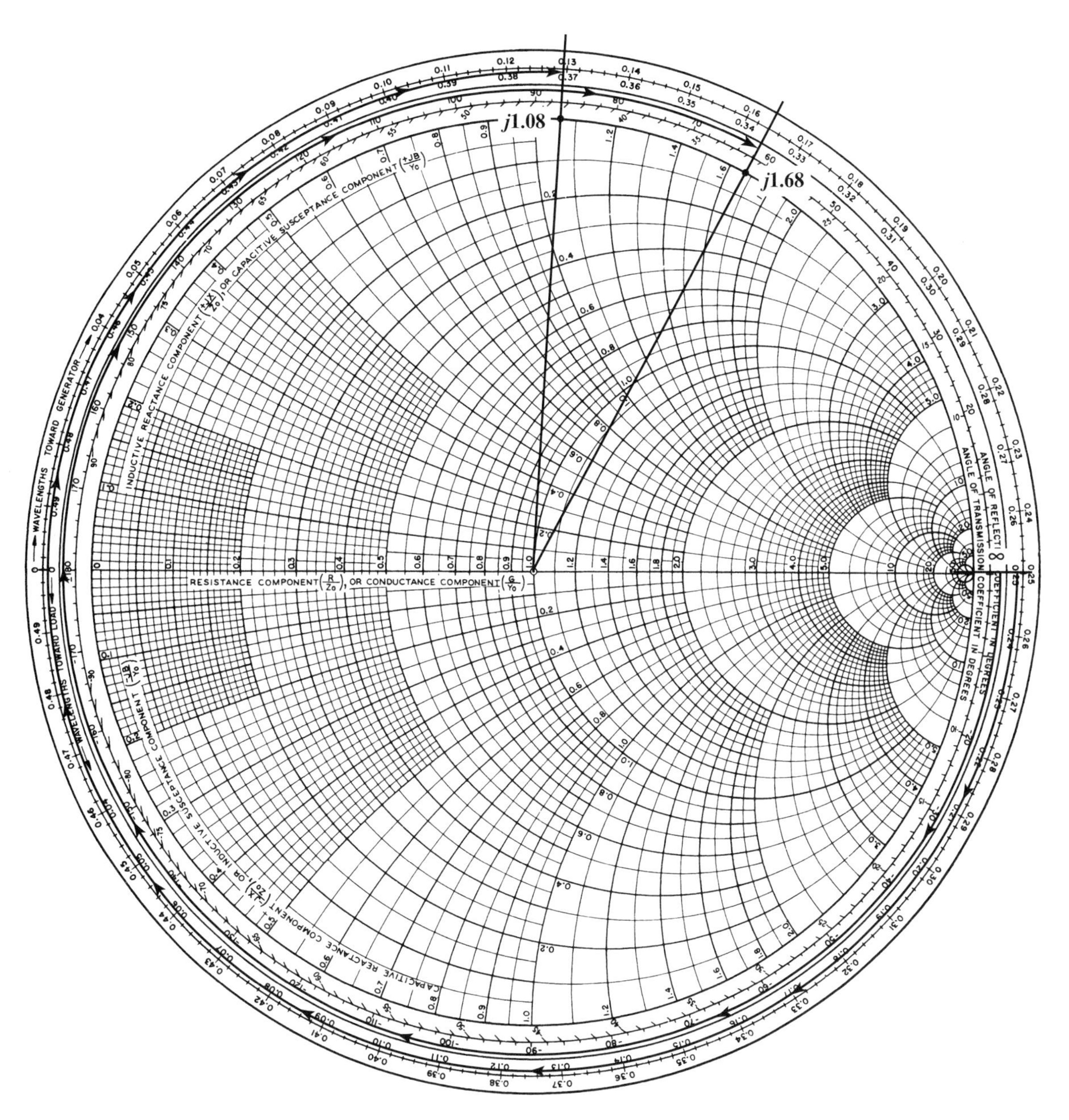

Figure 2-28. Double shunt stub matching example: chart B.

Then the required stub 1 susceptance is

$$j\tilde{B}_1(-l_{1\lambda}) = \tilde{B} \text{ at point } ⑤ - \tilde{B} \text{ at point } ④ = +j1.0 - (-j0.08) = +j1.08$$

To obtain l_1, we use another chart to avoid too much clutter on our main chart. Thus on Chart B, we rotate line passing through $\tilde{G} = \infty$ toward the generator until we obtain a reading on the rim of $+j1.08$. Stub length $l_{1\lambda} = (0.25 + 0.13)\lambda$, or

$$l_{1\lambda} = 0.38\lambda$$

This stub is then placed on the line and corresponds to location $a - a'$ in Figure 2-24.

5. Back to Chart A, rotate point ⑤ at constant radius a distance d_λ (to the $\tilde{G} = 1$ circle). This is at point ⑥.

$$\tilde{Y}_{⑥} = 1 - j1.68$$

Required stub 2 susceptance is then $j\tilde{B}_2(-l_2) = -(-j1.68) = j1.68$ to cancel remaining susceptance on the line. To get stub length, again use Chart B and rotate from $\tilde{G} = \infty$ toward the generator to the point on the chart rim where $j\tilde{B}_2(-l_2) = j1.68$. Adding on stub 2 will move point ⑥ to the chart center along constant conductance $G = 1$ since shorted stub can add only susceptance. Point ⑦ is the matched line.

Example 2-11

Limitation on Double Stub Matching

Suppose we have the system in Figure 2-29 given and we are asked to find stub lengths $l_{1\lambda}$ and $l_{2\lambda}$ required to match.

Figure 2-29. System for which double stub matching is not possible.

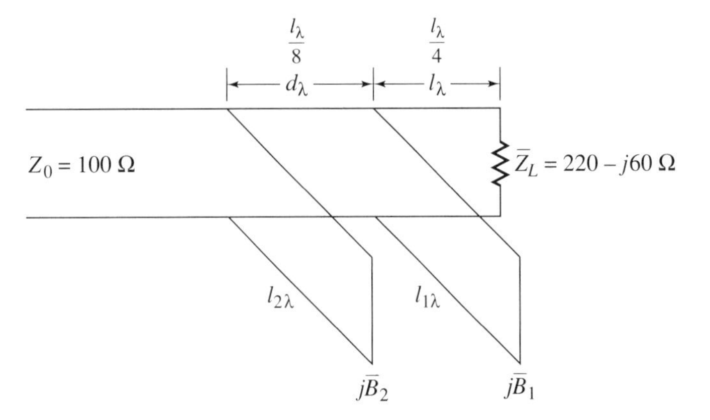

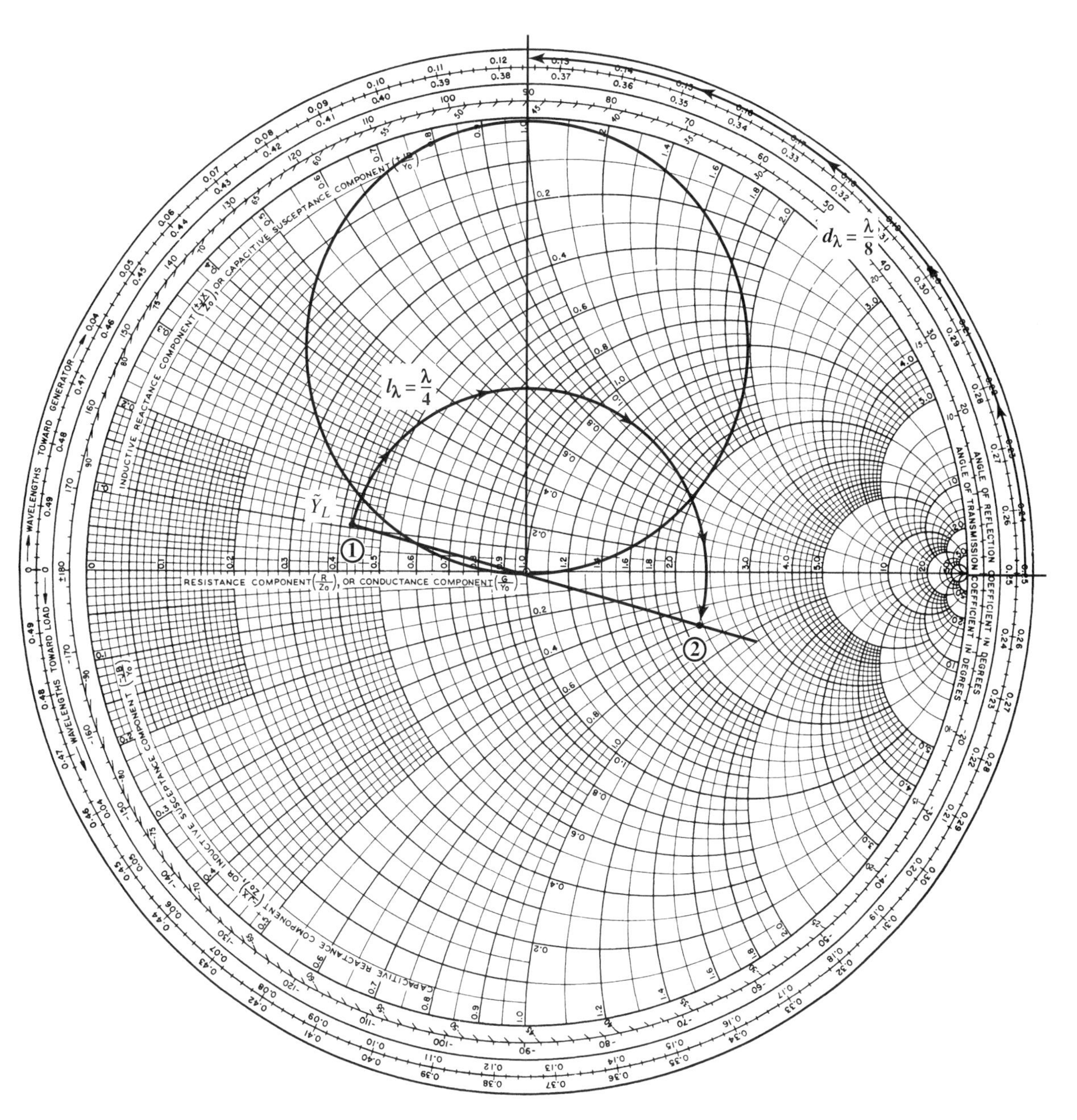

Figure 2-30. Second double stub matching example.

Rotate the $\tilde{G} = 1$ circle toward the load by $d_\lambda = \lambda_l/8$. See chart in Figure 2-30. Plot

$$\tilde{Y}_L = \frac{1}{\tilde{Z}_L} = \frac{1}{\left(\frac{220 - j60}{100}\right)} = 0.423 + j0.115$$

at point ①. Rotate $\tilde{Y}_L$ toward the generator at constant radius by amount $l_\lambda = \lambda_l/4$ to point ②, halfway around the chart.

Now the double stub technique requires that we add a susceptance at point ② that will put the point on the *rotated* $\tilde{G} = 1$ circle. However, since we can add only susceptance with a shorted stub, we would be constrained to move on the constant conductance circle $\tilde{G} = 2.2$, which would never intercept the rotated circle. Thus no match can be achieved with the system as shown. One obvious way to avoid this is to reduce d_λ so that the $\tilde{G} = 1$ circle is not rotated so far, but in many systems the distance d_λ is fixed, and the location l_λ is also specified.

Note also that there is an entire ''dead'' zone wherein no match can be effected. This is the region within the constant conductance circle that is tangent to the rotated $\tilde{G} = 1$ circle, in this case the $\tilde{G} = 2$ circle.

Figure 2-31. Stub tuner systems: (*a*), (*b*), (*c*) Single, double, and triple stub coaxial tuners, respectively; (*d*) Slide screw single stub tuner for waveguide (Chapter 9)

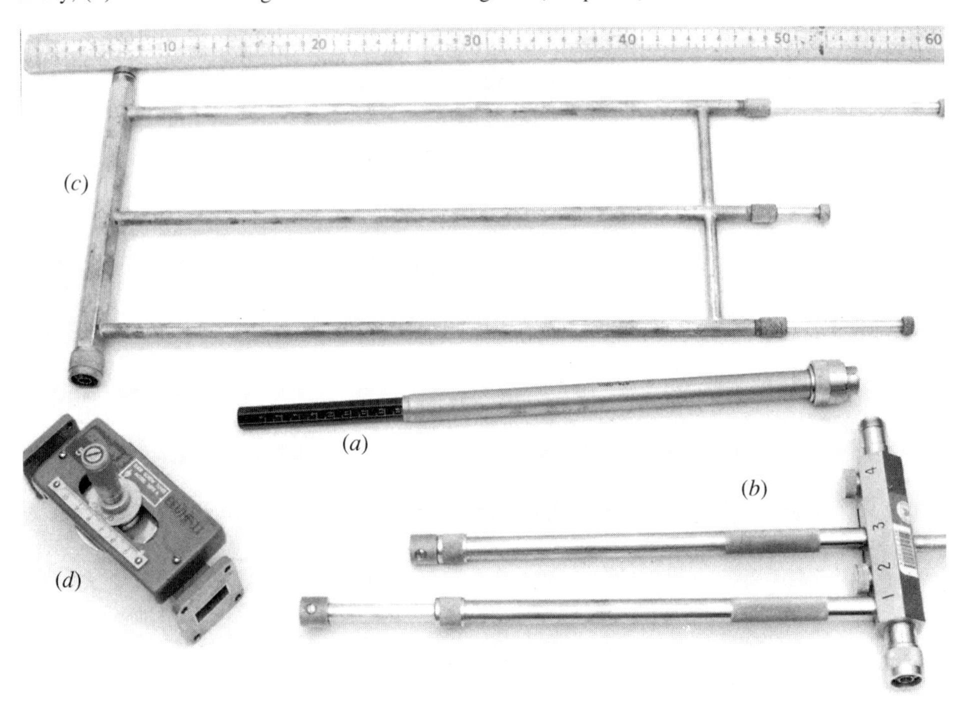

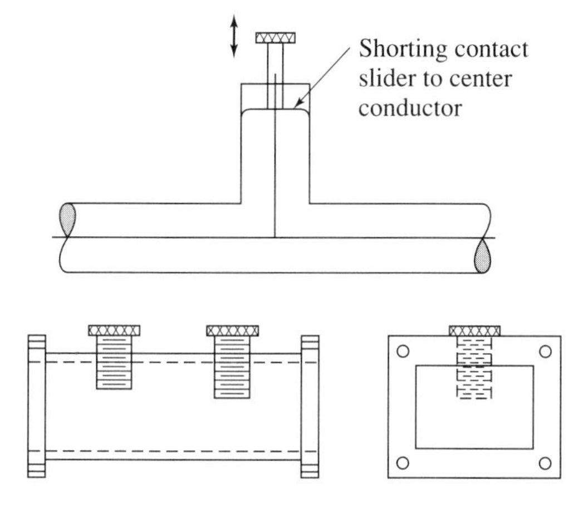

Figure 2-32. Schematic views of coaxial and waveguide tuners (stubs).

If the system is constrained as shown, an alternate solution is to place a third stub between the load and stub 1. This third stub in effect moves to the load to a new point such that stub 1 can then move point (2) to the rotated circle.

Figures 2-31 and 2-32 are pictures and operational cross sections of coaxial and waveguide stubs (tuners) used in practice.

Exercises

2-4.2a A shunt double stub tuner has a spacing of $\frac{1}{8}\lambda$ and is located 10λ from the load for which $\tilde{Y}_L = 0.5 + j1$ (normalized). Find the stub susceptances that will match this load. Use shorted stubs. Repeat for open circuited stubs and compare the two solutions (stub lengths, bandwidths, etc.).

2-4.2b An alternate series double stub tuner is shown in Figure 2-33. Develop a step-by-step Smith chart design procedure for determining the lengths l_1 and l_2 if the distances d and l are fixed. Describe any dead zone for this type of system.

2-4.2c A hybrid double stub tuner is shown in Figure 2-34. There is zero transmission line distance between the two stubs. Develop a design procedure for determining the required stub lengths l_1 and l_2. Describe any dead zone existing for this system. Distance l is fixed.

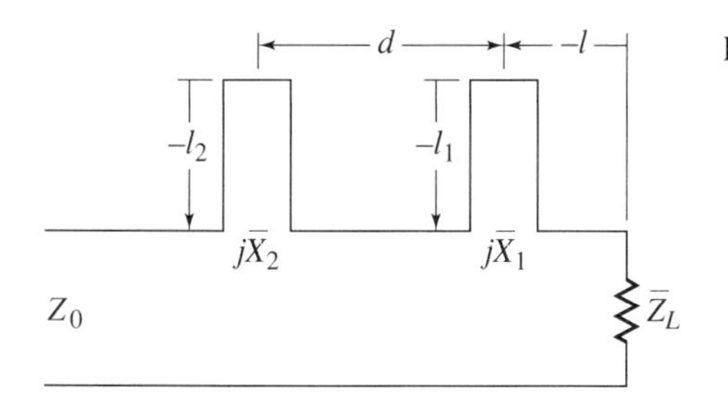

Figure 2-33.

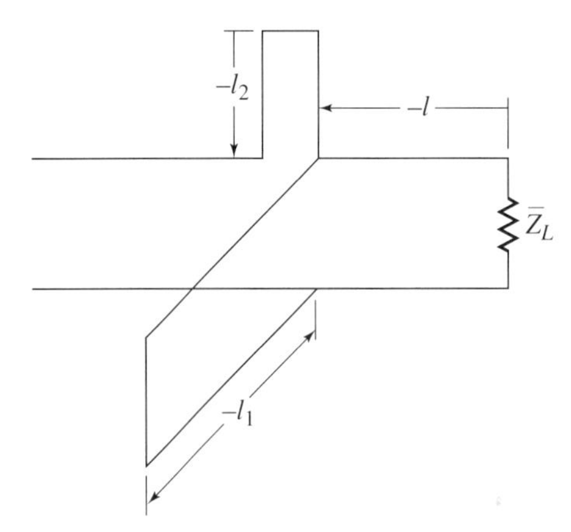

Figure 2-34.

2-4.2d Most applications of matching involve a design process for determining the stub lengths for a given, known load. Suppose we have an analysis problem where one is given the stub lengths, stub separation, and stub distance from the load as given in Figure 2-35. Determine the load impedance. Next suppose that a measurement error exists in all given values except stub spacing so they are all 10% high in value. What is the percentage error in the impedance magnitude with respect to the true value? The line wavelength is 15 cm, and the characteristic impedance of the line (assumed lossless) is 125 Ω. (Hint: Begin at a matched condition and remove stubs as you work back toward $\bar{Z}_L$.) (Partial answer: $114 - j84.4$ Ω)

2-4.2e Develop a design procedure for the triple stub tuner shown in Figure 2-36. Your design procedure should include a test to determine if stub 3 is even needed (i.e., whether a two stub match with only l_1 and l_2 will suffice).

2-4.2f Apply the triple shunt stub design developed in Exercise 2-4.2e to the specific cases $\bar{Z}_L = 220 - j60$ Ω, $Z_0 = 100$ Ω, $l_{L\lambda} = 9.11\lambda$, $l_\lambda = \lambda/8$, $d_\lambda = \lambda/8$. (Answer: $l_{3\lambda} = 0.184\lambda$, $l_{2\lambda} = 0.3032\lambda$, $l_{1\lambda} = 0.2092\lambda$)

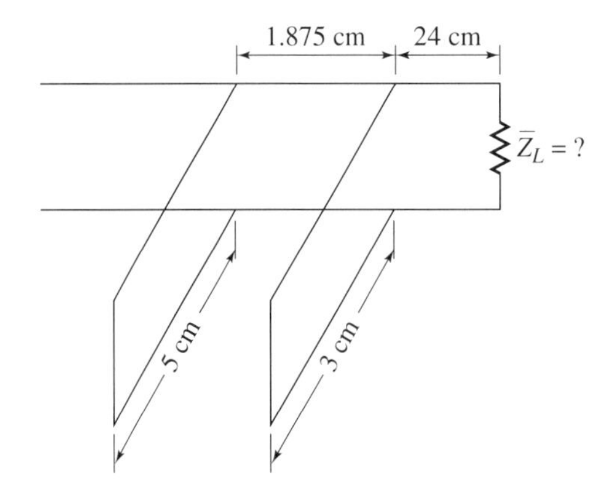

Figure 2-35.

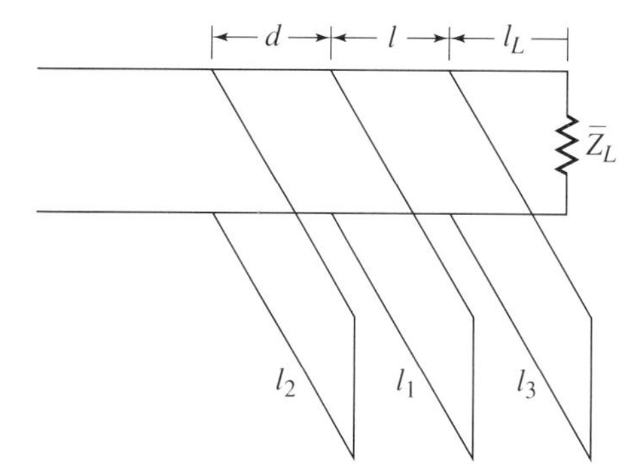

Figure 2-36.

2-5 The Smith Chart as an Impedance-Admittance Calculator for Lumped Element Systems

Up to this point the Smith chart has been used to work with impedances and admittances on distributed parameter transmission lines. Since it is possible to move around on the chart and hold either resistance (conductance) or reactance (susceptance) constant, we may use the same principles to combine lumped elements. This application makes it possible to take a given impedance and connect other elements to produce a desired input impedance. For example, if one wishes to connect a cable to the input of a bipolar transistor, a mismatch will generally exist. If we could add lumped elements to the input of the transistor to produce Z_0 of the line, it would not be necessary to use stubs on the line. Networks of the L, T and pi types may be designed using the Smith chart to provide the lumped element match. The chart has the advantage of avoiding complex arithmetic required when one uses a calculator, pencil, and paper—a difficult process even if the calculator will work directly with complex numbers.

Four basic principles apply when making lumped element computations with the Smith chart. These are:

1. When adding resistance (conductance) to a given impedance (admittance) point on the chart, move along the constant reactive (susceptive) component circles, since the imaginary component of the impedance (admittance) is not affected.
2. When adding a reactance (susceptance) to a given impedance (admittance) point on the chart, move along the constant resistive (conductive) component circles, since the real component of the impedance (admittance) is not affected.
3. As a network is being designed or analyzed there may be both series and parallel components. To convert from impedance to admittance (or vice versa) simply transfer the point 180° on the chart. It is a good idea to label each point on the chart with an appropriate Y or Z so you do not lose track of its meaning.
4. Since most networks will have impedance values in hundreds or thousands of ohms, it is necessary to scale the values to avoid the inaccurate right-hand portion of the Smith chart. To normalize the impedance values; arbitrarily select a Z_0 and divide all impe-

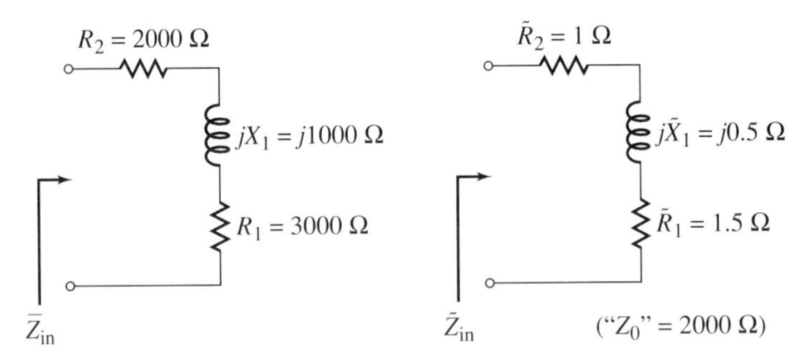

Figure 2-37. Circuit example for using the Smith chart as a lumped element impedance calculator.

dances by that number. It is obviously easiest to pick a purely real value for the normalization, and values in the range of $10^4\ \Omega$ are not uncommon. The value selected should be written on the chart for reference.

Since the theory of the principles has already been discussed earlier, the process will be presented using examples.

Example 2-12

We begin by considering the most elementary combination, a series circuit. The answer is obvious by inspection since impedances in series simply add. Our problem is to find the input impedance of the circuit shown in Figure 2-37 using the Smith chart.

We first arbitrarily select a normalizing impedance "Z_0" of 2000 Ω. The normalized circuit obtained by dividing all impedances by 2000 shown in Figure 2-37.

First, plot $\tilde{R}_1$, shown as point ① on the Smith chart of Figure 2-38.

Next the inductive reactance $j\tilde{X}_1$ is added in series by moving along the constant resistance circle in the positive reactance direction until the reactance value at point ① (zero in this case) is *increased* by $j0.5$. This is point ② in Figure 2-38.

The resistance $\tilde{R}_2$ is now added by moving along the constant reactance arc of point ③ to increase the resistance component by 1. This results in final $\tilde{Z}_{in}$ at point ③.

The actual value $\bar{Z}_{in}$ is then computed as

$$\bar{Z}_{in} = \tilde{Z}_{in} \times 2000 = (2.5 + j0.5)(2000) = 5000 + j1000\ \Omega$$

Example 2-13

We next consider the use of the Smith chart to compute the input impedance of a series-parallel lumped element circuit shown in Figure 2-39. The corresponding Smith chart is in Figure 2-40. Using a normalizing impedance of 1000 Ω results in the circuit in Figure 2-39.

For a normalizing impedance level of $Z_0 = 1000\ \Omega$,

$$\tilde{R}_1 = \tfrac{3000}{1000} = 3$$

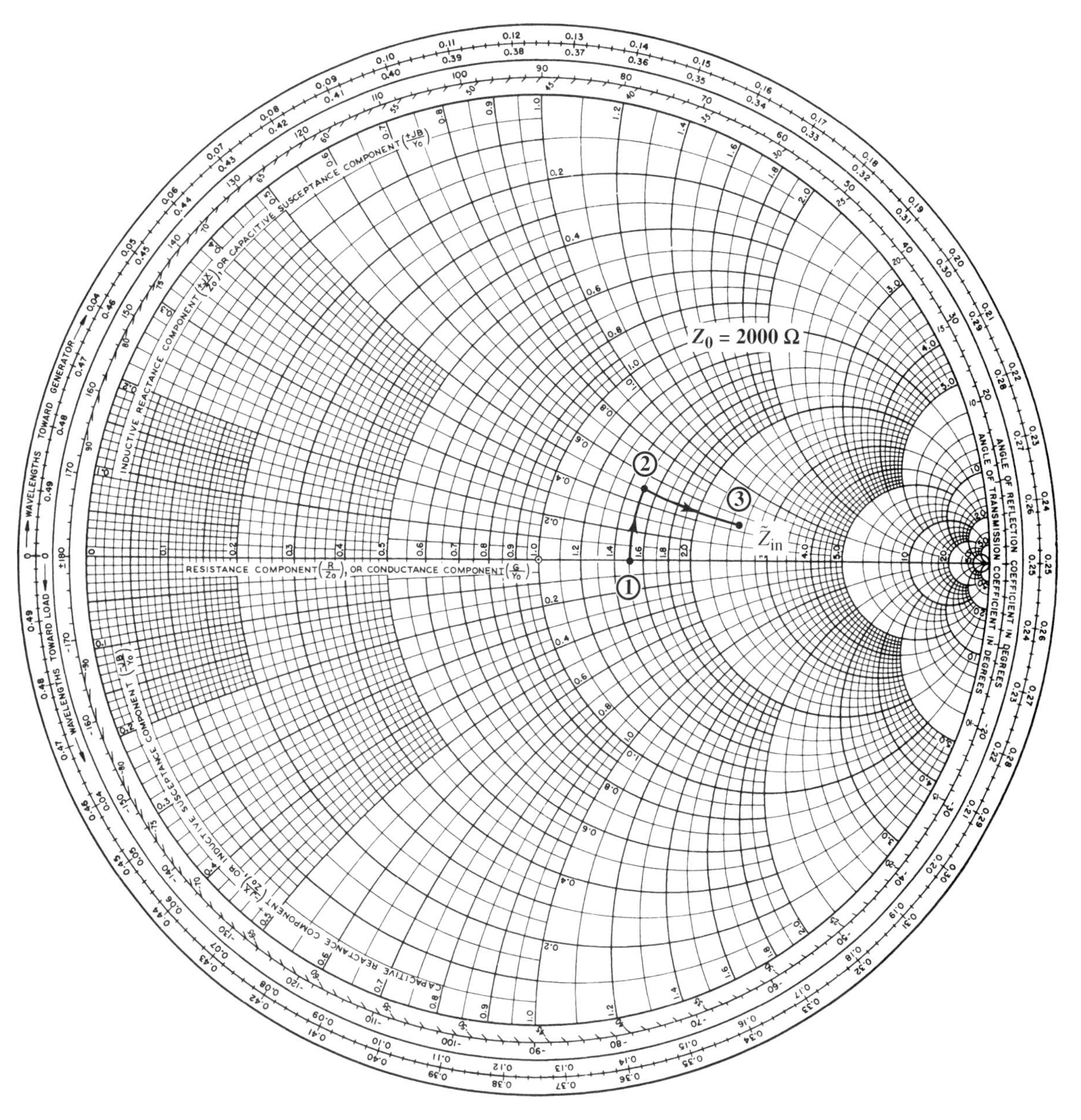

Figure 2-38. Example 2-12: Using the Smith chart as a lumped element series impedance calculator.

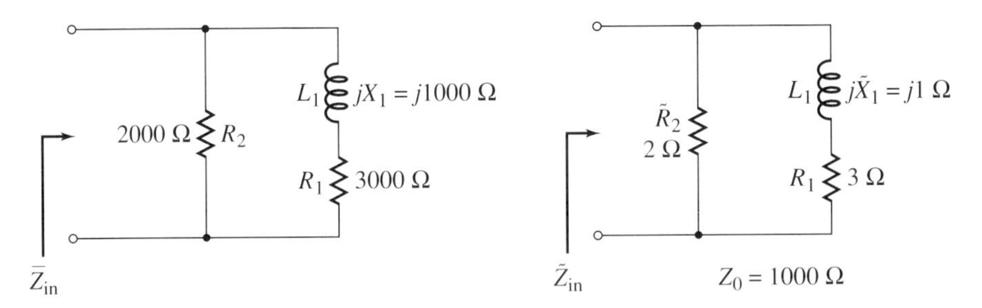

Figure 2-39. Circuit example for using the Smith Chart as a lumped element series-parallel calculator.

This is plotted as point ① on the chart. Note we have labeled it with a Z so we can keep track of the nature of the value the point represents (i.e., impedance or admittance).

Next

$$j\tilde{X}_1 = j\tfrac{1000}{1000} = j1$$

This is added in series with R_1 by moving along the constant-resistance (3-Ω) circle until we have increased the reactance by an amount $j1$.

This is point ②. (We could, of course, have plotted $3 + j1$ initially.) To parallel the 2000 Ω we first convert point ② to admittance by rotation 180° on the chart to ②′. This will be the admittance of the series L_1R_1 network.

Now parallel R_2:

$$\tilde{R}_2 = \frac{2000}{1000} = 2 \qquad \text{for which} \quad \tilde{G}_2 = \frac{1}{\tilde{R}_2} = 0.5$$

This is placed in parallel with the series network L_1R_1 by moving along constant susceptance circle ($-j0.095$) until the conductance is increased by an amount 0.5. This is point ③. Point ③ is the input admittance of the network. To obtain the impedance, we rotate 180° to point ③′.

From point ③′: $\bar{Z}_{in} = (1000)(1.24 + j0.155) = 1240 + j155\ \Omega$.

Exercises

2-5.0a Determine the input impedance of the circuit in Figure 2-41 using a Smith chart. What is the input admittance of the same network? (Answer: $625 - j200\ \Omega$; $0.00145 + j0.000464$ S)

2-5.0b Using the Smith chart with admittance coordinates, determine the input admittance of the circuit in Figure 2-42. We use the same process as example 1 substituting conductance for resistance, susceptance for reactance and Y_0 for Z_0.

2-5.0c Using the Smith chart determine the input admittance of the circuit in Figure 2-43. From the $\bar{Y}_{in}$ found above determine $\bar{Z}_{in}$ and draw a series circuit for this $\bar{Z}_{in}$ giving values (in ohms) for the components you show.

2-5.0d Determine the input impedances of the circuits in Figure 2-44.

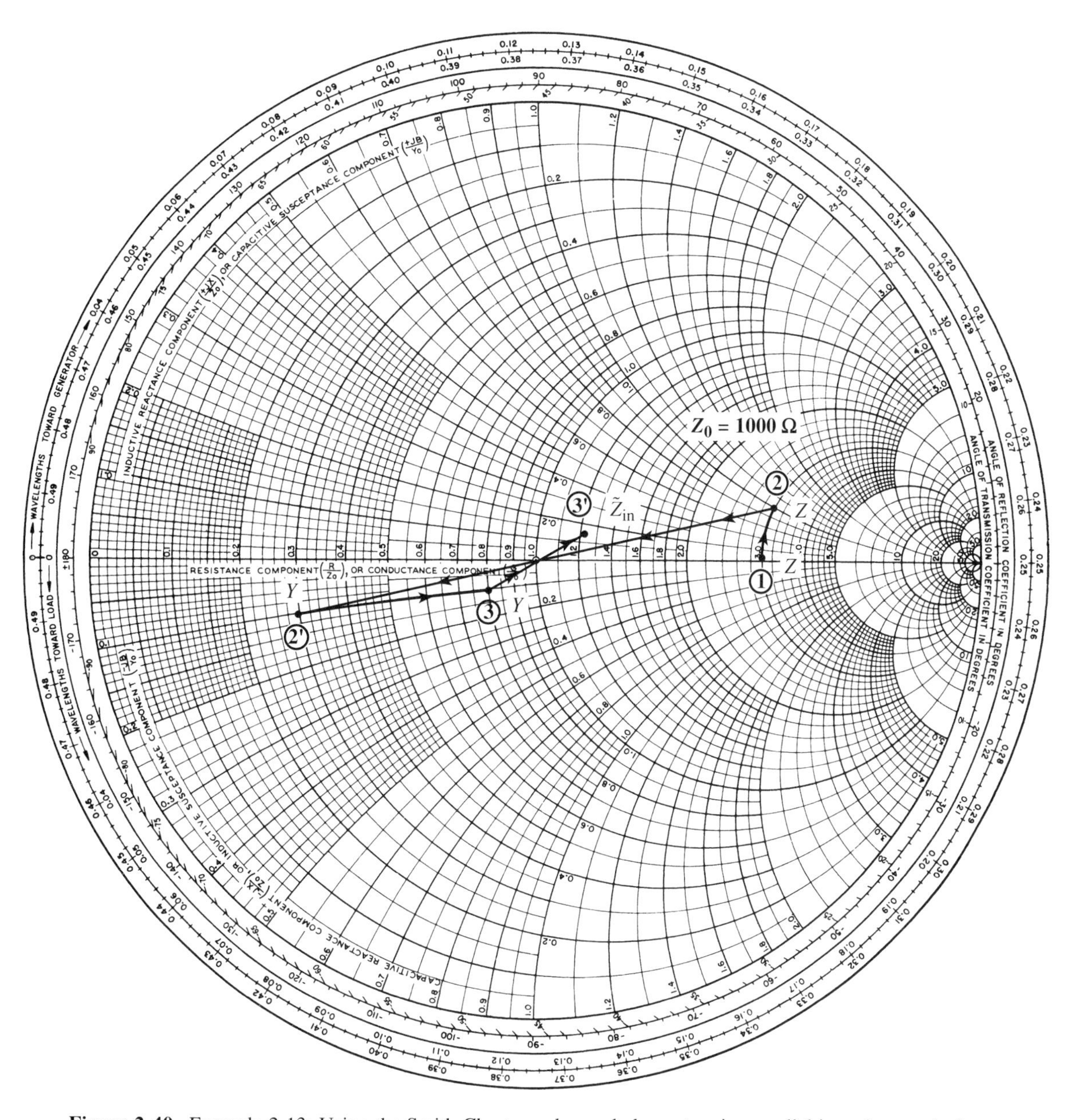

Figure 2-40. Example 2-13: Using the Smith Chart as a lumped element series-parallel impedance calculator.

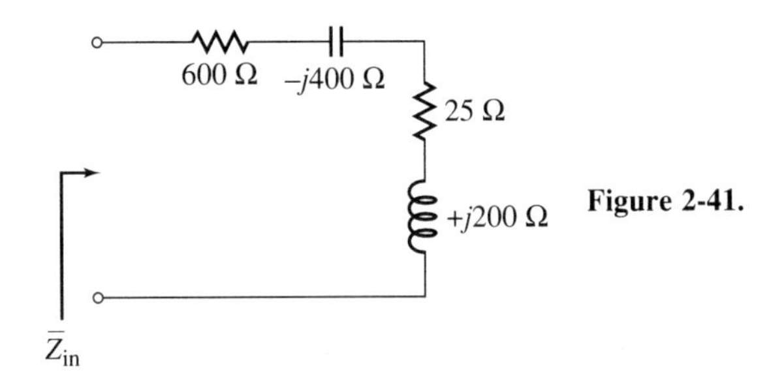

Figure 2-41.

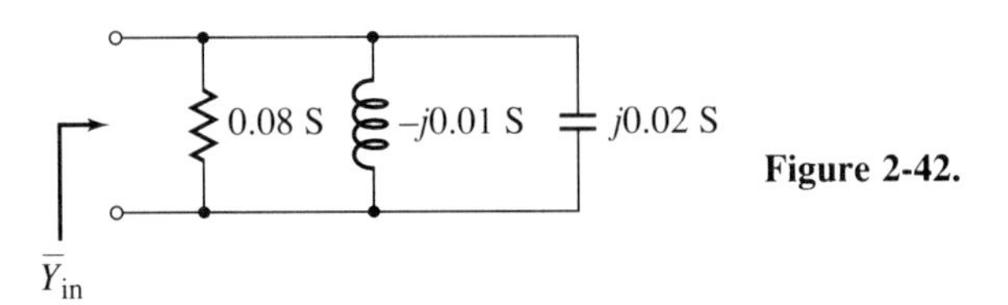

Figure 2-42.

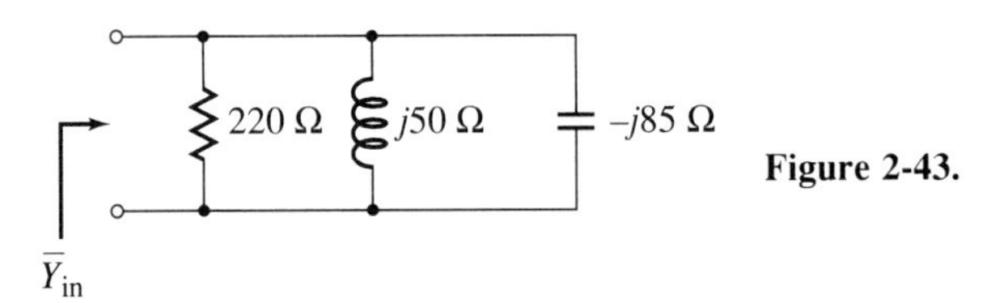

Figure 2-43.

Figure 2-44.

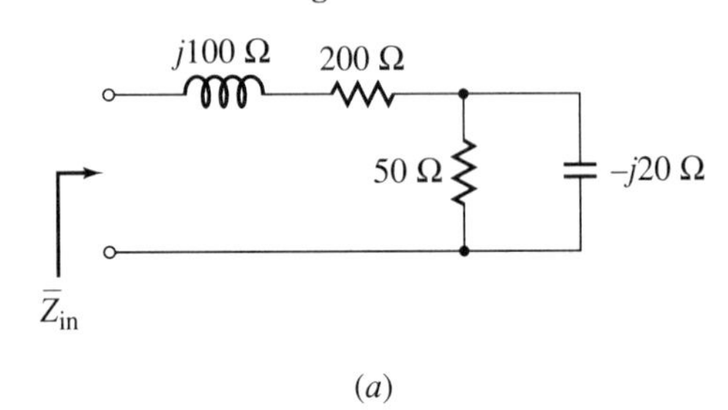

(*a*)

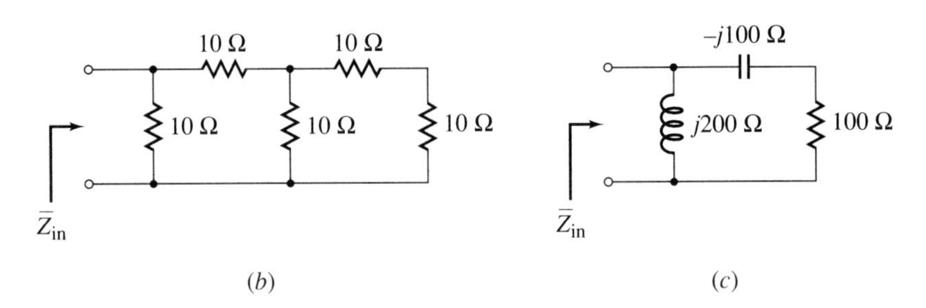

(*b*) (*c*)

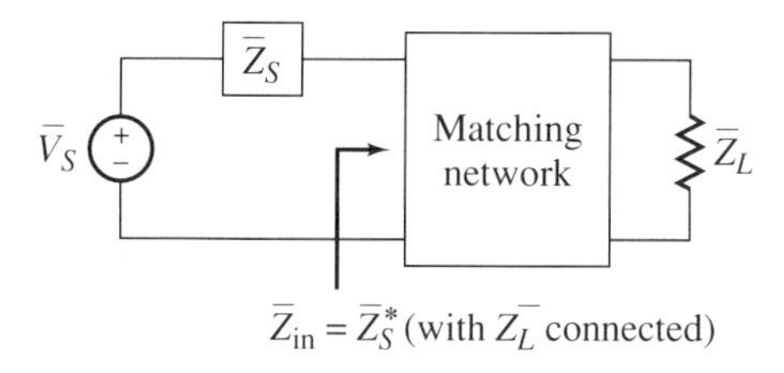

Figure 2-45. Matching network design requirements for producing a conjugate match.

2-5.1 Impedance Matching with Reactive Elements

Now that the basic principles of lumped element computations using the Smith chart have been presented we can consider some important applications. One of the most frequent design problems encountered is that of providing maximum power transfer, especially in high-frequency (above 10 MHz) electronic circuits. From circuit theory we recall that maximum power is transferred if the load impedance is the conjugate of the Thevenin impedance of the source driving the load. This is often referred to as *conjugate matching*. However, if both the load and source impedances are given we are faced with the problem of designing a network that can be placed between the source and load that will effectively produce a conjugate match. This problem is shown schematically in Figure 2-45.

To reduce power losses we use reactive elements. Since the reactance value is frequency sensitive, this matching will be exact at only one frequency; however, there is typically some bandwidth over which a certain amount of mismatch may be acceptable.

The technique is best presented by considering a specific example and pointing out the general principles as they are encountered. As with many design techniques there are several acceptable networks, and often other factors govern the particular configuration selected. Capacitors can often be used to block dc in addition to providing the matching function, for example.

Example 2-14

Suppose it is desired to add reactive (lossless) components to a load of 100-Ω resistance to produce an impedance level of $500 + j200\ \Omega$, as shown in Figure 2-46.

Many networks can perform this impedance transformation. We consider only one possibility. (See Smith [50] for other network topologies.)

Figure 2-46. One method for using reactive components to provide an impedance match.

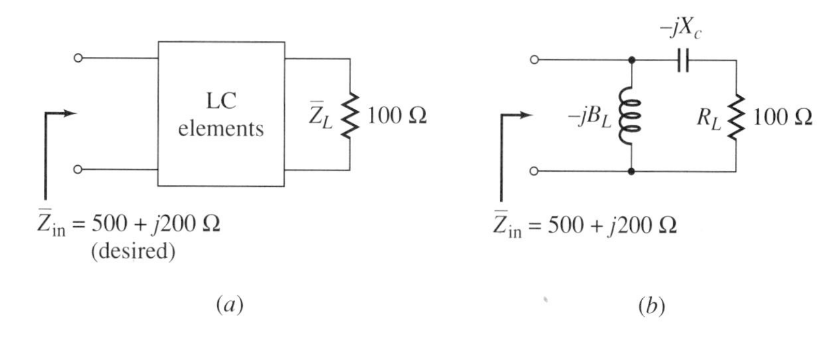

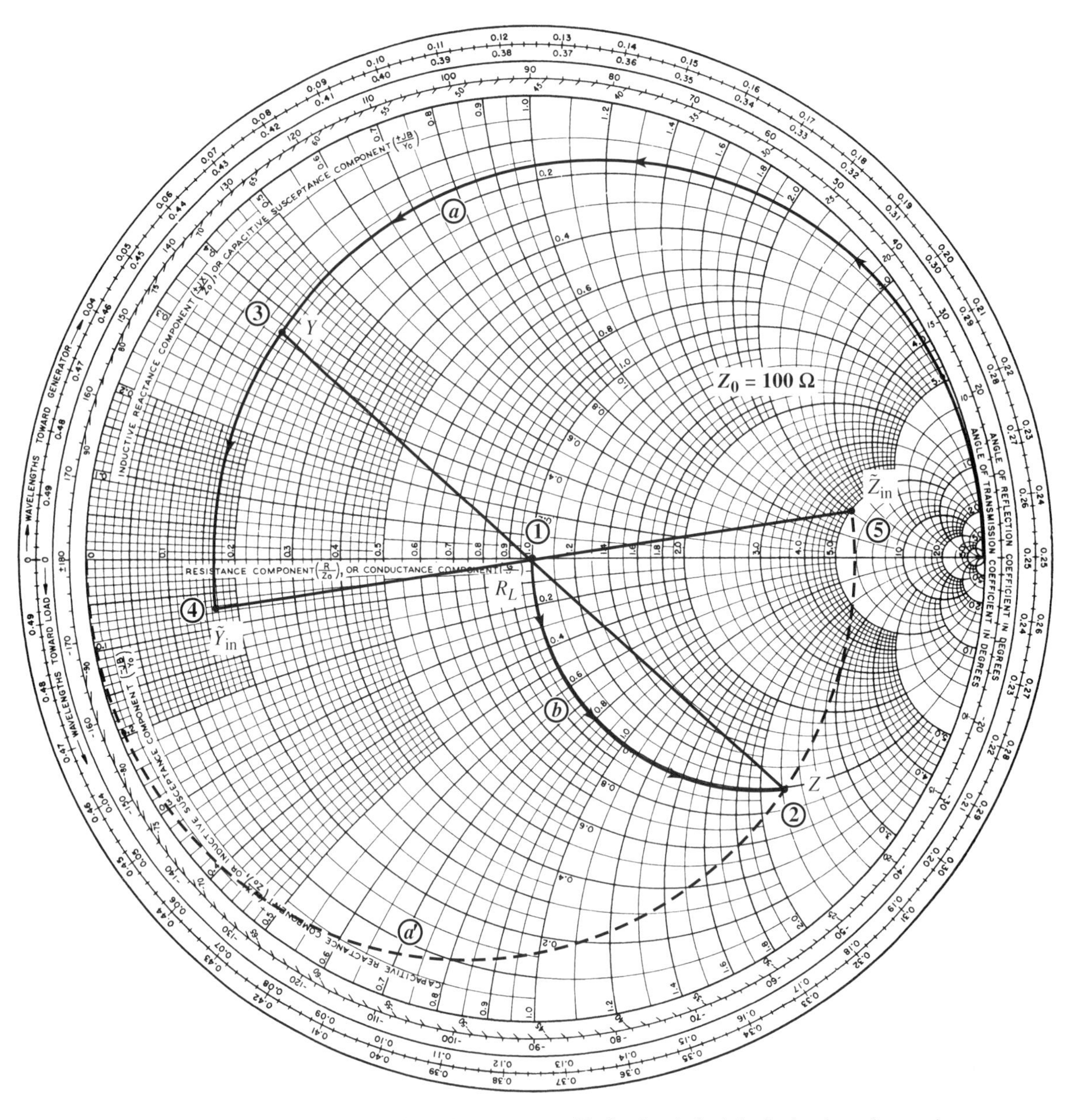

Figure 2-47. Changing a given impedance to a specified value (principle for lossless elements).

For convenience select $Z_0 = 100\ \Omega$ to normalize. $\tilde{Z}_L = 100/100 = 1$. This is plotted as point ① in Figure 2-47.

To complete the design we need to examine what happens as we add the inductor and capacitor. We first discuss the principles involved and then give the step-by-step design procedure. The analysis basically involves looking ahead to the point where one wants to end and set up the chart to yield that desired result.

Ultimately we want to end up at

$$\tilde{Z}_{\text{in}} = \frac{500 + j200}{100} = 5 + j2$$

(point ⑤ in Figure 2-47).

This $\tilde{Z}_{\text{in}}$ must be the value obtained by inverting the admittance $\tilde{Y}_{\text{in}}$ that resulted when the inductor was placed in parallel with the R_LC combination. This is $\tilde{Y}_{\text{in}}$ at point ④ on the chart, 180° from $\tilde{Z}_{\text{in}}$. Adding the inductance $-jB_L$ in parallel with the R_LC combination, which is a negative susceptance, could have only moved the admittance along the constant real (conductance) circle of value $\tilde{G}_{\text{in}} = 0.17$ along the solid path shown as ⓐ to get to point ④. The starting point on path ⓐ is not initially known, but must be at that point representing the *admittance* of the R_LC series combination.

The addition of the capacitor in *series* with the load must move along the constant-resistance (1-Ω) circle in the direction of negative reactance path ⓑ. Now the point at which we stop on path ⓑ will represent the impedance of the series R_LC combination. The reciprocal (admittance) of this series combination must end up on path ⓐ since the inductor can cause change only along that path to get to $\tilde{Y}_{\text{in}}$.

Thus to obtain the locus of all impedance points whose reciprocals lie on the required constant conductance path ⓐ we reciprocate all points on that constant conductance circle to obtain the dashed impedance circle ⓐ'. The intersection of the locus ⓐ' with the locus ⓑ uniquely determines point ②, and hence the required capacitive reactance.

The proper capacitance to add in series with R_L would be the value that moves the impedance from point ① to point ② along path ⓑ, since the reciprocal of point ② will then lie on path ⓐ (at point ③) as required. *If the dashed circle did not intersect the R_L circle some other type matching network must be used.*

Design Procedure

The design procedure will be given using the preceding example. See Smith chart in Figure 2-48.

1. Plot the desired impedance level $\tilde{Z}_{\text{in}}$ (point ① on chart, $\tilde{Z}_{\text{in}} = (500 + j200)/100 = 5 + j2$). (We select, for convenience, $Z_0 = 100\ \Omega$.)
2. Convert to $\tilde{Y}_{\text{in}}$ by moving 180° at constant radius to point ②.
3. Draw the reciprocal circle of the constant conductance value given by the real part of point ② ($\tilde{Y}_{\text{in}}$). This yields circle ⓐ. ($\tilde{G}_{\text{in}} = 0.17$ in this example.) (Note this passes through point ①.)
4. Plot the load $\tilde{Z}_L$ at point ③. (In this case $\tilde{Z}_L = 100/100 = 1$.)
5. Locate the intersection of circle ⓐ with the constant resistance circle corresponding to the real part of $\tilde{Z}_L$ (In this example $R_L = 1$, and the desired intersection is point ④ since we are using a series capacitor.) If there is no intersection a different network

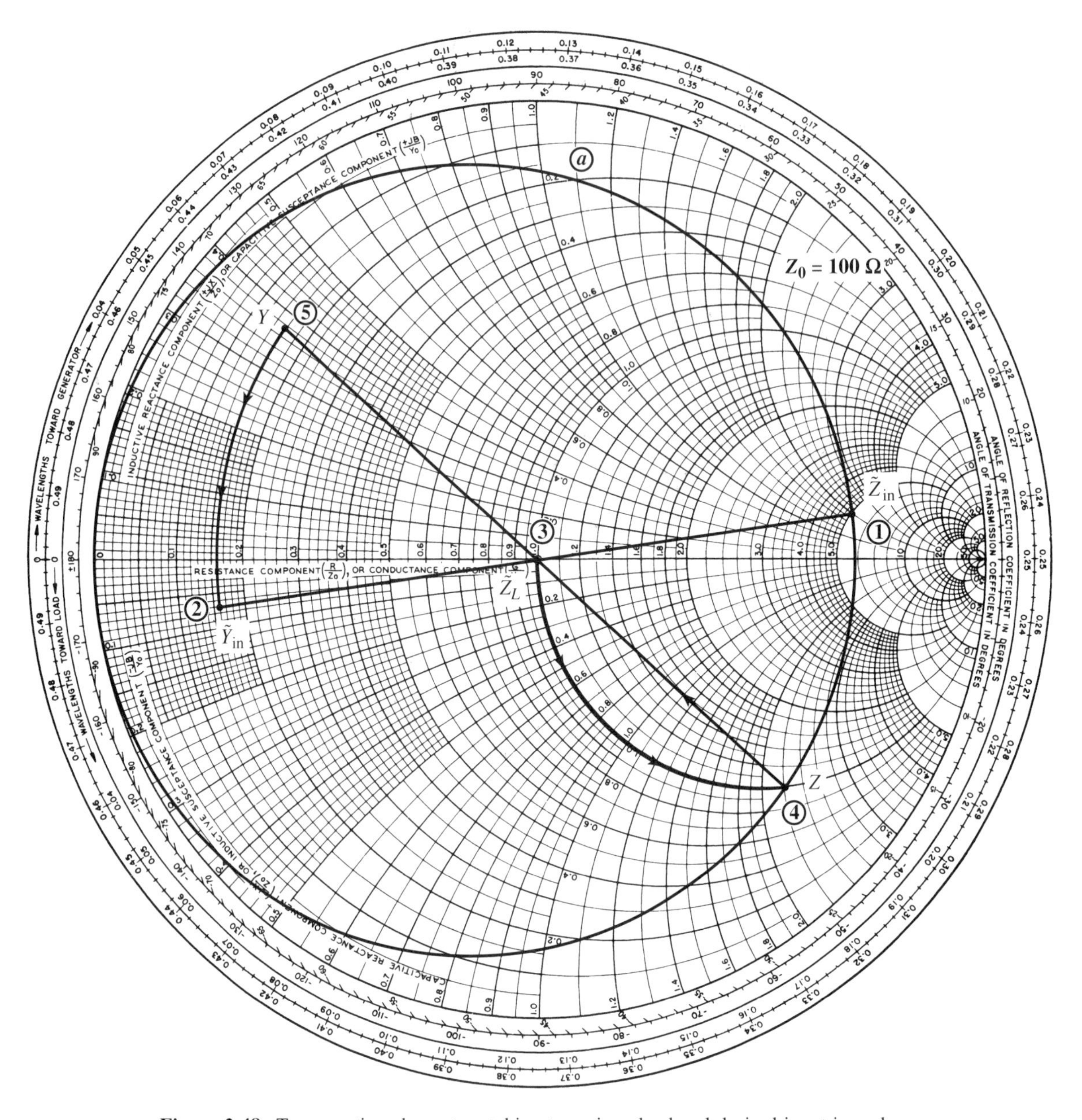

Figure 2-48. Two reactive element matching to a given load and desired input impedance.

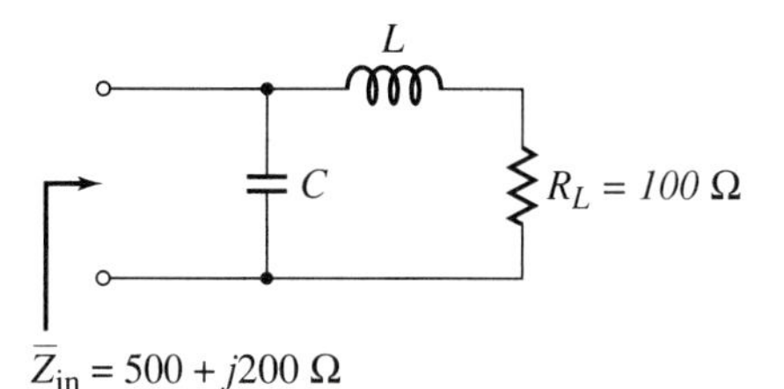

Figure 2-49.

must be used. Required $-j\bar{X}_c$ is the change in reactance needed to go from point ③ to point ④. In this case $-j\tilde{X}_c = -j2.20$ or $-j\bar{X}_c = (100)(-j2.20) = -j220\ \Omega$. If we let f = operating frequency in hertz, then

$$C = \frac{1}{(2\pi f)(\bar{X}_c)} = \frac{1}{(2\pi f)(220)}\ \text{F}$$

(Note: There is no guarantee the value of C will be practically obtainable. If not, some other network topology must be used.)

6. Convert point ④ to admittance by moving 180° on the chart to point ⑤, which by prior construction ends up on the correct real part of the desired Y_{in}.
7. Since this point ⑤ is admittance an inductor will move an amount $-j\tilde{B}_{L_1}$ to the required $\tilde{Y}_{in}$ at point ② along a constant conductance circle. In this example, the conductance circle is $\tilde{G} = 0.17$. The required inductive susceptance is (normalized) $-j\tilde{B}_{L_1} = j[-0.07 - (+0.38)] = -j0.45$. Thus

$$-j\bar{B}_{L_1} = (0.01)(-j0.45) = -j0.0045\ \text{S} \qquad \left(Y_0 = \frac{1}{Z_0} = \frac{1}{100} = 0.01\right)$$

$$L_1 = \frac{1}{(2\pi f)(\bar{B}_{L_1})} = \frac{1}{(2\pi f)(0.0045)}\ \text{H}$$

Again, there is no guarantee of a practical value of L, so it might be necessary to use some other network configuration.

As a general rule, if $|\bar{Z}_L| > |\bar{Z}_{in}|$ the element nearest $\bar{Z}_{in}$ will be a parallel L or C as a first guess. For $|\bar{Z}_L| < |\bar{Z}_{in}|$, a series element should be tried first.

Exercises

2-5.1a In the preceding example for which Figure 2-48 was used, show the locations of points ④ and ⑤ on the same chart for the network topology shown in Figure 2-49. Also give equations for L and C for a general frequency f.

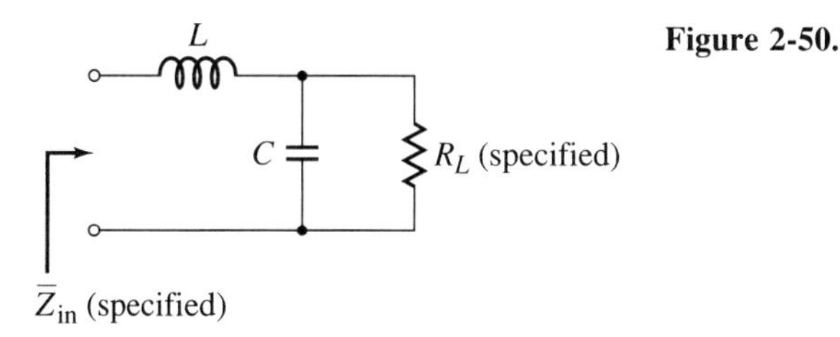

Figure 2-50.

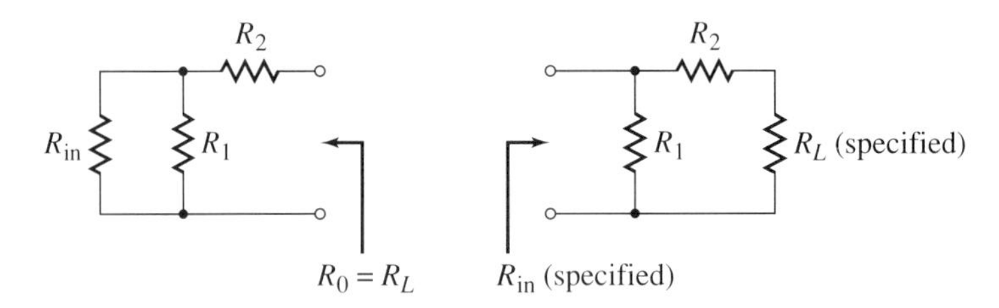

Figure 2-51.

Answer: $L = 2.2 \times \dfrac{100}{2\pi f}, \qquad C = \dfrac{(0.37 - 0.075)(0.01)}{2\pi f}$

2-5.1b A resistive load of 50 Ω is to be transformed to a value of 250 + j120 Ω. For an operating frequency of 200 MHz, determine the values of L and C which will provide this match using the network configuration of the form discussed in the example. Are these values physically reasonable?

2-5.1c Develop a design procedure for transforming a resistance load of value R_L to the impedance $\overline{Z}_{\text{in}} = R_{\text{in}} + jX_{\text{in}}$ using the configuration in Figure 2-50. Can you determine any $\overline{Z}_{\text{in}}$ for which the network will not work?

2-5.1d To obtain a frequency-independent network for an impedance transformation one must use resistances. Develop a design procedure for transforming a given load R_L to a given resistance R_{in} for the networks shown in Figure 2-51. An additional restriction is that the output impedance of the network is R_L. Are there any restrictions on the value of R_{in}? (See also General Exercise GE2.6 at the end of this chapter.)

2-5.1e Convert a load of value 50 Ω to a value of input impedance 200 + j100 Ω using

Figure 2-52.

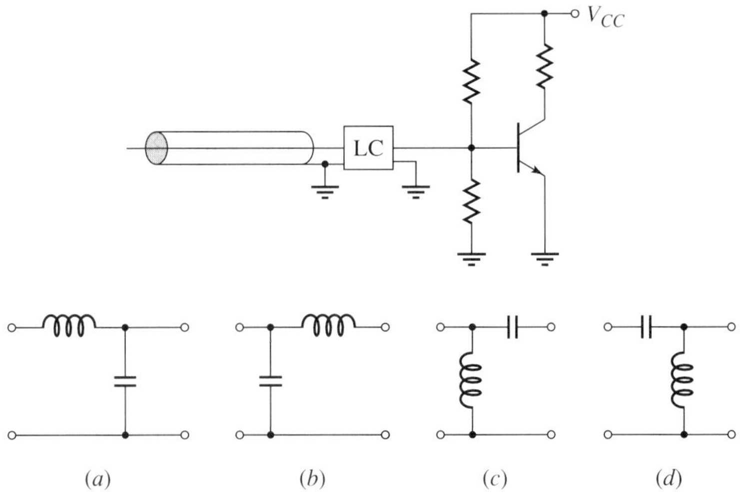

(*a*) (*b*) (*c*) (*d*)

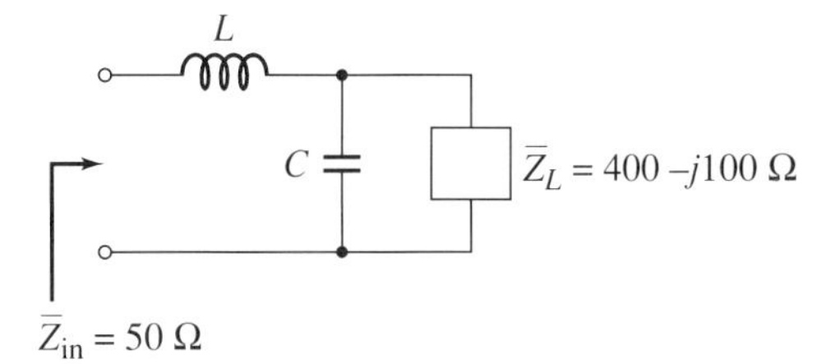

Figure 2-53.

reactive components at a frequency of 100 MHz. How are the component values for realizability?

2-5.1f For the design example presented in this section, what is the impedance seen by the load when a source having a series impedance of $\bar{Z}^*_{in}$ is connected at the input. Use the Smith chart and determine the impedance at the load terminals with the load removed. (Answer: 100 Ω)

2-5.1g A transmission line is to be matched to the input of a bipolar transistor as shown in Figure 2-52. Discuss the practical implications of each of the four possibilities shown. Consider the DC conditions.

2-5.1h Determine the values of L and C at 300 MHz that will provide the input impedance indicated for the circuit of Figure 2-53. (Answer: $C = 3.2$ pF, $L = 71.6$ nH)

2-5.2 Impedance Matching to Pure Real Values Using Reactive Elements

An important special case of the preceding matching technique is that of changing the resistive level of a resistive load using reactive components. As an example, suppose it is required to change a 100-Ω load resistance to a 200-Ω effective resistance as shown in Figure 2-54*a*, and the topology of Figure 2-54*b* is to be used. The result, of course, will only be valid at one frequency. The Smith chart which accompanies the following solution is in Figure 2-55. For convenience let us select a normalizing impedance of 100 Ω.

1. Plot $\tilde{R}_{in}$; point ①. $\tilde{R}_{in} = 200/100 = 2$.
2. Invert the point ① to point ② for conductance, $\tilde{G}_{in}$.
3. Draw reciprocal of the circle corresponding to $\tilde{G}_{in}$, circle ⓐ. In this case $\tilde{G}_{in} = 0.5$.

Figure 2-54. Changing resistance level of a resistive load using reactive components.

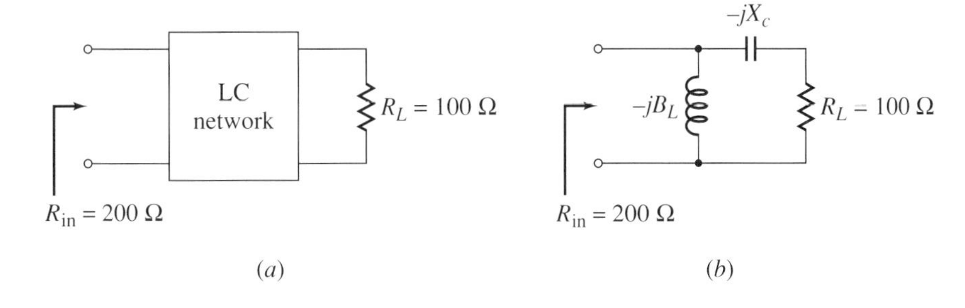

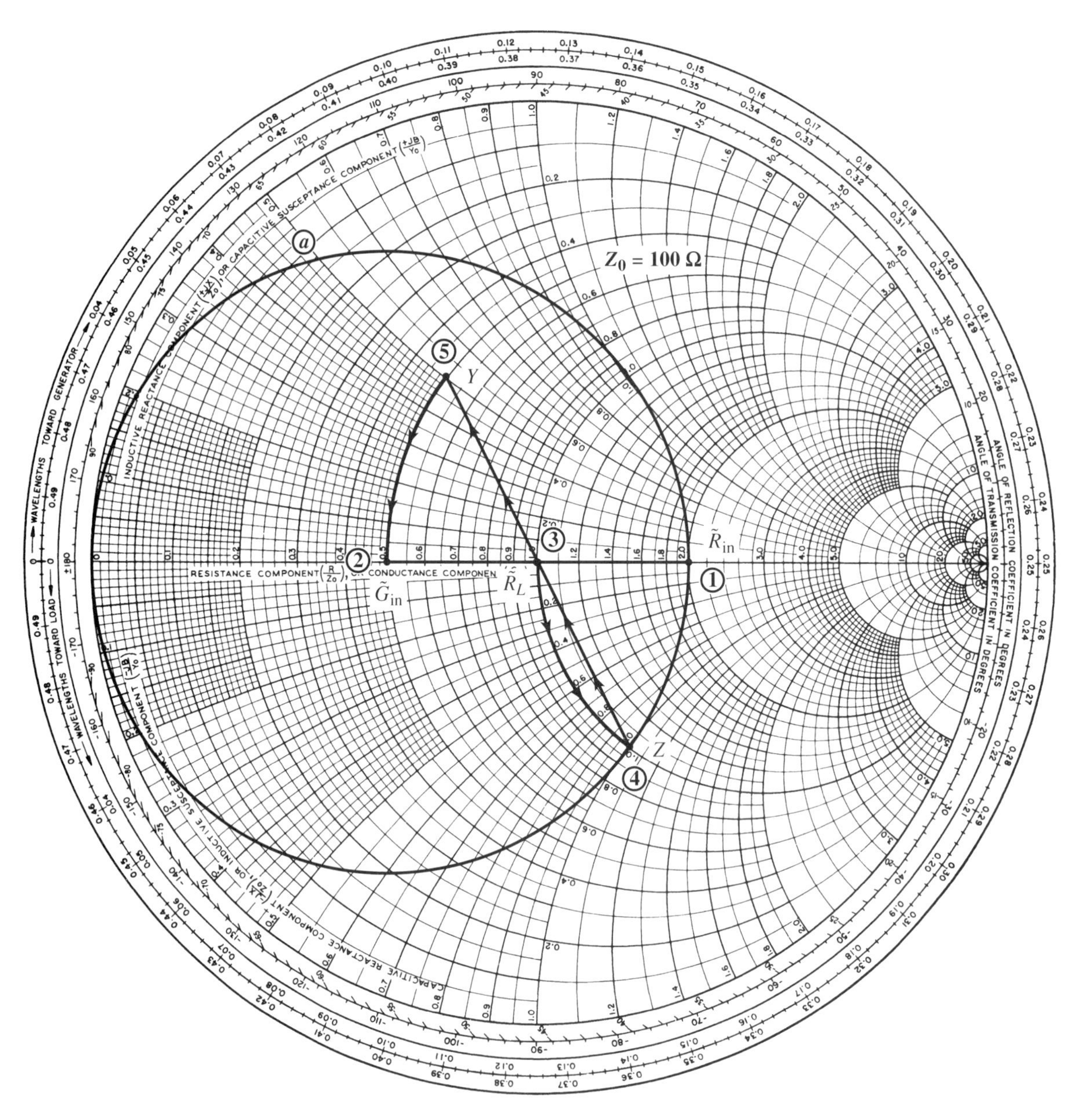

Figure 2-55. Changing resistance level using reactive elements only.

4. Plot $\tilde{R}_L$ at point ③, $\tilde{R}_L = 100/100 = 1$.
5. Locate point ④ intersection. (Note there is a second intersection above near the point $(1,j1)$ that could also be used if the L_1 and C_1 were interchanged.) The required capacitive reactance is

$$-j\overline{X}_c = (100)(-j1.0) = -j100\ \Omega$$

or

$$C_1 = \frac{1}{2\pi \overline{X}_{C_1}} = \frac{1}{(2\pi f)(100)}\ \text{F}$$

6. Convert point ④ to admittance at point ⑤.
7. Add $-j\tilde{B}_{L_1}$ to move point ⑤ down along constant conductance (0.5 in this case) to intersect at point ② ($\tilde{G}_{\text{in}}$).

$$-j\overline{B}_{L_1} = [0 - (+j0.5)](0.01) = -j0.005\ \text{S} \qquad \left(Y_0 = \frac{1}{Z_0} = \frac{1}{100} = 0.01\right)$$

The value of L_1 then is

$$L_1 = \frac{1}{2\pi f \overline{B}_{L_1}} = \frac{1}{(2\pi f)(0.005)}\ \text{H}$$

Exercises

2-5.2a Develop a design procedure for changing a resistance load, R_L, to an input resistance, R_{in}, for the case $R_{\text{in}} < R_L$, using reactive elements only. As an example, select $R_{\text{in}} = 50\ \Omega$ and $R_L = 100\ \Omega$.

2-5.2b Design an LC network that will match a 600-Ω load resistor to a transmission line having $Z_0 = 125\ \Omega$ at a frequency of 10 MHz.

2-5.2c Design an LC network that will match a 125-Ω load resistor to a transmission line having $Z_0 = 600\ \Omega$ at a frequency of 10 MHz.

2-5.3 Impedance Matching with Three Reactive Elements

As pointed out earlier, there may be cases where the two-element configuration will not allow a match due either to impractical component values or to the impossibility of any solution at all. Many times the system is to be designed for a given configuration that requires high-pass or low-pass properties and a match is to be obtained. The solution to the paradox is analogous to the three stub matching technique described in Exercises 2-4.2e and 2-4.2f; that is, we introduce a third element to allow us to remove the nonmatch condition and then proceed to complete the match using the two remaining elements. Samples of typical configurations are shown in Figure 2-56.

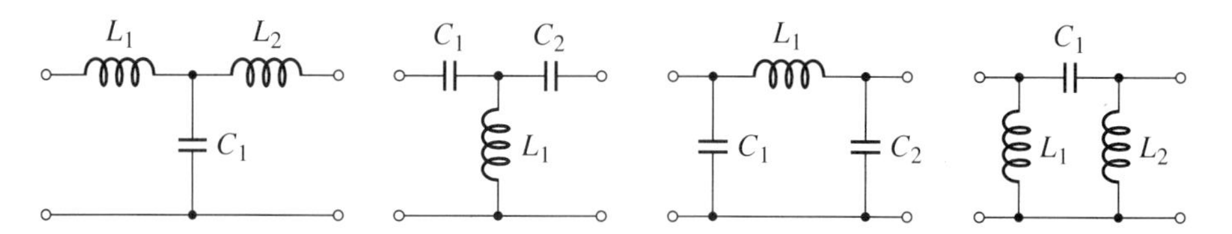

Figure 2-56. Examples of three-element reactive matching networks.

Example 2-15

This example is a case of system that has no two-element solution and for which a three-element match is required.

Suppose we are designing a low-pass system and desire to have the final configuration shown in Figure 2-57*a* (at least, this is what we desire initially).

We would proceed as with the two-element match process and plot $R_{in} = 50/100 = 0.5$ (using $Z_0 = 100\ \Omega$) as shown on the Figure 2-58 Smith chart (point ①).

Then reciprocate this point to point ② to obtain conductance since the capacitance is in parallel. Point ② is thus where we must end up when the capacitance is added (positive susceptance). To end at ② means the capacitive susceptance must have moved from some point on the lower $\tilde{G} = 2$ circle to point ② (all such points which lie on the portion labeled ⓐ). Now all points on arc ⓐ represent the reciprocal of the series impedance combination of the load with inductor L_1. Thus to obtain the locus of all possible impedance values that represent arc ⓐ (siemens) we reciprocate arc ⓐ to arc ⓑ. We next plot the load $\tilde{Z}_L = (250 - j100)/100 = 2.5 - j1$ as point ③. Now we would add an inductor in series with the load that must move along constant resistance ($\tilde{R} = 2.5$) circle in the positive reactance direction, arc ⓒ. But herein lies the problem: No matter what value of L_1 we use, we cannot get to the arc ⓑ! In fact, we can easily see that there is a "dead zone" or "no match zone" for this two-element configuration, which is the $\tilde{R} = 0.5$ circle, shown dotted in for ease of visualization. Any load value within this circle cannot be matched with this configuration.

As explained earlier, the technique is to add a third element, which effectively moves the load point ③ outside the no-match circle. Since we want to reduce the real part, we

Figure 2-57. Low-pass three reactive element impedance matching. (*a*) No match possible. (*b*) Three-element solution to system (*a*).

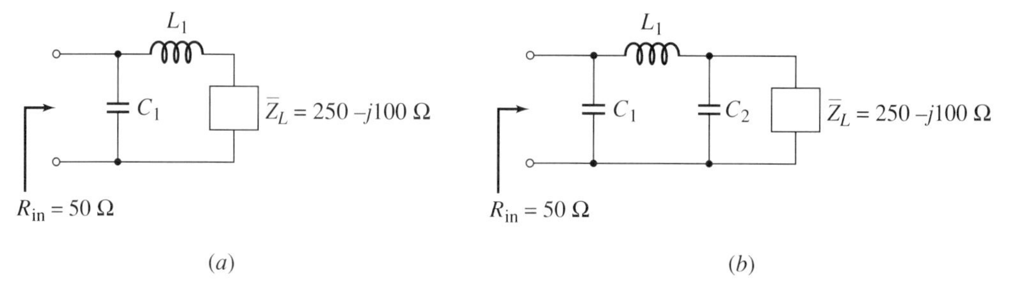

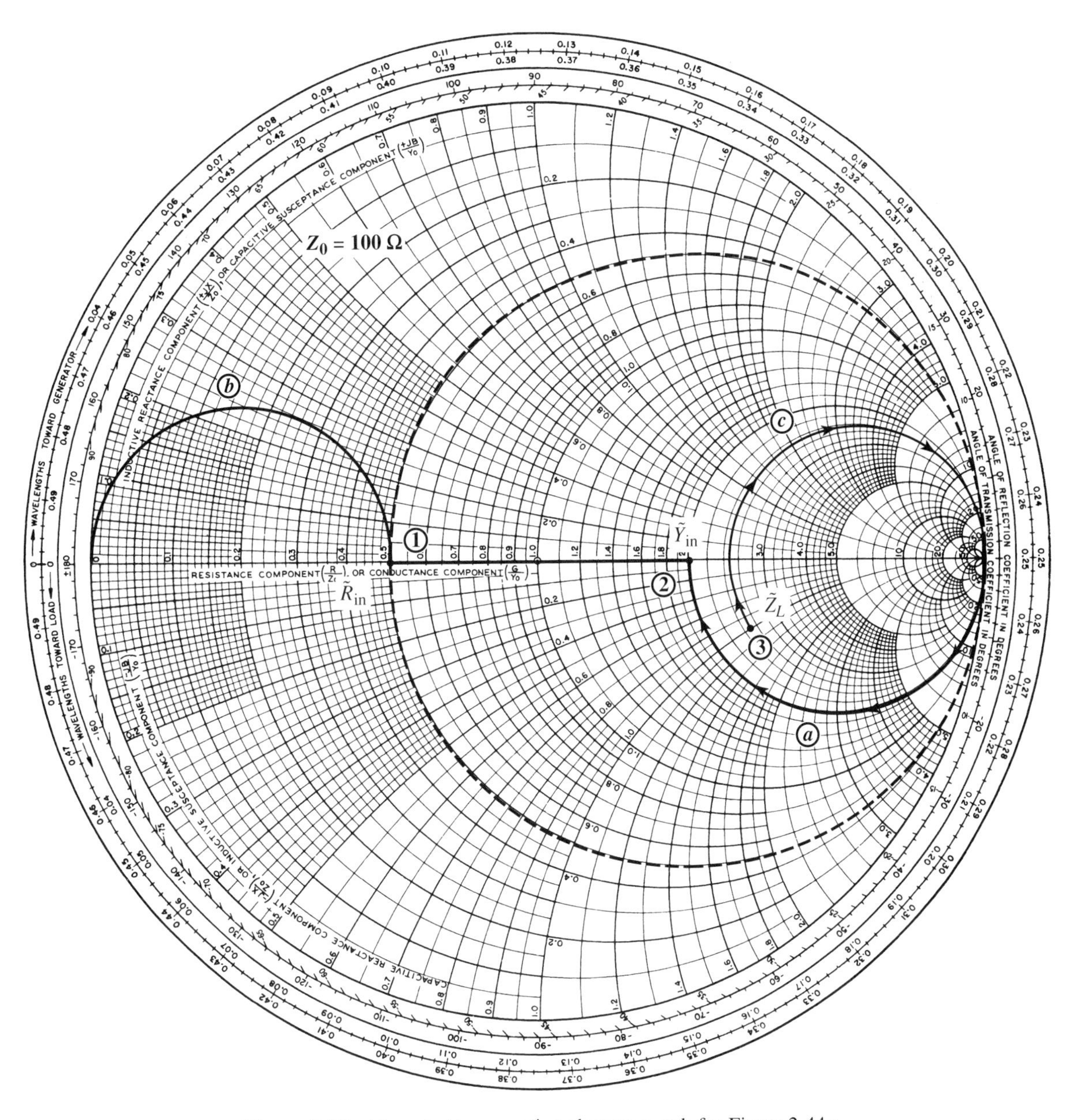

Figure 2-58. Attempted two reactive element match for Figure 2-44a.

may qualitatively think of paralleling with something ''shunting-down'' as it were. We then try the configuration shown in Figure 2-57*b*.

For this configuration, we obtain the match as follows: We know that the parallel combination of the load $\overline{Z}_L$ and the capacitor C_2 must lie outside the $\tilde{R} = 0.5$ circle. This means that the admittance of all these values must lie outside of the reciprocal of the $\tilde{R} = 0.5$ circle, which is shown as the dotted circle on the Figure 2-59 Smith chart. We then first plot $\tilde{Y}_L$ as point ①.

Next we add enough capacitive susceptance to move point ① outside the circle, to (arbitrary) point ②, say, along the constant conductance circle. The required susceptance is then $j\tilde{B}_{C_2} = j(1.0 - 0.14) = j0.86$:

$$j\overline{B}_{C_2} = j\tilde{B}_{C_2} \cdot Y_0 = \frac{j\tilde{B}_{C_2}}{100} = j0.0086 = j\omega C_2$$

Then

$$C_2 = \frac{0.0086}{\omega} \text{ F}$$

where ω is the operating frequency of the system.

We then reciprocate point ② (which is admittance) to point ③ to obtain the impedance equivalent of the parallel combination. Now draw arc ⓑ, which passes through $\tilde{R}_{\text{in}}$ and which represents the locus of points to which the inductor L_1 must take us on the chart. L_1 must thus be of the value that takes us from point ③ to arc ⓑ along the constant resistance circle to point ④.

Thus

$$j\tilde{X}_{L_1} = \underset{\text{point ④}}{j[0.25} - \underset{\text{point ③}}{(-0.92)]} = j1.17$$

From this

$$j\overline{X}_{L_1} = j\tilde{X}_{L_1} \cdot Z_0 = j117 = j\omega L_1$$

or

$$L_1 = \frac{117}{\omega} \text{ H}$$

Next reciprocate point ④ to point ⑤ so we can put the C_1 capacitor in parallel. C_1 will take point ⑤ to ⑥ which is the matched location desired, $\tilde{Y}_{\text{in}} = 2$, or $\tilde{Z}_{\text{in}} = 1/2 = 50/100$.

We need

$$j\tilde{B}_{C_1} = \underset{\text{point ⑤}}{j(1.6} - \underset{\text{point ⑥}}{0)} = j1.6 \text{ S}$$

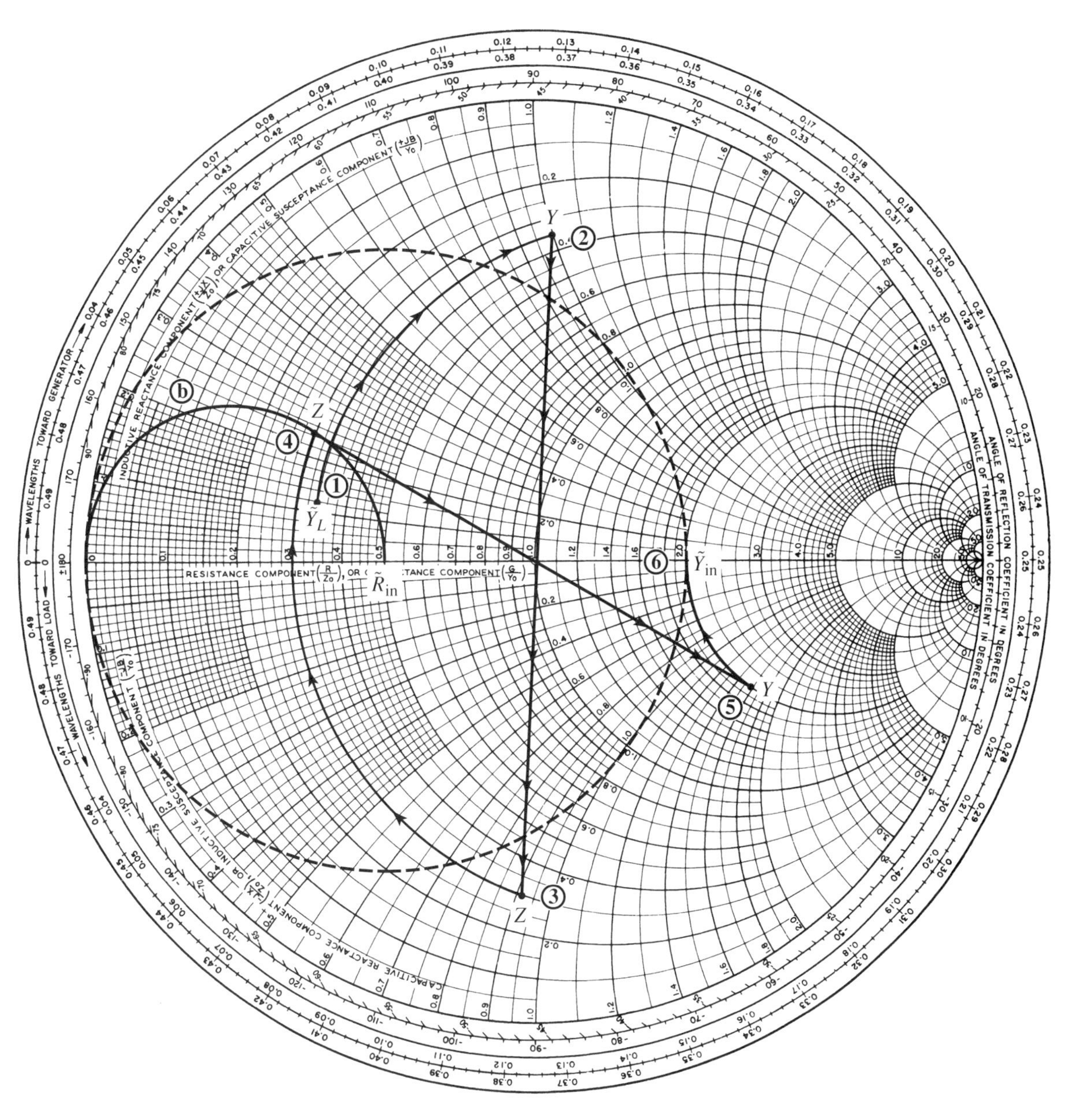

Figure 2-59. Example of a three-element (reactive) match procedure.

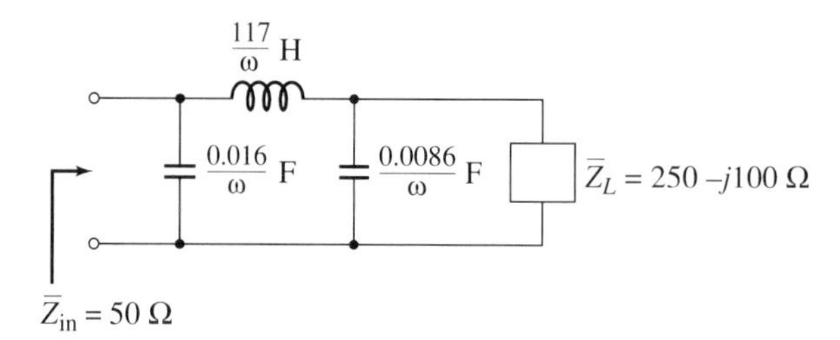

Figure 2-60. Final network for three reactive element match.

from which

$$jB_{C_1} = j\tilde{B}_{C_1} \cdot Y_0 = j\frac{B_{C_1}}{Z_0} = j0.016 = j\omega C_1$$

Thus

$$C_1 = \frac{0.015}{\omega}\text{ F}$$

Our final design would then be as shown in Figure 2-60.

As a final note, it may have been apparent that a two-element match could have been obtained by interchanging L_1 and C_1 as shown in Figure 2-61. However, there are occasions when restrictions on element placement dictate the basic configuration so that one is forced to a three-element configuration. Additionally, some configurations may yield impractical component values. The feed-through capacitor often used in high-frequency electronic circuits is a good example of a system constraint. The use of a coupling capacitor is another one that may function as both a DC blocking element as well as a part of a matching network.

Since there are so many configurations for matching networks, it is difficult to give a single procedure for all cases, but there are some general guidelines and processes that will assist in the design of three-element networks. This is given in the following section.

Step-by-Step Procedure for a Three-Element Match

1. If the resistive component of the load is larger than the resistive component of the desired impedance level, a shunt element is often required first nearest to load (either

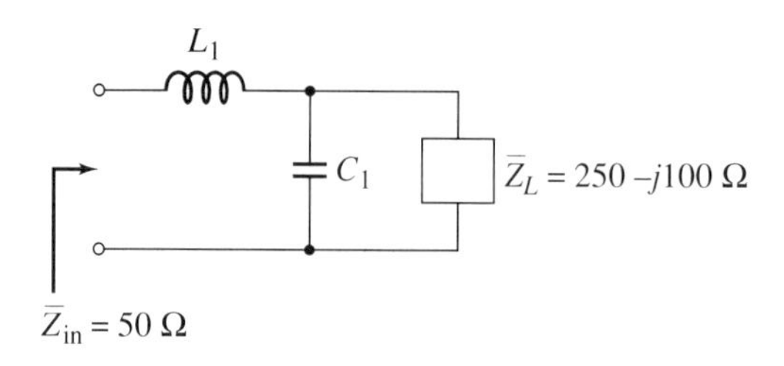

Figure 2-61. Possible two-element reactive matching network for reducing impedance values.

L or C). A series L or C may be appropriate for the first element (nearest load) if R_L is less than R_{match}.

2. Using elements that are already a part of the system, determine the desired configuration of the matching network.
3. Plot the desired impedance value (input impedance or match value).
4. Using ohms or siemens as appropriate, determine the locus of all points that are covered by the last element in the matching network (element farthest from the load) that end at the desired impedance value (let element value vary from 0 to ∞ to do this). This locus will be some portion of a constant $\tilde{R}$ or $\tilde{G}$ circle.
5. Draw the reciprocal of the locus values obtained in step 4. This new locus of points represents the locus to which the second element must move the point representing the combination of the load and the element nearest the load (again by moving along constant real part circles on the Smith chart).
6. Arbitrarily select any one of the constant real part circles that pass through the locus drawn in step 5. It is usually wise to avoid those near the edge of the chart or near the ends of the locus. Draw in the locus on this selected constant real part circle that represents all values obtainable by the second element as it assumes values that end on the locus of step 5.
7. Reciprocate the locus obtained in step 6 and plot the load ($\tilde{Z}_L$ if the element nearest the load is in series with the load; otherwise plot $\tilde{Y}_L$). Determine the element value nearest the load required to move the load point to the locus. If no value is possible, return to step 6, select a new network configuration, or continue to the next step.
8. Reciprocate the point on the locus obtained in step 7. By prior steps this should end up on the locus of step 6. Determine the value of the second element required to move this point to the locus of step 5.
9. Reciprocate the point obtained in step 8. By prior steps this should end up on the locus of step 4. Determine the value of the final element that will move this point to the desired match point.

Exercises

2-5.3a For the transmission line amplifier system shown in Figure 2-62, design a high-pass T matching network for an operating frequency of 100 MHz. (Answer: C_1 = 16 pF, C_2 = 32 pF, L_1 = 83 nH)

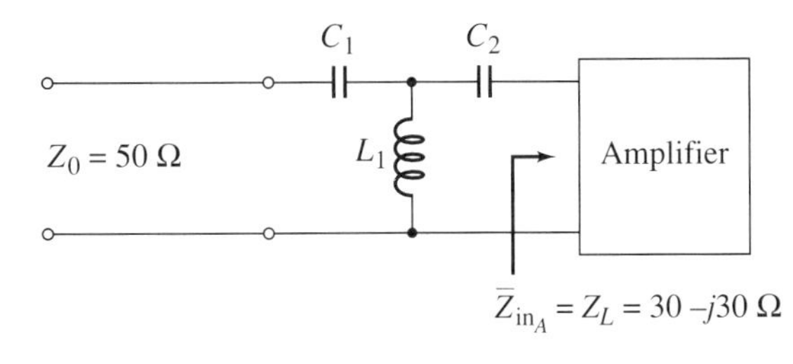

Figure 2-62.

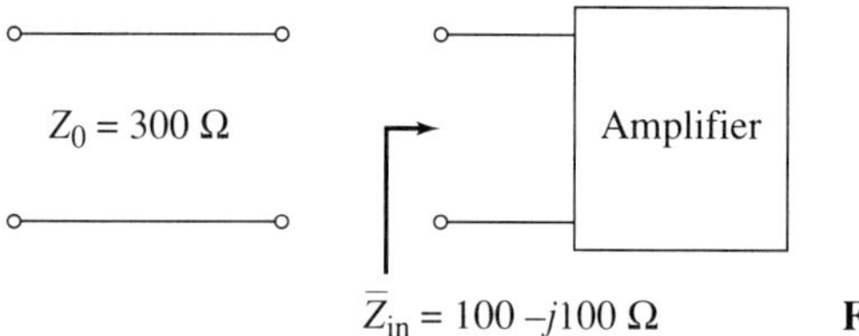

Figure 2-63.

2-5.3b A given system requires a low-pass configuration, and the engineer is required to design a system that will match the two elements shown in Figure 2-63. It is dictated by system constraints that three elements be used in the matching network. For an operating frequency of 200 MHz, determine the network configuration and its values. (Hint: Consider the general solution procedure described in the text.)

2-5.3c For the two systems of matching shown in Figure 2-64, determine the ''no-match'' zones and the relationships between the load resistive component and the desired input resistance value in terms of greater than or less than.

2-5.3d For the system shown in Figure 2-65*a*, determine the component values required to obtain a match at a frequency of 300 MHz. Now suppose that the system configuration required you to use the π network in Figure 2-65*b*. Determine the component values at the same frequency. Next we would like to compare the frequency response characteristics of the two matching networks you designed. To do this obtain a plot of the magnitude of the input impedance as a function of the source frequency for each of the systems. Comment on your results.

2-5.3e For the system shown in Figure 2-61, design the matching network L_1 and C_1 at a frequency of 85 MHz. Now draw the circuit of Figure 2-60 at the same frequency. Then plot the magnitude of the input impedance of each of the two circuits you designed for a frequency range that includes 85 MHz and for which $|\bar{Z}_{in}|$ differs, at most, 10% from the design value of 50 Ω. If we define the bandwidth of the matching network as the frequency range for which $|\bar{Z}_{in}|$ is within 10% of the design value, how do the designs compare?

Figure 2-64.

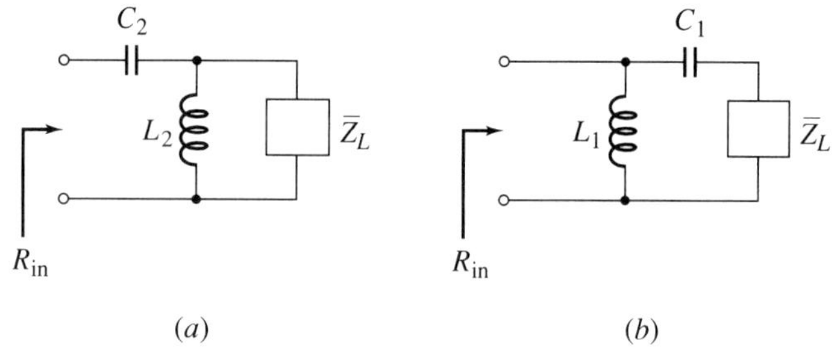

C_2 L_2 $\bar{Z}_L$ R_{in}

(*a*)

C_1 L_1 $\bar{Z}_L$ R_{in}

(*b*)

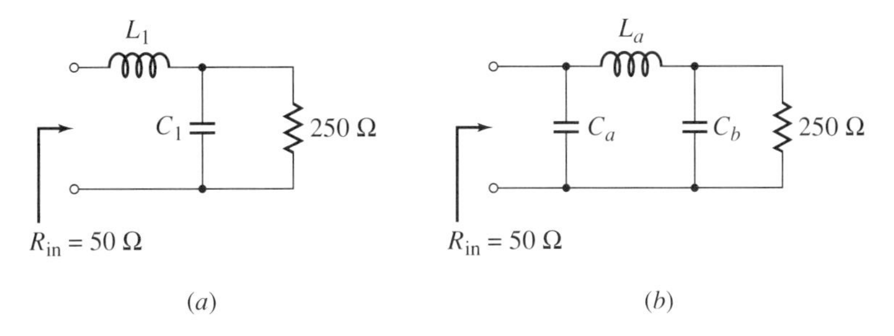

Figure 2-65.

Summary, Objectives, and General Exercises

Summary

We have seen that it is possible to greatly simplify the steady-state solutions to the transmission line equations by using the Smith chart. Four primary applications of importance in practice were presented:

1. Impedance, standing wave ratio, and voltage and power reflection coefficient determination
2. Admittance formulation for graphical analysis based on normalized impedance values
3. Impedance matching with transmission line systems
4. Lumped element computations and impedance matching

As is evident from the examples and exercises, the Smith chart has considerable practical significance. In engineering, the Smith chart is used to design high-frequency amplifiers, design matching components for both hybrid and monolithic systems and to analyze the performance of those systems.

From this chapter, the primary concepts that should be understood are both theoretical and practical, as the following list illustrates.

Objectives

1. Know the derivation of the Smith chart and know how to apply the chart to lossy and lossless transmission lines to:
   (a) Determine the reflection coefficient
   (b) Determine the standing wave ratio
   (c) Determine the return and reflected losses
   (d) Determine the input impedance and admittance at any point on a lossy or lossless line
   (e) Determine the minimum and maximum impedances and their location on the line
   (f) Determine the attenuation on a lossy line
   (g) Provide a single-stub match
   (h) Provide a double-stub match and determine the dead zone
   (i) Determine an unknown load using slotted line short circuit data measurements

2. Know how to use the Smith chart as a simple lumped-element series-parallel impedance calculator

3. Know how to use the Smith chart to compute lumped-element values for impedance matching networks that change impedance level.
   (a) Lossless (LC) components
      (i) Two-element matching
      (ii) Three-element matching
   (b) Resistive (frequency independent) matching

General Exercises

GE2-1 The vast majority of practical applications of the Smith chart assume that the characteristic impedance is real. For such cases, a match is obtained by ending at $\tilde{Z} = 1$ (center) on the chart. Derive the equations for the circles of the Smith chart for constant $\tilde{R}$ and $\tilde{X}$ and plot the generalized Smith chart for the case where both $\overline{Z}_L$ and Z_0 are complex, which allows $|\Gamma(-l)|$ to be greater than 1. See General Exercise GE1-35. (Answer: The chart will be somewhat like a figure eight with maximum and minimum values $\pm(1 + \sqrt{2})$)
Next, develop a design procedure for a single shorted stub of the same Z_0.

GE2-2 An engineering student wants to impress a date and purchases the best antenna available for a new frequency made available by a new broadcast station at 200 MHz. Some lead-in coaxial cable is also purchased to go from the antenna to the receiver. After paying for the equipment and returning home everything is set up. By this time it is Saturday afternoon and it is discovered that the antenna is 300 Ω, but the cable is only 50 Ω. The shop is now closed so no new cable can be bought, but a match of the antenna to the transmission line in the house is desired. The distance from the antenna terminals to the house location of the window where the tuner is located is 6.03 m. The tuners available have distances between stubs of $\lambda/8$, and the lead-in coax has a capacitance of 100 pF/m. Design the matching stub system for the student. Unfortunately, the receiver has an input impedance of 300 Ω and can't be connected to the 50-Ω cable system. Design a matching section (LC) that can be used to get 300 Ω at the input.

GE2-3 An RG34A/U cable has $Z_0 = 75\ \Omega$, $C = 67.6$ pF/m, and an attenuation of 0.125 dB/m. A generator having a peak voltage of 50 V drives a 6-m length of this cable on which the load is 200 Ω. Determine the power delivered by the generator and the power dissipated by the load. What fraction of the generator power is dissipated by the line? $f = 500$ MHz.

GE2-4 A transistor has an input impedance of $150 - j75\ \Omega$ at 175 MHz. Design a low-pass T and a low pass Π that will match this transistor to 50-Ω transmission line. Now for each design

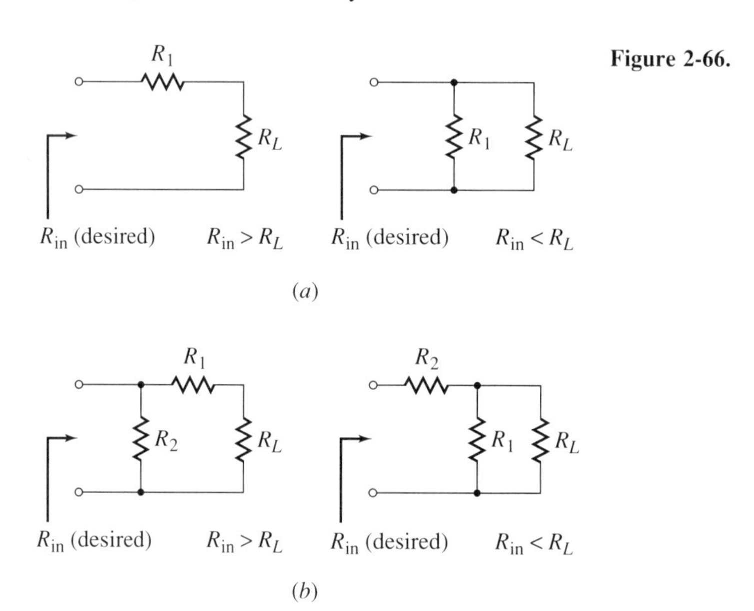

Figure 2-66.

plot $|\overline{Z}_{\text{in}}(\omega)|$ looking into the matching network with the transistor connected but the 50-Ω line disconnected. Compare the bandwidths of the two circuits at frequencies where $|\overline{Z}_{\text{in}}(\omega)|$ is within 10% of 50 Ω (i.e., 50 ± 5 Ω).

GE2-5 Repeat GE2.4 for a transistor having an input impedance of $40 - j35$ Ω.

GE2-6 In Exercise 2-5.1d, frequency-independent resistive matches were investigated. One might observe that one single series or parallel resistor would suffice to match a load to another system as shown in Figure 2-66*a*, where a given load R_L is to be changed to a desired (given) R_{in}. Suppose we have $R_L = 300$ Ω, which is to be changed to an R_{in} of 75 Ω. Determine R_1. For your network, next connect a resistor of value 75 Ω at the input terminals and remove the load. What is the resistance seen by the load at the load terminals (Thevenin equivalent resistance)? For the same load and input resistances, use the Smith chart to design a two-element match of the appropriate form shown in Figure 2-66*b*. Now determine the resistance (Thevenin) seen by the load. Comment on the results as compared with the single resistor match. Also compute the power efficiency of each of your designs ($P_{\text{out}}/P_{\text{in}}$).

3 Transients on Transmission Lines

3-1 Introduction

The preceding chapters have emphasized the sinusoidal steady-state performance of transmission lines. However, the importance of digital systems and signals makes the study of the propagation characteristics of signals that have discontinuities an essential consideration. One could, of course, take the given signal and express it in the frequency domain and then, assuming a linear system, sum the responses of each frequency component. There are, fortunately, better ways to approach such problems, which give good physical insights into the transient performance of lines.

This chapter begins with a study of the general solution of the lossless wave equation and then gives applications of the special case of resistively terminated lines. This is an important class of problems since in many practical cases the lines have very small losses and the study of resistive terminations contains much physical interpretation that carries over to general loads.

Following a study of step-function line response for resistive terminations, the basic concepts of time domain reflectometry (TDR) are presented. Time domain reflectometry can be used to determine whether a line load or discontinuity is inductive or capacitive. Similar concepts are also used to investigate the performance of fiber optic transmission systems.

A natural extension of pulse propagation is a study of the charging and discharging of currents and voltages on lines. For example, if a digital system uses level logic, a line may be charged to a logic one and then be discharged to produce a zero. We shall see that this process takes what can be a significant time in high-data-rate systems. Whereas the pulse studies are aimed primarily at an understanding of the traveling wave components on the line, this study develops insights into the manner in which *total* line current and line voltage build up and decay under switching conditions. Also, charged transmission lines can be used to generate precise pulse widths and amplitudes, and high-voltage DC power lines exhibit the same characteristics.

This chapter concludes with a discussion of the Laplace transform solution of the lossless transmission line equations. Although these solutions have a fair amount of detail, the results are applicable to transmission lines that have any Laplace-transformable signal applied and any linear component termination. From this approach some of the details of transmission line transient analysis, which may seem a bit heuristic, can be put on more solid theoretical grounds.

3-2 General Solutions of the Transmission Line Equations

We begin by examining the general forms of solutions of the lossless transmission line equations that were derived in Section 1-3 with $R = G = 0$. In summary these are:

$$\left.\begin{aligned} \frac{\partial v}{\partial z} &= -L\,\frac{\partial i}{\partial t} &\text{(a)}\\ \frac{\partial i}{\partial z} &= -C\,\frac{\partial v}{\partial t} &\text{(b)}\\ \frac{\partial^2 v}{\partial z^2} &= LC\,\frac{\partial^2 v}{\partial t^2} &\text{(c)}\\ \frac{\partial^2 i}{\partial z^2} &= LC\,\frac{\partial^2 i}{\partial t^2} &\text{(d)} \end{aligned}\right\} \tag{3-1}$$

General solutions of these equations, particularly (c) and (d), were developed by Jean le Rond d'Alembert and were derived as

$$f_1(at - bz) \qquad \text{and} \qquad f_2(at + bz) \tag{3-2}$$

where the constants a and b are determined by substituting either of these into the wave equation. For example, using the chain rule from calculus,

$$\frac{\partial f_1(at - bz)}{\partial t} = \frac{\partial f_1(at - bz)}{\partial(at - bz)} \cdot \frac{\partial(at - bz)}{\partial t} = f_1'(at - bz) \cdot a$$

where the prime denotes differentiation with respect to the entire argument as though it were replaced by a single variable, say $u = at - bz$. Using this same technique, we also evaluate the second-order partials. The results are:

$$\left.\begin{aligned} \frac{\partial^2 f_1(at - bz)}{\partial t^2} &= a^2 f_1''(at - bz) &\text{(a)}\\ \frac{\partial^2 f_1(at - bz)}{\partial z^2} &= b^2 f_1''(at - bz) &\text{(b)}\\ \frac{\partial^2 f_2(at + bz)}{\partial t^2} &= a^2 f_2''(at + bz) &\text{(c)}\\ \frac{\partial^2 f_2(at + bz)}{\partial z^2} &= b^2 f_2''(at + bz) &\text{(d)} \end{aligned}\right\} \tag{3-3}$$

The remarkable property of these results is that they are valid for any function having an argument variable $(at - bz)$ or $(at + bz)$ such as

$$e^{(at-bz)}, \qquad \log(at + bz), \qquad \sin(at - bz)$$

To find values of a and b that satisfy the transmission line equations substitute Eqs. 3-3 (a) and (b) into the voltage wave Eq. 3-1 (c). The result is

$$b^2 f_1''(at - bz) = LCa^2 f_1''(at - bz)$$

from which

$$\frac{b}{a} = \pm\sqrt{LC} \tag{3-4}$$

Using the function f_2 yields an identical result.

To develop the physical significance of the preceding result, consider the plot of a function $f_1(at - bz)$ at two different values of time, say $t = 0$ and $t = t_1 > 0$. Figure 3-1*a* shows the function for $t = 0$. At time $t = t_1$ this function must move to the right on the z axis, since one would have to have a larger value of z to obtain the same value of f_1. This is shown in Figure 3-1*b*. We then conclude that $f_1(at - bz)$ represents a function propagating in the positive z direction. For this reason it is convenient to select the plus sign for Eq. 3-4 above, so we have

$$\frac{b}{a} = \sqrt{LC} \tag{3-5}$$

Similar analysis shows that $f_2(at + bz)$ represents a function propagating in the negative z direction.

The next important property we need is the velocities of propagation of the functions f_1 and f_2. If we consider f_1 as plotted in Figure 3-1 and imagine standing at the peak of

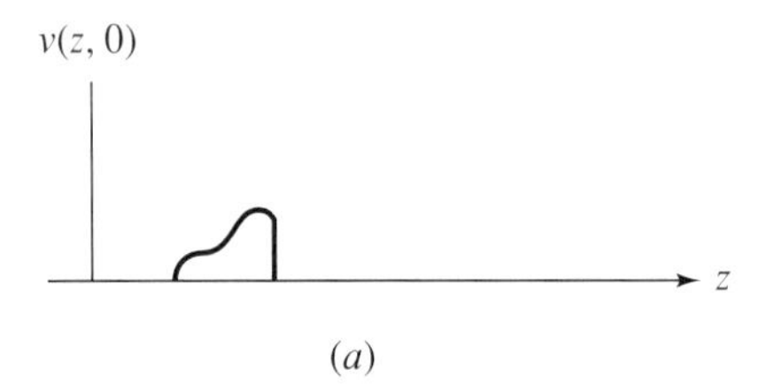

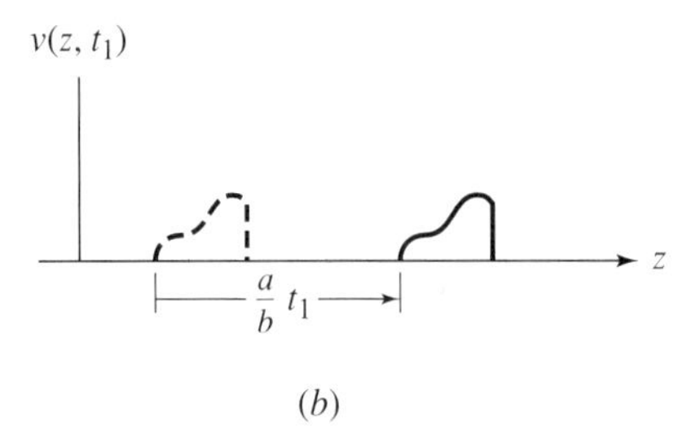

Figure 3-1. The physical effect of time on the d'Alembert solution $f_1(at - bz)$. (*a*) D'Alembert solution of f_1 for $t = 0$. (*b*) D'Alembert solution for f_1 for $t = t_1 > 0$.

the pulse and running along to stay there, we would then have to require $(at - bz)$ to be a constant so the function value would remain constant. We then set

$$at - bz = K$$

Differentiating with respect to time and recognizing that dz/dt is velocity, we have

$$V_p = \frac{dz}{dt} = \frac{a}{b}$$

Comparing this with Eq. 3-5 we have

$$\frac{a}{b} = V_p = \frac{1}{\sqrt{LC}} \tag{3-6}$$

which is the same as the phase velocity for sinusoids! Thus any waveform on a lossless line propagates at the phase velocity. With this result, we may write the d'Alembert solutions in a slightly different form by setting

$$(at \pm bz) = a\left(t \pm \frac{z}{a/b}\right) = a\left(t \pm \frac{z}{V_p}\right)$$

and then expressing the functions f_1 and f_2 as

$$\left.\begin{matrix} f_1\left(t - \dfrac{z}{V_p}\right) \\ f_2\left(t + \dfrac{z}{V_p}\right) \end{matrix}\right\} \tag{3-7}$$

Exercises

3-2.0a Show that the following functions satisfy the wave equation $\partial^2 v/\partial z^2 = A\partial^2 v/\partial t^2$, where A is a constant, and give the required values of A:
(a) $v(z, t) = 10\cos(10^9 t - 12z)$
(b) $v(z, t) = 8e^{j(10^{11}t + 1800z)}$
(c) $v(z, t) = 10\sin(t - 5 \times 10^{-9}z)$

3-2.0b By drawing appropriate figures, develop a physical interpretation of the d'Alembert solution $f_2(t + z/V_p)$ for z being the plot-independent variable. Consider two times $t = 0$ and $t = t_1 > 0$.

3-2.0c Suppose that in the d'Alembert solutions we hold z fixed at two different values, say $z = 0$ and $z = z_1 > 0$ and use t as the plot variable. Sketch the results and give physical interpretations for both solutions.

3-2.0d A function that satisfies a wave equation is given by $g(z, t) = 10\cos(6.28 \times 10^{10}t - 209.44z)$.

Plot the function $g(z, t_0)$ for the two cases $t_0 = 0.05$ ns and $t_0 = 0.06$ ns using the range $0 \le z \le 0.09$ m. Plot both curves on the same set of axes. Notice this amounts to holding time fixed and plotting the space (coordinate) variation. What do the results demonstrate? Next plot the function $g(z_0, t)$ for the two cases $z_0 = 0$ and $z_0 = 0.00375$ m using the range $0 \le t \le 0.3$ ns. Again, use the same set of axes for the plots and explain the results.

3-2.0e For the function $h(z, t) = 10 \cos(6.28 \times 10^{10}t + 209.44z)$, repeat Exercise 3-2.0d.

3-2.0f Derive the expression for the phase velocity of the d'Alembert solution $f_2(at + bz)$ and explain the result physically.

3-2.0g Compute the phase velocities of the functions given in Exercises 3-2.0a, 3-2.0d, and 3-2.0e.

With the d'Alembert solutions we are in a position to obtain general expression for current and voltage. Since the sum of solutions to a linear differential equation is also a solution, the voltage wave equation [3-1(c)] has the general form

$$v(z, t) = Af_1\left(t - \frac{z}{V_p}\right) + Bf_2\left(t + \frac{z}{V_p}\right) \tag{3-8}$$

where A and B are real constants (see Exercise 3-2.0j for another required constant). Then substituting this solution into Eq. 3-1(a), we obtain the equation for current as follows:

$$-\frac{A}{V_p}f_1'\left(t - \frac{z}{V_p}\right) + \frac{B}{V_p}f_2'\left(t + \frac{z}{V_p}\right) = -L\frac{\partial i}{\partial t}$$

Solving for the partial derivative of the current and then integrating partially with respect to time t yields

$$i(z, t) = \frac{A}{LV_p}f_1\left(t - \frac{z}{V_p}\right) - \frac{B}{LV_p}f_2\left(t + \frac{z}{V_p}\right) + g(z) \tag{3-9}$$

where $g(z)$ is the "constant" of integration required when integrating with respect to *time*. Recall that since we take the partial of $i(z, t)$ with respect to time, $g(z)$ would differentiate to zero.

Now if we substitute Eqs. 3-8 and 3-9 into Eq. 3-1(b), we would find that it is necessary to require

$$g(z) = \text{constant} = C_1 \tag{3-10}$$

We also have, using Eq. 3-6,

$$LV_p = \frac{L}{\sqrt{LC}} = \sqrt{\frac{L}{C}} = Z_0 \tag{3-11}$$

which is again the lossless line characteristic impedance! Using these last two equations our current general solution may be written

$$i(z, t) = \frac{A}{Z_0}f_1\left(t - \frac{z}{V_p}\right) - \frac{B}{Z_0}f_2\left(t + \frac{z}{V_p}\right) + C_1 \tag{3-12}$$

This last equation shows that C_1 is simply a constant added to a time signal and thus represents a superimposed DC level. In many problems we do not care about the DC level, so it is common practice to set $C_1 = 0$. Actually, we do this frequently in the laboratory by using the AC coupling position on the oscilloscope vertical amplifier since the series capacitor blocks the DC level and we view the time-varying component. Later, however, when we wish to plot total line voltage and current, we need to put this constant back into the solution. This will be shown in detail later.

Exercises

3-2.0h Beginning with Eqs. 3-8 and 3-1(a), obtain the equation for the partial derivative of current and then perform the partial integration to obtain Eq. 3-9.

3-2.0i Prove Eq. 3-10 by carrying out the steps indicated in that paragraph.

3-2.0j Show that if we express the *current* as the sum of the two functions similar to Eq. 3-8 the voltage would be given by

$$v(z, t) = Z_0 A f_1\left(t - \frac{z}{V_P}\right) - Z_0 B f_2\left(t + \frac{z}{V_P}\right) + C_2$$

This shows that in general both v and i must have constants in the solutions.

3-3 Reflection and Transmission Coefficients and Applications to General Line Waveforms

We can now apply the preceding results to determine the reflection and transmission at a discontinuity. Since we are investigating general time variations the sinusoidal concepts of impedance and the $j\omega$ operator cannot be applied. For our initial investigations we will assume resistive terminations only, and reserve the case of more general loads for later. The basic system is shown in Figure 3-2.

To determine the voltage reflection and transmission parameters we use the current and voltage solutions and Ohm's law. From the current solution (Eq. 3-12) we have, with $C_1 = 0$ as explained,

$$i_L = i(0, t) = \frac{1}{Z_0} A f_1(t) - \frac{1}{Z_0} B f_2(t)$$

Using Ohm's law and the voltage solution (Eq. 3-8) we obtain another expression for the load current,

$$i_L = i(0, t) = \frac{v_L}{R_L} = \frac{v(0, t)}{R_L} = \frac{A}{R_L} f_1(t) + \frac{B}{R_L} f_2(t)$$

Equating these expressions for load current and solving for the ratio of the forward and reverse traveling components, we obtain

$$\frac{B f_2(t)}{A f_1(t)} = \frac{R_L - Z_0}{R_L + Z_0} = \Gamma_L \tag{3-13}$$

This is the reflection coefficient defined in Chapter 1 for resistive loads.

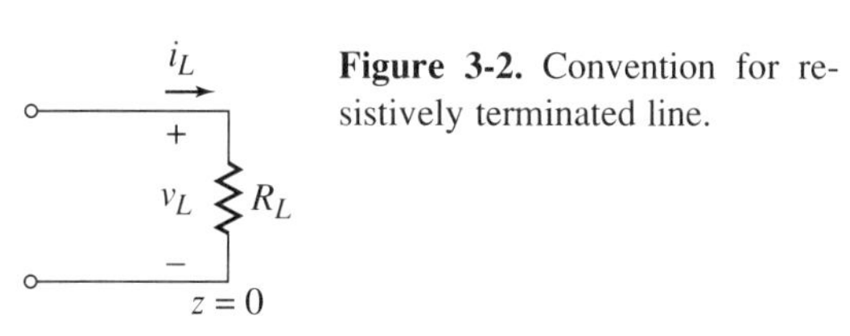

Figure 3-2. Convention for resistively terminated line.

The voltage transmission coefficient follows easily as

$$\tau_L = \frac{v_L}{v_{\text{incident}}} = \frac{v(0, t)}{Af_1(t)} = \frac{Af_1(t) + Bf_2(t)}{Af_1(t)} = 1 + \Gamma_L = \frac{2R_L}{R_L + Z_0} \tag{3-14}$$

Using similar definitions, the load current reflection and transmission coefficients are

$$\Gamma_L^i = \frac{\text{reflected current}}{\text{incident current}} = \frac{-Bf_2(t)}{Af_1(t)} = -\Gamma_L \tag{3-15}$$

$$\tau_L^i = \frac{\text{transmitted current}}{\text{incident current}} = \frac{Af_1(t) - Bf_2(t)}{Af_1(t)} = 1 + \Gamma_L^i = 1 - \Gamma_L \tag{3-16}$$

Example 3-1

For the system given in Figure 3-3*a*, determine the reflected current and voltage waveforms and the load voltage and current waveforms.

Figure 3-3. Reflection and load current and voltage for triangular pulse: (*a*) system and incident voltage; (*b*) reflected and lead waveforms.

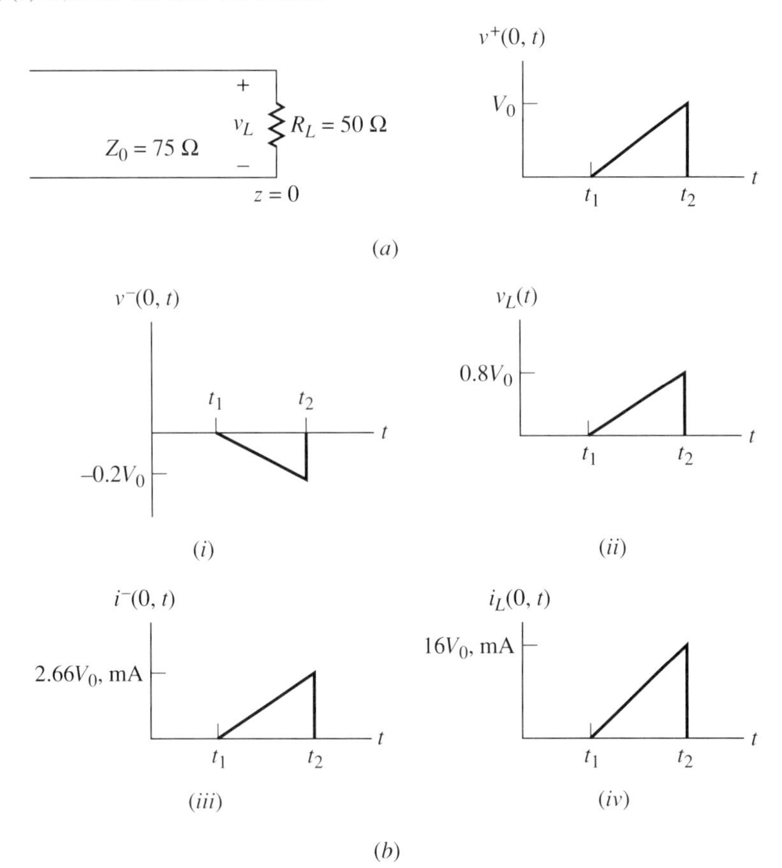

The waveform given as $v^+(0, t)$ corresponds to $Af_1(0, t)$. Thus the negative or reflected waveform will be

$$v^-(0, t) = \Gamma_L v^+(0, t) = \frac{50 - 75}{50 + 75} v^+(0, t) = -0.2v^+(0, t)$$

This is plotted as waveform (*i*) in Figure 3-3*b*. The total load voltage can now be obtained either by adding the incident and reflected components (or by using the transmission coefficient $\tau_L = 0.8$). This waveform is plotted in (*ii*) of Figure 3-3*b*.

From Eq. 3-12 we see that the reflected current is the negative of the reflected voltage divided by Z_0; that is,

$$i^-(0, t) = \frac{-Bf_2(0, t)}{Z_0} = \frac{-v^-(0, t)}{Z_0} = 2.66v^+(0, t) \text{ mA}$$

Thus we simply negate $v^-(0, t)$ and divide by Z_0. This is plotted in (*iii*) of Figure 3-3b.

The load current can be obtained by using Ohm's law for total load voltage v_L or by adding the incident and reflected currents. Since the incident current was not computed, Ohm's law is used directly to give the plot (*iv*) in the figure.

Exercises

3-3.0a For the preceding example, plot the incident current waveform. Also, explain why the reflected current is positive. (Consider which way the reflected current is flowing and what the current direction convention is in Figure 3-2.)

3-3.0b For the case of a resistively terminated lossless line develop equations for the power reflection coefficient (return loss) and the power transmission coefficient (reflected loss). Write the results in terms of R_L and Z_0. Write the power reflection coefficient in terms of Γ_L and the power transmission coefficient in terms of τ_L.

3-3.0c The example in the text determined the quantities at the load. Suppose that we move to the left of the load to location $z = -l. = -[(t_2 - t_1)/4)]V_p$. Plot the positive and negative traveling voltage waveforms and the total voltage waveform at $z = -l$. These are to be time plots. (Partial answer: See Figure 3-4)

3-3.0d For the example given in the text, suppose that the waveform in Figure 3-3*a* is the load voltage rather than the incident voltage as indicated. Determine the following:

(a) Load current waveform.

(b) Incident voltage and current waveforms.

(c) Reflected voltage and current waveforms.

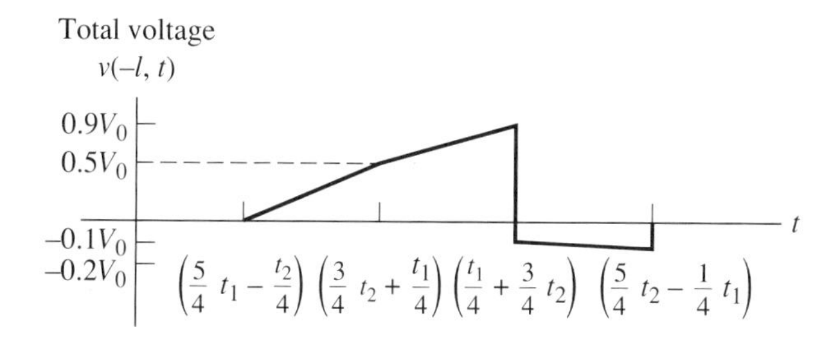

Figure 3-4.

3-4 Pulses and Step Functions on Lossless Lines

The problem of pulses on lines can be reduced to one of switching batteries (DC sources) on the lines using the superposition theorem, which applies to linear systems. In this case we use the theorem stated as follows: For a linear system the response due to the sum of signals is equal to the sum of the responses of each individual signal. The component signals that generate a pulse are illustrated in Figure 3-5. At z_a on the line the positive traveling pulse has been decomposed into the sum of two step-function signals. Each step is easily generated by switching a DC source onto the line at the appropriate time. However, due to the assumed linear time-invariance of the system we really need to solve for only *one* of the step responses. For the example in the figure we would first solve for the response due to v_1^+ and obtain the response due to v_2^+ by inverting the response due to v_1^+ and shifting it $(t_2 - t_1)$ seconds in time. The sum of the two resulting responses will then be the pulse response. Our first task is thus to determine how steps of voltage and current propagate on lines.

3-4.1 Propagation of Step Functions on an Infinite Line

We first consider a DC voltage source switched onto an infinite length line. Since there is no line discontinuity there will be no reflected component. This situation, along with the pertinent waveforms that will be discussed next, is shown in Figure 3-6. Note that we have two distinct types of diagrams (waveforms) to investigate, as shown in (*c*) and (*d*); namely, one where time is the variable and location is fixed and the other where location is the variable and time is fixed. Two examples of each are shown.

1. *Where time is the variable and we look at a fixed location on the line.* This is the waveform one would obtain on an oscilloscope connected to the line at a particular location. The left two diagrams (*c*) are of this type, for $z = 0$ and $z = z_1 > 0$.

   Note that at $z = z_1$ there is no voltage when the switch is first closed since there is a finite velocity of propagation. No voltage would be seen at z_1 until a time $t_1 = z_1/V_p$ after the closing of the switch.
2. *Where time is fixed and we examine the voltage at all points along the line*. Physically, this is a snapshot of the transmission line voltage at one instant. A sequence of such

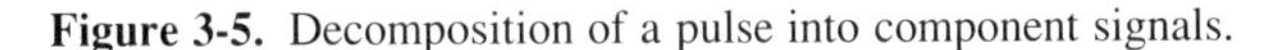

Figure 3-5. Decomposition of a pulse into component signals.

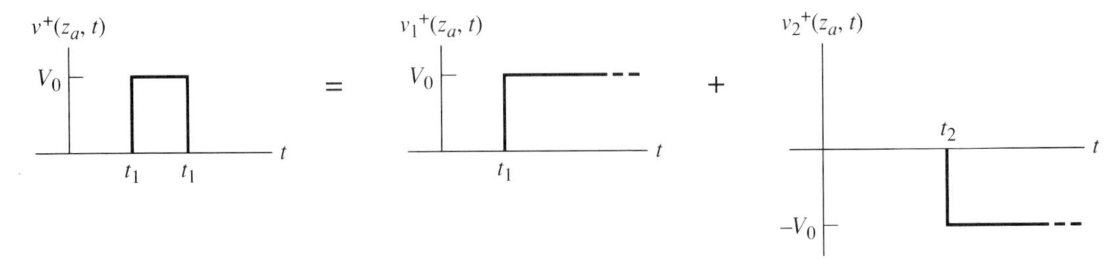

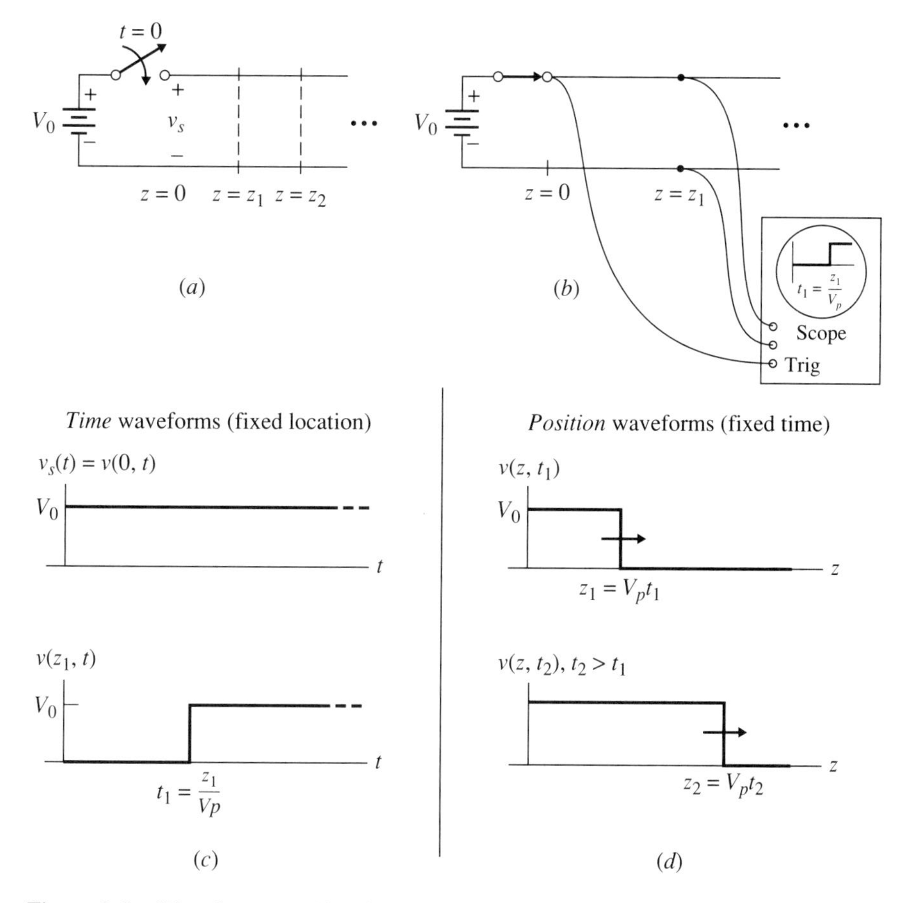

Figure 3-6. Waveforms resulting from a DC source switched onto an infinite length line.

snapshots reveals the traveling-wave nature of the voltage on the line. The two right waveforms (*d*) are of this type. The voltage discontinuity of value V_0 moves to the right with velocity V_p as time progresses. Note that as time increases the front of the step moves to the right, leaving behind a line charged to a voltage V_0. Crudely, we could think of charging a circuit of capacitors sequentially as shown in Figure 3-7.

Figure 3-7. Qualitative model for explaining the finite propagation time of a DC voltage switched onto a lossless line.

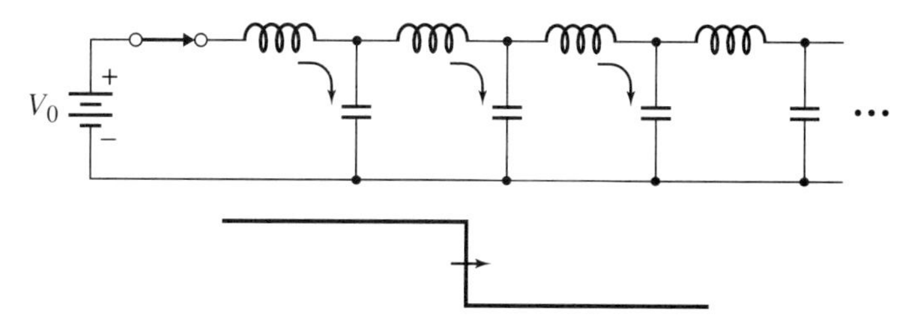

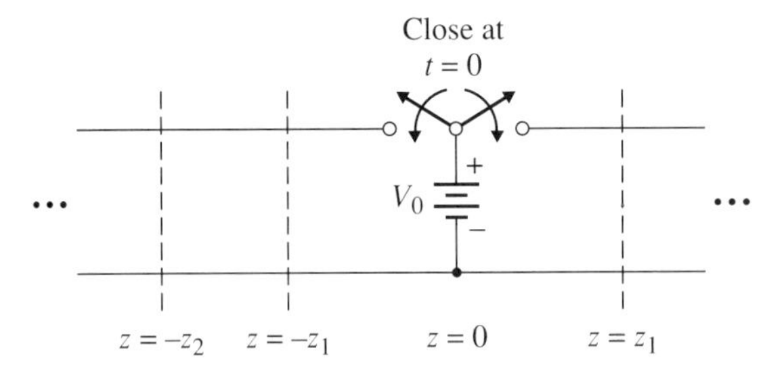

Figure 3-8.

Exercises

3-4.1a Suppose we connect a DC voltage in the center of a transmission line and call that location $z = 0$. See Figure 3-8. Make plots of the time waveforms at locations $-z_1$, $+z_1$, and $-z_2 < -z_1$. Also make plots of the position waveforms (voltage along the line) at $t = 0$, t_1, t_2, $t_1 < t_2$.

3-4.1b Since the preceding discussion has assumed lines of infinite length, we needed only consider the positive traveling voltage because there was never a reflected component. For the four waveforms drawn in the text obtain the corresponding waveforms for the forward traveling currents. (Hint: What is the relationship between V^+ and I^+?) Give amplitudes in terms of V_0 and other line constants as required.

3-4.2 Propagation of Step Functions on Finite-Length Lines

We next consider what happens on a finite length of shorted line. The basic configuration is shown in Figure 3-9. Before getting involved in the details of the calculation, however, let's look at the problem from a qualitative, physical point of view. Since the line termination is a short circuit we would expect that eventually the current in the source would become infinite (subject, obviously, to the real-world constraints that the line is either circuit-breaker protected or that the source would groan and burn out). However, since the actual short circuit is located at a point remote from the generator, we might be led to expect that the current would not rise instantly to the infinite value but somehow build up

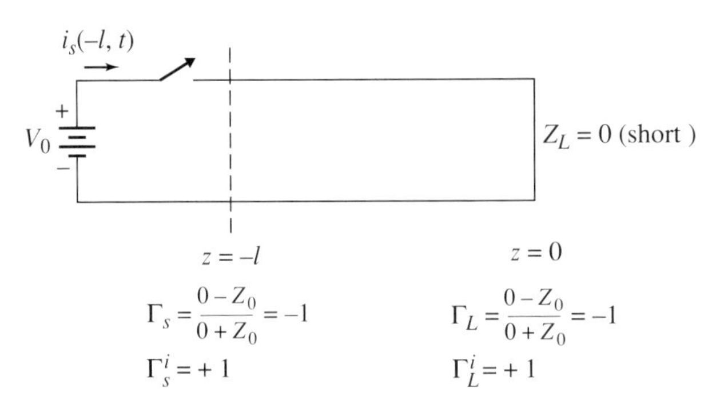

Figure 3-9. DC source switched onto a finite shorted line.

to that as a final value. Additionally, since there is no reflected component at the first instant of switch closure the line initially appears as an infinite line so that the source would see Z_0 as an impedance (here assumed real for a lossless line). Thus

$$i_s(-l, 0) = \frac{v^+(-l, 0)}{Z_0} - \frac{v^-(-l, 0)}{Z_0} \rightarrow \frac{v^+(-l, 0)}{Z_0}$$

From the preceding discussion $v^+(-l, 0)$ is V_0 so we would then expect that the source current would start at a value of

$$i_s(-l, 0) = \frac{V_0}{Z_0}$$

and end up finally at a value $i_s(-l, \infty) = \infty$ A. The mechanism of buildup between these values in time can be obtained using the equations derived earlier in this section. This will be investigated next:

Exercise

3-4.2a By considering the configuration of the shorted-line switching system, determine qualitatively what you would expect the plots of the voltages at $z = -l$ and $z = 0$ to look like (careful, now). Next discuss what you would expect the short-circuit load current at $z = 0$ to be like at $t = 0$, and $t = \infty$. (Answer: $v(-l, t) = V_0, t > 0$; $v(0, t) = 0$ all t; $i(0, 0) = 0$; $i(0, \infty) = \infty$)

We are now in a position to take a detailed look at the system performance. Remember also that $\Gamma_L^i = -\Gamma_L$. The solution will be presented as a series of steps with the results plotted in Figure 3-10.

1. The first step in all problems involving switching is to calculate the current and voltage reflection coefficients at both ends of the transmission line. At the source we use the Thevenin impedance of the generator as the ''load'' impedance in the reflection coefficient formula. The reason for this is two-fold: (1) the system has been tacitly assumed to be a linear system so that superposition is applicable, so we set the ideal source voltage to zero which leaves the source impedance; (2) any wave traveling to the *left* (i.e., a component reflected from the short) sees the generator as a ''load.'' These reflection values are given in Figure 3-9. Since $v(-l, t)$ is constant after switch closure $(= V_0)$, we plot $i_s(-l, t)$ instead.

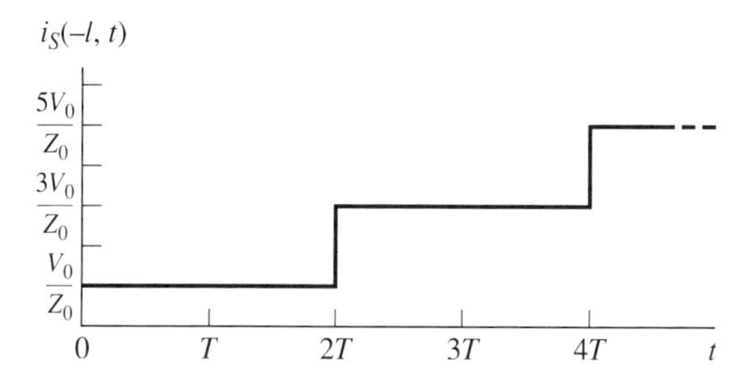

Figure 3-10. Source current resulting from a DC source switched onto a shorted line.

2. **At $t = 0$.**
There is no reflected wave at this time so that the constant B in the current and voltage is zero. The constants C_1 and C_2 are also zero since there were no current or voltage values on the line at $t < 0$. Then

$$v(-l, 0) = Af_1(-l, 0) = V_0 \equiv V^+$$

and

$$i_s(-l, 0) = \frac{Af_1(-l, 0)}{Z_0} = \frac{V_0}{Z_0} \equiv I^+$$

This i_s value will not change until a reflected component returns from the shorted load; thus it appears as DC on the line. If the phase velocity is V_p, then the one-way travel time is $T = l/V_p$. Thus it will be $2T$ seconds before the source can see any change (wave must travel down and back). The first part of the plot is started Figure 3-10, for $0 \leq t \leq 2T$.

3. **At short, $t = T$.**
V^+ has now reached the short:

$$\text{Reflected voltage} = Bf_2(0, T) = \Gamma_L V^+ = (-1)V_0 \equiv V^-$$

$V_L = V^+ + V^- = V_0 - V_0 = 0$ as it should for a short circuit. The V^- travels back toward the source.
Using Eq 3-15:

$$\text{Reflected current} = I^- = \Gamma_L^i I^+ = \frac{V_0}{Z_0}$$

$$I_L = I^+ + I^- = \frac{V_0}{Z_0} + \left(+\frac{V_0}{Z_0}\right) = \frac{2V_0}{Z_0} = i_L(0, T)$$

This I^- travels back to the source.

4. **At $t = 2T$ (back at source).**
The preceding quantities V^- and I^- of step 3 are now interpreted as the quantities *incident* at the source (moving toward source), since the source is an effective load to the incoming wave. So we use these as the *incident* terms in Γ_s and Γ_s^i. Remember that V^+ and I^+ are always the components moving in the $+z$ direction. Thus at the source

$$V^+ = \Gamma_S V^- = (-1)(-V_0) = V_0$$

This is the amount reflected from the source and sent back toward the short since it is the amplitude of the component moving in the positive direction.
Similarly,

$$I^+ = \Gamma_S^i I^- = (1)\left(\frac{V_0}{Z_0}\right) = \frac{V_0}{Z_0}$$

Since we are investigating transient phenomena we must now include the constant term, C_1, in the solution (see Eq. 3-12). This is the current that began to flow at $t = 0$. Thus

$$i_s(-l, 2T) = I^+ + I^- + \text{current already in line at source } (I_{DC})$$

$$= \frac{V_0}{Z_0} + \frac{V_0}{Z_0} + \frac{V_0}{Z_0} = \frac{3V_0}{Z_0}$$

This added to plot in Figure 3-10 at time $t = 2T$. (Note also that $v_S(-l, t) = V^+ + V^- + V_{DC} = V_0 - V_0 + V_0 = V_0$, as it should.)

5. **At t = 3T (back to short)**

$$V^- = \Gamma_L V^+ = (-1)(V_0) = -V_0$$

$$I^- = \Gamma_L^i I^+ = (1)\left(\frac{V_0}{Z_0}\right) = \frac{V_0}{Z_0}$$

Note that the incident quantities here (V^+ and I^+) are the values that were reflected from the source and sent back toward the short.

6. **At t = 4T (return to generator).**
Don't forget that the *incident* quantities are V^- and I^- here.

$$V^+ = \Gamma_S V^- = (-1)(-V_0) = V_0$$

$$I^+ = \Gamma_S^i I^- = (1)\left(\frac{V_0}{Z_0}\right) = \frac{V_0}{Z_0}$$

$$i_S(-l, 4T) = I^+ + I^- + \text{current already in line at source (at time } 2T).$$

$$= \frac{V_0}{Z_0} + \frac{V_0}{Z_0} + \frac{3V_0}{Z_0} = \frac{5V_0}{Z_0}$$

This current value is added to plot Figure 3-10.

Source current finally builds to infinite value as it should in the steady state, by increasing an amount $2V_0/Z_0$ amperes every $2T$ seconds.

Exercises

3-4.2b For the example above obtain a plot of the current in the short circuit as a function of time, that is, $i_L(0, t)$.

3-4.2c Remove the short circuit of the preceding example and plot the source current and load end voltage for $0 \le t \le 6T$. (Answer: See Figure 3-11)

3-4.2d For the system shown in Figure 3-12 obtain plots of $i_S(-l, t)$, $v_S(-l, t)$, and $i_L(0, t)$, $v_L(0, t)$. As far as the source is concerned, how long does the line appear to be? What are the steady values of load current, source current, and line voltage?

Figure 3-11.

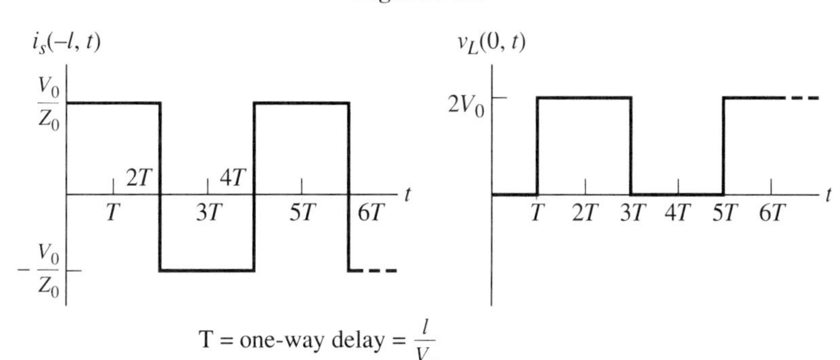

T = one-way delay = $\frac{l}{V_p}$

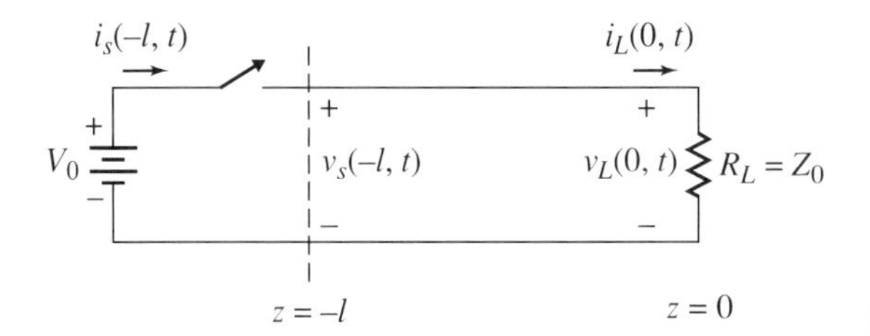

Figure 3-12.

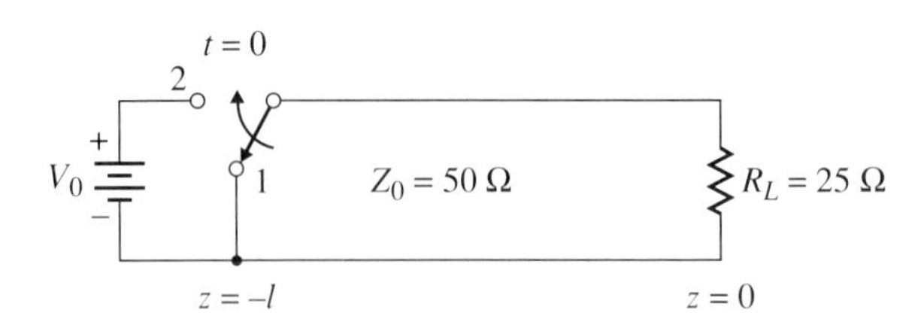

Figure 3-13.

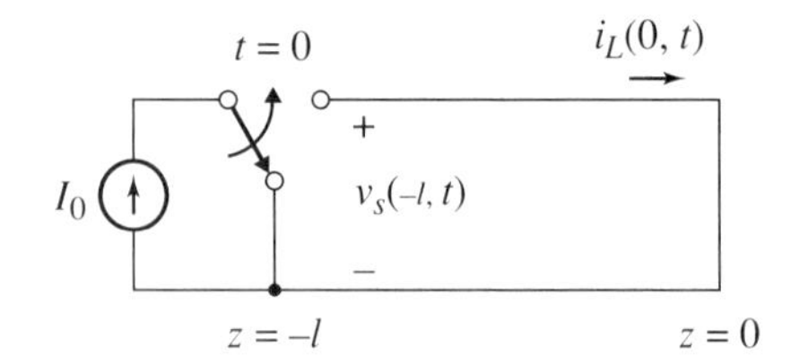

Figure 3-14.

3-4.2e Obtain a plot of the load voltage for the system in Figure 3-13. Steady state exists initially in position 1. Plot for $0 \leq t \leq 6T$. What is the limit value for $t \to \infty$?

3-4.2f For the system shown in Figure 3-14 obtain plots of source voltage and the shorted load current for $0 \leq t \leq 6T$. What are the limit values as $t \to \infty$?

3-4.2g In Figure 3-14, replace the short by a resistor of value 25 Ω and let Z_0 be 50 Ω. Obtain plots of source voltage, load voltage and load current, using $0 \leq t \leq 6T$ and give limit values as $t \to \infty$.

3-4.2h Repeat Exercise 3-4.2g for a load resistor value of 100 Ω. Plot the resistance value seen by the current source at its terminals as a function of time also.

For the cases where the source has an internal impedance, the most difficult problem is determining the line voltage at $t = 0$. Once it has been determined, the reflections can be handled just as in the preceding examples. Consider first the case where the source has a Thevenin resistance Z_0, the line characteristic impedance. The source is switched onto a shorted line as shown in Figure 3-15. Let's determine the voltage and current waveforms at the switch. First, we determine the current reflection coefficient and voltage reflection coefficient at each end as explained previously. These are shown in the figure.

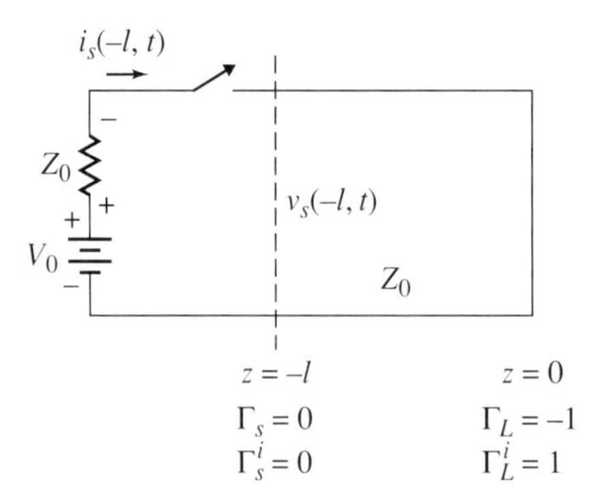

Figure 3-15. Matched DC source switched onto a shorted line.

At $t = 0$

$v_S(-l, 0) = Af_1(-l, 0) = V^+$ since there is no reflected term initially. Note also that now $V^+ \neq V_0$ since there will be a voltage drop across the source resistance. Also,

$$i_S(-l, 0) = \frac{Af_1(-l, 0)}{Z_0} = \frac{V^+}{Z_0} = I^+$$

Using Kirchhoff's voltage law around the source with the switch closed we obtain

$$V_0 = i_S(-l, 0)Z_0 + v_S(-l, t) = \frac{V^+}{Z_0} \cdot Z_0 + V^+ = 2V^+$$

From this result, $V^+ = 1/2V_0 = v_S(l, 0)$. Note this result would also be obtained if we used as the equivalent circuit at $t = 0^+$ that circuit shown in Figure 3-16. From this circuit the simple voltage divider formula gives the result directly. The equivalent circuit is also valid for any source resistance, not just Z_0. Simply use the source resistance actually present and connect it in series with Z_0.

It is emphasized that this equivalent is only valid at the instant of switching. Using the V^+ obtained we also have

$$I^+ = \frac{V^+}{Z_0} = \frac{V_0}{2Z_0} = i_S(-l, 0)$$

Using these two positive traveling components, we begin the plots shown in Figure 3-17. Since there will be no reflection returns for $2T$ seconds ($T = 1/V_p$) these values remain constant for $0 \leq t \leq 2T$.

At $t = T$ (at the short)

V^+ and I^+ have reached the short circuit, and the reflected components are

$$V^- = \Gamma_L V^+ = (-1)\left(\frac{V_0}{2}\right) = -\frac{V_0}{2}$$

$$I^- = \Gamma_L^i I^+ = (1)\left(\frac{V_0}{2Z_0}\right) = \frac{V_0}{2Z_0}$$

$$v(0, T) = V^+ + V^- = \frac{V_0}{2} - \frac{V_0}{2} = 0 \qquad \text{as required at the short}$$

$$i(0, T) = I^+ + I^- = \frac{V_0}{2Z_0} + \frac{V_0}{2Z_0} = \frac{V_0}{Z_0}$$

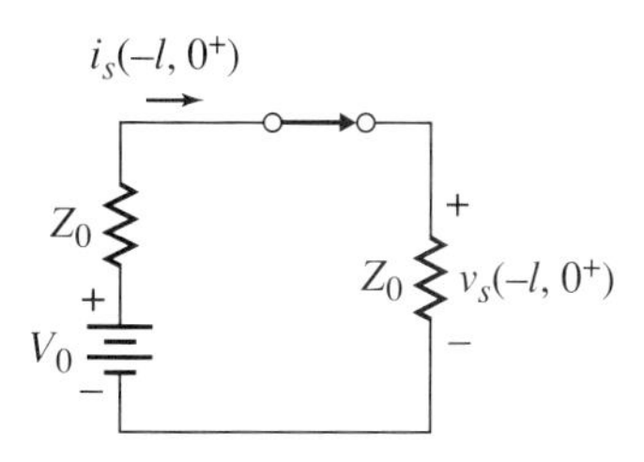

Figure 3-16. Equivalent circuit at $t = 0^+$ for matched DC source switched onto a shorted line.

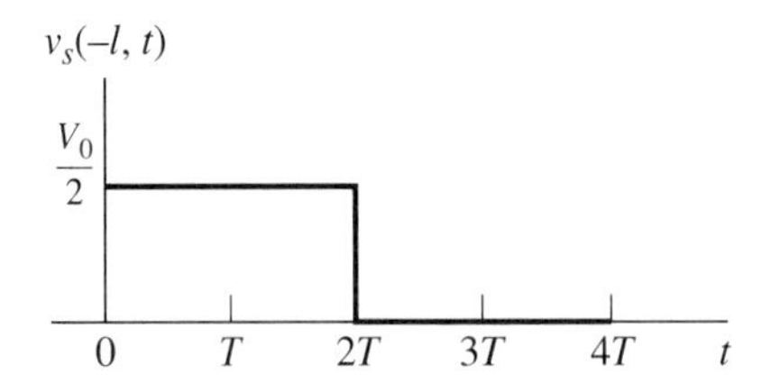

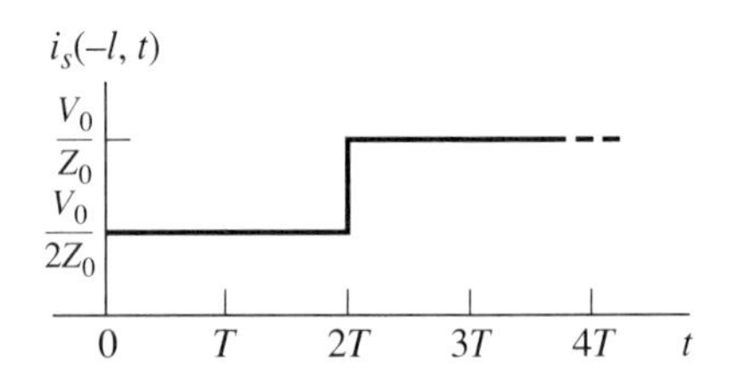

Figure 3-17. Source (sending end) current and voltage for a DC source having internal resistance Z_0 switched onto a shorted line of characteristic impedance Z_0.

The V^- and I^- travel back to the source end at $z = l$. Thus these are now considered the *incident* quantities at that location.

At $t = 2T$ (at source)

$$V^+ = \Gamma_S V^- = (0)\left(\frac{-V_0}{2}\right) = 0$$

$$I^+ = \Gamma_S^i I^- = (0)\left(\frac{V_0}{2Z_0}\right) = 0$$

Thus

$$\begin{aligned} v_S(-l, 2T) &= V^+ + V^- + \text{voltage already on the line} \\ &= 0 + \left(\frac{-V_0}{2}\right) + \frac{V_0}{2} = 0 \\ i_S(-l, 2T) &= I^+ + I^- + \text{current already in the line} \\ &= 0 + \frac{V_0}{2Z_0} + \frac{V_0}{2Z_0} = \frac{V_0}{Z_0} \end{aligned}$$

These are plotted on the graph. Since there are no reflections, the transient is completed. Note the result is physically correct since in the steady state, the equivalent circuit would be a short circuit around the source for $t > 2T$ as shown in Figure 3-18.

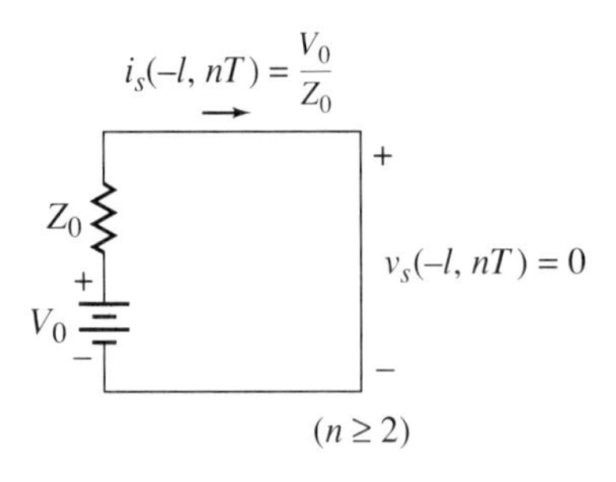

Figure 3-18. Steady-state equivalent circuit for a matched source switched onto a shorted line.

Exercises

3-4.2i For the preceding example, plot the current in the short at the load, $i(0, t)$. (Answer: 0 for $0 \le t < T$ and is V_0/Z_0 for $t > T$)

3-4.2j For the preceding example, plot the current and voltage at the center of the line, $z = -l/2$. (Answer: Figure 3-19)

3-4.2k Obtain plots for the voltage at the load and at the source terminals, $z = 0$ and $z = -l$, for the system shown in Figure 3-20. Are the results physically plausible. Explain.

3-4.2l Obtain plots for the voltages at the load and at the source terminals, $z = 0$ and $z = -l$, $K = \frac{1}{2}$ and 2 (two problems). See Figure 3-21.

3-4.2m Obtain plots for the voltages at the load and at the source terminals, $z = 0$ and $z = -l$, and compare to those in Exercise 3-4.2l. $K = \frac{1}{2}$ and 2. See Figure 3-22 for the configuration.

3-4.2n Obtain plots for the voltages at the load and at the source terminals, $z = 0$ and $z = -l$. See Figure 3-23.

3-4.2o The transistor shown in Figure 3-24 is assumed to act as an ideal switch. Obtain a plot of the load voltage.

3-4.2p Prove that at $t = 0$ the equivalent circuit of a source V_0 with internal resistance R switched onto a line of characteristic impedance Z_0 is a simple voltage divider consisting of R and Z_0. (Hint: Follow the initial step of the preceding example.)

Figure 3-19.

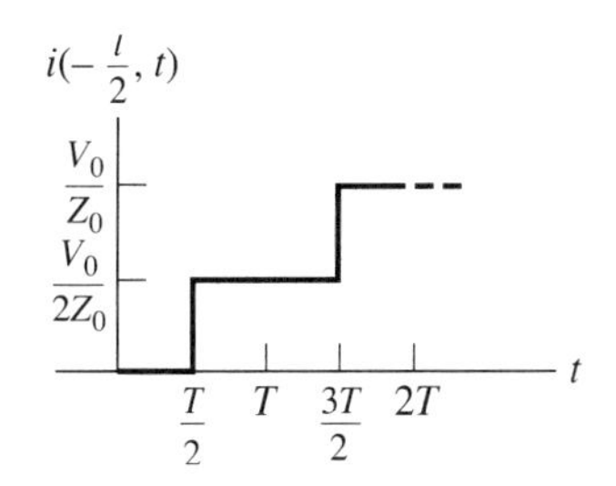

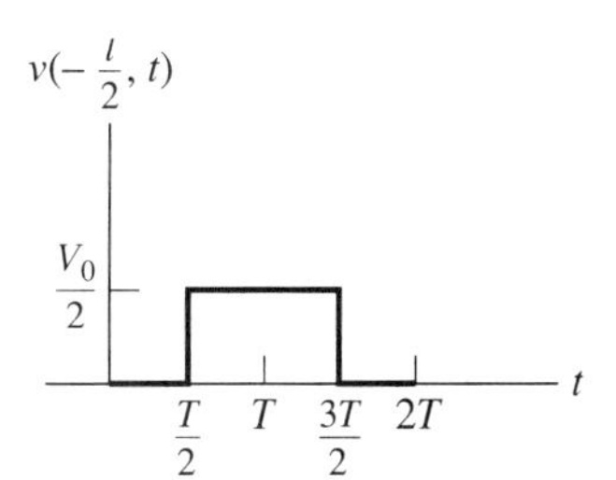

Figure 3-20.

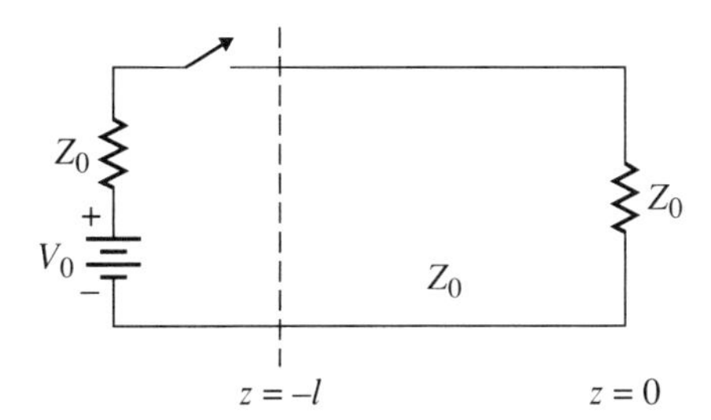

Figure 3-21.

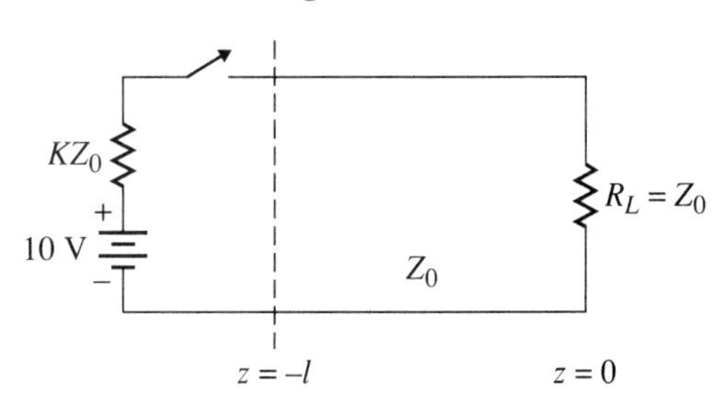

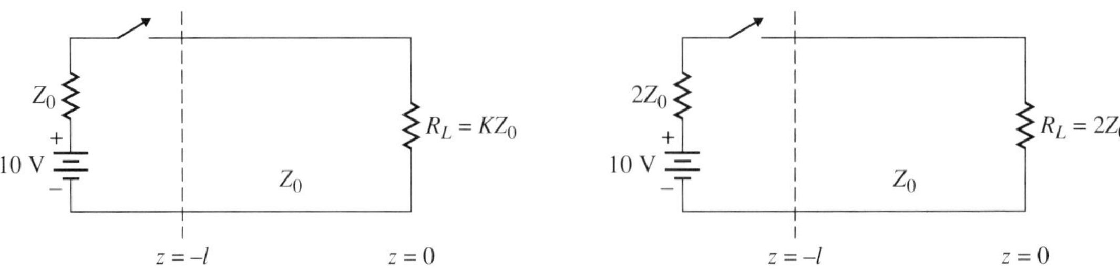

Figure 3-22.

Figure 3-23.

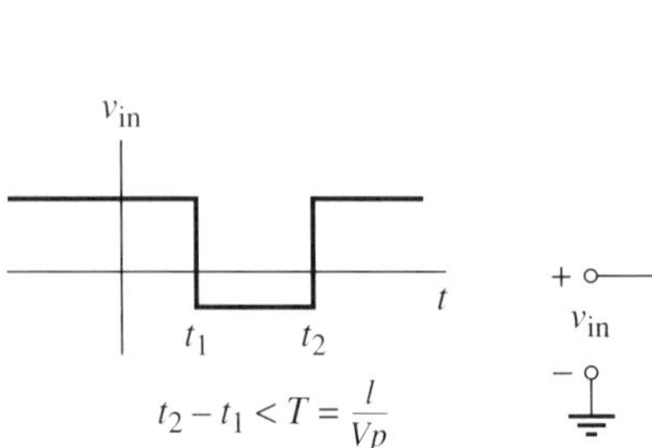

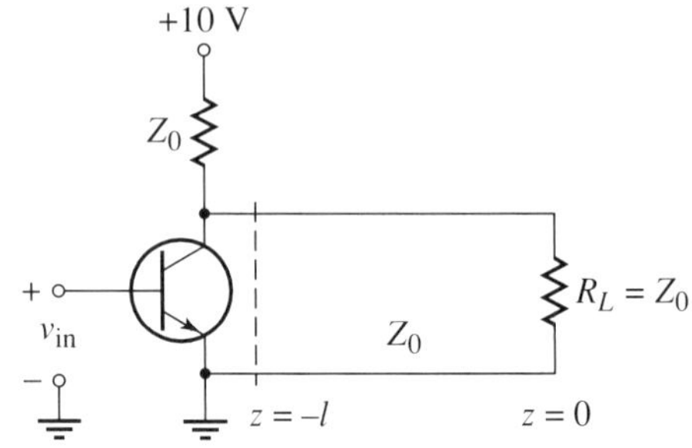

Figure 3-24.

3-5 Transients Produced by Discharging an Initially Charged Line

In digital systems, state transitions occur from low to high and from high to low. The latter case is equivalent to discharging a charged line, so it is important to understand the discharge mechanism. On high-voltage DC power transmission lines, the same action occurs when the power system is deenergized.

The examples presented to this point have all assumed that the lines had zero initial currents and voltages. Now suppose that the line were initially charged to a voltage V_0 and then connected to a resistance R at $t = 0$. The configuration is shown in Figure 3-25. The voltage and current reflection coefficients are also shown in the figure. We wish to obtain plots of $i_R(l, t)$ and $v_R(-l, t)$. Of course, we need to compute only one of the quantities since Ohm's law can be used to obtain the other as

$$i_R(-l, t) = -\frac{v_R(-l, t)}{R} \tag{3-17}$$

Note that the minus sign is required since the current flow is, by convention, up through R.

For purposes of this example, we'll select $R = \frac{1}{2}Z_0$ whenever specific values are to be computed.

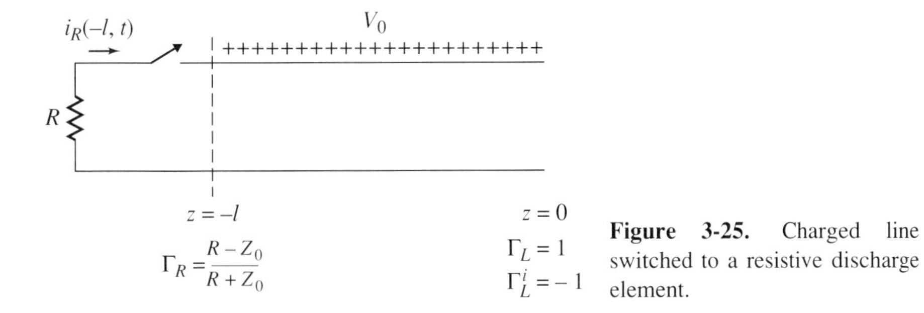

Figure 3-25. Charged line switched to a resistive discharge element.

At $t = 0$ (At the Resistance)

Since there is no reflected voltage yet the coefficient B in Eq. 3-8 is zero, but there is an initial DC component so that

$$v_R(-l, 0) = Af_1(-l, 0) + C_v = V^+ + V_0$$

Also

$$i_R(-l, 0) = \frac{Af_1(-l, 0)}{Z_0} + \text{current already in line}$$

$$= \frac{V^+}{Z_0} + 0 = \frac{V^+}{Z_0} = -\frac{v_R(-l, 0)}{R}$$

From the last equality, $v_R(-l, 0) = (R/Z_0)\, V^+$.

Using this result in the first equation for $v_R(-l, 0)$ we obtain

$$-\frac{R}{Z_0} V^+ = V^+ + V_0$$

from which

$$V^+ = -\frac{Z_0}{R + Z_0} V_0$$

The first equation for $v_R(-l, 0)$ above then becomes

$$v_R(-l, 0) = -\frac{Z_0}{R + Z_0} V_0 + V_0 = \frac{R}{R + Z_0} V_0$$

Again, this result shows that at the instant of switching the system can be replaced by the simple series circuit shown in Figure 3-26 and voltage division used to obtain this same result. The charged line has a simple Thevenin equivalent circuit, initially.

For $R = Z_0/2$ we obtain the specific values

$$V^+ = -\frac{2}{3} V_0, \qquad v_R(-l, 0) = \frac{V_0}{3}, \qquad \text{and} \quad i_R(-l, 0) = -\frac{2V_0}{3Z_0}$$

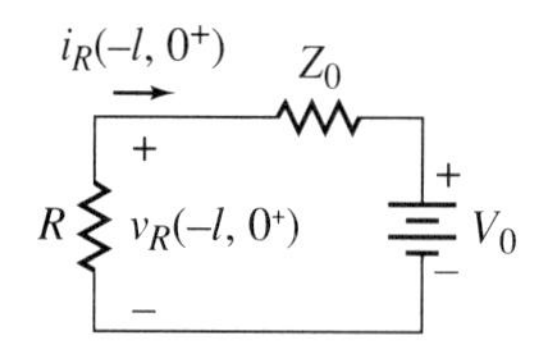

Figure 3-26. Equivalent circuit at $t = 0^+$ for a charged line switched onto a discharge resistor.

These values will persist until a reflection returns from the open end at $t = 2T$ seconds. (T is still the one-way line delay). The plot of $v_R(-l, t)$ is then started in Figure 3-27*a*.

At $t = T$ (At Open End)

$$V^- = \Gamma_L V^+ = (1)(-\tfrac{2}{3}V_0) = -\tfrac{2}{3}V_0$$

This travels back to the resistor.

At $t = 2T$ (At Resistor)

For our particular case,

$$\Gamma_S = \frac{\frac{1}{2}Z_0 - Z_0}{\frac{1}{2}Z_0 + Z_0} = -\frac{1}{3}$$

Then $V^+ = \Gamma_S V^- = (-\tfrac{1}{3})(-\tfrac{2}{3}) = \tfrac{2}{9}V_0$

$$\begin{aligned} v_R(-l, 2T) &= V^+ + V^- + \text{voltage already on line} \\ &= \tfrac{2}{9}V_0 - \tfrac{2}{3}V_0 + \tfrac{1}{3}V_0 = -\tfrac{1}{9}V_0 \end{aligned}$$

This voltage is added to the plot in the figure for $2T \le t \le 4T$. The V^+ from this step travels back to the open.

At $t = 3T$ (At Open End)

$$V^- = \Gamma_L V^+ = (1)(\tfrac{2}{9}V_0) = \tfrac{2}{9}V_0$$

This returns to the resistor.

At $t = 4T$ (At Resistor)

$$V^+ = \Gamma_S V^- = (-\tfrac{1}{3})(\tfrac{2}{9}V_0) = -\tfrac{2}{27}V_0$$

$$\begin{aligned} v_R(-l, 4T) &= V^+ + V^- + \text{voltage already there} \\ &= -\tfrac{2}{27}V_0 + \tfrac{2}{9}V_0 - \tfrac{1}{9}V_0 = \tfrac{1}{27}V_0 \end{aligned}$$

This is placed on the figure for $4T \le t \le 6T$.

The trend of the voltage waveform should now be clear. During each time interval, the amplitude is reduced by $-\frac{1}{3}$. Note that the waveform has the appearance of a damped square wave of period $4T$.

Using Ohm's law, the current $i_R(-l, t)$ is obtained by dividing $v_R(-l, t)$ by $-R$ $(= -\frac{1}{2}Z_0)$ as explained earlier. This waveform is plotted in Figure 3-27*b*.

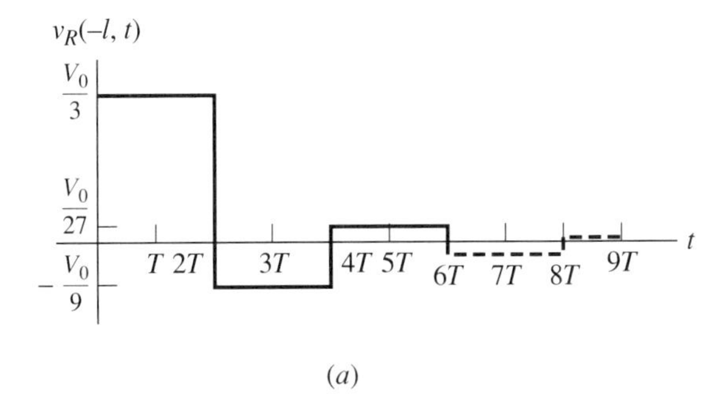

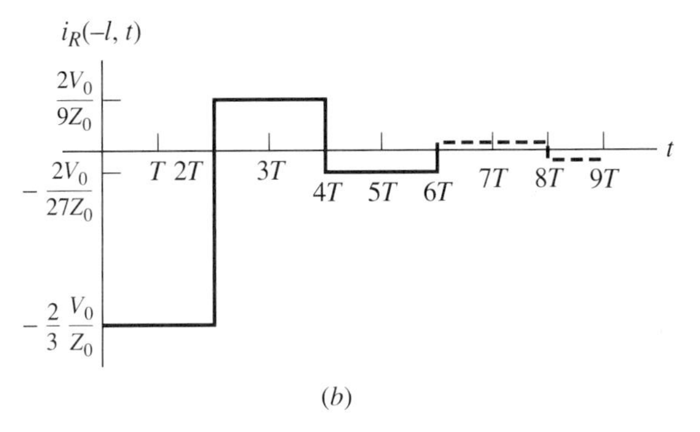

Figure 3-27. Discharge waveforms for a resistor switched onto a charged line; $R = \frac{1}{2}Z_0$.

Exercises

3-5.0a For the preceding example, compute $i_R(-l, t)$ using the current reflection coefficients directly and computing I^+ and I^- at each interval of time T. The answer, of course, should be Figure 3-27*b*. Also plot $v(0, t)$.

3-5.0b Repeat the preceding example for $R = 2Z_0$. Can you identify a decay characteristic from $v_R(-l, t)$ plot?

3-5.0c The circuit in Figure 3-28 is in the steady state initially. When the switch is moved to position 2, current will continue to flow in R_L since there is $V_0 = 10$ V on the line. The current will be supplied by the charge that exists on the line until the line has been discharged. (See Exercise 1-6.4e for properties of the cable, which we assume lossless for this problem.) What is the total energy stored in the length of line just before the switch is thrown? (Hint: What are the total cable capacitance and inductance?) Obtain the plot of the current through R_L and determine the heat given off by the resistor. (Hints: At all times

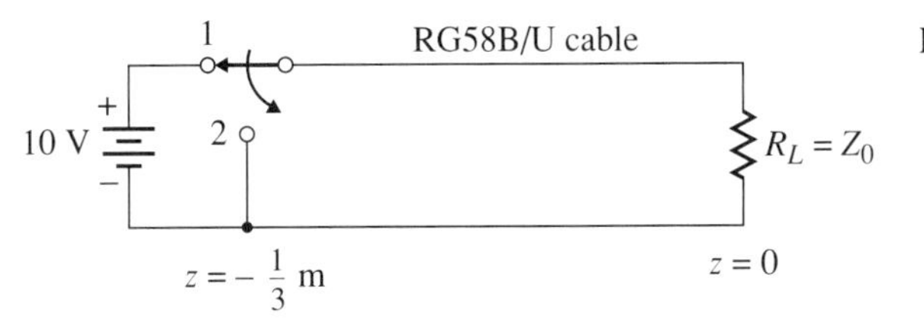

Figure 3-28.

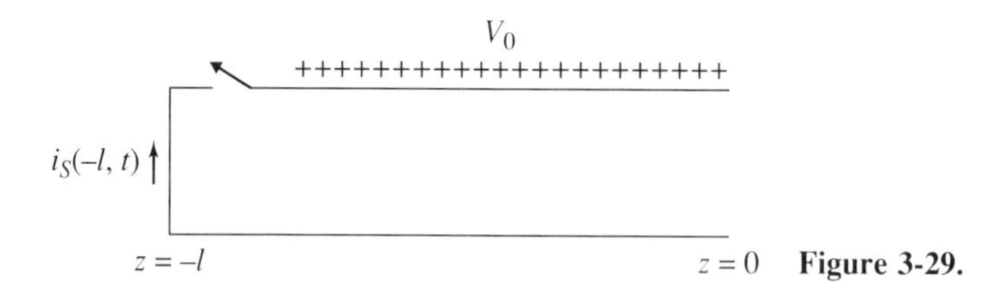

Figure 3-29.

and locations $v(z, t) = V^+ + V^- + V_{DC}$. At $z = -l$, $t = 0$, we have $V_{DC} = V_0$ and $V^- = 0$ since there is no negative traveling component, and $v(-l, t) = 0$. It might be good to find V^+ first and describe physically what will happen before you try to solve the problem.)

3-5.0d Given the charged line of length l as shown in Figure 3-29. The switch is closed at $t = 0$. Plot $i_S(-l, t)$ for the interval $0 \leq t \leq 6T$ where $T = l/V_P = l\sqrt{LC}$. The line characteristic impedance is Z_0.

3-5.0e A 2-m section of RG58B/U cable is charged to 100 V. The cable is discharged into a standard value 47-Ω resistor. Sketch the voltage across the resistance, giving voltage and time values numerically, and assuming a lossless cable. (See cable properties in Exercise 1-6.4e.) Compare the energy dissipated by the 47-Ω resistance to that stored in the cable initially.

3-5.0f The coaxial cable of Exercise 3-5.0e is discharged into a 100-Ω resistance instead of the 47-Ω resistance. Sketch the voltage across the 100-Ω resistance and compare.

3-5.0g For the lossless system shown in Figure 3-30, obtain plots of the load (short) current and the input current and voltage. The switch is in position 1 for a long time prior to opening.

3-5.0h A current source of constant value I_0 is switched onto the shorted transmission line as shown in Figure 3-31. Obtain a qualitative description of what you would expect from the system in terms of initial and final values of current and voltage and the short and the source. Then complete a detailed analysis of the configuration to obtain plots of the currents and voltages at the short and at the source.

3-5.0i An equivalent circuit of a transistor used to drive a transmission line, as shown in Figure 3-32, has a current source in parallel with an impedance that we assume to be resistive for our purposes here. Obtain plots of the current at the load and the voltage at the switch.

Figure 3-30.

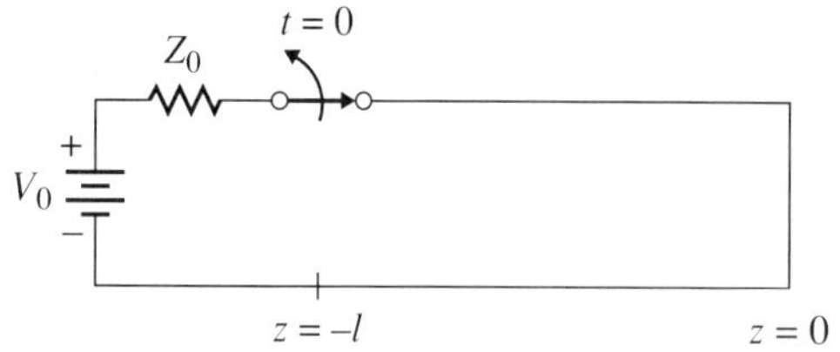

$t = 0$, Z_0, V_0, $z = -l$, $z = 0$

Figure 3-31.

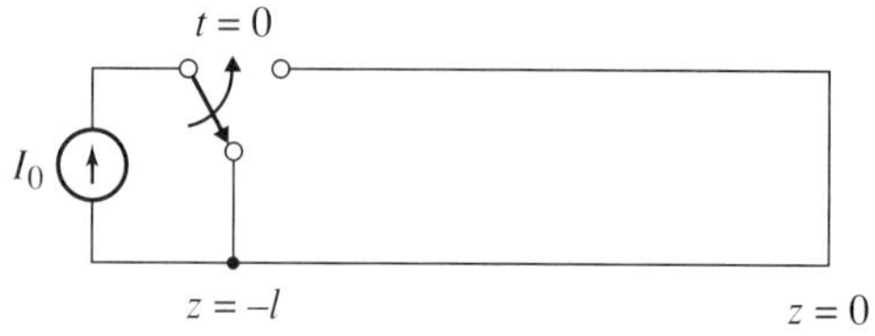

$t = 0$, I_0, $z = -l$, $z = 0$

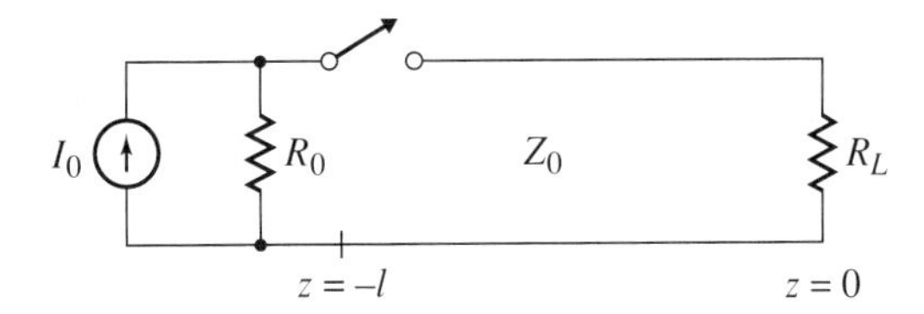

Figure 3-32.

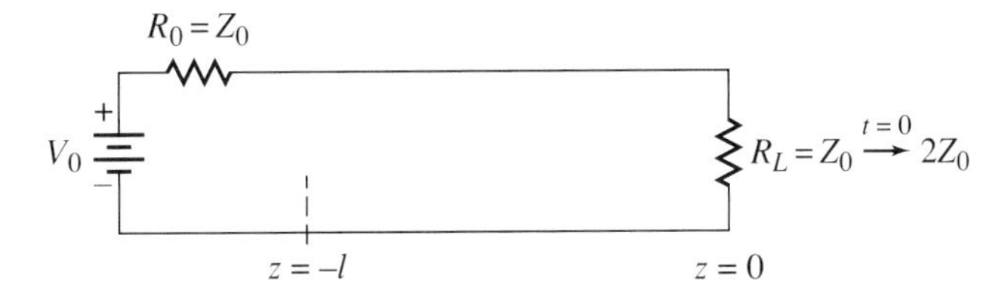

Figure 3-33.

Assume that $R_0 = Z_0$ and that R_L is $2Z_0$. (Hint: Can a source conversion be used to reduce this problem to one that has already been solved?)

3-5.0j For the system of Exercise 3-5.0i suppose that the transistor is operated so that a pulse is generated rather than a simple step by opening the switch again after a short time but leaving an R_0 attached across the line at the switch. Plot the load voltage that results for two cases: (1) switch open in a time less than $T = l/V_P$; and (2) switch open in a time greater than T but less than $2T$. (Hint: This system is linear and time-invariant, so the superposition theorem applies, and we can imagine the pulse to be composed of two step functions applied at two different times. Note how the use of the theorem avoids our having to trace a myriad of pulses back and forth on the line.)

3-5.0k The transmission system in Figure 3-33 is in the steady state initially. A disturbance that changes the load resistance to a value of $2Z_0$ occurs on the line. Obtain plots of the new load current and voltage and the generator end current and voltage.

3-6 Line Switching with Complex Terminations

In this section we consider line switching where the line is terminated in L, RL, C, and RC loads. Most of the discussion here will be based upon heuristic principles since exact reflection coefficient values are possible only for resistive loads for transient sources, unless one uses the Laplace transform, which is covered in the next section. We leave to the next section the task of obtaining exact solutions for these cases and for cases having general source functions and loads. As in the preceding sections we assume a lossless line. We consider only sources that have an internal (Thevenin) resistance of Z_0 since this is the most practical situation in time-domain reflectometry (TDR). With the basic principles

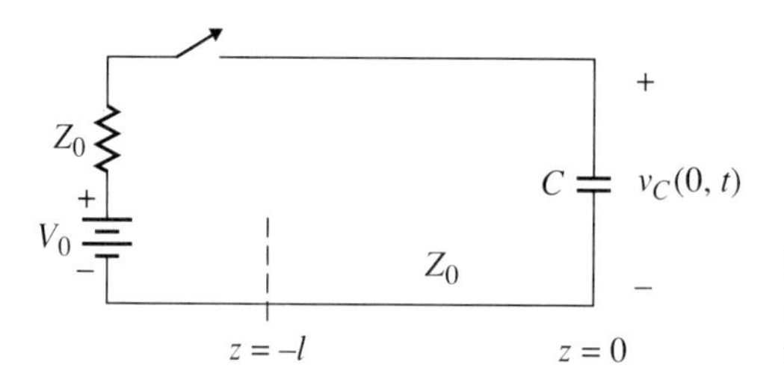

Figure 3-34. DC source switched onto a capacitively terminated line.

covered in this section we will be able to determine from time waveforms whether a given load or discontinuity is inductive or capacitive.

Suppose we consider a matched DC source (source impedance equal to the line characteristic impedance) switched onto a line terminated in a capacitor as shown in Figure 3-34. We wish to determine the capacitor voltage.

A good first step is to determine the desired quantity values at $t = 0$ and at $t = \infty$. Since the capacitor is initially at zero volts the plot of $v_C(0, t)$ will start at zero and remain at zero for the first T seconds (one way delay from switch). At $t = \infty$, the system will be in the steady-state DC condition, so there will be no capacitor current, $i_C = C dv_C/dt = 0$, since v_C is constant. This means the capacitor appears as an open circuit; so the line voltage and also the capacitor voltage $v_C(0, t = \infty)$ will be V_0. These two values are first placed on the plot in Figure 3-35a.

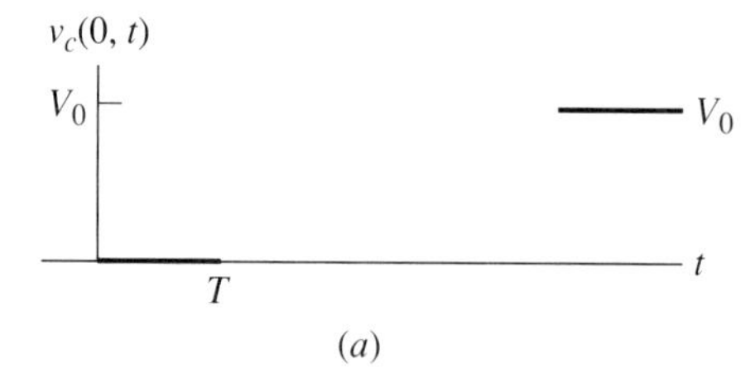

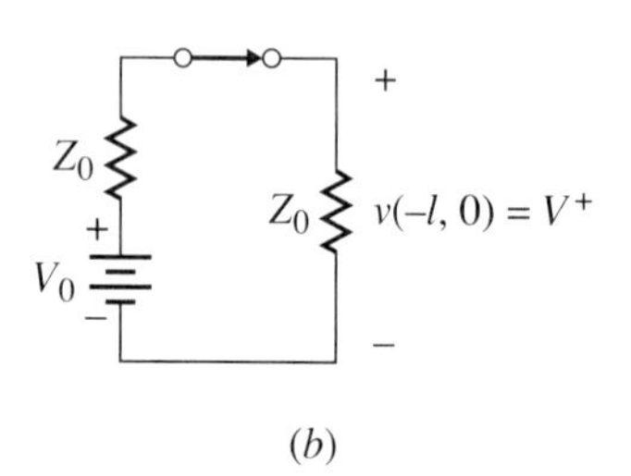

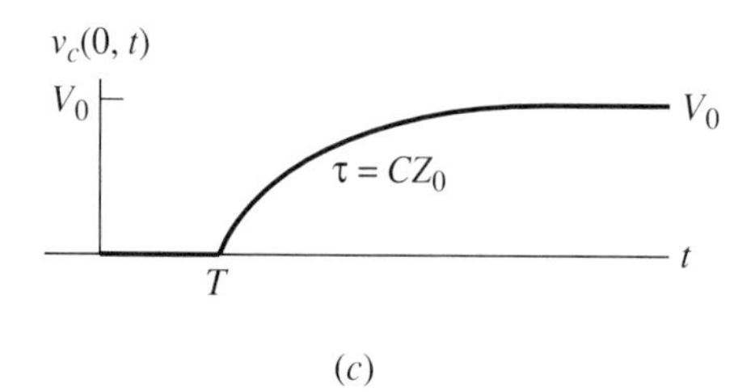

Figure 3-35. Capacitor voltage resulting from a matched DC source switched onto a capacitively terminated line. (a) Partial plot: initial and final values. (b) Equivalent circuit at source end at $t = 0$. (c) Complete capacitor waveform.

Answer:

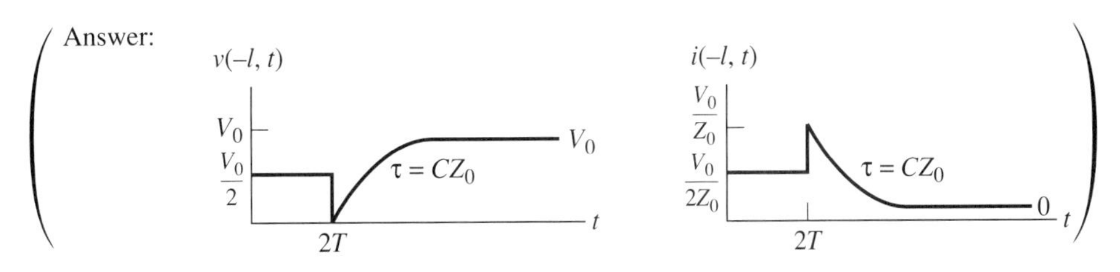

Figure 3-36.

At the source end, at $t = 0$, the equivalent circuit is as shown in Figure 3-35*b*. Thus $V^+ = \frac{1}{2}V_0$. This travels down to the capacitor, where the voltage must remain zero since the capacitor voltage cannot change instantaneously for finite currents. Zero volts is an effective short, so the reflection coefficient at that instant appears to be -1 and $V^- = -\frac{1}{2}V_0$. This V^- travels back to the generator. Since the source is matched, however, there will be no voltage reflection sent back to the capacitor. However, the line current now flows into the capacitor and charges the capacitor toward V_0 exponentially. Since the characteristic impedance of the line is resistive the Thevenin impedance of the line system appears to be Z_0, so one might expect that the time constant would be $\tau = Z_0C$ seconds. This is, in fact, the case as will be shown in Section 3.7. The exponential nature of the response is then drawn as shown in part (*c*) of the figure.

Exercises

3-6.0a For the preceding example, make a plot of $i_C(0, t)$. (Hint: Use the equation that describes a capacitor volt-ampere characteristic.)

3-6.0b For the preceding example, plot the voltage at the right of the switch and the current leaving the source. (Hints: What are the initial and final values, and what happens when the V^- returns from the capacitor? $v_{\text{total}} = V^+ + V^- + V_{DC}$). (Answer: Fig. 3-36)

3-6.0c Suppose that in the preceding example the line and capacitor are initially at voltage $\frac{1}{2}V_0$. Plot the capacitor voltage and current. Also plot the voltage at the right of the switch.

3-6.0d A capacitively terminated line having a characteristic impedance of 150 Ω is charged initially to 10 V. The system is shown in Figure 3-37. The system is to be discharged into a resistance of value Z_0. Plot the capacitor voltage and current and the voltage across the resistor Z_0.

As a second example, let's terminate the line with a series RL circuit and again switch on a matched (Z_0) DC source as shown in Figure 3-38. The voltage across the RL com-

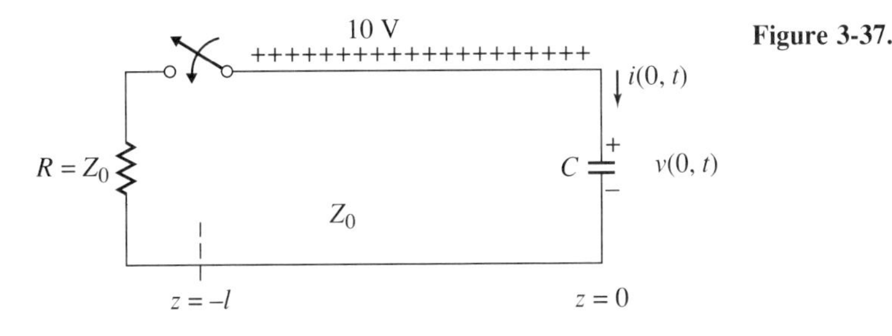

Figure 3-37.

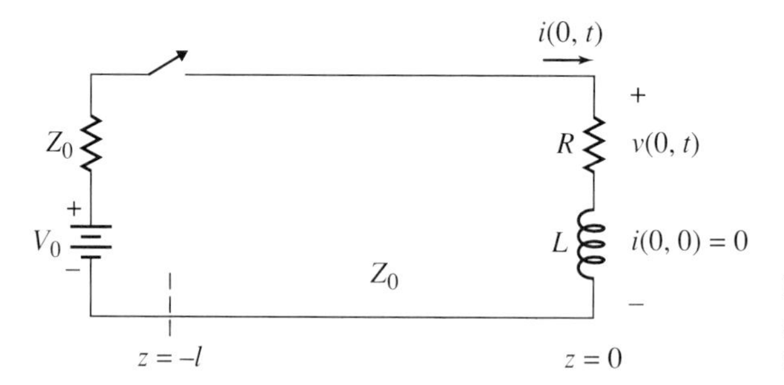

Figure 3-38. Matched DC source switched onto an RL terminated line.

bination is to be plotted. The developments of the results are shown in Figure 3-39. Steady state exists for $t < 0$.

Since the initial inductor current is zero, $v(0, 0) = 0$ also. After a very long time the equivalent circuit will be as shown in Figure 3-39*a* from which $v(0, \infty) = RV_0/(R + Z_0)$ (the inductor will be a short circuit then). These two results are used to begin the plot shown in part (*b*) of the figure.

As in the previous cases, at $t = 0$ the line appears to be a resistance of Z_0 Ω at the

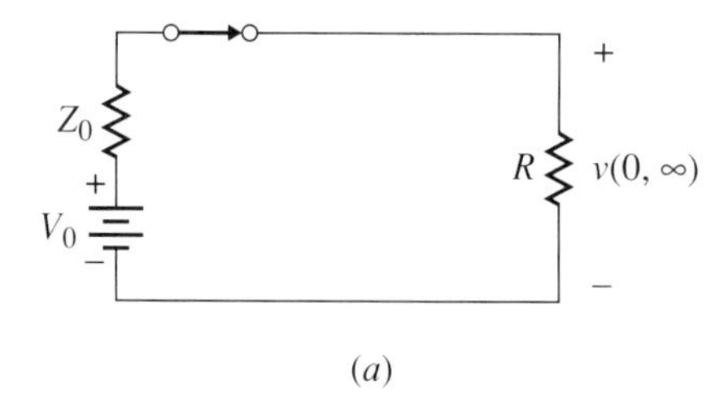

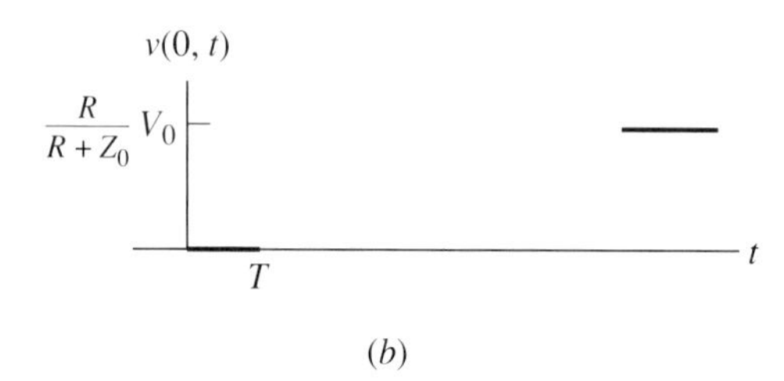

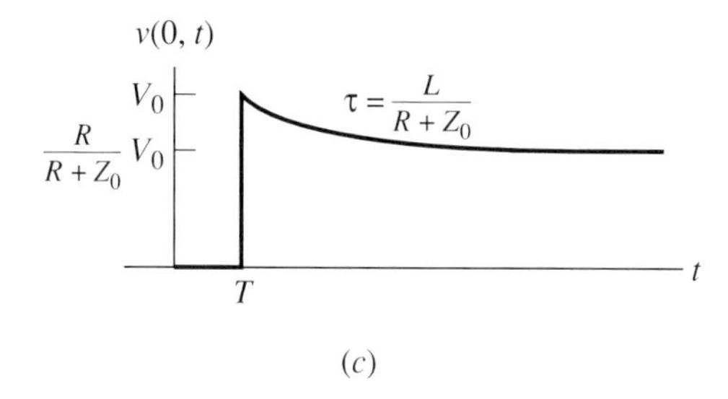

Figure 3-39. Load response for a matched DC source switched onto an RL terminated line. (*a*) Circuit at $t = \infty$. (*b*) Initial and final value plot. (*c*) Complete load voltage.

switch. Then the voltage divider effect gives $V^+ = \frac{1}{2}V_0$, which travels to the right to the load end. At the load the inductor current cannot change instantaneously, so the load appears to be an open circuit the moment V^+ reaches the load so the reflection coefficient is $+1$ at that moment. This makes $V^- = V^+ = \frac{1}{2}V_0$ so that

$$\begin{aligned} v(0, T) &= V^+ + V^- + \text{voltage on line} \\ &= \tfrac{1}{2}V_0 + \tfrac{1}{2}V_0 + 0 = V_0 \end{aligned}$$

This is a valid result since the inductor *voltage* can change instantaneously.

Last, the voltage must reach the final value already determined, and since there will be no voltage reflections from the source one might conclude that the decrease to the final value would be exponential. This is the actual result, and since the line impedance is Z_0 ohms and is in series with the resistor R, the total inductor ''charging'' resistance is $R + Z_0$. The time constant would then be

$$\tau = \frac{L}{R + Z_0} \text{ seconds}$$

The total load response is given in Figure 3-39*c*.

To obtain the current in the load, we simply use the volt-ampere characteristic for the inductor. This, of course, means that we must use the equation for the voltage across the L component only. Since the voltage across L goes ultimately to *zero* rather than to $[R/(R + Z_0)]V_0$, we use the equation for $t > T$ for the inductor voltage

$$v_L(0, t) = V_0 e^{-(t-T)/\tau} \qquad \text{where} \quad \tau = \frac{L}{R + Z_0}$$

We then obtain

$$\begin{aligned} i_L(0, t) &= \frac{1}{L}\int_T^t v(0, t)\, dt + i_L(0, T) = \frac{1}{L}\int_T^t V_0 e^{-(t-T)/\tau}\, dt + 0 \\ &= \frac{V_0}{L}(-\tau)e^{-(t-T)/\tau}\bigg|_T^t = -\frac{V_0\tau}{L}\left(e^{-(t-T)/\tau} - e^{-(T-T)/\tau}\right) \\ &= -\frac{V_0 \dfrac{L}{R + Z_0}}{L}\left(e^{-(t-T)/\tau} - 1\right) = \frac{V_0}{R + Z_0}\left(1 - e^{-(t-T)/\tau}\right) \text{ A} \end{aligned}$$

This is what one would expect on physical grounds since for $t \to \infty$ the steady-state current would be (replacing the inductor by a short)

$$i_L(0, t \to \infty) = \frac{V_0}{R + Z_0}$$

Thus the source Z_0 and resistance R would simply be in series across the source voltage effectively (L is a short for $t \to \infty$). See Figure 3-40.

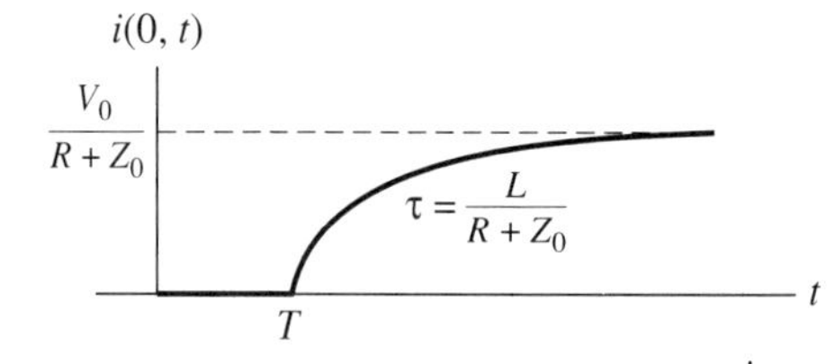

Figure 3-40. Load current for a matched DC source switched onto an RL terminated line.

Exercises

3-6.0e For the preceding example, plot the voltage across the resistor R and add the result to the equation written for the inductor voltage to show that the sum gives the solution $v(0, t)$ in Figure 3-39c.

3-6.0f For the example above plot the current and voltage waveforms at $z = -l$. (Answer: Figure 3-41)

3-6.0g For the system shown in Figure 3-42 plot $v(0, t)$, $i_C(0, t)$ and $v(-l, t)$. (Hint: What is the Thevenin resistance seen by the capacitor? Answer: 37.5 Ω. Why?)

3-6.0h For the system shown in Figure 3-43 plot the load voltage, capacitor voltage, and load current. (Hint: $\tau = (R + Z_0)C$.)

3-6.0i For the system shown in Figure 3-44 plot the load voltage, inductor current, and the load current as indicated.

3-6.0j In all of the examples and exercises in this section, the source internal resistance was shown to the left of the switch. Note that all of the results would be the same if the

Figure 3-41.

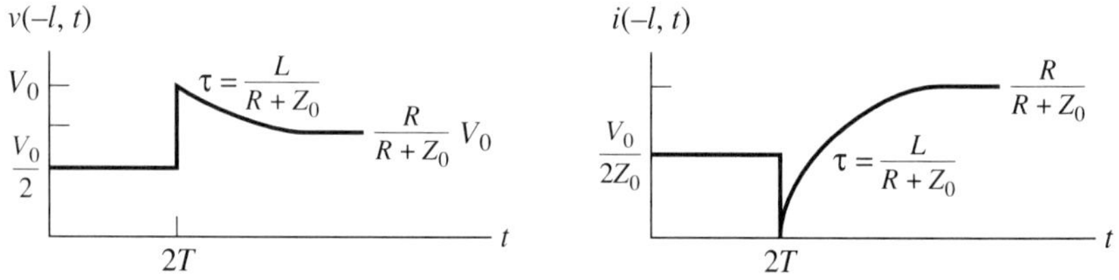

Figure 3-42.

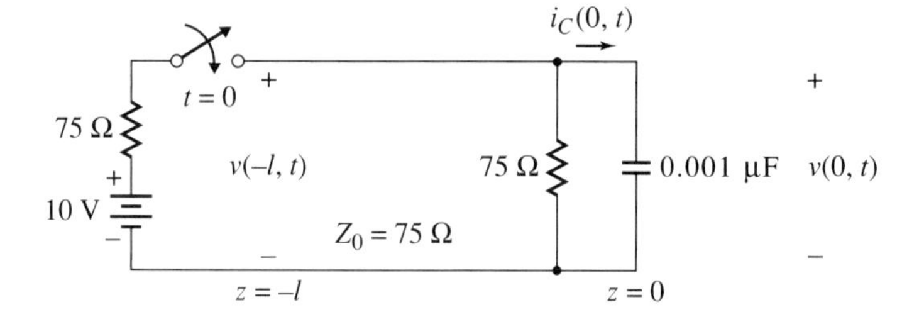

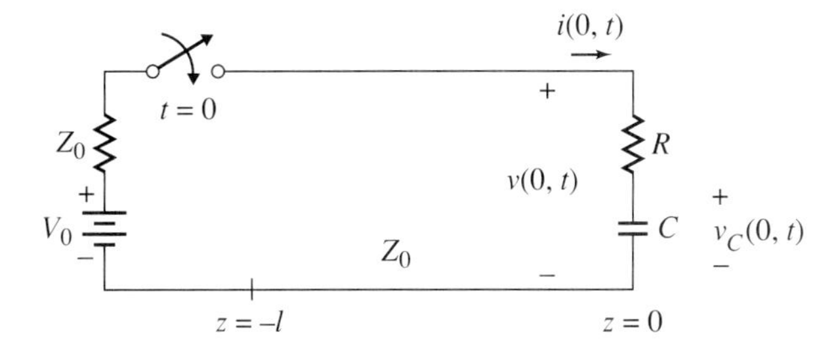

Figure 3-43.

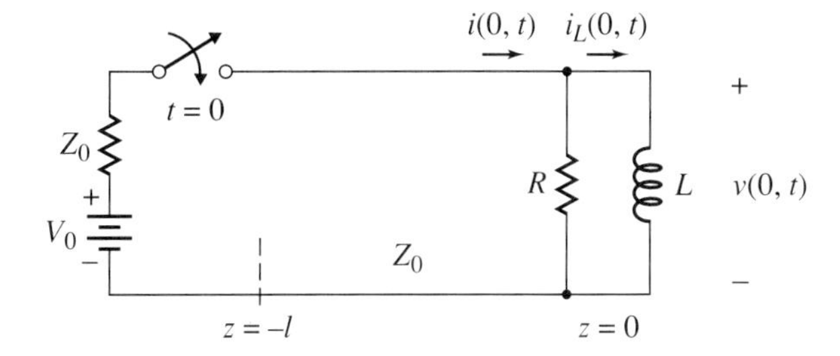

Figure 3-44.

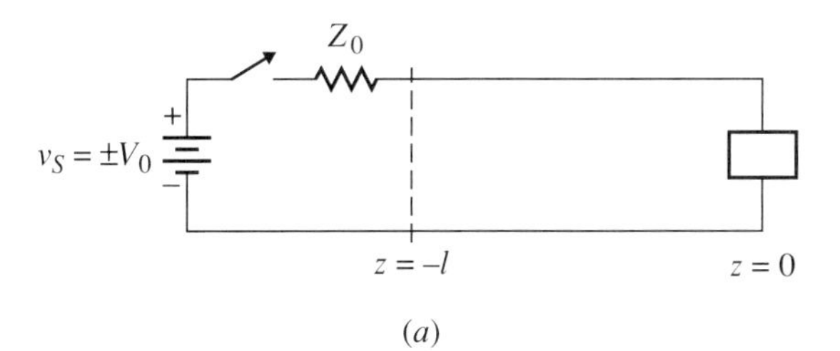

(a)

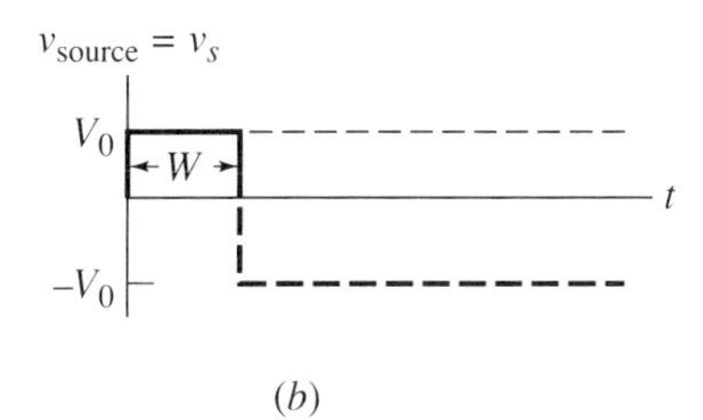

(b)

Figure 3-45.

resistances had been shown at the immediate right of the switch as shown in Figure 3-45*a*. Now if we wish to generate a pulse on the line we could think of applying two step functions separated by the pulse width W, as shown in Figure 3-45*b*. The sum of the two step functions generates the pulse. Using the superposition theorem, $V_0 = 10$ V, and $W = 50$ ns, determine the pulse response $v(0, t)$ for the system of Exercise 3-6.0g. Repeat for the system of Exercise 3-6.0i where $V_0 = 10$ V, $Z_0 = R = 150\ \Omega$, and $L = 1\ \mu$H.

3-7 Laplace Transform Solution of the Wave Equation

In the preceding section the types of systems that could be solved were quite limited. The source impedance was limited to the line characteristic impedance and only step functions were used. A pulse response could be generated by summing appropriate step function responses, but the effects of finite pulse rise time could not be obtained. In this section we develop the Laplace transform method and enlarge the number of problems that can be solved considerably. The price we pay for this generality is the introduction of considerable algebra and rather involved integrations. This additional complexity was probably evident in your study of ordinary differential equations using Laplace transforms. McLachlan [40] has essentially all of the material required. Our approach here follows that technique closely.

3-7.1 Laplace Transforms of the Transmission Line Equations and the Q Function Representations of Current and Voltage

Since the lossless approximation proves to be a very good one in engineering practice, we will make that assumption again. Actually, it is not too much more difficult to include the line losses, but the main goal here is to make our study an introductory one.

We begin with the lossless transmission line equations from Section 3-2 (Eqs. 3-1(a) and 3-1(b)); namely,

$$\frac{\partial v(z, t)}{\partial z} = -L\frac{\partial i(z, t)}{\partial t} \quad \text{and} \quad \frac{\partial i(z, t)}{\partial z} = -C\frac{\partial v(z, t)}{\partial t}$$

Taking the Laplace transform of each side of the left equation, we have the following results:

$$\begin{aligned} \mathcal{L}\left\{\frac{\partial v(z, t)}{\partial z}\right\} &= \int_{0^-}^{\infty} \frac{\partial v(z, t)}{\partial z} e^{-st}\, dt \\ &= \frac{\partial}{\partial z}\int_{0^-}^{\infty} v(z, t)e^{-st}\, dt = \frac{\partial V(z, s)}{\partial z} = \frac{dV(z, s)}{dz} \end{aligned}$$

and

$$\begin{aligned} \mathcal{L}\left\{-L\frac{\partial i(z, t)}{\partial t}\right\} &= -L\int_{0^-}^{\infty} \frac{\partial i(z, t)}{\partial t} e^{-st}\, dt = -L[sI(z, s) - i(z, 0^-)] \\ &= -sLI(z, s) + Li(z, 0^-) \end{aligned}$$

We can then write the original two differential equations as

$$\frac{dV(z, s)}{dz} = -sLI(z, s) + Li(z, 0^-) \tag{3-18}$$

$$\frac{dI(z, s)}{dz} = -sCV(z, s) + Cv(z, 0^-) \tag{3-19}$$

The preceding set of equations is said to be a *coupled set* since both of the functions we want to obtain, $V(z, s)$ and $I(z, s)$, appear in each. It would be much better if there were only one function. The following technique will show how a single equation can be obtained.

First multiply the second equation by a constant K (at least constant as far as the coordinate variable z is concerned) and add the result to the first equation. This gives, upon using the derivative of a sum law,

$$\frac{d[V(z, s) + KI(z, s)]}{dz} = -sCK\left[V(z, s) + \frac{L}{CK} I(z, s)\right] + CK\left[v(z, 0^-) + \frac{L}{CK} i(z, 0^-)\right] \tag{3-20}$$

Next we observe that the functions within the brackets of the derivative and in the left brackets in the right-hand member will be identical in form if we set

$$K = \frac{L}{CK}$$

from which

$$K^2 = \frac{L}{C}$$

or

$$K = \pm \sqrt{\frac{L}{C}}$$

Note, in this case, the right-bracket term on the right-hand side also has the same form.

From previous work we recognize that for a lossless line

$$\sqrt{\frac{L}{C}} \equiv Z_0$$

Thus we have

$$K = \pm Z_0$$

Using this value of K the terms in brackets become

$$V(z, s) \pm Z_0 I(z, s)$$

and

$$v(z, 0^-) \pm Z_0 i(z, 0^-)$$

We recognize the first result as a function of s for which the inverse Laplace transform would give the time response and the second result to be the initial condition function required to obtain the particular solution.

To simplify Eq. 3-20 we define a new function and its initial condition as

$$Q^{\pm}(z, s) \equiv V(z, s) \pm Z_0 I(z, s) \tag{3-21}$$

$$q^{\pm}(z, 0^-) \equiv v(z, 0^-) \pm Z_0 i(z, 0^-) \tag{3-22}$$

In what follows we must remember that in all equations and results, the top signs are all taken together and the bottom signs all taken together.

The differential Eq. 3-20 may now be written in terms of the new Q function as (leaving off the argument list of variables for easier writing):

$$\frac{dQ^{\pm}}{dz} = \mp sCZ_0 Q^{\pm} \pm CZ_0 q^{\pm}$$

Since $CZ_0 = C\sqrt{L/C} = \sqrt{LC} = 1/V_P$, where V_P is the phase velocity, this may be reduced to

$$\frac{dQ^{\pm}}{dz} = \mp \frac{1}{V_P} sQ^{\pm} \pm \frac{1}{V_P} q^{\pm} \tag{3-23}$$

We must now solve this equation for $Q^{\pm}$. We are, however, faced with obtaining $V(z, s)$ and $I(z, s)$ from this. From our definitions we have

$$Q^+ = V(z, s) + Z_0 I(z, s)$$

$$Q^- = V(z, s) - Z_0 I(z, s)$$

If we add these two equations and solve for $V(z, s)$ we find

$$V(z, s) = \frac{Q^+ + Q^-}{2} \tag{3-24}$$

If we subtract the two equations and solve for $I(z, s)$ we find

$$I(z, s) = \frac{Q^+ - Q^-}{2Z_0} \tag{3-25}$$

The solution process can now be easily given:

1. Solve Eq. 3-23 for $Q^{\pm}$.
2. Write expressions for $V(z, s)$ and $I(z, s)$ using Eqs. 3-24 and 3-25.
3. Take the inverse Laplace transforms to obtain $v(z, t)$ and $i(z, t)$.

Exercises

3-7.1a The work preceding assumed a lossless transmission line. Following the preceding discussion, use a lossy transmission line system of equations to determine the constant K (which will now be a function of s) and the single variable differential equation similar to Eq. 3-23.

3-7.1b In circuit theory the impedance is defined as the transform of the voltage divided by the transform of the current. Derive an expression for the line impedance in terms of Q^+ and Q^-. Which way would this impedance be looking, to the right or left at a location z on the line? What is the impedance looking the other direction? (Answer: $Z(z, s) = [(Q^+ + Q^-)/(Q^+ - Q^-)]Z_0$. The impedance would be that looking to the right since by convention the current in the impedance enters the positive sign of the voltage across it.)

3-7.1c When the Laplace transform of the partial derivative with respect to z was evaluated the partial derivative ultimately became an ordinary or total derivative. Explain why this is possible.

3-7.1d Evaluate the Laplace transform of the general voltage wave equation (Eq. 1-7). Identify the required initial conditions. If $v(z, 0^-)$ is 0 does it follow that $v'(z, 0) = 0$ also? (Prime indicates time derivative.) Give an example that illustrates your answer. If you were given $v(z, 0^-)$ and $i(z, 0^-)$ explain how you would obtain the time derivative initial conditions $v'(z, 0^-)$ and $i'(z, 0^-)$.

3-7.2 Solutions for the Q Functions and Initial Conditions

We turn our attention to the solution of the differential equation for $Q^{\pm}$. In standard form the differential equation (Eq. 3-23) is

$$\frac{dQ^{\pm}}{dz} \pm \frac{1}{V_P} sQ^{\pm} = \pm \frac{1}{V_P} q^{\pm} \tag{3-26}$$

which can be recognized as an inhomogeneous ordinary differential equation. The solution is obtained by finding the homogeneous solution, $Q_h^{\pm}$, of the homogeneous differential equation

$$\frac{dQ_h^{\pm}}{dz} \pm \frac{s}{V_P} Q_h^{\pm} = 0 \tag{3-27}$$

and adding it to the particular integral solution, $Q_{pi}^{\pm}$, to yield

$$Q^{\pm}(z, s) = Q_h^{\pm}(z, s) + Q_{pi}^{\pm}(z, s) \tag{3-28}$$

The solution of the first-order homogeneous differential equation with constant coefficients (s is constant and z is the independent variable here) is the well-known exponential function

$$Q_h^{\pm}(z, s) = Q_0^{\pm}(s)e^{\mp(s/V_P)z} \tag{3-29}$$

where $Q_0^{\pm}(s)$ is the constant required in the solution. You should reconstruct this solution from your previous study of ordinary differential equations.

The particular integral solution is obtained by assuming a functional form for $Q_{pi}^{\pm}$, which is basically of the form of the source term, in this case the right hand member, of the original differential equation (Eq. 3-26). Now $q^{\pm}$ is a function only of z, $q^{\pm}(z, 0) = v(z, 0) \pm Z_0 i(z, 0)$, so we expect there will be a component of $Q_{pi}^{\pm}$ that is a function of z only, call it $F^{\pm}(z)$. However, $Q_{pi}^{\pm}$ must also contain a part that has an s in it since that is the only way the s can be removed when $Q_{pi}^{\pm}$ is substituted into the left-hand side. Also, note that on the left-hand side we have the sum of the function and its derivative. To have two such terms combine to give a single $q^{\pm}$, the function $Q_{pi}^{\pm}$ and its derivative must have the same functional form. Thus we are led to use an exponential component involving s also in the function $Q_{pi}^{\pm}$. Summarizing the discussion of this paragraph we would then try a particular integral of the form

$$Q_{pi}^{\pm}(z, s) = F^{\pm}(z)e^{\mp(s/V_P)z} \tag{3-30}$$

Substituting this into the inhomogeneous differential equation (Eq. 3-26) we obtain

$$F^{\pm}(z)\left(\mp \frac{s}{V_P} e^{\mp(s/V_P)z}\right) + e^{\mp(s/V_P)z} \frac{dF^{\pm}(z)}{dz} \pm \frac{s}{V_P} F^{\pm}(z)e^{\mp(s/V_P)z} = \pm \frac{1}{V_P} q^{\pm}$$

or

$$\frac{dF^{\pm}(z)}{dz} = \pm \frac{1}{V_P} q^{\pm} e^{\pm(s/V_P)z}$$

Integrating this last result

$$F^{\pm}(z) = \pm \frac{1}{V_P} \int_0^z q^{\pm}(z, 0^-)e^{\pm(s/V_P)z}\, dz$$

With this the particular integral solution becomes

$$Q_{pi}^{\pm}(z, s) = \pm \frac{e^{\mp(s/V_P)z}}{V_P} \int_0^z q^{\pm}(z, 0^-)e^{\pm(s/V_P)z}\, dz \tag{3-31}$$

The general solution is then, using Eq. 3-28,

$$Q^{\pm}(z, s) = Q_0^{\pm} e^{\mp(s/V_P)z} \pm \frac{e^{\mp(s/V_P)z}}{V_P} \int_0^z q^{\pm}(x, 0^-)e^{\pm(s/V_P)x}\, dx \tag{3-32}$$

where we have replaced the integration variable z by x so that the integral upper limit z may be more clearly identified as the coordinate of interest along the line and confusion avoided. All we have to do, of course, is replace z by x in the given initial condition functions. As usual, there is an unknown constant $Q_0^{\pm}(s)$ which must be obtained by applying boundary conditions on the function $Q^{\pm}(z, s)$. This will be discussed in more detail later.

Now from the solution $Q^{\pm(z,s)}$ we could write the solutions for $I(z, s)$ and $V(z, s)$ using Eqs. 3-24 and 3-25. As a matter of fact, we must often write down these two solutions before $Q_0^{\pm}(s)$ is obtained since we frequently know $I(z, s)$ and $V(z, s)$ at points along the line rather than $Q^{\pm}(z, s)$ directly. This will be illustrated next.

Exercises

3-7.2a Verify by direct substitution that the homogeneous solution (Eq. 3-29) is a solution of the homogeneous Eq. 3-27.

3-7.2b Show by direct substitution that the particular integral solution (Eq. 3-31) does indeed satisfy the inhomogeneous differential equation (Eq. 3-26). (Hints: It is helpful to replace the variable of integration z in the integrand by the variable x. Also, you will need Leibniz' rule for differentiating an integral.)

3-7.2c When the differential equation for $F^{\pm}(z)$ was integrated, the lower limit was taken as zero. Show that this does not result in any loss of generality by assuming a lower limit L and substituting into the inhomogeneous differential equation. (Hint: $\int_a^c = \int_a^b + \int_b^c$.)

To obtain the solution to a particular problem we need to obtain the constant function $Q_0^{\pm}(s)$. Four types of initial conditions may be used to complete the solution. These are:

1. *Voltage conditions* (two required to obtain both Q_0^+ and Q_0^-). Voltages may be known as functions $v(z_i, t)$ at two points on the line. Once $V(z, s)$ has been constructed using Eq. 3-24 we may apply these voltage conditions by evaluating $\mathscr{L}\{v(z_i, t)\}$ and equating them to $V(z, s)$ evaluated at z_i. Note that this process yields two equations in the two unknowns Q_0^+ and Q_0^-. For example, at a short circuit $v(z_s, t) = 0$ for all t so that $V(z_s, s) = 0$. Ohm's law and Kirchhoff's law may be applied at a point on the line.
2. *Current conditions* (two required to obtain both Q_0^+ and Q_0^-). Currents may be known as functions $i(z_i, t)$ at two points on the line. Once $I(z, s)$ has been constructed using Eq. 3-25 we may apply these current conditions by evaluating $\mathscr{L}\{i(z_i, t)\}$ and equating them to $I(z, s)$ evaluated at z_i. Note that this gives two equations in the two unknowns Q_0^+ and Q_0^-. For example, at an open circuit $i(z_0, t) = 0$ for all t so $I(z_0, s) = 0$.
3. *Impedance conditions* (two required to obtain both Q_0^+ and Q_0^-). The impedances may be known as functions of s at two points. Thus once we have constructed $Z(z, s)$, see Exercise 3-7.1b, we can evaluated it at z_i and equate to the known functions and solve for Q_0^+ and Q_0^-. For example, if a transmission line were terminated in a capacitance then $Z(0, s) = 1/sC$. One must be a little careful with this approach, since if an element were placed at a location other than $z = 0$ (end of line) the impedance there is the element impedance in parallel or series with the rest of the line.
4. *Mixed conditions* (two required to obtain both Q_0^+ and Q_0^-). This case is usually a combination of any two of types 1, 2, or 3. Note that the conditions need not be at the same point on the line. This often may require writing KVL or KCL at the source end or some other point on the line.

At this point we can give a detailed solution procedure as our summary before presenting some examples.

1. Write expressions for $v(z, 0^-)$ and $i(z, 0^-)$ and then form

$$q^{\pm}(z, 0^-) = v(z, 0^-) \pm Z_0 i(z, 0^-)$$

2. Evaluate the integral

$$\int_0^z q^{\pm}(x, 0^-) e^{\pm(s/V_P)x}\, dx$$

3. Now $V(z, s) = (Q^+ + Q^-)/2$, $I(z, s) = (Q^+ - Q^-)/2Z_0$, and $Z(z, s) = [V(z, s)]/[I(z,s)]$, where we have shown that

$$Q^{\pm}(z, s) = Q_0^{\pm} e^{\mp(s/V_P)z} \pm \frac{e^{\mp(s/V_P)z}}{V_P} \int_0^z q^{\pm}(x, 0^-) e^{\pm(s/V_P)x}\, dx$$

We use any available boundary conditions on $V(z, s)$, $I(z, s)$, and/or the impedance $Z(z, s)$ to solve for $Q_0^{\pm}$. Two conditions are required to obtain both Q_0^+ and Q_0^-. Writing KVL and KCL at any switches also provides a mixed-boundary condition.

4. Having found $Q_0^{\pm}$ we form the solutions $Q^{\pm}(z, s)$, Eq. 3-32, and then obtain $V(z, s)$ and $I(z,s)$ using Eqs. 3-24 and 3-25.
5. Obtain $v(z, t)$ and $i(z, t)$ by taking the inverse Laplace transforms.

3-7.3 Some Examples of the Laplace Transform Solution

We now have at our disposal all of the tools required to solve transmission problems for complex load and source impedances. Along the way we will obtain exact solutions of the TDR problems presented in Section 3-6.

Example 3-2

For the first example we consider a DC source voltage switched onto a shorted transmission line shown in Figure 3-46. When the switch is closed, $v(z, 0)$ will be $v(z, 0^-) = 0$. Also $i(z, 0^-) = 0$.

Note that we use values just before the switch is closed as our initial conditions for solving differential equations. From Eq. 3-22, $q^{\pm}(z, 0^-) = 0$. This makes the integral part of the solution in Eq. 3-32 equal to zero. Equation 3-32 reduces to

$$Q^{\pm}(z, s) = Q_0^{\pm}(s)e^{\mp(s/V_P)z}$$

or

$$Q^{+}(z, s) = Q_0^{+}(s)e^{-(s/V_P)z} \qquad \text{and} \quad Q^{-}(z, s) = Q_0^{-}(s)e^{(s/V_P)z}$$

From Eq. 3-24

$$V(z, s) = \frac{Q_0^{+} e^{-(s/V_P)z} + Q_0 e^{-(s/V_P)z}}{2}$$

From Eq. 3-25

$$I(z, s) = \frac{Q_0^{+} e^{-(s/V_P)z} - Q_0 e^{-(s/V_P)z}}{2Z_0}$$

Using voltage boundary conditions $v(-l, t) = V_0U(t)$ and $v(0, t) = 0$ we have $V(-l, s) = V_0/s$ and $V(0, s) = 0$. We next apply these conditions to the preceding voltage solution.

At $z = -l$:

$$V(-l, s) = \frac{V_0}{s} = \frac{Q_0^{+} e^{sl/V_P} + Q_0^{-} e^{-sl/V_P}}{2}$$

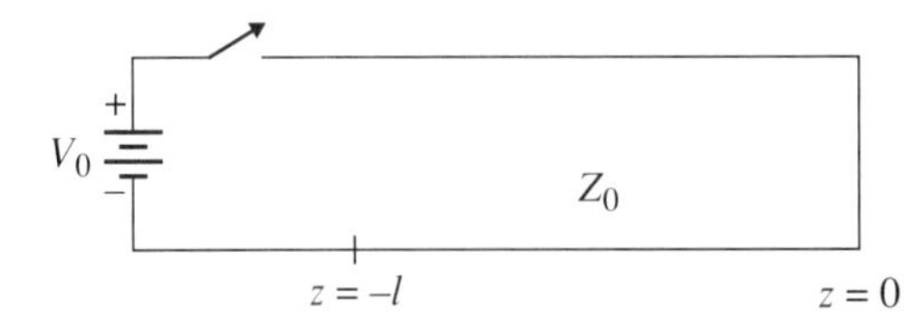

Figure 3-46. DC source switched onto a shorted line.

At $z = 0$:

$$V(0, s) = 0 = \frac{Q_0^+ + Q_0^-}{2} \qquad \text{from which} \quad Q_0^- = -Q_0^+$$

(Notice this corresponds to the fact that the reflection coefficient for voltage is -1!)

Using this last result in the voltage equation at $z = -l$,

$$\frac{V_0}{s} = Q_0^+ \frac{e^{sl/V_P} - e^{-sl/V_P}}{2}$$

from which

$$Q_0^+ = \frac{2V_0}{(e^{sl/V_P} - e^{-sl/V_P})s}$$

Then

$$Q_0^- = -Q_0^+ = \frac{-2V_0}{(e^{sl/V_P} - e^{-sl/V_P})s}$$

The solutions become:

$$V(z, s) = \frac{2V_0}{(e^{sl/V_P} - e^{-sl/V_P})s} \cdot \frac{e^{-(s/V_P)z} - e^{(s/V_P)z}}{2} = \frac{e^{-(s/V_P)z} - e^{(s/V_P)z}}{e^{sl/V_P} - e^{-sl/V_P}} \frac{V_0}{s}$$

$$I(z, s) = \frac{2V_0}{2Z_0(e^{sl/V_P} - e^{-sl/V_P})} \cdot \frac{e^{-(s/V_P)z} + e^{(s/V_P)z}}{s} = \frac{V_0}{sZ_0} \frac{e^{-(s/V_P)z} + e^{(s/V_P)z}}{e^{sl/V_P} - e^{-sl/V_P}}$$

If we are interested only in source quantities at $z = -l$ we have

$$V(-l, s) = \frac{V_0}{s}$$

Taking the inverse Laplace transform, $v(-l, t) = V_0U(t)$ as it should.

For the source current,

$$I(-l, s) = \frac{V_0}{sZ_0} \frac{e^{(l/V_p)s} + e^{-(l/V_p)s}}{e^{(l/V_p)s} - e^{-(l/V_p)s}} = \frac{V_0}{sZ_0} \coth\left(\frac{l}{V_p} s\right)$$

Recognizing that l/V_P is the line one-way delay T this may be written as

$$I(-l, s) = \frac{V_0}{sZ_0} \coth(Ts)$$

From a Laplace transform table we find that (see Appendix C)

$$\mathcal{L}^{-1}\left\{\frac{1}{s} \coth(Ts)\right\} = 2n - 1,$$

$$\text{for} \quad 2(n - 1)T < t < 2nT, \qquad \text{and} \quad n = 1, 2, 3, \ldots$$

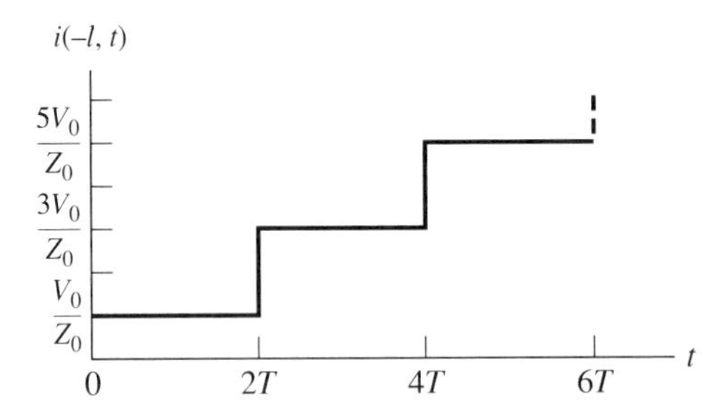

Figure 3-47. Current response in DC voltage source switched onto a shorted line.

Notice that the total line for a discontinuity to travel from the source to the short and back, $2T$, is automatically obtained in the function. We finally have

$$i(-l, t) = \frac{V_0}{Z_0}(2n - 1) \qquad \text{and} \quad 2(n - 1)T < t < 2nT, \quad n = 1, 2, 3, \ldots$$

This result is plotted in Figure 3-47. The source current is the same as that obtained by the reflection method in Section 3-4.2 (see particularly Figure 3-10).

Example 3-3

The next example will be that of a matched DC voltage source switched onto a capacitively terminated line as shown in Figure 3-48. We wish to determine $v(0, t)$.

If we assume that the line is initially discharged we will have

$$v(z, 0^-) = i(z, 0^-) = 0$$

Then $q^{\pm}(z, 0^-) = 0$ from Step 1 of the general procedure. This also means that the integral portion of the solution is zero (see Step 2 of the general procedure). We then construct the solution as

$$Q^{\pm}(z, s) = Q_0^{\pm} e^{\mp(s/V_P)z}$$

From this we have

$$V(z, s) = \frac{Q_0^{+} e^{-(s/V_P)z} + Q_0^{-} e^{(s/V_p)z}}{2}$$

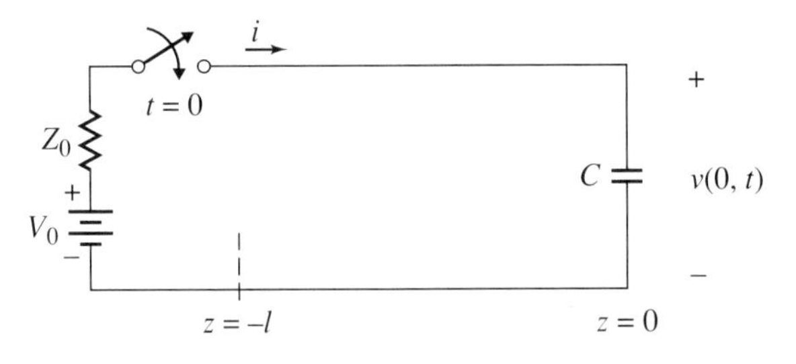

Figure 3-48. Matched DC source switched onto a capaci?tively terminated line.

and

$$I(z, s) = \frac{Q_0^+ e^{-(s/V_P)z} - Q_0^- e^{(s/V_P)z}}{2Z_0}$$

The impedance at any point on the line is then obtained by dividing these:

$$Z(z, s) = \frac{Q_0^+ e^{-(s/V_P)z} + Q_0^- e^{(s/V_P)z}}{Q_0^+ e^{-(s/V_P)z} - Q_0^- e^{(s/V_P)z}} Z_0$$

At $z = 0$ the impedance is $1/sC$ so that we require the impedance condition

$$Z(0, s) = \frac{Q_0^+ + Q_0^-}{Q_0^+ - Q_0^-} Z_0 = \frac{1}{sC} \quad \text{or} \quad \frac{Q_0^+ + Q_0^-}{Q_0^+ - Q_0^-} = \frac{1}{sZ_0C} \tag{3-33}$$

At the switch, $z = -l$, Kirchhoff's voltage law requires

$$v(-l, t) = V_0 - Z_0 i(-l, t)$$

Taking the Laplace transform, we obtain the voltage condition

$$V(-l, s) = \frac{V_0}{s} - Z_0 I(-l, s)$$

Using the expressions for $V(-l, s)$ and $I(-l, s)$ next, this becomes

$$\frac{Q_0^+ e^{(l/V_P)s} + Q_0^- e^{-(l/V_P)s}}{2} = \frac{V_0}{s} - Z_0 \frac{Q_0^+ e^{(l/V_P)s} - Q_0^- e^{-(l/V_p)s}}{2Z_0}$$

Again recognizing that l/V_P is the line one way delay T, and combining terms, the preceding reduces to

$$Q_0^+ = \frac{V_0}{s} e^{-Ts} \tag{3-34}$$

Substituting this into the impedance Eq. 3-33 for $Z(0, s)$ above and solving for Q_0^- we have

$$Q_0^- = \frac{V_0}{Z_0C} \frac{e^{-Ts}}{s\left(s + \dfrac{1}{Z_0C}\right)} - V_0 \frac{e^{-Ts}}{s + \dfrac{1}{Z_0C}} = \frac{V_0}{s} e^{-Ts} - 2V_0 \frac{e^{-Ts}}{s + \dfrac{1}{Z_0C}} \tag{3-35}$$

The second equality in the preceding equation was obtained by expanding the left term in the middle expression in a partial fraction series and combining terms.

Now that we have expressions for Q_0^+ and Q_0^-, the voltage equation becomes

$$V(z, s) = \frac{\dfrac{V_0}{s} e^{-Ts} e^{-(s/V_P)z} + \dfrac{V_0}{s} e^{-Ts} e^{(s/V_P)z} - 2V_0 \dfrac{e^{-Ts} e^{(s/V_P)z}}{s + \dfrac{1}{Z_0C}}}{2}$$

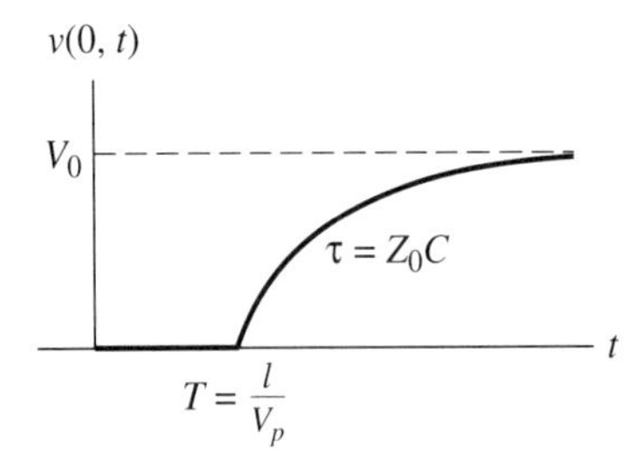

Figure 3-49. Capacitor voltage for a DC matched source switched onto a capacitively loaded line.

At the capacitor, $z = 0$, this reduces to

$$V(0, s) = \frac{V_0}{s} e^{-Ts} - \frac{V_0}{s + \dfrac{1}{Z_0C}} e^{-Ts} = V_0\left(\frac{1}{s} - \frac{1}{s + \dfrac{1}{Z_0C}}\right)e^{-Ts}$$

Taking the inverse Laplace transform (which is simplified by recognizing that the exponential e^{-Ts} is a time shift or delay),

$$v(0, t) = V_0(1 - e^{-(t-T/Z_0C)})U(t - T)$$

This result is plotted in Figure 3-49, and should be compared with the result of Section 3-6, especially Figure 3-35.

Exercises

3-7.3a For the shorted-line example discussed in the preceding section obtain plots of the current in the short and the voltage and current at $z = -l/2$. Use the results given in the discussion as the starting point.

3-7.3b For the capacitively terminated line example discussed in the preceding section determine equations for and plot:

(a) Capacitor current. (Hint: How does one differentiate in the Laplace domain?)
(b) Voltage at $z = -l$.
(c) Current at $z = -l$.

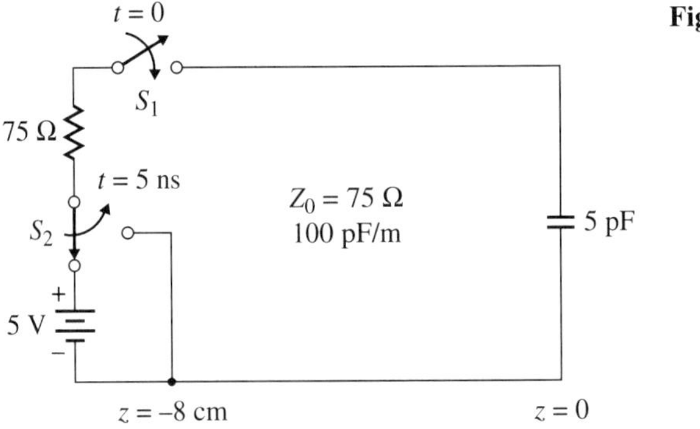

Figure 3-50.

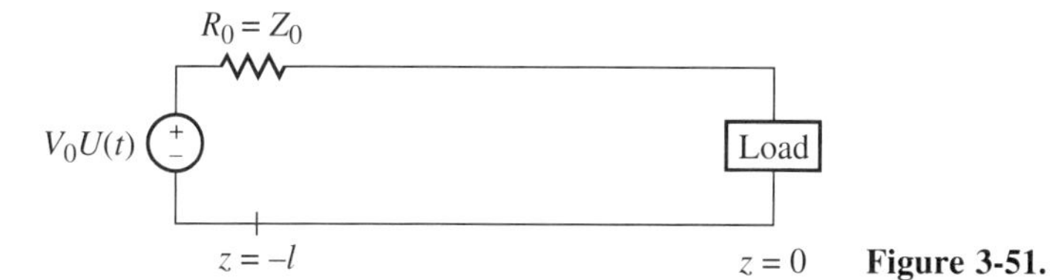

Figure 3-51.

3-7.3c In the system of Figure 3-48, change the source resistance to $Z_0/2$ and obtain a plot of $v(0, t)$. Be sure to mark the time constants on the plot.

3-7.3d The effect of a pulse applied to a transmission line can be simulated by removing the voltage component of the DC source switched onto a line and shorted the input. The source would be removed at a time equal to the pulse width. Figure 3-50 is the equivalent model. Compute the capacitor voltage. (Hint: Note that when the switch S_2 closes the result is a new problem of a charged line. Sum the two waveforms obtained when S_1 closes and then S_2 changes positions.)

If the pulse is assumed to arrive at the capacitor when the voltage is 2.5 V, what is the pulse delay?

3-7.3e A unit step-function voltage generator with internal impedance Z_0 is connected at $z = -l$ to a lossless line. At $z = 0$ the line is terminated in an inductance L_1. Find an expression for and plot $v(0,t)$. The line is initially uncharged.

3-7.3f In a transmission system it is often sufficient to know whether a given load is inductive or capacitive. In such cases, it is only necessary to obtain an oscilloscope waveform of the load or input waveform to make such a determination. This technique is called *time-domain reflectometry*, and makes use of a step generator whose Thevenin impedance is matched to the line. The basic configuration is shown in Figure 3-51. Determine the waveforms of voltage that appear at $z = 0$ and $z = -l$ for the following four cases (make plots):

(a) Pure inductor load.
(b) Pure capacitor load.
(c) Load consisting of series R and L.
(d) Load consisting of parallel R and C.

After you have completed the problem (or before if you can), see if you can devise a method for obtaining the results by considering time constants in the resulting equations and the exponential nature of the plots, including the values at key values of time. This, of course, is to be a qualitative discussion, but it should help you to appreciate the results.

3-7.3g A line is charged to 5 V and is in the steady state as shown in Figure 3-52. At $t = 0$ a resistor is connected to the other end of the line. Obtain the plots for the voltage across

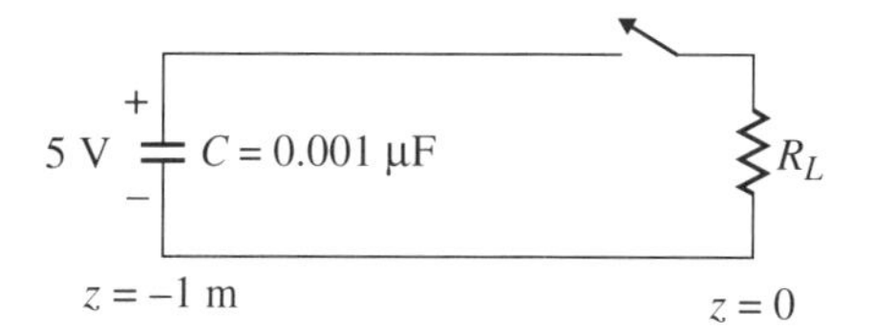

Figure 3-52.

the capacitor and resistor as functions of time for the cases $R_L = 25\ \Omega$, $50\ \Omega$ and $100\ \Omega$. The Z_0 of the line is $50\ \Omega$. Could you have obtained the plots from a qualitative consideration of the system configuration? The transmission line has a capacitance of 95 pF/m. Discuss qualitatively the effects of changing the value of the capacitor.

Example 3-4

The voltages applied to the transmission lines have been assumed to have vertical edges (zero rise time). This, of course, is a practical impossibility, so we next consider a ramp applied to a transmission line. We can construct pulses with finite rise and fall times using combinations of ramps and step functions and apply the superposition theorem for linear systems to obtain pulse responses. The basic system is shown in Figure 3-53. For simplicity we assume that R_S, Z_0, and R_L are purely real (resistive), so that the effects of finite slope may be emphasized more clearly. Also, the line is initially uncharged. Following the steps developed earlier, we have:

1. $v(z, 0^-) = i(z, 0^-) = 0$. Then

$$q^{\pm}(z, 0^-) = 0$$

2. $\int_0^z q^{\pm}(x, 0^-)e^{\pm(s/V_P)x}\,dx = 0$
3. $Q^{\pm}(z, s) = Q_0^{\pm} e^{\mp(s/V_P)z}$

$$V(z, s) = \frac{Q_0^+ e^{-(s/V_P)z} + Q_0^- e^{(s/V_P)z}}{2}, \qquad I(z, s) = \frac{Q_0^+ e^{-(s/V_P)z} - Q_0^- e^{(s/V_P)z}}{2Z_0}$$

At $z = 0$:

$$Z(0, s) = \frac{V(0, s)}{I(0, s)} = R_L = \frac{Q_0^+ + Q_0^-}{Q_0^+ - Q_0^-} Z_0$$

From the last equality,

$$\frac{Q_0^-}{Q_0^+} = \frac{R_L - Z_0}{R_L + Z_0} \qquad \text{(Does this look familiar?)} \tag{3-36}$$

Figure 3-53. Ramp applied to a transmission line.

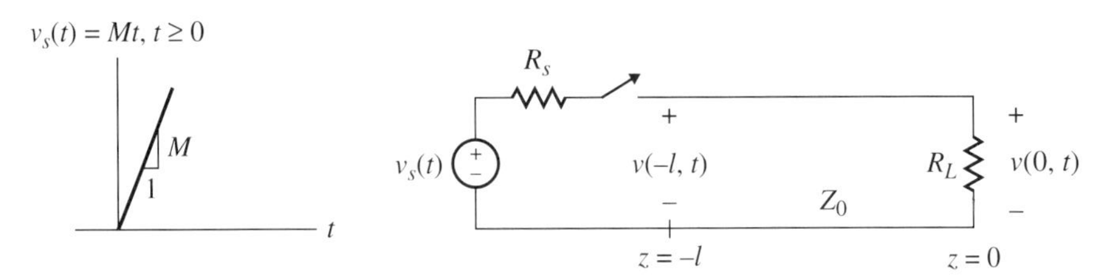

At $z = -l$:
KVL gives $V(-l, s) = V_s(s) - I(-l, s)R_S$, with $v_s(t) = Mt$. (M is the slope of the ramp)
Substituting the preceding $V(z, s)$ and $I(z, s)$ into this KVL equation:

$$\frac{Q_0^+ e^{(l/V_P)s} + Q_0^- e^{-(l/V_P)s}}{2} = \frac{M}{s^2} - \frac{Q_0^+ e^{(l/V_P)s} - Q_0^- e^{-(l/V_p)s}}{2Z_0} R_S$$

Recognizing l/V_P as the delay time T and combining terms we have

$$Q_0^+ e^{Ts} - Q_0^- e^{-Ts}\left(\frac{R_S - Z_0}{R_S + Z_0}\right) = \frac{2MZ_0}{s^2(R_S + Z_0)}$$

Substituting for Q_0^- from Eq. 3-36 and solving for Q_0^+:

$$Q_0^+ = \frac{2MZ_0}{R_S + Z_0}\frac{e^{-Ts}}{s^2}\left(\frac{1}{1 - \Gamma_S\Gamma_L e^{-2Ts}}\right)$$

where

$$\Gamma_S = \frac{R_S - Z_0}{R_S + Z_0}$$

and

$$\Gamma_L = \frac{R_L - Z_0}{R_L + Z_0}$$

Then

$$Q_0^- = \frac{2MZ_0\Gamma_L}{R_S + Z_0}\frac{e^{-Ts}}{s^2}\left(\frac{1}{1 - \Gamma_S\Gamma_L e^{-2Ts}}\right)$$

4. At the load, $z = 0$, the voltage is then

$$V(0, s) = \frac{MZ_0(1 + \Gamma_L)}{R_S + Z_0}\frac{e^{-Ts}}{s^2}\left(\frac{1}{1 - \Gamma_S\Gamma_L e^{-2Ts}}\right)$$

Since the general inverse of this voltage function would not readily reveal the reflection effects, we consider the particular case where the values are $R_S = 15\ \Omega$, $Z_0 = 150\ \Omega$, $M = 10^8$ V/s, and $R_L = 5$ kΩ. The corresponding reflection coefficients are $\Gamma_S = -0.818$ and $\Gamma_L = 0.942$. The line length corresponds to a delay, T, of 10 ns. The voltage expression for this case is

$$V(0, s) = 1.76 \times 10^8 \frac{e^{-10^{-8}s}}{s^2}\left(\frac{1}{1 + 0.77e^{-2\times 10^{-8}s}}\right)$$

This particular transform does not appear explicitly in Appendix C. Since the expressions for $V(z, s)$ and $I(z, s)$ invariably involve denominator exponential terms, it is worth the time to develop a procedure for obtaining the inverse Laplace transform. The numerator exponential is merely a time shift (see Transform 13 in the appendix) so it can be suppressed and then accounted for in the inverse transform of the remainder by simply replacing t by $t - 10^{-8}$. Let's examine the form in general terms as

$$F(s) = \frac{1}{s^2(1 + ae^{-2Ts})} = \left(\frac{1}{s^2}\right)\left(\frac{1}{1 + ae^{-2Ts}}\right)$$

Expanding the binomial term in an infinite series, we have, for $a \le 1$ and $T > 0$ so that convergence is assured,

$$\begin{aligned} F(s) &= \frac{1}{s^2}(1 - ae^{-2Ts} + a^2e^{-4Ts} - a^3e^{-6Ts} + a^4e^{-8Ts} - \cdots) \\ &= \frac{1}{s^2} - a\,\frac{e^{-2Ts}}{s^2} + a^2\,\frac{e^{-4Ts}}{s^2} - a^3\,\frac{e^{-6Ts}}{s^2} + a^4\,\frac{e^{-8Ts}}{s^2} - \cdots \end{aligned}$$

Taking the inverse transform:

$$\begin{aligned} f(t) = {} & tu(t) - a(t - 2T)u(t - 2T) + a^2(t - 4T)u(t - 4T) \\ & - a^3(t - 6T)u(t - 6T) + a^4(t - 8T)u(t - 8T) - \cdots \end{aligned}$$

For the general interval $2nT < t < 2(n + 1)T$ there would be a finite number of terms since all step functions after that range of t are zero. We then write

$$\begin{aligned} f_n(t) &= t - a(t - 2T) + a^2(t - 4T) - a^3(t - 6T) + \cdots \\ &\quad + (-1)^n a^n(t - 2nT), \qquad n = 0, 1, 2, \ldots \qquad \text{for} \quad 2nT \le t \le 2(n + 1)T \\ &= t[1 - a + a^2 - a^3 + \cdots + (-1)^n a^n] \\ &\quad + 2aT[0 + 1 - 2a + 3a^2 - \cdots + (-1)^{n+1} na^{n-1}] \end{aligned}$$

The first term in brackets is a finite geometric series with common ratio $-a$, first term 1, and last term $(-1)^n a^n$. (The second term in brackets increases without bound as we would expect since the input increases without bound also.) We finally obtain

$$f_n(t) = \frac{1 - (-1)^n a^n}{1 + a}\, t + 2aT[0 + 1 - 2a + 3a^2 - \cdots + (-1)^n(n - 1)a^{n-2}]$$

$$\text{for} \quad 2(n - 1)T \le t \le 2nT \qquad \text{with} \quad n = 1, 2, 3, \ldots$$

Note that the index n now starts at 1 rather than 0 since that is required by geometric series closed-form sum. Of course, the index n in the right-bracket term had to be changed to allow starting at $n = 1$, also.

The inverse transform of our particular $V(0, s)$ is then, putting in the time shift,

$$\begin{aligned} v(0, t) = 1.76 \times 10^8 \Bigg\{ & \frac{1 - (-1)^n(0.77)^n}{1.77}(t - 10^{-8}) + 1.54 \times 10^{-8}[0 + 1 - 2 \\ & \times 0.77 + 3 \times 0.77^2 - \cdots + (-1)^n(n - 1)(0.77)^{(n-2)}]\Bigg\} \\ & \text{for} \quad 2(n - 1) \times 10^{-8} \le t - 10^{-8} \le 2n \times 10^{-8} \\ & [\text{or } 2n \times 10^{-8} \le t \le (2n + 1) \times 10^{-8}] \quad \text{for} \quad n = 1, 2, 3, \ldots \end{aligned}$$

A plot of this result is given in Figure 3-54. The shifted source voltage is shown dotted for comparison.

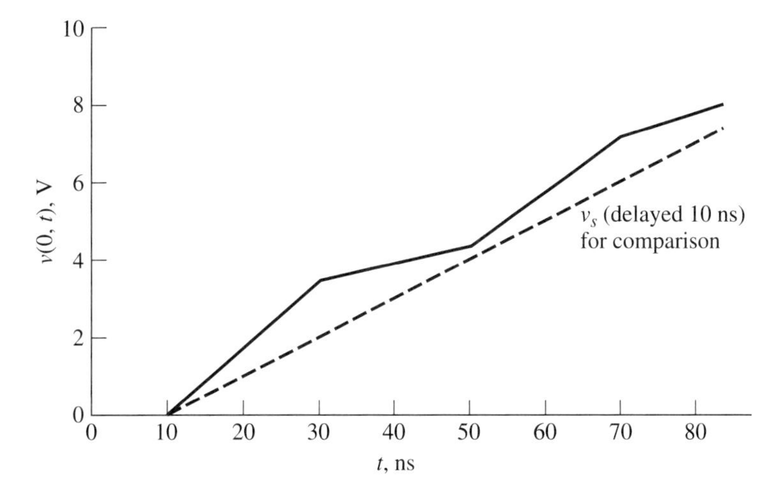

Figure 3-54. Ramp response of a resistively loaded transmission line.

Example 3-5

Once the ramp function response is available and understood, it is a simple matter to compute the line response to step or pulse functions with nonzero rise and fall times. We use the superposition theorem for linear systems. This will be illustrated by the final example, the source waveform for which is shown in Figure 3-55, which is a step function that has nonzero rise time. The desired source voltage is generated by adding two ramp functions.

For simplicity we will assume the same system and parameters as given in Figure 3-53, with $R_S = 14\ \Omega$, $Z_0 = 150\ \Omega$, $R_L = 5000\ \Omega$, and $M = 10^8$ V/s. The response is obtained by adding the responses due to each ramp individually. The negative ramp response is obtained by taking the positive ramp response and negating it and then shifting it by 50

Figure 3-55. Generation of a nonzero rise time step from two ramp functions.

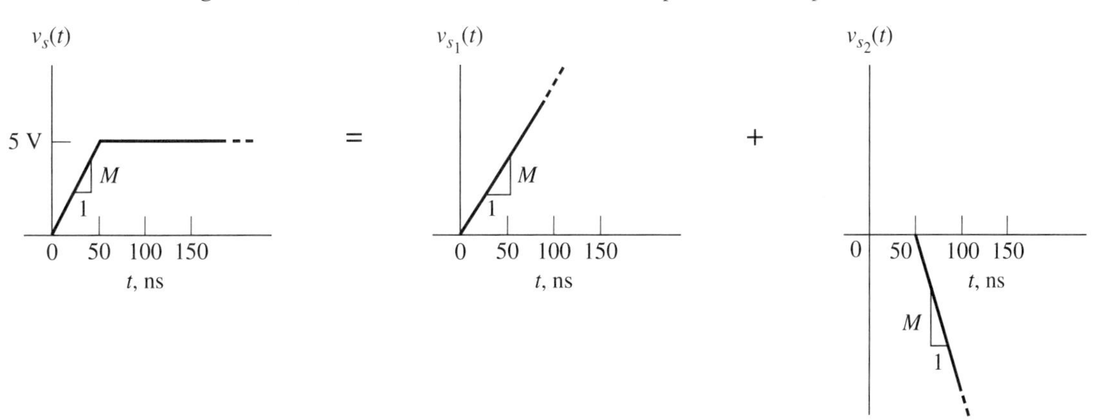

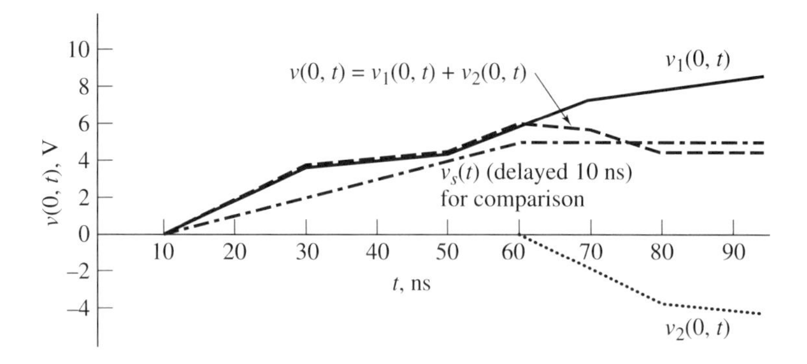

Figure 3-56. Finite rise time step response obtained by adding two ramp responses.

ns. These steps are shown in Figure 3-56. The two dashed curves are the individual ramp responses and solid curve is the actual input signal response. For comparison, the applied step with nonzero rise time is shown as the circle-marked curve.

Exercises

3-7.3h For the single ramp example, use the expression for $i(0, t)$ derived and write the specific expressions that apply to each section of the waveform. Answer:

$$\begin{pmatrix} 0, & 0 \le t \le 10^{-8} \\ 1.76 \times 10^8(t - 10^{-8}), & 10^{-8} \le t \le 3 \times 10^{-8} \\ 4.05 \times 10^8(t - 10^{-8}) + 2.71, & 3 \times 10^{-8} \le t \le 5 \times 10^{-8} \\ 1.45 \times 10^8(t - 10^{-8}) - 1.46, & 5 \times 10^{-8} \le t \le 7 \times 10^{-8} \\ 6.45 \times 10^8(t - 10^{-8}) + 3.36, & 7 \times 10^{-8} \le t \le 9 \times 10^{-8} \end{pmatrix}$$

3-7.3i To determine the effect of each parameter on the output—for the single ramp—compute the outputs corresponding to the change of only one variable to the values indicated (use values assumed in the example as base values):
(a) $R_S = 5\ \Omega$, $R_S = 30\ \Omega$
(b) $Z_0 = 75\ \Omega$, $Z_0 = 300\ \Omega$
(c) $M = 0.5 \times 10^8$ V/s, $M = 2 \times 10^8$ V/s
(d) $R_L = 2.2$ kΩ, $R_L = 10$ kΩ

Figure 3-57.

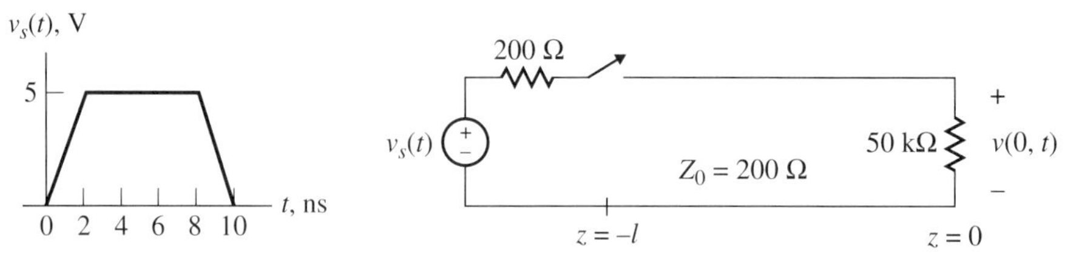

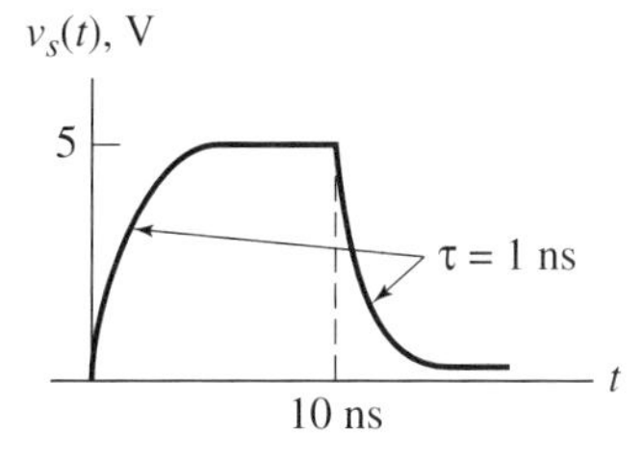

Figure 3-58.

3-7.3j Compute the load voltage for the system shown in Figure 3-57. (Hint: Use superposition. Also, the input source must be zero for $t > 10$ ns.)

3-7.3k In the preceding examples, the rise and fall portions of the waveforms were approximated by straight lines. Since those edges are in fact exponentials, a better approximation for a rise portion would be of the form $A(1 - e^{-t/\tau})$. For the system given in Figure 3-53, change the source to the function

$$5(1 - e^{-t/1\text{ns}}) \qquad \text{and plot} \quad v(0, t)$$

For system component values use those given in Example 3-4. (Hint: When you get to $V(0, s)$ there will be a term $(s + a)$ in one denominator. To get a more tractable form, change the Laplace variable by using $s' = s + a$.)

3-7.3l Using the results of Exercise 3-7.3k, determine the response of the system of Figure 3-53 to the waveform in Figure 3-58. Use system parameter values given in Example 3-4. (Hint: Use the superposition theorem. Be sure your functions cut off appropriately when you obtain the equation for $v_s(t)$.)

Summary, Objectives, and General Exercises

Summary

In this chapter we have generalized the solutions of the transmission line equations to include nonsinusoidal driving functions. Once the general properties of these solutions were derived, the responses of lines to step functions were investigated using lossless lines and resistive terminations. This special class of problems made possible the demonstration of physical insights that are of value in determining the responses when loads are complex. The concept of reflection coefficient was shown in those special cases to have the same mathematical form and interpretation as for the sinusoidal steady state. Discharge characteristics of lines were shown to form an important subclass of pulsed line problems that arise in level logic digital systems and DC power systems.

The current and voltage responses for reactively terminated lines driven by sources that are matched to the characteristic impedance of the line were next investigated. It was shown that, although some heuristic ideas were invoked, the responses were easily obtained using basic reflection coefficient and time constant ideas.

For unmatched driving sources and for waveforms other than zero rise time pulses or steps, the Laplace transform provides straightforward, although algebraically cumbersome, solutions. The important cases of finite rise time pulses and steps of importance in digital systems were presented.

Objectives

1. Be able to identify functions of the d'Alembert form that satisfy the lossless transmission line equations and show by substitution they are solutions.
2. Be able to compute the phase velocity of a d'Alembert solution.
3. Be able to plot time and distance graphs for step functions applied to infinite length lossless lines for both current and voltage at any point on the line.
4. Be able to solve for the transient response for a lossless transmission line:
   (a) Step function tracing of reflections (including pulses as superposition of step responses).
   (b) Charging and discharging of transmission lines.
   (c) Be able to sketch, qualitatively, the response to step functions (driving source matched to line) of RL and RC terminations.
   (d) Use the Laplace transform solution method for general sources and loads.

General Exercises

GE3-1 In the circuit of Figure 3-59 the switch is closed at $t = 0$, the transmission line having initially been charged as shown. Make plots of load current and voltage, and current and voltage on the right switch terminal.

GE3-2 Plot the voltage across the load resistance as a function of time for the circuit of Figure 3-60. Assume the diode is ideal.

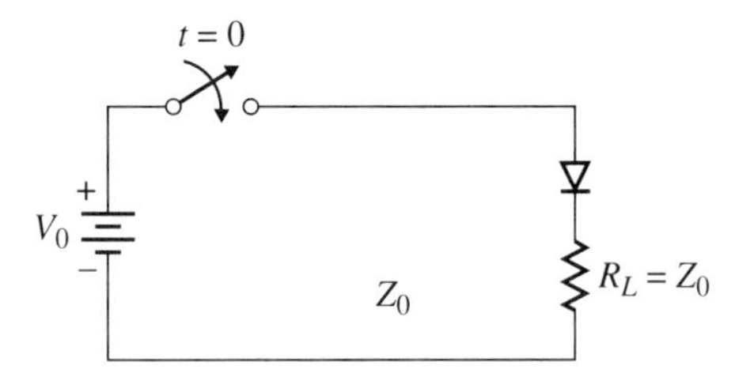

Figure 3-60.

GE3-3 Repeat GE3-2 for $R_L = 2Z_0$.

GE3-4 Repeat GE3-2 with the diode reversed.

GE3-5 Repeat GE3-2 for $R_L = 2Z_0$ and the diode reversed.

GE3-6 In Figure 3-60, suppose we replace the source V_0 by a resistor of value Z_0 and charge the line to +5 V. Plot the load voltage for the diode as shown and repeat for the diode reversed. Diode is ideal.

GE3-7 In Figure 3-60, add a series source resistance of value Z_0 and plot load and input line voltages and currents. Repeat for the diode reversed. Assume the diode is ideal.

GE3-8 Repeat Exercise GE3-7 for the cases where the source resistance is $\frac{1}{5}Z_0$ and $5Z_0$.

GE3-9 Without using the Laplace transform (see Section 3-6), obtain time plots of the inductor voltage and current for the system shown in Figure 3-61. $i_L(0, 0) = 0$. R_1 is directly across R_2 and L so there is zero transmission-line distance between them.

GE3-10 Repeat Exercise GE3-9 where the inductor is replaced by a capacitor.

GE3-11 Solve Exercise GE3-9 using the Laplace transform method.

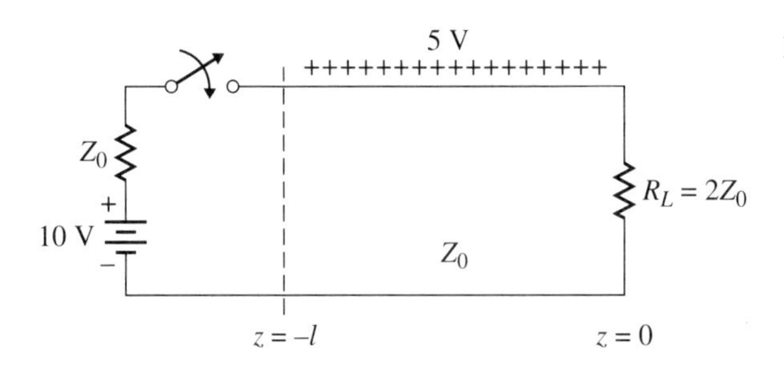

Figure 3-59.

GE3-12 Solve Exercise GE3-10 using the Laplace transform method.

GE3-13 A rectangular pulse may be thought of as the superposition of two step functions. The system in Figure 3-62 is in the steady state for $t < 0$, and we assume an ideal switching transistor. Sketch the waveforms at the load and at the emitter, both current and voltage. Assume the cable length is $\frac{1}{3}$ m and lossless. Next, remove the cable and 220-Ω resistance and compute the emitter current and voltage waveforms and compare with the previous results.

GE3-14 Since our systems have been assumed linear, we may use the superposition theorem and obtain the output of the system of Exercise GE3-13 for multiple input pulses. For the transistor base waveform of Figure 3-63, find the plots of emitter and load voltages.

GE3-15 As digital logic speeds increase, the frequency components of the pulses have wavelengths that are of the same order as the lengths of the logic interconnecting lines. Because of this, the circuit lines on printed circuit boards must be treated as transmission lines to account for delay and reflection problems. One obvious solution is to terminate the lines in the characteristic impedance. Two problems make this impractical. First, there are so many lines that the number of resistors would be prohibitive and second, the small values of Z_0 (around 100 Ω) would cause excessive wasteful current drain on the power supplies. Figure 3-64 shows an alternative way of using diodes to terminate a line, the so-called bilateral clamp. Assuming ideal diodes, compute the voltages at both ends of the *line* and compare with the voltages that would be present without the diodes. (These diodes can be an integral part of the transistor ICs.) At the input the two transistors are switched to provide a 200-ns pulse. The on transistor has a resistance

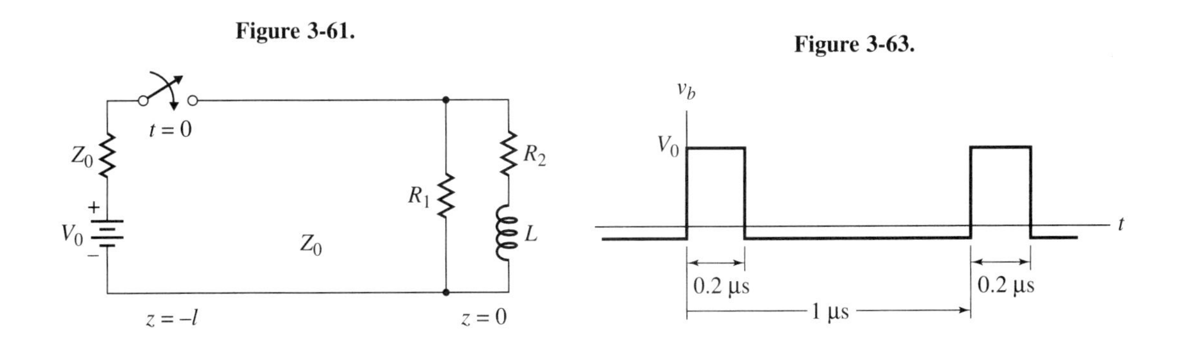

Figure 3-61.

Figure 3-63.

Figure 3-62.

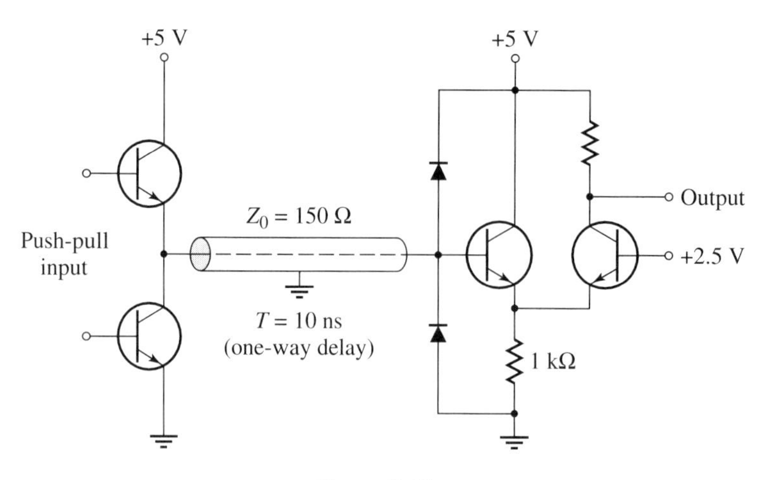

Figure 3-64.

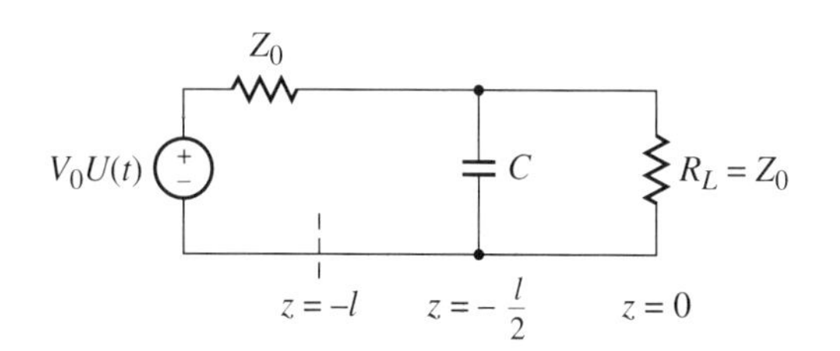

Figure 3-65.

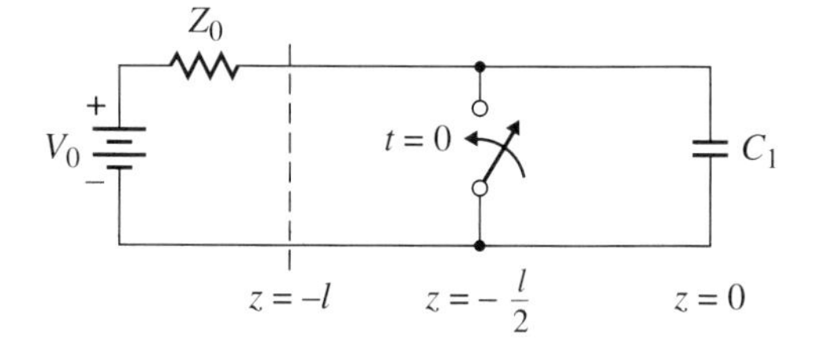

Figure 3-67.

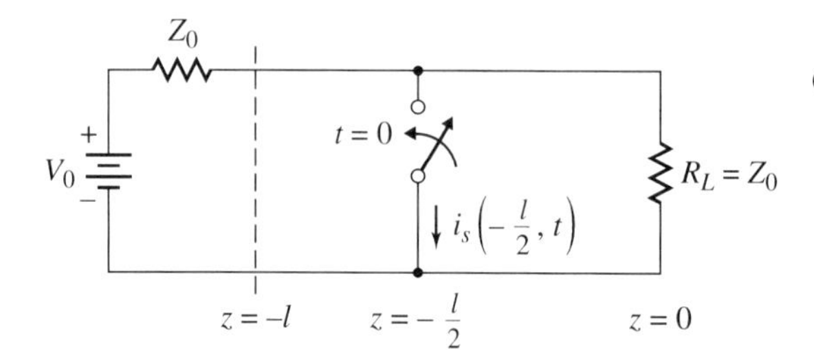

Figure 3-66.

at the emitter of 10 Ω. All transistors have $\beta = 150$. Repeat for a diode on voltage of 0.7 V.

GE3-16 Develop a Laplace transform solution procedure for the lossless line voltage wave equation. What initial conditions are needed?

GE3-17 A unit step-function voltage generator with internal impedance Z_0 is connected at $z = -l$ to a lossless line as shown in Figure 3-65. Find the expression for and plot $v(-l, t)$, $v_c(-l/2, t)$, $v_L(0, t)$. This situation could arise by placing a logic gate on a matched line, where C represents the gate input capacitance.

GE3-18 The circuit shown in Figure 3-66 is in the steady state for $t < 0$. At $t = 0$ the switch is closed. Obtain plots for $v(-l, t)$, $i_s(-l/2, t)$, and $v(0, t)$.

GE3-19 The system shown in Figure 3-67 is in the steady state for $t < 0$. At $t = 0$ the switch is closed. Obtain plots for $i(-l, t)$, $v(0,t)$.

4 Scalars, Vectors, Coordinate Systems, Vector Operations, and Functions

4-1 Introduction

This chapter is intended primarily as a review of coordinate system principles and the vector operations and computations used in field theory. The discussions are in sufficient detail, however, so that modest additional careful study should make the material accessible to those who have not had a formal mathematics course in vector analysis. Emphasis is given to the physical interpretations and spatial visualizations of the vector operations. Some results of importance from calculus are also given along with a short presentation of the use of *discontinuity* functions, specifically the step and impulse functions and their derivatives. These latter functions are of considerable utility for obtaining equations of physical systems that have abrupt value changes in their parameters as the independent variables cover given ranges of values. At this point time is not used extensively so one may concentrate on the properties of the coordinate systems and the vector operations. Since sinusoidal steady-state systems dominate initial exposure and applications, we use the steady state directly. (Those needing a review of steady-state concepts should use Appendix A.) Enough examples are given with time functions so that the more general applications are evident. Also, proofs are not given for all of the identities used in vector analysis and applications, but only those of more immediate interest. Most proofs are left as exercises, so that all identities appear either there or in the body of the text.

4-2 Types of Physical Quantities: Scalars and Vectors

The simplest physical quantity is the scalar. As a formal definition:

Definition. A *scalar* is a quantity characterized by a magnitude and phase at a point in space, where the *values* of magnitude and phase *at the given point* are independent of any *spatial* coordinate system.

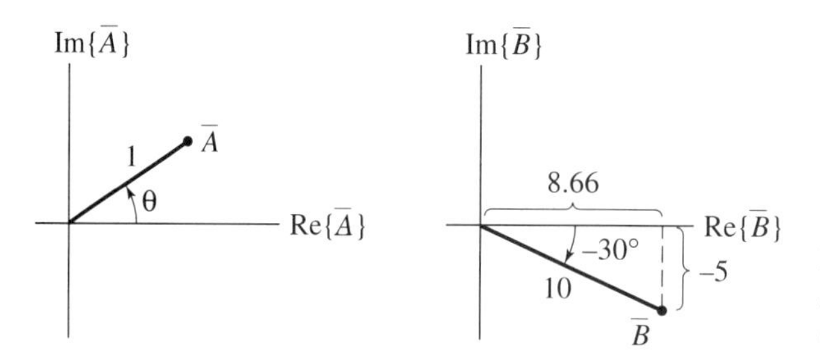

Figure 4-1. Representation of complex scalars on an argand (phasor) diagram.

The phase comes from the fact that we are considering sinusoidal steady-state time variations. In the terminology of circuit theory the scalar would thus be a *phasor*. Of course, for general time functions phase would have no meaning, so one would just have a magnitude that may vary with time. Examples of *complex scalars* (phasors), which we identify by an overbar, are

$$\bar{A} = e^{j\theta} = \cos\theta + j\sin\theta$$

$$\bar{B} = 10e^{j30^\circ} = 10\cos 30^\circ - j\sin 30^\circ = 8.66 - j5$$

Pictorially, these scalars may be drawn on what is called an *Argand diagram*, or, for the sinusoidal steady state, a *phasor diagram*. These complex scalars are shown in Figure 4-1. The operators Im{·} and Re{·} refer to the imaginary part and real part of the quantity within the braces, when the quantity is written in rectangular form.

Although some axes are used, they are *not* spatial or coordinate axes. The axes simply provide a way of representing what are called the *real* and *imaginary* components of the scalar. Note also that the value of a scalar may be different at different points in space, but at a given point the value is independent of the spatial coordinate system. The Argand diagram does not tell us the physical location of either $\bar{A}$ or $\bar{B}$, but such a location could be written on the diagram if it should need to be so identified. Another type of scalar is the *real scalar*, which is actually a subclass consisting of complex scalars with the angle (see $\bar{A}$ above) either 0° or 180° and general time functions. We write, for example,

$$C = -6 = 6\angle 180^\circ = 6e^{j\pi}$$

$$D = 4\cos 6t$$

$$b = 10 - 10\angle 0^\circ = 10e^{j0^\circ}$$

$$g = -9t$$

Note that overbars are not used on real scalars, and small letters are often used. As with complex scalars, the value of a real scalar may be different at two different points in some region of space, but at a *given* point the *value* is independent of the spatial coordinate system used to locate the point.

Example 4-1

As a first example, consider two parallel metal plates with a 6-V battery connected between them as shown in Figure 4-2. Here we have superimposed two different coordinate systems,

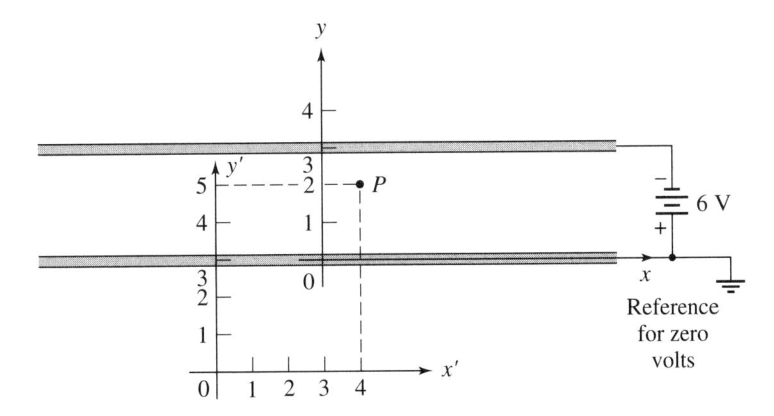

Figure 4-2. Example of a region that has a scalar value description.

a primed one and an umprimed one. Let V be the voltage at any point between the two plates. The equations for the voltage at a general point P between the plates are:

Primed system:

$$V(x', y') = -2(y' - 3)$$

Unprimed system:

$$V(x, y) = -2y$$

In this case V is a scalar (real) since at the same physical point P we obtain

Primed system: $V(P) = V(4, 5) = (-2)(5 - 3) = -4$ V

Unprimed system: $V(P) = V(1, 2) = (-2)(2) = -4$ V

Although the *equations* depend upon the coordinate systems, the *value* of voltage is not dependent upon the system.

Example 4-2

Another example is the density of a material in a box. The density may vary from point to point within the box. For one location in the material the density has one value independent of the coordinate system used to locate the point within the box.

Example 4-3

An example of a complex scalar comes from the study of transmission lines in Chapter 1. The voltage at any line location is given by Eq. 1-42, for the lossless line, as

$$\overline{V}(z) = V^+e^{-j\beta z} + V^-e^{j\beta z} \tag{4-1}$$

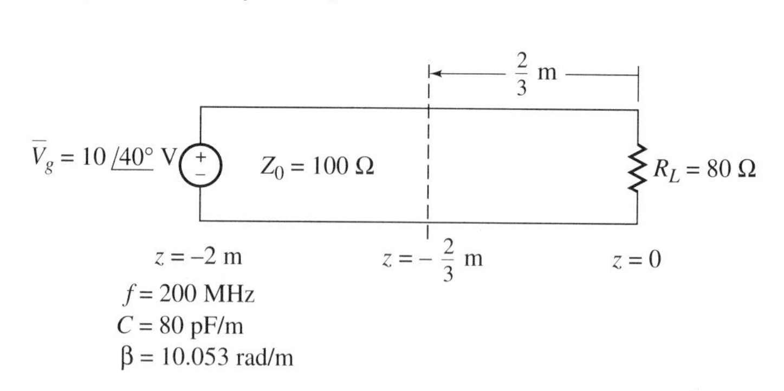

Figure 4-3. Transmission line example of a complex scalar that has spatial variation.

Suppose we draw a particular system configuration as given in Figure 4-3, with the parameters indicated. The corresponding current is

$$\overline{I}(z) = \frac{V^+}{Z_0} e^{-j\beta z} - \frac{V^-}{Z_0} e^{j\beta z} \tag{4-2}$$

Note that although the *equation* for $\overline{V}(z)$ (and $\overline{I}(z)$ also) depends upon a spatial coordinate z, the *value* of $\overline{V}$ at a given point on the line is independent of any particular coordinate system since we may select $z = 0$ at any place we wish. For example, if we select the load location to be $z = 0$, the voltage equation is

$$\overline{V}(z) = 9.16e^{-j35.4^\circ}e^{-j10.053z} - 1.02e^{-j35.4^\circ}e^{j10.053z}$$

However, if we select a coordinate system with $z' = 0$ at the generator the result is

$$\overline{V}(z') = 9.16e^{j36.6^\circ}e^{-j10.053z'} + 1.02e^{j72.4^\circ}e^{j10.053z'}$$

Evaluating these at point a which is $\frac{2}{3}$ m from the load and $\frac{4}{3}$ m from the source yields

$$\overline{V}(z = -\tfrac{2}{3}\text{ m}) = 9.16e^{-j35.4^\circ}e^{j6.702} - 1.02e^{-j35.4^\circ}e^{-j6.702} = 8.51\angle{-6.3^\circ}\text{ V}$$

$$\overline{V}(z' = \tfrac{4}{3}\text{ m}) = 9.16e^{j36.6^\circ}e^{-j13.404} + 1.02e^{j72.4^\circ}e^{j13.404} = 8.51\angle{-6.3^\circ}\text{ V}$$

(The exponents with no degree marks are in radian measure.) Thus even though the *equations* for $\overline{V}$ are dependent upon the coordinate system the *value* of the voltage at a *given* location is independent of the coordinate system. The value of the voltage will be different in general at another point in the region along a line.

Exercises

4-2.0a Devise a region where the temperature varies with location. Temperature is a scalar since the value at a given point is independent of any coordinate system. For your region, draw two coordinate systems with nonparallel axes and give the equations for temperature in the region with respect to both coordinate systems.

4-2.0b For the complex scalar example (Example 4-3), obtain the two corresponding equations for the current along the line and show that current is a complex scalar. When you evaluate the expressions, watch the degrees and radians carefully. (Partial answer: $\bar{I}(z) = 0.0916e^{-j35.4^\circ}e^{-j10.053z} + 0.0102e^{-j35.4^\circ}e^{j10.053z}$)

The next physical quantity to be described is a vector, defined as follows:

Definition. A *vector* is a quantity characterized by a *scalar* and a direction in *space*.

Notice that to specify direction in space some spatial reference is required. Thus the value of a vector depends upon the coordinate system for specification. Also note that the scalar part may be complex as defined earlier. The essential difference between a scalar and a vector is readily given. If one is told that the temperature at a point is 78° C, this is all the information necessary. However, if one is told that the force at a point is 5 N at an angle (direction) of 60°, one needs to inquire as to the spatial axis with respect to which the direction is measured. Vectors are identified by an arrow over the symbols to distinguish them from scalars. As an example, consider a constant magnitude force of 5 N, shown in Figure 4-4 with respect to two different coordinate systems. With respect to the unprimed system the force vector is

$$\vec{F} = 5\angle 45^\circ\text{N}$$

but with respect to the primed system, the force is

$$\vec{F}' = 5\angle 0^\circ\text{N}$$

Notice that the angle designation here is quite different from the angles used to describe complex scalars (phasors), even though they are symbolically (visually) identical. The arrow over the symbol is required to properly identify the vector. Note that if the scalar part of the vector (the 5 in the example above) were complex, we have the awkward situation of having two angles in the vector expression, one angle being a sinusoidal phase and the other a spatial orientation. This will be resolved fully shortly. For the more general case where the scalar part is a time function we would have forms like

$$\vec{F} = 5e^{-2t}\angle 45^\circ$$
$$\vec{F}' = 5\cos(2t - 30^\circ)\angle 45^\circ$$

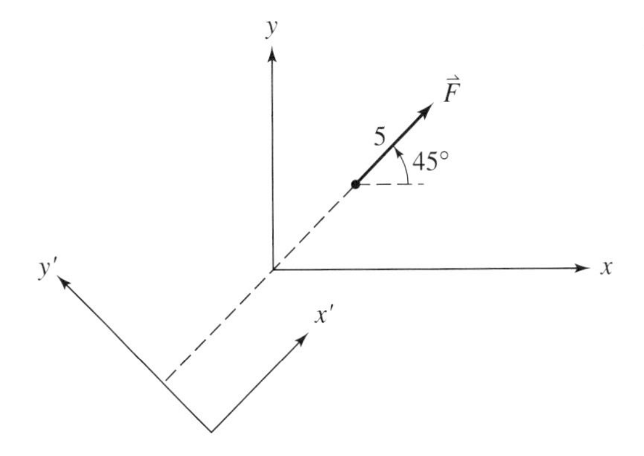

Figure 4-4. A force vector shown with respect to two coordinate systems.

This latter vector could be written in the sinusoidal steady state in the difficult (for now) form

$$\vec{F} = (5\angle -30°)\angle 45°$$

Other rather obvious examples of vector quantities are acceleration, velocity, and the electric field around a charge as described in physics texts.

Exercises

4-2.0c As with scalars, vectors may also vary from point to point in a region. Both the scalar part and the direction vary in general. Given an example of such a vector in a region and describe the system using two different coordinate systems. Show that the scalar part of your vectors do not depend upon the coordinate systems but the direction angles do.

4-2.0d Suppose that the force vector described in Figure 4-4 had a sinusoidally varying scalar part and were given as $5\angle 20°$. The complete vector representation would then be $\vec{F} = (5\angle 20°)\angle 45°\text{N}$ in the unprimed system. Discuss the admissibility of simplifying this to $\vec{F} = 5\angle 65°\text{N}$.

4-3 Scalar and Vector Fields

If a region of space is selected and at each point some quantity is defined (may be zero at some points) the quantity is said to form a *field* within the region. For the case where the quantity at each point of the region is a scalar the field is called a *scalar field*. When the quantity at each point in the region is a vector the field is called a *vector field*. In electromagnetic theory, vectors are so prevalent that we often use the term *field* alone with the vector nature being understood. Also, vectors may be further described as *free* or *fixed* vectors, but we do not need this distinction so no definitions are given. Some examples of fields are shown in Figure 4-5. The left portion of the figure is a region containing electric charge. The charged region may be described as having a certain charge density at each point in the region, denoted by $\rho(x, y, z)$, in coulomb/cubic meter. This is a scalar quantity

Figure 4-5. Examples of scalar and vector fields. (*a*) Charge density. (*b*) Force on charges.

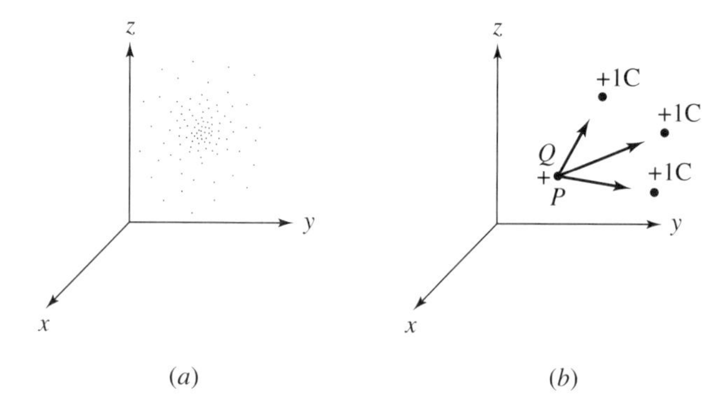

(since the value at a given point is independent of the coordinate system used to locate that point) that varies from point within the region. A scalar field results.

Figure 4-5*b* shows a point charge Q located at P and other 1-C charges in the neighborhood. From physics we know that the charge at P will exert a force on each of the other charges. The force will be different at each location of a 1-C charge, and at a given point, the force value will depend upon the coordinate system for direction specification. Thus the region around the point charge would be a *vector* force field.

Exercises

4-3.0a State whether the following systems describe scalar or vector fields, and substantiate your selections with diagrams and discussions.

(a) A can is filled with soil and tamped down from the top. Describe the density of the soil.

(b) A tub is filled with water and the pressure varies with depth. Describe the pressure.

(c) Magnetic north and south poles are brought in close proximity. Describe the region between the poles.

(d) As water drains down the sink a vortex is formed. Describe the motion of the water around the drain.

4-3.0b Suppose that in Figure 4-5*b* a charge on Q_1 is placed in the region. How would you obtain the total force on Q_1? (Hint: Consider the superposition theorem and the linearity problem.)

4-4 Addition of Vectors

This concept is probably known from your work in physics, but it is given here for completeness. Suppose we select a point in space and measure the values of two vectors $\vec{A}$ and $\vec{B}$ there. The vectors, of course, must have the same units just as with scalars. We cannot add force to angular momentum, for example. By convention we use arrows to represent a vector and the terminology is described in Figure 4-6. The vector is usually assumed to describe the value at the location of the initial end of the vector, labeled P.

The sum of two vectors $\vec{A}$ and $\vec{B}$ is defined as the vector $\vec{C}$ obtained by placing the initial end of vector $\vec{B}$ at the terminal end of vector $\vec{A}$ and drawing a line from the initial end of $\vec{A}$ to the terminal end of $\vec{B}$. Note that this is the same as completing the parallelogram described by the two vectors $\vec{A}$ and $\vec{B}$ both having initial ends at P with the sum vector $\vec{C}$ the diagonal that has the initial point also at P. These combinations are shown in Figure 4-7. These methods, of course, are graphical. Later, vector addition will be described from a computational viewpoint.

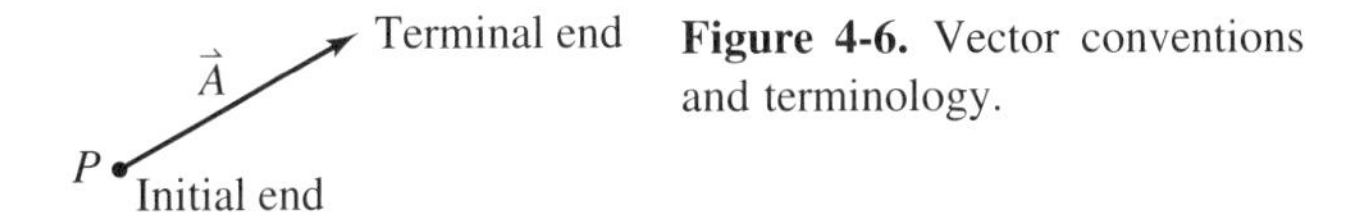

Figure 4-6. Vector conventions and terminology.

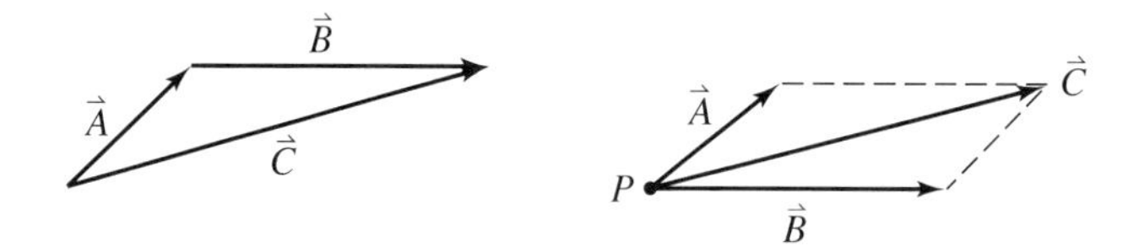

Figure 4-7. Two methods for graphical addition of two vectors.

Example 4-4

Given vectors $\vec{A}$ and $\vec{B}$ measured with respect to the x axis and the vectors are assumed measure at point P. See Figure 4-8. Physically, we could imagine the vectors as being generated by two different sources and we wish to find the combined effect as a vector $\vec{C}$. The first step is to decide upon an appropriate scale. Here we select 1 unit to be 5 mm. Beginning at point P we then draw $\vec{A}$ to scale (including the angle). Draw $\vec{B}$ to scale beginning at the terminal point of $\vec{A}$. Connect the initial point of $\vec{A}$ with the terminal point of $\vec{B}$. Scale this vector as $\vec{C}$ from the diagram. Note that $\vec{C}$ is a vector since it depends upon the coordinate axes for the direction angle. The scale solution is

$$\vec{C} = 18\angle 15°$$

Figure 4-8. Graphical solution for vector addition. (*a*) Problem statement. (*b*) Solutions.

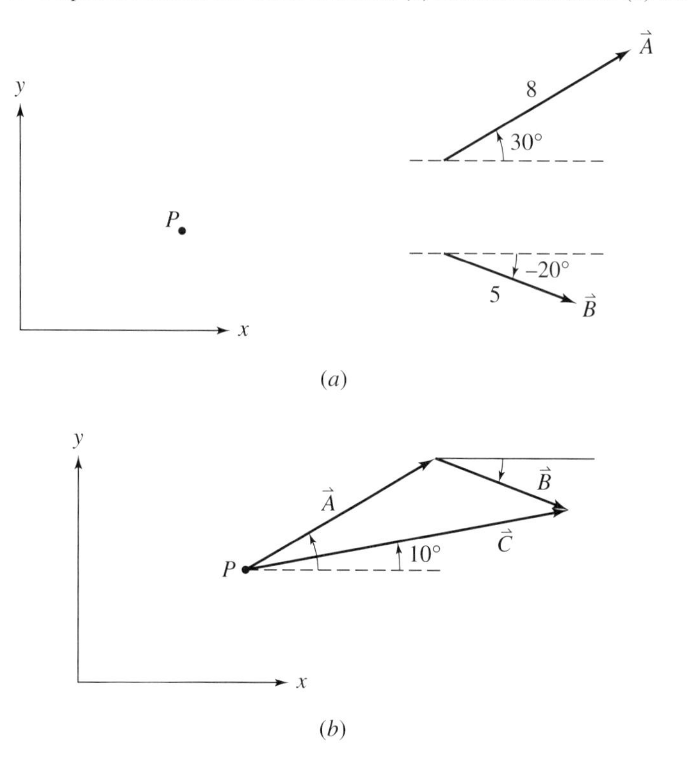

Exercises

4-4.0a Using a scale of 1 unit = 1 cm, obtain the vector sum of the following:
(a) $\vec{A} = 6\angle 120°$, $\vec{B} = 4\angle -90°$.
(b) $\vec{A} = 5\angle 135°$, $\vec{B} = 8\angle -135°$, $\vec{C} = 3\angle 0°$.
(c) $\vec{A} = 2\angle 60°$, $\vec{B} = -7$.

4-4.0b Use a graphical proof to prove the vector associative and commutative laws; that is,
(a) $(\vec{A} + \vec{B}) + \vec{C} = \vec{A} + (\vec{B} + \vec{C})$
(b) $\vec{A} + \vec{B} + \vec{C} = \vec{A} + \vec{C} + \vec{B}$

4-4.0c A vector $\vec{D} = 20\angle 45°$ is added to a vector $\vec{E}$ to give the sum vector $\vec{F} = 30\angle -30°$. What is the vector $\vec{E}$?

4-4.0d Given three vectors $\vec{A}$, $\vec{B}$, and $\vec{C}$, show how you could determine, graphically, whether or not the sum of any two of them gives the third. Is the solution unique?

4-4.0e Given only the magnitudes (lengths) of three vectors $\vec{A}$, $\vec{B}$, and $\vec{C}$, show how you could determine whether or not the sum of any two of them gives the third. Is the solution unique? (Answer: Use a compass and determine whether the three lengths form a closed triangle. The solution is not unique.)

4-5 Multiplication of a Vector by a Scalar

The multiplication of a vector $\vec{B}$ by a scalar $\overline{A}$ results in a vector $\vec{C}$ whose scalar part is $\overline{A}$ times the scalar part of $\vec{B}$ and whose direction is the same as $\vec{B}$. As an equation we denote this by

$$\vec{C} = \overline{A}\vec{B} \tag{4-3}$$

Multiplication by a real scalar is straightforward since the result is a simple increase in vector length if the scalar is positive and simple increase in vector length with a 180° change in vector direction if the scalar is negative. For example,

$$(6)(4\angle -160°) = 24\angle -160°$$

or

$$(-6)(4\angle -160°) = 24\angle -340° = 24\angle 20°$$

Note that in general we assume that $\overline{A}$ is complex or a time function. If the scalar part of $\vec{B}$ is also complex, the product of the scalars follows the usual rules of phasor algebra as described earlier.

Example 4-5

For the first example, consider the scalar $\overline{A} = 6e^{j65°}$ and a vector $\vec{B} = 2\angle -65°$. Right away we can see one problem with the angle form of representation. Unless the scalar and vector are clearly identified (in this case by overbar and arrow) one would have no way of knowing whether we had two scalars or, in fact, which was the vector. Shortly we will develop a method of avoiding this situation. To continue with the example, we would have the following product:

$$\vec{C} = \overline{A}\vec{B} = (6e^{j25°})(2\angle -65°) = (12e^{j25°})\angle -65°$$

Note that due to the awkward nature of the angle representation one might make the mistake of combining the angles to yield the *incorrect* vector

$$\vec{C} = 12\angle{-40°}$$

This would be wrong because the 25° is a phase of a sinusoid whereas the −65° is a spatial angle with respect to a coordinate axis. The difficulty is resolved in the next section.

Example 4-6

As a second example, and one which really shows the need for developing a better method for representing vectors, consider the scalar and vector given by

$$\overline{A} = 4e^{-j42°} \vec{B} = (9e^{j15°})\angle{-35°}$$

Using the definition of a scalar times a vector we have

$$\vec{C} = \overline{A}\vec{B} = (4e^{-j42°})(9e^{j15°})\angle{-35°} = (36e^{-j27°})\angle{-35°}$$

Note that the spatial direction of $\vec{B}$ is unchanged, but that the electrical phase has changed from −42° to −27°. Frequently the scalar is a real scalar b for which $\vec{C} = b\vec{B}$ represents (for positive b) a simple change in length of the vector $\vec{B}$. The direction is of course unchanged. The problem of negative values of b is handled either by letting the spatial direction change by 180° or letting the electrical phase change by 180°. As far as the physical result of a vector in space is concerned there is no difference, since phasors arise from sinusoidal steady state responses and changing the phase 180° can be accounted for by reversing the way a vector points in space.

4-6 Vector Decomposition: Coordinate System Representations and Unit Vectors

We next develop a system for representing vectors that eliminates the awkward double angle problem noted in the previous section. The process of addition gives us a two-dimensional example of the process of decomposition of a single vector into other vectors

Figure 4-9. Decomposition of a vector $\vec{C}$ into arbitrary components $\vec{A}$ and $\vec{B}$.

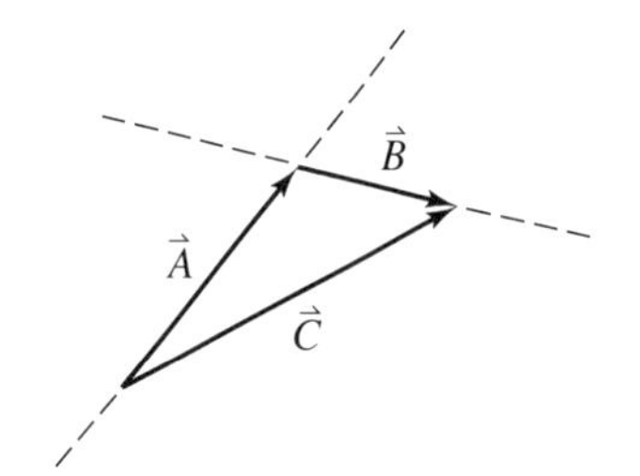

Figure 4-10. Decomposition of a vector $\vec{C}$ into orthogonal components $\vec{A}$ and $\vec{B}$.

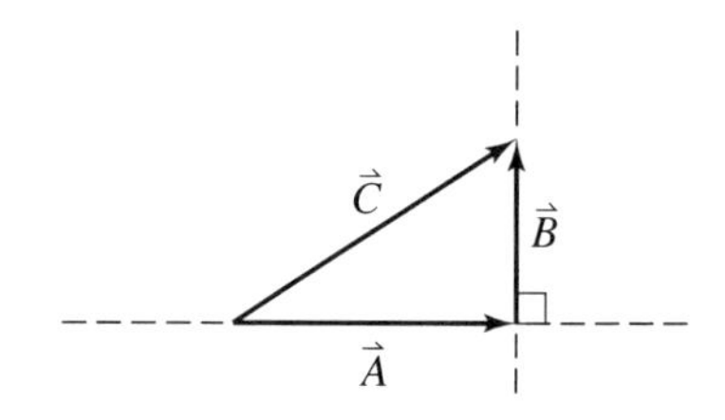

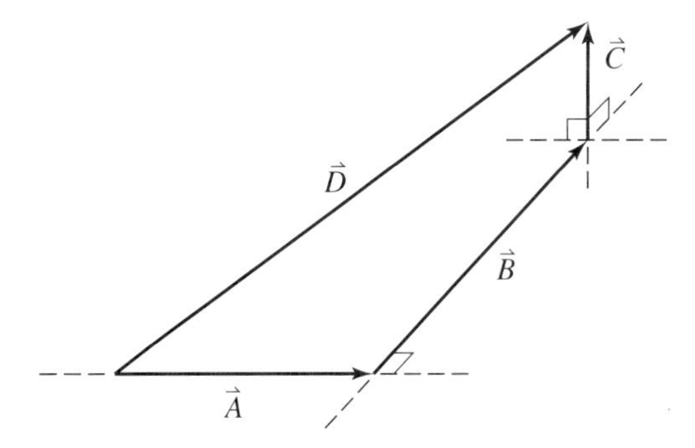

Figure 4-11. Decomposition of a vector $\vec{D}$ into three mutually orthogonal components. $\vec{A}$ and $\vec{B}$ define a plane to which $\vec{C}$ is normal.

whose sum is the original vector. Given a vector $\vec{C}$ we may, for example, decompose it into two vectors having arbitrary directions: Simply draw two lines, one each through initial and terminal ends of $\vec{C}$. The intersection of the two lines defines the two vectors that constitute the decomposition of the original vector. This is shown in Figure 4-9. Here $\vec{C}$ has been decomposed into $\vec{A}$ and $\vec{B}$. Note in particular, that the two lines may be selected to be orthogonal as shown in Figure 4-10. $\vec{A}$ and $\vec{B}$ are said to be *rectangular*, or *orthogonal*, components of the vector $\vec{C}$ in this case.

The extension of the decomposition to three dimensions follows easily. Figure 4-11 illustrates the three-dimensional decomposition of a vector $\vec{D}$ into the three components $\vec{A}$, $\vec{B}$, and $\vec{C}$, which are mutually perpendicular. As an equation we would have

$$\vec{D} = \vec{A} + \vec{B} + \vec{C}$$

With such a representation we can now take any vector in space and decompose it into orthogonal components, with the two-dimensional case as a special class. All we need to do is obtain the three orthogonal projections of the vector onto three mutually orthogonal lines called *axes*.

Exercises

4-6.0a Given four vectors $\vec{A}, \vec{B}, \vec{C}$, and $\vec{D}$ in a three-dimensional rectangular coordinate system. How can one determine whether or not three of them could possibly constitute the orthogonal vector decomposition of the fourth? (Hint: There is a famous theorem for right triangles.) Assume all scalars real.

4-6.0b A vector in a plane is decomposed into the two orthogonal components $6\angle 15°$ and $9\angle 105°$, where the angles are measured with respect to the positive horizontal. The initial ends of the original vector and the smaller orthogonal component are coincident. What is the original vector and what angle does it make with the smaller component? (Answer $10.82\angle 71.3°$, $56.3°$)

4-6.0c A two-dimensional orthogonal coordinate system is defined by two axes that intersect at right angles. One line (axis) is the X and the other, Y. A vector $\vec{E} = 45\angle 20°$ lies in the plane of the axes and passes through the origin of intersection. The angle is measured from the y axis. Give two possible vector decompositions which give $\vec{E}$. Suppose that the scalar part of $\vec{E}$ were $45e^{-j40°}$. Would this change either answer? (Partial answer: No)

4-6.0d A three-dimensional coordinate system is defined by three mutually orthogonal lines passing through a common point. A vector $\vec{E}$ having a scalar part of 60 makes angles of 20° and 80° with two of the lines. Find the possible decompositions of $\vec{E}$ into three components. (Hint: The sum of the squares of the direction cosines is unity.)

Stating a vector, for instance $\vec{A} = -10\angle 40°$, requires some given axis or coordinate system. Also, the use of the angle may cause some confusion (except for the arrow symbol) with usual phasor (scalar) quantities as described in the preceding sections. It is also desirable that in the expression of a vector the coordinate system be easily identifiable. This is done by decomposing the vector into components along the desired coordinate system axes and by specifying positive directions along the axes. For the majority of electromagnetic theory we are able to use a three-dimensional spatial decomposition. We furthermore specialize the three-dimensional coordinate systems to be orthogonal ones. The most common three-dimensional orthogonal system is the rectangular coordinate system. The three lines, or axes, are called the x, y, and z axes, respectively, as shown in Figure 4-12. The small half arrows on the axes tip represents the positive or increasing coordinate directions.

From the figure we may write the decomposition mathematically as

$$\vec{A} = \vec{A}_x + \vec{A}_y + \vec{A}_z$$

Another common way of drawing the diagram is to draw each of the *component* vectors as emanating from the common intersection of the three axes as illustrated in Figure 4-13.

If the original vector $\vec{A}$ did not describe the field at the common intersection of the

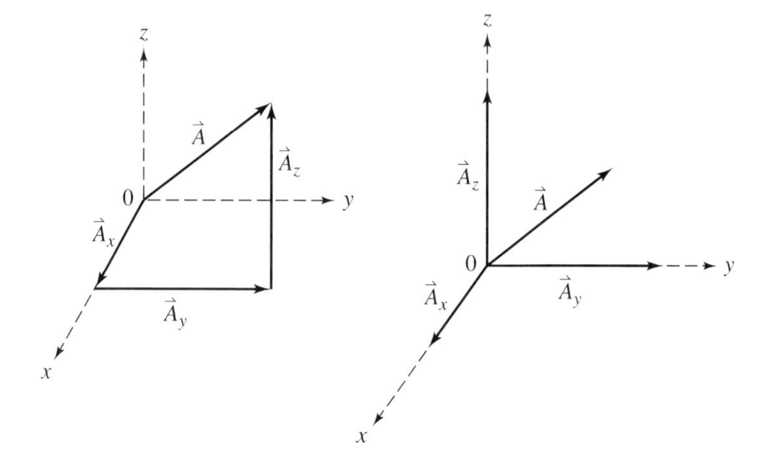

Figure 4-12. Definition of the orthogonal rectangular coordinate system.

Figure 4-13. Representation of an orthogonal decomposition as copunctual components.

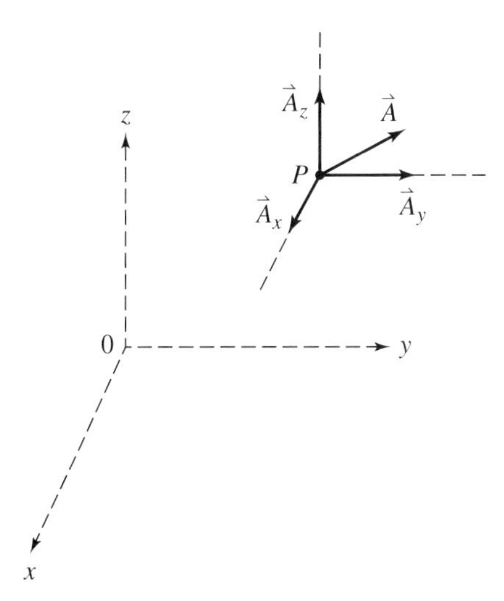

Figure 4-14. Vector decomposition components of a vector at a point other than the origin of coordinates.

axes we could draw axes *parallel* to the original ones at the point of interest. We would thus draw the system as shown in Figure 4-14.

Some 11 different orthogonal coordinate systems can be used to represent vector fields that are of importance in electromagnetic theory. Cylindrical and spherical are two of these. Examples will be given in detail later.

Exercises

4-6.0e A vector passes through the origin of a rectangular coordinate system and has a scalar part (magnitude) of 15 and makes angles of 30°, 80°, and 62° with the x, y, and z axes, respectively. Determine the orthogonal decomposition components and be sure to describe them completely.

4-6.0f Suppose that the vector $\vec{A}$ in Figure 4-13 has the x, y, and z components of lengths 3, 4, and 5, respectively. Now suppose that the coordinate system is rotated 20° about the z axis clockwise as viewed looking down along the z axis toward the xy plane, to produce x' and y' axes. What are the lengths of the three components in the new coordinate system? (Answer: 1.45, 4.78, 5)

Although our decomposition has helped in that the three vectors now contain information as to the coordinate system, it is still not a form that is best for detailed work. To obtain other forms we use the definition of a vector given earlier, which states that a vector can be expressed as a scalar and a spatial direction. Recalling that in general the scalar may be complex, we thus write the vector in what is called the *general three-component form* (rectangular in this case):

$$\vec{A} = \vec{A}_x + \vec{A}_y + \vec{A}_z = \overline{A}_x\hat{x} + \overline{A}_y\hat{y} + \overline{A}_z\hat{z} \tag{4-4}$$

In this last expression the quantities $\hat{x}$, $\hat{y}$, and $\hat{z}$ are defined as vectors directed parallel to the respective coordinate axis *in the direction of increasing coordinate value and of magnitude 1*. These are called *unit vectors* and are dimensionless. If the scalars are *real scalars*, we have the form

$$\vec{A} = A_x\hat{x} + A_y\hat{y} + A_z\hat{z} \tag{4-5}$$

For example: $\vec{A} = 4\hat{x} - 9\hat{y} + 5\hat{z}$. The -9 means a distance of 9 units measured in the negative y direction; that is, $\vec{A} = 3\hat{x} + 9(-\hat{y}) + 5\hat{z}$. The minus could also be interpreted as a 180° phase on the 9. The coefficients $\overline{A}_x$, $\overline{A}_y$, $\overline{A}_z$ (A_x, A_y, and A_z) are called *components* of the vector $\vec{A}$.

We are now in a position to give an accurate physical picture for the representation of vectors. Consider a vector given in its real time form as

$$\vec{C} = 6\cos(3 \times 10^8 t + 45°)\hat{x} + 10\cos(3 \times 10^8 t - 60°)\hat{y}$$

Using the phasor notation this could be written

$$\vec{C} = 6e^{j45°}\hat{x} + 10e^{-j60°}\hat{y}$$

Note that in field theory we use peak values rather than RMS as in circuit theory. This is a result of long-standing practice. If we draw each of the components in the xy plane of space it would appear as in Figure 4-15. The vector magnitude would be correct only if both components had the same phase, and then only at one particular instant of time when they were both maximum. Phase does not appear directly on the vector diagram. We also assumed a single frequency. Otherwise, *phase* has no meaning.

Note that as far as time and phase are concerned the diagram does not give any information. If we took a snapshot at various instants of time and superimposed them we would see component vectors of different lengths, and unless each corresponding value of time were shown there would still be no way of determining phase. This is why a minus sign or 180° can be interpreted either as a phase of 180° on the vector amplitude or as a spatial direction change of 180°. The time sequence would yield the configuration shown in Figure 4-16.

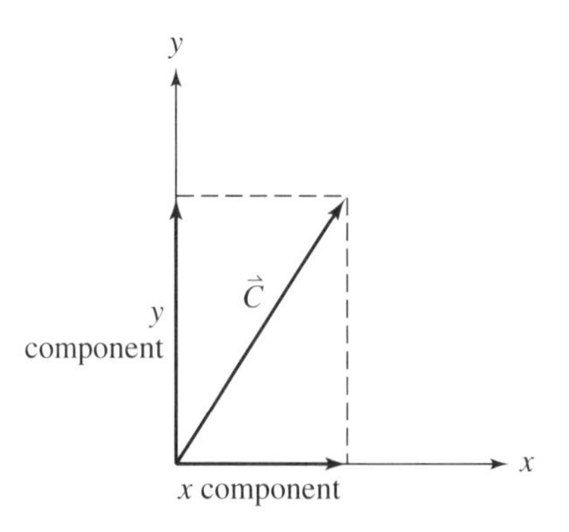

Figure 4-15. A vector with complex component scalar parts.

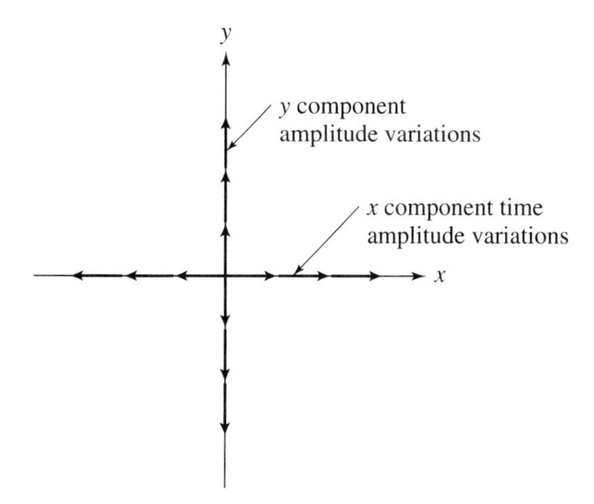

Figure 4-16. Time sequence of vector component amplitudes.

Exercises

4-6.0g One way of determining the effect of time on the vector $\vec{C}$ in the preceding example is to compute the components of the vector at different times and then plot the locus of the tip of the total vector obtained by combining the components at each instant of time. For the vector of the example above, sketch the locus for $0 \leq t \leq 20$ ns. (Answer: An ellipse whose major axis is approximately 18° from the y axis)

4-6.0h (a) Sketch the vector $\vec{E}_1 = 6\hat{x} - 2\hat{y} - \hat{z}$ on three axes. (There are two possibilities, constant vector or sinusoidal steady state.)

(b) Sketch the vector $\vec{E}_2 = 6e^{j80^\circ}\hat{x} - 2e^{j65^\circ}\hat{y} - 3e^{-j20^\circ}\hat{z}$ where $\vec{E}_2$ is a sinusoidal steady-state vector in phasor form. Discuss the differences between $\vec{E}_1$ and $\vec{E}_2$. What is the phase difference between E_{2x} and E_{2y}? Express the real time form of each component of $\vec{E}_2$ if the frequency is 1 GHz.

4-6.0i Although we have not intended to develop computer programming ability, thinking in terms of an algorithm often clarifies a concept. Develop a program that inputs a sinusoidal steady-state vector given as

$$\vec{A} = \overline{A}_x\hat{x} + \overline{A}_y\hat{y} + \overline{A}_z\hat{z}$$

and returns the following information:

(a) Magnitude of the vector $\vec{A}$ (as complex scalar) assuming all components have the same phase angle in the exponential. There may still be negative coefficients.

(b) The direction of the vector $\vec{A}$ with respect to each of the coordinate axes. Watch minus signs carefully!

(c) The magnitude of the vector $\vec{A}$ (having complex components) as a function of time.

(d) Apply your program to $\vec{A} = 8e^{j30^\circ}\hat{x} - 4e^{j30^\circ}\hat{y} + e^{j30^\circ}\hat{z}, f = 1$ GHz.

4-6.0j Using the program developed in Exercise 4-6.0i, apply it to the vector.

$$\vec{H} = -4e^{j20^\circ}\hat{x} + 3e^{j20^\circ}\hat{y} - 7e^{j20^\circ}\hat{z}$$

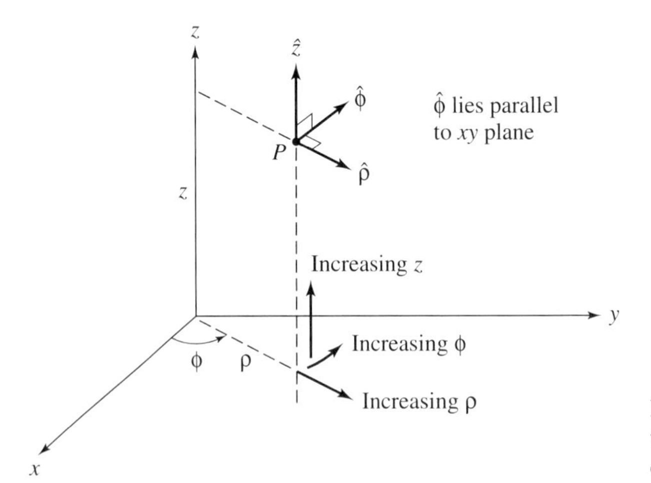

Figure 4-17. Determination of the unit vectors in the cylindrical coordinate system.

and obtain a plot of the quantity under output (c) of Exercise 4-6.0i using $f = 1$ GHz. Make a time plot of each component of $\vec{H}$. Do these plots give results consistent with answers (a) and (c) of Exercise 4-6.0i? (Note these plots do not give vector directions.)

The unit vectors for the other coordinate systems may not be as obvious as for the rectangular case. The definition of a unit vector, however, allows one to determine the correct directions easily. Consider the cylindrical coordinate system shown in Figure 4-17. A point P is located by specifying two lengths and one angle in the order ρ (normal to z

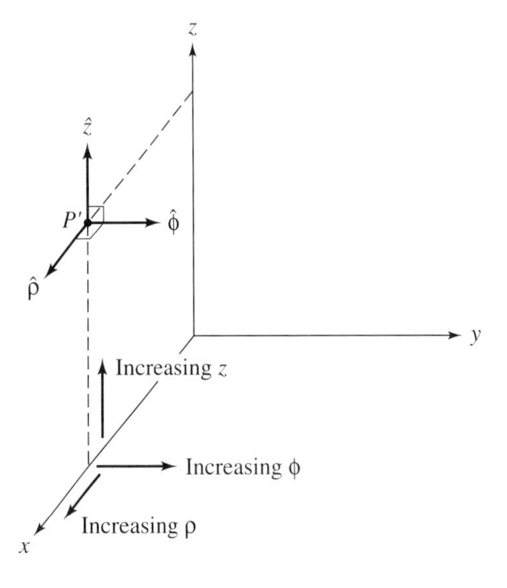

Figure 4-18. Unit vector direction change as a function of spatial location.

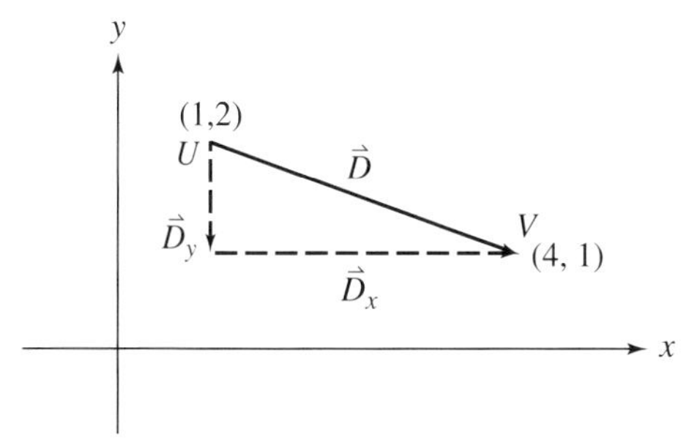

Figure 4-19. Vector and components of a vector defined by two points in space.

axis), ϕ (from positive x axis), and z (normal to xy plane). The unit vectors, according to the definition, are directed to increasing coordinate value.

Thus at point P the unit vectors are drawn in the direction of increasing coordinate values at P. Also notice that even though the ϕ coordinate describes an arc, we still draw a straight line to represent the unit vector $\hat{\phi}$ at P directed in the way ϕ increases. Another interesting feature of this coordinate system, and a feature of most other systems as well, is that at different locations in space the unit vectors may change direction. This is quite unlike the rectangular system where the unit vectors always point the same direction regardless of the location P. An example of this feature is shown in Figure 4-18. At the new point P' the directions of increasing coordinate value are different than in the case of the first location of P.

We may obtain the expression for a spatial vector by first obtaining the components and then constructing the vector in component form. Let's take a two-dimensional example in Figure 4-19. We desire to construct a vector $\vec{D}$ directed from U to V.

First imagine that $\vec{D}$ is decomposed into its two orthogonal components as shown in the figure. The x component $= 4 - 1 = 3 = D_x$. The y component $= 1 - 2 = -1 = D_y$. (Note the vector is directed from U to V so we subtract the U coordinates (initial end, or reference) from the terminal end coordinates V.)

We then obtain $\vec{D} = D_x\hat{x} + D_y\hat{y} = 3\hat{x} - 1\hat{y}$.

From this form it is an easy matter to obtain the distance from U to V (the magnitude of the vector $\vec{D}$) using the Pythagorean theorem:

$$|\vec{D}| = \sqrt{3^2 + (-1)^2} = \sqrt{10}$$

Exercises

4-6.0k Write in rectangular component form the vectors that extend (a) from the origin to $(-4, 1, 7)$, (b) from $(3, -8, 11)$ to $(-2, 5, 9)$, and (c) from $(-2, 5, 9)$ to $(3, -8, 11)$. Also determine the magnitude of each vector. Discuss the spatial relationship between the vector drawn between the two given points and the vector given in three-component form.

4-6.0l The preceding example assumed that the scalar parts of the vector were real. Under what conditions will the process describe there also work for complex scalar parts? (Answer: If the angles of all scalars are the same. Why?)

4-7 Mathematical Addition of Vectors and Their Properties in Three-Component Form

Earlier we had to combine vectors using a graphical process. Our component formulation allows a mathematical procedure. Thus if we have vectors of real scalar components we proceed as follows:

$$\vec{A} + \vec{B} = \overline{A}_x\hat{x} + \overline{A}_y\hat{y} + \overline{A}_z\hat{z} + \overline{B}_x\hat{x} + \overline{B}_y\hat{y} + \overline{B}_z\hat{z}$$

Combining like components since they are parallel gives:

$$\vec{A} + \vec{B} = (\overline{A}_x + \overline{B}_x)\hat{x} + (\overline{A}_y + \overline{B}_y)\hat{y} + (\overline{A}_z + \overline{B}_z)\hat{z}$$

Subtraction is, as usual, defined as the addition of the negative of the vector being subtracted. For example:

$$\vec{A} - \vec{B} \equiv \vec{A} + (-\vec{B}) = \overline{A}_x\hat{x} + \overline{A}_y\hat{y} + \overline{A}_z\hat{z} + (-\overline{B}_x\hat{x} - \overline{B}_y\hat{y} - \overline{B}_z\hat{z})$$
$$= \overline{A}_x\hat{x} + \overline{A}_y\hat{y} + \overline{A}_z\hat{z} - \overline{B}_x\hat{x} - \overline{B}_y\hat{y} - \overline{B}_z\hat{z}$$
$$\vec{A} - \vec{B} = (\overline{A}_x - \overline{B}_x)\hat{x} + (\overline{A}_y - \overline{B}_y)\hat{y} + (\overline{A}_z - \overline{B}_z)\hat{z}$$

As a numerical example if $\vec{A} = 6\hat{x} - 4\hat{y} + 3\hat{z}$ and $\vec{B} = 9\hat{x} + 7\hat{y} - 2\hat{z}$ we would have

$$\vec{A} - \vec{B} = (6 - 9)\hat{x} + (-4 - 7)\hat{y} + (3 + 2)\hat{z} = -3\hat{x} - 11\hat{y} + 5\hat{z}$$

Having a vector expressed in orthogonal component form allows us to reduce a vector equation to three scalar equations. This means that we could solve for as many as three unknowns from the original vector equation depending upon where the unknowns appear. To obtain the maximum number of unknowns requires at least one unknown in each component. (Why?) Suppose three vectors are related by $\vec{A} + \vec{B} = \vec{C}$. Expressing the vectors in component form:

$$(A_x + B_x)\hat{x} + (A_y + B_y)\hat{y} + (A_z + B_z)\hat{z} = C_x\hat{x} + C_y\hat{y} + C_z\hat{z}$$

Now, if we have a vector given in terms of its orthogonal coordinate components, we may change any one of its components without changing the others. Thus the coordinate components are independent. The preceding equation then requires

$$A_x + B_x = C_x$$
$$A_y + B_y = C_y$$
$$A_z + B_z = C_z$$

Note we could have only one new unknown in each equation, so we could obtain at most three unknowns. This could be extended to the sum of any number of vectors on each side of the equality. The magnitude of a vector can be expressed in convenient mathematical form for any orthogonal coordinate system vector form by using the distance formula from analytic geometry. For real scalars in rectangular coordinates:

$$\text{Magnitude of } \vec{A} \equiv |\vec{A}| = \sqrt{A_x^2 + A_y^2 + A_z^2} \qquad \text{(4-6)}$$

Exercises

4-7.0a Using the mathematical approach, prove the commutative and associative laws for vector addition

$$\vec{A} + \vec{B} = \vec{B} + \vec{A} \qquad (\vec{A} + \vec{B}) + \vec{C} = \vec{A} + (\vec{B} + \vec{C})$$

4-7.0b Suppose three vectors are related by the equation $\vec{A} - \vec{B} = \vec{C}$. If $\vec{A} = A_x\hat{x} + 3\hat{y} - 8\hat{z}$, $\vec{B} = -2\hat{x} + B_y\hat{y} + 5\hat{x}$ and $\vec{C} = \hat{x} - 9\hat{y} + C_z\hat{z}$, find A_x, B_y, and C_z.

4-8 Multiplication of Vectors—I: The Dot Product

In Section 4-5 we gave the definition for multiplying a vector by a scalar. We now inquire as to the meaning of multiplying a vector by another vector. One soon discovers that three types of multiplication are possible, the first of which is discussed in this section. The others are presented in subsequent sections.

The first vector combination is called the *dot product*. This is also often called the *scalar* or *inner* product in its more general applications. This name is derived from the fact that the result of this type of multiplication is a scalar. As a formal definition of the dot or scalar product we have:

Definition. The *dot product* of two general vectors existing at the same point in space in an orthogonal coordinate system is defined as the sum of the products of the corresponding components. The result is a scalar and the operation is represented by $\vec{\mathcal{A}}(t) \cdot \vec{\mathcal{B}}(t)$.

Example 4-7

1. $\vec{A} = A_\rho\hat{\rho} + A_\phi\hat{\phi} + A_z\hat{z}, \qquad \vec{B} = B_\rho\hat{\rho} + B_\phi\hat{\phi} + B_z\hat{z}$

$$\vec{A} \cdot \vec{B} = A_\rho B_\rho + A_\phi B_\phi + A_z B_z$$

2. $\vec{A} = 6\hat{x} - 4\hat{y} + 3\hat{z}, \qquad \vec{B} = 9\hat{x} + 7\hat{y} - 2\hat{z}$

$$\vec{A} \cdot \vec{B} = (6)(9) + (-4)(7) + (3)(-2) = 20$$

3. $\vec{\mathcal{A}}(t) = A_x \cos(\omega t + \alpha_y)\hat{y} \quad \vec{\mathcal{B}}(t) = B_x \cos(\omega t + \beta_x)\hat{x} + B_y \cos(\omega t + \beta_y)\hat{y}$

$$\vec{\mathcal{A}}(t) \cdot \vec{\mathcal{B}}(t) = A_x B_x \cos(\omega t + \alpha_x)\cos(\omega t + \beta_x) + A_y B_y \cos(\omega t + \alpha_y)\cos(\omega t + \beta_y)$$

In the preceding examples we have not used any complex (phasor) components. These require special care as will be shown shortly.

Exercises

4-8.0a Find the dot products of the following vector pairs:

(a) $\vec{E} = -3\hat{r} + 2\hat{\theta} \quad \vec{\mathrm{H}} = 2\hat{r} - 8\hat{\theta}$ (Answer: -22)

(b) $\vec{\mathrm{B}} = 2\hat{r} + 4\hat{\phi} \quad \vec{\mathrm{H}} = 4\hat{r} - 3\hat{\theta}$ (Answer: 8)

(c) $\vec{\mathscr{E}} = \cos\omega t\hat{\phi} + \cos\omega t\hat{z}$ $\vec{\mathscr{D}} = 3\cos\omega t\hat{\phi} - 2\cos\omega t\hat{z}$ (Answer: $\cos^2\omega t$)
(d) $\vec{\mathrm{B}} = 3\hat{x} - 7\hat{y} + \hat{z}$ $\vec{\mathrm{H}} = -2\hat{x} + \hat{y}$ (Answer: -13)

4-8.0b Prove that for the scalar (dot) product the commutative and distributive laws are valid; that is, prove:

$$\vec{A}\cdot\vec{B} = \vec{B}\cdot\vec{A}$$

$$\vec{A}\cdot(\vec{B}+\vec{C}) = \vec{A}\cdot\vec{B} + \vec{A}\cdot\vec{C}$$

(Hint: Use the definitions of vector sum and dot product along with the laws of scalar algebra.)

The dot product has some important properties. The first of these results from taking the dot product of a vector with itself. Suppose we have a vector $\vec{A}$ in rectangular coordinates with real scalar parts. Using the three components we obtain

$$\vec{A}\cdot\vec{A} = A_xA_x + A_yA_y + A_zA_z = A_x^2 + A_y^2 + A_z^2$$

Comparing this with Eq. 4-6 we see that

$$\vec{A}\cdot\vec{A} = |\vec{A}|^2 \tag{4-7}$$

which is the square of the magnitude of the vector.

Suppose next that

$$\vec{\mathscr{E}} = \cos\omega t\hat{x} + \sin\omega t\hat{y}$$

$$|\vec{\mathscr{E}}| = \sqrt{\vec{\mathscr{E}}\cdot\vec{\mathscr{E}}} = \sqrt{\cos^2\omega t + \sin^2\omega t} = 1$$

That this is a reasonable result can be shown by plotting the locus of the tip of the vector $\vec{E}$ as a function of time in the xy plane. The locus is a circle in the xy plane, and the vector $\vec{E}$ is rotating counterclockwise. The radius of the circle is 1.

A second important property of the dot product is that we may easily obtain the angle between two vectors measured at the same point in space. For simplicity we consider the two-dimensional system shown in Figure 4-20. This is really no severe restriction since any two vectors that intersect at a point will define a plane.

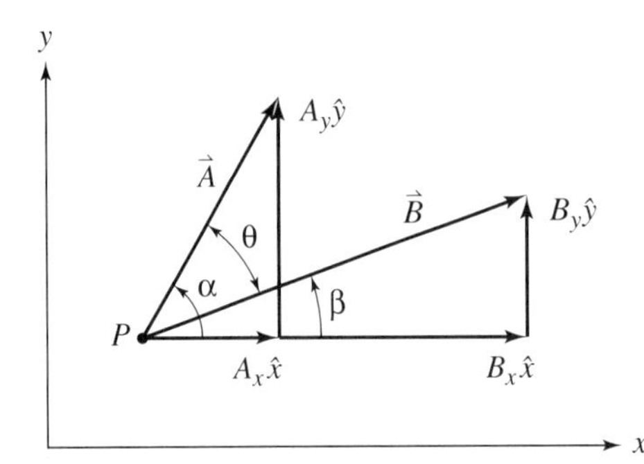

Figure 4-20. System identification for two vectors located at the same point in space.

Using the definition of the dot product we have the general result

$$\vec{A} \cdot \vec{B} = A_x B_x + A_y B_y$$

Without altering the results, we may multiply each term on the right by the unity ratio

$$\frac{|\vec{A}||\vec{B}|}{|\vec{A}||\vec{B}|} = \frac{\sqrt{A_x^2 + A_y^2}\,\sqrt{B_x^2 + B_y^2}}{\sqrt{A_x^2 + A_y^2}\,\sqrt{B_x^2 + B_y^2}}$$

to obtain

$$\vec{A} \cdot \vec{B} = A_x \frac{\sqrt{A_x^2 + A_y^2}}{\sqrt{A_x^2 + A_y^2}} B_x \frac{\sqrt{B_x^2 + B_y^2}}{\sqrt{B_x^2 + B_y^2}} + A_y \frac{\sqrt{A_x^2 + A_y^2}}{\sqrt{A_x^2 + A_y^2}} B_y \frac{\sqrt{B_x^2 + B_y^2}}{\sqrt{B_x^2 + B_y^2}}$$

But from the figure

$$\frac{A_x}{\sqrt{A_x^2 + A_y^2}} = \cos x \quad \frac{A_y}{\sqrt{A_x^2 + A_y^2}} = \sin x, \qquad \text{etc.}$$

So we may write

$$\begin{aligned}\vec{A} \cdot \vec{B} &= \sqrt{A_x^2 + A_y^2}\,\sqrt{B_x^2 + B_y^2} \cos \alpha \cos \beta + \sqrt{A_x^2 + A_y^2}\,\sqrt{B_x^2 + B_y^2} \sin \alpha \sin \beta \\ &= |\vec{A}||\vec{B}|(\cos \alpha \cos \beta + \sin \alpha \sin \beta) = |\vec{A}||\vec{B}| \cos(\alpha - \beta)\end{aligned}$$

Since $\cos(\alpha - \beta) = \cos \theta$ from the figure, we finally obtain

$$\vec{A} \cdot \vec{B} = |\vec{A}||\vec{B}| \cos \theta \tag{4-8}$$

From this result we can find the angle between two vectors as

$$\cos \theta = \frac{\vec{A} \cdot \vec{B}}{|\vec{A}||\vec{B}|} \tag{4-9}$$

A physical interpretation of this result is also possible. Since the product in Eq. 4-8 are ordinary arithmetic products of scalars we may group terms as follows

$$\vec{A} \cdot \vec{B} = |\vec{B}|(|\vec{A}| \cos \theta)$$

Now $|\vec{A}| \cos \theta$ is simply a measure of the amount of the vector $\vec{A}$ lying in the direction of $\vec{B}$. Three important special cases are:

1. If $\theta = 90°$, $\vec{A} \cdot \vec{B} = 0$, since $\vec{A}$ and $\vec{B}$ are orthogonal. This means, of course, that there is no part of $\vec{A}$ lying in the direction of $\vec{B}$ (or of $\vec{B}$ lying in the direction of $\vec{A}$).
2. If $\theta = 0°$, $\vec{A} \cdot \vec{B} = |\vec{A}||\vec{B}|$, since $\vec{A}$ and $\vec{B}$ are parallel. This also means that all of $\vec{A}$ is in the direction of $\vec{B}$.
3. If $\theta = 180°$, $\vec{A} \cdot \vec{B} = -|\vec{A}||\vec{B}|$, since $\vec{A}$ and $\vec{B}$ are antiparallel. This means that all of $\vec{A}$ is in the direction opposite to the direction of $\vec{B}$.

Since work (energy) is force times the distance moved *in the direction of the force*, we may write, for Figure 4-21,

$$W = (|\vec{F}| \cos \theta)\,|\vec{D}| = |\vec{F}||\vec{D}| \cos \theta = \vec{F} \cdot \vec{D}$$

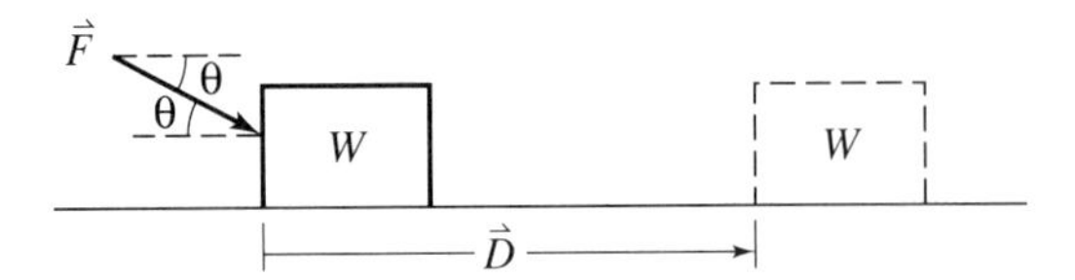

Figure 4-21. Application of the dot product in a mechanical system.

The preceding results provide a way for constructing a unit vector in the direction of a given vector. Suppose the given vector is

$$\vec{A} = A_x\hat{x} + A_y\hat{y} + A_z\hat{z}$$

and we desire to construct a unit vector $\hat{a}$ having the same direction as $\vec{A}$. Since dividing a vector by a positive real scalar does not change the direction of the result compared with the original vector we proceed as follows:

$$\hat{a} = \frac{\vec{A}}{|\vec{A}|} = \frac{\vec{A}}{\sqrt{\vec{A}\cdot\vec{A}}} = \frac{\vec{A}}{\sqrt{A_x^2 + A_y^2 + A_z^2}}$$

$$= \frac{A_x}{\sqrt{A_x^2 + A_y^2 + A_z^2}}\,\hat{x} + \frac{A_y}{\sqrt{A_x^2 + A_y^2 + A_z^2}}\,\hat{y} + \frac{A_z}{\sqrt{A_x^2 + A_y^2 + A_z^2}}\,\hat{z}$$

That this is a unit vector is easily verified by evaluating $\hat{a}\cdot\hat{a}$, which will be one. Following are some examples of various calculations using the dot product.

Example 4-8

Given vectors in orthogonal x, y, z coordinate system:

$$\vec{A} = 6\hat{x} + 4\hat{y} - 9\hat{z}$$

$$\vec{B} = 4\hat{x} - 2\hat{y} + 3\hat{z}$$

$$\vec{C} = 30\angle 18^\circ \quad \vec{D} = 46\angle{-40^\circ}$$

$\vec{C}$ and $\vec{D}$ are coplanar in the xy plane (angle from $+x$ axis).

Find $|\vec{A}|$:

$$|\vec{A}| = \sqrt{6^2 + 4^2 + 9^2} = \sqrt{36 + 16 + 81} = \sqrt{133} = 11.5$$

Find $|\vec{C}|$:

$$|\vec{C}| = 30 = (\vec{C}\cdot\vec{C})^{1/2}$$

Find $\vec{A}\cdot\vec{B}$:

$$\vec{A}\cdot\vec{B} = 6\cdot 4 + 4\cdot(-2) + 3\cdot(-9) = 24 - 8 - 27 = -11$$

(What does minus sign mean? It means the component of $\vec{A}$ along the line of $\vec{B}$ is in the direction opposite to the direction of $\vec{B}$.)

Find $\vec{C}\cdot\vec{D}$:

$$\vec{C}\cdot\vec{D} = |\vec{C}||\vec{D}|\cos\theta_{CD} = (30)(46)\cos(18 - (-40))$$

$$= 1380\cos 58^\circ = 731$$

Find $\vec{A} \cdot \vec{C}$:

$$\vec{C} = 30\angle 18° = 28.5\hat{x} + 9.3\hat{y}$$

$$\vec{A} \cdot \vec{C} = (6)(28.5) + (4)(9.3) + (-9)(0)$$
$$= 171 + 37.2 = 208.2$$

Construct a unit vector in the direction of $\vec{A}$:

$$\hat{a} = \frac{\vec{A}}{|\vec{A}|} = \frac{6\hat{x} + 4\hat{y} + 9\hat{z}}{11.5}$$
$$= 0.522\hat{x} + 0.348\hat{y} - 0.782\hat{z}$$

Construct a unit vector in the direction of $\vec{D}$:

$$\hat{d} = \frac{\vec{D}}{|\vec{D}|} = \frac{46\angle -40°}{46} = 1\angle -40°$$

Find the angle between $\vec{A}$ and $\vec{B}$:

$$\vec{A} \cdot \vec{B} = |\vec{A}||\vec{B}| \cos\theta$$

$$|\vec{A}| = 11.5 \qquad \text{from the previous calculation}$$

$$|\vec{B}| = \sqrt{4^2 + 2^2 + 3^2} = 5.38$$

$$\vec{A} \cdot \vec{B} = 11 \qquad \text{from the previous calculation}$$

$$\cos\theta = \frac{-11}{(11.5)(5.38)} = \frac{-11}{62} = -0.1776$$

$$\theta = 100.2°$$

Example 4-9

As an example using time functions, suppose we wish to find the spatial angle between the two vectors

$$\vec{\mathcal{A}}(t) = \cos\omega t\hat{x} + \cos\omega t\hat{y} \qquad \text{and} \qquad \vec{\mathcal{B}}(t) = \sin\omega t\hat{x} + \sin\omega t\hat{y}$$

$$\vec{\mathcal{A}}(t) \cdot \vec{\mathcal{B}}(t) = \cos\omega t \sin\omega t + \cos\omega t \sin\omega t = 2\cos\omega t \sin\omega t$$

$$\cos\theta = \frac{\vec{\mathcal{A}} \cdot \vec{\mathcal{B}}}{|\vec{\mathcal{A}}||\vec{\mathcal{B}}|} = \frac{2\cos\omega t \sin\omega t}{\sqrt{\cos^2\omega t + \cos^2\omega t}\sqrt{\sin^2\omega t + \sin^2\omega t}} = 1$$

or

$$\theta = 0°$$

To show this physically, one can plot the loci of the tips of the vectors $\vec{\mathcal{A}}$ and $\vec{\mathcal{B}}$ as functions of time and one will find that *both* describe lines in the xy plane at angles of 45° with respect to the $+x$ axis.

Exercises

4-8.0c (a) Write in rectangular component form the expression for the vector that extends from the origin to $(+4, -1, -7)$; call this vector $\vec{E}_1$.

(b) Repeat (a) for the vector that extends from $(-3, +8, 11)$ to $(+2, 5, -9)$; call this vector $\vec{E}_2$.

(c) Determine the magnitude of each vector and the unit vectors in the direction of each vector.

(d) Find the dot product of the two vectors and the angle between them. Draw a diagram showing the two vectors.

(e) Obtain the vector $\vec{E} = \vec{E}_2 - \vec{E}_1$ and also the unit vector in the direction of $\vec{E}$.

4-8.0d Given two real vectors $\vec{A}$ and $\vec{B}$, whose angles with respect to x, y, and z axes are α_x, α_y, α_z for $\vec{A}$, and β_x, β_y, and β_z for $\vec{B}$. Prove that if γ is the angle between the vectors $\vec{A}$ and $\vec{B}$,

$$\cos\gamma = \cos\alpha_x \cos\beta_x + \cos\alpha_y \cos\beta_y + \cos\alpha_z \cos\beta_z$$

4-8.0e What values of the constant β will make the vectors $\vec{H} = 2\beta\hat{x} - 4\hat{y} + \frac{1}{2}\hat{z}$ and $\vec{E} = \beta\hat{x} + \frac{1}{2}\beta\hat{y} + \frac{1}{2}\hat{z}$ orthogonal? Parallel? (Answer: 0.8536, 0.1464, none)

4-8.0f Find the magnitude of $\vec{P} = 2\hat{x} - 3\hat{y} + 6\hat{z}$ and determine how much (what percentage) of that value lies in the direction of the vector $\vec{Q} = \hat{x} + 2\hat{y} + 3\hat{z}$. (Hint: Convert $\vec{Q}$ to $\hat{q}$) (Answer: 7, 53.4%)

4-8.0g What angles do the vectors $\vec{P}$ and $\vec{Q}$ of Exercise 4-8.0f make with the x, y, and z axes? (Answer: For $\vec{P}$: 73.4°, 115.4°, 0.31°)

The properties of coordinate systems may be conveniently described in terms of the dot product. Consider two vectors $\vec{A}$ and $\vec{B}$ in the rectangular coordinate system:

$$\vec{A}\cdot\vec{B} = (A_x\hat{x} + A_y\hat{y} + A_z\hat{z})\cdot(B_x\hat{x} + B_y\hat{y} + B_z\hat{z})$$

Using the distributive property of the dot product (see Exercise 4-8.0b) and removing real scalars from the resulting dot products we have

$$\begin{aligned}\vec{A}\cdot\vec{B} &= (A_x\hat{x} + A_y\hat{y} + A_z\hat{z})\cdot B_x\hat{x} + (A_x\hat{x} + A_y\hat{y} + A_z\hat{z})\cdot B_y\hat{y} \\ &\quad + (A_x\hat{x} + A_y\hat{y} + A_z\hat{z})\cdot B_z\hat{z} \\ &= A_xB_x\hat{x}\cdot\hat{x} + A_yB_x\hat{y}\cdot\hat{x} + A_zB_x\hat{z}\cdot\hat{z} + A_xB_y\hat{x}\cdot\hat{y} + A_yB_y\hat{y}\cdot\hat{y} \\ &\quad + A_zB_y\hat{z}\cdot\hat{y} + A_xB_z\hat{x}\cdot\hat{z} + A_yB_z\hat{y}\cdot\hat{z} + A_zB_z\hat{z}\cdot\hat{z}\end{aligned}$$

Now since the unit vectors $\hat{x}$, $\hat{y}$, and $\hat{z}$ have magnitude one, and since any vector is parallel to itself ($\theta = 0°$) we have

$$\hat{x}\cdot\hat{x} = \hat{y}\cdot\hat{y} = \hat{z}\cdot\hat{z} = 1$$

For the mixed dot products $\hat{x}\cdot\hat{y}$, etc., the angle between the vectors is $\theta = 90°$. Thus all mixed dot products are zero:

$$\hat{y}\cdot\hat{x} = \hat{z}\cdot\hat{x} = \hat{z}\cdot\hat{y} = \cdots = 0$$

Then

$$\vec{A}\cdot\vec{B} = A_xB_x + A_yB_y + A_zB_z \qquad \text{as before}$$

The orthogonality property of unit vectors is an important property of many coordinate systems; namely

Definition. Any coordinate system for which the dot products between the different unit vectors of the system are zero is called an *orthogonal* coordinate system.

Cylindrical and spherical coordinate systems are orthogonal coordinate systems.

Exercise

4-8.0h Sketch the cylindrical and spherical coordinate systems and, using the definition of unit vector given earlier, show that each coordinate system is an orthogonal system.

We next have to address the problem of how to handle phasors that are components (scalar parts) of vectors. To show that something additional is required consider the example given earlier where we showed that for $\vec{\mathscr{E}} = \cos \omega t\hat{x} + \sin \omega t\hat{y}$,

$$|\vec{\mathscr{E}}| = 1$$

Now if we use phasor notation we would write the vector as

$$\vec{E} = 1\angle 0°\hat{x} + 1\angle 90°\hat{y} = \hat{x} + j\hat{y}$$

Then if we use the usual dot product definition we would obtain

$$|\vec{E}| = \sqrt{\vec{E} \cdot \vec{E}} = \sqrt{(1)(1) + (j)(j)} = \sqrt{1 - 1} = 0$$

which is clearly an error. To make the result agree with the time domain we would need to do the following to obtain the correct value of one:

$$1 = \frac{\sqrt{2}}{\sqrt{2}} = \frac{\sqrt{1 + 1}}{\sqrt{2}} = \frac{\sqrt{1 + (j)(-j)}}{\sqrt{2}}$$

The numerator is what is obtained for phasors if we use $\vec{E} \cdot \vec{E}^*$ rather than $\vec{E} \cdot \vec{E}$. Thus we suspect that the conjugate operation will need to be involved along with a constant.

Consider a general case where

$$\vec{\mathscr{E}}(t) = E_x \cos(\omega t + \phi_x)\hat{x} + E_y \cos(\omega t + \phi_y)\hat{y} + E_z \cos(\omega t + \phi_z)\hat{z}$$

Taking the dot product of the vector with itself we have

$$\begin{aligned}|\vec{\mathscr{E}}(t)|^2 &= E_x^2 \cos^2(\omega t + \phi_x) + E_y^2 \cos^2(\omega t + \phi_y) + E_z^2 \cos^2(\omega t + \phi_z)\\ &= E_x^2 \frac{[\cos(2\omega t + 2\phi_x) + 1]}{2} + E_y^2 \frac{[\cos(2\omega t + 2\phi_y) + 1]}{2}\\ &\quad + E_z^2 \frac{[\cos(2\omega t + 2\phi_z) + 1]}{2}\\ &= \tfrac{1}{2}[E_x^2 + E_y^2 + E_z^2 + (E_x^2 \cos 2\phi_x + E_y^2 \cos 2\phi_y + E_z^2 \cos 2\phi_z) \cos 2\omega t]\\ &\quad - (E_x^2 \sin 2\phi_x + E_y^2 \sin 2\phi_y + E_z^2 \cos 2\phi_z) \sin 2\omega t]\end{aligned}$$

Since the time averages of sine and cosine functions over a period are zero we have, using the long overbar to represent the time average:

$$\overline{|\vec{\mathscr{E}}(t)|^2} = \frac{E_x^2 + E_y^2 + E_z^2}{2} = \left(\frac{E_x}{\sqrt{2}}\right)^2 + \left(\frac{E_y}{\sqrt{2}}\right)^2 + \left(\frac{E_z}{\sqrt{2}}\right)^2$$

which might have been anticipated from circuit theory as the mean square value. Now for the phasor representation we write

$$\vec{\mathscr{E}}(t) \rightarrow \vec{E} = E_x e^{j\phi_x}\hat{x} + E_y e^{j\phi_y}\hat{y} + E_z e^{j\phi_z}\hat{z}$$

Using the definition of the dot product for complex numbers we have

$$\begin{aligned}\vec{E} \cdot \vec{E}^* &= (E_x e^{j\phi_x})(E_x e^{-j\phi_x}) + (E_y e^{j\phi_y})(E_y e^{-j\phi_y}) + (E_z e^{j\phi_z})(E_z e^{-j\phi_z}) \\ &= (E_x^2 + E_y^2 + E_z^2)\end{aligned}$$

Thus, as far as a single vector is concerned, if we take the square root of the dot product of the complex vector and its conjugate we obtain the square root of the time-average value of the squared vector magnitude multiplied by $\sqrt{2}$, or, formally,

$$\vec{E} \cdot \vec{E}^* = \sqrt{2}\,\overline{|\vec{\mathscr{E}}(t)|^2} \tag{4-10}$$

Note particularly that for complex vectors the dot product may be zero, and yet the vector is not necessarily zero. Only the time average *length* is zero.

Now for two different vectors $\mathscr{A}(t)$ and $\mathscr{E}(t)$, for example,

$$\begin{aligned}\vec{\mathscr{A}}(t) &= A_x \cos(\omega t + \phi_x)\hat{x} + A_y \cos(\omega t + \phi_y)\hat{y} + A_z \cos(\omega t + \phi_z)\hat{z} \\ \vec{\mathscr{B}}(t) &= B_x \cos(\omega t + \theta_x)\hat{x} + B_y \cos(\omega t + \theta_y)\hat{y} + B_z \cos(\omega t + \theta_z)\hat{z} \\ \vec{\mathscr{A}} \cdot \vec{\mathscr{B}} &= A_x B_x \cos(\omega t + \phi_x)\cos(\omega t + \theta_x) + A_y B_y \cos(\omega t + \phi_y)\cos(\omega t + \theta_y) \\ &\quad + A_z B_z \cos(\omega t + \phi_z)\cos(\omega t + \theta_z) \\ &= \tfrac{1}{2}A_x B_x[\cos(2\omega t + \phi_x + \theta_x) + \cos(\phi_x - \theta_x)] + \tfrac{1}{2}A_y B_y[\cos(2\omega t + \phi_y + \theta_y) \\ &\quad + \cos(\phi_y - \theta_y)] + \tfrac{1}{2}A_z B_z[\cos(2\omega t + \phi_z + \theta_z) + \cos(\phi_z - \theta_z)] \\ &= \tfrac{1}{2}[A_x B_x \cos(\phi_x - \theta_x) + A_y B_x \cos(\phi_y - \theta_y) + A_z B_z \cos(\phi_z - \theta_z] \\ &\quad + \tfrac{1}{2}[A_x B_x \cos(2\omega t + \phi_x + \theta_x) + A_y B_y \cos(2\omega t + \phi_y + \theta_y) \\ &\quad + A_z B_z \cos(2\omega t + \phi_z + \theta_z]\end{aligned}$$

Again, taking the time average

$$\overline{\vec{\mathscr{A}} \cdot \vec{\mathscr{B}}} = \tfrac{1}{2}[A_x B_x \cos(\phi_x - \theta_x) + A_y B_y \cos(\phi_y - \theta_y) + A_z A_z \cos(\phi_z - \theta_z)]$$

This result is one-half the time average of the sum of the projections of the components of one vector onto the corresponding components of the other, or, in other words, one-half the time average of the projection of one vector on the other. Since the difference of angles can be obtained by the products of conjugates, we begin with the phasor representations of $\vec{\mathscr{A}}(t)$ and $\vec{\mathscr{B}}(t)$; namely,

$$\begin{aligned}\vec{A} &= A_x e^{j\phi_x}\hat{x} + A_y e^{j\phi_y}\hat{y} + A_z e^{j\phi_z}\hat{z} \\ \vec{B} &= B_x e^{j\theta_x}\hat{x} + B_y e^{j\theta_y}\hat{y} + B_z e^{j\theta_z}\hat{z}\end{aligned}$$

Then

$$\vec{A} \cdot \vec{B}^* = A_x B_x e^{j(\phi_x - \theta_x)} + A_y B_y e^{j(\phi_y - \theta_y)} + A_z B_z e^{j(\phi_z - \theta_z)}$$

One-half of the real part of this dot product is then the same as the time average of the projection of one vector on another as shown above. Formally,

$$\text{Re}\{\vec{A} \cdot \vec{B}^* = 2\overline{\vec{\mathcal{A}}(t) \cdot \vec{\mathcal{B}}(t)} \tag{4-11}$$

Exercises

4-8.0i Using the results given in the preceding discussion, show that the time average angle between two sinusoidal vectors is given by

$$\cos\,\theta = \frac{\text{Re}\{\vec{A} \cdot \vec{B}^*\}}{|\vec{A}||\vec{B}|}$$

4-8.0j Given two vectors $\vec{E} = \hat{x} + \hat{j}$ and $\vec{H} = \hat{x} - j\hat{y}$, plot the angle between the two vectors as a function of time, and hence demonstrate that $\text{Re}\{\vec{E} \cdot \vec{H}^*\}$ satisfies the result derived in the text above.

4-8.0k Sketch the time loci of the tips of the vectors whose phasor representations are:

(a) $\vec{E} = j3\hat{x} + 4\hat{y}$

(b) $\vec{H} = \hat{x} + \hat{y}$

(c) $\vec{D} = j\hat{x} + j\hat{y}$

4-8.0l For the vectors $\vec{E} = 4e^{j30^\circ}\hat{x} - 5e^{j30^\circ}$ and $\vec{H} = 2e^{j30^\circ}\hat{x} + e^{j30^\circ}\hat{y}$

(a) Find the dot product of the two vectors (time average).

(b) Find the time average magnitude of each.

4-9 Multiplication of Vectors—II: The Cross Product

The second vector product combination we will investigate is called the *cross* product. This is also often called the *vector* product because the result of the operation is another vector. We represent the operation by $\vec{A} \times \vec{B}$. As a formal definition we use the following:

Definition. The *vector* (cross) product of two vectors is defined as a vector whose *magnitude* is equal to the product of the magnitudes of the two vectors and the sine of the smaller angle between them and whose *direction* is the direction a right-hand screw would move if the first vector were rotated toward the second through the smaller angle.

The direction follows what is commonly called the right-hand rule; that is, if one points the fingers of the right hand in the direction that would rotate (cross) the first vector toward the second, the extended thumb points in the direction of the cross product vector. Rotation, as required by the definition, is through the smaller angle between the two vectors. Two examples are given in Figure 4-22. From these examples three important properties can be seen:

1. The vector (cross) product is *not* commutative, and, in fact,

$$\vec{A} \times \vec{B} = -\vec{B} \times \vec{A} \tag{4-12}$$

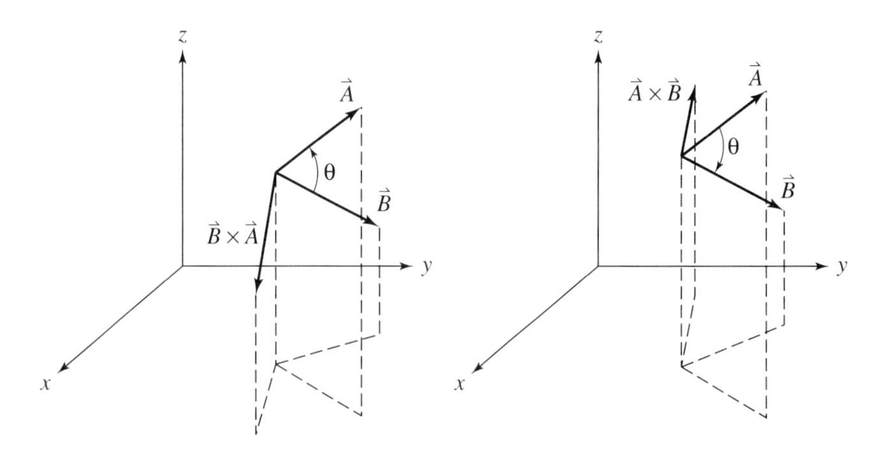

Figure 4-22. Examples of the cross (vector) product.

2. The vector product is normal to the plane defined by the two vectors. Thus we may represent the cross product algebraically as

$$\vec{A} \times \vec{B} = |\vec{A}||\vec{B}| \sin \theta_{A,B}\hat{n} \tag{4-13}$$

where $\hat{n}$ is a unit vector normal to the plane of $\vec{A}$ and $\vec{B}$, in the right-hand sense of $\vec{A}$ to $\vec{B}$ through the smaller angle. From this result we also have, by taking the magnitudes of both sides:

$$|\sin \theta_{A,B}| = \frac{|\vec{A} \times \vec{B}|}{|\vec{A}||\vec{B}|}$$

Since this yields only magnitude we have the angle ambiguity.

3. The quantity $|\vec{B}| \sin \theta_{A,B}$ represents the amount of $\vec{B}$ that is normal to $\vec{A}$. This fact will be used shortly.

Torque provides an application of the vector product. *Torque* is defined as the product of the lever arm normal to the axis of rotation and the component of the force normal to the lever arm and the axis of rotation. The direction of the torque vector is normal to the lever arm and force vectors. This is illustrated in Figure 4-23. As an equation we would then write

$$\vec{T} = \vec{L} \times \vec{F} = |\vec{L}||\vec{F}| \sin \theta_{L,F}\hat{t}$$

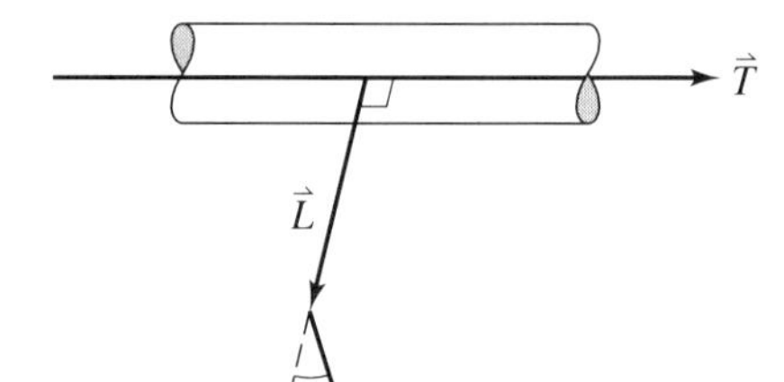

Figure 4-23. Example of torque as an application of the cross product.

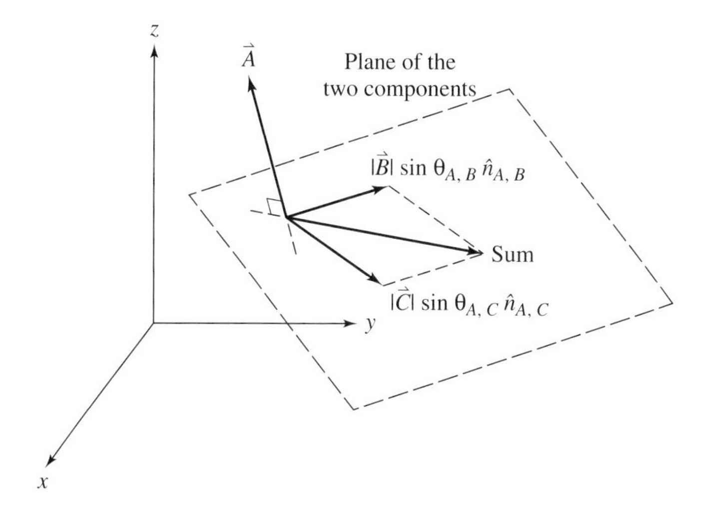

Figure 4-24. Configuration defining the sum terms in the distributive cross product theorem.

To develop alternatives for evaluating the cross product and to prove some useful theorems we need to prove the distributive property

$$\vec{A} \times (\vec{B} + \vec{C}) = \vec{A} \times \vec{B} + \vec{A} \times \vec{C} \tag{4-14}$$

We give a heuristic exposition of the proof. Using the definition of the curl the identity is written

$$\begin{aligned} |\vec{A}||\vec{B} + \vec{C}| \sin \theta_{A,B+C}\hat{n}_{A,B+C} &= |\vec{A}||\vec{B}| \sin \theta_{A,B}\hat{n}_{A,B} + |\vec{A}||\vec{C}| \sin \theta_{A,C}\hat{n}_{A,C} \\ &= |\vec{A}|[|\vec{B}| \sin \theta_{A,B}\hat{n}_{A,B} + |\vec{C}| \sin \theta_{A,C}\hat{n}_{A,C}] \end{aligned}$$

By definition all of the unit $\hat{n}$ vectors are normal to the vector $\vec{A}$. This means that $\hat{n}_{A,B}$ and $\hat{n}_{A,C}$ are coplanar and define a plane normal to $\vec{A}$. Thus the terms in the brackets on the right side represent the two vectors in a plane normal to $\vec{A}$. This situation is described in Figure 4-24. Also, $|\vec{B}| \sin \theta_{A,B}$ and $|\vec{C}| \sin \theta_{A,C}$ represent the amounts of vectors $\vec{B}$ and $\vec{C}$ normal to $\vec{A}$, which shows that the bracket term is a vector whose value is the sum of the components of vectors $\vec{B}$ and $\vec{C}$ normal to $\vec{A}$. On the left side of the equation the expression $|\vec{B} + \vec{C}| \sin \theta_{A,B+C}\hat{n}_{A,B+C}$ is a vector normal to $\vec{A}$ whose value is the amount of the sum of vectors $\vec{B}$ and $\vec{C}$ normal to $\vec{A}$, which is the interpretation, in essence, of the right-hand side of the equation.

Exercise

4-9.0a Prove the following: $\vec{A} \times (\bar{b}\vec{B}) = \bar{b}(\vec{A} \times \vec{B})$ where $\bar{b}$ is a complex scalar. (Hints: Use the definition of cross product. Also, the magnitude of the product of two quantities is the same as the product of the separate magnitudes.)

There is a convenient way to evaluate the cross product of vectors expressed in component form. The development uses the distributive property of the cross product. Consider two vectors $\vec{A}$ and $\vec{B}$ given in the rectangular coordinate system (the generalization to other coordinate systems is given later—it turns out to have the same form):

$$\vec{A} \times \vec{B} = (A_x\hat{x} + A_y\hat{y} + A_z\hat{z}) \times (B_x\hat{x} + B_y\hat{y} + B_z\hat{z})$$

Using the distributive law

$$\begin{aligned}
\vec{A} \times \vec{B} &= (A_x\hat{x} + A_y\hat{y} + A_z\hat{z}) \times B_x\hat{x} + (A_x\hat{x} + A_y\hat{y} + A_z\hat{z}) \times B_y\hat{y} \\
&\quad + (A_x\hat{x} + A_y\hat{y} + A_z\hat{z}) \times B_z\hat{z} \\
&= A_x\hat{x} \times B_x\hat{x} + A_y\hat{y} \times B_x\hat{x} + A_z\hat{z} \times B_x\hat{x} \\
&\quad + A_x\hat{x} \times B_y\hat{y} + A_y\hat{y} \times B_y\hat{y} + A_z\hat{z} \times B_y\hat{y} \\
&\quad + A_x\hat{x} \times B_z\hat{z} + A_y\hat{y} \times B_z\hat{z} + A_z\hat{z} \times B_z\hat{z}
\end{aligned}$$

Since scalars may be removed from the cross product operation (see Exercise 4-8.0a):

$$\begin{aligned}
\vec{A} \times \vec{B} = {} & A_xB_x(\hat{x} \times \hat{x}) + A_yB_x(\hat{y} \times \hat{x}) + A_zB_x(\hat{z} \times \hat{x}) \\
& + A_xB_y(\hat{x} \times \hat{y}) + A_yB_y(\hat{y} \times \hat{y}) + A_zB_y(\hat{z} \times \hat{y}) \\
& + A_xB_z(\hat{x} \times \hat{z}) + A_yB_z(\hat{y} \times \hat{z}) + A_zB_x(\hat{z} \times \hat{z})
\end{aligned}$$

Now the cross products depend upon the physical orientations of the unit vectors, we must adopt some convention to locate a point in rectangular coordinates. It is customary to write $P(x, y, z)$ and to locate axes as shown in Figure 4-25*a*.

From this convention we have that the unit vectors are orthogonal ($\theta = 90°$) so that Eq. 4-13 applied to the unit vectors yields

$$\hat{x} \times \hat{x} = \hat{y} \times \hat{y} = \hat{z} \times \hat{z} = 0 \qquad (\sin 0° = 0)$$

$$\hat{x} \times \hat{y} = -\hat{y} \times \hat{x} = \hat{z} \qquad (\hat{z} \text{ normal to plane of } \hat{x} \text{ and } \hat{y})$$

$$\hat{z} \times \hat{x} = -\hat{x} \times \hat{z} = \hat{y} \qquad (\hat{y} \text{ normal to plane of } \hat{x} \text{ and } \hat{z})$$

$$\hat{y} \times \hat{z} = -\hat{z} \times \hat{y} = \hat{x} \qquad (\hat{x} \text{ normal to plane of } \hat{y} \text{ and } \hat{z})$$

Figure 4-25. Three examples of physical orientations of the rectangular coordinate axes in the order *x*-*y*-*z* that define a right-hand system.

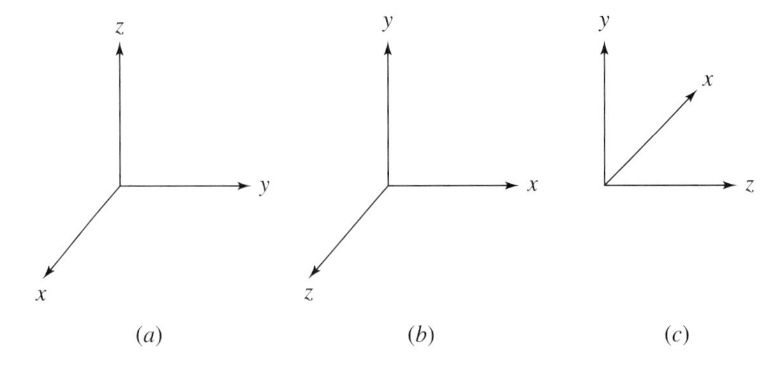

Thus we obtain

$$\vec{A} \times \vec{B} = (A_y B_z - A_z B_y)\hat{x} + (A_z B_x - A_x B_z)\hat{y} + (A_x B_y - A_y B_x)\hat{z}$$

This result can be recognized as the expansion of the determinant:

$$\vec{A} \times \vec{B} = \begin{vmatrix} \hat{x} & \hat{y} & \hat{z} \\ A_x & A_y & A_z \\ B_x & B_y & B_z \end{vmatrix} \tag{4-15}$$

Note that a simple proof of the result $\vec{A} \times \vec{B} = -\vec{B} \times \vec{A}$ can be given by merely interchanging the last two rows of the determinate which changes the sign of the result:

$$\vec{A} \times \vec{B} = \begin{vmatrix} \hat{x} & \hat{y} & \hat{z} \\ A_x & A_y & A_z \\ B_x & B_y & B_z \end{vmatrix} = -\begin{vmatrix} \hat{x} & \hat{y} & \hat{z} \\ B_x & B_y & B_z \\ A_x & A_y & A_z \end{vmatrix} = -\vec{B} \times \vec{A}$$

It should also be pointed out that for points located as $P(x, y, z)$ the vector products are $\hat{x} \times \hat{y} = \hat{z}$, $\hat{y} \times \hat{z} = \hat{x}$, and $\hat{z} \times \hat{x} = \hat{y}$ for the choice of axes shown. The unit vector cross products are taken in the same (cyclic) order as the coordinates are given in the argument list for the location of a point P. As a definition we have:

Definition. When the order of coordinates (in this case x, y, z) is the same as the vector product result $\hat{x} \times \hat{y} = \hat{z}$ (or cyclic permutations $\hat{y} \times \hat{z} = \hat{x}$ and $\hat{z} \times \hat{x} = \hat{y}$) the coordinate system is said to be a *right-hand orthogonal* coordinate system.

Physically the same three axes may be oriented in many other ways and still be a right-hand system. Two other common orientations are shown in Figure 4-25. A quick application of the right-hand rule reveals the equivalency of these systems in the cyclic order x-y-z -x-y-z, etc., as required for points located by the convention $P(x, y, z)$.

We conclude with some examples.

Example 4-10

Given vectors in the orthogonal x, y, z coordinate system:

$$\vec{A} = 6x + 4y - 9z$$

$$\vec{B} = 4x - 2y + 3z$$

$$\vec{C} = 30\angle 18^\circ \quad \vec{D} = 46\angle{-40^\circ}$$

$\vec{C}$ and $\vec{D}$ are coplanar in the xy plane (angles from the $+x$ axis).

Find $\vec{A} \times \vec{B}$:

$$\vec{A} \times \vec{B} = \begin{vmatrix} \hat{x} & \hat{y} & \hat{z} \\ 6 & 4 & -9 \\ 4 & -2 & 3 \end{vmatrix}$$

$$= (4 \cdot 3 - (-2)(-9))\hat{x} + (4 \cdot (-9) - (3)(6))\hat{y} + (6 \cdot (-2) - 4 \cdot 4)\hat{z}$$

$$= (12 - 18)\hat{x} + (-36 - 18)\hat{y} + (-12 - 16)\hat{z}$$

$$= -6\hat{x} - 54\hat{y} - 28\hat{z}$$

Find $\vec{C} \times \vec{D}$:

$$\vec{C} \times \vec{D} = |\vec{C}||\vec{D}| \sin \theta_{C \to D}\hat{n} = (30)(46) \cdot \sin(-40^\circ - 18^\circ)\hat{n}$$
$$= 1380 \sin(-58^\circ)\hat{n} = -1170\hat{n}$$

Since $\vec{C}$ and $\vec{D}$ are in the xy plane, $\vec{C} \times \vec{D}$ is normal to that plane so $\hat{n} = \hat{z}$

$$\vec{C} \times \vec{D} = -1170\hat{z}$$

($\vec{C} \times \vec{D}$ is in the negative z direction.) Find the angle between $\vec{A}$ and $\vec{B}$:

$$\vec{A} \times \vec{B} = |\vec{A}||\vec{B}| \sin \theta\hat{n}$$
$$|\vec{A} \times \vec{B}| = |\vec{A}||\vec{B}||\sin\theta|$$
$$|\sin \theta| = \frac{|\vec{A} \times \vec{B}|}{|\vec{A}||\vec{B}|} = \frac{\sqrt{36 + 2916 + 784}}{\sqrt{36 + 16 + 81} \cdot \sqrt{16 + 4 + 9}}$$
$$= \frac{\sqrt{3726}}{133 \cdot 29} = \sqrt{0.968} = 0.983$$
$$\theta = \pm 79^\circ, \pm 101^\circ$$

Note there is an ambiguity of 180° due to the fact we have taken the magnitude of $\sin \theta$. Thus when using the vector operations to find the angle between two vectors, it is best to use the dot product.

Exercises

4-9.0b Using the matrix evaluation of the cross product, Eq. 4-15, prove the distributive law of the cross product

$$\vec{A} \times (\vec{B} \times \vec{C}) = \vec{A} \times \vec{B} + \vec{A} \times \vec{C}$$

4-9.0c (a) Determine the magnitude of the cross product using the definition of cross product defined by the vector extending from the origin to (−4, 1, 7); vector $\vec{E}_1$.
(b) Repeat (a) for the vector extending from (3, −8, 11) to (−2, 5, 9); vector $\vec{E}_2$.
(c) Explain how one would obtain the direction for the cross product, and draw a sketch.
(d) Find the angle between the vectors $\vec{E}_1$ and $\vec{E}_2$. Is the solution unique?
(e) Obtain the cross product using the determinant formulation, call it $\vec{E}$, and obtain a vector in the direction of the cross product.
(f) What are the angular relationships between the unit vectors in the directions of $\vec{E}_1$, $\vec{E}_2$, and $\vec{E}$?

4-9.0d Develop computer programs that will evaluate $\vec{A} \cdot \vec{B}$ and $\vec{A} \times \vec{B}$ and give the angle between $\vec{A}$ and $\vec{B}$. Apply these routines to the following vector pairs:
(a) $\vec{E}_1 = 6\hat{x} - 3\hat{y}$, $\vec{H}_1 = -8\hat{x} + 5\hat{y} + 3\hat{z}$.
(b) $\vec{E}_2 = 9e^{j26^\circ}\hat{x} + 13e^{j26^\circ}\hat{y}$, $\vec{H}_2 = e^{j26^\circ}\hat{x} - 10e^{j26^\circ}\hat{y}$.

As with the dot product, special considerations are required of the cross product when phasors (complex vectors) are used. Consider first two sinusoidal steady-state vectors in the time domain,

$$\vec{\mathscr{E}}(t) = E_x \cos(\omega t + \phi_x)\hat{x} + E_y \cos(\omega t + \phi_y)\hat{y} + E_z \cos(\omega t + \phi_z)\hat{z}$$

and

$$\vec{\mathscr{H}}(t) = H_x \cos(\omega t + \theta_x)\hat{x} + H_y \cos(\omega t + \theta_y)\hat{y} + H_z \cos(\omega t + \theta_z)\hat{z}$$

Then

$$\begin{aligned}\vec{\mathscr{E}}(t) \times \vec{\mathscr{H}}(t) = {} & [E_y H_z \cos(\omega t + \phi_y)\cos(\omega t + \theta_z) - E_z H_y \cos(\omega t + \phi_z)\cos(\omega t + \theta_y)]\hat{x} \\ & + [E_z H_x \cos(\omega t + \phi_z)\cos(\omega t + \theta_x) - E_x H_z \cos(\omega t + \phi_x)\cos(\omega t + \theta_z)]\hat{y} \\ & + [E_x H_y \cos(\omega t + \phi_x)\cos(\omega t + \theta_y) - E_y H_x \cos(\omega t + \phi_y)\cos(\omega t + \theta_x)]\hat{z}\end{aligned}$$

Using the trig identity for the product of two cosines this becomes

$$\begin{aligned}\vec{\mathscr{E}}(t) \times \vec{\mathscr{H}}(t) = {} & \tfrac{1}{2}[E_y H_z \cos(2\omega t + \phi_y + \theta_z) + E_y H_z \cos(\phi_y - \theta_z) - E_z H_y \cos(2\omega t + \phi_z + \theta_y) \\ & - E_z H_y \cos(\phi_z - \theta_y)]\hat{x} + \tfrac{1}{2}[E_z H_x \cos(2\omega t + \phi_z + \theta_x) + E_z H_x \cos(\phi_z - \theta_x) \\ & - E_x H_z \cos(2\omega t + \phi_x + \theta_z) - E_x H_z \cos(\phi_x - \theta_z)]\hat{y} + \tfrac{1}{2}[E_x H_y \cos(2\omega t + \phi_x + \theta_y) \\ & + E_x H_y \cos(\phi_x - \theta_y) - E_y H_x \cos(2\omega t + \phi_y + \theta_x) - E_y H_x \cos(\phi_y - \theta_x)]\hat{z}\end{aligned}$$

If we take the *time* average of this expression and use the fact that $4 < \cos(n\omega t + \alpha)4 \geq 0$ over a period, we obtain

$$\begin{aligned}\overline{\vec{\mathscr{E}}(t) \times \vec{\mathscr{H}}(t)} = {} & \tfrac{1}{2}\{[E_y H_z \cos(\phi_y - \theta_z) - E_z H_y \cos(\phi_z - \theta_y)]\hat{x} \\ & + [E_z H_x \cos(\phi_z - \theta_x) - E_x H_z \cos(\phi_x - \theta_z)]\hat{y} \\ & + [E_x H_y \cos(\phi_x - \theta_y) - E_y H_x \cos(\phi_y - \theta_x)]\hat{z}\}\end{aligned}$$

Now if we try to decide what the equivalent operation would be using phasor notation, we first notice that we need to subtract the angles. When we multiply phasors the angles can be made to subtract if we conjugate one of the quantities first. Thus, for the phasor representations of $\vec{\mathscr{E}}$ and $\vec{\mathscr{H}}$,

$$\vec{E} = E_x e^{j\phi_x}\hat{x} + E_y e^{j\phi_y}\hat{y} + E_z e^{j\phi_z}\hat{z}$$

$$\vec{H} = H_x e^{j\theta_x}\hat{x} + H_y e^{j\theta_y}\hat{y} + H_z e^{j\theta_z}\hat{z}$$

we would try

$$\begin{aligned}\vec{E} \times \vec{H}^* & = (E_x e^{j\phi_x}\hat{x} + E_y e^{j\phi_y}\hat{y} + E_z e^{j\phi_z}\hat{z}) \times (H_x e^{-j\theta_x}\hat{x} + H_y e^{-j\theta_y}\hat{y} + H_z e^{-j\theta_z}\hat{z}) \\ & = [E_y H_z e^{j(\phi_y - \theta_z)} - E_z H_y e^{j(\phi_z - \theta_y)}]\hat{x} + [E_z H_x e^{j(\phi_z - \theta_x)} - E_x H_z e^{j(\phi_x - \theta_z)}]\hat{y} \\ & \quad [E_x H_y e^{j(\phi_x - \theta_y)} - E_y H_x e^{j(\phi_y - \theta_x)}]\hat{z}\end{aligned}$$

If we take the real part of this ($\mathrm{Re}\{e^{j\alpha}\} = \cos\alpha$) we obtain

$$\begin{aligned}\mathrm{Re}\{\vec{E} \times \vec{H}^*\} = {} & [E_y H_z \cos(\phi_y - \theta_z) - E_z H_y \cos(\phi_z - \theta_y)]\hat{x} \\ & [E_z H_x \cos(\phi_z - \theta_x) - E_x H_z \cos(\phi_x - \theta_z)] \\ & [E_x H_y \cos(\phi_x - \theta_y) - E_y H_x \cos(\phi_y - \theta_x)]\end{aligned}$$

Comparing this with the result of the cross product time domain average we conclude

$$\overline{\vec{\mathscr{E}}(t) \times \vec{\mathscr{H}}(t)} = \tfrac{1}{2}\,\mathrm{Re}\{\vec{E} \times \vec{H}^*\} \tag{4-16}$$

This result will prove to be important when we want to consider power flow in electromagnetic wave propagation. This will also be used to determine the wave impedance seen by a propagating wave. Although the proof has made use of rectangular coordinates, the result is also valid for any right-hand orthogonal system, since one can always convert a vector in any orthogonal system to rectangular form as will be shown in the next section. Let's next look at a few examples for some insight.

Example 4-11

$$\vec{E} = 3\angle 20°\hat{x} - 5\angle 40°\hat{y}$$

Compute $\vec{E} \times \vec{E}^*$:

$$\begin{aligned}\vec{E} \times \vec{E}^* &= (3\angle 20°\hat{x} - 5\angle 40°\hat{y}) \times (3\angle{-20°}\hat{x} - 5\angle{-40°}\hat{y}) \\ &= [-15\angle{-20°} - (-15\angle 20°)]\hat{z} = (15\angle 160° + 15\angle 20°)\hat{z} \\ &= 10.26\angle 90°\hat{z}\end{aligned}$$

or

$$\vec{E} \times \vec{E}^* = j10.26\hat{z}$$

This may seem to be an incorrect result, for certainly a vector is parallel to itself so that the cross product should be zero. But $\vec{E}^*$ is obviously not $\vec{E}$. If we look at the time domain we obtain, using Eq. 4-16,

$$\overline{\vec{\mathscr{E}}(t) \times \vec{\mathscr{E}}(t)} = \tfrac{1}{2}\,\mathrm{Re}\{j10.26\hat{z}\} = 0$$

so at least on the average (time) the cross product is zero. As a matter of fact, if one sets $\vec{\mathscr{H}}(t) = \vec{\mathscr{E}}(t)$ in the general cross product expansion $\vec{\mathscr{E}}(t) \times \vec{\mathscr{H}}(t)$ developed earlier, the result would be zero for every instant of time as well.

Example 4-12

In cylindrical coordinates $\vec{D} = 4e^{-j2z}\hat{\rho} + 3e^{-j2z}\hat{\phi}$. Compute the magnitude of $\vec{D}$.

$$|\vec{D}| = \sqrt{\vec{D} \cdot \vec{D}^*} = \sqrt{(4e^{-j2z})(4e^{j2z}) + (3e^{-j2z})(3e^{j2z})} = 5$$

In the time domain, $\vec{\mathscr{D}}(t) = 4\cos(\omega t - 2z)\hat{\rho} + 3\cos(\omega t - 2z)\hat{\phi}$. From this,

$$\begin{aligned}|\vec{\mathscr{D}}(t)|^2 &= 16\cos^2(\omega t - 2z) + 9\cos^2(\omega t - 2z) = 25\cos^2(\omega t - 2z) \\ &= \tfrac{25}{2}[\cos(2\omega t - 4z) + 1]\end{aligned}$$

Then

$$\overline{|\vec{\mathscr{D}}(t)|^2} = \tfrac{25}{2}$$

from which

$$\sqrt{\overline{|\vec{\mathscr{D}}(t)|^2}} = \frac{5}{\sqrt{2}}$$

which is in agreement with the general result developed earlier.

Example 4-13

Determine the spatial relationship of the two vectors

$$\vec{\mathscr{E}}(t) = 4\sin\pi x\cos(\omega t - 3z)\hat{y} \quad \text{and}$$

$$\vec{\mathscr{H}}(t) = 0.01\sin\pi x\cos(\omega t - 3z)\hat{x} + 0.003\cos\pi x\cos(\omega t - 3z)\hat{z}$$

Converting to phasor representations we first have

$$\vec{E} = 4 \sin \pi x e^{-j3z}\hat{y} \qquad \text{and} \qquad \vec{H} = 0.01 \sin \pi x e^{-j3z}\hat{x} + 0.003 \cos \pi x e^{-j3z}\hat{z}$$

Then

$$\vec{E} \cdot \vec{H}^* = 0$$

so the vectors are on time-average orthogonal. They are also orthogonal at each instant of time as well. Although this result is easy to see by inspection of the given vector quantities, the example shows the utility of the phasor representation.

Example 4-14

Given the two vectors of the preceding example. Evaluate $\vec{E} \times \vec{H}^*$ and show that $|\vec{E} \times \vec{H}^*| = |\vec{E}||\vec{H}|$. What would this mean?

Example 4-15

$$\vec{U} = 2\hat{\rho} + 2j\phi \qquad \text{and} \qquad \vec{V} = 2j\hat{\rho} - 2\hat{\phi}$$

What is the spatial relationship of these vectors?

$$\vec{U} \cdot \vec{V}^* = (2)(-j2) + (2j)(-2) = -j8$$

Then

$$\overline{\vec{\mathcal{U}}(t) \cdot \vec{\mathcal{V}}(t)} = \tfrac{1}{2}\text{Re}\{\vec{U} \cdot \vec{V}^*\} = \tfrac{1}{2}\text{Re}\{-j8\} = 0$$

This means that on the time average $\vec{\mathcal{U}}$ is orthogonal to $\vec{\mathcal{V}}$.

The preceding examples have shown that we may use the phasor representations in the vector products, but special care must be exercised in the physical interpretation of the results. We close this section with a short presentation of two common coordinate systems so that the ideas presented in the preceding section may be applied. The exercises that follow give an opportunity for analyzing a new coordinate system.

Example 4-16

Cylindrical Coordinate System

A point is located by the coordinate convention $P(\rho, \phi, z)$ as shown in Figure 4-26. The unit vectors $\hat{\rho}$, $\hat{\phi}$, and $\hat{z}$ are also shown at P. Note the unit vectors are directed to *increasing* coordinate values.

This is an *orthogonal* system since

$$\hat{\rho} \cdot \hat{\phi} = \hat{\phi} \cdot \hat{z} = \hat{z} \cdot \hat{\rho} = 0$$

This is a *right-handed* system since

$$\hat{\rho} \times \hat{\phi} = \hat{z}, \qquad \hat{\phi} \times \hat{z} = \hat{\rho}, \qquad \text{and} \quad \hat{z} \times \hat{\rho} = \hat{\phi}$$

in cyclic order of coordinate location, ρ–ϕ–z–ρ–ϕ.

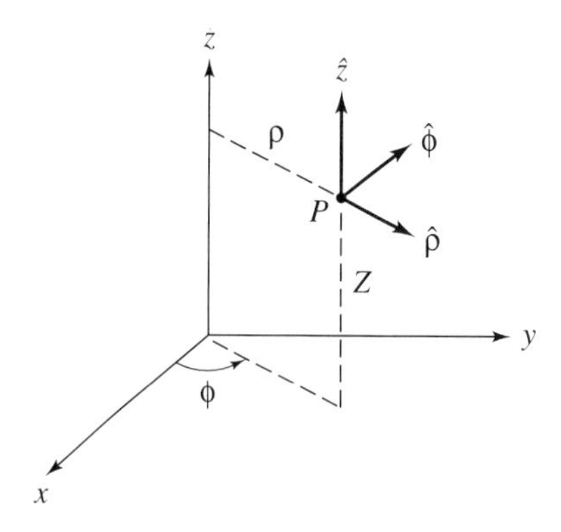

Figure 4-26. The cylindrical coordinate system describing $P(\rho, \phi, z)$.

Example 4-17

Spherical Coordinate System

A point is located by the coordinate convention $P(r, \theta, \phi)$ as shown in Figure 4-27. The unit vectors $\hat{r}$, $\hat{\theta}$, and $\hat{\phi}$ are also shown at P. As usual, the unit vectors are directed to increasing coordinate values.

This is an *orthogonal* system since

$$\hat{r} \cdot \hat{\theta} = \hat{\theta} \cdot \hat{\phi} = \hat{\phi} \cdot \hat{r} = 0$$

This is a *right-hand* system since

$$\hat{r} \times \hat{\theta} = \hat{\phi}, \qquad \hat{\theta} \times \hat{\phi} = \hat{r}, \qquad \text{and} \quad \hat{\phi} \times \hat{r} = \hat{\theta}$$

are in cyclic coordinate order r–θ–ϕ–r–θ. Eleven important coordinate systems have the orthogonal right-hand property. See the references [42] and [55] for additional discussions.

A word of caution is in order with respect to vector operations with vectors expressed in these other coordinate systems. There would seem to be such a close analogy in the systems (x, y, z), (ρ, ϕ, z), and (r, θ, ϕ) that one would be tempted to conclude that all

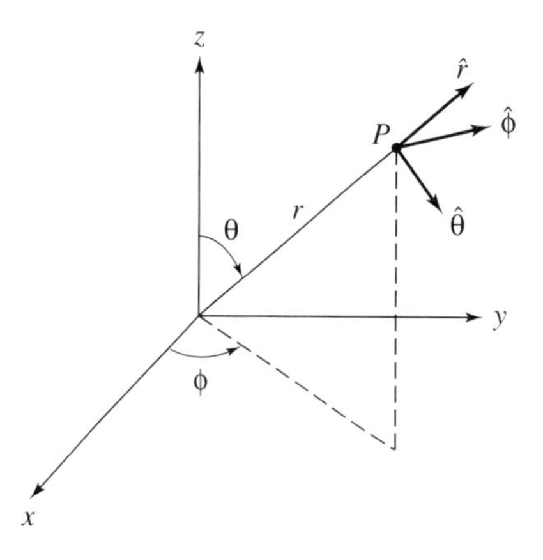

Figure 4-27. The spherical coordinate system describing $P(r, \theta, \phi)$.

the previous vector operations given in terms of components hold also by simply substituting ρ for x, etc., in the results. In general this is *not* true. Two examples will illustrate this.

Example 4-18

Consider an element of volume dV. In rectangular coordinates $dV = dxdydz$, but if we attempt to write, in cylindrical coordinates, $dV = d\rho d\phi dz$ we immediately see an error since dimensionally this would be *area*, not volume.

Example 4-19

Consider the sum of two vectors. Let $\vec{A} = 2\hat{x} + 3\hat{y}$ and $\vec{B} = 4\hat{x} + 8\hat{y}$. Then $\vec{C} = \vec{A} + \vec{B} = (2 + 4)\hat{x} + (3 + 8)\hat{y} = 6\hat{x} + 11\hat{y}$ is true. Now suppose we have, in cylindrical coordinates, $\vec{A} = 2\hat{\rho} + 3\hat{\phi}$ and $\vec{B} = 4\hat{\rho} + 8\hat{\phi}$, as shown in Figure 4-28, in the xy plane.

1. Vectors $\vec{A}$ and $\vec{B}$ are *not* at same point in space. Plotting these at P_A and P_B, we have (looking down the z axis)

$$\vec{C} = \vec{A} + \vec{B} \neq (2 + 4)\hat{\rho} + (3 + 8)\hat{\phi}$$

   The process fails since the $\hat{\rho}$ and $\hat{\phi}$ unit vectors are not parallel in both vectors. Although this is not a common operation, that is, adding two vectors at two different points in space, it points out clearly the need to use caution if one is tracking vectors while moving in space.

2. If the vectors $\vec{A}$ and $\vec{B}$ were at the *same* point, it would be permissible to use direct substitution:

$$\vec{C} = \vec{A} + \vec{B} = (2 + 4)\hat{\rho} + (3 + 8)\hat{\phi} = 6\hat{\rho} + 11\hat{\phi}$$

   since the corresponding unit vectors are now parallel.

Exercises

4-9.0e For a spherical coordinate system, sketch the three surfaces defined by (a) r = constant, (b) θ = constant, and (c) ϕ = constant.

4-9.0f The parabolic cylindrical coordinate system consists of intersecting parabolic sheets. The sheets extend to $\pm\infty$ in the z direction. Some of the sheet cross sections in the xy plane are shown in Figure 4-29. On the system at points P and Q, draw the unit vectors

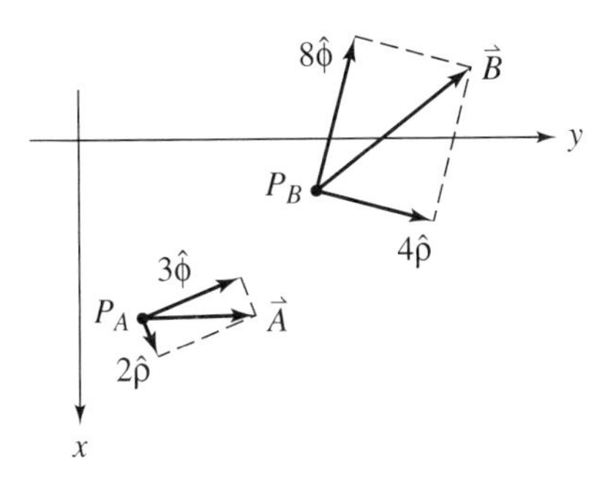

Figure 4-28. Two vectors represented in cylindrical coordinates in the xy plane.

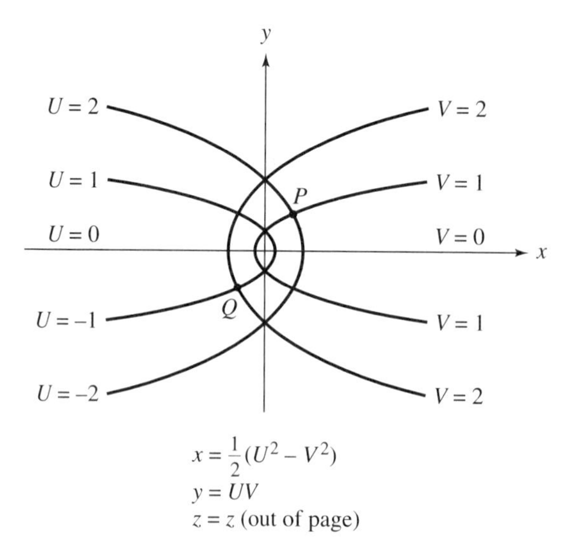

Figure 4-29.

and show that it is a right-hand orthogonal system if the coordinates of a point are located as $P(u, v, z)$. (The parabolas are also confocal.)

4-10 Multiplication of Vectors—III: Dyadic Product

This type of vector multiplication is better presented after a discussion of vector operators. The definitions for this kind of operation are in Section 4-17. Appendix G is an introductory discussion of the dyadic concept. Applications of this operation are not considered in this text.

4-11 Coordinate System Conversion

In the solution of electromagnetic problems it is often necessary to convert from one coordinate system to another. Here we will be concerned only with the more common rectangular, cylindrical, and spherical coordinate systems, but the principles apply to any coordinate conversion process. Since a vector consists of a magnitude (scalar) part and a direction, two steps are required to convert a vector from one coordinate system to another:

1. Change of coordinate variables to new system.
2. Change of components to new unit vectors.

The principle will be applied by converting a vector originally given in spherical coordinates into rectangular form; that is, convert $\vec{A} = A_r\hat{r} + A_\theta\hat{\theta} + A_\phi\hat{\phi}$ into rectangular form

$$\vec{A} = A_x\hat{x} + A_y\hat{y} + A_z\hat{z}$$

We will assume a particular $\vec{A}$ to demonstrate the process, and yet do some of the manipulations in general terms so that the process will be clear.

Suppose we have the specific vector given by the expression in spherical coordinates:

$$\vec{A} = A_r\hat{r} + A_\theta\hat{\theta} + A_\phi\hat{\phi} = \frac{\sin\theta}{r^2}\hat{r} + \frac{\cos\phi}{r^2}\hat{\theta} + \frac{\sin\theta}{r^2}\hat{\phi}$$

1. Change coordinate variables. Draw a diagram that locates a point P in both coordinate systems. For our specific example, this is shown in Figure 4-30.

   From the figure, we have the coordinate variable relations

$$r^2 = (\sqrt{x^2 + y^2})^2 + z^2 = x^2 + y^2 + z^2$$

$$\phi = \tan^{-1}\frac{y}{x}$$

and

$$\cos\phi = \frac{x}{\sqrt{x^2 + y^2}}$$

$$\sin\phi = \frac{y}{\sqrt{x^2 + y^2}}$$

$$\theta = \tan^{-1}\frac{\sqrt{x^2 + y^2}}{z}$$

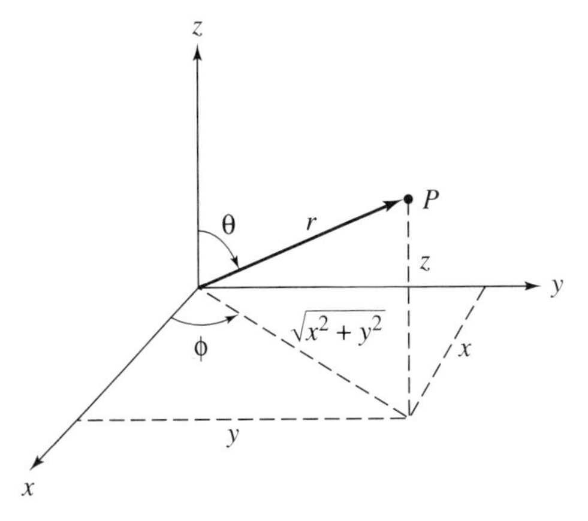

Figure 4-30. Coordinate variable conversion between spherical and rectangular systems.

and
$$\cos\theta = \frac{z}{\sqrt{x^2 + y^2 + z^2}}$$
$$\sin\theta = \frac{\sqrt{x^2 + y^2}}{\sqrt{x^2 + y^2 + z^2}}$$

(Note: If we were transforming from rectangular to spherical, we would initially write the relations for x, y, and z in terms of spherical coordinate variables.)

Thus our vector $\vec{A}$ to be transformed becomes, for the first step:

$$\vec{A} = A_r\hat{r} + A_\theta\hat{\theta} + A_\phi\hat{\phi} = \frac{\sqrt{x^2 + y^2}}{(x^2 + y^2 + z^2)\sqrt{x^2 + y^2 + z^2}}\,\hat{r} + \frac{x}{\sqrt{x^2 + y^2}(x^2 + y^2 + z^2)}\,\hat{\theta} + \frac{\sqrt{x^2 + y^2}}{(x^2 + y^2 + z^2)\sqrt{x^2 + y^2 + z^2}}\,\hat{\phi}$$

Note that the components are still A_r, A_θ, and A_ϕ, but they are in terms of the rectangular coordinates:

$$\vec{A} = \underbrace{\frac{\sqrt{x^2 + y^2}}{(x^2 + y^2 + z^2)^{3/2}}}_{A_r}\,\hat{r} + \underbrace{\frac{x}{\sqrt{x^2 + y^2}\,(x^2 + y^2 + z^2)}}_{A_\theta}\,\hat{\theta} + \underbrace{\frac{\sqrt{x^2 + y^2}}{(x^2 + y^2 + z^2)^{3/2}}}_{A_\phi}\,\hat{\phi} \tag{4-17}$$

2. Change the components to the new unit vectors $\hat{x}$, $\hat{y}$, and $\hat{z}$. Now A_x is the component of $\vec{A}$ in the direction of $\hat{x}$. Using the interpretation of the dot product given earlier as a component in a given direction:

$$\begin{aligned} A_x &= \vec{A}\cdot\hat{x} = (A_r\hat{r} + A_\theta\hat{\theta} + A_\phi\hat{\theta})\cdot\hat{x} \\ &= A_r\hat{r}\cdot\hat{x} + A_\theta\hat{\theta}\cdot\hat{x} + A_\phi\hat{\phi}\cdot\hat{x} \end{aligned} \tag{4-18}$$

We draw three unit-vector diagrams that show the relationship of each of the spherical unit vectors to the rectangular unit vectors. See Figure 4-31. It is best to draw such diagrams with point P in the principal octant of the rectangular system. From these figures we obtain expressions for the three dot products required in Eq. 4-18.

Since the angle between the unit vectors $\hat{x}$ and $\hat{r}$ is not known directly, two trigonometric projections are required to find the amount of $\hat{x}$ in the direction of $\hat{r}$ (Similarly for $\hat{y}$ on $\hat{r}$ and $\hat{\theta}$ on $\hat{x}$.) Thus, from Figure 4-31*a* we have, remembering unit vectors have unit magnitude,

$$\begin{aligned} \hat{r}\cdot\hat{x} = \hat{x}\cdot\hat{r} &= [(|\hat{x}|\cos\phi)\cos(90-\theta)]|\hat{r}| = |\hat{x}||\hat{r}|\cos\phi\sin\theta \\ &= \cos\phi\sin\theta = \frac{x}{\sqrt{x^2 + y^2}}\cdot\frac{\sqrt{x^2 + y^2}}{\sqrt{x^2 + y^2 + z^2}} = \frac{x}{\sqrt{x^2 + y^2 + z^2}} \end{aligned}$$

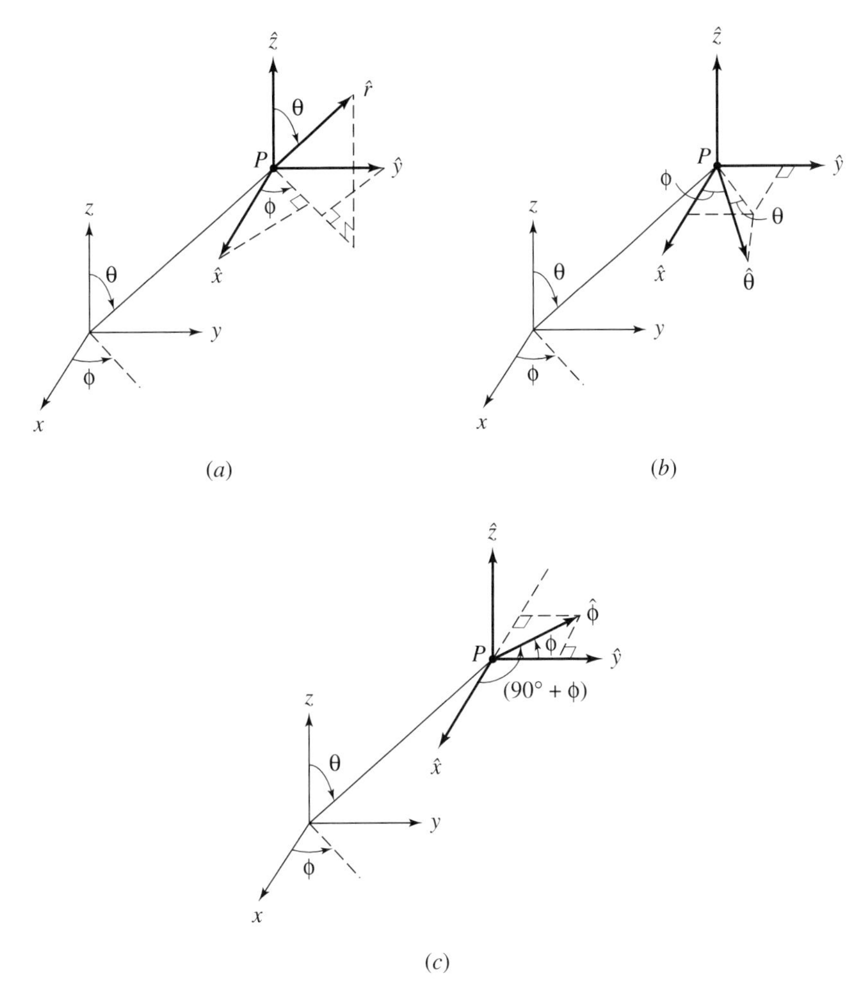

Figure 4-31. Spherical coordinate system unit vectors separately drawn with respect to rectangular coordinate unit vectors.

Similarly from Figure 4-31*b* we obtain

$$\hat{\theta} \cdot \hat{x} = \hat{x} \cdot \hat{\theta} = ((|\hat{x}| \cos \phi) \cos \theta)|\hat{\theta}| = \cos \phi \cos \theta$$

$$= \frac{xz}{\sqrt{x^2 + y^2}\sqrt{x^2 + y^2 + z^2}}$$

From Figure 4-31*c*

$$\hat{\phi} \cdot \hat{x} = \hat{x} \cdot \hat{\phi} = \cos(90^\circ + \phi) = -\sin \phi = -\frac{y}{\sqrt{x^2 + y^2}}$$

Using these three unit-vector dot products in Eq. 4-18

$$A_x = A_r \frac{x}{\sqrt{x^2 + y^2 + z^2}} + A_\theta \frac{xz}{\sqrt{x^2 + y^2}\,\sqrt{x^2 + y^2 + z^2}} + A_\phi\left(-\frac{y}{\sqrt{x^2 + y^2}}\right)$$

We next substitute the expressions for the spherical components as evaluated in step 1, Eq. 4-17) to obtain the final result for A_x:

$$A_x = \frac{\sqrt{x^2 + y^2}}{(x^2 + y^2 + z^2)^{3/2}} \cdot \frac{x}{\sqrt{x^2 + y^2 + z^2}} + \frac{x}{\sqrt{x^2 + y^2}\,(x^2 + y^2 + z^2)} \cdot \frac{xz}{\sqrt{x^2 + y^2}\,\sqrt{x^2 + y^2 + z^2}} + \frac{\sqrt{x^2 + y^2}}{(x^2 + y^2 + z^2)^{3/2}} \cdot \frac{-y}{\sqrt{x^2 + y^2}}$$

$$A_x = \frac{x\sqrt{x^2 + y^2}}{(x^2 + y^2 + z^2)^2} + \frac{x^2 z}{(x^2 + y^2)(x^2 + y^2 + z^2)^{3/2}} - \frac{y}{(x^2 + y^2 + z^2)^{3/2}}$$

Following a similar procedure, the y component is

$$\begin{aligned} A_y &= \vec{A} \cdot \hat{y} = A_r\hat{r} \cdot \hat{y} + A_\theta\hat{\theta} \cdot \hat{y} + A_\phi\hat{\phi} \cdot \hat{y} \\ &= A_r \frac{y}{\sqrt{x^2 + y^2 + z^2}} + A_\theta \frac{yz}{\sqrt{x^2 + y^2}\,\sqrt{x^2 + y^2 + z^2}} + A_\phi \frac{x}{\sqrt{x^2 + y^2}} \\ &= \frac{y\sqrt{x^2 + y^2}}{(x^2 + y^2 + z^2)^2} + \frac{xyz}{(x^2 + y^2)(x^2 + y^2 + z^2)^{3/2}} + \frac{x}{(x^2 + y^2 + z^2)^{3/2}} \end{aligned}$$

Also,

$$\begin{aligned} A_z &= \vec{A} \cdot \hat{z} = A_r\hat{r} \cdot \hat{z} + A_\theta\hat{\theta} \cdot \hat{z} + A_\phi\hat{\phi} \cdot \hat{z} \\ &= A_r \frac{z}{\sqrt{x^2 + y^2 + z^2}} - A_\theta \frac{\sqrt{x^2 + y^2}}{\sqrt{x^2 + y^2 + z^2}} + A_\phi \cdot 0 \\ &= \frac{z\sqrt{x^2 + y^2}}{(x^2 + y^2 + z^2)^2} - \frac{x}{(x^2 + y^2 + z^2)^{3/2}} \end{aligned}$$

The transformed vector is then constructed as

$$\vec{A} = A_x\hat{x} + A_y\hat{y} + A_z\hat{z}$$

where A_x, A_y, and A_z are as evaluated previously.

Exercises

4-11.0a By evaluating the dot products $A_y = \vec{A} \cdot \hat{y}$ and $A_z = \vec{A} \cdot \hat{z}$ as done in step 2 in the preceding discussion, obtain the two equations for A_y and A_z given above in terms of A_r, A_θ, A_ϕ and the rectangular coordinates.

4-11.0b Obtain an expression for the unit *vector* $\hat{\theta}$ (spherical coordinates) in terms of the rectangular coordinate system unit *vectors* $\hat{x}$, $\hat{y}$, $\hat{z}$.

Table 4-1. Rectangular, Cylindrical, Spherical Coordinate Conversion Relationships

	Rectangular to Cylindrical		Cylindrical to Rectangular
Variable change	$x = \rho\cos\phi$, $y = \rho\sin\phi$, $z = z$	Variable change	$\rho = \sqrt{x^2 + y^2}$; $\phi = \tan^{-1}\left(\frac{y}{x}\right)$ $\left\{ \sin\phi = \frac{y}{\sqrt{x^2 + y^2}},\ \cos\phi = \frac{x}{\sqrt{x^2 + y^2}} \right.$; $z = z$
Component change	$A_\rho = A_x\cos\phi + A_y\sin\phi$; $A_\phi = -A_x\sin\phi + A_y\cos\phi$; $A_z = A_z$	Component change	$A_x = A_\rho\frac{x}{\sqrt{x^2 + y^2}} - A_\phi\frac{y}{\sqrt{x^2 + y^2}}$; $A_y = A_\rho\frac{y}{\sqrt{x^2 + y^2}} + A_\phi\frac{x}{\sqrt{x^2 + y^2}}$; $A_z = A_z$

	Rectangular to Spherical		Spherical to Rectangular
Variable change	$x = r\sin\theta\cos\phi$, $y = r\sin\theta\sin\phi$, $z = r\cos\theta$	Variable change	$r = \sqrt{x^2 + y^2 + z^2}$; $\theta = \cos^{-1}\frac{z}{\sqrt{x^2 + y^2 + z^2}}$ $\left\{ \cos\theta = \frac{z}{\sqrt{x^2 + y^2 + z^2}},\ \sin\theta = \frac{\sqrt{x^2 + y^2}}{\sqrt{x^2 + y^2 + z^2}} \right.$; $\phi = \tan^{-1}\left(\frac{y}{x}\right)$ $\left\{ \cos\phi = \frac{x}{\sqrt{x^2 + y^2}},\ \sin\phi = \frac{y}{\sqrt{x^2 + y^2}} \right.$
Component change	$A_r = A_x\sin\theta\cos\phi + A_y\sin\theta\sin\phi + A_z\cos\theta$; $A_\theta = A_x\cos\theta\cos\phi + A_y\cos\theta\sin\phi - A_z\sin\theta$; $A_\phi = -A_x\sin\phi + A_y\cos\phi$	Component change	$A_x = \frac{A_r x}{\sqrt{x^2 + y^2 + z^2}} + \frac{A_\theta xz}{\sqrt{(x^2 + y^2)(x^2 + y^2 + z^2)}} - \frac{A_\phi y}{\sqrt{x^2 + y^2}}$; $A_y = \frac{A_r y}{\sqrt{x^2 + y^2 + z^2}} + \frac{A_\theta yz}{\sqrt{(x^2 + y^2)(x^2 + y^2 + z^2)}} + \frac{A_\phi x}{\sqrt{x^2 + y^2}}$; $A_z = \frac{A_r z}{\sqrt{x^2 + y^2 + z^2}} - \frac{A_\theta\sqrt{x^2 + y^2}}{\sqrt{x^2 + y^2 + z^2}}$

4-11.0c For the parabolic cylindrical coordinate system described in Exercise 4-9.0f, evaluate the unit-vector dot products $\hat{u} \cdot \hat{x}$, $\hat{v} \cdot \hat{x}$, $\hat{u} \cdot \hat{y}$, $\hat{v} \cdot \hat{y}$, $\hat{u} \cdot \hat{z}$, and $\hat{v} \cdot \hat{z}$. These are to be in terms of x, y, and z.

To avoid having to go through all the general steps each time, tables that eliminate steps that would yield the same results each time have been prepared. Even when the tables are used, both steps must be completed. The tables compile the drearier details of the unit-vector dot product computations of step 2. Table 4-1 is a complete table for conversion between the more common cylindrical and spherical systems and the rectangular system. See also Appendix E.

Exercises

4-11.0d Develop a table similar to Table 4-1 that will allow a coordinate conversion between the rectangular coordinate system and the parabolic cylindrical coordinate system of Exercise 4-9.0f.

4-11.0e Develop a computer program that has as the input a spherical coordinate system vector and outputs a rectangular coordinate system vector. Allow for complex scalar parts all having the same angle.

Example 4-20

We conclude this section with an example that shows how Table 4-1 is used.

The following vector is in cylindrical coordinates and is to be converted to rectangular coordinates:

$$\vec{A} = 6\rho \cos \phi\hat{\rho} + 2z\rho\hat{\phi} - 3z \sin \phi\hat{z}$$

We think of this vector as the form $\vec{A} = A_\rho\hat{\rho} + A_\phi\hat{\phi} + A_z\hat{z}$.

Note that each of the vector components is a function of several coordinate variables ρ, ϕ, z.

1. Change coordinate variables. From Table 4-1 we have

$$\rho = \sqrt{x^2 + y^2} \qquad \cos \phi = \frac{x}{\sqrt{x^2 + y^2}} \qquad \sin \phi = \frac{y}{\sqrt{x^2 + y^2}} \qquad z = z$$

Thus making these substitutions in the given vector,

$$\vec{A} = \frac{6\sqrt{x^2 + y^2}\, x}{\sqrt{x^2 + y^2}}\hat{\rho} + 2z\sqrt{x^2 + y^2}\,\hat{\phi} - 3z\frac{y}{\sqrt{x^2 + y^2}}\hat{z}$$

$$\vec{A} = \underbrace{6x}_{A_\rho}\hat{\rho} + \underbrace{2z\sqrt{x^2 + y^2}}_{A_\phi}\,\hat{\phi} - \underbrace{\frac{3yz}{\sqrt{x^2 + y^2}}}_{A_z}\hat{z}$$

2. Change components to rectangular components. Again using Table 4-1 we calculate each of the components A_x, A_y, and A_z.

$$A_x = A_\rho \frac{x}{\sqrt{x^2 + y^2}} - A_\phi \frac{y}{\sqrt{x^2 + y^2}} = \frac{6x \cdot x}{\sqrt{x^2 + y^2}} - \frac{2z\sqrt{x^2 + y^2}\, y}{\sqrt{x^2 + y^2}}$$

$$= \frac{6x^2}{\sqrt{x^2 + y^2}} - 2zy$$

$$A_y = A_\rho \frac{y}{\sqrt{x^2 + y^2}} + A_\phi \frac{x}{\sqrt{x^2 + y^2}} = \frac{6xy}{\sqrt{x^2 + y^2}} + 2z\sqrt{x^2 + y^2} \frac{x}{\sqrt{x^2 + y^2}}$$

$$= \frac{6xy}{\sqrt{x^2 + y^2}} + 2xz$$

$$A_{z_{\text{rect}}} = A_{z_{\text{cyl}}} = \frac{-3yz}{\sqrt{x^2 + y^2}}$$

We now form the final transformed vector:

$$\vec{A} = A_x\hat{x} + A_y\hat{y} + A_z\hat{z} = \left(\frac{6x^2}{\sqrt{x^2 + y^2}} - 2zy\right)\hat{x}$$

$$+ \left(\frac{6xy}{\sqrt{x^2 + y^2}} + 2xz\right)\hat{y} - \frac{3yz}{\sqrt{x^2 + y^2}}\hat{z}$$

Exercises

4-11.0f Make the following coordinate system conversions:

(a) $\frac{y}{x}\hat{x}$ rectangular → cylindrical.

(b) $\rho\hat{\rho} + \theta\hat{\theta}$ cylindrical → rectangular.

(c) $y\hat{x} + x\hat{y} + z\hat{z}$ rectangular → spherical.

(d) $\frac{1}{r \sin \theta}\hat{\phi}$ spherical → rectangular.

4-11.0g Develop a computer program that will take a given vector in rectangular coordinates and convert it to cylindrical coordinates. Apply the subroutine to the vector given by

$$\vec{E}(x, y, z) = 2xy\hat{x} - 3yz\hat{y} + \hat{z} \qquad \text{at the point} \quad P(x, y, z) = (4, -5, 6)$$

4-12 Generalized Curvilinear Coordinates

Before continuing the study of vector operations, we digress to develop ideas that will allow us to derive our vector operations in forms applicable to more than just rectangular, cylindrical, and spherical coordinate systems. Here we limit the work to right-hand orthogonal coordinate systems since these will be the only types we shall use. A few preliminary comments will set the stage for what follows:

1. In rectangular coordinates as only one coordinate changes value, the locus traced out by a point is a straight line as shown in Figure 4-32a. However, if we consider cylin-

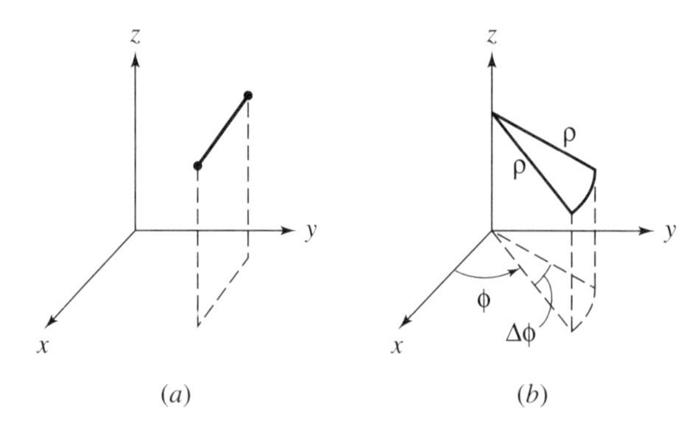

Figure 4-32. Paths generated by allowing only one coordinate to vary. (*a*) Rectangular, x only. (*b*) Cylindrical, ϕ only.

drical coordinates and let ϕ only increase in value, the locus of a point is an arc as shown in Figure 4-32*b*. Additionally, even though ρ is a constant, its position in space changes.

We conclude, then, that in a generalized coordinate system, we must allow for curvature of one coordinate varying contours (paths), even though unit vectors are always orthogonal at a given point.

2. Now rather than let only one coordinate vary suppose we hold only one coordinate constant. This generates a surface. In rectangular coordinates, holding x constant generates a plane surface. See Figure 4-33*a*. Now if we take cylindrical coordinates and hold ρ constant, we obtain a cylinder or curved surface as shown in figure 4-33*b*. Thus to have a generalized coordinate system, we need to allow curved surfaces to be generated when one coordinate is held constant.
3. When we transfer from one coordinate system to a rectangular system, we find that the new coordinate system variables may depend upon more than one rectangular coordinate. For instance, in spherical coordinates r is a function of x, y, and z as $r = \sqrt{x^2 + y^2 + z^2}$. In cylindrical coordinates, the radial coordinate is given by $\rho = \sqrt{x^2 + y^2}$. It follows then that in generalized coordinates our coordinate variables must

Figure 4-33. Surfaces generated by holding one coordinate constant. (*a*) Rectangular, x constant. (*b*) Cylindrical, ρ constant.

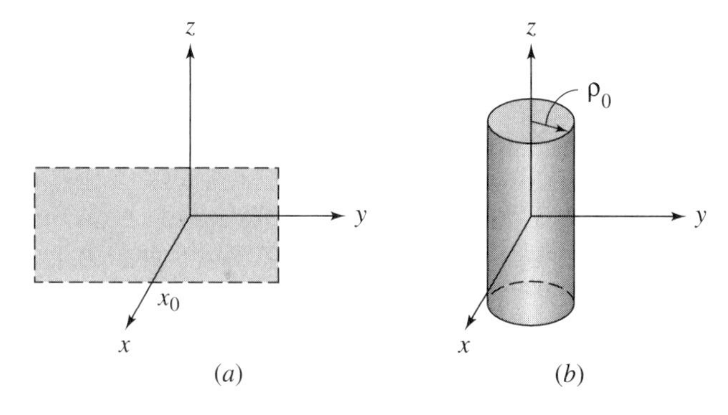

be allowed to depend upon combinations of x, y, and z. Of course, in specific coordinate systems, all three rectangular coordinates may not be involved.

With the foregoing comments in mind we then set up a right-hand, orthogonal, generalized coordinate system using the three coordinates u_1, u_2, and u_3 for which a point is located as $P(u_1, u_2, u_3)$. We associate a unit vector with each coordinate, directed toward increasing coordinate values as before, and identify them as $\hat{u}_1$, $\hat{u}_2$, and $\hat{u}_3$. To be a right-hand system requires $\hat{u}_1 \times \hat{u}_2 = \hat{u}_3$. As specific examples we would have:

Cylindrical coordinates: $u_1 = \rho \quad u_2 = \phi \quad u_3 = z$

Spherical coordinates: $u_1 = r \quad u_2 = \theta \quad u_3 = \phi$

In general, the coordinates will be functions of x, y, and z as discussed earlier so we may write

$$\left.\begin{aligned} u_1 &= u_1(x, y, z) \\ u_2 &= u_2(x, y, z) \\ u_3 &= u_3(x, y, z) \end{aligned}\right\} \tag{4-19}$$

Assuming that each point is unique (single-valued) in its representation (i.e., only one set u_1, u_2, u_3 for a given (x, y, z)) we could solve these equations to obtain the inverse relationships:

$$\left.\begin{aligned} x &= x(u_1, u_2, u_3) \\ x &= y(u_1, u_2, u_3) \\ z &= z(u_1, u_2, u_3) \end{aligned}\right\} \tag{4-20}$$

From these equations if we hold, say, u_1 and u_3 constant so that only u_2 varies, we observe that in general a curve will be traced in space since x and y and z are all (generally) dependent upon u_2.

The example in cylindrical coordinates given above illustrates this. For cylindrical coordinates

$$\begin{aligned} x &= \rho \cos \phi \\ y &= \rho \sin \phi \\ z &= z \end{aligned}$$

Holding ρ and z constant results in both x and y varying with ϕ $(\sim u_2)$ so that a circular arc is generated as shown in Figure 4-32*b*.

Exercises

4-12.0a For the following coordinate systems draw the curves described by *varying* the one coordinate indicated (give equations):

(a) spherical coordinates: θ.

(b) Cylindrical coordinates: ρ.

(c) Parabolic cylindrical coordinates: v (see Exercise 4-9.0f).

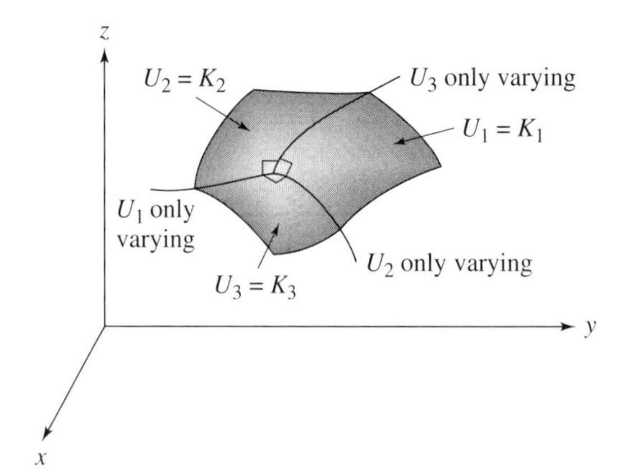

Figure 4-34. General surfaces and contours in a generalized coordinate system.

4-12.0b For the following coordinate systems draw the surfaces described by holding *constant* the coordinate indicated:

(a) Spherical coordinates: θ.

(b) Cylindrical coordinates: ρ.

(c) Parabolic cylindrical coordinates: v (Exercise 4-9.0f).

From the results of the preceding discussion, we would then draw the generalized coordinate configuration shown in Figure 4-34. Shown are general contours (curves) for which only one coordinate varies and the general surfaces for which only one coordinate is constant. Note that the curves intersect at right angles since we limit our discussion to orthogonal systems.

If we select a point $P(u_1, u_2, u_3)$ we can show the unit vectors at that point as given in Figure 4-35. The unit vectors $\hat{u}_1$, $\hat{u}_2$, and $\hat{u}_3$ are given to define a right-hand system. Also, the unit vectors are tangent to the respective coordinate contours at P.

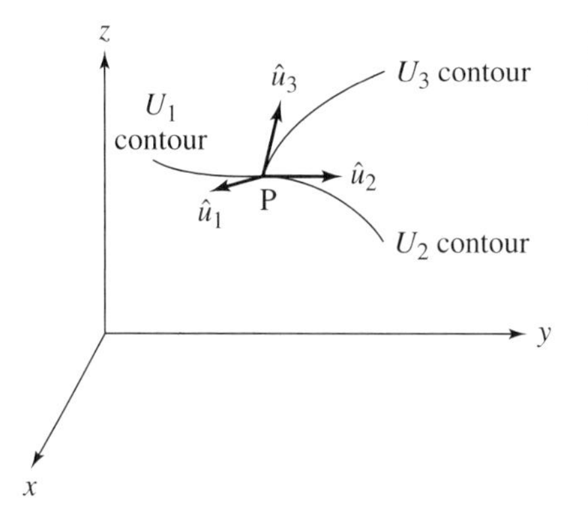

Figure 4-35. Unit vector representations in general curvilinear coordinates.

4-12.1 Generalized Curvilinear Coordinate Representation of Length

We first consider the representation of lengths. Some of our coordinates u_1, u_2, or u_3 may be angles. In rectangular coordinates, x, y, z, we can represent lengths as dx, dy, and dz. In cylindrical coordinates—ρ, ϕ, z—following this analogy would yield $d\rho$, $d\phi$, and dz. However, we see that $d\phi$ is not dimensionally length so we need some multiplying scale factor h to preserve the length dimension. Note that the h values need not be constants. For example, in the cylindrical coordinate system of Figure 4-36, the correct expression for a length generated by an angle change $d\phi$ is $dl = \rho d\phi$. In this case, the scale factor for the ϕ coordinate is ρ Thus we set length representations as

$$\left.\begin{aligned} dl_1 &= h_1 du_1 \\ dl_2 &= h_2 du_2 \\ dl_3 &= h_3 du_3 \end{aligned}\right\} \tag{4-21}$$

In some specific coordinate systems some of the hs may be unity, and in fact they all are for the rectangular system. Since distance is a vector we write in general:

$$\left.\begin{aligned} d\vec{l}_1 &= dl_1\hat{u}_1 = h_1 du_1\hat{u}_1 \\ d\vec{l}_2 &= dl_2\hat{u}_2 = h_2 du_2\hat{u}_2 \\ d\vec{l}_3 &= dl_3\hat{u}_3 = h_3 du_3\hat{u}_3 \end{aligned}\right\} \tag{4-22}$$

The total differential distance $d\vec{l}$ would then be the sum of the components:

$$\begin{aligned} d\vec{l} &= d\vec{l}_1 + d\vec{l}_2 + d\vec{l}_3 = dl_1\hat{u}_1 + dl_2\hat{u}_2 + dl_3\hat{u}_3 \\ &= h_1 du_1\hat{u}_1 + h_2 du_2\hat{u}_2 + h_3 du_3\hat{u}_3 \end{aligned} \tag{4-23}$$

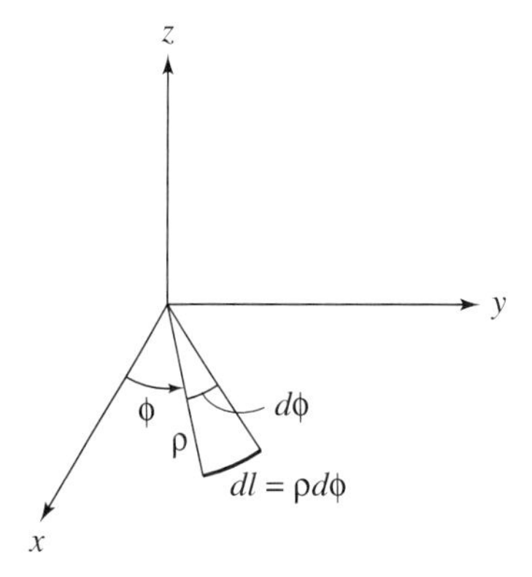

Figure 4-36. Demonstration of the need for a scale factor for the ϕ coordinate in the cylindrical system.

All we need now is a way of getting the hs. From partial differential calculus, we calculate the total differential from the partial derivatives as

$$d\vec{l} = \frac{\partial \vec{l}}{\partial u_1} du_1 + \frac{\partial \vec{l}}{\partial u_2} du_2 + \frac{\partial \vec{l}}{\partial u_3} du_3 \tag{4-24}$$

Comparing the coefficients of the two equations for $d\vec{l}$, Eqs. 4-23 and 4-24, we find, upon equating differential coefficients, which is legitimate for orthogonal systems,

$$\left.\begin{aligned} \frac{\partial \vec{l}}{\partial u_1} &= h_1 \hat{u}_1 \\ \frac{\partial \vec{l}}{\partial u_2} &= h_2 \hat{u}_2 \\ \frac{\partial \vec{l}}{\partial u_3} &= h_3 \hat{u}_3 \end{aligned}\right\} \tag{4-25}$$

Or, since the magnitude of a unit vector is one (by definition),

$$\left.\begin{aligned} h_1 &= \left|\frac{\partial \vec{l}}{\partial u_1}\right| \\ h_2 &= \left|\frac{\partial \vec{l}}{\partial u_2}\right| \\ h_3 &= \left|\frac{\partial \vec{l}}{\partial u_3}\right| \end{aligned}\right\} \tag{4-26}$$

This doesn't seem to have helped much since to find $\vec{l}$ we need to have the scale factors h, anyway, but it's a start. To continue, in rectangular coordinates a length $\vec{l}$ may be written

$$\vec{l} = x\hat{x} + y\hat{y} + z\hat{z}$$

Now from Eq. 4-20 x, y, and z can be expressed in terms of u_1, u_2, and u_3. Thus taking the partial derivative of $\vec{l}$ with respect to u_1 we obtain

$$\frac{\partial \vec{l}}{\partial u_1} = x \frac{\partial \hat{x}}{\partial u_1} + \hat{x} \frac{\partial x}{\partial u_1} + y \frac{\partial \hat{y}}{\partial u_1} + \hat{y} \frac{\partial y}{\partial u_1} + z \frac{\partial \hat{z}}{\partial_1} + \hat{z} \frac{\partial z}{\partial u_1}$$

The unit vectors $\hat{x}$, $\hat{y}$, and $\hat{z}$ are always directed the same way in space regardless of the location of the vector in the region. Thus these unit vectors do not vary with u_1, u_2, or u_3. Then

$$\frac{\partial \hat{x}}{\partial u_1} = \frac{\partial \hat{y}}{\partial u_1} = \frac{\partial \hat{z}}{\partial u_1} = \cdots = 0$$

for which the length derivative becomes

$$\frac{\partial \vec{l}}{\partial u_1} = \frac{\partial x}{\partial u_1}\hat{x} + \frac{\partial y}{\partial u_1}\hat{y} + \frac{\partial z}{\partial u_1}\hat{z}$$

Now the magnitude of a vector is the square root of the vector dotted with itself. It is independent of the coordinate system as long as it is orthogonal. Then from the preceding derivative:

$$\left|\frac{\partial \vec{l}}{\partial u_1}\right| = \left(\frac{\partial \vec{l}}{\partial u_1}\cdot\frac{\partial \vec{l}}{\partial u_1}\right)^{1/2} = \left[\left(\frac{\partial x}{\partial u_1}\right)^2 + \left(\frac{\partial y}{\partial u_1}\right)^2 + \left(\frac{\partial z}{\partial u_1}\right)^2\right]^{1/2}$$

Similarly,

$$\left|\frac{\partial \vec{l}}{\partial u_2}\right| = \left[\left(\frac{\partial x}{\partial u_2}\right)^2 + \left(\frac{\partial y}{\partial u_2}\right)^2 + \left(\frac{\partial x}{\partial u_2}\right)^2\right]^{1/2}$$

$$\left|\frac{\partial \vec{l}}{\partial u_3}\right| = \left[\left(\frac{\partial x}{\partial u_3}\right)^2 + \left(\frac{\partial y}{\partial u_3}\right)^2 + \left(\frac{\partial z}{\partial u_3}\right)^2\right]^{1/2}$$

Combining these with Eq. 4-26 we may write the general scale factor result

$$h_i = \left[\left(\frac{\partial x}{\partial u_i}\right)^2 + \left(\frac{\partial y}{\partial u_i}\right)^2 + \left(\frac{\partial z}{\partial u_i}\right)^2\right]^{1/2}, \qquad i = 1, 2, 3 \tag{4-27}$$

Thus if we can find the equations for x, y, and z as functions of the new coordinate system variables u_1, u_2, and u_3, we can determine the scale factors h_1, h_2, and h_3 required by Eqs. 4-22 and 4-23.

Example 4-21

Consider cylindrical coordinates, where we have $u_1 = \rho$, $u_2 = \phi$, $u_3 = z$. (See Figure 4-37.) From the diagram:

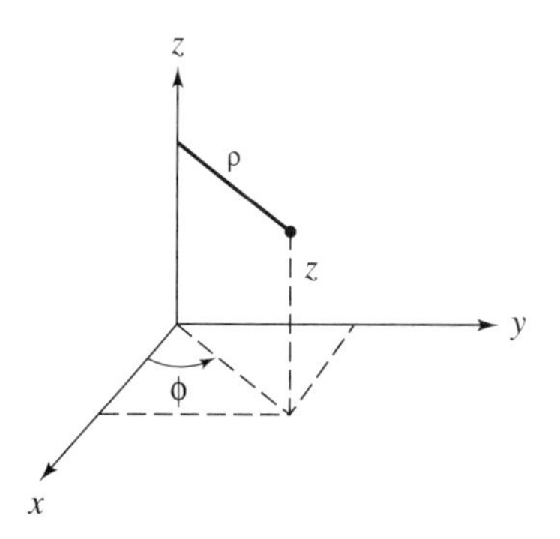

Figure 4-37. Cylindrical coordinate example.

$$x = \rho \cos \phi$$
$$y = \rho \sin \phi$$
$$z = z$$

Using Eq. 4-27:

$$h_1 = \sqrt{(\cos \phi)^2 + (\sin \phi)^2 + (0)^2} = 1$$
$$h_2 = \sqrt{(-\rho \sin \phi)^2 + (\rho \cos \phi)^2 + (0)^2} = \rho$$
$$h_3 = \sqrt{(0)^2 + (0)^2 + (1)^2} = 1$$

Thus, from Eqs. 4-21, since $dl_i = h_i du_i$, $i = 1, 2, 3$,

$$dl_1 = d\rho$$
$$dl_2 = \rho d\phi$$
$$dl_3 = dz$$

which agree with known results in cylindrical coordinates. Then

$$d\vec{l} = d\rho\hat{\rho} + \rho d\phi\hat{\phi} + dz\hat{z}$$

Rather than having to repeatedly calculate each h_i for all the coordinate systems, these have been tabulated, and they are given in Morse and Feshbach [42] and Spiegel [55].

Exercises

4-12.1a The parabolic cylindrical coordinate system consists of parabolic sheets parallel to the z axis as shown in Figure 4-38. The location of a point P is denoted by $P(u, v, z)$. The z axis is out of the page. With respect to the xyz axes (rectangular coordinates) the defining equations are

$$x = \tfrac{1}{2}(u^2 - v^2) \qquad -\infty \leq u \leq \infty$$
$$y = uv \qquad \text{for)} \qquad v \geq 0$$
$$z = z \qquad z = z$$

(a) At point P(2, $\frac{5}{2}$, 0) show the unit vectors.
(b) Find the scale factors h_i for the system.
(c) Express a general vector $\vec{A}$ in its component form. (See Eq. 4-4.)
(d) Give expressions for differential lengths $d\vec{l}$, dl_1, dl_2, dl_3.

4-12.1b The bipolar coordinate system consists of cylinders parallel to the z axis (out of the page in Figure 4-39). The location of a point P is denoted by $P(u, v, z)$. Surfaces u = constant are cylinders whose axes are on the y axis. Surfaces v = constant are cylinders having axes on the x axis. The defining equations are

$$y = \frac{a \sin u}{\cosh v - \cos u}, \qquad z = z, \qquad x = \frac{a \sinh v}{\cosh v - \cos u}$$

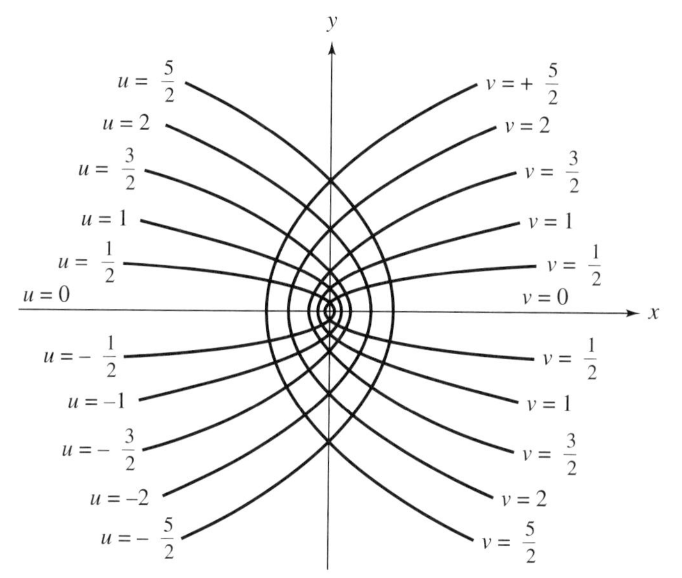

Figure 4-38.

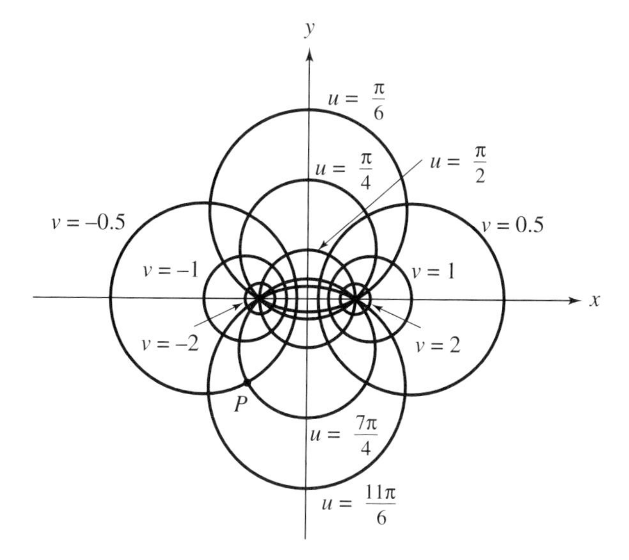

Figure 4-39.

$v = \pm\infty$ are points on the x axis at locations $(-a, 0)$ and $(a, 0)$ where a is a constant. $\pm\, a$ are where the constant u circles intersect on the x axis.

(a) At point P shown, $P(7\pi/4, -5, 0)$, draw the unit vectors $\hat{u}$ and $\hat{v}$.

(b) Find the scale factors for this system and simplify using appropriate hyperbolic and trigonometric identities.

(c) Give the expressions for the differential lengths $d\vec{l}$, dl_1, dl_2, dl_3.

(d) What are the contours for $u = 0$ and $u = \infty$?

4-12.2 Generalized Curvilinear Coordinate Representation of Area

We next consider the representation of area in generalized coordinates. We remember that areas are vectors with the direction normal to the surface. There are three combinations of the dl_i, which yield areas

$$\left.\begin{aligned} \vec{da}_1 &= dl_2 dl_3 \hat{u}_1 = h_2 h_3 du_2 du_3 \hat{u}_1 \\ \vec{da}_2 &= dl_3 dl_1 \hat{u}_2 = h_3 h_1 du_3 du_1 \hat{u}_2 \\ \vec{da}_3 &= dl_1 dl_2 \hat{u}_3 = h_1 h_2 du_1 du_2 \hat{u}_3 \end{aligned}\right\} \tag{4-28}$$

Dimensionally, these must work out to be square meters.

Example 4-22

Consider the cylindrical coordinate system with $u_1 = \rho$, $u_2 = \phi$, and $u_3 = z$. We earlier found $h_1 = h_3 = 1$ and $h_2 = \rho$.

Then from Eq. 4-28 $\vec{da}_1 = \rho d\phi dz \hat{\rho}$, $\vec{da}_2 = dz d\rho \hat{\phi}$, $\vec{da}_3 = \rho d\rho d\phi \hat{z}$. These are sketched in Figure 4-40.

4-12.3 Generalized Curvilinear Coordinate Representation of Volume

Finally, consider the representation of volume in generalized coordinates. This is a scalar.

$$dv = dl_1 dl_2 dl_3 = h_1 h_2 h_3 du_1 du_2 du_3 \tag{4-29}$$

Example 4-23

In cylindrical coordinates, $u_1 = \rho$, $u_2 = \phi$, $u_3 = z$, $h_1 = h_3 = 1$, $h_2 = \rho$.

$dv = \rho d\rho d\phi dz$ as shown from other methods. See Figure 4-41.

Now that we feel at home with the generalized coordinates, we will work with them in subsequent sections and as examples we will specialize to a particular system.

Figure 4-40. Generation of three incremental surfaces in cylindrical coordinates using the generalized curvilinear coordinate method.

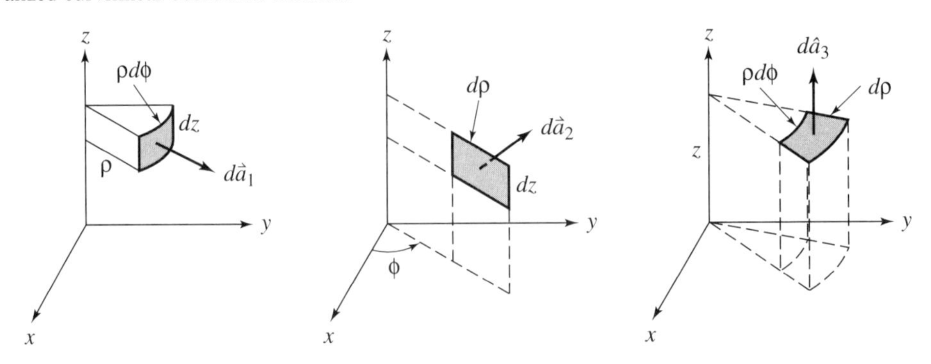

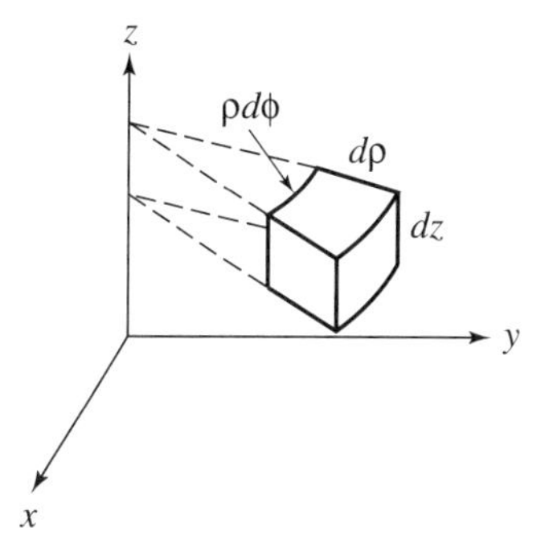

Figure 4-41. Generation of incremental volume in cylindrical coordinates using the generalized curvilinear coordinate method.

Exercises

4-12.3a For the coordinate systems given in Exercises 4-12.1a and 4-12.1b, give expressions for the $\vec{da}_i$ and dv and sketch each.

4-12.3b Determine the scale factors h_1, h_2, and h_3 for the spherical coordinate system and write expressions for differential lengths, areas, and volume. Sketch these on a coordinate system. Use three sets for clarity.

4-13 Dot and Cross Products in Generalized Curvilinear Coordinates

So the results obtained may be applied to any right-hand orthogonal coordinate system, we use generalized curvilinear coordinates in the derivations. First consider the dot (inner, scalar) product of two vectors expressed in generalized coordinate component form.

Let

$$\vec{A} = \overline{A}_1\hat{u}_1 + \overline{A}_2\overline{u}_2 + \overline{A}_3\hat{u}_3$$

and

$$\vec{B} = \overline{B}_1\hat{u}_1 + \overline{B}_2\hat{u}_2 + \overline{B}_3\hat{u}_3$$

Using the distributive property of the dot product:

$$\begin{aligned}\vec{A} \cdot \vec{B}^* = {} & \overline{A}_1\overline{B}_1^*\hat{u}_1 \cdot \hat{u}_1 + \overline{A}_1\overline{B}_2^*\hat{u}_1 \cdot \hat{u}_2 + \overline{A}_1\overline{B}_3^*\hat{u}_1 \cdot \hat{u}_3 + \overline{A}_2\overline{B}_1^*\hat{u}_2 \cdot \hat{u}_1 \\ & + \overline{A}_2\overline{B}_2^*\hat{u}_2 \cdot \hat{u}_2 + \overline{A}_2\overline{B}_3^*\hat{u}_2 \cdot \hat{u}_3 + \overline{A}_3\overline{B}_1^*\hat{u}_3 \cdot \hat{u}_1 \\ & + \overline{A}_3\overline{B}_2^*\hat{u}_3 \cdot \hat{u}_2 + \overline{A}_3\overline{B}^*{}_3\hat{u}_3 \cdot \hat{u}_3\end{aligned}$$

Now if the spatial locations at which the vectors $\vec{A}$ and $\vec{B}$ are given are the *same* point, then

$$\hat{u}_1 \cdot \hat{u}_1 = \hat{u}_2 \cdot \hat{u}_2 = \hat{u}_3 \cdot \hat{u}_3 = 1$$

Also, for the case of orthogonal systems all *mixed* dot products of unit vectors will be 0 for both vectors located at the same point, so that $\hat{u}_i \cdot \hat{u}_j = 0$, $i \neq j$. Then

$$\vec{A} \cdot \vec{B}^* = \overline{A}_1\overline{B}_1^* + \overline{A}_2\overline{B}_2^* + \overline{A}_3\overline{B}_3^* \tag{4-30}$$

Note that this is actually what the basic definition would have given. If the vectors $\vec{A}$ and $\vec{B}$ are not complex (components real numbers) this becomes:

$$\vec{A} \cdot \vec{B} = A_1B_1 + A_2B_2 + A_3B_3 \tag{4-31}$$

Thus the result derived earlier is valid for any orthogonal coordinate system if the two vectors are at the same point in space. This is usually the case in engineering practice. To evaluate the cross (vector) product of two vectors we proceed basically the same as in Section 4-9. For the same vectors $\vec{A}$ and $\vec{B}$ as above:

$$\begin{aligned}\vec{A} \times \vec{B} = {} & \overline{A}_1\overline{B}_1(\hat{u}_1 \times \hat{u}_1) + \overline{A}_1\overline{B}_2(\hat{u}_1 \times \hat{u}_2) + \overline{A}_1\overline{B}_3(\hat{u}_1 \times \hat{u}_3) \\ & + \overline{A}_2\overline{B}_1(\hat{u}_2 \times \hat{u}_1) + \overline{A}_2\overline{B}_2(\hat{u}_1 \times \hat{u}_2)\overline{A}_2\overline{B}_3(\hat{u}_2 \times \hat{u}_3) \\ & + \overline{A}_3\overline{B}_1(\overline{u}_3 \times \hat{u}_1) + \overline{A}_3\overline{B}_2(\overline{u}_3 \times \hat{u}_2) + \overline{A}_3\overline{B}_3(\hat{u}_3 \times \hat{u}_3)\end{aligned}$$

(Recall from Section 4-9 that to relate the sinusoidal steady-state cross product to the time domain we need to evaluate $\vec{A} \times \vec{B}^*$.) Again we make the restriction that the two vectors under consideration be at the same spatial location so that their corresponding unit vectors will be parallel. Then, since we have a right-hand orthogonal system in the order 1–2–3–1–2 . . . ,

$$\begin{aligned}&\hat{u}_1 \times \hat{u}_1 = \hat{u}_2 \times \hat{u}_2 = \hat{u}_3 \times \hat{u}_3 = 0 \\ &\hat{u}_1 \times \hat{u}_2 = -\hat{u}_2 \times \hat{u}_1 = \hat{u}_3 \\ &\hat{u}_2 \times \hat{u}_3 = -\hat{u}_3 \times \hat{u}_2 = \hat{u}_1 \\ &\hat{u}_3 \times \hat{u}_1 = -\hat{u}_1 \times \hat{u}_3 = \hat{u}_2\end{aligned}$$

Thus

$$\begin{aligned}\vec{A} \times \vec{B} &= \overline{A}_1\overline{B}_2\hat{u}_3 - \overline{A}_1\overline{B}_3\hat{u}_2 - \overline{A}_2\overline{B}_1\hat{u}_3 + \overline{A}_2\overline{B}_3\hat{u}_1 + \overline{A}_3\overline{B}_1\hat{u}_2 - \overline{A}_3\overline{B}_2\hat{u}_1 \\ &= (\overline{A}_2\overline{B}_3 - \overline{A}_3\overline{B}_2)\hat{u}_1 + (\overline{A}_3\overline{B}_1 - \overline{A}_1\overline{B}_3)\hat{u}_2 + (\overline{A}_1\overline{B}_2 - \overline{A}_2\overline{B}_1)\hat{u}_3\end{aligned}$$

In determinant form:

$$\vec{A} \times \vec{B} = \begin{vmatrix} \hat{u}_1 & \hat{u}_2 & \hat{u}_3 \\ \overline{A}_1 & \overline{A}_2 & \overline{A}_3 \\ \overline{B}_1 & \overline{B}_2 & \overline{B}_3 \end{vmatrix} \tag{4-32}$$

which is the same as the result for the rectangular coordinate system if the two vectors are located at the same point.

4-14 Vector Operators—I: The Gradient and Its Physical Interpretation

It is often desirable to describe vector and scalar fields by special properties they may possess. For example, the field lines representing a vector field may form closed loops or paths in a given region or the lines may terminate on a source. A scalar field may change value more rapidly in one direction than another. A vector operator that allows us to quantify the properties of such fields is called the *del*, or nabla, operator denoted by the symbol ∇. We next develop some concepts that allow us to describe scalar and vector fields using this operator.

The first concept is that of the *gradient of a scalar* function defined in some region of space. Before developing the computational expression, we begin with the physical meaning of the gradient. Consider a hill that has been marked with the contours of constant elevation on the top and side views as shown in Figure 4-42. Elevation is a scalar, and the contour lines represent constant scalar (elevation) values. Now suppose you imagine someone standing at point 1, and the person wants to get to the top by the most direct route. It should be clear that such a path would be normal to the contour lines in the top view. This specifies the direction of the gradient; namely, the direction of greatest increase of scalar value (elevation) at the point of interest. For another person starting at point 2, the gradient direction would be different from that at 1. Because of this directional property, the gradient of a scalar function is a vector quantity.

The next question is how we compare paths 1 and 2. Obviously, a hiker starting at point 2 has a much steeper climb than a hiker at 1. We describe this by saying that the gradient (slope) of path 2 is greater than path 1. Physically, one can see that the closer the scalar contour lines are in the top view the steeper the hill will be, that is, the larger the gradient. In the near limit as the contour lines almost touch, one would have a cliff on the mountain, and walking (climbing?) normal to the contours at that point would be a

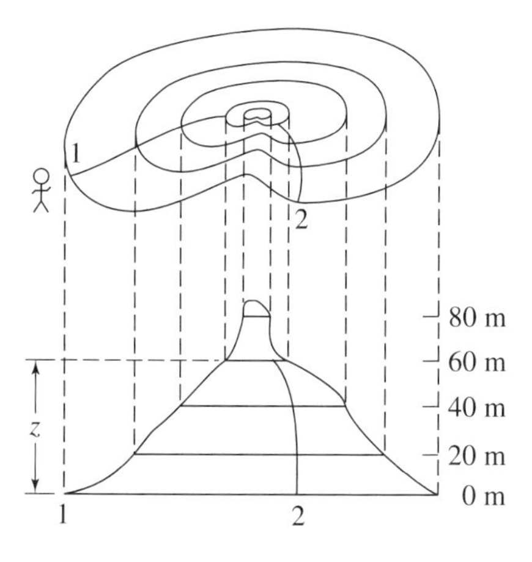

Figure 4-42. Top and side views of hill with contour (constant elevation) lines marked and two gradient paths indicated.

real challenge due to the nearly vertical rise (large gradient). In any case we denote these paths in words as:

$$\text{Steepest ascent slope of hiking path at a given point} \\ = \text{gradient of potential energy} = \nabla\ (\text{P.E.})$$

This is called a *point function* since the slope and direction depend upon the point (location) at which one starts, and the result is a vector. Another point of view may be taken. Suppose one places a large boulder at some point on the mountain and then releases it. The boulder will take the path of steepest descent which is normal to the scalar contour lines. Thus we would write:

$$\text{Steepest descent path} = -\nabla\ (\text{P.E.})$$

which, of course, means that the potential energy of the boulder is decreasing. Summarizing the results of the preceding discussion we then have the following formal definition: The gradient of a scalar field function, Φ, is given symbolically by

$$\text{Gradient } \Phi = \text{Grad}\Phi \equiv \nabla\Phi$$

and is a vector representing the magnitude and direction of the greatest increase of the scalar function value at the point of interest in the scalar field. This is often stated verbally as ''del phi.'' From the discussion we must remember that the gradient will be normal to the contours of constant scalar value Φ. To determine explicit expressions for the del operator and the gradient, one could derive a result for each coordinate system. The rectangular is most easily visualized and this is often used as a starting point. However, to avoid having to repeat the derivation for each coordinate system we use the results of our generalized right-hand orthogonal coordinate systems discussion to obtain the result once and for all. Since $\nabla\Phi$ is a vector, it may have components in all three of the coordinate directions. Thus we let the gradient be expressed by

$$\nabla\Phi = f_1(u_1, u_2, u_3)\,\hat{u}_1 + (u_1, u_2, u_3)\,\hat{u}_2 + f_3(u_1, u_2, u_3)\,\hat{u}_3 \qquad (4\text{-}33)$$

where f_1, f_2, and f_3 are to be determined.

For a general differential path length Eq. 4-23 requires

$$d\vec{l} = h_1 du_1 \hat{u}_1 + h_2 du_2 \hat{u}_2 + h_3 du_3 \hat{u}_3 \qquad (4\text{-}34)$$

Since the gradient is the maximum change of the scalar with distance, we have, using the dot product of these last two equations,

$$d\Phi = \nabla\Phi \cdot \vec{dl} = f_1 h_1 du_1 + f_2 h_2 du_2 + f_3 h_3 du_3 \qquad (4\text{-}35)$$

There are no cosines in this expression since we require that the path $d\vec{l}$ lie in the direction of the gradient, that is, $\cos\,\theta_i = 1$. From differential calculus we have the mathematical expression for the total differential of a function as

$$d\Phi = \frac{\partial\Phi}{\partial u_1}\,du_1 + \frac{\partial\Phi}{\partial u_2}\,du_2 + \frac{\partial\Phi}{\partial u_3}\,du_3 \qquad (4\text{-}36)$$

Comparing these last two equations term by term, which is valid for these scalar equations due to the fact that coordinates are orthogonal:

$$\left.\begin{aligned} f_1h_1 &= \frac{\partial\Phi}{\partial u_1} \quad \text{or} \quad f_1 = \frac{1}{h_1}\frac{\partial\Phi}{\partial u_1} \\ f_2h_2 &= \frac{\partial\Phi}{\partial u_2} \quad \text{or} \quad f_2 = \frac{1}{h_2}\frac{\partial\Phi}{\partial u_2} \\ f_3h_3 &= \frac{\partial\Phi}{\partial u_3} \quad \text{or} \quad f_3 = \frac{1}{h_3}\frac{\partial\Phi}{\partial u_3} \end{aligned}\right\} \tag{4-37}$$

Putting these results into Eq. 4-33 we finally obtain the general form

$$\nabla\Phi = \frac{1}{h_1}\frac{\partial\Phi}{\partial u_1}\hat{u}_1 + \frac{1}{h_2}\frac{\partial\Phi}{\partial u_2}\hat{u}_2 + \frac{1}{h_3}\frac{\partial\Phi}{\partial u_3}\hat{u}_3 \tag{4-38}$$

(The h_i are determined using Eq. 4-27.) We must now show that this is in fact the path giving the great rate of change. Physically, we would expect that the path of steepest change would be orthogonal to the constant elevation (potential) contours, as described earlier. Consider one of the constant elevation contours. We first find a vector that is tangent to the contour and is parallel to the contour in the horizontal plane as shown in Figure 4-43.

On the contour the potential Φ is a constant by definition. Let us select the vector $d\vec{l}_T$ to be in the direction of the tangent $\vec{T}$. Then if we move in the direction $d\vec{l}_T$ the potential will not change. Thus from Eq. 4-35,

$$d\Phi = 0 = \nabla\Phi \cdot d\vec{l}_T$$

But if the dot product is zero for all $\nabla\Phi$ and $d\vec{l}_T$, the vectors are orthogonal. Thus $\nabla\Phi$ is *normal to the constant potential contours*. It is also important to understand the dimensional effect of the ∇ operator on the scalar function. The terms in the operator are of the form

$$\frac{2}{h_i}\frac{\partial}{\partial u_i}, \qquad i = 1, 2, 3$$

From Eq. 4-21 the unit of $h_i du_i$ is the meter. Thus the operator is dimensionally 1/meter. For example, suppose we have a scalar function $V(u_1, u_2, u_3)$, which represents the voltage

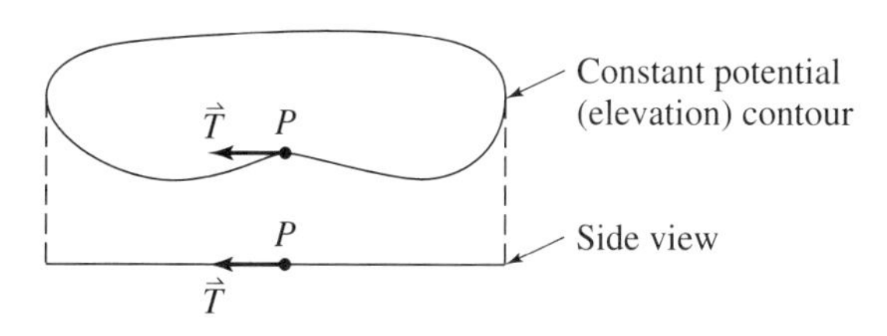

Figure 4-43. Tangent vector at a constant potential contour.

(potential) in a region, measured in volts. If we take the gradient of that function, the units of ∇V would then be volt/meter for the resulting vector field.

Example 4-24

Suppose we take a large cylinder and fill it with soil. We then tamp the soil at the top. This will lead to an uneven density of the soil, being most dense near top and center and decreasing down and toward the outside edge. See Figure 4-44.

From the interpretation of the gradient as the direction of increase of the scalar (density in this problem) we would expect that the vector representing the gradient would be directed toward the center (negative ρ) and toward the top (positive z). Now suppose that the equation describing density is given by the crude reasonable form

$$d = 8(K_2 + z)(K_1 - \rho)\ \frac{\text{kg}}{\text{m}^3} \qquad \text{(increases with } z\text{, decreases with } \rho\text{)},\ K_1 \text{ and } K_2 \text{ constants}$$

For cylindrical coordinates $h_1 = h_3 = 1$, and $h_2 = \rho$. Equation 4-38 gives

$$\nabla d = \frac{\partial d}{\partial \rho}\hat{\rho} + \frac{1}{\rho}\frac{\partial d}{\rho\phi}\hat{\phi} + \frac{\partial d}{\partial z}\hat{z}$$

Thus

$$\nabla d = -8(K_2 + z)\hat{\rho} + 8(K_1 - \rho)\hat{z} = 8(K_2 + z)(-\hat{\rho}) + 8(K_1 - \rho)\hat{z}$$

From this last result we see that the gradient is directed toward negative ρ and positive z as we expected physically, since the density increases most rapidly toward the top and center of the can.

Expressions for ∇V in common coordinate systems for a scalar function $V(u_1, u_2, u_3)$ are given here for future reference. See also Appendix D.

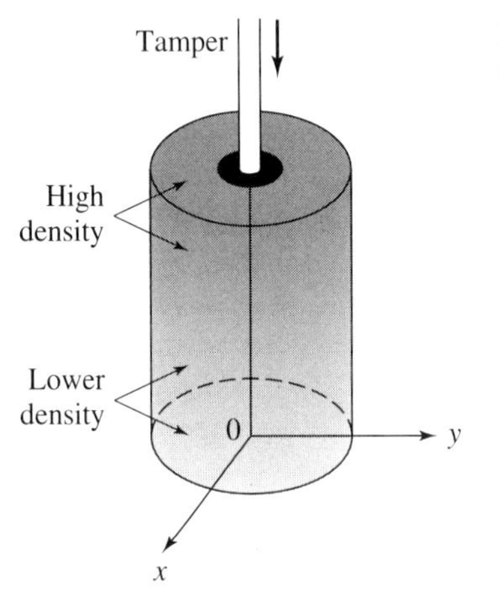

Figure 4-44. System that has nonzero gradient of density.

1. Rectangular: $h_1 = h_2 = h_3 = 1$, $u_1 = x$, $u_2 = y$, $u_3 = z$

$$\nabla V = \frac{\partial V}{\partial x}\hat{x} + \frac{\partial V}{\partial y}\hat{y} + \frac{\partial V}{\partial z}\hat{z} \equiv \left(\frac{\partial}{\partial x}\hat{x} + \frac{\partial}{\partial y}\hat{y} + \frac{\partial}{\partial z}\hat{z}\right) V$$

To obtain this last expression we have imagined that we have *factored out* a V. This results in the so-called operator form where we define $(\partial/\partial x)V = \partial V/\partial x$, etc. Note that $(\partial/\partial x)V \neq V(\partial/\partial x)$. The operator in parentheses is a vector and is given the symbolic form

$$\nabla \equiv \frac{\partial}{\partial x}\hat{x} + \frac{\partial}{\partial y}\hat{y} + \frac{\partial}{\partial z}\hat{z} \qquad \text{for rectangular coordinates}$$

2. Cylindrical: $h_1 = h_3 = 1$, $h_2 = \rho$, $u_1 = \rho$, $u_2 = \phi$, $u_3 = z$

$$\nabla V = \frac{\partial V}{\partial \rho}\hat{\rho} + \frac{1}{\rho}\frac{\partial V}{\partial \phi}\hat{\phi} + \frac{\partial V}{\partial z}\hat{z}$$
$$\equiv \left(\frac{\partial}{\partial \rho}\hat{\rho} + \frac{1}{\rho}\frac{\partial}{\partial \phi}\hat{\phi} + \frac{\partial}{\partial z}\hat{z}\right) V$$

3. Spherical: $h_1 = 1$, $h_2 = r$, $h_3 = r \sin\theta$, $u_1 = r$, $u_2 = \theta$, $u_3 = \phi$

$$\nabla V = \frac{\partial V}{\partial r}\hat{r} + \frac{1}{r}\frac{\partial V}{\partial \theta}\hat{\theta} + \frac{1}{r\sin\theta}\frac{\partial V}{\partial \theta}\hat{\phi}$$
$$\equiv \left(\frac{\partial}{\partial r}\hat{r} + \frac{1}{r}\frac{\partial}{\partial \theta}\hat{\theta} + \frac{1}{r\sin\phi}\frac{\partial}{\partial \phi}\hat{\phi}\right) V$$

Although the del operator is defined in terms of the rectangular coordinate system, it is common practice to use the del symbol to represent the gradient operation in any other system with the understanding that the gradient is evaluated using the result in Eq. 4-38.

Exercises

4-14.0a Obtain the expression for the gradient of a scalar function $V(u, v, z)$ in parabolic cylindrical coordinates. (See Exercises 4-9.0f and 4-12a.)

4-14.0b In spherical coordinates suppose we have the scalar function $V(r, \theta, \phi) = 4r \sin\theta$. Obtain the gradient of this function. Next devise some physical example for which this V is a reasonable scalar function and show that your ∇V agrees with physical expectations.

4-14.0c The two metal plates shown in Figure 4-45 have DC voltages applied to them. The

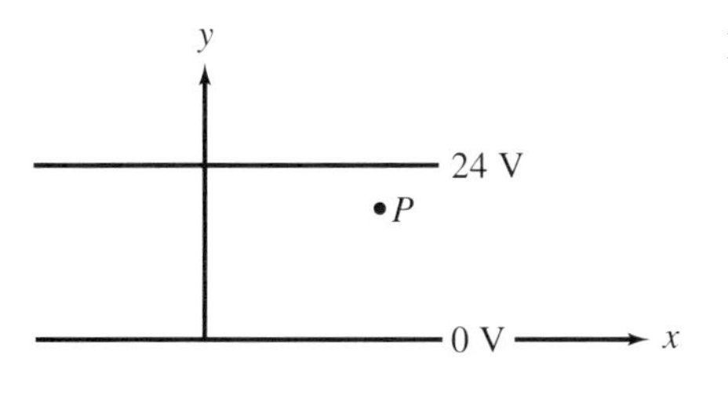

Figure 4-45.

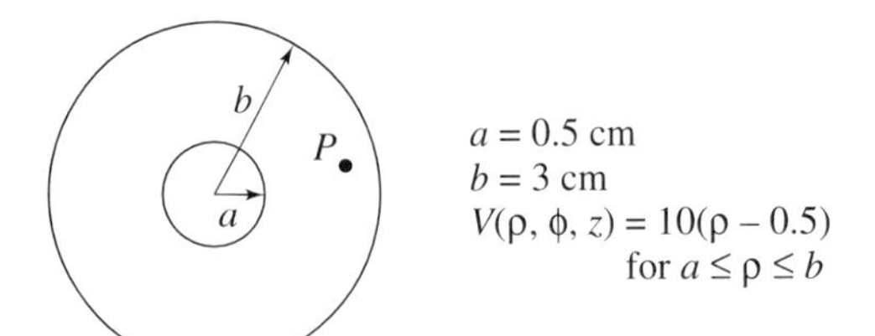

Figure 4-46.

equation for the voltage (potential) everywhere between the plates is given by $V(x, y, z) = 6y$. Draw the figure to scale and sketch on the figure four lines of constant potential value (y = constant). Now determine the gradient of the potential and sketch four of these vector lines; be sure to use arrows on the appropriate vector quantities. Redraw the figure and plot the same four vector lines representing the *negative* of the gradient of the potential. If a positive charge were placed at point P between the plates, what do you think the path would be if it were allowed to move? (Answer: $\nabla V = 6\hat{y}$. Charge would move $-\ \hat{y}$)

4-14.0d Repeat Exercise 4-14.0c for the configuration shown in Figure 4-46, which is the cross section of a coaxial cable.

4-14.0e What is the equation for the vector whose field line or path is normal to the constant potential contours of the scalar function $V(x, y, z) = x^2y - 2xz$? What is the value at the point $P(2, -4, 6)$? Now what is the unit vector in the direction of the normal at point P? (Hint: The gradient is normal to the constant potential contours. How do you convert a vector to a unit vector?)

(a) In what direction from the point (2, 1, 2) does the potential function $V = x^2yz^2$ have its maximum rate of change?

(b) What is the magnitude of this maximum?

(c) Determine the rate of change of the potential function in some other direction at the same point and calculate the magnitude of the rate of change. Is this in fact less than the answer you obtained in part (b)? (Hint: Don't forget to use the dot product in this exercise, since the dot product is defined also as the portion of one vector in the direction of another.)

4-14.0f Evaluate the following gradient functions:

(a) $\nabla(r^3)$ in spherical coordinates and in cylindrical coordinates. (Replace r by ρ for cylindrical.)

(b) $\nabla(re^{-r})$ in spherical and cylindrical coordinates. (Replace r by ρ for cylindrical.)

(c) $\nabla(z(u^2 + v^2)^{3/2})$ in parabolic cylindrical coordinates (see Exercise 4-14.0a).

4-15 Vector Operators—II: The Divergence and Its Physical Interpretation

The second concept to be described is the divergence of a vector function defined in some region of space. As before we begin with the physical meaning of divergence. Consider a region of space which contains gas molecules at a temperature T_1. Now in this region

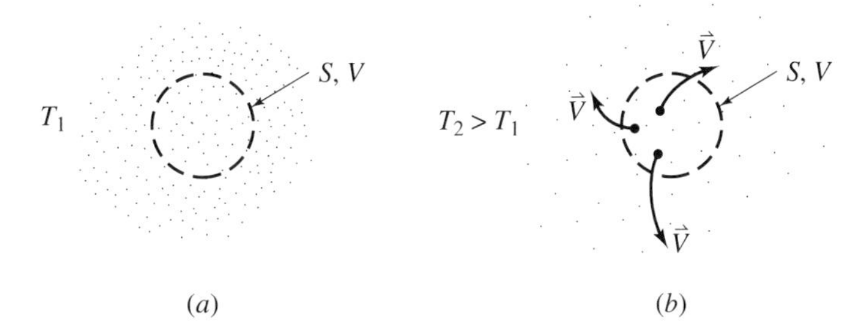

Figure 4-47. Effect of temperature on gas molecules. (*a*) Original state is temperature T_1. (*b*) Final state is temperature $T_2 > T_1$.

picture an imaginary surface S that encloses a volume v of the region. This is shown in Figure 4-47*a*.

Next suppose the temperature is increased to $T_2 > T_1$. The molecules will develop velocity vectors that will on average cause a decrease of the density of the molecules as one would expect on physical grounds. Thus the molecular velocity vectors will be directed out of the volume passing through the surface S as indicated in Figure 4-47*b*. We may state this another way by saying that the velocity vectors leave or diverge from within the volume. It is also of interest and important to observe that the vector field (velocity) is produced by scalar quantities, in this case temperature change and molecular motion. Furthermore, the divergence of the vector quantity is a measure of what is happening within the volume to the scalar quantities. Before giving a formal statement of divergence one more point should be noted. The velocity vectors of molecules outside the volume might be directed to cause a molecule to pass through the volume. What about the effect of such velocity vectors? Specifically, consider the path of the velocity vector labeled ① in Figure 4-48. It should be evident that the contribution to divergence of vectors through surface S of this velocity vector is zero since the vector entered one side of the volume and left another side. The net effect is zero. We conclude, then, that only a consideration of quantities inside the volumes are of any importance in the determination of divergence. Summarizing the results of the discussion we would describe the divergence as follows:

Definition. The *divergence* of a vector field is a measure of the net outflow (or inflow if negative) of the vector field lines at a given location in space. The value of the divergence is a function of the scalar quantity within the region that generates the vector quantity.

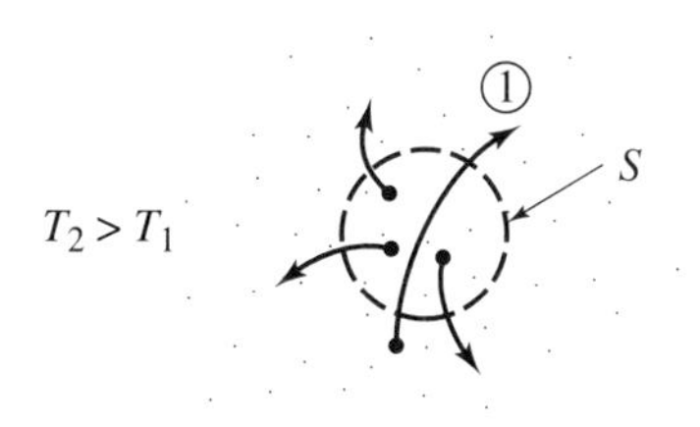

Figure 4-48. Molecular velocity vector beginning and ending outside region S of interest.

We represent the required operation symbolically as an operation on a vector field, say $\vec{A}$, by

$$\text{Divergence } \vec{A} = \text{Div } \vec{A}$$

As with the gradient, we are now faced with the problem of getting some mathematical way of calculating the divergence. We will show, in fact, that we can write the divergence in terms of the same operator, ∇, that was used in conjunction with the gradient. It should also be noted that the divergence of a vector field is a scalar quantity, the scalar quantity being the source (generator) of $\vec{A}$.

We next develop the formal mathematical definition of divergence. In our previous example we considered a finite volume of space. However, since the density change may in general vary from point to point within the region, we are led to consider a very small (infinitesimal) volume. This will of course lead us to what is called a *point function*, since as we let $\Delta v \to 0$ we would have the divergence at a point. The other quantity involved consists of the vector flow lines that cross the surface S, which would now become ΔS for an infinitesimal volume. We then define divergence as the lines that pass through the surface ΔS per unit volume Δv. Symbolically, this may be written (see Figure 4-49) as

$$\text{Div } \vec{A} \equiv \lim_{\Delta v \to 0} \frac{\oint_{\Delta S} \vec{A} \cdot d\vec{a}}{\Delta v} \tag{4-39}$$

In general, a surface integral (not necessarily closed) of a vector field $\vec{A}$ is called the *flux* of $\vec{A}$ over a surface S. The numerator in Eq. 4-39 is then the flux of the field $\vec{A}$ out of a closed surface ΔS. In the definition, $\oint$ indicates a closed surface integration over ΔS that encloses the volume Δv, $d\vec{a}$ is an element of area (a vector) of ΔS, and $\vec{A} \cdot d\vec{a}$ is the component of $\vec{A}$ passing through surface element $d\vec{a}$ so that the closed surface integral represents the net lines (or flux of $\vec{A}$) passing through the surface. We apply this definition to a volume in general orthogonal curvilinear coordinates shown in Figure 4-50.

The vector $\vec{A}$, being a vector field, exists throughout the region, but for clarity we show only one field line of $\vec{A}$. From Eq. 4-22 we have, in incremental form, the lengths of the sides of the incremental volume. These are

$$\Delta l_1 = h_1 \Delta u_1$$
$$\Delta l_2 = h_2 \Delta u_2$$
$$\Delta l_3 = h_3 \Delta u_3$$

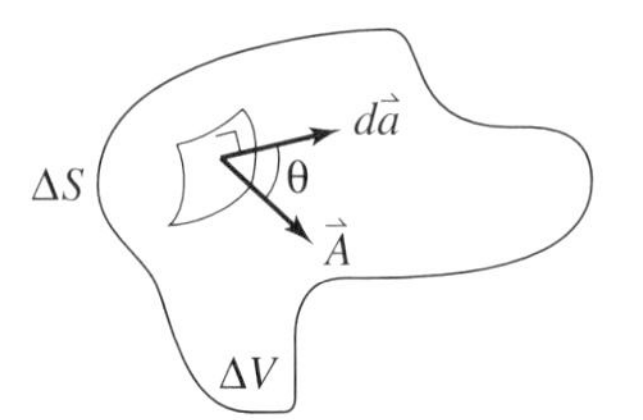

Figure 4-49. Relationships among the quantities used to define divergence.

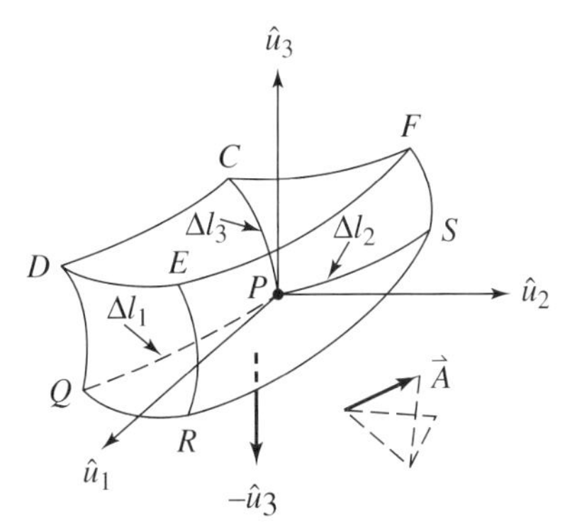

Figure 4-50. Generalized volume used to develop the divergence formula.

As the volume shrinks to zero, it compresses toward point P, which is the point at which the divergence will then be evaluated. We first evaluate the closed surface integral part of the definition of divergence over the six faces of the curvilinear parallelopiped volume. Note in the limit of small volume the unit vectors will be parallel or antiparallel to the surface vector directions (recall that the surface vector is normal to and outward from the physical surface area). We also assume that the vector $\vec{A}$ is directed to increasing coordinate values, $\vec{A} = \overline{A}_1\hat{u}_1 + \overline{A}_2\hat{u}_2 + \overline{A}_3\hat{u}_3$, the $\overline{A}_i$s being functions of the u_is generally. Now for the surface $PQRS$ we have to evaluate the integral

$$\int\int_{PQRS} \vec{A} \cdot d\vec{a}$$

which is at the bottom of the volume. However, since we are considering an incremental area that shrinks to zero as Δv goes to zero we may assume $\vec{A}$ is constant over this small area. Thus $\vec{A}$ may be removed from the integrand, and we proceed as follows:

$$\int\int_{PQRS} \vec{A} \cdot d\vec{a} = \vec{A} \cdot \int\int_{PQRS} d\vec{a} = \vec{A} \cdot \Delta\vec{a} = \vec{A} \cdot [\Delta l_1 \Delta l_2(-\hat{u}_3)]$$
$$= (\overline{A}_1\hat{u}_1 + \overline{A}_2\hat{u}_2 + \overline{A}_3\hat{u}_3) \cdot [h_1\Delta u_1 h_2 \Delta u_2(-\hat{u}_3)]$$

Note there is a minus sign on the unit vector for the area since the surface vector is directed away from the volume and normal to Δa; the direction is down ($-\hat{u}_3$ direction) in this case. Expanding the dot product

$$\int\int_{PQRS} \vec{A} \cdot d\vec{a} = -h_1 h_2 \Delta u_1 \Delta u_2 (\overline{A}_1\hat{u}_1 \cdot \hat{u}_3 + \overline{A}_2\hat{u}_2 \cdot \hat{u}_3 + \overline{A}_3\hat{u}_3 \cdot \hat{u}_3)$$

Since our coordinate system is an orthogonal one

$$\hat{u}_1 \cdot \hat{u}_3 = \hat{u}_2 \cdot \hat{u}_3 = 0 \qquad \text{and} \quad \hat{u}_3 \cdot \hat{u}_3 = 1$$

Thus for the bottom surface, which has the point P as one of its corners

$$\int\int_{PQRS} \vec{A} \cdot d\vec{a} = -h_1 h_2 \overline{A}_3 \Delta u_1 \Delta u_2$$

This represents the flux of $\vec{A}$ out of the bottom surface. Next we evaluate the flux of $\vec{A}$ out of the top surface given by

$$\int\int_{CDEF} \vec{A}_{\text{top}} \cdot d\vec{a}_{\text{top}}$$

One major difference with this integral evaluation is that $d\vec{a}$ and $\hat{u}_3$ are now in the same direction so the dot product is positive. Also, we can obtain the value of this integral by using the fact that this top surface is only an incremental distance from the bottom one. Thus the flux out of the top surface can be computed by adding an increment of flux to the bottom surface value (with the sign change because of the vector directions described above). Mathematically, this is

$$\begin{aligned}\int\int_{CDEF} \vec{A}_{\text{top}} \cdot d\vec{a}_{\text{top}} &= -\text{ flux out bottom } + \text{ flux increment for } \Delta u_3 \\ &\quad \text{coordinate change} \\ &= h_1 h_2 \overline{A}_3 \Delta u_1 \Delta u_2 + \frac{\partial(h_1 h_2 \overline{A}_3 \Delta u_1 \Delta u_2)}{\partial u_3} \Delta u_3\end{aligned}$$

Since the coordinates are orthogonal, Δu_1 and Δu_2 do not change value with u_3 so they may be removed from the derivative. The result is

$$\int\int_{CDEF} \vec{A}_{\text{top}} \cdot d\vec{a}_{\text{top}} = h_1 h_2 \overline{A}_3 \Delta u_1 \Delta u_2 + \frac{\partial(h_1 h_2 \overline{A}_3)}{\partial u_3} \Delta u_1 \Delta u_2 \Delta u_3$$

The net flux passing out through the top and bottom surfaces is the sum of the integrals:

$$\int\int_{CDEF} + \int\int_{PQRS} = \frac{\partial(h_1 h_2 \overline{A}_3)}{\partial u_3} \Delta_1 \Delta_2 \Delta u_3$$

Similarly, for the left and right faces we obtain the net flux as

$$\int\int_{PQDC} + \int\int_{ERSF} = \frac{\partial(h_1 h_3 \overline{A}_2)}{\partial u_2} \Delta u_1 \Delta u_2 \Delta u_3$$

For the back and front faces the net flux is

$$\int\int_{PCFS} + \int\int_{REDQ} = \frac{\partial(h_2 h_3 \overline{A}_1)}{\partial u_1} \Delta u_1 \Delta u_2 \Delta u_3$$

From the preceding results, the net outward flux from the closed surface is then the sum

$$\begin{aligned}\oint_{\Delta S} \vec{A} \cdot d\vec{a} &= \frac{\partial(h_2 h_3 \overline{A}_1)}{\partial u_1} \Delta u_1 \Delta u_2 \Delta u_3 + \frac{\partial(h_1 h_3 \overline{A}_2)}{\partial u_2} \Delta u_1 \Delta u_2 \Delta u_3 \\ &\quad + \frac{\partial(h_1 h_2 \overline{A}_3)}{\partial u_3} \Delta u_1 \Delta u_2 \Delta u_3\end{aligned}$$

The incremental volume is $\Delta v = \Delta l_1 \Delta l_2 \Delta l_3 = h_1 h_2 h_3 \Delta u_1 \Delta u_2 \Delta u_3$. Using the limit expression for divergence we finally obtain the general expression:

$$\text{Div } \vec{D} = \lim_{\Delta v \to 0} \frac{\oint_{\Delta S} \vec{A} \cdot d\vec{a}}{\Delta v}$$

$$= \lim_{\Delta v -> 0} \left\{ \frac{1}{h_1 h_2 h_3 \Delta u_1 \Delta u_2 \Delta u_3} \left[\frac{\partial (h_2 h_3 \overline{A}_1)}{\partial u_1} \Delta u_1 \Delta u_2 \Delta u_3 \right. \right.$$

$$\left. \left. + \frac{\partial (h_1 h_3 \overline{A}_2)}{\partial u_2} \Delta u_1 \Delta u_2 \Delta u_3 + \frac{\partial (h_1 h_2 \overline{A}_3)}{\partial u_3} \Delta u_1 \Delta u_2 \Delta u_3 \right] \right\}$$

$$\text{Div } \vec{D} = \frac{1}{h_1 h_2 h_3} \left[\frac{\partial (h_2 h_3 \overline{A}_1)}{\partial u_1} + \frac{\partial (h_1 h_3 \overline{A}_2)}{\partial u_2} + \frac{\partial (h_1 h_2 \overline{A}_3)}{\partial u_3} \right] \tag{4-40}$$

Exercises

4-15.0a This exercise is designed to help you visualize some of the concepts involved in the derivation of the divergence formula. Spherical coordinates will be used as a practical example. First, draw an incremental volume in spherical coordinates in the first octant of space (x, y, z all positive). Draw it fairly large.

(a) Identify the corner P. (This is the corner that has the smallest values of coordinates r, θ, and ϕ.)

(b) Identify Δl_1, Δl_2, and Δl_3 and give expressions for them in spherical coordinates.

(c) Identify the unit vectors for the six surfaces in terms of $\hat{r}$, $\hat{\phi}$, and $\hat{\theta}$. (Answers are + or − the coordinate unit vectors)

(d) Identify the pairs of parallel surfaces that correspond to Δr only, $\Delta\theta$ only and $\Delta\phi$ only varying respectively. Now identify these surfaces as top and bottom, front and back and left and right sides as used in the generalized coordinate derivation. Place letters at the corresponding corners to make this identification.

4-15.0b Carry out in detail the steps required to obtain the right and left and front and back contributions to the closed surface integral, the results of which are given above.

4-15.0c Obtain the expression for divergence for a vector $\vec{A}$ given in bipolar coordinates. See Exercise 4-12.1b.

Since rectangular, cylindrical, and spherical coordinate systems are widely used we give expressions for these for future reference. In particular, the rectangular coordinate result allows us to introduce the ∇ operator into the divergence operation symbolism.

1. Rectangular $h_1 = h_2 = h_3 = 1$, $u_1 = x$, $u_2 = y$, $u_3 = z$

$$\text{Div } \vec{A} = \frac{\partial A_x}{\partial x} + \frac{\partial A_y}{\partial y} + \frac{\partial A_z}{\partial z}$$

Note in rectangular coordinates we may think of the result here as the dot product

$$\text{Div } \vec{A} = \left(\frac{\partial}{\partial x} \hat{x} + \frac{\partial}{\partial y} \hat{y} + \frac{\partial}{\partial z} \hat{z} \right) \cdot (A_x \hat{x} + A_y \hat{y} A_z \hat{z})$$

$$= \left(\frac{\partial}{\partial x} \hat{x} + \frac{\partial}{\partial y} \hat{y} + \frac{\partial}{\partial z} \hat{z} \right) \cdot \vec{A}$$

Now the term in parentheses is the *rectangular* expression for ∇ (see rectangular expression for $\nabla\phi$) so in rectangular coordinates we may write exactly:

$$\text{Div } \vec{A} = \nabla \cdot \vec{A} = \frac{\partial A_x}{\partial x} + \frac{\partial A_y}{\partial y} + \frac{\partial A_z}{\partial z}$$

Although the operation $\nabla \cdot \vec{A}$ can be performed only in rectangular coordinates to obtain the expression for the divergence, it is still common practice to use the notation to represent divergence in general with the understanding that expressions in other coordinate systems are obtained using the derived result, Eq. 4-40.

2. Cylindrical $h_1 = h_3 = 1$, $h_2 = \rho$, $u_1 = \rho$, $u_2 = \phi$, $u_3 = z$

$$\text{Div } \vec{A} = \frac{1}{\rho}\left[\frac{\partial(\rho A_\rho)}{\partial \rho} + \frac{\partial A_\phi}{\partial \phi} + \frac{\partial(\rho A_z)}{\partial z}\right]$$

Since ρ does not vary with z, it may be removed from the partial derivative with respect to z. Then

$$\text{Div } \vec{A} = \frac{1}{\rho}\frac{\partial(\rho A_\rho)}{\partial \rho} + \frac{1}{\rho}\frac{\partial A_\phi}{\partial \phi} + \frac{\partial A_z}{\partial z}$$

Now the operator that works on the three components of $\vec{A}$ here is *not* the same as the ∇ operator expression in the cylindrical form of the gradient. It is still common practice, however, to use the dot product *symbolism* although it is not evaluated formally that way.

$$\text{Div } \vec{A} = \nabla \cdot \vec{A} = \frac{1}{\rho}\frac{\partial(\rho A_\rho)}{\partial \rho} + \frac{1}{\rho}\frac{\partial A_\phi}{\partial \phi} + \frac{\partial A_z}{\partial z}$$

3. Spherical $h_1 = 1$, $h_2 = r$, $h_3 = r \sin\theta$, $u_1 = r$, $u_2 = \theta$, $u_3 = \theta$. Using the same dot product symbolism

$$\nabla \cdot \vec{A} = \frac{1}{r^2 \sin\theta}\left[\frac{\partial(r^2 \sin\theta\, A_r)}{\partial r} + \frac{\partial(r \sin\theta\, A_\theta)}{\partial \theta} + \frac{\partial(r\, A_\phi)}{\partial \phi}\right]$$

$$= \frac{1}{r^2 \sin\theta}\left[\sin\theta \frac{\partial(r^2\, A_r)}{\partial r} + r\frac{\partial(\sin\theta\, A_\theta)}{\partial \theta} + r\frac{\partial(A_\phi)}{\partial \phi}\right]$$

$$\nabla \cdot \vec{A} = \frac{1}{r^2}\frac{\partial(r^2\, A_r)}{\partial r} + \frac{1}{r \sin\theta}\frac{\partial(\sin\theta\, A_\theta)}{\partial \theta} + \frac{1}{r\sin\theta}\frac{\partial A_\phi}{\partial \phi}$$

Exercises

4-15.0d By working directly with the cylindrical coordinate system, use the limit definition, Eq. 4-39 and determine $\nabla \cdot \vec{A}$ in cylindrical coordinates. (Hint: Draw a careful figure showing a volume ∇V.)

4-15.0e For the vectors $\vec{A}$ and $\vec{B}$ and a scalar function b prove that (a) $\nabla \cdot (\vec{A} + \vec{B}) = \nabla \cdot \vec{A} + \nabla \cdot \vec{B}$ and (b) $\nabla \cdot (b\vec{B}) = (\nabla b) \cdot \vec{B} + b(\nabla \cdot \vec{B})$. (Use rectangular coordinates.)

4-15.0f For the scalar function $\Phi(\rho, \phi, z) = 9\rho^2 z \cos\phi$, evaluate the following:
(a) $\nabla\phi$ at the point $P(3, \pi, 1)$. (Answer: $-54\hat{\rho} - 81\hat{z}$)
(b) $\nabla \cdot (\nabla\phi)$ at the point $(1, \pi/2, -1)$. (Answer: 0)

4-15.0g In spherical coordinates evaluate the expressions (a) $\nabla \cdot \left(\frac{1}{r}\hat{r}\right)$ and (b) $\nabla \cdot \left(\frac{\hat{r}}{r^2}\right)$.

4-15.0h If the vector $\vec{D} = x^2 z\hat{x} - 2y^3 z\hat{y} + x^2 y^2 z\hat{z}$, find $\nabla \cdot \vec{D}$ at the point $P(1, -1, 1)$. Also explain physically what you think the lines representing the vector $\vec{D}$ are doing at P.

4-15.0i Since the gradient of a scalar is a vector, we would suppose that we could take the divergence of the gradient. Symbolically we would have such operations denoted by

$$\text{Div (Grad } V) = \nabla \cdot \nabla V$$

where V is a scalar function of coordinates. If we have $V(x, y, z) = 2xy^2z^3$ evaluate the divergence of the gradient. If $V(r, \theta, \phi) = 3r \sin\theta \cos\phi$, evaluate $\nabla \cdot \nabla V$ at the point $P(2, 0.8, 0.15)$.

Two final comments conclude the discussion on divergence. Since the vectors are functions of coordinates the divergence of the vector field will be a function of coordinates. To obtain a particular numerical value one must substitute the coordinates of a particular point $P(u_1, u_2, u_3)$. For this reason the divergence is called a *point function*. This is also the case for the gradient. Last, vector fields for which $\nabla \cdot \vec{B} = 0$ are called *solenoidal* fields. From our physical description of divergence this means that there is no scalar source at that point so that no net lines leave the volume. Thus any line entering must leave that point, which implies that each field line closes on itself at that point.

4-16 Vector Operators—III: The Curl and Its Physical Interpretation

The third concept to be described is the curl of a vector function defined in some region of space. We begin with a physical description of the concept.

Suppose we have a viscous liquid flowing in an open channel and place a paddle wheel in the liquid. The small solid arrows represent the velocity vectors of the fluid, being zero at the boundary, increasing upward, as indicated in Figure 4-51.

The top of the paddle wheel will have greater force on it than the bottom due to different fluid velocities. This will cause a torque on the paddle wheel, which is a vector along the paddle wheel axis. Torque is *normal* to the velocity vector $\vec{V}$, and in this case is directed into the page, following the right-hand rule. The velocity vector field is said to have curl,

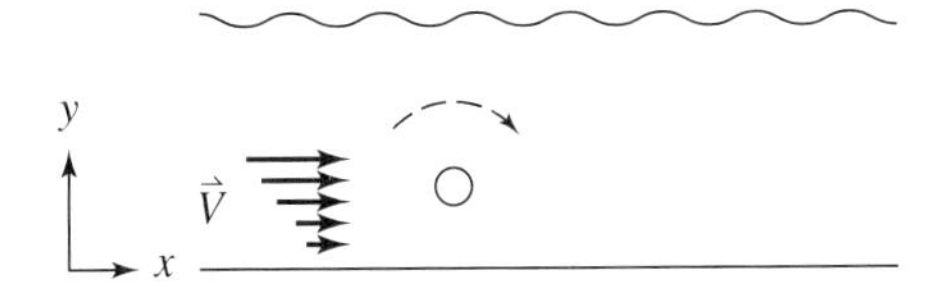

Figure 4-51. Illustration of a vector field $\vec{V}$ that has the property of curl.

the curl being the torque vector normal to the vector field. The curl of a vector is thus a vector. Note the analogy of the fact that the cross product of two vectors is a vector

$$\vec{A} \times \vec{B} = \vec{C}$$

$\vec{C}$ being normal to the plane defined by $\vec{A}$ and $\vec{B}$. Since the magnitude of the torque depends upon where the paddle wheel is placed in the channel the curl is also a point function, having different values, in general, at different points in the vector field. Figure 4-51 also shows one of the key properties that a vector field has if it has curl. If one imagines walking in a direction perpendicular to the field vectors, it is observed that the amplitude of the field vector varies. Note this does *not* mean to say that the converse is true; namely, if amplitude changes normal to the lines that the field has curl.

As another example, let's take a paper and float it on a channel of water. This example will show us that *surface* is also involved in the concept of curl. See Figure 4-52.

Consider the top view and examine the velocity vectors in opposite pairs. The paper occupies an area, and the velocity vectors exist on the boundaries and on the water surface under the paper. Examining first the x-directed velocities we assume for definiteness that $\partial V_x/\partial z$ is positive so that the vector amplitudes are as indicated by arrow lengths. This pair of vectors will tend to make the paper rotate *counterclockwise* to produce a $+\hat{y}$ torque. For the z-directed velocities the pair will tend to produce a small net *clockwise* rotation to produce a $-\hat{y}$ torque. Thus the net curl (which is a y-directed vector) will be the difference of the two effects, which will be proportional to the velocity vector and the lever arm about point P.

$$\text{Magnitude } + \hat{y} \text{ component} \sim \left(V_x + \frac{\partial V_x}{\partial z}\Delta z\right)\frac{\Delta z}{2} - V_x\frac{\Delta z}{2} = \frac{\partial V_x}{\partial z}\frac{(\Delta z)^2}{2}$$

$$\text{Magnitude } - \hat{y} \text{ component} \sim \left(V_z + \frac{\partial V_z}{\partial x}\Delta x\right)\frac{\Delta x}{2} - V_z\frac{\Delta x}{2} = \frac{\partial V_z}{\partial x}\frac{(\Delta x)^2}{2}$$

$$\text{Net curl} \sim (+\hat{y} \text{ component}) \text{ minus } (-\hat{y} \text{ component})$$

$$\sim \frac{\partial V_x}{\partial z}\frac{(\Delta z)^2}{2} - \frac{\partial V_z}{\partial x}\frac{(\Delta x)^2}{2}$$

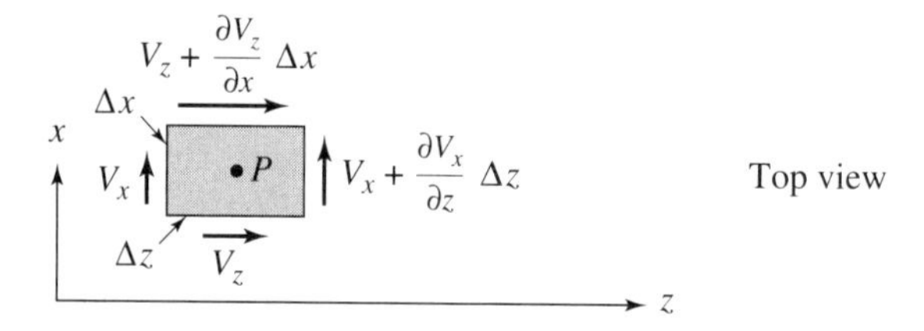

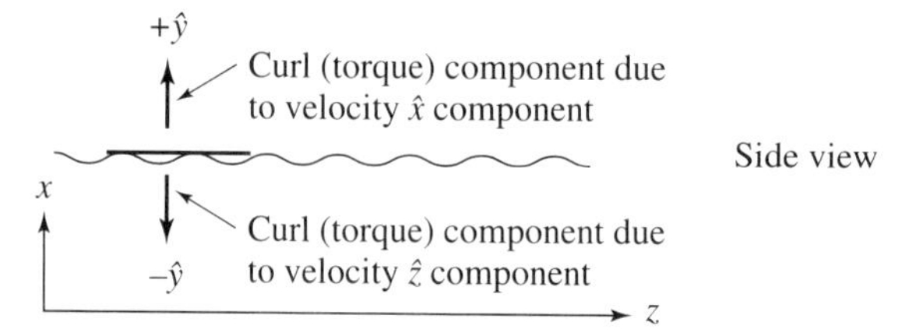

Figure 4-52. Curl as a surface phenomenon.

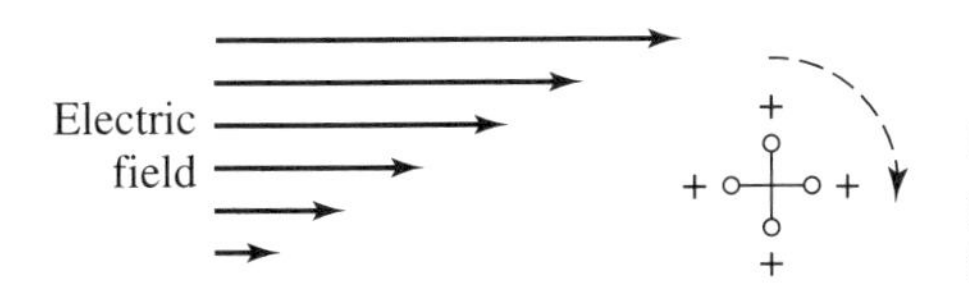

Figure 4-53. Testing for the presence of curl in an electric field.

For a square sheet of side Δl this reduces to

$$\text{Net curl} \sim \left(\frac{\partial V_x}{\partial z} - \frac{\partial V_z}{\partial x}\right)\frac{(\Delta l)^2}{2}$$

This is intended to serve only as a physical explanation so it is qualitative, but this result does show that one important feature is that the vector field components change values in directions normal to those components. Since the torque or curl is at a point we would actually have to imagine that the area of the paper shrinks to zero. This will again give us a vector function or value at a point similar to divergence. This idea will be explored more precisely later.

For a third example, recall from physics that an electric field produces a force on a charge. The stronger the electric field, the greater the force. Suppose we have an electric field with a spatial distribution as shown in Figure 4-53. We construct a very small electrostatic paddle wheel consisting of two crossed insulating bars with small positive charges at each end.

From the figure one can see that the topmost charge will have the largest force so that the paddle wheel will rotate as indicated, which shows the presence of curl, which is normal to and into the plane of the figure.

We can now formalize the physical concept of curl as follows:

Definition. The *curl of a vector field* is the measure of the tendency of the field to produce a rotation, the rotation being described by a vector normal to the vector field. One property that the vector field has if it has nonzero curl is variation in magnitude or direction as one travels normal to the field lines. Symbolically we represent this property by

$$\text{Curl of a vector} = \text{Curl } \vec{V} = \text{Rot } \vec{V}$$

Rot stands for rotation and is not widely used today, although it does have more physical significance perhaps than the term *curl.*

The next step is to develop a way to evaluate the curl mathematically. Some observations are appropriate first, however. That a vector field has amplitude variation normal to the vector lines does not guarantee it will have curl. All that has been said is that *if* curl is present, the vector lines must vary in magnitude or direction in a direction normal to the field lines. This can be seen by referring back to Figure 4-52. If the V_x and V_z components were of the same value there would have been no torque even though V_x varies in magnitude as one moves in the z direction.

Exercises

4-16.0a For which of the vector fields in Figure 4-54 could a curl exist? The lengths of the vectors indicate field magnitude (intensity). Describe where curls could exist within the field. Are there isolated points of zero or nonzero curl?

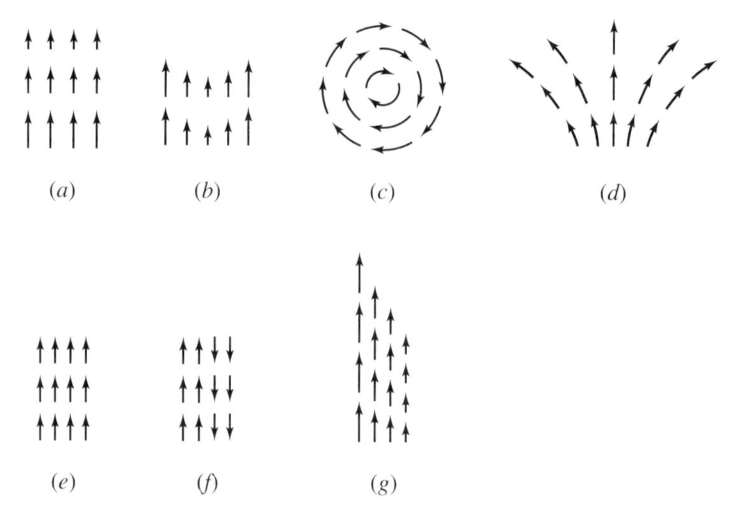

Figure 4-54.

4-16.0b A standard experiment for displaying the magnetic field around a bar magnet is to place a sheet of paper over the magnet and carefully sprinkle iron filings on the paper. The iron filings line up along the field lines. Devise a magnetic device which could be used to determine whether or not a magnetic field has curl. (Hint: See the electrostatic example in the preceding discussion.) Does the field around the bar magnet have curl, or could it have it qualitatively?

In the preceding examples we observe that the vector field was expressible in some plane, and that in the paper on the water example, the path around the paper was important since there is where we examined the values of the vector field. The boundary of the paper enclosed the area of the sheet, and thus we need to somehow enter the area as a part of our formal definition. Additionally, we are interested in the value of the curl at each *point* in the region so we would like to consider a small (infinitesimal) region. Thus we end up with a *point function* again, as with the gradient and divergence, and we have a vector result.

Consider an increment of area ΔS (see Figure 4-55) that has as its bounding contour ΔC. Let $\hat{n}$ be a (right-hand) unit normal vector to the surface ΔS. The vector field $\vec{A}$ exists in the region of the surface, but not necessarily all in the plane of ΔS. Also, since the

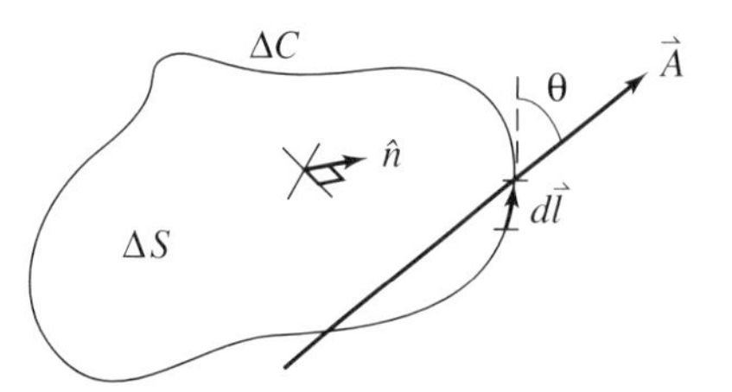

Figure 4-55. Configuration for defining curl.

component of a vector in a given direction is the dot product of the vector with a unit vector in that direction, we define the component of curl normal to the surface ΔS as

$$\text{Component of curl in direction of } \hat{n} = (\text{Curl } \vec{A})_n = (\text{Curl } \vec{A}) \cdot \hat{n} \equiv \lim_{\Delta S \to 0} \frac{\oint_{\Delta C} \vec{A} \cdot d\vec{l}}{\Delta S} \tag{4-41}$$

The integral term in this definition is often called the *circulation* of $\vec{A}$, or *vorticity*. The curl of $\vec{A}$ would then be the circulation per unit area. The circulation integral is a measure of the vector field parallel to the contour ΔC since $d\vec{l}$ is a differential of ΔC. Compare this idea with Figure 4-47. In the derivation that follows we will not use the overbar to indicate complex components of the vector, but the results do hold for complex values. We next apply this definition to a surface ΔS (having contour ΔC) in general orthogonal curvilinear coordinates. Since Curl $\vec{A}$ is a vector, it will have the component form:

$$\text{Curl } \vec{A} = (\text{Curl } \vec{A})_{u_1}\hat{u}_1 + (\text{Curl } \vec{A})_{u_2}\hat{u}_2 + (\text{Curl } \vec{A})_{u_3}\hat{u}_3 \tag{4-42}$$

where $(\text{Curl } \vec{A})_{u_i}$ is a component of a Curl $\vec{A}$ in the $\hat{u}_i$ direction. Now since the component of a vector in a given direction is the dot product of that vector with a unit vector in the given direction we evaluate each of the three orthogonal components of Curl $\vec{A}$ using Eq. 4-41 with $\hat{n} = \hat{u}_1$, then $\hat{u}_2$ and then $\hat{u}_3$. Figure 4-56 shows the configuration for finding the component $(\text{Curl } \vec{A})_{u_1}$. Note particularly the direction of the closed path ΔC_1 with respect to the surface and the surface normal. It follows the right-hand rule. From Eq. 4-22, we have for the sides of the surface

$$\Delta\vec{l}_2 = h_2 \Delta u_2 \hat{u}_2 \qquad \text{and} \qquad \Delta\vec{l}_3 = h_3 \Delta u_3 \hat{u}_3$$

In the limit the surface shrinks down to point P.

For the surface shown,

$$\oint_{\Delta C_1} \vec{A} \cdot d\vec{l} = \int_{PQ} \vec{A}_P \cdot d\vec{l}_2 + \int_{QR} \vec{A}_Q \cdot d\vec{l}_3 + \int_{RS} \vec{A}_R \cdot d\vec{l}_2 + \int_{SP} \vec{A}_S \cdot d\vec{l}_3$$

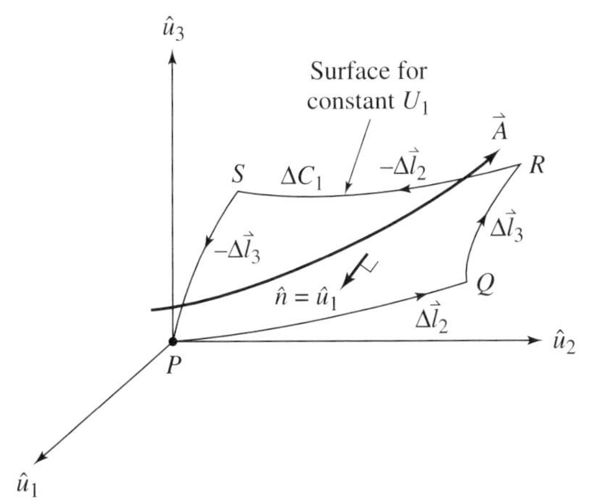

Figure 4-56. General curvilinear coordinate surface of constant U_1 at point P.

Evaluating the first integral, using the fact that for infinitesimal $\Delta\vec{l}_2$ the value $\vec{A}_P$ will be constant along that length PQ,

$$\int_{PQ} \vec{A}_P \cdot d\vec{l}_2 = \vec{A}_P \cdot \int_{PQ} d\vec{l}_2 = \vec{A}_P \cdot \Delta\vec{l}_2 = (A_1\hat{u}_1 + A_2\hat{u}_2 + A_3\hat{u}_3) \cdot (h_2\Delta u_2\hat{u}_2)$$
$$= h_2A_2\Delta u_2$$

We next evaluate the integral for path RS. Now $\vec{A}$ at point R will not be the same as along path PW since the path RS is displaced along coordinate u_3 direction. Also, the path of integration for RS is in the negative $\hat{u}_2$ direction, so the dot product of this path with $\vec{A}$ will be negative. (Notice how similar this discussion is with the divergence derivation.) Since the sides are incremental, we may take the value of the integral along path RS to be the value along PQ plus an increment of value change due to moving from path PQ up to path RS. Mathematically this is expressed as

$$\int_{RS} \vec{A}_R \cdot d\vec{l}_2 = -\left(\int_{PQ} \vec{A}_P \cdot d\vec{l}_2 + \frac{\partial(\int_{PQ}\vec{A}_P \cdot d\vec{l}_2)}{\partial u_3}\Delta u_3\right)$$

For incremental path $\vec{A}_P$ may be considered constant and $\vec{A}_P \cdot d\vec{l}_2$ is the A_2 component of $\vec{A}_P$. Then

$$\int_{RS} \vec{A}_R \cdot d\vec{l}_2 = -\left[h_2A_2\Delta u_2 - \frac{\partial(h_2A_2\Delta u_2)}{\partial u_3}\Delta u_3\right]$$

The minus sign, as described above, is due to the fact that the path and vector $\vec{A}$ are opposite directions as compared with path PQ. Also, since the coordinate system is an orthogonal one Δu_2 will not change with respect to u_3 so it may be removed from the derivative. We then obtain

$$\int_{RS} \vec{A}_R \cdot d\vec{l}_2 = -h_2A_2\Delta u_2 - \frac{\partial(h_2A_2)}{\partial u_3}\Delta u_2\Delta u_3$$

Using similar arguments we have for the other two sides:

$$\int_{SP} \vec{A}_S \cdot d\vec{l}_3 = (A_1\hat{u}_1 + A_2\hat{u}_2 + A_3\hat{u}_3) \cdot [h_3\Delta u_3(-\hat{u}_3)]$$
$$= -h_3A_3\Delta u_3$$

and

$$\int_{QR} \vec{A}_Q \cdot d\vec{l}_3 = h_3A_3\Delta u_3 + \frac{\partial(h_3A_3)}{\partial u_2}\Delta u_2\Delta u_3$$

Adding the four component integrals

$$\oint_{\Delta C_1} \vec{A} \cdot d\vec{l} = \left[\frac{\partial(h_3A_3)}{\partial u_2} - \frac{\partial(h_2A_2)}{\partial u_3}\right]\Delta u_2\Delta u_3$$

From the figure $\Delta S_1 = \Delta l_2\Delta l_3 = h_2h_3\Delta u_2\Delta u_3$. Applying the definition

$$(\text{Curl}\ \vec{A})_{u_1} = (\text{Curl}\ \vec{A}) \cdot \hat{u}_1 = \lim_{\Delta S_1 \to 0} \frac{\oint_{\Delta C_1} \vec{A} \cdot d\vec{l}}{\Delta S_1} = \frac{1}{h_2h_3}\left[\frac{\partial(h_3A_3)}{\partial u_2} - \frac{\partial(h_2A_2)}{\partial u_3}\right]$$

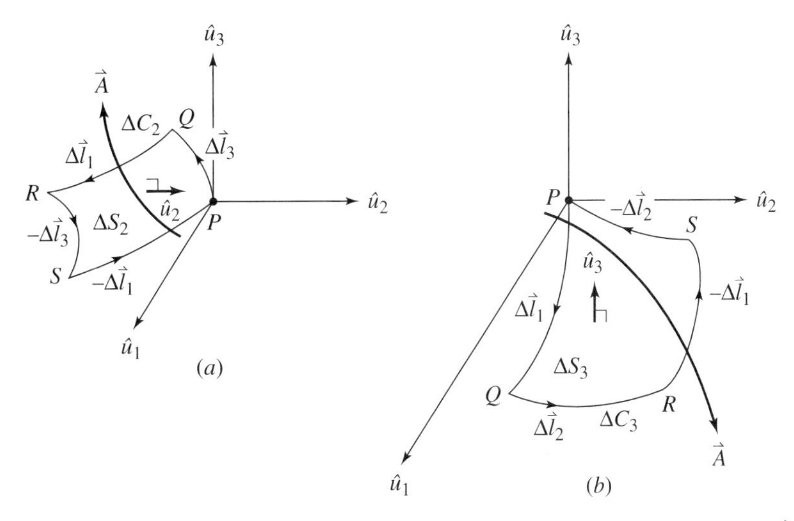

Figure 4-57. Configurations for determining the $\vec{u}_2$ and $\vec{u}_3$ components of Curl $\vec{A}$.

By exactly analogous procedures on the two surfaces shown in parts (*a*) and (*b*) of Figure 4-57 we obtain the other two components of the curl. The results are

$$(\text{Curl } \vec{A})_{u_2} = (\text{Curl } \vec{A}) \cdot \hat{u}_2 = \frac{1}{h_1 h_3}\left[\frac{\partial(h_1 A_1)}{\partial u_3} - \frac{\partial(h_3 A_3)}{\partial u_1}\right]$$

$$(\text{Curl } \vec{A})_{u_3} = (\text{Curl } \vec{A}) \cdot \hat{u}_3 = \frac{1}{h_1 h_2}\left[\frac{\partial(h_2 A_2)}{\partial u_1} - \frac{\partial(h_1 A_1)}{\partial u_2}\right]$$

Putting these results into Eq. 4-42 we obtain the general form:

$$\begin{aligned}\text{Curl } \vec{A} = &\frac{1}{h_2 h_3}\left[\frac{\partial(h_3 A_3)}{\partial u_2} - \frac{\partial(h_2 A_2)}{\partial u_3}\right]\hat{u}_1 \\ &+\frac{1}{h_1 h_3}\left[\frac{\partial(h_1 A_1)}{\partial u_3} - \frac{\partial(h_3 A_3)}{\partial u_1}\right]\hat{u}_2 \\ &\frac{1}{h_1 h_2}\left[\frac{\partial(h_2 A_2)}{\partial u_1} - \frac{\partial(h_1 A_1)}{\partial u_2}\right]\hat{u}_3\end{aligned} \tag{4-43}$$

Factoring out $1/h_1h_2h_3$ we could write

$$\text{Curl } \vec{A} = \frac{1}{h_1 h_2 h_3}\left\{\left[\frac{\partial(h_3 A_3)}{\partial u_2} - \frac{\partial(h_2 A_2)}{\partial u_3}\right] h_1\hat{u}_1 + \left[\frac{\partial(h_1 A_1)}{\partial u_3} - \frac{\partial(h_3 A_3)}{\partial u_1}\right] h_2\hat{u}_2 + \left[\frac{\partial(h_2 A_2)}{\partial u_1} - \frac{\partial(h_1 A_1)}{\partial u_2}\right] h_3\hat{u}_3\right\}$$

Now if we interpret the partial derivatives as operators written as

$$\frac{\partial(h_j A_j)}{\partial u_i} = \frac{\partial}{\partial u_i}(h_j A_j)$$

and then recognize the term in braces as the expansion of a determinant, we could write the preceding curl in a very nice determinant form:

$$\text{Curl } \vec{A} = \frac{1}{h_1 h_2 h_3} \begin{vmatrix} h_1 \hat{u}_1 & h_2 \hat{u}_2 & h_2 \hat{u}_2 \\ \dfrac{\partial}{\partial u_1} & \dfrac{\partial}{\partial u_2} & \dfrac{\partial}{\partial u_3} \\ h_1 A_1 & h_2 A_2 & h_3 A_3 \end{vmatrix} \tag{4-44}$$

Care must be exercised in using this format, however, to ensure that the scale factors appear at the proper places with respect to the partial derivatives. For example, the major diagonal product would be written

$$\begin{aligned}(h_1 \hat{u}_1)\left(\frac{\partial}{\partial u_2}\right)(h_3 A_3) &= (h_1 \hat{u}_1)\,\frac{\partial(h_3 A_3)}{\partial u_2} \\ &= h_1\,\frac{\partial(h_3 A_3)}{\partial u_2}\,\hat{u}_1\end{aligned}$$

Note that h_1 remains outside the partial derivative operator.

Exercises

4-16.0c This exercise is designed to help you visualize some of the concepts involved in the derivation of the curl formula. Spherical coordinates will be used as a practical example. First, draw three separate coordinate diagrams showing each of the three surfaces from which the three components of curl can be obtained. Use the first octant of space (x, y, and z positive).

(a) Identify the point P in each diagram. (This is the corner with the smallest values of coordinates r, θ, and ϕ.)

(b) Identify the contours ΔC_1, ΔC_2, and ΔC_3 and the corresponding surfaces ΔS_1, ΔS_2, and ΔS_3 (these latter three as vectors). Give expressions for the sections Δl of each contour in terms of spherical coordinates.

(c) Identify the path pairs that correspond to only one coordinate varying, as used in the generalized coordinate derivation. Place letters at the corresponding corners to make the identification on each path.

4-16.0d Expand Eq. 4-44 to obtain the form of Eq. 4-43.

4-16.0e Obtain the determinant form and the explicit equation form for Curl $\vec{A}$ in the parabolic cylindrical coordinates. (See Exercise 4-9.0f.)

4-16.0f By working directly with the rectangular coordinate system, use the limit definition, Eq. 4-41, and determine Curl $\vec{A}$ in rectangular coordinates.

Since rectangular, cylindrical, and spherical coordinate systems are widely used, we give expressions for these for future reference. In particular, the rectangular coordinate result allows us to introduce the ∇ operator into the curl operation symbolism.

1. Rectangular: $h_1 = h_2 = h_3$, $u_1 = x$, $u_2 = y$, $u_3 = z$. Using these in the general determinant form:

$$\text{Curl } \vec{A} = \begin{vmatrix} \hat{x} & \hat{y} & \hat{z} \\ \dfrac{\partial}{\partial x} & \dfrac{\partial}{\partial y} & \dfrac{\partial}{\partial z} \\ A_x & A_y & A_z \end{vmatrix}$$

Expanding the determinant yields the result:

$$\text{Curl } \vec{A} = \left(\frac{\partial A_z}{\partial y} - \frac{\partial A_y}{\partial z}\right)\hat{x} + \left(\frac{\partial A_x}{\partial z} - \frac{\partial A_z}{\partial x}\right)\hat{y} + \left(\frac{\partial A_y}{\partial x} - \frac{\partial A_x}{\partial y}\right)\hat{z}$$

Note here that in rectangular coordinates we may examine the cross product

$$\vec{C} \times \vec{B} = \begin{vmatrix} \hat{x} & \hat{y} & \hat{z} \\ C_x & C_y & C_z \\ B_x & B_y & B_z \end{vmatrix}$$

and compare it with the above determinant to identify the vector $\vec{C}$ as ∇. Thus we often write

$$\text{Curl } \vec{A} = \nabla \times \vec{A}$$

Although this exact identification cannot be made in other coordinate systems we still use the symbol $\nabla\times$ to represent the curl, realizing that to expand the operation we use the appropriate scale factors in the general result.

2. Cylindrical: $h_1 = h_3 = 1$, $h_2 = \rho$, $u_1 = \rho$, $u_2 = \phi$, $u_3 = z$.

$$\text{Curl } \vec{A} = \nabla \times \vec{A} = \frac{1}{\rho}\begin{vmatrix} \hat{\rho} & \rho\hat{\phi} & \hat{z} \\ \dfrac{\partial}{\partial \rho} & \dfrac{\partial}{\partial \phi} & \dfrac{\partial}{\partial z} \\ A_\rho & \rho A_\phi & A_z \end{vmatrix}$$

$$= \frac{1}{\rho}\left(\frac{\partial A_z}{\partial \phi} - \frac{\partial(\rho A_\phi)}{\partial z}\right)\hat{\rho} + \frac{\rho}{\rho}\left(\frac{\partial A_\rho}{\partial z} - \frac{\partial A_z}{\partial \rho}\right)\hat{\phi} + \frac{1}{\rho}\left(\frac{\partial(\rho A_\rho)}{\partial \phi} - \frac{\partial A_\rho}{\partial \phi}\right)\hat{z}$$

Since ρ is independent of z it may be taken outside of the partial in the $\hat{\rho}$ component:

$$\nabla \times \vec{A} = \left(\frac{1}{\rho}\frac{\partial A_z}{\partial \phi} - \frac{\partial(A_\phi)}{\partial z}\right)\hat{\rho}$$
$$+ \left(\frac{\partial A_\rho}{\partial z} - \frac{\partial A_z}{\partial \rho}\right)\hat{\phi} + \frac{1}{\rho}\left(\frac{\partial(\rho A_\rho)}{\partial \rho} - \frac{\partial A_\rho}{\partial \phi}\right)\hat{z}$$

3. Spherical: $h_1 = 1, h_2 = r, h_3 = r \sin\theta, u_1 = r, u_2 = \theta, u_3 = \phi$.

$$\nabla \times A = \frac{1}{r^2 \sin\theta} \begin{vmatrix} \hat{r} & r\hat{\theta} & r\sin\theta\hat{\phi} \\ \frac{\partial}{\partial r} & \frac{\partial}{\partial \theta} & \frac{\partial}{\partial \phi} \\ A_r & rA_\theta & r\sin\theta\, A_\phi \end{vmatrix}$$

Expanding the determinant and taking advantage of the fact that some coordinate variables may be taken outside other partial derivatives:

$$\nabla \times A = \frac{1}{r\sin\theta}\left[\frac{\partial(\sin\theta A_\phi)}{\partial\theta} - \frac{\partial A_\theta}{\partial\phi}\right]\hat{r} + \left[\frac{1}{r\sin\theta}\frac{\partial A_r}{\partial\phi} - \frac{1}{r}\frac{\partial(rA_\phi)}{\partial r}\right]\hat{\theta}$$
$$+ \frac{1}{r}\left[\frac{\partial(rA_\theta)}{\partial r} - \frac{\partial A_r}{\partial\theta}\right]\hat{\phi}$$

Exercises

4-16.0g Using the rectangular coordinates prove the following identities:

(a) $\nabla \cdot (\vec{A} \times \vec{B}) = B \cdot (\nabla \times \vec{A}) - \vec{A} \cdot (\nabla \times \vec{B})$.

(b) $\nabla \times (\vec{A} + \vec{B}) = \nabla \times \vec{A} + \nabla \times \vec{B}$.

(c) $\nabla \times (a\vec{A}) = (\nabla a) \times \vec{A} + a(\nabla \times \vec{A})$, a = scalar function.

(d) $\nabla \times (\nabla V) \equiv 0$, V is a scalar function.

(d) $\nabla \cdot (\nabla \times \vec{A}) \equiv 0$.

4-16.0h Evaluate the curls of the following vector fields in the coordinate systems indicated and at the points in space specified:

(a) $\vec{E} = 6xy\hat{x} - \hat{y} - 2zx\hat{z}$ (rectangular), (1,1,1).

(b) $\vec{E} = (\cos\phi/\rho^2)\hat{\rho} + \sin\phi\hat{\phi} + 6z\cos\phi\hat{z}$ (cylindrical), (2,1.5,6).

(c) $\vec{E} = (\cos\phi/r^2)\hat{r}$ (spherical), (4,1,2).

4-16.0i In a mechanical rotating system, the velocity $\vec{V}$ of a point on the body rotating at angular velocity $\vec{\omega}$ depends upon its distance $\vec{r}$ from the center of rotation. The vector relationship is

$$\vec{V} = \vec{\omega} \times \vec{r}$$

If ω is a constant vector, show that $\vec{\omega}$ = Curl $\vec{V}/2$ by expanding and using rectangular coordinates.

Two final comments conclude the discussion of curl. Since the vectors are functions of coordinates, the curl of the vector field will also be a function of coordinates. To obtain a particular numerical value, one must substitute the coordinates of a particular point $P(u_1, u_2, u_3)$. For this reason the curl is called a *point function*. Last, vector fields for which $\nabla \times \vec{A} = 0$ are called *irrotational*, or *conservative*, fields. From our physical description of curl this means that there would be no rotation of a small test paddle wheel used to probe the field. One consequence of this field type is that a line integral between two points in such a field is independent of the path taken between the points. Conclusion: select the easy path.

There is an important theorem about vector fields called the *vector Helmholtz theorem*, which involves both the divergence and curl operations. In essence it states that a continuous vector field is completely specified if its divergence and curl are specified. The proof of this theorem is not important here, so it is given in Appendix F, along with other background details required.

4-17 Second-Order Vector Operators: Scalar and Vector Laplacian Operators and Unit-Vector Multiplication

In many engineering applications there are cases where the ∇ operator may be applied more than once on a given field structure. The two most important are the divergence of the gradient, denoted by $\nabla \cdot (\nabla\Phi)$, and the curl of the curl, given symbolically by $\nabla \times \nabla \times \vec{A}$. The first combination is of considerable importance in obtaining solutions for electrostatic field systems and is called the Laplacian of a scalar; it is usually given the abbreviated form $\nabla^2\Phi$. It will be shown that the Laplacian of a scalar yields another scalar, as the symbol suggests. The double curl operation is important in the study of radio-wave propagation. This operation allows us to derive wave equations for electric and magnetic fields that have their counterparts in the voltage and current wave equations for transmission lines. All of the transmission line ideas transfer directly to wave propagation. For the evaluation of the Laplacian, we simply use the general results for the gradient and divergence developed earlier. From Eq. 4-38 we have, in generalized curvilinear coordinates,

$$\nabla\Phi = \frac{1}{h_1}\frac{\partial\Phi}{\partial u_1}\hat{u}_1 + \frac{1}{h_2}\frac{\partial\Phi}{\partial u_2}\hat{u}_2 + \frac{1}{h_3}\frac{\partial\Phi}{\partial u_3}\hat{u}_3$$

which is a vector. Thus

$$\nabla^2\Phi = \nabla \cdot \nabla\Phi = \nabla \cdot \left(\frac{1}{h_1}\frac{\partial\Phi}{\partial u_1}\hat{u}_1 + \frac{1}{h_2}\frac{\partial\Phi}{\partial u_2}\hat{u}_2 + \frac{1}{h_3}\frac{\partial\Phi}{\partial u_3}\hat{u}_3\right)$$

Using the generalized divergence relation, Eq. 4-40, we obtain the general form

$$\nabla^2\Phi = \frac{1}{h_1h_2h_3}\left[\frac{\partial\left(\frac{h_2h_3}{h_1}\frac{\partial\Phi}{\partial u_1}\right)}{\partial u_1} + \frac{\partial\left(\frac{h_1h_3}{h_2}\frac{\partial\Phi}{\partial u_2}\right)}{\partial u_2} + \frac{\partial\left(\frac{h_1h_2}{h_3}\frac{\partial\Phi}{\partial u_3}\right)}{\partial u_3}\right] \tag{4-45}$$

This result yields specific expressions for the rectangular, cylindrical, and spherical coordinate systems for later reference. The rectangular form in particular extends the ∇ operator to the Laplacian operator ∇^2.

1. Rectangular: $h_1 = h_2 = h_3 = 1$, $u_1 = x$, $u_2 = y$, $u_3 = z$

$$\nabla^2\Phi = \frac{\partial^2\Phi}{\partial x^2} + \frac{\partial^2\Phi}{\partial y^2} + \frac{\partial^2\Phi}{\partial z^2}$$

If we factor out Φ we may write the preceding in operator notation

$$\nabla^2\Phi = \left(\frac{\partial^2}{\partial x^2} + \frac{\partial^2}{\partial y^2} + \frac{\partial^2}{\partial z^2}\right)\Phi$$

Now if we dot a vector $\vec{A}$ with itself we have, from Eq. 4-6 and 4-7, $\vec{A} \cdot \vec{A} = A_x^2 + A_y^2 + A_z^2$, so we are led to try to get a form similar to this for the operator.

Using an analogous idea, we could write, for example,

$$\frac{\partial^2}{\partial x^2} = \frac{\partial\left(\dfrac{\partial}{\partial x}\right)}{\partial x} = \left(\frac{\partial}{\partial x}\right)\left(\frac{\partial}{\partial x}\right) = \left(\frac{\partial}{\partial x}\hat{x}\right) \cdot \left(\frac{\partial}{\partial x}\hat{x}\right)$$

We could then write the second partial operator as

$$\begin{aligned}\nabla \cdot (\nabla\Phi) &= \left(\frac{\partial}{\partial x}\hat{x} + \frac{\partial}{\partial y}\hat{y} + \frac{\partial}{\partial z}\hat{z}\right) \cdot \left(\frac{\partial}{\partial x}\hat{x} + \frac{\partial}{\partial y}\hat{y} + \frac{\partial}{\partial z}\hat{z}\right)\Phi \\ &= \left(\frac{\partial^2}{\partial x^2} + \frac{\partial^2}{\partial y^2} + \frac{\partial^3}{\partial z^2}\right)\Phi = \nabla^2\Phi\end{aligned}$$

The rectangular coordinate system is the only system for which this equality is strictly true. In other coordinate systems the dot product of the expression for ∇ with itself does *not* yield the correct result for ∇^2. Nevertheless, it is common practice to use the *symbolism* $\nabla \cdot \nabla$ with the understanding that the correct coordinate system expression is to be used for ∇.

2. Cylindrical: $h_1 = h_3 = 1$, $h_2 = \rho$, $u_1 = \rho$, $u_2 = \phi$, $u_3 = z$.
   You should justify each step in the following expansion:

$$\begin{aligned}\nabla^2\Phi &= \frac{1}{\rho}\left[\frac{\partial\left(\rho\dfrac{\partial\Phi}{\partial\rho}\right)}{\partial\rho} + \frac{\partial\left(\dfrac{1}{\rho}\dfrac{\partial\Phi}{\partial\phi}\right)}{\partial\phi} + \frac{\partial\left(\rho\dfrac{\partial\Phi}{\partial z}\right)}{\partial z}\right] \\ &= \frac{1}{\rho}\left[\frac{\partial\left(\rho\dfrac{\partial\Phi}{\partial\rho}\right)}{\partial\rho} + \frac{1}{\rho}\frac{\partial\left(\dfrac{\partial\Phi}{\partial\phi}\right)}{\partial\phi} + \rho\frac{\partial\left(\dfrac{\partial\Phi}{\partial z}\right)}{\partial z}\right]\end{aligned}$$

$$\nabla^2\Phi = \frac{1}{\rho}\frac{\partial\left(\rho\dfrac{\partial\Phi}{\partial\rho}\right)}{\partial\rho} + \frac{1}{\rho^2}\frac{\partial^2\Phi}{\partial\phi^2} + \frac{\partial^2\Phi}{\partial z^2}$$

3. Spherical: $h_1 = 1$, $h_2 = r$, $h_3 = r\sin\theta$, $u_1 = r$, $u_2 = \theta$, $u_3 = \phi$.
   Can you justify each step in the following expansion?

$$\nabla^2\Phi = \frac{1}{r^2 \sin\theta}\left[\frac{\partial\left(r^2 \sin\theta \dfrac{\partial\Phi}{\partial r}\right)}{\partial r} + \frac{\partial\left(\sin\theta \dfrac{\partial\Phi}{\partial\theta}\right)}{\partial\theta} + \frac{\partial\left(\dfrac{1}{\sin\theta}\dfrac{\partial\Phi}{\partial\phi}\right)}{\partial\phi}\right]$$

$$= \frac{1}{r^2 \sin\theta}\left[\sin\theta \frac{\partial\left(r^2 \dfrac{\partial\Phi}{\partial r}\right)}{\partial r} + \frac{\partial\left(\sin\theta \dfrac{\partial\Phi}{\partial\theta}\right)}{\partial\theta} + \frac{1}{\sin\theta}\frac{\partial\left(\dfrac{\partial\Phi}{\partial\phi}\right)}{\partial\phi}\right]$$

$$\nabla^2\Phi = \frac{1}{r^2}\frac{\partial\left(r^2 \dfrac{\partial\Phi}{\partial r}\right)}{\partial r} + \frac{1}{r^2\sin\theta}\frac{\partial\left(\sin\theta \dfrac{\partial\Phi}{\partial\theta}\right)}{\partial\theta} + \frac{1}{r^2\sin^2\theta}\frac{\partial^2\Phi}{\partial\phi^2}$$

Exercises

4-17.0a Evaluate $\nabla^2 V$ for the scalar function $V = 1/\rho^2$, cylindrical coordinates, first using rectangular coordinates and then using the cylindrical coordinates. This shows the importance of using the coordinate system that is natural to a given problem. (Hint: $\rho = \sqrt{x^2 + y^2}$.

4-17.0b Obtain the Laplacian of the function $V = \ln(1/r)$ using spherical coordinates.

4-17.0c Obtain the Laplacian of the function $V = a \cos bxe^{cx}$.

4-17.0d Given the bipolar coordinate system shown in Figure 4-39, (u, v, z), with $v = \pm\infty$ at $x = -a$, $x = +a$.

(a) On the diagram, show an element of area and give its vector expression.

(b) Suppose V is a scalar field. Give the expression for the Laplacian in this coordinate system. Simplify where possible.

As derivations are made in vector notation, using the symbol ∇ as a formal operator, we often encounter the form

$$\nabla \times (\nabla \times \vec{A})$$

This is an important term in the derivation of the wave equation for propagating fields which is of importance in radio wave studies. The result is the vector equivalent of the transmission line equation studied earlier. One obvious way to determine the expansion of this vector function would be to apply the generalized coordinate expression for the $\nabla\times$ twice. This leads to an extremely complex result and, in practice, does not yield an easily interpreted expansion. We recall, however, that when we reduced our generalized coordinate expressions to rectangular coordinates, we were able to treat ∇ as if it were a vector in that $\nabla \cdot \vec{A} \sim \vec{B} \cdot \vec{A}$ and $\nabla \times \vec{A} \sim \vec{B} \times \vec{A}$ where $\nabla \sim \vec{B}$. It was then noted that even though we could not make the same analogy in other coordinate systems, it was still convenient to retain the symbolism $\nabla \cdot$ and $\nabla \times$. Thus we are led to expand $\nabla \times \nabla \times \vec{A}$ in rectangular coordinates. Using the determinant form for the curl we obtain

$$\nabla \times \nabla \times \vec{A} = \nabla \times \begin{vmatrix} \hat{x} & \hat{y} & \hat{z} \\ \dfrac{\partial}{\partial x} & \dfrac{\partial}{\partial y} & \dfrac{\partial}{\partial z} \\ A_x & A_y & A_z \end{vmatrix}$$

$$- \nabla \times \left[\left(\frac{\partial A_z}{\partial y} - \frac{\partial A_y}{\partial z} \right) \hat{x} + \left(\frac{\partial A_x}{\partial z} - \frac{\partial A_z}{\partial x} \right) \hat{y} + \left(\frac{\partial A_y}{\partial x} - \frac{\partial A_x}{\partial y} \right) \hat{z} \right]$$

$$= \begin{vmatrix} \hat{x} & \hat{y} & \hat{z} \\ \dfrac{\partial}{\partial x} & \dfrac{\partial}{\partial y} & \dfrac{\partial}{\partial z} \\ \left(\dfrac{\partial A_z}{\partial y} - \dfrac{\partial A_y}{\partial z} \right) & \left(\dfrac{\partial A_x}{\partial z} - \dfrac{\partial A_z}{\partial x} \right) & \left(\dfrac{\partial A_y}{\partial x} - \dfrac{\partial A_x}{\partial y} \right) \end{vmatrix}$$

$$= \left[\frac{\partial}{\partial y} \left(\frac{\partial A_y}{\partial x} - \frac{\partial A_x}{\partial y} \right) - \frac{\partial}{\partial z} \left(\frac{\partial A_x}{\partial z} - \frac{\partial A_z}{\partial x} \right) \right] \hat{x}$$

$$+ \left[\frac{\partial}{\partial z} \left(\frac{\partial A_z}{\partial y} - \frac{\partial A_y}{\partial z} \right) - \frac{\partial}{\partial x} \left(\frac{\partial A_y}{\partial x} - \frac{\partial A_x}{\partial y} \right) \right] \hat{y}$$

$$+ \frac{\partial}{\partial x} \left(\frac{\partial A_x}{\partial_z} - \frac{\partial A_z}{\partial x} \right) - \frac{\partial}{\partial y} \left(\frac{\partial A_z}{\partial y} - \frac{\partial A_y}{\partial z} \right) \Bigg] \hat{z}$$

Expanding the derivatives and combining all mixed derivative terms on separate unit vectors:

$$\nabla \times \nabla \times \vec{A} = \left(-\frac{\partial^2 A_x}{\partial y^2} - \frac{\partial^2 A_x}{\partial z^2} \right) \hat{x} + \left(-\frac{\partial^2 A_y}{\partial x^2} - \frac{\partial^2 A_y}{\partial z^2} \right) \hat{y} + \left(-\frac{\partial^2 A_z}{\partial x^2} - \frac{\partial^2 A_z}{\partial y_2} \right) \hat{z}$$

$$+ \left(\frac{\partial^2 A_y}{\partial x \partial y} + \frac{\partial^2 A_z}{\partial x \partial z} \right) \hat{x} + \left(\frac{\partial^2 A_z}{\partial y \partial z} + \frac{\partial^2 A_x}{\partial x \partial y} \right) \hat{y} + \left(\frac{\partial^2 A_x}{\partial x \partial z} + \frac{\partial^2 A_y}{\partial y \partial z} \right) \hat{z}$$

Now the scalar components in the first three terms begin to look like the Laplacians of scalars since the A_i are scalar functions. Thus we add and subtract the appropriate terms to complete the Laplacian of each. Whatever terms remain are placed in with the other unit vectors.

$$\nabla \times \nabla \times \vec{A} = - \left(\frac{\partial^2 A_x}{\partial x^2} + \frac{\partial^2 A_x}{\partial y^2} + \frac{\partial^2 A_x}{\partial z^2} \right) \hat{x} - \left(\frac{\partial^2 A_y}{\partial x^2} + \frac{\partial^2 A_y}{\partial y^2} + \frac{\partial^2 A_y}{\partial z^2} \right) \hat{y}$$

$$\left(\frac{\partial^2 A_z}{\partial x^2} + \frac{\partial^2 A_z}{\partial y^2} + \frac{\partial^2 A_z}{\partial z^2} \right) \hat{z} + \left(\frac{\partial^2 A_x}{\partial x^2} + \frac{\partial^2 A_y}{\partial x \partial y} + \frac{\partial^2 A_z}{\partial x \partial z} \right) \hat{x}$$

$$\left(\frac{\partial^2 A_x}{\partial x \partial y} + \frac{\partial^2 A_y}{\partial y^2} + \frac{\partial^2 A_z}{\partial y \partial z} \right) \hat{y} + \left(\frac{\partial^2 A_x}{\partial x \partial z} + \frac{\partial^2 A_y}{\partial y \partial z} + \frac{\partial^2 A_z}{\partial z^2} \right) \hat{z}$$

Since the A_x, A_y, and A_z are scalars, we now recognize the Laplacians and write the first three terms in operator notation and remove a common partial derivative from each of the last three terms:

$$\nabla \times \nabla \times \vec{A} = -\left(\frac{\partial^2}{\partial x^2} + \frac{\partial^2}{\partial y^2} + \frac{\partial^2}{\partial z^2}\right)(A_x\hat{x} + A_y\hat{y} + A_z\hat{z}) + \frac{\partial}{\partial x}\left(\frac{\partial A_x}{\partial x} + \frac{\partial A_y}{\partial y} + \frac{\partial A_z}{\partial z}\right)\hat{x}$$

$$+ \frac{\partial}{\partial y}\left(\frac{\partial A_x}{\partial x} + \frac{\partial A_y}{\partial y} + \frac{\partial A_z}{\partial z}\right)\hat{y} + \frac{\partial}{\partial z}\left(\frac{\partial A_x}{\partial x} + \frac{\partial A_y}{\partial y} + \frac{\partial A_z}{\partial z}\right)\hat{z}$$

$$= -\nabla^2\vec{A} + \left(\frac{\partial}{\partial x}\hat{x} + \frac{\partial}{\partial y}\hat{y} + \frac{\partial}{\partial z}\hat{z}\right)\left(\frac{\partial A_x}{\partial x} + \frac{\partial A_y}{\partial y} + \frac{\partial A_z}{\partial z}\right)$$

$$= -\nabla^2\vec{A} + \left(\frac{\partial}{\partial x}\hat{x} + \frac{\partial}{\partial y}\hat{y} + \frac{\partial}{\partial z}\hat{z}\right)(\nabla \cdot \vec{A})$$

$$\nabla \times \nabla \times \vec{A} = -\nabla^2\vec{A} + \nabla(\nabla \cdot \vec{A}) \tag{4-46}$$

Although this applies strictly only to rectangular coordinates, we use this result to *define* $\nabla^2\vec{A}$ in other systems. Note, we have the Laplacian of a vector, defined now (from the preceding equation) as

$$\nabla^2\vec{A} \equiv \nabla(\nabla \cdot \vec{A}) - \nabla \times \nabla \times \vec{A} \tag{4-47}$$

Since the members of the right-hand side can be expanded in any orthogonal coordinate system using previous results, we can determine the left side.

Exercises

4-17.0e Given the two vector equations (called Maxwell's equations)

$$\nabla \times \vec{E} = -\mu \frac{\partial \vec{H}}{\partial t}$$

$$\nabla \times \vec{H} = +\epsilon \frac{\partial \vec{E}}{\partial t}$$

where t is time and ϵ and μ are constants.

By taking the curl of the first equation (to obtain $\nabla \times \nabla \times \vec{E}$ form) and realizing that partial derivatives on time and coordinates may be interchanged, obtain an equation involving $\vec{E}$ only and use Eq. 4-47 to obtain an equation for $\vec{E}$ that involves the vector Laplacian operation. Compare the result to the wave equation developed in the transmission line study and make any comparisons or analogies that are evident.

4-17.0f Give a physical description of the vector $\vec{E} = 10\cos(2\pi 10^6 t - 2\pi \times 10^6\sqrt{\mu\epsilon}\, z)\hat{y}$ (Hint: Recall the transmission line equation solution.) Show that this vector satisfies the equation developed in Exercise 4-17.0e.

4-17.0g Prove the following identities using rectangular coordinates:

(a) $\nabla(\vec{A} \cdot \vec{B}) = (\vec{A} \cdot \nabla)\vec{B} + (\vec{B} \cdot \nabla)A + \vec{A} \times (\nabla \times \vec{B}) + \vec{B} \times (\nabla \times \vec{A})$

(b) $(\vec{A}\nabla) \cdot \vec{B} = \vec{A}(\nabla \cdot \vec{B})$

(c) $\nabla \times (\vec{A} \times \vec{B}) = A\nabla \times \vec{B} - \vec{B}\nabla \cdot \vec{A} + (\vec{B} \cdot \nabla)\vec{A} - (\vec{A} \cdot \nabla)\vec{B}$

The final second-order operation to be described in this section is the ordinary multiplication of unit vectors. This is a different operation from both the dot and cross products. In this type of operation, we write the product of unit vectors in the traditional way without any special symbols. For example, if we wish to take $\hat{x}$ times $\hat{z}$ we write

$$\hat{x} \times \hat{z} = \hat{x}\hat{z} \tag{4-48}$$

In this result, the combination $\hat{x}\hat{z}$ is thought of as a single operator, and is called a *unit dyad*. It so happens that the unit vector order in the dyad is critical since the components of the dyad are not commutative; that is,

$$\hat{x}\hat{z} \neq \hat{z}\hat{x}$$

Since this operator is not of interest for this text, the details are given in a short Appendix G for completeness.

4-18 Vector Identities

Many vector identities involving both vector combinations and vector operator (∇) combinations are important in electromagnetic theory. In this section we list the more common ones encountered in practice. They are given without proof but are easily verified either by using vectors expressed in rectangular, three-component form ($\vec{A} = \overline{A}_x\hat{x} + \overline{A}_y\hat{y} + \overline{A}_z\hat{z}$) or by using theorems proved in earlier exercises. One may wonder whether proofs using the special rectangular form would suffice to be a proof for other coordinate systems. However, since one can always make a coordinate conversion to rectangular, which leaves the actual field structure unchanged, the proof using rectangular coordinates suffices. (See the General Exercises at the end of this chapter.)

1. Null identities:

$$\nabla \cdot (\nabla \times \vec{A}) \equiv 0 \text{ for any vector field } \vec{A}$$

$$\nabla \times (\nabla\Phi) \equiv 0 \text{ for any scalar field } \Phi$$

$$\vec{A} \cdot (\vec{A} \times \vec{C} \equiv 0$$

2. General identities (a, b scalar fields):

$$\nabla \times \nabla \times \vec{A} = \nabla(\nabla \cdot \vec{A}) - \nabla^2\vec{A}$$

$$\nabla \cdot (\vec{A} \times \vec{B}) = \vec{B} \cdot (\nabla \times \vec{A}) - \vec{A} \cdot (\nabla \times \vec{B})$$

$$\nabla(ab) = a\nabla b + b\nabla a$$

$$\nabla \cdot (a\vec{B}) = \vec{B} \cdot \nabla a + a\nabla \cdot \vec{B}$$

$$\nabla \cdot (a\nabla b) = \nabla a \cdot \nabla b + a\nabla^2 b$$

$$\nabla \times (a\vec{B}) = (\nabla a) \times \vec{B} + a(\nabla \times \vec{B})$$

$$\nabla \times (\vec{A} \times \vec{B}) = \vec{A}\nabla \cdot \vec{B} - \vec{B}\nabla \cdot \vec{A} + (\vec{B} \cdot \nabla)\vec{A} - (\vec{A} \cdot \nabla)\vec{B}$$

$$\nabla \cdot (\vec{A} \cdot \vec{B}) = (\vec{A} \cdot \nabla)\vec{B} + (\vec{B} \cdot \nabla)\vec{A} + \vec{A} \times (\nabla \times \vec{B}) + \vec{B} \times (\nabla \times \vec{A})$$

3. Distributive and associative identities (a, b, Φ scalar fields):

$$\nabla(a + b) = \nabla a + \nabla b$$

$$\nabla \cdot (\vec{A} + \vec{B}) = \nabla \cdot \vec{A} + \nabla \cdot \vec{B}$$

$$\nabla \times (\vec{A} + \vec{B}) = \nabla \times \vec{A} + \nabla \times \vec{B}$$

$$\vec{A} \times (\vec{B} \times \vec{C}) = \vec{B}(\vec{A} \cdot \vec{C}) - \vec{C}(\vec{A} \cdot \vec{B}) \text{ (vector triple product)}$$

$$(\vec{A} \times \vec{B}) \times \vec{C} = \vec{B}(\vec{A} \cdot \vec{C}) - \vec{A}(\vec{B} \cdot \vec{C})$$

$$\vec{A} \cdot (\vec{B} \times \vec{C}) = \begin{vmatrix} \bar{A}_1 & \bar{A}_2 & \bar{A}_3 \\ \bar{B}_1 & \bar{B}_2 & \bar{B}_3 \\ \bar{C}_1 & \bar{C}_2 & \bar{C}_3 \end{vmatrix} \text{ (scalar triple product)}$$

$$\vec{A} \cdot (\vec{B} \times \vec{C}) = (\vec{A} \times \vec{B}) \cdot \vec{C}$$

$$(\vec{A} \times \vec{B}) \cdot (\vec{C} \times \vec{D}) = (\vec{A} \cdot \vec{C})(\vec{B} \cdot \vec{D}) - (\vec{A} \cdot \vec{D})(\vec{B} \cdot \vec{C})$$

$$\begin{aligned}(\vec{A} \times \vec{B}) \times (\vec{C} \times \vec{D}) &= \vec{B}(\vec{A} \cdot \vec{C} \times \vec{D}) - \vec{A}(\vec{B} \cdot \vec{C} \times \vec{D}) \\ &= \vec{C}(\vec{A} \cdot \vec{B} \times \vec{D}) - \vec{D}(\vec{A} \cdot \vec{B} \times \vec{C})\end{aligned}$$

$$\nabla \cdot (a\nabla b - b\nabla a) = a\nabla^2 b - b\nabla^2 a$$

$$(\vec{A} \times \nabla)\Phi = \vec{A} \times (\nabla\Phi)$$

$$\vec{A} \times (\vec{B} \times \vec{C}) \neq (\vec{A} \times \vec{B}) \times \vec{C}$$

$$a(\vec{A} \times \vec{B}) = (a\vec{A}) \times \vec{B} = \vec{A} \times (a\vec{B})$$

4. Commutative identities:

$$\vec{A} \times \vec{B} = -\vec{B} \times \vec{A}$$

$$\vec{A} \cdot \vec{B} = \vec{B} \cdot \vec{A}$$

4-19 Integral Theorems for Vector Fields and their Interpretations

In the preceding sections we have emphasized the differential forms of vector analysis. All of the operations were point functions. This section is devoted to the subject of evaluating integrals of vector fields over finite regions of space. The evaluations include volume, surface, and line (contour) integrations of importance in electromagnetic theory in addition to the theorems that require the evaluations of such integrals. As in previous sections, one of the main goals here is to develop physical insights into the meanings of the integrals and the interpretations of the results.

Formal proofs of these theorems will not be given here since the electromagnetic equivalents of these result naturally from a consideration of fundamental experiments. We instead concentrate on the interpretation of the results.

Suppose that we have a *closed* surface S (see Figure 4-58) that surrounds a volume V. Also, suppose that permeating this volume we have vector field $\vec{A}$, which has its lines passing through the surface S. Let $\hat{n}$ be the vector direction of the area increment $d\vec{a}$ (outward normal to da as usual). The divergence theorem states:

$$\oint_S \vec{A} \cdot d\vec{a} = \int_V \nabla \cdot \vec{A} dv \tag{4-49}$$

In words, this theorem states that the net amount of a vector field crossing a closed surface is the same as the amount of divergence of the vector field within the enclosed volume. The closed surface integral of a vector field is called the *flux* of the vector field over surface S. If one considers the definition of divergence discussed earlier, this can readily be given a physical interpretation. As before, if we let the vector field $\vec{A}$ be the velocity vectors of the molecules (see Figure 4-47), then the divergence of the velocity vector field would be represented by the molecules as they enter and leave a small region of space. Thus if the closed surface integral of the velocity vector field were positive, there would be a divergence (leaving) of molecules from the enclosed volume. In a slightly different manner, one could say that there is a scalar quantity within the volume V (molecular density), which is a measure of the divergence of some vector field (velocity) that it produces. The properties of the scalar quantity (density, number, temperature, etc.) can be determined by calculating the flux of the vector field produced by the scalar.

A rather crude qualitative proof can be given for this theorem by considering the definition of divergence.

$$\nabla \cdot \vec{A} = \lim_{\Delta v \to 0} \frac{\oint_{\Delta S} \vec{A} \cdot d\vec{a}}{\Delta v} = \frac{\oint_{dS} \vec{A} \cdot d\vec{a}}{dv}$$

$$\nabla \cdot \vec{A} dv = \oint_{dS} \vec{A} \cdot d\vec{a}$$

We next add up (integrate) the effects due to all of the incremental volumes and surfaces. Consider, however, the vector $\vec{A}$ at the common face between the two incremental

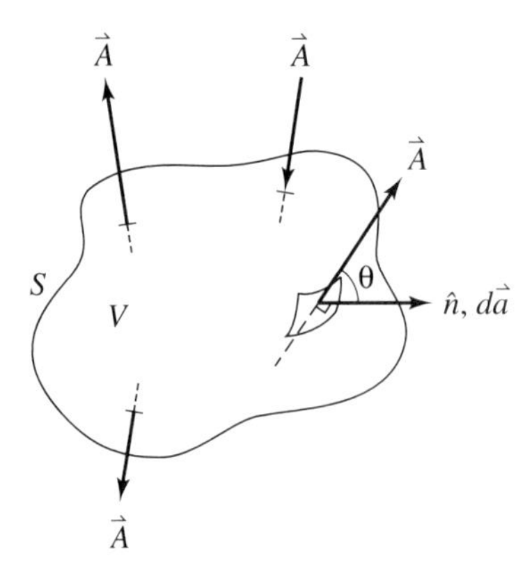

Figure 4-58. Illustration of the quantities used in the divergence theorem.

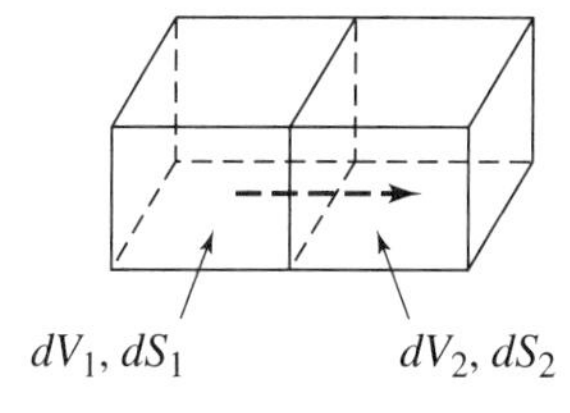

Figure 4-59. Flux at the interface of two contiguous increments of volume.

volumes shown in Figure 4-59. The amount of flux leaving dS_1 will enter dS_2 so that these contributions will cancel when added together since the dot product of $\vec{A} \cdot \vec{dS}_1$ will be the negative of $\vec{A} \cdot d\vec{S}_2$ (remember that the unit normal is directed away from the volume of the surface being considered). Therefore, all that is important is the surface outside (i.e., which encloses) the volume. Thus integrating both sides of the differential expression "yields" the divergence theorem, Eq. 4-49.

Example 4-25

A vector field is given by the expression

$$\vec{D} = r \sin\phi \cos\theta \hat{r} \qquad \text{(spherical coordinates)}$$

Show that the divergence theorem is satisfied by this vector field, for a sphere centered at the origin having a radius of 5 cm.

We first evaluate the flux of $\vec{D}$ (the left side of Eq. 4-49). In spherical coordinates, $d\vec{a} = r^2 \sin\theta d\phi d\theta \hat{r}$ so that

$$\vec{D} \cdot d\vec{a} = r \sin\phi \cos\theta \hat{r} \cdot r^2 \sin\theta d\theta d\phi \hat{r} = r^3 \sin\phi \sin\theta \cos\theta d\theta d\phi$$

Then

$$\oint_S \vec{D} \cdot d\vec{a} = \int_{\phi=0}^{2\pi} \int_{\theta=0}^{\pi} r^3 \sin\phi \sin\theta \cos\theta d\theta d\phi = r^3 \int_{\phi=0}^{2\pi} \sin\phi d\phi \int_{\theta=0}^{\pi} \sin\theta \cos\theta d\theta$$

$$= (0.05)^3(-\cos\phi)\Big|_0^{2\pi} \left(\tfrac{1}{2}\sin^2\theta\right)\Big|_0^{\pi}$$

$$= (1.25) \times 10^{-4})(-1 - (-1))(\tfrac{1}{2} \cdot 0 - \tfrac{1}{2} \cdot 0) = 0$$

(This is not a very exciting result, but correct nonetheless.) For the right-hand side of Eq. 4-49 we write

$$dv = r^2 \sin\theta d\theta d\phi dr$$

and

$$\nabla \cdot \vec{D} = \frac{1}{r^2}\frac{\partial(r^2 D_r)}{\partial r} + \frac{1}{r \sin\theta}\frac{\partial(\mathrm{D}_\theta \sin\theta)}{\partial\theta} + \frac{1}{r\sin\theta}\frac{\partial D_\phi}{\partial\phi}$$

Since D_θ and D_ϕ are zero

$$\nabla \cdot \vec{D} = \frac{1}{r^2}\frac{\partial(r^3 \sin\phi \cos\theta)}{\partial r} = 2 \sin\phi \cos\theta$$

Then

$$\int_V \nabla \cdot \vec{D} dv = \int_{r=0}^{0.05} \int_{\phi=0}^{2\pi} \int_{\theta=0}^{\pi} 2 \sin \phi \cos \theta r^3 \sin \theta d\theta d\phi dr$$

$$= 2 \int_0^{0.05} r^2 dr \int_0^{2\pi} \sin \phi d\phi \int_0^{\pi} \sin \theta \cos \theta d\theta$$

$$= 2(\tfrac{1}{3} r^3)\Big|_0^{0.05} (-\cos \phi)\Big|_0^{2\pi} (\tfrac{1}{2} \sin^2 \theta)\Big|_0^{\pi}$$

$$= (2)(4.16\overline{6} \times 10^{-5})(-1 - (-1))(\tfrac{1}{2} \cdot 0 - \tfrac{1}{2} \cdot 0) = 0$$

This is the same value as obtained from the surface integral, so the theorem is verified for this case.

The subject of vector integration is covered in more detail later. You will next have the opportunity of completing a guided exercise where each step is described and you fill in the details.

Example 4-26

Guided Exercise

Suppose we have a vector field given by the equation $\vec{D} = \rho z \cos \phi \hat{\rho} + \rho^2 z \hat{z}$ in cylindrical coordinates. Verify the divergence theorem for this field over the cylinder shown in Figure 4-60 by completing the steps given.

1. On the closed, capped cylinder S, draw a differential area and indicate the unit normal for each of the three surfaces which define the closed surface S.
2. Write an expression for each of the differential areas. Put the unit normal vectors in terms of the unit vectors of the cylindrical coordinate system as you do this.
3. Evaluate the dot product $\vec{D} \cdot \vec{da}$ for each of the three differential area vectors.
4. What are the limits of integration of each coordinate on each of the three surfaces? For each surface give the values of any constant coordinates.

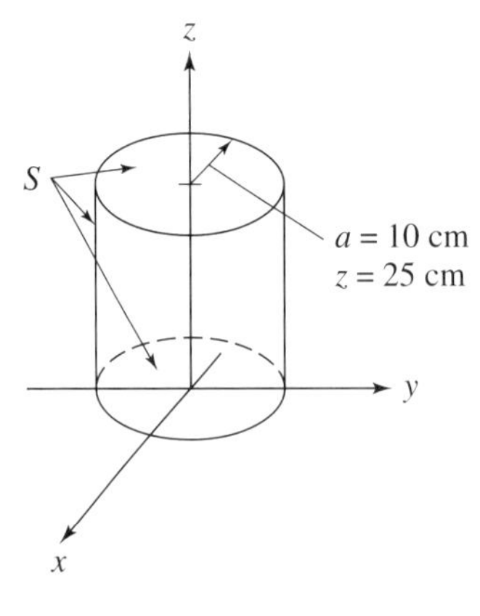

Figure 4-60. Configuration for divergence theorem computations.

5. Evaluate each of the three integrals and sum them to obtain a value for the flux of $\vec{D}$. This is the left member of Eq. 4-49.
6. In the cylinder, draw a differential volume and write an expression for that element.
7. Calculate the divergence of $\vec{D}$.
8. What are the limits of integration for each of the coordinate variables in dv?
9. Evaluate the volume integral term of the divergence theorem. Compare the results of the volume and surface integrations.

Exercises

4-19.0a A cube having edge of length 2 has one corner at the origin and is in the first octant of space ($x, y, z > 0$). The vector field $\vec{B} = -4xy\hat{x} + 8yz\hat{y} + (4yz - 4z^2)\hat{z}$ exists in the region. Verify the divergence theorem for this field. Suppose the cube is located at a general region and rotated 45° about a diagonal. Now if one were interested in determining the flux of $\vec{B}$ passing through the six faces of such a cube, of what value would the divergence theorem be in making the computation?

4-19.0b In the parabolic cylindrical coordinate system (see Exercise 4-9.0f) suppose we have a cylinder extending from $z = 2$ to $z = 5$, whose cross section is the football-shaped region defined by the intersections of the curves $v = 2$ and $u = 2, -2$. In the region there is a field $\vec{D} = 3\hat{u} - 4z\hat{z}$. Verify the divergence theorem for this system. What would you do if the integrals could not be evaluated in closed form?

4-19.0c Using the divergence theorem, prove that

$$\oint_S \vec{D} \times d\vec{a} = -\int_V \nabla \times \vec{D} dv$$

(Hint: In the divergence theorem let $\vec{A} = \vec{D} \times \vec{K}$, where $\vec{K}$ is a constant vector.)

The second important integral theorem is known as Stokes' theorem. This theorem relates surface and line (path) integrations, and the basic configuration is shown in Figure 4-61.

Suppose we have a closed-line path (contour) C that is the contour around the edge of an open surface S. Also, suppose there is a vector field $\vec{A}$ having components that pass through the surface. Let $\hat{n}$ be the normal at the surface increment $d\vec{a}$ and $d\vec{l}$ be an increment of the contour C.

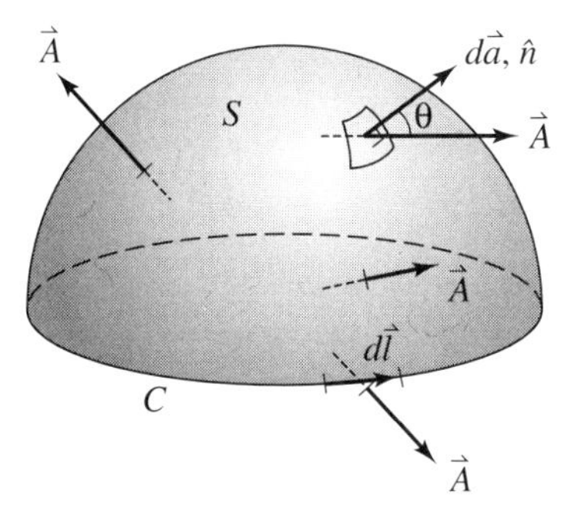

Figure 4-61. Illustration of quantities used in Stokes' theorem.

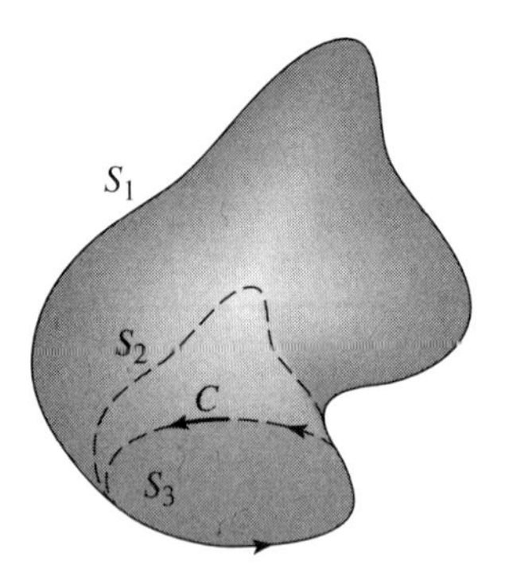

Figure 4-62. Multiple possibilities for surfaces which have the same contour.

Stokes' theorem states:

$$\oint_C \vec{A} \cdot d\vec{l} = \int_S \nabla \times \vec{A} \cdot d\vec{a} \tag{4-50}$$

In words, this says that the flux of the curl of the vector field passing through the surface is the same as the component of the vector around the contour, or rim, of the surface. The contour is traversed so that as one imagines walking the contour, the surface is on the left hand (right-hand rule where the thumb points in the general direction of $d\vec{a}$). This theorem relates a surface integral to a line integral, and the line (path) integral is often called the *circulation* of the vector field $\vec{A}$. There is a very important consequence of this theorem. Note that the shape of the *surface* is immaterial as long as it has the same contour on its edge. See Figure 4-62. Also, the surface S_3 as a special case could be the plane area within the contour, and the contour C is assumed to be simple; that is, it does not cross over itself as in a figure eight, for example. The surface S_2 is inside S_1. Think of one balloon inside another but only partly inflated. S_3 would be the mouth area of the balloon.

Exercises

4-19.0d Develop a qualitative proof of Stokes' theorem by following the steps that parallel those given in the proof of the divergence theorem. Begin with the definition of curl, Eq. 4-41.

Example 4-27

A vector field exists in a region and is described by the equation

$$\vec{E} = r \cos \theta \sin \phi \hat{r} + r^2 \sin \theta \hat{\phi}$$

Verify that Stokes' theorem is satisfied and evaluate the surface integral using both the hemispherical shell and the equation cross-sectional circle area. The sphere radius is 0.1 m. See Figure 4-63.

For the contour C we observe physically that $d\vec{l}$ is in the $\hat{\phi}$ direction in the spherical coordinate system (r, θ, ϕ) so that $d\vec{l} = h_3 du_3 \hat{u}_3 = r \sin \theta d\phi \hat{\phi}$. Thus $\vec{E} \cdot d\vec{l} = r^3 \sin^2 \theta d\phi$. Then

$$\oint_C \vec{E} \cdot d\vec{l} = \int_{\phi=0}^{2\pi} r^3 \sin^2\theta d\phi = r^3 \sin^2 \theta \int_{\phi=0}^{2\pi} d\phi = 2\pi r^3 \sin^2 \theta$$

Now on the contour $r = 0.1$ m and $\theta = 90°$ so that

$$\oint_C \vec{D} \cdot d\vec{l} = 2\pi \times 10^{-3}$$

For the curl:

$$\nabla \times \vec{E} = \frac{1}{r \sin \theta}\left[\frac{\partial(r^2 \sin^2 \theta)}{\partial\theta} - 0\right]\hat{r} + \left[\frac{1}{r \sin \theta}\frac{\partial(r \cos \theta \sin \phi)}{\partial\phi} - \frac{1}{r}\frac{\partial(r^3 \sin \theta}{\partial r}\right]\hat{\theta}$$
$$+ \frac{1}{r}\left[0 - \frac{\partial(r \cos \theta \sin \phi)}{\partial\theta}\right]\hat{\phi}$$

$$\nabla \times \vec{E} = 2r \cos \theta\hat{r} + (\cos \phi \cot \theta - 3r \sin \theta)\hat{\theta} + \sin \theta \sin \phi\hat{\phi}$$

On the surface S_1, $d\vec{a} = r^2 \sin \theta d\phi d\theta\hat{r}$ so that

$$\nabla \times \vec{E} \cdot d\vec{a} = 2r^3 \cos \theta \sin \theta d\phi d\theta$$

Then

$$\int_{S_1} \nabla \times \vec{E} \cdot d\vec{a} = \int_{\phi=0}^{2\pi}\int_{\theta=0}^{\pi/2} 2r^3 \cos \theta \sin \theta d\theta d\phi = 2r^3 \int_0^{\pi} \cos \theta \sin \theta d\theta \int_0^{2\pi} d\phi$$
$$= 2r^3(\tfrac{1}{2} \sin^2 \theta) \int_0^{\pi/2} 2\pi = 2\pi r^3(1 - 0) = 2\pi r^3$$

For $r = 0.1$ m, $\int_{S_1} \nabla \times \vec{E} \cdot d\vec{a} = 2\pi \times 10^{-3}$ as before. To evaluate the surface integral over the surface S_2 in the xy plane, we first observe that for S_2, $\theta = \pi/2$. Then

$$\nabla \times \vec{E}|_{\theta=\pi/2} = -3r\hat{\theta} + \sin \phi\hat{\phi}$$

Now since the circle represents a cylinder cross section it would be better to evaluate the integral in terms of cylindrical coordinates (ρ, ϕ, z). In the xy plane the unit vector $\hat{\theta}$ is oriented downward, or in the negative $\hat{z}$ direction. To see this remember that the unit vector

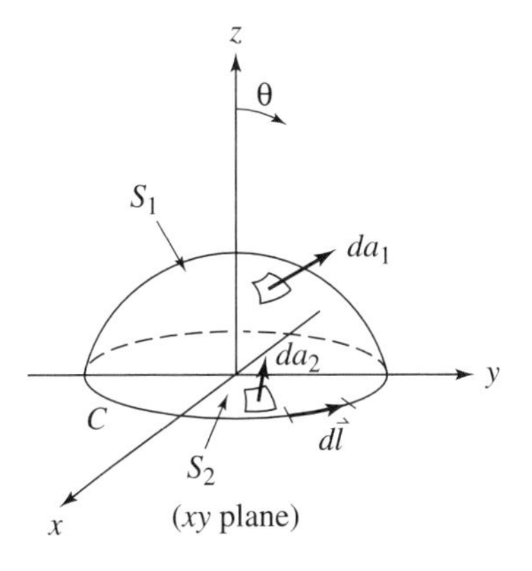

Figure 4-63. Example configuration for the application of Stokes' theorem.

$\hat{\theta}$ points in the direction of increasing coordinate value which is in $-\hat{z}$ on contour C and everywhere on S_2. Thus in cylindrical coordinates

$$\nabla \times \vec{E}|_{xy \text{ plane}} = -3\rho(-\hat{z}) + \sin\phi\hat{\phi} = 3\rho\hat{z} + \sin\phi\hat{\phi}$$

Notice also that spherical coordinate r is the same as cylindrical coordinate ρ *in the plane*, both being $\sqrt{x^2 + y^2}$. In cylindrical coordinates, for S_2, $d\vec{a} = \rho d\rho d\phi\hat{z}$ so that

$$\int_{S_2} \nabla \times \vec{E} \cdot d\vec{a} = \int_0^{2\pi}\int_0^{0.1} 3\rho^2 d\rho d\phi = \int_0^{2\pi} d\phi \int_0^{0.1} 3\rho^2 d\rho$$

$$= 2\pi \Big|_0^{0.1} = 2\pi \times 10^{-3}$$

which again agrees with the earlier evaluations. The following guided exercise is intended to have you carry out the steps in a verification of Stokes' theorem by giving direction to the order of computation.

Example 4-28

Guided Exercise

Suppose we have a vector field given by the equation $\vec{D} = \rho z \cos\phi\hat{\rho} + \rho^2 z\hat{z}$ in cylindrical coordinates. Verify Stokes' theorem for this field for each of the two surfaces S_1 and S_2 indicated in Figure 4-64. The contour C is to be the top rim, in both instances, as shown.

1. On the contour C show the incremental path $d\vec{l}$. (Note: $d\vec{l}$ is *always* in the direction of increasing coordinate values.)
2. Compute $\vec{A} \cdot d\vec{l}$ and state the limits of integration of the differential coordinate in $d\vec{l}$ in the form from ______ to ______. Don't forget to adhere to the convention for traversing the contour. (Answer: From 2π to 0)
3. Evaluate the closed-path integral part of Stokes' theorem.
4. On the surface S_1 draw the differential areas and give expressions for them. Note there will be two, one for the curved cylindrical side and one for the bottom. (Answer: $0.1 d\phi dz\hat{\rho}$ and $\rho d\phi d\rho(-\hat{z})$)

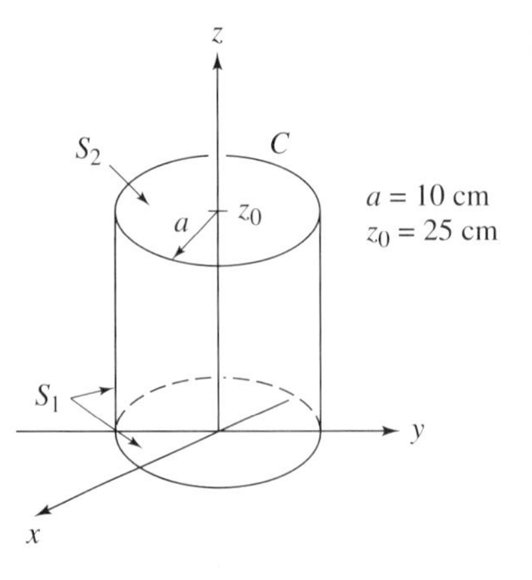

Figure 4-64.

5. Compute $\nabla \times \vec{A}$ and $\nabla \times \vec{A} \cdot d\vec{a}$ for each part of S_1, and state the limits for all coordinate differentials. (Answer: 0 to 2π and 0 to 0.25; 0 to 0.1 and 0 to 2π)
6. Evaluate the area integral part of Stokes' theorem. Note you will have to add two integrals to cover S_1. How does this value compare with the answer to (3)?
7. Repeat the preceding steps for the surface S_2 bounded by contour C. (Do not recompute those previous items that are unchanged by selecting new surface S_2. Simply relist them. (Answer: differential area $= \rho d\phi d\rho(-\hat{z})$; 0 to 0.1 and 0 to 2π. Also observe that z is constant on S_2, $z_0 = 0.1$.)

Exercises

4-19.0e A cube having edge of length 2 has one corner at the origin and is in the first octant of space (x, y, z). A vector field $\vec{B} = -4xy\hat{x} + 8yz\hat{y} + (4yz - 4z^2)\hat{z}$ exists in the region. Verify Stokes' theorem where the contour consists of the edges in the xy plane and for two cases of surfaces:

(a) The upper five cube faces.

(b) The face in the xy plane.

4-19.0f In the elliptic cylindrical coordinate system (see General Exercise 4-10 at the end of this chapter) suppose that we have the surface corresponding to $u = 3/2$ bounded top and bottom by $z = 2$ and -1, respectively, with the focus a located at $x = \frac{1}{2}$. In this region there exists a field $\vec{D} = 4 \cosh u \sin v\hat{u} - 2uv\hat{v} + 7 \cos v\hat{z}$. Verify Stokes' theorem for the case where the contour is the elliptical rim at $z = -1$. You may pick any surface you wish. (Which is the best?)

4-19.0g For a scalar field ϕ use Stokes' theorem to show that

$$\oint_C \phi d\vec{l} = -\int_S \nabla\phi \times d\vec{a}$$

(Hint: Define a vector $\vec{A} = \phi\vec{K}$ where $\vec{K}$ is a constant vector and ϕ is a scalar function.)

Although the divergence theorem and Stokes' theorem are the most widely used in introductory electromagnetics, other integral theorems have sufficiently frequent application that they need to be included. The first two results are called Green's theorems and are derived using the divergence theorem. We begin by using the general identity involving two scalar fields, listed in the second part of Section 4-18; namely, if a and b are two scalar fields,

$$\nabla \cdot (a\nabla b) = \nabla a \cdot \nabla b + a\nabla^2 b$$

Integrating over a volume,

$$\int_V \nabla \cdot (a\nabla b)dv = \int_V (\nabla a \cdot \nabla b + a\nabla^2 b)dv$$

Using the divergence theorem on the left side

Green's First Identity or Theorem:
$$\oint_S a\nabla b \cdot d\vec{a} = \int_V (\nabla a \cdot \nabla b + a\nabla^2 b)dv \qquad (4\text{-}51)$$

Next interchange the scalar fields a and b in the first divergence identity above to obtain

$$\nabla \cdot (b\nabla a) = \nabla b \cdot \nabla a + b\nabla^2 a = \nabla a \cdot \nabla b + \nabla^2 a$$

Subtracting the original identity from this result yields

$$\nabla \cdot (b\nabla a) - \nabla \cdot (a\nabla b) = b\nabla^2 a - a\nabla^2 b$$

Integrating this over a volume V (after using the distributive property of divergence)

$$\int_V \nabla \cdot (b\nabla a - a\nabla b)dv = \int_V (b\nabla^2 a - a\nabla^2 b)dv$$

Now applying the divergence theorem to the left side of the equation we finally have

Green's Second Identity or Theorem:
$$\oint_S (b\nabla a - A\nabla b) \cdot d\vec{a} = \int_V (b\nabla^2 a - a\nabla^2 b)dv \qquad (4\text{-}52)$$

The practical value of these results may appear questionable, especially since the integrands in both identities do not seem to have reduced the computational effort much. However, in many practical problems the Laplacian (∇^2) terms are frequently zero in value so some simplification may result. Also, if the situation arises where an integral evaluation is required, it may turn out that the equivalent integral may be easier to evaluate. Additional vector integral theorems that find applications in electromagnetic theory are (without proof here)

$$\left.\begin{aligned} \int_V \nabla \times \vec{A}dv &= -\oint_S \vec{A} \times d\vec{a} \\ \int_V \nabla\Phi dv &= \oint_S \Phi d\vec{a} \qquad \Phi \text{ a scalar field function} \\ \int_S \nabla\Phi \times d\vec{a} &= -\oint_C \Phi d\vec{l} \end{aligned}\right\} \qquad (4\text{-}53)$$

The physical spatial relationships of the contours, areas, and volumes in the above integrals are the same as those described for the divergence and Stokes' theorems. Incidentally, the proofs of these last results make use of the divergence and Stokes' theorems and are briefly outlined in Exercises 4-19.0c, GE4-30, and 4-19.0g, respectively. Although vector integration is usually covered in a calculus course we summarize the main ideas and present some examples here, sufficient to give meaning to the terms of the integrals. The general procedure for evaluating such integrals is:

1. Identify carefully the paths (contours), surfaces, and/or volumes of interest, and state the important coordinate values pertaining to these elements. Usually some coordinates are constant.
2. In the case of the paths and surfaces, write the sum of the integrals required to cover the entire path or surface.
3. Write an expression for each factor that appears in the integral expression, and the corresponding limits.

4. Substitute the expressions into the integrals and perform any vector operations required.
5. Evaluate the integrals using appropriate limits and path or surface functions.

Example 4-29

Suppose we are given the vector field $\vec{E} = 2xy\hat{x} + yz\hat{y} - 5xz\hat{z}$ and asked to evaluate $\int \vec{E} \cdot d\vec{l}$ along the path shown in Figure 4-65, path $a \rightarrow b \rightarrow c$.

The path has already been identified, but we should furthermore specify that for $a \rightarrow b$, x has the constant value 2 and for $b \rightarrow c$, y has the constant value 0. For both paths, z has the value 1. The path from $a \rightarrow c$ will thus need to be written as the sum of the two integrals:

$$\int_a^c \vec{E} \cdot d\vec{l} = \underset{\substack{x=2\\ z=1}}{\int_a^b \vec{E} \cdot d\vec{l}} + \underset{\substack{y=0\\ z=1}}{\int_b^c \vec{E} \cdot d\vec{l}}$$

For both integrals, $\vec{E} = 2xy\hat{x} + yz\hat{y} - 5xz\hat{z}$. For the path $a \rightarrow b$, $d\vec{l} = dy\hat{y}$. It is important to note that we always write the incremental path as being in the direction of *increasing* coordinate. The limits of integration take care of the fact that the path is in the direction of decreasing y. To put a minus sign with the $d\vec{l}$ would account for the path direction *twice*, thus leading to an error. The general expression for $d\vec{l}$ as given in Eq. 4-23 can be used in every case. We set a portion of that expression to zero if a coordinate is constant, in this case $dx = 0$ and $dz = 0$. For path section $b \rightarrow c$: $d\vec{l} = dx\hat{x}(dy = dz = 0$ in the general expression). Thus

$$\begin{aligned}\int_a^c \vec{E} \cdot d\vec{l} &= \int_a^b (2xy\hat{x} + yz\hat{z} - 5xz\hat{z}) \cdot dy\hat{y} + \int_b^c (2xy\hat{x} + yz\hat{y} - 5xz\hat{z}) \cdot dx\hat{x} \\ &= \int_3^0 yzdy + \int_2^1 2xydx = \int_3^0 (y)(1)dy + \int_2^1 2(x)(0)dx = \int_3^0 ydy \\ &= \frac{1}{2}y^2\bigg|_3^0 = \frac{1}{2}(9) = 4.5\end{aligned}$$

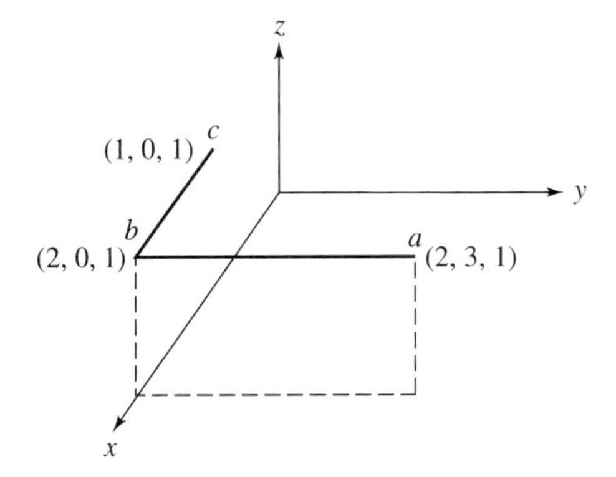

Figure 4-65. Path for evaluation of a line integral.

Example 4-30

Next suppose we want to evaluate the same integral as in the preceding example, but the path is from a to c directly as shown in Figure 4-66.

The path is described by the equation $y = 3(x - 1)$ and for this path $z = 1$, and the ranges (limits) of the variables are

$$x : 2 \rightarrow 1$$

$$y : 3 \rightarrow 0$$

In this case, one integral is the complete path. Now $\vec{E} = 2xy\hat{x} + yz\hat{y} - 5xz\hat{z}$ is given and we have the general $d\vec{l} = h_1 du_1 \hat{u}_1 + h_2 du_2 \hat{u}_2 + h_3 du_3 \hat{u}_3 = dx\hat{x} + dy\hat{y} + dz\hat{z}$. But z is constant so $dz = 0$, which reduces $d\vec{l}$ to

$$d\vec{l} = dx\hat{x} + dy\hat{y}$$

Putting these into the integral:

$$\int_a^c \vec{E} \cdot d\vec{l} = \int_a^c (2xy\hat{x} + yz\hat{y} - 5xz\hat{z}) \cdot (dx\hat{x} + dy\hat{y})$$

$$= \int_a^c (2xydx + yzdy) = \underset{z=0}{\int_2^1 2xydx} + \underset{z=1}{\int_3^0 yzdy} = \int_2^1 2xydx + \int_3^0 ydy$$

Now in the first (dx) integral set $y = 3(x - 1)$ to obtain the result

$$\int_a^c \vec{E} \cdot d\vec{l} = \int_2^1 (6x^2 - 6x)dx + \int_3^0 ydy = \left(\frac{6}{3}x^3 - \frac{6}{2}x^2\right)\Bigg|_2^1 + \frac{1}{2}y^2\Bigg|_3^0$$

$$= 2 - 3 - 16 + 12 - 4.5 = -9.5$$

This is not the same value we obtained for the other path, but nothing has been derived indicating that they should be the same. The condition for which the two values would be the same is that the vector field is conservative, that is, $\nabla \times \vec{E} = 0$. See the end of the discussion of curl in Section 4-16.

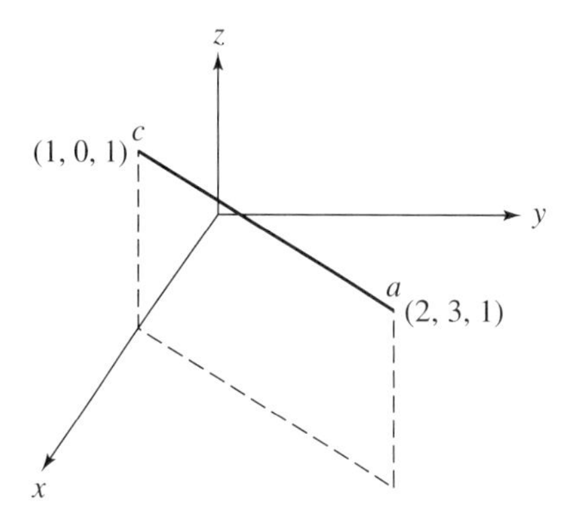

Figure 4-66. Configuration for direct path integration.

Exercises

4-19.0h For the vector $\vec{E} = 6x^2\hat{x} + 3xy\hat{y} - z\hat{z}$, evaluate $\int_a^d \vec{E} \cdot d\vec{l}$ for the paths *a-b-c-d* and *a-d* shown in Figure 4-67.

4-19.0i In cylindrical coordinates, we have $\vec{E} = 6/\rho \cos\phi\hat{\phi} + z\sin\phi\hat{\rho} + 6\hat{z}$. Evaluate $\int_a^c \vec{E} \cdot d\vec{l}$ for the paths shown in Figure 4-68 in the directions *a-b-c* and *a-c*. Don't forget the scale factors h_i on the expressions for $d\vec{l}$ as appropriate.

4-19.0j Develop a computer software unit that will evaluate $\int_a^b \vec{E} \cdot d\vec{l}$. Use rectangular coordinates. Solve Exercise 4-14.0h with your program. As in circuit theory when we program integrals, we use finite increments so that $d\vec{l} \rightarrow \nabla\vec{l}$, and so on.

4-19.0k As mentioned in the preceding example, there are some cases for which the line integral value is independent of the path between the two points. This is a valuable piece of information since if a given vector integration is independent of the path, it may be possible to select a path for which the integration is much easier to perform. In this exercise we will show that if a vector field $\vec{A}$ is derivable from a scalar field by $\vec{A} = \nabla\Phi$, where Φ is a function of coordinates, a path integral is independent of the path between the two points. You are asked to fill in the steps of the proof. You will also develop a test for determining if $\vec{A}$ can be obtained as the gradient of a scalar function. The theorem statement to be proved is:

If a vector field $\vec{A}$ can be derived as the gradient of a scalar function, or if $\nabla \times \vec{A} = 0$, then a path integral between two points is independent of the path taken between the points. Additionally, if the path is a closed contour the result is

$$\oint \vec{A} \cdot d\vec{l} = 0$$

Prove this by proceeding as follows:

(a) Express the vector $\vec{A}$ in generalized coordinate three-component form; that is, three components with the three generalized unit vectors $\hat{u}_1$, $\hat{u}_2$, and $\hat{u}_3$.

(b) Write the generalized coordinate expression for $d\vec{l}$.

(c) Evaluate the dot product $\vec{A} \cdot d\vec{l}$.

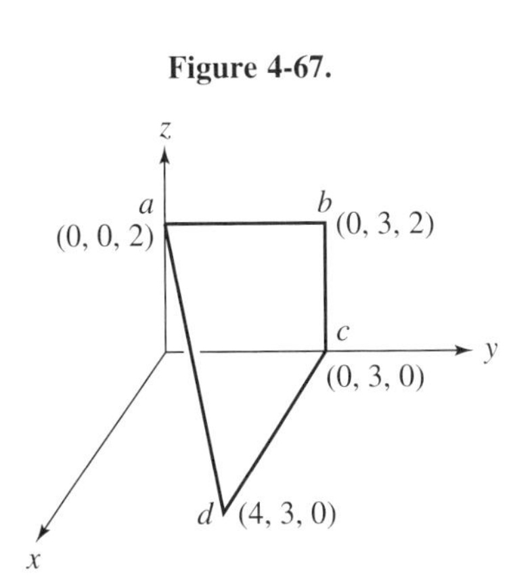

Figure 4-67.

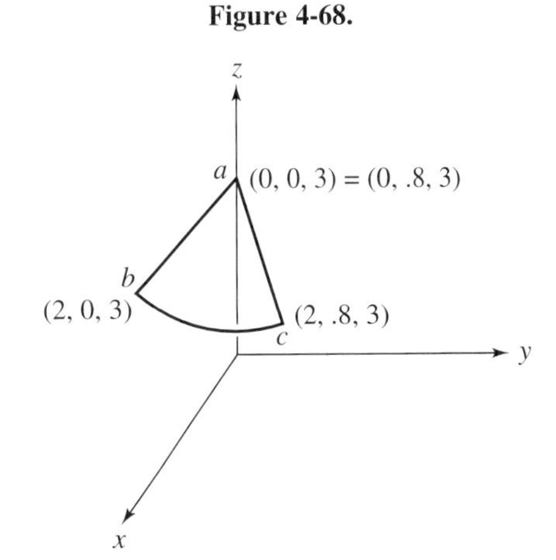

Figure 4-68.

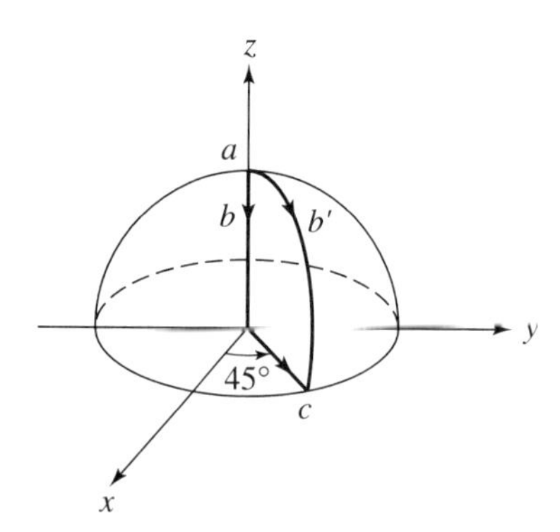

Figure 4-69.

(d) Write the generalized coordinate expression for $\nabla\Phi$ and identify A_1, A_2, and A_3 such that $\vec{A} = \nabla\Phi$. Substitute these components into the dot product of (c) above.

(e) Using your knowledge of partial differential calculus, write the expression for the total differential $d\Phi$ in terms of partial derivatives with respect to u_1, u_2, and u_3.

(f) By substituting your results of (d) and (e) show that

$$\int_a^b \vec{A} \cdot d\vec{l} = \Phi_b - \Phi_a$$

which shows that the value of the line integral depends only upon the end points and not on the path between. From this result show that

$$\oint_C \vec{A} \cdot d\vec{l} = 0$$

(Hint: a and b are the same point.)

(g) The question arises as to how we know whether or not $\vec{A}$ can be obtained as the gradient of a scalar function. Show that if $\vec{A} = \nabla\Phi$ then $\nabla \times \vec{A} = 0$. Use a vector identity involving multiple ∇ operations.

4-19.0l A particle is subject to a force $\vec{F} = (2x + z)\hat{x} + (x - z^2)\hat{y} + (3x + y - 4z)\hat{z}$. What is the total work done in moving the particle around the circumference of a circle of radius 10 cm, the circle being centered at the origin in the xy plane? (Hint: $dw = \vec{F} \cdot d\vec{l}$, and a change of coordinates may be helpful.)

4-19.0m In the region shown in Figure 4-69 the field is given by $\vec{E} = r\cos\phi\hat{r} + \sin\theta\hat{\theta}$. Evaluate the integral

$$\int_a^c \vec{E} \cdot d\vec{l}$$

for the paths indicated:

a-b-c

a-b'-c

The evaluation of surface integrals also requires special care in the determination of expressions for the area vectors. However, the general procedure given earlier still serves as a guide if the steps are carefully interpreted. Examples that illustrate the concepts follow.

Example 4-31

Evaluate $\oint_S \vec{B} \cdot d\vec{a}$ where $\vec{B}$ is the vector field $\vec{B} = 3x^2\hat{x} - xy\hat{y} + 5z^2y\hat{z}$ and S is the closed surface of the parallelopiped having dimensions as shown in Figure 4-70.

The closed surface in this case consists of the six sides of the box. The unit normal to each surface is also shown, the convention being that the unit normal is directed away from the volume enclosed by the surface.

Now to obtain the closed surface integral, we write

$$\oint_S \vec{B} \cdot d\vec{a} = \int\int_R \vec{B} \cdot d\vec{a} + \int\int_L \vec{B} \cdot d\vec{a} + \int\int_T \vec{B} \cdot d\vec{a} + \int\int_B \vec{B} \cdot d\vec{a} + \int\int_A \vec{B} \cdot d\vec{a} + \int\int_F \vec{B} \cdot d\vec{a}$$

We next write expressions for each term, noting the expression for $\vec{B}$ is the same for each integral:

$$\begin{aligned}
&R: \quad d\vec{a} = dxdz\hat{y} && y = 2 \\
&L: \quad d\vec{a} = dxdz(-\hat{y}) = -dxdz\hat{y} && y = 0
\end{aligned} \Bigg\} \; 0 \le x \le 1, \quad 0 \le z \le 3$$

$$\begin{aligned}
&T: \quad d\vec{a} = dxdy\hat{z} && z = 3 \\
&B: \quad d\vec{a} = dxdy(-\hat{z}) = -dxdy\hat{z} && z = 0
\end{aligned} \Bigg\} \; 0 \le x \le 1, \quad 0 \le y \le 2$$

$$\begin{aligned}
&F: \quad d\vec{a} = dydz\hat{x} && x = 1 \\
&A: \quad d\vec{a} = dydz(-\hat{x}) = -dydz\hat{x} && x = 0
\end{aligned} \Bigg\} \; 0 \le y \ge 2, \quad 0 \le z \le 3$$

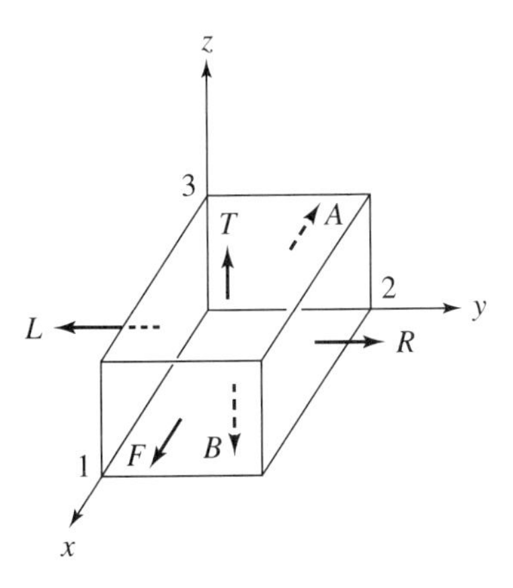

Figure 4-70. Configuration for the evaluation of a closed surface integral.

Substituting these we obtain, placing under each integral any constant coordinate values,

$$\oint_S \vec{B} \cdot d\vec{a} = \int_0^3 \int_0^1 (3x^2\hat{x} - xy\hat{y} + 5z^2y\hat{z}) \cdot dxdz\hat{y}$$
$$y = 2$$
$$+ \int_0^3 \int_0^1 (3x^2\hat{x} - xy\hat{y} + 5z^2y\hat{z}) \cdot (-dxdz\hat{y})$$
$$y = 0$$
$$+ \int_0^2 \int_0^1 (3x^2\hat{x} - xy\hat{y} + 5z^2y\hat{z}) \cdot dxdy\hat{z}$$
$$z = 3$$
$$+ \int_0^2 \int_0^1 (3x^2\hat{x} - xy\hat{y} + 5z^2y\hat{z}) \cdot (-dxdy\hat{z})$$
$$z = 0$$
$$+ \int_0^3 \int_0^2 (3x^2\hat{x} - xy\hat{y} + 5z^2\hat{z}) \cdot dydz\hat{x}$$
$$x = 1$$
$$+ \int_0^3 \int_0^2 (3x^2\hat{x} - xy\hat{y} + 5z^2y\hat{z}) \cdot (-dydz\hat{x})$$
$$x = 0$$

Performing the vector operations and substituting the coordinate value that is constant for the particular integral:

$$\oint_S \vec{B} \cdot d\vec{a} = -2\int_0^3 \int_0^1 xdxdz + 0 + 45\int_0^2 \int_0^1 ydxdy$$
$$+ 0 + 3\int_0^3 \int_0^2 dydz - 3\int_0^3 \int_0^2 dydz$$

Evaluating the integrals:

$$\oint_S \vec{B} \cdot d\vec{a} = (-2)\left(\frac{1}{2}x^2\right)\Bigg|_0^1 (z)\Bigg|_0^3 + (45)(x)\Bigg|_0^1 \left(\frac{1}{2}y^2\right)\Bigg|_0^2 = -(1 - 0)(3 - 0)$$
$$+ \left(\frac{45}{2}\right)(1 - 0)(4 - 0)$$
$$= 87$$

Example 4-32

Our next example indicates the care that must be taken to identify the unit vectors of the surfaces. Evaluate $\oint_S \vec{B} \cdot d\vec{a}$ for the semicylinder shown in Figure 4-71 for $\vec{B} = 2\rho \cos \phi\hat{\rho} - \sin \phi\hat{\phi} + 6z\hat{z}$ and where

Radius of cylinder = 1

Cylinder height = 2

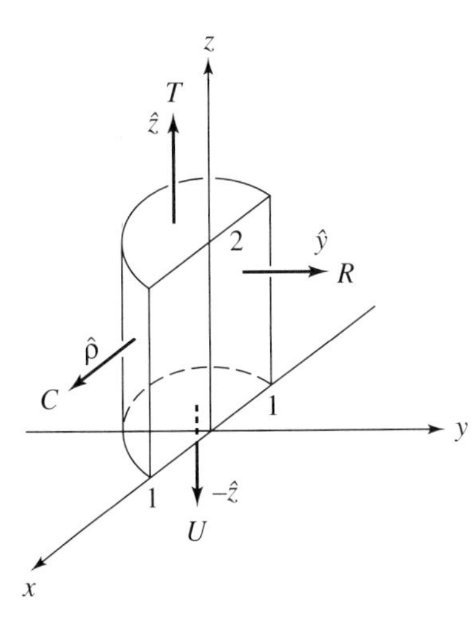

Figure 4-71. Semicylinder for closed surface integration example.

We use the general procedure given earlier. The surfaces and the corresponding unit vectors are indicated in the figure. For each surface we have the following:

For surface T: $\quad z = 2$

For surface U: $\quad z = 0$

For surface R: $\quad y = 0$

For surface C: $\quad \rho = 1$

$$\oint_S \vec{B} \cdot d\vec{a} = \int_T \vec{B} \cdot d\vec{a} + \int_U \vec{B} \cdot d\vec{a} + \int_C \vec{B} \cdot d\vec{a} + \int_R \vec{B} \cdot d\vec{a}$$

For surface T, since we have cylindrical coordinates,

$$d\vec{a} = h_1 h_2 du_1 du_2 \hat{u}_3 = 1 \cdot \rho \cdot d\rho d\phi \hat{z} = \rho d\rho d\phi \hat{z}, \qquad z = 2$$

For surface U,

$$d\vec{a} = h_1 h_2 du_1 du_2(-\hat{u}_3) = 1 \cdot \rho d\rho d\phi(-\hat{z}) = -\rho d\rho d\phi \hat{z}, \qquad z = 0$$

Surface R is a rectangular surface for which

$$d\vec{a} = dxdz\hat{y}, \qquad y = 0$$

For surface C,

$$d\vec{a} = h_2 h_3 du_2 du_3 \hat{u}_1 = \rho \cdot 1 d\phi dz\hat{\rho}z = \rho d\phi dz\hat{\rho}, \qquad \rho = 1$$

Note that the expressions for $\vec{B}$ will need to be in cylindrical coordinates for surfaces T, U, and C, but in rectangular coordinates for surface R. We next use the techniques learned earlier for vector conversion to rectangular coordinates.

1. Change variable:

$$\vec{B} = 2\sqrt{x^2 + y^2} \cdot \frac{x}{\sqrt{x^2 + y^2}}\,\hat{\rho} - \frac{y}{\sqrt{x^2 + y^2}}\,\hat{\phi}$$

$$+ 6z\hat{z} = 2x\hat{\rho} - \frac{y}{\sqrt{x^2 + y^2}}\,\hat{\phi} + 6z\hat{z}$$

2. Change components:

$$B_x = (2x)\frac{x}{\sqrt{x^2 + y^2}} + \left(-\frac{y}{\sqrt{x^2 + y^2}}\right)\frac{y}{\sqrt{x^2 + y^2}} = \frac{2x^2}{\sqrt{x^2 + y^2}} + \frac{y^2}{x^2 + y^2}$$

$$B_y = (2x)\frac{y}{\sqrt{x^2 + y^2}} + \left(-\frac{y}{\sqrt{x^2 + y^2}}\right)\frac{x}{\sqrt{x^2 + y^2}} = \frac{2xy}{\sqrt{x^2 + y^2}} - \frac{xy}{x^2 + y^2}$$

$$B_x = 6z$$

Then

$$\vec{B} = \left(\frac{2x^2}{\sqrt{x^2 + y^2}} + \frac{y^2}{x^2 + y^2}\right)\hat{x} + \left(\frac{2xy}{\sqrt{x^2 + y^2}} - \frac{xy}{x^2 + y^2}\right)\hat{y} + 6z\hat{z}$$

We now have:

$$\oint_S \vec{B} \cdot d\vec{a} = \int_{\pi}^{2\pi}\int_0^1 (2\rho\cos\phi\,\hat{\rho} - \sin\phi\,\hat{\phi} + 6z\hat{z}) \cdot (\rho d\rho d\phi\,\hat{z}) \quad z = 2$$

$$+ \int_{\pi}^{2\pi}\int_0^1 (2\rho\cos\phi\,\hat{\rho} - \sin\phi\,\hat{\phi} + 6z\hat{z}) \cdot (-\rho d\rho d\phi\,\hat{z}) \quad z = 2$$

$$+ \int_0^2\int_{-1}^1 \left[\left(\frac{2x^2}{\sqrt{x^2 + y^2}} + \frac{y^2}{x^2 + y^2}\right)\hat{x} + \left(\frac{2xy}{\sqrt{x^2 + y^2}} - \frac{xy}{x^2 + y^2}\right)\hat{y} + 6z\hat{z}\right] \cdot dxdz\hat{y} \quad y = 0$$

$$+ \int_0^2\int_{\pi}^{2\pi} (2\rho\cos\phi\rho\hat{\rho} - \sin\phi\,\hat{\phi} + 6z\hat{z}) \cdot \rho d\phi dz\hat{\rho} \quad \rho = 1$$

$$= \int_{\pi}^{2\pi}\int_0^1 6z\rho d\rho d\phi \Big|_{z=2} - \int_{\pi}^{2\pi}\int_0^1 6z\rho d\rho d\phi \Big|_{z=0} + \int_0^2\int_{-1}^1 \left(\frac{2xy}{\sqrt{x^2 + y^2}} - \frac{xy}{x^2 + y^2}\right) dxdz \Big|_{y=0}$$

$$+ \int_0^2\int_{\pi}^{2\pi} 2\rho^2\cos\phi d\phi dz \Big|_{\rho=1}$$

Substituting in the coordinate values that are constant:

$$\oint_S \vec{B} \cdot d\vec{a} = \int_{\pi}^{2\pi}\int_0^1 12\rho d\rho d\phi - 0 + 0 + \int_0^2\int_{\pi}^{2\pi} 2\cos\phi d\phi dz$$

$$= 12(\tfrac{1}{2}\rho^2)\Big|_0^1 (\phi)\Big|_{\pi}^{2\pi} + 2(\sin\phi)\Big|_{\pi}^{2\rho} (z)\Big|_0^2$$

$$= 12(\tfrac{1}{2} - 0)(2\pi - \pi) + 4(0 - 0) = 12 \times \tfrac{1}{2} \times \pi = 6\pi$$

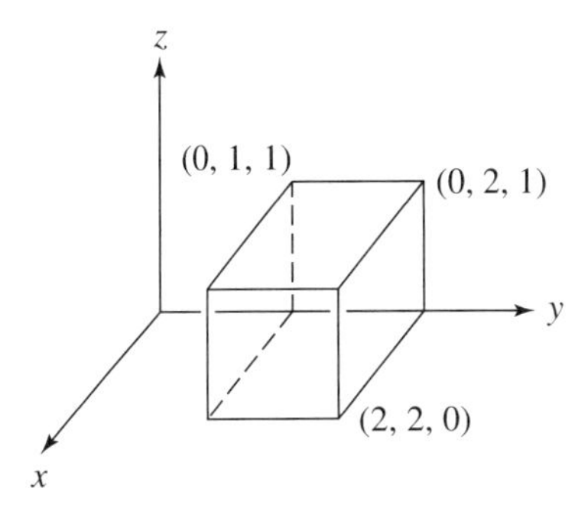

Figure 4-72.

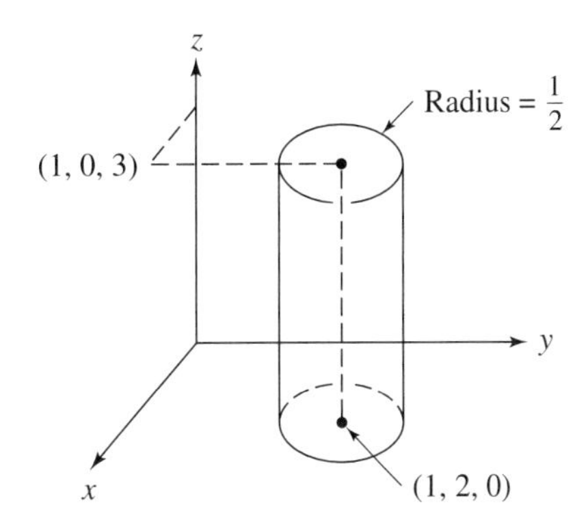

Figure 4-73.

Exercises

4-19.0n Evaluate the integral $\oint_S (\nabla \times \vec{B}) \cdot d\vec{a}$ for the surface shown in Figure 4-72 and for $\vec{B} = xy\hat{x} + yz\hat{y} + zx\hat{z}$.

4-19.0o The surface integral $\int_S \vec{B} \cdot d\vec{a}$ may be put in a form amenable to computer evaluation by writing the differential surface area as $d\vec{a} = da\hat{n}$ where $\hat{n}$ is the unit vector direction that is normal to the surface increment da. We then have

$$\int_S \vec{B} \cdot d\vec{a} = \int_S \vec{B} \cdot (da\hat{n}) = \int_S \vec{B} \cdot \hat{n} da$$

For this form, develop a summation approximation to the integral using rectangular coordinates and then write a computer program that will evaluate the integral. Use the program to solve Exercise 4-19n.

4-19.0p Evaluate $\int_S \vec{B} \cdot d\vec{a}$ where S is the surface consisting of the *top* and *bottom* of the region shown in Figure 4-73. For this region $\vec{B} = xyz\hat{x} + xyz\hat{z}$.

4-19.0q Determine $\oint_S \vec{B} \cdot d\vec{a}$ for the hemispherical configuration shown in Figure 4-74 and for $\vec{B} = r \sin\theta\hat{r} + 3 \sin\phi\hat{\theta} + 8 \cos\theta \sin\phi\hat{\phi}$.

4-19.0r For $\vec{B} = 6 \cos\phi\hat{\rho} - \rho z \sin\phi\hat{\phi}$ evaluate $\oint_S \vec{B} \cdot d\vec{a}$ for the surface shown in Figure 4-75. (Watch your surface unit vectors carefully.)

4-19.0s Verify the divergence theorem for the vector field and volume shown in Exercise 4-19r.

4-19.0t Suppose that the bottom of the hemisphere in Exercise 4-19q is open so that the circle in the xy plane is the contour sounding the hemispherical surface. Verify Stokes' theorem for the vector $\vec{B}$ given in the exercise for two cases:

(a) Surface S_1 = hemispherical surface.

(b) Surface S_2 = area of circle in the xy plane.

Since the volume integrals encountered are simple scalar integrals evaluated in calculus courses, no examples are given of these. The only thing that needs to be remembered is that

$$dv = h_1 h_2 h_3 du_1 du_2 du_3$$

which can be evaluated using the results derived earlier for the scale factors. A few exercises that give a review practice for such integrals follow.

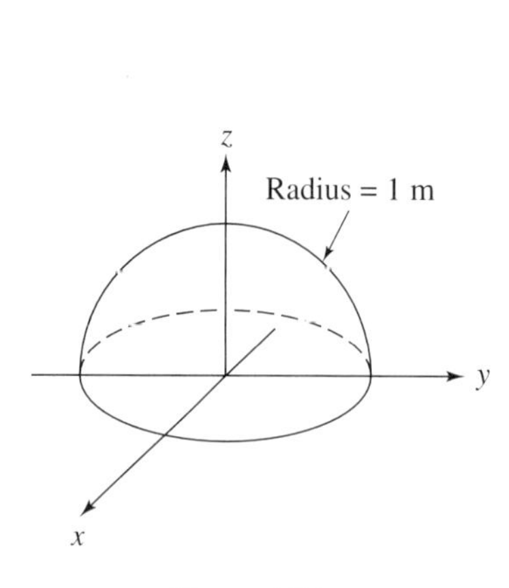

Figure 4-74.

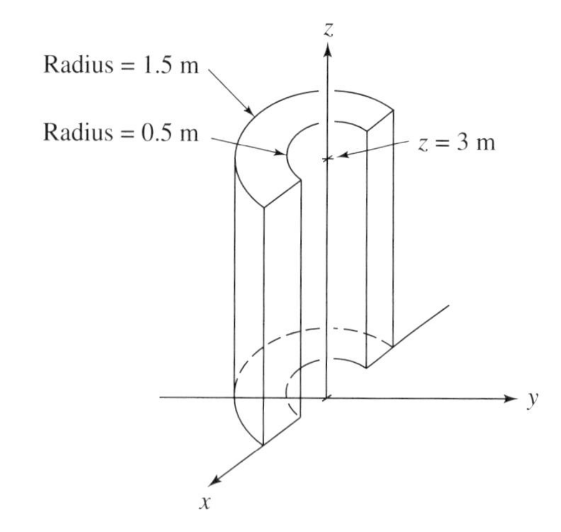

Figure 4-75.

Exercises

4-19.0u Given the scalar functions $U(x, y, z) = 6xyz$ and $V(x, y, z) = x + y + z$ verify by evaluation, both of Green's integral identities for a unit cube having one corner at the coordinate system origin and lying wholly in the positive octant.

4-19.0v For the scalar functions given in Exercise 4-19u, verify Green's identities for the spherical shell volume shown in Figure 4-76.

4-19.0w Write a computer program that will perform volume integrations for scalar functions in

(a) Rectangular coordinates.

(b) Cylindrical coordinates.

(c) Spherical coordinates.

(d) Generalized orthogonal curvilinear coordinates. For this case, make provisions for entering the scale factors so that it may be easily adapted to any coordinate system.

Write the program so that it will handle a FUNCTION subprogram for the integrand.

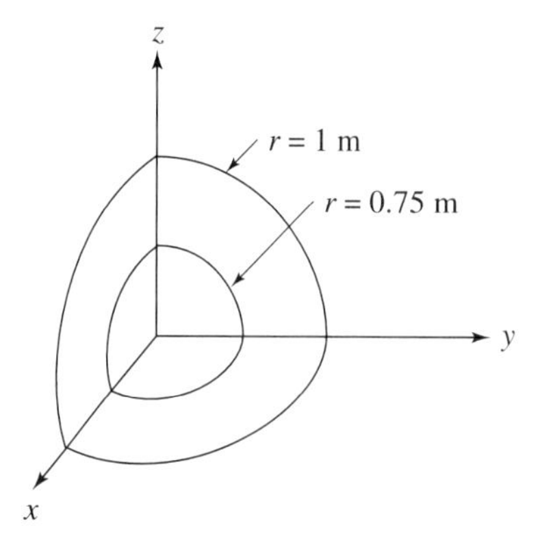

Figure 4-76.

4-20 Differentiation of Integrals with Variable Limits

In the problems considered to this point, the various contours, surfaces, and volumes were considered fixed in space. However, in many practical applications these may be moving in the region of interest. As an example, consider a sinusoidal voltage generator. The loop is usually of fixed dimensions, but it is rotating in a magnetic field, so the contour of the loop is not fixed with respect to the earth-referenced coordinate system. For cases like the generator, what usually arises is an integral that represents the flux of a vector field (see discussion following Eq. 4-39.) Since the boundary may be in motion then the limits of the integral are time varying. If the time rate of change of the flux of the field is desired, one is faced with the problem of differentiating an integral that has time-varying limits. We will use the results of this section in a later chapter. To develop the results desired we need to borrow four theorems from calculus, the formal statements of which do not concern us at this point. These are:

1. If $g(x) = \int_a^x f(y)dy$, $dg/dx = g'(x) = f(x)$. In words, this result states that the derivative of an integral with respect to the upper limit, which is a variable, is simply the integrand evaluated at that upper limit.
2. $\int_a^b f(x)dx = -\int_b^a f(x)dx$. In words, this result states that reversing the limits of integration negates the value of the integral.
3. If $g(x) = \int_a^b f(x, y)dy$, $dg/dx = \int_a^b [\partial f(x, y)/\partial x]dy$ (Leibniz' rule) (a and b constants).
4. For $g[x(t), y(t), u(t), \cdots]$, $dg/dt = (\partial g/\partial x)(dx/dt) + (\partial g/\partial y)(dy/dt) + (\partial g/\partial u)(du/dt) + \cdots$.

The first theorem we derive involves a single integral that has both limits depending upon a parameter t, which we write as

$$g[a(t), b(t)] = \int_{a(t)}^{b(t)} f(x, t)dx$$

We wish to evaluate dg/dt. First, change the variables with the substitutions $u = a(t)$ and $v = b(t)$. this gives

$$g(u, v, t) = \int_u^v f(x, t)dx$$

Using Theorem 4:

$$\frac{dg}{dt} = \frac{\partial g}{\partial u}\frac{du}{dt} + \frac{\partial g}{\partial v}\frac{dv}{dt} + \frac{\partial g}{\partial t}\frac{dt}{dt}$$

Since $dt/dt = 1$,

$$\frac{dg}{dt} = \frac{\partial g}{\partial u}\frac{du}{dt} + \frac{\partial g}{\partial v}\frac{dv}{dt} + \frac{\partial g}{\partial t}$$

Substituting the integral for $g(u, v, t)$

$$\frac{dg}{dt} = \frac{\partial[\int_u^v f(x, t)dx]}{\partial u}\frac{du}{dt} + \frac{\partial[\int_u^v f(x, t)dx]}{\partial v}\frac{dv}{dt} + \frac{\partial[\int_u^v f(x, t)dx]}{\partial t}$$

Using Theorem 2 on the first term on the right side, Theorem 1 on the middle term (remembering that the partial derivative means we hold one limit constant), and Theorem 3 on the third term this becomes

$$\frac{dg}{dt} = \frac{\partial[-\int_v^u f(x,t)dx]}{\partial u}\frac{du}{dt} + f(v,t)\frac{dv}{dt} + \int_u^v \frac{\partial f(x,t)}{\partial t}dx$$

Next use Theorem 1 on the first term, where now v is considered constant since the partial derivative is with respect to u,

$$\frac{dg}{dt} = -f(u,t)\frac{du}{dt} + f(v,t)\frac{dv}{dt} + \int_u^v \frac{\partial f(x,t)}{\partial t}dx$$

Finally, changing back to the original variables $a(t)$ and $b(t)$ we obtain

$$\frac{d}{dt}\int_{a(t)}^{b(t)} f(x,t)dx = \int_{a(t)}^{b(t)} \frac{\partial f(x,t)}{\partial t}dx + f(b(t),t)\frac{db}{dt} - f(a(t),t)\frac{da}{dt} \quad (4\text{-}54)$$

This is Leibniz' theorem for single integrals. Using the same technique, we may extend the theorem to double integrals. The result is (see Exercise 4-20o):

$$\begin{aligned}\frac{d}{dt}\int_{a_2(t)}^{b_2(t)}\int_{a_1(t)}^{b_1(t)} f(x,y,t)dxdy = &\int_{a_2(t)}^{b_2(t)}\int_{a_1(t)}^{b_1(t)} \frac{\partial f(x,y,t)}{\partial t}dxdy \\ &+ \int_{a_1(t)}^{b_1(t)} f(x,b_2,t)dx\frac{db_2}{dt} \\ &- \int_{a_1(t)}^{b_1(t)} f(x,a_2,t)dx\frac{da_2}{dt} \\ &+ \int_{a_2(t)}^{b_2(t)} f(b_1,y,t)dy\frac{db_1}{dt} \\ &- \int_{a_2(t)}^{b_2(t)} f(a_1,y,t)dy\frac{da_1}{dt}\end{aligned} \quad (4\text{-}55)$$

This is Leibniz' theorem for double integrals. Although the preceding theorems may appear somewhat formidable, in many cases some of the units are constants so some terms vanish.

Example 4-33

Suppose we have a vector field

$$\vec{\mathcal{B}} = 2\sin x\cos\frac{\pi}{4}y\cos 5t\hat{z}$$

in a region of space and we wish to evaluate the time derivative of the flux of the field through the surface S, shown in Figure 4-77. The boundary edge $x = c$ moves with velocity $\vec{V}_x$.

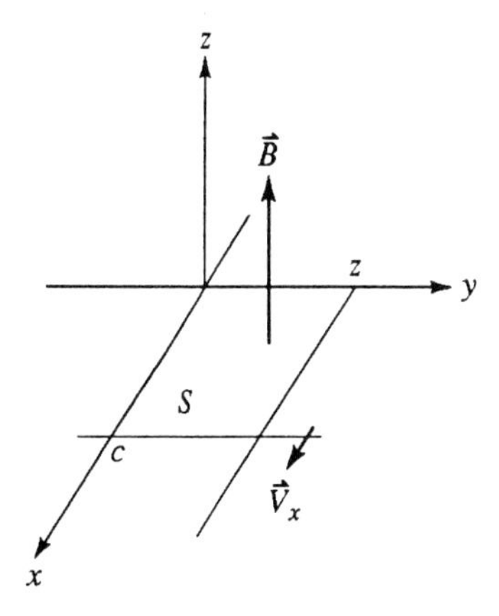

Figure 4-77. Moving boundary system.

This means c is a function of time. By definition, the flux of $\vec{\mathcal{B}}$ through the area S is

$$\Phi = \int_S \vec{\mathcal{B}} \cdot d\vec{a} = \int_0^{c(t)} \int_0^2 2 \sin x \cos \frac{\pi}{4} y \cos 5t\hat{z} \cdot dydx\hat{z}$$

$$= \int_0^{c(t)} 2 \sin x \cos 5t \left[\int_0^2 \cos \frac{\pi}{4} ydy \right] dx = \int_0^{c(t)} 2 \sin x \cos 5t \cdot \frac{4}{\pi} dx$$

$$= \int_0^{c(t)} \frac{8}{\pi} \sin x \cos 5tdx$$

Using Leibniz' rule for single integrals, the time rate of change of the flux of $\vec{\mathcal{B}}$ is (noting that $a(t) = 0$ in this example):

$$\frac{d\Phi}{dt} = \frac{d}{dt} \int_0^{c(t)} \frac{8}{\pi} \sin x \cos 5tdx = \int_0^{c(t)} - \frac{40}{\pi} \sin x \sin 5tdx$$

$$+ \frac{8}{\pi} \sin [c(t)] \cos 5t \frac{dc}{dt}$$

Now dc/dt is the velocity with which the boundary is moving, so we have

$$\frac{d\Phi}{dt} = -\frac{40}{\pi} \sin 5t \int_0^{c(t)} \sin xdx + \frac{8}{\pi} V_x \cos 5t \sin[c(t)]$$

$$= -\frac{40}{\pi} \sin 5t(-\cos x)\Big|_0^{c(t)} + \frac{8}{\pi} V_x \cos 5t \sin[c(t)]$$

$$= \frac{40}{\pi} \sin 5t(\cos[c(t)] - 1) + \frac{8}{\pi} V_x \cos 5t \sin[c(t)]$$

For the special case where the velocity is constant, $c(t) = V_x t$ and the final result would be

$$\frac{d\Phi}{dt} = \frac{40}{\pi} (\cos(V_x t) - 1) \sin 5t + \frac{8}{\pi} V_x \cos 5t \sin(V_x t)$$

Exercises

4-20.0a Prove Leibniz' theorem for double integrals. (Hint: In Theorems 1 and 3 the function f is the inner integral of dx over the range $a_1(t)$ to $b_1(t)$.)

4-20.0b In the text example, suppose that the edge shown at $y = 2$ is not fixed but is also moving to the right at constant velocity $\vec{\mathbf{V}}_y$. Denote the position of that boundary by $d(t)$ and determine the time rate of change of the flux of $\vec{\mathcal{B}}$ through the surface S. Use Leibniz' theorem for double integrals.

4-20.0c For the example in the text suppose that the field $\vec{\mathcal{B}} = 2\hat{z}$. For the velocity $\vec{V}_x$ constant find the time rate of change of the flux of $\vec{\mathcal{B}}$ and plot as a function of time for $\vec{V}_x = 1\hat{x}$ m/S. Plot also for $\vec{V}_x = 2\hat{x}$ m/S.

4-21 Dirac Delta and Unit Step Functions with Applications to Systems Having Discontinuities

In the evaluations of integrals discussed in Section 4-19 it was assumed that the integrands were defined for all space and that there were no discontinuities in the fields. In most practical situations the fields have only finite extent in the modeling of the system. This section is devoted to developing methods for writing equations for field and scalar quantities as a single equality without the necessity of specifying ranges of variables over which the equations apply. In other words, the effects of discontinuities are built into the equations directly. Some examples will show the ideas involved.

In circuit theory it is often convenient to be able to express a voltage that has a nonzero value only at one instant of time as though it were a continuous function of time. Figure 4-78 shows this ''function'' of voltage and the normal method of describing it. Thus rather than having to write time constraints, it would be nice to be able to have some way of symbolically representing this function with a single line equation form.

Another case that occurs frequently in engineering is when the voltage $v(t)$ is nonzero over a short span of time as shown in Figure 4-79, along with an equation that might be used to express it functionally.

Figure 4-78. Example of discontinuous voltage function.

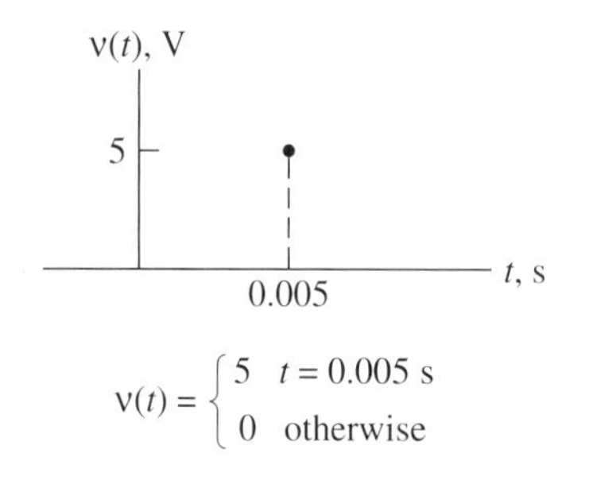

$$v(t) = \begin{cases} 5 & t = 0.005 \text{ s} \\ 0 & \text{otherwise} \end{cases}$$

Figure 4-79. Example of a voltage function that is nonzero on a finite interval.

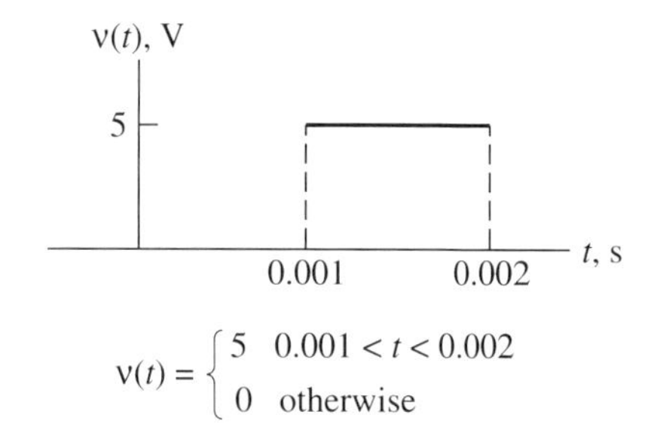

$$v(t) = \begin{cases} 5 & 0.001 < t < 0.002 \\ 0 & \text{otherwise} \end{cases}$$

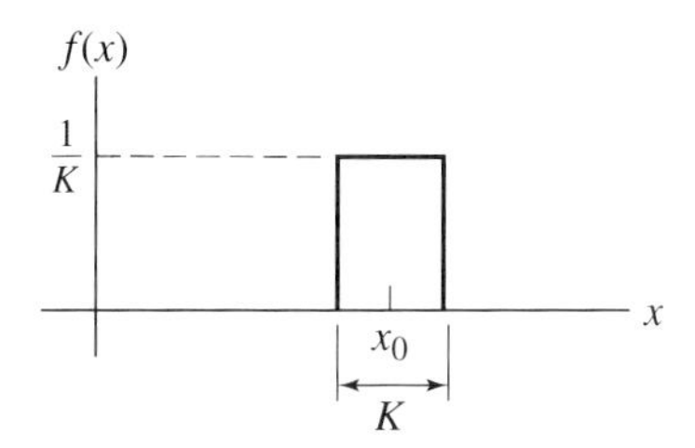

Figure 4-80. Rectangular pulse having constant area.

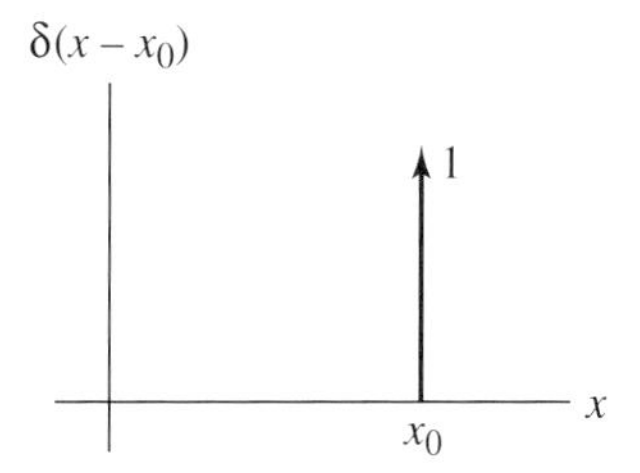

Figure 4-81. Limit function that describes the Dirac delta function.

The next discussions will develop ways of writing a single line equation form that has the desired feature of giving the correct form as it is plotted point by point without requiring the explanatory time interval specifications. Our immediate interest is the application to coordinate systems so we simply use coordinate values in place of the time t.

We begin with the description of the Dirac delta function [19], denoted by $\delta(x)$. This function provides us with a means of describing quantities that have values at isolated points. The most we can do here is give a heuristic presentation of this special function. Take a rectangular pulse of width K and amplitude $1/K$, centered at $x = x_0$ as shown in Figure 4-80.

$$\text{Area under pulse} = \int_{-\infty}^{\infty} f(x)dx = \frac{1}{K} \cdot K = 1 \qquad \text{independent of } K$$

Now suppose $K \to 0$. The pulse width goes to zero, the amplitude becomes infinite, but the area remains 1. Crudely, this is our delta function $\delta(x - x_0)$. We place a 1 beside the *curve* to indicate its area or, more formally its *weight*. This ''function'' can be thought of as having value zero everywhere except at $x = x_0$ as shown in Figure 4-81.

Now suppose the amplitude of the original pulse were given as W/K, where W is a scalar. We would generate an impulse (delta function) of weight w as shown in Figure 4-82.

It is important to notice that the discussion of the delta function involved area, which is an integration in the most general sense. We would expect then that expressions involving the delta function should be looked at basically as an implied integration. More will be covered on this next.

Figure 4-82. Impulse of arbitrary weight W.

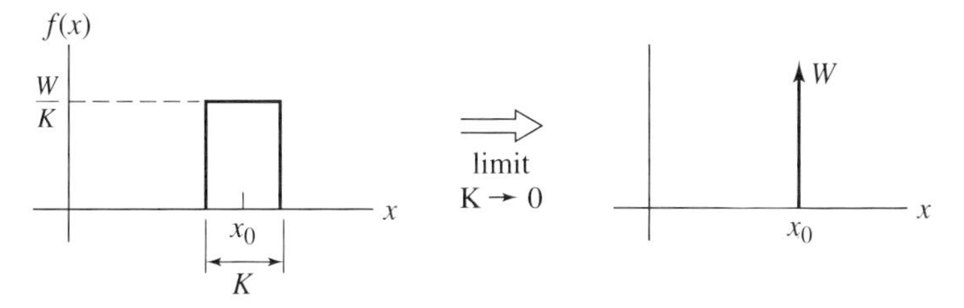

(1)

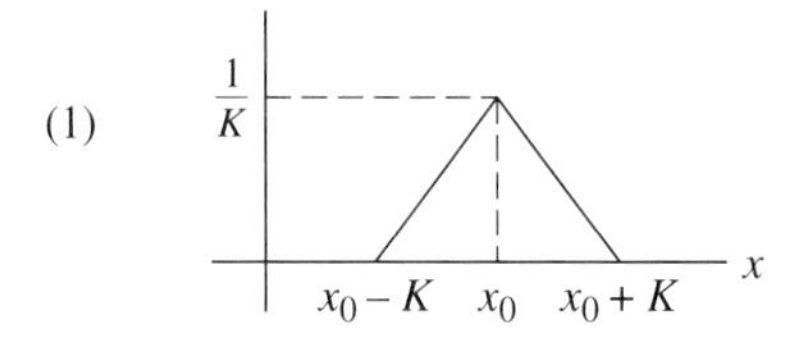

(2)

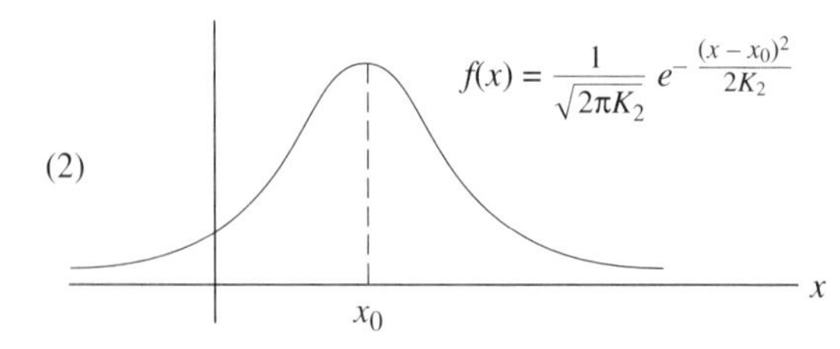

Figure 4-83.

Exercises

4-21.0a Show that the unit impulse may also be described as the limit of the functions shown in Figure 4-83 (for $K \to 0$).

4-21.0b For the triangular approximation shown in Exercise 4-21a form the function

$$g(x) = f(x) \cdot \text{triangular function}$$

where $f(x) = \sin x$. Evaluate $\int_{-\infty}^{\infty} g(x)dx = \int_{-\infty}^{\infty} f(x) \cdot \text{triangular function} \cdot dx$. Now let $K \to 0$. Comment on your answer; that is, describe how you could have obtained the answer without performing integration. Could you conjecture as to the integral value for any continuous function $f(x)$?

We are now in a position to give the formal definition of the delta function and develop some of its properties. The delta function is defined as

$$\int_a^b \delta(x - x_0)dx \equiv \begin{cases} 1 \text{ if } x_0 \in (a, b) \\ 0 \text{ if } x_0 \notin (a, b) \end{cases} \tag{4-56}$$

Suppose we have a function $f(x)$ that is continuous at $x = x_0$ and that we multiply the delta function by $f(x)$ and then integrate over some interval (a, b). Then we have the situation shown in Figure 4-84, for which we write

$$\int_a^b f(x)\delta(x - x_0)dx$$

Now, since $\delta(x - x_0)$ is zero at every point except x_0, we write

$$\int_{x_0^-}^{x_0^+} f(x)\delta(x - x_0)dx$$

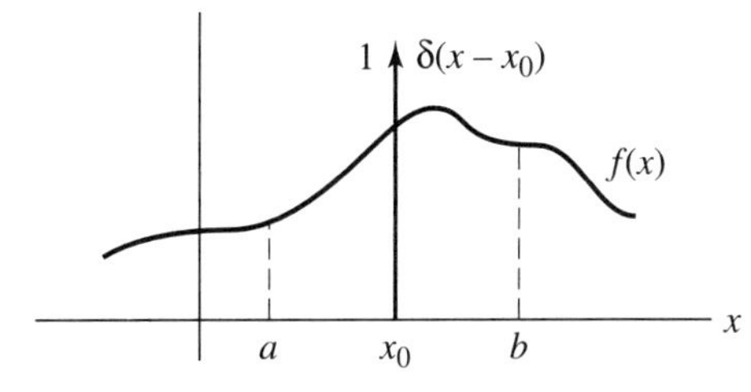

Figure 4-84. Multiplication of a continuous function and a delta function.

If $x_0 \notin (a, b)$, this integral will be 0. If $x_0 \in (a, b)$, we say that the interval (x_0^-, x_0^+) is so small that $f(x)$ will be constant over this interval and have value $f(x_0)$ (a rather crude application of the mean value theorem for integrals). Then

$$\int_a^b f(x)\delta(x - x_0)dx = \int_{x_0^-}^{x_0^+} f(x_0)\delta(x - x_0)dx$$

$$= f(x_0)\int_{x_0^-}^{x_0^+} \delta(x - x_0)dx = f(x_0)$$

Thus

$$\int_a^b f(x)\delta(x - x_0)dx = \begin{cases} f(x_0) & x_0 \in (a, b) \\ 0 & \text{otherwise} \end{cases} \tag{4-57}$$

Note that what the delta function effectively does is sample one point of the function $f(x)$. Although the delta function has significance only when appearing as part of an integrand, we often write the equations without the integral sign to obtain the symbolic relationship

$$f(x)\delta(x - x_0) = f(x_0)$$

Example 4-34

$e^{-x}\delta(x - 3) = e^{-3} = 0.0498$. This is thought of as e^{-x} evaluated at $x = 3$.

$$\cos x \sin 3x\delta(x - 2) = \cos 2 \sin 6 = (-0.42)(-0.28) = 0.116$$

$$3xe^{-2x}\delta(x + 1) = (3)(-1)e^{-}2(-1) = -3e^2 = -22.17$$

There are two general ways for interpreting these examples. One is that when a function is multiplied by a delta function, the effect is simply to weight the delta function by a constant $f(x_0)$. The second more general interpretation is to say that when a function is multiplied by a delta function the result is the value of the function at the point or points where the argument list of the delta function is zero; that is, for $\delta(Y)$ the points where $Y = 0$, for Y any function. Note that $Y = x - x_0$ is a special case.

Example 4-35

As an example consider the evaluation of the expression

$$e^{-x}\delta(\sin 2x)$$

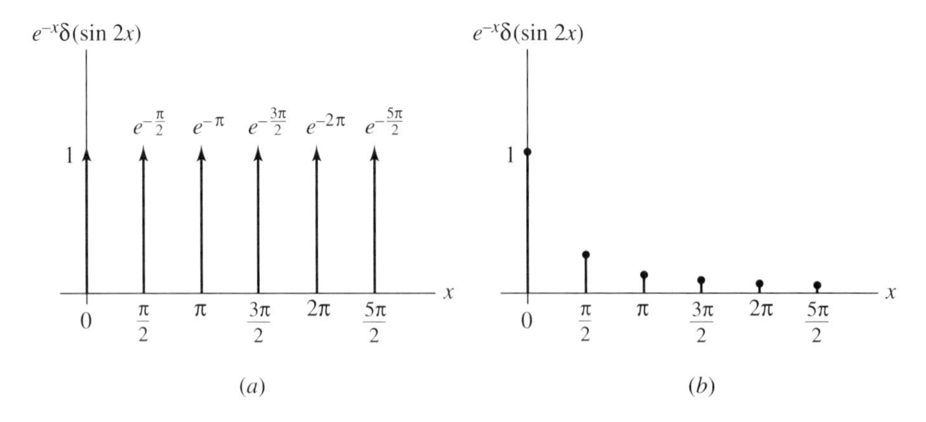

Figure 4-85. Plots of $e^{-x}\delta(\sin 2x)$ for different interpretations. (*a*) Weighted delta functions. (*b*) Function evaluation.

The trigonometric function $\sin 2x$ is zero for $x = n\pi/2$, $n = 0, 1, \ldots$. Two plots corresponding to the two interpretations given above are shown in Figure 4-85. The implied integration discussed earlier applies.

Exercises

4-21.0a Suppose we multiply a continuous function by two delta functions, $f(x, y)\delta(x - x_0)\delta(y - y_0)$. What would you conjecture the meaning of this product to be? Discuss fully in terms of integration and so on.

4-21.0b Give numerical values for the following:

(a) $\cosh(x - 3)\delta(x - 3)$

(b) $(e^{jx} - e^{-jx})\delta(x + 2)$

(c) $6x \cos x\delta(x)$

4-21.0c Plot the following functions using both interpretations discussed in the text.

(a) $4 \sin 3x\delta(\sin x)$

(b) $e^{x}\delta(x - 3)$ over the interval $-3 \leq x \leq 2$

(c) $3x^2\delta[\cos \pi(x + 2)]$

Our original purpose was to devise a way to write discontinuous systems as a single line equation, so we next give an example of an application. Suppose we have a point mass of 10 kg situated on the x axis as shown in Figure 4-86. We write the equation for the mass as a function of x as

$$m(x) = 10\delta(x - 2)$$

This seems at least plausible since the delta function (and hence also $m(x)$) is zero for all x except at $x = 2$ where the delta function occurs. However, at $x = 2$ the delta function has "∞" value, which may seem a bit awkward. Since the real meaning of such functions is tied to integration, we check to see if consistency exists for such integration. Now the total mass on the line is given by

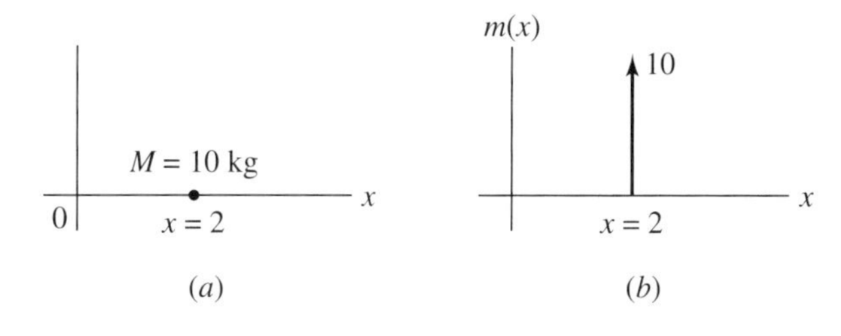

Figure 4-86. Discontinuous system and its functional plot. (*a*) Functional plot. (*b*) Original system.

$$M = \int_{-\infty}^{\infty} m(x)dx = \int_{-\infty}^{\infty} 10\delta(x - 2)dx$$

$$= 10 \int_{-\infty}^{\infty} \delta(x - 2)dx = (10)(1) = 10 \text{ kg}$$

which agrees with the physical situation. One additional point should be made. M is in kilograms and dx in meters, so the integrand must be in kilograms per meter. Since 10 is the mass at $x = 2$, then, in context, the δ function must have effectively 1/meter for its unit. Thus multiplying by the δ function has changed our mass *point* to a line mass *density* as $10\delta(x - 2)$ kg/m.

Exercises

4-21.0d Suppose we have a point charge of Q_0 C located at $y = 6$ on the y axis. Express this as a function of y and plot $q(y)$. What are the units on $q(y)$?

4-21.0e Delta functions are also used to represent point time functions using the form $\delta(t - a)$. What would the effective unit of the delta function be in the time domain?

Since we have defined the delta function we next inquire about the meaning of derivatives of the delta function. Suppose we again consider our rectangular area concept, but let's differentiate before we let $K \to 0$. Since the sides have infinite slope, we have the results shown in Figure 4-87 for $K \to 0$.

This is called the *unit doublet*, and is denoted symbolically by

$$\delta'(x - x_0) \tag{4-58}$$

Figure 4-87. Generation of the derivative of the delta function.

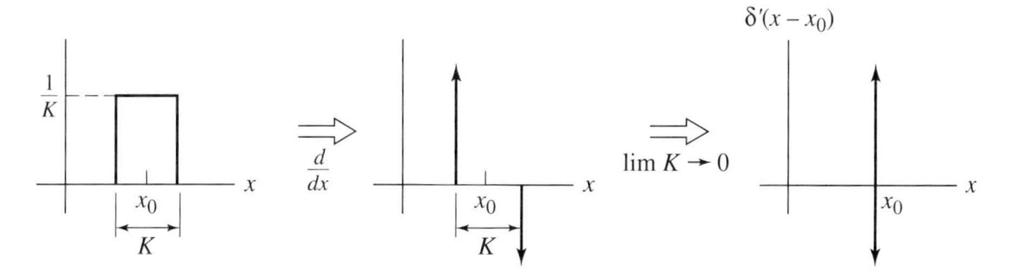

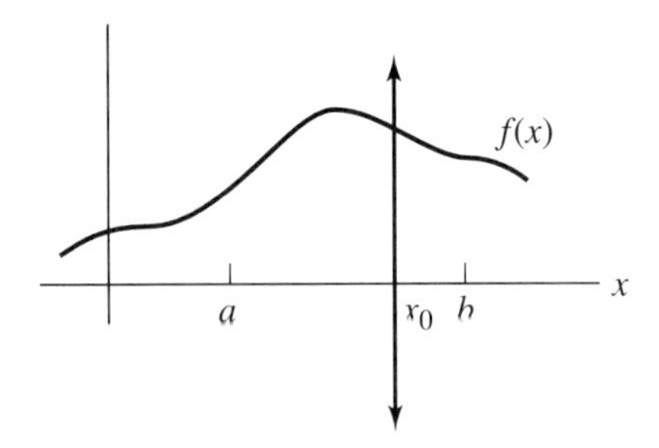

Figure 4-88. Multiplication of a continuous function by a unit doublet.

Next we multiply a function $f(x)$ by this doublet and then integrate over an interval that includes the doublet as indicated in Figure 4-88. Symbolically, we write the function as

$$g(x) = f(x)\delta'(x - x_0)$$

and the integral as

$$\int_a^b f(x)\delta'(x - x_0)dx$$

Integrating by parts:

$$\int_a^b f(x)\delta'(x - x_0)dx = [f(x)\delta(x - x_0)]_a^b - \int_a^b f'(x)\delta(x - x_0)dx$$

$$= f(b)\delta(b - x_0) - f(a)\delta(a - x_0) - \int_a^b f'(x)\delta(x - x_0)dx$$

Since $x_0 \neq a$ and $x_0 \neq b$, $\delta(a - x_0) = 0$ and $\delta(b - x_0) = 0$ and our integral becomes

$$\int_a^b f(x)\delta'(x - x_0)dx = -\int_a^b f'(x)\delta(x - x_0)dx = \begin{cases} -f'(x_0) & x_0 \in (a, b) \\ 0 & \text{otherwise} \end{cases} \tag{4-59}$$

Thus the doublet evaluates the negative of the derivative of the function at x_0.

Example 4-36

For $f(x) = \sin 3x$,

$$\int_6^{10} \sin 3x\delta'(x - 8)dx = -\left.\frac{d(\sin 3x)}{dx}\right|_{x=8} = -3 \cos 3x\Big|_{x=8}$$

$$= -3 \cos 24 = -1.273$$

For $f(x) = \sin 3x$ again, $\int_6^1 0 \sin 3x\delta'(x - 5)dx = 0$ (5 is outside integration limits so that $\delta'(x - 5) = 0$ everywhere on (6, 10)).

By repeated application of integration by parts we may show

$$\int_a^b f(x)\delta^{(n)}(x - x_0)dx = \begin{cases} (-1)^n f^{(n)} & x_0 \in (a, b) \\ 0 & \text{otherwise} \end{cases} \tag{4-60}$$

It is now a relatively easy matter to extend the preceding ideas to several dimensions. The idea behind the extended definition is shown in Figure 4-89. We use the vector locator $\vec{r}$ terminology as a simple generic shorthand way of indicating several dimensions or coordinates are used generally.

Let $\vec{r}$ denote the location of a point in the xy plane as shown in the figure and let $\vec{r}_0$ be the location of the delta function.. Note that there is no z coordinate here, the vertical arrow at (x_0, y_0) is simply to show where the impulse occurs. The extended definition is

$$\int_V \delta(\vec{r} - \vec{r}_0)dv = \begin{cases} 1 & \vec{r}_0 \text{ inside } V \\ 0 & \text{otherwise} \end{cases} \tag{4-61}$$

where V is some region of space.

Using reasoning similar to that for the one-dimensional case, we may show that

$$\int_V f(\vec{r})\delta(\vec{r} - \vec{r}_0)dv = \begin{cases} f(\vec{r}_0) & \vec{r}_0 \text{ inside } V \\ 0 & \text{otherwise} \end{cases} \tag{4-62}$$

and

$$\int_V f(\vec{r})\delta^{(n)}(\vec{r} - \vec{r}_0)dv = \begin{cases} (-1)^n f^{(n)}(\vec{r}_0) & \vec{r}_0 \text{ inside } V \\ 0 & \text{otherwise} \end{cases} \tag{4-63}$$

In both of the preceding equations $f(\vec{r})$ does not mean a vector function, but merely that the result is in general a function of more than one coordinate.

We now need to develop expressions for various coordinate systems. Whatever symbolism we use to write the δ function in a specific coordinate system it must be consistent with the basic definition, Eq. 4-61. In generalized coordinates, suppose we set

$$\delta(\vec{r} - \vec{r}_0) = A\delta(u_1 - u_{10})\delta(u_2 - u_{20})\delta(u_3 - u_{30})$$

where the delta function is located at (u_{10}, u_{20}, u_{30}). The constant A is to be determined so that the function satisfies the definition. Integrating over all space (the delta function has to be somewhere):

$$\int_{u_{3\min}}^{u_{3\max}} \int_{u_{2\min}}^{u_{2\max}} \int_{u_{1\min}}^{u_{1\max}} A\delta(u_1 - u_{10})\delta(u_2 - u_{20})\delta(u_3 - u_{30})h_1h_2h_3du_1du_2du_3 = 1$$

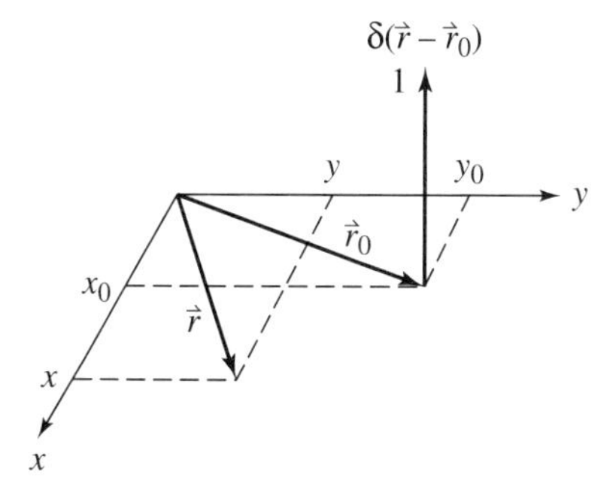

Figure 4-89. Representation of a two-dimensional delta function.

Since A is a constant and u_1, u_2, and u_3 are independent (orthogonal coordinate systems), this may be written

$$A \int_{u_{3\min}}^{u_{3\max}} \delta(u_3 - u_{30}) \int_{u_{2\min}}^{u_{2\max}} \delta(u_2 - u_{20}) \int_{u_{1\min}}^{u_{1\max}} \delta(u_1 - u_{10}) h_1 h_2 h_3 du_1 du_2 du_3 = 1$$

The scale factors cannot be taken out of the integrals since they are functions of coordinates. Also, the delta functions are the one-dimensional cases now so we use Eq. 4-57 on the u_1 integration to obtain $[f(x) \sim h_1 h_2 h_3]$:

$$A \int_{u_{3\min}}^{u_{3\max}} \delta(u_3 - u_{30}) \int_{u_{2\min}}^{u_{2\max}} \delta(u_2 - u_{20}) h_1(u_{10}) h_2(u_{10}) h_3(u_{10}) du_2 du_3 = 1$$

Two more applications of Eq. 4-57 result in

$$A h_1(u_{10}, u_{20}, u_{30}) h_2(u_{10}, u_{20}, u_{30}) h_3(u_{10}, u_{20}, u_{30}) = 1$$

This means that all three scale factors are evaluated at the point where the delta function occurs. As before, we may simplify this by writing $\vec{r}_0$ in place of the specific coordinates to obtain

$$A = \frac{1}{h_1(\vec{r}_0) h_2(\vec{r}_0) h_3(\vec{r}_0)}$$

With this, the generalized delta function becomes

$$\delta(\vec{r} - \vec{r}_0) = \frac{\delta(u_1 - u_{10})}{h_1(\vec{r}_0)} \cdot \frac{\delta(u_2 - u_{20})}{h_2(\vec{r}_0)} \cdot \frac{\delta(u_3 - u_{30})}{h_3(\vec{r}_0)} \tag{4-64}$$

It is important to note that if there were no delta function in one of the coordinate variables, the *scale factor* for that variable is removed as well. This also means that the remaining scale factors may have coordinate *variables* remaining, which would not be evaluated at a particular point for the missing delta function. Some examples will be given shortly. In the three most common coordinate systems we have

Rectangular: $$\delta(\vec{r} - \vec{r}_0) = \delta(x - x_0)\delta(y - y_0)\delta(z - z_0) \tag{4-65}$$

Cylindrical: $$\delta(\vec{r} - \vec{r}_0) = \delta(\rho - \rho_0) \frac{\delta(\phi - \phi_0)}{\rho_0} \delta(z - z_0) \tag{4-66}$$

Spherical: $$\delta(\vec{r} - \vec{r}_0) = \delta(r - r_0) \frac{\delta(\theta - \theta_0)}{r_0} \frac{\delta(\phi - \phi_0)}{r_0 \sin \theta_0} \tag{4-67}$$

To show that the form for $\delta(\vec{r} - \vec{r}_0)$ in cylindrical coordinates is correct according to Eq. 4-61, we substitute in the $\delta(\vec{r} - \vec{r}_0)$ as follows:

$$\int_V \delta(\vec{r} - \vec{r}_0)\, dv = \int_0^\infty \int_{-\infty}^\infty \int_0^{2\pi} \frac{\delta(\rho - \rho_0)\delta(\phi - \phi_0)\delta(z - z_0)}{\rho_0} (\rho d\phi dz d\rho)$$

Taking out terms that do not involve ϕ from the first ($d\phi$) integration, this becomes

$$\int_V \delta(\vec{r} - \vec{r}_0)\, dv = \int_0^\infty \int_{-\infty}^\infty \frac{\delta(\rho - \rho_0)}{\rho_0} \delta(z - z_0) \left[\int_0^{2\pi} \delta(\phi - \phi_0) d\phi \right] \rho dz d\rho$$

The integral of $\delta(\phi - \phi_0)$ is 1 since ϕ_0 is somewhere in $(0, 2\pi)$.

$$\int_V \delta(\vec{r} - \vec{r}_0)\, dv = \frac{1}{\rho_0}\int_0^\infty \int_{-\infty}^\infty \delta(\rho - \rho_0)\delta(z - z_0)\rho dz d\rho$$

$$= \frac{1}{\rho_0}\int_0^\infty \rho\delta(\rho - \rho_0)\left[\underbrace{\int_{-\infty}^\infty \delta(z - z_0)dz}_{=1}\right] d\rho$$

$$= \frac{1}{\rho_0}\int_0^\infty \rho\delta(\rho - \rho_0)d\rho$$

$$= \frac{1}{\rho_0}\rho_0 = 1 \qquad \text{as required by the definition}$$

Example 4-37

As an example of a two-dimensional delta function, consider spherical coordinates where there is no delta function in the θ coordinate. Thus we leave out the corresponding delta function and scale factor and use θ in place of θ_0. Thus we would write, using Eq. 4-67,

$$\delta(\vec{r}\ \vec{r}_0) = \delta(r - r_0)\frac{\delta(\phi - \phi_0)}{r_0 \sin\theta}$$

The dimensional effects of the multidimensional delta functions are also important. We must remember that *each* component delta function has a 1/meter effect. Thus each of the expressions in Eqs. 4-65, 4-66, and 4-67 would have $1/\text{m}^3$ as the dimension. The expression for the two-dimensional spherical delta function would have dimension $1/\text{m}^2$.

Example 4-38

As an example of the writing of equations for field quantities, suppose we have a current flowing along a very thin cylindrical sheet as shown in Figure 4-90. Let the current be

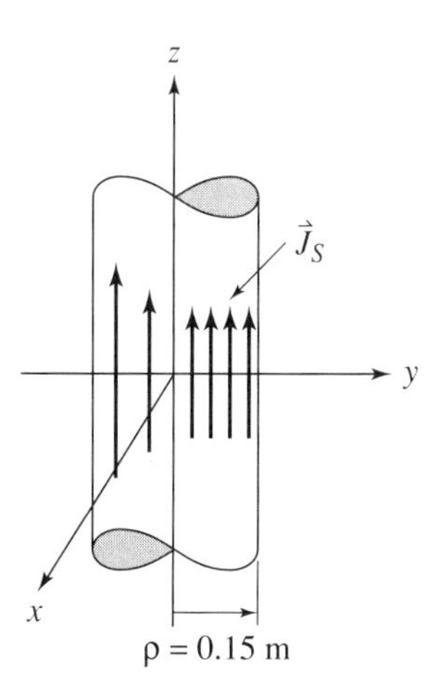

Figure 4-90. Current flow on the surface of a cylindrical sheet.

$J_s = 3$ A/m where the meter dimension is normal to the direction of current flow; that is, amperes per meter around the cylindrical surface.

For later comparison, note that the total current flow along the sheet is

$$I = J_S \cdot \text{length across which current flows}$$
$$= 3 \times 2\pi(0.15) = 0.9\pi \text{ A}$$

The equation for the current flow valid anywhere in space is seen to contain a delta function in ρ since the current is nonzero only at $\rho = 0.15$ m. Using the appropriate parts of Eq. 4-66 we write

$$\vec{J} = 3\delta(\rho - 0.15)\hat{z} \text{ A/m}^2$$

Note the dimensional change in the current expression. Integrating in the xy plane, which is normal to $\vec{J}$, the total current passing through the plane is

$$I = \int_0^\infty \int_0^{2\pi} 3\delta(\rho - 0.15)\rho d\phi d\rho = 3\int_0^\infty \rho\delta(\rho - 0.15)d\rho \int_0^{2\pi} d\phi$$
$$= 3 \cdot 0.15 \cdot 2\pi = 0.9\pi \text{ A}$$

as before. When we use δ functions, if there are no constraints on the coordinate variable in the δ function we merely integrate over the entire range of that variable in the coordinate system.

Exercises

4-21.0f Show that in spherical coordinates the expression for the delta function

$$\delta(\vec{r} - \vec{r}_0) = \frac{\delta(r - r_0)\delta(\theta - \theta_0)\delta(\phi - \phi_0)}{r_0^2 \sin \theta_0}$$

is consistent with the definition of the three dimensional delta function. The subscript 0 means that quantity is a constant.

Figure 4-91.

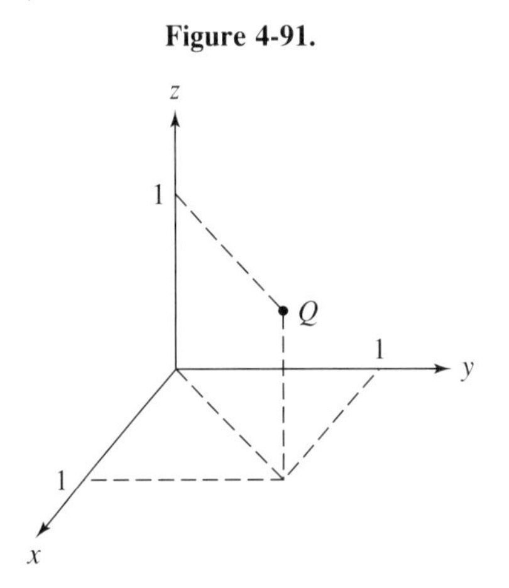

Figure 4-92.

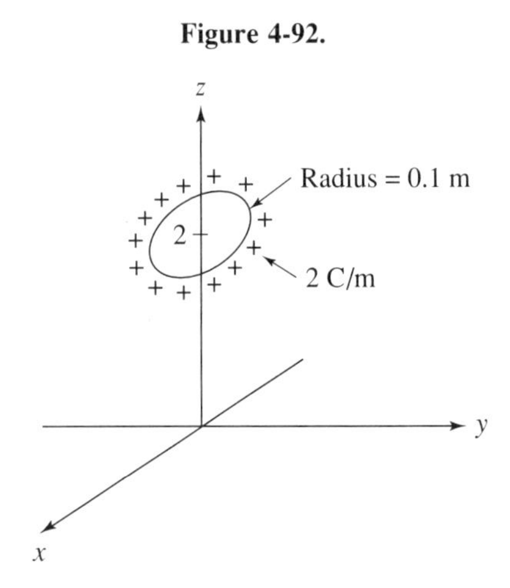

4-21.0g Obtain the expression for the three-dimensional delta function for parabolic cylindrical coordinates, and show that it satisfies the defining equation. (See Exercise 4-12.1a.)

4-21.0h For the bipolar coordinate system (see Exercise 4-12.1b), obtain the delta function for the coordinates u and z and show that the delta function is consistent with the integral definition.

4-21.0i For a point charge of Q coulombs located as shown in Figure 4-91, express the system as a charge density in coulombs/meter^3 in (a) rectangular coordinates, (b) cylindrical coordinates, and (c) spherical coordinates.

4-21.0j In Figure 4-92 is shown a ring of charge that is parallel to the xy plane. Express the charge density everywhere in space as a single equation. (Answer: $\rho_v = 2\delta(\rho - 0.1)\delta(z - 2)$ C/m^3)

Now integrate the equation over all space ($0 \leq \rho \leq \infty, 0 \leq \phi \leq 2\pi, -\infty \leq z \leq \infty$) and obtain the total charge. (Answer: 0.4π C)

The last topic in this section is a presentation of a method for writing single line equations for finite extent (in space or time) systems. Since we desire a functional form that is zero for some range of the coordinate variable and nonzero for some other range, it has been found convenient to define a unit step function as follows:

$$U[f(x)] = \begin{cases} 1 & f(x) > 0 \\ 0 & f(x) < 0 \end{cases} \tag{4-68}$$

Example 4-39

Consider the step function $U(x - 3)$. The argument list is greater than zero for $x > 3$, and so has value 1 for $x > 3$. For $x < 3$, the argument is negative so the function is zero. This is shown as a plot in Figure 4-93.

For the function $U(5 - x)$ the argument is positive for $x < 5$ and negative for $x > 5$. This would be plotted as in Figure 4-94.

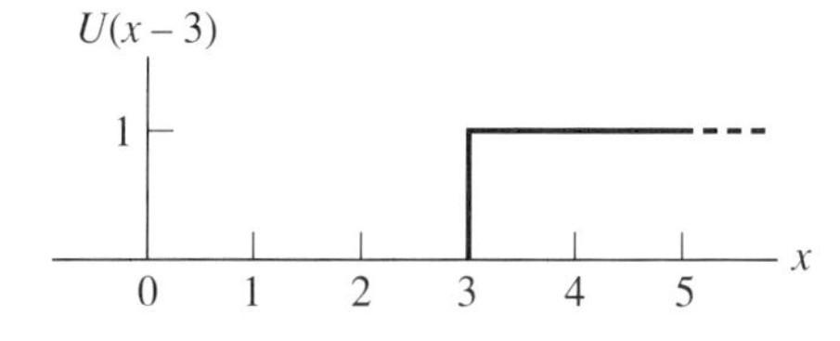

Figure 4-93. Fundamental unit step function.

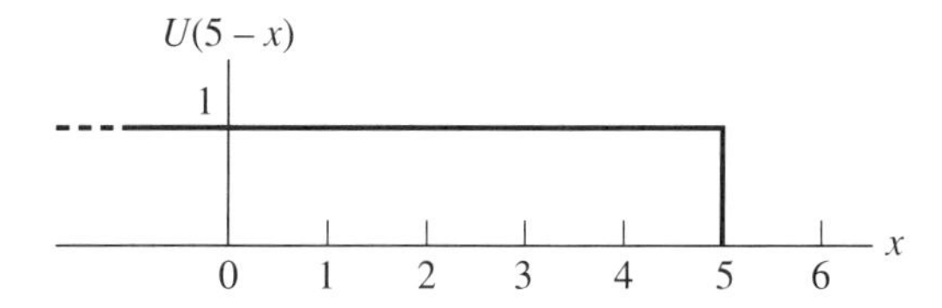

Figure 4-94. Step function that is nonzero for negative x.

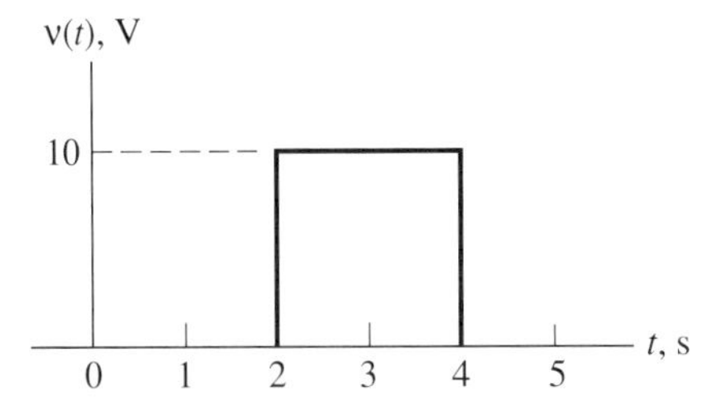

Figure 4-95. Finite duration voltage signal.

Example 4-40

For the next example consider a finite time duration function. Express the voltage function shown in Figure 4-95 as a *continuous* single expression equation.

Method 1: Step function subtraction. $V(t) = 10[U(t-2) - U(t-4)]$ volts. To show that this is correct, we plot each piece of the equation and combine them as indicated in Figure 4-96.

Method 2: $V(t) = 10U(t-2)U(4-t)$ volts. To show that this is correct, plot each factor and combine by multiplying the plots point-by-point as shown in Figure 4-97.

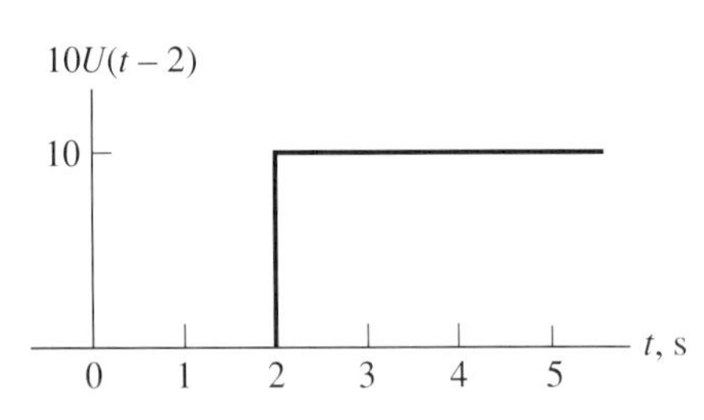

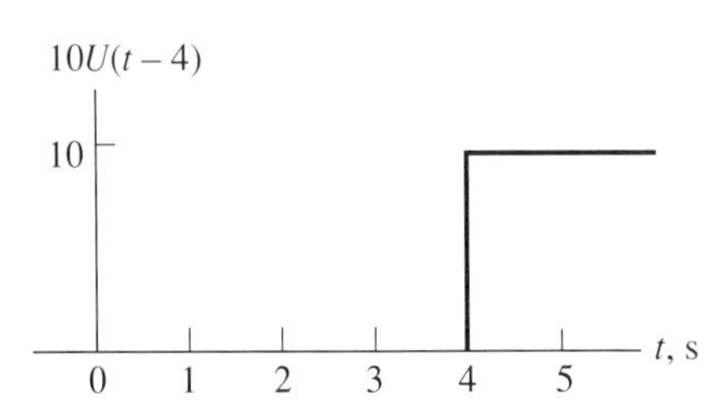

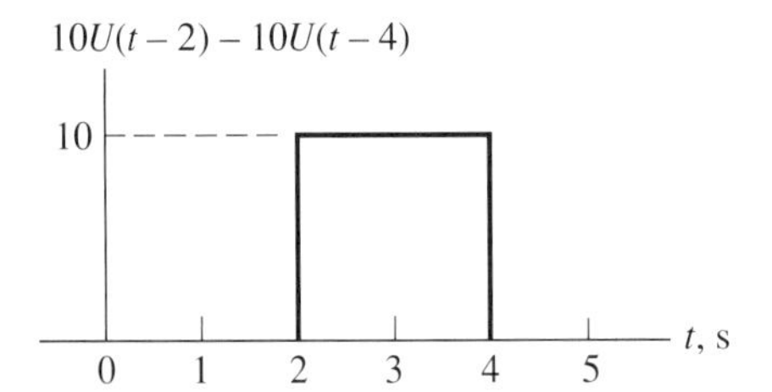

Figure 4-96. Generation of a pulse by step function subtraction.

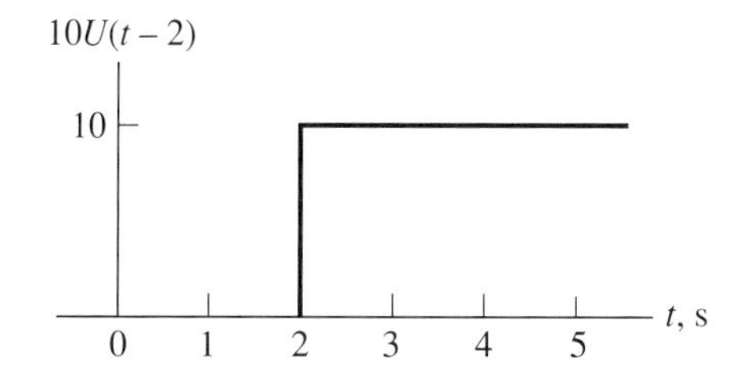

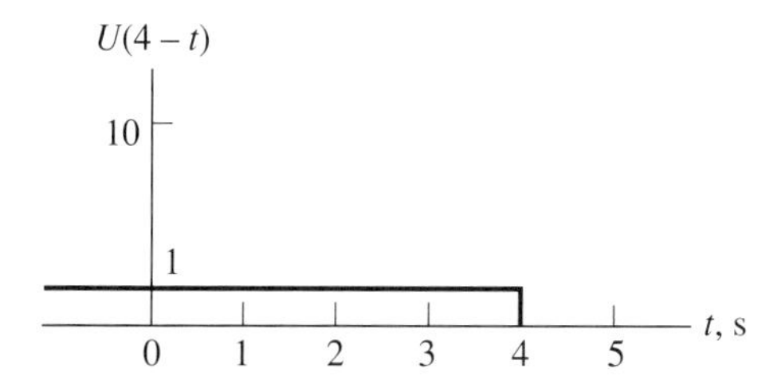

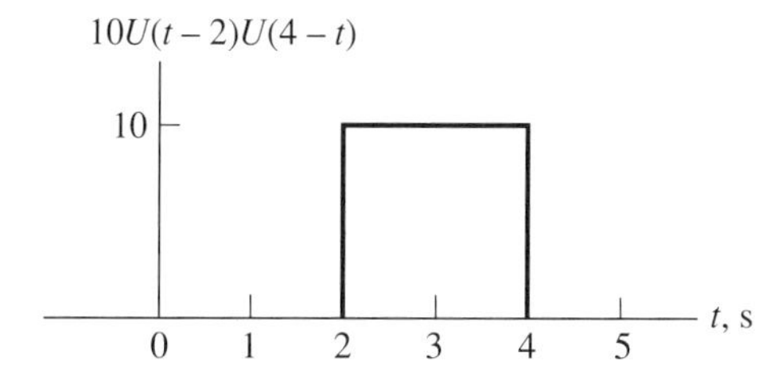

Figure 4-97. Generation of a pulse-by-step function multiplication.

Finally, suppose we have a fictitious mass that is distributed along a line and given as $M(x) = xe^{-x/2}$ kilograms/meter for x values $1 \leq x \leq 4$. See Figure 4-98.

To write an expression valid for all x we could use either

$$m(x) = xe^{-x/2}[U(x - 1) - U(x - 4)]$$

or

$$m(x) = xe^{-x/2}[U(4 - x)U(x - 1)]$$

This is plotted in Figure 4-99. Here we have used the same techniques learned for time functions using x in place of time t, so we make no additional comment.

Note that the multiplication by *step* functions doesn't change the dimensionality since to find the total mass, we evaluate by multiplying by dx, but $M(x)$ is already in kilograms/

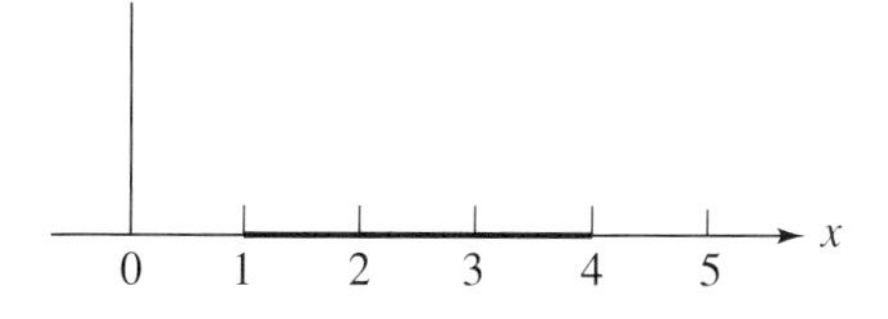

Figure 4-98. Mass distributed along a portion of a line.

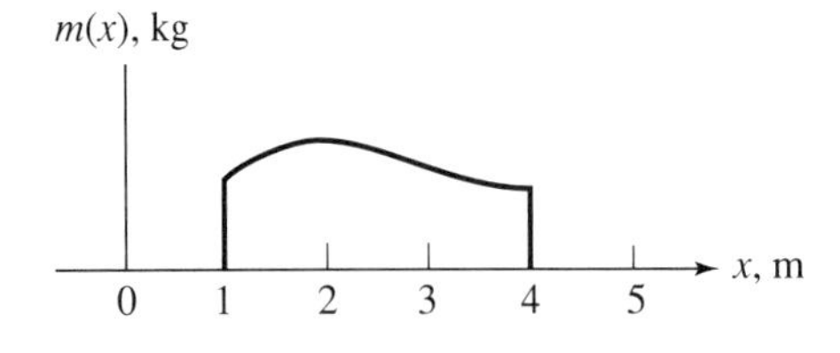

Figure 4-99. Plot of finite extent mass distribution.

meter, so $M(x)U(4 - x)U(x - 1)dx$ is dimensionally kilograms. The total mass can be obtained as

$$M = \int_{-\infty}^{\infty} xe^{-x/2}U(4 - x)U(x - 1)dx = \int_{1}^{4} xe^{-x/2}dx = \left(-\frac{x}{2} - 1\right) 4e^{-x/2}\bigg|_{1}^{4}$$

Note the integrand is 0 outside the limits 1 and 4.

$$f = 4e^{-2}(-2 - 1) - 4e^{-0.5}(-0.5 - 1) = -1.62 + 3.64 = 2.02 \text{ kg}$$

Exercises

4-21.0k On the z axis we have a small section of wire that has an electric charge on it distributed as $Q(z) = 3 \cos z$ coulomb/meter over the interval $-1 \leq z \leq 1$. Express this as a function valid for all coordinates and determine the total charge Q on this wire by performing a volume integration. (Answer: $q(x, y, z) = 3 \cos z\delta(x)\delta(y)[u(z + 1) - u(z - 1)]$ coulombs/meter^3)

4-21.0l Evaluate $3e^{-4x}U(x - 3)U(6 - x)\delta(x - 4)$.

4-21.0m Plot the function $f(x) = U(\cos x)$.

4-21.0n (a) Express the line charge shown in Figure 4-100 as a function of rectangular coordinates such that the expression is valid for all x, y, z, values of coordinates. (Hint: You will have both delta functions and step functions in the equation.)

(b) What are the units of your resulting expression? What is the total charge on the line as obtained directly from the given line charge and also as obtained by integrating your expression over all space coordinates $(-\infty, \infty)$.

4-21.0o Suppose we have a sphere of charge as shown in Figure 4-101. Express the charge density, in coulombs/meter^3, which is valid for all values of coordinates.

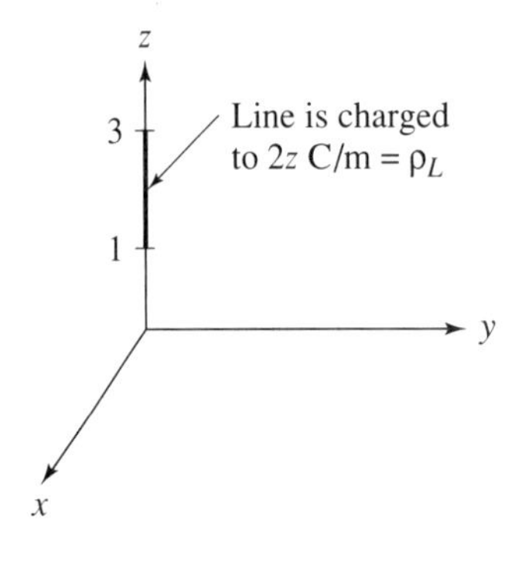

Figure 4-100.

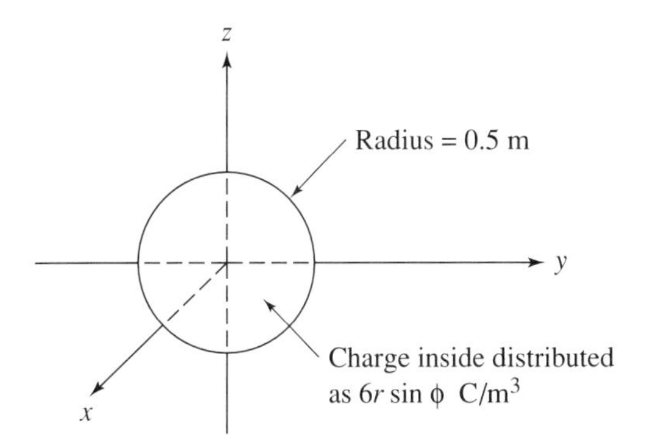

Figure 4-101.

4-21.0p A region of space has a charge within it given by the expression

$$q(x, y, z) = 3 \times 10^{-4} xyzU(x - 2)U(6 - x)U(y + 3)$$
$$U(y - 1)\delta(z - 2) \text{ coulomb/meter}^3$$

Integrate this charge equation over all space and determine the total charge present in coulombs. It might be helpful to sketch roughly the region covered by the charge to aid in visualizing the nonzero regions for the integration.

Summary, Objectives, and General Exercises

Summary

To this point we have been developing insights to the physical interpretations of coordinate systems and the fundamental vector operations in those systems. This study, along with the ability to solve basic problems that involve space and time coordinates discussed in Chapter 1, prepares us to apply the ideas to electromagnetic systems. At this juncture we could delve right into a general study of time-varying vector fields. One would simply postulate a basic set of field equations (Maxwell's equations) and develop solutions. Pedagogically, however, it is best to apply some of the concepts individually and build to the more difficult situations. We choose this latter approach, along a little more historical line, and hope to solidify some physical understanding of electric and magnetic fields and material properties. This is easier if one first looks briefly at static fields, and this is the approach taken. However, we use static fields only as a steppingstone to the time-varying case and do not spend time solving static problems, which is left to later work. It is felt that most generally the time-varying cases are more prevalent in practice, so that after a short study of static electric and magnetic fields, familiarity with the applications of vector operations can quickly carry over to the time-varying problems.

Objectives

1. Know and be able to give examples of real and complex scalars and scalar fields.
2. Be able to define and give examples of vectors, vector fields, and vector components and to write a vector in general component form. Be

able to obtain the expression for a vector between two given points.

3. Be able to add and subtract vectors and to decompose a given vector into any orthogonal or nonorthogonal coordinate component form using unit vector notation and the dot product.
4. Know the definition of a unit vector and be able to determine a unit vector in the direction of any given vector.
5. Know how to multiply a vector by a scalar.
6. Know the definition of a right-hand orthogonal coordinate system and be able to write the location of a point in that system by giving coordinates in proper sequence. Be able to give three examples of such coordinate systems in pictorial form.
7. Be able to determine whether or not a given coordinate system is a right-hand system using the dot and cross product definitions.
8. Be able to reduce a given vector equation to a set of three scalar equations to solve for unknown vector components.
9. Be able to perform the dot (scalar) product of two given vectors and be able to find the angle between the vectors. Also be able to state formally the definition of the dot product, and to explain the physical significance.
10. Be able to state the definition of the cross (vector) product and to perform the cross product of two vectors and to find the angle between them. Be able to explain the physical significance of the cross product.
11. Be able to convert a vector given in one coordinate system to its representation in another right-hand orthogonal system.
12. Be able to state the definition of a right-hand orthogonal general curvilinear coordinate system and to write expressions for incremental length, surface, and volume in terms of the scale factors h_i.
13. Be able to state the definition of a coordinate system scale factor and to explain why scale factors are needed by giving examples in cylindrical and spherical coordinates.
14. Know how to calculate the scale factors for a given right-hand orthogonal coordinate system and to write expressions for $d\vec{l}$, $d\vec{a}$, and dv in that system.
15. Be able to write the definitions of gradient, divergence, and curl in mathematical form including a picture showing the meaning of each symbol that appears in the definitions. Be able to evaluate these for given scalar and vector functions.
16. Be able to describe the physical significance of the terms gradient, divergence, and curl and know the meaning of each symbol that appears in the general orthogonal curvilinear system expansion of these three operations and the Laplacian.
17. Be able to derive expressions for divergence and curl for a given coordinate system using the definitions of each.
18. Know the definitions of fields that are irrotational (conservative) and fields that are solenoidal. State these definitions in terms of divergence and curl operations and give their physical significance.
19. Know the definitions of the Laplacian of a scalar and the Laplacian of a vector and be able to evaluate these for given scalar and vector fields.
20. Be able to apply any of the vector identities involving the ∇ operator (∇^2, $\nabla \times \nabla \times$, etc.) to given scalar and vector fields as appropriate.
21. Be able to state the divergence theorem and Stokes' theorem and to give the physical relationships between the contours, surfaces, and volumes appearing in these theorems. Be able to evaluate each integral appearing in these theorems. Understand the qualitative relationships between these theorems and the corresponding definitions of divergence and curl.
22. Be able to apply and know how to evaluate each term appearing in Green's first and second identities.
23. Be able to evaluate line (contour) integrals and to determine whether or not the values of such integrals are dependent upon the path between points.

24. Be able to evaluate surface and volume integrals that involve scalar and vector functions.
25. Be able to differentiate an integral that has variable limits and a multiple variable function integrand (Liebniz theorem).
26. Be able to state the formal mathematical definition of the multidimensional delta function and know how to express the delta function in any right-hand orthogonal coordinate system. Know the physical interpretation of the delta function and the meaning of *weight of a delta function*.
27. Be able to apply the sampling and derivative properties of the delta function as they appear in integrals.
28. Be able to state the formal mathematical definition of the unit step function and give examples of its use.
29. Be able to express a discontinuous function given either as a plot or a function in variable range limited form in terms of delta functions and step functions.
30. Be able to evaluate integrals that have delta functions and/or step functions in the integrands.

General Exercises

GE4-1 Obtain the dot product of the following two vectors. Convert the two vectors to phasor form and repeat.

$$\vec{\mathscr{E}} = 2\cos(10^9 t - 20°)\,\hat{x} - 4\sin(10^9 t - 20°)\,\hat{y}$$

$$\vec{\mathscr{H}} = 6\cos(10^9 t - 20°)\,\hat{x} + 3\sin(10^9 - 20°)\,\hat{y} + 2\cos(10^9 t - 20°)\,\hat{z}$$

GE4-2 Using the cosine as the basis function, express two vectors of GE4-1 in phasor form and compute the dot product.

GE4-3 The vector $\vec{B} = -3\angle 70°\hat{x} + 2\angle 70°\hat{y} + 2\angle 70°\hat{z}$ has a frequency of 100 MHz. At $t = 2.5$ ns, what angles does $\vec{B}$ make with the x, y, and z axes? (Hint: Convert $\vec{B}$ to the time domain and compute the components at that instant. Then use the dot product.) (Answer: 36.7°, 122.3°, 105.5°)

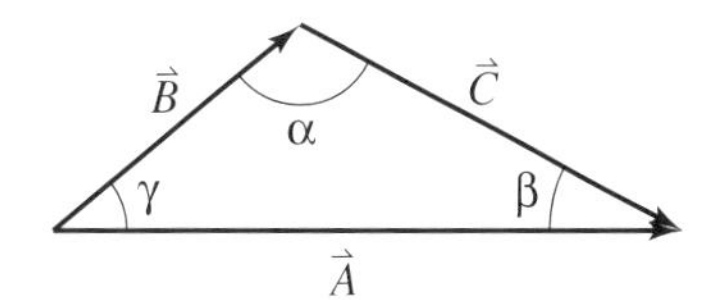

Figure 4-102.

GE4-4 Given three vectors that form a triangle shown in Figure 4-102, prove the law of cosines. (Hint: $\vec{C} = \vec{A} - \vec{B}$. Dot this with itself.)

GE4-5 Using Figure 4-102, prove the law of sines. (Hint: $\vec{A} + \vec{B} + \vec{C} = 0$. Cross this equation in turn by $\vec{A}$, $\vec{B}$, and $\vec{C}$ and then use the definition of the cross product.)

GE4-6 Prove that the area of a parallelogram having sides $\vec{A}$ and $\vec{B}$ with θ the smaller angle between them is given by $|\vec{A} \times \vec{B}|$.

GE4-7 Given $\vec{E} = 2\hat{x} - 3\hat{y} + \hat{z}$ and $\vec{H} = -4\hat{x} + 6\hat{y} - 8\hat{z}$. Find a unit vector normal to the plane defined by $\vec{E}$ and $\vec{H}$. (Hint: What is the direction of the vector product?)

GE4-8 Prove $\vec{A} \times (\vec{B} \times \vec{C}) = (\vec{A} \cdot \vec{C})\vec{B} - (\vec{A} \cdot \vec{B})\vec{C}$. (Hint: Express the vectors in rectangular component form and use the determinant form of the cross product.)

GE4-9 A force $\vec{F} = -3\hat{x} - 2\hat{y} + 4\hat{z}$ exists at a point $P(-1, 1, -2)$. Find the moment (torque) about the point $(-2, 1, -3)$.

GE4-10 The coordinate system shown in Figure 4-103 is called the *elliptic cylindrical* coordinate system. It consists of concentric confocal elliptic cylinders (tubes) for the u coordinate constant and confocal hyperbolas, which are sheets in the z (outward) coordinate when the v coordinate is constant. Determine the expression for the gradient for this coordinate system.

For a practical application of this seemingly odd coordinate system, consider the following. In an integrated circuit package, two metal tabs are bent near the chip. Since the circuit is very

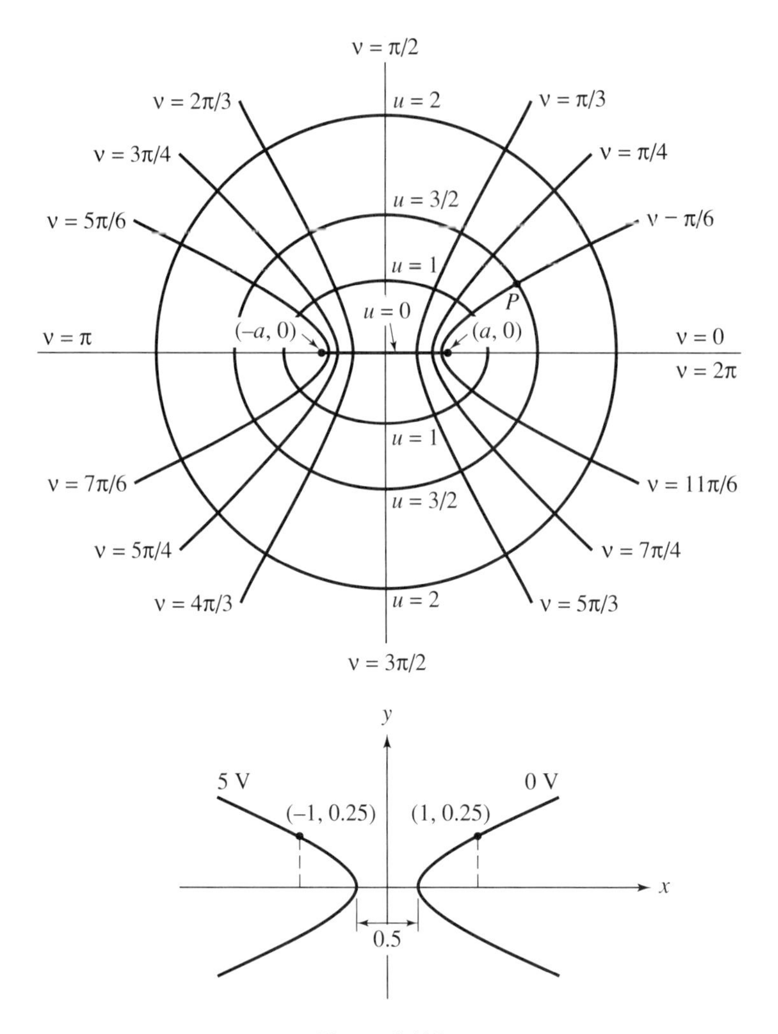

Figure 4-103.

small the two tabs may be considered as two folded flat sheets and the corners considered to approximate hyperbolas as shown in the figure above. Some dimensions are measured in millimeters and are also given on the diagram. The two tabs are in close proximity as indicated and are at two different potentials. The equation for the potential everywhere between the two hyperbolic sheets is

$$V(u, v, z) = A(v - B)$$

where A and B are constants to be determined as outlined below. All coordinates are in millimeters. Answer the following:

(a) What are the equations for the two hyperbolas in x, y coordinates? (Use some formulas from analytic geometry.)

(b) Find the constant a in the equations that relate the rectangular and elliptic cylindrical coordinates. Note that a is the focus.

(c) Using the coordinate system transformation equations, find the values of v that define the two parabolas. (Hint: Some point locations are known on each parabola. Find the corresponding values of v, using simultaneous equations for the known points.) Sketch the hyperbolas to some reasonable scale. What is the value of u that corresponds to the rectangular point $(-1, 0.25)$?

(d) For the values of v that define the hyperbolic sections in the positive y half-plane, find the constants A and B in $V(u, v, z)$ by using the known voltages at these two surfaces. This process will yield two simultaneous equations for determining A and B.

(e) Determine the negative gradient of the function of $V(u, v, z)$ you now have, and sketch the direction of the resulting vector at various points along the u = constant curve value you obtained in (c).

GE4-11 For the elliptic cylindrical coordinate system shown in the preceding exercise, plot constant coordinate contours for the cases $a = 0.25$, 0.5, 1, 2. This will show you the effect of the parameter a on the physical appearance of the particular system.

GE4-12 Show that in rectangular coordinates:
(a) $\vec{A} \cdot \nabla V = (\vec{A} \cdot \nabla)V$
(b) $\vec{A} \cdot \nabla V \neq (\nabla \cdot \vec{A})V$
(c) $\vec{A} \cdot \nabla \neq \nabla \cdot \vec{A}$

GE4-13 Prove the following vector identities involving the gradient where A and B are *scalar* functions of position $A(u_1, u_2, u_3)$, $B(u_1, u_2, u_3)$.
(a) $\nabla(A + B) = \nabla A + \nabla B$
(b) $\nabla(AB) = A\nabla B + B\nabla A$

GE4-14 A vector field for which $\nabla \cdot \vec{B} = 0$ is said to be a solenoidal field. Suppose that the field is given by the expression $\vec{B} = (x+y)\hat{x}-(y-z)\hat{y}+(x+y+az)\hat{z}$ where the constant a is to be determined. What is the value of a so that the $\vec{B}$ field is solenoidal?

GE4-15 Using the expression for the divergence derived for general curvilinear coordinates, develop expressions for the divergence in the following coordinate systems:
(a) Parabolic cylindrical coordinates (see Exercise 4-12.1a).

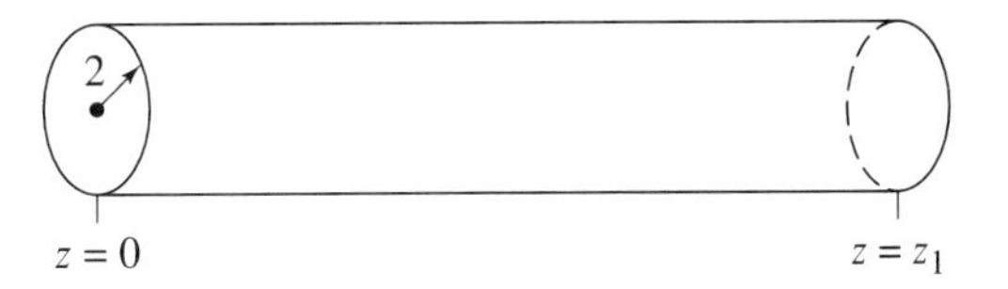

Figure 4-104.

(b) Bipolar coordinates (see Exercise 4-12.1b).
(c) Elliptic cylindrical coordinates (see Exercise GE4-10).

GE4-16 In the cylindrical pipe shown in Figure 4-104 a compressible fluid is flowing. Let's consider two velocity fields given by
(a) $\vec{V} = 6z(2 - \rho)\hat{z}$
(b) $\vec{V} = (2 - \rho)/(6z)\hat{z}$
Calculate the divergence of the velocity vectors in each case and describe physically what you would expect to be happening to the fluid as z increases down the pipe. (Hint: Consider density.)

Also, for each case draw a pipe and indicate the vectors on each diagram as a function of several z locations. Use different lengths of arrows to indicate magnitude.

GE4-17 A vector field $\vec{B}$ is said to be irrotational if $\nabla \times \vec{B} = 0$. Find the constants a, b, and c in order that the following fields be irrotational:

$$\vec{B} = (2x + y + 3az)\hat{z} + (x - 2by + 6z)\hat{y} + (cx - 2y - 7z)\hat{z}$$

(Hint: One vector equation can yield three scalar equations.)

GE4-18 For the rotating system of Exercise 4-15i show that the divergence of $\vec{V}$, $\nabla \cdot \vec{V}$, is zero. What does this tell you about the material from which the body is constructed? (Hint: Consider the physical significance of divergence applied to a small section of the body.)

GE4-19 Since the curl of a vector is also a vector, we would expect that we could take the curl of the new vector. Symbolically, we would write

$$\text{Curl Curl } \vec{E} = \nabla \times \nabla \times \vec{E}$$

Evaluate this operation for the vector

$$\vec{E} = K_1 \sin\left(\frac{3\pi}{a} x\right) e^{-j\beta z} \hat{y}$$

where K_1, a, and β are constants. Physically, this vector represents an electric field propagating in space between two metal plates. Such vector fields satisfy a vector equation that involves this double curl operation.

GE4-20 The field vectors that describe the electromagnetic radiation from a short antenna are given by

$$\vec{\mathscr{E}} = \mathscr{E}_r \hat{r} + \mathscr{E}_\theta \hat{\theta} + \mathscr{E}_\phi \hat{\phi} \quad \text{and} \quad \vec{\mathscr{H}} = \mathscr{H}_r \hat{r} + \mathscr{H}_\theta \hat{\theta} + \mathscr{H}_\phi \hat{\phi}$$

where

$$\left.\begin{aligned}
\mathscr{E}_r &= \frac{2Idl \cos\theta}{4\pi\epsilon}\left(\frac{\cos \omega t'}{r^2 v} + \frac{\sin \omega t'}{\omega r^3}\right) \\
\mathscr{E}_\theta &= \frac{Idl \sin\theta}{4\pi\epsilon}\left(\frac{-\omega \sin \omega t'}{r v^2} + \frac{\cos \omega t'}{r^2 v} + \frac{\sin \omega t'}{\omega r^3}\right) \\
\mathscr{E}_\phi &= 0,\ H_\theta = 0,\ H_r = 0 \\
\mathscr{H}_\phi &= \frac{Idl \sin\theta}{4\pi}\left(\frac{-\omega \sin \omega t'}{r v} + \frac{\cos \omega t'}{r^2}\right)
\end{aligned}\right\} \quad \text{where } t' = \left(t - \frac{r}{v}\right)$$

In the preceding, I, dl, v, ω, ϵ are constants, with $v = 1/\sqrt{\mu\epsilon}$. Show that the $\vec{\mathscr{E}}$ and $\vec{\mathscr{H}}$ vectors satisfy the equation

$$\nabla \times \vec{\mathscr{E}} = -\mu \frac{\partial \vec{\mathscr{H}}}{\partial t}$$

GE4-21 Prove the null identities (use rectangular coordinates)

(a) $\nabla \cdot (\nabla \times \vec{A}) \equiv 0$

(b) $\nabla \times (\nabla \Phi) = 0$

GE4-22 Given two vectors $\vec{A}$ and $\vec{B}$ that exist at the same point in space. Prove that the area of the parallelogram having $\vec{A}$ and $\vec{B}$ as sides is $|\vec{A} \times \vec{B}|$.

GE4-23 Prove that $|\vec{A} \times \vec{B}|^2 + |\vec{A} \cdot \vec{B}|^2 = |\vec{A}|^2 |\vec{B}|^2$. (Hint: Use the trigonometric forms of the vector and scalar products.)

GE4-24 For a vector in general curvilinear coordinates, say, $\vec{A} = \overline{A}_1 \hat{u}_1 + \overline{A}_2 \hat{u}_2 + \overline{A}_3 \hat{u}_3$, show that

$$\vec{A} = (\vec{A} \cdot \hat{u}_1)\hat{u}_1 + (\vec{A} \cdot \hat{u}_2)\hat{u}_2 + (\vec{A} \cdot \hat{u}_3)\hat{u}_3$$

GE4-25 Prove that the scalar triple product of three vectors in generalized curvilinear coordinates may be expressed as the determinant

$$\vec{A} \cdot (\vec{B} \times \vec{C}) = \begin{vmatrix} A_1 & A_2 & A_3 \\ B_1 & B_2 & B_3 \\ C_1 & C_2 & C_3 \end{vmatrix}$$

GE4-26 Expand the following in rectangular coordinates to illustrate the difference between $(\vec{A} \cdot \nabla)\vec{B}$ and $\vec{A} \cdot (\nabla \vec{B})$. (Hint: Unit dyads are involved. See Appendix G.)

GE4-27 Evaluate $\nabla^2 (1/r)$ in spherical coordinates. Now using $r = \sqrt{x^2 + y^2 + z^2}$, evaluate $\nabla^2(1/\sqrt{x^2 + y^2 + z^2})$ in rectangular coordinates. Compare the amount of work involved in the two computations. What does this tell you about selecting a coordinate system?

GE4-28 Show that if a scalar field satisfies Laplace's equation $\nabla^2 \Phi = 0$, the vector field $\nabla \Phi$ is both solenoidal and irrotational. (Hint: $\nabla^2 = \nabla \cdot \nabla$.)

GE4-29 Suppose we have a cylindrical region of space defined in the bipolar coordinate system (Exercise 4-12.1b) defined by the surface $v = 1$ and extending from $z = -1$ to $z = +3$. A field in the region is given by $\vec{A} = 4uvz\vec{u}$. Verify the divergence theorem for this system.

GE4-30 Using the divergence theorem prove that for a scalar field Φ

$$\int_V \nabla \Phi dv = \oint_S \Phi d\vec{a}$$

(Hint: In the divergence theorem let $\vec{A} = \Phi\vec{K}$ where $\vec{K}$ is a constant vector function.)

GE4-31 A force field is given by $\vec{F} = 6xy\hat{x} + 2y^2\hat{y}$ newtons in the xy plane. Evaluate the path integral

$$\int_C \vec{F} \cdot d\vec{l}$$

where C is the curve in the plane given by $y = 3x^2$, from the point (0, 0) to (2, 12). What is the work involved in this path?

GE4-32 A particle in a force field $\vec{F} = 2x\hat{x} + (3xy - z)\hat{y} - 8\hat{z}$ newtons is moved along the following three paths. Determine the work done in each case.

(a) Straight line from (1, 1, 1) to (1, 2, 3).

(b) Curve defined by $x = t^2$, $y = 2t$, $z = 3t^2 + t$ from $t = 0$ to $t = 1$. The variable t is time.

(c) The curve given by $x^2 = 9y$ and $4x^3 = \frac{1}{2}z$ from $x = 0$ to $x = 3$.

GE4-33 Compute the circulation of the vector field $\vec{H} = (y - 2x)\hat{x} + (3x - 2y)\hat{y}$ around the circle in the xy plane centered at the origin and of radius 3. (Hint: You might try a change of coordinate system.)

GE4-34 Suppose we have a vector field $\vec{\mathcal{B}}(t) = (4t^2 - t)\hat{x} + (4 - 12t)\hat{y} + 8t\hat{z}$. Evaluate

$$\int_3^5 \vec{\mathcal{B}}(t)dt$$

(There is no dot product here and the integral of a vector is still a vector.)

GE4-35 In the elliptic cylindrical coordinate system (see General Exercise GE4.10) there is a vector field $\vec{E} = 4 \cos v\hat{v} - 2z \cosh u\hat{u} + 9\hat{z}$. Compute the closed path integral of $\vec{E}$ around the curve $u = \frac{3}{2}$ in the xy plane for the case $a = \frac{1}{3}$. If a closed form of an integral does not result, use numerical evaluation.

GE4-36 In the elliptical cylindrical coordinate system (see General Exercise GE4-10), evaluate the closed surface integral for the vector field in General Exercise GE4-35. The closed surface is defined by the intersecting surfaces $u = 1$, $v \in (0, \pi)$ with $z \in (-1, 1)$. Use $a = \frac{1}{4}$.

GE4-37 For the bipolar coordinate system, determine the volume of the cylinder enclosed by the surfaces $u = \pi/6$ and $z = 2$ m. Use $a = 1$. Also obtain the value of the cross-sectional area of the cylinder in the xy plane and from this obtain the cylinder radius. (See Exercise 4-12.1b for the coordinate system.)

GE4-38 Plot the following function for $0 < x < 20$

$$f(x) = 25 \sin xU (\sin x)$$

Suggest an application for your result.

GE4-39 Using the results given in Appendix E, take the partial derivative with respect to ϕ of the following vector field function:

$$\vec{F}(\rho, \phi, z) = 3z \cos \phi\hat{\rho} - 2\rho z\hat{\phi} + 3 \sin \phi \cos \phi\hat{z}$$

(Hint: Some unit vectors change direction as coordinates change, so you must use the product rule and think of the unit vectors as functions.)

5 Theory, Physical Description, and Basic Equations of Electric Fields

5-1 Introduction

Once the basic ideas and mathematical representations of voltage and current propagation on transmission lines are understood, the next step is to become familiar with the propagation of electric and magnetic fields. However, the mathematics of fields (time-varying as well as static) is somewhat more involved and this is the reason we needed to make a mathematical diversion in Chapter 4 to develop that second tool. Except for techniques for solving partial differential equations, which are covered in later chapters, our mathematical preparation to this point can take us a long way toward our goal.

The third tool area we need to develop is that of the basic concepts and mathematical representations of electric and magnetic fields. This chapter and Chapter 6 will be devoted to those studies. For the initial study we will consider primarily static fields so that the encumbrance of time variation will not detract from the basic ideas; however, we will not spend time solving many static problems but will move quickly to the time-varying cases in Chapter 7. This is not meant to imply that static field problems are not important, for many significant applications exist. As examples, consider electrostatic precipitators used to clean particles from gases, electrostatic image processes used in duplicating machines, noncontact printers, which use charged ink droplets, and the ionization processes used to generate ultra-low vacuums.

In this chapter many new symbols and names are introduced. It is important to understand the physical meanings of the symbols and the basic electrical and mechanical units. One should also try to form mental images insofar as possible. This will help one feel at ease with the myriad symbols required to describe electric field quantities. If this understanding is not developed, there is a danger that field theory reduces to mere symbol manipulation.

5-2 Coulomb's Law and Applications

When one considers the fact that early workers in electromagnetic theory did not have spectrum analyzers, 250-MHz oscilloscopes, and digital transducers and meters, one marvels at the ingenuity they possessed. The extreme care they exercised in both their equipment construction and their observations gave us laws that have stood the test of decades. (Could it be that we too often let our instruments come between us and our tests or that we blindly rely on those instruments to tell us *all* we need to know?) Maxwell has presented a concise review of these experiments in his treatise [39].

The experiments that Charles Augustin Coulomb performed with the torsion balance (and similar experiments by Cavendish and Townsend) produced the law of force between two electrically charged particles. This law is known as Coulomb's law, which he reported in 1785, and is (see Figure 5-1):

$$\vec{F}_2 = K \frac{q_1 q_2}{|\vec{R}|^2} \hat{R} \tag{5-1}$$

where $\vec{F}_2$ = force on point charge q_2, Newtons (N)
K = experimentally determined constant
q_1, q_2 = point charges, coulombs (C)
$\vec{R}$ = vector distance directed from charge q_1 to charge q_2, meters
$\hat{R}$ = unit vector in direction of $\vec{R}$, $\vec{R}/|\vec{R}|$ (unitless)

The units on the constant K would be square meters · newton/square coulomb, but a simpler set of units will be developed shortly.

The law suggests that there is some field quantity that exists at q_2, produced by q_1, which gives the force on q_2. This will be defined later.

Another experiment that demonstrates the existence of some field quantity that exerts a force on a charge is the cathode ray tube deflection observation. This is shown schematically in Figure 5-2. As the voltage across the two plates increases, the deflection of an electron increases, as evidenced by the movement of the spot on the tube-face phosphor. In the figure, the value of V for Path 2 is greater than that for Path 1. The amount of

Figure 5-1. Coulomb's Law configuration for the force of charge q_1 on charge q_2.

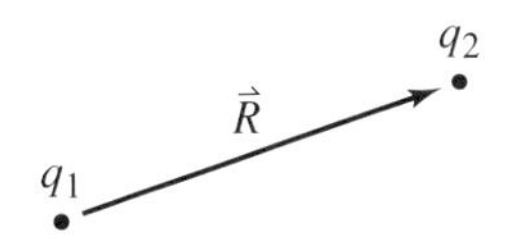

Figure 5-2. Deflection of an electron in a cathode ray tube.

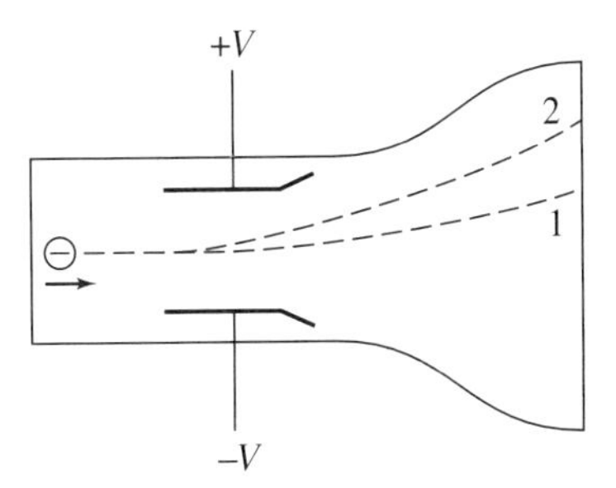

deflection is observed to vary linearly with V. Thus the voltage V must produce some field quantity between the plates which exerts a force on the charge.

In investigations with Coulomb's law, one finds that the force is maximum in air (vacuum) and decreases as other materials are placed between the two charges. Specifically, if nonconducting materials such as porcelain, bakelite, and Teflon—called *dielectrics*—are used, the force between the charges is different for each material. To account for this material-dependent force, the constant K in the rationalized MKS system of units (i.e., the SI system) is written as

$$K = \frac{1}{4\pi\epsilon} \tag{5-2}$$

where ϵ is called the *permittivity* of the material. The factor 4π is used because it simplifies the forms of equations that are used in time-varying field calculations. The "constant" ϵ is further written as the product of two quantities as

$$\epsilon = \epsilon_0 \epsilon_r \tag{5-3}$$

where ϵ_0 is the permittivity of vacuum and ϵ_r is called the *relative permittivity* or *dielectric constant* of the material. The constant ϵ_r is what one obtains from tables of dielectric materials and is in the range of 1.5 to 15 in value for most materials, although in special cases the value may be in the hundreds or thousands. The constant ϵ_r is dimensionless, so it is is a direct measure of how much greater the permittivity of a material is than vacuum. Since the forces decrease with the materials present, $\epsilon_r \geq 1$. There are anomalous cases, but they will not be discussed here. Most gasses have ϵ_r close to unity. The description of the physical mechanism in materials that determine ϵ_r values will be covered later in the discussion of polarization.

Experiments in vacuum ($\epsilon_r = 1$) give a value of ϵ_0 as

$$\epsilon_0 = 8.854 \times 10^{-12} \approx \frac{1}{36\pi} \times 10^{-9} \text{ farad/meter} \tag{5-4}$$

The units of ϵ_0 will be explained later as field computations and applications to capacitance are made. Coulomb's law now has the form in most common use:

$$\vec{F}_2 = \frac{q_1 q_2}{4\pi\epsilon|\vec{R}|^2} \hat{R} \tag{5-5}$$

where, as before, $\hat{R}$ is directed from q_1 to q_2.

For linear dielectrics (i.e., those dielectrics whose ϵ_r values are not functions of any fields or forces present) the superposition theorem may be applied to obtain the total force that several charges produce on one particular charge. Suppose we have n point charges in a system and wish to compute the force on the jth charge as shown in Figure 5-3. Summing the forces on the jth charge yields

$$\vec{F}_j = \sum_{i=1}^{n} \frac{q_i q_j}{4\pi\epsilon|\vec{r}_{ij}|^2} \hat{R}_j, \qquad i \neq j \tag{5-6}$$

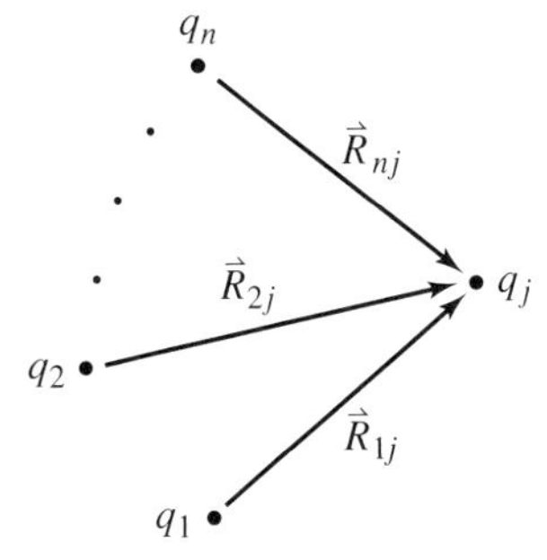

Figure 5-3. System of n point charges with the force on charge q_j to be determined.

The expressions for the vectors and unit vectors are obtained using the techniques of Chapter 4.

Example 5-1

Figure 5-4 is a three-point charge configuration for which we want to compute the force on the 1-C charge. The medium around the charges is vacuum. From the figure and using the methods of Chapter 4, we proceed as follows:

$$\vec{R}_{21} = (7 - 1)\hat{x} + (2 - 1)\hat{y} = 6\hat{x} + \hat{y}$$

$$|\vec{R}_{21}|^2 = \vec{R}_{21} \cdot \vec{R}_{21} = 36 + 1 = 37$$

$$\hat{R} = \frac{\vec{R}_{21}}{|\vec{R}_{21}|} = \frac{6\hat{x} + \hat{y}}{\sqrt{37}} = 0.9863\hat{x} + 0.1644\hat{y}$$

Then, with $\epsilon = \epsilon_0$,

$$\vec{F}_{21} = \frac{(-2)(1)}{4\pi + 8.854 \times 10^{-12} \times 37}(0.9863\hat{x} + 0.1644\hat{y})$$
$$= -4.79 \times 10^8\hat{x} - 7.99 \times 10^8\hat{y} \text{ N}$$

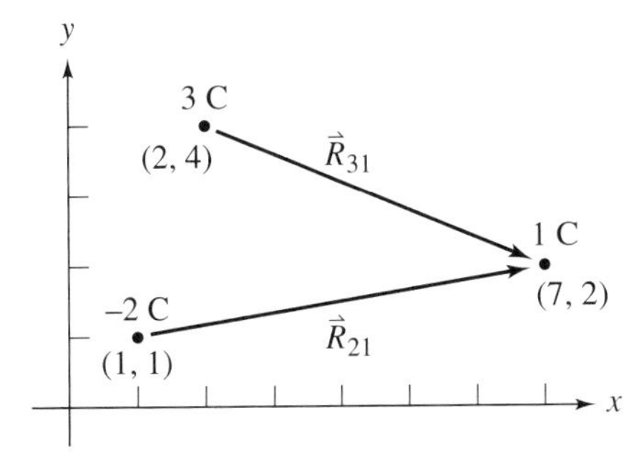

Figure 5-4. A three-point charge system.

Following the same procedure we have

$$\vec{R}_{31} = 5\hat{x} - 2\hat{y}$$
$$|\vec{R}_{31}|^2 = 29$$
$$\hat{R}_{31} = 0.9285\hat{x} - 0.3714\hat{y}$$

Then, for the force of the 3-C charge on the 1-C charge:

$$\vec{F}_{31} = 8.63 \times 10^8\hat{x} - 3.45 \times 10^8\hat{y} \text{ N}$$

Using the superposition theorem,

$$\vec{F}_1 = \vec{F}_{21} + \vec{F}_{31} = 3.84 \times 10^8\hat{x} - 11.44 \times 10^8\hat{y} \text{ N}$$

From this example, one can see that the forces resulting from charges in the coulomb value range are very large. In practice, the amount of charge is frequently in the range of 10^{-3} to 10^{-6} C (except for large energy-storage systems for high-energy plasma generation). To avoid having to write the powers of ten, a system of prefixes has been standardized for use. Table 5-1 is a summary of these prefixes. Notice particularly that, when written out, all prefixes start with lowercase letters, even if the symbol is a capital letter.

Exercises

5-2.0a Devise an experiment that can be used to demonstrate that the force between two charged particles decreases as a dielectric is placed between them.

5-2.0b For Example 5-1, determine the angle that the force $\vec{F}_1$ makes with the positive x axis. (Answer: $-71.4°$)

5-2.0c Assuming that the charge configuration shown in Figure 5-5 exists in vacuum, determine the force on the 2-C charge and sketch the force direction on the diagram. Convert your answer into cylindrical and spherical coordinates. Repeat for the 1-C charge.

Table 5-1. Prefixes for Power-of-Ten Presentations of Large and Small Quantities

Factor by Which Unit Is Multiplied	Prefix	Symbol	Comments
10^{18}	exa	E	pronounced ĕx′ŭ
10^{15}	peta	P	pronounced pā′tŭ
10^{12}	tera	T	pronounced tĕ′rŭ
10^{9}	giga	G	pronounced jĭg′ŭ
10^{6}	mega	M	
10^{3}	kilo	k	
10^{-2}	centi	c	not recommended
10^{-3}	milli	m	
10^{-6}	micro	μ	
10^{-9}	nano	n	pronounced năn′ō
10^{-12}	pico	p	pronounced pē̄′cō
10^{-15}	femto	f	pronounced fĕm′tō
10^{-18}	atto	a	pronounced ăt′ō

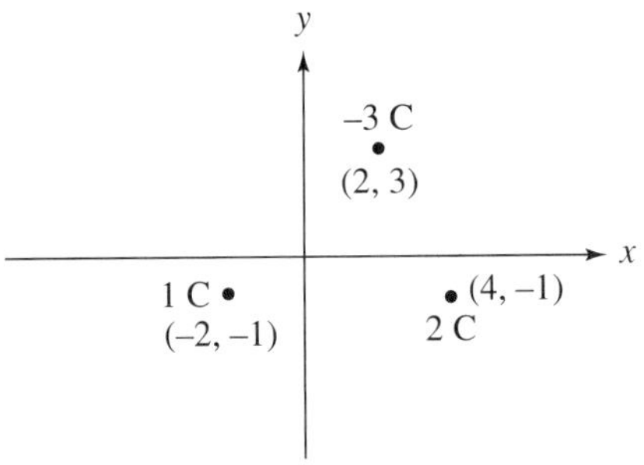

Figure 5-5.

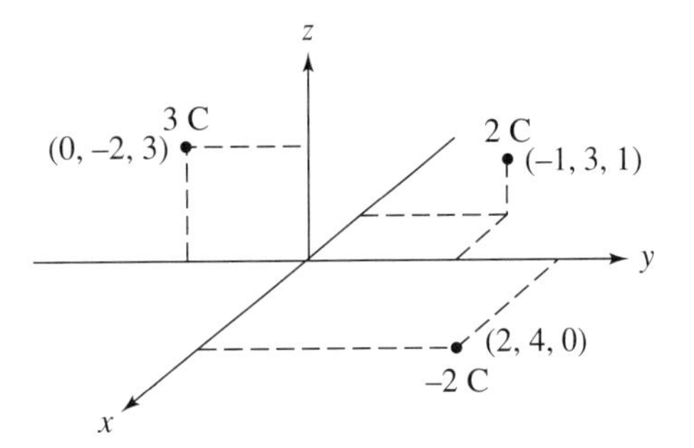

Figure 5-6.

5-2.0d Assuming that the charge configuration shown in Figure 5-6 exists in vacuum, determine the force on the 3-C charge and sketch the force components at that charge. Convert your answer into cylindrical and spherical coordinates and sketch these components at the charge. Use separate diagrams for the three sketches.

5-2.0e Develop a computer program that will compute the force on a selected charge in a given point charge configuration. If a graphics terminal is available, set up a program that will plot the configuration and show the force at the selected charge. Solve Exercises 5-2.0c and 5-2.0d using your program.

5-2.0f Suppose that, instead of vacuum, the charge configurations in Exercises 5-2.0c and 5-2.0d were in a dielectric that has $\epsilon_r = 4$. What are the forces now (give values)? (Answer: Forces are decreased by a factor of 4)

5-2.0g Two charged spheres, each of mass 0.001 kg and +2-C charge, are oriented as shown in Figure 5-7, with the z axis being perpendicular to the earth. The surrounding medium is vacuum. At what separation, D, could equilibrium theoretically occur where the upper charge is at rest if the spheres are considered point charges? Now suppose we had a hypothetical dielectric with $\epsilon_r = 6$, in which the upper charge could move. Now what is the equilibrium distance D? To get an idea of permissible sphere size, what would the allowable sphere radius of the upper sphere be if the equilibrium distance D is not to be more than 1% different from the distance from the center of the lower sphere to the lower edge of the upper sphere for the two cases? Repeat for sphere charges of 2 μC.

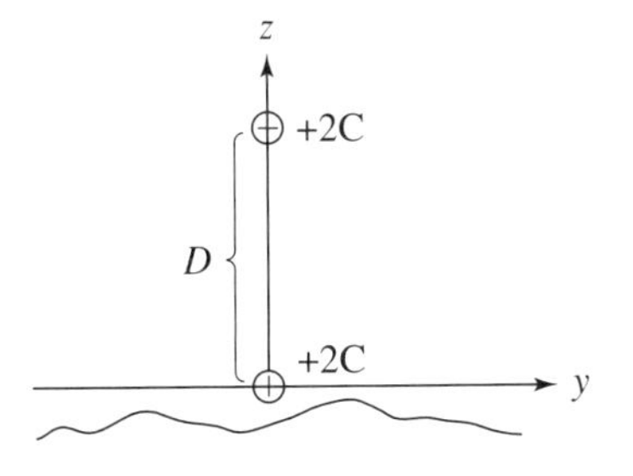

Figure 5-7.

A qualitative explanation for the fact that the force between charges decreases in the presence of materials can be developed from a consideration of the effect of an isolated charge on an atom or molecule. The sequence of the discussion is shown in Figure 5-8. In Figure 5-8*a* the atom has a circular orbit so that neutrality exists at points in the region. When a positive charge is brought near the atom, the nucleus is repelled and the orbit of the electron is pulled toward the charge and is elongated as shown in Figure 5-8*b*. This orbit elongation results in a net negative region to the right of the nucleus. This effect can be represented by a small dipole as shown in Figure 5-8*c*, where the charges are average values over the orbit and the distance between them an average displacement. The displacement and charge values will, of course, be functions of the magnitude of $+q$ and the distance of the atom from $+q$. Such atoms are said to be *polarized*. (This will be quantified in a later section.) A similar qualitative view can be given for molecules that are polarized.

Now suppose that we embed the charge $+q$ in a dielectric material where the atoms (molecules) become polarized as shown in Figure 5-8*d*. Each polarized unit (dipole) becomes polarized to an extent that depends upon the proximity of the charge to the atom, those closer being ''stretched'' more than those farther removed. Notice also that the boundary surface of the material has dipoles that form an effective surface charge of the same polarity as the embedded charge. This is a *bound* charge, however, as opposed to the $+q$ charge, which is mobile. These bound charges may move in a small region at the site of the atom but generally not through the material as a free charge. (Some atoms may

Figure 5-8. Interaction between an atom and a point charge and the effect within a dielectric. (*a*) Isolated atom. (*b*) Orbit perturbation by a point charge. (*c*) Net dipole of elongated orbit. (*d*) Point charge surrounded by many atoms.

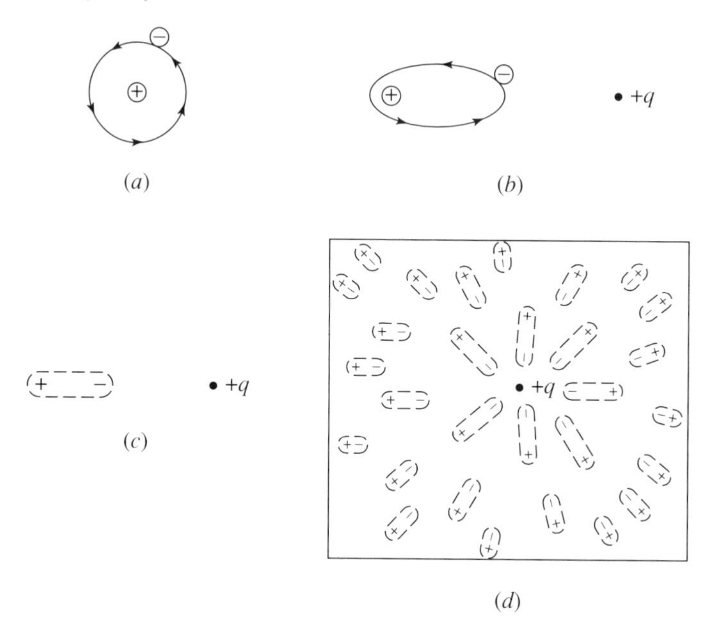

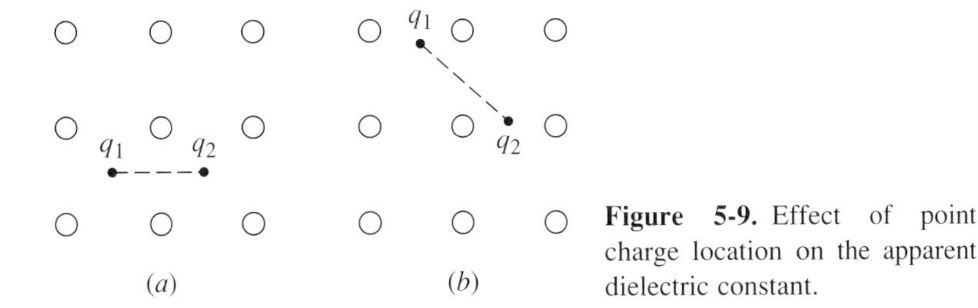

Figure 5-9. Effect of point charge location on the apparent dielectric constant.

have the capability of giving up a charge (electron) that can move through the material as a conduction electron, but this does not increase the net charge of the system.) The effect of the bound charges will be considered in more detail in a following section.

To return to the main item for discussion, we observe that although the material as a whole has a net $+q$ charge, the effect of the adjacent dipoles is to diffuse the point charge and ''neutralize'' it locally due to the close negative charge. Now if a second charge q_2 were placed into the material, the force between the charges would be reduced due to this diffusion effect. This force reduction is accounted for by the relative permittivity ϵ_r.

Most dielectrics do not have a single value of ϵ_r, and there are materials that behave differently from the simple picture given here. The dielectric ''constant'' is usually a function of the frequency of an electric field and also temperature. The variation of ϵ_r is caused by the fact that the ability of an atom (molecule) to polarize is a function of these parameters. If we imagine, for example, that the value of q were varied with time, the amount of polarization of the adjacent molecules would change. If the value of q changed very rapidly, the polarization may not even be able to follow the fluctuations of q due to the inertia (mass) of the dipole components. In this case, the material may appear nearly as vacuum since it cannot polarize rapidly enough, and ϵ_r would be near unity!

An additional dielectric effect in materials can be explained by considering two point charges placed in a latticelike dielectric structure, as shown in Figure 5-9. It should seem intuitively reasonable that the force between the two charges in Figure 5-9*a* would be different from the force between the same two charges oriented as in Figure 5-9*b*. Effectively, the two charges are ''looking'' at each other through different dimensions of the crystal, and hence through different polarization capabilities. Such materials are said to be *anisotropic*, and to represent ϵ_r we must give a different value for each direction. We use a matrix form as

$$[\epsilon_r] = \begin{bmatrix} \epsilon_{11} & \epsilon_{21} & \epsilon_{31} \\ \epsilon_{12} & \epsilon_{22} & \epsilon_{32} \\ \epsilon_{13} & \epsilon_{23} & \epsilon_{33} \end{bmatrix} \tag{5-7}$$

This is called the *tensor* or *matrix* relative permittivity. As before, each of the elements ϵ_{ij} is a function of frequency, etc., as with the scalar ϵ_r. Such materials may still be linear, of course, since doubling the value of q may still double the force in a given direction.

Thus the superposition theorem still applies. For rectangular coordinates, the subscript 1 is x, 2 is y and 3 is z.

Example 5-2

Find the ratio of the forces between two charges in vacuum and in a dielectric material. (Assume material is isotropic.) Use constant separation between charges.

$$\vec{F}_0 = \frac{q_1 q_2}{4\pi\epsilon_0|\vec{R}_{12}|^2}\hat{R}_2 \qquad \vec{F}_d = \frac{q_1 q_2}{4\pi\epsilon|\vec{R}_{12}|^2}\hat{R}_2$$

Then

$$\frac{|\vec{F}_0|}{|\vec{F}_d|} = \frac{\epsilon}{\epsilon_0} = \frac{\epsilon_0\epsilon_r}{\epsilon_0} = \epsilon_r$$

as expected, since this is how we entered the dielectric effect into the force equation. If ϵ_r were 10, the force between the charges in the dielectric would be reduced by a factor of 10 compared with the force in vacuum.

Example 5-3

Assuming that electrostatic forces only are significant and that a dipole may be idealized by equal-value positive and negative electron charges as shown in Figure 5-10, determine the equilibrium separation distances. The symbol e is the electron charge magnitude 1.6×10^{-19} C. Assume free space.

Since both distance vectors are directed along the x axis, we have

$$\vec{R}_1 = R_1\hat{x} \quad \text{and} \quad \vec{R}_2 = R_2(-\hat{x}) = -R_2\hat{x} = -(10^{-3} - R_1)\hat{x}$$

Then

$$\vec{R}_1 = \hat{x} \quad \text{and} \quad \hat{R}_2 = -\hat{x}$$

The total force on the $-e$ charge is then

$$\vec{F}_{-e} = \vec{F}_{q,-e} + \vec{F}_{e,-e} = \frac{-1.6 \times 10^{-19} \times 9 \times 10^{-18}}{4\pi\epsilon_0(10^{-3} - R_1)^2}(-\hat{x})$$

$$+ \frac{-1.6 \times 10^{-19} \times 1.6 \times 10^{-19}}{4\pi\epsilon_0 R_1^2}\hat{x}$$

$$= \left[\frac{1.296 \times 10^{-26}}{(10^{-3} - R_1)^2} - \frac{2.304 \times 10^{-28}}{R_1^2}\right]\hat{x}$$

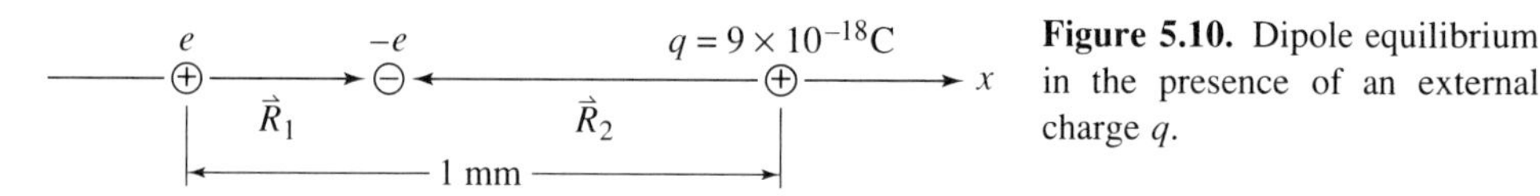

Figure 5.10. Dipole equilibrium in the presence of an external charge q.

For equilibrium the force must be zero, so that

$$\frac{1.296 \times 10^{-26}}{(10^{-3} - R_1)^2} = \frac{2.304 \times 10^{-28}}{R_1^2}$$

Rearranging results in

$$R_1^2 + 3.62 \times 10^{-5} R_1 - 1.81 \times 10^{-8} = 0$$

from which

$$R_1 = 0.118 \text{ mm}$$

and

$$R_2 = 0.882 \text{ mm}$$

Exercises

5-2.0h For Example 5-3, compute the forces on $-e$ due to each of the other charges. (They are equal and opposite.)

5-2.0i Determine the force on charge q for Example 5-3 for the case where $-e$ is removed, distances remaining the same. Next determine the force on q due to the $-e$ alone with e removed.

What is the effect of the dipole on the forces seen by q due to each separate charge? What is the value of a single charge located midway between e and $-e$, which would replace the dipole as far as the force experienced by q is concerned?

5-2.0j Suppose that we have two charges $+q$ of value 9×10^{-18} C, each located 3 mm apart as shown in Figure 5-11. The dipole e, $-e$ is placed between the colinear with the q and the dipole center is $\frac{1}{3}$ the distance from the right q. Determine the equilibrium distance between the dipole charges.

Calculate the forces on the right-hand charge q with and without the dipole present. For this hypothetical situation, what would be the value of ϵ_r, the relative permittivity? Repeat for the case where the two dipole $\pm e$ charges are reversed in position. Comment on this computation.

5-2.0k For Example 5-3, suppose that the charge q moves *vertically* past the dipole starting at $-\infty$ and moving to $+\infty$, passing through the point shown in Figure 5-10. Explain the effect of this motion on the dipole charge separation, dipole orientation, and effective single charge equivalent of the dipole.

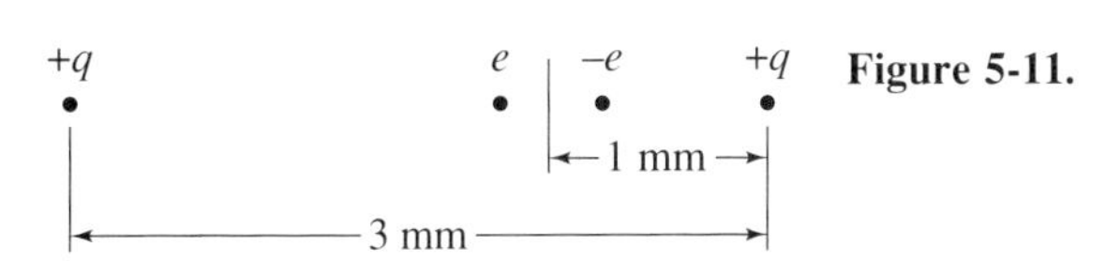

Figure 5-11.

5-3 Electric Field Intensity: Introduction

In both of the experiments considered in the first part of the preceding section, force was the main effect described. The forces, however, were produced by the charges, so we would like to define a field quantity produced by the charges that can account for those forces. Furthermore, since force is a vector, we shall let the field quantity have the same spatial vector direction as the force at the point or points of interest. For purposes of being able to visualize the field quantity and to draw physical systems, we use lines that describe the direction of the force at each point in the system.

The electric field intensity (also called the *electric field strength*) at a point is defined as the force per unit positive charge at that point. As described earlier, since force is a vector the electric field intensity is also a vector in that direction. As a simple definition equation, we could express this by

$$\vec{E} \triangleq \frac{\text{force}}{+\text{unit charge}}, \qquad \text{in units of newtons/coulomb} \tag{5-8}$$

Obviously we can't put 1 C of charge at the point since that would disturb the field. A more precise definition would be

$$\vec{E} \triangleq \lim_{+q \to 0} \frac{\vec{F}}{q} \text{ N/C} \tag{5-9}$$

Note that this definition also gives us a visual way of making a qualitative plot of the field lines. One simply imagines wearing a positively charged coat and then jumping into the field. Then the path one is pushed along is the field line.

Example 5-4

Determine the direction of the electric field at point P within the two plates shown in Figure 5-12*a*.

Figure 5-12. Qualitative determination of the electric field at a point. (*a*) Basic system. (*b*) Test charge at *P*.

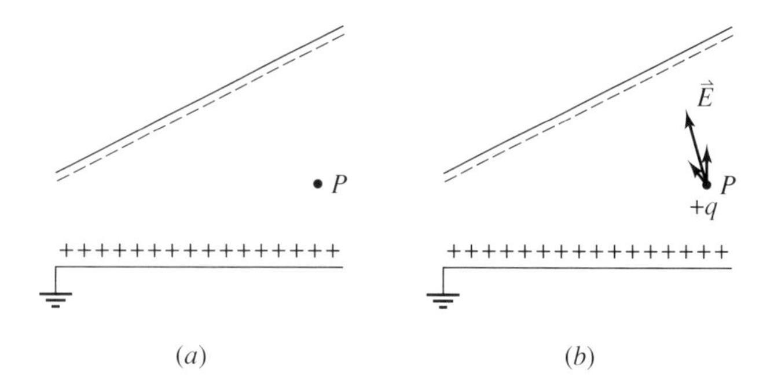

We imagine a positive q placed at point P as shown in Figure 5-12b. The charge will be attracted to the negatively charged plate and pushed from the positively charged plate (ground plate) in the two directions shown. Adding these two forces gives the resulting direction of $\vec{E}$ at P. Note that the vertical force due to the positive plate is drawn larger than the other force because q is closer to the positive plate. (Here we have assumed that the charges per unit area on the plates are the same.)

Example 5-5

For the system shown in Figure 5-13a, use the definition of $\vec{E}$ to obtain a qualitative sketch of the field line that passes through point P. The charge $+Q$ and the charged plate constitute the system.

The technique is to imagine a small positive test charge at P and then trace its path as it is pushed by the charge system at successive points, as it moves incrementally in the force produced by the field.

Beginning at P, we indicate the effects of the sheet and point charge Q. The sum of these indicates the direction of motion of the small test charge. Note that the dominant force will be due to the $+Q$ when P is near that charge. This is shown in Figure 5-13b. We then move a small distance in that direction as shown in Figure 5-13c and repeat the process. Then connect all these locations and the result defines one field line of the system as shown in Figure 5-13f.

Figure 5-13. Qualitative determination of a field line in a charge system. (a) Basic system. (f) Solution for one $\vec{E}$ field line.

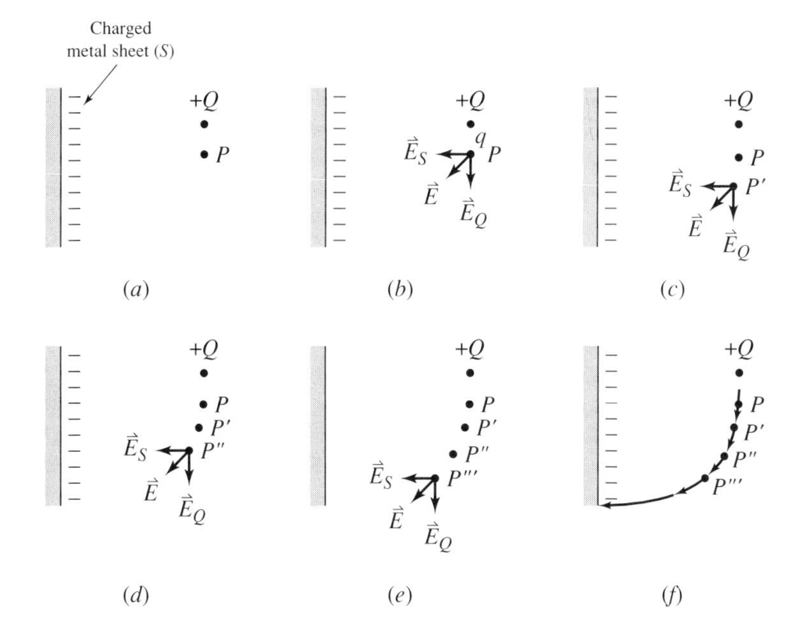

We next develop some quantitative results for computing values of electric field intensity. From the basic definition (Eq. 5-9) it is an easy matter to compute $\vec{E}$ for a point charge. Suppose we have a point charge Q located at $\vec{r}\,'$ from the origin as shown in Figure 5-14. The distance vector $\vec{r}\,'$ is called the *source point vector* since it locates the point in space at which the source that causes the field is found. The distance vector $\vec{r}$ is called the *field point vector* since it locates the point in space P at which the field is to be computed. Note also that $\vec{r}$ and $\vec{r}\,'$ are *independent* since any point P can be selected for a given point charge location $\vec{r}\,'$. We now place a positive test charge q at location $P(\vec{r}\,)$ and use Coulomb's law to obtain the force on q as

$$\vec{F}_q = \frac{qQ}{4\pi\epsilon|\vec{R}|^2}\hat{R} = \frac{qQ}{4\pi\epsilon|\vec{R}|^3}\vec{R} \qquad \left(\hat{R} = \frac{\vec{R}}{|\vec{R}|}\right) \tag{5-10}$$

Since $\vec{R} = \vec{r} - \vec{r}\,'$ this may be written

$$\vec{F}_q = \frac{qQ}{4\pi\epsilon|\vec{r} - \vec{r}\,'|^3}(\vec{r} - \vec{r}\,')$$

Then

$$\vec{E} \text{ at P} \equiv \vec{E}(\vec{r}\,) = \lim_{q\to 0}\frac{\vec{F}_q}{q} = \frac{Q}{4\pi\epsilon|\vec{r} - \vec{r}\,'|^3}(\vec{r} - \vec{r}\,') \tag{5-11}$$

Note that for Q positive $\vec{E}$ is directed from Q toward P. In spite of the presence of the cube in the denominator, this is basically an inverse square function since one of the magnitude quantities was used to generate the unit vector $\hat{R}$. Also, the symbolism $\vec{E}(\vec{r}\,)$ is a generic way of indicating that $\vec{E}$ is a function of location in space and can represent any coordinate system used to locate P. For example, in rectangular and cylindrical coordinates we would write the arguments more explicitly as

$$\vec{E}(x, y, z) \qquad \text{and} \qquad \vec{E}(\rho, \phi, z)$$

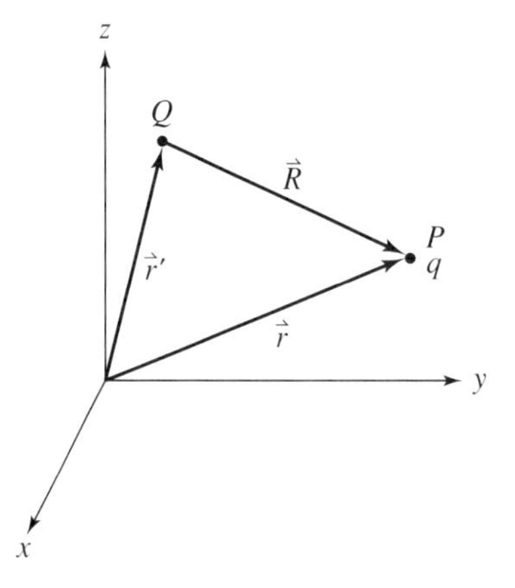

Figure 5-14. Configuration for determining the electric field of a point charge Q.

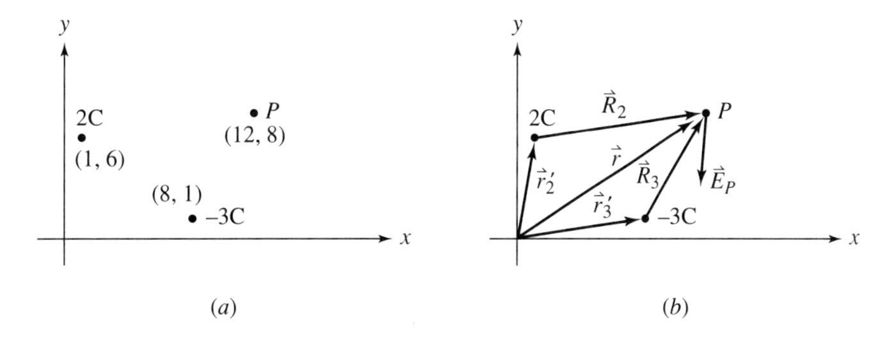

Figure 5-15. Charge configuration for Example 5-6.

As with forces in linear materials, the electric field intensity satisfies the superposition theorem so that the total electric field (force per unit positive charge) at a point is the sum of the electric fields produced by each individual point charges. This, of course, must be a vector sum. Thus for n point charges,

$$\vec{E}_{\text{total}} = \sum_{i=1}^{n} \vec{E}_i \tag{5-12}$$

where all of the $\vec{E}_i$ are evaluated at the same point P in space.

The effect of the dielectric constant (relative permittivity) can be determined by writing ϵ specifically in terms of ϵ_r as

$$\vec{E}(\vec{r}) = \frac{Q}{4\pi\epsilon_0\epsilon_r|\vec{r} - \vec{r}'|^3}(\vec{r} - \vec{r}')$$

Thus for a given charge Q the electric field intensity will be smaller at a given point P than the value in vacuum. This is true since the dielectric has the effect of reducing the value of Q as explained earlier. The other point of view is that since $\vec{E} \propto \vec{F}$, then $\vec{E}$ is reduced since $\vec{F}$ is reduced.

Example 5-6

Charges of $+2$ C and -3 C are located as shown in Figure 5-15*a*. Find the total electric field intensity at point P. Also, indicate the approximate direction of $\vec{E}$ at P.

We first draw the location vectors to point P as indicated in Figure 5-15*b*. Using Eqs. 5-11 and 5-12, we have

$$\vec{E}(\vec{r}) = \vec{E}(12, 8) = \vec{E}_2(12, 8) + \vec{E}_3(12, 8) = \frac{2}{4\pi\epsilon|\vec{R}_2|^3}\vec{R}_2 + \frac{-3}{4\pi\epsilon|\vec{R}_3|^3}\vec{R}_3$$

$$= \frac{1}{4\pi\epsilon}\left[\frac{2}{|\vec{r} - \vec{r}_2'|^3}(\vec{r} - \vec{r}_2') - \frac{3}{|\vec{r} - \vec{r}_3'|^3}(\vec{r} - \vec{r}_3')\right]$$

$$= \frac{1}{4\pi\epsilon}\left[\frac{2((12\hat{x} + 8\hat{y}) - (1\hat{x} + 6\hat{y}))}{|(12\hat{x} + 8\hat{y}) - (1\hat{x} + 6\hat{y})|^3} - \frac{3((12\hat{x} + 8\hat{y}) - (8\hat{x} + 1\hat{y}))}{|(12\hat{x} + 8\hat{y}) - (8\hat{x} + 1\hat{y})|^3}\right]$$

$$= \frac{1}{4\pi\epsilon}\left[\frac{22\hat{x} + 4\hat{y}}{1397.5} - \frac{12\hat{x} + 21\hat{y}}{524}\right]$$

$$= \frac{1}{4\pi\epsilon}(-0.007\hat{x} - 0.037\hat{y}) = E_x\hat{x} + E_y\hat{y}$$

The direction of $\vec{E}$ is indicated on the diagram, a minus $\hat{x}$ component and a minus $\hat{y}$ component. Physically, this means that a positive charge at P would move in the direction of $\vec{E}$.

Exercises

5-3.0a Charges of -2 C and $+3$ C are located as indicated in Figure 5-16*a*. Find the total electric field intensity (field strength) at $P(2, -5)$. Also indicate, roughly, the direction of $\vec{E}$ at P. Assume vacuum. If a charge of $\frac{1}{2}$ C were placed at P, what force would it experience? (Get this value from $\vec{E}$.) What would be the force on $-\frac{1}{2}$ C at P?

5-3.0b Charges of $+2$ C and -3 C are located as shown in Figure 5-17. Find the total electric field at $P(-4, -4)$, and sketch the direction of $\vec{E}$ there. Assume vacuum.

5-3.0c For the systems shown in Figure 5-18 determine, qualitatively, the direction and path of the field lines passing through the points P.

5-3.0d Since the electric field intensity is defined as the force per unit charge, work is expended (or absorbed, depending upon the sign of the charge) in moving a charge in such a field. Suppose we have the charge configuration shown in Figure 5-19 and we move a 1-C charge from A to B directly. Find the total work done on the charge for that path. Repeat for the path $A \to C \to B$. From Chapter 4, what is the condition that must be satisfied if a line integral value is indepedent of the path? Make the test. (Hints: First write the equation for $\vec{E}(x, y)$ for a general point in the first quadrant. Then $dw = \vec{F} \cdot dl = q\vec{E} \cdot dl$ where q is the moving charge. Then use the integration techniques of Chapter 4.)

Figure 5-16.

Figure 5-17.

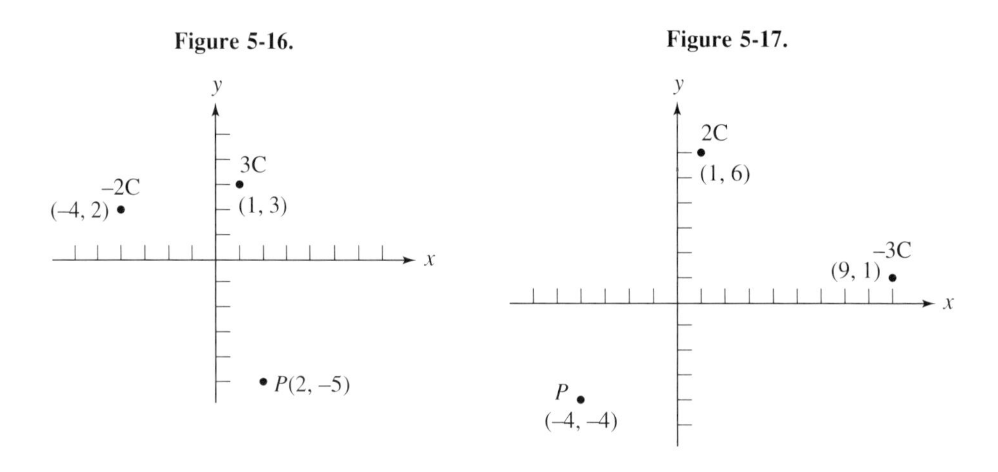

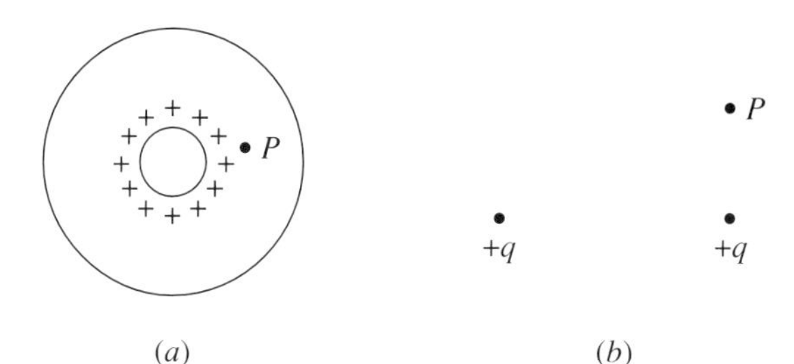

Figure 5-18. (*a*) Concentric cylinders. (*b*) Two point charges.

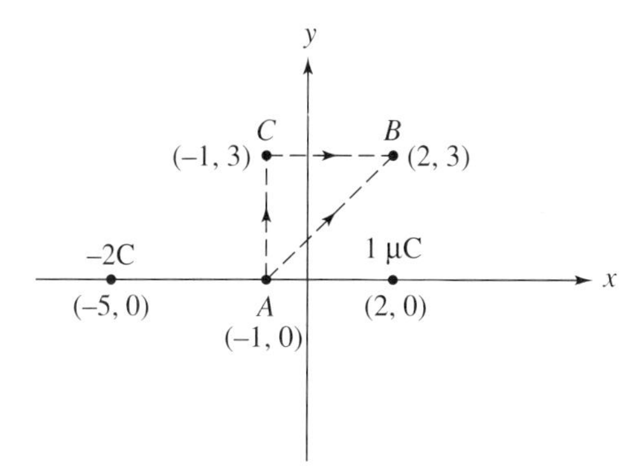

Figure 5-19.

5-4 Electric Field Intensity: Continuous Charge Distributions

In many systems of engineering interest we deal with continuous distributions of charge rather than with point charges, so we need some way to express the distributions as equations. To this end we define the following:

Line charge density: $$\rho_l(\vec{r}) = \lim_{\Delta l \to 0} \left(\frac{\Delta q}{\Delta l}\right) \text{ coulomb/meter} \qquad (5\text{-}13)$$

Surface charge density: $$\rho_s(\vec{r}) = \lim_{\Delta s \to 0} \left(\frac{\Delta q}{\Delta s}\right) \text{ coulomb/square meter} \qquad (5\text{-}14)$$

Volume charge density: $$\rho_v(\vec{r}) = \lim_{\Delta v \to 0} \left(\frac{\Delta q}{\Delta v}\right) \text{ coulomb/cubic meter} \qquad (5\text{-}15)$$

In these we again have used the argument list symbol $\vec{r}$ to represent, generically, the fact that the charge may be a function of coordinate variables as discussed in the preceding section.

These three types of charge ''density'' would seem to require us to develop three different expressions for $\vec{E}$, but very frequently we can use the delta function described in Chapter 4 to express all of them as equivalent volume charge densities. For example, a surface charge density may have a delta function in one dimension (coordinate) since the charge is zero everywhere except on the sheet or surface. Thus we may express an equivalent volume charge density as, for the coordinate surface $u_i = u_{i0}$ (constant),

$$\rho_v = \rho_s \frac{\delta(u_i - u_{i0})}{h_i(u_{i0})} \text{ C/m}^3 \tag{5-16}$$

(See Sections 4-12 and 4-21. $h_i(u_{i0})$ is the scale factor (metric coefficient) of coordinate u_i evaluated at the coordinate position u_{i0}.) This is a plausible equation since it yields zero charge value except on the coordinate surface $u_i = u_{i0}$, which is on the sheet. As an example, suppose that in the rectangular coordinate system the plane, parallel to the xy plane and passing through the z axis at $z = 3$, has a surface charge density given by

$$\rho_s(\vec{r}) = \rho_s(x, y, z) = 3x^2 + y - 2z \text{ C/m}^2$$

This could be expressed as a volume charge density in the form for rectangular coordinates with $h_z = 1$:

$$\rho_v(\vec{r}) = \rho_v(x, y, z) = \rho_s(x, y, z)\ \delta(z - 3) = (3x^2 + y - 2z)\ \delta(z - 3) \text{ C/m}^3$$

One might also wonder about the units on this equation and Eq. 5-16 since the left sides are coulomb per cubic meter, whereas the surface charge is coulomb per square meter. In Chapter 4 it was shown that multiplying by a delta function (and the associated scale factor) of space coordinates has the effect of increasing the dimensionality by 1/meter, thus making both sides consistent.

Now suppose we have a volume charge density which is a function of coordinates, $\rho_v(\vec{r}')$, as shown in Figure 5-20. Actually, all one has to do is write the equation for ρ_v in the usual (''unprimed'') coordinate system and then simply put a prime on each coordinate. As before, we use the primes to denote a general point location in the region that contains the source (in this case charge) of the external field (i.e., source points) and the unprimes to locate a general point at which the field is to be computed (i.e., field point). We imagine the differential volume to contain an increment of charge

$$dq = \rho_v(\vec{r}')dv'$$

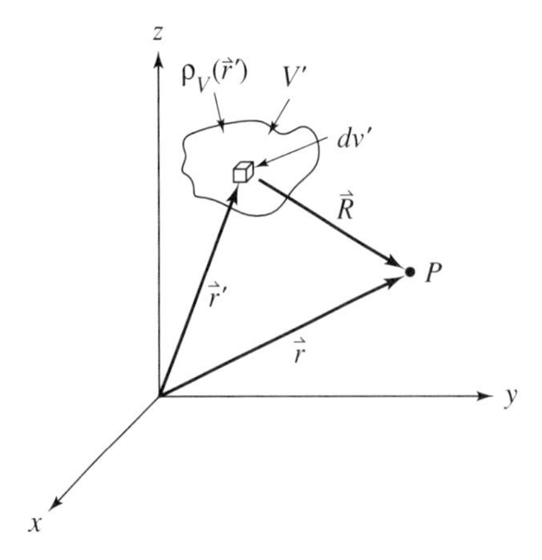

Figure 5-20. General conventions for determining the electric field intensity external to a volume charge density.

where dv' is a differential volume written in prime coordinate notation. (See Equation 4-29 with primes placed on all coordinate variables in the expression, including coordinates in the h's.) Then, using Equation (5-10) for this point charge,

$$d\vec{E}(\vec{r}) = \frac{dq}{4\pi\epsilon|\vec{R}|^2}\hat{R} = \frac{\rho_v(\vec{r}\,')dv'\hat{R}}{4\pi\epsilon|\vec{R}|^2}$$

Integrating, we obtain the total field as

$$\vec{E}(\vec{r}) = \int_{V'} \frac{\rho_v(\vec{r}\,')\hat{R}}{4\pi\epsilon|\vec{R}|^2}\,dv', \qquad \text{with} \quad \vec{R} = \vec{r} - \vec{r}\,' \tag{5-17}$$

It is important to notice that the unit vector $\hat{R}$ is *inside* the integral since it will, in general, change direction as the primed coordinates move over the volume V'. One should also keep in mind that the *unprimed* coordinates are considered *constants*.

The examples that follow present ideas that will also be useful in solving problems for time-varying fields. For the first example, all of the steps will be carried out so that the fine points of the general solution can be developed.

Example 5-7

A line of charge on the z axis extends from $-L$ to $+L$ and has a uniform (constant) charge density ρ_{l0} C/m along its length. Find the electric field intensity at point P as indicated in Figure 5-21.

Although we could use the line integration easily in this case, we will use Eq. 5-17 to show the application of the delta function, step function, and primed coordinates. Also we could, without loss in generality, take the point P to be in either the xz or yz plane, but, again, we want to emphasize some of the ideas involved in evaluating the integrals.

Using the techniques of Chapter 4, we write

$$\rho_v(\vec{r}) = \rho_v(x, y, z) = \rho_{l0}\ \delta(x)\ \delta(y)U(L - z)U(L + z)\ \text{C/m}^3$$

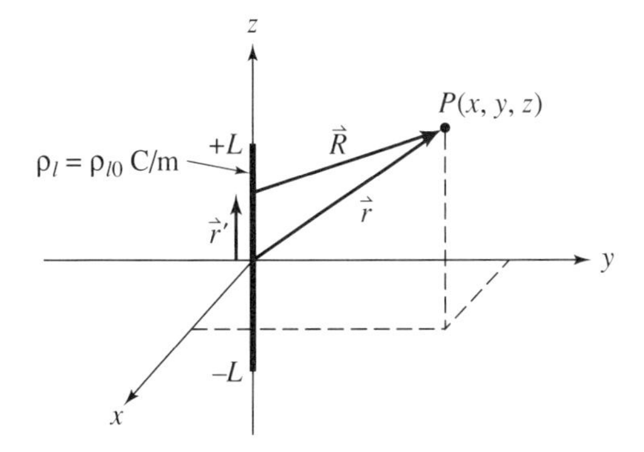

Figure 5-21. Configuration for the determination of the electric field intensity of a finite length of charged line.

In primed coordinates we have then

$$\rho_v(\vec{r}\,') = \rho_v(x', y', z') = \rho_{l0}\ \delta(x')\ \delta(y')U(L - z')U(L + z')\ \text{C/m}^3$$

Since each delta function puts a 1/m dimension onto ρ_l, the dimension is per unit volume. Note that the expression for ρ_v is correct for any values of x', y', and z', so we may integrate over all space.

From the figure

$$\vec{r} = x\hat{x} + y\hat{y} + z\hat{z}$$

which locates any point in space, and

$$\vec{r}\,' = z'\hat{z}$$

which locates a general point in the charge distribution. (Actually, one could have used $\vec{r}\,' = x'\hat{x}' + y'\hat{y}' + z'\hat{z}'$ and still be correct because the delta functions will automatically reduce this to the simpler expression.) From these last two expressions

$$\vec{R} = \vec{r} - \vec{r}\,' = x\hat{x} + y\hat{y} + z\hat{z} - z'\hat{z}'$$

Now rectangular coordinates is a particularly nice coordinate system since no matter where the points are taken, $\hat{z}$ and $\hat{z}'$ always point the same direction, even if the prime and umprimed are two different points, so we may replace $\hat{z}'$ by $\hat{z}$. The vector $\vec{R}$ then becomes

$$\vec{R} = x\hat{x} + y\hat{y} + (z - z')\hat{z}$$

Then

$$|\vec{R}|^2 = \vec{R}\cdot\vec{R} = x^2 + y^2 + (z - z')^2$$

and

$$\hat{R} = \frac{\vec{R}}{|\vec{R}|} = \frac{x\hat{x} + y\hat{y} + (z - z')\hat{z}}{\sqrt{x^2 + y^2 + (z - z')^2}}$$

Also, for rectangular coordinates

$$dv' = dx'dy'dz'$$

with these, Eq. 5-17 becomes

$$\vec{E}(\vec{r}) = \vec{E}(x, y, z)$$
$$= \frac{1}{4\pi\epsilon}\int_{-\infty}^{\infty}\int_{-\infty}^{\infty}\int_{-\infty}^{\infty}\frac{\rho_{l0}\ \delta(x')\ \delta(y')U(L + z')U(L - z')[x\hat{x} + y\hat{y} + (z - z')\hat{z}]}{[x^2 + y^2 + (z - z')^2]\sqrt{x^2 + y^2 + (z - z')^2}}\,dx'dy'dz'$$

Expanding the binomial square terms and recognizing that $x^2 + y^2 + z^2$ is r^2, the preceding becomes (for ρ_{l0} constant)

$$\vec{E}(x, y, z) = \frac{\rho_{l0}}{4\pi\epsilon}\int_{-\infty}^{\infty}\int_{-\infty}^{\infty}\int_{-\infty}^{\infty}\frac{\delta(x')\ \delta(y')U(L + z')U(L - z')[x\hat{x} + y\hat{y} + (z - z')\hat{z}]}{(r^2 - 2zz' + z'^2)^{3/2}}\,dx'dy'dz'$$

Next we evaluate the x' and y' integrations using the concepts of Section 4-21 for delta functions:

$$\vec{E}(x, y, z) = \frac{\rho_{l0}}{4\pi\epsilon} \int_{-\infty}^{\infty} \frac{U(L + z')U(L - z')[x\hat{x} + y\hat{y} + z\hat{z} - z'\hat{z}]}{(r^2 - 2zz' + z'^2)^{3/2}}\, dz'$$

From the properties of step functions, $U(L + z')$ is zero for $z' < -L$ and $U(L - z')$ is zero for $z' > L$, so the integral reduces finally to

$$\vec{E}(x, y, z) = \frac{\rho_{l0}}{4\pi\epsilon} \int_{-L}^{L} \frac{x\hat{x} + y\hat{y} + z\hat{z} - z'\hat{z}}{(r^2 - 2zz' + z'^2)^{3/2}}\, dz'$$

This is what one may have expected since a line of charge should be integrated over the extent of the charge. We next write the preceding equation as the sum of four integrals and use the fact that all unprimed variables are constants (this includes unprimed unit vectors) so they may be removed from the integrals:

$$\vec{E}(x, y, z) = \frac{\rho_{l0}}{4\pi\epsilon} \left[x \int_{-L}^{L} \frac{1}{(r^2 - 2zz' + z'^2)^{3/2}}\, dz'\hat{x} + y \int_{-L}^{L} \frac{1}{(r^2 - 2zz' + z'^2)^{3/2}}\, dz'\hat{y} \right.$$
$$\left. + z \int_{-L}^{L} \frac{1}{(r^2 - 2zz' + z'^2)^{3/2}}\, dz'\hat{z} - \int_{-L}^{L} \frac{z'}{(r^2 - 2zz' + z'^2)^{3/2}}\, dz'\hat{z} \right]$$

Factoring out the common integral in the first three terms:

$$\vec{E}(x, y, z) = \frac{\rho_{l0}}{4\pi\epsilon} \left[(x\hat{x} + y\hat{y} + z\hat{z}) \int_{-L}^{L} \frac{dz'}{(r^2 - 2zz' + z'^2)^{3/2}} - \int_{-L}^{L} \frac{z'dz'}{(r^2 - 2zz' + z'^2)^{3/2}}\, \hat{z} \right] \tag{5-18}$$

Since the unprimed variables are treated as constants, we use integral tables and look for integrals having the cube of a quantity $\sqrt{a + bz' + z'^2}$ in the denominator (i.e., 3/2 power). Then

$$\vec{E}(x, y, z) = \frac{\rho_{l0}}{4\pi\epsilon(r^2 - z^2)} \left[(x\hat{x} + y\hat{y} + z\hat{z}) \left(\left. \frac{z' - z}{\sqrt{r^2 - 2zz' + z'^2}} \right|_{-L}^{L} \right) + \left(\left. \frac{r^2 - zz'}{\sqrt{r^2 - 2zz' + z'^2}} \right|_{-L}^{L} \right) \hat{z} \right]$$

Putting in the limits of z', we have finally

$$\vec{E}(x, y, z) = \frac{\rho_{l0}}{4\pi\epsilon(r^2 - z^2)} \left[(x\hat{x} + y\hat{y} + z\hat{z}) \left(\frac{L - z}{\sqrt{r^2 - 2zL + L^2}} + \frac{L + z}{\sqrt{r^2 + 2zL + L^2}} \right) + \left(\frac{r^2 - zL}{\sqrt{r^2 - 2zL + L^2}} - \frac{r^2 + zL}{\sqrt{r^2 + 2zL + L^2}} \right) \hat{z} \right] \text{N/C} \tag{5-19}$$

An interesting result is obtained if we evaluate the electric field at points very far away from the line (i.e., for $r \gg L$). In this case

$$\frac{L \pm z}{\sqrt{r^2 \pm 2zL + L^2}} \to \frac{L \pm z}{r}$$

and

$$\frac{r^2 \pm zL}{\sqrt{r^2 \pm 2zL + L^2}} \to r$$

Then for large r, $x\hat{x} + y\hat{y} + z\hat{z} = \vec{r}$

$$\vec{E}(x, y, z) \to \frac{\rho_{l0}}{4\pi\epsilon(r^2 - z^2)}\left[(x\hat{x} + y\hat{y} + z\hat{z})\left(\frac{L - z}{r} + \frac{L + z}{r}\right)\right]$$

$$= \frac{2\rho_{l0}L}{4\pi\epsilon(r^2 - z^2)}\frac{\vec{r}}{r}$$

Since $2\rho_{l0}L$ is the total charge on the line and $\vec{r}/r$ is the unit vector $\hat{r}$, this may be written (for $r^2 = x^2 + y^2 + x^2 \gg z^2$)

$$\vec{E}(x, y, z) = \frac{Q_T}{4\pi\epsilon r^2}\hat{r} \tag{5-20}$$

which means that for distances large compared to the dimensions of the charge distribution, the system looks like a point charge equal to the net total charge on the configuration.

There is one exception, or pathological case, in the general result, Eq. 5-19, that occurs at $x = y = 0$; that is, for field points on the z axis for $z > L$. For such points

$$r^2 - z^2 = x^2 + y^2 + z^2 - z^2 \to 0$$

so that the denominator of the term in front of the brackets becomes zero. It turns out, however, that the terms within the brackets also give zero, so the indeterminant form 0/0 results. To resolve this dilemma we must return to the original solution to the point just before integration, Eq. 5-18. If we set x and y to zero in that equation and then integrate, we obtain

$$\vec{E}(0, 0, z) = \frac{\rho_{l0}}{4\pi\epsilon}\left[z\left(-\frac{1}{-2(z - z')^2}\right) - \left(\frac{-1}{z - z'} + \frac{z}{2(z - z')^2}\right)\right]\Bigg|_{z'=-L}^{L}\hat{z}$$

Putting in the limits gives

$$\vec{E}(0, 0, z) = \frac{2\rho_{l0}L}{4\pi\epsilon(z^2 - L^2)}\hat{z} = \frac{Q_T}{4\pi\epsilon(z^2 - L^2)}\hat{z} \qquad (|z| > L) \tag{5-21}$$

This equation, along with Eq. 5-19, completes the description of the static electric field intensity around a charged line.

Although Eq. 5-17 handles all static situations, it is sometimes convenient to have forms that apply specifically to surface or line distributions of charge, especially when the surfaces or lines do not coincide with a constant coordinate surface or single-variable coordinate line. The equations are

$$\vec{E}(\vec{r}) = \int_{s'} \frac{\rho_s(\vec{r}\,')\vec{R}da'}{4\pi\epsilon|\vec{R}|^2} \tag{5-22}$$

$$\vec{E}(\vec{r}) = \int_{c'} \frac{\rho_l(\vec{r}\,')\vec{R}dl'}{4\pi\epsilon|\vec{R}|^2} \tag{5-23}$$

In these expressions we define the following:

$\vec{r}\,'$:	Vector expressing the distance from the origin to a general point on the charge distribution (source point)
$\vec{r}$:	Vector expressing the distance from the origin to a general point in the region at which the field is to be computed (field point)
$\vec{R}$:	$\vec{r} - \vec{r}\,'$ is the vector expressing the distance from the general source point to the general field point
$\hat{R}$:	Unit vector of $\vec{R}$
dl', da':	General scalar expressions for differential length and area, respectively, of the charge configuration
c', s':	Integration limits for the contour (line) or surface, which includes all the charge; contour is single integration and surface is double integration

---

Example 5-8

This example illustrates how one handles unit vectors in a prime coordinate system which cannot be taken outside of the integration. The configuration is shown in Figure 5-22 and consists of a charged circular plate in the xy plane on which the surface charge density varies with the radial coordinate, ρ. We evaluate the electric field intensity on the z axis.

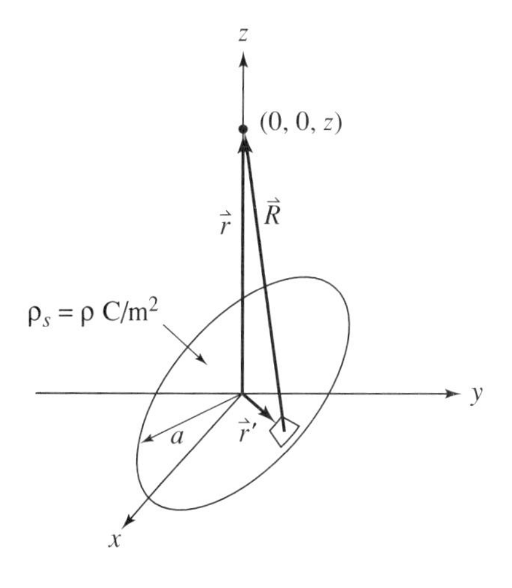

Figure 5-22. Configuration for determining $\vec{E}$ on the central axis of a charged circular sheet.

We could use either Eq. 5-17 or Eq. 5-22. Using the former, we identify the various components of the integral from the figure as follows:

$$\rho_v(\vec{r}\,') = \rho' \ \delta(z')U(a - \rho') \ \frac{\text{C}}{\text{m}^3}$$

$$dv' = \rho' d\rho' d\phi' dz' \text{ (cylindrical; see Chapter 4)}$$

$$\vec{r} = z\hat{z}$$

$$\vec{r}\,' = \rho'\hat{\rho}'$$

$$\vec{R} = \vec{r} - \vec{r}\,' = z\hat{z} - \rho'\hat{\rho}'$$

$$\hat{R} = \frac{\vec{R}}{|\vec{R}|} = \frac{z\hat{z} - \rho'\hat{\rho}'}{\sqrt{z^2 + \rho'^2}}$$

Substituting into Eq. 5-17,

$$\vec{E}(\vec{r}) = \vec{E}(0, 0, z)$$

$$= \int_{-\infty}^{\infty} \int_{0}^{2\pi} \int_{0}^{\infty} \frac{\rho' \ \delta(z')U(a - \rho')\left(\dfrac{z\hat{z} - \rho'\hat{\rho}'}{\sqrt{z^2 + \rho'^2}}\right)}{4\pi\epsilon(z^2 + \rho'^2)} \rho' d\rho' d\phi' dz'$$

Evaluating for the delta function and step function using the methods of Example 5-7 and Chapter 4, this reduces to

$$\vec{E}(0, 0, z) = \frac{1}{4\pi\epsilon} \int_0^{2\pi} \int_0^a \frac{\rho'^2(z\hat{z} - \rho'\hat{\rho}')}{(z^2 + \rho'^2)^{3/2}} d\rho' d\phi'$$

This is what one would have anticipated from Eq. 5-22.

$$\vec{E}(0, 0, z) = \frac{1}{4\pi\epsilon} \int_0^{2\pi} \int_0^a \frac{z\rho'^2\hat{z}}{(z^2 + \rho'^2)^{3/2}} d\rho' d\phi' - \frac{1}{4\pi\epsilon} \int_0^{2\pi} \int_0^a \frac{\rho'^3\hat{\rho}'}{(z^2 + \rho')^{3/2}} d\rho' d\phi'$$

In the left integral, as ϕ' varies from 0 to 2π; z and ρ' do not change values since they are scalars. Also, as ϕ' varies, $\hat{z}$ always has the same direction. Thus in the left integral all of the inner integral may be removed from the ϕ' integration.

In the right integral, as ρ' varies from 0 to a, $\hat{\rho}'$ always points the same direction and has magnitude 1 so $\hat{\rho}'$ may be removed from the ρ' integration.

With these observations we obtain the form

$$\vec{E}(0, 0, z) = \frac{1}{4\pi\epsilon} \int_0^a \frac{z\rho'^2\hat{z}}{(z^2 + \rho'^2)^{3/2}} d\rho' \int_0^{2\pi} d\phi' - \frac{1}{4\pi\epsilon} \int_0^{2\pi} \int_0^a \frac{\rho'^3}{(z^2 + \rho'^2)^{3/2}} d\rho' \hat{\rho}' d\phi'$$

Since the unprimed quantities are considered constants as far as the prime integrations are concerned, some of the integrations are easily evaluated. Then

$$\vec{E}(0,0,z) = \frac{1}{4\rho\epsilon} z\left[\frac{-\rho'}{\sqrt{z^2+\rho'^2}} + \ln(\rho' + \sqrt{z^2+\rho'^2})\right]_0^a \hat{z}\cdot 2\pi$$

$$- \frac{1}{4\pi\epsilon}\int_0^{2\pi}\left[\sqrt{z^2+\rho'^2} + \frac{z^2}{\sqrt{z^2+\rho'^2}}\right]_0^a \hat{\rho}' d\phi'$$

$$= \frac{z}{2\epsilon}\left[-\frac{a}{\sqrt{z^2+\rho'^2}} + \ln\left(\frac{a+\sqrt{z^2+a^2}}{z}\right)\right]\hat{z}$$

$$- \frac{\sqrt{z^2+a^2} + \dfrac{z^2}{\sqrt{z^2+a^2}} - 2z}{4\pi\epsilon}\int_0^{2\pi}\hat{\rho}' d\phi'$$

In the remaining integral, $\hat{\rho}'$ changes direction as ϕ' changes so it cannot be removed from the integrand. This is shown in Figure 5-23. From the figure, one might intuitively conclude (correctly) that the result of such an integration (summation) would be zero over 2π range since for each ϕ' value there would be another unit vector diametrically opposite, which would cancel the one at ϕ'. (If we had only half of the circular sheet, this would not be true, of course.) This integral may be evaluated in two ways. One way is to convert the integral to the more tractable rectangular coordinates. Thus using coordinate conversion (see Section 4-9),

$$\hat{\rho}' = \frac{x'}{\sqrt{x'^2+y'^2}}\hat{x}' + \frac{y'}{\sqrt{x'^2+y'^2}}\hat{y}'$$

$$\phi' = \tan^{-1}\left(\frac{y'}{x'}\right) \rightarrow d\phi' = \frac{-y'dx' + x'dy'}{x'^2+y'^2}$$

Because the technique for evaluating integrals of this type is important for later work as well, we give the details for this basic case.

The change on the coordinate ϕ' is not a one-to-one transformation since there is a tangent function. For the tangent function the principal values for ϕ' are $-\pi/2$ to $\pi/2$, or,

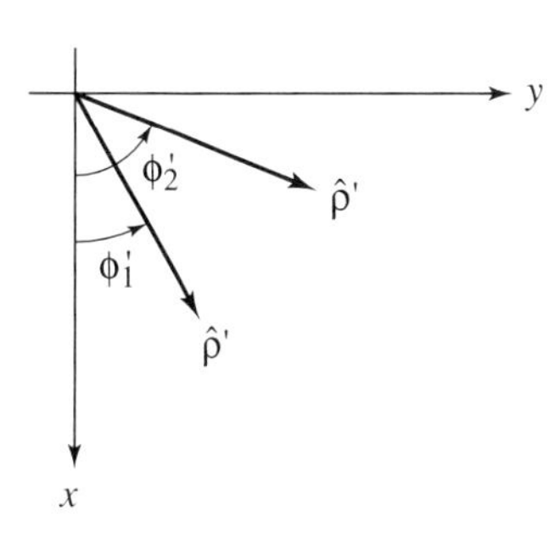

Figure 5-23. Unit vector movement with coordinate ϕ' change.

equivalently, $3\pi/2$ to 2π and 0 to $\pi/2$. Thus we need to write the original integral as the sum of integrals that include the principal values as follows:

$$\int_0^{2\pi} \hat{\rho}' d\phi' = \int_0^{\pi/2} \hat{\rho}' d\phi' + \int_{\pi/2}^{3\pi/2} \hat{\rho}' d\phi' + \int_{3\pi/2}^{2\pi} \hat{\rho}' d\phi'$$

The magnitude of the unit vector $\hat{\rho}'$ is 1 so we may impose the constraints $x'^2 + y'^2 = 1$; thus

$$\left.\begin{aligned} \hat{\rho}' &= x'\hat{x}' + y'\hat{y}' \\ x' &= \pm\sqrt{1 - y'^2} \\ y' &= \pm\sqrt{1 - x'^2} \\ d\phi' &= -y'dx' + x'dy' \end{aligned}\right\}$$

Replacing the primed rectangular unit vectors by unprimed ones (see Example 5-7):

$$\hat{\rho}' d\phi' = -x'y'dx'\hat{x} + x'^2dy'\hat{x} - y'^2dx'\hat{y} + x'y'dy'\hat{y}$$

Now, for the first ingegral, 0 to $\pi/2$, y' is positive so we use $+\sqrt{1 - x'^2}$, as is x', so we use $+\sqrt{1 - y'^2}$. For limits: $\phi' = 0 \rightarrow x' = 1$, $y' = 0$; $\phi' = \pi/2 \rightarrow x' = 0$, $y' = 1$. The first integral becomes, using the $\hat{\rho}' d\phi'$ expression given earlier,

$$\begin{aligned} \int_0^{\pi/2} \hat{\rho}' d\phi' &= \int_1^0 - x'\sqrt{1 - x'^2}\, dx'\hat{x} + \int_0^1 (1 - y'^2)dy'\hat{x} \\ &\quad - \int_1^0 (1 - x'^2)dx'\hat{y} + \int_0^1 y'\sqrt{1 - y'^2}\, dy'\hat{y} \\ &= \hat{x} + \hat{y} \end{aligned}$$

For the second integral, $\pi/2$ to $3\pi/2$, y' is positive over $(\pi/2, \pi)$ but negative over $(\pi, 3\pi/2)$, so we need to consider this integral as the sum of two others by

$$\int_{\pi/2}^{3\pi/2} \hat{\rho}' d\phi' = \int_{\pi/2}^{\pi} \hat{\rho}' d\phi' + \int_{\pi}^{3\pi/2} \hat{\rho}' d\phi'$$

In this result, for the $(\pi/2, \pi)$ portion, y' is positive so we use $+\sqrt{1 - x'^2}$ where needed, and x' is negative so we use $-\sqrt{1 - y'^2}$ where needed. The limits are $\phi' = \pi/2 \rightarrow x' = 0$, $y' = 1$ and $\phi' = \pi \rightarrow x' = -1$, $y' = 0$. Then

$$\begin{aligned} \int_{\pi/2}^{\pi} \hat{\rho}' d\phi' &= \int_0^{-1} -x'\sqrt{1 - x'^2}\, dx'\hat{x} + \int_1^0 (1 - y'^2)dy'\hat{x} \\ &\quad - \int_0^{-1} (1 - x'^2)dx'\hat{y} + \int_1^0 - y'\sqrt{1 - y'^2}\, dy'\hat{y} = -\hat{x} - \tfrac{1}{3}\hat{y} \end{aligned}$$

For the second half, $(\pi, 3\pi/2)$, y' is negative so we use $-\sqrt{1 - x'^2}$ where needed and x' is also negative so we use $-\sqrt{1 - y'^2}$ where needed. The corresponding limits are now $\phi' = \pi \rightarrow x' = -1$, $y' = 0$ and $\phi' = 3\pi/2 \rightarrow x' = 0$, $y' = -1$.

Then

$$\int_{\pi}^{3\pi/2} \hat{\rho}' d\phi' = \int_{-1}^{0} + x'\sqrt{1 - x'^2}\, dx'\hat{x} + \int_{0}^{-1} (1 - y'^2) dy'\hat{x}$$
$$- \int_{-1}^{0} (1 - x'^2) dx'\hat{y} + \int_{0}^{-1} - y'\sqrt{1 - y'^2}\, dy'\hat{y} = -\hat{x} + \tfrac{1}{3}\hat{y}$$

With these two integrations we obtain the result for the second part of our original integration as

$$\int_{\pi/2}^{3\pi/2} \hat{\rho}' d\phi' = -\hat{x} - \tfrac{1}{3}\hat{y} - \hat{x} + \tfrac{1}{3}\hat{y} = -2\hat{x}$$

Finally, the third part of the original integration over the interval $(3\pi/2, 2\pi)$ is evaluated using the same techniques to yield

$$\int_{3\pi/2}^{2\pi} \hat{\rho}' d\phi' = \hat{x} - \hat{y}$$

Now we sum the three results to obtain

$$\int_{0}^{2\pi} \hat{\rho}' d\phi' = \hat{x} + \hat{y} - 2\hat{x} + \hat{x} - \hat{y} = 0$$

as expected.

(Although the evaluation of the integral has resulted in some digression, this technique will prove to be very helpful for integrals where the unit vectors cannot be removed from the integrand and also where the integral value is not obvious by inspection.)

The expression for the electric field intensity now has the final form

$$\vec{E}(0, 0, z) = \frac{z}{2\epsilon}\left[\ln\left(\frac{a + \sqrt{z^2 + a^2}}{z}\right) - \frac{a}{\sqrt{z^2 + a^2}}\right]\hat{z} \text{ N/C}$$

Exercises

5-4.0a The entire z axis is a line charged to a constant value ρ_{l0} C/m as shown in Figure 5-24. Derive the expression for the electric field intensity everywhere.

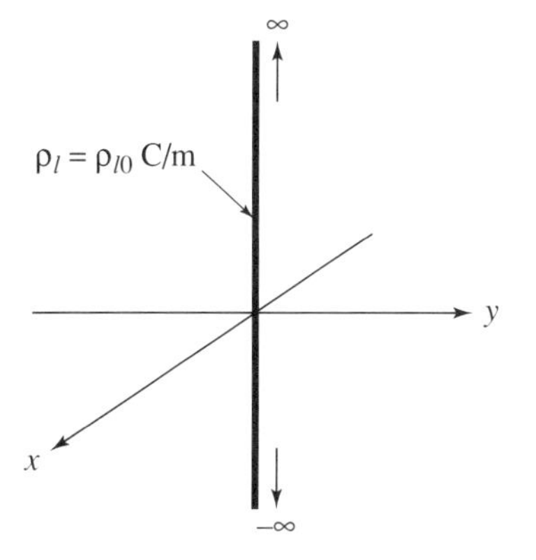

Figure 5-24.

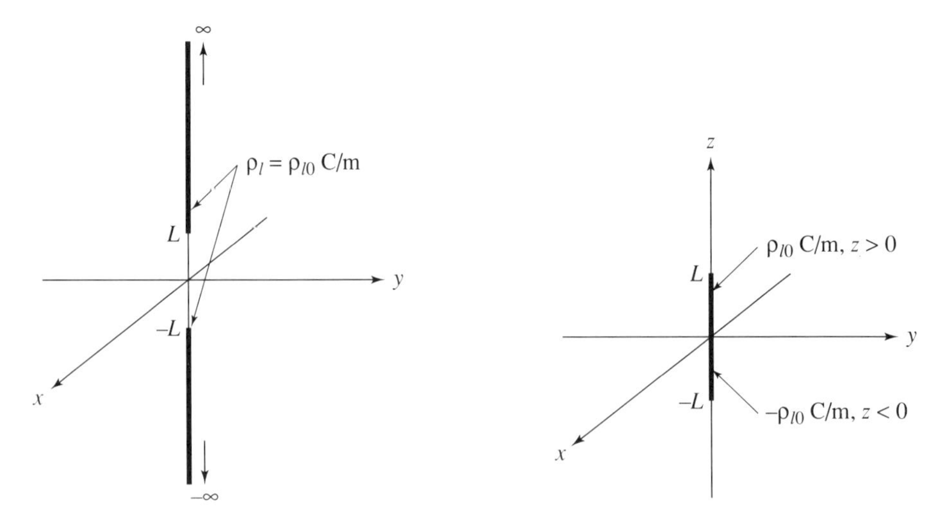

Figure 5-25.

Figure 5-26.

5-4.0b The z axis from $+L$ to $+\infty$ and $-L$ to $-\infty$ is charged to a constant value ρ_{l0} C/m as shown in Figure 5-25. Determine the expression for the electric field *everywhere*. (You may work the problem one of two ways: direct integration or since linearity is assumed, by superposition (adding the negative of Example 5-7 with the results of Exercise 5-4.0a).

5-4.0c Compute the electric field intensity everywhere for the charge system shown in Figure 5-26. What is $\vec{E}$ for distances far from the origin? (Partial answer: For large distance, $\vec{E} \rightarrow 0$ since the net charge of the system is zero)

5-4.0d Determine the electric field intensity on the z axis for the xy plane half disk shown in Figure 5-27. The surface charge density is constant, ρ_{s0} C/m^2. For $a = 1$ cm, what is the force on a $+2$-μC charge placed on the z axis at 10 cm?

5-4.0e The spherical charge region shown in Figure 5-28 has a varying charge density $\rho_v = 1/r$ C/m^3. Determine the electric field intensity for $r > a$. (Even though ρ_v is infinite at $r = 0$, the field values will be finite. The charge distribution is said to have an integrable singularity.) Is there a largest value of n for which $\rho_v = 1/r^n$ C/m^3 will not give an acceptable solution for $\vec{E}$?

5-4.0f An infinite sheet of charge lies in the xy plane and has a constant surface charge density ρ_{s0} C/m^2. Determine the electric field for $z > 0$. Note that, since the sheet is infinite in extent, we can take the field point to be on the z axis without loss of generality. Either rectangular or cylindrical coordinates will work. (Answer: $\vec{E} = (\rho_{s0})/(2\epsilon)\hat{z}$ N/C)

5-4.0g A square sheet of side dimension $2a$ has a constant surface charge density ρ_{s0} C/m^2 and lies in the xy plane with its center at the origin. Find the electric field intensity for all points in space. Repeat for $\rho_s(\vec{r}) = xy$ C/m^2.

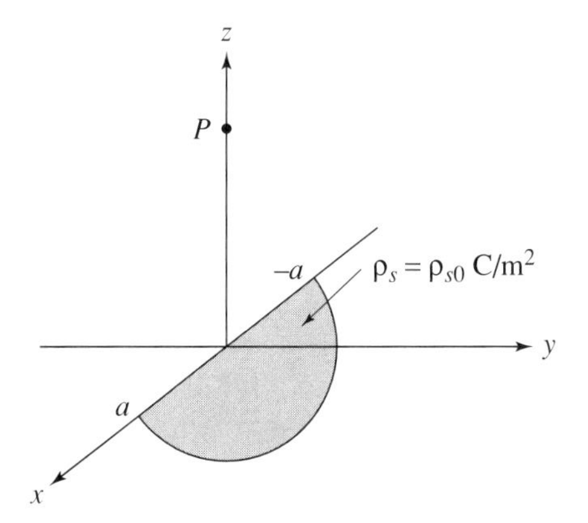

Figure 5-27.

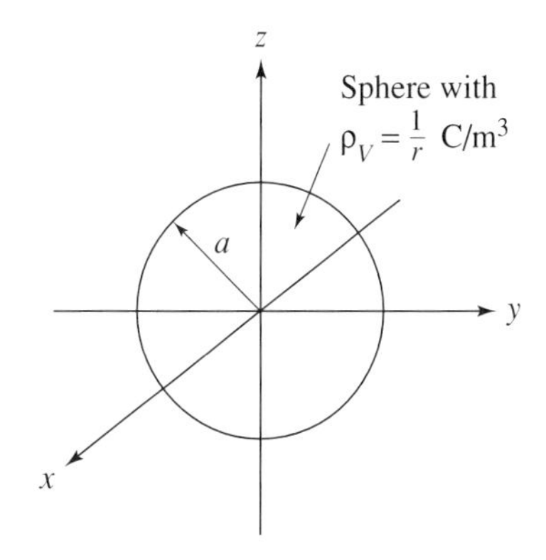

Figure 5-28.

5-5 Electric Field Intensity: Field Patterns

The examples in the preceding section have shown how to develop equations for the field intensity, but it is just as important to be able to obtain from those equations some understanding of what is happening in space around the charge distribution. One way of obtaining some insights is to make what are called *field pattern* plots. To obtain the field pattern, what one does is imagine being at a fixed distance from the origin and then walking around the system at that fixed distance and determining the amplitude of $\vec{E}$ at each location (and also phase of the field in case of sinusoidal time variations) along the circular path. A polar plot is obviously the best way to present these values, and we use the scaled distance from the plot origin to represent the field strength at each angle. Since there are many circular paths one can take around the system and an infinite number of radii, it is important to indicate on such plots the plane of the circular path and the corresponding radius.

Example 5-9

Obtain the field patterns of the charged line of Example 5-7 in the yz plane for distances $2L$ and $4L$ from the origin using $L = 0.1$ m.

Using the expression developed in Example 5-7 and setting $x = 0$ to obtain the yz plane field, we have, for $r^2 = y^2 + z^2$,

$$\vec{E}(0, y, z) = \frac{\rho_{l0}}{4\pi\epsilon y^2}\left[(y\hat{y} + z\hat{z})\left(\frac{L - z}{\sqrt{y^2 + z^2 - 2zL + L^2}} + \frac{L + z}{\sqrt{y^2 + z^2 + 2zL + L^2}}\right) + \left(\frac{y^2 + z^2 - zL}{\sqrt{y^2 + z^2 - 2zL + L^2}} - \frac{y^2 + z^2 + zL}{\sqrt{y^2 + z^2 + 2zL + L^2}}\right)\hat{z}\right]$$

In engineering practice it is usual to *normalize* the field values by dividing out any constants in the field equation. For this case we normalize by dividing by $(\rho_{l0})/(4\pi\epsilon)$. The

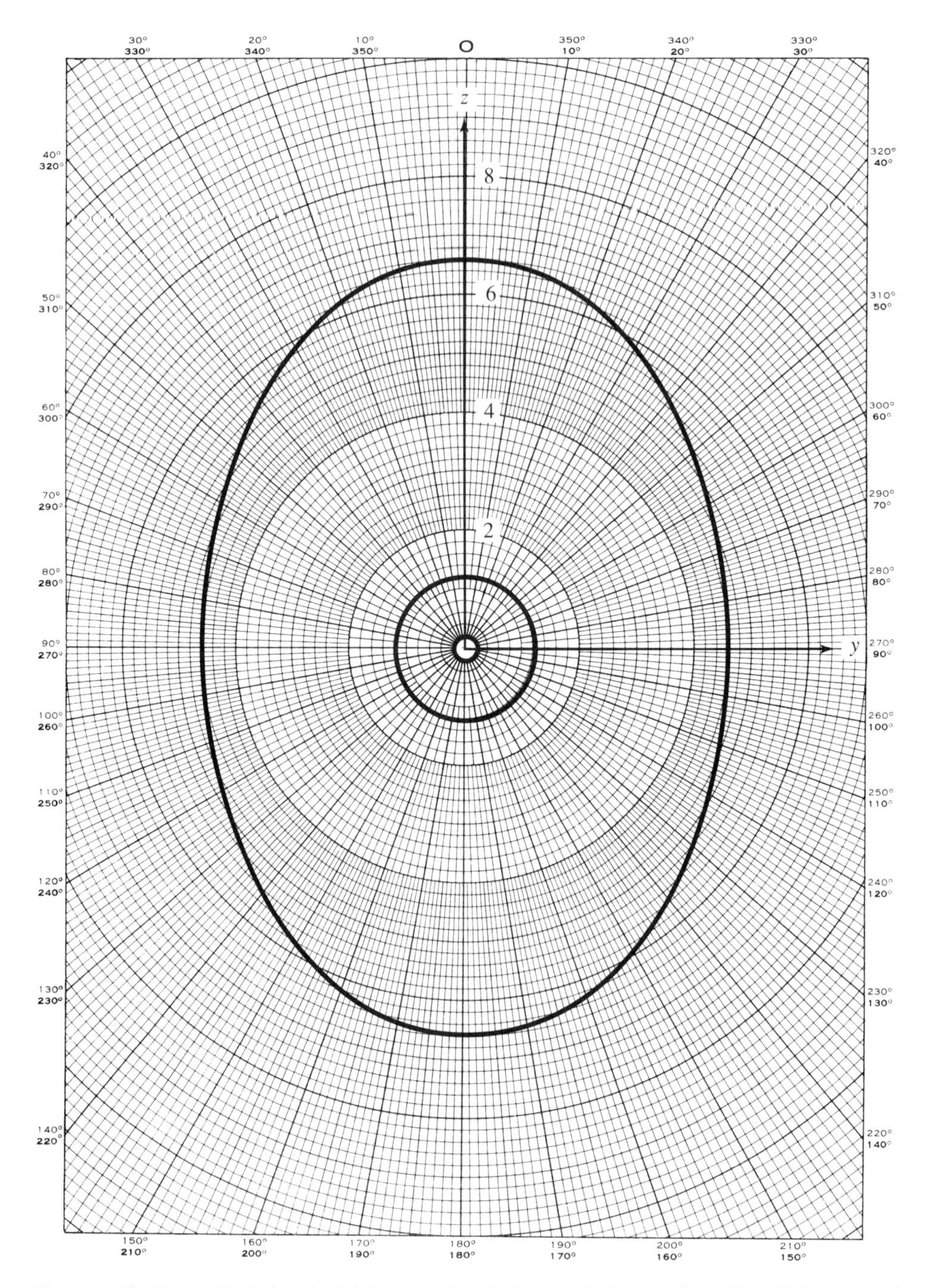

Figure 5-29. Normalized electric field pattern for a finite length charged line of length $2L$ centered on the z axis. Larger pattern is for distance $2L$ from the origin, smaller pattern at distance $4L$.

variational or functional expression remaining is often called the *pattern factor*. Also, in this case, to stay at a constant radius from the origin requires

$$y^2 + z^2 = (KL)^2 = (K \times 0.1)^2 = 0.01K^2$$

where $K = 2$ and 4 for the specified distances.

The field equation—with y removed using the preceding relation between y and z, normalizing by $(\rho_{l0})/(4\pi\epsilon)$, and combining the $\hat{z}$ components—reduces to

$$\vec{E}_{\mathrm{N}}(0, y, z) = \frac{1}{\sqrt{0.1K^2 - z^2}}\left(\frac{0.1 - z}{\sqrt{0.01(K^2 + 1) - 0.2z}} + \frac{0.1 + z}{\sqrt{0.01(K^2 + 1) + 0.2z}}\right)\hat{y}$$
$$+ \left(\frac{1}{\sqrt{0.01(K^2 + 1) - 0.2z}} - \frac{1}{\sqrt{0.01(K^2 + 1) + 0.2z}}\right)\hat{z}$$
$$\text{for} \quad -0.1K \le z \le 0.1K \qquad (K = 2, 4)$$

The angle measured clockwise from the positive z axis (looking into the yz plane from a point on the $+x$ axis) is

$$\phi = \tan^{-1}\left(\frac{y}{z}\right) = \tan^{-1}\left(\frac{\sqrt{0.01K^2 - z^2}}{z}\right)$$

The field pattern is the magnitude of $\vec{E}(0, y, z)$, which is the square root of the sum of the squares of the spatial components. Thus the procedure is to select a value of coordinate z in the specified range and compute the angle ϕ from the z axis and the magnitude of $\vec{E}$. Then plot the magnitude along the line from the origin that makes angle ϕ with the z axis. This is the usual polar plot procedure.

The results were programmed on a hand calculator and a polar plot made as shown in Figure 5-29. In the plot each major circle has a value of 2 for magnitude plotting. The plot reveals that the electric field intensity is strongest on the z axis (directly above the charged line) and weakest on the y axis (broadside to the line).

One word of caution regarding the programming of patterns is that there may be singular values of z where division by zero is called for in the computation. One must be alert for these possibilities. In this case, problems arise at $z = 0$ for ϕ and $z = 0.1K$ for $|\vec{E}|$. If such points are at important areas of interest in the pattern, one can either make computations close to the singular values of coordinate or derive a separate expression at those singular values as was done in Example 5-7.

As expected, the magnitude of electric field is smaller for the case of distance $4L$ from the configuration. Notice also that the pattern is almost circular (a point charge) at the $4L$ distance that is only 0.4 m from the origin for $L = 0.1$ m.

Exercises

5-5.0a Plot the field pattern for the line charge configuration of Exercise 5-4.0c for the case $L = 0.1$ m and for distances $2L$ and $4L$ from the origin.

5-5.0b Plot the field patterns for the semicircular disk of Exercise 5-4.0d at a distance $2a$ from the origin, where $a = 5$ cm, in the xy plane, yz plane and xz plane.

5-5.0c Two point charges of $+1\ \mu$C are located on the z axis, one at $z = -0.1$ m and one at $z = 0.1$ m. Plot the field pattern in the yz plane at a distance of 0.5 m.

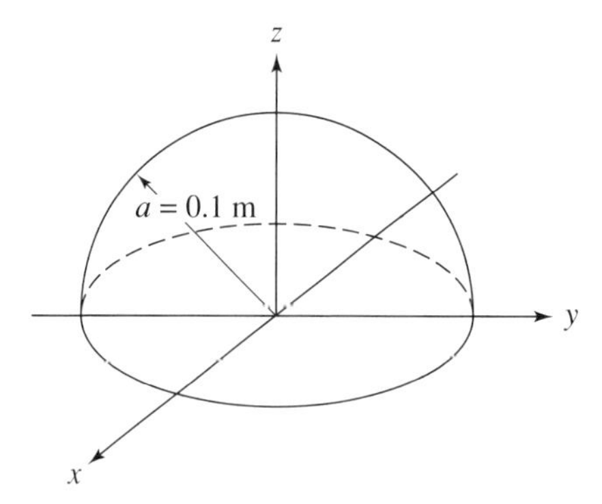

Figure 5-30.

5-5.0d Repeat Exercise 5-5.0c if the charge at $z = -0.1$ m is -1 μC.

5-5.0e A hemispherical shell, shown in Figure 5-30, is charged uniformly to -30 μC/m^2 on the curved surface (bottom is open). Plot the field patterns in the xy and yz planes at a distance of 0.2 m.

5-6 Electric Field Intensity: Field Lines

Field patterns described in the preceding section only give the magnitude of the field intensity at a given distance from the charge distribution. It is often necessary to determine the direction of a path (i.e., a field line) that passes through a given point as an equation rather than qualitatively as was done in Section 5-3. The field line is the path a small positive charge would follow in the region of the charge distribution if it were placed at the point in the field and released. These lines have other names such as *stream* lines, *flow* lines, or *flux* lines.

The simplest case is a planar field structure as shown in Figure 5-31. We wish to determine the equation and the plot of the field line passing through the point $P(x_1, y_1)$.

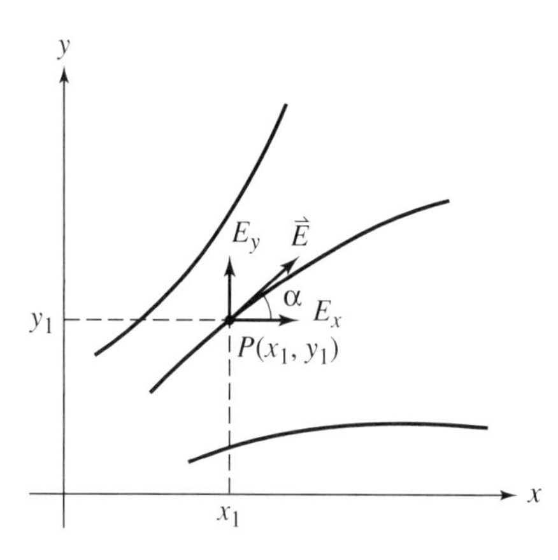

Figure 5-31. Definition of terms used to determine the equation of a field line passing through a given point.

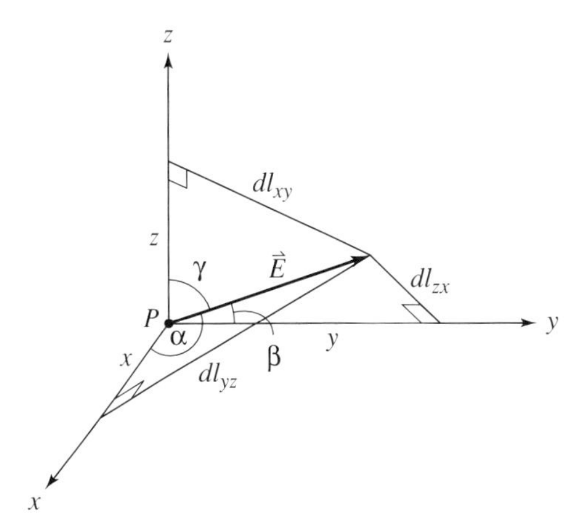

Figure 5-32. Direction angles and dimension definition for determining the equation of a three-dimensional field line.

Since the force is along the field line, $\vec{E}$ will be tangent to the field line at all points. The slope is then given by

$$\tan\alpha = \frac{E_y}{E_x} = \frac{dy}{dx} \tag{5-24}$$

Thus once we have solved for the function $\vec{E}(\vec{r})$ we then need to solve the preceding differential equation for $y(x)$.

For three-dimensional systems we need to compute the direction angles at each point using the configuration shown in Figure 5-32 and results from solid geometry. The pertinent equations are

$$\left.\begin{aligned} \tan\alpha &= \frac{dl_{yz}}{dx} = \frac{\sqrt{E_y^2 + E_z^2}}{E_x} \\ \tan\beta &= \frac{dl_{zx}}{dy} = \frac{\sqrt{E_z^2 + E_x^2}}{E_y} \\ \tan\gamma &= \frac{dl_{xy}}{dz} = \frac{\sqrt{E_x^2 + E_y^2}}{E_x} \end{aligned}\right\} \tag{5-25}$$

To aid in the computations we may use the following differential equivalents:

$$\left.\begin{aligned} dl_{yz} &= \sqrt{dy^2 + dz^2} \\ dl_{zx} &= \sqrt{dz^2 + dx^2} \\ dl_{xy} &= \sqrt{dx^2 + dy^2} \end{aligned}\right\} \tag{5-26}$$

Unfortunately, except for a few simple cases, the differential equations that result cannot be solved explicitly, so some numerical procedure must be used. Here we will not concern ourselves with questions of best method, errors, convergence, singularities, etc., but simply give an example of field-line computation.

Example 5-10

Suppose we wish to find the equation and its plot of the field line for the system of Figure 5-21 that passes through the point (0, 0.02, 0.02) for the case $L = 0.2$ m and $\rho_{l0} = 4\pi\epsilon$ C/m (so constants cancel for convenience). Repeat for the point (0, 0.02, 0.15). Since the system is symmetric around the line charge, the field line will be in the yz plane. In that plane ($x = 0$), the slope of the field lines with the y axis generally will be (for $\vec{E}$ of Example 5-7 with $L = 0.2$ m, $x = 0$, $\rho_{l0} = 4\pi\epsilon$):

$$\tan\beta = \frac{dl_{zx}}{dy} = \frac{\sqrt{dx^2 + dz^2}}{dy} = \frac{dz}{dy} = \frac{E_z}{E_y}$$

Note that x is constant in the yz plane. Then:

$$\frac{dz}{dy} = \frac{\dfrac{y^2 + z^2 - 0.2z}{\sqrt{y^2 + z^2 - 0.4z + 0.04}} - \dfrac{y^2 + z^2 + 0.2z}{\sqrt{y^2 + z^2 + 0.4z + 0.04}} + \dfrac{0.2z - z^2}{\sqrt{y^2 + z^2 + 0.4z + 0.04}} + \dfrac{0.2z + z^2}{\sqrt{y^2 + z^2 + 0.4z + 0.04}}}{\dfrac{0.2y - yz}{\sqrt{y^2 + z^2 - 0.4z + 0.04}} + \dfrac{0.2y + yz}{\sqrt{y^2 + z^2 + 0.4z + 0.04}}}$$

Combining terms where possible and rearranging:

$$\frac{dz}{dy} = \frac{y\sqrt{y^2 + z^2 + 0.4z + 0.04} - y\sqrt{y^2 + z^2 - 0.4z + 0.04}}{(0.2 - z)\sqrt{y^2 + z^2 + 0.4z + 0.04} + (0.2 + z)\sqrt{y^2z^2 - 0.4z + 0.04}}$$

This is a very nonlinear differential equation so we resort to some form of numerical solution. The most straightforward methods are the Euler and Runge-Kutta. For this case let us begin by evaluating dz/dy at the point through which the field line is to pass, in this case, the first line is to pass through (0, 0.02, 0.02). Then use the incremental approximation of the derivative to evaluate the coordinates of the next point, using

$$\left.\frac{dz}{dy}\right|_{\substack{y=y_0=0.02\\ z=z_0=0.02}} \approx \frac{\Delta z}{\Delta y} = \frac{z_1 - z_0}{y_1 - y_0} = \frac{z_1 - 0.02}{y_1 - 0.02}$$

Select an increment Δy (which amounts to selecting y_1 since y_0 is known) which is just a little removed from y_0. Just how large this can be depends upon how much curvature the field line has, and a qualitative plot can often reveal this. The values of y_1 and z_1 are then computed from

$$\Delta y = y_1 - y_0$$

or

$$y_1 = y_0 + \Delta y$$

and

$$z_1 - z_0 = \left(\frac{\Delta z}{\Delta y}\right)(y_1 - y_0) \approx \left.\frac{dz}{dy}\right|_{y_0, z_0} (y_1 - y_0)$$

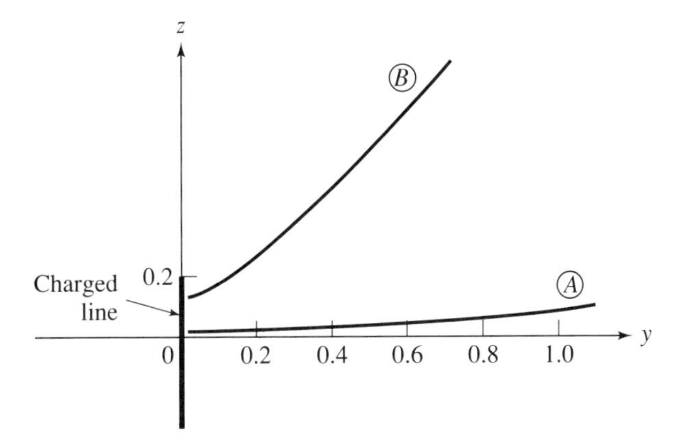

Figure 5-33. Field lines from a charged line passing through prescribed points. See example 5-10.

from which

$$z_1 = \left.\frac{dz}{dy}\right|_{y_0,z_0} (y_1 - y_0) + z_0$$

Next evaluate dz/dy at (y_1, z_1) and estimate y_2 and z_2 using similar expressions. Generalizing this procedure we obtain

$$z_{i+1} = z_i + \left.\frac{dz}{dy}\right|_{y_i,z_i} \cdot (y_{i+1} - y_i), \qquad i = 0, 1, 2, \ldots \tag{5-27}$$

This last equation provides us with an iterative solution process that is ideally suited to computer techniques (hand-held or PC). Equation 5-27 was programmed for the two different points given in the original problem statement. The results are plotted in Figure 5-33. Note that it is possible to use negative Δy values to extrapolate back toward the charge system if desired. It is interesting to note that for large y values, a tangent to the field line, if extended back toward the charged line will pass through the origin. This agrees with the fact that, at large distances, the line appears as a point charge, from which the field lines eminate radially (see Eq. 5-20).

Exercises

5-6.0a Calculator solutions simply give point values but not arrow directions on the field lines. Explain how you would determine the arrow direction using physical principles. (Hint: What is the definition of $\vec{E}$?) Apply it to Example 5-10 and draw the arrows.

5-6.0b Write a computer program that will plot the electric field lines in the region of two point charges. Apply your program to the configuration shown in Figure 5-34 and plot six field lines. Repeat if the right charge is changed to 1 μC, 3 μC, and -3 μC. If graphics

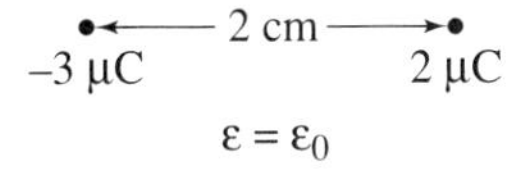

•← 2 cm →•
−3 μC   2 μC
$\varepsilon = \varepsilon_0$

Figure 5-34.

capability is available, set up the program to plot some field lines and put direction on them.

5-6.0c For the charge configuration of Exercise 5-4.0b, determine and plot the field line passing through the point (0, 0, 0.05) for the case $L = 0.1$ m and ρ_{l0} such that the total coefficient in the solution for $\vec{E}$ is equal to unity. Take the plot in the yz plane.

5-6.0d For the charge configuration of Exercise 5-4.0c, determine and plot the field line passing through the point (0.02, 0, 1.5) for the case $L = 0.8$ m. Use the xy plane. Does this field line seem correct physically? Why or why not?

5-7 Current, Current Density, Resistance, and the Continuity Equation

The current flow of interest in circuit theory arises from the flow of free charges, that is, charges that can move to any point within the material or in the evacuated region of a beam of charges. In circuit theory the convention adopted is that, for an arrow drawn on a diagram, the current is the flow of positive charge in the direction of the arrow. For example, in the circuit of Figure 5-35 a current of 2 A means that at the particular instant of time under consideration +2 C/s is the charge flow in that direction. A current of −2 A means that at that instant positive charge is flowing in a direction opposite to the arrow. We represent the current in basic form as

$$i(t) = \frac{dq}{dt} \text{ A} \tag{5-28}$$

If one now looks at the wire cross section through which $i(t)$ flows, it would be reasonable to assume that the charge distribution over the cross section may not be uniform. All one could really say is that the total charge passing the cross section gives $i(t)$. Referring to Figure 5-36, we then define a current density function that provides this variation as

$$\vec{\mathcal{J}}(t) = \lim_{\Delta a \to 0} \frac{\Delta i(t)}{\Delta a} \hat{n} \text{ A/m}^2 \tag{5-29}$$

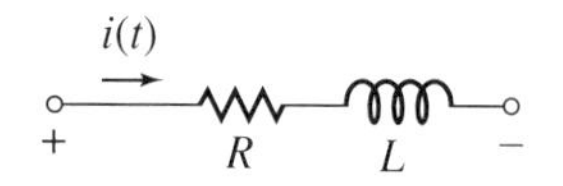

Figure 5-35. Current convention.

Figure 5-36. Configuration for defining current density.

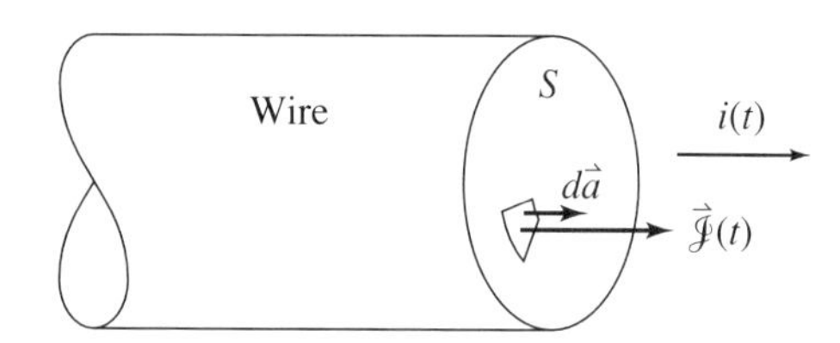

where $\hat{n}$ is the normal unit vector at Δa. We now have a function that varies over the cross section and produces total current $i(t)$ according to

$$i(t) = \int_S \vec{\mathcal{J}}(t) \cdot d\vec{a} \tag{5-30}$$

Here S is the surface through which the current is being computed. One may also wonder why the dot product is used since, in the figure, the current density vector and area are parallel so one could use scalars throughout. This more general expression allows one to draw any surface cutting across a current density distribution and compute the current crossing or passing through the surface. Note the use of script $\vec{\mathcal{J}}$ rather than J. This script nomenclature is used to indicate that the current distribution over the cross section may vary with time in general. The block letters have been used to denote time invariant quantities or sinusoidal steady state.

Now if free charges are subject to an electric field intensity $\vec{\mathcal{E}}(t)$, the force on the charges causes them to move and thus generates a current. If the charges are in vacuum, they will continue to accelerate as long as they are in the dielectric field. On the other hand, if the charges are inside a material, their motion and acceleration are impeded by the stationary material. Thus the ability of electrons to conduct through the material and, hence, establish a current flow depends upon the material properties. The relationship was developed by George Ohm, who reported it in 1827. In field form it is given as a *conduction current density* by the formula

$$\vec{\mathcal{J}}_{\text{cond}}(t) = \sigma\vec{\mathcal{E}}(t) \text{ A/m}^2 \tag{5-31}$$

In this expression σ is called the *conductivity* of the material in units of siemen/meter (S/m). The unit siemen is taken from the family name of four brothers (of a family of ten children): Werner, Carl, William, and Friedrich Siemens of Germany, who all worked in electrical applications (principally telegraphy and heavy machinery) in the early 1800s. If we examine the *units* of $\mathcal{E}$ from the preceding Ohm's law, we obtain

$$\mathcal{E} \approx \frac{\text{A/m}^2}{\text{S/m}} = \frac{\text{A/S}}{\text{m}} = \frac{\text{V}}{\text{m}}$$

where A/S is the volt (V) as used in circuit theory. Thus we have two sets of units for the electric field intensity, which are of course equivalent, namely, N/C and V/m. The relationship between the electric field and voltage will be developed in detail in a later section, but for a simple result we use the configuration shown in Figure 5-37.

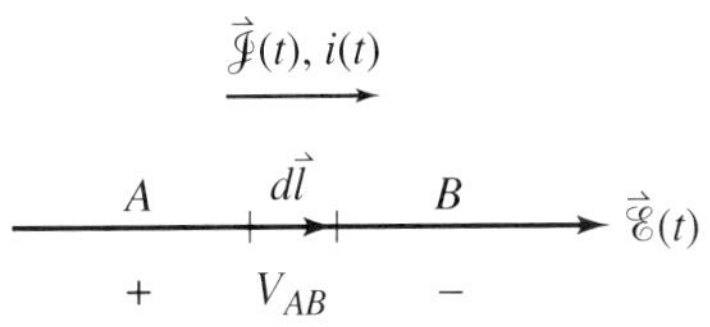

Figure 5-37. Configuration for determining a voltage drop in an electric field.

Since the electric field is in V/m, we would expect to compute a voltage drop (from dimensional considerations) as

$$dv_{AB} = \vec{\mathscr{E}}(t, \vec{r}) \cdot d\vec{l} = \mathscr{E}(t, \vec{r})dl$$

because, in the present cases, $\vec{E}$ and $d\vec{l}$ will be taken as parallel. Then the voltage drop from A to B would be

$$v_{AB}(t) = \int_A^B \mathscr{E}(t, \vec{r})dl = -\int_B^A \mathscr{E}(t, \vec{r})dl \tag{5-32}$$

where we take B as the reference for 0 V and point A would then be $v_{AB}(t)$ V above reference.

Example 5-11

An electric field intensity of $0.1\hat{x}$ V/m(N/C) exists in bar of material 5 cm in length. What is the voltage drop across the bar? The field is uniform over the cross section.

Since the given electric field is time invariant, the voltage will be a DC value of $(dl \to dx)$:

$$V_{AB} = \int_0^{0.05} 0.1dx = (0.1)(0.05) = 5 \text{ mV}$$

We can now develop a general formula for the resistance of a piece of material. The system is shown in Figure 5-38. The coordinates u_1 and u_2 are cross-sectional coordinates.

The process will be to compute the resistance of a section of length dz and then sum the results since the sections are in series from A to B. The resistance of the incremental section dz from A' to B' is the parallel combination of all the small "bars" that make up the volume between A' and B'. The resistance of one of the small bars is

$$\delta R_{A'B'} = \frac{dv_{A'B'}}{di} = \frac{\int_z^{z+dz} \mathscr{E}_z(u_1, u_2, z', t)dz'}{\int_{ds_z} \mathscr{J}(u_1', u_2', z, t)da_z'} = \frac{\int_z^{z+dz} \mathscr{E}_z(u_1, u_2, z', t)dz'}{\int_{u_1}^{u_1+du_1} \int_{u_2}^{u_2+du_2} \mathscr{J}_z(u_1', u_2', z, t)da_z'}$$

Figure 5-38. Configuration for the determination of the resistance of a section of material.

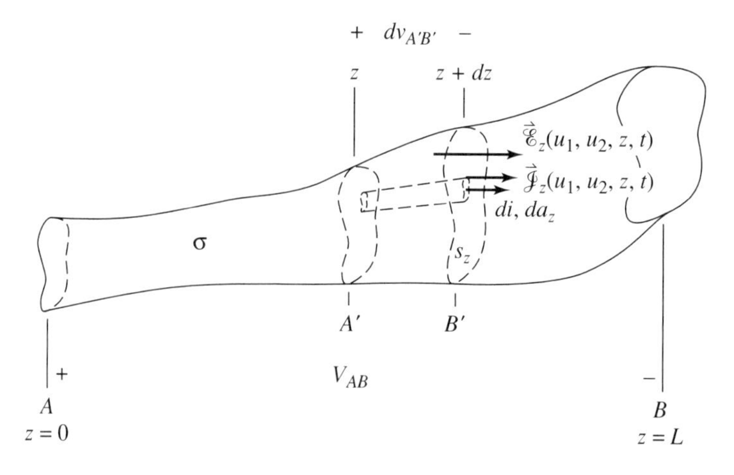

where, as explained in Chapter 4, $da'_z = h_1 h_2 du'_1 du'_2$. However, over incremental coordinate changes, $\mathscr{E}_z$ and $\mathscr{J}_z$ may be considered constant so they may be removed from the integrals. The integrals become dz in the numerator and da_z in the denominator so we have for the small bar

$$\delta R_{A'B'} = \frac{\mathscr{E}_z(u_1, u_2, z, t)dz}{\mathscr{J}_z(u_1, u_2, z, t)da_z} = \frac{dz}{\sigma(u_1, u_2, z, t)da_z}$$

Since we must parallel the small bars in order to cover all the region between dz and $z + dz$, we convert to conductance so we can add as in circuit theory. Then

$$\delta G_{A'B'} = \frac{\sigma(u_1, u_2, z, t)da_z}{dz}$$

We have allowed for varying conductivity within the conductor (material is *inhomogeneous*).

Next, paralleling the bars by adding (integrating) over the cross section (assumed normal to z, so z and u_1, u_2 are independent):

$$dG_{A'B'} = \int_{S'_z} \frac{\sigma(u'_1, u'_2, z, t)}{dz} da'_z = \frac{\int_{S'_z} \sigma(u'_1, u'_2, z, t)da'_z}{dz}$$

From this last result,

$$dR_{A'B'} = \frac{dz}{\int_{S'_z} \sigma(u'_1, u'_2, z, t)da'_z}$$

The total resistance is the sum (integral) of the series resistances between A and B so that

$$R_{AB} = \int_0^L \frac{dz}{\int_{S'_z} \sigma(u'_1, u'_2, z, t)da'_z} \tag{5-33}$$

Note that the cross section may vary with z so the denominator integral in general cannot be removed from the integration on z. Special cases of interest are:

1. Constant conductivity (independent of coordinates):

$$R_{AB} = \frac{1}{\sigma} \int_0^L \frac{dz}{\int_{S'_z} da'_z} \tag{5-34}$$

2. Constant conductivity and constant cross section of area A (often called the *DC resistance*):

$$R_{AB} = \frac{L}{\sigma A} \tag{5-35}$$

3. Constant cross section and σ independent of z only:

$$R_{AB} = \frac{L}{\int_S \sigma(u'_1, u'_2, t)da'_z} \tag{5-36}$$

On an ohms *per meter* basis, as required for *uniform* transmission lines, we have from the last two equations

$$R = \frac{R_{AB}}{L} = \frac{1}{\sigma A} \quad \text{or} \quad \frac{1}{\int_S \sigma(u_1', u_2', t)\,da_z'} \ \Omega/\text{m} \tag{5-37}$$

Suppose there exists a free charge density $\rho_v(t)$ C/m^3 that, under the influence of an electric field, moves with a velocity $\vec{u}$. This results in a *convection current density*, expressible as

$$\vec{\mathcal{J}}_{\text{conv}}(t) = \rho_v(t)\vec{u}(t) \tag{5-38}$$

For free charge density existing in a material, the velocity would be proportional to the electric field so that

$$\vec{u}(t) = \mu\vec{\mathcal{E}}(t) \tag{5-39}$$

where μ is called the *mobility* of free charge in the material. Using this expression in Eq. 5-38, we have an alternate expression for *conduction current* in a material of constant ρ_v as

$$\vec{\mathcal{J}}_{\text{conv}}(t) \rightarrow \vec{\mathcal{J}}_{\text{cond}}(t) = \rho_v\mu\vec{\mathcal{E}}(t)$$

Comparing this last result with Eq. 5-31, we have

$$\sigma = \rho_v\mu \tag{5-40}$$

This result is not too surprising since the more mobile the charges are, the higher the conductivity of the material.

Example 5-12

A glass tube has free electrons and free positively charged atoms inside, and an electric field is impressed axially so that the velocities are $\vec{u}_n$ and $\vec{u}_p$ as shown in Figure 5-39. The density of the negative electrons is ρ_{vn} and that of the positive atoms ρ_{vp}. Compute the current I as shown in the figure.

Since the positive charges are moving in the direction for which I is being computed, Eq. 5-38 applies directly. The negative charge flowing to the left is equivalent to an identical amount of positive charge moving to the right in the direction of the desired I,

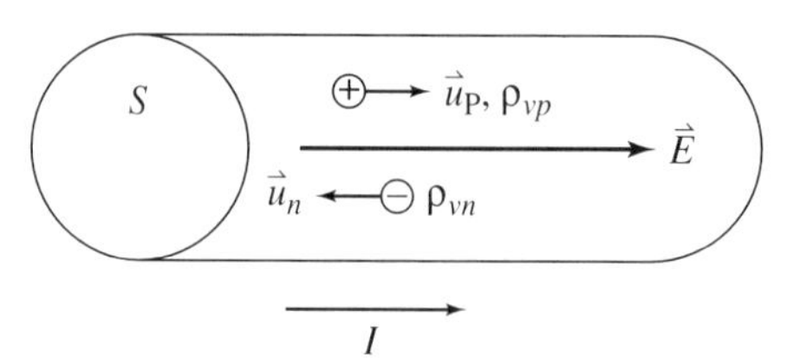

Figure 5-39. Computation of positive and negative contributions to total current flow.

so we can again use Eq. 5-38, as it is given, directly. Then, upon summing the two current contributions, we have, in scalar form,

$$J_{\text{conv}} = \rho_{vp}u_p + \rho_{vn}u_n \text{ A/m}^2$$

Note that ρ_{vn} is negative and u_n will be negative with respect to the direction of I, so the product will be positive, as required physically. If we assume this current is constant across the glass tube cross section, then

$$I = J_{\text{conv}}S = (\rho_{vp}u_p + \rho_{vn}u_n)S \text{ A}$$

Example 5-13

Suppose the preceding example is modified so that the charge densities and velocities are inside of a material rather than a hollow tube. This is analogous to charged carriers in a semiconductor as electrons and holes. Compute the conductivity of the material.

In a material, the convection current becomes the conduction current for which the velocities are proportional to the electric field intensity as given by Eq. 5-39. For the electrons, however, we need to negate the equation since the electron velocity will be opposite the electric field. (By convention, the mobilities are positive numbers.) The conduction current is then

$$J_{\text{cond}} = \rho_{vp}\mu_p E + \rho_{vn}(-\mu_n E) = (\rho_{vp}\mu_p - \rho_{vn}\mu_n)E$$

Comparing this with Eq. 5-31 gives

$$\sigma = \rho_{vp}\mu_p - \rho_{vn}\mu_n \tag{5-41}$$

Remember that ρ_{vn} is negative since it is an electron volume charge density.

Example 5-14

Compute the DC resistance of the cylindrical shell shown in Figure 5-40. Compute the resistance per meter that would correspond to the outer conductor R parameter of a coax idealized to a solid conductor, as discussed in Chapter 1.

For DC conditions and uniform cross section, Eq. 5-35 applies. For this case, A is the annular ring area shown shaded in the figure.

$$A = \pi a^2 - \pi b^2 = \pi(a^2 - b^2)$$

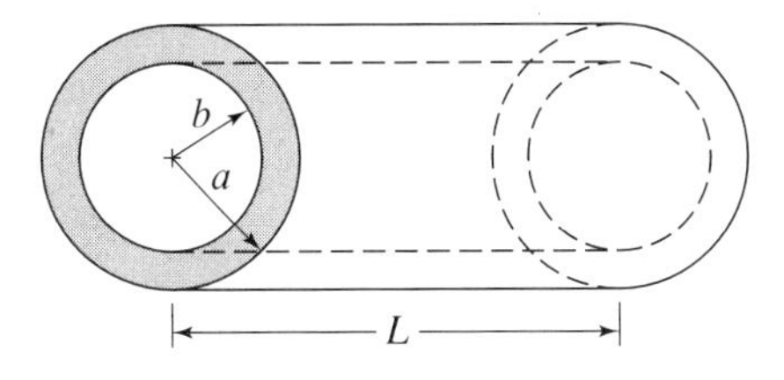

Figure 5-40. Dimensions for computing the resistance of a cylindrical shell.

Then

$$R_{DC} = \frac{L}{\sigma\pi(a^2 - b^2)}\ \Omega$$

From this we obtain

$$R = \frac{R_{DC}}{L} = \frac{1}{\sigma\pi(a^2 - b^2)}\ \Omega/\text{m}$$

Example 5-15

Suppose that in Example 5-14 the conductivity is not constant over the cross section but varies with the cylindrical coordinates as

$$\sigma(\rho, \phi) = \frac{\sigma_0}{(a-b)^2}\left(\rho - \frac{a+b}{2}\right)^2$$

where σ_0 is a constant. Using Eq. 5-36,

$$\begin{aligned} R_{\text{shell}} &= \frac{L}{\displaystyle\int_b^a \int_0^{2\pi} \frac{4\sigma_0}{(a-b)^2}\left(\rho' - \frac{a+b}{2}\right)^2 \rho' d\phi' d\rho'} \\ &= \frac{(a-b)^2 L}{8\pi\sigma_0} \frac{1}{\left.\left(\dfrac{1}{4}\rho'^4 - \dfrac{a+b}{3}\rho'^3 + \dfrac{(a+b)^2}{8}\rho'^2\right)\right|_b^a} \\ &= \frac{3L}{\pi\sigma_0(a^2 - b^2)}\ \Omega \end{aligned}$$

Then

$$R = \frac{R_{\text{shell}}}{L} = \frac{3}{\pi\sigma_0(a^2 - b^2)}\ \Omega/\text{m}$$

Example 5-16

In Figure 5-41 is shown a solid metal truncated cone made of a material of constant conductivity σ_0. Determine the resistance of the element.

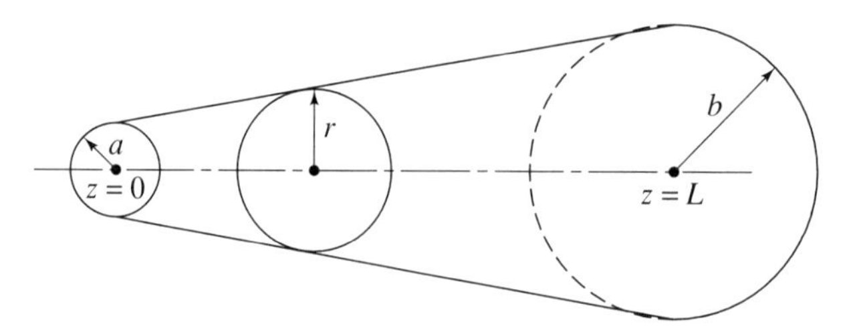

Figure 5-41. Configuration for the determination of the resistance of a truncated right circular cone.

The radius r of a general cross section as a function of z (the cone axis) is

$$r = a + \frac{b-a}{L} z$$

Using Eq. 5-34,

$$R = \frac{1}{\sigma_0} \int_0^L \frac{dz}{\int_0^r \int_0^{2\pi} r' d\phi' dr'} = \frac{1}{2\pi\sigma_0} \int_0^L \frac{dz}{\int_0^{a+\frac{b-a}{L}z} r' dr'}$$

$$= \frac{L^2}{\pi\sigma_0(b-a)^2} \int_0^L \frac{dz}{\frac{a^2L^2}{(b-a)^2} + \frac{2aL}{b-a} z + z^2}$$

$$= \frac{L^2}{\pi\sigma_0(b-a)^2} \int_0^L \frac{dz}{\left[\frac{aL}{(b-a)} + z\right]^2} = \frac{L}{\pi ab\sigma_0} \ \Omega$$

Example 5-17

In pure semiconductor materials (called *intrinsic semiconductors*), charged carriers are produced by thermal ionization or excitation. For each electron excited a positive charge (hole) is also created, so the number of densities of the two will be equal. At 300 K the number density is $2.4 \times 10^{19}/\text{m}^3$. The mobility of electrons at that temperature is $0.39 \text{ m}^2/\text{Vs}$ and for holes $0.19 \text{ m}^2/\text{Vs}$. Compute the conductivity of germanium.

$$\rho_{Vn} = -1.6 \times 10^{-19} \times 2.4 \times 10^{19} = -3.84 \text{ C/m}^3$$

$$\rho_{Vp} = -\rho_{Vn} = 3.84 \text{ C/m}^3$$

Using Eq. 5-41,

$$\sigma_{\text{Ge}} = 3.84 \times 0.39 - (-3.84) \times 0.19 = 1.04 \text{ S/m}$$

Exercises

5-7.0a For the intrinsic germanium of Example 5-17, suppose a small square bar 1 mm on a side and 10 mm long carries 10 mA distributed uniformly over the cross section. Determine the current density, electric field intensity within the bar, and the voltage drop across the bar. (Answer: 10 kA/m^2, 9.6 kV/m, 96 V)

5-7.0b For the result of Example 5-14, show that the resistance of the element for $b \to 0$ is consistant with that obtained by using Eq. 5-35 for a cylindrical conductor. Repeat for Example 5-16 where $b \to a$.

5-7.0c At 300 K the intrinsic carrier densities of silicon and gallium arsenide are 1.45×10^{16} and $1.79 \times 10^{12}/\text{m}^3$, respectively. The corresponding mobilities for silicon are $\mu_n = 0.135$ and $\mu_p = 0.048$ and for gallium arsenide, $\mu_n = 0.68$ and $\mu_p = 0.068$. Compute the conductivities of these materials.

5-7.0d A solid, uniform cross section, cylindrical conductor whose axis is the z axis carries a current density of $3\rho^2\hat{x} \text{ A/m}^2$. For a wire diameter of 3 mm compute the total current flow.

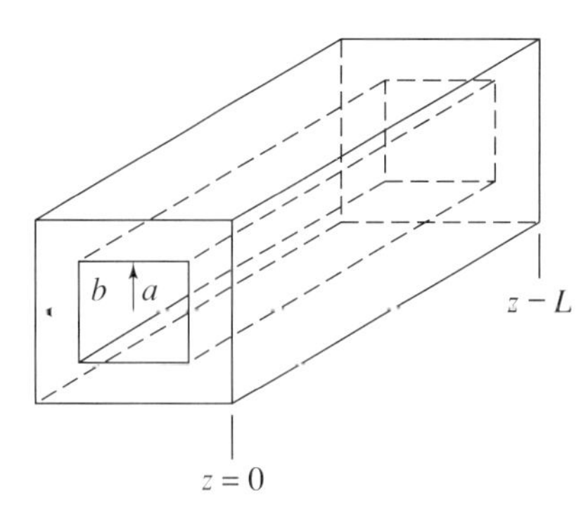

Figure 5-42.

5-7.0e Determine the resistance of the section of square shell which is shown in Figure 5-42. Conductivity is constant, σ_0. Show that a correct result is obtained for the case $a \to 0$.

5-7.0f For the configuration of Exercise 5-7.0e suppose the conductivity varies as $\sigma_0 + z$. Determine the resistance of the shell.

Since the conservation of charge is of fundamental importance in physics, we close this section with the derivation of the relationship between current density and charge density. Consider a closed surface S that has a current density $\vec{\mathcal{J}}$ flowing through it. We use a script letter here to denote that the current density may vary with time in general. Applying Eq. 5-30 to this configuration, we would expect that if the closed surface integral were not zero, there must be a *decrease* of charge within that surface. As an equation we would write:

$$\oint_S \vec{\mathcal{J}} \cdot d\vec{a} = -\frac{dq}{dt}$$

The minus sign might seem incorrect since the derivative is already current flow. The sign depends upon the convention we use. In the case of current $i(t)$ given by Eq. 5-28, it is the current coming out of the surface as we stand outside the region. This is analogous to Kirchhoff's current law in the form of current leaving a wire tied at a node. In the preceding equation we are standing inside the closed region and asking what the charge is doing there that causes current to flow out of the region.

Now the charge inside the closed surface S is given by

$$q(t) = \int_V \rho_v(\vec{r}, t) dv$$

so the preceding becomes

$$\oint_S \vec{\mathcal{J}} \cdot d\vec{a} = \frac{-d\int_V \rho_v(\vec{r}, t) dv}{dt} = -\int_V \frac{\partial \rho_v}{\partial t} dv$$

The last equality is obtained by applying Leibniz' theorem (see Chapter 4) to the volume integral where the surface is stationary so that the limits are constants. Next, applying the divergence theorem to the closed surface integral, we obtain

$$\int_V \nabla \cdot \vec{\mathcal{J}} dv = \int_V -\frac{\partial \rho_v}{\partial t} dv$$

This must be true for any closed surface (volume), so we may equate the integrands to obtain

$$\nabla \cdot \vec{\mathcal{J}} = -\frac{\partial \rho_v}{\partial t} \tag{5-42}$$

This is the *continuity equation*, which, in words, says that if current is diverging at a given point in space, the charge there must be decreasing. Such a result is not unexpected, but has been placed on a theoretical basis for computation. Obviously, it follows that if there is no charge at a given point no current can diverge from the point.

Exercises

5-7.0g In cylindrical coordinates a current density, given by $\vec{\mathcal{J}}(\vec{r}, t) = 6\rho \cos \omega_0 t \rho + 2e^{-3z}\rho\hat{z}$ A/m^2, flows inside a metal tube of radius 0.1 m. Compute the current $i(t)$ flowing out of a surface of radius 0.05 m and length 0.5 m, extending in the positive z direction from $z = 0$. Next determine how the volume charge density is changing with time.

5-7.0h Suppose we have a time invariant current given by the expression

$$\vec{J}(\vec{r}) = 3xy\hat{x} - 2yz\hat{y} + 5z^2\hat{z} \text{ A/m}^2$$

Compute the current i flowing out of a cube that is 0.5 m on a side with the center of the cube at the origin of coordinates. Determine the rate at which the volume charge density is changing at the point (0.25, 0.3, 0.4). What is the time rate of change of total charge inside the cube? Is the charge within the cube increasing or decreasing?

5-8 Electric Flux and Electric Displacement (Flux Density); Gauss' Law

In earlier sections it was shown that the forces and electric field intensity in a medium were dependent upon the medium dielectric constant. Michael Faraday reported in 1834 the results of his experiments, which indicated another electric phenomenon that does not depend upon the dielectric. Although Faraday did not use spheres in his original experiements, we will do so in our discussion here since the results happen to be independent of the shapes of the parts of the system. The various steps and measurements are given in Figure 5-43.

We begin by charging a metal sphere as shown in Figure 5-43*a*. Then a dielectric is carefully placed around the sphere and a second metal shell is placed (using insulated tools) around the system (Figure 5-43*b*), and the charge on the outer sphere is measured.

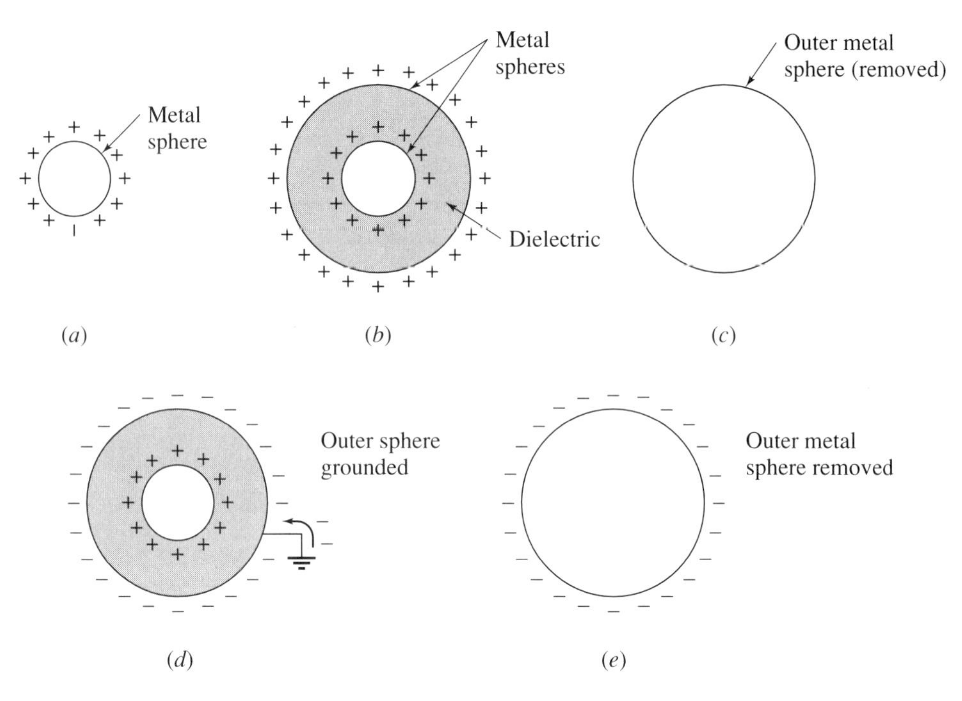

Figure 5-43. Idealization of Faraday's experiments.

The outer sphere is then carefully removed as shown in Figure 5-43*c* and the charge is measured on the isolated larger metal sphere. Next the system is assembled as before except the outer sphere is grounded through a conductor as indicated in Figure 5-43*d*. Finally, the shell is removed again and its charge measured. The following observations result from these experiments:

1. For Figure 5-43*b*, the charge on the outer sphere is exactly equal to the charge on the inner sphere in magnitude and sign, and independent of the dielectric.
2. For Figure 5-43*c* the removed sphere has zero charge.
3. For Figure 5-43*d* the outer sphere has a charge equal in magnitude but of opposite sign to that on the inner sphere, and the result is again independent of the dielectric.
4. For Figure 5-43*e* the removed sphere retains the same charge that it had in Figure 5-43*d*, indicating that charge was obtained through the ground line.

Faraday concluded that there was some type of electric displacement between the two spheres, the inner sphere displacing charge onto the outer sphere by either induction for Figure 5-43*b* or by conduction for Figure 5-43*d*. Furthermore, the result depends only upon the charge inside and not upon the dielectric.

Faraday performed two other experiments that shed additional light on these conclusions. First, he observed that the same results were obtained no matter where the inside charged metal sphere was placed within the larger shell. Thus symmetry played no part in the results. Second, when he placed two charged bodies within the shell, the charge on the outer shell (Figure 5-43*b*) was equal to the algebraic sum of the charges within, so that only the net free charge inside was of importance.

From these conclusions we are led to define an electric flux, which we'll call Ψ, which has the ability to displace charge. Since the results depend only upon the net free charge enclosed, we define

$$\Psi = Q_{\text{enclosed}} \text{ C} \tag{5-43}$$

This is called *Gauss' law*.

The total charge enclosed within a region can be obtained by integrating the charge distribution within the region, whether it is a line, surface, or volume configuration. Thus the most general form of the preceding would be

$$\Psi = \int_V \rho_v(\vec{r})dv \text{ C} \tag{5-44}$$

Since the electric flux passes to a surface, we next define a surface density of these flux lines, which is called the *electric displacement* density, or *electric flux* density, by

$$\vec{D} = \lim_{\Delta S \to 0} \frac{\Delta\Psi}{\Delta S} \hat{n} \text{ C/m}^2 \tag{5-45}$$

where $\hat{n}$ is the unit normal to the surface ΔS. With this result we can then compute the electric flux passing through any surface S drawn in the region containing $\vec{D}$ by integrating as follows:

$$\Psi_S = \int_S \vec{D} \cdot d\vec{a}$$

However, to use this relationship along with Eq. 5-44, the surface must enclose the volume V so we have

$$\oint_S \vec{D} \cdot d\vec{a} = \int_V \rho_v(\vec{r})dv \tag{5-46}$$

which is called the integral form of Gauss' law.

Now if we apply the divergence theorem (see Chapter 4), we may write the preceding as

$$\int_V \nabla \cdot \vec{D}dv = \int_V \rho_v(\vec{r})dv$$

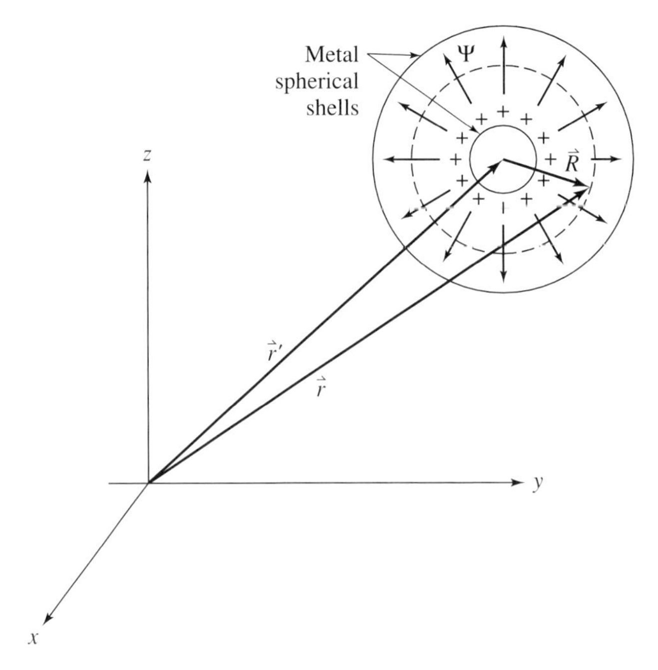

Figure 5-44. Configuration for determining the electric displacement (flux density) for a spherically symmetric region.

Since this result must hold for *any* volume, the integrands must be equal. Thus

$$\nabla \cdot \vec{D} = \rho_v(\vec{r}) \tag{5-47}$$

which is the differential form of Gauss' law. It is important to remember that in all of the preceding results the ρ_v is the *free* (moveable) charge within the region. The importance of this fact will be explored in detail in the next section. Incidentally, Eq. 5-47 is one of Maxwell's equations and is valid even when ρ_v and $\vec{D}$ are time-varying functions! This equation is called the *point* form of Guass' law since it will be the value of charge density at points in the region of $\vec{D}$.

Let's next apply Guass' law to a nice, spherically symmetric system shown in Figure 5-44. Suppose we look at an imaginary spherical surface of radius R between the two metal shells. All of the flux Ψ passes through this surface and will do so uniformly due to symmetry. Thus

$$\vec{D} = \frac{\Psi}{\text{area}}\hat{R} = \frac{Q_{\text{enclosed}}}{4\pi|\vec{R}|^2}\hat{R} = \frac{Q_{\text{enclosed}}}{4\pi|\vec{r} - \vec{r}'|^2}\frac{(\vec{r} - \vec{r}')}{|\vec{r} - \vec{r}'|}$$

For a point charge (inner sphere shrinks to a point) we would have $Q_{\text{enclosed}} \rightarrow q$, so that

$$\vec{D} = \frac{q}{4\pi|\vec{r} - \vec{r}'|^3}(\vec{r} - \vec{r}')$$

Note that this is independent of the medium. Now if we compare this with the equation for the electric field intensity of a point charge, Eq. 5-11, we have

$$\vec{D} = \epsilon\vec{E} \tag{5-48}$$

This is called the *constitutive relationship for electric fields* and is valid also when $\vec{D}$ and $\vec{E}$ are time varying. Although this was derived for a point charge, it is used for any region where an electric field is present. The result is also reasonable since it tells us that the ability of a field to displace charge is proportional to its ability to apply a force to a charge.

This last equation also provides us with a way to get the units on the permittivity ϵ. Since $\vec{D}$ is in C/m^2 and $\vec{E}$ in V/m, we have

$$\epsilon \stackrel{\delta}{=} \frac{\text{C/m}^2}{\text{V/m}} = \frac{\text{C/V}}{\text{m}} = \frac{\text{farad}}{\text{meter}}$$

which should be compared to the result given in Eq. 5-4. As a final note, since $\vec{D}$ and $\vec{E}$ are related by a constant, all of the integral results for $\vec{E}$ derived earlier also apply to obtaining $\vec{D}$ if we simply leave out the ϵ in those results.

Example 5-18

A point charge q is located outside a closed surface S. Find the flux passing through the surface as shown in Figure 5-45.

From Gauss' law (Eq. 5-43), we simply observe that the flux passing through the surface is equal to the enclosed charge. Since there is no charge *inside* S, then Q_{enclosed} is zero. Thus

$$\Psi = 0$$

This result may seem to contradict the figure, but if we recall that Gauss' law gives us the net flux, the result is justified because any flux line entering one side of the surface passes out another side.

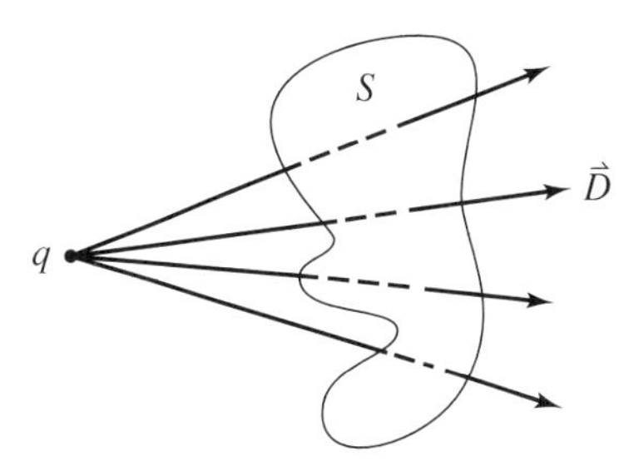

Figure 5-45. Configuration for Example 5-18.

Example 5-19

Consider a region of space (vacuum) where the charge density is given by the spherically symmetric distribution

$$\rho_v(\vec{r}) = \rho_v(r) = \frac{1}{r}\ \text{C/m}^2$$

as shown in Figure 5-46. Determine the electric flux density at a distance r from the origin, and show that the differential form of Gauss' law is satisfied.

Since the system is symmetric (independent of θ and ϕ), we observe that $\vec{D}$ will be in the $\hat{r}$ direction and a function r only. Thus in component form,

$$\vec{D} = D_r(r)\hat{r}$$

From the right side of Eq. 5-46 we have

$$Q_{\text{enclosed}}(r_0) = \int_0^{r_0}\int_0^{2\pi}\int_0^{\pi} \frac{1}{r}\, r^2 \sin\,\theta d\theta d\phi dr = 2\pi r_0^2$$

From the left side of Eq. 5-46 we have

$$\begin{aligned}\Psi(r_0) &= \int_0^{2\pi}\int_0^{\pi} D_r(r_0)\hat{r} \cdot r_0^2 \sin\,\theta d\theta d\phi \hat{r} \\ &= r_0^2 D_r(r_0) \int_0^{2\pi}\int_0^{\pi} \sin\,\theta d\theta d\phi = 4\pi r_0^2 D_r(r_0)\end{aligned}$$

Equating the two results yields

$$D_r(r_0) = \tfrac{1}{2}$$

from which

$$\vec{D} = \tfrac{1}{2}\hat{r}$$

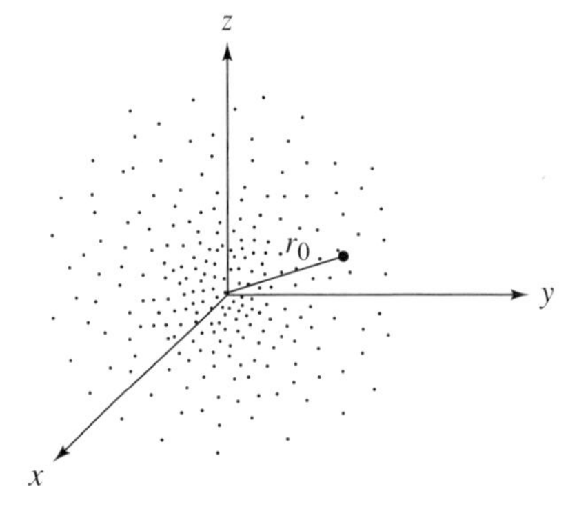

Figure 5-46. Configuration for Example 5-19.

Using the spherical expansion of the divergence operator of the point form of Gauss' law, Eq. 5-47, we have the following expansion since D_θ and D_ϕ are zero:

$$\nabla \cdot \vec{D} = \frac{1}{r^2}\frac{\partial(r^2 D_r)}{\partial r} = \frac{1}{r^2}\frac{d(r^2\frac{1}{2})}{dr} = \frac{1}{r^2}\cdot r = \frac{1}{r} = \rho_v(r)$$

This agrees with the originally given charge density.

Example 5-20

A coaxial line consists of a center conductor, two dielectrics, and an outer conductor as shown in Figure 5-47. The center conductor is charged to $+\rho_l$ C/m length and the outer conductor to $-\rho_l$ C/m (same magnitude). Determine equations for and make plots of $\vec{D}$ and $\vec{E}$ for $a < \rho < c$, and $\epsilon_2 > \epsilon_1$. Here we suppose ρ_l is a constant.

For a long coax (or for locations near the center of the length removed from the ends), the symmetry of the system dictates that $\vec{D}$ will vary only with the radial coordinate ρ and that $\vec{D}$ will only have a $\hat{\rho}$ component. We then have

$$\vec{D} = D_\rho(\rho)\hat{\rho}$$

If we take a length L of cable, Gauss' law (Eq. 5-43) requires, for any location ρ between a and c, that only the inner conductor charge contributes to net outward flux, since it is the only charge in the volume between a and c. Also note that it does not matter whether coordinate ρ is in medium 1 or 2 since $\vec{D}$ is independent of the dielectric. We then have

$$Q_{\text{enclosed}} = \rho_l L \text{ C}$$

$$\Psi = \int_S \vec{D}\cdot d\vec{a} = \int_0^L\int_0^{2\pi} D_\rho(\rho)\hat{\rho}\cdot\rho d\phi dz\hat{\rho} = \int_0^L\int_0^{2\pi} D_\rho(\rho)\rho d\phi dz$$

$$= \rho D_\rho(\rho)\int_0^L\int_0^{2\pi} d\phi dz = 2\pi L\rho D_\rho(\rho) \text{ C}$$

Equating these, we obtain

$$D_\rho(\rho) = \frac{\rho_l}{2\pi\rho}$$

Figure 5-47. Two-dielectric coax configuration for Example 5-20.

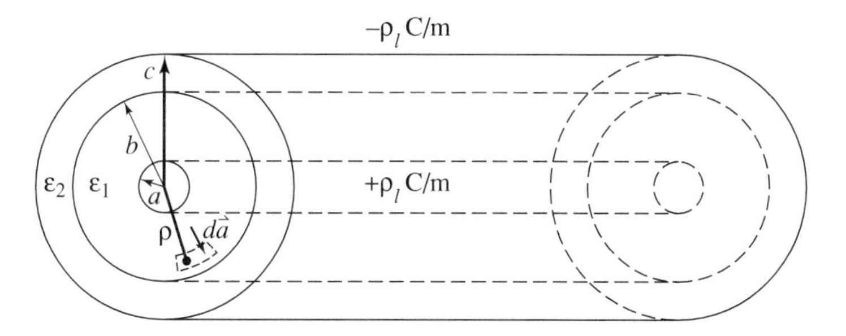

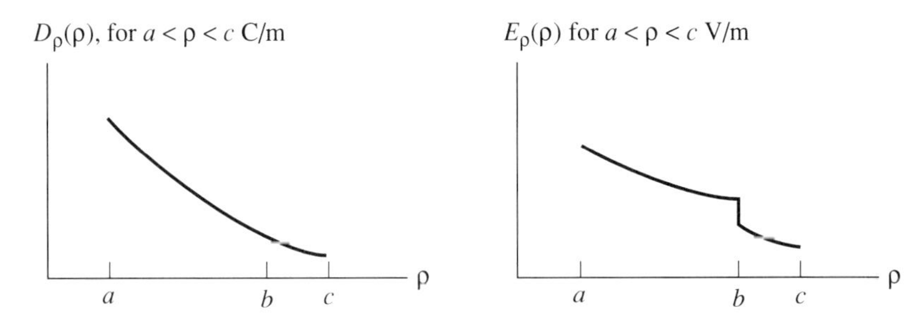

Figure 5-48. Electric displacement and electric field intensity for the interior of the coax of Figure 5-57.

from which

$$\vec{D} = D_\rho(\rho)\hat{\rho} = \frac{\rho_l}{2\pi\rho}\hat{\rho} \qquad a < \rho < c$$

Using the constitutive relationship for electric fields:

$$\vec{E}_1 = \frac{\vec{D}}{\epsilon_1} = \frac{\rho_l}{2\pi\epsilon_1\rho}\hat{\rho} \qquad \text{and} \quad \vec{E}_2 = \frac{\vec{D}}{\epsilon_2} = \frac{\rho_l}{2\pi\epsilon_2\rho}\hat{\rho}$$

Since $\epsilon_2 > \epsilon_1$, $|\vec{E}_1| > |\vec{E}_2|$ at $\rho = b$. The plots of the field components are given in Figure 5-48.

Exercises

5-8.0a For the system of Figure 5-49, determine the value of electric flux for each of the closed surfaces 1, 2, 3, and 4. (Answer: 0, q, $-q$, 0, respectively)

5-8.0b For example 5-19, suppose the volume charge density were generalized to

$$\rho_v(r) = \frac{1}{r^\alpha}\ \text{C/m}^3$$

For what values of α does the solution for $\vec{D}$ exist? (Answer: $\alpha \leq 2$)

5-8.0c In Example 5-19, why was the partial derivative replaced by the total derivative? (Answer: $\vec{D}$ was a function of only one variable r)

5-8.0d Make plots of D_ρ and E_ρ for Example 5-20 for the case $\epsilon_2 < \epsilon_1$ for the same ρ values

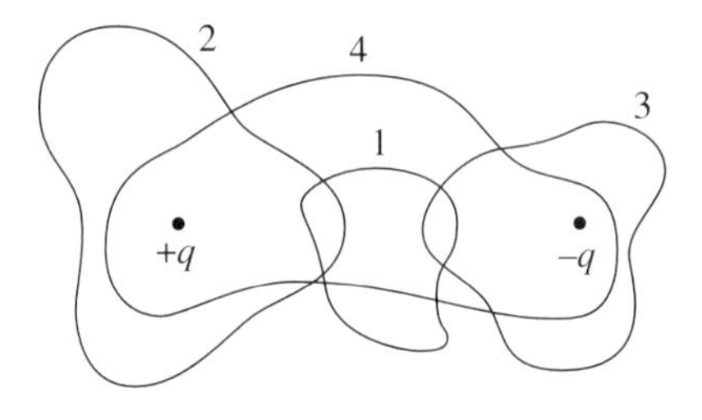

Figure 5-49.

(i.e., $a < \rho < c$). Also show that the point form (differential) of Gauss' law is satisfied. (Answer: $\nabla \cdot \vec{D} = 0$)

5-8.0e Develop equations for $\vec{D}$ and $\vec{E}$ for the case $\rho > c$ in the example above. (Answer: $\vec{D} = \vec{E} = 0$. Why?)

5-8.0f A perfectly conducting spherical shell of radius 3 m is charged uniformly to $\frac{1}{4}$ C/m^2. Around this sphere is placed a dielectric having $\epsilon_r = 5$ and thickness 1 m. Vacuum exists beyond the dielectric. Plot $D_r(r)$ and $E_r(r)$ for $0 < r < 6$ m. (Hint: Compute $\vec{D}$ for $0 < r < 3$ and then for $r \geq 3$.)

5-8.0g Find the electric field intensity external to a metal sphere of radius a in vacuum, and charged uniformly to ρ_s C/m^2. (Hint: Find the displacement first. How does this process compare with finding $\vec{E}$ by integration?) (Answer: $\rho_s a^2/\epsilon_0 r^2)\hat{r}$)

5-8.0h Obtain equations for the electric flux density and electric field intensity external to an infinite metal cylinder of radius a charged uniformly to ρ_s C/m^2. Let the cylinder lie along the z axis and thus the fields will have only radial components. (Answer: $(a\rho_s/\rho)\hat{\rho}$)

5-8.0i Suppose it has been determined that the electric displacement in a region of space has the form

$$\vec{D} = 3xyz^2\hat{x} - 8xz\hat{y} + 2(x^2 + y^2 + z^2)\hat{z} \text{ C/m}^2$$

What is the charge density ρ_v at the point (1, 2, 3)? (Answer: 66 C/m^3)

5-9 Polarization, Bound Charge, and Electric Susceptibility

In previous sections references to free charges have been made. Such charge was taken to be that which was free to move through a material when an electric field was present. This implies the presence of another kind of charge, and we call this a *bound* charge because its motion is limited to a local region. Such charge is the result of polarization, which will be described next. Following that, the relationship between the polarization and the dielectric constant will be developed.

Suppose we have an atom or molecule and immerse it into an electric field as shown in Figure 5-50. We let $\vec{E}_0$ be the original electric field existing in vacuum as indicated. Now the positively charged components of the atom or molecule will be pushed in the direction of $\vec{E}_0$ and the negative components opposite to $\vec{E}_0$. The result is the generation of an electric dipole with equal but opposite sign charges at the ends since the original unit was electrically neutral. Note that within the dipole there will be an electric field $\vec{E}_d$ as shown. (Can you verify this from the definition of electric field intensity?) It is easy to see that the net electric field $\vec{E}$ will be the difference between $\vec{E}_0$ and $\vec{E}_d$. (This also gives us another qualitative reason why the electric field intensity is reduced by the presence of

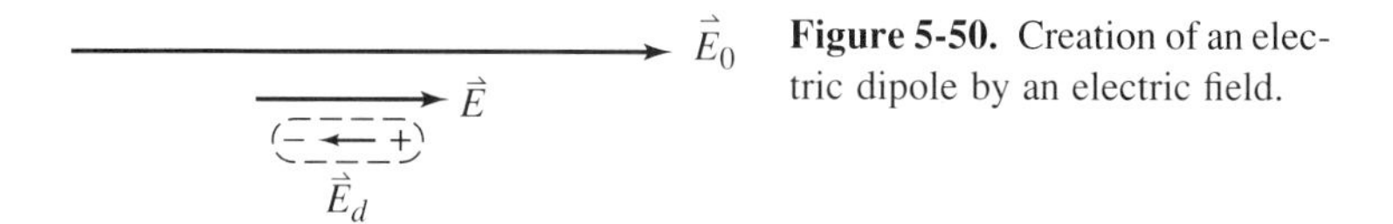

Figure 5-50. Creation of an electric dipole by an electric field.

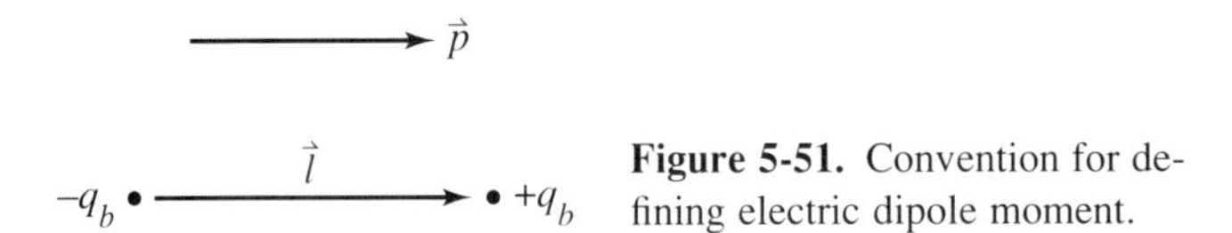

Figure 5-51. Convention for defining electric dipole moment.

a dielectric.) To quantify the phenomenon, we defined the *dipole moment* as shown in Figure 5-51 by the following equation:

$$\vec{p} = (q_b l)\hat{l}_{-+} \text{ C} \cdot \text{m} \tag{5-49}$$

In this equation q_b is the magnitude of the charge, l is the distance between the charges and $\hat{l}_{-+}$ is a unit vector directed from the negative charge to the positive charge. There are some important features of this definition:

1. The vector $\vec{p}$ is opposite the direction of $\vec{E}_d$.
2. The value of $\vec{p}$ depends upon $\vec{E}_0$ or $\vec{E}$ since the distance l will change as the force on the dipole changes.
3. The charge present is a bound charge since it can only move within the dipole region.
4. The charges cannot be indefinitely ''stretched'' apart, so a saturation might occur. Also, as the applied field changes magnitude, the increase and decrease in l may not follow the same size at equal E values, leading to a hysteresis effect. Such materials are called *ferroelectrics*.

Materials may have three main sources of dipoles. The first is *electronic polarization* (or atomic polarization) wherein the electrons in their atomic orbits are displaced with respect to the nucleus by an elongation of the orbital path to an elliptical pattern. The nucleus will be near one focus of the path and the average electron location at the other focus. A second source is *ionic polarization*, normally associated with lighter inorganic molecules. Such molecules are ''stretched'' by the electric field so the positive ion is at one end and the negative at the other. Another example would be sulfuric acid H_2SO_4, where the negative sulfate ion and hydrogen nuclei produce the end charges.

The final source of dipoles is called *molecular polarization*, usually associated with overall molecular effects or those molecules that have a polarization without the presence of an external field analogous to a small permanent magnet (called *polar* molecules). For the nonpolar case, the overall effect of an applied field is to pull electrons to one side and nuclei to another much like electronic polarization, but now with many atoms of the molecule in action to produce an effective net negative location at one point and positive at another. (Note also that electronic polarization occurs within the individual atoms of the molecule.) The polar case is exemplified by the H_2O molecule where the permanent dipoles are at random orientation and an applied electric field tends to align the molecules and to stretch them even further. This is often called *orientational polarization*. There are, incidentally, a few polar materials in which the permanent dipoles are not in completely random orientation so that a piece of the material has a permanent positive end and negative end similar to bar magnet. These materials are called *electrets*.

Irrespective of the source of the dipoles, the same definition and physical configuration may be used to represent the dipoles. Note also that an equivalent bound charge could be identified independent of the type of polarization.

In many cases we are not concerned about the state of polarization of each individual atom or molecule but rather with the state of a large piece of the material. Thus we need a quantity that can represent this state. We first consider a volume Δv of material and determine the overall dipole moment by vector-summing the individual dipole moments to obtain

$$\Delta\vec{p} = \sum_{i=1}^{\Delta n} \vec{p}_i \qquad \text{C} \cdot \text{m} \tag{5-50}$$

where Δn is the number of dipoles in Δv. From this we define the *polarization vector* as a dipole moment per unit volume by

$$\vec{P} = \lim_{\Delta v \to 0} \frac{\sum_{i=1}^{\Delta n} \vec{p}_i}{\Delta v} \quad \text{C/m}^2 \tag{5-51}$$

One could also define an average dipole moment and use

$$\vec{P} = \lim_{\Delta v \to 0} \left(\frac{\vec{p}_{\text{ave}}\, \Delta n}{\Delta v} \right) = \vec{p}_{\text{ave}} N \qquad \text{C/m}^2 \tag{5-52}$$

where N is the dipole density in dipoles/m^3. It is also interesting to note that the units of $\vec{P}$ are the same as $\vec{D}$.

Now consider a block of material immersed in a field $\vec{E}_0$ as pictured in Figure 5-52. As the dipoles line up, the positive ends move to the right and the negative ends to the

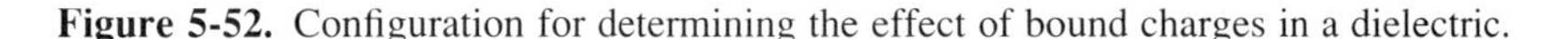

Figure 5-52. Configuration for determining the effect of bound charges in a dielectric.

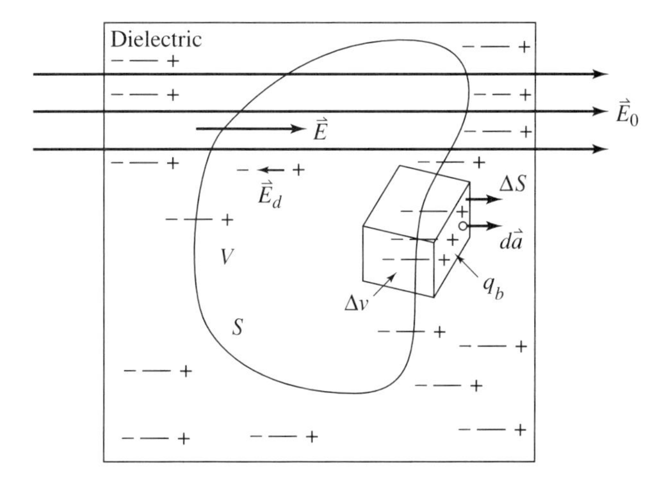

left. For an arbitrary volume V of the block and a volume Δv containing part of the surface S of V, we see that the positive bound charges q_b pass through ΔS almost as a current flow during the alignment process. This current is, in fact, a bound current, or more properly, a momentary displacement current since the q_b have displaced through the surface Δv, and equilibrium will be established when static conditions prevail with the q_b outside of Δv. Now, the bound charge came from the dipoles and, since the polarization $\vec{P}$ is the charge per unit area (dipole moment per unit volume), we obtain the total charge passing out of Δv as

$$\Delta Q_b = \oint_{\Delta S} \vec{P} \cdot d\vec{a}$$

Now the bound charge remaining *within* the volume Δv is the *negative* of this, so we may define a bound volume charge density as

$$\rho_{vb} = \lim_{\Delta v \to 0} \frac{-(\Delta Q_b)}{\Delta v} = \lim_{\Delta v \to 0} \frac{-\oint_{\Delta S} \vec{P} \cdot d\vec{a}}{\Delta v} = -\nabla \cdot \vec{P}$$

where we have used the definition of divergence (see Equation 4-39). We then have

$$\nabla \cdot \vec{P} = -\rho_{vb} \tag{5-53}$$

Let us compute the total volume charge density and seek to relate all of the parameters and field quantities. We express this total as the sum of the free and bound charge densities

$$\rho_{vT} = \rho_v + \rho_{vb}$$

Using Eqs. 5-47 and 5-53, we have

$$\rho_{vT} = \nabla \cdot \vec{D} - \nabla \cdot \vec{P} = \nabla \cdot (\vec{D} - \vec{P})$$

Now if we take $\vec{D}$ and subtract the polarization $\vec{P}$, the result must be the electric displacement corresponding to $\vec{E}$ without the dielectric (not $\vec{E}_0$), namely, $\epsilon_0\vec{E}$. Thus we write

$$\nabla \cdot (\epsilon_0 \vec{E}) = \rho_{vT} \tag{5-54}$$

Then

$$\vec{D} - \vec{P} = \epsilon_0 \vec{E}$$

or

$$\vec{D} = \epsilon_0 \vec{E} + \vec{P} \tag{5-55}$$

Now, as mentioned earlier, the dipole moment $\vec{p}$ is proportional to the electric field which it experiences, and thus $\vec{P}$ will also be proportional to the field $\vec{E}$, which is the net field. We then define

$$\vec{P} = \epsilon_0 \chi_e \vec{E} \tag{5-56}$$

where χ_e is called the *electric susceptibility*, a dimensionless number and a property of the material since some materials may polarize easier than others. Substituting this into Eq. 5-55, we obtain

$$\vec{D} = \epsilon_0\vec{E} + \epsilon_0\chi_e\vec{E} = \epsilon_0(1 + \chi_e)\vec{E} \tag{5-57}$$

Comparing this last result with Eq. 5-48, we have the important result

$$\epsilon = \epsilon_0(1 + \chi_e) \tag{5-58}$$

Using Eq. 5-3, we obtain

$$\epsilon_r = 1 + \chi_e \tag{5-59}$$

An important conclusion of this section is that, if we use the constitutive relationship $\vec{D} = \epsilon\vec{E}$, we need to concern ourselves only with free charges since the effect of the bound charges has already been included in the permittivity ϵ.

As a final note, there may seem to be a contradiction in Eq. 5-55 since it appears that $\vec{D}$ is a function of the dielectric. However, when a dielectric is placed in an electric field, $\vec{E}$ decreases to offset the added polarization component $\vec{p}$.

Example 5-21

For the results of Example 5-20, obtain a qualitative plot of the polarization $\vec{P}$ as a function of radius ρ for $a < \rho < c$.

Using Eqs. 5-56 and 5-59:

$$P_\rho(\rho) = \epsilon_0\chi_e E_\rho(\rho) = \epsilon_0(\epsilon_r - 1)E_\rho(\rho)$$

Thus, since ϵ_r changes value at $\rho = b$, the polarization also has a discontinuity at $\rho = b$. The plot is given in Figure 5-53. This is physically correct since as one moves away from the center conductor the electric field intensity decreases, which means the dipole moments are smaller so $\vec{P}$ also decreases.

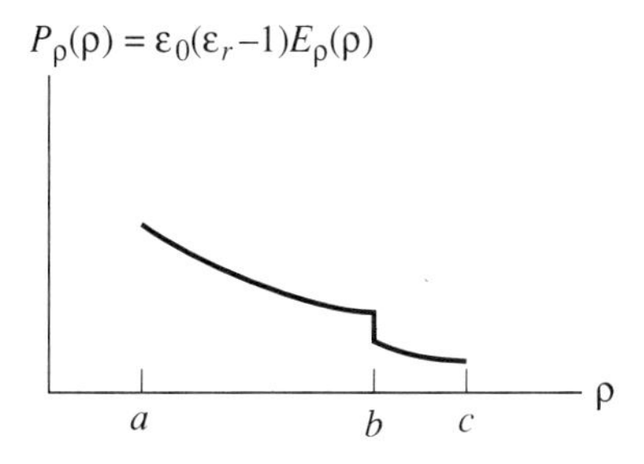

Figure 5-53. Polarization for the dielectrics of Example 5-20.

Example 5-22

A dielectric system has a relative permittivity (dielectric constant) of 2.7. The electric field is known to be

$$\vec{E} = 3xyz\hat{x} - 8xz\hat{y} + 2(x^2 + y^2 + z^2)\hat{z} \text{ V/m (N/C)}$$

Compute the total, free, and bound charge densities within the material.

$$\vec{D} = \epsilon\vec{E} = \epsilon_0\epsilon_r\vec{E} = 2.7\epsilon_0\vec{E} = 71.6xyz\hat{x} - 191xz\hat{y} + 47.7(x^2 + y^2 + z^2)\hat{z} \text{ pC/m}^2$$

$$\epsilon_0\vec{E} = 26.5xyz\hat{x} - 71xz\hat{y} + 17.7(x^2 + y^2 + z^2)\hat{z} \text{ pC/m}^2$$

Using Eq. 5-54:

$$\rho_{vT}(\vec{r}) = \nabla \cdot (\epsilon_0\vec{E}) = 26.5yz + 35.4z \text{ pC/m}^3$$

Using Eq. 5-47:

$$\rho_v(\vec{r}) = \nabla \cdot \vec{D} = 71.6yz + 95.4z \text{ pC/m}^3$$

Using Eqs. 5-53 and 5-55 (there are other choices)

$$\rho_{vb} = \nabla \cdot \vec{P} = -\nabla \cdot (\vec{D} - \epsilon_0\vec{E}).$$

Substituting the previous expressions for $\vec{D}$ and $\epsilon_0\vec{E}$,

$$\begin{aligned}\rho_{vb} &= -\nabla \cdot [45.1xyz\hat{x} + 30(x^2 + y^2 + z^2)\hat{z}] \times 10^{-12} \\ &= -45.1yz - 60z \text{ pC/m}^3\end{aligned}$$

Note that this last result is consistent with $\rho_{vb} - \rho_{vT} - \rho_v$.

Exercises

5-9.0a Show that the units of the left and right sides of Eq. 5-56 are equal.

5-9.0b A point charge of 5 μC is embedded in polystyrene, which has a dielectric constant of 2.5. Develop an equation for $\vec{P}$ of the dielectric. (Answer: $(2.39 \times 10^{-7}/r^2)\hat{r}$ C/m^2)

5-9.0c For Example 5-19, compute ρ_{vb} and ρ_{vT}. Note that the charge is in vacuum. (Answer: 0, $1/r$ C/m^3)

5-9.0d For the system of Exercise 5-8.0f, plot the polarization for all values of r ($0 \le r < \infty$).

5-9.0e Table salt, NaCl, has a dielectric constant (relative permittivity) of 6.1, a density of 2 g/cm^3, and a molecular weight of 58.5. If a solid section of salt is measured to have an electric field intensity of 20 kV/m (kN/C), compute the polarization, the average dipole moment, and the bound charge density. (Hint: You will need to compute the number of molecules (dipoles) per unit volume. Avogadro's number is 6.01×10^{23} molecules/g-mole.) (Answer: $P = 902$ nC/m^2, $p_{\text{ave}} = 4.4 \times 10^{-35}$ C · m, $\rho_{vb} = 0$)

We conclude this section with a short discussion of adjectives used to describe materials. Although these are described for electric fields, the same terms hold for conductivity or other material parameters used to relate two field quantities. These adjectives (and their opposites) are:

Linear (nonlinear)
Homogeneous (inhomogeneous)
Time-invariant (time-varying)
Isotropic (anisotropic)
Reciprocal (nonreciprocal)

1. A medium is said to be *linear* if the permittivity is independent of the magnitude of $\vec{D}$ or $\vec{E}$. A linear medium could, for example, have $\epsilon_r = 3$, whereas a nonlinear medium might have

$$\epsilon_r = \frac{3}{\left(1 - \left|\frac{\vec{E}}{10^5}\right|\right)^2}$$

2. A medium is said to be *homogeneous* if the permittivity has the same value at two different physical locations in the material. If ϵ_r is a function of coordinate values, the material is inhomogeneous.
3. A medium is said to be *time-invariant* if the value of permittivity is independent of time. If a material were subject to pressure, the value of ϵ_r might change so that if the pressure fluctuated with time, ϵ_r would also, thus giving a time-varying medium.
4. A medium is said to be *isotropic* if the value of permittivity is independent of the *direction* of a field within the material. For example, in a crystalline material as one "looks" different directions from a given point within the material, the lattice structure presents a different "view" and hence a different value of ϵ_r. This would be an anisotropic material.
5. A material is said to be *reciprocal* if the value of permittivity is the same looking in directions 180° from each other at a given location. This is a particular case of 4, but of sufficient importance to justify a separate classification. It is particularly important for fields that are functions of time or, more particularly, sinusoidal.

An anisotropic medium gives us particular concern because, in order to use the electric field constitutive relation between $\vec{D}$ and $\vec{E}$, we need different ϵ_r values for the x, y, and z directions. The usual practice is to express ϵ as a matrix and it is called the tensor permittivity. The matrix would contain nine numbers, which would account for all possibilities as follows:

$$[\epsilon] = \begin{bmatrix} \epsilon_{11} & \epsilon_{12} & \epsilon_{13} \\ \epsilon_{21} & \epsilon_{22} & \epsilon_{23} \\ \epsilon_{31} & \epsilon_{32} & \epsilon_{33} \end{bmatrix}$$

We would then write the constitutive relationship as

$$[\vec{D}] = [\epsilon][\vec{E}] \tag{5-60}$$

It turns out that one important effect of an anisotropic medium is to make $\vec{D}$ and $\vec{E}$ have different directions in space, as the next example shows.

Example 5-23

Suppose we have a dielectric constant tensor given by

$$[\epsilon_r] = \begin{bmatrix} 2 & 0.6 & 0.6 \\ 0.6 & 1.7 & 0.3 \\ 0.6 & 0.3 & 1.5 \end{bmatrix}$$

and that measurements show that, in the material, the electric field is $\vec{E} = 2 \times 10^{-3}\hat{x} + 3 \times 10^{-3}\hat{y}$ N/C. Find the electric displacement (flux density).

Using Eq. 5-60,

$$\begin{bmatrix} D_x \\ D_y \\ D_z \end{bmatrix} = \epsilon_0[\epsilon_r]\begin{bmatrix} E_x \\ E_y \\ E_z \end{bmatrix} = \epsilon_0\begin{bmatrix} 2 & 0.6 & 0.6 \\ 0.6 & 1.7 & 0.3 \\ 0.6 & 0.3 & 1.5 \end{bmatrix}\begin{bmatrix} 2 \times 10^{-3} \\ 3 \times 10^{-3} \\ 0 \end{bmatrix} = \begin{bmatrix} 5.1 \times 10^{-14} \\ 5.6 \times 10^{-14} \\ 1.8 \times 10^{-14} \end{bmatrix} \text{C/m}^2$$

Thus

$$\vec{D} = 5.1 \times 10^{-14}\hat{x} + 5.6 \times 10^{-14}\hat{y} + 1.8 \times 10^{-14}\hat{z} \text{ C/m}^2$$

Note that $\vec{D}$ has a $\hat{z}$ component but $\vec{E}$ does not, so $\vec{D}$ and $\vec{E}$ are not parallel.

Exercises

5-9.0f Using the dot product, find the angle between $\vec{D}$ and $\vec{E}$ in Example 5-23. (Answer: 15.4°)

5-9.0g In a vacuum, $\vec{D}$ is $3 \times 10^{-12}\hat{x} - 7 \times 10^{-12}\hat{y}$. Determine the force on an electron placed in this flux density. Suppose now that an anisotropic material is placed in this field ($\vec{D}$ does not change) for which the relative permittivity tensor is

$$[\epsilon_r] = \begin{bmatrix} 2 & 0.6 & 0.6 \\ 0.6 & 1.7 & 0.3 \\ 0.6 & 0.3 & 1.5 \end{bmatrix}$$

Now compute the force on the electron. Compare force directions. (Hint: You will have to invert a matrix.)

5-9.0h For the relative permittivity tensor of Exercise 5-9.0g, obtain the electric susceptibility tensor $[\chi_e]$. (Hint: Replace the 1 by the identity or unit matrix $[I]$ in Eq. 5-59.)

5-10 Potential and Equipotential Contours

Previous sections have dealt with vector field quantities and the vector formulation tends to divert our attention from the problems being investigated. Care must be exercised in manipulating the various quantities to preserve correct spatial orientation as well as magnitude. It would be convenient if we could obtain a scalar quantity that makes the com-

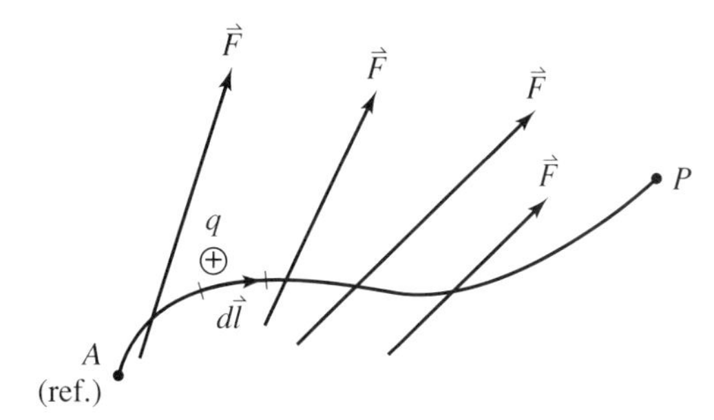

Figure 5-54. Configuration for the determination of the work required to move a positive test charge in a force field.

putations simpler but from which we could obtain the vector field quantities. For electric fields this is called the *electric scalar potential*, or potential, for brevity.

By definition, the electric scalar potential at a point in an electric field is the negative of the energy required to move a unit positive charge from any selected reference point to the desired point in the electric field. The negative sign is required because to increase the potential energy of a positive charge, one would have to push the charge against the electric field vector. The next discussion will demonstrate this. From this definition, it is usual to say that the potential is with respect to the selected reference.

For a positive test charge in a force field as pictured in Figure 5-54, the differential work (energy) to move the test charge along dl is

$$dw = \vec{F} \cdot d\vec{l}$$

where $d\vec{l}$ is the differential path in the direction of increasing *coordinate* values. It need not point from the reference to the final point. Integration limits will account for the actual direction one wishes to move on the path. (See discussion in Chapter 4.) This means that we may always use the expression for $d\vec{l}$ developed in Chapter 4; namely,

$$dl = h_1 du_1 \hat{u}_1 + h_2 du_2 \hat{u}_2 + h_3 du_3 \hat{u}_3$$

If any of the coordinates happen to be constant over a particular path, then those corresponding du_is are simply set to zero in the preceding general expression. Now, if we let V represent the scalar potential, the definition of potential results in

$$dV = -\frac{dw}{q} = -\frac{\vec{F} \cdot d\vec{l}}{q} = -\vec{E} \cdot d\vec{l} \tag{5-61}$$

From this result, we know that the units of potential are work per unit positive charge, or joule/coulomb (J/C). This is called the *volt* in common terminology. Also, this verifies the units of electric field intensity as volt/meter used in earlier sections.

In a battery chemical energy separates the charges of the system. When the first positive charge is ''sent'' to the positive terminal, it produces an electric field. To get a second positive charge there, that charge must be pushed against the field set up by the first. This also shows why we need the minus sign in the definition. To increase the potential (i.e., make it positive), the path $d\vec{l}$ is opposite that of $\vec{E}$ so that the dot product gives a negative sign. The negative sign in the definition cancels that one, yielding a positive result.

Integrating the preceding equation, we have the expression for potential as

$$V = -\int_{\text{ref}}^{P} \vec{E} \cdot d\vec{l} \tag{5-62}$$

This is the integral form. What we want, though, is some way to compute $\vec{E}$ if we first determine V. From calculus, we write the total differential as (assuming V is not a function of time so there is no time differential)

$$\begin{aligned} dV &= \frac{\partial V}{\partial x} dx + \frac{\partial V}{\partial y} dy + \frac{\partial V}{\partial z} dz \\ &= \left(\frac{\partial V}{\partial x} \hat{x} + \frac{\partial V}{\partial y} \hat{y} + \frac{\partial V}{\partial z} \hat{z} \right) \cdot (dx\hat{x} + dy\hat{y} + dz\hat{z}) \end{aligned}$$

In the last expression we identify the rectangular operation and path:

$$\nabla V = \frac{\partial V}{\partial x} \hat{x} + \frac{\partial V}{\partial y} \hat{y} + \frac{\partial V}{\partial z} \hat{z}$$

$$d\vec{l} = dx\hat{x} + dy\hat{y} + dz\hat{z}$$

The voltage differential becomes

$$dV = \nabla V \cdot d\vec{l}$$

Comparing this with Eq. 5-61, we then have

$$\vec{E} = -\nabla V \qquad \text{(time invariant fields)} \tag{5-63}$$

Some important observations should be made. First, V is a function of coordinates, so the values of V form a scalar field. To obtain the vector field $\vec{E}$ we simply expand the preceding equation in the appropriate coordinate systems as explained in Chapter 4. Second, V is a point function, so we may obtain the potential at any location by merely substituting the coordinates of that location.

Example 5-24

Find the electric field between the two metal plates for the system shown in Figure 5-55. Also compute the energy required to push a positive charge from the lower plate to the upper plate vertically.

We note that the voltage on the bottom plate at $y = 0$ is $V(x, 0) = 0$ and on the top plate at $y = 1$ m is $V(x, 1) = 50$ V. This suggests a potential function $V(y) = 50y$. Using rectangular coordinates, Eq. 5-63 gives

$$\vec{E} = -\nabla V = -\left(\frac{\partial(50y)}{\partial x} \hat{x} + \frac{\partial(50y)}{\partial y} \hat{y} + \frac{\partial(50y)}{\partial z} \hat{z} \right) = -50\hat{y} \text{ V/m}$$

To check this answer qualitatively, put a positive test charge between the plates. Since the top plate is at +50 V, the test charge would move down ($-\hat{y}$) as predicted by the expression for $\vec{E}$.

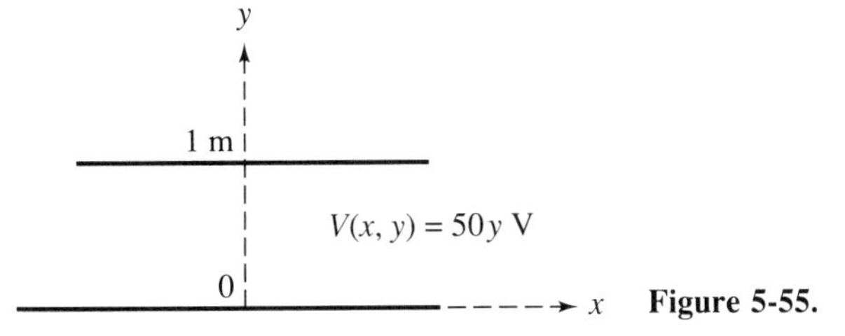

Figure 5-55.

Next, using the value of $\vec{E}$, we compute energy:

$$W = \int_{y=0}^{y=1} dw = \int_{y=0}^{y=1} \vec{F} \cdot d\vec{l}$$

Now, to move the charge up, we are pushing against the electric field, that is, against the force the field exerts on the charge. Thus

$$\vec{F} = -(q\vec{E}) = -(1)(-50\hat{y}) = 50\hat{y}$$

Since x and y do not change, dx and dz are zero so that

$$d\vec{l} = dx\hat{x} + dy\hat{y} + dz\hat{z} \rightarrow dy\hat{y}$$

Then

$$W = \int_0^1 50\hat{y} \cdot dy\hat{y} = 50 \text{ J}$$

Exercises

5-10.1a The coaxial cable shown in Figure 5-56 has a voltage applied to it such that the potential distribution is given by

$$V(\rho, \phi, z) = 25000(10^{-3} - \rho) + 100 \text{ V}$$

(a) What is the potential on each conductor of the coax?
(b) Find the equation of the electric field between the conductors.
(c) Explain why your solution for $\vec{E}$ is physically reasonable.
(d) If the dielectric constant (relative permittivity) is 3, what is the equation for the electric flux density (electric displacement)? What is ρ_V? (Answer: $V(1 \text{ mm}) = 100$ V;

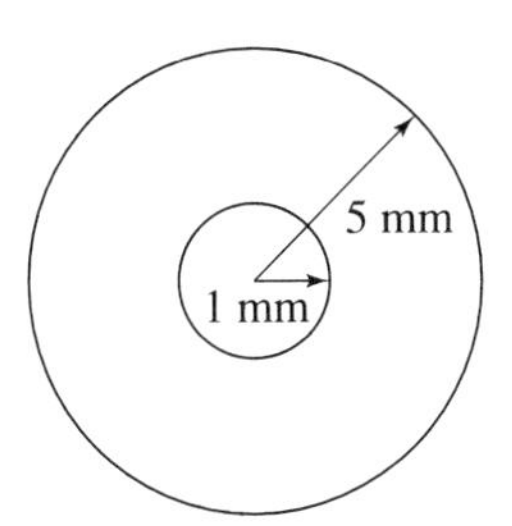

Figure 5-56.

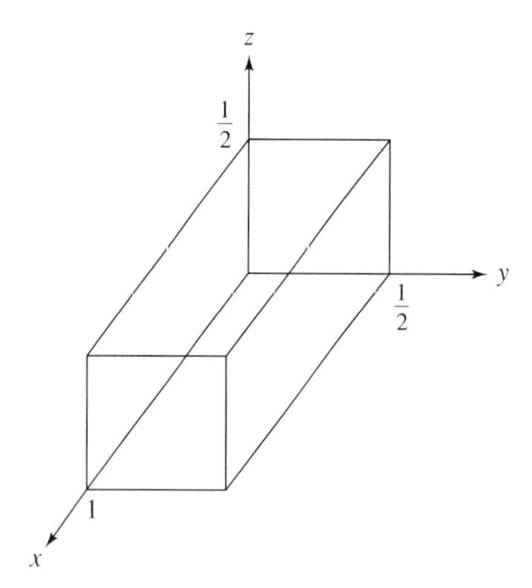

Figure 5-57.

$V(5 \text{ mm}) = 0$ V; $25{,}000\hat{\rho}$ V/m; positive charge will be pushed away from the 100-V conductor; $0.66\hat{\rho}\ \mu\text{C/m}^2$; $0.66/\hat{\rho}\ \text{C/m}^3$)

5-10.0b For the system of Exercise 5-10.0a, what energy (work) is required to move $a + 1$ C charge from 1 mm to 5 mm radially?

5-10.0c For the closed metal box shown in Figure 5-57, the potential within is

$$V(x, y, z) = 10 \sin \pi x \sin 2\pi y \sin 2\pi z \text{ V}$$

(a) What is the potential on each box surface?
(b) Find the electric field everywhere inside the box.
(c) Find a unit vector in the direction of $\vec{E}$ in the center of the box.
(d) Find the equation for the volume charge density inside the box.

5-10.0d A point charge is located at the origin as shown in Figure 5-58. Compute the potential at point B with respect to point A (reference) by direct integration. How could you tell

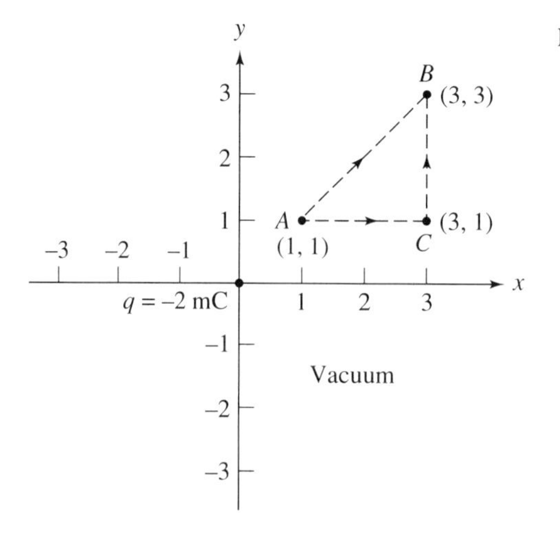

Figure 5-58.

ahead of time whether or not the integration is path independent? (See Example 4-30 of Section 4-19.) (Answer: 8.48 MV; 8.48 MV; $\nabla \times \vec{E} = 0$)

Since the determination of potential from the electric field intensity requires a selection of a path between the two points, the criterion of "best path selection" would be helpful. Equation 5-62 tells us in general that

$$V = -\int_{\text{ref}}^{P} \vec{E} \cdot d\vec{l}$$

Using Eq. 5-63 and the equation just preceding it (which apply to non–time-varying fields only), we have

$$V = -\int_{\text{ref}}^{P} -\nabla V \cdot d\vec{l} = \int_{\text{ref}}^{P} dV = V(P) - V(\text{ref})$$

We conclude from this that the potential in the *static* (time-invariant) case is only a function of the locations of the end points of the path, so we may select any path between them we wish. An interesting result is obtained if we start at a point A, follow any path C, and return to A. The preceding equation yields

$$V_{AA} = -\int_{A}^{A} \vec{E} \cdot d\vec{l} = -\oint_{C} \vec{E} \cdot d\vec{l} = V(A) - V(A) = 0$$

Applying Stokes' theorem (Chapter 4) to the closed path integral yields

$$\oint_{C} \vec{E} \cdot d\vec{l} = \int_{S} \nabla \times \vec{E} \cdot d\vec{a} = 0 \tag{5-64}$$

Since this must be true for any path C (or surface S), the integrand must be zero. Thus

$$\nabla \times \vec{E} = 0 \qquad \text{(time invariant)} \tag{5-65}$$

As discussed in Chapter 4, this defines a conservative field. In conclusion, if a field is conservative, the contour integration between two points in the field is independent of the contour (path) between the points.

Example 5-25

Find the equation for the potential at a position $\vec{R}$ from a point charge as shown in Figure 5-59. Use the location $R = \infty$ as reference.

As in earlier work, we let $\vec{r}'$ locate the source point and $\vec{r}$ the field point P at which the voltage is to be determined. From Eq. 5-10, we can determine the electric field intensity as

$$\vec{E} = \frac{q}{4\pi\epsilon|\vec{R}|^2}\hat{R}, \qquad \vec{R} = \vec{r} - \vec{r}'$$

Now, if one were to evaluate $\nabla \times \vec{E}$ in spherical coordinates, the result would be zero, so that $\vec{E}$ describes a conservative field. Thus we may select any path we wish from ∞ in to

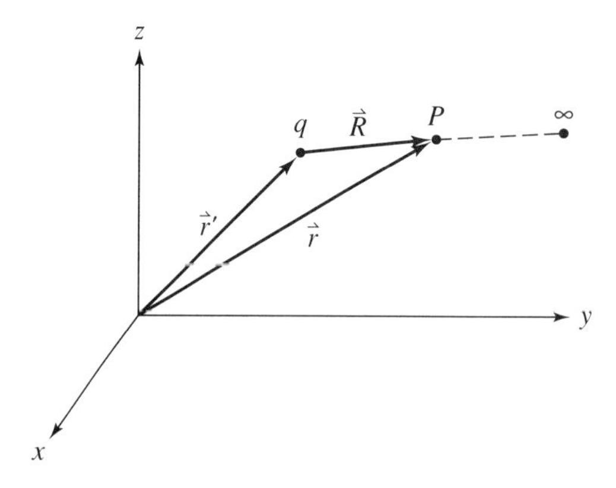

Figure 5-59. Configuration used to determine the potential around a point charge q.

P. Let us start at ∞ and enter along a line that is the extension of the line q to P (i.e., in the $\hat{R}$ direction). We then have, using R as $|\vec{R}|$,

$$V = -\int_{\infty}^{R} \vec{E} \cdot dl = -\int_{\infty}^{R} \frac{q}{4\pi\epsilon R^2} \hat{R} \cdot dR\hat{R} = -\frac{q}{4\pi\epsilon}\int_{\infty}^{R} \frac{dR}{R^2} = \frac{q}{4\pi\epsilon R} \qquad (5\text{-}66)$$

Note that the integration path is actually in the negative (incoming) direction. This is taken care of by the limits.

It might be appropriate to point out here that it may not always be possible to select ∞ as reference. This is particularly true when the charge distribution extends to ∞, such as a line or cylinder of charge. In such cases we pick any convenient location as reference. Physically, this may be interpreted as follows: Suppose we have two metal plates, one at 200 V and one at 75 V. If we place a voltmeter between the two, we measure 125 V with respect to the 75-V plate as reference. Note that all we have assumed is that the potentials of the two plates are initially given with respect to the same reference (often called *ground*).

Exercises

5-10.0e Show that $\nabla \times \vec{E}$ for the point charge is zero. Remember $\vec{r}\,'$ is a constant when taking derivatives with respect to unprimed coordinates. Use rectangular coordinates where $\vec{r} = x\hat{x} + y\hat{y} + z\hat{z}$, etc.

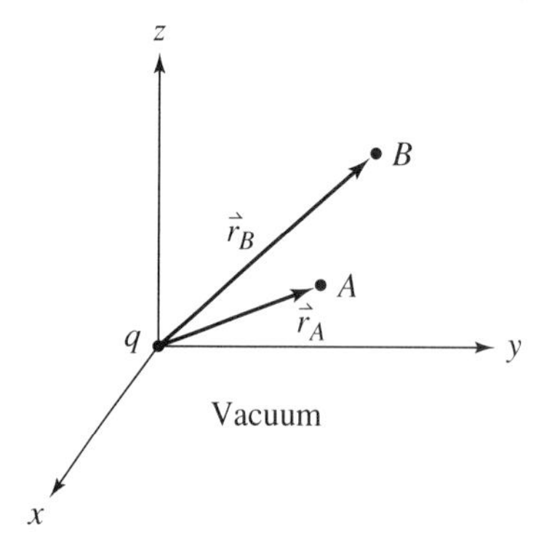

Figure 5-60.

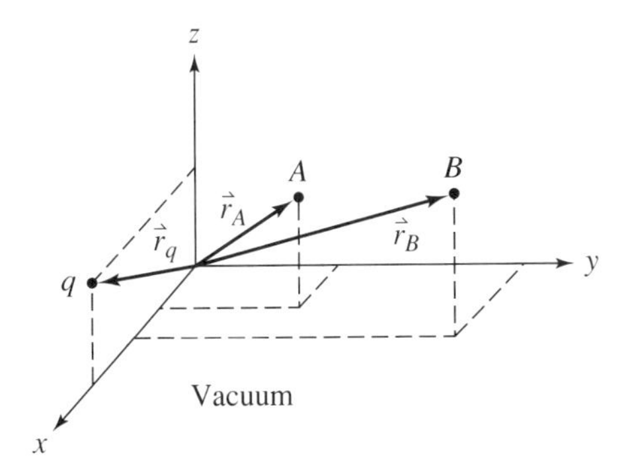

Figure 5-61.

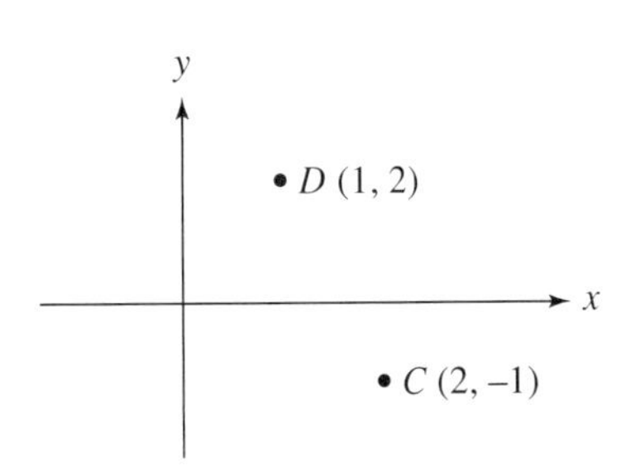

Figure 5-62.

5-10.0f For the point charge shown in Figure 5-60, determine the potential of A with respect to B. $q = 1\ \mu\text{C}$, $|\vec{r}_A| = 2$ cm, $|\vec{r}_B| = 3$ cm. (Answer: 150 kV)

5-10.0g For the point charge shown in Figure 5-61, determine the voltage at A with respect to B. $q = 2\ \mu\text{C}$, $\vec{r}_q = 3\hat{x} + 2\hat{z}$, $\vec{r}_A = \hat{x} + 2\hat{y} + 2\hat{z}$, and $\vec{r}_B = 2\hat{x} + 5\hat{y} + 3\hat{z}$.

5-10.0h An electric field is given by $\vec{E} = 2x\hat{x} + 3y\hat{y}$. What is the potential of point D with respect to C, as shown in Figure 5-62?

5-10.0i For the system of Exercise 5-10.0f, what would the potential be if the dielectric were not vacuum? Explain the physical reason for your answer. (Answer: $150/\epsilon_r$ kV. For a dielectric the field intensity decreases, so the force per unit charge decreases. Thus it requires less energy (work) to move a charge between two points.)

Since in engineering we deal more with distributions of charge rather than point charges, we now need a general formula for determining the potential external to such a region. Since we are assuming linear media, we may use the superposition theorem and merely add the potentials produced by the individual charges. The process is much nicer than for the electric field computation since we are now dealing with scalars. What we do is first obtain an expression for the potential at a general location $V(\vec{r})$, and then obtain the static electric field intensity using the gradient relationship, Eq. 5-63.

In Figure 5-63 the conventions are shown. As before, the primes locate points in the charge (source) region and the unprimes locate points at which V is to be determined (field point). The argument list $(\vec{r})$ is a generic way of representing a function or coordinates and does not mean the function is a vector function. From the figure and using Eq. 5-66:

$$dq = \rho_v(\vec{r}\,')dv'$$

$$dV = \frac{dq}{4\pi\epsilon R} = \frac{\rho_v(\vec{r}\,')dv'}{4\pi\epsilon R}$$

Integrating, we have

$$V(\vec{r}) = \int_{v'} \frac{\rho_v(\vec{r}\,')}{4\pi\epsilon R}\,dv' \qquad \text{reference at } \infty \tag{5-67}$$

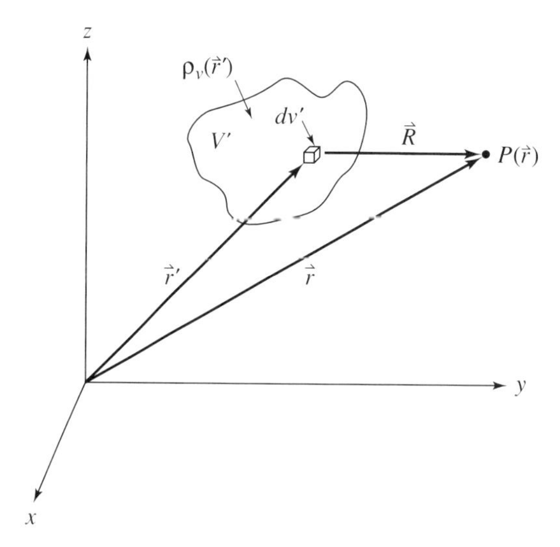

Figure 5-63. Definitions of terms used to develop the general potential formula.

Although this can be used for surface and line charge densities by using delta functions as done in earlier sections, we may use surface and line integrals directly as

$$V(\vec{r}) = \int_{S'} \frac{\rho_s(\vec{r}\,')}{4\pi\epsilon R}\, da' \qquad \text{reference at } \infty \tag{5-68}$$

$$V(\vec{r}) = \int_{C'} \frac{\rho_l(\vec{r}\,')}{4\pi\epsilon R}\, dl' \qquad \text{reference at } \infty \tag{5-69}$$

Incidentally, the quantity $1/4\pi\epsilon R$ appearing in all of the above is the potential of a positive 1-C point charge, which corresponds to a charge density $\rho_v = 1\delta_{u_1}\delta_{u_2}\delta_{u_3}$ C/m^3, and would be the "impulse" response of the system, that is, the impulse potential distribution in the region. Such a response function is called the Green's function, $G(R)$, of the system. From linear system theory we recall that if we have the impulse response of a system, we can obtain the response to any input (charge distribution) by convolving the input with the impulse response. Convolution is a folding-shifting-integrating process (Faltung integration). We so frequently use time functions in other studies that we might overlook the analogous operation with coordinate variable functions. This same convolution operation can be identified in the integrals that yield the electric field intensities resulting from arbitrary charge distributions as given by Eqs. 5-17, 5-22, and 5-23. It is instructive to identify the Green's function in those equations (a vector function this time).

Example 5-26

For the two-charge configuration shown in Figure 5-64, determine the potential at points P far from the two charges, that is, for $r \gg l$. (This is called the *far field*, or Fraunhofer zone, result.) Compute and plot $|\vec{E}|$.

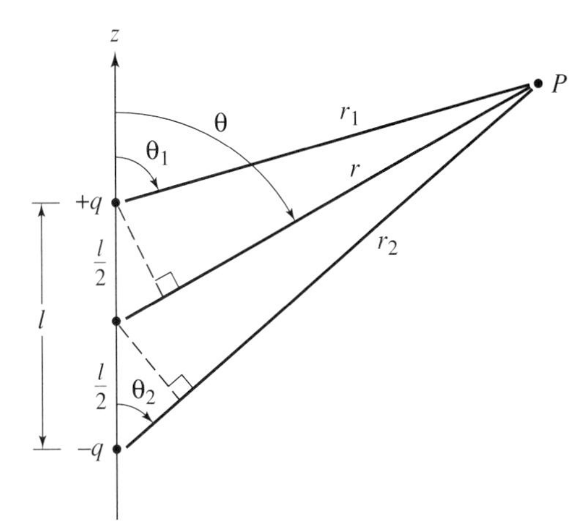

Figure 5-64. Electric dipole of two charges separated by an arbitrary distance l.

Using the superposition of the two potentials, we begin by writing

$$V(\vec{r}) = V_{+q} + V_{-q} = \frac{q}{4\pi\epsilon r_1} + \frac{-q}{4\pi\epsilon r_2} = \frac{1}{4\pi\epsilon}\left(\frac{1}{r_1} - \frac{1}{r_2}\right)$$

Now, at large distances the lines r_1, r, and r_2 will be essentially parallel so that $\theta_1 = \theta_2 = \theta$. Then, using the law of cosines,

$$V(\vec{r}) \approx \frac{q}{4\pi\epsilon}\left(\frac{1}{\sqrt{\left(\frac{l}{2}\right)^2 + r^2 - 2\frac{l}{2}r\cos\theta}} - \frac{1}{\sqrt{\left(\frac{l}{2}\right)^2 + r^2 - 2\frac{l}{2}r\cos(180^\circ - \theta)}}\right)$$

Since $r \gg l$, we may neglect $(l/2)^2$ and then factor an r^2 out of each radical to obtain

$$V(\vec{r}) \approx \frac{q}{4\pi\epsilon r}\left(\frac{1}{\sqrt{1 - \frac{l}{r}\cos\theta}} - \frac{1}{\sqrt{1 + \frac{l}{r}\cos\theta}}\right)$$

Now l/r is also very small, so we may use the following approximation from algebra:

$$(1 \pm x)^n \approx 1 \pm nx \qquad \text{for} \quad x \ll 1$$

For $n = -\frac{1}{2}$ the voltage expression becomes

$$V(\vec{r}) \approx \frac{q}{4\pi\epsilon r}\left[\left(1 + \frac{1}{2}\frac{l}{r}\cos\theta\right) - \left(1 - \frac{1}{2}\frac{l}{r}\cos\theta\right)\right] = \frac{ql}{4\pi\epsilon r^2}\cos\theta$$

From the discussion of polarization in Section 5-9.0, we can recognize ql as the magnitude of the dipole moment. We finally have

$$V(\vec{r}) \approx \frac{|\vec{p}|\cos\theta}{4\pi\epsilon r^2}$$

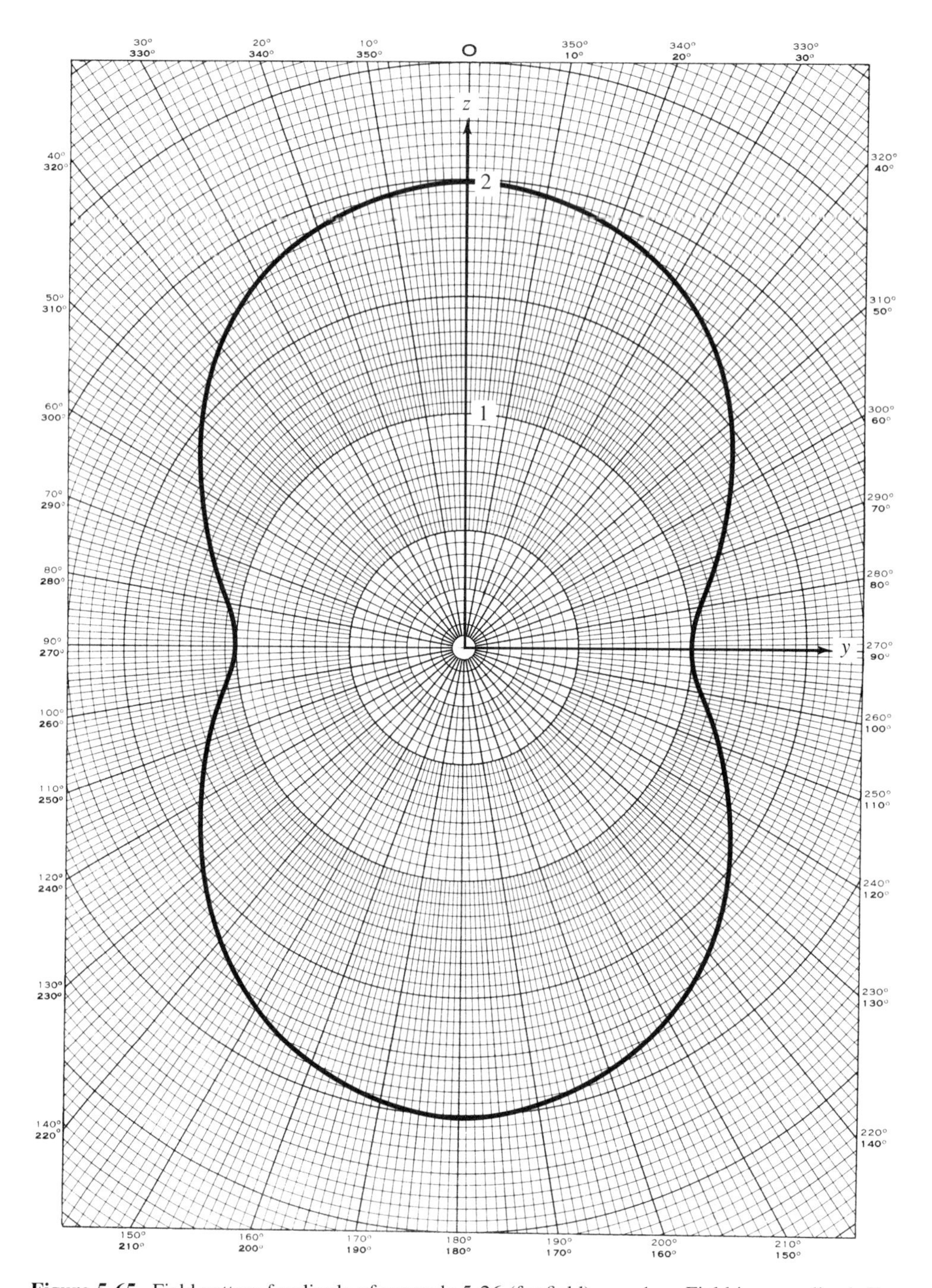

Figure 5-65. Field pattern for dipole of example 5-26 (far field), $r = 1$ m. Field is normalized, E_N.

Using polar coordinates, we have the far-field electric field as

$$\vec{E} = -\nabla V \approx \frac{2|\vec{p}|\cos\theta}{4\pi\epsilon r^3}\hat{r} + \frac{|\vec{p}|\sin\theta}{4\pi\epsilon r^3}\hat{\theta}$$

Now, to plot the field pattern $|\vec{E}|$ we hold r constant, as described earlier, and then normalize $\vec{E}$ by dividing out any constants. This results in

$$|\vec{E}|_N = \sqrt{4\cos^2\theta + \sin^2\theta} = \sqrt{4 - 3\sin^2\theta}$$

This is plotted in Figure 5-65.

Example 5-27

Obtain the equations for the potential (voltage) and electric field intensity for the finite length line charged to ρ_{l0} C/m (constant) shown in Figure 5-66.

Using the techniques presented in Chapter 4 and previous examples in this chapter:

$$\begin{aligned} R &= |\vec{r} - \vec{r}\,'| = |x\hat{x} + y\hat{y} + z\hat{z} - x'\hat{x}' - y'\hat{y}' - z'\hat{z}'| \\ &= |(x - x')\hat{x} + (y - y')\hat{y} + (z - z')\hat{z}| \end{aligned}$$

$$\rho_v(\vec{r}\,') = \rho_{l0}\delta(x')\delta(y')U(H + z')U(H - z')\ \text{C/m}^3$$

Using Eq. 5-67:

$$V(x, y, z) = \int_{-\infty}^{\infty}\int_{-\infty}^{\infty}\int_{-\infty}^{\infty} \frac{\rho_{l0}\delta(x')\delta(y')U(H + z')U(H - z')}{4\pi\epsilon\sqrt{(x - x')^2 + (y - y')^2 + (z - z')^2}}\,dx'dy'dz'$$

Using the properties of delta and step functions and converting unprimed to polar:

$$\begin{aligned} V(r, \theta, \phi) &= \frac{\rho_{l0}}{4\pi\epsilon}\int_{-H}^{H} \frac{dz'}{\sqrt{x^2 + y^2 + z^2 - 2zz' + z'^2}} \\ &= \frac{\rho_{l0}}{4\pi\epsilon}\int_{-H}^{H} \frac{dz'}{\sqrt{r^2 - 2r\cos\theta z' + z'^2}} \end{aligned}$$

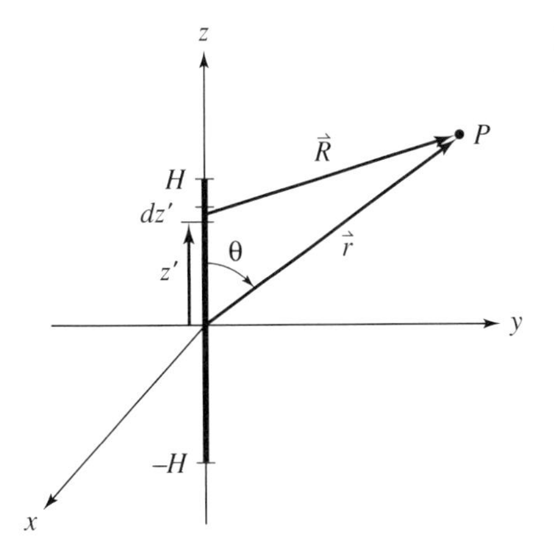

Figure 5-66. Finite charged line for Example 5-27.

Since the unprimed coordinates are constants as far as z' is concerned, we look up in an integral table the form $(a + bz' + z'^2)^{-1/2}$. Then:

$$V(r, \theta, \phi) = \frac{\rho_{l0}}{4\pi\epsilon} \ln\left(\frac{\sqrt{H^2 + r^2 - 2rH\cos\theta} - r\cos\theta + H}{\sqrt{H^2 + r^2 + 2rH\cos\theta} - r\cos\theta - H}\right)$$

Using this potential function, the electric field is obtained using

$$\vec{E} = -\nabla V$$

in spherical coordinates. This differentiation will not be done here.

Example 5-28

This example will show a case for which ∞ cannot be used as the reference for zero potential. This is usually the case when any dimension of the system is infinite. Consider the case of the infinite length charged line, which is Figure 5-66, with $H \to \infty$. If we let H become infinite in $V(r, \theta, \phi)$ of Example 5-27, we obtain infinite value for $V(r, \theta, \phi)$.

Since the line is infinite, we may take a point in the xy plane because the result will be independent of z. For the line, the field intensity, using Gauss' law (see General Exercise GE5-21), is (see Figure 5-67)

$$\vec{E}(\rho) = \frac{\rho_{l0}}{2\pi\epsilon\rho}\hat{\rho}$$

One can easily verify that $\nabla \times \vec{E}$ is zero in cylindrical coordinates, so the vector field is conservative. We may then select any path, so we select a path in the ρ coordinate (constant ϕ). Using Eq. 5-62:

$$V(\rho) = -\int_{\text{ref}}^{\rho} \vec{E}\cdot d\vec{l} = -\int_{\text{ref}}^{\rho} \frac{\rho_{l0}}{2\pi\epsilon\rho}\hat{\rho}\cdot d\rho\hat{\rho} = -\frac{\rho_{l0}}{2\pi\epsilon}\ln\rho\Big|_{\text{ref}}^{\rho}$$

Note that we cannot select reference at $\rho = \infty$ since the logarithm is undefined there. Arbitrarily select $\rho = 1$ m as the reference. Since $\ln 1 = 0$, the potential is

$$V(\rho) = -\frac{\rho_{l0}}{2\pi\epsilon}\ln\rho \text{ V}$$

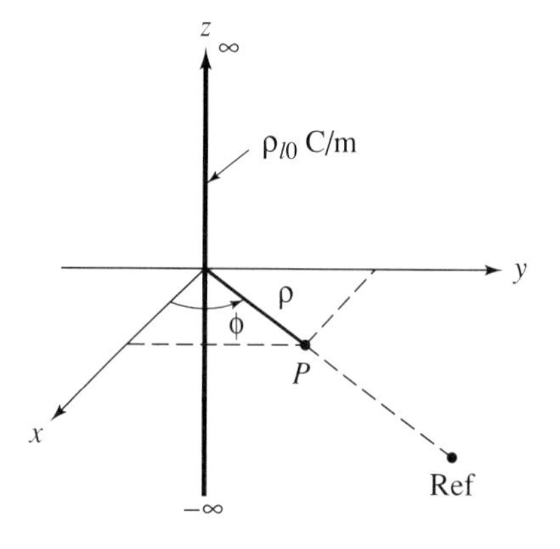

Figure 5-67. Finite charged line for Example 5-28.

Note that for any ρ less than 1 m the potential will be the same sign as ρ_{l0} since the point would be closer to the line than the reference, but it will be the opposite sign if ρ is greater than 1 m.

Example 5-29

There are many problems that have simple configurations but for which the integrations are difficult or impossible. In such cases, plots of potential are obtained by numerical techniques. Figure 5-68 is such a system. From the figure we write expressions for the various parts of the integrand using the surface form, Eq. 5-68.

$$\rho_s(\vec{r}\,') = \frac{10^{-6}}{\rho}\ \text{C/m}^2 \qquad \text{(function of coordinate)}$$

$$da' = \rho' d\rho' d\phi' \qquad \text{(note this is not a vector)}$$

$$\begin{aligned}
\vec{R} &= \vec{r} - \vec{r}\,' = x\hat{x} + y\hat{y} + z\hat{z} - x'\hat{x}' - y'\hat{y}' \\
&= (x - x')\hat{x} + (y - y')\hat{y} = z\hat{z} \\
R &= |\vec{R}| = \sqrt{(x - x')^2 + (y - y')^2 + z^2} \\
&= \sqrt{x^2 - 2xx' + x'^2 + y^2 - 2yy' + y'^2 + z^2} \\
&= \sqrt{r^2 - 2xx' - 2yy' + \rho'^2} \\
&= \sqrt{r^2 - 2x\rho' \cos\phi' - 2y\rho' \sin\phi' + \rho'^2} \\
&= \sqrt{r^2 - 2(x\cos\phi' + y\sin\phi')\rho' + \rho'^2}
\end{aligned}$$

From these,

$$V(\vec{r}) = \int_0^{2\pi}\int_0^{a} \frac{\dfrac{10^{-6}}{\rho'}}{4\pi\epsilon\sqrt{r^2 - 2(x\cos\phi' + y\sin\phi')\rho' + \rho'^2}}\,\rho' d\rho' d\phi'$$

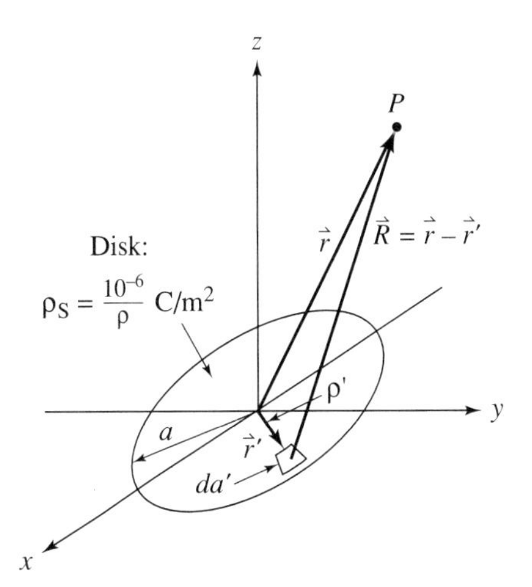

Figure 5-68. Configuration for determining the potential at a general point around a charged circular disk in the xy plane (Example 5-29).

Using an integral table, we obtain the ρ' integration, remembering that r^2, x, and y are all constants now, as is ϕ' for the first integration:

$$V(\vec{r}) = \frac{10^{-6}}{4\pi\epsilon}\int_0^{2\pi}\left[\ln\left(\sqrt{r^2 - 2(x\cos\phi' + y\sin\phi')\rho' + \rho'^2} + \rho' + \frac{-2(x\cos\phi' + y\sin\phi')}{2}\right)\right]_{\rho'=0}^{\rho'=a} d\phi'$$

$$V(\vec{r}) = \frac{10^{-6}}{4\pi\epsilon}\int_0^{2\pi}\ln\left[\frac{\sqrt{r^2 - 2a(x\cos\phi' + y\sin\phi') + a^2} + a - (x\cos\phi' + y\sin\phi')}{r - (x\cos\phi' + y\sin\phi')}\right]d\phi'$$

The remaining ϕ' integration is clearly not a simple one. Thus to plot $V(\vec{r})$ one selects a distance r and then any two of the coordinates x, y, z (since any two give the third because r^2 is $x^2 + y^2 + z^2$). Choosing x and y is probably best since they already appear explicitly in the equation. Then put these values in the integrand and numerically integrate $\phi' \in [0, 2\pi]$. For a specific case, suppose we plot V in the yz plane ($x = 0$) using $a = 2$ cm, $r = 4$ cm and $\epsilon = \epsilon_0$. Then the preceding equation reduces to

$$V(r = 0.04, x = 0, y) = 9000\int_0^{2\pi}\ln\left(\frac{\sqrt{0.002 - 0.04y\sin\phi'} + 0.02 - y\sin\phi'}{0.04 - y\sin\phi'}\right)d\phi'$$

for $y \in (0, 0.04)$.

As pointed out in earlier examples, when plotting quantities, one needs to be aware of singular points of the function. In many cases the z axis is a problem. For this case, $y = 0$ (with x already zero) does not yield a singularity. However, for $y = 0.04$ and $\phi' = \pi/2$, the logarithm argument is 0/0. So, a problem lies there. Using l'Hospital's rule, we find that the limit is 0 for the integrand at that point. Thus one either must select increments in the numerical integration that avoid that point or use 0 for the value at that point. The results are plotted in Figure 5-69, using $\Delta\phi' = 0.06283$ rad (100 increments).

Exercises

5-10.0j The charge configuration of Figure 5-70 is called an *electric quadripole*. Determine the potential at points P that are far from the charges, $r \gg l$. (See Example 5-26 and use additional terms in the expansion approximation.) Also, find $\vec{e}(\vec{r})$. l is the charge separation.

5-10.0k In Example 5-26 approximations were made to obtain the expressions for $V(\vec{r})$ and $\vec{E}(\vec{r})$. To check if the electric field is still a static field, determine whether or not $\vec{E}$ is conservative.

5-10.0l Two charges are located in the xy plane, a −2-C charge at (1, 6) and a +3-C charge at (8, 1).

(a) Find the potential $V(x, y)$ at a general point P, and $V(12, 8)$.

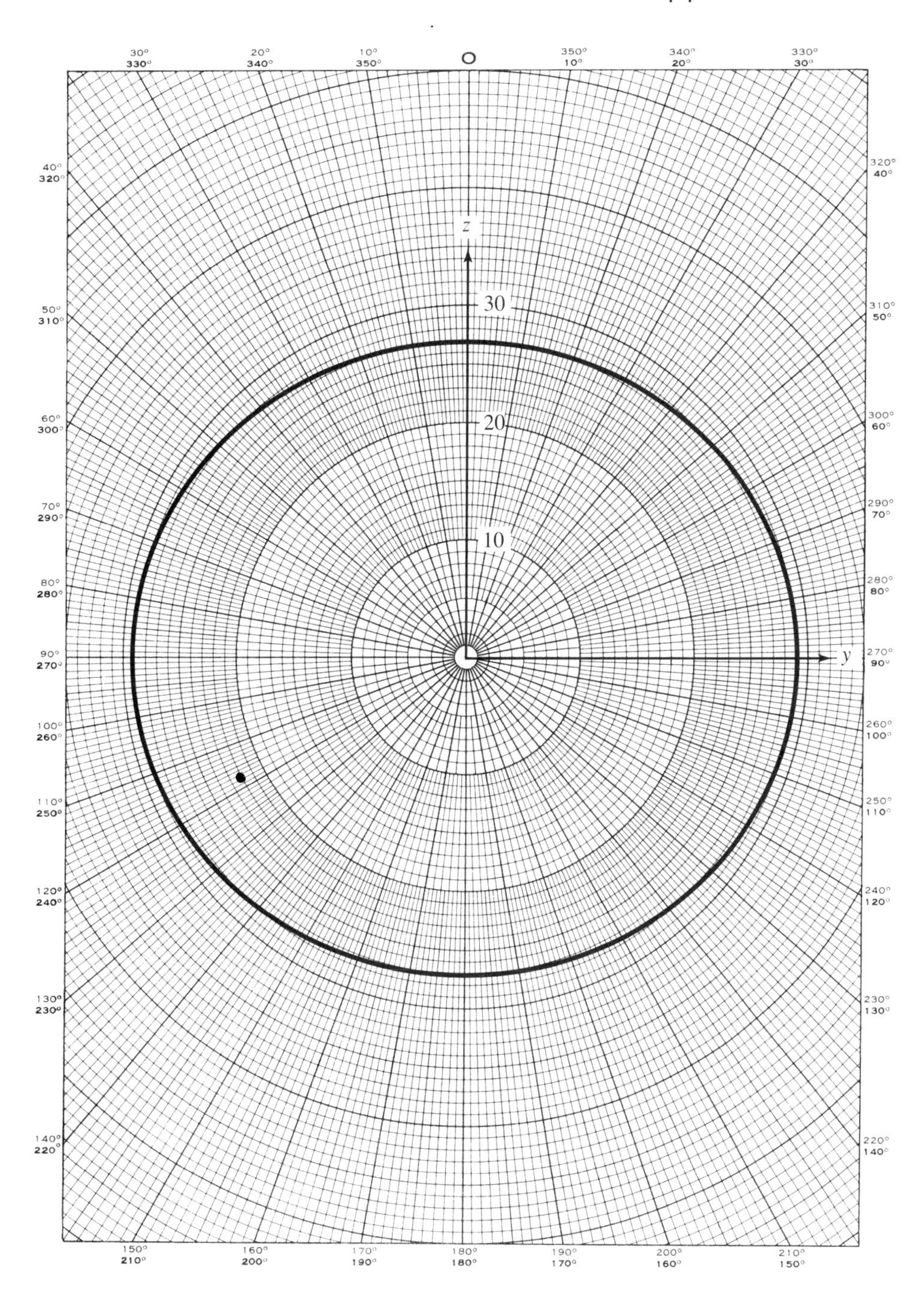

Figure 5-69. Potential distribution around a flat charged disk at 4 cm from disk center (Example 5-29). Units of kilovolts.

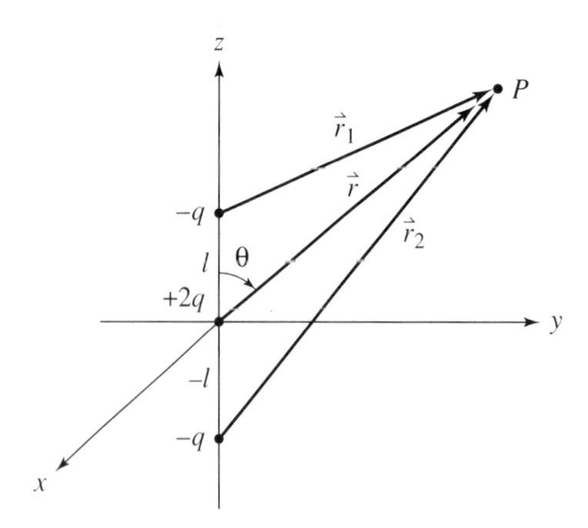

Figure 5-70.

(b) Find $\vec{E}(x, y)$ and evaluate it and plot it at (12, 8). Can you justify the direction qualitatively?

5-10.0m Repeat Exercise 5-10.01 for charge locations $(-4, 2)$ for the -2 C and (1, 3) for the 3 C.

5-10.0n For the equation for $\rho_v(\vec{r}\,')$ of Example 5-27, show by integration over all space that the total charge is $2H\rho_{l0}$.

5-10.0o The thin wire ring shown in Figure 5-71 is charged uniformly to ρ_{l0} C/m. Find the potential at a general point $V(\vec{r})$ and on the ring axis $V(0, 0, z)$. Simplify the integrals as much as possible, but do not evaluate unless a closed form results. Next, obtain expressions for $\vec{E}(\vec{r})$ and $\vec{E}(0, 0, z)$ using the two V results. Does $\vec{E}$ for the general point reduce to the z axis $\vec{E}$ by using $x = y = 0$? Explain. (Hint: When taking the partial derivatives with respect to unprimed coordinates, one may take the derivatives inside the integrals on primed coordinates.)

5-10.0p A sheet of charge is located in the xy plane as shown in Figure 5-72. The charge is constant at ρ_{so} C/m^2. Obtain the equations for $V(x, y, z)$ and $V(0, 0, z)$. Simplify to the extent possible.

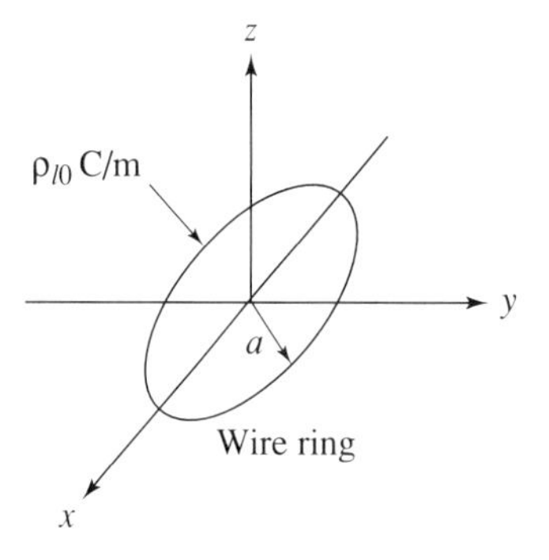

Figure 5-71.

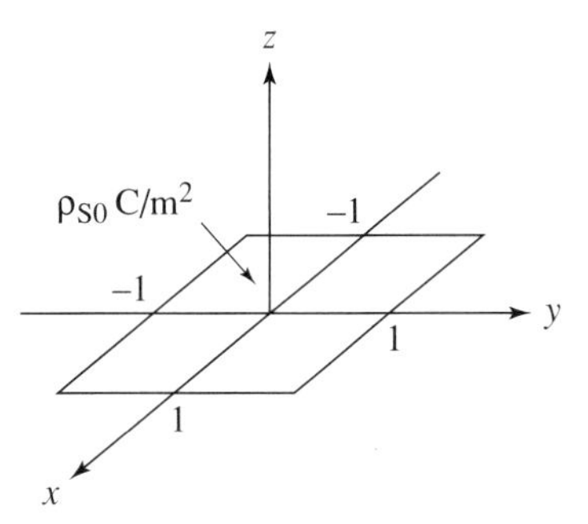

Figure 5-72.

The final concept of this section is that of equipotential contours. An *equipotential contour* (line or surface) is a contour on which the potential function is a constant. Note this is quite different from a field plot where we keep the distance from the system constant and plot the value. For example, a point charge would have a spherical equipotential surface since the potential function

$$\frac{q}{4\pi\epsilon R} = V_0 \qquad \text{(constant)}$$

requires

$$R = \frac{q}{4\pi\epsilon V_0} = R_0 \qquad \text{(constant)}$$

As one might suspect, the most difficult part of the problem is often the solving of the potential equation to obtain the equation relating the coordinate variables.

Since the gradient of a scalar function is normal to constant scalar (potential) contours (see discussion of gradient in Chapter 4), it follows that the field lines of $\vec{E}$ are normal to the equipotential contours because $\vec{E}$ is the negative gradient of V. This is easily seen for a point charge, as shown in Figure 5-73, since $\vec{E}$ has only an $\hat{r}$ component, as Eq. 5-11 shows, and equipotentials are at constant R_0 values.

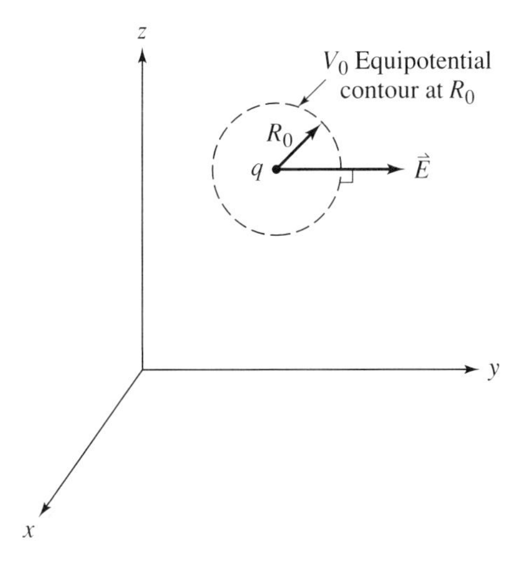

Figure 5-73. Demonstration of the orthogonality of $\vec{E}$ and the equipotential contour for a point charge.

Example 5-30

A 5-pC point charge is located in the yz plane at (4, 5) as shown in Figure 5-74. The y and z axes are marked in 1-cm values with the lines in 0.5-cm increments and the solid circles are equipotential contours in the yz plane. Estimate $\vec{E}$ at thc points P and T in rectangular coordinates. The medium is vacuum.

In rectangular coordinates we may replace the partial derivatives by increments as follows:

$$\vec{E} = -\left(\frac{\partial V}{\partial y}\,\hat{y} + \frac{\partial V}{\partial z}\,\hat{z}\right) \approx -\left(\frac{\Delta V_y}{\Delta y}\,\hat{y} + \frac{\Delta V_z}{\Delta z}\,\hat{z}\right)$$

By definition (see Chapter 4), the gradient is in the direction of greatest space rate of *increase* of potential. However, since $\vec{E}$ is the *negative* gradient, we move in the direction of greatest *decrease* of potential. Also, since $\vec{E}$ is perpendicular to equipotential contours at point P, we would move as shown in the figure to some close, but arbitrary, lower potential contour. The same goes for point T. Thus, from point P to the 1-V contour, we estimate the various increments by interpolating between the marked values as follows:

$$\Delta V_y = (1 - 1.18) = -0.18 \text{ V}$$

$$\Delta V_z = (1.18 - 1.25) = -0.07 \text{ V}$$

Figure 5-74. Estimation of electric field intensity from equipotential contours; $q = 5$ pC.

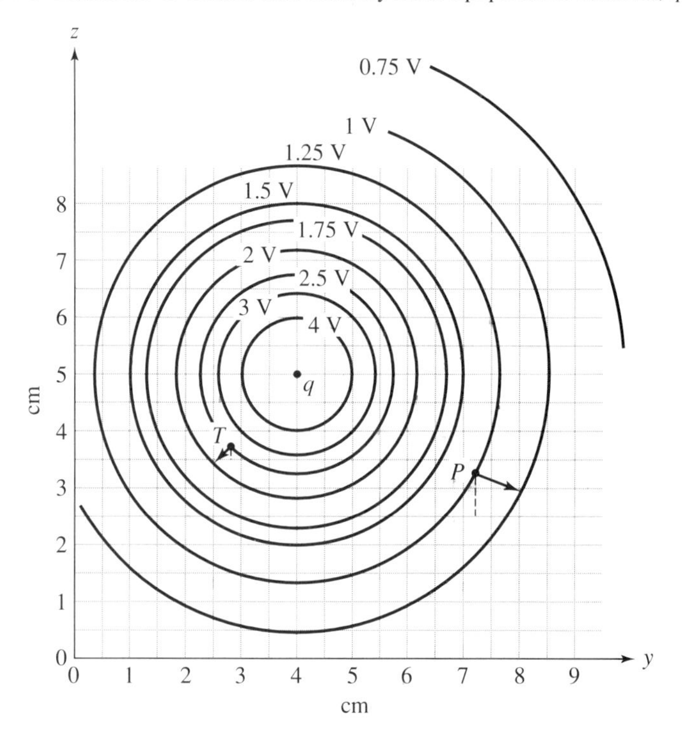

$$\Delta y \approx (8 - 7.20) = 0.8 \text{ cm} = 0.008 \text{ m}$$

$$\Delta z \approx (3 - 3.5) = -0.5 \text{ cm} = -0.005 \text{ m}$$

Then

$$\vec{E}(P) \approx -\left(\frac{-0.18}{0.008}\hat{y} + \frac{-0.07}{-0.005}\hat{z}\right) = 23\hat{y} - 14\hat{z} \text{ V/m}$$

The exact value is $33\hat{y} - 16\hat{z}$ V/m. A rather large error exists in the $\hat{y}$ component. For point T:

$$\Delta V_y = (2 - 2.25) = -0.25 \text{ V}$$

$$\Delta V_z = (2.25 - 2.5) = -0.25 \text{ V}$$

$$\Delta y = (2.6 - 2.8) \text{ cm} = -0.002 \text{ m}$$

$$\Delta z = (3.5 - 3.8) \text{ cm} = -0.003 \text{ m}$$

$$\vec{E}(T) = -\left(\frac{-0.25}{-0.002}\hat{y} + \frac{-0.25}{-0.003}\hat{z}\right) = -120\hat{y} - 83\hat{z} \text{ V/m}$$

The exact value is $-101\hat{y} - 101\hat{z}$ V/m (about $\pm 20\%$ error in each component).

Example 5-31

Two lines of infinite length are charged to ρ_{l0} and $-\rho_{l0}$ cm as shown in Figure 5-75. Obtain a plot of the equipotential contours in the xy plane.

The first step is to obtain the general expression for the potential in the xy plane, a portion of which is shown dashed in the figure. For convenience we will use the distances from lines r_a and r_b as the variables rather than the distance from the origin, as would be the usual case. Using Eq. 5-69 we have

$$V(P) = V_a + V_b = \int_{-\infty}^{\infty} \frac{\rho_{l0}}{4\pi\epsilon R_a}\, dz'_a + \int_{-\infty}^{\infty} \frac{-\rho_{l0}}{4\pi\epsilon R_b}\, dz'_b$$

$$= \frac{\rho_{l0}}{4\pi\epsilon}\left[\int_{-\infty}^{\infty} \frac{dz'_a}{\sqrt{r_a^2 + z_a'^2}} - \int_{-\infty}^{\infty} \frac{dz'_b}{\sqrt{r_a^2 + z'^2}}\right]$$

Since z'_a and z'_b are variables of integration in definite integrals, we may replace them by z' for convenience. Also, since the integrands are even functions, we may integrate from 0 to ∞ and double the result. Then

$$V(P) = \frac{\rho_{l0}}{2\pi\epsilon}\left[\int_0^{\infty} \frac{dz'}{\sqrt{r_a^2 + z'^2}} - \int_0^{\infty} \frac{dz'}{\sqrt{r_b^2 + z'^2}}\right] = \frac{\rho_{l0}}{2\pi\epsilon} \ln\left(\frac{z' + \sqrt{r_a^2 + z'^2}}{z' + \sqrt{r_b^2 + z'^2}}\right)\Bigg|_0^{\infty}$$

$$= \frac{\rho_{l0}}{2\pi\epsilon} \ln\left(\frac{r_a}{r_b}\right)$$

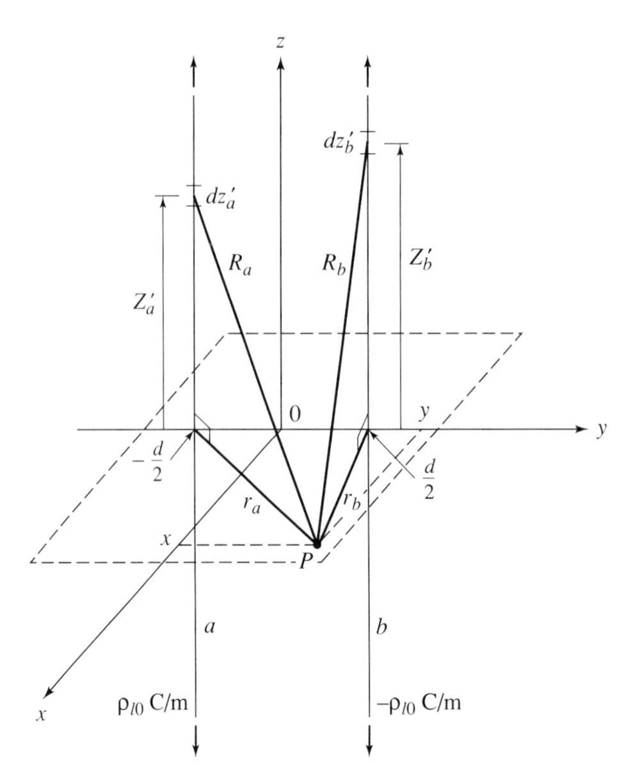

Figure 5-75. Definitions of terms used to compute the potential at a point in the *xy* plane.

The second step is to set V to a constant and solve for the coordinate (r_a and r_b) variation:

$$V_0 = \frac{\rho_{l0}}{2\pi\epsilon} \ln\left(\frac{r_a}{r_b}\right)$$

In the xy plane,

$$r_a = x^2 + \left(y + \frac{d}{2}\right)^2, \qquad r_b^2 = x^2 + \left(y - \frac{d}{2}\right)^2$$

$$V_0 = \frac{\rho_{l0}}{2\pi\epsilon} \ln\left[\frac{x^2 + \left(y + \frac{d}{2}\right)^2}{x^2 + \left(y - \frac{d}{2}\right)^2}\right]^{1/2} = \frac{\rho_{l0}}{2\pi\epsilon} \ln\left[\frac{x^2 + \left(y + \frac{d}{2}\right)^2}{x^2 + \left(y - \frac{d}{2}\right)^2}\right]^{1/2}$$

Using the definition of logarithm and its properties

$$e^{(4\pi\epsilon V_0/\rho_{l0})} = \frac{x^2 + \left(y + \frac{d}{2}\right)^2}{x^2 + \left(y - \frac{d}{2}\right)^2}$$

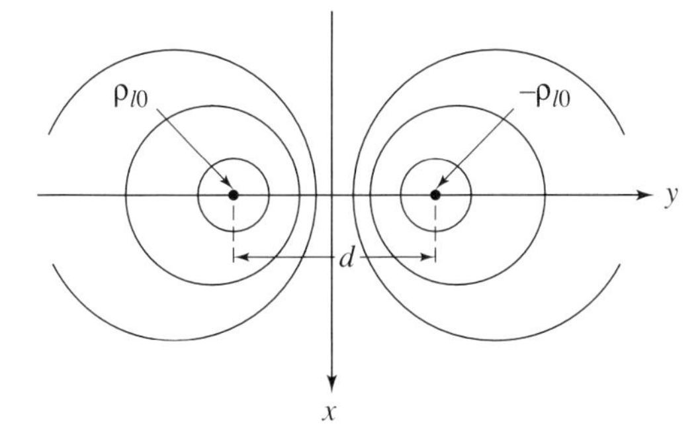

Figure 5-76. Typical equipotential contours in the xy plane for the charged lines of Figure 5-75.

Let the left side be denoted by the constant K_0^2. Then,

$$K_0^2 x^2 + K_0^2\left(y - \frac{d}{2}\right)^2 = x^2 + \left(y + \frac{d}{2}\right)^2$$

Expanding the y binomials and collecting terms,

$$x^2 + y^2 - d\,\frac{1 + K_0^2}{1 - K_0^2}\,y + \frac{d^2}{4} = 0$$

Completing the square on y, we finally obtain

$$x^2 + \left(y + \frac{d}{2}\,\frac{K_0^2 + 1}{K_0^2 - 1}\right)^2 = \left(\frac{K_0^2 d}{K_0^2 - 1}\right)^2$$

These are circles with centers at

$$\left(0, \pm\frac{d}{2}\,\frac{K_0^2 + 1}{K_0^2 - 1}\right)$$

and radii

$$\frac{K_0 d}{K_0^2 - 1}$$

From this we see that, as V_0 is set to larger values (K_0 larger), the radius decreases and the centers move to $\pm d/2$. Typical equipotential contours are plotted in Figure 5-76.

Exercises

5-10.0q For the charged wire ring of Exercise 5-10.0o, plot three equipotential contours in the yz plane where the medium is vacuum; $a = 10$ cm, $\rho_{l0} = 1$ nC/m, $V = 1$ V, 10 V, 100 V.

5-10.0r For the quadripole of Exercise 5-10.0j, plot three equipotential contours for $q = 10$ nC, $l = 5$ cm, using $V = 1$ V, 10 V, 100 V.

5-10.0s Figure 5-77 is a plot of some equipotentials around two positive point charges located at ± 1 cm. Estimate the electric field intensity at points P, Q, and R.

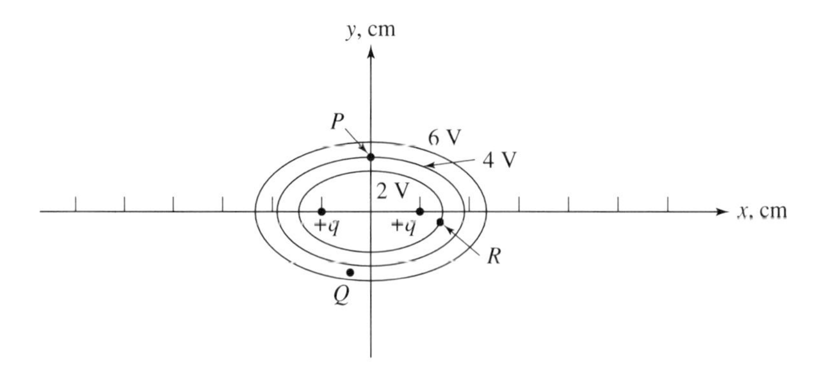

Figure 5-77.

5-11 Energy and Energy Density

We again follow the general outline considered by Maxwell in his *Treatise on Electricity and Magnetism*, volume 1 [39], and imagine all charge has been removed to infinity. Consider a local point P, at which the electric field will be zero since all charge is at infinity as shown in Figure 5-78*a*. Now bring in one charge from infinity to point P. This does not require any work (energy) since it is not pushing against any field, as shown in Figure 5-78b. Thus

$$W_1 = 0 \text{ J}$$

Now bring in a second charge from infinity as shown in Figure 5-78*c*. This requires energy to push against q_1. The potential at q_2 due to q_1 is

$$V_{21} = \frac{q_1}{4\pi\epsilon R_{12}} \text{ V or J/C}$$

The energy required to bring in q_2 is then

$$W_{21} = q_2 V_{21} = \frac{q_2 q_1}{4\pi\epsilon R_{12}} \text{ J}$$

Now, to bring in the ith charge requires pushing against all the preceding $i - 1$ charges. Then

$$\begin{aligned} W_i &= W_{i1} + W_{i2} + W_{i3} + \cdots + W_{i(i-1)} \\ &= q_i V_{i1} + q_i V_{i2} + q_i V_{i3} + \cdots + q_i V_{i(i-1)} = q_i \sum_{j=1}^{i-1} V_{ij} \end{aligned}$$

Thus for N charges, the total electric energy to assemble the distribution is, starting the sum at 2 since $W_1 = 0$,

$$W_e = \sum_{i=2}^{N} W_i = \sum_{i=2}^{N} \left(q_i \sum_{j=2}^{i-1} V_{ij} \right) = \sum_{i=2}^{N} \left(q_i \sum_{j=1}^{i-1} \frac{q_j}{4\pi\epsilon R_{ji}} \right)$$

This result has a nice physical interpretation. The terms in parentheses tell us that for charge q_i we add up all the potentials that the remaining charges produce at q_i and then

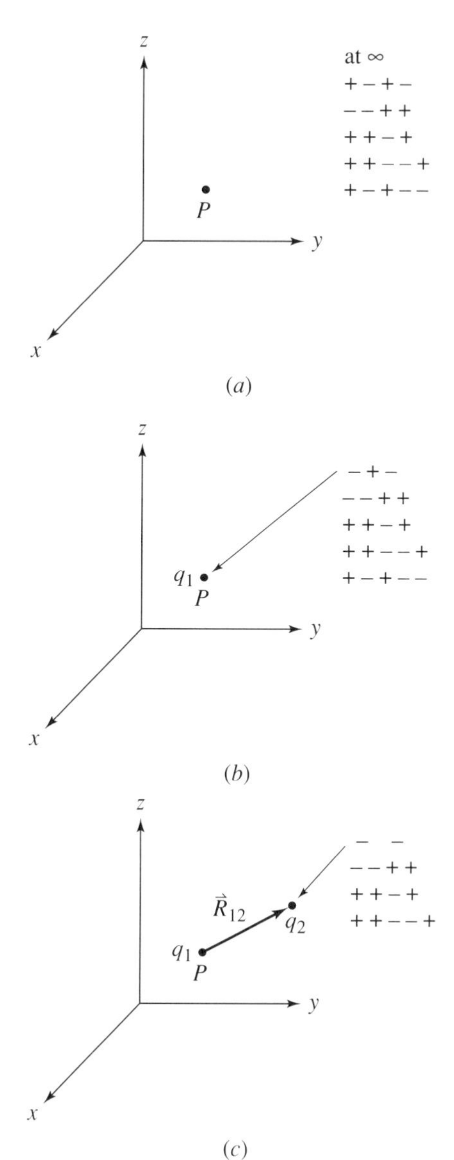

Figure 5-78. Process for determining energy to assemble point charges.

multiply by q_i to get the energy to place q_i at its location. Since the q_i charge does not interact with its *own* potential, j is never equal to i as the upper limit on the inner sum shows; that is, j is always one less than i.

To have the lower limits symmetric, we could start the i sum at one also and then note that $j \neq i$ or, equivalently, require $V_{mm} = 0$ for all integers m:

$$W_e = \sum_{i=1}^{N}\left(q_i \sum_{j=1}^{i-1} V_{ij}\right) = \sum_{i=1}^{N}\left(q_i \sum_{j=1}^{i-1} \frac{q_j}{4\pi\epsilon R_{ji}}\right), \qquad j \neq i \; (V_{mm} = 0) \qquad \text{(5-70)}$$

From the individual energy terms W_{ij} we use the fact that $R_{ij} = R_{ji}$ to obtain

$$W_{ij} = q_i V_{ij} = q_i \frac{q_j}{4\pi\epsilon R_{ji}} = q_j \frac{q_i}{4\pi\epsilon R_{ij}} = q_j V_{ji} = W_{ji} \tag{5-71}$$

Thus in Eq. 5-70 we may interchange i and j and obtain the same result.

Expanding a few terms in Eq. 5-70 we have

$$\begin{aligned} W_e = &\, q_2 V_{21} + q_3(V_{31} + V_{32}) + q_4(V_{41} + V_{42} + V_{43}) \\ &+ q_5(V_{51} + V_{52} + V_{53} + V_{54}) + \cdots \end{aligned}$$

Multiplying out and interchanging the subscripts according to Eq. 5-71,

$$\begin{aligned} W_e = &\, q_1 V_{12} + q_1 V_{13} + q_2 V_{23} + q_1 V_{14} + q_2 V_{24} + q_3 V_{34} \\ &+ q_1 V_{15} + q_2 V_{25} + q_3 V_{35} + q_4 V_{45} + \cdots \end{aligned}$$

Adding the preceding two expressions for W_e and combining terms on the q_i, there results:

$$\begin{aligned} 2W_e = &\, q_1(V_{12} + V_{13} + V_{14} + V_{15} + \cdots) + q_2(V_{21} + V_{23} + V_{24} + V_{25} + \cdots) \\ &+ q_3(V_{31} + V_{32} + V_{34} + V_{35} + \cdots) + q_4(V_{41} + V_{42} + V_{43} + V_{45} + \cdots) \\ &+ q_5(V_{51} + V_{52} + V_{53} + V_{54} + \cdots) + \cdots \end{aligned}$$

Note that each sum contains all voltages except the self voltage V_{ii} quantity. Thus the preceding equation for N charges may be written

$$2W_e = \sum_{i=1}^{N} \left(q_i \sum_{j=1}^{N} V_{ij} \right), \qquad j \neq i$$

or

$$W_e = \frac{1}{2} \sum_{i=1}^{N} \left(q_i \sum_{j=1}^{N} \frac{q_j}{4\pi\epsilon R_{ij}} \right), \qquad i \neq j \tag{5-72}$$

Example 5-32

Three charges are located in the xy plane as shown in Figure 5-79. Compute the energy required to form the charge configuration. Note that this could also be identified as the energy contained in the distribution.

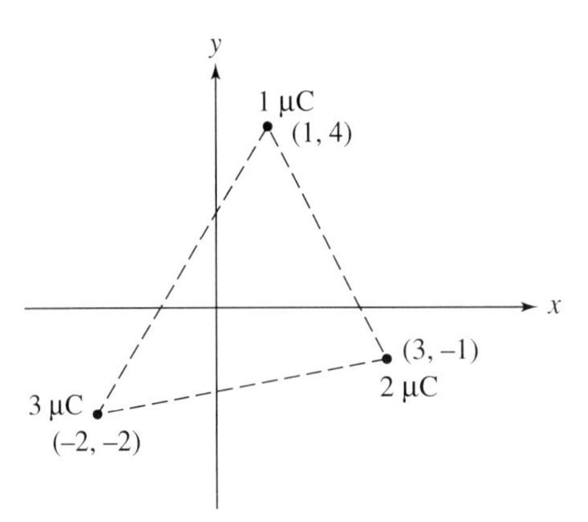

Figure 5-79. Configuration for the computation of electric energy.

Using Eq. 5-72 with $N = 3$ charges, $i \neq j$, and $\epsilon = \epsilon_0$,

$$W_e = \frac{1}{2}\sum_{i=1}^{3}\left(q_i \sum_{j=1}^{3} \frac{q_j}{4\pi\epsilon_0 R_{ij}}\right) = \frac{1}{8\pi\epsilon_0}\sum_{i=1}^{3} q_i \left(\frac{q_1}{R_{i1}} + \frac{q_2}{R_{i2}} + \frac{q_3}{R_{i3}}\right)$$

$$= \frac{1}{8\pi\epsilon_0}\left[q_1\left(\frac{q_2}{R_{12}} + \frac{q_3}{R_{13}}\right) q_2\left(\frac{q_1}{R_{21}} + \frac{q_3}{R_{23}}\right) + q_3\left(\frac{q_1}{R_{31}} + \frac{q_2}{R_{32}}\right)\right]$$

Since $R_{ij} = R_{ji}$,

$$W_e = \frac{1}{8\pi\epsilon_0}\left[\frac{2q_1q_2}{R_{12}} + \frac{2q_1q_3}{R_{13}} + \frac{2q_2q_3}{R_{23}}\right]$$

$$= \frac{1}{4\pi \times 1/36\pi \times 10^{-9}}\left(\frac{1 \times 10^{-6} \times 2 \times 10^{-6}}{\sqrt{(3-1)^2 + (-1-4)^2}} + \frac{1 \times 10^{-6} \times 3 \times 10^{-6}}{\sqrt{(-2-1)^2 + (-2-4)^2}}\right.$$

$$\left. + \frac{2 \times 10^{-6} \times 3 \times 10^{-6}}{\sqrt{(-2-3)^2 + (-2-(-1))^2}}\right)$$

$$W_e = 0.018 \text{ J}$$

Exercises

5-11.0a Repeat Example 5-32 using Eq. 5-70.

5-11.0b In Figure 5-79, change the $+3$-μC charge to -3 μC and compute the electric energy. Explain the physical significance of negative energy ($\epsilon = \epsilon_0$).

5-11.0c Four charges of value -20 μC are at the corners of a square of side length 1 m. Compute the electric energy in the system. Does the solution depend upon the selection of a coordinate system? Explain why or why not.

5-11.0d Three charges of values 10 μC, 50 μC, -80 μC are located at (2, 3, 4), $(-4, 6, -3)$, and $(2, -8, 1)$, respectively, in rectangular coordinates. Compute the electric energy for $\epsilon = \epsilon_0$.

5-11.0e In Example 5-32, compute W_e for the case $\epsilon = 6\epsilon_0$. Explain the reason for the difference between the two values corresponding to ϵ_0 and $6\epsilon_0$.

Since most engineering systems consist of accumulations of charge densities rather than discrete point charges, we need formulations for those configurations. Figure 5-80 is the system definition.

To use Eq. 5-72, we make the following identifications:

$$q_i \sim \rho_v(\vec{r}'_i)\,\Delta v'_i$$

$$q_j \sim \rho_v(\vec{r}_j)\,\Delta v_j$$

$$R_{ij} \sim R = |\vec{R}| = |\vec{r}_j - \vec{r}'_i|$$

Equation 5-72 then becomes

$$W_e = \frac{1}{2}\sum_{i=1}^{N} \rho_v(\vec{r}'_i)\,\Delta v'_i\left(\sum_{j=1}^{N} \frac{\rho_v(\vec{r}_j)\,\Delta v_j}{4\pi\epsilon|\vec{r}_j - \vec{r}'_i|}\right), \qquad i \neq j$$

$$= \sum_{i=1}^{N} \frac{1}{2}\rho_v(\vec{r}'_i)\left(\sum_{j=1}^{N} \frac{\rho_v(\vec{r}_j)\,\Delta v_j}{4\pi\epsilon|\vec{r}_j - \vec{r}'_i|}\right)\Delta v'_i, \qquad i \neq j$$

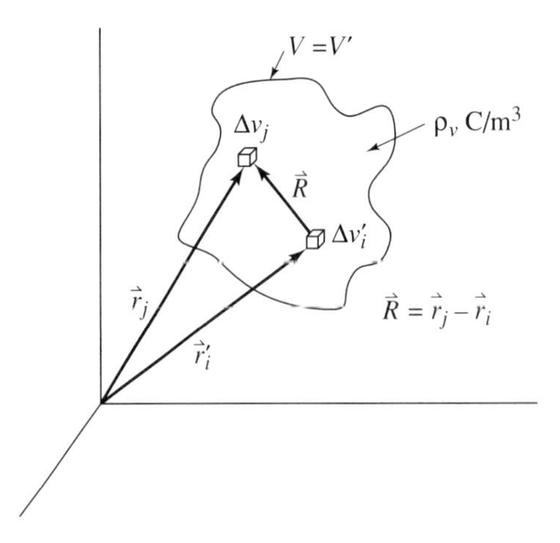

Figure 5-80. Configuration and definitions for determining the electric energy of continuous charge distributions.

Taking the limits as dv_j and dv'_i approach zero, the sums become integrals:

$$W_e = \int_{V'} \frac{1}{2}\,\rho_v(\vec{r}\,')\left(\int_V \frac{\rho_v(\vec{r})dv}{4\pi\epsilon|\vec{r} - \vec{r}\,'|}\right)dv' \tag{5-73}$$

Now in the limit, as Δv_j and $\Delta v'_i$ go to zero, the charge also becomes zero and the energy to form those charges is zero, so we do not need the specification $i \neq j$ any more. This can be explained quantitatively by observing that even though $|\vec{r}_j - \vec{r}\,'_i| \rightarrow 0$ for $i \rightarrow j$, the product $\Delta v'_i\ \Delta v_j \rightarrow 0$ in a higher order (i.e., faster) way since volume is the cube of linear dimension, which results in zero value of the integral.

Using Eq. 5-67 on the inner integral of the preceding equation we have:

$$W_e = \int_{V'} \frac{1}{2}\,\rho_v(\vec{r}\,')V(\vec{r}\,')dv' \ \text{J} \tag{5-74}$$

A final form in terms of field quantities only can be obtained by using the identity for the divergence of a scalar times a vector; namely,

$$\nabla \cdot (V\vec{D}) = V\,\nabla \cdot \vec{D} + \vec{D} \cdot \nabla V$$

Using Eqs. 5-46 and 5-63 this becomes

$$\nabla \cdot (V\vec{D}) = V\rho_v(\vec{r}) - \vec{D} \cdot \vec{E}$$

from which

$$\rho_v(\vec{r})V = \nabla \cdot (V\vec{D}) + \vec{D} \cdot \vec{E}$$

Substituting this into the integrand of the preceding equation for W_e we have

$$W_e = \int_{V'} \frac{1}{2}\,\nabla \cdot (V\vec{D})dv' + \int_{V'} \frac{1}{2}\,\vec{D} \cdot \vec{E}dv'$$

From Eqs. 5-45 and 5-66 we see that, since ∇ has dimension 1/m,

$$\nabla \cdot (V\vec{D}) \text{ will have dimension } \left(\frac{1}{\text{m}}\right)\left(\frac{1}{\text{m}}\right)\left(\frac{1}{\text{m}^2}\right) \sim \frac{1}{\text{m}^4}$$

Thus in the left integral, the entire expression will vary as 1/m since volume is dimensionally m^3. If we include all space so that $\vec{r}$ recedes to ∞, the integral goes to zero. We then have

$$W_e = \int_{\text{all space}} \frac{1}{2} \vec{D} \cdot \vec{E} dv' \text{ J} \tag{5-75}$$

This last result tells us that we may imagine the energy to be stored in the *field* quantities $\vec{D}$ and $\vec{E}$ between the charges rather than in the charges themselves.

Note also that the integrands of Eqs. 5-74 and 5-75 may be interpreted as energy densities since the volume integrals yield energy. Thus we write

$$w_e = \tfrac{1}{2}\rho_v(\vec{r})V(\vec{r}) = \tfrac{1}{2}\vec{D} \cdot \vec{E} \text{ J/m}^3 \tag{5-76}$$

The constitutive relationship for electric fields, Eq. 5-48, allows the preceding equation to be written

$$w_e = \tfrac{1}{2}\epsilon\vec{E} \cdot \vec{E} = \tfrac{1}{2}\epsilon|\vec{E}|^2 \text{ J/m}^3 \tag{5-77}$$

Example 5-33

A spherical region of charge varies according to

$$\rho_v(r) = r[U(r) - U(r-2)] \text{ C/m}^3 \text{ (zero charge density for } r \geq 2)$$

Compute the energy stored using the three alternate expressions for energy density:

1. *Electric field intensity formulation.* Since the charged region has spherical symmetry (no θ or ϕ variations), $\vec{D}$ and $\vec{E}$ will have only radial ($\hat{r}$) components so that

$$\vec{D} = \epsilon\vec{E} = \epsilon E_r \hat{r}$$

For $0 \leq r \leq 2$ m, Gauss' law requires

$$\oint_S \vec{D} \cdot d\vec{a} = Q_{\text{encl}} = \int_V \rho_v dv$$

$$\int_0^{\pi}\int_0^{2\pi} \epsilon E_r \hat{r} \cdot r^2 \sin\theta d\phi d\theta \hat{r} = \int_0^{\pi}\int_0^{2\pi}\int_0^{r} (r)(r^2 \sin\theta dr d\phi d\theta)$$

from which

$$E_r = \frac{r^2}{4\epsilon} \qquad 0 \leq r \leq 2$$

For $r \geq 2$ m, Gauss' law becomes (remember $\rho v = 0$ for $r \geq 2$)

$$\int_0^{\pi}\int_0^{2\pi} \epsilon E_r \hat{r} \cdot r^2 \sin\theta d\phi d\theta \hat{r} = \int_0^{\pi}\int_0^{2\pi}\int_0^{2} (r)(r^2 \sin\theta dr d\phi d\theta)$$

From which

$$E_r = \frac{4}{\epsilon r^2} \qquad r \geq 2$$

Then

$$\vec{E} = \begin{cases} \dfrac{r^2}{4\epsilon}\hat{r} & 0 \leq r \leq 2 \\ \dfrac{4}{\epsilon r^2}\hat{r} & r \geq 2 \end{cases}$$

Using Eq. 5-77,

$$w = \begin{cases} \dfrac{r^4}{32\epsilon} & 0 \leq r \leq 2 \\ \dfrac{8}{\epsilon r^4} & r \geq 2 \end{cases}$$

Then

$$W_e = \int_V \frac{1}{2}\epsilon|\vec{E}|^2 dv = \underset{0 \leq r \leq 2}{\int_V \frac{1}{2}\epsilon E_r^2 dv} + \underset{r \geq 2}{\int_V \frac{1}{2}\epsilon E_r^2 dv}$$

$$= \int_0^{\pi}\int_0^{2\pi}\int_0^{2}\left(\frac{r^4}{32\epsilon}\right) r^2 \sin\theta dr d\phi d\theta + \int_0^{\pi}\int_0^{2\pi}\int_2^{\infty}\left(\frac{8}{\epsilon r^4}\right) r^2 \sin\theta dr d\phi d\theta$$

$$= \frac{16\pi}{7\epsilon} + \frac{16\pi}{\epsilon} = \frac{57.4}{\epsilon}\ \text{J}$$

2. *Displacement-intensity formulation.* Using the preceding results for $\vec{E}$, we have

$$\vec{D} = \epsilon\vec{E} = \begin{cases} \dfrac{r^2}{4}\hat{r} & 0 \leq r \leq 2 \\ \dfrac{4}{r^2}\hat{r} & r \geq 2 \end{cases}$$

Then

$$w_e = \frac{1}{2}\vec{D} \cdot \vec{E} = \begin{cases} \dfrac{r^4}{32\epsilon} & 0 \leq r \leq 2 \\ \dfrac{8}{\epsilon r^4} & r \geq 2 \end{cases}$$

Since this is identical with w_e of the previous method, the total energy will be the same.

3. *Charge density-potential formulation.* Using Eq. 5-32 and using $r = \infty$ as reference B, the voltage is (using $d\vec{l} = d\vec{r} = dr\hat{r}$)

$$V(r) = \begin{cases} -\int_{\infty}^{2} \vec{E} \cdot d\vec{r} - \int_{2}^{r} \vec{E} \cdot d\vec{r} & 0 \le r \le 2 \\ -\int_{\infty}^{r} \vec{E} \cdot d\vec{r} & r \ge 2 \end{cases}$$

$$= \begin{cases} 0 - \int_{\infty}^{2} \frac{4}{\epsilon r^2}\, dr - \int_{2}^{r} \frac{r^2}{4\epsilon}\, dr & 0 \le r \le 2 \\ -\int_{\infty}^{r} \frac{4}{\epsilon r^2}\, dr & r \ge 2 \end{cases}$$

$$= \begin{cases} \frac{8}{3\epsilon} - \frac{r^3}{12\epsilon} & 0 \le r \le 2 \\ \frac{4}{\epsilon r} & r \ge 2 \end{cases}$$

Using the middle equality in Eq. 5-76:

$$w_e = \frac{1}{2}\rho_v(\vec{r})V(\vec{r}) = \frac{1}{2}\rho_v(r)V(r) = \frac{1}{2}\begin{cases} rV(r) & 0 \le r \le 2 \\ 0 \cdot V(r) & \mathrm{r} \ge 2 \end{cases}$$

$$= \frac{1}{2} rV(r) \qquad 0 \le r \le 2$$

$$= \frac{4r}{3\epsilon} - \frac{r^4}{24\epsilon} \qquad 0 \le r \le 2$$

Then

$$W_e = \int_0^{\pi}\int_0^{2\pi}\int_0^{2} \left(\frac{4r}{3\epsilon} - \frac{r^4}{24\epsilon}\right) r^2 dr \sin\, d\phi d\theta = \frac{57.4}{\epsilon}\ \mathrm{J}$$

Exercises

5-11.0f Using the units of the quantities in the expressions for energy density in Eqs. 5-76 and 5-77, show that each is in fact joule/cubic meter.

5-11.0g Suppose that a region of spherical charge has a density given by

$$\rho_v(r) = \frac{1}{r}\,[U(r) - U(r - 3)]\ \mathrm{C/m^3}$$

Determine the energy required to form this configuration, that is, the energy stored. Do this using the $\vec{E}$ field only and then the charge density-voltage product.

5-11.0h A 1-m length of coaxial cable, shown in Figure 5-81, has the following potential distribution:

$$V(\rho) = 62.13 \ln\left(\frac{5 \times 10^{-3}}{\rho}\right)\ \mathrm{V}$$

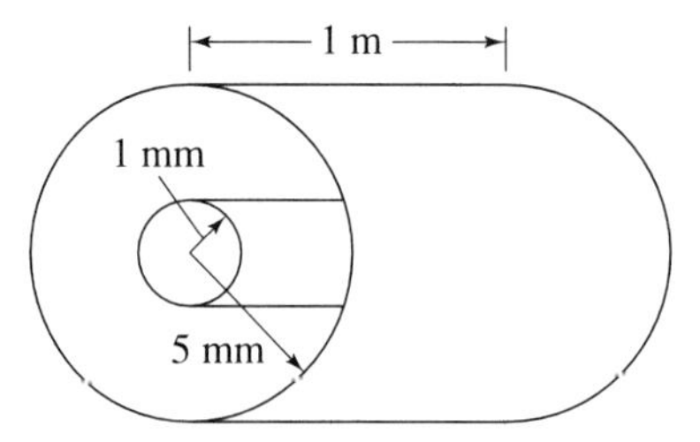

Figure 5-81.

Compute the electric field between the conductors of the coax and then compute the energy stored in this section of coax. Use $\epsilon = 3\epsilon_0$.

5-12 Boundary Conditions of Electric Field Quantities at Dielectric-Dielectric and Dielectric-Conductor Interfaces and Physical Constraints on Electric Field Values

Most practical systems contain discontinuities in material properties primarily because we usually work with finite-extent systems. It is thus important that we know what the effects of discontinuities or material changes are on electric field quantities at those interface boundaries. Additionally, in subsequent work we will be solving differential equations that require boundary conditions to obtain a complete solution. These correspond to initial conditions in the solution of differential equations having time as the independent variable.

Along with boundary conditions, there are some physical constraints on electric field quantities in terms of allowed values and variations in time and space. A knowledge of these constraints often simplifies the solutions of many electromagnetic systems. Although some of these will not be used until later chapters, they are included for completeness of ''boundary value'' conditions lists.

The general boundary configuration is shown in Figure 5-82*a*. Only one field vector arrow is shown for clarity, even though the entire region may be filled with a field quantity. We use the convention that a unit normal vector $\hat{n}$ is directed from Medium 1 into Medium 2. We also allow for the possibility that there may be a surface charge density ρ_s on the boundary. Since vectors may be decomposed into components, we show in Figure 5-82*b* decompositions into components tangent to the boundary at the point of the field vector (denoted by the subscript t) and components normal to the boundary (denoted by the subscript n), so that the sums yield the original field vectors. Material properties are identified by ϵ and σ. The dotted regions will be explained later.

Consider first the tangential components on which is shown a path (contour) *ABCD*. From Eq. 5-64 we have

$$\oint \vec{E} \cdot d\vec{l} = \int_A^B \vec{E} \cdot d\vec{l} + \int_B^C \vec{E} \cdot d\vec{l} + \int_C^D \vec{E} \cdot d\vec{l} + \int_D^A \vec{E} \cdot d\vec{l} = 0$$

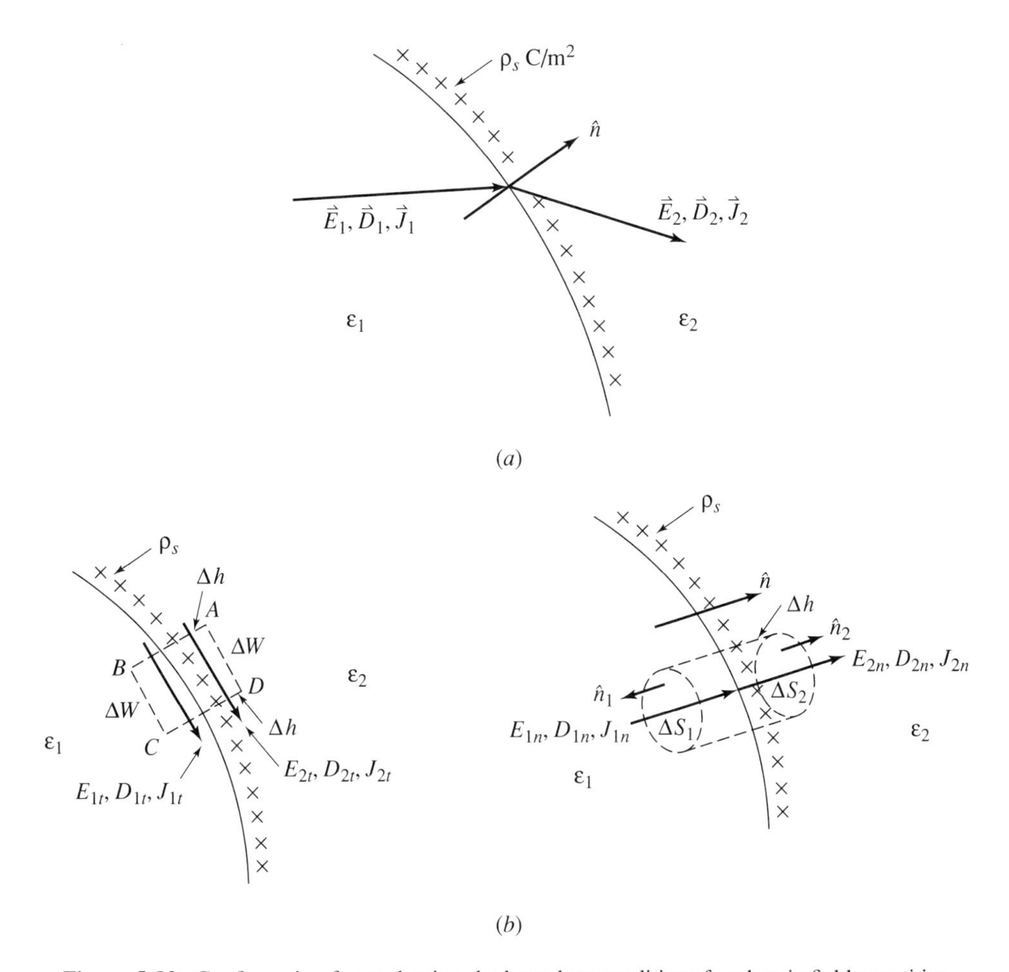

Figure 5-82. Configuration for evaluating the boundary conditions for electric field quantities.

For path components A to B and C to D the dot products in the integrals are zero for small path length Δh (since we must let Δh approach zero to get values at the boundary) because $\vec{E}$ and $d\vec{l}$ are orthogonal. For path B to C, the differential $d\vec{l}$ and $\vec{E}$ are parallel ($\cos 0° = 1$) whereas for path D to A they are antiparallel ($\cos 0° = -1$). Thus the preceding equation reduces to

$$\int_B^C E_{1t} dl + \int_D^A (-E_{2t} dl) = 0$$

To get to one point on the boundary we will also need to let Δw approach zero. Then for small path lengths, E_{1t} and E_{2t} may be considered constants and removed from the integrands to yield

$$E_{1t} \int_B^C dl - E_{2t} \int_D^A dl = E_{1t}\, \Delta w - E_{2t}\, \Delta w = 0$$

or, dividing by Δw,

$$E_{1t} = E_{2t} \tag{5-78}$$

Using the relationship between $\vec{D}$ and $\vec{E}$:

$$\frac{D_{1t}}{\epsilon_1} = \frac{D_{2t}}{\epsilon_2} \quad \text{or} \quad D_{2t} = \frac{\epsilon_2}{\epsilon_1} D_{1t} \tag{5-79}$$

Qualitatively, these results show that tangential components of electric field intensity are continuous (equal) whereas electric displacement components are discontinuous by the ratio of the material properties. Note that given any one tangential component we can compute all the others!

Next consider the normal component configuration of Figure 5-82*b* on which is shown a small incremental volume Δv enclosed by the cylindrical surface, Δs on the ends and Δh in length. Unit vectors on the ends are also shown (remember they are directed out of the volume normal to the surface). There is no unit vector shown on the cylindrical side since Δh will go to zero anyway to get on the boundary and no electric flux passes through the side wall. Also, for small Δh the surfaces Δs_1 and Δs_2 will be equal. We begin with Gauss' law, Eq. 5-46,

$$\Psi_e = \oint_S \vec{D} \cdot d\vec{a} = \int_V \rho_v dv = Q_{\text{enclosed}}$$

Since the only charge present is on the boundary when $\Delta h \to 0$, the charge enclosed by the cylinder is $\rho_s \, \Delta s$. Then

$$\int_{\Delta s} \vec{D}_1 \cdot d\vec{a}_1 + \int_{\Delta s} \vec{D}_2 \cdot d\vec{a}_2 = \rho_s \, \Delta s \tag{5-80}$$

In Medium 1 the field and $\hat{n}_1$ are antiparallel ($\cos 180° = -1$) and in Medium 2 they are parallel ($\cos 0° = 1$) so that the preceding becomes

$$\int_{\Delta s_1} - D_{1n} da_1 + \int_{\Delta s_2} D_{2n} da_1 = \rho_s \, \Delta s$$

Now as Δs becomes small, the field may be considered constant over the area and thus removed from the integrals to give (using Δs_1 and $\Delta s_2 = \Delta s$)

$$-D_{1n} \, \Delta S D_{2n} \, \Delta S = \rho_s \, \Delta S$$

or

$$D_{2n} - D_{1n} = \rho_s \tag{5-81}$$

Thus the electric displacement is discontinuous by the amount of surface charge. This is physically reasonable since the lines of $\vec{D}$ from the surface charge will be in the same direction as D_{2n} but opposing (reducing) D_{1n}. Note $\vec{D}$ is continuous ($D_{1n} = D_{2n}$) if $\rho_s = 0$.

Again, using the constitutive relationship, we can obtain boundary conditions on normal electric field intensity as

$$\epsilon_2 E_{2n} - \epsilon_1 E_{1n} = \rho_s. \tag{5-82}$$

One important special case of engineering interest is that of having Medium 1 a perfect conductor, $\sigma_1 = \infty$. In practice, the conductivities of metals in common use are of high conductivity or the base material is often silver- or gold-plated to give a high-conductivity surface. One consequence of this is that the electric field intensity in the material is zero. If it were not so Ohm's law (Eq. 5-30) would yield infinite current density. Thus we would have in Medium 1

$$\vec{E}_1 = 0$$
$$\vec{D}_1 = \epsilon_1 \vec{E}_1 = 0$$
$$\vec{J}_1 = \sigma_1 \vec{E}_1 = \infty \cdot 0 = \text{finite limit value}$$

The current density need not be zero since an indeterminant form can have a nonzero, finite limit! Substituting these values into the preceding boundary conditions we obtain for $\sigma_1 = \infty$,

$$E_{2t} = 0 \quad \text{perfect conductor,} \quad \sigma_1 = \infty \tag{5-83}$$

$$D_{2t} = 0 \quad \text{perfect conductor,} \quad \sigma_1 = \infty \tag{5-84}$$

$$D_{2n} = \rho_s \quad \text{perfect conductor,} \quad \sigma_1 = \infty \tag{5-85}$$

$$\epsilon_2 E_{2n} = \rho_s \quad \text{perfect conductor,} \quad \sigma_1 = \infty \tag{5-86}$$

These results tell us that the electric field is normal at a perfect conductor boundary. Other field conditions will be given following some examples and exercises.

Example 5-34

A metal with $\sigma = \infty$ interfaces with a dielectric which has $\epsilon_r = 2.1$ and $\sigma = 0$ S/m. An electric field is directed normal to the boundary as shown in Figure 5-83. The electric field must be normal to the boundary since there cannot be a tangential component (Eq. 5-83).

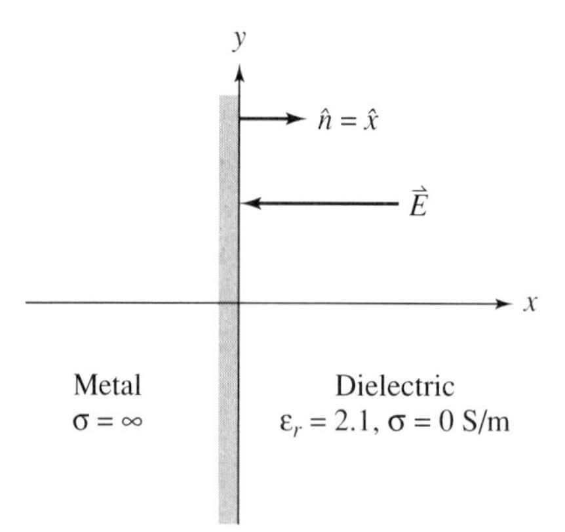

Figure 5-83.

The electric field is given by

$$\vec{E} = E_x(x, y)\hat{x} = -\frac{6}{3 + y^2}\,\hat{x}$$

For convenience we take the metal to be Medium 1 and the dielectric as Medium 2 so that the unit normal will be into the dielectric, and in this case the $\hat{x}$ direction.

Compute the surface charge density on the boundary.

Using Eq. 5-86 we have

$$\rho_s = \epsilon_2 E_{2n}\big|_{x=0} = \epsilon_r\epsilon_0 E_{2x}\big|_{x=0} = (2.1)\left(\frac{1}{36\pi}\times 10^{-9}\right)\left(-\frac{6}{3 + y^2}\right)$$

$$= -\frac{1.11 \times 10^{-10}}{3 + y^2}\ \mathrm{C/m^2}$$

Example 5-35

A closed metal box of conductivity $\sigma = \infty$ is shown in Figure 5-84. The potential within the box is given by the equation

$$V(x, y, z) = 10 \cos\frac{\pi}{2}x \sin 2\pi y \sin 2\pi z\ \mathrm{V}$$

We are to compute the distribution of charge on the side of the box located at $y = \frac{1}{2}$. The dielectric within the box has zero conductivity.

If we select the metal to be Medium 1 and the interior of the box as Medium 2, then $\hat{n} = -\hat{y}$ as shown.

Using Eq. 5-86 we have

$$\rho_s(x, \tfrac{1}{2}, z) = \epsilon_2 E_{2n} = \epsilon_2(-E_y)\big|_{y=1/2} = -\epsilon_2 E_y\big|_{y=1/2}$$

For static fields in rectangular coordinates,

$$\vec{E} = E_x\hat{x} + E_y\hat{y} + E_z\hat{z} = -\nabla V = -\frac{\partial V}{\partial x}\hat{x} - \frac{\partial V}{\partial y}\hat{y} - \frac{\partial V}{\partial z}\hat{z}$$

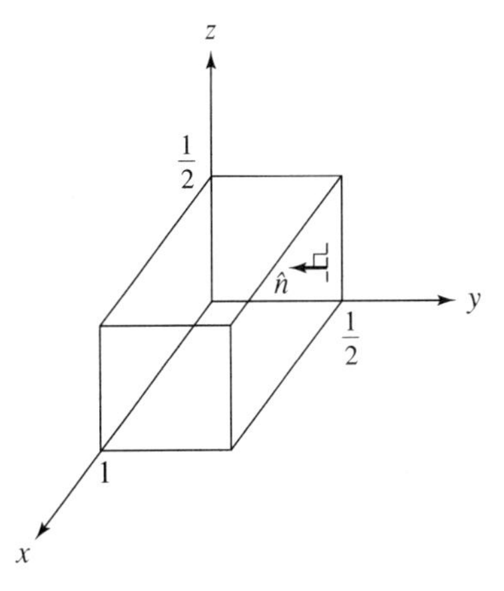

Figure 5-84.

Therefore

$$E_y = -\frac{\partial V}{\partial y} = -10 \cos \frac{\pi}{2} x(2\pi \cos 2\pi y) \sin 2\pi z$$

$$= -20\pi \cos \frac{\pi}{2} x \cos 2\pi y \sin 2\pi z$$

Then

$$E_y|_{y=1/2} = 20\pi \cos \frac{\pi}{2} x \sin 2\pi z$$

We finally obtain

$$\rho_s\left(x, \frac{1}{2}, z\right) = -20\pi\epsilon_2 \cos \frac{\pi}{2} x \sin 2\pi z \text{ C/m}^2$$

Note that even though the potential of the wall at $y = \frac{1}{2}$ is 0 V there is still charge there.

Example 5-36

Suppose we have a dielectric-dielectric boundary with perfect dielectrics, that is, $\sigma = 0$ for both. There is a charge of -2 C/m^2 at the interface, as shown in Figure 5-85. Compute the normal component of $\vec{E}_2$ in terms of $|\vec{E}_1|$.

Using Eq. 5-82 we obtain

$$\epsilon_2 E_{2x} - \epsilon_1 E_{1x} = -2$$

Then

$$E_{2n} = E_{2x} = \frac{\epsilon_1 E_{1x} - 2}{\epsilon_2}$$

From the figure

$$E_{1x} = |\vec{E}_1| \sin 30° = \tfrac{1}{2}|\vec{E}_1|$$

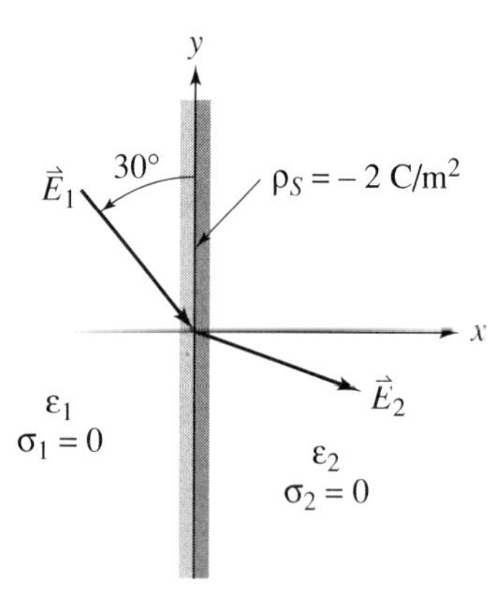

Figure 5-85.

to give

$$E_{2n} = \frac{\frac{1}{2}\epsilon_1|\vec{E}_1| - 2}{\epsilon_2}$$

Exercises

5-12.0a A metal sheet is oriented so it is in the xy plane. The sheet is charged to a constant value of 7 C/m^2. Find the values of $\vec{D}$ and $\vec{E}$ at the boundary on top and bottom of the sheet. Remember these are vectors. Air is the medium near the sheet.

5-12.0b Given the perfect dielectric-dielectric boundary shown in Figure 5-86 with $\rho_s = 0$ and $\vec{E}_1 = 6\hat{x} - 3\hat{y}$. Determine the following:

(a) Sketch $\vec{E}_1$ at (0, 0)
(b) E_{1t}
(c) E_{1n}
(d) Angle between $\vec{E}_1$ and the negative x axis
(e) E_{2t}
(f) E_{2n}
(g) Sketch $\vec{E}_2$ at (0, 0)
(h) $\vec{E}_2$
(i) Angle between $\vec{E}_2$ and the positive x axis
(j) $\vec{D}_2$

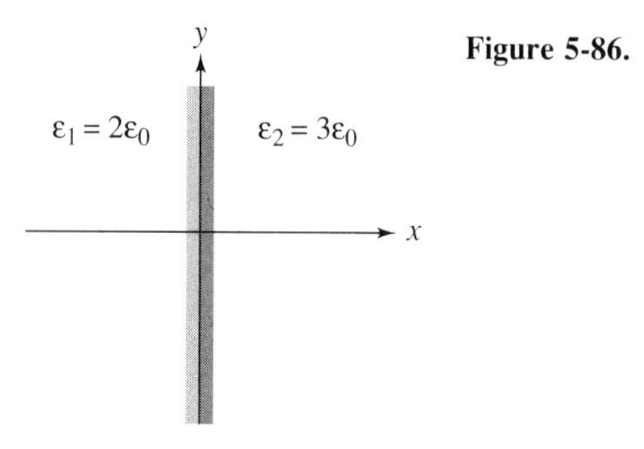

Figure 5-86.

5-12.0c A metal sphere of perfect conductivity and radius 1 m is centered at the origin and has a charge of -3 C/m^2. Give values for $\vec{D}$ and $\vec{E}$ outside the sphere if $\epsilon = 3\epsilon_0$.

5-12.0d Two concentric, perfectly conducting metal spheres have a potential distribution between them given by

$$V(r) = 50\,\frac{b - r}{b - a} \qquad a \leq r \leq b$$

where a is the radius of the inner sphere and b the radius of the outer sphere. Compute $\vec{D}$, $\vec{E}$, and ρ_s on both spheres, and the total charge on each sphere.

5-12.0e Explain why it is not permissible to use the substitutions $\vec{E}_1 = \vec{J}_1/\sigma_1$ and $\vec{E}_2 = \vec{J}_2/\sigma_2$ in the boundary conditions described by Eqs. 5-78 and 5-82. (Answer: When two materials of different conductivity are brought into contact a barrier (contact) potential develops. This is expressible as a dipole layer that is not accounted for in the derivations. Consider a *PN* junction diode.)

The remaining conditions satisfied by electric fields and currents are often easier to apply even though some may be equivalent to boundary conditions obtained earlier. Others of the conditions to be stated apply at special coordinate values, often at points where a coordinate becomes zero or infinite.

Condition 1

The first condition states that for finite dimensional systems with specified charges and/or voltages the potential at infinite distance from the system remains finite (including zero). For example, if one had an equation for potential outside of a charged object of the form

$$V(x, y, z) = f(x, y) + K_1 e^{cz}$$

and if infinite z coordinate were in the region of the potential, we would have to require either the constant K_1 or the constant c to be zero.

Condition 2

For finite dimensional systems in which a coordinate value of zero is in the region of interest and no point charge is at that point, the potential must remain finite or zero. For example, if the potential is of the form

$$V(r, \theta, \phi) = 10 \cos\theta \sin\phi + \frac{K_2}{r}$$

and if $r = 0$ is in the region of interest, then K_2 must be zero.

Condition 3

The potential is continuous across a charge-free boundary between two materials. Suppose the situation is as pictured in Figure 5-87, where there is a discontinuity of potential at the boundary. Figure 5-87*b* is a plot of such a function. If this were the case and we compute $\vec{E}$ using the gradient function of V, we would obtain infinite $\vec{E}$ at the boundary since ∇V represents the derivative of V (slope) with respect to coordinate and the slope of the curve

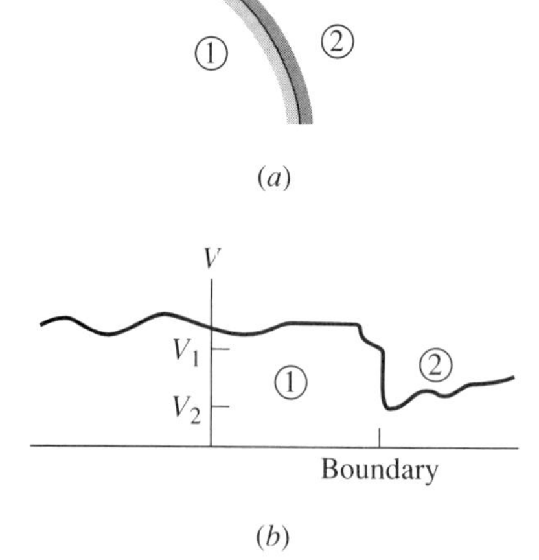

Figure 5.87. Boundary configuration for assumed potential discontinuity. (*a*) Boundary definition. (*b*) Potential distribution.

V is infinite at the boundary. Thus for $\vec{E}$ to remain finite, V would have to be continuous. If this condition is used, the tangential electric field boundary condition will not yield any new information about the system.

Condition 4

Suppose we have a uniform field that is very large in extent as shown in Figure 5-88*a*. If we immerse an object into the field there will be distortion of the field as shown in Figure 5-88*b*. However, at large distances from the object the field will maintain its original uniform configuration (usually for infinite coordinate values).

Condition 5

For finite field sources, the energy stored in a given volume remains constant. This is called the Meixner edge condition [41]. This is frequently used for systems in which time-varying fields impinge on an edge. Physically, we know that charge concentrations and electric field values become very large at sharp edges and corners. If a solution produces infinite energy in a region near a corner, the terms that produce this large value must be

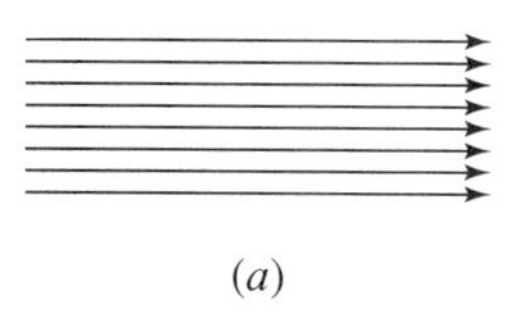

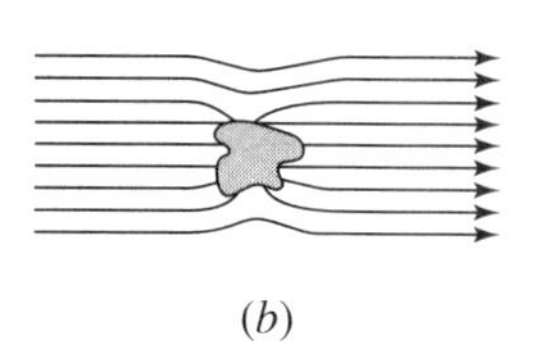

Figure 5-88. Illustration of the constancy of fields of large extent. (*a*) Original field. (*b*) Object immersed in field.

set to zero or exponents made of appropriate value to yield finite energy. This particular condition will not be used in this course.

Condition 6

This condition also applies to time-varying fields and will be used in a later chapter. We may use our experience from transmission line theory to assist us with a physical picture. For a sinusoidal source the voltage at a point down the line must lag the source in a prescribed way; specifically, of the form $e^{-j\beta z}$, where $z = 0$ is at the source.

Analogous to this is a finite dimension radiating source radiating outward. Thus for spherical coordinates, we expect a propagation term of the form $e^{-j\beta r}$ to appear in the solution. There is an additional feature with an antenna. As the energy moves away from the source it is distributed over a larger and larger surface (volume) so that at a given point the energy density and $|\vec{E}|$ are smaller than at points closer to the antenna. We would expect a $1/r^{\alpha}$ type variation of amplitude. This *radiation condition* is given formally as

$$\lim_{r\to\infty}\left[r\left(\frac{\epsilon\vec{E}}{\partial r} + j\beta\vec{E}\right)\right] = 0$$

and is called the Sommerfeld radiation condition [53].

5-13 Laplace's and Poisson's Equations

In Section 5-10 it was shown that if we could find the potential function $V(\vec{r})$ we could easily obtain all of the other field quantities using the gradient to obtain $\vec{E}$ and then using the constitutive relations to compute $\vec{D}$ and $\vec{J}$. This requires some way to obtain V.

We begin with Eq. 5-47 and use the constitutive relationship between $\vec{D}$ and $\vec{E}$ for an *isotropic* material:

$$\nabla \cdot \vec{D} = \rho_v$$

and then

$$\nabla \cdot \epsilon\vec{E} = \rho_v$$

If ϵ is not a function of coordinates (material also *homogeneous*), it may be removed from the derivative to obtain

$$\epsilon\, \nabla \cdot \vec{E} = \rho_v$$

or

$$\nabla \cdot \vec{E} = \frac{\rho_v}{\epsilon}$$

For *static* fields

$$\vec{E} = -\nabla V$$

so that the preceding equation becomes

$$\nabla \cdot \nabla V = -\frac{\rho_v}{\epsilon}$$

Using the definition $\nabla \cdot \nabla \equiv \nabla^2$ (see Chapter 4),

$$\nabla^2 V = -\frac{\rho_v}{\epsilon} \tag{5-87}$$

This is Poisson's equation.

If the region of interest has no volume charge density,

$$\nabla^2 V = 0 \tag{5-88}$$

This is Laplace's equation.

The problem now becomes one of solving these two partial differential equations in the appropriate coordinate system (see Chapter 4) and applying the boundary conditions to the potential and field quantities determined from the potential.

Exercises

5-13.0a Suppose that a region has an ϵ which is a function of coordinate (material is inhomogeneous). Derive the form of Poisson's equation for $\nabla^2 V$ for this case. If one were examining a region that is free of charge, what is the form of the equation? Compare this charge-free region equation with Eq. 5-87 and comment on the effect of inhomogeneity.

5-13.0b For the dipole far field expression developed in Example 5-26, show that Laplace's equation is satisfied. Can you account for the fact that Laplace's equation yields $\rho_v = 0$ and yet charges that form the dipole are present?

5-13.0c Suppose that the potential in the region between two concentric metal cylinders ($\sigma = \infty$) is given by $V(\rho) = 50\rho$ V. What is the charge density between the cylinders? The dielectric has $\epsilon_r = 2.4$. What are the surface charge densities on each cylinder if the inner radius is 0.05 m and the outer radius is 0.5 m?

5-13.0d A potential region has the form

$$V(r) = \frac{1}{r}$$

Show that this satisfies Laplace's equation.

Our problem now is to solve either Laplace's or Poisson's equation—whichever applies to a given problem—and use the gradient relationship to obtain the electric field intensity. At this point we will not discuss such solutions in detail but will defer them to a later chapter in the context of time-varying fields. In the next section we give a few examples having to do with the computation of capacitance for special configurations.

An important question, however, is that if we obtain a solution, how do we know it is the correct one? The *uniqueness theorem* gives the answer, which in rough form states that if a solution satisfies the differential equation and the boundary conditions it is the only solution. To show this we suppose a region of charge density $\rho_v(\vec{r})$ (can be zero) in a volume V enclosed by a surface S. Assume there are two potential solutions $V_1(\vec{r})$ and $V_2(\vec{r})$ that satisfy both Poisson's equation and the boundary conditions on surface S. Then

$$\nabla^2 V_1 = -\frac{\rho_v}{\epsilon} \quad \text{and} \quad \nabla^2 V_2 = -\frac{\rho_v}{\epsilon}$$

Subtracting,

$$\nabla^2 V_1 - \nabla^2 V_2 = \nabla^2(V_1 - V_2) = 0$$

If we set $V_1 - V_2 = V_0$, a new function, this becomes

$$\nabla^2 V_0 = 0$$

From Chapter 4 we have the vector identity

$$\nabla \cdot (V_0 \nabla V_0) = V_0 \nabla^2 V_0 + \nabla V_0 \cdot \nabla V_0 = V_0 \nabla^2 V_0 + |\nabla V_0|^2$$

Since V_0 satisfies Laplace's equation as shown previously, this reduces to

$$\nabla \cdot (V_0 \nabla V_0) = |\nabla V_0|^2$$

Integrating over the volume V,

$$\int_V \nabla \cdot (V_0 \nabla V_0) dv = \int_V |\nabla V_0|^2 dv$$

Using the divergence theorem on the left member

$$\oint_S V_0 \nabla V_0 \cdot d\vec{a} = \int_V |\nabla V_0|^2 dv$$

Now $\nabla V_0 \cdot d\vec{a}$ is the component of ∇V_0 in the direction of $d\vec{a}$, and $d\vec{a}$ is normal to the surface S. Then we may write

$$\nabla V_0 \cdot d\vec{a} = \frac{\partial V_0}{\partial n}, \qquad n = \text{normal coordinate}$$

Then

$$\oint_S V_0 \frac{\partial V_0}{\partial n} da = \int_V |\nabla V_0|^2 dv \tag{5-89}$$

Three types of boundary conditions are applicable to the surface S, any one of which makes the left integral zero.

1. Potential specified (given) on all of S. This is the *Dirichlet boundary condition*. Let V_S be the potential distribution on S. Since both V_1 and V_2 satisfy the same boundary conditions (by assumption)

   $$V_0(S) = V_1(S) - V_2(S) = V_S - V_S = 0$$

   Then

   $$\int_V |\nabla V_0|^2 dv = 0$$

Since the integrand is always positive we must have

$$|\nabla V_0|^2 = 0$$

or

$$\nabla V_0 = \nabla(V_1 - V_2) = 0 \qquad \text{everywhere in the system}$$

This requires

$$V_1 - V_2 = K \qquad \text{everywhere}$$

However, since the boundary is part of the system, and we have already shown that $V_1 - V_2 = 0$ there, the constant K is also zero. Thus we have

$$V_1 = V_2$$

2. Suppose that at the boundary the normal derivative of potential is specified; that is, $\partial V/\partial n$ is given at the boundary. This is called the *Neumann boundary condition*. Now $\partial V/\partial n$ is the same as the normal component of $\vec{E}$, namely, E_n. Now since both V_1 and V_2 satisfy the same boundary conditions we have over the surface

$$\frac{\partial V_1}{\partial n} - \frac{\partial V_2}{\partial n} = 0$$

Then

$$\frac{\partial(V_1 - V_2)}{\partial n} = \frac{\partial V_0}{\partial n} = 0$$

Thus in this case the left integral is also zero so that we again are led to set

$$\nabla(V_1 - V_2) = 0$$

or

$$V_1 - V_2 = K$$

Since the potential has not been specified on the surface, the most we can say is that two solutions can differ at most by a constant. Physically, this simply means a shift of DC reference or level of the system potential function.

3. The last condition is called a *mixed boundary condition*, and states that the potential is specified on part of the boundary and the normal derivative specified on the remainder. The conclusions would thus be the same as for 1 and 2 preceding for the appropriate parts of the system.

5-14 Capacitance

Capacitance between two conductors is an important electrical circuit element. In transmission line theory a key parameter was shown to be the capacitance per meter of line

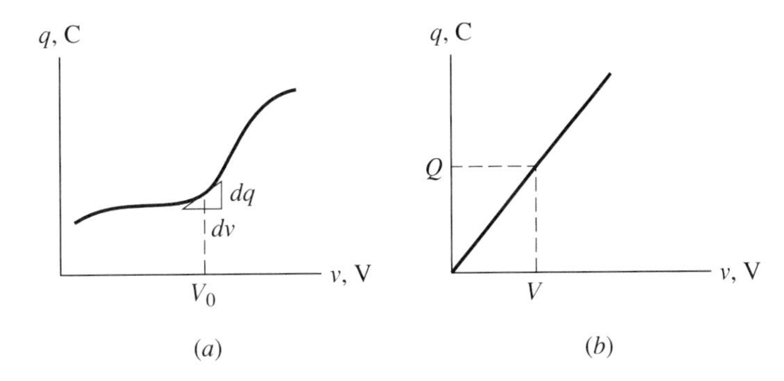

Figure 5-89. Charge-voltage characteristics for defining capacitance. (*a*) General variation. (*b*) Linear variation.

between the two conductors. That capacitance determined the velocity of propagation of signals and the phase shift of sinusoidal signals on the line.

Capacitance is defined in terms of the charge-voltage (Q-V) characteristic of a configuration. Two Q-V plots are shown in Figure 5-89.

Formally, capacitance is defined for Figure 5-89*a*, which holds in general. For a charge change dq accompanying a voltage change dv at a point V_0 on the Q-V characteristic we define capacitance as

$$C(V_0) \equiv \left.\frac{dq}{dv}\right|_{v=V_0} \tag{5-90}$$

This, of course, is the slope of the characteristic at V_0. This definition produces units of coulomb/volt, which is called the *farad*, denoted by F. See Example 5-37.

In many practical cases the charge on two conductors of a system is linearly related to the voltage between the conductors as shown in Figure 5-89*b*. This is the linear (constant) capacitor (Q-V is a straight line, through the origin) for which the slope is simply the ratio of any corresponding Q and V values. Thus

$$C = \frac{Q}{V} \qquad \text{linear (constant) capacitor} \tag{5-91}$$

For the majority of our work we will assume linear characteristic capacitors.

Example 5-37

A variable capacitance whose value is a function of the voltage across it is obtained by placing reverse bias on a PN junction diode. The capacitance is that of the depletion layer at the junction. The Q-V characteristic may be approximated by

$$q(V) = \frac{K}{(V_A - V)^{m-1}}$$

where K, V_A, and m (>1) are constants.

This is the variable capacitance diode, called a *varicap*. The capacitance is then

$$C(V_0) = \left.\frac{dq}{dv}\right|_{V=V_0} = \left.\frac{(m-1)K}{(V_A - V)^m}\right|_{V=V_0} \quad \text{F}$$

Note that simply taking the ratio of q to v would give quite a different result and that the capacitance is a function of the DC bias (V_0) on the diode.

A linear Q-V characteristic is given by

$$q = 10^{-6}v$$

or

$$Q = 10^{-6}V$$

Then

$$C = \frac{Q}{V} = 10^{-6} = 1\ \mu\text{F}$$

There are occasions when it is convenient to know how much charge can be placed on a single, isolated conductor. In such cases the second "plate" is assumed to be at infinity.

If conductors are of infinite extent we usually calculate the capacitance per unit length, such as for transmission lines.

Three techniques are used to compute capacitance using Eq. 5-90. These are first developed and then followed by examples that apply these techniques.

Method I

This method assumes a charge on the conductors. It is applicable to systems that have a high degree of symmetry (cylindrical and spherical, for example), which allows one to write expressions for the charge on the surfaces, usually a constant ρ_s. These usually allow a straightforward application of Gauss' law to obtain the field structure.

Procedure:

1. Assume that the surface charge density is uniform (constant) of value ρ_s C/m^2. (A symbol may be used; one does not need a particular numerical value.)
2. Use Gauss' law to obtain $\vec{D}$ as a function of coordinates and then use $\vec{E} = \vec{D}/\epsilon$.
3. Compute the charge on one conductor using

$$Q = \left| \int_S \rho_s da \right|$$

4. Compute the potential between the two conductors using

$$V = \left| \int_{\text{cond } 1}^{\text{cond } 2} \vec{E} \cdot d\vec{l} \right|$$

5. Compute C using Eq. 5-91.

Method II

This method assumes potentials on the conductors. One is usually selected as 0 V and one as V_0 V. The method is used for systems not having obvious symmetry, but for which the two conductors lie along coordinate system surfaces so that the selected potentials can be identified at one or two coordinate values. For example, $V(6, y, z) = V_0$. Such systems are said to have *regular boundaries.*

Procedure:

1. Solve Laplace's equation subject to the assumed boundary potentials to obtain $V(\vec{r})$.
2. Compute $\vec{D} = \epsilon\vec{E} = -\epsilon\,\nabla V$.
3. Using the boundary condition at a metal boundary, Eq. 5-85, obtain $\rho_s(\vec{r})$ as

$$\rho_s(\vec{r}) = D_n \qquad \text{(evaluated at the boundary coordinate)}$$

4. Evaluate $Q = \left|\int_S \rho_s\, da\right|$
5. Compute C using $V = |V_0 - 0| = V_0$ (positive value in Eq. 5-91.

Method III

This is for systems that do not have regular boundaries or have regions for which the solution of Laplace's equation directly is difficult. Additionally, this method applies only to those systems having a uniform cross section, such as a transmission line. This is a two-dimensional numerical technique and gives a numerical value in farads/meter rather than an equation involving general dimensional symbols. It is called the method of *curvilinear squares.*

Procedure:

1. Draw the cross section of the system to scale. The system may be expanded or reduced in size as long as the relative proportions and lengths are maintained.
2. Sketch in a series of equipotential contours or determine them experimentally. Generally, 5 to 10 such contours that divide the region into smaller sections are adequate. It is best to start in areas where $\vec{E}$ is most likely to be uniform so that the equipotential contours are easily visualized (away from corners or irregularities). While drawing these, keep in mind that the $\vec{E}$ field lines must be normal to these since $\vec{E}$ is the gradient of V.
3. Sketch in the electric field lines beginning normal from one conductor, passing normal to the equipotential lines and ending normal to the other conductor. As you draw subsequent lines try to produce squares as far or as close as possible. If there are extremely irregular ''squares'' it may be necessary to subdivide such squares by drawing finer equipotential contours therein and then additional field lines. A typical section of a region is shown in Figure 5-90.
4. Let Δp be the distance between equipotential lines in whatever length is being used. Let Δf be the distance between field lines. Now for the dimensioned cell shown in Figure 5-90 we think of V_1 and V_2 as the potentials on two small capacitor plates of width Δf and length 1 m into the page. For this small capacitor the potential across it is

$$V = V_1 - V_2$$

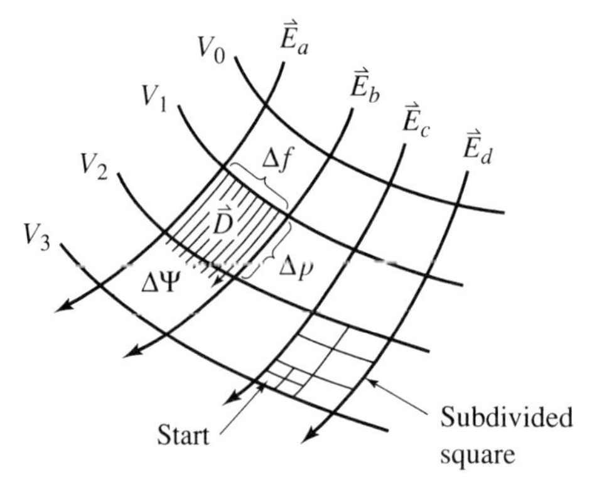

Figure 5-90. Sample region divided for application of curvilinear squares technique.

The electric displacement $\vec{D}$ is then of magnitude

$$D = \epsilon E = \epsilon \frac{V_1 - V_2}{\Delta p}$$

The equivalent charge on one ''plate'' of the cell is then obtained using the boundary condition as follows:

$$Q = \rho_s(\Delta f \times 1) = D\,\Delta f = \epsilon \frac{V_1 - V_2}{\Delta p} \Delta f$$

From these the capacitance per cell (per meter) is

$$C_{\text{cell}} = \frac{Q}{V} = \epsilon \frac{\Delta f}{\Delta p} \text{ F/m} \tag{5-92}$$

If we have approximate squares then Δf equals Δp so that

$$C_{\text{cell}} = \epsilon \text{ F/m} \tag{5-93}$$

Now if we stay between flux lines $\vec{E}_a$ and $\vec{E}_b$ we would have capacitors in *series*, so that if we draw squares the capacitance in one tube of flux would be

$$C_{\text{tube}} = \frac{\epsilon}{N_p}$$

where N_p is one less than the total number of equipotential lines. On the other hand, if we visualize staying between equipotential contours, say V_1 and V_2, the cell ''plates'' would be connected together, which parallels the cells. If there are N_f tubes of flux, the parallel capacitance would be

$$C_{\text{equi}} = N_f \epsilon$$

Thus the technique is to start with the smallest subdivided unit (identified by ''Start'' in Figure 5-90) and placing cells in series and parallel and working outward until all cells have been accounted for. The result is the capacitance of the system per meter of depth.

Example 5-38

Two parallel metal plates are separated by a distance d and have width $2a$ in the y dimension and b in the x dimension as shown in Figure 5-91. Solve for the capacitance of the system, neglecting fringing effects, using the three methods outlined in the preceding discussion.

Method I

We assume uniform surface charge on the plates ρ_s. The charge on one plate is then

$$Q = \rho_s \times \text{area of plate} = 2ab\rho_s$$

Next draw a small volume that includes the bottom plate. Since we neglect flux outside the plates, all the charge is on the inner surface of the bottom plate. From Gauss' law the flux passing out of the volume (note none goes out the sides) is of value

$$\Psi_e = \text{enclosed charge} = (-\rho_s)(2ab)$$

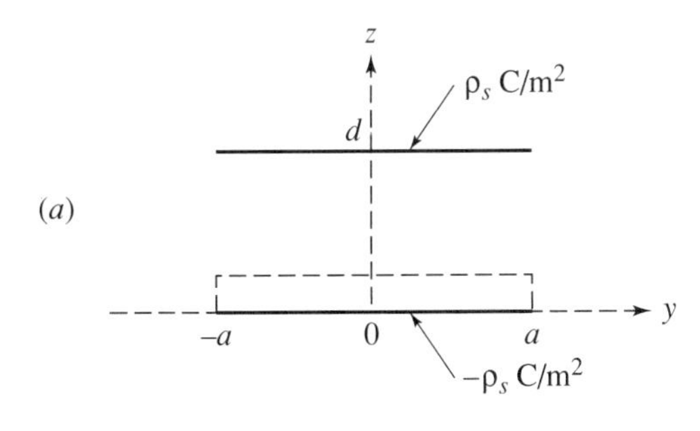

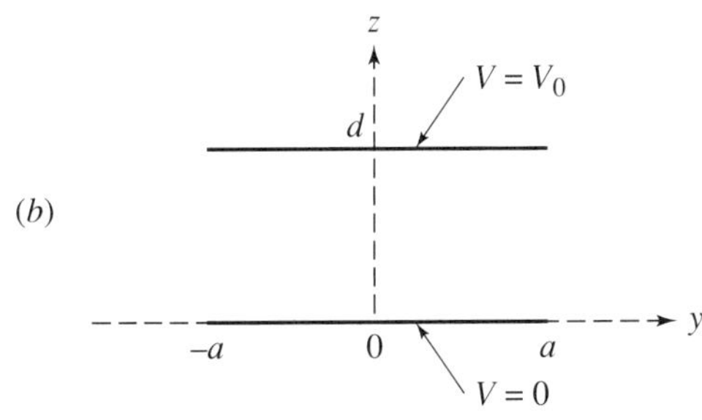

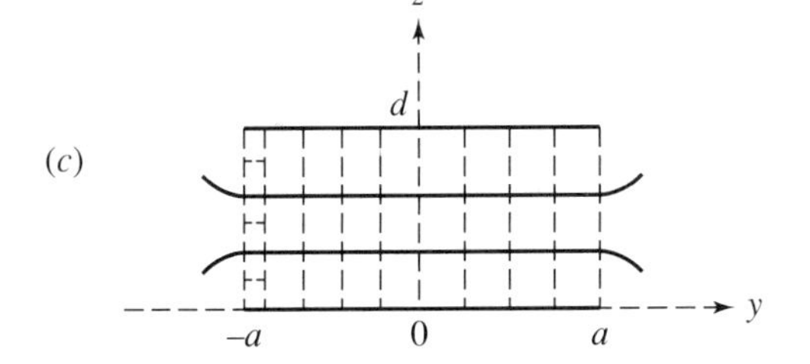

Figure 5-91. Solution of parallel plate capacitance problem using three methods. (*a*) Assumed charge. (*b*) Assumed voltage. (*c*) Curvilinear squares.

Thus

$$\vec{D} = \frac{\Psi}{\text{area}}\hat{z} = \frac{(-\rho_s)(2ab)}{2ab}\hat{z} = -\rho_s\hat{z}$$

as expected. Then

$$\vec{E} = \frac{\vec{D}}{\epsilon} = -\frac{\rho_s}{\epsilon}\hat{z}$$

From this,

$$V = -\int_0^d \vec{E}\cdot d\vec{l} = -\int_0^d -\frac{\rho_s}{\epsilon}\hat{z}\cdot dz\hat{z} = \frac{\rho_s}{\epsilon}d$$

Finally,

$$C = \frac{Q}{V} = \frac{2ab\epsilon}{d}$$

Method II

Assume that the top plate is at V_0 volts and the bottom at 0 V. First we expand Laplace's equation in rectangular coordinates

$$\frac{\partial^2 V}{\partial x^2} + \frac{\partial^2 V}{\partial y^2} + \frac{\partial^2 V}{\partial z^2} = 0$$

Since we have assumed no variation with x or y (all field quantities $\hat{z}$ directed) the partials (changes) with respect to x and y are zero. Laplace's equation reduces to

$$\frac{\partial^2 V}{\partial z^2} = \frac{dV^2}{dz^2} = \frac{d\left(\frac{dV}{dz}\right)}{dz} = 0$$

This requires

$$\frac{dV}{dz} = K_1$$

Integrating this,

$$V(z) = K_1 z + K_2$$

Applying the known plate voltages

$$V(0) = 0 = K_1 \cdot 0 + K_2$$

Then

$$K_2 = 0$$

$$V(d) = V_0 = K_1 d$$

Then

$$K_1 = \frac{V_0}{d}$$

We finally have

$$V(z) = \frac{V_0}{d} z$$

Then

$$\vec{E} = -\nabla V = -\frac{dV}{dz}\hat{z} = -\frac{V_0}{d}\hat{z}$$

and

$$\vec{D} = \epsilon\vec{E} = -\frac{\epsilon V_0}{d}\hat{z}$$

From the metal-dielectric boundary condition, Eq. 5-85,

$$\rho_s = D_n = -\frac{\epsilon V_0}{d}$$

$Q = \rho_s \times$ area of plate $= (-\epsilon V_0/d)(2ab)$. (We use the magnitude of this.) Finally,

$$C = \frac{Q}{V} = \frac{2ab\epsilon}{d}$$

as before.

Method III

Draw two equipotential lines between the plates. In practice, we would often draw more of these to get a better result. These are the solid horizontal lines between the plates. Next, draw in some dotted electric field lines beginning at the right on one plate, passing normal to the equipotentials and ending on the other plate. Note that successive lines are drawn so that approximate squares are formed. At the left side of the plate the last field line does not produce squares so we subdivide those cells to get closer squares.

Using the technique developed above the capacitance of the left vertical set of six squares is (per meter) the series result

$$C_6 = \frac{\epsilon}{6}$$

For each of the other eight vertical sets of three cells,

$$C_3 = \frac{\epsilon}{3}$$

Paralleling the cells we obtain

$$C = 8 \times \frac{\epsilon}{3} + \frac{\epsilon}{6} = \frac{17}{6}\epsilon \text{ F/m}$$

For depth b,

$$C = \frac{17b}{6}\epsilon = 2.83b\epsilon \text{ F}$$

As a check on the preceding results, the ratio of a to d scaled from the figure is about 1. From the results of the other two methods we would have

$$C = 2b\epsilon \text{ F}$$

This is an error of about 40%, which may seem high, but we used only three equipotential contours, so it is not a bad estimate.

Example 5-39

Using the method of assumed charge, compute the capacitance per meter of coaxial metal cylinders having two different concentric dielectrics between them as shown in Figure 5-92. Assume a charge ρ_l C/m on the center conductor and $-\rho_l$ C/m on the outer conductor.

Using Gauss' law as in Example 5-20 we have

$$\vec{D} = \frac{\rho_l}{2\pi\rho}\hat{\rho} = D_\rho(\rho)\hat{\rho}$$

From the boundary condition at $\rho = a$ we have

$$\rho_s = D_n\big|_{\rho=a} = D_\rho\big|_{\rho=a} = \frac{\rho_l}{2\pi a}$$

in both media. Then for 1-m length

$$Q = \frac{\rho_l}{2\pi a}(2\pi a) \times 1 = \rho_l \text{ C/m}$$

$$\begin{aligned} V &= -\int_a^c \vec{E}\cdot d\vec{l} = -\int_a^b \vec{E}_1\cdot d\vec{l} - \int_b^c \vec{E}_2\cdot d\vec{l} \\ &= -\int_a^b \frac{\vec{D}_1}{\epsilon_1}\cdot d\vec{l} - \int_b^c \frac{\vec{D}_2}{\epsilon_2}\cdot d\vec{l} \\ &= -\int_a^b \frac{\rho_l}{2\pi\epsilon_1\rho}\hat{\rho}\cdot d\rho\hat{\rho} - \int_b^c \frac{\rho_l}{2\pi\epsilon_2\rho}\hat{\rho}\cdot d\rho\hat{\rho} \\ &= -\frac{\rho_l}{2\pi\epsilon_1}\ln\left(\frac{b}{a}\right) - \hat{\rho}\cdot d\rho\hat{\rho} - \frac{\rho_l}{2\pi\epsilon_2}\ln\left(\frac{c}{d}\right) \end{aligned}$$

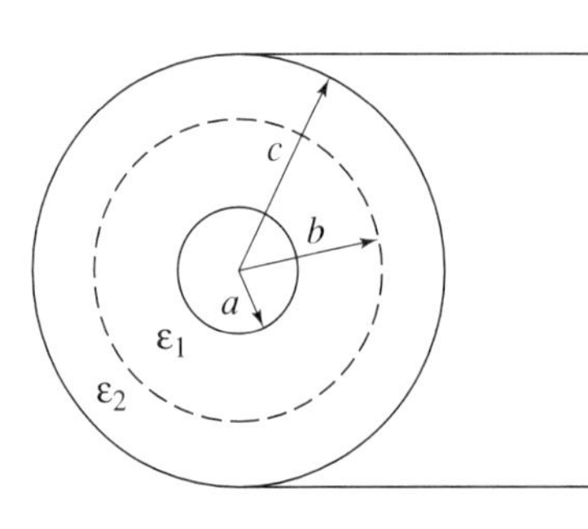

Figure 5-92. Multiple dielectric coaxial cable cross section.

As before, we use magnitudes. Thus

$$C = \frac{Q}{V} = \frac{1}{\dfrac{1}{2\pi\epsilon_1}\ln\left(\dfrac{b}{a}\right) + \dfrac{1}{2\pi\epsilon_2}\ln\left(\dfrac{c}{b}\right)} \text{ F/m}$$

Example 5-40

Derive the expression for the capacitance between two metal concentric spheres separated by two dielectric shells as shown in Figure 5-93. Use the method of assumed potentials, with the inner sphere at V_0 and the outer sphere at zero.

Using the spherical coordinate expansion of Laplace's equation and noting that there is symmetry (no variation) with the θ and ϕ coordinates so that only the r terms remain, we have

$$\frac{1}{r^2}\frac{\partial}{\partial r}\left(r^2\frac{\partial V}{\partial r}\right) = \frac{1}{r^2}\frac{d}{dr}\left(r^2\frac{dV}{dr}\right) = 0$$

For $r \neq 0$ we must have the term in parentheses constant so that

$$r^2\frac{dV}{dr} = K_1$$

Dividing by r^2 and integrating this becomes

$$V(r) = -\frac{K_1}{r} + K_2$$

From this

$$\vec{E} = -\nabla V = -\frac{dV}{dr}\hat{r} = -\frac{K_1}{r^2}\hat{r}$$

and

$$\vec{D} = \epsilon\vec{E} = -\frac{\epsilon K_1}{r^2}\hat{r}$$

Since there are two dielectrics, we need to use different constants in the expressions above to account for this. For the voltage solution use A_1 and B_1 for Medium 1 and F_2 and G_2 for Medium 2.

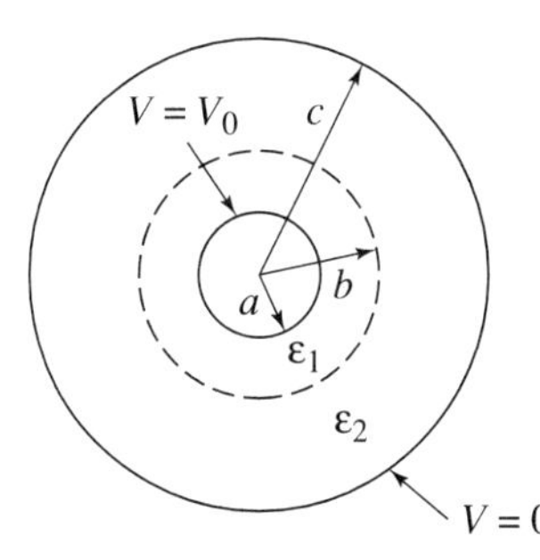

Figure 5-93. Configuration for determining the capacitance between concentric spheres using assumed potentials.

For Medium 1	For Medium 2
$V_1(r) = -\frac{A_1}{r} + B_1$	$V_2(r) = -\frac{F_2}{r} + G_2$
$V_1(a) = V_0 = -\frac{A_1}{a} + B_1$	$V_2(c) = 0 - -\frac{F_2}{c} + G_2$
or $B_1 = V_0 + \frac{A_1}{a}$	or $G_2 = F_2$
Then	Then
$V_1(r) = -\frac{A_1}{r} + V_0 + \frac{A_1}{a}$	$V_2(r) = -\frac{F_2}{r} + \frac{F_2}{c}$
and	and
$\vec{D}_1(r) = -\frac{\epsilon_1 A_1}{r^2}\hat{r}$	$\vec{D}_2(r) = -\frac{\epsilon_2 F_2}{r^2}\hat{r}$

At the common boundary $r = b$, the normal components of $\vec{D}$ must be equal since there is no surface charge there (see Eq. 5-81 for $\rho_s = 0$). Since the normal component of $\vec{D}$ is the $\hat{r}$ component, we require

$$-\frac{\epsilon_1 A_1}{b^2} = -\frac{\epsilon_2 F_2}{b^2}$$

or

$$F_2 = \frac{\epsilon_1}{\epsilon_2} A_1$$

Using this result the potential and flux density become

For Medium 1	For Medium 2
$V_1(r) = -\frac{A_1}{r} + V_0 + \frac{A_1}{a}$	$V_2(r) = -\frac{\epsilon_1 A_1}{\epsilon_2 r} + \frac{\epsilon_1}{A_1}$
$\vec{D}_1(r) = -\frac{\epsilon_1 A_1}{r^2}\hat{r}$	$\vec{D}_2(r) = -\frac{\epsilon_1 A_1}{r^2}\hat{r}$

At $r = b$ the potential must be continuous (see Condition 3 at the end of Section 5-12). Thus

$$-\frac{A_1}{b} + V_0 + \frac{A_1}{a} = -\frac{\epsilon_1}{\epsilon_2}\frac{A_1}{b} + \frac{\epsilon_1 A_1}{\epsilon_2 c}$$

Then

$$A_1 = \frac{V_0}{\frac{1}{b} - \frac{1}{a} + \frac{\epsilon_1}{\epsilon_2}\frac{1}{c} - \frac{\epsilon_1}{\epsilon_2}\frac{1}{b}}$$

With this the expressions for $\vec{D}$ become

$$\vec{D}_1(r) = \vec{D}_2(r) = \frac{\epsilon_1 V_0}{\left(\frac{1}{a} - \frac{1}{b} - \frac{\epsilon_1}{\epsilon_2}\frac{1}{c} + \frac{\epsilon_1}{\epsilon_2}\frac{1}{b}\right) r^2}\hat{r} = D_r(r)\hat{r}$$

Now at $r = a$ (could use $r = b$) the metal-dielectric boundary condition gives

$$\rho_s(r = a) = D_n(r = a) = D_r(r = a) = \frac{\epsilon_1 V_0}{\left(\frac{1}{a} - \frac{1}{b} - \frac{\epsilon_1}{\epsilon_2}\frac{1}{c} + \frac{\epsilon_1}{\epsilon_2}\frac{1}{b}\right) a^2} \text{ C/m}^2$$

Total charge on inner sphere is then, since ρ_s is a constant,

$$Q = \rho_s(r = a) \times (\text{area at } r = a) = \frac{4\pi a^2 \epsilon_1 V_0}{\left(\frac{1}{a} - \frac{1}{b} - \frac{\epsilon_1}{\epsilon_2}\frac{1}{c} + \frac{\epsilon_1}{\epsilon_2}\frac{1}{b}\right) a^2}$$

Thus the capacitance is

$$C = \frac{Q}{V} = \frac{Q}{V_0 - 0} = \frac{4\pi\epsilon_1}{\frac{1}{a} - \frac{1}{b} + \frac{\epsilon_1}{\epsilon_2}\left(\frac{1}{b} - \frac{1}{c}\right)}$$

Exercises

5-14.0a Using the Method I, show that the capacitance of two metal cylinders separated by a single dielectric is

$$C = \frac{2\pi\epsilon}{\ln\left(\frac{b}{a}\right)} \text{ F/m}$$

where a is the inner radius and b the outer. Using this result, show that the results of Example 5-39 could be obtained by visualizing that configuration as two capacitors in series with the dielectric interface as a common "plate."

5-14.0b Solve Exercise 5-14.0a using Method II, Laplace's equation.

5-14.0c For the coaxial system of Exercies 5-14.0a, suppose the inner radius is 15 mm and the outer is 45 mm. If the material between the cylinders is polyethylene, compute the capacitance per meter using curvilinear squares and compare the result with the exact value from Exercise 5-14.0a equation.

5-14.0d (a) For two concentric spheres separated by a single dielectric, use Method I to show that the capacitance is

$$C = \frac{4\pi ab\epsilon}{b - a}$$

(b) Using this result, show that the solution of Example 5-40 can be obtained by putting two capacitors in series with the boundary between the dielectrics visualized as the common "plate."

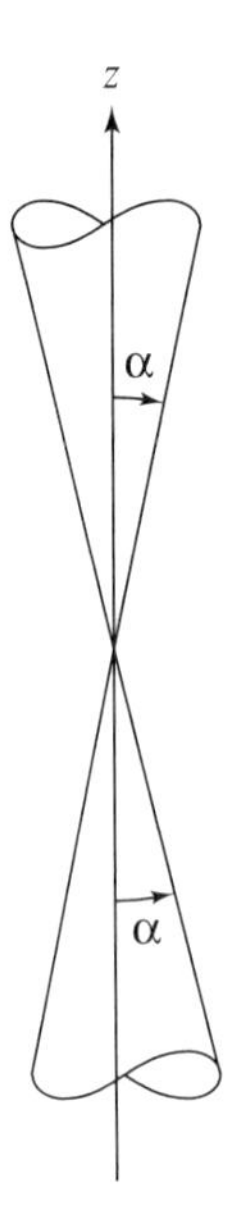

Figure 5-94.

(c) Find the capacitance of an isolated sphere by letting b recede to infinity. This is the capacitance of the sphere on a Van de Graff generator.

5-14.0e Obtain the equation for the capacitance per meter length between the two infinite cones shown in Figure 5-94. (Hints: Use the method of assumed potentials, spherical coordinates, and the fact that potential does not vary with ϕ or r.) There is a small gap at the points.

5-14.0f Repeat Exercise 5-14.0e for the case of a lower cone angle $\beta \neq \alpha$. From this result, obtain the capacitance per meter of a cone over a flat metal plate.

5-14.0g For the parallel plate Example 5-38, compute the energy density between the two plates and then compute the total stored energy. Show that this equals $\frac{1}{2}CV_0^2$.

The case of an imperfect, lossy dielectric ($\sigma \neq 0$) produces an interesting result of practical importance. In Chapter 1 the equivalent circuit of the transmission line contained a parallel G and C combination. The capacitance is the capacitance per meter length between the conductors and G the conductance per meter between them. There is a simple relationship between these two parameters of the line. The conductance G, of course, results in conductive current flow between conductors.

Using the definition of capacitance and the constitutive relationships between field quantities we have

$$C = \frac{Q}{V} = \frac{\int_S \rho_s da}{V} = \frac{\int_S D_n da}{V} = \frac{\epsilon \int_S E_n da}{V} = \frac{\frac{\epsilon}{\sigma} \int_S J_n da}{V}$$

Since J_n is A/m^2, the last integral is the current flow leaving the conductor (we assume 1-m length as one dimension of the surface). Thus

$$C = \frac{\epsilon}{\sigma}\frac{I}{V} = \frac{\epsilon}{\sigma} G$$

From this we conclude that if we can find the capacitance equation or value using any of the techniques of this section, we can obtain the conductance equation or value using

$$G = \frac{\sigma}{\epsilon} C \text{ S or S/m} \tag{5-94}$$

This result applies *in general* to any capacitance system, although we have referred to the transmission line as an example.

Another important relationship was derived in Chapter 1 relating C and L for the transmission line, assuming small losses R and G. This is the phase velocity, Eq. 1-77:

$$V_p = \frac{1}{\sqrt{L\,C}}$$

We will show in a later chapter that for the field configurations existing in two-conductor transmission systems with low losses

$$V_p = \frac{1}{\sqrt{\mu\epsilon}}$$

From these two equations the line inductance is

$$L = \frac{\mu\epsilon}{C} \text{ H/m} \tag{5-95}$$

This applies to transmission lines only.

5-15 Pressure on Charged Surfaces

Since a surface that has charge on it also has an electric field intensity there one might expect a force on the surface since $\vec{E}$ has units of N/C. This is in fact the case. We will now derive the equation for the pressure on a charged surface.

In Figure 5-95 is shown a surface of charge and a local force given to a small area Δs. The small volume contains energy of amount

$$\Delta W = w_e\,\Delta v = w_e\,\Delta l\,\Delta S$$

The work done is

$$\Delta W = F\,\Delta l = (P_e\,\Delta S)\,\Delta l$$

where P_e is the electric pressure per unit area measured in pascals (Pa). From these two equations we have, upon equating,

$$P_e = w_e \text{ Pa} \tag{5-96}$$

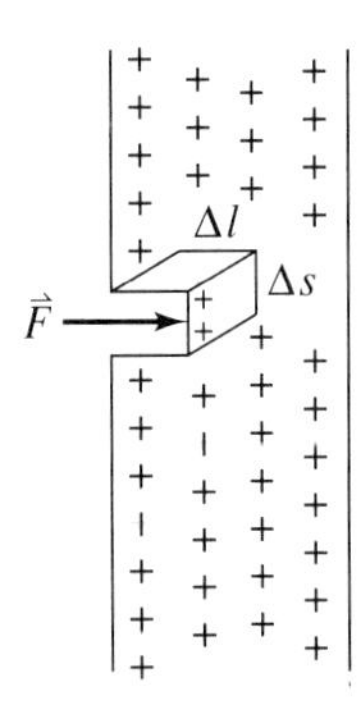

Figure 5-95. Configuration for determining the force on a charged surface.

To obtain the result in terms of field quantities we use Equations (5-77) and (5-85) to obtain the pressure on a charged metal surface:

$$P_e = \frac{1}{2}\epsilon|\vec{E}|^2 = \frac{1}{2}\epsilon E_n^2 = \frac{1}{2\epsilon}D_n^2 = \frac{\rho_s^2}{2\epsilon}\ \text{Pa} \tag{5-97}$$

One possible solution procedure for obtaining the force on a charged metal surface would be as follows:

1. Assume the potentials on the conductors and solve Laplace's or Poisson's equation as appropriate.
2. Evaluate $\vec{D} = \epsilon\vec{E} = -\epsilon\,\nabla V$.
3. Evaluate, on the conductor $\rho_s = D_n$.
4. $P_e = (1/2\epsilon)\rho_s^2$
5. $F = \int_{\text{surface}} P_e da$

Note that this yields only the magnitude of the force. Can you develop a method for determining direction?

Example 5-41

Neglecting fringing effects, determine the force on the top capacitor plate of the system shown in Figure 5-96. Expanding Laplace's equation and observing that neglecting fringing means V will vary only with z, there results

$$\frac{d^2V}{dz^2} = 0$$

This requires

$$\frac{dV}{dz} = K_1$$

Integrating:

$$V(z) = K_1 z + K_2$$

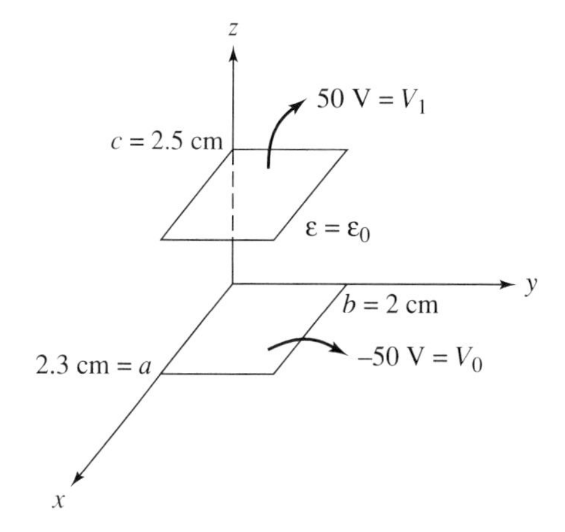

Figure 5-96.

Applying the assumed potentials:

$$V(z = 0) = -50 = K_1 \cdot 0 + K_2 = K_2$$

Then

$$V(z) = 4000z - 50$$
$$V(z = 0.025) = 50 = K_1(0.025) - 50$$
$$K_1 = 4000$$

Finally,

$$V(z) = 4000z - 50$$

$$\vec{D} = -\epsilon_0 \nabla V = -\frac{1}{36\pi} \times 10^{-9} \times 4000\hat{z} = -3.54 \times 10^{-8}\hat{z}$$

$$\rho_s = |D_n| = |D_z| = 3.54 \times 10^{-8} \text{ C/m}^2$$

$$P_e = \frac{1}{2\epsilon_0}\rho_s^2 = 7.086 \times 10^{-5} \text{ N/m}^2$$

$$F = \int_0^{.02}\int_0^{.023} 7.086 \times 10^{-5} dxdy = 3.26 \times 10^{-8} \text{ N}$$

The direction of this force would be down since the upper plate is attracted toward the lower one.

Exercises

5-15.0a From a unit analysis of the quantities of the rightmost equality of Eq. 5-97, show that the result is in fact Pa or N/m^2.

5-15.0b Two concentric metal spheres of radii a and b ($a < b$) have voltages 100 V and 0 V on the inner and outer spheres, respectively. For $a = 2$ cm and $b = 4$ cm, compute the force on the inner sphere for air dielectric and for $\epsilon_r = 4$. Is the force an explosive or implosive force?

5-15.0c An electron is located a distance h from an infinite metal plane. Determine

(a) Total force on the plane and the direction
(b) Energy stored in the system
(c) If one were to use a finite-dimensional plane, what would the dimensions be so that the force would be within (i) 10% and (ii) 1% of the true value obtained in (a)? (Hints: Place an image charge of opposite sign a distance h below the plane and then remove the plane and compute the $\vec{D}$ at the plane location. Then replace the plane and compute ρ_s. This is the method of images.)

5-15.0d Determine the pressure on the inner and outer conductors of a coaxial cable of inner radius a and voltage V_0, and outer radius b and voltage zero.

Summary, Objectives, and General Exercises

Summary

This chapter has been devoted to the basic concepts and computations of electric field quantities and applications to basic systems. Although Laplace's and Poisson's equations were developed, a detailed study of solutions to systems having more than one coordinate variation is deferred to propagation studies with time-varying field components. Sufficient examples and exercises were given to develop a facility for working with vector operations and integrations without the encumbrance of time. For this chapter the key ideas and computations one should be able to perform are listed in the following.

Objectives

1. Be able to give the definition of electric current measured in the unit of ampere (A).
2. Be able to give the definition of electric charge measured in the unit of coulomb (C).
3. Be able to state Coulomb's law for the force between two charges and to give the physical significance of each symbol in the expression.
4. Be able to compute the force on a given charge produced by several other point charges.
5. Be able to give a qualitative description of the effect of a dielectric on the field around a point charge.
6. Be able to state the definition of electric field intensity and know the symbol and units corresponding to the field.
7. Be able to determine qualitatively the electric field lines in a system by using the definition of electric field.
8. Be able to compute the total electric field intensity at a given point in space due to several point charges and due to a distribution of charge density. Be able to plot the field patterns of charge configurations for constant radius around the distribution.
9. Be able to explain qualitatively the reason why the electric field intensity varies inversely with dielectric constant.
10. Know the constitutive relationship for electric fields and be able to explain the physical description of the electric flux density (also called *electric displacement* density) $\vec{D}$.
11. Know the definitions, units, and physical significance of:
    (a) Electric dipole moment
    (b) Electric susceptibility
    (c) Polarization
    (d) Relative dielectric constant

(e) Dielectric constant (permittivity)
(f) Dielectric tensor

12. Be able to describe and distinguish among the following types of dielectric materials:
(a) Isotropic
(b) Nonlinear
(c) Inhomogeneous
(d) Anisotropic
(e) Time-varying

13. State Gauss' law in differential (point) form and in integral form and give a physical interpretation of both.

14. Be able to use Gauss' law in integral form to compute the electric flux density $\vec{D}$ for a symmetric charge distribution.

15. Distinguish between electric flux and net electric flux as applicable to Gauss' law.

16. Be able to make computations for the $\vec{D}$, $\vec{P}$, and $\vec{E}$ fields and to distinguish between bound and free charge densities.

17. Know and be able to state the definition of potential (voltage) along with its units and equation in integral form.

18. Know and be able to make computations using both the differential and integral forms of the relationships between V and $\vec{E}$ for static (time-invariant) fields.

19. Be able to compute the potential at a given point in space due to point charges and due to a charged region, given as line, surface or volume charge. Be able to plot the equipotential contours.

20. Be able to write the integral and differential relations that describe the conservative property of static electric fields.

21. Be able to state the boundary conditions for $\vec{D}$ and $\vec{E}$ at a metal-dielectric boundary and a dielectric-dielectric boundary, and also to apply those conditions.

22. Be able to derive Poisson's equation and Laplace's equation as they apply to static (time invariant) potential distributions beginning with $\nabla \cdot \vec{D} = \rho$.

23. Be able to state the uniqueness theorem for solutions to Poisson's equation and Laplace's equation and the three types of boundary conditions that yield unique solutions.

24. Be able to express point, line, and surface charge regions as volume charge densities using unit step and delta functions, and be able to evaluate potential functions and electric-field functions from integrals having such densities in the integrand.

25. Be able to solve Poisson's equation and Laplace's equation for simple directly integrable charge distributions and potential distributions.

26. Be able to use the solution to Laplace's equation to compute the electric field intensity, electric displacement density, and the surface charge density at a boundary.

27. Be able to describe physically and to apply to a given problem the following supplementary boundary conditions:
(a) Finiteness of potential
(b) Continuity of potential across a charge-free boundary
(c) Finiteness of energy at an edge (Meixner edge condition)
(d) Sommerfeld radiation condition
(e) Constancy of fields having infinite dimensional extent (perturbation condition)

28. Be able to solve one-dimensional electrostatic boundary value problems that have multiple dielectrics in the region of interest for $\vec{E}$, $\vec{D}$, and $\vec{J}$.

29. Be able to state the definition of capacitance and to examine the units of the quantities involved. Additionally, be able to describe and apply the steps for computing capacitance (and capacitance per meter) using:
(a) Symmetry considerations
(b) Solutions of Laplace's equation
(c) Method of curvilinear squares

30. Be able to compute the energy stored in an electrostatic system using:
(a) System capacitance and voltage
(b) Summation formulation for point charge configurations
(c) The integration formulation for continuous

charge regions and the associated potential distribution

31. Be able to compute the energy stored in a given electrostatic field configuration and the associated energy density function.
32. Be able to compute the force on a charged conductor in an electrostatic configuration.
33. Be able to give the physical significance of the current continuity equation and to state the equation.
34. Be able to compute conduction and convection current densities.
35. Be able to compute the resistance of a given conductor configuration.
36. Know and be able to use the relationship between G and C parallel elements on transmission lines.

General Exercise

GE5-1 Two small charged balls are suspended by threads (assumed massless) as shown in Figure 5-97. Determine the equilibrium angles that the threads make with the vertical.

GE5-2 For the configuration shown in Figure 5-98, derive the equation for $\vec{E}(x, y)$ and show that the resulting field is a conservative one. Suppose you were asked to evaluate the line integral from A to B. What path would you suggest?

GE5-3 Find the electric field intensity at point P shown in Figure 5-99. What force would be exerted on a +1-C charge placed at P?

GE5-4 The finite circular cylindrical surface section shown in Figure 5-100 is charged to a con-

Figure 5-97.

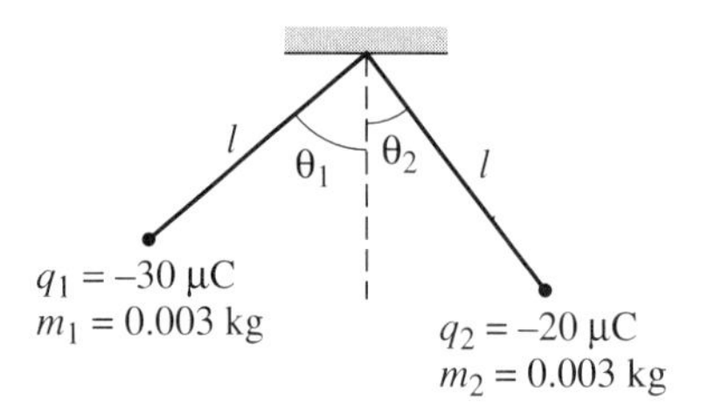

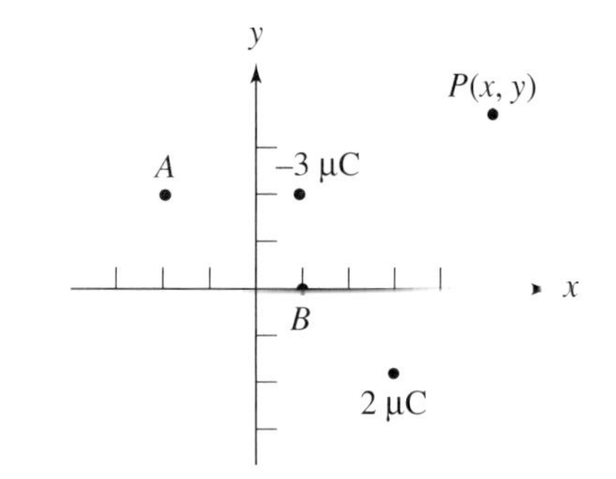

Figure 5-98.

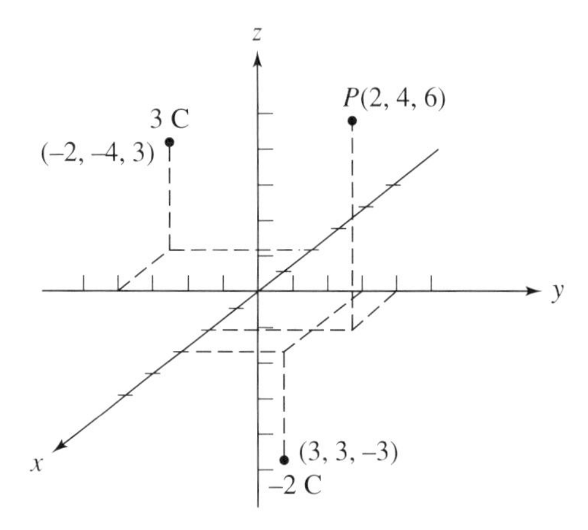

Figure 5-99.

Figure 5-100.

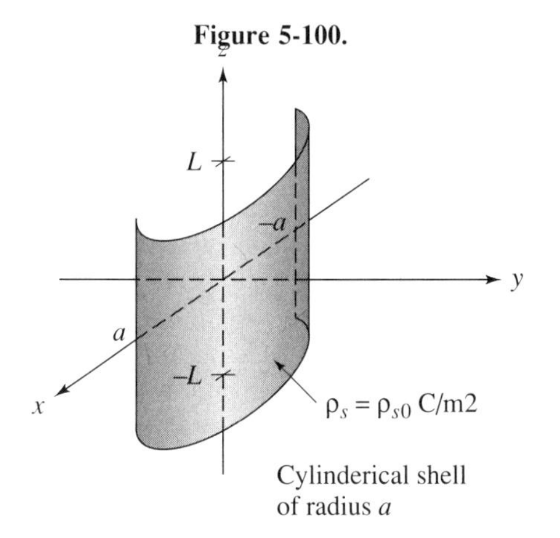

stant value charge density ρ_{s0} C/m^2. Compute the electric field intensity around the configuration. Are there any singular points to watch out for where $|\vec{E}| \to \infty$?

GE5-5 The elliptic cylindrical coordinate system is shown in Figure 5-101 (see Chapter 4 for a discussion of coordinate systems). The hyperbolae describe hyperbolic cylindrical surfaces that are uniform in the z direction. Suppose there is a charged hyperbolic cylinder corresponding to the surface $v = \pi/6$ and $v = 11\pi/6$ and it is charged to a surface charge density of constant value 30 μC/m^2. The constant a of the coordinate system is 0.5 cm. Determine the electric field intensity for the region of space defined by $\pi/6 < v < 11\pi/6$.

GE5-6 The configuration of General Exercise GE5-5 is modified by having the portion of the surface in the region $y < 0$ charged to -30 μC/m^2 (upper part still positive). Determine the electric field intensity for the system for $\pi/6 < v < 11\pi/6$. Determine the equation for and plot the field line that passes through the point (3/2, π/2, 0).

GE5-7 For the system of General Exercise GE5-2, plot the field line that passes through the point (5, 4). Plot the line so it approaches the two charges, and also mark it with an arrow.

GE5-8 For the system of General Exercise GE5-3, plot the field line that passes through the point (2, 4, 6), and indicate the direction of

Figure 5-101.

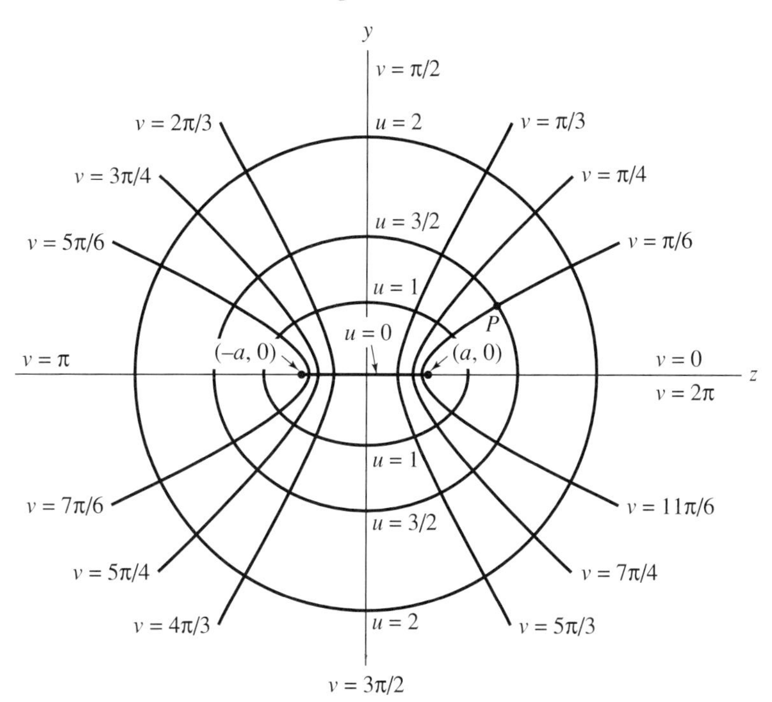

$x = a \cosh u \cos v,\ y = a \sinh u \sin v,\ z = z$
where $u \geq 0,\ 0 \leq v < 2\pi,\ -\infty < z < \infty$
$h_u = h_v = a\sqrt{\sinh^2 u + \sin^2 v},\ h_z = 1$

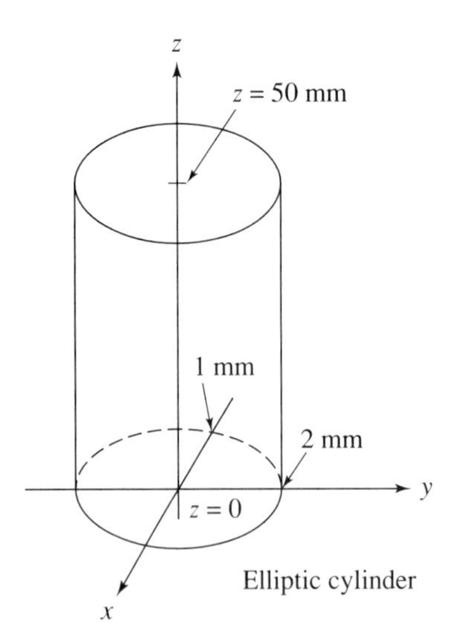

Figure 5-102.

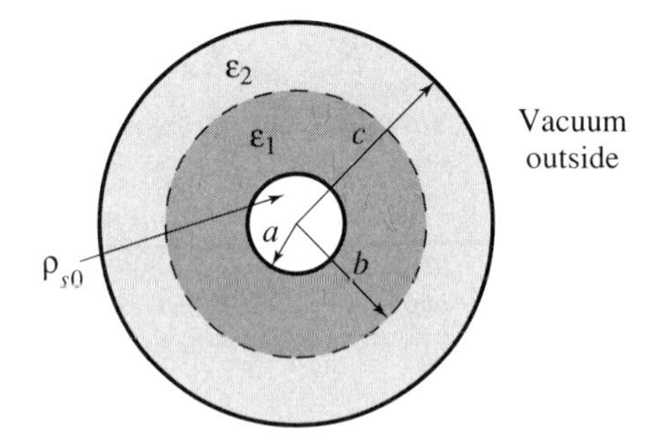

Figure 5-103.

the line. Plot the line so it approaches the two charges.

GE5-9 The elliptic cylindrical metal bar has a constant cross section, as shown in Figure 5-102, and constant conductivity of 6×10^7 S/m. Compute the resistance of the bar. Use elliptic cylindrical coordinates (see Figure 5-101) and find the value of a by substituting the two known points, which are the major and minor axes points shown at the bottom of the figure.

GE5-10 Suppose we have oxygen gas enclosed in a container and then the gas is ionized. Is there any flux line passing through a small imaginary closed surface inside the container? Is there any *net* flux passing through the surface? Explain your reasoning.

GE5-11 Suppose we have a metal sphere of radius a which has a uniform surface charge density ρ_{s0} C/m^2. Around this shell we place two different dielectric shells concentric with the metal sphere as shown in Figure 5-103. The dielectrics have relative permittivities ϵ_{r1} and ϵ_{r2} with $\epsilon_{r1} > \epsilon_{r2}$. There is no metal outer sphere. For this system find equations for and plot $\vec{D}$ and $\vec{E}$ for radius $r \geq 0$.

GE5-12 Repeat GE5-11 for the case where a second metal sphere charged to $-\rho_{s0}$ C/m^2 is placed at $r = c$. Describe the physical effect of this second shell.

GE5-13 In an oscilloscope a beam of electrons is used to obtain the visible trace on the screen. The moving electrons constitute a current flow (which direction?). The beam has a finite, nonzero cylindrical shape of radius a. Suppose we denote the beam current as a constant I_0 amperes and that the electrons have velocity $\vec{\mu}_0$. The electrons are furthermore assumed uniformly distributed over the cross section of the beam. We may treat this as a static problem since the number of charges in a given region is constant with as many electrons entering one side as leaving the other. Using Gauss' law, find $\vec{D}$ and $\vec{E}$ for all radii ρ for $0 \leq \rho \leq \infty$. If we consider some $\rho_0 < a$, what is the effect of $\vec{E}$ on the electrons at points $\rho_0 < \rho < a$? What is the result of that effect on the beam structure?

GE5-14 Suppose we have a point charge of value Q located at the origin of a rectangular coordinate system. Now, centered around this point charge, imagine a cylinder of height h in the z direction and radius b. Find the total flux passing through this cylinder by integrating $\vec{D}$ over the closed surface. (Hint: You will need to convert a vector in spherical coordinates into one in cylindrical coordinates. See Chapter 4.) Does your result agree with Gauss' law?

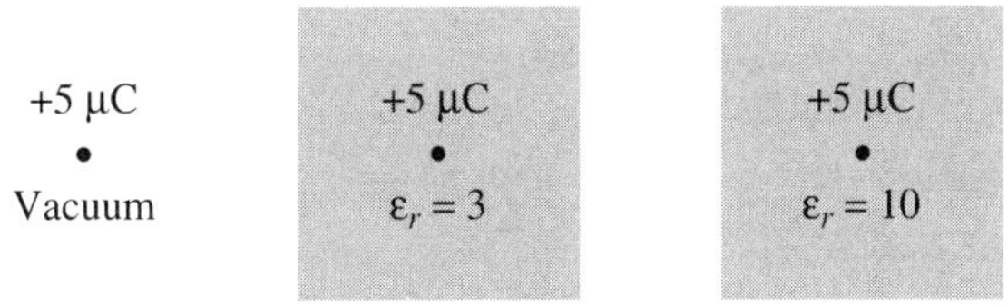

+5 μC • Vacuum

+5 μC • $\varepsilon_r = 3$

+5 μC • $\varepsilon_r = 10$

Figure 5-104.

GE5-15 For the cases shown in Figure 5-104, determine the electric flux density (displacement) $\vec{D}$ and the electric field intensity $\vec{E}$ at a distance 1 m from the charge. What is the force on a 1-C charge at that same radius for each case?

GE5-16 Since the permittivity of anisotropic materials is dependent upon direction, the values of ϵ_{ij} depend upon the orientation of the coordinate axes with respect to crystal structure. Suppose we limit ourselves to a two-dimensional "crystal" shown in Figure 5-105*a* for which the relative permittivity tensor is

$$[\epsilon_r] = \begin{bmatrix} 2 & 0.6 \\ 0.6 & 1.5 \end{bmatrix}$$

Now suppose we rotate the axes through an angle ϕ_0 shown in Figure 5-105*b* to obtain the prime coordinate system.

(a) Determine the values of ϵ'_{ij} for the new coordinate axes tensor $[\epsilon'_r]$ in terms of ϕ.

(b) Find the angle ϕ_0 for which the relative permittivity tensor is diagonal. These particular axes are called the *principal axes.* What are the relationships between the corresponding components of $\vec{D}$ and $\vec{E}$ for this case?

Figure 5-105.

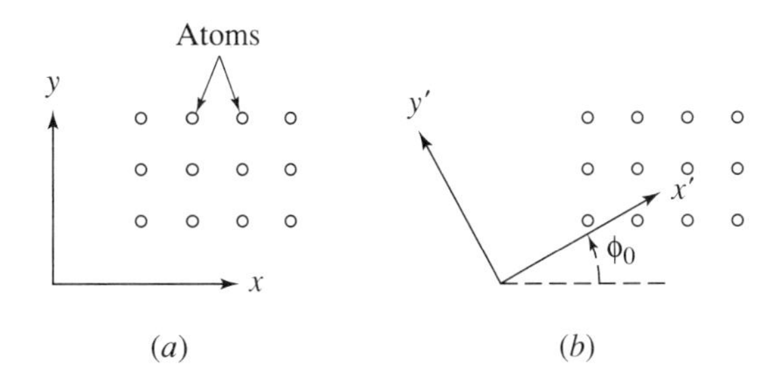

GE5-17 How much energy does it take to move 1 electron through 1 V of potential? This is called 1 electron volt of energy. (Answer: 1.6×10^{-19} J)

GE5-18 The intersection of two parabolic sheets defines a tube having a football-shaped cross section as shown in the heavy outline in Figure 5-106. The potential distribution within the cross section is

$$V(u, v) = 150(4 - u^2)(u^2 - v^2)\sin(2 - v) + 100 \text{ V}$$

(a) Find the potentials on each curved boundary and along the y axis within the cross section.

(b) Find the equation for the electric field intensity within the cross section.

(c) For a dielectric with $\epsilon_r = 4$, obtain the electric flux density, polarization, and bound charge density in the cross section.

GE5-19 A rectangular box in the positive octant of space has one corner at the origin. The lengths of the sides are a, b, and c along the x, y and z axes, respectively. The potential within the box is given by

$$V(x, y, z) = \frac{1600}{\pi^2} \sum_{n=1,3,5}^{\infty} \sum_{m=1,3,5}^{\infty} \frac{\sin\left(\frac{m\pi}{a}x\right)\sin\left(\frac{n\pi}{b}y\right)\sinh\sqrt{\left(\frac{m\pi}{a}\right)^2 + \left(\frac{n\pi}{b}\right)^2}\,z}{nm \sinh\left(\sqrt{\left(\frac{m\pi}{a}\right)^2 + \left(\frac{n\pi}{b}\right)^2}\,c\right)}$$

(a) Obtain the expression for the electric field intensity within the box. (The derivative of a sum is the sum of the derivatives.)

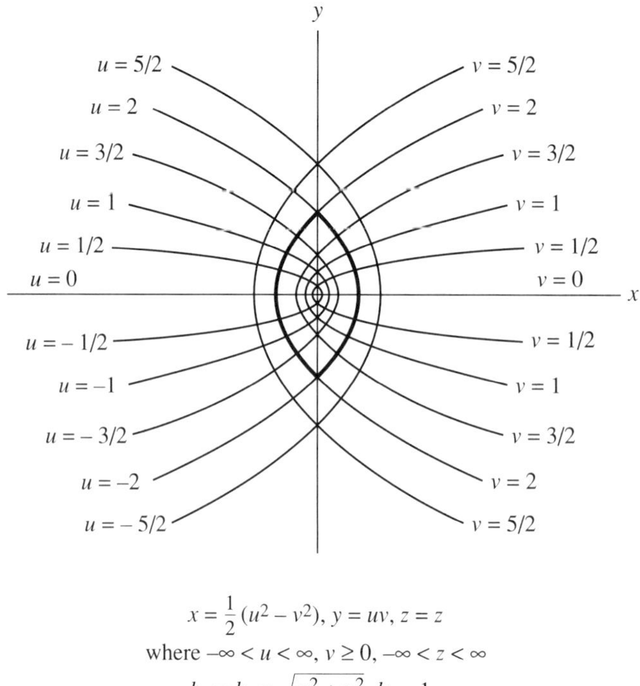

$$x = \frac{1}{2}(u^2 - v^2),\ y = uv,\ z = z$$
where $-\infty < u < \infty,\ v \geq 0,\ -\infty < z < \infty$
$$h_u = h_v = \sqrt{u^2 + v^2},\ h_z = 1$$

Figure 5-106.

(b) Draw the box in the first octant giving the potentials on each side wall and the dimensions along each axis, knowing that the potentials on each of the sides are constants. (Hint: Recognizing a double Fourier series will be necessary.)

GE5-20 An infinite rectangular metal channel has one corner along the z axis and the other corners are in the positive xy quadrant. The cross section has x dimension a and y dimension b. All sides are at constant (0 or nonzero) potentials. The potential within the channel is

$$V(x, y) = \frac{4500}{\pi} \sum_{m=1,3,5}^{\infty} \frac{\sinh\left(\frac{m\pi}{a} x\right)}{m \sinh\left(\frac{m\pi b}{a}\right)} \sin\left(\frac{m\pi}{a} y\right)$$

(a) Obtain the equation for $\vec{E}(x, y)$. (The derivative of the sum is the sum of the derivatives.)

(b) Draw the channel, properly dimensioned, and give the potentials (constants) on all of the sides. Look for a Fourier series on one of the sides.)

GE5-21 A charged line of infinite length is coincident with the z axis and is charged uniformly to ρ_{l0} C/m of constant value. Use Gauss' law to find $\vec{D} = D_\rho(\rho)\hat{\rho}$, then $\vec{D} = \epsilon\vec{E}$.

GE5-22 Write a computer program that will compute and plot the potential at a given point in space due to given point charges. Use this program to plot on a graphics terminal (or print out) the exact potential of the dipole of Example 5-26, using $q = 1$ nC, $l = 0.01$ m, for several values of r including values near l. Now, by comparing the exact values available from your program to approximate values, determine the smallest r value for which the maximum absolute error in V is 10%. Is this r dependent upon θ? q? The change in the potential values as the approximation begins to break down can be made visually by plotting both exact and approximate V values.

GE5-23 Compute $\vec{E}$ of Example 5-27 using the gradient relationship.

GE5-24 Make a polar plot of the potential around the disk of Example 5-29 for r = 2.5 cm and r = 3 cm. Describe what is happening, as r increases, to the plot and give the physical reason. (See discussion of Example 5-7, Eq. 5-20.

GE5-25 In this exercise we show how Gauss' law, applied to systems having a high degree of symmetry, yields results consistent with the potential integration used to ultimately determine $\vec{E}$. Consider the cylinder shown in Figure 5-107, of radius b and uniform surface charge ρ_{s0} C/m^2. Assume L is large compared with the distance from the cylinder to the point at which we want to determine V and $\vec{E}$.

(a) Find $\vec{E}$ at a general point P(ρ), $\rho > b$. Note we may use $\vec{D} = D_\rho(\rho)\hat{\rho}$.

Figure 5-107.

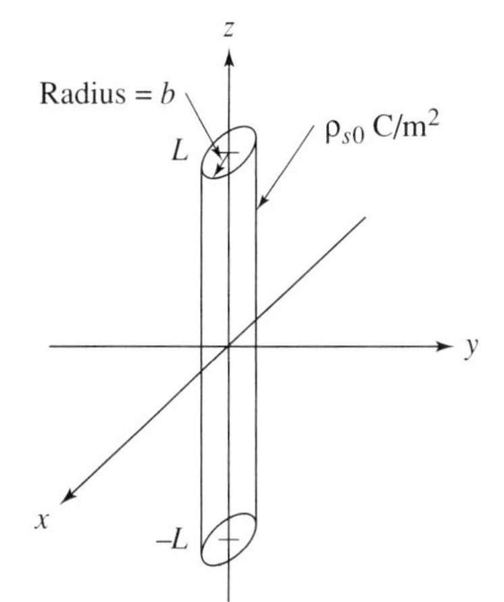

(b) Next compute V by integrating $\vec{E}$ using 1 m as reference point for 0 V ($b < 1$ m).

(c) Finally, compute V by direct integration of the charge distribution and compare.

GE5-26 The charged disk of Example 5-29 has a charge density $\rho_s(\rho) = \rho\ \mu$C/m^2. Obtain the integral expression for $V(\vec{r})$ by direct integration of the charge system. Answer: If we let $g(\phi') = x\cos\phi' + y\sin\phi'$, then $V(\vec{r})$ is

$$\frac{10^{-6}}{4\pi\epsilon}\int_0^{2\pi}\left\{\frac{[a+3g(\phi')]\sqrt{r^2-2ag(\phi')+a^2}-3rg(\phi')}{2}+\frac{3g^2(\phi')-r^2}{2}\ln\left[\frac{\sqrt{r^2-2ag(\phi')+a^2}+a-g(\phi')}{r-g(\phi')}\right]\right\}d\phi'$$

GE5-27 Suppose a metal sphere of radius b is centered at the origin of coordinates and has a uniform (constant) surface charge ρ_{s0} C/m^2. Find the potential at a distance $r > b$. Use integration formula and then repeat using Gauss' Law to find $\vec{E}$ and then V. (See GE5-25.)

GE5-28 For the infinite, uniformly charged line shown in Figure 5-108, determine the potential at any point in the xy plane.

GE5-29 A very long metal strip of width $2b$ is uniformly charged to ρ_{s0} C/m^2. See Figure 5-109. Find the potential in the xy plane. Work the prob-

Figure 5-108.

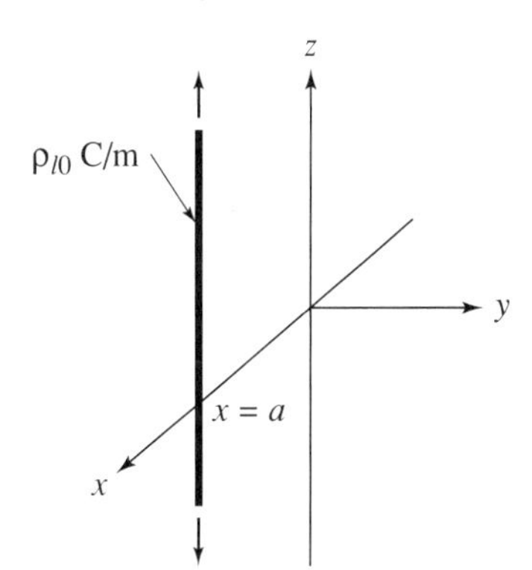

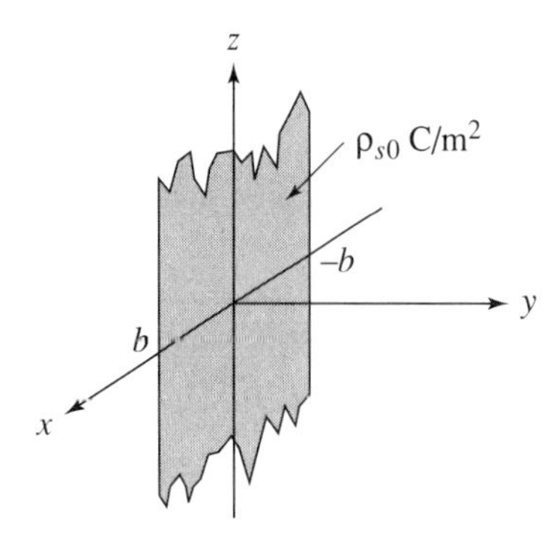

Figure 5-109.

lem two ways: (1) direct integration of the charge distribution; (2) divide the strip into widths dx' and located at x' and use the result of GE5-28 to express dV for one of the strips and then integrate over $(b, -b)$.

GE5-30 It is possible to compute the far field of a charge distribution directly by expanding Eq. 5-67 into a series from which the *first* nonzero term is the far field. Using Figure 5-110, derive the expansion using the technique of Example 5-26. Begin by expressing $R = |\vec{R}|$ using the law of cosines, factoring out r^2 and then using the first three terms in $(1 \pm x)^2$ for small x and $n = -1/2$.

Figure 5-110.

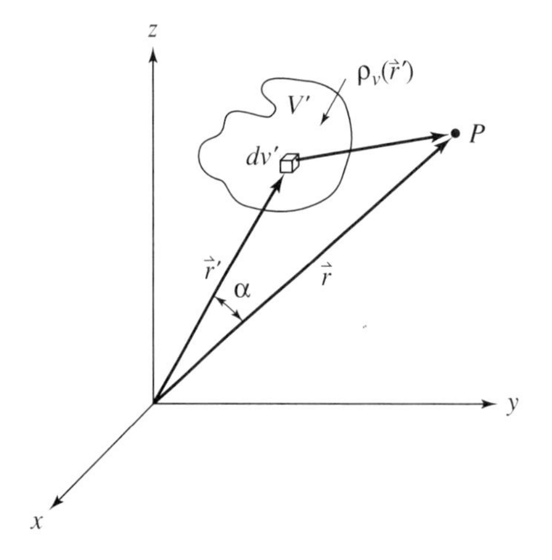

Also use the fact that $r'r \cos\alpha = \vec{r}' \cdot \vec{r}$. Answer:

$$\frac{1}{4\pi\epsilon r}\int_{V'} \rho_v(\vec{r}')dv' + \frac{1}{4\pi\epsilon r^2}\int_{V'} \rho_v(\vec{r}')(\vec{r}' \cdot \hat{r})dv' + \frac{1}{\pi\epsilon r^3}\int_{V'} \rho_v(\vec{r}')\left[\frac{3}{2}(\vec{r}' \cdot \hat{r})^2 - \frac{r'^2}{2}\right]dv' + \cdots$$

GE5-31 Using the result of GE5-30, obtain the far field potential $V(\vec{r})$ for the following (first nonzero term):

(a) Point charge
(b) Dipole of Figure 5-64
(c) Quadripole of Figure 5-70
(d) Wire ring of Figure 5-71
(e) Square sheet of Figure 5-72
(f) Finite line of Figure 5-66
(g) Figure 5-66 with the modification $\rho = +\rho_{l0}$: $0 < z < H$; $-\rho_{l0}$: $-H < z < 0$

GE5-32 For the two charged lines shown in Figure 5-75, compute the potential at a general point P in space (not xy plane) for finite length lines, $z = \pm L$. Use r_a and r_b distances to P as the variables. Reduce the result to points in the xy plane. Let θ_a and θ_b be the angles from the positive z part of each line to the corresponding lengths r_a and r_b. Answer:

$$V = \frac{\rho_{l0}}{4\pi\epsilon}\ln\left[\frac{L - r_a\cos\theta_a + \sqrt{L^2 + r_a^2 - 2r_aL\cos\theta_a}}{L - r_b\cos\theta_b + \sqrt{L^2 + r_b^2 - 2r_bL\cos\theta_b}} \cdot \frac{-L - r_b\cos\theta_b + \sqrt{L^2 + r_b^2 + 2r_bL\cos\theta_b}}{-L - r_a\cos\theta_a + \sqrt{L^2 + r_a^2 + 2r_aL\cos\theta_a}}\right]$$

$$\Rightarrow \frac{\rho_{l0}}{2\pi\epsilon}\ln\left[\frac{L + \sqrt{L^2 + r_a^2}}{L + \sqrt{L^2 + r_b^2}} \cdot \frac{r_b}{r_a}\right]$$

GE5-33 Obtain the general potential functions and the far field potential functions for the two configurations shown in Figure 5-111. Also plot two equipotential contours in the yz plane for each system, using $Q = 1\ \mu$C, $L = 20$ cm, for $V = 10$ V and 100 V.

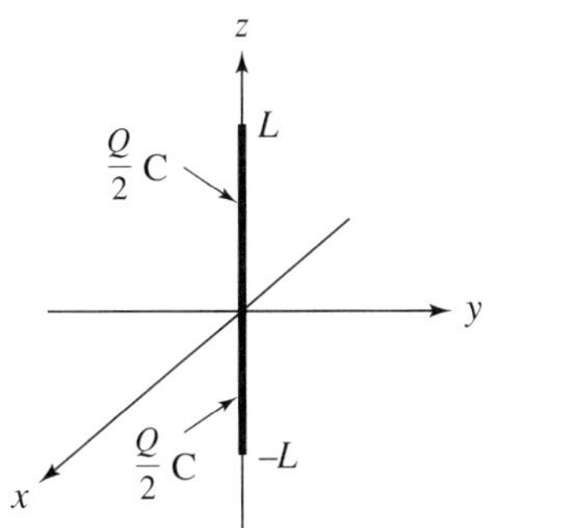

Figure 5-111.

GE5-34 Suppose an electron is positioned at a distance h above a horizontal, perfectly conducting, infinite metal plane. What is the energy in this configuration? (Hint: Use the image charge to replace the metal plane as done in physics.)

GE5-35 A spherically symmetric charged region has a density given by

$$\rho_v(\vec{r}) = \rho_0 r \text{ C/m}^3$$

(a) Compute $\vec{E}$ within the charge region.

(b) Show that, for a rectangular path in the yz plane (centered at the origin) having corners at (1, 2), (−1, 2), (−1, −2), and (1, −2)

$$\oint_C \vec{E} \cdot d\vec{l} = 0$$

Integrate the path counterclockwise.

Figure 5-112.

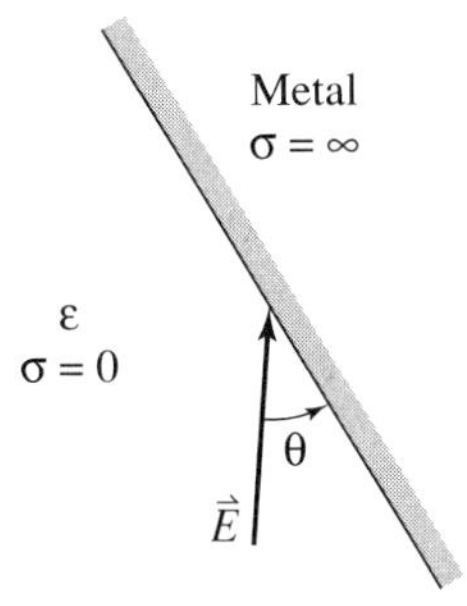

GE5-36 For the configuration of Figure 5-112, determine the angle θ.

GE5-37 For the configuration of Figure 5-113, determine $\vec{E}$ and $\vec{E}_3$. $\vec{E}_1 = 4\hat{x} - 2\hat{y}$.

GE5-38 Show that the far field potential of the quadripole of Exercise 5-10.0j satisfies Laplace's equation.

GE5-39 Suppose we have two hinged metal plates as shown in Figure 5-114. The hinge is an insulator, and the plates are very long vertically. The outside edge is a distance b from the hinge pivot. Find the expression for the capacitance per meter length of these plates. Remember we have capacitances between the inside surfaces and

Figure 5-113.

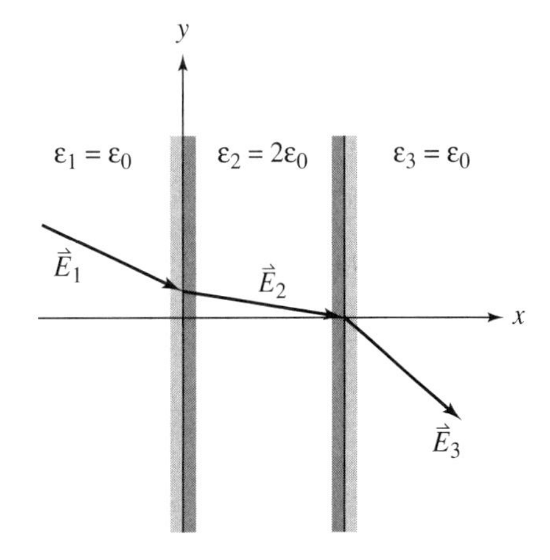

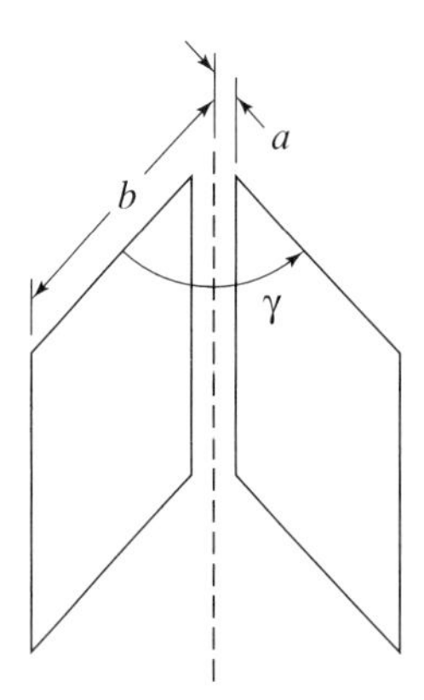

Figure 5-114.

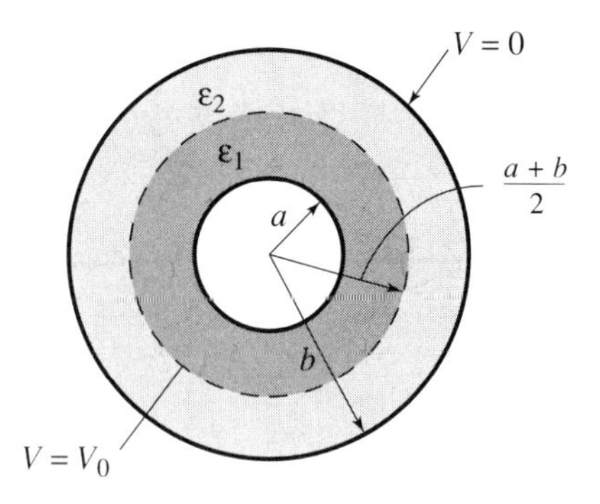

Figure 5-115.

also between the outside surfaces. Are they in series or parallel? Neglect fringing and solve Laplace's equation as though the plates were infinite in extent in the radial direction, and then cut if off at b.

GE5-40 Determine the capacitance per meter between the two elliptic cylinders $u = 0$ and $u = 1$ as shown in Figure 5-101. The potential will not vary with z or v. Explain this qualitatively.

GE5-41 A spherical capacitor is formed of two concentric metal spheres of radii a and b ($a < b$) and is charged so the inner sphere is at V_0 and the outer sphere is at 0 V. Compute the energy stored between the spheres using both the energy density approach and the capacitor-voltage approach. What is the energy stored on an isolated sphere of radius a?

GE5-42 For the system shown in Figure 5-115, obtain the equation for the potential distribution between the infinite length concentric metal cylinders and determine the pressure on each cylinder.

GE5-43 In GE5-42, suppose the system is interpreted as concentric spheres. Repeat GE5-42 for this case.

GE5-44 A metal sphere of radius b is connected to ground (0 V). Surrounding the sphere is a charge density $\rho_v = e^{-r}$, with charge density zero for $r > a (a > b)$. Find the total force on the metal sphere. (Note that the surface charge on the sphere is unknown and will need to be determined.)

6 Theory, Physical Description, and Basic Equations of Magnetic Fields

6-1 Introduction

Magnetic phenomena have captured the imagination and interest of young and old, beginning centuries ago when scientists and philosophers of Greece and China discovered stones that exhibited peculiar attractive and repulsive properties. (The word *magnet* appears to have been derived from the town Magnesia, an ancient city of Asia Minor.) Those investigators observed that small bars of the material oriented themselves along the earth north–south direction when properly suspended, which led to the development of the compass. This property of the magnet gave rise to the term *lodestone*, a derivative of *lead stone* or *way stone*. The end of the bar that pointed toward the North Pole of the earth was called the north end of the magnet.

It was not until nearly three millenniums later that the relationship of magnetic effects to current flow was investigated and quantified. Later experiments revealed that currents which are functions of time produce time-varying magnetic fields. Ultimately, it was Maxwell who unified electric and magnetic theories by developing a consistent set of equations relating electric and magnetic fields. In this chapter we examine some of the basic experiments and develop appropriate equations that describe magnetic fields.

6-2 Basic Experiments and Observations Involving Magnetic Fields: Magnetic Field Lines and Magnetic Flux Density

For the first experiment we place a compass over a horizontal wire, the wire being connected through a switch to a battery and current limiting resistor and the switch is open. The wire is turned so that it parallels the compass needle, which has come to rest in its north-south position as shown in Figure 6-1a. (This experiment was demonstrated by Hans C. Örsted about 1820.) When S_1 is closed, the north end of the needle moves to the right

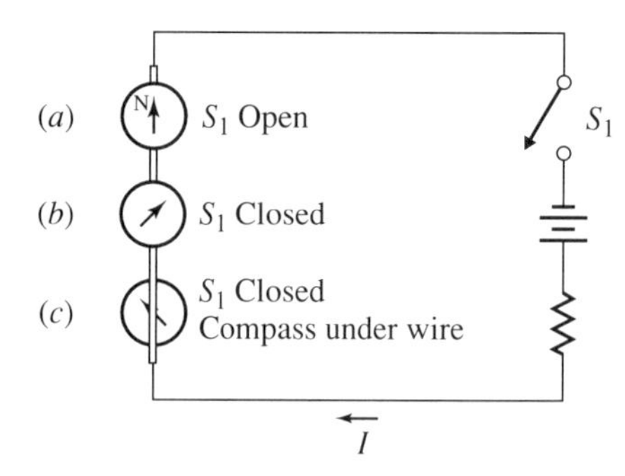

Figure 6-1. Compass deflection by a current in a wire.

as shown in Figure 6-1*b*. If the compass were initially under the wire, the north end would deflect to the left as illustrated in Figure 6-1*c*. Thus the current flow in the wire produces a field quantity outside the wire that produces a force. By definition, the direction of this magnetic force vector is the direction the north end moved. This convention is analogous to determining the direction of an electric field by agreeing to use a positive test charge. Although an isolated north pole has not been observed, we can still use the notion mentally to probe a magnetic field to determine its direction. Using this convention we observe that if we point the thumb of the right hand in the direction of the current flow with the wire gripped inside the fingers, the fingers point in the direction of the north-end deflection. The field will then form closed lines around the wire. This is the right-hand rule for the magnetic field. Additionally, one observes that the larger the current value, the greater the deflection.

Symbolic representations of the magnetic field lines are given by the Greek letter ϕ, which is called the *magnetic flux* (a scalar) measured in webers (Wb), and by $\vec{B}$, which is the magnetic flux density measured in webers/square meter, or tesla (T). The flux density is thus a measure of the number of lines per unit area. The relationship is shown in Figure 6-2, where we are looking at the end of the wire with the current flowing out of the page.

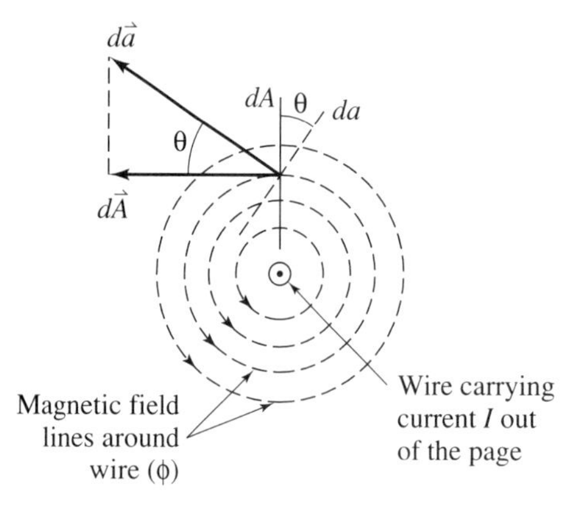

Figure 6-2. Magnetic field lines and areas in the vicinity of a current-carrying conductor.

Note that to this point the units of weber and tesla do not give us a relationship to force. That relationship comes from other experiments to be described later.

Imagine that a small area da is placed in the magnetic field so that the edge of the area is what is seen looking along the wire axis as shown in Figure 6-2. The area vector is represented by $d\vec{a}$. Since da can be placed at any arbitrary angle within the field, to compute the density of lines passing through an area we need to project da to an area that is physically normal to the magnetic field lines. This new area is denoted by dA, with a vector direction $d\vec{A}$ as shown. If we denote the unit vector of $d\vec{A}$ as $\hat{A}$, then the flux density can be written as

$$\vec{B} = \frac{d\phi}{dA}\hat{A} \tag{6-1}$$

Note that $\vec{B}$ is in the direction of $d\vec{A}$.

For a general area da we would then write

$$\vec{B} = \frac{d\phi}{da\cos\theta}\hat{A}$$

from which the magnitude of the vector would be

$$B = |\vec{B}| = \frac{d\phi}{da\cos\theta}$$

or

$$d\phi = Bda\cos\theta = \vec{B}\cdot d\vec{a} \tag{6-2}$$

From the preceding equation one can then compute the magnetic flux through a general area in a magnetic field using

$$\phi = \int_S \vec{B}\cdot d\vec{a} \tag{6-3}$$

Example 6-1

A region of space has a magnetic flux density given by

$$\vec{B} = 3\cos\frac{\pi}{2a}x\hat{z} - 2\sin\frac{\pi}{b}y\hat{x}\ \mathrm{T}$$

Find the magnetic flux that passes through the area shown in Figure 6-3, using the area vector as the positive z direction. Using the math techniques of earlier chapters we evaluate Eq. 6-3 as follows:

$$\Phi = \int_S \vec{B}\cdot d\vec{a} = \int_0^b\int_0^a\left(3\cos\frac{\pi}{2a}x\hat{z} - 2\sin\frac{\pi}{b}y\hat{x}\right)\cdot(dxdy\hat{z})\bigg|_{z=0}$$

$$= \int_0^b 0\int_0^a 3\cos\frac{\pi}{2a}xdxdy = \frac{6ab}{\pi}\ \mathrm{Wb}$$

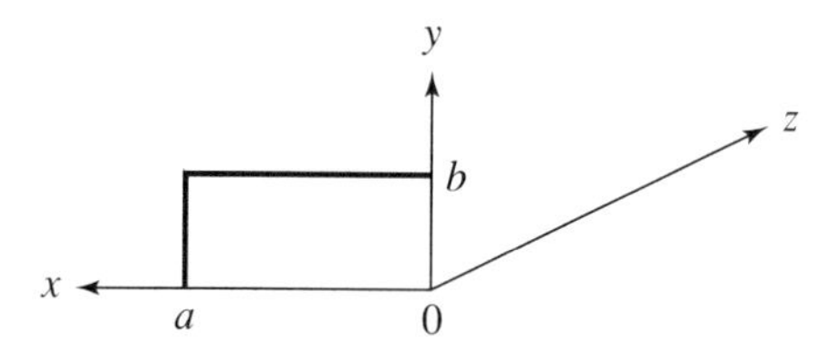

Figure 6-3.

Exercises

6-2.0a Suppose there is a region of space where there exists a magnetic flux density

$$\vec{B} = 3 \cos y\hat{x} + 2xz\hat{y} \text{ T}$$

What is the total flux passing through a square 1 m on a side located in the positive quadrant of the xz plane and has one corner at the origin? Is your answer unique? (Remember area vectors have two possible directions.) (Answer: $\pm\frac{1}{2}$ Wb, area vector can be $\pm\hat{y}$)

6-2.0b In cylindrical coordinates, a magnetic field is given by

$$\vec{B} = 3\rho\hat{\rho} - 2z\hat{\phi} + \cos z\hat{z}$$

An imaginary cylindrical surface of length L and radius a is placed so its axis is coincident with the z axis and lying along the positive z axis. Compute the flux passing outward through each of the three surfaces that make up the closed cylinder.

6-2.0c A straight cylindrical wire that is very long has a magnetic flux density around it (outside wire radius)

$$\vec{B} = \frac{\mu_0 I}{2\pi\rho}\hat{\phi}$$

where the wire is coincident with the z axis, ρ is the distance from the z axis, μ_0 is the constant $4\pi \times 10^{-7}$. Number 22 wire has a diameter of 25.35 mils (1 mil = 0.001 in.). What is the magnetic flux density around the wire at a distance 2 cm from the wire center if the current is 20 mA DC? What is the flux passing through the area of the square shown in Figure 6-4?

6-2.0d Suppose each component of the magnetic flux density $\vec{B}$ of Exercise 6-2.0a is multiplied by the time function $\cos 10^6 t$. Now what is the total flux through the square described in

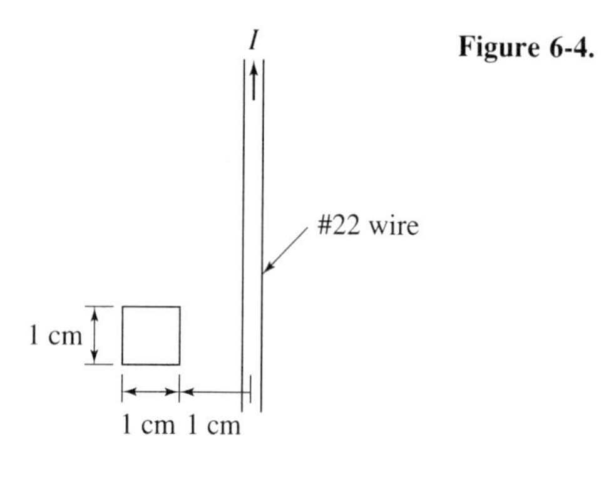

Figure 6-4.

that exercise? Next compute $d\phi/dt$ and determine the value of time for which this derivative is maximum. (Answer: $\pm\frac{1}{2}\cos 10^6 t$, since the coordinate system is assumed stationary the time function may be removed from the area integration. $\pm\frac{1}{2} \times 10^6 \sin 10^6 t$ Wb/s $\cdot\ t = 2n\pi \times 10^{-6}$, $n = 0, 1, \ldots$)

6-2.0e The magnetic flux density was defined in Eq. 6-1 as the ratio of differential flux to differential area. Under what conditions can B be computed as the ratio ϕ/A? (Use Eq. 6-3). (Answer: $\vec{B}$ parallel to $d\vec{A}$, and $\vec{B}$ independent of coordinate values)

The second experiment is a very basic one familiar to many. If we take a bar magnet for which the north and south poles have been identified by its alignment in the earth's magnetic field and place it under a sheet of paper, iron filings carefully sprinkled on the paper show the field lines around the bar. This result is shown in Figure 6-5. This, however, does not give us the *direction* of the magnetic flux density vector. To determine the direction we use the convention suggested from the earlier experiment, namely, the direction of the field lines is the direction a north "test" pole experiences a force. Thus if we imagine an isolated north pole inserted into the region, the direction the pole would move is the field direction. Since we observe that north poles of two magnets repel, then our small test pole N would be pushed away from the north pole and attracted toward the south pole, which allows us to draw arrows on the field lines as indicated on the line passing by the test north pole in the figure. The larger the value of $\vec{B}$, the greater the force on the test pole.

6-3 Forces Produced by Magnetic Flux Density on Current Distributions

For the next experiment we place a current-carrying conductor in a magnetic field as illustrated in Figure 6-6*a*. If the experiment is set up so that the force on the wire can be measured, one finds that the force is directed into the page as shown by the tail of the arrow in the figure. Additionally, it is observed that the magnitude of the force varies as the sine of the angle between the magnetic field and the conductor as shown. Also, as the current increases the magnitude of the force increases.

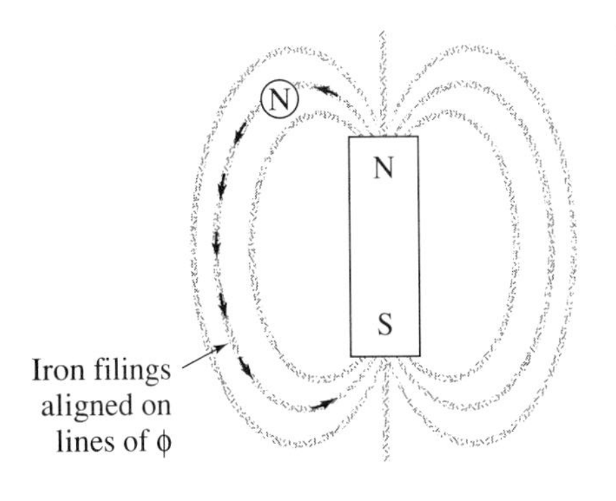

Figure 6-5. Pattern produced by iron filings sprinkled on a sheet placed over a magnet.

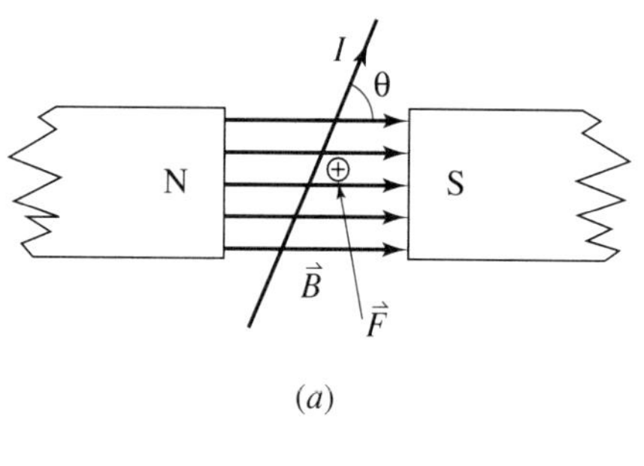

(a)

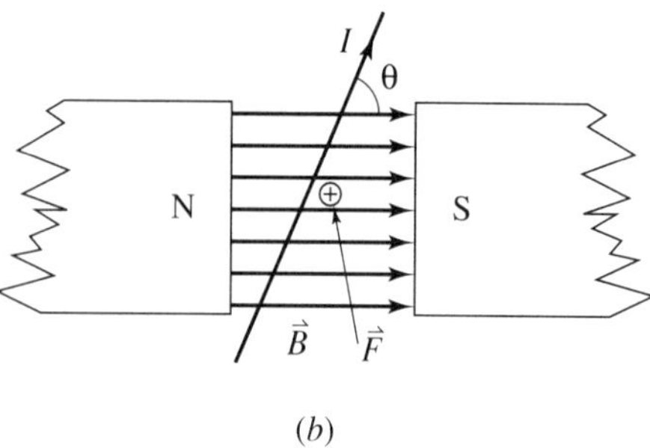

(b)

Figure 6-6. Current-carrying conductor in a magnetic field.

In Figure 6-6*b* the sizes of the magnetic pole faces have been increased ($\vec{B}$ remaining the same) so that the field interacts with a longer segment of wire. It is found that the force is a linear function of the conductor length l over which the field is impressed (i.e., if we double the effective interaction length, the force doubles).

From the experimental results of the last two paragraphs we would write

$$F = KIlB \sin \theta$$

which can be written in differential form as

$$dF = KIdlB \sin \theta$$

The constant K is a proportionality constant. In the MKS (SI) system of units K is set to unity so that

$$dF = IdlB \sin \theta$$

Recognizing the last result as a cross product we can write

$$d\vec{F} = Id\vec{l} \times \vec{B} \tag{6-4}$$

which is also easily remembered by the right-hand rule. This equation allows us to obtain other units for the magnetic flux density as follows:

$$\text{tesla} \equiv \frac{\text{weber}}{\text{square meter}} = \frac{\text{newton}}{\text{ampere} \cdot \text{meter}}$$

From this the unit of magnetic flux ϕ would be

$$\text{weber} = \frac{\text{newton} \cdot \text{meter}}{\text{ampere}} = \frac{\text{joule}}{\text{ampere}}$$

There is a close analogy between magnetic flux ϕ and potential V introduced in Chapter 5. Charges were the source of potential and potential was defined as the energy per unit charge. In the present case we have seen that currents are the source of magnetic field and the magnetic field is the energy per unit ampere.

Example 6-2

A circular ring of radius a carries a constant current I_0 and is in a constant magnetic field of value $\vec{B} = B_0\hat{z}$. Determine the net force on the portion of the ring in the positive y portion of the xy plane.

From Figure 6-7, we identify the terms in Eq. 6-4 as

$$I d\vec{l} \Rightarrow (I_0)(\rho d\phi\hat{\phi})\big|_{\rho=a} = I_0 a d\phi\hat{\phi}$$

$$\vec{B} \Rightarrow B_0\hat{z}$$

$$\vec{F}_+ = \int_0^{\pi} d\vec{F}_+ = \int_0^{\pi} I_0 a d\phi\hat{\phi} \times B_0\hat{z} = aB_0 I_0 \int_0^{\pi} \hat{\rho} d\phi$$

Since $\hat{\rho}$ changes direction with ϕ, it cannot be removed from the integral, so we proceed by making a coordinate conversion to rectangular. Using the techniques described in Chapter 5, Example 5-8, and the fact that $\sqrt{x^2 + y^2} = a$ for this example,

$$\hat{\rho} = \frac{x}{\sqrt{x^2 + y^2}}\hat{x} + \frac{y}{\sqrt{x^2 + y^2}}\hat{y} = \frac{x}{a}\hat{x} + \frac{y}{a}\hat{y}$$

and

$$d\phi = d\left[\tan^{-1}\left(\frac{y}{x}\right)\right] = \frac{1}{1 + \frac{y^2}{x^2}}\left(-\frac{y}{x^2}\right)dx + \frac{1}{1 + \frac{y^2}{x^2}}\left(\frac{1}{x}\right)dy = -\frac{y}{a^2}dx + \frac{x}{a^2}dy$$

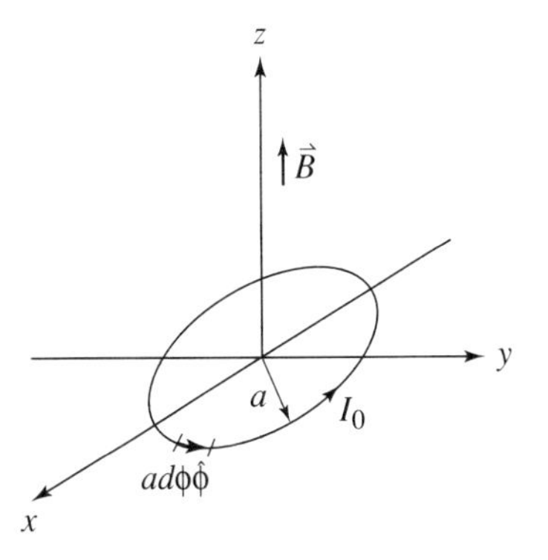

Figure 6-7.

From these,

$$\hat{\rho}d\phi = -\frac{xy}{a^3}dx\hat{x} + \frac{x^2}{a^3}dy\hat{x} - \frac{y^2}{a^3}dx\hat{y} + \frac{xy}{ra^3}dy\hat{y}$$

with

$$\begin{cases} x^2 + y^2 = a^2 \\ x = \pm\sqrt{a^2 - y^2} \\ y = \pm\sqrt{a^2 - x^2} \end{cases}$$

Since the principle $\tan^{-1}$ values are in the interval $(-\pi/2, \pi/2)$, we break up and evaluate the integral as in Example 5-8 and obtain

$$\int_0^{\pi} \hat{\rho}d\phi = \int_0^{\pi/2} \hat{\rho}d\phi + \int_{\pi/2}^{\pi} \hat{\rho}d\phi = (\hat{x} + \hat{y}) + (-\hat{x} + \hat{y}) = 2\hat{y}$$

With this result the force $\vec{F}_+$ becomes

$$\vec{F}_+ = 2aB_0I_0\hat{y} \text{ N}$$

Exercises

6-3.0a For Example 6-2, determine the force $\vec{F}_-$ on the portion of the ring in the negative y portion of the xy plane. (Answer: $-2aB_0I_0\hat{y}$ N)

6.03b For Example 6-2, determine the force on the negative x portion of the ring. On the basis of this result, what is the effect of the magnetic field intensity on the ring? (Answer: $-2aB_0I_0\hat{x}$ N; to stretch or explode the ring outward in the plane of the loop)

6-3.0c In Example 6-2, explain what changes would be necessary in the equations if the direction of the loop current were reversed. Explain how you would in general determine the correct sign on the term $I\vec{dl}$ for a given configuration. (*Answer:* Negate I_0 wherever it appears in the equations. I is considered positive if it is flowing in the direction of $\vec{dl}$.)

6-3.0d For the configuration shown in Figure 6-8, compute the net force on the loop and explain what would happen if the loop were free to move. $\vec{B} = 2xz\hat{y}$ T, $I = 1$ A.

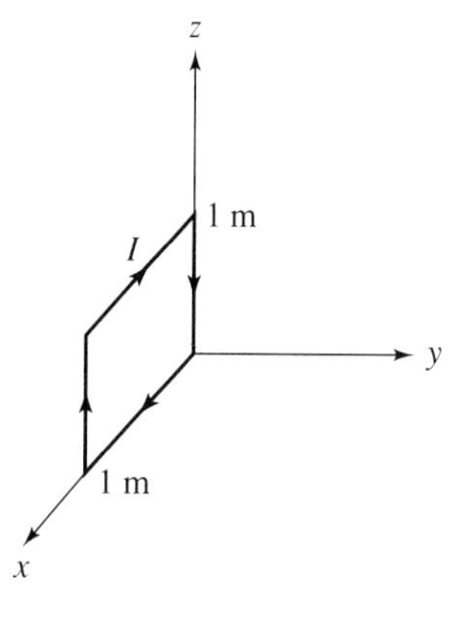

Figure 6-8.

In many engineering systems currents flow not only in filamentary structures but on sheets and in volumes. It is then desirable to cast Eq. 6-4 into forms applicable to these other systems. Figure 6-9*a* is a picture of current flow on a surface. The current flow is of zero thickness into the surface and is described by the surface current density vector $\vec{J}_s$ in units of A/m. In this current the meter dimension is normal to the current flow, as illustrated in Figure 6-9*b*, Line 1. The current flow across dw would then be (for Line 1),

$$dI = J_s dw$$

from which

$$I_{ab} = \int_a^b J_s dw$$

If, however, we consider Line 2 of the figure we would first have to project the differential path length onto the dotted normal line and then write

$$dI = J_s d\lambda \cos \theta = \vec{J}_s \cdot d\vec{\boldsymbol{\lambda}}$$

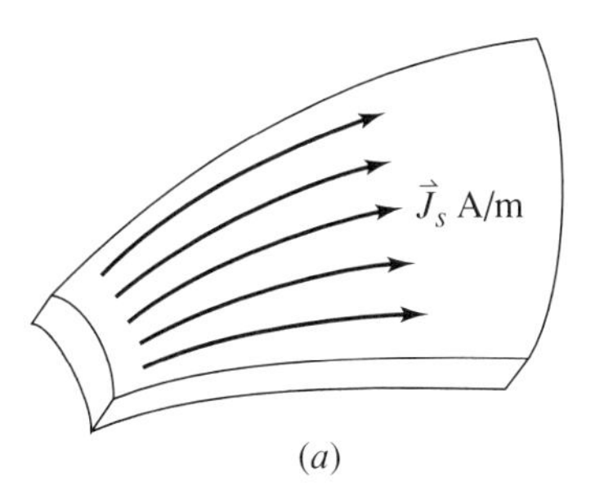

(*a*)

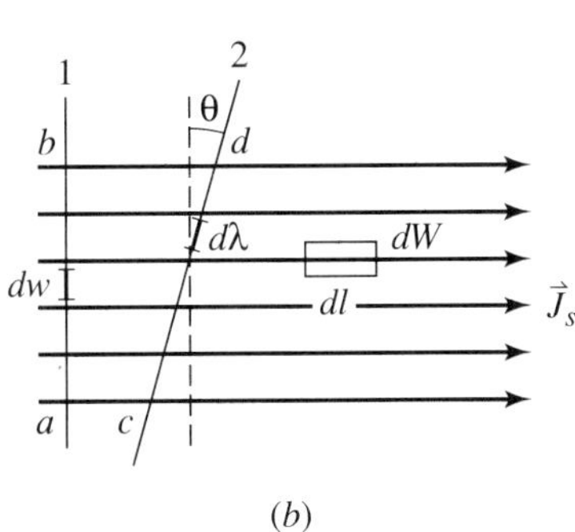

(*b*)

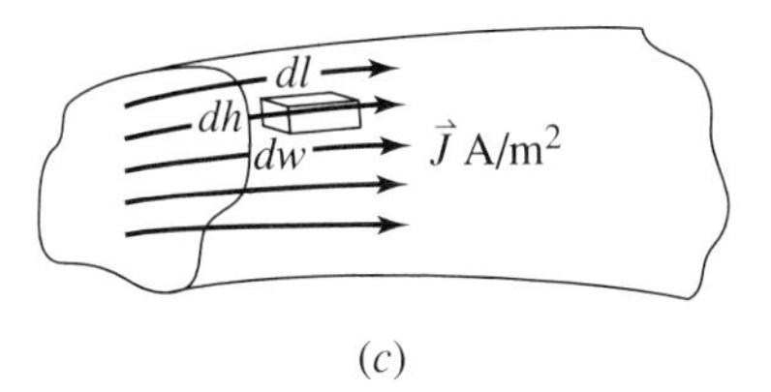

(*c*)

Figure 6-9. Currents on surfaces and in volumes. (*a*) Surface current vector. (*b*) Surface current flowing across lines having different orientations. (*c*) Current flowing in a volume.

Then

$$I_{cd} = \int_c^d \vec{J}_s \cdot d\vec{\lambda} \tag{6-5}$$

Now to obtain an equivalent current element $Id\vec{l}$ we first draw an incremental surface area on the current density as shown in Figure 6-9*b*. We then have, since $\vec{J}_s$ is parallel to $d\vec{l}$,

$$Id\vec{l} = (J_s dw)d\vec{l} = \vec{J}_s dwdl = \vec{J}_s da$$

Where da is an increment of surface over which $\vec{J}_s$ flows. Substituting this last result into Eq. 6-4 there results

$$d\vec{F} = \vec{J}_s da \times \vec{B}$$

Since scalars may be removed from the cross product we obtain

$$d\vec{F} = (\vec{J}_s \times \vec{B})da \tag{6-6}$$

Figure 6-9*c* shows a current density $\vec{J}$ A/m^2 flowing in a volume. This is the same current density discussed in Section 5-7 of Chapter 5 (specifically, see Eqs. 5-29 and 5-30). Here we use the block letter rather than the script to emphasize time-invariant currents. From the figure, and again using the fact that $d\vec{l}$ is parallel to $\vec{J}$, we write

$$Id\vec{l} = (Jdhdw)d\vec{l} = \vec{J}dhdwdl = \vec{J}dv$$

Substituting this into Eq. 6-4 yields

$$d\vec{F} = \vec{J} \times \vec{B}dv \tag{6-7}$$

Actually, this last equation can always be used because any current distribution can be expressed as a current density $\vec{J}$ by using the delta and step functions as explained in Chapters 4 and 5. The next example illustrates this.

Example 6-3

A sheet of current on which $\vec{J}_s = -3xy\hat{y}$ is shown in Figure 6-10. (How this is generated is a subject for later discussion). The magnetic flux density is given by $\vec{B} = 2x\hat{x} - 5(z-4)\hat{z}$. Compute the force on the sheet.

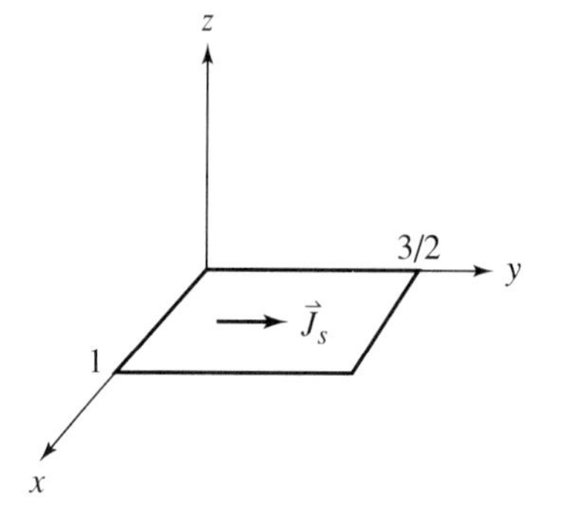

Figure 6-10.

Using step and delta functions as in Chapter 5 we have

$$\begin{aligned}\vec{J} &= \vec{J}_s\ \delta(z)U(y)U(\tfrac{3}{2} - y)U(x)U(1 - x) \\ &= -3xy\ \delta(z)U(y)U(\tfrac{3}{2} - y)U(x)U(1 - x)\hat{y}\ \text{A/m}^2\end{aligned}$$

Using Eq. 6-7:

$$\vec{F} = \int_V d\vec{F} = \int_{-\infty}^{\infty}\int_{-\infty}^{\infty}\int_{-\infty}^{\infty} [6x^2y\ \delta(z)U(y)U(\tfrac{3}{2} - y)U(x)U(1 - x)\hat{z}$$
$$+ 15xy(z - 4)\ \delta(z)U(y)U(\tfrac{3}{2} - y) \cdot U(x)U(1 - x)\hat{x}]dxdydz$$

Using the properties of the delta and step functions this becomes

$$\vec{F} = \int_0^{3/2}\int_0^1 6x^2y\hat{z}dxdy + \int_0^{3/2}\int_0^1 - 60xy\hat{x}dxdy = -\frac{135}{4}\hat{x} + \frac{9}{4}\hat{z}\ \text{N}$$

Exercises

6-3.0e A semicylindrical sheet shown in Figure 6-11 carries a current $\vec{J}_s = 2\sin 2\phi\hat{z}$ and is in a magnetic field having $\vec{B} = -(3/\rho)\hat{\rho}$. Determine the force on the sheet. Describe how the right-hand rule can qualitatively verify your result.

6-3.0f A solenoid can be approximated by a circumferential current sheet on a cylindrical surface as shown in Figure 6-12. Suppose the solenoid is tightly (closely) wound with #26 wire and the wire current is 20 mA. Neglecting the magnetic field generated by the current, determine (a) The effective (equivalent) surface current $\vec{J}_s$ (Answer: 49.4 $U(l + z)U(l - z)\hat{\phi}$ A/m) and (b) the total force on the solenoid surface.

6-3.0g A beam of electrons can be approximated by a cylindrical stream as modeled in Figure 6-13. If the total beam current is 10 mA and assumed uniform over the beam cross section of radius 0.5 mm, determine:

(a) $\vec{J}$ (Answer: $\vec{J} = -12.7U(a - \rho)\hat{z}$ kA/m^2) and

(b) the total force per unit length (m) on the beam in a magnetic field of value $\vec{B} = 0.04\hat{x}$ T.

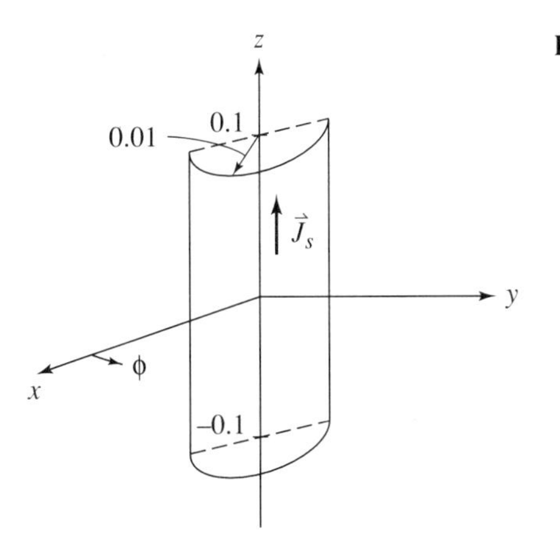

Figure 6-11.

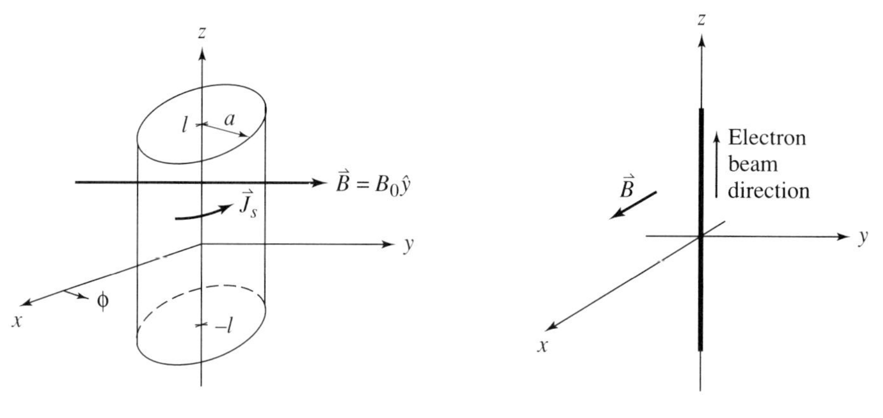

Figure 6-12.

Figure 6-13.

6-4 The Biot-Savart Law and Magnetic Field Intensity

From the experiment with the compass and wire with a current we deduced that a magnetic field that interacted with the compass north pole was produced. Since magnetic fields also produce forces on current distributions, let us replace the compass by another current-carrying conductor. The general configuration is shown in Figure 6-14. Ignore $d(d\vec{F})$ and $d\vec{H}$ for the moment; they will be described in detail later.

In the figure the prime system is the original wire and the unprimed system is the new wire system. For convenience we have placed a coordinate system in the configuration to locate points on the primed and unprimed wire systems just as was done in the preceding

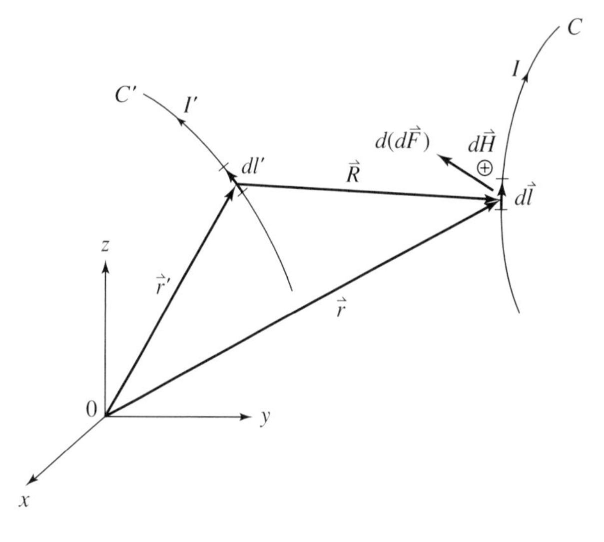

Figure 6-14. Magnetic interaction system of two current-carrying conductors.

chapter for electric fields. As before, the primed system is the source system and the unprimed is the field system, and $\vec{R} = \vec{r} - \vec{r}\,'$.

Since the wires (described by contours or paths C and C') must form closed paths for time-invariant currents, we are led to consider differential elements of the paths and determine the interaction of these elements. This is analogous to considering point charges in Coulomb's law of electrostatics. As expected, there is a force on contour C, particularly on each incremental element $d\vec{l}$. The source of the force is also a differential element, so we would expect that the force would be a second-order differential force. To obtain the representation of a quantity that depends upon two differential variables, we could examine briefly the derivatives of a function of two independent variables, namely

$$y = g(x, z)$$

Then

$$\frac{dy}{dx} = \frac{\partial g}{\partial x}\frac{dx}{dx} + \frac{\partial g}{\partial z}\frac{dz}{dx} = \frac{\partial g}{\partial x}$$

z does not change with x since they are independent, so $(dz/dx) = 0$. Also,

$$\frac{d\left(\dfrac{dy}{dx}\right)}{dz} = \frac{\partial\left(\dfrac{\partial g}{\partial x}\right)}{\partial z}\frac{dz}{dz} + \frac{\partial\left(\dfrac{\partial g}{\partial x}\right)}{\partial x}\frac{\partial x}{\partial z} = \frac{\partial\left(\dfrac{\partial g}{\partial x}\right)}{\partial z} = \frac{\partial^2 g}{\partial x \partial z}$$

Putting the left member in terms of differentials we have

$$\frac{d(dy)}{dxdz} = \frac{\partial^2 g}{\partial x \partial z}$$

from which

$$d(dy) = \frac{\partial^2 g}{\partial x \partial z}\, dxdz$$

In the present discussion, the force is a function of the two independent contours and currents therein so we expect the force to be represented by $d(d\vec{F})$.

Ampere performed a set of four experiments to determine the forces between current-carrying conductors, and he assumed that the force in the incremental section depended upon the straight line distance between the two incremental sections, denoted by $\vec{R}$ in Figure 6-14. (See Maxwell, volume two, pages 158–163 and 173 (articles 502–510 and 526) [39].) These experiments were initially mathematically quantified by Jean-Baptiste Biot and Felix Savart, who were colleagues of Ampere. Ampere later generalized their work even further. Now we have already observed that the current I' in $d\vec{l}\,'$ as shown produces some vector field quantity $d\vec{H}$, which is perpendicular to the page in Figure 6-14. The experiments showed that the force was in the plane of the two differential elements and that the force was proportional to the strength of the current I. The two quantities $d(d\vec{F})$ and $d\vec{H}$ are shown in the figure. The relationship of these variables is (using k for the proportionality constant)

$$d(d\vec{F}) = kId\vec{l} \times d\vec{H} \tag{6-8}$$

For the quantity $d\vec{H}$, it was found that the force increased directly with the current I' and varied inversely with R^2, and depended upon the sine of the angle between $d\vec{l}\,'$ and $\vec{R}$, so that we may write

$$dH = \frac{I' dl'}{4\pi R^2} \sin\theta_R$$

The inclusion of the 4π has the nice features of making the resulting force equation have the same appearance as Coulomb's law discussed in Chapter 5. In vector form, the preceding equation may be written as

$$d\vec{H} = \frac{I' d\vec{l}\,' \times \hat{R}}{4\pi R^2} \tag{6-9}$$

This last result is usually called the *Biot-Savart law*, but a few authors choose to call it Ampere's law. We will use the former name since that is the most common, and since the name Ampere's law is applied to another relationship.

The field quantity $\vec{H}$ is called the *magnetic field intensity* and can be seen from Eq. 6-9 to have the units of ampere/meter (A/m). Substituting Eq. 6-9 into Eq. 6-8 we have

$$d(d\vec{F}) = kI d\vec{l} \times \left(\frac{I' d\vec{l}\,' \times \hat{R}}{4\pi R^2}\right)$$

The constant k is customarily denoted by the symbol μ, which is called the *permeability*, the value of which is dependent upon the medium between the two conductors. Specifically, for the vacuum the symbol μ_0 is used. With this convention the preceding equation becomes

$$d(d\vec{F}) = \mu I d\vec{l} \times \left(\frac{I' d\vec{l}\,' \times \hat{R}}{4\pi R^2}\right) = I d\vec{l} \times \mu\left(\frac{I' d\vec{l}\,' \times \hat{R}}{4\pi R^2}\right)$$

Upon integrating this equation over the contour C' (i.e., with respect to $d\vec{l}\,'$) and noting that the primed and unprimed coordinates are independent as described in Chapter 5

$$d\vec{F} = I d\vec{l} \times \mu \int_{C'} \frac{I' d\vec{l}\,' \times \hat{R}}{4\pi R^2} = I d\vec{l} \times \mu\vec{H} \tag{6-10}$$

Comparing this result with Eq. 6-4 we obtain

$$\vec{B} = \mu \int_{C'} \frac{I' d\vec{l}\,' \times \hat{R}}{4\pi R^2} \tag{6-11}$$

or

$$\vec{B} = \mu\vec{H} \tag{6-12}$$

This is called the *constitutive relationship for magnetic fields*. The units for permeability μ are obtained from this relationship and those of $\vec{B}$ and $\vec{H}$:

$$\mu \sim \frac{\text{units of } \vec{B}}{\text{units of } \vec{H}} = \frac{\text{T}}{\text{A/m}} = \frac{\text{Wb/m}^2}{\text{A/m}} = \frac{\text{Wb}}{\text{A} \cdot \text{m}} = \frac{\text{J/A}}{\text{A} \cdot \text{m}}$$

$$= \frac{\dfrac{\text{J}}{\text{C/s}}}{\text{A} \cdot \text{m}} = \frac{\text{V} \cdot \text{s}}{\text{A} \cdot \text{m}}$$

Now from the circuit theory, $V = L(di/dt)$ for a constant inductance, from which the henry has units of v · s/A. Thus for permeability

$$\mu \sim \frac{\text{H}}{\text{m}}$$

This is analogous to ϵ which has units of F/m (see Chapter 5). Note that to obtain Eq. 6-12 we made use of the substitution

$$\vec{H} = \int_{C'} \frac{I' d\vec{l}' \times \hat{R}}{4\pi R^2} \tag{6-13}$$

Since for DC conditions current must have a closed path, it follows that actual experimental investigation of the Biot-Savart law requires C' to be a closed contour. Two important observations need to be made here:

1. In Eq. 6-10 note that every quantity except the permeability has defined quantitative measure, so that in essence we have *defined* the value of permeability of vacuum. This turns out to be

$$\mu_0 = 4\pi \times 10^{-7} \text{ H/m} \qquad \text{(exactly)}$$

   This will be shown later.
2. We should think of currents as being the sources which generate magnetic field intensity $\vec{H}$. The permeability μ is the quantity that yields the magnetic flux density, which produces the forces.

As in the preceding section, we need to be able to generalize Eq. 6-13 so that we can compute the magnetic field intensity for surface and volume currents, $\vec{J}_s$ and $\vec{J}$. Use exactly the same technique as in that section we replace $I'd\vec{l}$ by its equivalents to obtain:

$$\vec{H} = \int_{S'} \frac{\vec{J}_s(\vec{r}') \times \hat{R}}{4\pi R^2} da' \tag{6-14}$$

where da is an increment of S', and

$$\vec{H} = \int_{V'} \frac{\vec{J}(\vec{r}\,') \times \hat{R}}{4\pi R^2}\, dv' \tag{6-15}$$

where dv' is an increment of V'. In these results the symbols $\vec{r}\,'$, $\vec{R}$, da', and dv' are interpreted in exactly the same way as in the electrostatic case.

Example 6-4

Determine the magnetic field intensity on the axis of a loop of radius b, carrying a current I_0 as shown in Figure 6-15.

We use Eq. 6-13 and evaluate it using the identical techniques as in electric field computations.

$$I'd\vec{l}\,' \Rightarrow I_0\rho' d\phi' \hat{\phi}' = I_0 b\, d\phi' \hat{\phi}' \qquad \text{(on loop)}$$

$$\vec{R} = \vec{r} - \vec{r}\,' \Rightarrow z\hat{z} - \rho'\hat{\rho}'\big|_{\rho'=b} = z\hat{z} - b\hat{\rho}'$$

$$\hat{R} = \frac{\vec{R}}{|\vec{R}|} = \frac{\vec{R}}{R} \Rightarrow \frac{z\hat{z} - b\hat{\rho}'}{\sqrt{z^2 + b^2}}$$

$$R^2 = \vec{R} \cdot \vec{R} \Rightarrow z^2 + b^2$$

Then

$$\vec{H} = \int_0^{2\pi} \left[\frac{I_0 b\, d\phi' \hat{\phi}'}{4\pi(z^2 + b^2)} \times \frac{z\hat{z} - b\hat{\rho}'}{\sqrt{z^2 + b^2}}\right] = \frac{bI_0}{4\pi}\int_0^{2\pi} \frac{z\hat{\rho}' d\phi'}{(z^2 + b^2)^{3/2}}$$

$$+ \frac{bI_0}{4\pi}\int_0^{2\pi} \frac{b\hat{z}'}{(z^2 + b^2)^{3/2}}\, d\phi'$$

$$= \frac{zbI_0}{4\pi(z^2 + b^2)^{3/2}}\int_0^{2\pi} \hat{\rho}' d\phi' + \frac{b^2 I_0}{4\pi(z^2 + b^2)^{3/2}}\int_0^{2\pi} \hat{z}' d\phi'$$

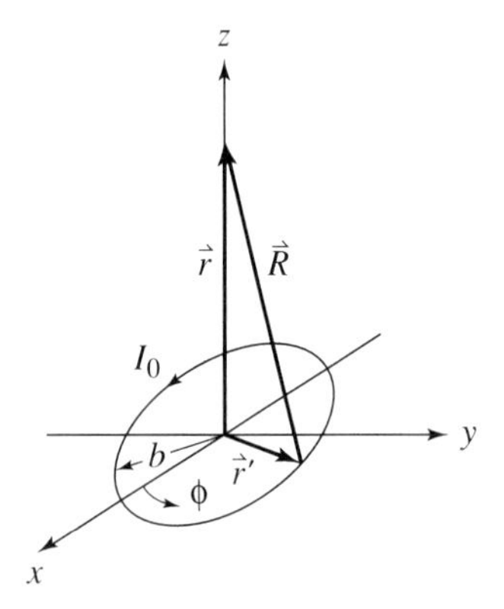

Figure 6-15.

The first integral was shown to be zero in Example 5-8, and primed rectangular coordinate unit vectors may be replaced by the unprimed unit vectors. Then

$$\vec{H} = \frac{b^2 I_0}{2(z^2 + b^2)^{3/2}}\,\hat{z}\ \text{A/m}$$

Example 6-5

This example shows how a modest change in a problem statement can result in considerably increased solution complexity, and it also indicates what can be done to handle these complexities. Suppose we take the loop of Example 5-4 but now compute the magnetic field intensity for general points in space as indicated in Figure 6-16. Since the system is symmetric in ϕ (independent of ϕ) we may take our point to be in the xz plane without unduly restricting the solution.

From the figure:

$$I'd\vec{l}\,' \Rightarrow I_0\rho' d\phi'\hat{\phi}'\big|_{\rho'=b} = bI_0 d\phi'\hat{\phi}'$$

$$\vec{r}\,' \Rightarrow \rho'\hat{\rho}'\big|_{\rho'=b} = b\hat{\rho}' \qquad \vec{r} \Rightarrow x\hat{x} + z\hat{z}\big|_{\phi=0} = \rho\hat{\rho} + z\hat{z}$$

$$\vec{R} = \vec{r} - \vec{r}\,' \Rightarrow \rho\hat{\rho} + z\hat{z} - b\hat{\rho}' \qquad \text{note } \hat{\rho} \neq \hat{\rho}'$$

$$R^2 = \vec{R}\cdot\vec{R} \Rightarrow \rho^2 - 2b\rho\hat{\rho}\cdot\hat{\rho}' + z^2 + b^2 = \rho^2 - 2b\rho\cos(\phi - \phi') + z^2 + b^2\big|_{\phi=0}$$
$$= \rho^2 + b^2 + z^2 - 2b\rho\cos\phi'$$

$$\hat{R} \Rightarrow \frac{\rho\hat{\rho} + z\hat{z} - b\hat{\rho}'}{\sqrt{\rho^2 + b^2 + z^2 - 2b\rho\cos\phi'}}$$

Using Eq. 6-13 and removing unprimed coordinates from integrands where possible:

$$\vec{H} = \frac{bI_0}{4\pi}\int_0^{2\pi} \frac{(\rho d\phi')\hat{\phi}' \times \hat{\rho} + (zd\phi')\hat{\phi}' \times \hat{z} - (bd\phi')\hat{\phi}' \times \hat{\rho}'}{(\rho^2 + b^2 + z^2 - 2b\rho\cos\phi')^{3/2}}$$

$$= \frac{bI_0}{4\pi}\int_0^{2\pi} \frac{-\rho d\phi'\sin\phi'\hat{z} + zd\phi'\hat{\rho}' + bd\phi'\hat{z}}{(\rho^2 + b^2 + x^2 - 2b\rho\cos\phi')^{3/2}}$$

$$= \frac{bI_0}{4\pi}\left[-\rho\int_0^{2\pi} \frac{\sin\phi' d\phi'}{(b^2 + \rho^2 + z^2 - 2b\rho\cos\phi')^{3/2}}\,\hat{z}\right.$$
$$+ z\int_0^{2\pi} \frac{\hat{\rho}' d\phi'}{(b^2 + \rho^2 + z^2 - 2b\rho\cos\phi')^{3/2}}$$
$$\left. + b\int_0^{2\pi} \frac{d\phi'}{(b^2 + \rho^2 + z^2 - 2b\rho\cos\phi')^{3/2}}\,\hat{z}\right]$$

Since the $\hat{\rho}'$ cannot be removed from the integrand of the second integral we convert prime coordinates to rectangular coordinates so that all integrals will have no unit vectors in the

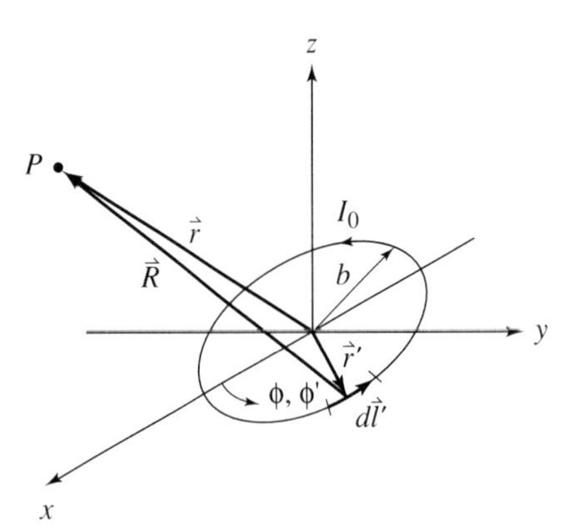

Figure 6-16.

integrands. Also, a shown in Example 5-8, the integrals that result must be carefully set up to account for principal values of the trigonometric functions. Using the same process the expression for $\vec{H}$ expands to the basic form

$$
\begin{aligned}
\vec{H} = \frac{bI_0}{4pi}\Bigg\{ & -\rho\int_0^{2\pi}\frac{\sin\phi' d\phi'}{(b^2+\rho^2+z^2-2b\rho\cos\phi')^{3/2}}\hat{z} + \frac{z}{b^3}\Bigg[-\int_b^0\frac{x'\sqrt{b^2-x'^2}}{(b^2+\rho^2+z^2-2\rho x')^{3/2}}dx'\hat{x} \\
& + \int_0^b\frac{b^2-y'^2}{(b^2+\rho^2+z^2-2\rho\sqrt{b^2-y'^2})^{3/2}}dy'\hat{x} - \int_b^0\frac{b^2-x'^2}{(b^2+\rho^2+z^2-2\rho x')^{3/2}}dx'\hat{y} \\
& + \int_0^b\frac{y'\sqrt{b^2-y'^2}}{(b^2+\rho^2+z^2-2\rho\sqrt{b^2-y'^2})^{3/2}}dy'\hat{y} - \int_0^{-b}\frac{x'\sqrt{b^2-x'^2}}{(b^2+\rho^2+z^2-2\rho x')^{3/2}}dx'\hat{x} \\
& + \int_b^0\frac{b^2-y'^2}{(b^2+p^2+z^2+2\rho\sqrt{b^2-y'^2})^{3/2}}dy'\hat{x} - \int_0^{-b}\frac{b^2-x'^2}{(b^2+p^2+z^2-2\rho x')^{3/2}}dx'\hat{y} \\
& - \int_b^0\frac{y'\sqrt{b^2-y'^2}}{(b^2+\rho^2+z^2+2\rho\sqrt{b^2-y'^2})^{3/2}}dy'\hat{y} + \int_{-b}^0\frac{x'\sqrt{b^2-x'^2}}{(b^2+\rho^2+z^2-2\rho x')^{3/2}}dx'\hat{x} \\
& + \int_0^{-b}\frac{b^2-y'^2}{(b^2+\rho^2+z^2+2\rho\sqrt{b^2-y'^2})^{3/2}}dy'\hat{x} - \int_{-b}^0\frac{b^2-x'^2}{(b^2+\rho^2+z^2-2\rho x')^{3/2}}dx'\hat{y} \\
& - \int_0^{-b}\frac{y'\sqrt{b^2-y'^2}}{(b^2+\rho^2+z^2+2\rho\sqrt{b^2-y'^2})^{3/2}}dy'\hat{y} + \int_0^b\frac{x'\sqrt{b^2-x'^2}}{(b^2+\rho^2+z^2-2\rho x')^{3/2}}dx'\hat{x} \\
& + \int_{-b}^0\frac{b^2-y'^2}{(b^2+\rho^2+z^2-2\rho\sqrt{b^2-y'^2})^{3/2}}dy'\hat{x} - \int_0^b\frac{b^2-x'^2}{(b^2+\rho^2+z^2-2\rho x')^{3/2}}dx'\hat{y} \\
& + \int_{-b}^0\frac{y'\sqrt{b^2-y'^2}}{(b^2+\rho^2+z^2-2\rho\sqrt{b^2-y'^2})^{3/2}}dy'\hat{y}\Bigg] + b\int_0^{2\pi}\frac{d\phi'}{(b^2+\rho^2+z^2-2b\rho\cos\phi')^{3/2}}\hat{z}\Bigg\}
\end{aligned}
$$

On examining the rectangular coordinate integrals we find that some may be combined over extended limits and some cancel so the preceding becomes

$$\vec{H} = \frac{bI_0}{4\pi}\left\{-\rho\int_0^{2\pi}\frac{\sin\phi' d\phi'}{(b^2+\rho^2+z^2-2b\rho\cos\phi')^{3/2}}\hat{z} + \frac{z}{b^3}\left[2\int_{-b}^{b}\frac{x'\sqrt{b^2-x'^2}}{(b^2+\rho^2+z^2-2\rho x')^{3/2}}dx'\hat{x}\right.\right.$$
$$+\int_{-b}^{b}\frac{b^2-y'^2}{(b^2+\rho^2+z^2-2\rho\sqrt{b^2-y'^2})^{3/2}}dy'\hat{x} + \int_{-b}^{b}\frac{y'\sqrt{b^2-y'^2}}{(b^2+\rho^2+z^2-2\rho\sqrt{b^2-y'^2})^{3/2}}dy'\hat{y}$$
$$\left.-\int_{-b}^{b}\frac{b^2-y'^2}{(b^2+\rho^2+z^2+2\rho\sqrt{b^2-y'^2})^{3/2}}dy'\hat{x} + \int_{-b}^{b}\frac{y'\sqrt{b^2-y'^2}}{(b^2+\rho^2+z^2+2\rho\sqrt{b^2-y'^2})^{3/2}}dy'\hat{y}\right]$$
$$\left.+\,b\int_0^{2\pi}\frac{d\phi'}{(b^2+\rho^2+z^2-2b\rho\cos\phi')^{3/2}}\hat{z}\right\}$$

Now the integrals in the $\hat{y}$ direction are both odd functions integrated over symmetric limits $(-b, b)$ so those integrals are both zero. We could expect this since in the xz plane we would expect no $\hat{y}$ component. Similarly, the $\hat{x}$ integral with the 2 multiplier will also be zero.

The remaining two integrals in $\hat{x}$ are even integrals over symmetric limits so we may take twice the half-range integration. The expression for $\vec{H}$ is now

$$\vec{H} = \frac{bI_0}{4\pi}\left\{-\rho\int_0^{2\pi}\frac{\sin\phi' d\phi'}{(b^2+\rho^2+z^2-2b\rho\cos\phi')^{3/2}}\hat{z}\right.$$
$$+\frac{2z}{b^3}\left[\int_0^{b}\frac{b^2-y'^2}{(b^2+\rho^2+z^2-2\rho\sqrt{b^2-y'^2})^{3/2}}dy'\hat{x}\right.$$
$$\left.-\int_0^{b}\frac{b^2-y'^2}{(b^2+\rho^2+z^2+2\rho\sqrt{b^2-y'^2})^{3/2}}dy'\hat{x}\right]$$
$$\left.+\,b\int_0^{2\pi}\frac{d\phi'}{(b^2+\rho^2+z^2-2b\rho\cos\phi')^{3/2}}\hat{z}\right\}$$

To compact this a little more we could combine the $\hat{z}$ integrals and the $\hat{x}$ integrals to finally obtain

$$\vec{H} = \frac{bI_0}{4\pi}\left\{\int_0^{2\pi}\frac{(b-\rho\sin\phi')d\phi'}{(b^2+\rho^2+z^2-2b\rho\cos\phi')^{3/2}}\hat{z} + \frac{2z}{b^3}\int_0^{b}\left[\frac{b^2-y'^2}{(b^2+\rho^2+z^2-2\rho\sqrt{b^2-y'^2})^{3/2}}\right.\right.$$
$$\left.\left.-\frac{b^2-y'^2}{(b^2+\rho^2+z^2+2\rho\sqrt{b^2-y'^2})^{3/2}}\right]dy'\hat{x}\right\}$$

This may seem like an unusual stopping point, but these remaining integrals are most easily evaluated numerically. We would need to be given the loop radius b. Then we would select a point in the xz plane by specifying ρ and z coordinates and perform the numerical integrations to obtain $\vec{H}$ at that point.

Plots of the field lines can be obtained by using the same technique described in Section 5-6. This will not be repeated here.

(The integrals above can be evaluated in terms of elliptic functions. These are treated briefly in Example 6-37. The integrals in the $\hat{x}$ component would be first converted to trigonometric forms.)

Example 6-6

For the current sheet shown in Figure 6-17, compute the magnetic field intensity on the x axis. $\vec{J}_s = -k(a^2 - y^2)\hat{z}$ A/m.

From the figure and using the symbols in Eq. 6-14:

$$\vec{J}_s(\vec{r}\,') \Rightarrow -k(a^2 - y'^2)\hat{z}' = -k(a^2 - y'^2)\hat{z}$$

$$\vec{R} = \vec{r} - \vec{r}\,' \Rightarrow x\hat{x} - (y'\hat{y}' + z'\hat{z}') = x\hat{x} - y'\hat{y}' - z'\hat{z}$$

$$R^2 = \vec{R} \cdot \vec{R} \Rightarrow x^2 + y'^2 + z'^2$$

$$\hat{R} = \frac{\vec{R}}{R} \Rightarrow \frac{x\hat{x} - y'\hat{y} - z'\hat{z}}{\sqrt{x^2 + y'^2 + z'^2}}$$

Then from Eq. 6-14

$$\vec{H} = -\frac{k}{4\pi}\int_{-b}^{b}\int_{-a}^{a}\frac{x(a^2 - y'^2)\hat{y} + y'(a^2 - y'^2)\hat{x}}{(x^2 + y'^2 + z'^2)^{3/2}}\,dy'dz'$$

$$= -\frac{k}{4\pi}\left[\int_{-b}^{b}\int_{-a}^{a}\frac{(a^2 - y'^2)}{(x^2 + y'^2 + z'^2)^{3/2}}\,dy'dz'\hat{y}\right.$$

$$\left.+ \int_{-b}^{b}\int_{-a}^{a}\frac{a^2y' - y'^3}{(x^2 + y'^2 + z'^2)^{3/2}}\,dy'dz'\hat{x}\right]$$

The $\hat{x}$ component integration is zero since the y' portion is an odd function integrated over symmetric limits. (One may have suspected no $\hat{x}$ component on physical grounds.) For the

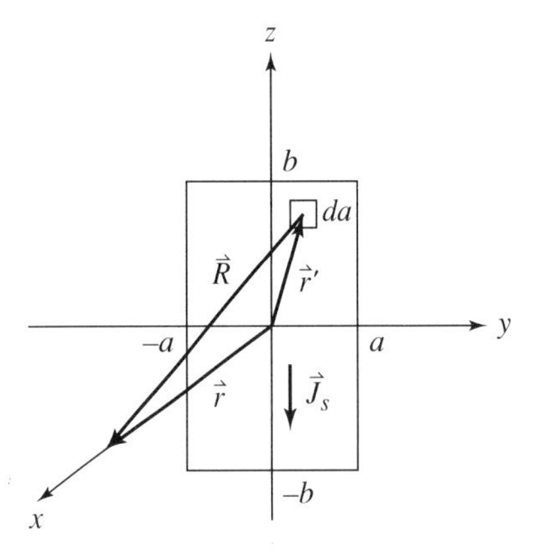

Figure 6-17. Configuration of sheet current for Example 6-6.

$\hat{y}$ component, since the y' portion is an even function over symmetric limits we may double the half-range integration. This is also true for the z' integration:

$$\vec{H} = -\frac{kx}{\pi}\left[\int_0^b \int_0^a \frac{a^2}{(x^2 + y'^2 + z'^2)^{3/2}}\, dy'dz' - \int_0^b \int_0^a \frac{y'^2}{(x^2 + y'^2 + z'^2)^{3/2}}\, dy'dz'\right]\hat{y}$$

$$\vec{H} = -\frac{kx}{\pi}\left[a^2 \int_0^b \frac{a}{(x^2 + z'^2)\sqrt{a^2 + x^2 + z'^2}}\, dz'\right.$$

$$\left. - \int_0^b \left[\frac{-a}{\sqrt{a^2 + x^2 + z'^2}} + \ln\left(\frac{a + \sqrt{a^2 + x^2 + z'^2}}{\sqrt{x^2 + z'^2}}\right)\right] dz'\right]\hat{y}$$

As in the preceding example, we are now forced to a numerical solution for a particular problem. We would need to have particualr dimensions a and b given and then apply numerical integration for a selected point x.

Example 6-7

A beam of current $\vec{J} = J_0\hat{z}$ flows along the z axis and has radius a. Compute the magnetic field intensity on the x axis as indicated in Figure 6-18. The beam is assumed to be of infinite length. We use the volume integral form, Eq. 6-15. From the figure we have

$$\vec{J}(\vec{r}') \Rightarrow J_0\hat{z}'$$

$$\vec{R} = \vec{r} - \vec{r}' \Rightarrow x\hat{x} - (\rho'\hat{\rho}' + z'\hat{z}') = x\hat{x} - \rho\hat{\rho}' - z'\hat{z}'$$

$$R^2 = \vec{R} \cdot \vec{R} \Rightarrow x^2 - 2x\rho'\hat{x} \cdot \hat{\rho}' + \rho'^2 + z'^2 = x^2 - x\rho' \cos\phi' + \rho'^2 + z'^2$$

$$\hat{R} = \frac{\vec{R}}{R} \Rightarrow \frac{x\hat{x} - \rho'\hat{\rho}' - z'\hat{z}'}{\sqrt{x^2 - 2x\rho' \cos\phi' + \rho'^2 + z'^2}}$$

$$dv' \Rightarrow \rho' d\phi' d\rho' dz'$$

$$\vec{H} = \int_{-\infty}^{\infty} \int_0^a \int_0^{2\pi} \frac{J_0\hat{z}' \times \left(\dfrac{x\hat{x} - \rho'\hat{\rho}' - z'\hat{z}'}{\sqrt{x^2 - 2x\rho' \cos\phi' + \rho'^2 + z'^2}}\right)}{4\pi(x^2 - 2x\rho' \cos\phi' + \rho'^2 + z'^2)}\, \rho' d\phi' d\rho' dz'$$

$$= \frac{J_0}{4\pi} \int_{-\infty}^{\infty} \int_0^a \int_0^{2\pi} \frac{x\hat{y} - \rho'\hat{\phi}'}{(x^2 - 2x\rho' \cos\phi' + \rho'^2 + z'^2)^{3/2}}\, \rho' d\phi' d\rho' dz'$$

$$= \frac{J_0}{4\pi}\left[x \int_{-\infty}^{\infty} \int_0^a \rho' \int_0^{2\pi} \frac{d\phi'}{(x^2 - 2x\rho' \cos\phi' + \rho'^2 + z'^2)^{3/2}}\, d\phi' d\rho' dz'\hat{y}\right.$$

$$\left. - \int_{-\infty}^{\infty} \int_0^a \rho'^2 \int_0^{2\pi} \frac{\hat{\phi}' d\phi'}{(x^2 - 2x\rho' \cos\phi' + \rho'^2 + z'^2)^{3/2}}\, d\rho' dz'\right]$$

The evaluation techniques (which may include numerical ones) have already been presented, so this example will not be carried out further. The point of this example was to show how the volume integrations are set up. Later, another technique that applies potential theory much like electrostatics of Chapter 5 will be used and will give us the closed-form solution.

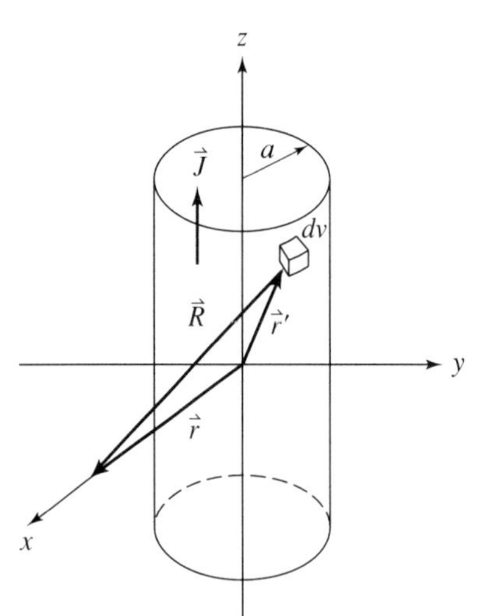

Figure 6-18. Beam current system for Excample 6-7.

Example 6-8

Determine the magnetic field intensity at a distance ρ from an infinite-length current filament shown in Figure 6-19. The z axis is the line current.

Since the system is symmetric in ϕ and is of infinite length, we may select the field point on the x axis without losing any generality.

$$Id\vec{l}' \Rightarrow I_0 dz'\hat{z}'$$

$$\vec{R} = \vec{r} - \vec{r}' \Rightarrow x\hat{x} - z'\hat{z}'$$

$$R^2 = \vec{R} \cdot \vec{R} \Rightarrow x^2 + z'^2$$

$$\hat{R} = \frac{\vec{R}}{R} \Rightarrow \frac{x\hat{x} - z'\hat{z}'}{\sqrt{x^2 + z'^2}}$$

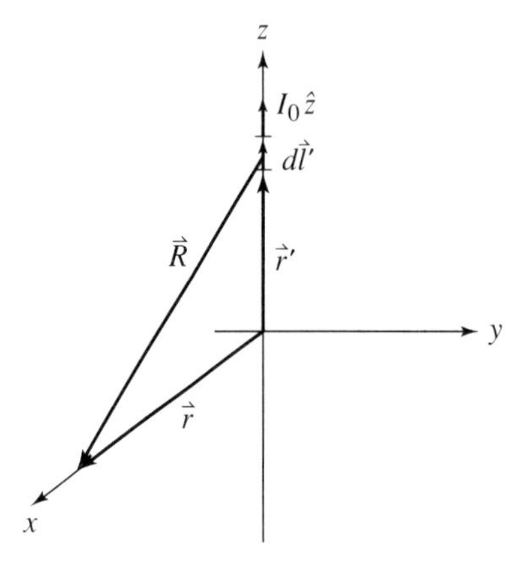

Figure 6-19. Infinite-length current line for Example 6-8.

Using Eq. 6-13:

$$\vec{H} = \int_{-\infty}^{\infty} \frac{xI_0 dz'\hat{y}'}{4\pi(x^2 + z'^2)^{3/2}} = \frac{I_0 x}{4\pi}\int_{-\infty}^{\infty} \frac{dz'}{(x^2 + z'^2)^{3/2}}\,\hat{y}$$

Since the integrand is an even function we may double the value of half-range integration and write

$$\vec{H} = \frac{I_0 x}{2\pi}\int_0^{\infty} \frac{dz'}{(x^2 + z'^2)^{3/2}}\,\hat{y} = \frac{xI_0}{2\pi}\left[\lim_{z_0\to\infty}\int_0^{z_0} \frac{dz'}{(x^2 + z'^2)^{3/2}}\right]\hat{y}$$

Evaluating this, remembering that x is considered constant,

$$\vec{H} = \frac{xI_0}{2\pi x^2}\lim_{z_0\to\infty}\left(\frac{z'}{\sqrt{x^2 + z'^2}}\right)\Bigg|_0^{z_0}\hat{y} = \frac{I_0}{2\pi x}\lim_{z_0\to\infty}\left(\frac{z_0}{\sqrt{x^2 + z_0^2}}\right)\hat{y} = \frac{I_0}{2\pi x}\,\hat{y}$$

A rectangular to cylindrical coordinate conversion yields

$$\vec{H} = \frac{I_0}{2\pi\rho}\,\hat{\phi}$$

Example 6-9

A wire is bent in a parabolic shape as shown in Figure 6-20 and carries a current I_0 and is of infinite length. Determine the magnetic field intensity at the focus of the parabola (origin of rectangular coordinates). The parabolic cylindrical coordinate system will be used as described in Chapter 4.

The line passes through the point y_0 and we need to determine the value of coordinate v_0. The rectangular coordinate point is $(0,y_0)$. Using the relationships for parabolic cylindrical coordinates (see Chapter 4, Exercise 4-8.0f), we have

$$0 = \tfrac{1}{2}(u_0^2 - v_0^2) \qquad y_0 = u_0 v_0$$

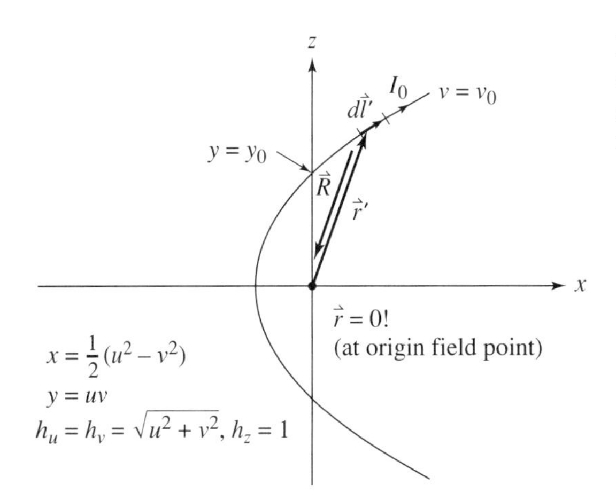

Figure 6-20. Infinite length parabolic wire carrying constant current I_0 for Example 6-9; coordinates: (u, v, z).

Solving for v_0,

$$v_0 = \sqrt{y_0} \qquad (v \geq 0 \text{ for this system})$$

For this coordinate system, $d\vec{l} = h_u du\hat{u} + h_v dv\hat{v} + h_z dz\hat{z}$. Since v and z are constant, $d\vec{l} = h_u du\hat{u} \Rightarrow \sqrt{u^2 + v_0^2}\, du\hat{u} = \sqrt{u^2 + y_0}\, du\hat{u}$. Thus

$$I' d\vec{l}' \Rightarrow I_0\sqrt{u'^2 + y_0}\, du'\hat{u}'$$

Since the field point is at the origin, $\vec{r} = 0$, and the line is in the $z' = 0$ plane,

$$\vec{R} = \vec{r} - \vec{r}' = -\vec{r}' \Rightarrow -(x'\hat{x}' + y'\hat{y}') = -[\tfrac{1}{2}(u'^2 - y_0)\hat{x}' + u'\sqrt{y_0}\,\hat{y}']$$

$$R^2 = \vec{R}\cdot\vec{R} \Rightarrow \tfrac{1}{4}(u'^2 - y_0)^2 + u'^2 y_0 = \tfrac{1}{4}(u'^2 + y_0)^2$$

$$\hat{R} = \frac{\vec{R}}{R} \Rightarrow \frac{-\frac{1}{2}(u'^2 - y_0)\hat{x}' - u'\sqrt{y_0}\,\hat{y}'}{\frac{1}{2}(u'^2 + y_0)}$$

Using Eq. 6-13, substitution yields

$$\vec{H} = \int_{-\infty}^{\infty} \frac{I_0[-\frac{1}{2}(u'^2 - y_0)du'\hat{u}' \times \hat{x}' - u'\sqrt{y_0}\, du'\hat{u}' \times \hat{y}']}{4\pi\frac{1}{8}(u'^2 + y_0)^{5/2}}$$

$$= \frac{-2I_0}{\pi}\left[\frac{1}{2}\int_{-\infty}^{\infty} \frac{(u'^2 - y_0)\hat{u}' \times \hat{x}'}{(u'^2 + y_0)^{5/2}}\, du' + \sqrt{y_0}\int_{-\infty}^{\infty} \frac{u'\hat{u}' \times \hat{y}'}{(u'^2 + y_0)^{5/2}}\, du'\right]$$

To evaluate the cross products we use the techniques of coordinate conversion discussed in Chapter 4. The spatial relationships are shown in Figure 6-21, where the angles of interest have been identified.

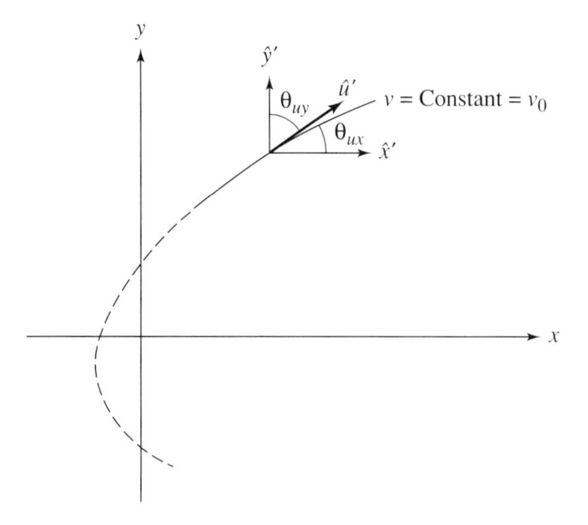

Figure 6-21. Relationship in space of unit vectors for parabolic cylindrical coordinate u' and rectangular coordinates x' and y'.

From Figure 6-21 and using the right-hand rule for cross-product direction:

$$\hat{u}' \times \hat{x}' = \sin\theta_{ux}(-\hat{z}') = -\sin[(\tan^{-1}(\text{slope of } v = v_0 \text{ curve})]\hat{z}'$$

$$= -\sin\left[\tan^{-1}\left(\left.\frac{\partial y}{\partial x}\right|_{v'=v_0}\right)\right]\hat{z}' = -\sin\left[\tan^{-1}\left(\left.\frac{\dfrac{\partial y}{\partial u}du + \dfrac{\partial y}{\partial v}dv}{\dfrac{\partial x}{\partial u}du + \dfrac{\partial x}{\partial v}dv}\right)\right|_{v'=v_0}\right]\hat{z}'$$

Since v is a constant, $dv = 0$ so that

$$\hat{u}' \times \hat{x}' = -\sin\left[\tan^{-1}\left(\left.\frac{\partial y'/\partial u'}{\partial x'/\partial u'}\right|_{v'=v_0}\right)\right]\hat{z}' = -\sin\left[\tan^{-1}\left(\frac{v_0}{u'}\right)\right]\hat{z}'$$

$$= -\frac{v_0}{\sqrt{u'^2 + v_0^2}}\hat{z}' = \frac{-\sqrt{y_0}}{\sqrt{u'^2 + y_0}}\hat{z}'$$

Similarly

$$\hat{u}' \times \hat{y}' = \sin\theta_{uy}\hat{z}' = \sin(90^\circ - \theta_{ux})\hat{z}' = \cos\theta_{ux}\hat{z}' = \frac{u'}{\sqrt{u'^2 + y_0}}\hat{z}'$$

The integral for the magnetic field then becomes

$$\vec{H} = -\frac{2I_0}{\pi}\left[-\frac{\sqrt{y_0}}{2}\int_{-\infty}^{\infty}\frac{u'^2 - y_0}{(u'^2 + y_0)^3}du'\hat{z} + \sqrt{y_0}\int_{-\infty}^{\infty}\frac{u'^2}{(u'^2 + y_0)^3}du'\hat{z}\right]$$

Combining u'^2 integrands after writing the left integral as the sum of two integrals:

$$\vec{H} = -\frac{I_0\sqrt{y_0}}{\pi}\left[\int_{-\infty}^{\infty}\frac{u'^2}{(u'^2 + y_0)^3}du' + y_0\int_{-\infty}^{\infty}\frac{1}{(u'^2 + y_0)^3}du'\right]\hat{z}$$

Since the integrands are even functions of u' we can double the half-range integrations:

$$\vec{H} = -\frac{2I_0\sqrt{y_0}}{\pi}\lim_{u_0\to\infty}\left[\int_0^{u_0}\frac{u'^2}{(u'^2 + y_0)^3}du' + y_0\int_0^{u_0}\frac{du'}{(u'^2 + y_0)^3}\right]\hat{z} = -\frac{I_0}{2y_0}\hat{z}$$

(These integrals were evaluated using the substitution $w = u'^2$ and Section 2-21 of [24].)

Exercises

6-4.0a For the result of Example 6-4 determine the equation for the magnetic flux density $\vec{B}$ on the axis of the loop. (Answer: $\vec{B} = \mu_0\vec{H} = 4\pi \times 10^{-7}\vec{H}$)

6-4.0b For the configuration of Example 6-6, set up the expression for $\vec{H}$ by initially writing an equation for $\vec{J}$ using step and delta functions and using Eq. 6-15 integrated over all space. Then reduce using the properties of the functions to obtain the double integral form over finite limits. Do not evaluate any further integrations.

6-4.0c From the last result in Example 6-5, show that the correct result is obtained for the special case $\rho = 0$ (which is Example 6-4).

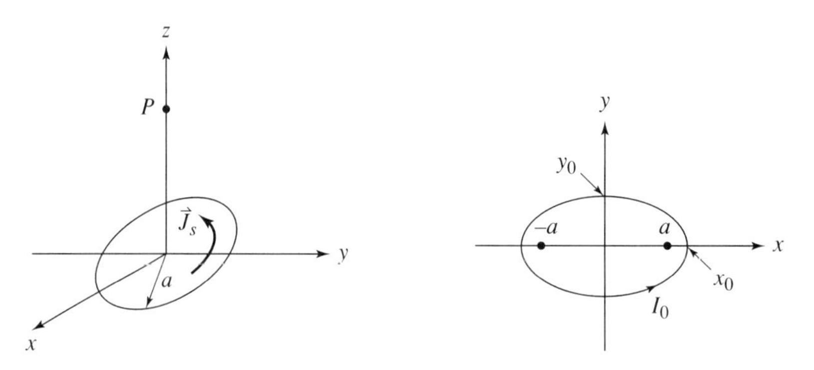

Figure 6-22. **Figure 6-23.**

6-4.0d In Example 6-5, verify that the expansion of $\int_{\pi/2}^{\pi} \hat{\rho}' d\phi'$ results in the fifth, sixth, seventh, and eighth integrals inside the square brackets.

6-4.0e In Example 6-6, suppose the current density is changed to $\vec{J}_s = -J_0\hat{z}$. Solve for the magnetic field intensity on the x axis.

6-4.0f For the circular sheet shown in Figure 6-22 compute $\vec{H}$ on the z axis. $\vec{J}_s = J_0\hat{\phi}$.

6-4.0g Solve Example 6-8 using cylindrical coordinates exclusively for $\phi = 0$ (field point still on x axis).

6-4.0h An elliptical wire loop carries a constant current I_0 as shown in Figure 6-23. Compute the current at the center of the ellipse ($x = y = z = 0$). For the elliptical coordinate system see Chapter 4, General Exercise GE 4-10. Where

$$x = a \cosh u \cos v \qquad y = a \sinh u \sin v \qquad z = z$$

$$u \geq 0, \qquad 0 \leq v \leq 2\pi, \qquad -\infty < z < \infty$$

$$h_u = h_v = a\sqrt{\sinh^2 u + \sin^2 v}, \qquad h_z = 1$$

6-4.0i A solenoid of infinite length in Figure 6-24 has n turns per meter and current I_0 in wire turns. Assume that the turns are closely wound so that $\vec{J}_s = nI_0\hat{\phi}$ A/m. Determine $\vec{H}$ at the center (at the origin). (Answer: $nI_0\hat{z}$)

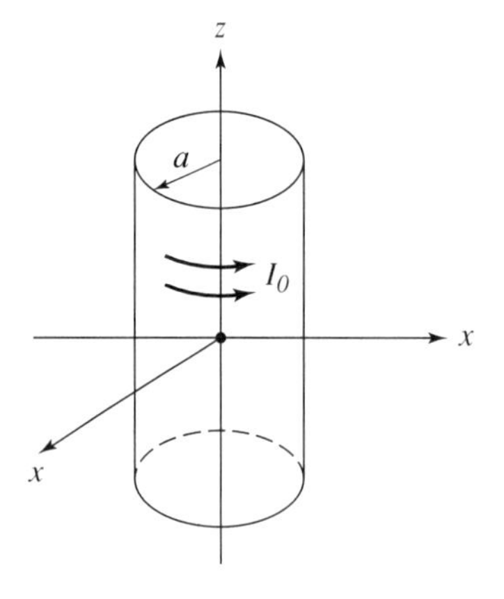

Figure 6-24.

6-4.0j Suppose the wire of Example 6-8 has finite length and extends from $-L$ to L. Compute the magnetic field intensity on the x axis.

It should be noted that since the computation of $\vec{H}$ depends only upon the current distribution, the systems are linear at the field-point locations. Thus the superposition theorem can be applied in cases where there are several different current distributions present. All that needs to be done is to compute the magnetic field intensity at a point due to each current system separately and add the results (vector addition, of course).

The initial discussion of this section concerned the determination of the force between two current systems. To this point all that has been done is to show how the magnetic field intensity can be computed from the primed system current distribution, which gives $\vec{H}$ as a function of unprimed coordinates. Referring to Eq. 6-10 we next must determine the force on the unprimed current distribution. If the umprimed system is either a surface current $\vec{J}_s$ or a volume current $\vec{J}$, we can replace $Id\vec{l}$ by the appropriate current increment as before to have the additional forms for force.

$$d\vec{F} = \vec{J}_s da \times \mu\vec{H} \tag{6-16}$$

$$d\vec{F} = \vec{J} dv \times \mu\vec{H} \tag{6-17}$$

The $\vec{H}$ in these equations and also Eq. 6-10, is computed at a general point on the unprimed current distribution; that is, $d\vec{F}$ has no primes in it. Then, to determine the force a second integration that involves techniques already presented is performed. Since the concepts are quite similar, only one example will be given.

Example 6-10

The ampere is now formally defined as that constant current which, if maintained in two parallel, infinite, straight, small-cross-section wires places 1 m apart, results in a force of 2×10^{-7} N/m of length in vacuum. The National Institute for Standards and Technology (NIST, formerly the National Bureau of Standards) maintains a system that is used to determine the ampere. Let us determine the force on two such wires, which carry 1 A as shown in Figure 6-25. We select one wire to be on the z axis and denote it by the primed coordinates and the other wire 1 m out on the y axis. We compute the force per meter on the right wire.

The first step is to locate a general point on the current distribution on which the force is to be computed. This is $d\vec{l}$ in the figure and corresponds to the field point location denoted by $\vec{r}$. We compute $\vec{H}$ at the field point produced by the primed current distribution.

From the figure, for $y = 1$ m, and $\hat{z}' = \hat{z}$ as usual,

$$I'd\vec{l}' \Rightarrow 1dz'\hat{z}' = dz'\hat{z}' = dz'\hat{z}$$

$$\vec{r}' \Rightarrow z'\hat{z}'$$

$$\vec{r} \Rightarrow y\hat{y} + z\hat{z}$$

$$\vec{R} = \vec{r} - \vec{r}' \Rightarrow y\hat{y} + z\hat{z} - z'\hat{z}' = y\hat{y} + (z - z')\hat{z}$$

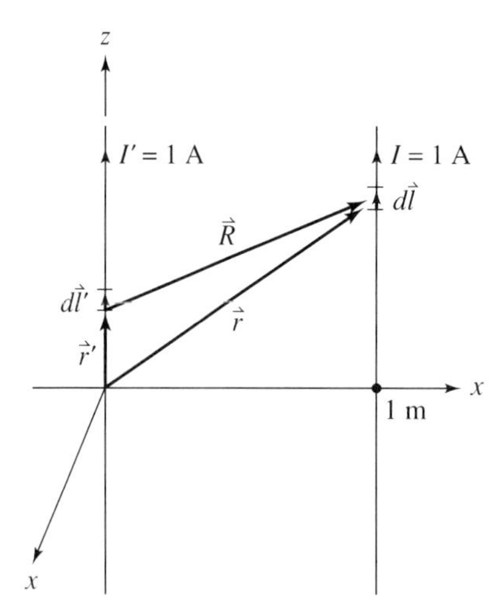

Figure 6-25. Two-wire system for Example 6-10.

$$R^2 = \vec{R} \cdot \vec{R} \Rightarrow 1 + (z - z')^2$$

$$\hat{R} = \frac{\vec{R}}{R} \Rightarrow \frac{\hat{y} + (z - z')\hat{z}}{\sqrt{1 + (z - z')^2}}$$

Then at the field point $d\vec{l}$, using Eq. 6-13,

$$\vec{H} = \int_{-\infty}^{\infty} \frac{-\hat{x}dz'}{4\pi[1 + (z - z')^2]^{3/2}} = -\frac{1}{2\pi}\int_0^{\infty} \frac{dz'}{(1 + z^2 - 2zz' + z'^2)^{3/2}}\,\hat{x}$$

Remembering that z is a constant in this integration we use integral tables to find the form $(a + bz' + z'^2)^{-3/2}$.

$$\vec{H} = -\frac{1}{2\pi}\lim_{z_0\to\infty}\left[\frac{z' - z}{\sqrt{1 + z^2 - 2zz' + z'^2}}\right]_0^{z'_0}\hat{x} = -\frac{1}{2\pi}\hat{x}$$

Using this result in Eq. 6-10 the force on 1 m of the right line is, for $\mu = \mu_0$ ($\vec{B} = \mu_0\vec{H}$ is the flux density which produces the force),

$$\vec{F} = \int_{-0.5}^{0.5} Id\vec{l} \times \mu_0\vec{H} = \int_{-0.5}^{0.5} 1dz\hat{z} \times \mu_0\left(-\frac{1}{2\pi}\hat{x}\right)$$

$$= -\frac{\mu_0}{2\pi}\int_{-0.5}^{0.5} \hat{y}dz = -\frac{\mu_0}{2\pi}\int_{-0.5}^{0.5} dz\hat{y} = -\frac{\mu_0}{2\pi}\hat{y}\text{ N}$$

Since this is to be 2×10^{-7} N/m, μ_0 must be $4\pi \times 10^{-7}$ as stated earlier.

Exercises

6-4.0k Repeat Example 6-10 using the right wire as the primed system and the left wire as the unprimed system, and determine the force per meter on the left wire.

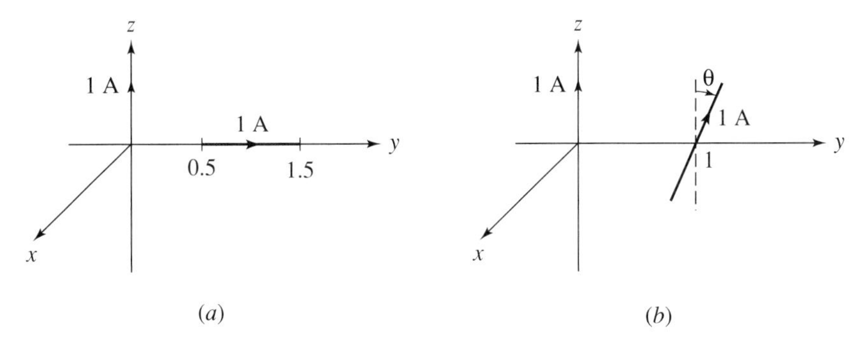

Figure 6-26.

6-4.0l Suppose by some stratagem we could obtain a finite length wire 1 m long carrying 1 A of current. Orient it along the y axis as shown in the Figure 6-26*a*. Compute the force on the 1 m wire. Compare this result with the result of Example 6-10, and suggest an equation for the force on a 1-m length line in the yz plane pivoted about the $y = 1$ m point, as a function of ϕ.

6-4.0m Repeat Exercise 6-4.01 for the configurations shown in Figure 6-27 where the wire is in the xy plane.

Figure 6-27.

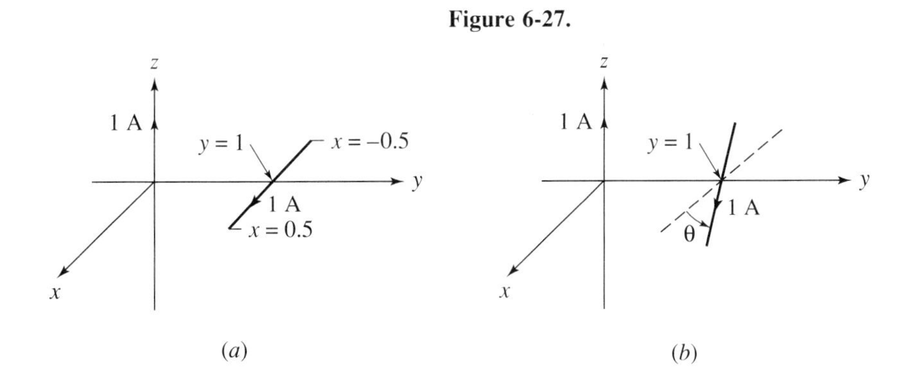

6-5 Forces on Charges: Lorentz Force Equation

There are many practical situations where it is necessary to compute the force on a charged particle. Electron beams in cathode ray tubes, particles in accelerators, and carriers in doped semiconductor materials are some examples. Such particles can be subject to both electric and magnetic fields so it is important to be able to compute the total force.

To obtain the magnetic force we begin with Eq. 6-4, and express the current as the time rate of flow of charge to yield

$$d\vec{F}_m = Id\vec{l} \times \vec{B} = \frac{dq}{dt}\, d\vec{l} \times \vec{B}$$

Since the charge is moving in the path $d\vec{l}$ this may be written

$$d\vec{F}_m = dq\,\frac{d\vec{l}}{dt} \times \vec{B} = dq\vec{V} \times \vec{B}$$

where $\vec{V}$ is the charge velocity. If we consider a quantity of charge q (zero volume) the force would be

$$\vec{F}_m = q\vec{V} \times \vec{B}$$

For the case where the charge q is subjected to both electric and magnetic fields we sum the two forces to obtain

$$\vec{F} = \vec{F}_e + \vec{F}_m = q\vec{E} + q\vec{V} \times \vec{B} \tag{6-18}$$

This is the Lorentz force equation.

Example 6-11

Suppose we have a p-type semiconductor material with an applied voltage as shown in Figure 6-28*a*. The holes will drift in the direction shown due to the force $\vec{F}_e$ of the electric field produced by the voltage source across the material. Since the material is electrically neutral and there is uniform distribution of charge a voltmeter placed across opposite faces of the material will not register any voltage. Now if a magnetic field $\vec{B}$ is simultaneously applied to the material as shown in Figure 6-28*b* the positive charges will tend to move across the width through the material due to force F_m, $\vec{V}$ being the hole velocity, which

Figure 6-28. Illustration of the Hall effect created by a magnetic flux density in p-type material.

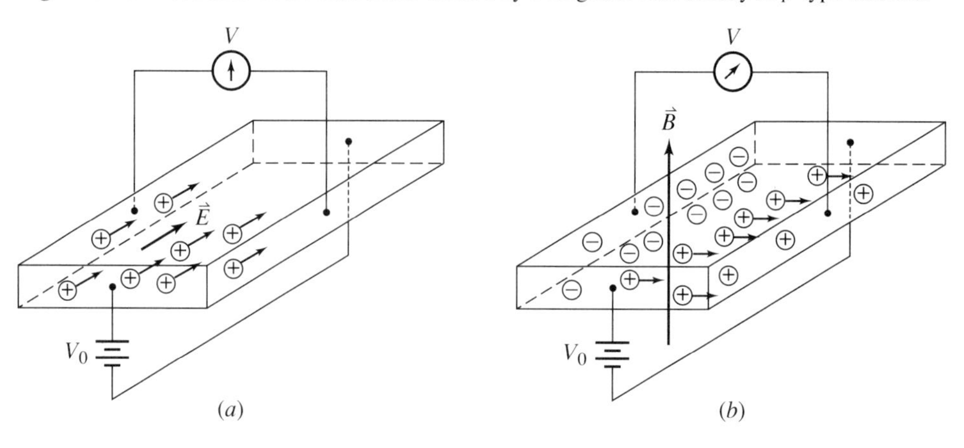

would tend to put excess plus charge on the *right* side and leave behind the negative lattice points. This tends to create an effective cross-bar electric field component that can be sensed by the voltmeter, positive on the right, negative on the left. This voltage is called the *Hall effect voltage*. Notice also that the diagonal motion produced by the sum of the two electric fields would have an effective current component across the bar in the direction between the voltmeter leads.

Another way of visualizing this effect is to attribute a cross-bar electric field generated by the magnetic field that causes the current to flow diagonally. The voltage sensed by the voltmeter would be this electric field component multiplied by the bar width.

Exercises

6-5.0a Suppose an electron passes by #20 wire that carries 50 mA DC. The electron moves parallel to the wire in the direction of current flow and is 5 mm from the wire center. The electron velocity is 9×10^6 m/s. What is the force on the electron and approximately how much will the electron path be deflected in 0.1 μs?

6-5.0b Using the system of Figure 6-28, determine the polarity of the Hall voltage for *n*-type material.

6-5.0c An MHD (magneto hydro dynamic) generator produces electric power by burning (ionizing) a gas at high velocity horizontally between two vertical parallel plates and applying a magnetic field parallel to the plates and normal to the ion flow. Draw a diagram showing how the system would be configured physically. Show the output current in a resistor connected between the plates. An ionized gas consists of electrons and ions.

6-5.0d A point charge of 2 C has a velocity given by $\vec{V} = -2\hat{x} + 3\hat{y} + 6\hat{z}$ m/s. Compute the force on the charge if it is in fields $2\hat{x} - 4\hat{y} - 3\hat{z}$ v/m and $9\hat{x} + 3\hat{y} - 4\hat{z}$ T simultaneously.

6-5.0e Prove that a particle of mass m, charge q, an velocity $\vec{V}$ moving in a magnetic field $\vec{B}$ only has constant kinetic energy; that is, a magnetic force cannot change the particle energy or $|\vec{V}|$.

6-6 Ampere's Circuital Law

In Section 6-4 we developed equations that allowed us to compute the magnetic field intensity $\vec{H}$ from a given current distribution. Suppose, however, that we have a magnetic field intensity and would like to determine something about the currents in the region.

The motivation for what we are about to do stems initially from the fact that the magnetic field lines always form closed paths since there has not been discovered an isolated north pole element to act as a source or sink of magnetic field lines which would be analogous to electric charge. We are then led to integrate the field intensity around a closed path. The result we obtain is known as Ampere's circuital law. This is one reason why the Biot-Savart law was so named rather than calling it Ampere's law, so that there would not be confusion as to which law was being used or discussed. Although the derivation is lengthy there are important techniques used that show applications of vector theorems and primed coordinates.

We begin, then, by integrating the general result Eq. 6-15 around a closed path (contour) C. The direction of the contour will be in the right-hand sense; that is, if one curls the fingers of the right hand along the contour, the thumb defined the enclosed area vector. Mathematically, we could also say we traverse the (regular) surface region so the area is on the left-hand side. From Eq. 6-15:

$$\oint_C \vec{H} \cdot d\vec{l} = \oint_C \int_{V'} \frac{\vec{J}(\vec{r}\,') \times \hat{R}}{4\pi R^2}\, dv'dl$$

Since scalars may be taken inside cross products we can write

$$\oint_C \vec{H} \cdot d\vec{l} = \oint_C \int_{V'} \frac{1}{4\pi} \vec{J}(\vec{r}\,') \times \left(\frac{\hat{R}}{R^2}\right) dv'dl$$

In rectangular coordinates

$$\begin{aligned} R &= |\vec{R}| = |\vec{r} - \vec{r}\,'| = |(x - x')\hat{x} + (y - y')\hat{y} + (z - z')\hat{z}| \\ &= \sqrt{(x - x')^2 + (y - y')^2 + (z - z')^2}. \end{aligned}$$

From this it is easy to show (see Appendix F) that the gradient with respect to the unprimed coordinates is

$$\nabla\left(\frac{1}{R}\right) = -\frac{\hat{R}}{R^2}$$

Then

$$\begin{aligned} \oint_C \vec{H} \cdot d\vec{l} &= \oint_C \int_{V'} \frac{1}{4\pi} \vec{J}(\vec{r}\,') \times \left[-\nabla\left(\frac{1}{R}\right)\right] dv'dl \\ &= \oint_C \int_{V'} \frac{1}{4\pi} \nabla\left(\frac{1}{R}\right) \times \vec{J}(\vec{r}\,') dv'dl \end{aligned} \qquad (6\text{-}19)$$

We next use the vector identity $\nabla \times (a\vec{A}) = \nabla a \times \vec{A} + a\nabla \times \vec{A}$ and write it in the form

$$\nabla a \times \vec{A} = \nabla \times (a\vec{A}) - a\nabla \times \vec{A}$$

From the preceding integrand if we associate $a \Rightarrow 1/R$ and $\vec{A} \Rightarrow \vec{J}(\vec{r}\,')$ this identity becomes

$$\nabla\left(\frac{1}{R}\right) \times \vec{J}(\vec{r}\,') = \nabla \times \left(\frac{\vec{J}(\vec{r}\,')}{R}\right) - \frac{1}{R} \nabla \times \vec{J}(\vec{r}\,')$$

However, since ∇ operates on unprimed coordinates and $\vec{J}(\vec{r}\,')$ is a function only of primed coordinates the right curl term is zero. Substituting the remaining term into the double integral of Eq. 6-19:

$$\oint_C \vec{H} \cdot d\vec{l} = \oint_C \int_{V'} \frac{1}{4\pi} \nabla \times \left(\frac{\vec{J}(\vec{r}\,')}{R}\right) dv'dl$$

Again, ∇ operates on umprimed coordinates so it may be removed from the first (inner) integral (note that R is a function of primes):

$$\oint_C \vec{H} \cdot d\vec{l} = \frac{1}{4\pi} \oint_C \nabla \times \left[\int_{V'} \frac{\vec{J}(\vec{r})}{R} dv' \right] dl$$

Next apply Stokes' theorem (see Chapter 4):

$$\oint_C \vec{H} \cdot d\vec{l} = \frac{1}{4\pi} \int_S \left\{ \nabla \times \nabla \times \left[\int_{V'} \frac{\vec{J}(\vec{r}\,')}{R} dv' \right] \right\} \cdot d\vec{a} \qquad (C \text{ encloses } S)$$

Now use the vector identity for the double curl $\nabla \times \nabla \times \vec{A} = \nabla(\nabla \cdot \vec{A}) - \nabla^2 \vec{A}$, where we associate $\vec{A}$ with the V' integral in brackets. Then

$$\oint_C \vec{H} \cdot d\vec{l} = \frac{1}{4\pi} \int_S \nabla \left\{ \left[\nabla \cdot \left(\int_{V'} \frac{\vec{J}(\vec{r}\,')}{R} dv' \right) \right] - \nabla^2 \left(\int_{V'} \frac{\vec{J}(\vec{r}\,')}{R} dv' \right) \right\} \cdot d\vec{a} \qquad (6\text{-}20)$$

Working with the term that has the Laplacian we may take ∇^2 inside the integral since it operates on unprimed coordinates. Thus

$$\nabla^2 \left(\int_{V'} \frac{\vec{J}(\vec{r}\,')}{R} dv' \right) = \int_{V'} \nabla^2 \left[\frac{\vec{J}(\vec{r}\,')}{R} \right] dv' = \int_{V'} \vec{J}(\vec{r}\,')\, \nabla^2 \left(\frac{1}{R} \right) dv'$$

Now $\nabla^2(1/R) = -4\pi\, \delta(R)$ (see Appendix F) so this becomes

$$\nabla^2 \left(\int_{V'} \frac{\vec{J}(\vec{r}\,')}{R} dv' \right) = \int_{V'} \vec{J}(\vec{r}\,')[-4\pi\, \delta(R)] dv' = -4\pi \int_{V'} \vec{J}(\vec{r}\,')\, \delta(R) dv'$$

The delta function requires us to evaluate the integrand at the point where the delta function occurs (see Chapter 4), that is, $R = 0$. Now R is zero for $x' = x$, $y' = y$, $z' = z$, or $\vec{r}\,' = \vec{r}$. Then we have

$$\nabla^2 \left(\int_{V'} \frac{\vec{J}(\vec{r}\,')}{R} dv' \right) = -4\pi \vec{J}(\vec{r})$$

Substituting this result into Eq. 6-20:

$$\oint_C \vec{H} \cdot d\vec{l} = \frac{1}{4\pi} \int_S \left\{ \nabla \left[\nabla \cdot \left(\int_{V'} \frac{\vec{J}(\vec{r}\,')}{R} dv' \right) \right] + 4\pi \vec{J}(\vec{r}) \right\} \cdot d\vec{a} \qquad (6\text{-}21)$$

Using similar ideas on the remaining V' integration term, and then applying the divergence theorem

$$\nabla \left[\nabla \cdot \left(\int_{V'} \frac{\vec{J}(\vec{r}\,')}{R} dv' \right) \right] = \nabla \left[\int_{V'} \nabla \cdot \left(\frac{\vec{J}(\vec{r}\,')}{R} \right) dv' \right] = \nabla \left[\int_{S'} \frac{\vec{J}(\vec{r}\,')}{R} \cdot d\vec{a}\,' \right]$$

Now the divergence theorem requires that the surface S' encloses the entire volume V' in which the current $\vec{J}$ flows. Thus the surface S' must enclose all the current, so none of the current of V' can be on the boundary. Another way of looking at this is to imagine S' extended just a little beyond V'. But we cannot include any additional current since only the current in V' is to be considered, so $\vec{J}$ must be zero on S'. In either case, the surface integral is zero, which makes the left side of the preceding equation zero. With this, Eq. 6-21 evaluates to (since the 4πs cancel),

$$\oint_C \vec{H} \cdot d\vec{l} = \int_S \vec{J}(\vec{r}) \cdot d\vec{a} \tag{6-22}$$

This is Ampere's circuital law in integral form for static magnetic fields. Since the surface S is the surface within (bounded by) C, the right integral is customarily interpreted as the total current enclosed by C and an alternate form is

$$\oint_C \vec{H} \cdot d\vec{l} = I_{\text{enclosed}} \tag{6-23}$$

This form is similar to Gauss' law for electrostatics developed in Chapter 5.

Thus knowing $\vec{H}$ we can obtain the total current within a selected path C if that should be of interest.

Next we apply Stokes' theorem to the left side of Eq. 6-22 to obtain

$$\int_S \nabla \times \vec{H} \cdot d\vec{a} = \int_S \vec{J}(\vec{r}) \cdot d\vec{a}$$

Since the surfaces are the same and since we can select any contour (hence any surface) we wish, the integrands must be equal, so that

$$\nabla \times \vec{H} = \vec{J} \tag{6-24}$$

This is Ampere's circuital law in differential or point form for static fields. In words, this tells us that if there is a current density at a given point then there is also a magnetic field intensity at that same point whose curl is the current density. This again emphasizes that currents are sources for $\vec{H}$.

Equation 6-21 can also be used in some special cases to determine $\vec{H}$ for a given $\vec{J}$. This usually requires a judicious selection of the contour C. The following examples illustrate this.

Example 6-12

Consider a thin wire along the z axis which carries a current I_0 in the positive $\hat{z}$ direction. The top view of this configuration is shown is Figure 6-29, where we have selected a circular path around the wire as shown for the contour C at a radius ρ. The surface enclosed by this contour is taken to be the disk within C. We use Eq. 6-22. From symmetry the magnetic field intensity will not be a function of ϕ. Also, since the line is of infinite length the field will not vary with z. We assume the field has only a $\hat{\phi}$ component. Then we write

$$\vec{H} = H_\phi(\rho)\hat{\phi} \tag{6-25}$$

Since the area is taken in the positive sense,

$$d\vec{a} = \rho d\phi d\rho \hat{z}$$

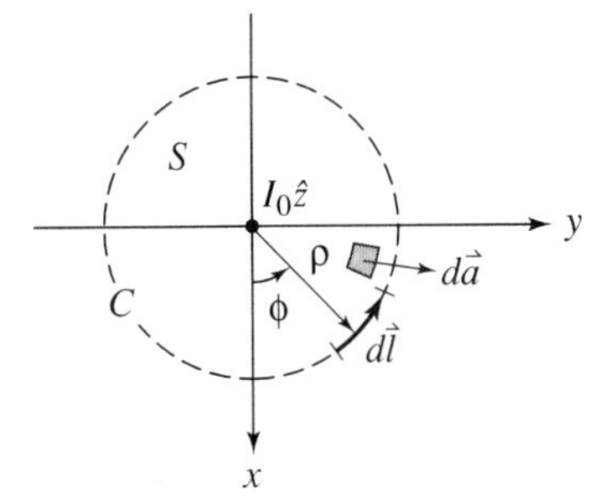

Figure 6-29. System for Example 6-12. Current is out of page along z axis.

and

$$d\vec{l} = \rho d\phi\hat{\phi}$$

Using the techniques of Chapters 4 and 5:

$$\vec{J}(\vec{r}) = I_0\ \delta(\rho)\ \frac{\delta(\phi)}{\rho}\ \hat{z}\ \text{A/m}^2$$

Substituting these components into Eq. 6-22:

$$\int_0^{2\pi} H_\phi(\rho)\hat{\phi}\cdot\rho d\phi\hat{\phi} = \int_0^\rho \int_0^{2\pi} I_0\ \delta(\rho)\ \frac{\delta(\phi)}{\rho}\ \hat{z}\cdot\rho d\phi d\rho\hat{z}$$

Carrying out the vector operations

$$\int_0^{2\pi} H_\phi(\rho)\rho d\phi = I_0 \int_0^\rho \int_0^{2\pi} \delta(\phi)\ \delta(\rho) d\phi d\rho$$

Since $\vec{H}$ is not a function of ϕ it may be removed from the integral sign. The right side is evaluated using the properties of the delta function (the integrand is the constant 1). The results are:

$$\rho H_\phi(\rho) 2\pi = I_0$$

$$H_\phi = \frac{I_0}{2\pi\rho}$$

Putting this result into Eq. 6-25:

$$\vec{H} = \frac{I_0}{2\pi\rho}\ \hat{\phi}$$

This agrees with the result obtained using the more complex integration of the current distribution as in Example 6-8.

Example 6-13

A wire of radius a carries a total current I_0 assumed to be uniformly distributed over the cross section. The top view of the system is shown in Figure 6-30. Determine the magnetic

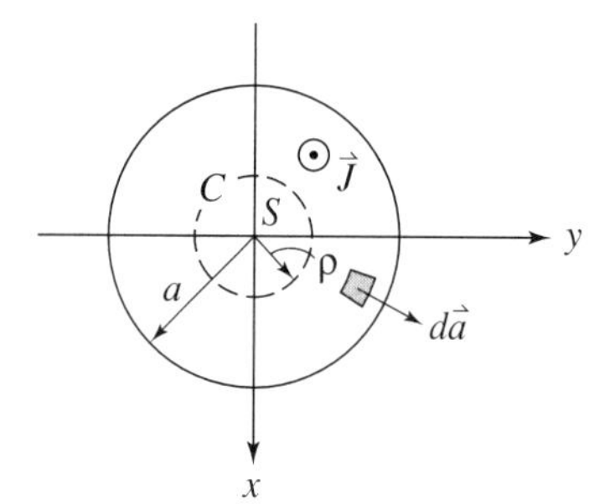

Figure 6-30. Wire carrying current out of page.

field for all values of ρ, and plot it. Using the same arguments as in the preceding example, we would have

$$\vec{H} = H_\phi(\rho)\hat{\phi}$$
$$d\vec{a} = \rho d\phi d\rho \hat{z}$$
$$d\vec{l} = \rho d\phi \hat{\phi}$$

For the uniform current:

$$\vec{J} = \frac{I_0}{\pi a^2} U(a - \rho)\hat{z}$$

Since the current density equation changes at $\rho = a(\vec{J} = 0)$ we will need to evaluate Ampere's law for each region.

For $0 \leq \rho \leq a$:

$$\int_0^{2\pi} H_\phi(\rho)\hat{\phi} \cdot \rho d\phi \hat{\phi} = \int_0^{\rho}\int_0^{2\pi} \frac{I_0}{\pi a^2} U(a - \rho)\hat{z} \cdot \rho d\phi d\rho \hat{z}$$

The step function is 1 for the range of ρ noted. Using the process of Example 6-12,

$$\int_0^{2\pi} H_\phi(\rho)\rho d\phi = \frac{I_0}{\pi a^2}\int_0^{\rho}\int_0^{2\pi} \rho d\phi d\rho$$

$$2\pi\rho H_\phi(\rho) = \frac{I_0}{\pi a^2} \cdot 2\pi \cdot \frac{1}{2}\rho^2$$

$$H_\phi(\rho) = \frac{I_0}{2\pi a^2}\rho$$

For $a \leq \rho \leq \infty$:

$$\int_0^{2\pi} H_\phi(\rho)\hat{\phi} \cdot \rho d\phi \hat{\phi} = \int_0^{a}\int_0^{2\pi} \frac{I_0}{\pi a^2}\hat{z} \cdot \rho d\phi d\rho \hat{z} + \int_a^{\rho}\int_0^{2\pi} 0 \cdot \rho d\phi d\rho \hat{z}$$

$$2\pi\rho H_\phi(\rho) = \frac{I_0}{\pi a^2} \cdot 2\pi \cdot a^2 = I_0$$

$$H_\phi(\rho) = \frac{I_0}{2\pi\rho}$$

These results are plotted in Figure 6-31.

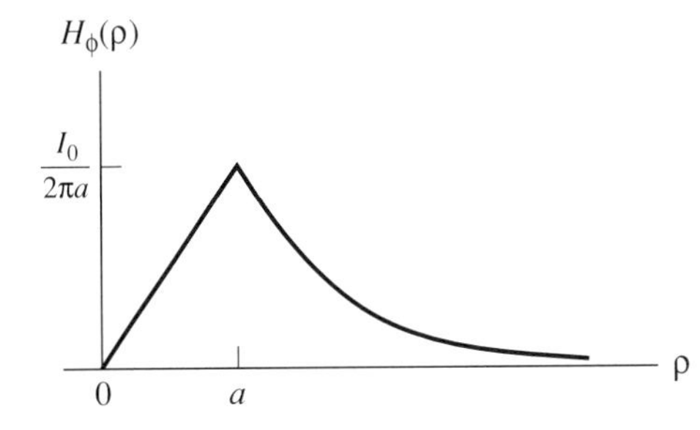

Figure 6-31. Magnetic field intensity for a wire of radius a carrying uniformly distributed current I_0.

Example 6-14

Determine and plot the magnetic field intensity around a coaxial cable carrying current I_0 assumed uniformly distributed in both conductor cross sections shown in Figure 6-32. For this configuration:

$$\vec{J} = \frac{I_0}{\pi a^2} U(a - \rho)\hat{z} - \frac{I_0}{\pi c^2 - \pi b^2} U(c - \rho)U(\rho - b)\hat{z}$$

In this problem there are four regions to consider. Since $\vec{J}$ is constant within each region we will not do the formal integration but simply multiply by the corresponding areas to obtain the *enclosed* current. Note that currents *outside* are not used. For each region the general expression for the magnetic field is

$$\vec{H} = H_\phi(\rho)\hat{\phi}$$

with corresponding path

$$d\vec{l} = \rho d\phi\hat{\phi}$$

so that

$$\oint \vec{H} \cdot d\vec{l} = \int_0^{2\pi} H_\phi(\rho)\rho d\phi = 2\pi\rho H_\phi(\rho) = I_{\text{enclosed}}$$

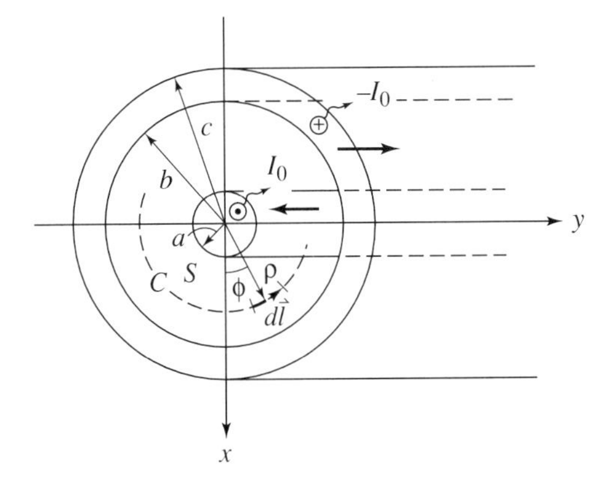

Figure 6-32. Configuration for determining the magnetic field intensity around a coaxial cable.

For $0 \le \rho \le a$:

$$I_{\text{enclosed}} = \left(\frac{I_0}{\pi a^2}\right)\pi\rho^2 = I_0\,\frac{\rho^2}{a^2}$$

Then

$$2\pi\rho H_\phi(\rho) = I_0\,\frac{\rho^2}{a^2}$$

$$H_\phi(\rho) = \frac{I_0}{2\pi a^2}\,\rho$$

For $a \le \rho \le b$:

$$I_{\text{enclosed}} = \left(\frac{I_0}{\pi a^2}\right)\pi a^2 + 0 \cdot (\pi\rho^2 - \pi a^2) = I_0$$

Then

$$2\pi\rho H_\phi(\rho) = I_0$$

$$H_\phi(\rho) = \frac{I_0}{2\pi\rho}$$

For $b \le \rho \le c$:

$$I_{\text{enclosed}} = \left(\frac{I_0}{\pi a^2}\right)(\pi a^2) + 0 + \frac{-I_0}{\pi c^2 - \pi b^2}\,(\pi\rho^2 - \pi b^2)$$

$$= I_0 - I_0\left(\frac{\rho^2 - b^2}{c^2 - b^2}\right) = I_0\left(\frac{c^2 - \rho^2}{c^2 - b^2}\right)$$

Then

$$2\pi\rho H_\phi(\rho) = I_0\left(\frac{c^2 - \rho^2}{c^2 - b^2}\right)$$

$$H_\phi = \frac{I_0}{2\pi(c^2 - b^2)}\left(\frac{c^2 - \rho^2}{\rho}\right)$$

For $c \le \rho \le \infty$:

$$I_{\text{enclosed}} = \left(\frac{I_0}{\pi a^2}\right)(\pi a^2) + 0 - \frac{I_0}{\pi c^2 - \pi b^2}\,(\pi c^2 - \pi b^2) = 0$$

Then

$$2\pi\rho H_\phi(\rho) = 0$$

$$H_\phi(\rho) = 0$$

These results are plotted in Figure 6-33. This shows that ideally there is no magnetic field external to the coax to interact with an external circuit.

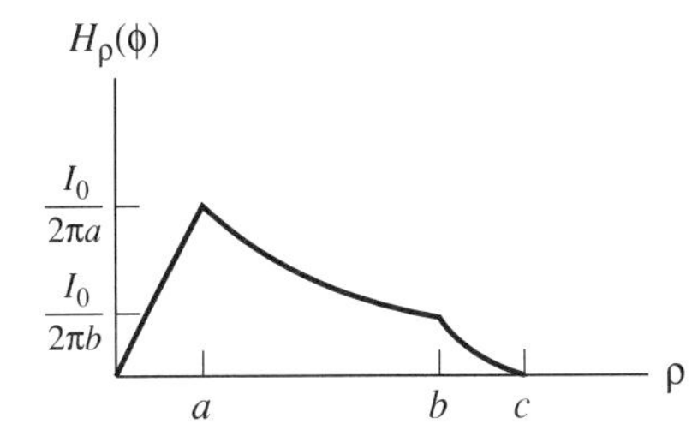

Figure 6-33. Magnetic field intensity around a coaxial cable.

Example 6-15

Determine the magnetic field intensity around a flat sheet of width $2a$, extending $\pm\infty$ in the z direction and carrying uniform (constant) surface current $\vec{J}_s$ A/m. (See Figure 6-17 of Example 6-6, with $b = a$ but with $\vec{J}_s = J_{s_0}\hat{z}$.) Put the strip width along the x axis for this example.

From the preceding examples we observe that what one tries to do is select a coordinate system that has the system current configuration as a coordinate surface. Specifically, in those examples ρ is a constant that defines the current surfaces. For this example, if we look down along the z axis we see a line between $x = \pm a$. If we examine the elliptic cylindrical coordinate system shown in Figure 6-34a we see that for $u = 0$ we have just such a strip.

Note that for $u = 0$ the current does not vary as v varies across the sheet (corresponding in cylindrical coordinates to current not varying as ϕ varies). Thus we assume that on a constant u path shown in Figure 6-34b the magnetic field intensity will not vary with v and will be directed along $\hat{v}$ (right-hand rule). We then write

$$\vec{H} = H_v(u)\hat{v}, \; \vec{J} = J_{s_0}\frac{\delta(u)}{h_u}\hat{z} = J_{s_0}\frac{\delta(u)}{a\sqrt{\sinh^2 u + \sin^2 v}}\hat{z}$$

From Chapter 2, with u and z constant for the path and z constant for the surface,

$$\begin{aligned} d\vec{l} &= h_u du\hat{u} + h_v dv\hat{v} + h_z dz\hat{z} = h_v dv\hat{v} \\ &= a\sqrt{\sinh^2 u + \sin^2 v}\, dv\hat{v} \\ d\vec{a} &= h_u h_v du dv\hat{z} = a^2(\sinh^2 u + \sin^2 v) du dv\hat{z} \end{aligned}$$

Substituting these into Ampere's circuital law, (Eq. 6-21),

$$\int_0^{2\pi} aH_v(u)\sqrt{\sinh^2 u + \sin^2 v}\, dv = \int_0^{\pi}\int_0^{u} aJ_{s_0}\sqrt{\sinh^2 u + \sin^2 v}\;\delta(u) du dv$$

Note that on the left integral the limits on v are 0 to 2π since we must make a closed path, whereas on the right integral limits are 0 to π since that range covers the sheet from a to $-a$.

$$aH_v(u)\int_0^{2\pi}\sqrt{\sinh^2 u + \sin^2 v}\, dv = aJ_{s_0}\int_0^{\pi}\sqrt{\sin^2 v}\, dv = aJ_{s_0}\int_0^{\pi}|\sin v| dv$$

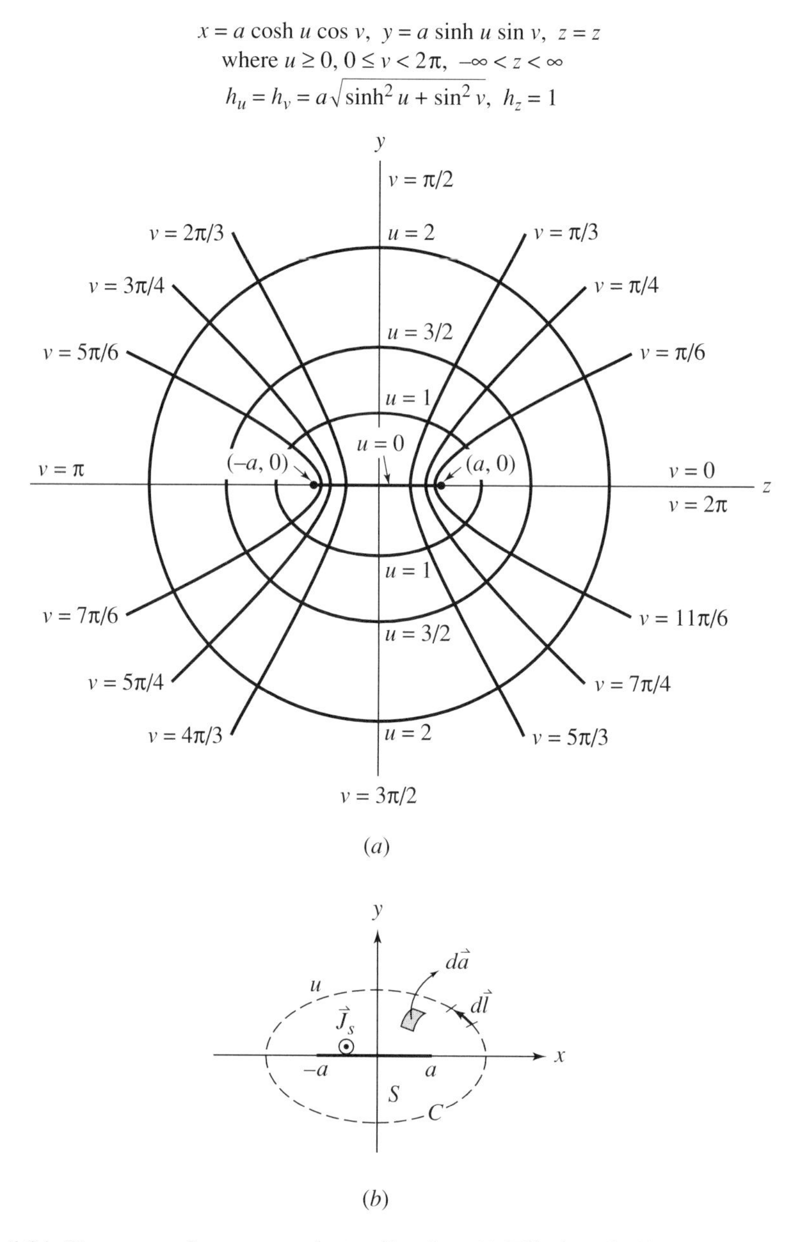

Figure 6-34. Flat current sheet, current in $+z$ direction. (*a*) Elliptic cylindrical coordinate system. (*b*) Current sheet $\pm a$ on x axis, extending along entire z axis.

Since the sine function is positive for $0 \leq v \leq \pi$, the absolute value bars may be removed and the right integral evaluated in the usual way. Then

$$aH_v(u) \int_0^{2\pi} \sqrt{\sinh^2 u + \sin^2 v}\, dv = 2aJ_{s_0}$$

(Note that the enclosed current in this case could have been obtained by inspection since the surface current $\vec{J}_s$ is constant across the sheet of width $2a$.)

Since u is a constant in the remaining integration we can factor out sinh u and then solve for $H_v(u)$ and then write the field vector as

$$\vec{H} = H_v(u)\hat{v} = \frac{2J_{s_0}}{\sinh u \int_0^{2\pi} \sqrt{1 + \operatorname{csch}^2 u \, \sin^2 v}\, dv} \hat{v}$$

The remaining integral cannot be expressed in terms of elementary functions, so one must either use a numerical evaluation by selecting a value of u at which to evaluate the field, or obtain a more advanced set of tables having special functions of this type. (In this case, see [24], formula 2.597 (2).)

This result has an interesting, recognizable form for large u (large distance from the sheet). Suppose we take a point on the y axis, where $v = \pi/2$. Now as u increases, sinh u increases and csch u approaches zero so the integral approached 2π in value. Also, x is zero and $y = a \sinh u$. With these results we have

$$\vec{H} \geq \frac{2J_{s_0}}{\left(\frac{y}{a}\right)(2\pi)} \hat{v} = \frac{2aJ_{s_0}}{2\pi y} \hat{v} = \frac{I_0}{2\pi y} \hat{v}$$

which says that a long way from the strip the strip appears just like a filament line current, I_0. (See Example 6-8.) From the coordinate system it can be seen that for large u the ellipses approach circles, so this result is reasonable.

Of course, for a general value of u, one could always make a coordinate conversion as explained in Chapter 4 to get the rectangular components. From these the equations for the lines can be obtained using the technique already presented in Chapter 5.

Example 6-16

Figure 6-35 is a picture of a thin sheet of current. The sheet is of large extent, and we assume that the return current path is a very large distance away. The current sheet is parallel to the xy plane.

Applying the right-hand rule, one would conclude that on top of the sheet the magnetic field would be to the left and below the sheet to the right as shown.

We select the contour C to be rectangular around the surface current with the plane of the contour S, normal to the current.

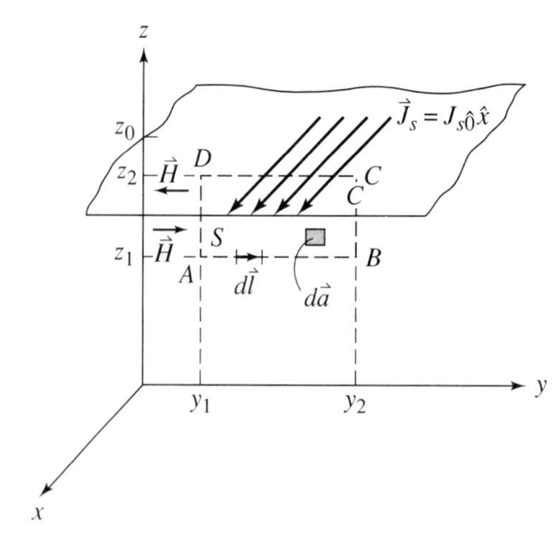

Figure 6-35. Uniform (constant) surface current sheet of large extent.

For this configuration,

$$\vec{J} = J_{s_0}\, \delta(z - z_0)\hat{x}$$
$$d\vec{l} = dy\hat{y} \text{ and } d\vec{l} = dz\hat{z}$$
$$d\vec{a} = dydz\hat{x}$$
$$\vec{H} = \begin{cases} H_y\hat{y} & z \leq z_0 \\ -H_y\hat{y} & z \geq z_0 \end{cases}$$

Using Ampere's circuital law,

$$\oint \vec{H} \cdot d\vec{l} = \int_A^B \vec{H} \cdot dy\hat{y} + \int_B^C \vec{H} \cdot dz\hat{z} + \int_C^D \vec{H} \cdot dy\hat{y}$$
$$+ \int_D^A \vec{H} \cdot dz\hat{z} = \int_{y_1}^{y_2} \int_{z_1}^{z_2} J_{s_0}\, \delta(z - z_0) dz dy$$

Now since there is no H_z (adjacent current lines will have opposing $\hat{z}$ components) the path integrals $B \rightarrow C$ and $D \rightarrow A$ will be zero. The preceding reduces to

$$\int_{y_1}^{Y_2} H_y\hat{y} \cdot dy\hat{y} + \int_{y_2}^{y_1} - H_y\hat{y} \cdot dy\hat{y} = J_{s_0}(y_2 - y_1)$$

Since H_y will not be a function of y it may be removed from the integrals. This yields

$$H_y(y_2 - y_1) - H_y(y_1 - y_2) = J_{s_0}(y_2 - y_1)$$
$$2H_y(y_2 - y_1) = J_{s_0}(y_2 - y_1)$$
$$H_y = \tfrac{1}{2} J_{s_0}$$

This can be put in a vector form by identifying a unit normal vector $\hat{N}$, which is away from the sheet on the side one is imagined to be standing; that is, $\hat{N} = \hat{z}$ for $z > z_0$ or $\hat{N} = -\hat{z}$ for $z < z_0$. This is a convenient right-hand rule again.

$$\vec{H} = \tfrac{1}{2}\vec{J}_s \times \hat{N} \qquad \text{single sheet} \tag{6-26}$$

Note that this result is independent of the distance from the sheet.

Example 6-17

Next we take two surface current sheets that are near each other as shown in Figure 6-36*a*. Both sheets are assumed to be very large in extent. It was pointed out earlier that superposition applies so that the magnetic fields of the separate sheets may be added. The field structure is shown in Figure 6-36*b*. Since the field due to a single sheet is independent of the distance from the sheet, all of the field magnitudes are the same. Thus outside the sheets there is no magnetic field, but the field intensity is doubled between the plates. We then have

$$\vec{H} = \vec{J}_s \times \hat{N} \qquad \text{double sheet} \tag{6-27}$$

Example 6-18

Using the result of Example 6-4 for the magnetic field intensity on the axis of a circular current filament, compute the field intensity on the axis of a very long (infinite) solenoid. The solenoid consists of tightly wound turns, n per meter, and the wire current is I_0 A. The configuration is shown in Figure 6-37. The surface current may be approximated by $nI_0\hat{\phi}$ A/m. For the differential ring of thickness dz, the equivalent ring current would be

Figure 6-36. Two-current sheet system. (*a*) System showing output and return currents. (*b*) Edge view of sheets showing fields produced by both currents.

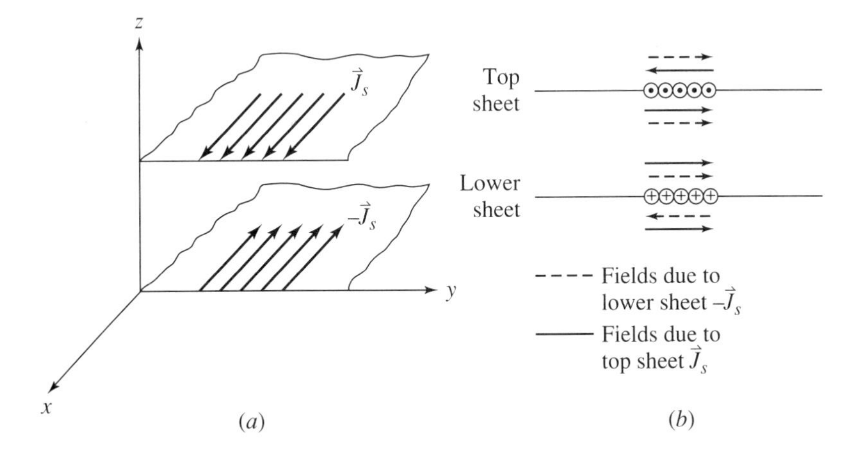

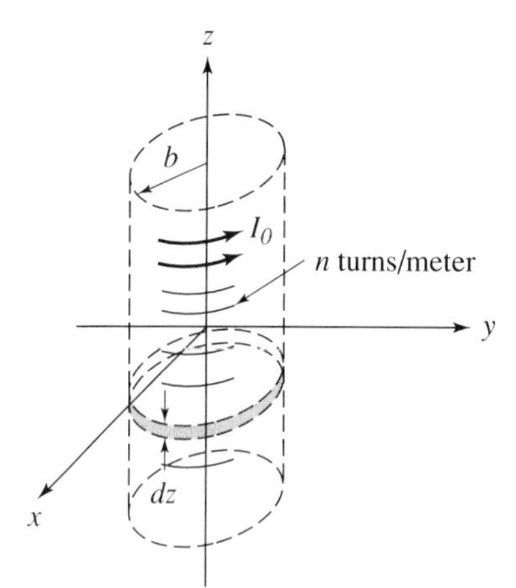

Figure 6-37. Infinite solenoid approximated by a current sheet.

nI_0dz ampere. Using the result of Example 6-4 the magnetic field intensity at the center is

$$dH_z = \frac{(nI_0dz)b^2}{2(z^2 + b^2)^{3/2}}$$

Summing up all the differential loops,

$$H_z = \int_{-\infty}^{\infty} \frac{nb^2I_0}{2(z^2 + b^2)^{3/2}}\,dz$$

$$= \frac{nb^2I_0}{2} \cdot 2\int_0^{\infty} \frac{1}{(z^2 + b^2)^{3/2}}\,dz = nb^2I_0 \lim_{z_0\to\infty}\left[\frac{z}{b^2(b^2 + z^2)}\right]_0^{z_0}$$

$$H_z = nI_0 \tag{6-28}$$

Example 6-19

Determine the magnetic field intensity everywhere inside and outside a very long solenoid for points far from the end. The solenoid configuration is the same as that given in Figure 6-37. The axial cross section is shown in Figure 6-38. To use Ampere's law, the required contour and surface are shown in the figure, where the path section $1 \to 2$ has been selected on the z axis since $\vec{H}$ is known along that section from Example 6-17. Applying Ampere's law:

$$\oint_C \vec{H} \cdot d\vec{l} = \int_1^2 \vec{H} \cdot dz\hat{z} + \int_2^3 \vec{H} \cdot d\rho\hat{\rho} + \int_3^4 \vec{H} \cdot dz\hat{z}$$
$$+ \int_4^1 \vec{H} \cdot d\rho\hat{\rho} = \int_S \vec{J} \cdot d\vec{a} = J_sL = nI_0L$$

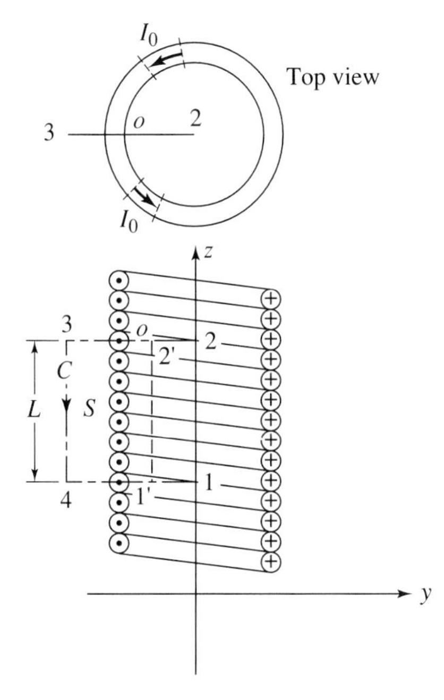

Figure 6-38. Axial cross section of long solenoid.

Now from physical considerations, there will not be a $\hat{\rho}$ component of $\vec{H}$ or a $\hat{\phi}$ component. That this is so can be seen by considering the top view of one turn as shown in the figure. For a typical point o on the turn, there will always be current elements on opposite sides of the point that will (or would) tend to produce $\hat{\rho}$ and $\hat{\phi}$ components, which would cancel since the currents flow opposite directions. Thus the integrals for paths $2 \rightarrow 3$ and $4 \rightarrow 1$ will be zero due to the dot product. From Example 6-18, the integral along the z axis would be

$$\int_1^2 nI_0\hat{z} \cdot dz\hat{z} = nI_0 \int_1^2 dz = nI_0L$$

Thus Ampere's law reduces to

$$nI_0L + \int_3^4 \vec{H} \cdot d\vec{l} = nI_0L$$

$$\int_3^4 \vec{H} \cdot d\vec{l} = 0$$

Since the actual path $3 \rightarrow 4$ has not been specified the preceding integral must be zero for any path. Thus

$$\vec{H}_{\text{outside}} \equiv 0$$

If we return to the original sum of integrals—now using the facts that for paths $2 \rightarrow 3$, $3 \rightarrow 4$ and $4 \rightarrow 1$ are all zero—we would have

$$\int_1^2 \vec{H} \cdot dz\hat{z} = nI_0L$$

Suppose we select a different path parallel to the z axis, say path $1' \rightarrow 2'$. The preceding equation must still hold since all the other integrals are zero. Since the solenoid is very long $\vec{H}$ will not be a function of z. We have already shown that $\vec{H}$ has no $\hat{\phi}$ or $\hat{\rho}$ components. Thus at most the field could be a function of ρ because of circular symmetry. Then we set

$$\vec{H}_{\text{inside}} = H_z(\rho)\hat{z}$$

Substituting this into the preceding integral,

$$\int_{1'}^{2'} H_z(\rho)\hat{z} \cdot dz\hat{z} = H_z(\rho) \int_{1'}^{2'} dz = nI_0L$$

$$H_z(\rho)L = nI_0L \Rightarrow H_z(\rho) = nI_0$$

Thus $\vec{H}$ is constant inside the solenoid everywhere

$$\vec{H}_{\text{inside}} = nI_0\hat{z} \ \frac{\text{ampere} - \text{turns}}{\text{meter}} \left(\frac{\text{A} \cdot \text{t}}{\text{m}}\right)$$

Exercises

6-6.0a (a) Derive Eq. 6-24 by taking the curl of Eq. 6-15 with respect to unprimed coordinates and then apply vector theorems and the techniques used in the first part of this section.
(b) Now integrate both sides of the result (Eq. 6-24) with respect to unprimed area (use the dot product). Then apply Stokes' theorem to obtain Eq. 6-22, Ampere's law.

6-6.0b A cylinder of radius a, whose axis coincides with the z axis, carries a surface current $\vec{J}_s = J_1\hat{z}$. Determine the magnetic field everywhere. Identify the contour as a circular path and the surface as the disk within. The cylinder is a thin shell.

6-6.0c A wire is bent into the shape of a parabola, and carries a current I_0 as shown in Figure 6-39. Using the parabolic cylindrical coordinate system, find the magnetic field intensity on the x axis (positive and negative). The wire passes through the points $x = 0$, $y = \pm 4$, and the wire opens (concave) to the right.

6-6.0d For Example 6-15, change the direction of the current and solve for the magnetic field intensity. Show the problem set-up carefully. (Answer: Negative of the result in the example)

6-6.0e For Example 6-17, change the direction of the current on the lower sheet and solve for the magnetic field intensity everywhere. (Answer: Zero between the sheets, J_s outside sheets)

6-6.0f The following magnetic fields are given. Determine the current densities which generate these fields. Use the point or differential form of Ampere's circuital law.
(a) $-3xy\hat{x} + 2xz\hat{y} + 7\hat{z}$ (rectangular)
(b) $(\sin\phi - 1)\hat{\rho} + 2(\rho - 1)\hat{\phi} + 3z\hat{z}$ (cylindrical)
(c) $\cos\theta\hat{\phi} - \cos\theta\hat{\theta}$ (spherical)

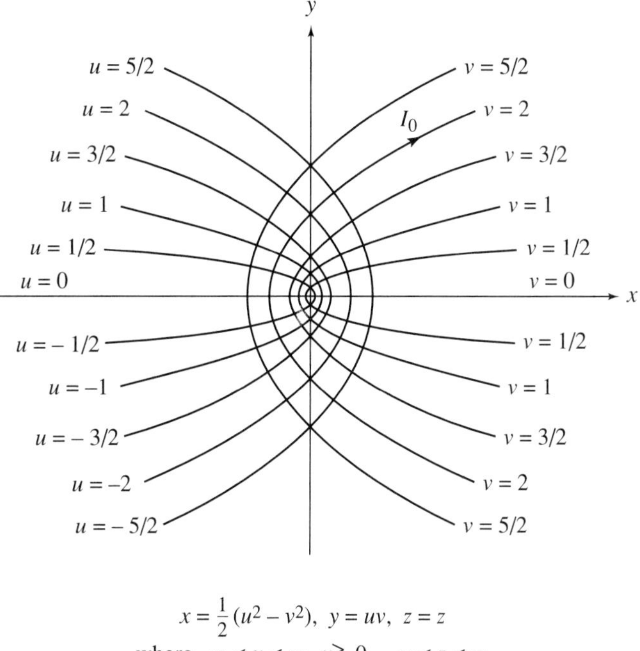

$$x = \frac{1}{2}(u^2 - v^2), \;\; y = uv, \;\; z = z$$
$$\text{where } -\infty < u < \infty, \;\; v \geqq 0, \;\; -\infty < z < \infty$$
$$h_u = h_v = \sqrt{u^2 + v^2}, \;\; h_z = 1$$

Figure 6-39.

6-7 Properties of Magnetic Materials: Magnetic Moment, Magnetic Susceptibility, Magnetization, Permeability, Types of Magnetic Phenomena

In Section 6-4, the constitutive relationship between $\vec{B}$ and $\vec{H}$, Eq. 6-12, was derived to include a parameter μ called the permeability. No mention was made of how this parameter arises physically, other than the fact that for free space (vacuum) the value was defined to be $4\pi \times 10^{-7}$ and denoted by μ_0. In this section we present some of the fundamental magnetic ideas that give a few insights into this parameter and also some additional practical significance to the constitutive relationship.

Most of the quantitative work having to do with the atomic level of magnetic phenomena is based on quantum mechanics and advanced experiments (see [17]). For purposes of gaining physical insights, we take a more qualitative approach, which for most applications serves our needs very well. To this end we return to the current carrying circular loop of wire presented in Example 6-4. If we multiply the numerator and denominator by π the field on the axis would be

$$\vec{H} = \frac{\pi b^2 I_0 \hat{z}}{2\pi(z^2 + b^2)^{3/2}} \text{ A/m}$$

Note that the numerator is a constant for the given parameters of the system and is the loop area times the current around the loop. This field has associated with it a magnetic flux density $\vec{B}$ that is μ times the preceding equation. This look flux density would interact with an external $\vec{B}$ field brought near to the loop. The axis of the loop effectively being a north–south magnet would tend to line up in the applied field. Thus the loop may be thought of as a magnetic dipole. By definition the dipole moment is the product of area and current, the vector direction being that of the area in the right hand sense; that is,

$$\vec{m} = IS\hat{n} \text{ A} \cdot \text{m}^2 \tag{6-29}$$

This is shown in Figure 6-40.

On an atomic scale there are four primary sources of magnetic dipole moment; namely, electron, nucleus, atom and molecule.

The electron magnetic dipole moment, $\vec{m}_e$, can be qualitatively pictured as being produced by an electron spinning on its own axis. One could imagine some equivalent filamentary current being produced by the spinning charge at some average radius. Although this concept does not produce quantitative values, it serves as a visual idea. The conventional current flow (positive charge) would be opposite the rotation direction of the negative charge. Now if this dipole moment were immersed in a field $\vec{B}$ it would tend to line up in the field so the total local field would be the sum (vector) of the two fields present. For an electron the spin magnetic moment is called the *Bohr magneton* and has a value of 9.27×10^{-24} A · m^2

As with the electron, the nucleus may also be spinning and produce a magnetic dipole $\vec{m}_n$. This effect is also very small.

The third source, the atomic magnetic dipole $\vec{m}_a$, arises from the electrons revolving in orbits around the nucleus. Since orbits may occur in spherical plane sections the net moment would have to be considered; that is, one must add the moments of the various orbital paths in vector sense. One would expect this to be the dominant effect since the area of the orbit would be larger than the equivalent electronic or nuclear radii areas.

Finally we have the molecular magnetic moment, $\vec{m}_m$, which is produced by rotating molecules or ions. Oxygen gas, O_2, is a good example of this source of moment. All

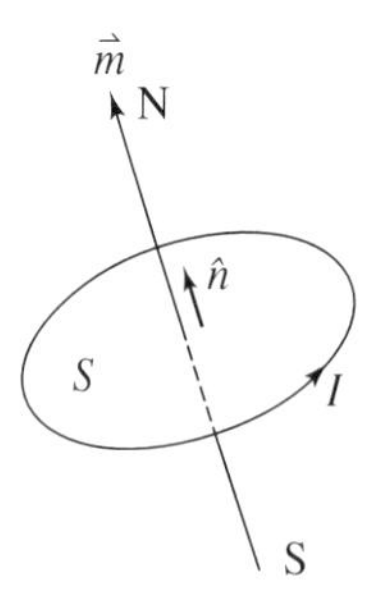

Figure 6-40. Illustration of the quantities used to define the magnetic dipole moment.

charges in the molecule are in rotation about a molecular axis and the net ''current flow'' in that rotation produces the moment.

From the preceding discussion we conclude that the total magnetic moment produced by a material molecule would be the sum of these effects:

$$\vec{m}_T = \left(\sum_{i=1}^{z} \vec{m}_e \right) + \vec{m}_n + \vec{m}_a + \vec{m}_m \tag{6-30}$$

In the summation of electron moments over the atomic number, z, the only electrons that actually need to be considered are those in the unfilled shells of the atom since those moments in filled shells cancel. This result is from quantum mechanics which predicts electrons in a shell have $\pm$ values of magnetic moment, and only unpaired ones contribute to net value of $\vec{m}$.

At this point we are not interested in the atomic level, but on the large scale or macroscopic level. Suppose that the number of dipole moments per unit volume is denoted by N. In a volume Δv there would be $N\,\Delta v$ magnetic dipoles. The net moment in Δv would be

$$\vec{m}_{\text{net}} = \sum_{i=1}^{N\,\Delta v} \vec{m}_{Ti}$$

We define the *magnetization* of a material as the magnetic moment per unit volume:

$$\vec{M} \equiv \lim_{\Delta v \to 0} \left[\frac{\left(\sum_{i=1}^{N\,\Delta v} \vec{m}_{Ti} \right)}{\Delta v} \right] \text{A/m} \tag{6-31}$$

The units A/m come from considering the units of the quantities in the definition

$$\frac{\vec{m}}{\Delta v} \Rightarrow \frac{\text{A} \cdot \text{m}^2}{\text{m}^3} = \frac{\text{A}}{\text{m}}$$

The units of magnetization are the same as $\vec{H}$. This is probably not too surprising since currents are sources of $\vec{H}$ and $\vec{M}$. Now suppose a material which has magnetization $\vec{M}$ is subjected to an external magnetic flux density field. Since the effect of the external field is to align the magnetic dipole moments, the vectors $\vec{m}_{Ti}$ in Eq. 6-31 would tend toward parallelism so that the value of the sum would change. Thus the total magnetic flux density would be the sum of the free space field plus the effects of magnetization. As an equation we may write, since $\vec{H}$ and $\vec{M}$ have the same units,

$$\vec{B} = \mu_0 \vec{H} + \mu_0 \vec{M} \tag{6-32}$$

In words this tells us that the total flux density is equal to the free space flux density produced by $\vec{H}$ plus the flux density contributed by the magnetic moments.

As just explained, the magnetization $\vec{M}$ changes with an applied $\vec{B}$, so one would expect a proportional relationship:

$$\mu_0 \vec{M} = \chi_m \vec{B} = \chi_m \mu_0 \vec{H}$$

Note we have used the free space value of $\vec{B}$ since this is in the space around the moments. Solving the outer equality:

$$\vec{M} = \chi_m \vec{H} \tag{6-33}$$

where we do not consider permanent magnetization. In this relationship χ_m is called the *magnetic susceptibility* and is analogous to the dielectric susceptibility discussed in Chapter 5. It is dimensionless, since $\vec{M}$ and $\vec{H}$ are the same. Substituting Eq. 6-33 into Eq 6-32 we obtain

$$\vec{B} = \mu_0 \vec{H} + \mu_0 \chi_m \vec{H} = \mu_0(1 + \chi_m)\vec{H} = \mu_0 \mu_r \vec{H} = \mu \vec{H} \tag{6-34}$$

The quantity μ_r is called the *relative permeability*, and this is what one usually finds in a table of material magnetic properties.

$$\mu_r = 1 + \chi_m = \frac{\mu}{\mu_0} \tag{6-35}$$

As with the permittivity ϵ, the permeability value may depend upon the direction that the field is applied in the material, so that μ may require a matrix representation—called *tensor permeability*—denoted by

$$[\mu] = \mu_0[\mu_r] = \mu_0(I + [\chi_m]) \tag{6-36}$$

where I is the unit or identity matrix. This describes an anisotropic medium.

Since $\vec{M}$ has the same units as $\vec{H}$ we now reexamine Ampere's circuital law. Solving Eq. 6-32 for $\vec{H}$ and substituting into Eq. 6-22 we obtain:

$$\oint \vec{H} \cdot d\vec{l} = \oint \frac{\vec{B}}{\mu_0} \cdot d\vec{l} - \oint \vec{M} \cdot d\vec{l} = \int_S \vec{J}(\vec{r}) \cdot d\vec{a} = I_{\text{enclosed}}$$

Thus in magnetic materials the enclosed current, which is a free flow of charge, may be considered equal to the total current passing through the surface minus the currents caused by charges bound in an orbit. We might represent this by replacing the integrals by currents:

$$I_{\text{total}} - I_{\text{bound}} = I_{\text{enclosed}} \qquad (\text{or } I_{\text{free}})$$

In terms of current densities,

$$\int_S \vec{J}_{\text{total}} \cdot d\vec{a} - \int_S \vec{J}_{\text{bound}} \cdot d\vec{a} = \int_S \vec{J} \cdot d\vec{a}$$

We could use the term $\vec{J}_{\text{free}}$, but this is not usual practice.

This situation is shown in Figure 6-41. The magnetic moments (which produce $\vec{M}$) have components $\vec{m} \cdot d\vec{l}$ along the contour C, and on the enclosed surface S have currents that are bound to the atom but still passing through the surface S. Note that the right-hand rule for the free current density $\vec{J}$ predicts that the dipole moments would be aligned around the contour as shown.

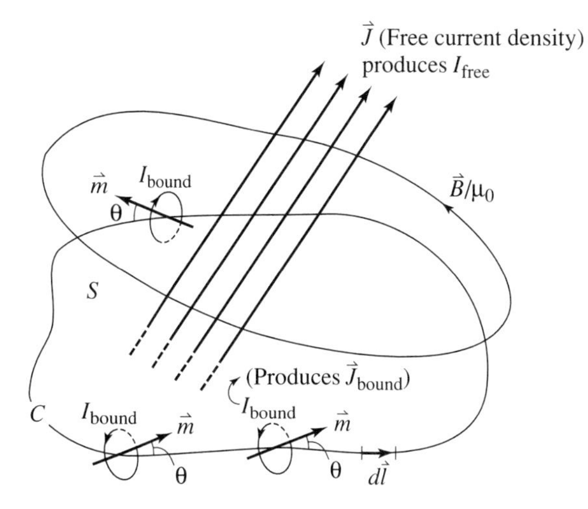

Figure 6-41. Components in Ampere's circuital law in magnetic materials.

Note particularly that around the contour C all the bound currents pass the same direction, in the case shown all up through the surface S.

Similar results may be obtained using the point form of Ampere's circuital law (6-24):

$$\nabla \times \vec{H} = \nabla \times \left(\frac{\vec{B}}{\mu_0} - \vec{M}\right) = \nabla \times \left(\frac{\vec{B}}{\mu_0}\right) - \nabla \times \vec{M} = \vec{J} \qquad \text{(free current density)}$$

$$\text{Then} \begin{cases} \nabla \times \left(\dfrac{\vec{B}}{\mu_0}\right) = \vec{J}_{\text{total}} \\ \nabla \times \vec{M} = \vec{J}_{\text{bound}} \end{cases} \tag{6-37}$$

The magnetic effects of materials due to the magnetic moments of the atoms and molecules are quite varied. They are also temperature dependent as one might expect due to the thermal agitation of the atoms. The most important of these effects are classified as

1. Diamagnetic and paramagnetic
2. Ferromagnetic
3. Antiferromagnetic
4. Ferrimagnetic

The materials in class 1 are those which have very little magnetization capability (i.e., $\mu_r \approx 1$). These materials are characterized by magnetic susceptibilities in the range

$$-10^{-5} < \chi_m < 10^{-5}$$

Note that for negative χ_m Eq. 6-34 would predict a reduced local value of $\vec{B}$, quantitatively corresponding to a permeability less than that of vacuum. Those materials having negative values are called *diamagnetic* materials, and those with positive values called *paramagnetic*. In the case of diamagnetic materials, the small line-up of the dipoles is accompanied by a precession of the magnetic moments about the magnetic moment axis that reduces the net dipole moment that accounts for negative susceptibility. The precession is called the Larmour precession of the electrons in the orbits. This is somewhat equivalent to Lenz's law from physics, which states that the induced current is in such a direction as to create a flux which opposes the excitation flux. This effect is observed in nontransition (closed shell) metals.

The *paramagnetic* effect is observed in many of the transition and some rare earth elements. These incomplete shell atoms have residual electron spin moments. Actually, according to quantum theory the orientations of these permanent magnetic moments are not initially completely random but exist in finite sets of orientations. The lining up of these dipoles yields the paramagnetic effect, much like that described for electric dipoles in dielectrics. The effect is small but measurable. The susceptibility of these materials varies approximately inversely with absolute temperatures in cryogenic ranges. This is called the Curie law, and the equation relating the susceptibility to inverse temperature is known as the Langevin equation. Details will not be explored in this book.

Materials that are capable of residual magnetic moment without any applied field constitute class 2. They have large values of magnetic susceptibility (up to several thousand or more) and are called *ferromagnetic* materials. These materials usually contain iron, cobalt, or nickel, or some of the lanthanide (rare earth) elements. They also exhibit nonlinear and hysteresis effects as the applied (external) field value changes. The theory of these materials was described by P. Weiss in 1907 on the basis of two hypotheses:

1. Magnetic material of macroscopic size contains small magnetized regions (called *magnetic* domains). These regions are spontaneously magnetized in different directions due to the tendency of the large atomic dipole moments to align with each other. This is shown diagramatically in Figure 6-42. The partitions between the domains are called *domain*, or *Bloch*, walls. The net spontaneous (permanent) magnetization is the vector sum of the domain magnetizations.
2. Within each domain the spontaneous magnetization is due to the existence of a molecular field (now called the *Weiss field*), which produces alignment of the microscopic dipoles.

The nonlinear and hysteretic properties of these materials is shown in Figure 6-43. The nonlinearity of the susceptibility is evident from the fact that μ is the slope of a line from the origin to a point on the curve. If there were a small varying magnetic component superimposed on a static (DC) level, the permeability would be called a *dynamic* permeability defined as the local slope (derivative) of the curve.

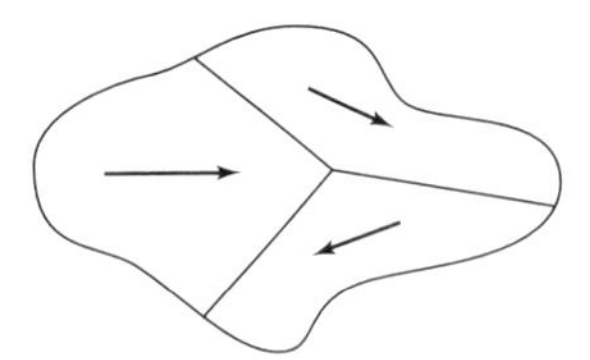

Figure 6-42. Illustration of permanent magnetic domains in a ferromagnetic material separated by Bloch walls.

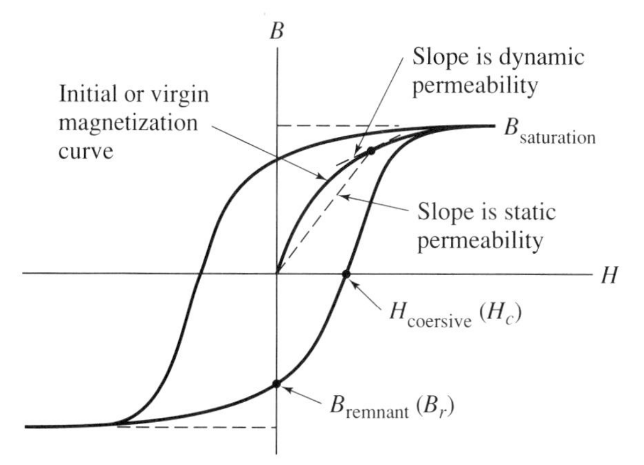

Figure 6-43. Magnetic curve for ferromagnetic material.

From this curve the definitions are:

1. *Initial or virgin magnetization curve.* The magnetization path the first time an external field is applied to the material.
2. $B_{\text{saturation}}$. The maximum magnetic flux density that can exist in the material when all domains are aligned. Called the *saturation point field.*
3. H_{coercive}. The magnetic field intensity required to bring the flux density to zero. Called *coercive force.*
4. B_{remnant}. The magnetic flux density remaining in the material in the absence of an applied $\vec{H}$ field.

As these materials are heated up there is a temperature reached above which there is no longer a residual or remnant field. The thermal agitation destroys the domains and hence the remnant field. This is called the *Curie temperature* and is 1043 K for iron, 1404 K for cobalt, 631 K for nickel, 230 K for terbium, and 20 K for erbium. Materials then become paramagnetic in nature.

The class 3 phenomenon of *antiferromagnetism* is primarily an effect observed in ionic compounds and occurs below a critical temperature called the *Néel temperature*. Below this temperature the magnetic moments line up antiparallel and are of equal moment value, which results in zero net moment. It is difficult to align these moments so χ_m is small. Néel temperatures range from 24 K for FeCl to 725 K for CrSb. This phenomenon has not been of significant engineering interest.

Class 4 *ferrimagnetism* is important in these materials used extensively in communications systems and circuits. This is called the *ferrite* group of materials. These are synthetic materials made by taking magnetite, $Fe^{2+}Fe_2^{3+}O_4$, and replacing the Fe^{2+} ion with another divalent metal ion such as Ni, Mn, Co, Cu, Mg, Zn, or Cd. These are usually sintered materials whose chief advantage is their high electrical resistivity as opposed to the low resistivity of iron materials. The term *ferrimagnetism* was suggested by Néel. These materials exhibit overall ferromagnetic behavior in that they possess relatively large valued of χ_m similar to irons and steels.

Néel's theory of ferrites accounts for this action and was based on a physical explanation of an experimental result. The experimental result was that the saturation value of magnetite has a much lower value than predicted by the presence of the three iron atoms. Néel observed that if the two Fe^{3+} atoms were antiferromagnetic (spins aligned opposite) the magnetic effects due to those would cancel leaving the Fe^{2+} only to require a field for saturation. Now when we replace the Fe^{2+} by some other divalent ion, this would have significant effect. It turns out that all the interactions of moments are antiferromagnetic, but due to incomplete cancellation of ionic moments a net field results, and we gain the advantage of higher resistivity by removing the divalent iron. This material class is also characterized by a Curie temperature, so when a ferrite is selected for use, the environment may be an important consideration.

The final property of magnetic flux density we consider is obtained by considering the fact that in all of the configurations presented in this chapter the flux lines formed closed paths. For the bar magnet the lines leaving the north pole reentered at the south pole and continued through the magnet back to the north pole. If a bar magnet is broken into smaller pieces, each piece contains north and south poles. This implies that there does not exist an isolated north (or south) pole. No such isolated poles have been found experimentally. For the field around a current-carrying conductor, the lines also formed closed paths. Thus if we select a point in a magnetic flux density field the flux line enters and leaves that point. On a larger scale one could take a volume in the field and conclude that there could be no net flux entering (or leaving) the surface around the volume since there are no isolated poles to terminate (south) or generate (north) lines of flux. As an equation we would then write

$$\oint_S \vec{B} \cdot d\vec{a} = 0 \tag{6-38}$$

This is Gauss' law for magnetic flux density (integral form). If we apply the divergence theorem to this equation we obtain

$$\int_V \nabla \cdot \vec{B} dv = 0$$

Since this must be true for any volume, the integrand must be zero. Thus

$$\nabla \cdot \vec{B} = 0 \tag{6-39}$$

This is the differential or point form of Gauss' law for magnetic flux density. Thus, $\vec{B}$ is a solenoidal field.

The preceding result could also be obtained from Eq. 6-15 as follows, using the ideas of primed and unprimed vector operations as in Section 6-6:

$$\nabla \cdot \vec{B} = \nabla \cdot (\mu\vec{H}) = \nabla \cdot \left[\mu \int_{V'} \frac{\vec{J}(\vec{r}\,') \times \hat{R}}{4\pi R^2} dv'\right] = \mu\nabla \cdot \int_{V'} \frac{\vec{J}(\vec{r}\,') \times \hat{R}}{4\pi R^2} dv'$$

$$= \frac{\mu}{4\pi} \int_{V'} \nabla \cdot \left(\frac{\vec{J}(\vec{r}\,') \times \hat{R}}{R^2}\right) dv' = \frac{\mu}{4\pi} \int_{V'} - \nabla \cdot \left[\vec{J}(\vec{r}\,') \times \nabla\left(\frac{1}{R}\right)\right] dv'$$

Using the vector identity for $\nabla \cdot (\vec{A} \times \vec{B})$

$$\nabla \cdot \vec{B} = -\frac{\mu}{4\pi}\left[\int_{V'} \nabla\left(\frac{1}{R}\right) \cdot \nabla \times \vec{J}(\vec{r}')dv' - \int_{V'} \vec{J}(\vec{r}') \cdot \nabla \times \left(\nabla\left(\frac{1}{R}\right)\right)dv'\right]$$

The left integrand is zero since ∇ operates only on unprimed coordinates and $\vec{J}$ is a function only of primed coordinates. The right integrand is zero because the curl of a gradient is always zero. Then we have

$$\nabla \cdot \vec{B} = 0$$

as before.

Example 6-20

The long current filament of Example 6-12 is inserted into a material that has a relative permeability of 50. Determine the following:

1. Magnetic susceptibility of the material.
2. The magnetization $\vec{M}$ of the material around the wire.
   (a) Since $\mu_r = 1 + \chi_m$,

$$\chi_m = \mu_r - 1 = 49$$

   (b) From Example 6-12 the magnetic field intensity around the wire is

$$\vec{H} = \frac{I_0}{2\pi\rho}\,\hat{\phi} \text{ A/m}$$

Then

$$\vec{B} = \mu\vec{H} = \mu_0\mu_r\vec{H} = 4\pi \times 10^{-7} \times 50 + \frac{I_0}{2\pi\rho}\,\hat{\phi} = \frac{10^{-5}I_0}{\rho}\,\hat{\phi} \text{ T}$$

Using Eq. 6-32

$$\vec{M} = \frac{\vec{B}}{\mu_0} - \vec{H} = \frac{25I_0}{\pi\rho}\,\hat{\phi} - \frac{I_0}{2\pi\rho}\,\hat{\phi} = \frac{7.8I_0}{\rho}\,\hat{\phi} \text{ A/m}$$

This is what one would expect physically since at locations of strong magnetic field (close to wire) more magnetic dipoles line up, which produces larger $\vec{M}$.

Example 6-21

An engineer has been asked to devise a system that will generate a magnetic intensity pattern (in vacuum)

$$\vec{H} = 3x\hat{x} + 3y\hat{y} \text{ A/m}$$

What are the chances for success?

$$\vec{B} = \mu_0\vec{H} = 3.77x\hat{x} + 3.77y\hat{y}\ \mu\text{T}$$

$$\nabla \cdot \vec{B} = \frac{\partial B_x}{\partial x} + \frac{\partial B_y}{\partial y} = 7.54\ \mu\text{T/m}$$

Since Eq. 6-39 requires this to be zero, the chances for an engineer to generate $\vec{H}$ exactly are very small (zero).

Exercises

6-7.0a In Example 6-21, what change in $\vec{H}$ would make it a theoretical possibility to generate the $\vec{H}$? What is the current density in the region of the $\vec{H}$? (Answer: Negate either one component of $\vec{H}$ so $\nabla \cdot \vec{B} = 0$. From Eq. 6-24, $\vec{J} = 0$.)

6-7.0b For Example 6-20, compute the bound current density in the magnetic material.(Answer: $\vec{J}_{\text{bound}} = \nabla \times \vec{M} = 0$)

6-7.0c A magnetic material has a relative permeability of 20, and a field $\vec{B} = -10x\hat{z}$ μT. Compute χ_m, μ, $\vec{J}$, $\vec{J}_{\text{bound}}$, $\vec{J}_{\text{total}}$, $\vec{H}$, and $\vec{M}$. (Answer: 19, 25.1×10^{-6} H/m, $398\hat{y}$ A/m^2, 7.56 $\hat{y}$ A/m^2, $7.96\hat{y}$ A/m^2, $-398x\hat{z}$ A/m, $-7.56x\hat{z}$ A/m)

6-7.0d Show that the magnetic field intensity for the straight current-carrying wire satisfies the divergence Eq. 6-39.

6-8 Vector Magnetic Potential

In the preceding section it was shown that $\vec{B}$ is a solenoidal field

$$\nabla \cdot \vec{B} = 0$$

From vector calculus there is an identity that states

$$\nabla \cdot \nabla \times \vec{A} \equiv 0$$

Thus we may set

$$\vec{B} \equiv \nabla \times \vec{A} \tag{6-40}$$

The vector $\vec{A}$ is called the *vector magnetic potential* since it serves a role similar to the electric potential developed in Chapter 5. We learned there that it was easier to determine the electric potential V and then obtain the electric field intensity by a derivative calculation:

$$\vec{E} = -\nabla V$$

By analogy, if we could determine the vector magnetic potential by a simpler computation than the integrals for $\vec{H}$, then we could compute the $\vec{B}$ field from the preceding equation and then use the constitutive relation to obtain $\vec{H}$.

To get the equation for $\vec{A}$, we again return to the basic equation for $\vec{H}$ (Eq. 6-15) and obtain $\vec{B}$ as follows:

$$\vec{B} = \mu \int_{V'} \frac{\vec{J}(\vec{r}\,') \times \hat{R}}{4\pi R^2}\, dv' = \frac{\mu}{4\pi} \int_{V'} \vec{J}(\vec{r}\,') \times \left(\frac{\hat{R}}{R^2}\right) dv'$$

Using $\nabla(1/R) = -\hat{R}/R^2$,

$$\vec{B} = \frac{\mu}{4\pi}\int_{V'} -\vec{J}(\vec{r}\,') \times \nabla\left(\frac{1}{R}\right)dv' = \frac{\mu}{4\pi}\int_{V'} \nabla\left(\frac{1}{R}\right) \times \vec{J}(\vec{r}\,')dv'$$

We next use the vector identity $\nabla \times (f\vec{F}) = \nabla f \times \vec{F} + f\nabla \times \vec{F}$ and rearrange it to the following form:

$$\nabla f \times \vec{F} = \nabla \times (f\vec{F}) - f\nabla \times \vec{F}$$

Next identifying $f \Rightarrow 1/R$ and $\vec{F} \Rightarrow \vec{J}$ the integral for $\vec{B}$ becomes

$$\vec{B} = \frac{\mu}{4\pi}\int_{V'} \nabla \times \left[\frac{1}{R}\vec{J}(\vec{r}\,')\right]dv' - \frac{\mu}{4\pi}\int_{V'} \frac{1}{R}\nabla \times \vec{J}(\vec{r}\,')dv'$$

The right integral is zero since ∇ operates on unprimed coordinates only. In the left integral the curl operator may be removed from the integrand (why?) and there results

$$\vec{B} = \nabla \times \frac{\mu}{4\pi}\int_{V'} \frac{\vec{J}(\vec{r}\,')}{R}\,dv'$$

Comparing this with Eq. 6-40 we finally obtain

$$\vec{A} = \mu \int_{V'} \frac{\vec{J}(\vec{r}\,')}{4\pi R}\,dv' \qquad (6\text{-}41)$$

This allows us to compute the vector potential $\vec{A}$ at points *outside* the current distribution $\vec{J}(\vec{r}\,')$. Note this integration is much simpler in principle than those for $\vec{H}$ and $\vec{B}$, so we have again the advantage of a potential. Note that the vector potential is *parallel* to $\vec{J}$.

The units of the vector potential are easily obtained from Eq. 6-40, remembering that ∇ is unit-wise 1/m,

$$\vec{A} \Rightarrow \mathrm{T\cdot m} = \frac{\mathrm{Wb}}{\mathrm{m}^2}\cdot \mathrm{m} = \frac{\mathrm{Wb}}{\mathrm{m}} = \frac{\mathrm{J}}{\mathrm{A\cdot m}} = \frac{\mathrm{N}}{\mathrm{A}}$$

To determine the vector magnetic potential *inside* a current distribution we begin with the point form derived earlier

$$\nabla \times \vec{H} = \vec{J}$$

where here $\vec{H}$ and $\vec{J}$ are at the same point. Multiplying both sides by μ and assuming the material is homogeneous (not a function of position),

$$\nabla \times \mu\vec{H} = \nabla \times \vec{B} = \nabla \times \nabla \times \vec{A} = \mu\vec{J}$$

Applying the identity for the double curl (see Chapter 4)

$$\nabla(\nabla \cdot \vec{A}) - \nabla^2\vec{A} = \mu\vec{J}$$

The vector Helmholtz theorem (see Appendix F) states that a vector field is completely specified when its curl and divergence are specified. So far we have only set the curl of $\vec{A}$ to be $\vec{B}$. Thus we can select the divergence. We set $\nabla \cdot \vec{A} = 0$ so the equation reduces to

$$\nabla^2 \vec{A} = -\mu \vec{J} \tag{6-42}$$

for points *inside* $\vec{J}$. This choice for $\nabla \cdot \vec{A}$ is called the *Coulomb* gauge.

Note that we cannot determine the divergence of $\vec{A}$ from the integral definition, Eq. 6-41, since that applies only outside the current density. In most cases the boundary conditions needed to solve this differential equation are those on $\vec{H}$ and $\vec{B}$. So, we solve the preceding equation, compute $\vec{B}$ and $\vec{H}$ with the constants in them and then apply boundary conditions. The boundary conditions are developed in Section 6-9. Direct integration may sometimes be used as shown next.

Example 6-22

Compute the vector magnetic potential at a point on the y axis for the finite-length line current shown in Figure 6-44. Using techniques already illustrated in previous examples we have:

$$\vec{J}(\vec{r}\,') = I_0\, \delta(x')\, \delta(y')U(L - z')\hat{z}' \text{ A/m}^2$$

$$\vec{R} = \vec{r} - \vec{r}\,' = y\hat{y} - z'\hat{z}'$$

$$R = |\vec{R}| = \sqrt{y^2 + z'^2}$$

$$dv' = dx'dy'dz'$$

Putting these into Eq. 6-41 and evaluating the step and delta functions as usual

$$\vec{A} = \frac{\mu}{4\pi}\int_{-\infty}^{\infty}\int_{-\infty}^{\infty}\int_{-\infty}^{\infty} \frac{I_0\, \delta(x')\, \delta(y')U(L - z')U(L + z')\hat{z}'}{\sqrt{y^2 + z'^2}}\, dx'dy'dz'$$

$$= \frac{\mu I_0}{4\pi}\int_{-L}^{L} \frac{1}{\sqrt{y^2 + z'^2}}\, dz'\hat{z}$$

$$= \frac{\mu I_0}{2\pi}\int_{0}^{L} \frac{dz'}{\sqrt{y^2 + z'^2}}\,\hat{z} = \frac{\mu I_0}{2\pi} \ln\left(\frac{L + \sqrt{y^2 + L^2}}{y}\right)\hat{z}$$

For $L \gg y$ (long wire), $L + \sqrt{y^2 + L^2} \approx 2L$, so that

$$\vec{A} \approx \frac{\mu I_0}{2\pi} \ln\left(\frac{2L}{y}\right)\hat{z}$$

Note $\vec{A}$ is parallel to $\vec{J}$ as expected.

In cylindrical coordinates y corresponds to ρ as the problem was initially set up. We may thus write $\vec{A}$ as

$$\vec{A} \approx \frac{\mu I_0}{2\pi} \ln\left(\frac{2L}{\rho}\right)\hat{z}$$

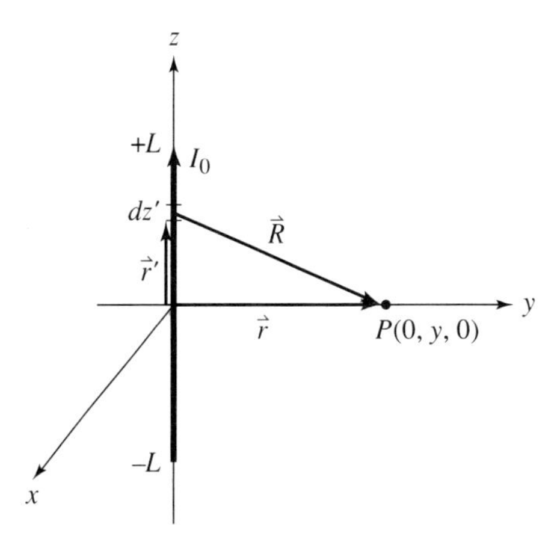

Figure 6-44.

Then

$$\vec{B} = \nabla \times \vec{A} = -\frac{\partial A_z}{\partial \rho}\,\hat{\phi} = \frac{\mu I_0}{2\pi\rho}\,\hat{\phi}$$

as before.

Example 6-23

Determine and plot the vector magnetic potential inside and outside the current beam of Example 6-7, Figure 6-18.

For $0 < \rho < a$ the second order differential Eq. 6-42 applies, and since $\vec{A}$ is parallel to $\vec{J}$, $\vec{A}$ will have a z component only, which will vary only with ρ by symmetry. We then set

$$\vec{A} = A_z(\rho)\hat{z}$$

Using $\mu = \mu_0$ in space:

$$\nabla^2\vec{A} = \nabla^2[A_z(\rho)\hat{z}] = [\nabla^2 A_z(\rho)]\hat{z} = -\mu_0 J_0 \hat{z}$$

Thus we have the scalar equation

$$\nabla^2 A_z = -\mu_0 J_0$$

Since A_z is a function only of ρ the scalar Laplacian in cylindrical coordinates (Appendix D) reduces to

$$\frac{1}{\rho}\frac{d}{d\rho}\left(\rho\,\frac{dA_z}{d\rho}\right) = -\mu_0 J_0$$

Multiplying by ρ and integrating

$$\left(\rho\,\frac{dA_z}{d\rho}\right) = -\frac{1}{2}\,\mu_0 J_0 \rho^2 + K_1$$

Dividing by ρ and integrating again,

$$A_z(\rho) = -\tfrac{1}{4}\mu_0 J_0 \rho^2 + K_1 \ln \rho + K_2$$

Since this is a potential we may select reference for zero at $\rho = a$. (This is arbitrary, but make a judicious choice.)

$$0 = -\tfrac{1}{4}\mu_0 J_0 a^2 + K_1 \ln a + K_2$$

Solving for K_2 and substituting it into the general equation:

$$A_z(\rho) = \frac{1}{4}\mu_0 J_0 (a^2 - \rho^2) + K_1 \ln\left(\frac{\rho}{a}\right)$$

To get K_1 we have to use this vector magnetic potential to find $\vec{H}$ and then apply Ampere's circuital law:

$$\vec{H} = \frac{\vec{B}}{\mu_0} = \frac{\nabla \times \vec{A}}{\mu_0} = \frac{1}{\mu_0}\left(-\frac{dA_z}{d\rho}\right)\hat{\phi} = -\frac{1}{\mu_0}\left(-\frac{1}{2}\mu_0 J_0 \rho + K_1 \frac{1}{\rho}\right)\hat{\phi}$$
$$= H_\phi(\rho)\hat{\phi}$$

$$\oint_C \vec{H} \cdot d\vec{l} = \int_0^{2\pi} H_\phi(\rho)\hat{\phi} \cdot \rho d\phi \hat{\phi} = I_{\text{enclosed}}$$

Since H_ϕ is independent of ϕ, it may be removed from the integral. Also, if we select $\rho = a$ (we are solving for $\rho \geq a$) the enclosed current is $J_0 \pi a^2$. Then, at $\rho = a$, the preceding integral reduces to

$$2\pi\left(\frac{1}{2} J_0 a^2 - \frac{K_1}{\mu_0}\right) = J_0 \pi a^2$$

Solve for K_1:

$$\pi J_0 a^2 - \frac{2\pi}{\mu_0} K_1 = J_0 \pi a^2$$

$$K_1 = 0$$

Then

$$A_z(\rho) = \tfrac{1}{4}\mu_0 J_0 (a^2 - \rho^2)$$

from which

$$\vec{A} = \tfrac{1}{4}\mu_0 J_0 (a^2 - \rho^2)\hat{z} \text{ Wb/m} \quad \text{or} \quad \text{N/A} \qquad (\rho < a)$$

For $a < \rho$ we have a choice. Since the current is zero outside we could either use the integral (Eq. 6-41) or solve the second-order equation with $\vec{J} = 0$. Choosing the latter method,

$$\nabla^2 \vec{A} = 0$$

Expanding as before

$$\frac{1}{\rho}\frac{d}{d\rho}\left(\rho\frac{dA_z}{d\rho}\right) = 0$$

Multiply by ρ and integrate

$$\left(\rho\frac{dA_z}{d\rho}\right) = K_1$$

Divide by ρ and integrate

$$A_z(\rho) = K_1 \ln \rho + K_2$$

Setting the reference potential at $\rho = a$ again, since that was our choice before,

$$A_z(\rho = a) = 0 = K_1 \ln a + K_2$$

$$K_2 = -K_1 \ln a$$

Then

$$A_z(\rho) = K_1 \ln\left(\frac{\rho}{a}\right)$$

computing $\vec{H}$ as before

$$\vec{H} = \frac{1}{\mu_0}\vec{B} = \frac{1}{\mu_0}\nabla\times\vec{A} = -\frac{1}{\mu_0}\frac{dA_z}{d\rho}\hat{\phi} = -\frac{K_1}{\mu_0}\frac{1}{\rho}\hat{\phi}$$

Applying Ampere's circuital law at $\rho = a$ (or any $\rho \geq a$):

$$\int_0^{2\pi} -\frac{K_1}{\mu_0}\frac{1}{a}\hat{\phi}\cdot a\,d\phi\hat{\phi} = I_{\text{enclosed}} = \pi a^2 J_0$$

$$-\frac{2\pi K_1}{\mu_0} = \pi a^2 J_0$$

$$K_1 = -\frac{\mu_0 J_0 a^2}{2}$$

Then

$$\vec{A} = A_z(\rho)\hat{z} = -\frac{\mu_0 J_0 a^2}{2}\ln\left(\frac{\rho}{a}\right)\hat{z}\ \text{Wb/m} \qquad (\rho > a)$$

A plot of these results is given in Figure 6-45. Note also that we are able to get a solution for $\vec{H}$, which was very difficult in Example 6-7.

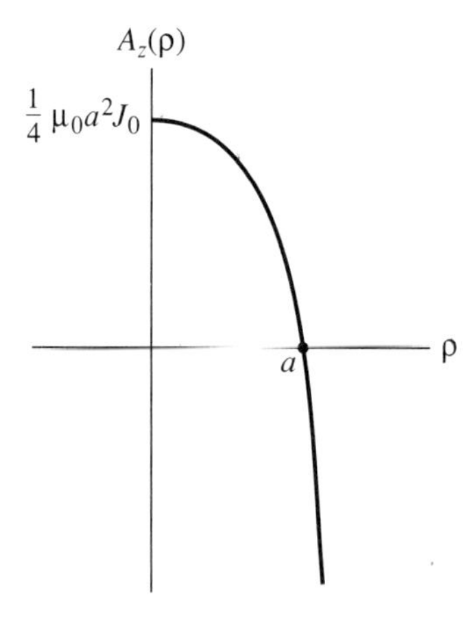

Figure 6-45. Plot of the vector magnetic potential of a uniform current beam.

Exercises

6-8.0a Determine the vector magnetic potential for the current loop of Example 6-5, Figure 6-16. Using this, compute the magnetic field intensity on the z axis. Assume the medium is free space. Note that any ∇ operations can be taken inside integrals if necessary.

6-8.0b Obtain the expression for the vector magnetic potential at any point in the xy plane for the parallel wire transmission line shown. See Figure 6-46. Use superposition and assume free space. Determine the magnetic field intensity from your solution.

6-8.0c A sheet of width a is located as shown in Figure 6-47. The sheet carries a total current I_0 and the current is uniform across the sheet. The sheet is in free space. Determine the vector magnetic potential and magnetic field intensity in the xy plane by two methods: (a) First writing an expression for $\vec{J}$ and use the general integral for $\vec{A}$.

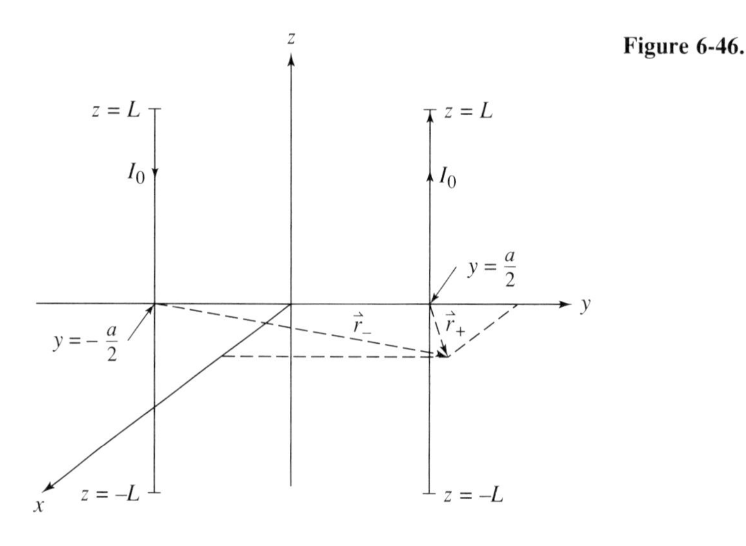

Figure 6-46.

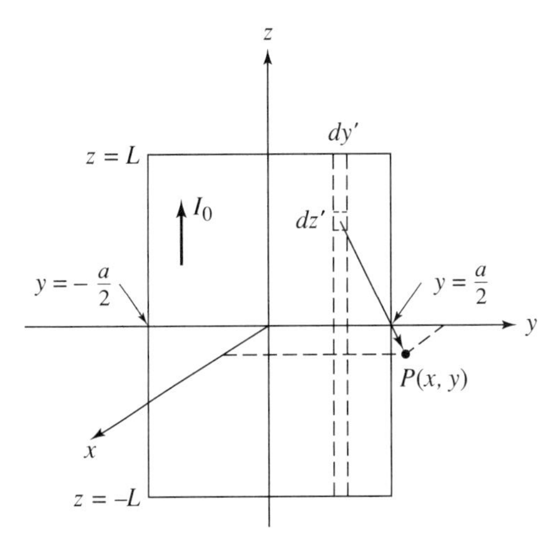

Figure 6-47.

(b) By using superposition of small line strips of width dy for $d\vec{A}$ and summing (integrating) the results to find $\vec{A}$.

As usual, ∇ operators work on unprimed coordinates.

6-8.0d Determine the vector magnetic potential for the wire of Example 6-9 in free space at a point ($u = u$, $v = v$, $z = 0$).

6-8.0e Show that the vector magnetic potential solution of Example 6-23 yields the correct magnetic field intensity. (Answer: See Example 6-7)

In many instances we are interested in the value of a field quantity at points that are very far from the actual current distribution. Fields at large distances ($\vec{R}$ large compared with the largest dimension of the current system) are called *far-field values*. Often the integrations may be simplified greatly so that the computations are much easier. Obviously, when an antenna on the earth is transmitting a signal to the moon, the size of the antenna is not perceptible to the receiver. The configuration is shown in Figure 6-48.

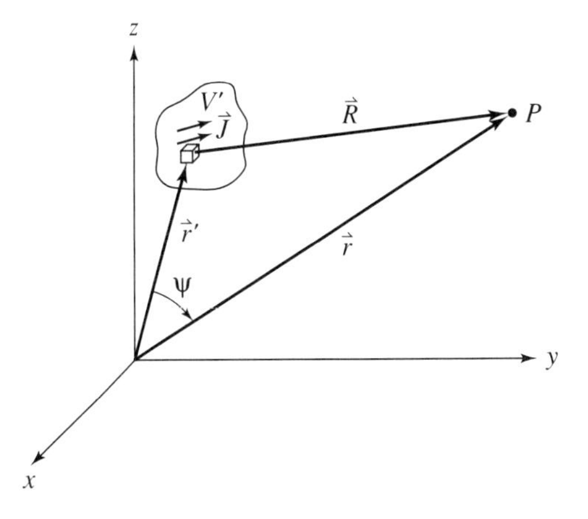

Figure 6-48. Configuration for determining the far-field expansion of the vector magnetic potential.

For the far-field expansion of the vector magnetic potential, we are interested in large distances $\vec{r}$. This term is located in the distance quantity R in the integral expression for $\vec{A}$. Using the law of cosines we have

$$\frac{1}{R} = \frac{1}{|\vec{r} - \vec{r}'|} = \frac{1}{\sqrt{r^2 + r'^2 - 2rr' \cos \psi}} = \frac{1}{r} \frac{1}{\sqrt{1 + \left(\frac{r'}{r}\right)^2 - 2\frac{r'}{r} \cos \psi}}$$

Since $\vec{r}' \cdot \vec{r} = rr' \cos \psi$, we may eliminate $\cos \psi$. Also, the squared term in the radicand is second-order smallness compared with r'/r so that term may be neglected. With these changes the preceding expression becomes

$$\frac{1}{R} \approx \frac{1}{r} \frac{1}{\sqrt{1 - \frac{2\vec{r}' \cdot \vec{r}}{r^2}}} = \frac{1}{r} \frac{1}{\sqrt{1 - 2\frac{\vec{r}' \cdot \hat{r}}{r}}}$$

From algebra we have the series binomial expansion ($x^2 < 1$)

$$(1 - x)^n = 1 - nx + \frac{n(n-1)}{2!} x^2 - \frac{n(n-1)(n-2)}{3!} x^3 + \frac{n(n-1)(n-2)(n-3)}{4!} x^4 - \cdots$$

For $n = -\frac{1}{2}$ we may use this to expand the reciprocal of the radical. Doing this we obtain

$$\frac{1}{R} \approx \frac{1}{r}\left[1 + \frac{\vec{r}' \cdot \hat{r}}{r} + \frac{3}{8}\frac{(\vec{r}' \cdot \hat{r})^2}{r^2} + \frac{5}{16}\frac{(\vec{r}' \cdot \hat{r})^3}{r^3} + \frac{35}{128}\frac{(\vec{r}' \cdot \hat{r})^4}{r^4} + \cdots\right]$$

Substituting this into Eq. 6-41 and removing the r terms from the integrand since they are independent of the prime coordinates we obtain

$$\vec{A} \approx \frac{\mu}{4\pi r}\int_{V'} \vec{J}(\vec{r}')dv' + \frac{\mu}{4\pi r^2}\int_{V'} \vec{J}(\vec{r}')(\vec{r}' \cdot \hat{r})dv' + \frac{3\mu}{32\pi r^3}\int_{V'} \vec{J}(\vec{r}')(\vec{r}' \cdot \hat{r})^2 dv' + \cdots \tag{6-43}$$

Note that the coefficients of the integrals get smaller and smaller for r very large. By definition, the far field is the first nonzero integral in this expansion. What we do then is begin with the simplest left integral and work each integral out and stop at the first nonzero one.

Example 6-24

Find the far field expression for $\vec{A}$ for a wire of length $2L$ that carries a current I_0 as shown in Figure 6-49. From the figure

$$dv' = dx'dy'dz'$$

$$\vec{r}' = z'\hat{z}' = z'\hat{z}$$

$$\vec{r} = x\hat{x} + y\hat{y} + z\hat{z}$$

$$\vec{r}\,' \cdot \hat{r} = \vec{r}\,' \cdot \frac{\vec{r}}{r} = z'\hat{z} \cdot \frac{x\hat{x} + y\hat{y} + z\hat{z}}{\sqrt{x^2 + y^2 + z^2}} = \frac{z'z}{\sqrt{x^2 + y^2 + z^2}} = \frac{z}{r}z'$$

$$\vec{J}(\vec{r}\,') = I_0\,\delta(x')\,\delta(y')U(L - z')U(L + z')\hat{z}'$$

Using the first integral in Eq. 6-43, replacing $\hat{z}'$ by $\hat{z}$ in the current density equation,

$$\vec{A} \approx \frac{\mu}{4\pi r}\int_{-\infty}^{\infty}\int_{-\infty}^{\infty}\int_{-\infty}^{\infty} I_0\,\delta(x')\,\delta(y')U(L - z')U(L + z')\hat{z}dx'dy'dz'$$

$$\approx \frac{\mu I_0}{4\pi r}\int_{-L}^{L} dz'\hat{z} = \frac{\mu I_0 L}{2\pi r}\hat{z}$$

Exercises

6-8.0f For the result of Example 6-24, compute the far-field magnetic field intensity in spherical coordinates. Reduce this result to points in the xy plane and compare with the result of Example 6-12.

6-8.0g Determine the equation for the far fields $\vec{A}$ and $\vec{B}$ for the circular current carrying loop of Figure 6-15, and identify the loop magnetic moment. (Answer: $B_r = (\mu m/2\pi r^3)\cos\theta$, $B_\theta = (\mu m/4\pi r^3)\sin\theta$, $\beta_\phi = 0$)

6-8.0h Modify Example 6-24 so that the current is up for $z > 0$ and down for $z < 0$. Develop an equation for the far field $\vec{A}$. In this case your first integral should be zero. Next plot $|\vec{A}|$, for constant distance from the origin, on a polar plot in the yz plane.

6-8.0i Derive the far-field expansion for the electric potential V of Chapter 5, Eq. 5-67.

6-9 Boundary Conditions for Magnetic Field Quantities

In many engineering applications there are boundary discontinuities or several materials in a given system. We need to know how magnetic fields behave at these system discontinuities to be able to describe the fields everywhere. Also, as mentioned in the preceding

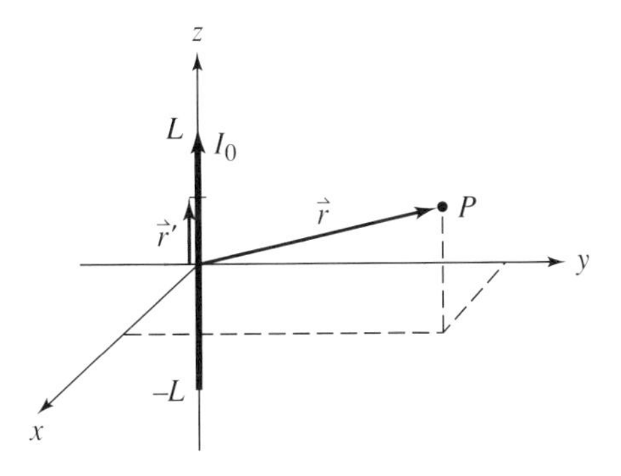

Figure 6-49. Far-field computation for line current.

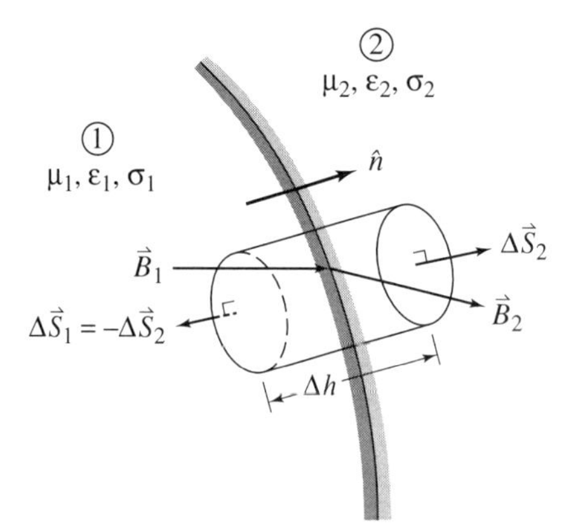

Figure 6-50. Configuration used to determine the normal boundary conditions for magnetic fields.

section, we often need to have boundary conditions on $\vec{B}$ and $\vec{H}$ to solve for the magnetic vector potential from the differential equations. These conditions are also used for time varying field solutions. They are also analogous to those obtained for electric field quantities in Chapter 5. The derivations are very similar, so some details are left out when they are the same.

Consider the small cylindrical can which cuts across a boundary as shown in Figure 6-50. We apply the integral form of Gauss' law for magnetic fields and expand it for the cylinder. Note our unit vector normal extends from Medium 1 into Medium 2. Thus, $\Delta\vec{S}_2$ is in the direction of $\hat{n}$ and $\Delta\vec{S}_1$ is antiparallel to $\hat{n}$. Then

$$\oint_S \vec{B} \cdot d\vec{a} = \int_{\Delta S_1} \vec{B} \cdot d\vec{A} + \int_{\Delta S_2} \vec{B} \cdot d\vec{a} + \int_{\text{side}} \vec{B} \cdot d\vec{a} = 0$$

Now at the boundary, Δh must be reduced to zero so there is no area on the side through which $\vec{B}$ can pass so that integral will be zero. The preceding reduces to

$$\int_{\Delta S_1} \vec{B}_1 \cdot d\vec{a}_1 + \int_{\Delta S_2} \vec{B}_2 \cdot d\vec{a}_2 = \int_{\Delta S_1} - B_{in} da_1 + \int_{\Delta S_2} B_{2n} da_2 = 0$$

where the subscript n denotes the component of the field normal to the boundary. For small ΔS_1 and ΔS_2 the fields may be considered constant so they may be removed from the integrals. Also, since $\Delta h \to 0$, $\Delta S_1 \to \Delta S_2$. Then

$$-B_{ln}\, \Delta S_2 + B_{2n}\, \Delta S_2 = 0$$

$$B_{ln} = B_{2n} \tag{6-44}$$

This result shows that the normal component of B is continuous across a boundary. Using the constitutive relationship for magnetic fields this yields

$$\mu_1 H_{ln} = \mu_2 H_{2n} \tag{6-45}$$

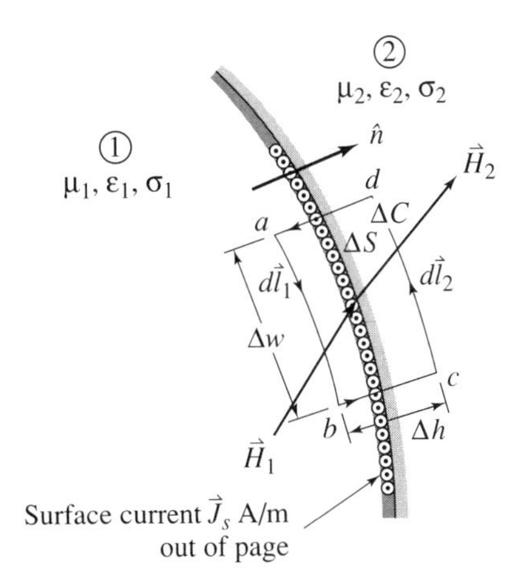

Figure 6-51. Configuration used to determine the tangential boundary conditions for magnetic fields.

Figure 6-51 is the configuration from which the tangential boundary conditions are obtained. We apply Ampere's circuital law to the contour ΔC, noting that for the path $a \rightarrow b$ the angle between $d\vec{l}$ and $\vec{H}_1$ is greater than 90° so the dot product of the two is negative. Also, in the limit at the boundary $\Delta h \rightarrow 0$ so there is no path contribution from the two ends. Thus

$$\oint_{\Delta C} \vec{H} \cdot d\vec{l} = \int_a^b \vec{H}_1 \cdot d\vec{l}_1 + \int_b^c \vec{H} \cdot d\vec{l} + \int_c^d \vec{H}_2 \cdot d\vec{l}_2$$

$$+ \int_d^a \vec{H} \cdot d\vec{l} = I_{\text{enclosed}} = J_s \, \Delta w$$

which reduces to

$$\int_a^b - H_{1t} \, dl_1 + \int_c^d H_{2t} \, dl_2 = J_s \, \Delta w$$

For small Δw the magnetic field intensity may be considered constant and removed from the integrals:

$$-\vec{H}_{1t} \, \Delta w + H_{2t} \, \Delta w = J_s \, \Delta w$$

$$H_{2t} - H_{1t} = J_s \tag{6-46}$$

This is a reasonable result on physical grounds since the right-hand rule applied to the current shows that the field produced by the current aids the tangential component of $\vec{H}_2$ and opposes the tangential component of $\vec{H}_1$. Thus H_{2t} is greater than H_{1t} for the current as shown.

Using the constitutive relationship we have the alternative form

$$\frac{B_{2t}}{\mu_2} - \frac{B_{1t}}{\mu_1} = J_s \tag{6-47}$$

Since the tangential component can be obtained by the vector operation $|\vec{H} \times \hat{n}|$ we could write Eq. 6-46 in vector form:

$$|\vec{H}_2 \times \hat{n}| - |\vec{H}_1 \times \hat{n}| = J_s$$

If we observe next that $\vec{H} \times \hat{n}$ is in the direction of $-\vec{J}_s$ we could write this in complete vector form as

$$\vec{H}_2 \times \hat{n} - \vec{H}_1 \times \hat{n} = -\vec{J}_s$$

Since the cross product is distributive but negatively commutative this becomes

$$\hat{n} \times (\vec{H}_2 - \vec{H}_1) = \vec{J}_s \tag{6-48}$$

To obtain boundary conditions on the magnetization $\vec{M}$ we use the fact that (see Eq. 6-33) $\vec{M} \times \chi_m \vec{H}$. Thus, for normal components of $\vec{M}$ Eq. 6-45 yields

$$\frac{\mu_1}{\chi_{m1}} M_{1n} = \frac{\mu_2}{\chi_{m2}} M_{2n} \tag{6-49}$$

For the tangential components of $\vec{M}$ Eqs. 6-46 and 6-48 give

$$\frac{M_{2t}}{\chi_{m2}} - \frac{M_{1t}}{\chi_{m1}} = J_s$$
$$\hat{n} \times \left(\frac{\vec{M}_2}{\chi_{m2}} - \frac{\vec{M}_1}{\chi_{m1}}\right) = \vec{J}_s \tag{6-50}$$

Exercises

6-9.0a Explain why if a current density $\vec{J}$ A/m^2 flows in the metal $H_{2t} - H_{1t} = 0$, whereas if a surface current flows the difference is not zero. Show in the derivation how this is justified. (Hint: What is ΔS as $\Delta h \geq 0$?)

6-9.0b For the system shown in Figure 6-52 determine $\vec{H}_2$ for the case $\vec{J}_s = 0$. (Answer: $(\sqrt{3}/2)\hat{x} + \hat{y}$)

6-9.0c Repeat Exercise 6-9.0b if $\vec{J}_s = -2\hat{z}$ A/m. (Answer: $(\sqrt{3}/2)\hat{x} - \hat{y}$

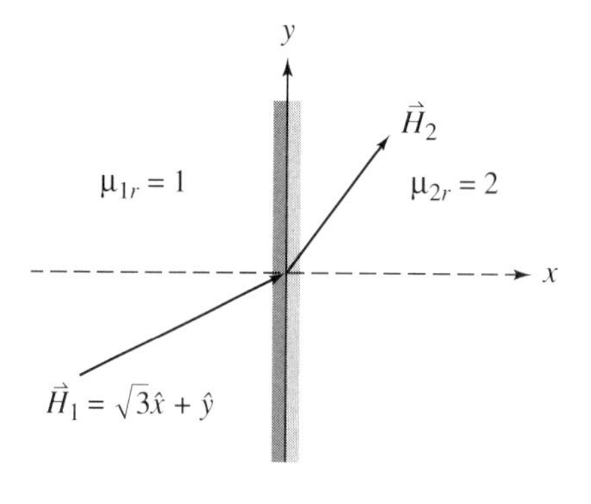

Figure 6-52.

6-10 Scalar Magnetic Potential for Regions with $\vec{J} = 0$

In Chapter 5 we found it convenient to define a scalar potential V because it was often an easier quantity to compute. We could then obtain the electric field intensity by taking the gradient of that scalar. The same situation arises for magnetic fields in regions where there is no current density $\vec{J}$. For such regions Eq. 6-24 reduces to

$$\nabla \times \vec{H} = 0$$

A vector theorem (see Chapter 4) states that the curl of the gradient is always zero. Thus we can define a scalar magnetic potential from which the magnetic field intensity can be computed as

$$\vec{H} = -\nabla V_m \tag{6-51}$$

The negative sign is used only to establish an analogy with the electrostatic case of Chapter 5. The inverse relation is then

$$V_m = -\int_{\text{ref}}^{A} \vec{H} \cdot d\vec{l} \tag{6-52}$$

This last equation has one important limitation on the path of integration. In the electrostatic potential case we have

$$\oint \vec{E} \cdot d\vec{l} = 0$$

but for the magnetostatic case,

$$\oint \vec{H} \cdot d\vec{l} = I_{\text{enclosed}}$$

Therefore, if we are to have Eq. 6-52 give a result that is independent of the path we must be certain that there is no current on or within the path. In practice we can insure this by never forming a closed path around any current.

To obtain an equation which we can use to solve for V_m we use the divergence equation for magnetic fields and proceed as follows:

$$\nabla \cdot \vec{B} = \nabla \cdot (\mu\vec{H}) = \mu\nabla \cdot (-\nabla V_m) = -\mu\nabla \cdot \nabla V_m = -\mu\, \nabla^2 V_m = 0$$

$$\nabla^2 V_m = 0 \tag{6-53}$$

The solution of Laplace's equation requires boundary conditions derived in Section 6-9 to determine the constants. One such condition is that we may select the location for zero reference at any point where the function V_m is finite when we have only one material. Other conditions are required when there are two materials and we use the previous boundary conditions on $\vec{H}$ and $\vec{B}$.

Consider the tangential components of $\vec{H}$ at a boundary

$$H_{2t} - H_{1t} = J_s$$

Note it is permissible to have J_s since we have only required $\vec{J}$ to be zero. Thus we have two cases to consider. For $\vec{J}_s = 0$:

$$H_{2t} - H_{1t} = 0$$

or

$$(-\nabla V_{m1})_t - (-\nabla V_{m2})_t = 0$$

Since the difference of the tangential components of vector is the same as the tangential component of the difference of the vectors we may write

$$(\nabla V_{m2} - \nabla V_{m1})_t = 0$$

Since the gradient operator is distributive,

$$\nabla(V_{m2} - V_{m1})|_t = 0, \qquad (\vec{J}_s = 0) \tag{6-54}$$

For $\vec{J}_s \neq 0$, using Eq. 6-48 and the gradient definition:

$$\hat{n} \times (\nabla V_{m1} - \nabla V_{m2}) = \vec{J}_s \tag{6-55}$$

Consider the normal components of $\vec{B}$ at a boundary that must satisfy Eq. 6-45. Substituting for $\vec{H}$ in terms of the scalar magnetic potential we obtain

$$\mu_1 \nabla V_{m1} = \mu_2 \nabla V_{m2}, \qquad (\text{any value of } \vec{J}_s) \tag{6-56}$$

Example 6-25

Determine the scalar magnetic potential between the conductors of the coax shown in earlier Figure 6-32. Find $\vec{H}$.

Since there is no current density $\vec{J}$ between the conductors,

$$\nabla^2 V_m = 0$$

Now V_m must be independent of ρ since the right-hand rule shows that $\vec{H}$ has only a ϕ component. Similarly, V_m will not be a function of z.

$$\nabla^2 V_m(\phi) = \frac{1}{\rho^2}\frac{d^2 V_m(\phi)}{d\phi^2} = 0$$

Multiply by ρ^2 and integrate twice to obtain

$$V_m(\phi) = K_1\phi + K_2$$

Using the first condition given in the text, we may select the reference location for $V_m = 0$. Pick $\phi = 0$. Then

$$0 = K_1 \cdot 0 + K_2 \Rightarrow K_2 = 0$$

Then we have

$$V_m(\phi) = K_1\phi$$

Now inside the center conductor there is no V_m since there is a current flow there. We may compute an equivalent surface current $\vec{J}_s$ around the center conductor as

$$\vec{J}_{\text{eq}} = \frac{I_0}{2\pi\alpha}\,\hat{z}$$

Using Eq. 6-55 with $V_{m2} = 0$, and $\hat{n} = -\hat{\rho}$ since the normal is into Medium 2:

$$-\hat{\rho} \times \nabla V_m(\phi)\big|_{\rho=a} = \frac{I_0}{2\pi\alpha}\,\hat{z}$$

$$-\rho \times \left(\frac{1}{\rho}\,K_1\hat{\phi}\right)\Bigg|_{\rho=a} = \frac{I_0}{2\pi\alpha}\,\hat{z}$$

$$-\frac{1}{a}\,K_1\hat{z} = \frac{I_0}{2\pi a}\,\hat{z} \Rightarrow K_1 = \frac{-I_0}{2\pi}$$

Then

$$V_m(\phi) = -\frac{I_0}{2\pi}\,\phi \qquad (0 \le \phi < 2\pi)$$

Note we cannot use 2π for ϕ or we enclose a current I_0. Finally,

$$\vec{H} = -\nabla V_m = -\left(\frac{1}{\rho}\frac{-I_0}{2\pi}\right)\hat{\phi} = \frac{I_0}{2\pi\rho}\,\hat{\phi}, \qquad a < \rho < b$$

More advanced techniques for solving Laplace's equation will be given in a later chapter.

Exercises

6-10.0a Using $\phi = 0$ as reference, evaluate the integral in Eq. 6-52 using the result of the preceding example. Use a constant radius path. (Answer: $-I_0/2\pi\phi$)

6-10.0b Determine V_m for the coax for $\rho > c$ for Figure 6-32. Note that the equivalent surface current is computed from the total current existing for $\rho < c$. (Answer: $V_m = 0$)

6-10.0c A cylinder of infinite length has a surface current $\vec{J}_s = J_0\hat{\phi}$ and radius b. Determine the scalar magnetic potential and magnetic field intensity for $\rho > b$. Justify that V_m will be a function only of z? (Answer: $J_0 z$, $-J_0\hat{z}$)

6-11 Permanent Magnets

Ferromagnetic materials below the Curie temperature possess remnant magnetization. Since there are no currents due to free charge flow, we characterize a permanent magnet by

$$\vec{J} = 0, \qquad \vec{J}_s = 0, \qquad \vec{M} \neq 0$$

This characterization allows us to use either the scalar magnetic potential or vector magnetic potential to solve for the magnetic field configurations both inside and outside the material. The application of both types of potential will be presented.

Since these magnets have a permanent magnetization the general technique is to separate the magnetization $\vec{M}$ from $\vec{B}$ and $\vec{H}$ using Eq. 6-32, repeated here for reference:

$$\vec{B} = \mu_0\vec{H} + \mu_0\vec{M}$$

A point of note is that in this section only μ_0 is used and never μ because we will separate out the polarization effects used to obtain μ by using $\vec{M}$ explicitly.

6-11.1 Solution by Scalar Magnetic Potential

Since there is no current density $\vec{J}$ the potential is exactly the same as that derived in the preceding section

$$\vec{H} = -\nabla V_m$$

Using the fact that the divergence of $\vec{B}$ is zero (Eq. 6-39)

$$\begin{aligned}\nabla \cdot \vec{B} &= \nabla \cdot (\mu_0\vec{H} + \mu_0\vec{M}) = \mu_0\nabla \cdot \vec{H} + \mu_0\nabla \cdot \vec{M} = \mu_0\nabla \cdot (-\nabla V_m) + \mu_0\nabla \cdot \vec{M} \\ &= -\mu_0\, \nabla^2 V_m + \mu_0\nabla \cdot \vec{M} = 0\end{aligned}$$

Solving for $\nabla^2 V_m$:

$$\nabla^2 V_m = \nabla \cdot \vec{M}$$

Now we can make this analogous to Poisson's equation of electrostatics if we define a volume magnetic charge density by

$$\nabla \cdot \vec{M} = -\frac{\rho_{vm}}{\mu_0}$$

or

$$\rho_{vm} \equiv -\mu_0\nabla \cdot \vec{M} \tag{6-57}$$

Then for the scalar potential

$$\nabla^2 V_m = -\frac{\rho_{vm}}{\mu_0} \tag{6-58}$$

We usually know the state of magnetization of an object so we can obtain ρ_{vm} from the divergence equation easily. The boundary conditions at the surface of the magnet are as follows, where Medium 1 is the magnet and 2 the surrounding medium:

1.

$$V_{m1} = V_{m2} \tag{6-59}$$

This must be the case since if there were a discontinuity in V_m the gradient would yield an infinite $\vec{H}$ at the boundary.

2. To obtain the second condition we use the general boundary condition that normal $\vec{B}$ is continuous. Using the same conventions as in Figure 6-50,

$$B_{1n} = B_{2n}$$

Then

$$\mu_0 H_{1n} + \mu_0 M_{1n} = \mu_0 H_{2n} + \mu_0 M_{2n}$$

Rearranging and using the fact that the normal component of a vector is the vector dotted with the unit normal vector,

$$\mu_0 H_{2n} - \mu_0 H_{1n} = \mu_0 \vec{M}_1 \cdot \hat{n} - \mu_0 \vec{M}_2 \cdot \hat{n}$$

In Chapter 5 we learned that the normal electric flux density is discontinuous by the amount of surface charge density. The left side of the equation above is the difference of normal magnetic flux densities so we are led to define an equivalent magnetic surface charge density as

$$\rho_{sm} \equiv \mu_0 \vec{M} \cdot \hat{n} \tag{6-60}$$

We then have the second boundary condition

$$\mu_0 H_{2n} - \mu_0 H_{1n} = \rho_{sm_1} - \rho_{sm_2}$$

or

$$\mu_0(-\nabla V_{m2})_n - \mu_0(-\nabla V_{m1})_n = \rho_{sm_1} - \rho_{sm_2} \tag{6-61}$$

Note particularly that for a permanent magnet in free space $\vec{M}_2 = 0$, since we must have molecules to produce polarization.

6-11.2 Solution by Vector Magnetic Potential

For this case we use the same potential function $\vec{A}$ as in Section 6-8 since the derivation starts from the same divergence of $\vec{B}$ equation. Therefore,

$$\vec{B} = \nabla \times \vec{A}$$

From this,

$$\nabla \times \nabla \times \vec{A} = \nabla \times \vec{B} = \nabla \times (\mu_0 \vec{H} + \mu_0 \vec{M}) = \mu_0 \nabla \times \vec{H} + \mu_0 \nabla \times \vec{M}$$

Since there is no current density $\vec{J}$, the curl of $\vec{H}$ is zero (see Eq. 6-24), so that

$$\nabla \times \nabla \times \vec{A} = \mu_0 \nabla \times \vec{M}$$

$$\nabla(\nabla \cdot \vec{A}) - \nabla^2 \vec{A} = \mu_0 \nabla \times \vec{M}$$

For the same reason as given in Section 6-8 (vector Helmholtz theorem) we set $\nabla \cdot \vec{A} = 0$, so that

$$\nabla^2 \vec{A} = -\mu_0 \nabla \times \vec{M}$$

Comparing this result with Eq. 6-42 we are then led to define an equivalent magnetic current density as

$$\vec{J}_m \equiv \nabla \times \vec{M} \text{ A/m}^2 \tag{6-62}$$

Finally,

$$\nabla^2 \vec{A} = -\mu_0 \vec{J}_m \tag{6-63}$$

As explained in Section 6-8, the easiest boundary conditions to apply would be to obtain $\vec{B}$ from $\nabla \times \vec{A}$, carrying along the constants to be determined, and then apply the continuity of normal $\vec{B}$. Then, obtain the equation for $\vec{H}$ using $\vec{H} = \dfrac{\vec{B}}{\mu_0} - \vec{M}$ for both media and making tangential $\vec{H}$ continuous since there is no true $\vec{J}_s$ (see Eq. 6-48).

Actually, we can derive a second boundary condition from the tangential field condition by expressing $\vec{H}$ in terms of the magnetization explicitly as follows:

$$\hat{n} \times (\vec{H}_2 - \vec{H}_1) = \vec{J}_s = 0$$

$$\hat{n} \times \left[\left(\frac{\vec{B}_2}{\mu_0} - \vec{M}_2\right) - \left(\frac{\vec{B}_1}{\mu_0} - \vec{M}_1\right)\right] = 0$$

Using the distributive property of the cross product and then rearranging:

$$\hat{n} \times \left(\frac{\vec{B}_2}{\mu_0} - \frac{\vec{B}_1}{\mu_0}\right) = \hat{n} \times \vec{M}_2 - \hat{n} \times \vec{M}_1 = \vec{M}_1 \times \hat{n} - \vec{M}_2 \times \hat{n}$$

We next define an equivalent magnetic surface current density

$$\vec{J}_{sm} \equiv \vec{M} \times \hat{n} \tag{6-64}$$

With this definition and substituting for $\vec{B}_2$ and $\vec{B}_1$ in terms of the vector magnetic potentials

$$\hat{n} \times \left(\frac{\nabla \times \vec{A}_2}{\mu_0} - \frac{\nabla \times \vec{A}_1}{\mu_0}\right) = \vec{J}_{sm_2} - \vec{J}_{sm_1} \tag{6-65}$$

In both methods of solution presented here, the technique is to solve the Poisson equation (Eqs. 6-58 or 6-60) for each medium in the system; that is, the magnet (Medium 1) and then the surrounding medium (Medium 2). Use subscripts on the potentials to identify the media. Then apply the appropriate boundary conditions to determine constants in the solutions. In many cases the surrounding Medium 2 is nonmagnetic so that $\vec{M}_2 = 0$. This simplifies the boundary conditions. At this point we have not developed methods for solving the Poisson equations, a subject for a later chapter. Thus in the examples that follow the solutions proceed to the equation to be solved for the particular problem.

Example 6-26

Using the scalar magnetic potential.

Suppose we have a cylindrical bar magnet with permanent uniform magnetization $\vec{M} = M_0\hat{z}$ with free space outside. See Figure 6-53. Inside the magnet

$$\rho_{vm_1} = -\mu_0 \nabla \cdot \vec{M} = 0$$

The Poisson equation (Eq. 6-58) in cylindrical coordinates is then (see expansions in Chapter 4)

$$\frac{1}{\rho}\frac{\partial}{\partial \rho}\left(\rho \frac{\partial V_{m_1}}{\partial \rho}\right) + \frac{1}{\rho^2}\frac{\partial^2 V_{m_1}}{\partial \phi^2} + \frac{\partial^2 V_{m_1}}{\partial z^2} = 0$$

Now within the magnet we would expect to have an $\vec{H}_1$ with at most a $\hat{\rho}$ component and $\hat{z}$ component. By symmetry $\vec{H}_1$ would not be a function of ϕ. Thus V_{m_1} would not be a function of ϕ since $H_{\phi_1} = 0$, ($\vec{H}_1 = -\nabla V_{m_1}$). (See Figure 6-53 for a couple of field lines). Laplace's equation reduces to

$$\frac{1}{\rho}\frac{\partial}{\partial \rho}\left(\rho \frac{\partial V_{m_1}}{\partial \rho}\right) + \frac{\partial^2 V_{m_1}}{\partial z^2} = 0, \qquad \rho \le a, \qquad |z| \le L$$

Using the same arguments for the air around the magnet with $\vec{M}_2 = 0$

$$\frac{1}{\rho}\frac{\partial}{\partial \rho}\left(\rho \frac{\partial V_{m_2}}{\partial \rho}\right) + \frac{\partial^2 V_{m_2}}{\partial z^2} = 0, \qquad \rho \ge a, \qquad |z| \ge L$$

As mentioned, techniques for solving these will be covered later. The solutions have constants in them that can be determined by applying boundary conditions on the top, bottom, and side of the magnet.

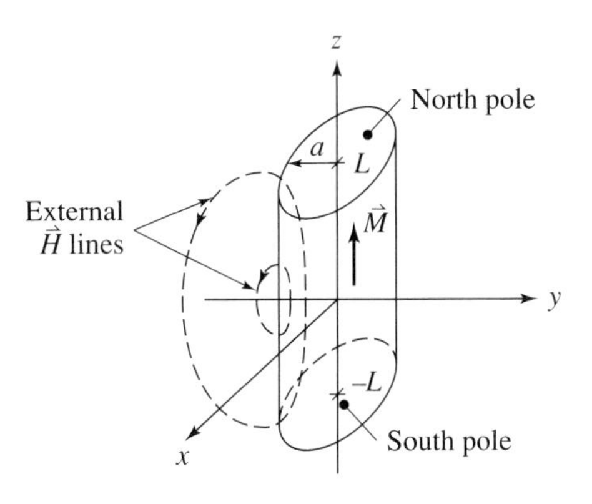

Figure 6-53. Cylindrical permanent bar magnet.

Example 6-27

Using the vector magnetic potential.

Next take a uniformly magnetized sphere of radius a and magnetization $\vec{M} = M_0\hat{z}$. The sphere is in free space, and is shown in Figure 6-54.

Using the techniques of Chapter 4 we convert the sphere magnetization to spherical coordinates

$$\vec{M}_1 = M_0\hat{z} = M_0 \cos\theta\hat{r} - M_0 \sin\theta\hat{\theta}$$

Within the sphere Eq. 6-62 yields

$$\vec{J}_{m_1} = \nabla \times \vec{M}_1 = \frac{-1}{r} M_0 \sin\theta\hat{\phi}$$

From physical considerations we expect $\vec{H}_1$ to have at most $\hat{r}$ and $\hat{\theta}$ components, and to be independent of ϕ by symmetry. Also, since the vector potential is parallel to $\vec{J}$, A_r and A_θ must be zero (see Eq. 6-41). Expanding the vector Laplacian using the identity for $\nabla \times \nabla \times \vec{A}$ given in Chapter 4 and imposing the constraints stated

$$\left(\nabla^2 A_{\phi_1} - \frac{1}{r^2}\csc^2\theta A_{\phi_1}\right)\hat{\phi} = -\frac{1}{r} M_0 \sin\theta\hat{\phi}$$

Substituting for the scalar Laplacian and equating vector components on the two sides of the preceding equation:

$$\frac{1}{r^2}\frac{\partial}{\partial r}\left(r^2\frac{\partial A_{\phi_1}}{\partial r}\right) + \frac{1}{r^2 \sin\theta}\frac{\partial}{\partial\theta}\left(\sin\theta\frac{\partial A_{\phi_1}}{\partial\theta}\right) - \frac{1}{r^2}\csc^2\theta A_{\phi_1} = -\frac{1}{r} M_0 \sin\theta$$

Outside the sphere $\vec{M}_2 = 0$, so $\vec{J}_{m_2} = 0$. Following similar steps as earlier, and multiplying through by r^2:

$$\frac{\partial}{\partial r}\left(r^2\frac{\partial A_{\phi_2}}{\partial r}\right) + \frac{1}{\sin\theta}\frac{\partial}{\partial\theta}\left(\sin\theta\frac{\partial A_{\phi_2}}{\partial\theta}\right) - \csc^2\theta A_{\phi_2} = 0$$

The preceding two equations would be solved for A_{ϕ_1}, and A_{ϕ_2} and the boundary conditions given in the text applied to find the constants.

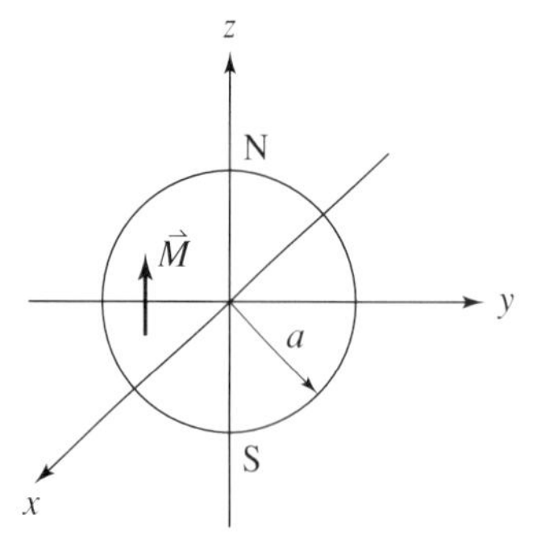

Figure 6-54. Uniformly magnetized sphere.

Exercises

6-11.2a Set up the solution for Example 6-26 using the vector magnetic potential.

6-11.2b Set up the solution to Example 6-27 using the scalar magnetic potential. (Answer: For both Medium 1 and Medium 2: $\frac{\partial}{\partial r}\left(r^2 \frac{\partial V_m}{\partial r}\right) + \frac{1}{\sin\theta}\frac{\partial}{\partial\theta}\left(\sin\theta \frac{\partial V_m}{\partial\theta}\right) = 0$)

6-12 Magnetic Energy, Energy Density, Pressure, and Force

In Chapter 5 the energy density in an electric field was derived and shown to be an important quantity in the calculation of capacitance by energy storage. Since the product of the electric flux (displacement) density and electric field intensity gave energy per unit volume, we are led to inquire whether a similar relationship holds for magnetic fields. If we consider an analysis of units we find (see Eq. 6-4):

$$\text{units of } B \cdot H = \text{T} \cdot \frac{\text{A}}{\text{m}} = \frac{\text{Wb}}{\text{M}^2} \cdot \frac{\text{A}}{\text{m}} = \frac{\text{N}}{\text{A} \cdot \text{m}} \cdot \frac{\text{A}}{\text{m}} = \frac{\text{N}}{\text{m}^2} = \frac{\text{N} \cdot \text{m}}{\text{m}^3} = \frac{\text{J}}{\text{m}^3}$$

Thus this product is also energy density. The next problem we face is how to obtain the exact functional relationship. The derivation of the magnetic field energy density is not as simple as that for electric field energy, and in fact involves a consideration of the buildup of the magnetic field so time is involved. However, the transient buildup of the field is taken to be arbitrarily slow so that we can still apply static field equations. The full derivation is given in Appendix H. The result is

$$dW_m = \int_V \vec{H} \cdot d\vec{B} dv \text{ J} \tag{6-66}$$

Where V is the volume in which the energy is being computed.

For the special case of a linear medium (μ not a function of $\vec{H}$) we may use the constitutive relationship and take the differential to obtain

$$d\vec{B} = \mu d\vec{H}$$

The energy becomes

$$dW_m = \int_V \vec{H} \cdot \mu d\vec{H} dv = \int_V \mu\vec{H} \cdot d\vec{H} dv = \int_V \mu \tfrac{1}{2} d(\vec{H} \cdot \vec{H}) dv$$

$$= \int_V \tfrac{1}{2}\mu d(H^2) dv$$

To obtain the total energy we integrate with respect to the magnetic field to obtain

$$W_m = \int_0^{H^2} \int_V \tfrac{1}{2}\mu d(H^2) dv = \int_V \tfrac{1}{2}\mu \int_0^{H^2} d(H^2) dv$$

$$W_m = \int_V \tfrac{1}{2}\mu H^2 dv \text{ J} \tag{6-67}$$

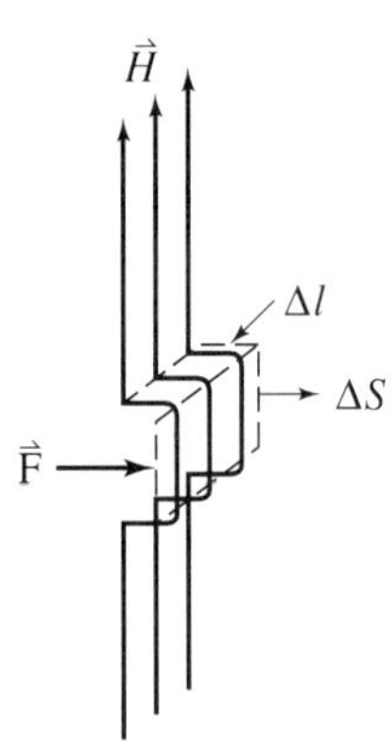

Figure 6-55. System for determining the force produced by a magnetic field.

The integrand is now *interpreted* as the energy density of the magnetic field in a medium of permeability μ:

$$w_m = \tfrac{1}{2}\mu H^2 \text{ J/m}^3 \tag{6-68}$$

To determine the pressure that a magnetic field exerts on a surface, we follow a procedure similar to that in Section 5-15 for electric field pressure. The basic structure is shown in Figure 6-55. Note that the force will be normal to the magnetic field lines. Note that to move the lines a permeability different from that of the medium in which $\vec{H}$ initially resides is required.

The principle of virtual work is applied, so we take the work done by the force and equate it to the energy in the displaced region. Then using the tangential component of $\vec{H}$

$$\Delta\text{work} = F_n\,\Delta l = (\tfrac{1}{2}\mu H_t^2)\,\Delta v = (\tfrac{1}{2}\mu H_t^2)\,\Delta S\,\Delta l$$

$$F_n = \tfrac{1}{2}\mu H_t^2\,\Delta S$$

The pressure on the surface defined by plane of $\vec{H}$ is then

$$P_m = \lim_{\Delta S \to 0} \frac{F_n}{\Delta S} = \frac{1}{2}\,\mu H_t^2 = \frac{1}{2}\frac{B_t^2}{\mu} \text{ Pa} \tag{6-69}$$

Example 6-28

Compute the magnetic energy per meter stored between the conductors of a coaxial cable. See Figure 6-32 where we want the energy in the region $a \le \rho \le b$. Use air medium.

From Example 6-14, the magnetic field intensity in that region is

$$H_\phi(\rho) = \frac{I_0}{2\pi\rho}$$

Using Eq. 6-67 with $\mu = \mu_0$

$$W_m = \int_0^l \int_a^b \int_0^{2\pi} \frac{1}{2}\mu_0 \frac{I_0^2}{4\pi^2\rho^2} \rho d\phi d\rho dz = 10^{-7} I_0^2 \ln\left(\frac{b}{a}\right) l \qquad \text{J}$$

For $l = 1$ m:

$$W_m = 10^{-7} I_0^2 \ln\left(\frac{b}{a}\right) \text{ J/m}$$

Example 6-29

A sheet of current is oriented as shown in Figure 6-56. Determine the magnetic force on the side facing the positive x axis. $\vec{J}_s$ is a constant over the sheet.

Assuming there is no $\vec{H}$ inside the metal sheet the boundary condition Eq. 6-46 or 6-48 yields the magnetic field intensity at the boundary ($\vec{H}_2 = 0$, $\hat{n} = -\hat{x}$):

$$\vec{H} = \vec{J}_s = \vec{J}_0 \hat{z}$$

$$H_t = J_0$$

Using Eq. 6-69, and $\mu = \mu_0$,

$$P_m = \tfrac{1}{2}\mu_0 J_0^2 \text{ Pa}$$

$$F_x = \int_{-b}^{b} \int_{-a}^{a} P_m dy dz = 2ab\mu_0 J_0^2 \qquad \text{N}$$

This force will be in the $-\hat{x}$ direction.

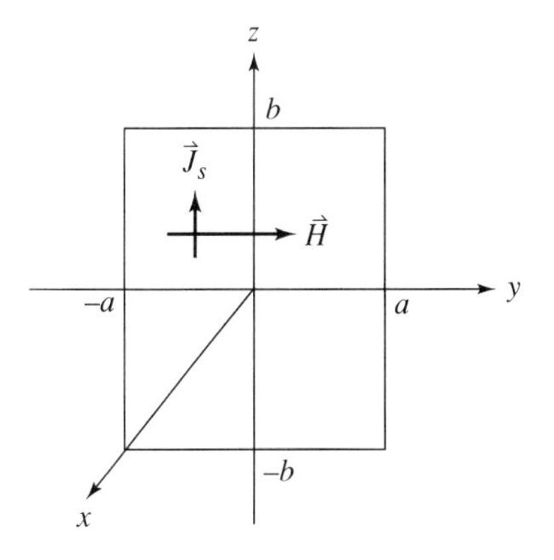

Figure 6-56. Current sheet.

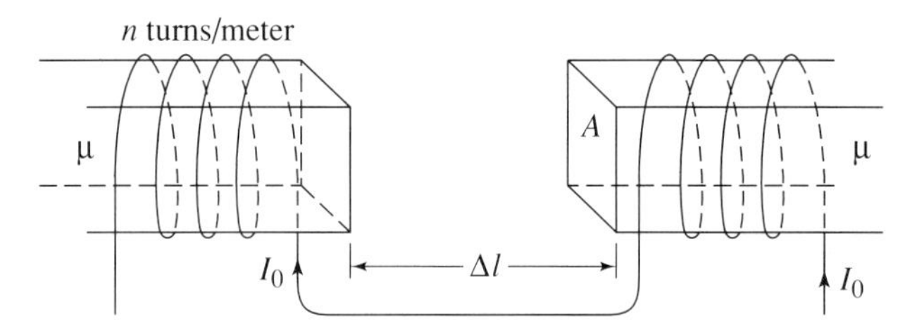

Figure 6-57.

Example 6-30

Determine the force between the two magnetic core materials shown in Figure 6-57. The core material has a permeability $\mu \neq \mu_0$, and the cross section area is A, with n turns per meter. Only the cores are moved. The turns are stationary. From Example 6-18 the magnetic field intensity inside the core is nI_0. Then

$$B_{\text{material}} = \mu H_{\text{material}} = \mu n I_0 \text{ T}$$

Now we suppose that the two bars initially touched and were then drawn apart a small distance dl. Neglecting fringing we assume the same flux ϕ crosses the gap so that $B_{\text{air}} = B_{\text{material}}$. The energy in the gap is given by Eq. 6-67 to be

$$\Delta W_m = \int_0^{\Delta l} \int_A \frac{1}{2} \mu_0 H_{\text{air}}^2 da dl = \frac{1}{2} \mu_0 \left(\frac{B_{\text{air}}}{\mu_0} \right)^2 A \, \Delta l = \frac{1}{2\mu_0} (B_{\text{material}})^2 A \, \Delta l$$

The work done is given by

$$\Delta W_m = F \, \Delta l$$

Comparing these last two equations:

$$F = \frac{1}{2\mu_0} (B_{\text{material}})^2 A = \frac{1}{2\mu_0} (\mu n I_0)^2 A = \frac{\mu^2 n^2 I_0^2 A}{2\mu_0} \text{ N}$$

See Exercise 6-12.0d on this result!

Exercises

6-12.0a For the coax configuration of Figure 6-32 derive expressions for the energy stored in the outer conductor and in the center conductor per meter of coax.

6-12.0b Determine the pressure on the outer conductor of the coax. What is the direction of the pressure?

6-12.0c Determine the energy stored in the interior region of an inductor of n turns per meter, I_0 ampere, and length L neglecting end effects.

6-12.0d Explain why the result of Example 6-30 would not be valid for a core material with $\mu = \mu_0$, such as wood. What does this tell you about permissible μ values? (Answer: Wooden cores with $\mu = \mu_0$ would not displace or change $\vec{H}$ so no energy would be required. This result is then valid only for $\mu \gg \mu_0$.)

As pointed out in Exercise 6-12.0d the result of Example 6-28 is a little deceptive because it would predict a force on the core for any μ whereas only magnetizable cores have forces on them. What we need is an expression for the forces in terms of the magnetization so the results more clearly give the true picture. To obtain this expression we need to return to the forces on dipole moments as shown in Figure 6-58. We follow the technique in [37], ¶34. The loop carrying a current I_0 is immersed in a magnetic flux density $\vec{B}$. The total force on the loop is the sum of the forces on each side. Using Eq. 6-4 we have, for the incremental lengths:

$$\begin{aligned}\Delta\vec{F} &= \Delta\vec{F}_1 + \Delta\vec{F}_2 + \Delta\vec{F}_3 + \Delta\vec{F}_4 \\ &= I_0[\Delta x\hat{x} \times \vec{B}(y) + \Delta y\hat{y} \times \vec{B}(x + \Delta x) + \Delta x(-\hat{x}) \times \vec{B}(y + \Delta y) \\ &\quad + \Delta y(-\hat{y}) \times \vec{B}(x)] \\ &= I_0[\Delta x(B_y(y)\hat{z} - B_z(y)\hat{y}) + \Delta y(B_x(x + \Delta x)(-\hat{z}) + B_z(x + \Delta x)\hat{x}) \\ &\quad + \Delta x(B_y(y + \Delta y)(-\hat{z}) + B_z(y + \Delta y) + \Delta y(B_x(x)\hat{z} - B_z(x)\hat{x})]\end{aligned}$$

Factoring out $\Delta x\,\Delta y$ and collecting terms on the unit vectors:

$$\begin{aligned}\Delta\vec{F} = I_0\,\Delta x\,\Delta y\Bigg[&\frac{B_z(x + \Delta x) - B_z(x)}{\Delta x}\hat{x} + \frac{B_z(y + \Delta y) - B_z(y)}{\Delta y}\hat{y} \\ &- \left(\frac{B_x(x + \Delta x) - B_x(x)}{\Delta x} + \frac{B_y(y + \Delta y) - B_y(y)}{\Delta y}\right)\hat{z}\Bigg]\end{aligned}$$

Now the product $I_0\,\Delta x\,\Delta y$ is the dipole moment of the loop (Eq. 6-29) in the $\hat{z}$ direction. Also, if we take the limit Δx and Δy going to zero, the incremental terms become partial derivatives. With these, the force becomes

$$\vec{F} = m_z\left[\frac{\partial B_z}{\partial x}\hat{x} + \frac{\partial B_z}{\partial y}\hat{y} - \left(\frac{\partial B_x}{\partial x} + \frac{\partial B_y}{\partial y}\right)\hat{z}\right] \text{ N}$$

The term in parentheses is almost the divergence of $\vec{B}$ given by

$$\nabla \cdot \vec{B} = \frac{\partial B_x}{\partial x} + \frac{\partial B_y}{\partial y} + \frac{\partial B_z}{\partial z} = 0$$

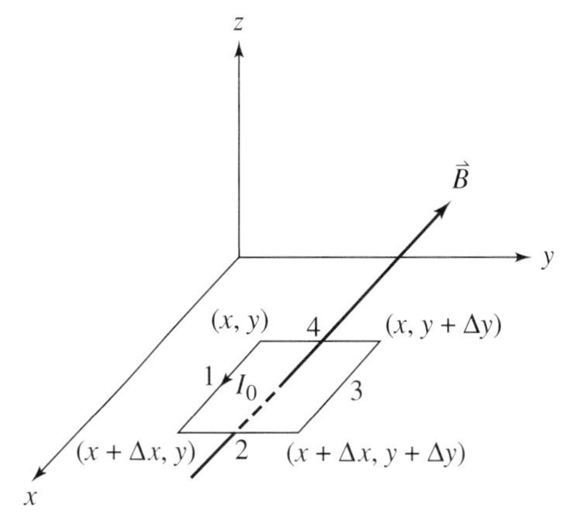

Figure 6-58. Configuration for determining the force on a magnetic dipole immersed in a magnetic field $\vec{B}$.

From this,

$$-\left(\frac{\partial B_x}{\partial x} + \frac{\partial B_y}{\partial y}\right) = \frac{\partial B_z}{\partial z}$$

So the force equation becomes

$$\vec{F} = m_z\left(\frac{\partial B_z}{\partial x}\hat{x} + \frac{\partial B_z}{\partial y}\hat{y} + \frac{\partial B_z}{\partial z}\hat{z}\right) \tag{6-70}$$

The first two components look like curl terms so we evaluate $\nabla \times \vec{H} = \vec{J}$, using Eq. 6-32 to replace $\vec{H}$:

$$\nabla \times \vec{H} = \nabla \times \left(\frac{\vec{B}}{\mu_0} - \vec{M}\right) = \frac{1}{\mu_0}\nabla \times \vec{B} - \nabla \times \vec{M} = \vec{J}$$

Solving:

$$\nabla \times \vec{B} = \mu_0\vec{J} + \mu_0\nabla \times \vec{M}$$

Using the magnetic current $\vec{J}_m$, Eq. 6-62 and expanding the left curl:

$$\left(\frac{\partial B_z}{\partial y} - \frac{\partial B_y}{\partial z}\right)\hat{x} + \left(\frac{\partial B_x}{\partial z} - \frac{\partial B_z}{\partial x}\right)\hat{y} + \left(\frac{\partial B_y}{\partial x} - \frac{\partial B_x}{\partial y}\right)\hat{z} = \mu_0\vec{J} + \mu_0\vec{J}_m = \mu_0(\vec{J} + \vec{J}_m)$$

Equating vector components

$$\frac{\partial B_z}{\partial y} - \frac{\partial B_y}{\partial z} = \mu_0(J_x + J_m x)$$

$$\frac{\partial B_x}{\partial z} - \frac{\partial B_z}{\partial x} = \mu_0(J_y + J_m y)$$

$$\frac{\partial B_y}{\partial x} - \frac{\partial B_x}{\partial y} = \mu_0(J_z + J_m z)$$

Solving the first equation for $\partial B_z/\partial y$ and the second for $\partial B_z/\partial x$ and substituting those results into Eq. 6-70 there results

$$\vec{F} = \left(m_z\frac{\partial B_x}{\partial z}\hat{x} + m_z\frac{\partial B_y}{\partial z}\hat{y} + m_z\frac{\partial B_z}{\partial z}\hat{z}\right) + \mu_0[-m_z(J_y + J_m y)\hat{x} + m_z(J_x + J_m x)\hat{y}]$$

The left part of this expression in parentheses could be obtained by the vector operation (see Chapter 4, Exercise 4-17.0g and Section 4-18 identities in part 2)

$$(m_z\hat{z}\cdot\nabla)\vec{B} = \left[m_z\hat{z}\cdot\left(\frac{\partial}{\partial x}\hat{x} + \frac{\partial}{\partial y}\hat{y} + \frac{\partial}{\partial z}\hat{z}\right)\right](B_x\hat{x} + B_y\hat{y} + B_z\hat{z})$$

$$= m_z\frac{\partial}{\partial z}(B_x\hat{x} + B_y\hat{y} + B_z\hat{z}) = m_z\frac{\partial B_x}{\partial z}\hat{x} + m_z\frac{\partial B_y}{\partial z}\hat{y} + m_z\frac{\partial B_z}{\partial z}\hat{z}$$

The right part of the expression for $\vec{F}$ in brackets could be obtained by

$$\mu_0 m_z \hat{z} \times (\vec{J} + \vec{J}_m) = \mu_0 \begin{bmatrix} \hat{x} & \hat{y} & \hat{z} \\ 0 & 0 & m_z \\ J_x + J_{mx} & J_y + J_{my} & J_z + J_{mz} \end{bmatrix}$$

$$= \mu_0[-m_z(J_y + J_{my})\hat{x} + m_z(J_x + J_{mx})\hat{y}]$$

For the more general case where the moment $\vec{m}$ has $\hat{x}$ and $\hat{y}$ components as well as $\hat{z}$ component we would replace $m_z\hat{z}$ by $m_x\hat{x} + m_y\hat{y} + m_z\hat{z} = \vec{m}$ and the equation for $\vec{F}$ becomes

$$\vec{F} = (\vec{m} \cdot \nabla)\vec{B} + \mu_0 \vec{m} \times (\vec{J} + \vec{J}_m) \text{ N} \tag{6-71}$$

The preceding equation is for a single dipole. If the density of these dipoles is n per unit volume the resultant force density would be, using $\vec{M} = \vec{m}n$ since $\vec{M}$ is the dipole moment per unit volume (Eq. 6-31) and replacing $\vec{J}_m$ by $\nabla \times \vec{M}$,

$$\vec{F}_d = (\vec{M} \cdot \nabla)\vec{B} + \mu_0 \vec{M} \times (\vec{J} + \nabla \times \vec{M}) \text{ N/m}^3 \tag{6-72}$$

Using Eq. 6-32 to replace $\vec{B}$

$$\begin{aligned} \vec{F}_d &= (\vec{M} \cdot \nabla)(\mu_0\vec{H} + \mu_0\vec{M}) + \mu_0\vec{M} \times (\vec{J} + \nabla \times \vec{M}) \\ &= \mu_0(\vec{M} \cdot \nabla)\vec{H} + \mu_0(\vec{M} \cdot \nabla)\vec{M} + \mu_0\vec{M} \times \vec{J} + \mu_0\vec{M} \times (\nabla \times \vec{M}) \end{aligned} \tag{6-73}$$

Using the vector identity (see Section 4-18):

$$\nabla(\vec{A} \cdot \vec{B}) = (\vec{A} \cdot \nabla)\vec{B} + (\vec{B} \cdot \nabla)\vec{A} + \vec{A} \times (\nabla \times \vec{B}) + \vec{B} \times (\nabla \times \vec{A})$$

with $\vec{A}$ and $\vec{B} \Rightarrow \vec{M}$ we obtain

$$\nabla(\vec{M} \cdot \vec{M}) = 2(\vec{M} \cdot \nabla)\vec{M} + 2\vec{M} \times (\nabla \times \vec{M})$$

or

$$\vec{M} \times (\nabla \times \vec{M}) = -(\vec{M} \cdot \nabla)\vec{M} + \tfrac{1}{2}\nabla(\vec{M} \cdot \vec{M})$$

Substituting this into Eq. 6-73 there results

$$\vec{F}_d = \mu_0(\vec{M} \cdot \nabla)\vec{H} + \mu_0\vec{M} \times \vec{J} + \tfrac{1}{2}\mu_0 \nabla(\vec{M} \cdot \vec{M})$$

The total force on the material is then

$$\vec{F} = \int_V \vec{F}_d dv = \int_V \mu_0(\vec{M} \cdot \nabla)\vec{H} dv + \int_V \mu_0\vec{M} \times \vec{J} dv + \tfrac{1}{2}\mu_0 \int_V \nabla(\vec{M} \cdot \vec{M}) dv$$

Using the theorem (see General Exercise GE4-30, Chapter 4)

$$\int_V \nabla\Phi dv = \oint_S \Phi d\vec{a}$$

the right volume integral becomes

$$\int_V \nabla(\vec{M} \cdot \vec{M}) dv = \oint_S (\vec{M} \cdot \vec{M})\, d\vec{a} = 0$$

This is zero since the surface S encloses all $\vec{M}$, so $\vec{M}$ is zero on the surface.

We finally obtain

$$\vec{F} = \int_V [\mu_0(\vec{M} \cdot \nabla)\vec{H} + \mu_0\vec{M} \times \vec{J}]dv \tag{6-74}$$

This result shows that if there is no magnetization, there is no magnetic attraction of two bodies in the absence of free current density $\vec{J}$.

Example 6-31

A sphere has a permanent, uniform magnetization given by $\vec{M} = M_0\hat{z}$. It is brought near the south pole of another magnet for which the magnetic flux density is $B_0\hat{z}$ as shown in Figure 6-59. Determine the force on the sphere. $\mu = \mu_0$ in the intervening region.

In the air, the field due to the magnet is

$$\vec{H} = \frac{\vec{B}}{\mu_0} = \frac{B_0}{\mu_0}\hat{z}$$

Converting to spherical coordinates as explained in Chapter 4:

$$\vec{H} = \frac{B_0}{\mu_0}\cos\theta\hat{r} - \frac{B_0}{\mu_0}\sin\theta\hat{\theta}$$

Similarly, for the sphere

$$\vec{M} = M_0\cos\theta\hat{r} - M_0\sin\theta\hat{\theta}$$

We will assume that $\vec{H}$ is small enough that it does not disturb the magnetization ($\vec{M}$) value, so it penetrates the sphere with minimal effect on $\vec{M}$.

In spherical coordinates the del operator is

$$\nabla = \frac{\partial}{\partial r}\hat{r} + \frac{1}{r}\frac{\partial}{\partial\theta}\hat{\theta} + \frac{1}{r\sin\theta}\frac{\partial}{\partial\phi}\hat{\phi}$$

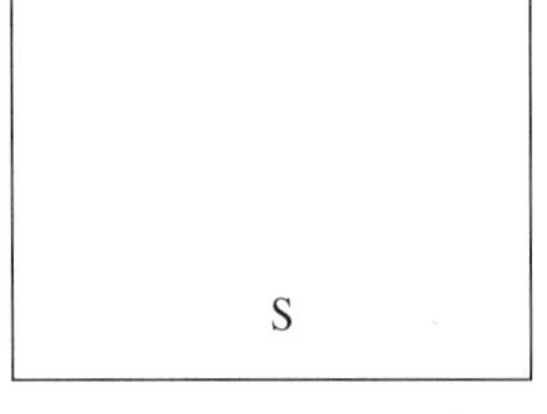

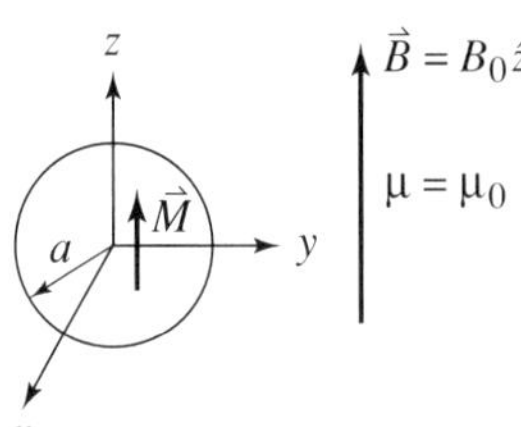

Figure 6-59. Permanently magnetized sphere in a uniform magnetic field.

Substitute these results into Eq. 6-74 using the fact also that $\vec{J}$ is zero in the sphere. Remember that unit vector must be differentiated, too (see Appendix E):

$$\vec{F} = \mu_0 \int_0^a \int_0^{2\pi} \int_0^{\pi} \left[M_0 \cos\theta \frac{\partial}{\partial r} - \frac{M_0 \sin\theta}{r} \frac{\partial}{\partial \theta} \right] \left(\frac{B_0}{\mu_0} \cos\theta \hat{r} - \frac{B_0}{\mu_0} \sin\theta \hat{\theta} \right) r^2 \sin\theta d\theta d\phi dr$$

$$= M_0 B_0 \int_0^a \int_0^{2\pi} \int_0^{\pi} - \frac{\sin\theta}{r} \left(\cos\theta \frac{\partial \hat{r}}{\partial \theta} - \sin\theta \hat{r} - \sin\theta \frac{\partial \hat{\theta}}{\partial \theta} - \cos\theta \hat{\theta} \right) r^2 \sin\theta d\theta d\phi dr$$

$$= M_0 B_0 \int_0^a \int_0^{2\pi} \int_0^{\pi} - \frac{\sin\theta}{r} [\cos\theta \hat{\theta} - \sin\theta \hat{r} + \sin\theta \hat{r} - \cos\theta \hat{\theta}] r^2 \sin\theta d\theta d\phi dr$$

$$= 0$$

This result should not be too surprising since the attraction of the magnet's south pole with the top (N) of the sphere is canceled by the downward force on the bottom of the sphere since $\vec{B}$ is constant.

Exercises

6-12.0e In Example 6-31 replace the sphere by a uniformly magnetized cylinder of radius a and length L. Magnetization is $M_0\hat{z}$. Prove that the force is zero.

6-12.0f For Example 6-31 replace the external field by $\vec{B} = [(a + z/a]B_0\hat{z}$ and center the sphere at the origin of coordinates.

6-13 Magnetomotive Force and Magnetic Circuit Theory

Since many of the ideas presented in this chapter have analogous results in Chapter 5 one might suspect that circuit theory concepts can be applied to magnetic systems. By way of summary, Table 6-1 identifies the analogies. The equations are numbered for cross reference and are written in time-invariant form for the static case using block letters for field quantities in place of the script as in earlier chapters.

The key analogies are flux ϕ corresponding to current I, magnetostatic potential (mmf) corresponding to electrostatic potential (emf), and permeability μ corresponding to conductivity σ. Since Ohm's law defines a resistance that relates emf to current, we define reluctance $\mathcal{R}$, which similarly relates mmf to flux. Reluctance is measured in the unit of rel, and its computation for a given configuration is similar to that for resistance, an example of which is given in the last entry in the table. On this basis, all of the DC circuit theory concepts using Kirchhoff's laws and Ohm's law apply directly to magnetic circuits. The notions of series and parallel elements is also very helpful.

As with the electric circuit case, the reluctance of a series path (same flux ϕ in all sections) is the sum of the reluctances of the elements in series. The use of this concept is best illustrated by example.

Table 6-1. Electric and Magnetic Analogies

Electrical	Magnetic
Current, ampere $I = \int_S \vec{J} \cdot d\vec{a}$ (Eq. 5-30)	Flux weber $\Phi = \int_S \vec{B} \cdot d\vec{a}$ (Eq. 6-3)
Voltage potential, or electromotive force (emf), volt $V_{A,\mathrm{ref}} = -\int_{\mathrm{ref}}^{A} \vec{E} \cdot d\vec{l}$ (Eq. 5-62)	Scalar magnetic potential or magnetomotive force (mmf), ampere (also ampere-turns, A · t) $V_m = -\int_{\mathrm{ref}}^{A} \vec{H} \cdot d\vec{l}$ (Eq. 6-52)
Ohm's law $\vec{J} = \sigma\vec{E}$ (Eq. 5-31) or $V_{\mathrm{AB}} = \mathrm{RI}$	Constitutive relation $\vec{B} = \mu\vec{H}$ (Eq. 6-34) or, by analogy, $V_m \equiv \mathcal{R}\Phi$ (new definition)
For constant cross section (A) and length L, resistance (ohm) $R = \dfrac{L}{\sigma A}$ (Eq. 5-35)	For constant cross section (A) and length L, reluctance (rel) $\mathcal{R} \equiv \dfrac{L}{\mu A}$ (new definition)

Example 6-32

Consider the magnetic circuit of Figure 6-60*a*. Here we neglect any flux outside the core area and suppose that in the air gap fringing of 20% of the area occurs. Determine the current required to produce 50 mT flux density in the gap and determine the flux density in the core. There are 300 turns on the coil. A magnetic circuit is shown in Figure 6-60*b*. For the gap the effective area is

$$A_g = (1 + 0.2)(1.2 \times 2)\ \mathrm{cm}^2 = 2.88 \times 10^{-4}\ \mathrm{m}^2$$

The gap reluctance is then ($\mu = \mu_0$ for air)

$$\mathcal{R}_g = \frac{L_g}{\mu_0 A_g} = \frac{0.0015}{4\pi \times 10^{-7} \times 2.88 \times 10^{-4}} = 4.145 \times 10^6\ \text{rels}$$

The gap flux desired (assuming uniform distribution) is

$$\phi_g = B_g \cdot A_g = (0.05)(2.88 \times 10^{-4}) = 1.44 \times 10^{-5}\ \mathrm{Wb}$$

The mmf drop across the air gap is then

$$V_{mg} = \mathcal{R}_g \phi_g = 59.7\ \mathrm{A \cdot t}$$

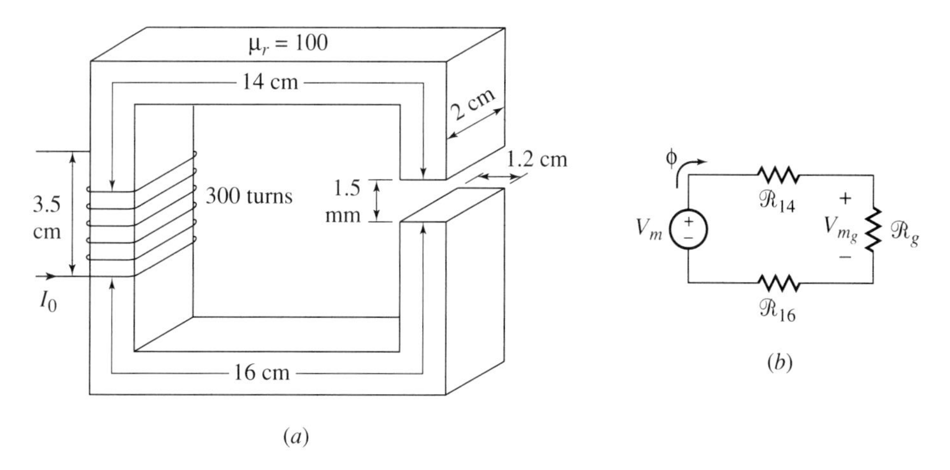

Figure 6-60. Magnetic network. (*a*) Physical configuration. (*b*) Magnetic circuit.

Flux density in the magnetic core (neglecting leakage):

$$B_c = \frac{\phi_c}{A_c} = \frac{\phi_g}{A_c} = \frac{1.44 \times 10^{-5}}{0.012 \times 0.02} = 60 \text{ mT}$$

Reluctance of the core outside the coil

$$\begin{aligned} \mathcal{R}_c = \mathcal{R}_{14} + \mathcal{R}_{16} &= \frac{0.14}{100 \times 4\pi \times 10^{-7} \times 2.4 \times 10^{-4}} \\ &+ \frac{0.16}{100 \times 4\pi \times 10^{-7} \times 2.4 \times 10^{-4}} \\ &= 9.95 \times 10^6 \text{ rels} \end{aligned}$$

The mmf drop around the core material is then

$$V_{mc} = \mathcal{R}_c \phi_c = 9.95 \times 10^6 \times 1.44 \times 10^{-5} = 143 \text{ A} \cdot \text{t}$$

Total mmf that must be provided by the coil, using the analog of Kirchhoff's voltage law,

$$V_m = V_{mg} + V_{mc} = 203 \text{ A} \cdot \text{t}$$

Making an assumption of a long coil so we can use the results of Example 6-19 (a little optimistic since the coil winding is short), the driving source is found from the field intensity

$$H_w = nI_0 = \frac{300}{0.035} I_0 = 8.57 \times 10^3 I_0 \text{ A} \cdot \text{t/m}$$

(Note n is turns per meter.)

Using the integral expression for V_m, Eq. 6-52 in Table 6-1 (using magnitudes only), the driving mmf source becomes

$$V_m = 203 = \int_0^{.035} 8.57 \times 10^3 I_0 dl = 300 I_0$$

$$I_0 = 0.68 \text{ A}$$

A practical point should be made at this time. Since the core is magnetic, one would have to check the nonlinear characteristic of the core to determine whether the core flux density of 60 mT corresponded with $\mu_r = 100$ since some saturation may reduce the actual μ_r value (see Figure 6-43, the static permeability). The next example illustrates this.

Example 6-33

The core configuration shown in Figure 6-61*a* is made of a material that has the trade name Supermendur. The material has the magnetic characteristic given in Figure 6-61*b*. Determine the flux density in the core.

This is analogous to a single resistor (core) connected to a source. Assuming a perfect solenoid for the windings, the magnetic field inside the windings (the driving source) is

$$H_w = nI_0 = \left(\frac{100}{0.05}\right)\left(\frac{1}{2}\right) = 1000 \text{ A} \cdot \text{t/m}$$

$$R_{\text{core}} = \frac{L}{\mu A} = \frac{0.3}{\mu 4 \times 10^{-4}} = \frac{750}{\mu}$$

The problem is that we have no value for μ because we don't know where the operating point on the curve is. For example, if the operation were at Point 1, the slope of the dotted line yields

$$\mu_1 = \frac{B_1}{H_1} \approx \frac{0.2}{12} = 0.017 \text{ H/m} \qquad (\mu_{r1} \approx 13{,}270)$$

However, at Point 2 we would have

$$\mu_2 = \frac{B_2}{H_2} \approx \frac{1.8}{26} = 0.069 \text{ H/m} \qquad (\mu_{r2} \approx 55{,}100)$$

Since the mmf drop in the core must equal the source we then have

$$V_m = R_{\text{core}}\phi = R_{\text{core}} B \cdot A = \frac{750}{\mu} B \times 4 \times 10^{-4} = 0.3 \frac{B}{\mu}$$

From H_ω:

$$V_m = \int_0^{0.05} H dl = 1000 \times 0.05 = 50 \text{ A} \cdot \text{t}$$

Equating two values of V_m:

$$0.3 \frac{B}{\mu} = 50$$

Then we require

$$\frac{B}{\mu} = \frac{50}{0.3} = 167 = H$$

From the material characteristic curves, this value of H has a corresponding magnetic flux density of

$$B = 2.35 \text{ T}$$

The static (DC) permeability at Point 3 is

$$\mu_3 \approx \frac{2.35}{167} = 0.014 \qquad (\mu_r \approx 10{,}960)$$

Exercises

6-13.0a For Example 6-32, increase the air gap by a factor of 2 to 3 mm and find the current required to produce the same flux density of 50 mT.

6-13.0b For Example 6-32, suppose there were 1500 turns wound to cover all the core but leave the gap exposed. This would eliminate the core reluctance terms and leave only the gap. Determine the current required to produce the 1.5 T flux density.

6-13.0c For Example 6-33, estimate the flux density in the coil region, assuming the characteristic (b) in Figure 6-61 is flat beyond 70 A/m.

6-13.0d Assuming linear material and $\mu_r = 300$ in Figure 6-62, compute the flux and flux density in the core of each leg. Assume that the center core winding covers its entire mean length. (Hint: The two parallel legs must have the same mmf across them.)

6-14 Inductance

Inductors are important energy storage devices, and their presence is manifest by the magnetic flux produced by current flow. From Chapter 1 we found that the characteristic impedance of a transmission line is a function of the inductance per meter, as is the signal propagation velocity. We next study methods for computing this important parameter.

Formally we define the inductance of a system as

$$L = \frac{\Lambda}{I} \tag{6-75}$$

where Λ is called the flux linkage of the system and is the total flux in the region under consideration, and I is the current that produces the flux from which Λ is computed. This definition gives inductance the unit of weber/ampere, which is called the *henry*, denoted

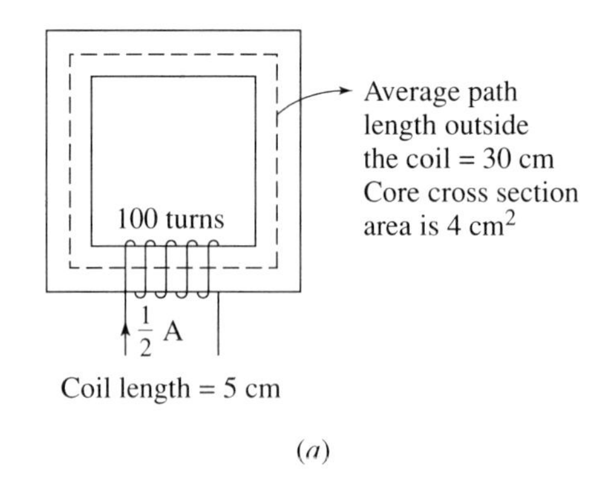

(*a*)

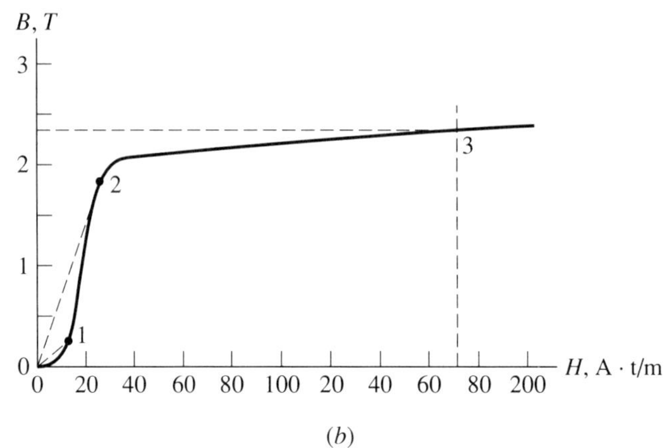

(*b*)

Figure 6-61. Magnetic circuit. (*a*) Configuration. (*b*) Properties of the Supermendur core material.

Figure 6-62.

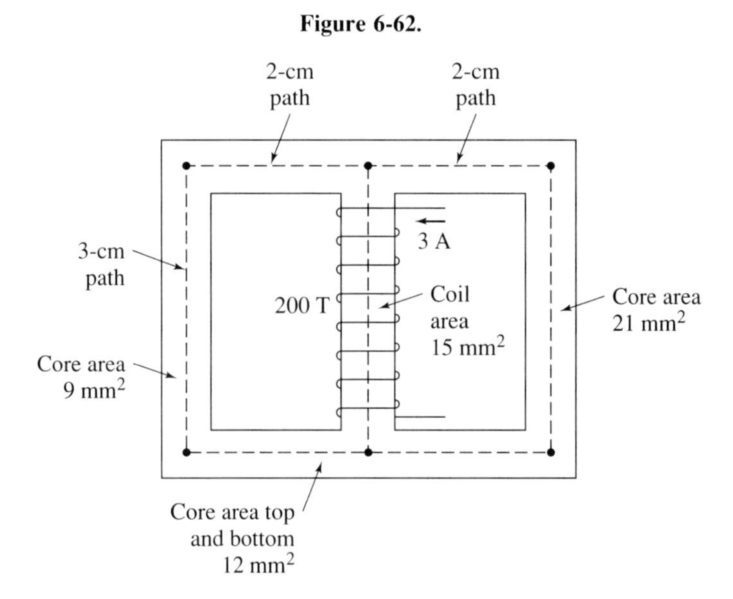

by H. In some systems the inductance is not a constant due to nonlinear effects of the materials in which the flux is computed. These cases are usually described by the equation

$$L = \left.\frac{d\Lambda}{dI}\right|_{I=I_{DC}} \tag{6-76}$$

Where I_{DC} is the current in the system (often called the *offset* or *bias* current) about which the small (differential) changes are being determined.

The inductance can also be computed using the energy relationship from circuit theory:

$$W_L = \tfrac{1}{2}LI^2$$

This yields

$$L = \frac{2W_L}{I^2} \tag{6-77}$$

In this context the energy stored is in the magnetic fields of the structure. This energy is computed using the integral relations given in Section 6-12. There is one important feature of inductance caused by the fact that *static* magnetic fields can exist inside a perfect conductor whereas electric fields cannot. Thus we need to define external and internal inductances. The external inductance is due to that flux linkage external to the conductor and internal inductance due to the flux linkage in the current region. Under Eq. 6-75 the relationship is

$$L_{\text{total}} = \frac{\Lambda_{\text{total}}}{I} = \frac{\Lambda_{\text{ext}} + \Lambda_{\text{int}}}{I} = L_{\text{ext}} + L_{\text{int}} \tag{6-78}$$

The most difficult part of the computation process is the proper identification of surfaces and contours, and care must be given to their selection. In many practical cases the internal inductance is small and only L_{ext} is used.

Example 6-34

Derive the expression for the external inductance of the coaxial cable shown in Figure 6-63.

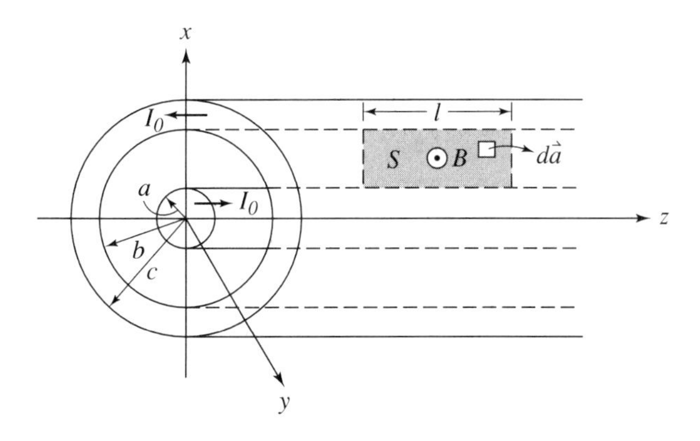

Figure 6-63. Computation configuration for coax external inductance.

The area between the conductors gives the external inductance. Using the result of Example 6-14 of Section 6-6 for the magnetic field between the conductors

$$\Lambda_{\text{ext}} = \phi_{\text{ext}} = \int_S \vec{B} \cdot d\vec{a} = \mu \int_S \vec{H} \cdot d\vec{a} = \mu \int_0^l \int_a^b \frac{I_0}{2\pi\rho} \hat{\phi} \cdot d\rho dz \hat{\phi}$$

$$= \frac{\mu I_0 l}{2\pi} \ln\left(\frac{b}{a}\right)$$

$$L_{\text{ext}} = \frac{\Lambda_{\text{ext}}}{I_0} = \frac{\mu l}{2\pi} \ln\left(\frac{b}{a}\right) \text{ H}$$

On a per-meter basis,

$$L = \frac{L_{\text{ext}}}{l} = \frac{\mu}{2\pi} \ln\left(\frac{b}{a}\right) \text{ H/m}$$

Example 6-35

Compute the internal inductance per meter for the center conductor of the coax of Figure 6-63, assumed to be copper with $\mu = \mu_0$.

The proper area now is *within* the wire from $0 \leq \rho \leq a$. The area would have the same basic placement as that shown, except within the wire. Using the field equation from Example 6-14 for that region:

$$\Lambda_{\text{int}} = \int_{S_{\text{int}}} \vec{B} \cdot d\vec{a} = \int_0^l \int_0^a \frac{\mu_0 I_0}{2\pi a^2} \rho \hat{\phi} \cdot d\rho dz \hat{\phi} = \frac{\mu_0 I_0 l}{4\pi}$$

$$L_{\text{int}} = \frac{\Lambda_{\text{int}}}{I_0 l} = \frac{\mu_0}{4\pi} \text{ H/m} = 10^{-7} \text{ H/m}$$

Note that this is the inductance (at low frequencies) of a single wire, and it is independent of wire radius.

Example 6-36

Compute the external inductance of a tightly wound circular solenoid shown in Figure 6-64. This time use the stored energy concept and use the distance a as the mean toroid radius. The core cross section has radius b.

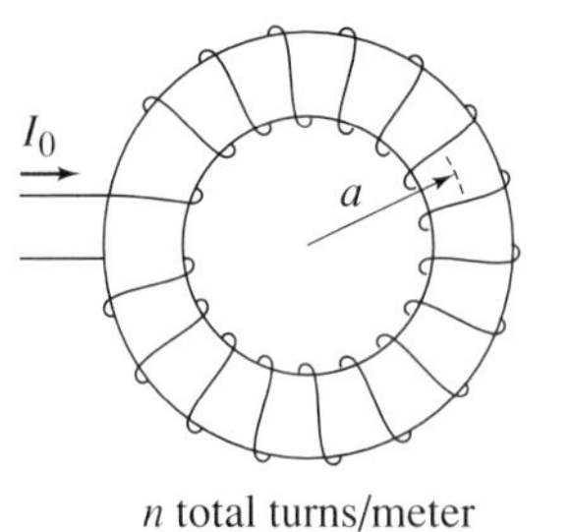

Figure 6-64. Definition of quantities used to compute toroid external inductance.

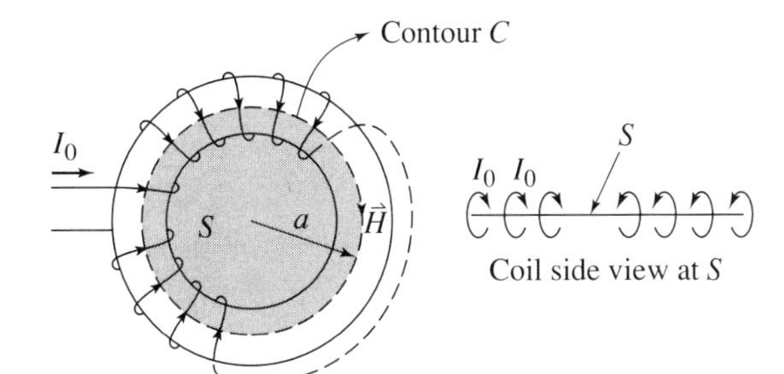

Figure 6-65. Identification of the surface and contour used to compute the magnetic field intensity in a toroid.

We first need to obtain the equation for the magnetic field, using Ampere's law. The surface and contour required by the law are shown in Figure 6-65. The surface S for Ampere's law is a disk of radius a in the plane of the toroid. Note that the current passes down through the disk. By the right hand rule the magnetic field intensity will be along the contour C which bounds S. Physically, $\vec{H}$ will not be a function of the coordinate ϕ, so we can write

$$\vec{H} = H_\phi(\rho)\hat{\phi}$$

Using Ampere's law, Eq. 6-23,

$$\oint \vec{H} \cdot d\vec{l} = \int_0^{2\pi} H_\phi(\rho)\hat{\phi} \cdot ad\phi\hat{\phi} = 2\pi a H_\phi(\rho)\big|_{\rho=a} = 2\pi a H_\phi(a)$$

We further assume (as was justified in Example 6-18 for a straight solenoid) that $\vec{H}$ is constant across the core so is also independent of the ρ coordinate.

$$I_{\text{enclosed}} = I \qquad \text{passing through} \qquad S = n \times 2\pi a \times I_0 \text{ A}$$

Equating this to the integral result we obtain

$$H_\phi = nI_0 \text{ A/m}$$

To obtain the energy in the core material we use Eq. 6-67 on Figure 6-64, using H_ϕ as a constant:

$$W_L = W_m = \int_{V_{\text{core}}} \tfrac{1}{2}\mu H^2 dv = (\tfrac{1}{2}\mu n^2 I_0^2)(\pi b^2 \cdot 2\pi a) = \mu\pi^2 ab^2 n^2 I_0^2$$

Using Eq. 6-77

$$L_{\text{ext}} = \frac{2 \times \mu\pi^2 ab^2 n^2 I_0^2}{I_0^2} = 2\mu\pi^2 ab^2 n^2 \text{ H}$$

Example 6-37

Determine the external inductance per meter of a parallel-plate transmission line shown in Figure 6-66. Assume the plates are close together so that $d \ll w$, which means that end effects are negligible.

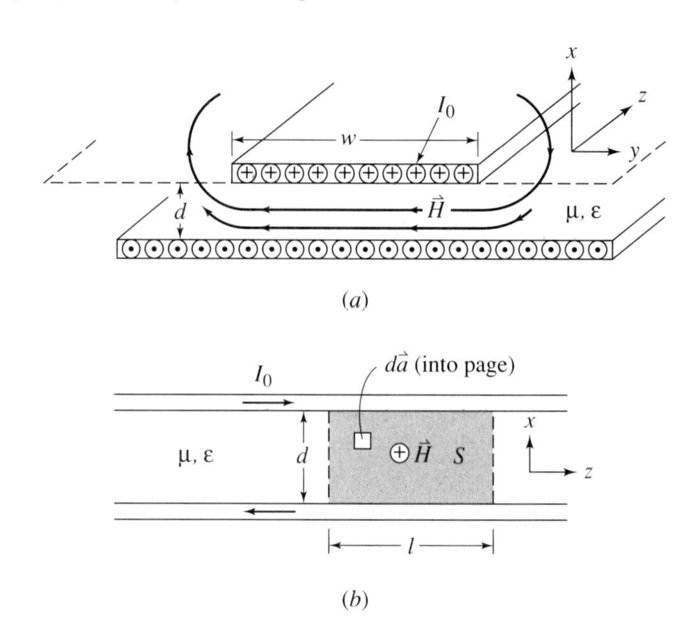

Figure 6-66. Strip transmission line magnetic field configuration for determining external inductance. (*a*) Perspective view (end). (*b*) Side view.

From Example 6-17 of Section 6-6 we have the magnetic field intensity $\vec{H}$ for parallel plates whose ends may be neglected:

$$H = J_s = \frac{I_0}{W} \text{ A/m}$$

Selecting the surface S as shown in Figure 6-66*b*:

$$\Lambda_{\text{ext}} = \int_S \vec{B} \cdot d\vec{a} = \int_0^l \int_0^d \mu \frac{I_0}{W} dxdz = \mu \frac{I_0}{W} ld$$

$$L_{\text{ext}} = \frac{\Lambda_{\text{ext}}}{I_0 l} = \mu \frac{d}{W} \text{ H/m}$$

Exercises

6-14.0a Determine the equation for the internal inductance of the outer conductor for the coax of Figure 6-63.

6-14.0b Using the stored field energy concept compute the external inductance of the coax of Figure 6-63.

6-14.0c In Example 6-17 of Section 6-6 it was shown that the two-current sheet configuration (Figure 6-36) had zero magnetic field outside the plates and J_s at the inner surface. In Figure 6-66 let the thickness of the top plate be t and assume the field inside decreases from J_s (which is I_0/W) to zero linearly and compute the internal inductance of the top plate per meter.

In circuit theory the concept of mutual inductance is introduced to account for the fact that a current flow in one part of a network can induce currents in other parts of a network by way of the magnetic field coupling. Although a time-varying current is required to induce the currents, the proportionality constant called *mutual inductance* can be estimated from static field structures. The symbol M is used for mutual inductance. This is unfortunately the same symbol used for magnetization, although magnetization is a vector quantity. In any event, the discussion in which the symbol occurs easily identifies which quantity is intended.

By definition, a current flow in Circuit 1 that generates a magnetic flux, a part of which links Circuit 2, has a mutual inductance:

$$M_{12} \equiv \frac{\Lambda_{12}}{I_1} = \frac{\phi_{12,\text{total}}}{I_1} \tag{6-79}$$

Expressing the flux in terms of $\vec{B}_1$ and then $\vec{A}_1$ followed by an application of Stokes' theorem

$$M_{12} = \frac{\int_{S_2}\vec{B}_1 \cdot d\vec{a}_2}{I_1} = \frac{\int_{S_2}\nabla \times \vec{A}_1 \cdot d\vec{a}_2}{I_1} = \frac{\oint_{C_2}\vec{A}_1 \cdot d\vec{l}_2}{I_1} \tag{6-80}$$

For a filament of current I_1 Eq. 6-41 reduces to

$$\vec{A}_1 = \mu \int_{C_1} \frac{I_1 d\vec{l}_1}{4\pi R} \Rightarrow \mu \oint_{C_1} \frac{I_1 d\vec{l}_1}{4\pi R}$$

The closed path integral results from the fact that for DC conditions the current must close on itself. Using this in M_{12}

$$M_{12} = \frac{\oint_{C_2}\left(\mu \oint_{C_1} \frac{I_1 d\vec{l}_1}{4\pi R}\right) \cdot d\vec{l}_2}{I_1} = \frac{\mu}{4\pi} \oint_{C_2} \oint_{C_1} \frac{1}{R} d\vec{l}_1 \cdot d\vec{l}_2.$$

Since the dot product commutes,

$$M_{12} = \frac{\mu}{4\pi} \oint_{C_1} \oint_{C_2} \frac{1}{R} d\vec{l}_2 \cdot d\vec{l}_1 = M_{21}$$

Thus the subscript is unnecessary on the mutual inductance so that

$$M = \frac{\mu}{4\pi} \oint_{C_1} \oint_{C_2} \frac{1}{R_{12}} d\vec{l}_2 \cdot d\vec{l}_1 \tag{6-81}$$

This is Neumann's formula for mutual inductance.

Example 6-38

This example also introduces elliptic functions. Determine the mutual inductance of two coaxial loops shown in Figure 6-67. The rectangular components of R_{12} are (assuming $b > a$)

$$R_{12x} = b - a\cos\phi$$
$$R_{12y} = a\sin\phi$$
$$R_{12z} = L$$

Then

$$R_{12} = \sqrt{L^2 + (a\sin\phi)^2 + (b - \alpha\cos\phi)^2} = \sqrt{a^2 + b^2 + L^2 - 2ab\cos\phi}$$

From the figure the general expressions for the differential lengths are

$$d\vec{l}_1 = ad\phi_1\hat{\phi}_1 \qquad d\vec{l}_2 = bd\phi_2\hat{\phi}_2$$

It is important to note here that the angle ϕ is the *relative* displacement angle between $d\vec{l}_1$ and $d\vec{l}_2$; that is,

$$\phi = \phi_1 - \phi_2$$

Thus

$$d\vec{l}_2 \cdot d\vec{l}_1 = abd\phi_1 d\phi_2 \hat{\phi}_1 \cdot \hat{\phi}_2 = abd\phi_1 d\phi_2 \cos\phi = ab\cos(\phi_1 - \phi_2)d\phi_1 d\phi_2$$

Using these results Neumann's formula (Eq. 6-81) gives

$$M = \frac{\mu}{4\pi}\int_0^{2\pi}\int_0^{2\pi}\frac{ab\cos(\phi_1 - \phi_2)d\phi_1 d\phi_2}{\sqrt{a^2 + b^2 + L^2 - 2ab\cos(\phi_1 - \phi_2)}}$$

The integration with respect to ϕ_1 requires holding ϕ_2 constant. Also, since *both* integrations are over $(0, 2\pi)$, we may start at any value of the variable ϕ_2 as long as the

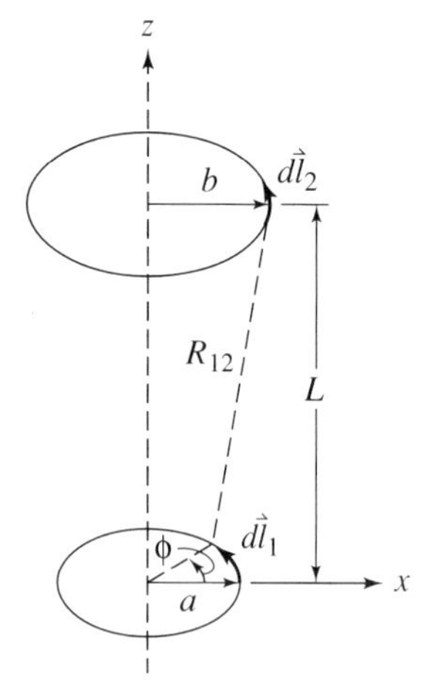

Figure 6-67. Coaxial loop system for computation of mutual inductance.

2π range is covered. Hold ϕ_2 = constant = 0 for the ϕ_1 integration. The expression for M becomes

$$M = \frac{\mu ab}{4\pi} \int_0^{2\pi} \int_0^{2\pi} \frac{\cos \phi_1 d\phi_1}{\sqrt{a^2 + b^2 + L^2 - 2ab \cos \phi_1}} d\phi_2$$

Since the inner integral is independent of ϕ_2 it may be taken outside the ϕ_2 integration,

$$M = \frac{\mu ab}{4\pi} \int_0^{2\pi} d\phi_2 \int_0^{2\pi} \frac{\cos \phi_1 d\phi_1}{\sqrt{a^2 + b^2 + L^2 - 2ab \cos \phi_1}}$$

$$= \frac{\mu ab}{2} \int_0^{2\pi} \frac{\cos \phi_1}{\sqrt{a^2 + b^2 + L^2 - 2ab \cos \phi_1}} d\phi_1$$

This integral does not appear explicitly in tables, so we have two alternatives:

1. Evaluate the integral numerically for given system dimensions.
2. Put the integral into a form identifiable in tables.

For the latter method, we observe in tables that the most frequent forms appearing have $\sqrt{1 \pm k^2 \sin^2 \phi}$, so we make the substitution

$$\phi_1 = 2\left(\frac{\pi}{2} - \theta\right) = \pi - 2\theta$$

where the $\pi/2$ converts a cosine to a sine and the 2 multiplier lets us use the double angle formula from trigonometry, which gives us squared functions. Then

$$\cos \phi_1 = \cos(\pi - 2\theta) = -\cos 2\theta = 2 \sin^2 \theta - 1$$

Along with this we also must use

$$d\phi_1 = -2d\theta$$

with the limits

$$\theta = \left.\frac{\pi - \phi_1}{2}\right|_{\phi_1 = 0, 2\pi} = \frac{\pi}{2}, -\frac{\pi}{2}$$

The expression for M becomes

$$M = \frac{\mu ab}{2} \int_{\pi/2}^{-\pi/2} \frac{2 \sin^2 \theta - 1}{\sqrt{L^2 + (a + b)^2 - 4ab \sin^2 \theta}} (-2d\theta)$$

Since the integrand is an even function over symmetric limits,

$$M = 2ab\mu \int_0^{\pi/2} \frac{2 \sin^2 \theta - 1}{\sqrt{L^2 + (a + b)^2 - 4ab \sin^2 \theta}} d\theta$$

$$= \frac{2ab\mu}{\sqrt{L^2 + (a + b)^2}} \int_0^{\pi/2} \frac{2 \sin^2 \theta - 1}{\sqrt{1 - \dfrac{4ab}{L^2(a + b)^2} \sin^2 \theta}} d\theta$$

If we let $k^2 = 4ab/[L^2 + (a + b)^2]$, this reduces to

$$M = \mu k\sqrt{ab}\int_0^{\pi/2}\frac{2\sin^2\theta - 1}{\sqrt{1 - k^2\sin^2\theta}}\,d\theta$$

$$= \mu k\sqrt{ab}\left(-\int_0^{\pi/2}\frac{d\theta}{\sqrt{1 - k^2\sin^2\theta}} + 2\int_0^{\pi/2}\frac{\sin^2\theta}{\sqrt{1 - k^2\sin^2\theta}}\,d\theta\right)$$

The left integral is from a function called an *elliptic integral of the first kind*, denoted by

$$F(\alpha, k) \equiv \int_0^{\alpha}\frac{d\theta}{\sqrt{1 - k^2\sin^2\theta}} \tag{6-82}$$

For our specific case, $\alpha = \pi/2$, and the function is called the *complete* elliptic integral of the first kind,

$$F\left(\frac{\pi}{2}, k\right) \equiv K(k) = \int_0^{\pi/2}\frac{d\theta}{\sqrt{1 - k^2\sin^2\theta}} \tag{6-83}$$

These functions are both given in tables. Thus, for given values of a,b, and L we evaluate k^2 and consult a table.

The right integral in the equation for M is found in [24] under Section 2.584 to be

$$\frac{1}{k^2}K(k) - \frac{1}{k^2}\int_0^{\pi/2}\sqrt{1 - k^2\sin^2\theta}\,d\theta$$

The integral in this result is from a function called an *elliptic function of the second kind*, denoted by

$$E(\alpha, k) = \int_0^{\alpha}\sqrt{1 - k^2\sin^2\theta}\,d\theta \tag{6-84}$$

In our case $\alpha = \pi/2$ and this becomes the *complete* elliptic function of the second kind:

$$E\left(\frac{\pi}{2}, k\right) \equiv E(k) = \int_0^{\pi/2}\sqrt{1 - k^2\sin^2\theta}\,d\theta \tag{6-85}$$

These functions are also tabulated.

The evaluation of the right integral in M is then

$$\frac{1}{k^2}K(k) - \frac{1}{k^2}E(k)$$

Putting these results into the expression for M and combining terms:

$$M = \mu\sqrt{ab}\left[\left(\frac{2 - k^2}{k}\right)K(k) - \frac{2}{k}E(k)\right]$$

Example 6-39

Determine the mutual inductance per meter between two coaxial solenoids as shown in Figure 6-68. Neglect end fields. Let the outer solenoid carry a current I_b. From Example 6-18,

$$\vec{H}_b = n_b I_b \hat{z}$$

Then $\vec{B}_b = \mu\vec{H}_b$.

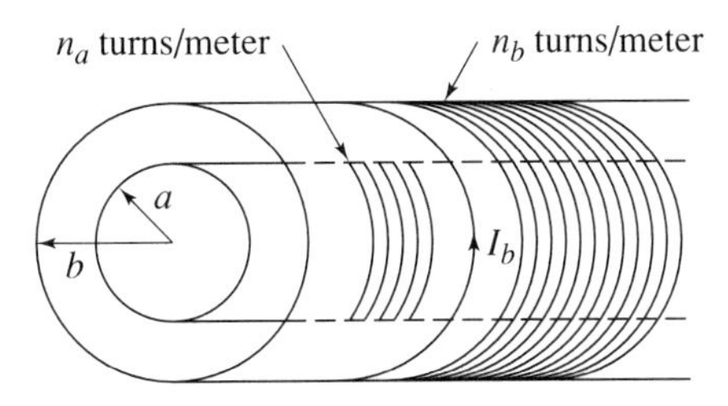

Figure 6-68. Configuration for determining the mutual inductance between coaxial solenoids.

Flux linking second (inner) solenoid:

$$\phi_{12} = \Lambda_{ba} \text{ (per meter)} = n_a \phi_{ba} = n_a \int_{S_a} \vec{B}_b \cdot d\vec{a}_a$$

$$= n_a \int_0^a \int_0^b \mu n_b I_b \hat{z} \cdot \rho d\phi d\rho \hat{z} = \mu n_a n_b \pi a^2 I_b$$

$$M = \frac{\Lambda_{ba}}{I_b} = \mu \pi n_a n_b a^2 \text{ H/m}$$

From the definitions of L and M we can obtain an expression for the tightness of magnetic coupling between two systems. If we take the total flux in System 1 and divide it into a flux component that links only *Circuit 1* (often called *leakage flux*) and a flux component that links Circuit 2 (the *mutual flux*) we would have

$$\Lambda_1 = \Lambda_{11} + \Lambda_{12}$$

Then

$$L_1 = \frac{\Lambda_1}{I_1} = \frac{\Lambda_{11} + \Lambda_{12}}{I_1} = \left(\frac{\Lambda_{11}}{\Lambda_{12}} + 1\right)\frac{\Lambda_{12}}{I_1}$$

$$= \left(\frac{\Lambda_{11} + \Lambda_{12}}{\Lambda_{12}}\right) M = \frac{\Lambda_1}{\Lambda_{12}} M$$

Similarly, since $M_{21} = M_{21} = M$ as proved earlier,

$$L_2 = \frac{\Lambda_2}{\Lambda_{21}} M$$

Multiplying these together and solving for M:

$$M = \sqrt{\frac{\Lambda_{21}\Lambda_{12}}{\Lambda_2\Lambda_1}} \sqrt{L_1 L_2}$$

Now since the mutual flux component is always less than the total flux the flux linkage radical term is always less than one. We define this as the coefficient of coupling:

$$k \equiv \sqrt{\frac{\Lambda_{21}\Lambda_{12}}{\Lambda_2\Lambda_1}} < 1$$

Then

$$M = k\sqrt{L_1 L_2} \qquad 0 \le k \le 1 \tag{6-86}$$

Example 6-40

Compute the coefficient of coupling for the two coupled inductors of Example 6-39.

If we use the results of Example 6-36 interpreted as a long, straight solenoid of length $2\pi a$ we have for the two solenoids in Figure 6-55, per meter,

$$L_a = \frac{L_{sa}}{2\pi a} = \frac{2\mu\pi^2 a \cdot a^2 n_a^2}{2\pi a} = \mu\pi a^2 n_a^2 \text{ H/m}$$

$$L_b = \frac{L_{sb}}{2\pi b} = \frac{2\mu\pi^2 b \cdot b^2 n_b^2}{2\pi b} = \mu\pi b^2 n_b^2 \text{ H/m}$$

Using the expression for M of Example 6-39 and Eq. 6-86:

$$k = \frac{M}{\sqrt{L_a L_b}} = \frac{a}{b}$$

This result is physically plausible since if the diameters of the two coils happen to be the same all the flux from one links the other (an ideal situation, of course).

Exercises

6-14.0d Derive the equation for the mutual inductance of Example 6-39 by assuming a current in solenoid a of value I_a. Assume flux outside a long solenoid is zero. Also note that the cross section area of a is also part of that of b. (Answer: $M = \mu\pi n_a n_b a^2$)

6-14.0e Derive the equation for the mutual inductance of the system shown in Figure 6-69. See Example 6-8, Section 6-4. Evaluate M for the specific case $a = 0.5$ cm, $b = c = 1$ cm.

6-14.0f Derive the equation for the mutual inductance between two square loops shown in Figure 6-70. The letters shown are coordinate values. Use Neumann's equation.

Figure 6-69.

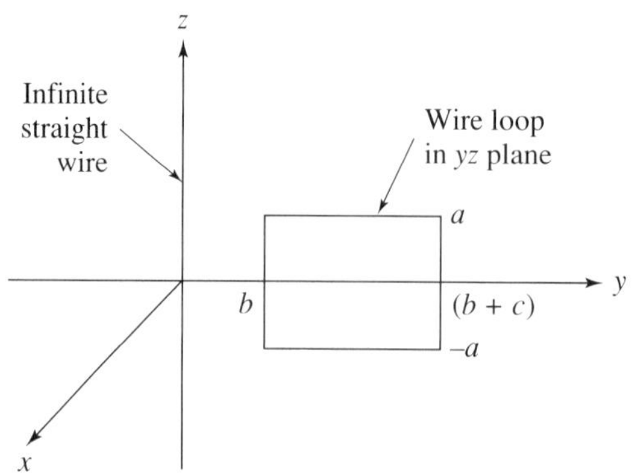

Figure 6-70.

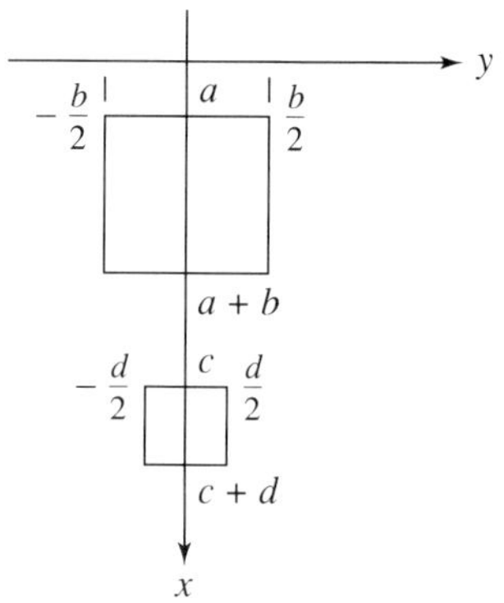

Summary, Objectives, and General Exercises

Summary

In this chapter we have learned the fundamentals of magnetic fields and the equations that govern their spatial properties and the forces they produce. The basic experiments have been presented, which led to the mathematic formulations that describe the fields. A short description of the properties of magnetic materials revealed the complex and varied matures of the response of particles, atoms, and molecules to magnetic fields.

The main applications given in this chapter included the determination of the magnetic field intensity produced by a given current distribution, the forces produced by magnetic flux density, the description of permanent magnets by scalar and vector magnetic potentials, magnetic circuit theory, and the computation of self and mutual inductances.

Objectives

1. Be able to give the names and units of and to define the following magnetic quantities: $\vec{B}\,\phi$, $\vec{J}_s$, μ, μ_0, μ_r, χ_m, $\vec{H}$, $\vec{m}$, $\vec{M}$, $\vec{A}$, V_m, ρ_{vm}, ρ_{sm}, W_m, $\mathcal{R}$, F, Λ, L, M, L_{ext}, L_{int}, k.
2. Know and be able to apply the integral relationship between the magnetic flux density and magnetic flux.
3. Describe how one determines the direction convention for a magnetic field in terms of a hypothetical isolated north pole.
4. Be able to compute the force on a current distribution immersed in a magnetic flux density field.
5. Be able to write the mathematical form of the Biot-Savart law and apply it to a given current distribution to compute the magnetic field intensity.
6. Know and be able to use the constitutive relationship for magnetic fields.
7. Be able to state and apply the Lorentz force equation.
8. Be able to describe the Hall effect.
9. Know and be able to apply Ampere's circuital law in both differential and integral forms to determine current density and magnetic field intensity.
10. Know the magnetic field intensity value in the space near a single current sheet and how to compute it, and also for two or more parallel current sheets using superposition.
11. Be able to define magnetic moment and magnetization (dipole moment per unit volume).
12. Know the relationships among the quantities $\vec{B}$, $\vec{H}$,χ_m, μ, μ_0, and μ_r.
13. Be able to compute and describe bound current and free current physically.
14. Be able to describe and define the following properties of magnetic materials:
    (a) Tensor permeability
    (b) Diamagnetic
    (c) Paramagnetic
    (d) Ferromagnetic
    (e) Curie law
    (f) Bloch wall
    (g) Antiferromagnetic
    (h) Ferrimagnetic
    (i) Weiss field
    (j) Hysteresis
    (k) Coercive field (force)
    (l) Remnant magnetization
    (m) Curie temperature
    (n) Néel temperature
    (o) Ferrite
15. Be able to state and apply Gauss' law for magnetic fields in differential and integral form.
16. Give the definition of vector magnetic potential and be able to compute it for points inside a current distribution and points outside a current distribution.
17. Know how to develop the far-field expansion for an integral expression of a field quantity.

Know how to apply the far-field expansion of the vector magnetic potential.

18. Be able to state and apply the tangential and normal boundary conditions for $\vec{B}$ and $\vec{H}$. Know how to obtain the magnetization ($\vec{M}$) boundary conditions from those of $\vec{B}$ and $\vec{H}$ using the relationship $\vec{M} = \chi_m \vec{H}$ and $\vec{B} = \mu_0 \vec{H} + \mu_0 \vec{M}$.
19. Be able to state and apply the definition of the scalar magnetic potential in differential and integral forms. Know the conditions under which the scalar magnetic potential can be used. ($\vec{J} = 0$, no encircling of currents). Be able to obtain the boundary conditions on V_m from the boundary conditions on $\vec{B}$ and $\vec{H}$.
20. Be able to compute the volume and surface magnetic charge densities from a given $\vec{M}$.
21. Be able to obtain the equation for magnetic current density from a given $\vec{M}$, and be able to state and use the relationship (differential) between $\vec{A}$ and $\vec{J}_m$.
22. Be able to compute the magnetic energy and magnetic energy density for a given magnetic field.
23. Know how to compute the pressure and force on a given surface current immersed in a magnetic field, and how to apply the force integral for magnetization $\vec{M}$, Eq. 6-74.
24. Be able to solve magnetic circuits using the concepts of reluctance and mmf, along with Kirchhoff's laws and Ohm's law.
25. Be able to write the defining equations for mutual and self inductances and to use them to compute expressions for M and L. Also be able to define external and internal inductances.
26. Be able to apply Neumann's formula for mutual inductance.
27. Know the definitions of the elliptic integrals of the first and second kinds.

General Exercises

GE6-1 At $t = 0$ a boundary is defined as shown in Figure 6-71. The flux density in the region is

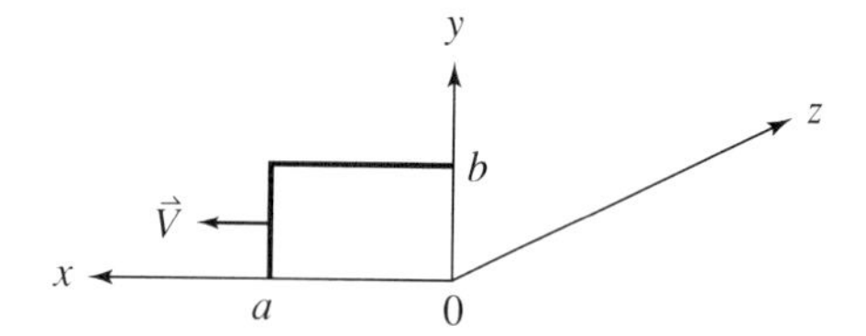

Figure 6-71.

$$\vec{B} = 3 \cos \frac{\pi}{2\alpha} x\hat{z} - 2 \sin \frac{\pi}{b} y\hat{x} \text{ T}$$

The left vertical edge then begins to move with constant velocity $\vec{V} = V_0\hat{x}$ so that the area increases. Determine the flux passing through the loop as a function of time. (Hint: The x boundary is at $a + V_0 t$). Why is it *not* permissible to use $dx = V_0 dt$ for the change of variable to integrate on time rather than x? (x defines the area not the boundary.)

GE6-2 A semicircular closed loop of radius 10 cm is in the positive y portion of the xy plane and carries a current of 200 mA. The magnetic field is $\vec{B} = -3\rho\hat{\rho} + 2\rho\hat{z}$ T. Compute the total force on the semi-circular loop if the current flows in the $+\phi$ direction.

GE6-3 A triangular system of wires carries 1 A of current as shown in Figure 6-72. $\vec{B} = -4\hat{z}$. Compute the force on each side of the system and the total force on the loop.

GE6-4 For the triangular loop shown in Figure 6-73, compute the force on side 1. Use $\vec{B} = -4\hat{z}$.

GE6-5 A semicircular plate lies in the negative y half of the xy plane and has radius b and carries

Figure 6-72.

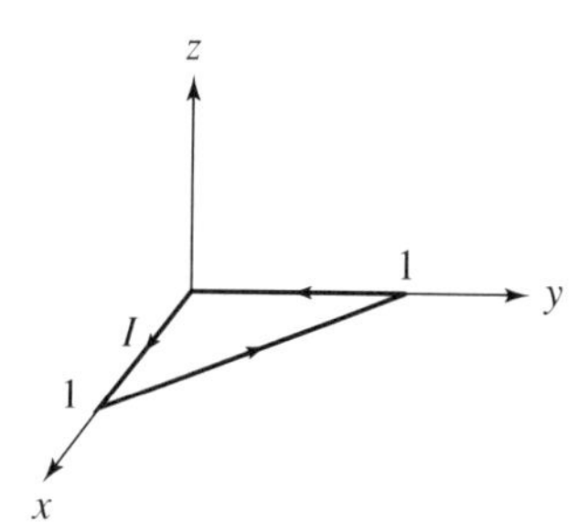

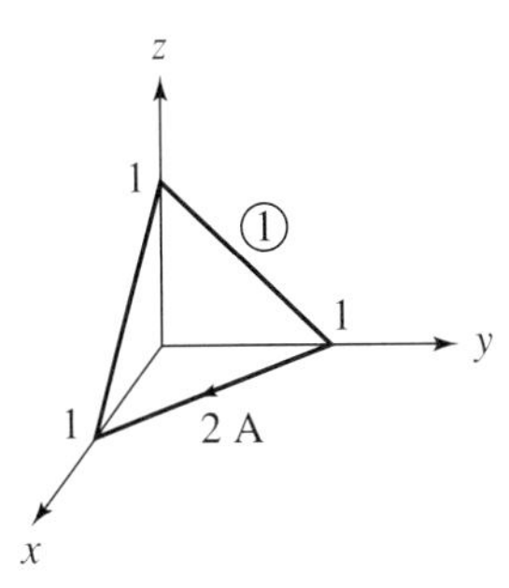

Figure 6-73.

a current $-0.02\hat{x}$ A/m. A magnetic field $\vec{B} = 0.05\hat{z}$ is also in the region. Determine the total force on the plate. The x axis is the plate diameter.

GE6-6 A square loop of wire of side length a is in the xy plane and centered at the origin. The wire carries I_0 amperes and there is a magnetic field $\vec{B} = B_0\hat{z}$. What is the torque on the loop about the x axis due to the interaction of the field and the current?

GE6-7 For the finite-length line shown in Figure 6-74, determine the magnetic field intensity on the x axis and express the answer in terms of β_1 and β_2.

Figure 6-74.

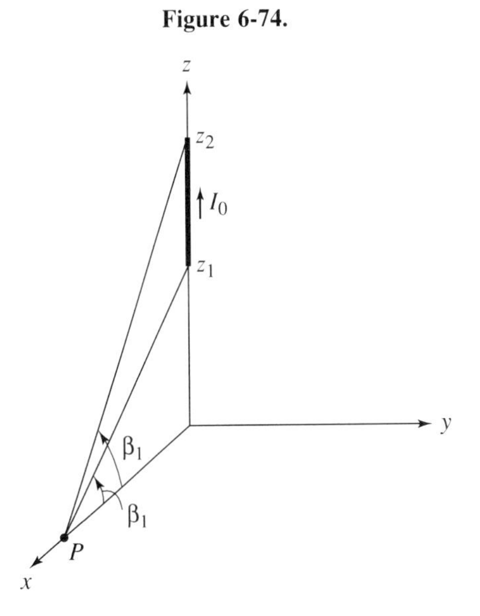

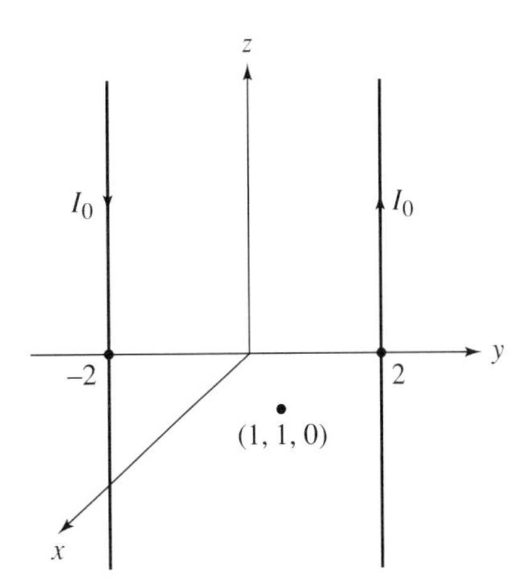

Figure 6-75.

GE6-8 Determine the magnetic field intensity at the point (1, 1, 0) for the infinite parallel lines shown in Figure 6-75. (Hint: Superposition applies. $I_0 = 1$ A.)

GE6-9 Compute the value of $\vec{H}$ at the origin for the semicircular filament ring of current shown in Figure 6-76.

GE6-10 For the system of Figure 6-76 compute $\vec{H}$ at a general point on the z axis.

GE6-11 Suppose the system in Figure 6-76 is a semicircular disk carrying current $\vec{J}_s = J_0\hat{\phi}$. Determine $\vec{H}$ on the z axis.

Figure 6-76.

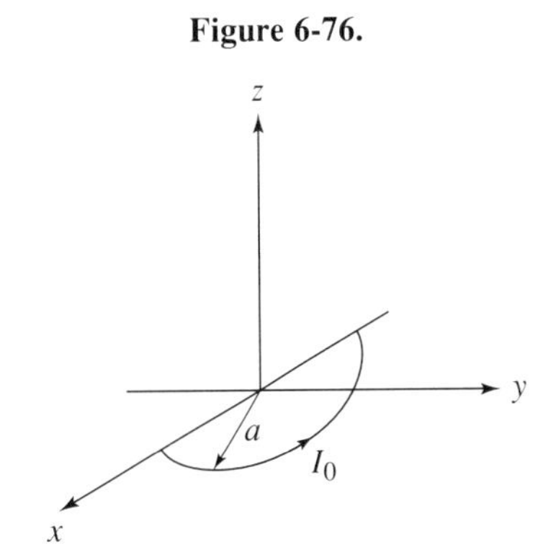

GE6-12 The positive x and positive y axes (0 to ∞) carry currents I_0 in the positive directions. Determine $\vec{H}$ for a general point $(x, y\ z)$ in space.

GE6-13 A spherical shell has a surface current $J_0\hat{\phi}$ flowing over it. For a shell of radius b, compute the magnetic field intensity at a point on the z axis. Repeat for a point on the x axis. Sphere is centered at the origin.

GE6-14 For the wires of length L shown in Figure 6-77, which carry constant currents I_0 amperes, determine the force on each wire caused by a magnetic flux density $\vec{B} = B_0\hat{z}$ T. From the results devise an equation that yields the force as a function of I_0, L, B_0, and θ, and suggest a vector form. (Partial answer: Force on y axis wire is $B_0I_0L\hat{x}$)

GE6-15 For Example 6-2 suppose the magnetic flux density were reoriented to $\vec{B} = B_0\hat{y}$. Find the forces on each half of the loop and describe the resultant motion. What is the total torque $\vec{T}$ on the loop about the loop diameter perpendicular to $\vec{B}$? The magnetic moment of a loop is defined as $\vec{m} = IA\hat{n}$, A = loop area, $\hat{n}$ = loop unit normal. Obtain $\vec{T}$ in terms of $\vec{m}$. If the loop started to turn, would $\vec{T}$ change? Suggest a general formula for $\vec{T}$ as a function of rotation angle.

GE6-16 A charge of 1 C and mass 1 kg is at the origin of a rectangular coordinate system moving with $\vec{v} = -3\hat{y}$ m/s. In the region there is an electric field $\vec{E} = 10\hat{x} - 3\hat{y}$ V/m and a magnetic field $\vec{B} = 1/2\hat{y} - \hat{z}/6$ T. Determine the position of the charge at $t = 3$ seconds. (Hint: $\vec{F} = m\vec{a}$. Obtain displacement from $\vec{a}$.)

GE6-17 For the Exercise 6-6.0b, suppose another thin cylindrical shell of radius $b > a$ is placed around the first and carries a current $-J_0\hat{z}$. Determine the magnetic field everywhere using superposition.

GE6-18 A toroidal surface carries a current $\vec{J}_s = J_0\hat{\rho}$ around its periphery as shown in Figure 6-78. Determine the magnetic field intensity inside and outside the toroid. (Answer: $J_s[(b - a)/\rho]\hat{\phi}$ inside, 0 outside). Take a cross-sectional cut by the xy plane, and use a distance ρ from the origin to a point inside the toroid to define the contour and surface as the ρ line sweeps around through the toroid.

GE6-19 The toroid of GE6-18 is constructed of closely wound fine wire n turns/meter and carries wire current I_0 ampere. Develop approximate formula for the magnetic field $\vec{H}$ inside and outside the toroid. (Answer: $(NI_0/2\pi\rho)\hat{\phi}$; 0)

GE6-20 Suppose that a #20 wire carries a current of 100 mA, assumed uniform over the wire cross section ($\vec{J}$ constant). What is the magnetic field intensity just outside the wire, obtained from Ampere's circuital law? Now if we assume the wire to be isolated from any other conductor, what is the equivalent surface current $\vec{J}_s$ along the wire surface using the tangential boundary condition? From this $\vec{J}_s$ calculate the total current flowing along the wire and compare with the original current. (Answer: $39.1\hat{\phi}$, $39.1\hat{z}$, 1 A)

GE6-21 In Chapter 4 it was pointed out that in Stokes' theorem the surface S was not unique. Any surface having C as its closed boundary could be used. All examples given in the text

Figure 6-77.

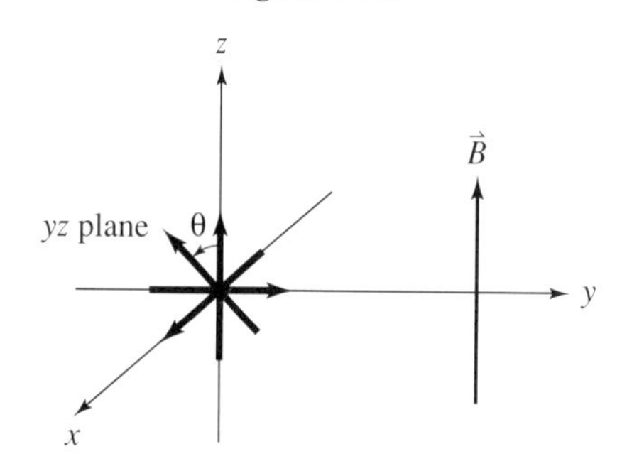

Figure 6-78.

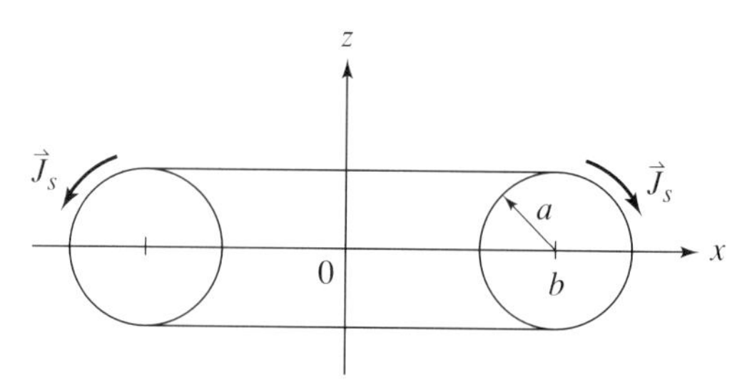

using Ampere's circuital law used the plane surface within contour C. Using the expression for $\vec{H}$ of a wire along the z axis carrying a current I_0 (see Example 6-12), verify Ampere's law in integral form for both surfaces shown in Figure 6-79. Surface S is the circular disk and surface S' is the hemispherical shell. A coordinate conversion to spherical may help in the latter case.

GE6-22 An electron beam of radius a has a current density given by $J_z = J_0[1 - (\rho/a)^3]$ A/m^2. Use Ampere's law to find the magnetic field intensity inside and outside the beam.

GE6-23 A wire carries a current I_0 along the positive z axis. A loop carrying a current I_2 as shown in Figure 6-80 is in vacuum (free space) near the straight wire. Determine (a) the force on each side of the loop, (b) the force on the entire loop, and (c) the torque on the loop around axis $0 - 0'$ and axis $1 - 1'$

GE6-24 Compute the vector magnetic potential of the disk current in Figure 6-22 of Section 6-4 for a general point in the xz plane. From this compute $\vec{H}$ on the disk axis.

GE6-25 Compute the vector magnetic potential at a general point in the xy plane for the elliptic current loop shown in Figure 6-23 of Section 6-4. Assume free space.

GE6-26 Derive the equation for the vector magnetic potential anywhere inside the solenoid of Figure 6-24 of Section 6-4. The solenoid is in air. Assume the solenoid also has an air core and is of length $\pm L$.

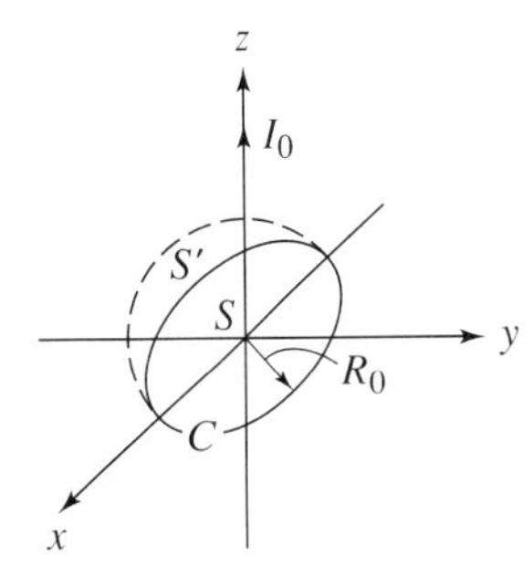

Figure 6-79.

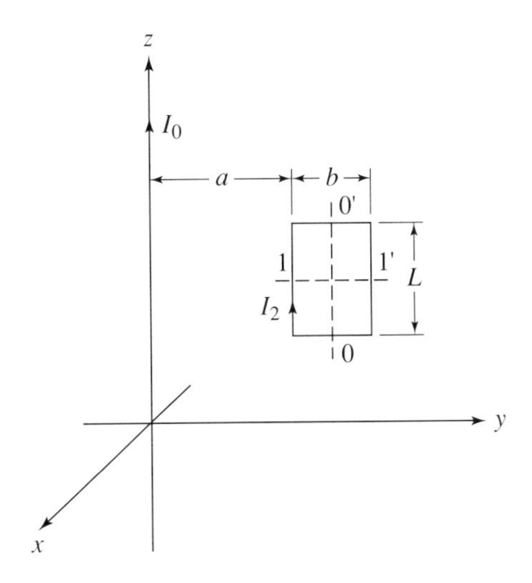

Figure 6-80.

GE6-27 Derive the equation for the vector magnetic potential for the uniform (constant) surface current sheet of large extent shown in Figure 6-35 for Example 6-16. From this compute $\vec{H}$. Assume $\mu = \mu_0$.

GE6-28 Develop the equation for the far-field vector magnetic potential for the beam current of Example 6-7, Figure 6-18.

GE6-29 Derive the far-field expansion for the electric field intensity $\vec{E}$, Eq. 5-17.

GE6-30 Determine the far-field vector magnetic potential for the current carrying disk of Figure 6-22 of Section 6-4.

GE6-31 Show that

$$\int_{-\infty}^{\infty} \frac{dz}{(az^2 + c)^{3/2}} = \frac{2}{c}$$

GE6-32 Measurement for a particular $\vec{B}$ field configuration reveals that $B_z = B_\rho = 0$. Using the divergence equation determine the possible coordinates of which B_ϕ may be a function.

GE6-33 An electron is injected into a magnetic field in the $\hat{x}$ direction and is observed to be deflected in the $\hat{z}$ direction on a phosphor screen. What components of $\vec{B}$ may be present and with which coordinate variables could these components vary? Consider the force equation and $\nabla \cdot$

$\vec{B} = 0$. Suppose that for the same field an electron is not deflected when injected along the $\hat{y}$ direction. What additional information can you obtain about the magnetic field?

GE6-34 On physical grounds it was concluded (Eq. 6-39) that $\nabla \cdot \vec{B} = 0$. Using Eq. 6-15, take the divergence with respect to unprimed coordinates and use the techniques of Section 6-6 to show that $\nabla \cdot \vec{H} = 0$. Now the constitutive relation $\vec{B} = \mu\vec{H}$ holds point by point in space and μ can also be a function of coordinates. Starting with $\nabla \cdot \vec{B} = 0$ and using the constitutive relation and vector identities, show that the gradient of μ is always normal to $\vec{H}$; that is, $(\nabla\mu) \cdot \vec{H} = 0$.

GE6-35 Prove that if $\vec{H}$ is zero, $\vec{B}$ need not be zero. (Hint: Consider Eq. 6-32.) Now starting at the upper right saturation point in Figure 6-43 and decreasing $\vec{H}$, tell which of the two paths would be followed down along the hysteresis characteristic.

GE6-36 Plot $|\vec{H}|$, $|\vec{B}|$, and $|\vec{M}|$ for the configuration of Figure 6-81, for $-\infty < z < \infty$

GE6-37 The two sections of the hysteresis curves shown in Figure 6-82 may be approximated by the following equations:

$$+ \text{ curve:} \qquad B_+ = B_{\text{sat}} \tanh\left(\frac{H + H_c}{H_c}\right)$$

$$- \text{ curve:} \qquad B_- = B_{\text{sat}} \tanh\left(\frac{H - H_c}{H_c}\right)$$

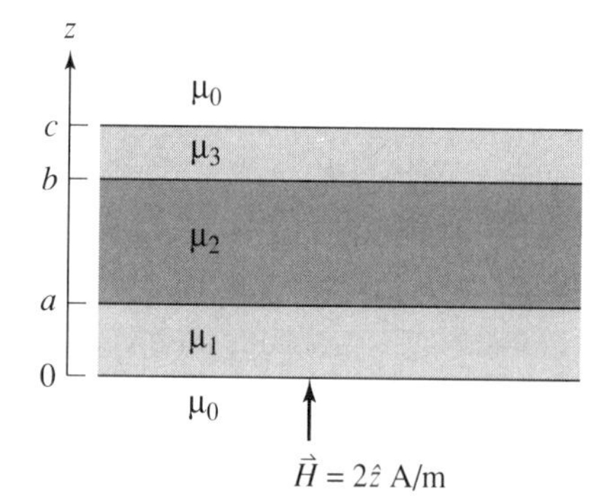

Figure 6-81.

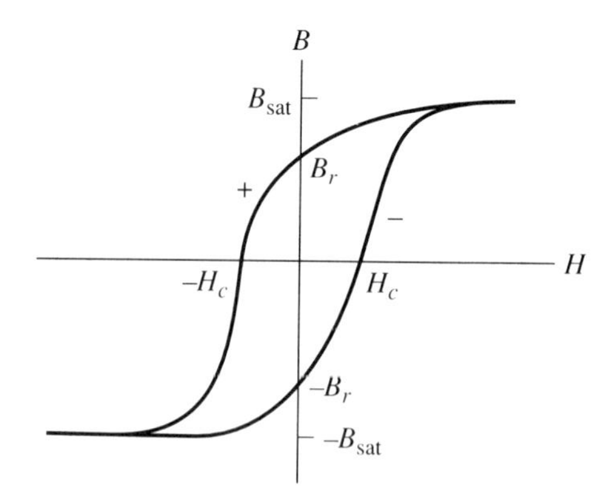

Figure 6-82.

(a) From the equations determine the expression for B_r.

(b) Compute the area of the hysteresis loop by integrating to $H = 3H_c$.

(c) Compute the approximate equations for μ if we define

$$\mu = \frac{B}{H} \quad \text{and} \quad \mu_d = \frac{dB}{dH}$$

for the case of the virgin or initial magnetization curve

$$\vec{B} = B_{\text{sat}} \tanh\left(\frac{H}{H_c}\right).$$

Plot both μ and μ_d versus H for $0 \le H \le 5H_c$.

GE6-38 Using the superposition theorem, compute and plot the magnetic field intensity $H_\phi(\rho)$ for the configuration shown in Figure 6-83. Current density $\vec{J}$ is uniform. (Hint: Let $-\vec{J}$ flow in the center region.)

GE6-39 Using the superposition of currents, obtain the equation for the magnetic field intensity inside the hollow region of radius a. See Figure 6-84. Assume the current density $\vec{J}$ is uniform in the large cylinder.

GE6-40 In Example 6-31 of Section 6-12 replace the sphere by a cylinder of radius a and length $2L$ centered on the origin of coordinates and hav-

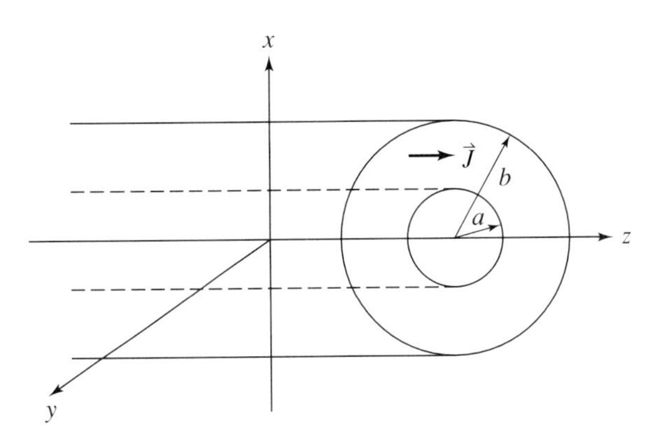

Figure 6-83.

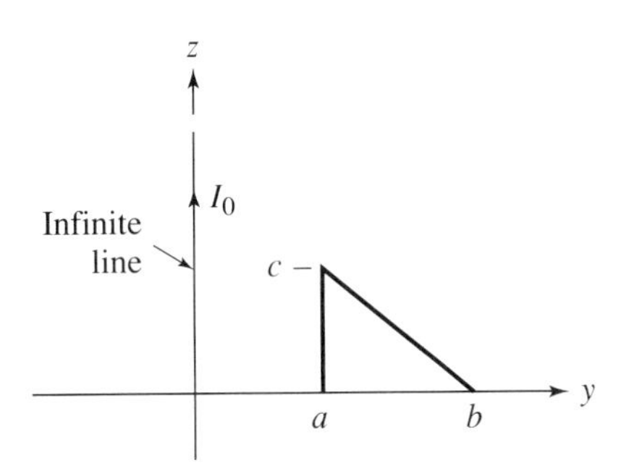

Figure 6-85.

ing magnetization $\vec{M} = M_0\hat{z}$. The external field is $\vec{B} = [(L + z)/L]B_0\hat{z}$. Compute the force on the cylinder.

GE6-41 Work Example 6-33 of Section 6-13 when the current is reduced to 60 mA. Compute the total flux in the core and in the coil. Discuss whether or not these have to be the same (They do.) Give a physical reason why ($\nabla \cdot \vec{B} = 0$). What approximations in the problem give errors?

GE6-42 For the coaxial cable of Figure 6-63, suppose that the material from $a < \rho < d$ ($d < b$) has permeability μ_1 and the remainder for $d < \rho < b$ has permeability μ_2. Determine the equation for the external inductance. Are the inductances for the two sections in parallel or series?

GE6-43 Using the external inductance from Example 6-37 of this chapter and the capacitance from Example 5-38, estimate the characteristic impedance of the parallel strip transmission line of Figure 6-66. Calculate this value for $\epsilon_r = 4$, $\mu_r = 1$, $w = 5$ mm, $d = 1$ mm. Neglect fringing fields.

GE6-44 Determine the mutual inductance for the configuration of line and triangular loop shown in Figure 6-85. Use flux linkages.

GE6-45 Determine the expression for the mutual inductance of two concentric wire rings shown in Figure 6-86, and evaluate it for the case $a = 5$ cm and $b = 10$ cm. Use Neumann's formula.

GE6-46 The top of the loop in Figure 6-69 of Exercise 6-14.0e in Section 6-14 is rotated and angle θ about the y axis with the top moving in the negative x direction. Determine the mutual inductance as a general function of θ, a, b, and c.

Figure 6-84.

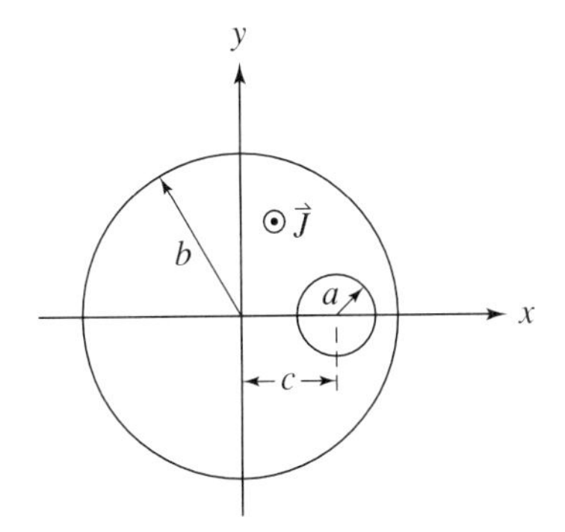

Figure 6-86.

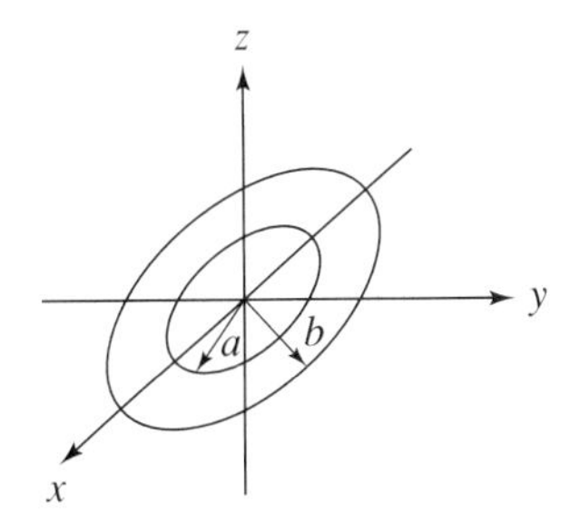

GE6-47 Determine the inductance of a very thin filament ring of radius a. Use the energy concept for computing inductance. Elliptic functions may be helpful. (See Example 6-38 and [24] for help.)

GE6-48 For the plate of Exercise GE6-5, find the torque on the plate about the plate straight edge.

GE6-49 Determine the vector magnetic potential at all values of radial coordinate ρ for the coaxial cable of Figure 6-32. Use the vector Laplacian equation for $\vec{A}$, and select $\rho = a$ as zero potential. Plot the results. From this $\vec{A}$, compute $\vec{H}$ for all values of ρ. (Note that $\vec{A}$ will be a function of ρ only. What components will $\vec{A}$ have?)

7 Theory, Physical Description, and Basic Equations of Time-Varying Electromagnetic Fields: Maxwell's Equations and Field Properties of Waves

7-1 Introduction

Time-varying electromagnetic fields constitute some of the more exciting applications of field theory. Radio waves, antennas, and wave guiding structures abound in engineering work. Nevertheless, a basic understanding of static fields is a significant asset to the interpretation of time-varying phenomena. The many equations in Chapters 5 and 6 often cause one to stop and ask just which ones are considered fundamental and which ones are the more important derived results. In Table 7-1 the fundamental equations of the preceding two chapters are summarized with the names associated with them. The equation numbers are also given for easy reference to earlier chapters. From the equations given in the table, several important auxiliary equations are obtained by applying vector differential and integral theorems. These are summarized in Table 7-2.

We begin this chapter by examining some additional experiments that were performed by Faraday and others and determine the modifications necessary to extend the tables to time-varying fields. Finally, we derive the general equations which describe propagating waves and give some of their properties. These results will be used extensively in the following chapters.

7-2 Faraday's Law of Induction and Lenz's Law

In 1831 Faraday performed a series of experiments with coils that were wound on the same core. He connected a galvanometer across the terminals of one coil and connected a battery and switch in series with the other coil. When the switch was opened and closed he observed that there were transient deflections of the galvanometer. Since magnetic fields

Table 7-1. Fundamental Equations of Static Fields

Maxwell's Equations for Static Fields			
Ampere's law:			
$\oint_C \vec{H} \cdot d\vec{l} = \int_S \vec{J} \cdot d\vec{a}$	(Eq. 6-22)	$\nabla \times \vec{H} = \vec{J}$	(Eq. 6-24)
Faraday's law:			
$\oint_C \vec{E} \cdot d\vec{l} = 0$	(Eq. 5-64)	$\nabla \times \vec{E} = 0$	(Eq. 5-65)
Gauss' laws:			
Electric fields:			
$\oint_S \vec{D} \cdot d\vec{a} = \int_V \rho_v dv$	(Eq. 5-46)	$\nabla \cdot \vec{D} = \rho_v$	(Eq. 5-46)
Magnetic fields:			
$\oint_S \vec{B} \cdot d\vec{a} = 0$	(Eq. 6-38)	$\nabla \cdot \vec{B} = 0$	(Eq. 6-39)

Companion Formulas			
Constitutive relationships:			
$\vec{D} = \epsilon\vec{E}$ (Eq. 5-48)	$\vec{B} = \mu\vec{H}$ (Eq. 6-12)		
$\vec{J}_{\text{conduction}} = \sigma\vec{E}$	(Eq. 5-31)		
$\vec{J}_{\text{convection}} = \rho_v\vec{v}$	(Eq. 5-38)		
Lorentz force equation:		Continuity equation (Kirchhoff's law)	
$\vec{F} = q\vec{E} + q\vec{v} \times \vec{B}$	(Eq. 6-17)	$\nabla \cdot \vec{J} = 0$	(Eq. 5-42 static ρ_v)
Ohm's laws:			
Electric fields:			
$V = IR$			
Magnetic fields:			
$V_m = \phi\mathcal{R}$	(Table 6-1)		

accompany currents, he concluded that time-changing flux linkages were responsible for inducing voltages in the first coil. Faraday also observed that if a loop of wire was moved in a steady magnetic field a voltage was also induced.

To put these experiments on a physical foundation, we consider the sequence of experiments portrayed in Figure 7-1. We will use an ammeter to be able to determine current

Table 7-2. Important Auxiliary Static Field Laws

Potential relations:

$$\vec{E} = -\nabla V \qquad \text{(Eq. 5-63)} \qquad\qquad V = -\int_{\text{ref}}^{P} \vec{E} \cdot d\vec{l} \qquad \text{(Eq. 5-62)}$$

$$\vec{H} = -\nabla V_m \qquad \text{(Eq. 6-51)}$$

$$\vec{B} = \nabla \times \vec{A} \qquad \text{(Eq. 6-34)} \qquad\qquad V_m = -\int_{\text{ref}}^{a} \vec{H} \cdot d\vec{l} \qquad \text{(Eq. 6-52)}$$

$$(\nabla \cdot \vec{A} = 0)$$

Poisson's Equation:

$$\nabla^2 V = -\frac{\rho_v}{\epsilon} \qquad \text{(Eq. 5-87)} \qquad\qquad \nabla^2 V_m = -\rho_{vm}/\mu_0 \qquad \text{(Eq. 6-58)}$$

$$\nabla^2 \vec{A} = \mu \vec{J} \qquad \text{(Eq. 6-42)} \qquad\qquad V(\vec{r}) = \int_{V'} \frac{\rho_v(\vec{r}\,')}{4\pi\epsilon R}\, dv' \qquad \text{(Eq. 5-67)}$$

$$\vec{A}(\vec{r}) = \int_{V'} \frac{\mu \vec{J}(\vec{r}\,')}{4\pi R}\, dv' \qquad \text{(Eq. 6-41)}$$

Boundary conditions:

Will not be repeated here. See Sections 6-9 and 6-11 for magnetic fields and Section 5-12 for electric fields.

Note: The normal components of *time-varying* magnetic fields are zero for perfect conductors. See Table 7-5.

directions in the loops. Figure 7-1*a* is the initial configuration showing a battery and resistance that produce a current I in the filamentary bar. Now when the switch is opened as in Figure 7-1*b* the induced current i_i flows momentarily in the direction shown. When the switch is closed, as in Figure 7-1*c*, current reverses as indicated.

Now rather than open and close the switch we move the loop with respect to the filament. When the loop is pulled away from the filament with velocity $\vec{V}$ (Figure 7-1*d*), the current flows counterclockwise. When moved toward the filament the current reverses as shown in Figure 71*e*.

Consider first case (*b*) Figure 7-1, where the switch is opened. The current I decreases, and since the flux ϕ is established by I, the value of ϕ will also decrease. However, the *direction* of the flux lines is still out of the page as shown in Figure 7-2 since I is decreasing, not changing direction. This flux is denoted by ϕ in this figure. Using the right-hand rule on the current i_i we find that the flux ϕ_i generated is also out of the page. We observe that ϕ_i adds to ϕ, and thus the effect of ϕ_i is to prevent ϕ from decreasing. This is generalized by saying that the induced current creates a flux that opposes the *change* of ϕ. This is Lenz's law (1834). See [39], volume 2, pages 179 and 190.

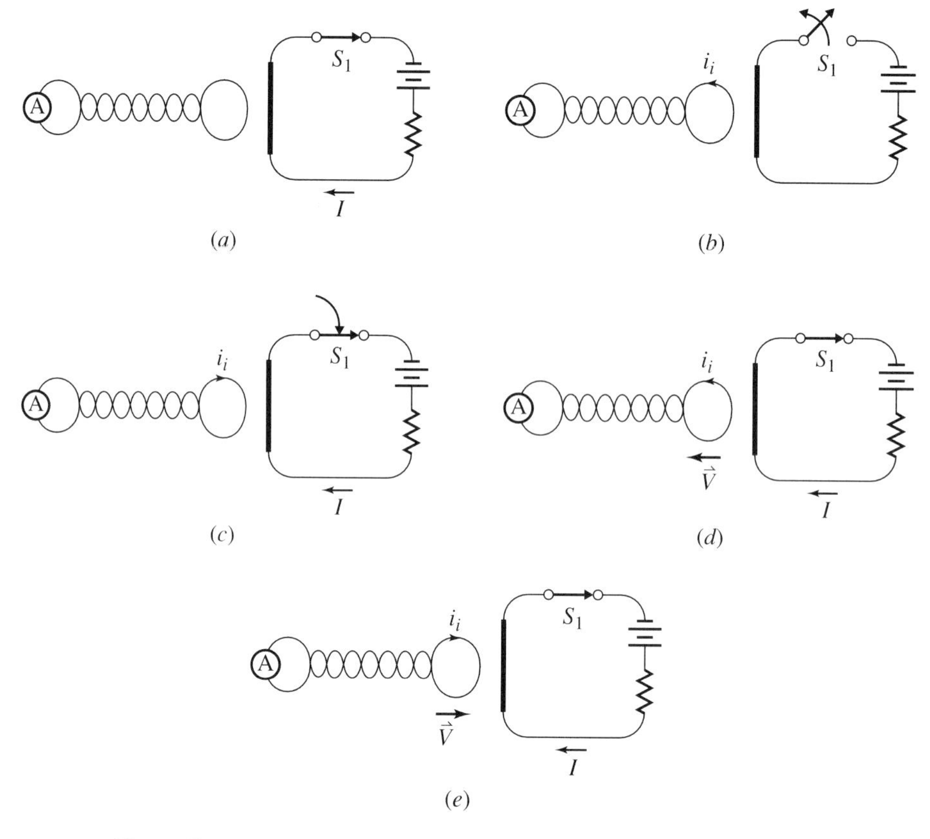

Figure 7-1. Series of experiments that lead to Faraday's law of induction.

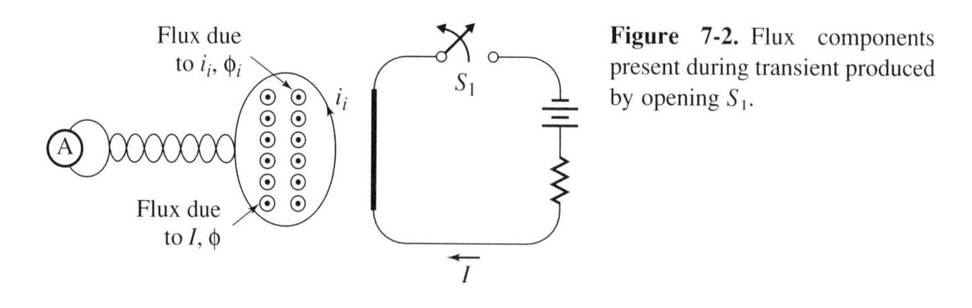

Figure 7-2. Flux components present during transient produced by opening S_1.

Exercises

7-2.0a Analyze Figure 7-1*c*, where S_1 closes and I increases from 0 to its final value, and show that the generalization stated above (Lenz's law) is satisfied.

7-2.0b From Example 6-12 of Section 6-6, the magnetic flux density around a current carrying conductor is given by

$$\vec{B} = \frac{I_0}{2\pi\rho}\,\hat{\phi}$$

Analyze Figure 7-1*d* qualitatively and show that the resulting current is consistent with Lenz's law. (Hint: Moving away from the filament has what effect on the loop flux?) Repeat for Figure 7-1*e*.

Lenz's law is physically reasonable since if, for example, ϕ were increasing and ϕ_i added to it there would be a resulting larger induced current i_i that increases ϕ_i ad infinitum.

Next we observe that current flow i_i in the loop implies that a voltage v_i exists in the loop. For the case where switch S_1 is opened (Figure 7-1*b*), the equivalent voltage source would be as shown in Figure 7-3.

The results of the preceding experiments and observations were stated by Maxwell [39], (volume 2, page 179, ¶ 531) formally:

> When the number of lines of magnetic induction (i.e., flux) which pass through the secondary circuit in the positive direction (i.e., right-hand–rule sense) is altered, an electromotive force acts around the circuit, which is measured by the rate of decrease of magnetic induction.

This is called Faraday's law and is written mathematically as

$$\text{induced emf} = -\frac{d\Lambda}{dt} \tag{7-1}$$

In this equation Λ is the flux linkage as used in Chapter 6.

The induced emf is a voltage that drives the induced current around the resistance wire loop. We may then suppose that there is a time-varying electric field intensity around the closed loop such that the integral of the field around the closed loop is the induced emf. Now as in Chapter 1 we will use the script letters to represent time-dependent quantities, so we write

$$\text{induced emf} = \oint_C \vec{\mathscr{E}} \cdot d\vec{l} \tag{7-2}$$

Although our discussions to this point have referred specifically to loops of resistance wire, there is nothing to prevent us from taking any closed path in space (or any material) and computing an induced emf for that path. Equation 7-2 defines the voltage drop around the contour C from a circuit theory point of view.

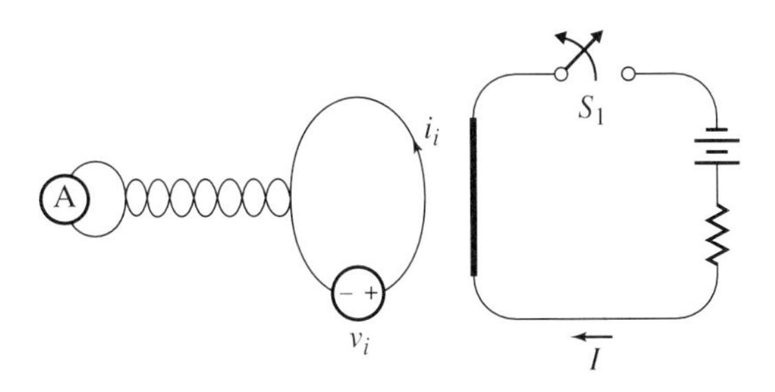

Figure 7-3. Representation of the induced voltage by a source.

From the preceding two equations we obtain

$$\oint_C \vec{\mathscr{E}} \cdot d\vec{l} = -\frac{d\Lambda}{dt}$$

Substituting for the total flux linkage within the contour in terms of magnetic flux density

$$\oint_C \vec{\mathscr{E}} \cdot d\vec{l} = -\frac{d}{dt}\int_S \vec{\mathscr{B}} \cdot d\vec{a} \tag{7-3}$$

This is integral form of Faraday's law for electromagnetic fields. Note that if there are no time variations this reduces to the static field case shown in Table 7-1.

To obtain the differential or point form of Faraday's law we apply Leibniz' rule for differentiating an integral (see Chapter 4) and assume that the surface S is not moving. Equation 7-3 becomes

$$\oint_C \vec{\mathscr{E}} \cdot d\vec{l} = \int_S -\frac{\partial \vec{\mathscr{B}}}{\partial t} \cdot d\vec{a} \tag{7-4}$$

Using Stokes' theorem on the contour integral:

$$\int_S \nabla \times \vec{\mathscr{E}} \cdot d\vec{l} = \int_S -\frac{\partial \vec{\mathscr{B}}}{\partial t} \cdot d\vec{a}$$

Since we have not restricted the size or shape of the surface the integrands must be equal and we obtain

$$\nabla \times \vec{\mathscr{E}} = -\frac{\partial \vec{\mathscr{B}}}{\partial t} \tag{7-5}$$

Note that $\vec{\mathscr{E}}$ and $\vec{\mathscr{B}}$ are orthogonal in space. As before, this reduces to the static case in Table 7-1 for no time variation. (The script letters, of course, would be replaced by the block letters in accordance with our convention.) This last result has a very important consequence for magnetic fields in perfect conductors. In Chapter 6 we saw that static magnetic fields can exist in any conductor, even a perfect one. Let us assume for the moment that we had a time-varying magnetic field in a perfect conductor. The preceding equation tells us that we also have a time-varying electric field intensity. But if the conductor were perfect ($\sigma = \infty$) this would result in infinite current density by Ohm's law. Thus there cannot be an electric field in the perfect conductor, which means our initial assumption of a time-varying magnetic field in the perfect conductor is false. *There cannot be a time-varying magnetic field in a perfect conductor*. Static magnetic fields, however, *can* exist in a perfect conductor.

Example 7-1

A circular loop of resistance wire 10 cm in diameter is placed so that the loop plane is normal to a magnetic field having $|\vec{\mathscr{B}}| = 0.3 \cos \omega t \ \mu T$, at a frequency of 100 KHz. The loop is constructed of #12 wire that has a resistance of 0.00525 Ω/m.

1. Determine the equation for the voltage induced in the wire as a function of time.
2. Determine the equation for the current flow induced in the wire (neglecting the loop inductance).

Since the loop plane is normal to the magnetic field and the area vector is normal to the loop plane we have

$$\vec{\mathcal{B}} \cdot d\vec{a} = |\vec{\mathcal{B}}| da = 0.3 \times 10^{-6} \cos 2\pi 10^5 t da$$

$$\begin{aligned} \Lambda = \phi_{\text{enclosed}} &= \int_S \vec{\mathcal{B}} \cdot d\vec{a} = \int_S 0.3 \times 10^{-6} \cos 2\pi 10^5 t da \\ &= (0.3 \times 10^{-6} \cos 2\pi 10^5 t) \times \text{area} \\ &= (0.3 \times 10^{-6} \cos 2\pi 10^5 t) \left[\pi \left(\frac{0.1}{2} \right)^2 \right] = 2.356 \times 10^{-9} \cos 2\pi 10^5 t \text{ Wb} \end{aligned}$$

Using Eq. 7-1,

$$v_i = -\frac{d\Lambda}{dt} = 0.00148 \sin 2\pi 10^5 t \text{ V}$$

Ohm's law yields

$$i_i = \frac{v_i}{R} = \frac{0.00148 \sin 2\pi 10^5 t}{(0.00525)(\pi \times 0.1)} = 0.898 \sin 2\pi 10^5 t \text{ A}$$

Example 7-2

This example illustrates that the boundary C need not be a metallic contour for a meaningful computation to be effected. A region of space has an electric field given by

$$\vec{\mathcal{E}} = 0.1 xze^{-2t}\hat{x} - 2xye^{-2t}\hat{y} \text{ V/m}$$

Determine the total flux and induced emf in the area of space shown dotted in Figure 7-4. Using the area vector to be in the $\hat{y}$ direction, we start at Point 1 and integrate counterclockwise around the defined contour:

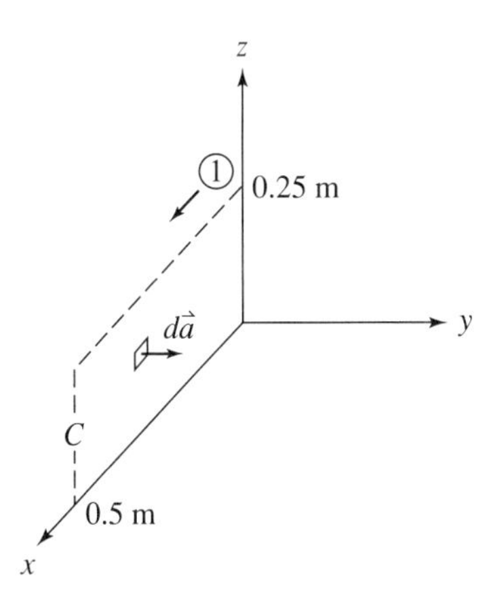

Figure 7-4. Contour in space that does not have a metal structure.

$$v_i = \oint_C \vec{\mathscr{E}} \cdot d\vec{l} = \underset{\substack{z = 0.25 \\ y = 0}}{\int_0^{0.5} \vec{\mathscr{E}} \cdot dx\hat{x}} + \underset{\substack{x = 0.5 \\ y = 0}}{\int_{0.25}^{0} \vec{\mathscr{E}} \cdot dz\hat{z}} + \underset{\substack{z = 0 \\ y = 0}}{\int_{0.5}^{0} \vec{\mathscr{E}} \cdot dx\hat{x}} + \underset{\substack{x = 0 \\ y = 0}}{\int_0^{0.25} \vec{\mathscr{E}} \cdot dz\hat{z}}$$

$$= \underset{\substack{z = 0.25 \\ z - 0}}{\int_0^{0.5} 0.1xze^{-2t}dx} + 0 + \underset{\substack{z = 0 \\ y = 0}}{\int_{0.5}^{0} 0.1xze^{-2t}dx} + 0$$

$$= \int_0^{0.5} 0.025xe^{-2t}dx$$

$$= 0.00313e^{-2t} \text{ V}$$

Since

$$\phi_{\text{total}} = \Lambda = \int_S \vec{\mathscr{B}} \cdot d\vec{a}$$

Eq. 7-3 yields

$$\oint_C \vec{\mathscr{E}} \cdot d\vec{l} = -\frac{d\phi_{\text{total}}}{dt}$$

Thus

$$0.00313e^{-2t} = -\frac{d\phi_{\text{total}}}{dt}$$

Integrating with respect to time

$$\phi_{\text{total}}(t) = 0.00156e^{-2t} + \text{constant Wb}$$

(Note that the "constant" could be a function of coordinates since the boundary is not moving with time.)

Example 7-3

In a particular region of space the electric field intensity has been measured to be

$$\vec{\mathscr{E}}(\vec{r}, t) = 6 \times 10^{-3} \cos(10^{10}t + 0.3z)\hat{y} \text{ V/m}$$

Obtain an expression for $\vec{B}(\vec{r}, t)$.

Expanding the curl side of Eq. 7-5 using the given $\vec{\mathscr{E}}$ we have

$$\nabla \times \vec{\mathscr{E}} = -\frac{\partial \vec{\mathscr{E}}_y}{\partial z}\hat{x} = 1.8 \times 10^{-3} \sin(10^{10}t + 0.3z)\hat{x}$$

Equation 7-5 then requires

$$-\frac{\partial \vec{\mathscr{B}}}{\partial t} = 1.8 \times 10^{-3} \sin(10^{10}t + 0.3z)\hat{x}$$

Thus as far as time-varying fields are concerned, $\vec{\mathscr{B}}$ has at most an $\hat{x}$ component that must satisfy

$$-\frac{\partial \mathscr{B}_x}{\partial t} = 1.8 \times 10^{-3} \sin(10^{10}t + 0.3z)$$

Integrating (neglecting the constant of integration),

$$\mathscr{B}_x = 1.8 \times 10^{-13} \cos(10^{10}t + 0.3z)$$

Then

$$\vec{\mathscr{B}}(\vec{r}, t) = 1.8 \times 10^{-13} \cos(10^{10}t + 0.3z)\hat{x} \text{ T}$$

Example 7-4

A circular disk of polystyrene is subjected to a $\hat{z}$-directed sinusoidal magnetic field

$$\vec{\mathscr{B}} = B_0 \cos 6 \times 10^8 t\hat{z} \text{ T}$$

The orientation is shown in Figure 7-5. The dielectric strength (breakdown) of polystyrene is $E_{\text{max}} = 47$ MV/m. What is the largest value of B_0 allowed?

Using Eq. 7-5 in cylindrical coordinates

$$\left(\frac{1}{\rho}\frac{\partial \mathscr{E}_z}{\partial \phi} - \frac{\partial \mathscr{E}_\phi}{\partial z}\right)\hat{\rho} + \left(\frac{\partial \mathscr{E}_\rho}{\partial z} - \frac{\partial \mathscr{E}_z}{\partial \rho}\right)\hat{\phi} + \left(\frac{1}{\rho}\frac{\partial(\rho \mathscr{E}_\phi)}{\partial \rho} - \frac{1}{\rho}\frac{\partial \mathscr{E}_\rho}{\partial \phi}\right)\hat{z}$$
$$= 6 \times 10^8 B_0 \sin 6 \times 10^8 t\hat{z}$$

From the symmetries of the problem, there will be no z variation and no ϕ variation of the electric field. The preceding reduces to

$$-\frac{\partial \mathscr{E}_z}{\partial \rho}\hat{\phi} + \frac{1}{\rho}\frac{\partial(\rho \mathscr{E}_\phi)}{\partial \rho}\hat{z} = 6 \times 10^8 B_0 \sin 6 \times 10^8 t\hat{z}$$

This result requires the $\hat{\phi}$ component to be zero, so if there is an $\mathscr{E}_z$ it cannot be a function of ρ, either. We then have

$$\frac{1}{\rho}\frac{\partial(\rho \mathscr{E}_\phi)}{\partial \rho} = 6 \times 10^8 B_0 \sin 6 \times 10^8 t$$

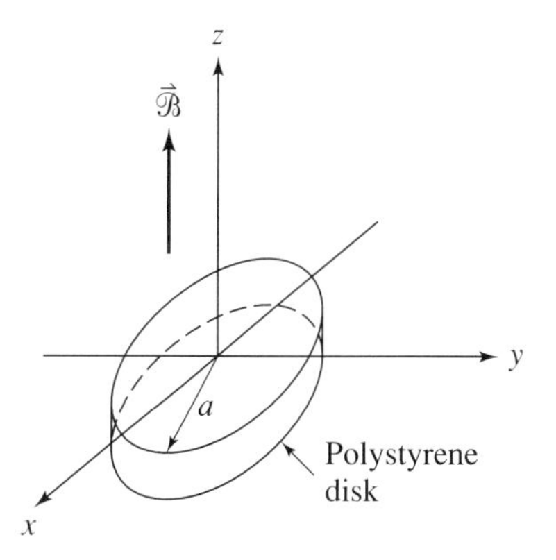

Figure 7-5. Polystyrene disk in a time-varying magnetic field.

Multiply by ρ and integrate with respect to ρ:

$$\rho \mathscr{E}_\phi = 3 \times 10^8 B_0 \rho^2 \sin 6 \times 10^8 t + f(z, \phi, t)$$

The function $f(z, \phi, t)$ is the ''constant'' of integration, whose partial derivative with respect to ρ is zero. However, we already know that $\mathscr{E}_\phi$ is not a function of z and ϕ, and that the value of the ''constant'' cannot depend on time, so the function f must be zero. Dividing by ρ we obtain

$$\mathscr{E}_\phi = 3 \times 10^8 B_0 \rho \sin 6 \times 10^8 t$$

This is maximum for $\rho = a$ and for unity sine function value. Therefore

$$\mathscr{E}_{\phi_{\max}} = 3 \times 10^8 a B_0$$

To avoid dielectric breakdown,

$$3 \times 10^8 a B_0 \leq 47 \times 10^6$$

from which

$$B_{0_{\max}} = \frac{0.157}{a} \text{ T}$$

Example 7-5

The betatron is a device used to accelerate particles by using a time-varying magnetic field. In Chapter 6 it was shown that a static magnetic field can only produce forces on charges that are already in motion. Now suppose we have a proton at rest in a magnetic field having cylindrical structure

$$\vec{\mathscr{B}} = B_0(t + k)^2 \rho^2 \hat{z}$$

as shown schematically in Figure 7-6. Determine the acceleration of the proton at $t = 0$ and also the emf around the path of radius b. Proton mass is 1.67×10^{-27} kg.

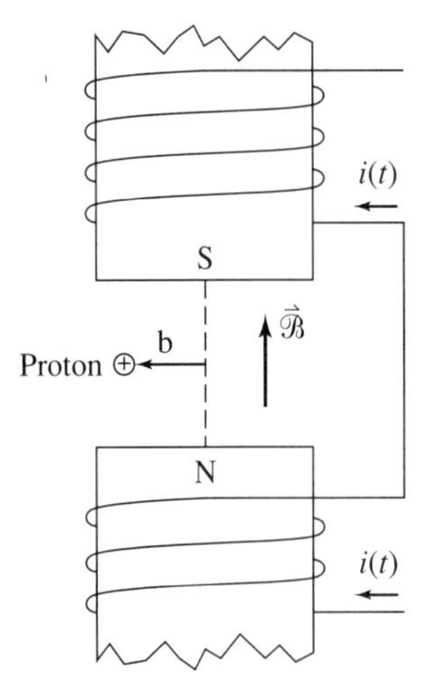

Figure 7-6. Schematic diagram of the betatron.

Expanding Faraday's law (Eq. 7-5) in point form and integrating once as explained in Example 7-4, we obtain (by symmetry, there cannot be variation with ϕ, so $\partial/\partial\phi = 0$, and also no z variation)

$$\mathscr{E}_\phi = -\tfrac{1}{2}B_0(t + k)\rho^3$$

The force on the proton is then

$$\vec{\mathscr{F}} = q\vec{\mathscr{E}} = 1.6 \times 10^{-19}[-\tfrac{1}{2}B_0(t + k)\rho^3\hat{\phi}]$$

At the first instant, $t = 0$, the radius is b so that

$$\vec{\mathscr{F}}(\rho = b, t = 0) = -8 \times 10^{-20}B_0kb^3\hat{\phi}$$

Using Newton's law,

$$-8 \times 10^{-20}B_0kb^3\hat{\phi} = m_p\vec{a}$$

Then

$$\vec{a} = \frac{-8 \times 10^{-20}B_0kb^3}{1.6 \times 10^{-27}}\hat{\phi} = -4.8 \times 10^7 B_0kb^3\hat{\phi}\ \text{m/s}^2$$

Flux inside radius b path:

$$\Lambda(t) = \phi_{\text{enclosed}}(t) = \int_0^b\int_0^{2\pi} \vec{\mathscr{B}} \cdot \rho d\phi d\rho\hat{z} = \int_0^b\int_0^{2\pi} B_0(t + k)^2\rho^3 d\phi d\rho$$

$$= \frac{\pi}{2}B_0(t + k)^2b^4\ \text{Wb}$$

$$\text{emf} = -\frac{d\Lambda}{dt} = -\pi B_0(t + k)b^4\ \text{V}$$

Note this emf is in agreement with the acceleration result being in the $-\hat{\phi}$ direction.

Example 7-6

A resistive wire loop is moving in a time-invariant magnetic field as shown in Figure 7-7. The magnetic field is given by the equation

$$\vec{B} = B_0y\hat{x}$$

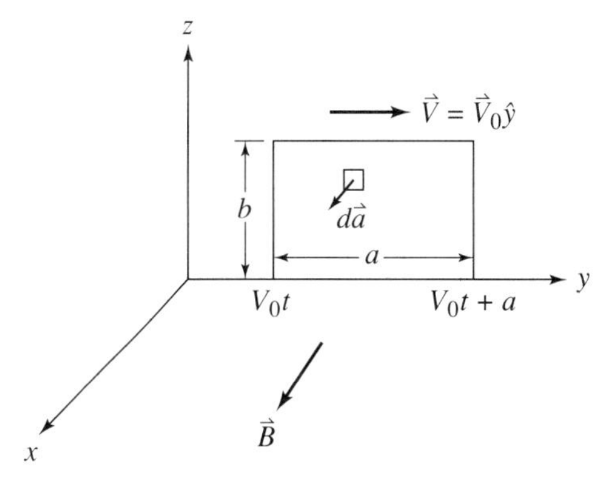

Figure 7-7. Wire loop moving in a constant magnetic field.

Derive the expression for the induced emf and determine the placement of the equivalent induced voltage source, v_i, in the loop:

$$\text{emf} = -\frac{d\Lambda}{dt} = -\frac{d\phi_{\text{enclosed}}}{dt} = -\frac{d\int_S \vec{B}\cdot d\vec{a}}{dt} = -\frac{d}{dt}\left[\int_0^b \int_{v_0 t}^{v_0 t+a} B_0 y\hat{x}\cdot dydz\hat{x}\right]$$
$$= -\frac{d}{dt}\left[bB_0 \int_{v_0 t}^{v_0 t+a} ydy\right]$$

Using Leibniz' rule (see Chapter 4),

$$\text{emf} = -bB_0\left[\int_{v_0 t}^{v_0 t+a} \frac{\partial y}{\partial t}dy + (v_0 t + a)\frac{d(v_0 t + a)}{dt} - v_0 t\frac{d(v_0 t)}{dt}\right]$$

Since the coordinate system is not moving, $\partial y/\partial t$ is zero. Then

$$\text{emf} = -bB_0[0 + (v_0 t + a)v_0 - v_0^2 t] = -bB_0 a v_0 \text{ V}$$

Note this could also have been obtained by performing the integration before differentiating:

$$\text{emf} = -bB_0\frac{d}{dt}\{\tfrac{1}{2}[(v_0 t + a)^2 - (v_0 t)^2]\} = -bB_0 a v_0$$

To determine the placement of the source, first determine the positive convention using the right-hand rule: since $d\vec{a}$ is in the $+\hat{x}$ direction, the positive direction of the contour C is counterclockwise around the loop. Now since the emf is negative, it forces current in a direction opposite the contour direction, so the source could be placed in the top wire with the positive sign of the source on the right and negative on the left. This result is consistent with Lenz's law since as the loop moves to the right the enclosed flux increases since $\vec{B}$ is increasing with y. The current flowing clockwise around the loop would produce a flux in the $-\hat{x}$ direction inside the loop, opposing the original increase.

Exercises

7-2.0c Show that the magnetic field derived in Example 7-3 satisfies Gauss's law in point form, $\nabla\cdot\vec{\mathcal{B}} = 0$.

7-2.0d A bar slides along a fixed parallel wire line as shown in Figure 7-8. The magnetic field is given by

$$\vec{B} = 4y\hat{x} \text{ T}$$

Derive the equation for the induced emf and show the placement of the voltage source in the left fixed end.

7-2.0e For example 7-5 compute the closed path integral of $\vec{\mathcal{E}}$ around the circular path of radius b. Compare this with the emf computed in the example.

7-2.0f For the fields of Example 7-3 obtain equations for $\vec{\mathcal{D}}$ and $\vec{\mathcal{H}}$ assuming free space. (Answer: $\vec{\mathcal{D}} = \epsilon_0\vec{\mathcal{E}}$, $\vec{\mathcal{H}} = (1/\mu_0)\vec{\mathcal{B}}$)

7-2.0g Suppose that the magnetic field intensity in free space is measured to be

$$\vec{\mathcal{H}}(\vec{r}, t) = 3\times 10^{-7}\cos(10^{11}t - 0.03z)\hat{x}$$

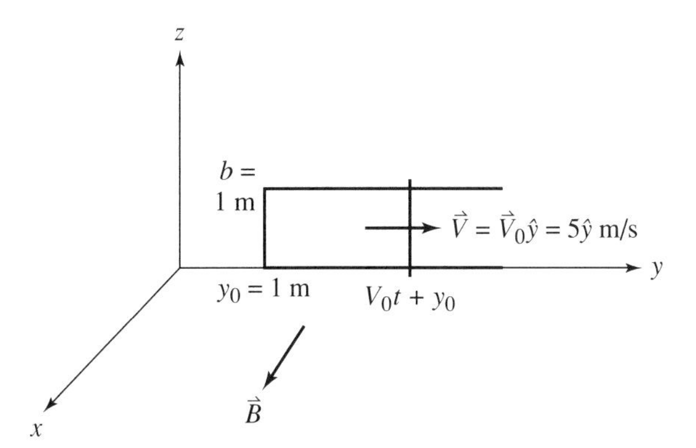

Figure 7-8.

Use the differential (point) form of Faraday's law to determine as many relationships as possible among the components of $\vec{\mathscr{E}}$ and obtain expressions for as many components of $\vec{\mathscr{E}}$ as possible. (Hint: Are there limits as to which coordinates the components can vary with?)

In some of the preceding examples, Leibniz' rule for differentiating an integral was used when the system boundaries were not stationary. Another form of Faraday's law that accounts for boundary motion directly can be developed. It should also be recalled that Eq. 7-4 was derived on the assumption that the boundary was stationary. If we use the Lorentz force equation—which, incidentally, is valid for time-varying fields—we may identify a motional electric field intensity as follows:

$$\vec{\mathscr{F}} = q\vec{\mathscr{E}} + q\vec{v} \times \vec{\mathscr{B}} = q\vec{\mathscr{E}} + q\vec{\mathscr{E}}_m = q\vec{\mathscr{E}}_{\text{equiv}} \tag{7-6}$$

If the flux is changing with time through the closed contour and the contour is moving the total emf would be the sum of the two effects, so that

$$\text{emf} = \oint_C \vec{\mathscr{E}}_{\text{equiv}} \cdot d\vec{l}$$

We could divide the integral into two pieces, expressed in words as emf due to time changing flux through surface plus emf due to boundary motion:

$$\text{emf} = -\int_S \frac{\partial \vec{\mathscr{B}}}{\partial t} \cdot d\vec{a} + \oint_C \vec{\mathscr{E}}_m \cdot d\vec{l} = -\int_S \frac{\partial \vec{\mathscr{B}}}{\partial t} \cdot d\vec{a} + \oint_C (\vec{v} \times \vec{\mathscr{B}}) \cdot d\vec{l}$$

Using Stokes' theorem on the contour integral,

$$\begin{aligned}\text{emf} &= -\int_S \frac{\partial \vec{\mathscr{B}}}{\partial t} \cdot d\vec{a} + \int_S \nabla \times (\vec{v} \times \vec{\mathscr{B}}) \cdot d\vec{a} \\ &= -\int_S \left[\frac{\partial \vec{\mathscr{B}}}{\partial t} - \nabla \times (\vec{v} \times \vec{\mathscr{B}})\right] \cdot d\vec{a}\end{aligned} \tag{7-7}$$

Applications of this equation may be difficult because any boundary curvature would require setting up an expression for a general increment of the boundary.

Example 7-7

A straight wire of length L moves with constant velocity $\vec{V} = V_0\hat{v}$ in a constant magnetic field $\vec{B} = B_0\hat{b}$. Compute the emf generated in the wire. Since there is no closed surface the only contribution is due to the motion. Thus

$$\text{emf} = \int_0^L (\vec{v} \times \vec{B}) \cdot d\vec{l}$$

For the particular case where $\vec{v}$, $\vec{B}$, and $d\vec{l}$ are mutually perpendicular, $\vec{v} \times \vec{B}$ will be parallel to $d\vec{l}$, so the dot product is just the product of the vector amplitudes. For this case,

$$\text{emf} = \int_0^L V_0 B_0 dl = V_0 B_0 L \qquad \text{V}$$

Example 7-8

A semicircular wire moves with constant velocity in a constant magnetic field as shown in Figure 7-9. Compute the emf induced in the wire. From the figure,

$$d\vec{l} = a d\phi \hat{\phi}$$

Converting the velocity to cylindrical coordinates (see Chapter 4):

$$\vec{V} = V_0 \sin \phi \hat{\rho} + V_0 \cos \phi \hat{\phi}$$

$$\text{emf} = \int_0^{\pi} [V_0 \sin \phi B_0(-\hat{\phi}) + V_0 \cos \phi B_0 \hat{\rho}] \cdot a d\phi \hat{\phi}$$

$$= -a V_0 B_0 \int_0^{\pi} \sin \phi d\phi = 2a V_0 B_0 \text{ V}$$

Since $2a$ is the loop diameter, this result says that the emf can be thought of as generated by the projection of the contour onto a line mutually perpendicular to $\vec{V}$ and $\vec{B}$.

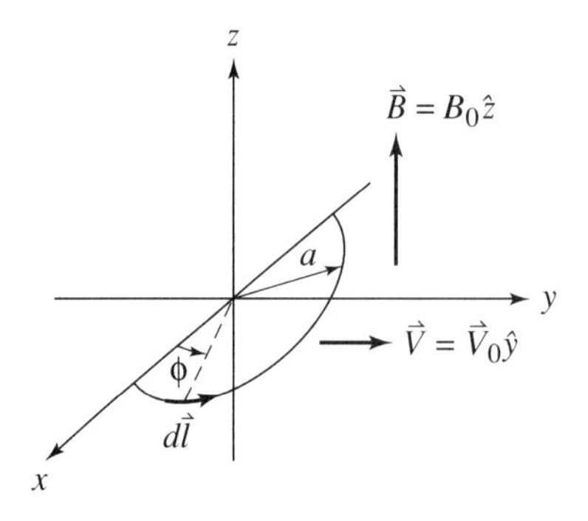

Figure 7-9. Semicircular wire moving with constant velocity in a constant magnetic field. The x axis is not part of the wire.

Exercises

7-2.0h Work Example 7-6 using the concept of motional emf applied to each side of the wire loop.

7-2.0i Work Exercise 7-2.0b using the concept of motional emf applied to the moving bar.

7-3 Displacement Current

We next examine Ampere's law to see if any modifications are required. From Table 7-1 the point form would normally be written in time form using the script letters as

$$\nabla \times \vec{\mathcal{H}} = \vec{\mathcal{J}}$$

Taking the divergence of this equation and using the vector identity that the divergence of the curl of any vector is identically zero results in

$$0 \equiv \nabla \cdot \vec{\mathcal{J}}$$

However, if charge is changing with time at a given point, there must be a divergence of current at that point. This statement is the continuity equation given in Chapter 5, Eq. 5-42:

$$\nabla \cdot \vec{\mathcal{J}} = -\frac{\partial \rho_v}{\partial t}$$

These last two equations are in conflict, for the first states that the divergence is always zero. The resolution of this dichotomy was proposed by James Maxwell, the result of which permanently attached his name to the set of equations brought into harmony by his genius. In his own words ([39], volume 2, page 253, article 610):

> One of the chief peculiarities of this treatise is the doctrine which it asserts, that the true electric current, that on which electromagnetic phenomena depend, is not the same thing as $\vec{\mathcal{J}}$, the current of conduction, but that the time variation of $\vec{\mathcal{D}}$, the electric displacement, must be taken into account in estimating the total movement of electricity, so that we must write,

$$\vec{\mathcal{J}}_{\text{total}} = \vec{\mathcal{J}} + \frac{\partial \vec{\mathcal{D}}}{\partial t} \tag{7-8}$$

This means that Ampere's law must be modified to

$$\nabla \times \vec{\mathcal{H}} = \vec{\mathcal{J}} + \frac{\partial \vec{\mathcal{D}}}{\partial t} \tag{7-9}$$

To show that this resolves the situation we take the divergence of the preceding equation:

$$\nabla \cdot \nabla \times \vec{\mathcal{H}} \equiv 0 = \nabla \cdot \vec{\mathcal{J}} + \nabla \cdot \left(\frac{\partial \vec{\mathcal{D}}}{\partial t}\right)$$

Interchanging the order of taking partial derivatives with respect to time and coordinate this becomes

$$\nabla \cdot \vec{\mathcal{J}} = -\frac{\partial}{\partial t}(\nabla \cdot \vec{\mathcal{D}})$$

From Table 7-1 the point form of Gauss' law for electric fields replaces the divergence of $\vec{\mathcal{D}}$ with the charge density ρ_v:

$$\nabla \cdot \vec{\mathcal{J}} = -\frac{\partial \rho_v}{\partial t}$$

This last result is exactly what is required, since it reduces to the proper form for static (time-invariant) quantities. The derivative of the displacement $\vec{\mathcal{D}}$ is dimensionally ampere per square meter, so this is called the *displacement current density* $\vec{\mathcal{J}}_d$:

$$\vec{\mathcal{J}}_d = \frac{\partial \vec{\mathcal{D}}}{\partial t} \tag{7-10}$$

We may now generalize the integral form called Ampere's circuital law by integrating Eq. 7-8 over a surface S and applying Stokes' theorem to the left side:

$$\int_S \nabla \times \vec{\mathcal{H}} \cdot d\vec{a} = \int_C \vec{\mathcal{H}} \cdot d\vec{l} = \int_S (\vec{\mathcal{J}} + \vec{\mathcal{J}}_D) \cdot d\vec{a} = I_{\text{enclosed}} \tag{7-11}$$

The result is basically the same except that the enclosed current consists of true current flow plus the contribution of displacement current.

The displacement current can also be "derived" by taking the static form of Ampere's law and adding an unknown vector quantity to it, and then determining what the vector function has to be to remove the inconsistency. We thus add an unknown current density (which it must be since $\vec{\mathcal{J}}$ is current density) to Ampere's law:

$$\nabla \times \vec{\mathcal{H}} = \vec{\mathcal{J}} + \vec{\mathcal{J}}_d$$

Taking the divergence and applying the vector identity

$$\nabla \cdot \nabla \times \vec{\mathcal{H}} \equiv 0 = \nabla \cdot \vec{\mathcal{J}} + \nabla \cdot \vec{\mathcal{J}}_d$$

or

$$\nabla \cdot \vec{\mathcal{J}} = -\nabla \cdot \vec{\mathcal{J}}_d$$

From the continuity Eq. 5-42 this requires

$$-\nabla \cdot \vec{\mathcal{J}}_d = -\frac{\partial \rho_v}{\partial t}$$

Using Gauss' law for electric fields, $\nabla \cdot \vec{\mathcal{D}} = \rho_v$,

$$\nabla \cdot \vec{\mathcal{J}}_d = \frac{\partial(\nabla \cdot \vec{\mathcal{D}})}{\partial t} = \nabla \cdot \left(\frac{\partial \vec{\mathcal{D}}}{\partial t}\right)$$

Therefore,

$$\vec{\mathcal{J}}_d = \frac{\partial \vec{\mathcal{D}}}{\partial t}$$

as before.

The classical example of the displacement current is that of a parallel plate capacitor, shown in Figure 7-10. Suppose that the applied source is a sinusoid, and that the resistance and inductance of the circuit may be neglected. For slow time variation (lower frequencies, often called the *quasi-static* approximation) we may use the result of Example 5-38 of Chapter 5, Section 5-14. Method II of that example gave the intermediate result that the electric field between two plates is equal to the voltage difference between the plates divided by the distance between the plates. In the present example we would write (just using magnitudes)

$$\mathcal{E} = \frac{v(t)}{d} = \frac{V_0 \cos \omega t}{d}$$

For a dielectric having permittivity ϵ between the plates the constitutive relation for electric fields gives

$$\mathcal{D} = \frac{\epsilon V_0}{d} \cos \omega t$$

The displacement current density is then

$$\mathcal{J}_D = \frac{\partial \mathcal{D}}{\partial t} = -\omega \frac{\epsilon V_0}{d} \sin \omega t$$

The total displacement current between the plates is then

$$i_D = \mathcal{J}_D A = -\omega \frac{\epsilon A}{d} V_0 \sin \omega t$$

Now if we use circuit theory the current $i(t)$ would be written for the parallel plate capacitor as

$$i(t) = C \frac{dv}{dt} = C(-\omega V_0 \sin \omega t) = -\omega C V_0 \sin \omega t$$

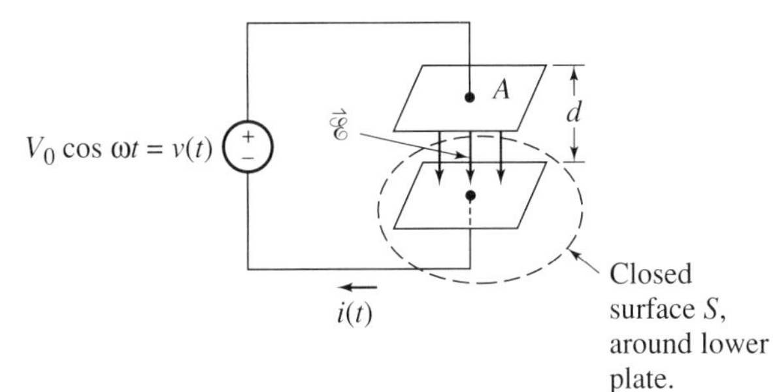

Figure 7-10. Capacitor as an illustration of displacement current.

Comparing the two expressions for current we would have

$$C = \frac{\epsilon A}{d}$$

which is the known result from physics (or Example 5-38 of Chapter 5). What we have actually done is generalized Kirchhoff's current law from circuit theory, namely, the sum of all currents (displacement and conduction) leaving a closed surface is zero. We are now no longer restricted to metal surfaces or wires.

Exercises

7-3.0a The magnetic field intensity in a region of free space is determined to be

$$\vec{\mathcal{H}}(\vec{r}, t) = 30 \cos(10^{10}t - \tfrac{100}{3}z)\hat{y} \text{ A/m}$$

(a) Obtain the equation for $\vec{\mathcal{B}}$.
(b) Obtain the equation for the displacement current if there are no free charges in the region.
(c) Obtain the equation for the electric flux (displacement) density $\vec{\mathcal{D}}$ by integration of (b). Neglect constant of integration.
(d) Obtain the equation for $\vec{\mathcal{E}}$.
(e) Show that the point form of Faraday's law is satisfied by $\vec{\mathcal{B}}$ and $\vec{\mathcal{E}}$. (Answer: $\vec{\mathcal{B}} = \mu_0\vec{\mathcal{H}} = 37.7 \cos[10^{10}t - (100/3)z]\hat{y}$ μT; $\vec{\mathcal{J}} = -10^3 \sin[10^{10}t - (100/3)z]\hat{x}$ A/m^2; $\vec{\mathcal{D}} = 10^{-7} \cos[10^{10}t - (100/3)z]\hat{x}$ C/m^2; $\vec{\mathcal{E}} = \vec{\mathcal{D}}/\epsilon_0 = 11{,}310 \cos[10^{10}t - (100/3)z]\hat{x}$ V/m)

7-3.0b Two circular disks form a parallel-plate capacitor. The plates are 5 cm in diameter and separated by 1 cm. The electric field intensity between them is

$$3000 \cos(10^8 t - \rho)\hat{z} \text{ V/m}$$

A dielectric with a relative permittivity of 9 is between the plates. Determine the equation for the displacement current, the current in the wires connected to the plates, the magnetic field between the plates and the voltage source attached to the capacitor. What is the phase relationship between the applied source and the line current? (Answer: $-23.87 \sin(10^8 t - \rho)\hat{z}$ A/m^2, $-0.04687 \sin 10^8 t$ A, $30 \cos(10^8 t - \rho)\hat{\phi}$ μT, $30 \cos 10^8 t$ V, i leads v by 90°)

7-3.0c Using the point forms of Faraday's law and Ampere's law, determine the spatial relationships (qualitatively) that exist between pairs of $\vec{\mathcal{J}}_D$, $\vec{\mathcal{D}}$, $\vec{\mathcal{E}}$, $\vec{\mathcal{B}}$, $\vec{\mathcal{H}}$, for linear, isotropic materials. (Answer: $\vec{\mathcal{J}}_D \| \vec{\mathcal{D}}$, $\vec{\mathcal{J}}_D \| \vec{\mathcal{E}}$, $\vec{\mathcal{E}} \perp \vec{\mathcal{B}}$, $\vec{\mathcal{E}} \perp \vec{\mathcal{H}}$, $\vec{\mathcal{D}} \| \vec{\mathcal{E}}$, $\vec{\mathcal{J}}_D \perp \vec{\mathcal{B}}$, etc.)

We conclude this section with a summary of the types of current flow that have been developed. The total current may be written as

$$\vec{\mathcal{J}}_{\text{total}} = \vec{\mathcal{J}} + \vec{\mathcal{J}}_D \tag{7-12}$$

The first term, often called the *free current* since it is the flow of free charges, can occur due to conduction or convection as discussed in Chapter 5. These are given by

$$\vec{\mathcal{J}} = \begin{cases} \rho_v \vec{v} & \text{(convection)} \\ \sigma \vec{\mathcal{E}} & \text{(conduction)} \end{cases} \tag{7-13}$$

The displacement current (which may be thought of as a radiation current) can be broken into components as well by using the constitutive relationship for electric fields and the polarization equations 5-56 through 5-59:

$$\begin{aligned}\vec{\mathcal{J}}_D &= \frac{\partial\vec{\mathcal{D}}}{\partial t} = \frac{\partial(\epsilon\vec{\mathcal{E}})}{\partial t} = \frac{\partial(\epsilon_0\epsilon_r\vec{\mathcal{E}})}{\partial t} = \frac{\partial[\epsilon_0(1 + \chi_e)\vec{\mathcal{E}}]}{\partial t} \\ &= \frac{\partial(\epsilon_0\vec{\mathcal{E}})}{\partial t} + \frac{\partial(\epsilon_0\chi_e\vec{\mathcal{E}})}{\partial t} = \frac{\partial(\epsilon_0\vec{\mathcal{E}})}{\partial t} + \frac{\partial\vec{P}}{\partial t} = \vec{\mathcal{J}}_{D0} + \vec{\mathcal{J}}_P\end{aligned} \tag{7-14}$$

Thus the displacement current actually consists of a free-space displacement current component plus a polarization current component. For time-varying fields, the material dipoles will be in motion since these dipoles consist of charges that have forces on them produced by the fields. This movement of charge constitutes a current flow, but since the charges do not move through the material this is often referred to as a *bound displacement current component.*

7-4 Maxwell's Equations: General and Sinusoidal Steady-State Forms

Fortunately, it turns out that the remainder of the field equations in Table 7-1 are unchanged. All we need to do is replace the block letters by script letters. The resulting four differential (point) field equations (and their integral counterparts) are referred to as Maxwell's equations. It is from these that we study wave propagation and other time-varying field effects, including radio frequency interference (RFI) and electromagnetic interference (EMI). A summary of the time-varying field equations is given in Table 7-3. Note that when the time derivatives are zero, they reduce to the results in Table 7-1.

Exercises

7-4.0a Determine which of the following fields satisfy Maxwell's equations:
(a) $\vec{\mathcal{D}} = yt^2\hat{x}$
(b) $\vec{\mathcal{H}} = -3e^{j(\omega t - z)}\hat{\phi}$
(c) $\vec{\mathcal{E}} = (E_0/r)\cos(\omega t - \beta r)\hat{\theta}$ (β constant)

7-4.0b Suppose you have discovered magnetic charge ρ_{vm} so that $\nabla \cdot \vec{\mathcal{J}}_m = -\partial(\rho_{vm})/\partial t$. What modification of the point form of Faraday's law, Eq. 7-5, is required so that Maxwell's equations are internally consistent?

The process for converting Maxwell's equations to the sinusoidal steady state is exactly the same as discussed in Chapter 1, Section 1-5. A review of this theory is given in Appendix A. Two steps were derived:

1. Replace each time derivative by $j\omega$.
2. Replace each script letter by the block phasor letter.

Table 7-3. Summary of Time-Varying Field Equations: General Form

Maxwell's Equations	
Ampere's law:	
$\oint_C \vec{\mathscr{H}} \cdot d\vec{l} = \int_S (\vec{\mathscr{J}} + \vec{\mathscr{J}}_D) \cdot d\vec{a}$ $\left(\vec{\mathscr{J}}_D = \frac{\partial \vec{\mathscr{D}}}{\partial t}\right)$	$\nabla \times \vec{\mathscr{H}} = \vec{\mathscr{J}} + \vec{\mathscr{J}}_D$
Faraday's law:	
$\oint_C \vec{\mathscr{E}} \cdot d\vec{l} = -\frac{d}{dt}\int_S \vec{\mathscr{B}} \cdot d\vec{a}$	$\times \vec{\mathscr{E}} = -\frac{\partial \vec{\mathscr{B}}}{\partial t}$
Gauss' law:	
Electric fields:	
$\oint_S \vec{\mathscr{D}} \cdot d\vec{a} = \int_V \rho_v dv$	$\nabla \cdot \vec{\mathscr{D}} = \rho_v$
Magnetic Field:	
$\oint_S \vec{\mathscr{B}} \cdot d\vec{a} = 0$	$\nabla \cdot \vec{\mathscr{B}} = 0$
Companion Formulas	
Constitutive relationships:	Lorentz force equation:
$\vec{\mathscr{D}} = \epsilon \vec{\mathscr{E}} = \mu \vec{\mathscr{H}}$	$\vec{\mathscr{F}} = q\vec{\mathscr{E}} + q\vec{v} \times \vec{\mathscr{B}}$
$\vec{\mathscr{J}}_{\text{conduction}} = \sigma \vec{\mathscr{E}}$	Continuity equation:
$\vec{\mathscr{J}}_{\text{convection}} = \rho_v \vec{v}$	$\nabla \cdot \vec{\mathscr{J}} = \frac{\partial \rho_v}{\partial t}$

The last step puts block letters back into the equations, which make them appear like the static forms somewhat. One can usually identify the sinusoidal steady state by the presence of an ω or a j, or else the context of a discussion will alert one to which is intended. When the sinusoidal steady-state equations have been solved, we obtain the time form using the process that was also described in Chapter 1. For example, if we have solved for the phasor representation of the magnetic field intensity, $\vec{H}$, then

$$\vec{\mathscr{H}}(\vec{r}, t) = \text{Re}\{\vec{\mathscr{H}} e^{j\omega t}\} \tag{7-15}$$

Using the steps given in the preceding paragraph we obtain the formulation summarized in Table 7-4.

Table 7-4 Summary of Time-Varying Field Equations: Sinusoidal Steady-State Form

Maxwell's Equations	
Ampere's law:	
$\oint_C \vec{H} \cdot d\vec{l} = \int_S (\vec{J} + \vec{J}_D) \cdot d\vec{a}$	$\nabla \times \vec{H} = \vec{J}$
$(\vec{J}_D = j\omega\vec{D})$	
Faraday's law:	
$\oint_C \vec{E} \cdot d\vec{l} = -j\omega \int_S \vec{B} \cdot d\vec{a}$	$\nabla \times \vec{E} = -j\omega\vec{B}$
Electric fields:	
$\oint_S \vec{D} \cdot d\vec{a} = \int_V \rho_v dv$	$\nabla \cdot \vec{D} = \rho_v$
Magnetic fields:	
$\oint_S \vec{B} \cdot d\vec{a} = 0$	$\nabla \cdot \vec{B} = 0$

Companion Formulas	
Constitutive relationships:	Lorentz force equation:
$\vec{D} = \epsilon\vec{E} \qquad \vec{B} = \mu\vec{H}$	$\vec{F} = q\vec{E} + q\vec{v} \times \vec{B}$
$\vec{J}_{\text{conduction}} = \sigma\vec{E}$	Continuity equation:
$\vec{J}_{\text{convection}} = \rho_v\vec{v}$	$\nabla \cdot \vec{J} = -j\omega\rho_v$

7-5 Boundary Conditions for Time-Varying Fields

To solve Maxwell's equations for specific systems we need to have some boundary conditions to obtain values of the constants that appear in the general solutions of the differential equations. Again we are fortunate because the normal and tangential components of the fields satisfy the same boundary conditions as those in the static field case. The derivations use the integral forms as was done in Chapter 5 and 6. Since time derivatives appear in Maxwell's equations, one boundary condition will be derived in detail to show how the derivatives are handled.

Consider the electric field at a boundary between two dielectrics shown in Figure 7-11. We wish to determine the boundary condition on the tangential electric field intensity. Applying Faraday's law in integral form we write

$$\oint_{\Delta C} \vec{\mathscr{E}} \cdot d\vec{l} = -\frac{d}{dt} \int_{\Delta S} \vec{\mathscr{B}} \cdot d\vec{a}$$

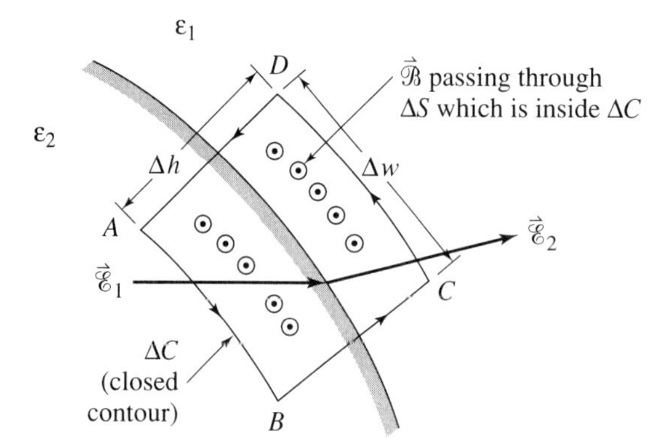

Figure 7-11. Determining the boundary condition of tangential components of a time-varying electric field.

Expanding around the contour

$$\oint_A^B \vec{\mathcal{E}}_1 \cdot d\vec{l} + \int_B^C \vec{\mathcal{E}} \cdot d\vec{l} + \int_C^D \vec{\mathcal{E}}_2 \cdot d\vec{l} + \int_D^A \vec{\mathcal{E}} \cdot d\vec{l} = -\frac{d}{dt}\int_0^{\Delta h}\int_0^{\Delta w} \vec{\mathcal{B}} \cdot d\vec{a}$$

To obtain values at the boundary we let $\Delta h \to 0$. This means that paths $B \to C$ and $D \to A$ are of zero length so two of the left integrals are zero. For the integral on the right, as $\Delta h \to 0$ there is no area so the flux is zero. The preceding equation reduces to

$$\int_A^B \vec{\mathcal{E}}_1 \cdot d\vec{l} + \int_C^D \vec{\mathcal{E}}_2 \cdot d\vec{l} = 0$$

The dot products select the tangential components on these remaining paths. Note that for path $C \to D$ the dot product is negative since the angle between the vectors is greater than 90°:

$$\int_A^B \mathcal{E}_{1t} dl + \int_C^D - \mathcal{E}_{2t} dl = 0$$

For small $\Delta \omega$ the field may be considered constant over that length

$$\mathcal{E}_{1t} \int_A^B dl - \mathcal{E}_{2t} \int_C^D dl = 0$$

or

$$\mathcal{E}_{1t}\, \Delta\omega - \mathcal{E}_{2t}\, \Delta w = 0$$

Thus

$$\mathcal{E}_{1t} = \mathcal{E}_{2t}$$

as before.

The general boundary conditions are summarized in Table 7-5. The equations in parentheses are obtained using the constitutive relationships. The table also gives the special case where Medium 1 is a perfect conductor $\sigma_1 = \infty$. To obtain the sinusoidal steady state form simply put in the block phasor letters.

Table 7-5 Summary of Boundary Conditions for Time-Varying Electromagnetic Fields

General Case	
$\mathscr{E}_{1t} = \mathscr{E}_{2t}$ $\left(\dfrac{\mathscr{D}_{1t}}{\epsilon} = \dfrac{\mathscr{D}_{2t}}{\epsilon}\right)$	$\mathscr{D}_{2n} - \mathscr{D}_{1n} = \rho_s$ $(\epsilon_2\mathscr{E}_{2n} - \epsilon_1\mathscr{E}_{1n} = \rho_s)$
$\mathscr{B}_{1n} = \mathscr{B}_{2n}$ $(\mu_1\mathscr{H}_{1n} = \mu_2\mathscr{H}_{2n})$	$\mathscr{H}_{2t} - \mathscr{H}_{1t} = \mathscr{J}_s$ (current density on surface, normal to $\vec{\mathscr{H}}$) $\left(\dfrac{\mathscr{B}_{2t}}{\mu_2} - \dfrac{\mathscr{B}_{1t}}{\mu_1} = \mathscr{J}_s\right)$ Alternative form: $\hat{n} \times (\vec{\mathscr{H}}_2 - \vec{\mathrm{H}}_1) = \vec{\mathscr{J}}_s$ ($\hat{n}$ is normal from medium 1 into medium 2)
Medium 1 a Perfect Conductor ($\sigma_1 = \infty$)	
$\mathscr{E}_t = 0$ $(\mathscr{D}_t = 0)$	$\mathscr{D}_{2n} = \rho_s$ $(\epsilon_2\mathscr{E}_{2n} = \rho_s)$
$\mathscr{B}_n = 0$ $(\mathscr{H}_n = 0)$	$\mathscr{B}_{2t} = \mathscr{H}_{2t}$ $\left(\dfrac{\mathscr{B}_t}{\mu_2} = \mathscr{J}_s\right)$
Note: For *static* fields the normal component of $\mathscr{B}$ is continuous, not necessarily zero, for perfect conductors. See Sections 6-9 and 6-11.	$\hat{n} \times \mathscr{H}_2 = \mathscr{J}_s$

In addition to the conditions listed in the table we must include the six physical auxiliary boundary conditions given at the end of Section 5-12. These are summarized as

1. For finite-dimension systems with specified charges and/or voltages and/or currents, the potentials and fields at infinite distances from the system remain finite (including zero).
2. For finite dimensional systems in which a coordinate value of zero is in the region of interest the fields and potentials must remain finite or zero.
3. Potentials are continuous across charge-free and current-free boundaries.
4. If an object of finite extent is immersed in a uniform field of large extent, the field at large distance from the object is unaltered.
5. For finite field sources the energy stored in a given region is finite (Meixner edge condition).

6. For propogating fields emanating from a finite source (Somerfeld radiation condition),

$$\lim_{r\to\infty}\left[r\left(\frac{\partial\vec{\mathscr{E}}}{\partial r}+j\beta\vec{\mathscr{E}}\right)\right]=0$$

The boundary conditions are applied in exactly the same manner as in Chapters 5 and 6. Also, the uniqueness of the solutions obtained using these results can be proved in exactly the same manner as in Chapter 5, Section 5-13.

Exercises

7-5.0a Derive the boundary condition for normal $\vec{\mathscr{D}}$.

7-5.0b Derive the boundary condition for normal $\vec{\mathscr{B}}$.

7-5.0c Derive the boundary condition for tangential $\vec{\mathscr{H}}$ using the configuration in Figure 6-51.

7-5.0d Suppose that medium 1 in Figure 7-11 is a perfect conductor. Explain how the lower half of Table 7-5 is obtained. (Hint: For time-varying fields *both* $\vec{\mathscr{E}}$ and $\vec{\mathscr{B}}$ are zero in a perfect conductor.) Why are there no numerical subscripts on tangential $\vec{\mathscr{E}}$ and normal $\vec{\mathscr{B}}$?

7-5.0e The electric field in the air to the left of the perfect conductor shown in Figure 7-12 has the form

$$\vec{\mathscr{E}}(\vec{r},t)=0.1\cos(10^{10}t-20z)\hat{y}+E_R\cos(10^{10}t+20z)\hat{y}$$

Determine the value of E_R. Note the boundary is at $z=0$. (Answer: $E_R=-0.1$)

7-5.0f Using the curl form of Faraday's law determine $\vec{\mathscr{H}}$ for the field given in Exercise 7-5.0e, using $E_r=-0.1$. From this $\vec{\mathscr{H}}$ determine the surface current on the metal and show it on the figure. Remember that inside the metal $\vec{\mathscr{H}}$ is zero.

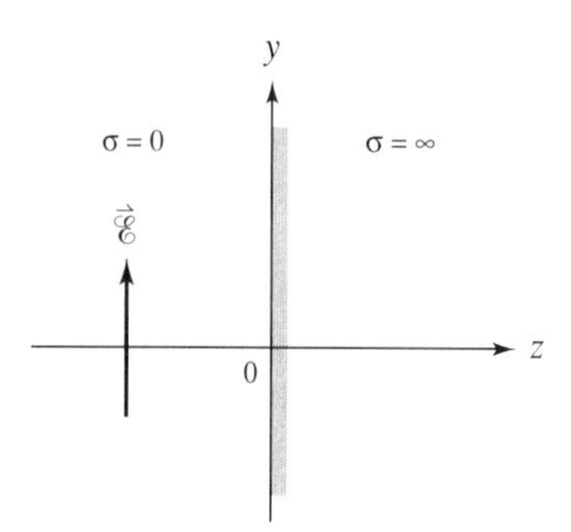

Figure 7-12.

7-6 Wave Equations for Electric Fields, Magnetic Fields, Scalar Electric Potential, and Vector Magnetic Potential; Retarded Potentials

In our study of transmission lines it was found convenient and direct to derive the general differential equations satisfied by voltage and current along the lines. It was then a relatively easy matter to specialize those results to specific cases. Such an approach made it possible to connect a derived particular result to its source. This same approach will be used for time-varying fields, which also makes it possible to obtain some physical insights into propagating fields by making analogies with the transmission line results.

For the electric field wave equation we begin by taking the curl of Eq. 7-5:

$$\nabla \times \nabla \times \vec{\mathscr{E}} = \nabla \times \left(- \frac{\partial \vec{\mathscr{B}}}{\partial t} \right)$$

Interchanging the order of taking partial derivatives on the right-hand side and using a vector identity on the left-hand side:

$$\nabla(\nabla \cdot \vec{\mathscr{E}}) - \nabla^2 \vec{\mathscr{E}} = - \frac{\partial}{\partial t} (\nabla \times \vec{\mathscr{B}})$$

Using the constitutive relationships from Table 7-3 we obtain

$$\nabla\left(\nabla \cdot \left(\frac{\vec{\mathscr{D}}}{\epsilon} \right) \right) - \nabla^2 \vec{\mathscr{E}} = - \frac{\partial}{\partial t} (\nabla \times (\mu \vec{\mathscr{H}}))$$

If we want the most general case we must allow ϵ and μ to be functions of coordinates (called *inhomogeneous materials*) and apply vector identities treating ϵ and μ as scalars, so that the del operations are on a scalar times a vector. The preceding equation becomes

$$\nabla\left[\vec{\mathscr{D}} \cdot \nabla\left(\frac{1}{\epsilon} \right) + \frac{1}{\epsilon} \nabla \cdot \vec{\mathscr{D}} \right] - \nabla^2 \vec{\mathscr{E}} = - \frac{\partial}{\partial t} [\nabla \mu \times \vec{\mathscr{H}} + \mu \nabla \times \vec{\mathscr{H}}]$$

In the left member replace the isolated $\vec{\mathscr{D}}$ by $\epsilon \vec{\mathscr{E}}$ and replace $\nabla \cdot \vec{\mathscr{D}}$ by ρ_v (Gauss' law for electric fields). In the right member replace $\nabla \times \vec{\mathscr{H}}$ by $\vec{\mathscr{J}} + (\partial \vec{\mathscr{D}} / \partial t)$ (Ampere's law). Then use the distributive property of differential operators, also assuming that μ is not varying with time (time-varying materials are usually nonlinear or anisotropic):

$$\nabla\left[\epsilon \vec{\mathscr{E}} \cdot \nabla\left(\frac{1}{\epsilon} \right) \right] + \nabla\left(\frac{1}{\epsilon} \rho_v \right) - \nabla^2 \vec{\mathscr{E}} = - \frac{\partial}{\partial t} (\nabla \mu \times \vec{\mathscr{H}}) - \mu \frac{\partial \vec{\mathscr{J}}}{\partial t} - \mu \frac{\partial^2 \vec{\mathscr{D}}}{\partial t^2}$$

Using the constitutive relationship for $\vec{\mathscr{D}}$ in the second partial with respect to time and rearranging the terms in the result we obtain (see Exercise 7-6.0a)

$$\nabla^2 \vec{\mathscr{E}} - \mu\epsilon \frac{\partial^2 \vec{\mathscr{E}}}{\partial t^2} = \nabla\left[\epsilon \vec{\mathscr{E}} \cdot \nabla\left(\frac{1}{\epsilon} \right) \right] + \nabla\left(\frac{1}{\epsilon} \rho_v \right) + \frac{\partial}{\partial t} (\nabla \mu \times \vec{\mathscr{H}}) + \mu \frac{\partial \vec{\mathscr{J}}}{\partial t} \qquad (7\text{-}16)$$

One of the important special cases of the preceding equation is that for which the medium is homogeneous (ϵ and μ not functions of position) for which the gradient terms are zero.

$$\nabla^2 \vec{\mathscr{E}} - \mu\epsilon \frac{\partial^2 \vec{\mathscr{E}}}{\partial t^2} = \frac{1}{\epsilon} \nabla \rho_v + \mu \frac{\partial \vec{\mathscr{J}}}{\partial t} \qquad (\mu, \epsilon \text{ constant}) \qquad (7\text{-}17)$$

In many cases we consider electromagnetic waves in a medium that has no net charge, no sources, and zero conductivity (called a *lossless, source-free* medium). These conditions make $\nabla\rho_v = 0$, $\vec{\mathcal{J}} = \vec{\mathcal{J}}_{\text{source}} + \vec{\mathcal{J}}_{\text{conduction}} = \vec{\mathcal{J}}_{\text{convection}} = \vec{\mathcal{J}}_{\text{source}} + \sigma\vec{\mathcal{E}} + \rho_v\vec{v} = 0$ so that the reduced wave equation is

$$\nabla^2\vec{\mathcal{E}} - \mu\epsilon\frac{\partial^2\vec{\mathcal{E}}}{\partial t^2} = 0 \quad (\mu,\ \epsilon \text{ constant; charge-free and source-free medium; } \sigma = 0) \tag{7-18}$$

This equation form is often called the *vector Helmholtz equation*. If we compare this last result with Eq. 1-7 for the transmission line lossless case with

$$R = G = 0$$

we see an immediate analogy in which $\vec{\mathcal{E}}$ is analogous to voltage and μ and ϵ analogous to L and C. Remember that the del-squared operator has second partial derivatives with respect to coordinates with units of $1/\text{m}^2$. Also, for transmission lines the phase velocity of a propagating voltage was obtained as

$$V_p = \frac{1}{\sqrt{LC}}$$

which suggests that for a propagating electric field there is *a* velocity corresponding to

$$\frac{1}{\sqrt{\mu\epsilon}}$$

This result, however, is true only for particular field configurations that will be described later. Using the techniques developed in Section 1-5.1, the three forms of the wave equation may be written in sinusoidal steady-state forms as

$$\nabla^2\vec{E} + \omega^2\mu\epsilon\vec{E} = \nabla\left[\epsilon\vec{E}\cdot\nabla\left(\frac{1}{\epsilon}\right)\right] + \nabla\left(\frac{1}{\epsilon}\rho_v\right) + j\omega(\nabla\mu \times \vec{H}) + j\omega\mu\vec{J} \tag{7-19}$$

$$\nabla^2\vec{E} + \omega^2\mu\epsilon\vec{E} = \frac{1}{\epsilon}\nabla\rho_v + j\omega\mu\vec{J} \qquad (\epsilon,\ \mu \text{ constants}) \tag{7-20}$$

$$\nabla^2\vec{E} + \omega^2\mu\epsilon\vec{E} = 0 \quad (\epsilon,\ \mu \text{ constants; charge-free and source-free medium; } \sigma = 0) \tag{7-21}$$

where

$$\vec{\mathcal{E}} = \text{Re}\{\vec{E}e^{j\omega t}\} \qquad \rho_v(t) = \text{Re}\{\rho_v e^{j\omega t}\}$$
$$\vec{\mathcal{H}} = \text{Re}\{\vec{H}e^{j\omega t}\} \quad \text{and} \quad \vec{\mathcal{J}} = \text{Re}\{\vec{J}e^{j\omega t}\} \tag{7-22}$$

Using a similar procedure we obtain the following wave equation forms for the magnetic field intensity (see Exercises 7-6.0b and 7-6.0c):

$$\nabla^2\vec{\mathcal{H}} - \mu\epsilon\frac{\partial^2\vec{\mathcal{H}}}{\partial t^2} = \nabla\left[\mu\vec{\mathrm{H}}\cdot\nabla\left(\frac{1}{\mu}\right)\right] - \nabla\times\vec{\mathcal{J}} - \frac{\partial(\nabla\epsilon\times\vec{\mathcal{E}})}{\partial t} \tag{7-23}$$

$$\nabla^2\vec{\mathcal{H}} - \mu\epsilon\frac{\partial^2\vec{\mathcal{H}}}{\partial t^2} = -\nabla\times\vec{\mathcal{J}} \qquad (\mu,\ \epsilon \text{ constants}) \tag{7-24}$$

$$\nabla^2\vec{\mathcal{H}} - \mu\epsilon\frac{\partial^2\vec{\mathcal{H}}}{\partial t^2} = 0 \qquad (\mu,\ \epsilon,\ \text{constants; source-free medium; } \sigma = 0) \tag{7-25}$$

$$\nabla^2\vec{H} + \omega^2\mu\epsilon\vec{H} = \nabla\left[\mu\vec{H}\cdot\nabla\left(\frac{1}{\mu}\right)\right] - \nabla\times\vec{J} - j\omega(\nabla\epsilon\times\vec{E}) \tag{7-26}$$

$$\nabla^2\vec{H} + \omega^2\mu\epsilon\vec{H} = -\nabla\times\vec{J} \qquad (\mu,\ \epsilon \text{ constants}) \tag{7-27}$$

$$\nabla^2\vec{H} + \omega^2\mu\epsilon\vec{H} = 0 \qquad (\mu,\ \epsilon,\ \text{constants; source-free medium; } \sigma = 0) \tag{7-28}$$

It may appear that we have set for ourselves an unenviable task of having to solve two vector equations (one from each of the preceding two sets) to obtain the field configuration. However, since the electric and magnetic fields are related by Maxwell's equations we really need only solve one equation and then use the appropriate curl equation to obtain the other field quantity. Although solving even one of the vector equations may seem challenging, we will use techniques similar to transmission line solutions by deriving reduced forms of the vector equations before proceeding with the solution. Much of the remainder of the text is devoted to developing various strategies for solving the vector equations.

Exercises

7-6.0a Eq. 7-16 contains a term having $\vec{\mathcal{H}}$ in it. Assuming that μ is time invariant, show that the term may be written as

$$-\frac{1}{\mu}\nabla\mu + (\nabla\times\vec{\mathcal{E}})$$

so that the wave equation will contain only the field $\vec{\mathcal{E}}$.

7-6.0b Derive Eqs. 7-23 through 7-28. Begin by taking the curl of the differential form of Ampere's law (Table 7-3) and then following the procedure of the text. Be sure to keep track of any constraints you place on μ, ϵ, and σ.

7-6.0c Equation 7-23 contains a term having $\vec{\mathscr{E}}$ in it. Assuming that ϵ is time invariant, show that the term may be written as

$$-\frac{1}{\epsilon}\nabla\epsilon \times (\nabla \times \vec{\mathscr{H}}) + \frac{1}{\epsilon}\nabla\epsilon \times \vec{\mathscr{J}}$$

so that the wave equation will contain only the field $\vec{\mathscr{H}}$.

7-6.0d Show that the field

$$\vec{\mathscr{E}}(\vec{r}, t) = E_0 \cos(\omega t - kz)\hat{y} \qquad E_0 \text{ constant}$$

will satisfy Eq. 7-18 if k is suitably defined. Identify this k by analogy with transmission line solution for $v(z, t)$ lossless case. (Answer: $k^2 = -\omega^2\mu\epsilon$)

7-6.0e Construct the sinusoidal steady-state equivalent for the $\vec{\mathscr{E}}$ of Exercise 7-6.0d and show that it satisfies Eq. 7-21.

In some problems it is more convenient and even easier to solve for the scalar electric and magnetic vector potentials and then $\vec{\mathscr{E}}(\vec{E})$ and $\vec{\mathscr{H}}(\vec{H})$ from them. This is particularly true when there are sources present ($\vec{\mathscr{J}}$ and ρ_v) since, as we shall see, the potentials can be derived directly from these. For the vector magnetic potential we begin with Gauss' law for magnetic fields

$$\nabla \cdot \vec{\mathscr{B}} = 0$$

Recalling the null vector identity

$$\nabla \cdot \nabla \times \vec{\mathscr{A}} = 0$$

we can define the vector potential by way of the formula

$$\vec{\mathscr{B}} = \nabla \times \vec{\mathscr{A}} \tag{7-29}$$

This is identical in form to the static case.

To obtain the scalar electric potential we use Faraday's law

$$\nabla \times \vec{\mathscr{E}} = \frac{\partial \vec{\mathscr{B}}}{\partial t}$$

Substituting the preceding equation for $\vec{\mathscr{B}}$,

$$\nabla \times \vec{\mathscr{E}} = -\frac{\partial(\nabla \times \vec{\mathscr{A}})}{\partial t} = -\nabla \times \frac{\partial \vec{\mathscr{A}}}{\partial t}$$

From this we obtain

$$\nabla \times \vec{\mathscr{E}} + \nabla \times \frac{\partial \vec{\mathscr{A}}}{\partial t} = 0$$

Using the distributive property of the ∇ operator,

$$\nabla \times \left(\vec{\mathscr{E}} + \frac{\partial \vec{\mathscr{A}}}{\partial t}\right) = 0$$

Another null vector identity states that

$$\nabla \times (-\nabla \mathcal{V}) = 0$$

where the minus sign is included so that the resulting potential function will reduce to the static case. Comparing the preceding two equations we can define the scalar electric potential as follows

$$-\nabla \mathcal{V} = \vec{\mathcal{E}} + \frac{\partial \vec{\mathcal{A}}}{\partial t}$$

or, in another form,

$$\vec{\mathcal{E}} = -\nabla \mathcal{V} - \frac{\partial \vec{\mathcal{A}}}{\partial t} \tag{7-30}$$

For static fields this reduces to the form given in Table 7-2, Eq. 5-63. Thus if we can solve for the two potential functions we can obtain $\vec{\mathcal{B}}(\vec{\mathcal{H}})$ and $\vec{\mathcal{E}}(\vec{\mathcal{D}})$ easily. To obtain the wave equation for the vector $\vec{\mathcal{A}}$ we begin with Ampere's law

$$\nabla \times \vec{\mathcal{H}} = \vec{\mathcal{J}} + \vec{\mathcal{J}}_D = \vec{\mathcal{J}} + \frac{\partial \vec{\mathcal{D}}}{\partial t}$$

Applying the constitutive relationships and assuming that ϵ is not varying with time

$$\nabla \times \frac{\vec{\mathcal{B}}}{\mu} = \vec{\mathcal{J}} + \epsilon \frac{\partial \vec{\mathcal{E}}}{\partial t}$$

Substituting for $\vec{\mathcal{B}}$ and $\vec{\mathcal{E}}$ from Eqs. 7-29 and 7-30:

$$\nabla \times \left(\frac{1}{\mu} \nabla \times \vec{\mathcal{A}}\right) = \vec{\mathcal{J}} + \epsilon \frac{\partial\left(-\nabla \vec{\mathcal{V}} - \frac{\partial \vec{\mathcal{A}}}{\partial t}\right)}{\partial t} = \vec{\mathcal{J}} - \epsilon \nabla\left(\frac{\partial \mathcal{V}}{\partial t}\right) - \epsilon \frac{\partial^2 \vec{\mathcal{A}}}{\partial t^2}$$

If μ is a function of coordinates, the left side is the curl of a scalar function times a vector $(\nabla \times \vec{\mathcal{A}})$. The vector expansion yields

$$\nabla\left(\frac{1}{\mu}\right) \times (\nabla \times \vec{\mathcal{A}}) + \frac{1}{\mu} \nabla \times \nabla \times \vec{\mathcal{A}} = \vec{\mathcal{J}} - \epsilon \nabla\left(\frac{\partial \mathcal{V}}{\partial t}\right) - \epsilon \frac{\partial^2 \vec{\mathcal{A}}}{\partial t^2}$$

Expanding the double curl

$$\nabla\left(\frac{1}{\mu}\right) \times (\nabla \times \vec{\mathcal{A}}) + \frac{1}{\mu} \nabla(\nabla \cdot \vec{\mathcal{A}}) - \frac{1}{\mu} \nabla^2 \vec{\mathcal{A}} = \vec{\mathcal{J}} - \epsilon \nabla\left(\frac{\partial \mathcal{V}}{\partial t}\right) - \epsilon \frac{\partial^2 \vec{\mathcal{A}}}{\partial t^2}$$

Multiplying through by μ and rearranging terms we obtain the general form

$$\nabla^2 \vec{\mathcal{A}} - \mu\epsilon \frac{\partial^2 \vec{\mathcal{A}}}{\partial t^2} = -\mu \vec{\mathcal{J}} + \mu\epsilon \nabla\left(\frac{\partial \mathcal{V}}{\partial t}\right) + \nabla(\nabla \cdot \vec{\mathcal{A}}) + \mu \nabla\left(\frac{1}{\mu}\right) \times (\nabla \times \vec{\mathcal{A}}) \tag{7-31}$$

For most cases μ and ϵ are constants (homogeneous medium) and Eq. 7-31 reduces to

$$\nabla^2\vec{\mathscr{A}} - \mu\epsilon\frac{\partial^2\vec{\mathscr{A}}}{\partial t^2} = -\mu\vec{\mathscr{J}} + \nabla\left(\mu\epsilon\frac{\partial\mathscr{V}}{\partial t}\right) + \nabla(\nabla\cdot\vec{\mathscr{A}})$$

Using the distributive property of the gradient we may combine the last two terms to obtain

$$\nabla^2\vec{\mathscr{A}} - \mu\epsilon\frac{\partial^2\vec{\mathscr{A}}}{\partial t^2} = -\mu\vec{\mathscr{J}} + \nabla\left(\mu\epsilon\frac{\partial\mathscr{V}}{\partial t} + \nabla\cdot\vec{\mathscr{A}}\right)$$

To this point we have really only specified what the curl of $\vec{\mathscr{A}}$ is, namely, Eq. 7-29. The vector Helmholtz theorem (Appendix F) requires that for a unique vector $\vec{\mathscr{A}}$ we must also specify its divergence. From the preceding equation it is beneficial if we define

$$\nabla\cdot\vec{\mathscr{A}} = -\mu\epsilon\frac{\partial\mathscr{V}}{\partial t} \qquad (7\text{-}32)$$

which removes the gradient term. This choice for the divergence of $\vec{\mathscr{A}}$ is called the *Lorentz gauge*. Note that for static fields it reduces to the *Coulomb gauge* $\nabla\cdot\vec{A} = 0$, which gave Eq. 6-42.

Applying the Lorentz gauge the wave equation reduces to

$$\nabla^2\vec{\mathscr{A}} - \mu\epsilon\frac{\partial^2\vec{\mathscr{A}}}{\partial t^2} = -\mu\vec{\mathscr{J}}(\vec{r}, t) \qquad (\mu, \epsilon \text{ constants}) \qquad (7\text{-}33)$$

where

$$\vec{\mathscr{J}} = \vec{\mathscr{J}}_{\text{source}} + \vec{\mathscr{J}}_{\text{conduction}} + \vec{\mathscr{J}}_{\text{convection}} = \vec{\mathscr{J}}_{\text{source}} + \sigma\vec{\mathscr{E}} + \rho_v\vec{v}$$

In most applications $\vec{\mathscr{J}}$ consists only of current sources for which the transmitted fields are to be determined. In the sinusoidal steady-state forms, Eqs. 7-31, 7-32, and 7-33 are

$$\nabla^2\vec{A} + \omega^2\mu\epsilon\vec{A} = -\mu\vec{J} + j\omega\mu\epsilon\,\nabla V + \nabla(\nabla\cdot\vec{A}) + \mu\nabla\left(\frac{1}{\mu}\right)\times(\nabla\times\vec{A}) \qquad (7\text{-}34)$$

$$\nabla\cdot\vec{A} = -j\omega\mu\epsilon V \qquad (7\text{-}35)$$

$$\nabla^2\vec{A} + \omega^2\mu\epsilon\vec{A} = -\mu\vec{J}(\vec{r}) \qquad (\mu, \epsilon \text{ constants}) \qquad (7\text{-}36)$$

The corresponding equations for the scalar electric potential are obtained by writing Gauss' law as

$$\nabla\cdot\vec{\mathscr{D}} = \nabla\cdot(\epsilon\vec{\mathscr{E}}) = \rho_v$$

and then using Eq. 7-30 to remove $\vec{\mathscr{E}}$. The details are left to Exercise 7-6.0f. The results are

$$\nabla^2 \mathcal{V} - \mu\epsilon \frac{\partial^2 \mathcal{V}}{\partial t^2} = -\frac{\rho_v}{\epsilon} - \frac{1}{\epsilon}\frac{\partial \vec{\mathcal{A}}}{\partial t} \cdot \nabla\epsilon - \frac{\nabla \mathcal{V} \cdot \nabla\epsilon}{\epsilon} \quad (\epsilon, \mu \text{ not functions of time}) \tag{7-37}$$

$$\nabla^2 \mathcal{V} - \mu\epsilon \frac{\partial^2 \mathcal{V}}{\partial t^2} = -\frac{1}{\epsilon} \rho_v(\vec{r}, t) \quad (\epsilon \text{ constant, } \mu \text{ not function of time}) \tag{7-38}$$

$$\nabla^2 \overline{V} + \omega^2 \mu\epsilon \overline{V} = -\frac{\rho_v}{\epsilon} - j\frac{\omega}{\epsilon}\vec{A} \cdot \nabla\epsilon - \frac{\nabla V \cdot \nabla\epsilon}{\epsilon} \tag{7-39}$$

$$\nabla^2 \overline{V} + \omega^2 \mu\epsilon \overline{V} = -\frac{1}{\epsilon} \rho_v(\vec{r}) \quad (\epsilon \text{ constant, } \mu \text{ not function of time}) \tag{7-40}$$

In the last two equations $\overline{V}$ is a phasor. Thus we would obtain in the usual way

$$\mathcal{V}(t) = \text{Re}\{\overline{V}e^{j\omega t}\}$$

The remaining task is to obtain solutions for the potential functions. Since our main interests will be for the case of homogeneous linear time-invariant media, in which ϵ and μ are constants, we will develop solutions only for Eqs. 7-33, 7-36, 7-38, and 7-40.

To obtain both solutions we write the generic form of the wave equations as

$$\nabla^2 \Phi - \mu\epsilon \frac{\partial^2 \Phi}{\partial t^2} = -ks(\vec{r}, t)$$
$$\left(k \Rightarrow \frac{1}{\epsilon} \text{ or } \mu;\ s \Rightarrow \text{source function: } \rho_v \text{ or component of } \vec{J}\right) \tag{7-41}$$

The general approach begins by solving this equation when $s(\vec{r},t)$ is a point function $\delta(t - t')\,\delta(\vec{r})$ to obtain the point solution $\Phi_p(\vec{r}, t) \equiv G(\vec{r}, t)$. We use $\delta(t - t')$ to allow the source to be turned on at an arbitrary time $t = t'$. This G is the impulse response or Green's function similar to that discussed in electrostatics following Eq. 5-69. We then generate the solution for given $s(\vec{r}, t)$ by expressing $s(\vec{r}, t)$ as a differential point function $s(\vec{r}\,',t')dv'$, dv' being a differential volume, and then integrating with G over the volume V'. This process is analogous to the scheme used in Chapter 5 where we first found the potential due to a point charge (Coulomb's law) and then expressed a general charge density distribution as an assembly of point charges $\rho_v(\vec{r}\,')dv'$. This is the same convolution process of convolving the source function with the impulse response. Although we have used a scalar in our generalized equation the results will apply equally well to vectors since we can apply the solution to each of the scalar components of the vector and then construct the vector by appending the unit vectors.

Here we give only an outline of the derivation of the solution G, since more detail

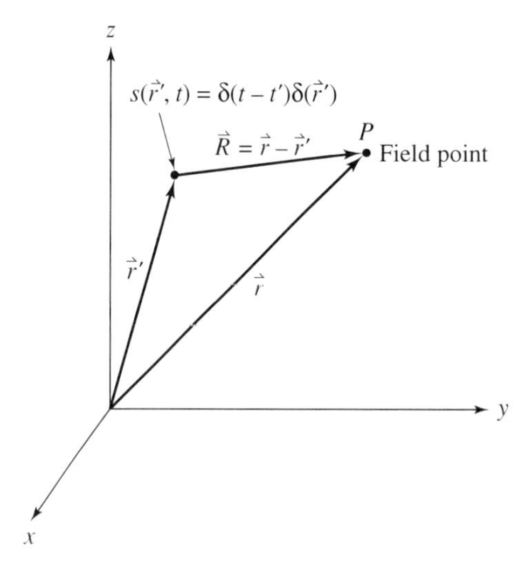

Figure 7-13. Configuration for the determination of the point solution of the wave equation.

would not help us in our goal. For a more complete discussion of the sinusoidal steady state see [56], pages 424–429.

The system configuration is shown in Figure 7-13. This is reminiscent of our static point charge. Since a point system is spherically symmetric we use a spherical coordinate expansion of Eq. 7-41 to obtain, using $\Phi_p = G$,

$$\frac{1}{r^2}\frac{\partial}{\partial r}\left(r^2\frac{\partial G}{\partial r}\right) - \mu\epsilon\frac{\partial^2 G}{\partial t^2} = -k\delta(t - t')\,\delta(\vec{r})$$

We change coordinate variables to R so that our particular system will have the point source at the origin $R = 0$ so that

$$\frac{1}{R^2}\frac{\partial}{\partial R}\left(R^2\frac{\partial G}{\partial R}\right) - \mu\epsilon\frac{\partial^2 G}{\partial t^2} = -k\delta(t - t')\,\delta(R)$$

For $R > 0$, $\delta(R) = 0$ and the equation reduces to

$$\frac{1}{R^2}\frac{\partial}{\partial R}\left(R^2\frac{\partial G}{\partial R}\right) - \mu\epsilon\frac{\partial^2 G}{\partial t^2} = 0 \qquad (R > 0)$$

If we multiply this equation through by R and then use the fact that

$$\frac{\partial^2(RG)}{\partial R^2} = \frac{1}{R}\frac{\partial\left(R^2\frac{\partial G}{\partial R}\right)}{\partial R} \tag{7-42}$$

We have

$$\frac{\partial^2(RG)}{\partial R^2} - \mu\epsilon\frac{\partial^2(RG)}{\partial t^2} = 0$$

We solve this for the product function RG.

This equation is identical in form to the transmission line wave Eq. 3-1 in Chapter 3. We can thus use the same d'Alembert solutions, for the product function RG in our case, to obtain

$$RG = K_1 f_1(t - \sqrt{\mu\epsilon}R) + K_2 f_2(t + \sqrt{\mu\epsilon}R) \tag{7-43}$$

(See Eqs. 3-2, 3-5, 3-6, and 3-7 where $\sqrt{LC} \Rightarrow \sqrt{\mu\epsilon}$.) From our transmission line study we know that f_1 represents a wave propagating in the $+R$ direction and f_2 a wave propagating in the $-R$ direction. Since our interest is in fields generated (transmitted) by the source $\delta(t - t')\,\delta(R)$, we use only the solution f_1. Thus our point source solution is, from Eq. 7-43

$$G(R, t) = \frac{f_1(t - \sqrt{\mu\epsilon}R)}{R} \qquad (R > 0) \tag{7-44}$$

As might have been expected, our solution shows that the potential at P is delayed in time due to the finite propagation velocity.

Now if we let $R \to 0$, then the potential would be observed immediately, and be of "infinite" value, which can be shown to be expressed as (see [56])

$$\lim_{R\to 0} G(R, t) = k\,\frac{\delta(t - t')}{4\pi R} \qquad (R = 0)$$

Using Eq. 7-44 as $R \to 0$, we obtain

$$k\,\frac{\delta(t - t')}{4\pi R} = \frac{f_1(t)}{R}$$

Therefore

$$f_1(t) = k\,\frac{\delta(t - t')}{4\pi}$$

Thus, in Eq. 7-44, we replace $f_1(t - \sqrt{\mu\epsilon}R)$ by its equivalent

$$k\,\frac{\delta[(t - \sqrt{\mu\epsilon}R) - t']}{4\pi}$$

to get

$$G(R,t) = \frac{\delta[(t - \sqrt{\mu\epsilon}R) - t']}{4\pi\epsilon R} \qquad \left(\text{scalar potential, } k = \frac{1}{\epsilon}\right) \tag{7-45}$$

or, similarly,

$$G(R, t) = \frac{\mu\delta[(t - \sqrt{\mu\epsilon}R) - t']}{4\pi R} \qquad (\text{vector potential, } k = \mu) \tag{7-46}$$

For a volume distribution of source, in either case,

$$\text{differential point source} = s(\vec{r}\,', t')dv'$$

where dv' is an incremental source volume as used in Chapters 5 and 6. Then

$$d\mathcal{V} = G(R, t)s(\vec{r}', t')dv' = \frac{\delta[-(t - \sqrt{\mu\epsilon}R) - t']s(\vec{r}', t')}{4\pi\epsilon R} dv'$$

$$= \frac{s(\vec{r}', t - \sqrt{\mu\epsilon}R)}{4\pi\epsilon R} dv' \qquad \text{(scalar potential)}$$

and

$$d\vec{\mathcal{A}} = G(R, t)\vec{s}(\vec{r}', t')dv' = \frac{\mu\delta[(t - \sqrt{\mu\epsilon}R) - t']\vec{s}(\vec{r}', t')}{4\pi R} dv'$$

$$= \frac{\mu\vec{g}(\vec{r}', t - \sqrt{\mu\epsilon}R)}{4\pi R} dv' \qquad \text{(vector potential)}$$

Using these we have our solutions to the wave equations for constant ϵ and μ as, after integrating over the source volume V':

$$\mathcal{V}(r, t) = \int_{V'} \frac{\rho_v[\vec{r}', (t - \sqrt{\mu\epsilon}R)]}{4\pi\epsilon R} dv' \tag{7-47}$$

and

$$\vec{\mathcal{A}}(r, t) = \int_{V'} \frac{\mu\vec{\mathcal{J}}[\vec{r}', (t - \sqrt{\mu\epsilon}R)]}{4\pi R} dv' \tag{7-48}$$

where $R = |\vec{R}| = |\vec{r} - \vec{r}'|$ from Figure 7-13.

These equations tell us that in the source functions ρ_v and $\vec{\mathcal{J}}$ we replace coordinate variables by the prime variables and replace time, t, by the quantity $(t - \sqrt{\mu\epsilon}R)$. See Exercise 7-6.0h. The quantity $(t = \sqrt{\mu\epsilon}R)$ is called the *retarded time* since any change in the source is not observed at point P until a time delay $\sqrt{\mu\epsilon}R$ later. Note we have assumed an infinite medium since any reflections have been rejected in our solution. For the sinusoidal steady state the solutions are obtained by substituting the phasor representations of ρ_v and $\vec{\mathcal{J}}$:

$$V(r) = \int_{V'} \frac{\rho_v(\vec{r}')e^{-j\omega\sqrt{\mu\epsilon}R}}{4\pi\epsilon R} dv' \tag{7-49}$$

$$\vec{A}(r) = \int_{V'} \frac{\mu\vec{J}(\vec{r}')e^{-j\omega\sqrt{\mu\epsilon}R}}{4\pi R} dv' \tag{7-50}$$

Exercises

7-6.0f Using the outline suggested in the text preceding Eq. 7-37, derive Eqs. 7-37 through 7-40. Assume ϵ and μ are not functions of time. You will also need to use the Lorentz condition.

7-6.0g A vector potential is given by $\vec{A} = 3x\hat{x} - yz\hat{y}$ in the sinusoidal steady state. Using the Lorentz condition determine V and $v(t)$. Use $f = 10^9$ Hz. Find expressions for $\vec{H}$ and $\vec{E}$, and show that Maxwell's equations are satisfied.

7-6.0h A charge distribution is given by

$$\rho_v(\vec{r}, t) = 4xyz \sin 10^6 t$$

What are the equations for $\rho_v(\vec{r}\,')$ and $\rho_v[\vec{r}\,', (t - \sqrt{\mu\epsilon}R)]$?

(Answer: $\rho_v(\vec{r}) = 4x'y'z'e^{-j90°}$, $\rho_v[\vec{r}\,', (t - \sqrt{\mu\epsilon}R)] = 4x'y'z' \sin[10^6(t - \sqrt{\mu\epsilon}\sqrt{(x - x')^2 + (y - y')^2 + (z - z')^2}])$

7-6.0i A point charge $2e^{-t}$ μC is at the location (0, 0, 0). Find the scalar electric potential at a general point r. (Hint: a point charge is conveniently expressed using a delta function, which makes the integration every easy.) Determine $\vec{\mathscr{E}}$ and $\vec{\mathscr{H}}$.

7-6.0j A wire of length h (h small compared with wavelength) has a sinusoidal current along it given by I_0. The wire is centered at the origin and oriented along the z axis. For this system write the expression for $\vec{J}$. Obtain the equation for $\vec{A}$ and express it in rectangular and spherical coordinates. Finally, find $\vec{E}$ and $\vec{H}$ in spherical coordinates (note $V = 0$). Show that $\vec{E}$ and $\vec{H}$ satisfy Maxwell's equations. Answer:

$$A_z = \frac{\mu h I_0}{4\pi r} e^{-j\omega\sqrt{\mu\epsilon}r}, \qquad A_r = A_z \cos\theta,\ A_\phi = -A_z \sin\theta$$

$$H_\phi = \frac{hI_0}{4\pi} e^{-j\omega\sqrt{\mu\epsilon}r} \left(\frac{j\omega\sqrt{\mu\epsilon}}{r} + \frac{1}{r^2}\right) \sin\theta$$

7-7 Power Flow in Waves: The Poynting Vector and the Poynting Theorem

In this section we develop the equations that describe the flow of power and energy in electromagnetic fields. We first develop the Poynting vector and then derive the Poynting theorem, both of which are named after John J. Poynting, an English physicist. Incidentally, the same results were also obtained the same year (1884) by Oliver Heaviside, but Poynting had apparently published his results first. It is fortunate, as we will discover shortly, that the Poynting vector actually points in the direction of the power flow! The development that follows is from some observations and suggestions by Booker [6].

The introduction of the power concept in circuit theory usually begins by considering a two-terminal box, with a current and a voltage defined at those terminals as shown in Figure 7-14*a*. The power is given by

$$p(t) = v(t)i(t)$$

Note that the configuration does not depend upon the exact nature of whatever might be connected to the left at the terminals since we have merely cut away everything else and placed at the terminals the current and voltage effects they produce.

In an analogous manner, we consider an incremental surface Δa, which is oriented so

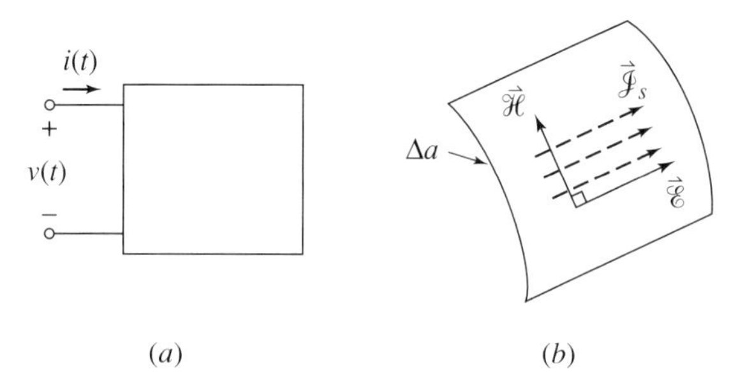

Figure 7-14. Configurations for defining power flow (*a*) in an electrical network (*b*) in a field structure.

that it contains $\vec{\mathscr{E}}$ and $\vec{\mathscr{H}}$ (which are at right angles by Faraday's law). This is shown in Figure 7-14*b*. Note that in order to cut away all fields *behind* the surface we simply need to place a source current $\vec{\mathscr{J}}_s$ A/m (normal to $\vec{\mathscr{H}}$ as the boundary condition requires) along Δa. Now in terms of the sources the power would be

$$|\vec{\mathscr{P}}(t)| = |\vec{\mathscr{J}}_s(t)||\vec{\mathscr{E}}(t)| \text{ A/m} \cdot \text{V/m or W/m}^2 \tag{7-51}$$

Since $|\vec{\mathscr{J}}_s| = |\vec{\mathscr{H}}|$ on Δa this may also be written

$$|\vec{\mathscr{P}}(t)| = |\vec{\mathscr{H}}(t)||\vec{\mathscr{E}}(t)| \text{ W/m}^2 \tag{7-52}$$

Now for circuit theory the power is imagined to flow from the source to the box, so for the field case we imagine power from the sources moving through and away from Δa. Since $\vec{\mathscr{E}}$ and $\vec{\mathscr{H}}$ are perpendicular, it is easy to see that a power flow through Δa—that is, parallel to the surface vector $\Delta\vec{a}$—would be appropriately written as

$$\vec{\mathscr{P}} = \vec{\mathscr{E}} \times \vec{\mathscr{H}} \text{ W/m}^2 \tag{7-53}$$

This is the Poynting vector. From this result, the power in watts through a surface S would be

$$p(t) = \int_S \vec{\mathscr{P}}(t) \cdot d\vec{a} \text{ W} \tag{7-54}$$

Since the sinusoidal steady state is an important special case we need the corresponding forms for Eqs. 7-53 and 7-54. In circuit theory the transformation is, using the overbar to represent complex numbers,

$$p(t) = v(t)i(t) \rightarrow \bar{S} = \tfrac{1}{2}\overline{VI^*} = P + jQ \tag{7-55}$$

where P and Q are the real and reactive components and where $\bar{S}$ is the complex power. The $\frac{1}{2}$ factor is required if $\bar{V}$ and $\bar{I}$ are represented using peak sinusoidal values. For field theory we must then use

$$\vec{\mathscr{P}} = \vec{\mathscr{E}} \times \vec{\mathscr{H}} \Rightarrow \bar{S} = \tfrac{1}{2}\vec{E} \times \vec{H}^* \text{ W/m}^2 \tag{7-56}$$

The integral form would then be

$$\overline{S} = \int_S \vec{S} \cdot d\vec{a} = \int_S \tfrac{1}{2}\vec{E} \times \vec{H}^* \cdot d\vec{a} \text{ W} \tag{7-57}$$

Note that we cannot simply replace $\vec{\mathscr{E}}$ by $\vec{E}$, and so on, because that process works only for linear combinations of field quantities, whereas power is a (cross) product. The real parts of $\vec{S}$ and $\overline{S}$ are average powers just as in circuit theory.

The Poynting theorem, which we next derive, is in essence a conservation-type law that will be seen to apply to a *closed* surface. Since a closed surface is what appears in the divergence theorem, we are led to consider the power that diverges from a volume, that is, passes out through its enclosing surface. Thus we begin by writing

$$\nabla \cdot \vec{\mathscr{P}} = \nabla \cdot (\vec{\mathscr{E}} \times \vec{\mathscr{H}}^*)$$

Applying a vector identity and substituting Maxwell's curl equations,

$$\nabla \cdot \vec{\mathscr{P}} = (\nabla \times \vec{\mathscr{E}}) \cdot \vec{\mathscr{H}} - (\nabla \times \vec{\mathscr{H}}) \cdot \vec{\mathscr{E}} = \left(-\frac{\partial \vec{\mathscr{B}}}{\partial t}\right) \cdot \vec{\mathscr{H}} - \left(\frac{\partial \vec{\mathscr{D}}}{\partial t} + \vec{\mathscr{J}}\right) \cdot \vec{\mathscr{E}}$$

Integrating over a volume and applying the divergence theorem,

$$\int_V \nabla \cdot \vec{\mathscr{P}} dv = \oint_S \vec{\mathscr{P}} \cdot d\vec{a} = -\int_V \left[\frac{\partial \vec{\mathscr{B}}}{\partial t} \cdot \vec{\mathscr{H}} + \frac{\partial \vec{\mathscr{D}}}{\partial t} \cdot \vec{\mathscr{E}}\right] dv - \int_V \vec{\mathscr{J}} \cdot \vec{\mathscr{E}} dv \tag{7-58}$$

This is the general form of the Poynting theorem. In qualitative terms it states that the power flow out of a closed region can be accounted for by decreases of field powers inside the region. Note that all terms in the right-hand member integrands are dimensionally W/m^3. The sinusoidal steady-state formulation is readily obtained by following the preceding steps on the expression

$$\nabla \cdot \vec{S} = \nabla \cdot (\tfrac{1}{2}\vec{E} \times \vec{H}^*)$$

The result is, allowing μ and ϵ to be complex (overbars) in general (covered in Chapter 8), and $\vec{J} = \sigma \vec{E}$,

$$\oint_S \vec{S} \cdot d\vec{a} = -j\omega \int_V [\tfrac{1}{2}\overline{\mu}\vec{H} \cdot \vec{H}^* + \tfrac{1}{2}\overline{\epsilon}^* \vec{E} \cdot \vec{E}^*] dv - \int_V \tfrac{1}{2}\sigma \vec{E} \cdot \vec{E}^* dv \tag{7-59}$$

If we express the complex permittivity and permeability as

$$\overline{\epsilon} = \epsilon' - j\epsilon'' \qquad \text{and} \qquad \overline{\mu} = \mu' - j\mu'' \tag{7-60}$$

we obtain

$$\begin{aligned}\oint_S \vec{S} \cdot d\vec{a} = &-\int_V [\tfrac{1}{2}\omega\mu''\vec{H} \cdot \vec{H}^* + \tfrac{1}{2}\omega\epsilon''\vec{E} \cdot \vec{E}^* + \tfrac{1}{2}\sigma\vec{E} \cdot \vec{E}^*] dv \\ &-j\int_V 2\omega[\tfrac{1}{4}\mu'\vec{H} \cdot \vec{H}^* - \tfrac{1}{4}\epsilon'\vec{E} \cdot \vec{E}^*] dv\end{aligned} \tag{7-61}$$

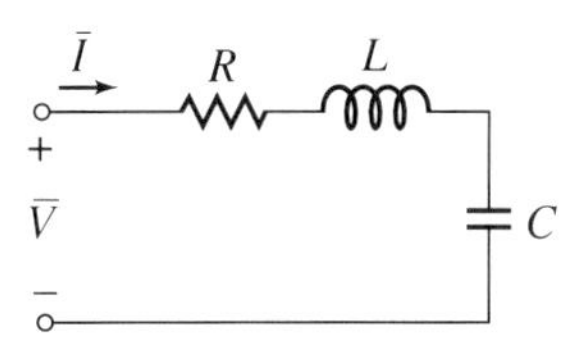

Figure 7-15. Circuit theory equivalent for identifying stored energy and dissipated energy.

From this last result it is easy to identify the heat loss components and the stored energy components. The analogy with circuit theory is obtained by considering the series RLC circuit of Figure 7-15. From the figure

$$\begin{aligned}\bar{S} &= \frac{1}{2}\,\bar{V}\bar{I}^* = \frac{1}{2}\,(\bar{I}\bar{Z})\bar{I}^* = \frac{1}{2}\,\bar{Z}\bar{I}\bar{I}^* = \frac{1}{2}\left(R + j\omega L - j\,\frac{1}{\omega C}\right)\bar{I}\bar{I}^* \\ &= \frac{1}{2}\,R\bar{I}\bar{I}^* + j2\omega\left(\frac{L\bar{I}\bar{I}^*}{4} - \frac{\bar{I}\bar{I}^*}{4\omega^2 C}\right) = P + j2\omega(W_L - W_C)\end{aligned} \tag{7-62}$$

Two special cases of particular interest lead to physically important results.

Case 1. μ and ϵ not functions of time, and both are real valued. Also, let $\vec{\mathcal{J}} = \sigma\vec{\mathcal{E}}$ (conduction current).

For the time derivative terms in Eq. 7-58:

$$\begin{aligned}\frac{\partial\vec{\mathcal{B}}}{\partial t}\cdot\vec{\mathcal{H}} &= \mu\,\frac{\partial\vec{\mathcal{H}}}{\partial t}\cdot\vec{\mathcal{H}} = \frac{1}{2}\,\mu\left(\frac{\partial\vec{\mathcal{H}}}{\partial t}\cdot\vec{\mathcal{H}} + \vec{\mathcal{H}}\cdot\frac{\partial\vec{\mathcal{H}}}{\partial t}\right) = \frac{1}{2}\,\mu\,\frac{\partial(\vec{\mathcal{H}}\cdot\vec{\mathcal{H}})}{\partial t} \\ &= \frac{\partial(\frac{1}{2}\mu\vec{\mathcal{H}}\cdot\vec{\mathcal{H}})}{\partial t}\end{aligned}$$

Similarly,

$$\frac{\partial\vec{\mathcal{D}}}{\partial t}\cdot\vec{\mathcal{E}} = \frac{\epsilon(\frac{1}{2}\epsilon\vec{\mathcal{E}}\cdot\vec{\mathcal{E}})}{\partial t}$$

These two results are identified with time rates of change of stored magnetic and electric energy (see Eqs. 5-77 and 6-68). The Poynting theorem then becomes

$$\oint_S \vec{\mathcal{P}}\cdot d\vec{a} = -\frac{\partial}{\partial t}\int_V [\tfrac{1}{2}\mu\vec{\mathcal{H}}\cdot\vec{\mathcal{H}} + \tfrac{1}{2}\epsilon\vec{\mathcal{E}}\cdot\vec{\mathcal{E}}]dv - \int_V \tfrac{1}{2}\sigma\vec{\mathcal{E}}\cdot\vec{\mathcal{E}}dv \tag{7-63}$$

This form can be physically interpreted as stating that the power passing out through a closed surface can be accounted for by the decrease in stored energy and heat dissipation (heat loss) within the volume.

Case 2. Representation in terms of electric and magnetic polarizations and polarization currents.

This form is obtained by writing the electric and magnetic flux densities in terms of the polarizations, Eqs. 5-55 and 6-32:

$$\vec{\mathcal{D}} = \epsilon_0\vec{\mathcal{E}} + \vec{\mathcal{P}}$$
$$\vec{\mathcal{B}} = \mu_0\vec{\mathcal{H}} + \mu_0\vec{\mathcal{M}}$$

(Note in this first equation that $\vec{\mathcal{P}}$ is the dielectric polarization, *not* the Poynting vector. Be sure to keep careful track of these two $\mathcal{P}$s since they may unfortunately both appear in the same equation.) For the electric field

$$\frac{\partial \vec{\mathcal{D}}}{\partial t} \cdot \vec{\mathcal{E}} = \left(\epsilon_0 \frac{\partial \vec{\mathcal{E}}}{\partial t} + \frac{\partial \vec{\mathcal{P}}}{\partial t} \right) \cdot \vec{\mathcal{E}} = \epsilon_0 \frac{\partial \vec{\mathcal{E}}}{\partial t} \cdot \vec{\mathcal{E}} + \frac{\partial \vec{\mathcal{P}}}{\partial t} \cdot \vec{\mathcal{E}}$$
$$= \frac{\partial}{\partial t}\left(\frac{1}{2} \epsilon_0 \vec{\mathcal{E}} \cdot \vec{\mathcal{E}} \right) + \frac{\partial \vec{\mathcal{P}}}{\partial t} \cdot \vec{\mathcal{E}} \equiv \frac{\partial}{\partial t}\left(\frac{1}{2} \epsilon_0 \vec{\mathcal{E}} \cdot \vec{\mathcal{E}} \right) + \vec{\mathcal{J}}_e \cdot \vec{\mathcal{E}}$$

Where $\vec{\mathcal{J}}_e$ is the electric polarization current density. Similarly,

$$\frac{\partial \vec{\mathcal{B}}}{\partial t} \cdot \vec{\mathcal{H}} = \frac{\partial}{\partial t}\left(\frac{1}{2} \mu_0 \vec{\mathcal{H}} \cdot \vec{\mathcal{H}} \right) + \mu_0 \frac{\partial \vec{\mathcal{M}}}{\partial t} \cdot \vec{\mathcal{H}} \equiv \frac{\partial}{\partial t}\left(\frac{1}{2} \mu_0 \vec{\mathcal{H}} \cdot \vec{\mathcal{H}} \right) + \vec{\mathcal{J}}_m \cdot \vec{\mathcal{H}}$$

With these results, the Poynting theorem may be written

$$\begin{aligned} \oint_S \vec{\mathcal{P}} \cdot d\vec{a} = &-\frac{\partial}{\partial t} \int_V \left[\frac{1}{2} \mu_0 \vec{\mathcal{H}} \cdot \vec{\mathcal{H}} + \frac{1}{2} \epsilon_0 \vec{\mathcal{E}} \cdot \vec{\mathcal{E}} \right] dv \\ &- \int_V (\vec{\mathcal{J}}_m \cdot \vec{\mathcal{H}} + \vec{\mathcal{J}}_e \cdot \vec{\mathcal{E}} + \vec{\mathcal{J}} \cdot \vec{\mathcal{E}}) dv \end{aligned} \tag{7-64}$$

Example 7-9

An electric field in vacuum has the sinusoidal form

$$\vec{E} = 10^{-5} e^{-j2.1z} \hat{x}$$

for a source frequency of 100 MHz. Determine the Poynting vector and compute the power passing through a 1 m^2 surface, oriented as shown in Figure 7-16, in the positive z direction.

Since $\nabla \times \vec{E} = -j\omega\mu_0 \vec{H}$ in vacuum,

$$\vec{H} = j \frac{1}{\omega\mu_0} \nabla \times \vec{E} = j \frac{1}{2\pi 10^8 \times 4\pi \times 10^{-7}} \frac{\partial Ex}{\partial z} \hat{y} = 2.66 \times 10^{-8} e^{-j2.1z} \hat{y}$$

Then from Eq. 7-56

$$\begin{aligned} \vec{S} &= \tfrac{1}{2} \vec{E} \times \vec{H}^* = \tfrac{1}{2} \times 10^{-5} e^{-j2.1z} \hat{x} \times 2.66 \times 10^{-8} e^{j2.1z} \hat{y} \\ &= 1.33 \times 10^{-13} \hat{z} \text{ W/m}^2 \end{aligned}$$

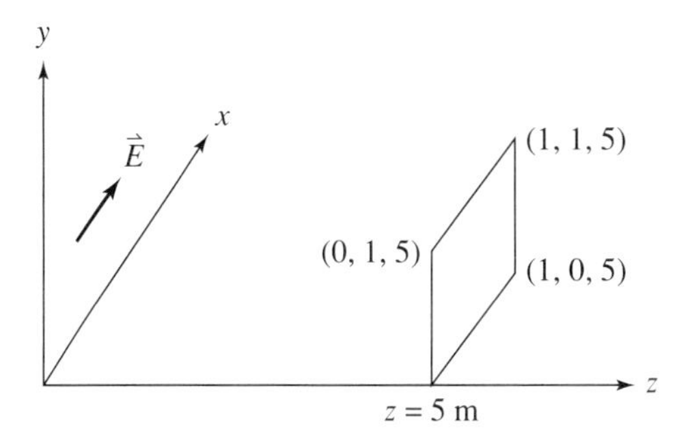

Figure 7-16.

For the positive z direction, $\vec{da} = dxdy\hat{z}$. Eq. 7-57 yields

$$\overline{S} = \int_0^1 \int_0^1 1.33 \times 10^{-13}\hat{z} \cdot dxdy\hat{z} = 1.33 \times 10^{-13} \text{ W}$$

Exercises

7-7.0a Carry out the steps that lead to Eq. 7-59.

7-7.0b Develop the sinusoidal steady-state equivalents of Eqs. 7-63 and 7-64. Collect the real and imaginary components so that the real and reactive "powers" may be identified.

7-7.0c For Example 7-9, convert the phasor $\vec{E}$ to the time domain and repeat the problem. This will demonstrate the utility of the sinusoidal steady-state method.

7-7.0d Express the electric and magnetic currents in sinusoidal steady-state form. Now Eq. 7-64 seems to identify the polarization currents with losses. What does the sinusoidal steady-state form imply?

7-7.0e Suppose in a dielectric that has $\epsilon_r = 3$ the total electric field is

$$\vec{\mathscr{E}} = 10^{-6} \cos(10 \times 10^6 t - 0.029z)\hat{y}$$

Determine the equation for the dielectric polarization current $\vec{\mathscr{J}}_e$. What is the phase relation between $\vec{E}$ and $\vec{J}_e$? (Hint: Relate $\vec{\mathscr{P}}$ to $\vec{\mathscr{E}}$ via the ϵ_r and χ_e.)

7-7.0f For the electric field of Exercise 7-7.0e determine the magnetic field intensity. Calculate $\vec{\mathscr{P}}$ and $\overline{S}$. Compute the average value of $\vec{\mathscr{P}}$ over one cycle of the field in time.

7-7.0g For the electric field of Exercise 7-7.0e determine expressions for the stored energies in the electric and magnetic fields. Plot their time derivatives over two cycles and determine their phase difference.

7-8 Wave Impedance as an Extension of Circuit Theory

One of the most useful concepts in circuit theory is that of impedance. The application of the impedance parameter is based on the definition of impedance as the ratio of the transform of the voltage divided by the transform of the current. The current and voltage are taken at the same point in the network. As an equation,

$$\overline{Z} = \frac{T(v(t))}{T(i(t))} = \frac{\overline{V}}{\overline{I}} \tag{7-65}$$

and the unit of ohm is used due to the similarity with Ohm's law for a resistor. For example, if the transform T is the Laplace transform $\mathscr{L}$ the impedance is expressed as

$$Z(s) = \frac{V(s)}{I(s)} \tag{7-66}$$

In the sinusoidal steady state, which can be developed from the Fourier transform $\mathscr{F}$, we write

$$\overline{Z}(\omega) = \frac{\overline{V}(\omega)}{\overline{I}(\omega)} \tag{7-67}$$

This impedance concept is also valuable in sinusoidal steady-state field theory. By analogy, we would like to define a wave impedance as the ratio of an electric field intensity to a magnetic field intensity so that we would have the ohm as the electrical unit. The difficulty, however, is that field quantities have spatial orientation as well as a phasor representation. Thus the spatial direction becomes an important factor. This impedance concept was first developed by Sergi A. Schelkunoff [47].

In circuit theory the direction one "looks" into two terminals within the network also affects the impedance value, that is, whether one looks to the left or right at a given pair of terminals. For field theory applications we first decide upon the direction we want to look and then use the field components normal to that direction.

Since we are going to deal with the Poynting vector we first examine impedance in circuit theory as a function of complex power. This is

$$\overline{S} = \tfrac{1}{2}\overline{VI^*} = \tfrac{1}{2}\overline{ZII^*}$$

Solving for impedance,

$$\overline{Z} = \frac{\tfrac{1}{2}\overline{VI^*}}{\tfrac{1}{2}\overline{II^*}} = \frac{\text{complex power}}{\text{current magnitude squared}} \tag{7-68}$$

To obtain the field theory counterpart we use the Poynting vector and take the component of it which is in the direction we want to look. If we let $\hat{\tau}$ be a unit vector in the desired location then the amount of the Poynting vector would be

$$\overline{S}_\tau = \vec{S} \cdot \hat{\tau} = \tfrac{1}{2}(\vec{E} \times \vec{H}^*) \cdot \hat{\tau}$$

Since we need the component of $\vec{H}$ that produces this $\overline{S}_\tau$, and since $\vec{H}$ would be normal to $\vec{S}$, the "current" component is $\hat{\tau} \times \vec{H}$. Using these last two results in Eq. 7-68 gives us the wave impedance in the direction $\hat{\tau}$:

$$Z_\tau^w = \frac{\tfrac{1}{2}(\vec{E} \times \vec{H}^*)\cdot \hat{\tau}}{\tfrac{1}{2}(\hat{\tau} \times \vec{H}) \cdot (\hat{\tau} \times \vec{H})^*} = \frac{(\vec{E} \times \vec{H}^*) \cdot \hat{\tau}}{(\hat{\tau} \times \vec{H}) \cdot (\hat{\tau} \times \vec{H})^*}\ \Omega \tag{7-69}$$

This is the result suggested by Booker [6]. An alternate form is given in Exercise 7-8.0a.

Example 7-10

An electric field in space is given by

$$\vec{E} = E_0 e^{-j\omega\sqrt{\mu\epsilon}z}\hat{y}$$

E_0 being a positive, real constant. Determine the wave impedance looking in the $+\hat{z}$ direction. From transmission line theory we know that terms of the form

$$e^{-j\beta z}$$

represent waves traveling in the $+z$ direction (see Section 1-5.3). We first compute $\vec{H}$ using Ampere's law from Maxwell's equations:

$$\vec{H} = \frac{1}{-j\omega\mu}\nabla \times \vec{E} - \sqrt{\frac{\epsilon}{\mu}}\, E_0 e^{-j\omega\sqrt{\mu\epsilon}z}\hat{x}$$

In our case $\hat{\tau} \Rightarrow \hat{z}$, so Eq. 7-69 gives

$$Z^w_{+z} = \frac{\left[E_0 e^{-j\omega\sqrt{\mu\epsilon}z}\hat{y} \times \left(-\sqrt{\frac{\epsilon}{\mu}}E_0 e^{-j\omega\sqrt{\mu\epsilon}z}\hat{x}\right)^*\right]\cdot\hat{z}}{\left[\hat{z} \times \left(-\sqrt{\frac{\epsilon}{\mu}}E_0 e^{-j\omega\sqrt{\mu\epsilon}z}\hat{x}\right)\right]\cdot\left[\hat{z} \times \left(-\sqrt{\frac{\epsilon}{\mu}}E_0 e^{-j\omega\sqrt{\mu\epsilon}z}\hat{x}\right)\right]^*} = \sqrt{\frac{\mu}{\epsilon}} \tag{7-70}$$

Exercises

7-8.0a Using the steps outlined in this section, show that the wave impedance may be expressed as

$$Z^w_\tau = \frac{(\hat{\tau} \times \vec{E}) \cdot (\hat{\tau} \times \vec{E}^*)}{(\vec{E} \times \vec{H}^*) \cdot \hat{\tau}}$$

(Hint: Begin by expressing the circuit theory complex power in terms of $\overline{V}$ and $\overline{Z}$ rather than $\overline{I}$ and $\overline{Z}$.)

7-8.0b Carry out the details of the vector operations in Eq. 7-70 to obtain the result given.

7-8.0c For the electric field in Example 7-10 derive the equation for the wave impedance looking in the $-\hat{z}$ direction. Be sure to identify the field components and $\hat{\tau}$ properly. (Answer: $Z^w_{-z} = -\sqrt{\mu/\epsilon}$)

7-8.0d Suppose we have a field given by

$$\vec{E} = E_0 e^{j\omega\sqrt{\mu\epsilon}z}\hat{y}$$

(a) What is the direction of propagation of this field?
(b) Compute the expressions for the two wave impedances, Z^w_{+z} and Z^w_{-z}.

7-9 General Equations for Waves in Systems that Have a *z* Coordinate

Following the derivations of the vector wave equations (Section 7-6) it was pointed out that there still remained the task of solving them. We also promised some reduced forms that greatly simplify the solutions. These two forms are particularly advantageous for wave propagation in coordinate systems that have a z coordinate. The results of solutions for these systems, while significant in their own right, also frequently provide many qualitative physical insights into the properties of propagating waves generally. In our work in this section we will restrict ourselves to the sinusoidal steady-state cases and for regions that are called source-free; that is, $\vec{J}_{\text{source}} = \rho_v = 0$. The appropriate equations are then Eqs. 7-20 and 7-27 with $\vec{J} = \vec{J}_{\text{source}} + \rho_v\vec{v} + \sigma\vec{E} \Rightarrow \sigma\vec{E}$ so that

$$\nabla^2\vec{E} + \omega^2\mu\epsilon\vec{E} = j\omega\mu\sigma\vec{E} \tag{7-71}$$

and

$$\nabla^2\vec{H} + \omega^2\mu\epsilon\vec{H} = j\omega\mu\sigma\vec{H} \tag{7-72}$$

Note that σ is the conductivity of the material medium. The particular combination $\omega^2\mu\epsilon$ appears frequently in wave propagation so it is customary to define the positive square root of it as

$$k = \omega\sqrt{\mu\epsilon} \tag{7-73}$$

which is a constant for the medium. This is usually called the wave number or propagation constant and has units of 1/m. With this definition the wave equations may be written as

$$\nabla^2\vec{E} + (k^2 - j\omega\mu\sigma)\vec{E} = 0 \tag{7-74}$$

$$\nabla^2\vec{H} + (k^2 - j\omega\mu\sigma)\vec{H} = 0 \tag{7-75}$$

The constant term may be cast into other important, physically descriptive forms. We define the complex term as

$$\bar{k}^2 = k^2 - j\omega\mu\sigma = \omega^2\mu\epsilon - j\omega\mu\sigma \tag{7-76}$$

One form is obtained by factoring out $\omega^2\mu$ to obtain

$$\bar{k}^2 = \omega^2\mu\left(\epsilon - j\frac{\sigma}{\omega}\right) \equiv \omega^2\mu(\epsilon' - j\epsilon'') = \omega^2\mu\bar{\epsilon} \tag{7-77}$$

This form defines an effective complex permittivity as postulated in Eqs. 7-59 and 7-60. The second form factors out $\omega^2\epsilon$ to obtain

$$\bar{k}^2 = \omega^2\epsilon\left(\mu - j\frac{\mu\sigma}{\omega\epsilon}\right) = \omega^2\epsilon(\mu' - j\mu'') = \omega^2\epsilon\bar{\mu} \tag{7-78}$$

This form defines an effective complex permeability.

Using the complex wave number, the wave equations assume a simple form

$$\nabla^2\vec{E} + \bar{k}^2\vec{E} = 0 \tag{7-79}$$

$$\nabla^2\vec{H} + \bar{k}^2\vec{H} = 0 \tag{7-80}$$

Now for coordinate systems that have a z coordinate, the Laplacian may be written as

$$\nabla^2 \equiv \nabla_t^2 + \nabla_z^2 = \nabla_t^2 + \frac{\partial^2}{\partial z^2} \tag{7-81}$$

where ∇_t^2 means that part of ∇^2 which involves only the derivatives with respect to the two coordinates that are transverse (or normal) to z, which we call t_1 and t_2. As examples,

rectangular: $t_1 \approx x,\quad t_2 \approx y$

cylindrical: $t_1 \approx \rho\quad t_2 \approx \phi$

elliptic cylindrical: $t_1 \approx u,\quad t_2 \approx v$

Using this the wave equation for $\vec{E}$ (or $\vec{H}$) assumes the form

$$\nabla_t^2 \vec{E} + \frac{\partial \vec{E}}{\partial x^2} + \bar{k}^2 \vec{E} = 0 \tag{7-82}$$

Now from our transmission line study we know that propagation in the z direction has the functional form

$$e^{-j\beta z} \tag{7-83}$$

Thus to allow the greatest generality, we let the z variation of the field quantities have the form

$$e^{-\gamma z} \tag{7-84}$$

where γ is a parameter to be determined for each system, and has the units of 1/m. It is also called the *propagation constant* but must not be confused with k defined earlier. In many cases γ will be a function of k. Note that there is no j in front of the γ, so to have propagation γ must have an imaginary part, meaning that γ is in general *complex* even though in practice we do not use an overbar to denote such. If we represent this complex propagation constant in rectangular form

$$\gamma = \alpha + j\beta \tag{7-85}$$

Equation 7-84 may be written as

$$e^{-\gamma z} = e^{-\alpha z} e^{-j\beta z} \tag{7-86}$$

From transmission line theory we recall that $e^{-\alpha z}$ is an attenuation factor and $e^{-j\beta z}$ is the phase or propagation constant. See Eqs. 1-25 and 1-31 and the corresponding discussion of Section 1-5.3. For Eq. 7-82 we may express a general field vector by the z direction form:

$$\vec{E}(t_1, t_2, z) = E_1(t_1, t_2)e^{-\gamma z}\hat{t}_1 + E_2(t_1, t_2)e^{-\gamma z}\hat{t}_2 + E_z(t_1, t_2)e^{-\gamma z}\hat{z} \tag{7-87}$$

$$= \vec{E}(t_1, t_2)e^{-\gamma z} \tag{7-88}$$

With the preceding form we can evaluate the Laplacian as

$$\nabla^2 \vec{E} = \left(\nabla_t^2 + \frac{\partial^2}{\partial z^2}\right)[\vec{E}(t_1, t_2)e^{-\gamma z}] = \nabla_t^2[\vec{E}(t_1, t_2)e^{-\gamma z}] + \frac{\partial^2[\vec{E}(t_1, t_2)e^{-\gamma z}]}{\partial z^2}$$

$$= [\nabla_t^2 \vec{E}(t_1, t_2)]e^{-\gamma z} + \gamma^2 \vec{E}(t_1, t_2)e^{-\gamma z}$$

Substituting this into the wave Eq. 7-79 and also using Eq. 7-88

$$[\nabla_t^2 \vec{E}(t_1, t_2)]e^{-\gamma z} + \gamma^2 \vec{E}(t_1, t_2)e^{-\gamma z} + \bar{k}^2 \vec{E}(t_1, t_2)e^{-\gamma z} = 0$$

Dividing out the exponents and combining the last two terms;

$$\nabla_t^2 \vec{E}(t_1, t_2) + (\gamma^2 + \bar{k}^2)\vec{E}(t_1, t_2) = 0 \tag{7-89}$$

Comparing this with Eq. 7-82, we see that all we ever need to do for z directed propagation is replace $\partial/\partial z$ by $-\gamma$, which means $\partial^2/\partial z^2 = (-\gamma)^2 = \gamma^2$. We will use this replacement in subsequent work.

To simplify writing we define a new constant:

$$k_c^2 = \gamma^2 + \bar{k}^2 \tag{7-90}$$

The constant k_c is also 1/m unit wise, and is called the cutoff or critical wave number. This name for k_c can be quickly justified by noting that there will be no propagation if the imaginary part of γ is zero since there will then be no exponential factor like Eq. 7-83. Thus propagation ceases when

$$\text{Im } \gamma = \text{Im}\sqrt{k_c^2 - \bar{k}^2} = 0 \tag{7-91}$$

For those cases where γ is purely imaginary, say $j\beta$, and where $\bar{k} = k$ for the lossless medium ($\sigma = 0$):

$$\text{Im } \gamma = \text{Im } j\beta = \beta = \text{Im}\sqrt{k_c^2 - k^2} = \text{Im } j\sqrt{k^2 - k_c^2} = \sqrt{k^2 - k_c^2} = 0$$

This requires

$$k^2 = k_c^2 \tag{7-92}$$

or, using Eq. 7-73 evaluated at the cutoff frequency,

$$\omega_c\sqrt{\mu\epsilon} = k_c^2$$

from which the cutoff frequency may be computed as

$$\omega_c = \frac{k_c^2}{\sqrt{\mu\epsilon}} \tag{7-93}$$

(γ purely real) Thus the determination of the value of k_c is an important part of the solution process. The wave Eq. 7-89 is then written as

$$\nabla_t^2\vec{E}(t_1, t_2) + k_c^2\vec{E}(t_1, t_2) = 0 \tag{7-94}$$

Similarly,

$$\nabla_t^2\vec{H}(t_1, t_2) + k_c^2\vec{H}(t_1, t_2) = 0 \tag{7-95}$$

We must not forget that in these forms the fields are functions only of *transverse* coordinates. The solution process in general would be to find solution of either of these equations, determine the equation for γ, and then construct the total field solution using Eq. 7-88.

As a side note to this point, notice that if we want to consider propagation in the $-z$ direction all we have to do is replace γ by $-\gamma$! To this point the only simplification we have obtained is the elimination of z from the vector wave equations. To show how the solution process can be further simplified we will work specifically with rectangular coordinates. We then leave the derivation of analogous results for other coordinate systems to exercises and general exercises. The process is identical to that which follows.

Again assuming the source-free case, Faraday's law in rectangular form is, for variations with x, y, and z,

$$\left(\frac{\partial E_z}{\partial y} - \frac{\partial E_y}{\partial z}\right)\hat{x} + \left(\frac{\partial E_x}{\partial z} - \frac{\partial E_z}{\partial x}\right)\hat{y} + \left(\frac{\partial E_y}{\partial x} - \frac{\partial E_x}{\partial y}\right)\hat{z} = -j\omega\mu H_x\hat{x} - j\omega\mu H_y\hat{y} - j\omega\mu H_z\hat{z}$$

Equating components:

$$\frac{\partial E_z}{\partial y} - \frac{\partial E_y}{\partial z} = -j\omega\mu H_x$$

$$\frac{\partial E_x}{\partial z} - \frac{\partial E_z}{\partial x} = -j\omega\mu H_y$$

$$\frac{\partial E_y}{\partial x} - \frac{\partial E_z}{\partial y} = -j\omega\mu H_z$$

Now for the z coordinate direction we may replace $\partial/\partial z$ by $-\gamma$ as explained earlier and express all field components as

$$E_i(x, y, z) = E_i(x, y)e^{-\gamma z}, \qquad i = x, y, z$$

and

$$H_i(x, y, z) = H_i(x, y)e^{-\gamma z}$$

Using these component expressions the preceding three equations are:

$$\frac{\partial E_z}{\partial y}e^{-\gamma z} + \gamma e^{-\gamma z}E_y(x, y) = -j\omega\mu H_x(x, y)e^{-\gamma z}$$

$$-\gamma e^{-\gamma z}E_x(x, y) - \frac{\partial E_z}{\partial x}e^{-\gamma z} = -j\omega\mu H_y(x, y)e^{-\gamma z}$$

$$\frac{\partial E_y}{\partial x}e^{-\gamma z} - \frac{\partial E_x}{\partial y}e^{-\gamma z} = -j\omega\mu H_z(x, y)e^{-\gamma z}$$

Cancelling the exponents and dropping the parentheses lists, which means we must remember that the field components are now functions of x and y (transverse) coordinates only,

$$\frac{\partial E_z}{\partial y} + \gamma E_y = -j\omega\mu H_x \tag{7-96}$$

$$-\gamma E_x - \frac{\partial E_z}{\partial y} = -j\omega\mu H_y \tag{7-97}$$

$$\frac{\partial E_y}{\partial x} - \frac{\partial E_x}{\partial y} = -j\omega\mu H_z \tag{7-98}$$

Similar steps for $\nabla \times \vec{H} = j\omega\epsilon\vec{E} + \sigma\vec{E} = (j\omega\epsilon + \sigma)\vec{E}$, give

$$\frac{\partial H_z}{\partial y} + \gamma H_y = (j\omega\epsilon + \sigma)E_x \tag{7-99}$$

$$-\gamma H_x - \frac{\partial H_z}{\partial x} = (j\omega\epsilon + \sigma)E_y \tag{7-100}$$

$$\frac{\partial H_y}{\partial x} - \frac{\partial H_x}{\partial y} = (j\omega\epsilon + \sigma)E_z \tag{7-101}$$

First solve Eq. 7-96 for H_x and substitute the result into Eq. 7-99:

$$-j\frac{\gamma}{\omega\mu}\frac{\partial E_z}{\partial x} - j\frac{\gamma^2}{\omega\mu}E_y - \frac{\partial H_z}{\partial x} = (j\omega\epsilon + \sigma)E_y$$

Multiply by $j\omega\mu$ and solve for E_y:

$$E_y = \frac{1}{\gamma^2 + \bar{k}^2}\left(-\gamma\frac{\partial E_z}{\partial y} + j\omega\mu\frac{\partial H_z}{\partial x}\right) \tag{7-102}$$

Similarly from Eqs. 7-97 and 7-99:

$$E_x = -\frac{1}{\gamma^2 + \bar{k}^2}\left(\gamma\frac{\partial E_z}{\partial x} + j\omega\mu\frac{\partial H_z}{\partial y}\right) \tag{7-103}$$

Using the same process for the $\vec{H}$ components we obtain

$$H_x = \frac{1}{\gamma^2 + \bar{k}^2}\left(j\omega\epsilon\frac{\partial E_z}{\partial y} - \gamma\frac{\partial H_z}{\partial x}\right) \tag{7-104}$$

$$H_y = -\frac{1}{\gamma^2 + \bar{k}^2}\left(j\omega\epsilon\frac{\partial E_z}{\partial x} + \gamma\frac{\partial H_z}{\partial y}\right) \tag{7-105}$$

These last four equations tell us that if a wave has E_z and/or H_z components, we can determine all other field components from them. Also, for propagation in the $-z$ direction, we simply replace γ by $-\gamma$.

Results similar to the preceding are obtained for any generalized coordinate system that has a z component (see Appendix J). Refer to Exercises 7-9.0d–g and General Exercise GE7-21.

All we need now are the equations from which we can obtain $E_z(t_1, t_2)$ and $H_z(t_1, t_2)$. For the E_z equation we begin with wave Eq. 7-94, and use the most general field structure Eq. 7-88 in transverse coordinates:

$$\vec{E}(t_1, t_2) = E_1(t_1, t_2)\hat{t}_1 + E_2(t_1, t_2)\hat{t}_2 + E_z(t_1, t_2)\hat{z}$$

or the shorthand form

$$\vec{E} = E_1\hat{t}_1 + E_2\hat{t}_2 + E_z\hat{z}$$

Substituting this into the wave Eq. 7-94:

$$\nabla_t^2(E_1\hat{t}_1 + E_2\hat{t}_2 + E_z\hat{z}) + k_c^2(E_1\hat{t}_1 + E_2\hat{t}_2 + E_z\hat{z}) = 0$$

Using the distributive property of the Laplacian to take out the z component:

$$\nabla_t^2(E_1\hat{t}_1 + E_2\hat{t}_2) + \nabla_t^2(E_z\hat{z}) + k_c^2(E_1\hat{t}_1 + E_2\hat{t}_2) + k_c^2(E_z\hat{z}) = 0$$

Now since the unit vector $\hat{z}$ does not change direction as coordinates t_1 and t_2 change, $\hat{z}$ may be removed from the transverse Laplacian, giving

$$\nabla_t^2(E_1\hat{t}_1 + E_2\hat{t}_2) + (\nabla_t^2 E_z + k_c^2 E_z)\hat{z} = 0$$

Since all vector components in the equation must be zero

$$\nabla_t^2 E_z + k_c^2 E_z = 0 \tag{7-106}$$

Similarly,

$$\nabla_t^2 H_z + k_c^2 H_z = 0 \tag{7-107}$$

These give us our goal of scalar equations in one field component.

As a general solution process for wave propagation looking in the z direction we now have:

1. Solve Eqs. 7-106 and 7-107 for $E_z(t_1, t_2)$ and $H_z(t_1, t_2)$.
2. Apply boundary conditions to determine as many of the constants as possible. It turns out that during this step we obtain k_c and γ!
3. Compute the transverse field components using Eqs. 7-102 to 7-105 or their equivalents in the appropriate coordinate system, Appendix J.
4. Construct the total field using Eq. 7-87. The same equation also applies to $\vec{H}$.

Exercises

7-9.0a Show that the wave number (propagation constant) k has the unit of reciprocal meters.

7-9.0b Carry out the derivation of Eq. 7-84, justifying carefully each step.

7-9.0c Complete the development of Eqs. 7-99, 7-100, and 7-101.

7-9.0d Beginning with Maxwell's two curl equations expanded in cylindrical coordinates, derive the equations that show that all the transverse field components E_ρ, E_ϕ, H_ρ, H_ϕ can be determined if E_z and H_z are known. Region is lossless and source-free. Use positive z only. Partial answer:

$$E_\rho = -\frac{1}{\gamma^2 + \overline{k}^2}\left(\gamma\frac{\partial E_z}{\partial \rho} + j\frac{\omega\mu}{\rho}\frac{\partial H_z}{\partial \phi}\right)$$

7-9.0e Repeat Exercise 7-9.0d for parabolic cylindrical coordinates.

7-9.0f Repeat Exercise 7-9.0d for elliptic cylindrical coordinates.

7-9.0g Repeat Exercise 7-9.0d for bipolar coordinates.

7-9.0h Derive Eq. 7-107.

7-9.0i Carry out the steps that reduce Eq. 7-27 to 7-72. You will have to use one of Maxwell's curl equations.

7-10 Mode Classification of Propagating Waves

The solution process given at the end of the preceding section gives us the total field structure. In practice it has been found convenient to identify specific types of field structures, called *modes*, which have properties that are very useful in the design of high-frequency systems. Since we have assumed linear media, more complex field configurations can usually be expressed as combinations of these basic types. In this section the basic modes are defined and solution processes are developed. Applications are given in following chapters.

One of the most important modes is the *transverse electromagnetic* (TEM) mode. This mode is defined as having both $\vec{E}$ and $\vec{H}$ perpendicular (transverse) to the direction one is interested in, usually the direction of wave propagation. The TEM mode is often called the *transmission-line* mode since this is the field structure most often found in our two-conductor transmission lines: coax, printec circuit boards, parallel line, and so on. For our particular case we are interested in looking in the z direction, so this mode requires $E_z = H_z = 0$. Now this leads to interesting conclusions if we try to use Eqs. 7-102 through 7-105 directly. These equations would predict that all field quantities are zero so no such solutions would exist. Notice, however, that the denominator of each of those equations is $\gamma^2 + \bar{k}^2$. If this term were zero, we would have the indeterminant form 0/0 for each component and a non zero field could result. Thus, we must set

$$\gamma_{\text{TEM}}^2 + \bar{k}^2 = 0$$

Therefore

$$\gamma_{\text{TEM}}^2 = -\bar{k}^2 = -(\omega^2\mu\epsilon - j\omega\mu\sigma) = j\omega\mu\sigma - \omega^2\mu\epsilon$$

or

$$\gamma_{\text{TEM}} = \sqrt{-\bar{k}^2} = \sqrt{j\omega\mu\sigma - \omega^2\mu\epsilon} \tag{7-108}$$

This also means that

$$k_{\text{c,TEM}}^2 = \gamma_{\text{TEM}}^2 + \bar{k}^2 = 0$$

so that Eqs. 7-94 and 7-95 reduce to

$$\nabla_t^2\vec{E}(t_1, t_2) = 0$$

$$\nabla_t^2\vec{H}(t_1, t_2) = 0$$

We next show that these equations can be replaced by scalar Laplacian equations. For example, expanding Faraday's law, and using $H_z = 0$ for TEM mode,

$$\nabla \times \vec{E} = (\nabla_t + \nabla_z) \times \vec{E} = \nabla_t \times \vec{E} + \nabla_z \times \vec{E} = -j\omega\mu\vec{H}_t - j\omega\mu H_z\hat{z} = -j\omega\mu\vec{H}_t$$

Now since the curl of a vector is normal to the vector, the term $\nabla_t \times \vec{E}$ must be the $\hat{z}$ component because $\vec{E}$ is only in the transverse plane for TEM waves. The other curl term must be in the transverse plane. Thus since there is no $\hat{z}$ component on the right-hand side we require

$$\nabla_t \times \vec{E} \equiv 0$$

Thus in the transverse plane (i.e., any plane where z is constant) the electric field is conservative. It follows that *in the transverse plane* the field $\vec{E}$ can be expressed as the gradient of a scalar function

$$\vec{E} = -\nabla_t \Phi \tag{7-109}$$

This is consistent with the vector identity

$$\nabla_t \times (-\nabla_t \Phi) \equiv 0$$

Equation 7-109 also means that

$$\Phi = -\int_a^b \vec{E} \cdot d\vec{l}_t \tag{7-110}$$

where $d\vec{l}_t$ is any path in the *transverse* plane. Using Gauss' law for charge-free region yields

$$\nabla_t \cdot \vec{D} = \epsilon \nabla_t \cdot \vec{E} = -\epsilon \nabla_t \cdot (\nabla_t \Phi) = -\epsilon \nabla_t^2 \Phi = \rho_v = 0$$

Thus for TEM modes

$$\nabla_t^2 \Phi = 0 \tag{7-111}$$

Actually, this result might have been obtained by observing that since $H_z = 0$ there is no magnetic flux in any closed path in the transverse plane, and hence no voltage induced in any contour. The conclusion is that *in the transverse plane* TEM waves obey *static* field laws even though there are time variations present in the transverse fields.

Since Eq. 7-111 is a second-order differential equation for voltage (potential) there must be at least two *separate* conductors on which the potentials are specified in order to have enough boundary conditions to solve for Φ uniquely. We then conclude that for TEM modes to exist we must have at least two separate conductors. This is why TEM modes are called *transmission-line* modes.

Consider, for example, the coaxial cable with an applied sinusoidal source as shown in Figure 7-17. Applying the right-hand rule for currents and the positive charge test for electric fields, the generated fields are as shown. Note that the power flow $\vec{E} \times \vec{H}^*$ is in the $+\hat{z}$ direction. This raises the interesting question as to where the power really resides, in the volt-ampere product or the Poynting vector. What is your response?

The solution process for TEM is then:

1. Solve $\nabla_t^2 \Phi = 0$ for Φ and apply boundary conditions.
2. Compute the electric field from

$$\vec{E} = -\nabla_t \Phi$$

(Note we use only the transverse parts of the del operator.)

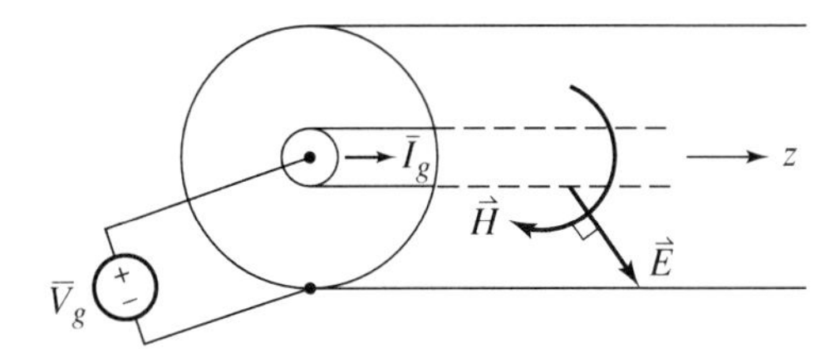

Figure 7-17. Illustration of the transmission line (TEM) mode in a coaxial cable.

3. Evaluate

$$\gamma_{\text{TEM}} = \sqrt{j\omega\mu\sigma - \omega^2\mu\epsilon}$$

4. Construct the total field vector

$$\vec{E}_{\text{total}} = \vec{E}_e^{-\gamma_{\text{TEM}}z}$$

5. Use Faraday's law to find $\vec{H}$

$$\vec{H}_{\text{total}} = j\frac{1}{\omega\mu}\nabla \times \vec{E}_{\text{total}}$$

(Total curl here, not transverse.)

The second mode is the *transverse electric* (TE) mode. This mode is defined as having total $\vec{E}$ perpendicular (transverse) to the direction one is interested in. This definition thus requires for z-directed propagation:

$$E_z = 0 \qquad H_z \neq 0$$

These modes are often referred to as H modes since only $\vec{H}$ has a field component in the direction of interest. The solution process for TE is then:

1. Solve Eq. 7-107 for H_z:

$$\nabla_t^2 H_z + k_c^2 H_z = 0$$

   Techniques for doing this are found in Appendix K, or from your mathematics courses. Many solutions will be carried out in detail in the following chapters.
2. Apply boundary conditions. For these modes it is frequently easier to first compute components of $\vec{E}$ using equations of the form given by Eqs. 7-102 and 7-103 or components of $\vec{H}$ from equations of the form given by Eqs. 7-104 and 7-105 and apply boundary conditions to them.
3. As the boundary conditions are applied, values of k_c and γ will be obtained ($k_c^2 = \gamma^2 + \overline{k}^2$).
4. Compute the remaining field components from the equations for transverse components.
5. Multiply solutions by

$$e^{-\gamma_{\text{TE}}z}$$

   to get total expressions. For these modes the Poynting vector is not in the z direction (see Exercise 7-10.0b).

The next mode type is the *transverse magnetic* (TM) mode. This mode is defined as having total $\vec{H}$ perpendicular (transverse) to the direction one is interested in. This definition thus requires for z-directed propagation

$$H_z = 0 \qquad E_z \neq 0$$

These modes are often referred to as E modes since only $\vec{E}$ has a field component in the direction of interest. The solution process for TM is then

1. Solve Eq. 7-106 for E_z:

$$\nabla_t^2 E_z + k_c^2 E_z = 0$$

2. Apply boundary conditions to E_z. As the boundary conditions are applied values for γ and k_c will be obtained ($k_c^2 = \gamma^2 + \bar{k}^2$).
3. Obtain solutions for the remaining components of $\vec{E}$ and $\vec{H}$ using the equations that give the transverse components in terms of E_z (Eqs. 7-102 through 7-105).
4. Multiply solutions by

$$e^{-\gamma_{\mathrm{TM}} z}$$

to obtain total field expressions.

The last mode class is the *hybrid* mode, which consists of E_z and H_z (combinations of TE and TM modes). These are of two types:

1. HE modes for which the H_z component determines the majority of the field amplitude.
2. EH modes for which the E_z component is the dominant component. These mode types occur frequently in fiber-optic systems and for propagation in nonmetallic boundary (dielectric) configurations. These problems are usually solved on a case-by-case basis.

Exercises

7-10.0a Draw a two-wire parallel (twin lead) transmission line and sketch the structures of $\vec{E}$ and $\vec{H}$. From these determine the direction of power flow.

7-10.0b For the TE modes, draw an appropriate figure and give physical arguments to show that the Poynting vector cannot be in the z direction (sketch total fields).

7-10.0c For the TM modes, draw an appropriate figure and give physical arguments to show that the Poynting vector cannot be in the z direction (sketch total fields). Since cutoff occurs when no energy (power) is transferred in the z direction, what would you conjecture would happen to the direction of the Poynting vector as the cutoff condition is approached?

Summary, Objectives, and General Exercises

Summary

The main purpose of this chapter has been to develop the wave equations for the electric and magnetic fields and for the scalar and vector potentials that produce fields. Although there have been no applications directly—these are covered in the following chapters—we now have at our disposal all the equations needed to solve for the vast majority of traveling wave configurations. The equations were developed for the general time-varying fields and for the sinusoidal steady state, and then reduced to important cases where μ and ϵ are constant and where the media are lossless ($\sigma = 0$) and source-free ($\rho_v = \vec{J} = 0$).

An important class of problems discussed was the case where the system had a z coordinate. The general wave equations and Maxwell's equations were put in reduced form for which solution processes could be developed. Along with this development, the various mode structures were defined and applied to z-directed systems.

The flow of power in propagating systems was presented and the Poynting vector and Poynting theorem developed. The flow of power in TE and TM modes was shown (in exercises) to be in a direction other than the z direction. Figure 7-18 shows one possible way to indicate how the theory was developed and how the results lead to specific applications.

Objectives

1. Be able to write the general time form and sinusoidal steady-state form of Maxwell's equations in both point (differential) and large-scale (integral) forms. Know the names and units of all symbols.
2. Be able to derive the integral forms of Maxwell's equations from the point forms.
3. Be able to write the constitutive relationships for time-varying fields and apply them to given fields and materials for linear, homogeneous, and isotropic media.
4. Be able to give a physical interpretation for each of the integral forms of Maxwell's equations and to identify each volume, surface, and contour. Make sketches showing terms.
5. Be able to evaluate the integral forms of Maxwell's equations for given field quantities and regions.
6. Be able to write the definition of displacement current and to explain the difference between displacement current, conduction current, and convection current.
7. Be able to derive the continuity equation and give a physical interpretation of it. This is for both point form and integral form.
8. Be able to state the boundary conditions for the normal and tangential components of the field quantities and to outline the steps involved in the proofs of these (three or four steps). Be able to state and describe qualitatively (physically) the six auxiliary boundary conditions.
9. Be able to apply the differential forms of Maxwell's equations to given field quantities.
10. Be able to describe the observation that led to the concept of displacement current. Be able to explain the types of current flow in a capacitor, both between and external to the plates.
11. Be able to compute the force exerted on a metal surface that carries a surface current $\vec{\mathcal{J}}_s$ (or has a tangential magnetic field at the boundary; see Chapter 6).
12. Be able to derive the wave equations for electric and magnetic fields beginning with Maxwell's equations in either sinusoidal steady-state form or general time-variation form. Be able to state any assumptions you make in the derivations.
13. Be able to write Maxwell's equations and the wave equations in the sinusoidal steady-state form if given the general time forms.
14. Be able to convert a phasor into its specific time equation.
15. Know and be able to compute the Poynting vector for both sinusoidal fields and general time-varying fields. Know the units of the Poynting vector and its physical interpretation.
16. Be able to compute the power loss and energy storage in a given material (μ and ϵ either or both complex) for given electric and magnetic fields for both the sinusoidal and general time-varying cases.
17. Be able to define electric polarization current and magnetic polarization current and to develop the units of both. Be able to compute

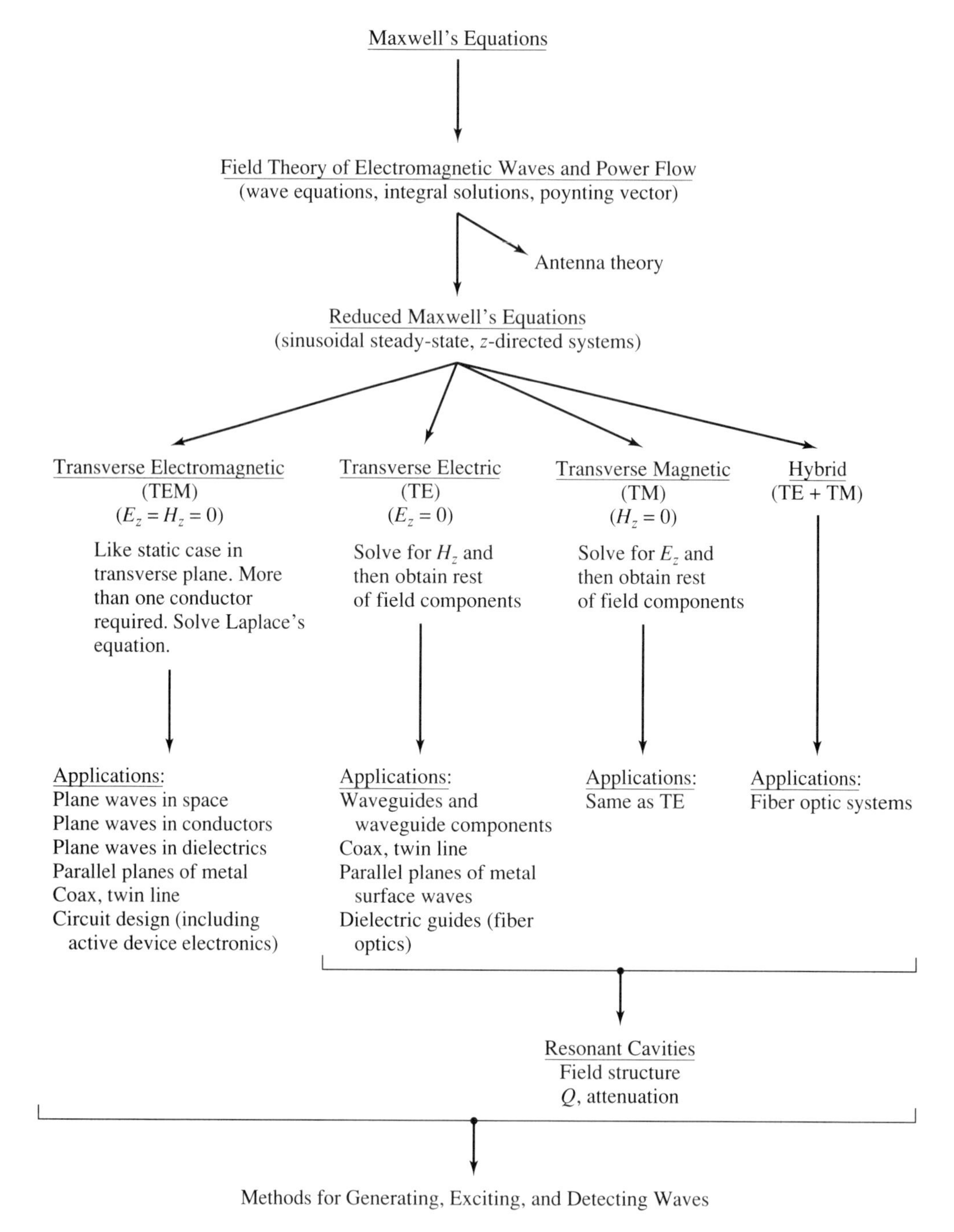

Figure 7-18. Chart showing what we have done and where we are going.

these for general time variations and sinusoidal steady state.

18. Know the conditions on μ and ϵ that are required to obtain both the general time form and sinusoidal steady-state form (differential) of Maxwell's equations.
19. Be able to identify the propagation constant (γ) in the wave equations in the sinusoidal steady state and know its units.
20. Be able to compute the wave impedance from the electric and magnetic field intensities.
21. Be able to give the physical interpretation of each term in the integral form of the Poynting theorem, particularly when μ and ϵ are not functions of time. Do this for both general time form and sinusoidal steady state.
22. Be able to reduce the wave equations for $\vec{\mathscr{E}}$ and $\vec{\mathscr{H}}$ to forms appropriate for
    (a) Charge-free regions ($\rho_v = 0$)
    (b) Lossless regions ($\sigma = 0$)
    (c) Charge-free and lossless regions.
23. For a right-hand orthogonal coordinate system (t_1, t_2, z) be able to express a general vector $\vec{E}(t_1, t_2, z)$ in three-component form and in a form that has only one transverse vector component and a z component (two-component form).
24. For any right-hand orthogonal coordinate system that has a z component, be able to take given expressions for ∇ and ∇^2 and write them in the forms $\nabla_t + \nabla_z$ and $\nabla_t^2 + \nabla_z^2$.
25. Be able to use the two-component form of a general field vector $\vec{E}(t_1, t_2, z)$ with ∇^2 to obtain the scalar wave equation for E_z.
26. Be able to tell how the z-variation exponential portion of $E_z(t_1, t_2, z)$ and $H_z(t_1, t_2, z)$ can be identified as a positive or negative traveling component.
27. Understand the difference between the total field $\vec{E}(t_1, t_2, z)$ and the transverse coordinate function component $\vec{E}(t_1, t_2)$ and how to construct $\vec{E}(t_1, t_2, z)$ from $\vec{E}(t_1, t_2)$. Similarly for $\vec{H}$.
28. Be able to take any one of Maxwell's equations in differential form and reduce it to an equation involving only the transverse coordinates t_1 and t_2 using the z variation of the form $e^{\pm j\beta z}$ and involving only the field quantities $\vec{E}(t_1, t_2)$ and/or $\vec{H}(t_1, t_2)$.
29. Be able to define and give vector diagrams for the TEM, TE, and TM modes.
30. Be able to begin with the reduced forms of Maxwell's equations (z coordinate) and write them in the forms appropriate for TEM, TE, and TM.
31. Be able to use Faraday's law and Lenz's law to determine currents and voltages induced in conductors.

General Exercises

GE7-1 Show qualitatively that the induced current in the twisted resistive wire loop of Figure 7-19 is zero. Can you suggest any applications for this idea in terms of magnetic field interference? Suppose there were two twists. Now what would happen?

GE7-2 For the generator shown in Figure 7-20, determine the induced emf as the loop rotates in the magnetic field. Also show the polarity signs on the equivalent voltage source. (Hint: $\theta(t) = \omega t$; compute flux through the area of the loop.)

GE7-3 Repeat Example 7-6 in Section 7-2 for the magnetic flux density

$$\vec{B} = \frac{B_0}{y + k}\,\hat{x}$$

Figure 7-19.

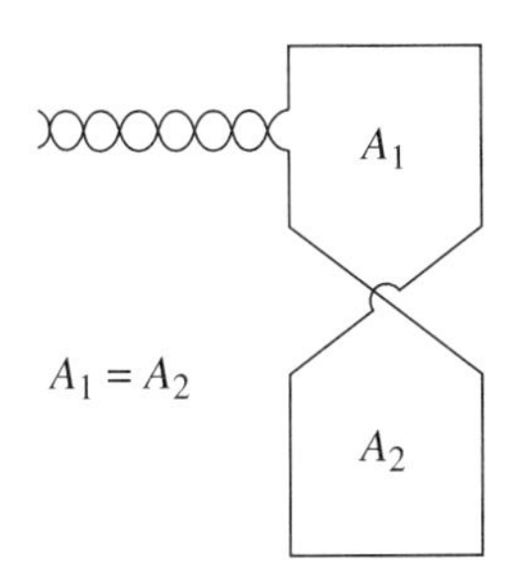

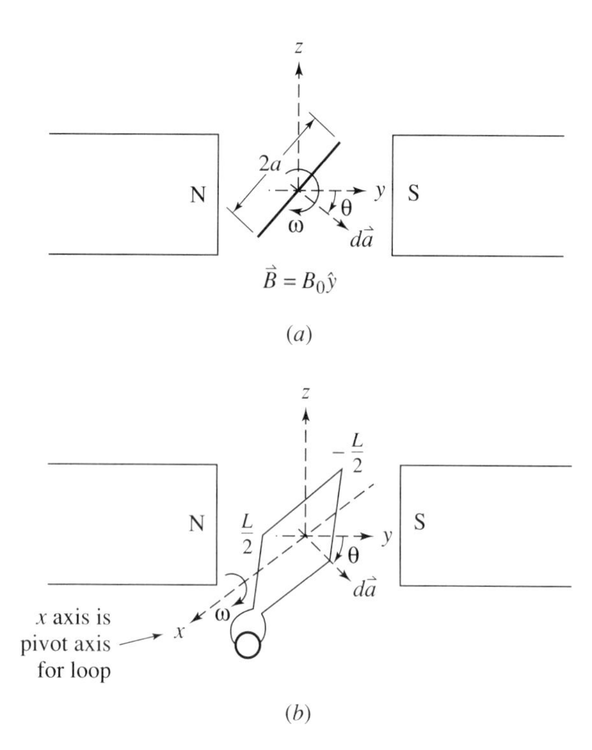

Figure 7-20. (*a*) Side view. (*b*) Perspective view.

GE7-4 A magnetic field is given by

$$\vec{\mathscr{B}}(\vec{r}, t) = 0.01 \cos 10^8 t\hat{x} - 0.06 \cos 10^8 t\hat{z} \text{ T}$$

A square resistive wire loop of side 0.1 m is in the xy plane and the center of the square is at the origin. Compute the induced emf in the loop and show the induced voltage source.

Figure 7-21.

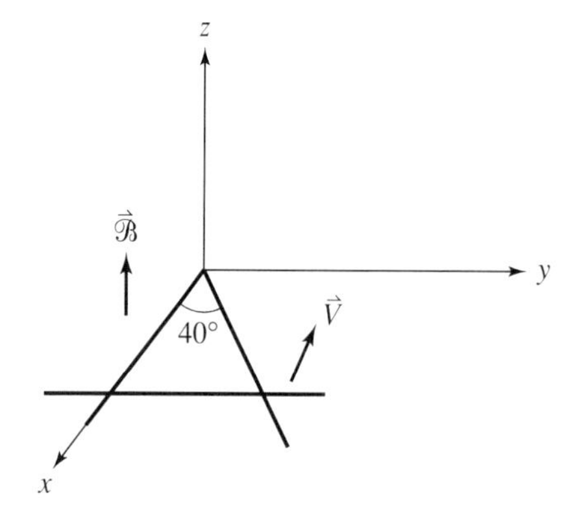

GE7-5 Compute the emf induced in the loop shown in Figure 7-21. $\vec{\mathscr{B}} = 6 \sin t\hat{z}$, and $\vec{V} = -10\hat{x}$.

GE7-6 For Example 7-1 in Section 7-2 compute the power dissipated by the loop $p(t)$ and the average power P.

GE7-7 The switch S_1 in the stationary loop is in Figure 7-22 opened and for a time-invariant magnetic field the voltmeter reads zero. Explain. (Answer: There is no relative motion between the conductors and the field.)

GE7-8 An electric field has an equation given by

Figure 7-22.

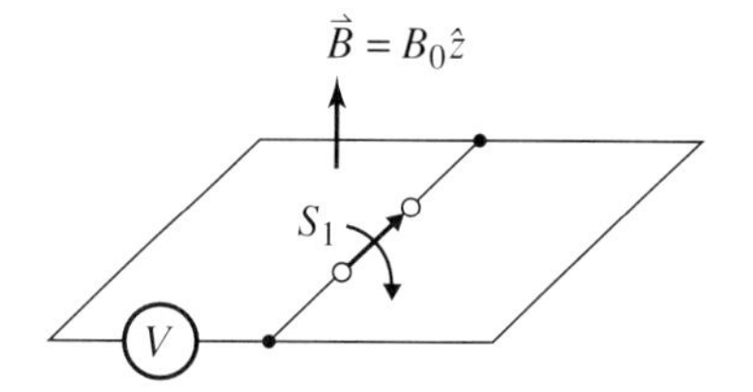

$$\vec{\mathscr{E}}(\vec{r}, t) = 0.1\cos(10^{10}t - 20z)\hat{y} - 0.1\cos(10^{10}t + 20z)\hat{y}$$

Determine the equation for $\vec{\mathscr{B}}$.

GE7-9 A magnetic field has the form (in a charge-free region and $\vec{\mathscr{J}} = 0$)

$$\vec{\mathscr{H}} = 10\cos(10^{10}t - \tfrac{200}{3}z)\hat{x} \text{ A/m}$$

and is in a material with $\mu = \mu_0$ and an unknown ϵ. Using Maxwell's curl equations determine ϵ as follows:

(a) Determine $\vec{\mathscr{D}}$ from the displacement current.

(b) Let $\vec{\mathscr{E}} = \vec{\mathscr{D}}/\epsilon$ and find the ϵ so the other curl equation is satisfied. What is the material ϵ_r?

GE7-10 Prove that for the scalar electric potential and vector magnetic potential superposition applies if there are multiple sources.

GE7-11 Prove by direct substitution that Eq. 7-47 for the scalar electric potential satisfies the wave Eq. 7-38. Watch the primes and unprimes carefully, and remember they are independent.

GE7-12 One disadvantage of the scalar electric potential and vector magnetic potential is that both must be found before $\vec{\mathscr{E}}(\vec{E})$ can be determined (Eq. 7-30). Hertz showed that it is possible to define a single potential function, called the Hertz vector, which is denoted by $\vec{\Pi}$. Since we must specify the curl and divergence of $\vec{\Pi}$ to make it unique, it is usual to define

$$\vec{A} = \mu\epsilon\frac{\partial\vec{\Pi}}{\partial t}$$

For μ and ϵ constant, show that this is the same as specifying the curl of $\vec{\Pi}$. (Hint: How do we define the vector magnetic potential?) Since we are free to define the divergence of $\vec{\Pi}$ (vector Helmholtz theorem), set

$$\nabla\cdot\vec{\Pi} = -\mathscr{V}$$

Determine expressions for $\vec{\mathscr{B}}$ and $\vec{\mathscr{E}}$ in terms of $\vec{\Pi}$ only. Beginning with Ampere's law in differential form, show that

$$\frac{\partial}{\partial t}\left(\nabla^2\vec{\Pi} - \mu\epsilon\frac{\partial^2\vec{\Pi}}{\partial t^2}\right) = -\frac{1}{\epsilon}\vec{\mathscr{J}}$$

and

$$\nabla^2\vec{\Pi} - \mu\epsilon\frac{\partial^2\vec{\Pi}}{\partial t^2} = 0$$

(for $\vec{\mathscr{J}} = 0$; i.e., source-free region)

How do you justify setting the constants of integration to zero when integrating partially with respect to time?

GE7-13 A point charge $2e^{-t}\mu$ C is located at (1, 2, 3). Determine the value of the scalar electric potential at the point (0, 0, 0).

GE7-14 The wave equation for $\vec{\mathscr{E}}$, $\vec{E}$, and $\vec{\mathscr{H}}$, $\vec{H}$ (Eqs. 7-17, 7-20, 7-24, and 7-27) are of the same forms as those for the scalar and vector potentials. By analogy with the integral solutions for the potentials, obtain the integral solutions for $\vec{\mathscr{E}}$, $\vec{E}$, and $\vec{\mathscr{H}}$, and $\vec{H}$. (Hint: Use prime coordinates and retarded time to generate the source functions in the integrands. Watch signs carefully, too.)

GE7-15 The derivation of Eq. 7-59 assumed that only conduction current was present. Derive the corresponding formula for $\vec{J} = \rho_v\vec{v}$.

GE7-16 An engineer proposes to extract energy (power) from the earth's magnetic field by placing a DC source across two parallel plates so that the resulting electric field is perpendicular to the magnetic field as shown in Figure 7-23. Discuss the feasibility of this proposal.

GE7-17 The equation for a magnetic field (in cylindrical coordinates) is given by

$$\vec{\mathscr{H}} = -\frac{10^{-8}}{\rho}\cos(10^9 t - 20\rho)\hat{\phi} \text{ A/m}$$

Figure 7-23.

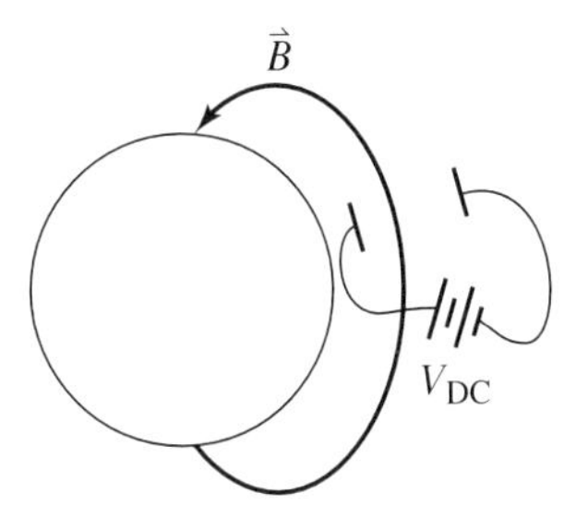

Determine the power flow through a cylinder of height 1 m and radius 1 m. Also determine whether Poynting's theorem is satisfied for a volume consisting of concentric cylinders of radii 0.5 m and 1 m, and of 1 m length. What does your result tell you about the field structure; that is, is it possible or not?

GE7-18 The field components of a wave are given by

$$\mathscr{E}_x = E_0 \cos(\omega t - \omega\sqrt{\mu\epsilon z})$$

$$\mathscr{H}_y = \sqrt{\frac{\epsilon}{\mu}} E_0 \cos(\omega t - \omega\sqrt{\mu\epsilon z})$$

(a) Show that these fields satisfy the wave equations for $\vec{\mathscr{E}}$ and $\vec{\mathscr{H}}$. Convert these to sinusoidal steady state and repeat.

(b) Obtain the equation for the Poynting vector $\vec{\mathscr{P}}$ and compute its time average (period $= T = 2\pi/\omega$). Compare this with $\overline{S}$.

GE7-19 The electric field intensity at a large distance from a finite radiating dipole is given by the spherical coordinate form

$$E_\theta = \sqrt{\frac{\mu}{\epsilon}} H_\phi = \frac{A}{r} e^{-j\omega\sqrt{\mu\epsilon r}} \sin\theta$$

where A is a constant, and the wire dipole is at the origin directed along the z axis. Find the average power $\overline{S}$ radiating through a sphere of radius r_0, r_0 large.

GE7-20 An electric field is given by

$$\vec{E} = (E_0 e^{-j\omega\sqrt{\mu\epsilon z}} + \tfrac{1}{3}E_0 e^{j\omega\sqrt{\mu\epsilon z}})\hat{y}$$

Obtain the equation for the wave impedance Z^w_{+z} and evaluate it for $f = 10^9$ Hz in vacuum at location $z = 1.2$ m.

GE7-21 For generalized curvilinear coordinate systems that have a z coordinate, show that the transverse field components of a $+z$-directed system can be obtained from E_z and H_z. Results are functions of transverse coordinates only. (Assume lossless, source-free region and begin by expanding the two curl equations) (Partial answer:

$$E_1 = -\frac{\gamma}{k_c^2 h_1}\frac{\partial E_z}{\partial t_1} - j\frac{\omega\mu}{k_c^2 h_2}\frac{\partial H_z}{\partial t_2}$$

where h_1 and h_2 are the scale factors.)

GE7-22 For the TEM modes the propagation constant was shown to be

$$\gamma_{\text{TEM}} = \sqrt{j\omega\mu\sigma - \omega^2\mu\epsilon}$$

Express this in the form $\alpha + j\beta$ and identify an attenuation using Eq. 7-86.

8 Propagation of Plane Waves

8-1 Introduction

Our first study in applications of propagating waves will be for plane waves, which are TEM-mode waves. These applications are important for three reasons:

1. Most of the radio frequency and microwave properties of materials are defined in terms of their response to plane wave fields.
2. For studies of the propagation of more complicated field configurations many helpful qualitative physical insights can often be obtained by examining some equivalent plane-wave properties.
3. There are many transmission-line analogies in plane wave propagation that make it much easier to visualize what is happening. Remember that TEM modes are also called *transmission line* modes. Some of the transmission line systems will be studied in Chapter 9.

A study of plane waves also provides a very good way to show how to apply the Poynting vector, use the wave impedance concept, define wavelength, and introduce wave or phase velocity. We shall see that the results are identical in form with the transmission line formulas of Chapter 1 and that the Smith chart can also be used. We will use the z direction as the basis for our work to make the analogies even more direct.

8-2 The Uniform Plane Wave: Definition, Solution Properties, and Description for Lossless, Source-Free Media

A *uniform plane wave* is defined as follows:

Definition. *Plane* means that $\vec{\mathscr{E}}$ and $\vec{\mathscr{H}}$ are in the same plane and both are perpendicular (transverse) to the direction of propagation.

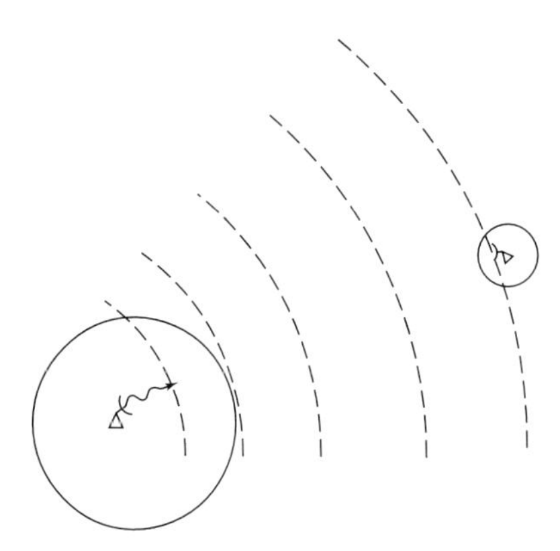

Figure 8-1. A receiving antenna on the moon appears to intercept a uniform plane wave.

Definition. *Uniform* means that the field vectors $\vec{\mathscr{E}}$ and $\vec{\mathscr{H}}$ do not vary in the plane containing them.

This definition may seem to be a little impractical since if $\vec{\mathscr{E}}$ and $\vec{\mathscr{H}}$ do not change in the plane they would have to extend to infinity. Also, since TEM modes require two or more conductors, the conductors are at infinity. However, in practice the portion of a wave being investigated may appear uniform locally. One extreme case is illustrated in Figure 8-1. A radio transmitter on the earth may appear as a point source radiating a spherical wave that expands as it propagates out. A receiving antenna on the moon is able to intercept only a small part of the wavefront passing across it, so the wave appears locally as a flat (uniform) plane wave.

If we consider the z direction of propagation, the electric field intensity will be in the xy plane by definition. We select $\vec{\mathscr{E}}$ to be in the $+\hat{x}$ direction as shown in Figure 8-2. Note that the x coordinate is into the page so the system is still a right-hand coordinate system. (This may appear to be an awkward orientation, but since it will be used throughout the remainder of the text it is best to introduce the convention now to avoid changing orientations later.)

Let's consider the sinusoidal steady-state lossless solution first ($\sigma = 0$). Since we do not have any boundaries for boundary conditions required to solve Laplace's equation for the TEM solution process (Section 7-10) we can proceed two ways. First, the general expression for electric field intensity (Eq. 7-87) reduces for this case to

$$\vec{E}(x, y, z) = E_x(x, y)E^{-\gamma_{\mathrm{TEM}}z}\hat{x}$$

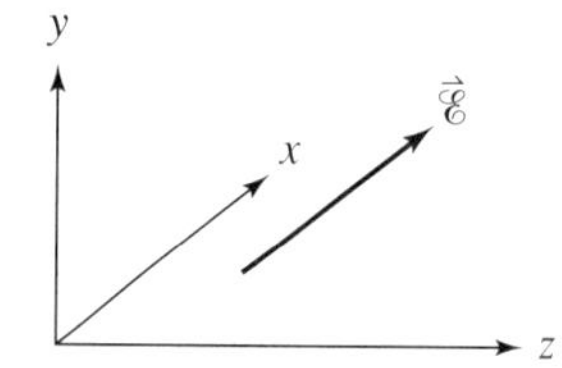

Figure 8-2. Convention for direction of the electric field intensity vector for a uniform plane wave.

By definition, however, $\vec{E}$ cannot vary in the plane, so

$$E_x(x, y) = E_{0x} \qquad \text{(constant)}$$

Also, from Eq. 7-108 with $\sigma = 0$ we obtain for TEM waves in lossless media

$$\gamma_{\text{TEM}} = j\omega\sqrt{\mu\epsilon} = jk$$

The solution for $\vec{E}$ is then

$$\vec{E} = E_{0x}e^{-j\omega\sqrt{\mu\epsilon}z}\hat{x} = E_{0x}e^{-jkz}\hat{x} \tag{8-1}$$

This is the result for a wave propagating in the $+\hat{z}$ direction. If the wave had both positive and negative traveling components we would have, analogous to transmission lines,

$$\vec{E} = (E_{0x}^{+}e^{-jkz} + E_{0x}^{-}e^{jkz})\hat{x} \tag{8-2}$$

Note that in general E_{0x}^{+} and E_{0x}^{-} may be complex, but it is usual practice not to use overbars.

An alternative method is to use the TEM wave definition and write

$$\vec{E} = E_x(z)\hat{x}$$

Next substitute this into the lossless wave equation obtained from Eq. 7-79 where $\bar{k} \to k$ (Eq. 7-76):

$$\nabla^2(E_x(z)\hat{x}) + k^2E_x(z)\hat{x} = 0$$

Since $\hat{x}$ does not vary with coordinates, it may be taken outside the derivative (del) operations

$$(\nabla^2E_x(z))\hat{x} + k^2E_x(z)\hat{x} = (\nabla^2E_x(z) + k^2E_x(z))\hat{x} = 0$$

Now since a vector is zero only if its components are zero the preceding equation requires that

$$\nabla^2E_x(z) + k^2E_x(z) = 0$$

Evaluating the Laplacian, this reduces to

$$\frac{d^2E_x(z)}{dz^2} + k^2E_x(z) = 0$$

This is a second-order ordinary differential equation with constant coefficients whose characteristic equation is

$$m^2 + k^2 = 0 \Rightarrow m = \pm jk$$

The solution then, using complex constants, is

$$E_x(z) = E_{0x}^{+}e^{-jkz} + E_{0x}^{-}e^{jkz}$$

for which the corresponding vector is

$$\vec{E} = E_x(z)\hat{x} = (E_{0x}^{+}e^{-jkz} + E_{0x}^{-}e^{jkz})\hat{x}$$

as before.

An important sidelight observation is that γ, our usual propagation constant, is the square root of the negative of the coefficient of E_x in the wave equation, when the derivative term has unit coefficient. The expression for $\vec{H}$ is obtained using Maxwell's curl equation (Faraday's law):

$$\vec{H} = \frac{1}{-j\omega\mu} \nabla \times \vec{E} = \left(\frac{k}{\omega\mu} E_{0x}^{+} e^{-jkz} - \frac{k}{\omega\mu} E_{0x}^{-} e^{jkz} \right) \hat{y}$$

$$= \left(\frac{E_{0x}^{+}}{\frac{\omega\mu}{k}} e^{-jkz} - \frac{E_{0x}^{-}}{\frac{\omega\mu}{k}} e^{jkz} \right) \hat{y}$$

Using the definition of k,

$$\frac{\omega\mu}{k} = \frac{\omega\mu}{\omega\sqrt{\mu\epsilon}} = \sqrt{\frac{\mu}{\epsilon}}$$

A dimensional analysis of this result yields

$$\sqrt{\frac{\mu}{\epsilon}} \Rightarrow \sqrt{\frac{\text{H/m}}{\text{F/m}}} = \sqrt{\frac{\text{V} \cdot \text{s/A}}{\text{A} \cdot \text{s/V}}} = \frac{\text{V}}{\text{A}} = \Omega$$

Since the term is dimensionally ohms and it is a property (characteristic) of the medium, it is called the *characteristic* or *intrinsic* impedance of the medium. We then define

$$\eta \equiv \sqrt{\frac{\mu}{\epsilon}} \tag{8-3}$$

Using this, our solution for the magnetic field is written

$$\vec{H} = \left(\frac{E_{0x}^{+}}{\eta} e^{-jkz} - \frac{E_{0x}^{-}}{\eta} e^{jkz} \right) \hat{y} \tag{8-4}$$

There are several transmission line analogies here. First, the intrinsic impedance has the same role as the characteristic impedance Z_0 in transmission-line theory. Second, μ and ϵ correspond to L and C of a transmission line (in fact, they have the same units). Third, the sign of the negative traveling component of $\vec{H}$ (current) is negated just as it was for transmission line current. These analogies show why we call TEM the transmission line mode. Additional analogies will be covered later.

The explicit time forms for $\vec{\mathscr{E}}$ and $\vec{\mathscr{H}}$ are obtained from Eqs. 8-2 and 8-4 using the sinusoidal steady-state technique discussed in earlier chapters.

$$\begin{aligned} \mathscr{E}(z, t) &= \text{Re}\{\vec{E}(z)e^{j\omega t}\} \\ &= [|E_{0x}^{+}| \cos(\omega t + \theta^{+} - kz) + |E_{0x}^{-}| \cos(\omega t + \theta^{-} + kz]\hat{x} \end{aligned} \tag{8-5}$$

$$\begin{aligned} \mathscr{H}(z, t) &= \text{Re}\{\vec{H}(z)e^{j\omega t}\} \\ &= \left[\frac{|E_{0x}^{+}|}{\eta} \cos(\omega t + \theta^{+} - kz) - \frac{|E_{0x}^{-}|}{\eta} \cos(\omega t + \theta^{-} + kz) \right] \hat{y} \end{aligned} \tag{8-6}$$

In these equations θ^+ and θ^- are the angles of the complex numbers E_{0x}^+ and E_{0x}^-, respectively.

These time waveforms are identical in form to those developed in Chapter 1 (except for the attenuation α assumed to be zero, or lossless, here). Thus the interpretations of the positive and negative traveling waves are identical with those of Eq. 1-30 and Figures 1-16 and 1-17.

For nonsinusoidal waveforms we must solve the general wave Eq. 7-18, which is applicable to the lossless case,

$$\nabla^2\vec{\mathscr{E}} - \mu\epsilon \frac{\partial^2 \vec{\mathrm{E}}}{\partial t^2} = 0$$

For a uniform plane wave we set

$$\vec{\mathscr{E}}(z, t) = \mathscr{E}_x(z, t)\hat{x}$$

and the wave equation becomes

$$\frac{\partial^2\mathscr{E}_x(z, t)}{\partial z^2} - \mu\epsilon \frac{\partial^2\mathscr{E}_x(z, t)}{\partial t^2} = 0 \tag{8-7}$$

This one-dimensional wave equation has the same form as the lossless transmission line wave equations of Eq. 3-1. The d'Alembert solution thus applies, so that (see Eq. 3-2)

$$\mathscr{E}_x(z, t) = \mathscr{E}_x^+ f_1(at - bz) + \mathscr{E}_x^- f_2(at + bz)$$

Thus any waveform will propagate with identifiable positive and negative traveling components in general. Specific solution forms for Eq. 8-7 may be obtained by the method of separation of variables (see Appendix K) using a product solution of the form.

$$\mathscr{E}_x(z, t) = Z(z)T(t) \tag{8-8}$$

(That this will work at all is assured by the uniqueness theorem stating that if we can get any solution that satisfies the differential equation and the boundary conditions we have the solution.) Then

$$\frac{\partial^2\mathscr{E}_x(z, t)}{\partial z^2} = \frac{\partial^2[Z(z)T(t)]}{\partial z^2} = T(t)\frac{d^2Z(z)}{dz^2} \equiv TZ''$$

Similarly,

$$\frac{\partial^2\mathscr{E}_x(z, t)}{\partial t^2} = T''Z$$

Substituting these into the wave Eq. 8-7,

$$TZ'' - \mu\epsilon T''Z = 0$$

Dividing by TZ,

$$\frac{Z''}{Z} - \mu\epsilon\frac{T''}{T} = 0$$

Rearranging,

$$\frac{T''}{T} = \frac{1}{\mu\epsilon}\frac{Z''}{Z}$$

Now since the left side is a function of t only and the right side a function of z only, they both must be a constant. That this must be so is explained by the fact that if z only were to change for a given time t, the right side could not change value since the left side is unchanged. Let this constant be a^2. The two solutions are then

$$\frac{T''}{T} = a^2 \qquad \frac{1}{\mu\epsilon}\frac{Z''}{Z} = a^2$$

$$T'' = a^2 T \qquad Z'' = a^2 \mu\epsilon Z$$

$$T(t) = C_1 e^{\pm at} \qquad Z(z) = C_2 e^{\pm a\sqrt{\mu\epsilon}}$$

Putting these solutions into the product form (Eq. 8-8), letting $C_1 C_2 = C$, yields

$$\mathscr{E}_x(z, t) = Ce^{\pm at \pm a\sqrt{\mu\epsilon z}}$$

The four possible exponents are

$$\begin{aligned} &at + a\sqrt{\mu\epsilon z} \\ -&at + a\sqrt{\mu\epsilon z} = -(at - a\sqrt{\mu\epsilon z}) \\ -&at - a\sqrt{\mu\epsilon z} = -(at + a\sqrt{\mu\epsilon z}) \\ &at - a\sqrt{\mu\epsilon z} \end{aligned}$$

Examining these possibilities we see that there are really only two distinct forms (i.e., $(at + a\sqrt{\mu\epsilon z})$ and $(at - a\sqrt{\mu\epsilon z})$) so that the solution can be written

$$\mathscr{E}_x(z, t) = \mathscr{E}_x^+ e^{(at - a\sqrt{\mu\epsilon z})} + \mathscr{E}_x^- e^{(at + a\sqrt{\mu\epsilon z})} \tag{8-9}$$

This is indeed of the form predicted by the d'Alembert solution.

The solution for H_y is obtained using Maxwell's equation (Faraday's law) on the vector expression using the preceding equation

$$\vec{\mathscr{E}} = \mathscr{E}_x(z, t)\hat{x}$$

and then integrating partially with respect to t and negating. The result is

$$\vec{\mathscr{H}} = \mathscr{H}_y(z, t)\hat{y} = \left[\frac{\mathscr{E}_x^+}{\eta} e^{(at - a\sqrt{\mu\epsilon z})} - \frac{\mathscr{E}_x^-}{\eta} e^{(at + a\sqrt{\mu\epsilon z})}\right]\hat{y} \tag{8-10}$$

Although the exponential form is most prevalent, other forms are more convenient for some classes of problems. These are reviewed in Appendix I.

We next look at some of the properties of these waves, beginning with the *phase velocity*. The phase velocity is obtained in exactly the same way as it was for transmission lines. Figure 8-3 is a plot of the forward traveling component of some $\mathscr{E}_x(z, t)$. To be able to stand at a given location on the waveform, one must move along with the waveform in such a way that the exponent remains constant (otherwise, the value of the function would

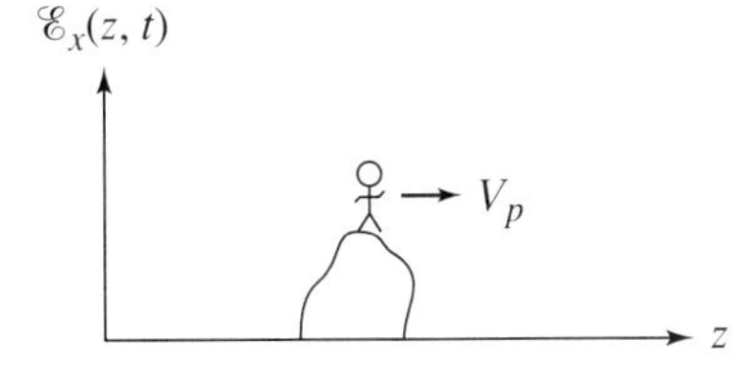

Figure 8-3. Physical illustration of the definition of phase velocity.

not be the same). On a sine wave this corresponds to staying at the same angle or constant phase point, hence the name *phase velocity*. Thus we set

$$at + a\sqrt{\mu\epsilon}z = \text{constant}$$

Differentiating with respect to time,

$$a + a\sqrt{\mu\epsilon}\,\frac{dz}{dt} = 0$$

The phase velocity is then

$$V_p \equiv \frac{dz}{dt} = \frac{1}{\sqrt{\mu\epsilon}}\ \text{m/s} \tag{8-11}$$

For the sinusoidal steady-state another form of V_p is obtained by multiplying the numerator by ω and then using earlier definitions to obtain

$$V_p = \frac{\omega}{\omega\sqrt{\mu\epsilon}} = \frac{\omega}{k} = \frac{\omega}{\text{Im}\{\gamma\}} \tag{8-12}$$

The last equality is actually more general than implied by our derivation, since the coefficient of z, which is Im γ, will always be the propagation component. See Eq. 7-85. For the negative traveling component see Exercise 8-2.0c.

Equation 8-11 shows another transmission line analogy for TEM waves. The material parameters μ and ϵ are again analogous to L and C; see Eq. 1-74.

The *wavelength* of a sinusoidal wave is defined as the distance one must move in a given direction to observe a phase change of 2π radians. If we evaluate the cosine function argument of Eq. 8-5 at two locations z_1 and z_2 we have

$$\omega t + \theta^+ - kz_1$$

$$\omega t + \theta^+ - kz_2$$

Subtracting these we obtain

$$k(z_2 - z_1)$$

Using the definition of wavelength, which we denote by λ_0 for the uniform plane wave, the difference term in the preceding expression must satisfy

$$k\lambda_0 = 2\pi$$

from which

$$\lambda_0 = \frac{2\pi}{k} = \frac{2\pi}{\mathrm{Im}\{\gamma\}} \tag{8-13}$$

The constant λ_0 is frequently called the *free medium*, *unbounded medium*, or *free space* wavelength; the word *space* is not restricted to vacuum here. A general form can be obtained by solving for k in terms of V_p and ω from Eq. 8-12 and substituting it into the preceding equation. This yields

$$\lambda_0 = \frac{V_p}{f} = \frac{1}{f\sqrt{\mu\epsilon}} \tag{8-14}$$

a familiar formula from physics.

To compute the *wave impedance* we use Eq. 7-69 and evaluate it in the $+\hat{z}$ direction, that is, using $\hat{\tau} \Rightarrow \hat{z}$. For the positive z components of the wave, we use the general expressions

$$\vec{E} = E_{0x}^{+}e^{-jkz}\hat{x} \qquad \vec{H} = H_{0y}^{+}e^{-jkz}\hat{y}$$

to obtain

$$Z_{z,\mathrm{TEM}}^{w} = \frac{(E_{0x}^{+}e^{-jkz}\hat{x} \times H_{0\,y}^{+*}e^{+jkz}\hat{y}) \cdot \hat{z}}{(\hat{z} \times H_{oy}^{+}e^{-jkz}\hat{y}) \cdot (\hat{z} \times H_{0\,y}^{+*}e^{jkz}\hat{y})} = \frac{E_{0x}^{+}}{H_{0y}^{+}}$$

From Eq. 8-4 we have

$$H_{0x}^{+} = \frac{E_{0x}^{+}}{\eta}$$

so that

$$Z_{z,\mathrm{TEM}}^{w} = \eta = \sqrt{\frac{\mu}{\epsilon}} \tag{8-15}$$

(see also Exercise 8-2.0d).

Again, the transmission line analogy is evident with $\mu, \epsilon \Rightarrow L,C$, giving the characteristic impedance for TEM waves.

The Poynting vector for the positive traveling wave components is

$$\begin{aligned} \vec{S} &= \tfrac{1}{2}(E_{0x}^{+}e^{-jkz}\hat{x}) \times (H_{0y}^{+}e^{-jkz}\hat{y})^{*} = \tfrac{1}{2}E_{0x}^{+}H_{0y}^{+*}\hat{z} \\ &= \frac{1}{2}E_{0x}^{+}\frac{E_{0x}^{+*}}{\eta^{*}}\hat{z} = \frac{1}{2}\frac{|E_{0x}^{+}|^{2}}{\eta}\hat{z} \qquad (\eta \text{ real}) \end{aligned} \tag{8-16}$$

or

$$= \tfrac{1}{2}\eta H_{0y}^{+}H_{0y}^{+*}\hat{z} = \tfrac{1}{2}|H_{0y}^{+}|^{2}\eta\hat{z} \tag{8-17}$$

These results look like the circuit theory resistor power equations

$$\frac{1}{2}\frac{|\bar{V}|^2}{R} \qquad \text{and} \qquad \frac{1}{2}|\bar{I}|^2 R$$

The main difference, however, is that the circuit theory equations represent a heat dissipation whereas the Poynting vector represents power flow across a surface. The Poynting power, of course, could ultimately be absorbed by a material.

Example 8-1

A uniform plane wave in vacuum has a frequency of 300 MHz and an electric field given by

$$\vec{E} = 10^{-5} e^{-j6.83z} \hat{x}$$

Determine the following: $\vec{H}$, V_p, λ_0, $Z^w_{z,\text{TEM}}$, $\vec{S}$.

Since the electric field vector is in the x direction the results of this section apply directly so that, using ϵ_0 and μ_0,

$$\vec{H} = \frac{E_{0x}}{\eta}\hat{y} = \frac{10^{-5}}{\sqrt{\dfrac{4\pi \times 10^{-7}}{1/36\pi \times 10^{-9}}}} e^{-j6.283z}\hat{y} = \frac{10^{-5}}{377} e^{-j6.283z}\hat{y}$$

$$= 2.65 \times 10^{-8} e^{-j6.283z}\hat{y}$$

$$V_p = \frac{1}{\sqrt{\mu_0 \epsilon_0}} \approx \frac{1}{\sqrt{4\pi \times 10^{-7} \times 1/36\pi \times 10^{-9}}} = 3 \times 10^8 \text{ m/s}$$

(This is the velocity of light in vacuum, so light waves obey Maxwell's equations.)

$$\lambda_0 = \frac{2\pi}{k} = \frac{2\pi}{6.283} = 1 \text{ m}$$

or

$$\lambda_0 = \frac{V_p}{f} = \frac{3 \times 10^8}{3 \times 10^8} = 1 \text{ m}$$

$$Z^w_{z,\text{TEM}} = \sqrt{\frac{\mu_0}{\epsilon_0}} \approx \sqrt{\frac{4\pi \times 10^{-7}}{1/36\pi \times 10^{-9}}} = 120\pi \approx 377\ \Omega$$

$$\vec{S} = \tfrac{1}{2}\vec{E} \times \vec{H}^* = \tfrac{1}{2} \times 10^{-5} e^{-j6.283z}\hat{x} \times 2.65 \times 10^{-8} e^{j6.283z}\hat{y}$$

$$= 0.133\hat{z} \text{ pW/m}^2$$

Example 8-2

The electric field of Example 8-1 is modified as to spatial orientation to be

$$\vec{E} = 10^{-5} e^{-j6.283z}\hat{y}$$

Determine the same quantities as in Example 8-1.

Since $\vec{E}$ is propagating in the $+\hat{z}$ direction, $\vec{E} \times \vec{H}^*$ must be in the $+\hat{z}$ direction. Using the right hand rule we find that $\vec{H}$ must be in the $-\hat{x}$ direction. Since the ratio of the two field components is $\eta = 377\ \Omega$,

$$\vec{H} = \frac{E_{0y}}{\eta} e^{-j6.283z}(-\hat{x}) = -2.65 \times 10^{-8} e^{-j6.283z}\hat{x}$$

(Maxwell's equation $\nabla \times \vec{E} = -j\omega\mu_0\vec{H}$ could also be used.) All the remaining quantities are identical.

Example 8-3

Although this example will be lengthy, the results and processes involved are important. This is a general xy plane orientation of fields.

A uniform plane wave propagates in the positive z direction, but is oriented off-axis in the xy plane as shown in Figure 8-4. The magnitude of the electric field is 10^{-4} V/m and is at 30° with respect to the x axis. The operating frequency is 1 GHz and the medium has $\mu = \mu_0$ and $\epsilon = 3\epsilon_0$. Determine $\vec{H}$, V_p, λ_0, $Z^w_{z,\text{TEM}}$, $\vec{S}$.

We first determine the vector expression for $\vec{E}$. The components of $\vec{E}$ are

$$E_{0x} = 10^{-4} \cos 30° = 8.66 \times 10^{-5} \text{ V/m}$$

$$E_{0y} = 10^{-4} \sin 30° = 5 \times 10^{-5} \text{ V/m}$$

$$k = \omega\sqrt{\mu\epsilon} = 2\pi 10^9 \sqrt{4\pi \times 10^{-7} \times 3 \times \frac{1}{36\pi} \times 10^{-9}} = 36.28 \text{ rad/m}$$

Thus

$$\vec{E} = (8.66 \times 10^{-5}\hat{x} + 5 \times 10^{-5}\hat{y})e^{-j36.28z} \text{ V/m}$$

which is the total field propagating in the $+z$ direction. The magnetic field can be determined by either of two methods as in Example 8-2. If we use the right-hand rule and remember that the two vectors are orthogonal, we conclude $\vec{H}$ must have a positive y component and a negative x component, and make an angle of 60° with respect to the negative x axis.

Since the ratio of positive propagating $\vec{E}$ to positive propagating $\vec{H}$ is η for TEM waves, we first obtain

$$\eta = \sqrt{\frac{4\pi \times 10^{-7}}{3 \times 1/36\pi + 10^{-9}}} = 218\ \Omega$$

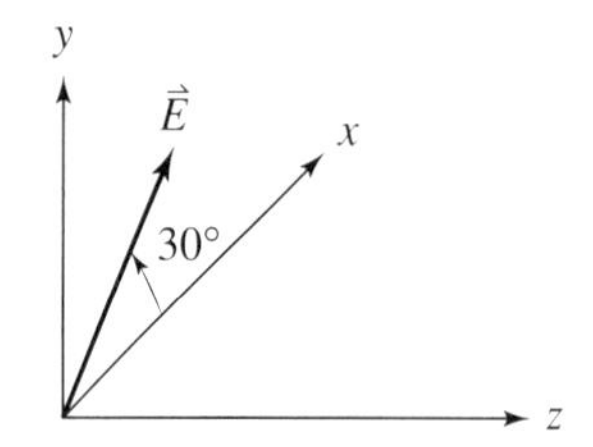

Figure 8-4. Electric field orientation.

Thus

$$|\vec{H}| = \frac{|\vec{E}|}{\eta} = \frac{10^{-5}}{218} = 4.59 \times 10^{-8} \text{ A/m}$$

The components are then

$$H_{0x} = -4.59 \times 10^{-8} \cos 60° = -2.29 \times 10^{-8}$$
$$H_{0y} = 4.59 \times 10^{-8} \sin 60° = 3.98 \times 10^{-8}$$

then

$$\vec{H} = (-2.29 \times 10^{-8}\hat{x} + 3.98 \times 10^{-8}\hat{y})e^{-j36.28z} \text{ A/m}$$

(Try generating this using Maxwell's equation on $\vec{E}$.)

$$V_p = \frac{1}{\sqrt{\mu\epsilon}} = \frac{1}{\sqrt{4\pi \times 10^{-7} \times 3 \times 1/36\pi \times 10^{-9}}} = 1.73 \times 10^8 \text{ m/s}$$

or

$$V_p = \frac{\omega}{k} = \frac{2\pi 10^9}{36.28} = 1.73 \times 10^8 \text{ m/s}$$

We know that the wave impedance is η, but it is instructive to use the impedance Eq. 7-69 on the general forms for $\vec{E}$ and $\vec{H}$ to substantiate the result. This will give confidence in using the equation in more complex cases.

$$Z^w_{z,\text{TEM}} = \frac{[(E_{0x}\hat{x} + E_{0y}\hat{y})e^{-jkz} \times (H_{0x}\hat{x} + H_{0y}\hat{y})^* e^{jkz}] \cdot \hat{z}}{[\hat{z} \times (H_{0x}\hat{x} + H_{0y}\hat{y})e^{-jkz}] \cdot [\hat{z} \times (H_{0x}\hat{x} + H_{0y}\hat{y})^* e^{jkz}]}$$
$$= \frac{[E_{0x}H^*_{0y}\hat{z} - E_{0y}H^*_{0x}\hat{z}] \cdot \hat{z}}{[H_{0x}\hat{y} - H_{0y}\hat{x}] \cdot [H^*_{0x}\hat{y} - H^*_{0y}\hat{x}]} = \frac{E_{0x}H_{0y} - E_{0y}H_{0x}}{H^2_{0x} + H^2_{0y}}$$

Next use Maxwell's equation on $\vec{E}$ to obtain $\vec{H}$:

$$\vec{H} = \frac{1}{-j\omega\mu} \nabla \times \vec{E} = \frac{1}{-j\omega\mu} [+jkE_{0y}e^{-jkx}\hat{x} - jkE_{0x}e^{-jkx}\hat{y}]$$
$$= -\frac{k}{\omega\mu} E_{0y}e^{-jkx}\hat{x} + \frac{k}{\omega\mu} E_{0x}e^{-jkx}\hat{y} = -\frac{E_{0y}}{\eta} e^{-jkx}\hat{x} + \frac{E_{0x}}{\eta} e^{-jkx}\hat{y}$$

Thus

$$H_{0x} = -\frac{E_{0y}}{\eta} \quad \text{and} \quad H_{0y} = \frac{E_{0x}}{\eta} \tag{8-18}$$

This states that the ratio of the orthogonal *components* of $\vec{E}$ and $\vec{H}$ are also related by η. Note that the minus sign is consistent with the fact that $E_{0y}\hat{y} \times H_{0x}\hat{x} = E_{0y}H_{0x}(-\hat{z})$. Placing the expressions for the components of $\vec{H}$ into the wave impedance formula above we obtain

$$Z^w_{z,\text{TEM}} = \frac{1}{\eta} \frac{E^2_{0x} + E^2_{0y}}{H^2_{0x} + H^2_{0y}}$$

This can be put in two different forms. Using Eq. 8-18 first to eliminate the electric field we have

$$Z^w_{z,\text{TEM}} = \frac{1}{\eta}\frac{(\eta H_{0y})^2 + (-\eta H_{0x})^2}{H^2_{0x} + H^2_{0y}} = \eta \tag{8-19}$$

Or, we can write the original sums of squares as radical products first and then use Eq. 8-18 to obtain the form

$$Z^w_{z,\text{TEM}} = \frac{1}{\eta}\frac{\sqrt{E^2_{0x} + E^2_{0y}}\sqrt{E^2_{0x} + E^2_{0y}}}{\sqrt{H^2_{0x} + H^2_{0y}}\sqrt{H^2_{0x} + H^2_{0y}}}$$

$$= \frac{1}{\eta}\frac{\sqrt{E^2_{0x} + E^2_{0y}}\sqrt{(\eta H_{0y})^2 + (\eta H_{0x})^2}}{\sqrt{H^2_{0x} + H^2_{0y}}\sqrt{H^2_{0x} + H^2_{0y}}}$$

$$= \frac{\sqrt{E^2_{0x} + E^2_{0y}}}{\sqrt{H^2_{0x} + H^2_{0y}}} = \frac{|\vec{E}|}{|\vec{H}|} \tag{8-20}$$

This last result shows that the ratio of *total* electric field to *total* magnetic field for the positive z direction is also η.

Another quantity asked for in this example is $\vec{S}$. Using Eq. 8-16, or 8-17 if desired, for *total* field values:

$$\vec{S} = \frac{1}{2}\frac{(10^{-4})^2}{218} = 22.9\ \mu\text{W/m}^2$$

Finally, from Eq. 8-13,

$$\lambda_0 = \frac{2\pi}{k} = 0.1732\ \text{m}$$

Example 8-4

The magnetic field intensity of a uniform plane wave (positive z propagation) has the time expression, for the medium, $\mu = \mu_0$,

$$\vec{\mathcal{H}}(\vec{r}, t) = -4 \times 10^{-9} \cos(10^{10}t - 45z)\hat{x} + 6 \times 10^{-9} \cos(10^{10}t - 45z)\hat{y}\ \text{A/m}$$

Determine $|\vec{H}|$, $|\vec{E}|$, ϵ_r, λ_0, f, $\vec{S}$.

From the given equation: $\omega = 10^{10}$, so $f = 1.59$ GHz.

From the given equation: $k = 45 = \omega\sqrt{\mu\epsilon} = 10^{10}\sqrt{4\pi \times 10^{-7} \times \epsilon}$ or $\epsilon = 1.611 \times 10^{-11}$. Then $\epsilon_r = \epsilon/\epsilon_0 = 1.82$:

$$|\vec{H}| = \sqrt{H^2_{0x} + H^2_{0y}} = 7.2 \times 10^{-9}\ \text{A/m}$$

$$\eta = \sqrt{\frac{\mu}{\epsilon}} = \sqrt{\frac{\mu_0}{\epsilon}} = \sqrt{\frac{4\pi \times 10^{-7}}{1.611 \times 10^{-11}}} = 279.3\ \Omega$$

Using Eq. 8-20, $|\vec{E}| = \eta|\vec{H}| = 2.01 \times 10^{-6}$ V/m.
Using Eq. 8-13, $\lambda_0 = 2\pi/k = 2\pi/45 = 0.1396$ m

Finally, since propagation is in the z direction,

$$\vec{S} = \tfrac{1}{2}|\vec{H}|^2\eta\hat{z} = 7.24 \times 10^{-15}\hat{z}\ \text{W/m}^2$$

Exercises

8-2.0a Beginning with the time-dependent wave equation, supply the steps that result in Eq. 8-7.

8-2.0b Carry out the steps that derive Eq. 8-10 from the electric field (Eq. 8-9).

8-2.0c Prove that the phase velocity of the negative traveling wave component is $-1/\sqrt{\mu\epsilon}$ and show that the dimensions are m/s.

8-2.0d Determine the wave impedance of the $-\hat{z}$ component of the uniform plane wave. (Answer: $-\eta$)

8-2.0e Work Example 8-1 for the case $\epsilon_r = 4$. Answer:

$$\vec{H} = 5.3 \times 10^{-8}e^{-j6.283z}\hat{y} \qquad V_p = 1.5 \times 10^8 \qquad \lambda = 0.5\ \text{m}$$

$$Z^w_{z,\text{TEM}} = 188\ \Omega \qquad \vec{S} = 0.266\ \text{pW/m}^2$$

8-2.0f Verify the value of $\vec{S}$ for Example 8-3 by evaluating the cross product definition of $\vec{S}$.

8-2.0g A uniform plane wave has an electric field intensity of 10 mV/m directed in the $-y$ direction and propagating in the $+z$ direction. The wavelength is 0.5 m and the material has $\mu = \mu_0$ and $\epsilon = 3\epsilon_0$. Determine $\vec{E}$, $\vec{H}$, $\vec{S}$, $\vec{\mathscr{E}}$, $\vec{\mathscr{H}}$, and $\vec{\mathscr{S}}$.

8-3 The Uniform Plane Wave: Properties and Description of Lossy, Source-Free Media

If the medium in which the wave is propagating is lossy, $\sigma \neq 0$, we must retain a current component

$$\vec{\mathscr{J}} = \sigma\vec{\mathscr{E}}$$

but still set $\rho_v = 0$. For this case the general wave Eq. 7-17 reduces to

$$\nabla^2\vec{\mathscr{E}} - \mu\epsilon\frac{\partial^2\vec{\mathscr{E}}}{\partial t^2} - \mu\sigma\frac{\partial\vec{\mathscr{E}}}{\partial t} = 0 \tag{8-21}$$

The sinusoidal steady-state form is obtained using the process developed in Chapter 1 and is

$$\nabla^2\vec{E} + \omega^2\mu\epsilon\vec{E} - j\omega\sigma\vec{E} = \nabla^2\vec{E} + (\omega^2\mu\epsilon - j\omega\sigma)\vec{E} = 0$$

Factoring $\omega^2\mu$ out of the coefficient we have

$$\nabla^2\vec{E} + \omega^2\mu\left(\epsilon - j\frac{\sigma}{\omega}\right)\vec{E} = 0 \tag{8-22}$$

Comparing this equation with Eq. 7-21, which is

$$\nabla^2\vec{E} + \omega^2\mu\epsilon\vec{E} = 0$$

one can see that all that needs to be done is to replace ϵ by a complex value given by

$$\bar{\epsilon} = \epsilon - j\frac{\sigma}{\omega} = \epsilon - j\epsilon'' \tag{8-23}$$

as suggested by Eq. 7-77. We can write the lossy wave equation as

$$\nabla^2\vec{E} + \omega^2\mu\bar{\epsilon}\vec{E} = 0 \tag{8-24}$$

In Chapter 7 it was shown that the loss could also be associated with a complex permeability $\bar{\mu}$. However, it is easy to show that it is in this instance more natural to use complex permittivity and call the medium a lossy dielectric. Using Ampere's law in differential form we have

$$\nabla \times \vec{H} = \vec{J} + j\omega D = \sigma\vec{E} + j\omega\epsilon\vec{E} = (\sigma + j\omega\epsilon)\vec{E} = j\omega\left(\epsilon - j\frac{\sigma}{\omega}\right)\vec{E} = j\omega\bar{\epsilon}\vec{E}$$

This result is of the same form as the lossless equation

$$\nabla \times \vec{H} = j\omega\vec{D} = j\omega\epsilon\vec{E}$$

with ϵ replaced by the complex permittivity $\bar{\epsilon}$.

Using the complex permittivity approach we describe a lossy dielectric by what is called the *loss tangent*, given by

$$\text{loss tangent} \equiv \tan\,\delta \equiv \frac{\epsilon''}{\epsilon'} \tag{8-25}$$

where δ is the loss angle. Using Eq. 8-23 this would yield

$$\tan\,\delta = \frac{\sigma}{\omega\epsilon} \tag{8-26}$$

for conduction losses only. The loss tangent is sometimes called the *dissipation factor*.

Other contributions to dielectric losses arise from the fact that electrons, ions, and molecules of materials absorb energy. See [56] for discussions of these. It is to be noted that ϵ' is not a constant either, but varies with frequency also. Thus ϵ' (here a direct replacement for ϵ) in the denominators of Eqs. 8-25 and 8-26 affects the loss tangent differently at each frequency. Table 8-1 is a sample of values taken from [45], Section 4.

The solution of wave Eq. 8-24 is obtained using steps similar to the lossless case. We have already shown that a uniform plane wave traveling in the positive z direction and which has only an x component is expressible as

$$\vec{E} = E_{0x}e^{-\gamma_{\text{TEM}}z}\hat{x} \qquad E_{0x} = \text{constant}$$

and where γ_{TEM}, the propagation constant, is the square root of the negative of the coefficient of $\vec{E}$ in the wave equation (the derivative term having unity coefficient). From Eq. 8-24 we then have

$$\gamma_{\text{TEM}} = \sqrt{-\omega^2\mu\bar{\epsilon}} = \sqrt{-\omega^2\mu(\epsilon' - j\epsilon'')} = \sqrt{j\omega^2\mu\epsilon'' - \omega^2\mu\epsilon'} \tag{8-27}$$

Table 8-1. Dielectric Constants and Loss Tangents of Representative Dielectrics

	Dielectric Constant (Relative Permittivity): $\epsilon'/\epsilon_0 : \epsilon_r$ (frequency, Hz)				Loss Tangent: Table Value Is $10^4 \times \epsilon''/\epsilon'$ (frequency, Hz)			
Material	10^3	10^6	10^8	3×10^9	10^3	10^6	10^8	3×10^9
Aluminum oxide	8.83	8.80	8.80	8.79	5.7	3.3	3	10
Barium titanate (values depends on $\lvert\vec{E}\rvert$)	1200	1143		600	4.4	2		23
Porcelain	5.36	5.08	5.04		140	75	78	
Silicon dioxide (fused quartz)	3.78	3.78	3.78	3.78	7.5	1	2	0.6
Epoxy resin (Araldite CN-501)	3.67	3.62	3.35	3.09	24	190	340	270
Epoxy resin (Epon resin RN-48)	3.63	3.52	3.32	3.04	38	142	264	210
Foamed polystyrene (25% filler)	1.03	1.03		1.03	<2	<1	<2	1
Phenol-formaldehyde								
(Bakelite BM120)	4.74	4.36	3.95	3.70	220	280	380	438
(50% paper luminate)	5.15	4.60	4.04	3.57	165	340	570	600
Polythylene	2.26	2.26	2.26	2.26	<2	<2	2	3.1
Polystyrene	2.56	2.56	2.55	2.55	<.5	.7	<1	3.3
Polytetrafluorethylene (Teflon)	2.1	2.1	2.1	2.1	<3	<2	<2	1.5
Sodium chloride (crystal)	5.9	5.9			<1	<2		<5
Water (distilled)		78.2	78	76.7		400	50	1570

Note that for the lossless case $\sigma = 0$ this reduces to the earlier result. This expression is in agreement with Eq. 7-108 for conductive dielectric losses only so that ϵ'' is given by Eq. 8-23.

Since γ_{TEM} is complex we need to determine α and β as suggested in Eqs. 7-85, 7-86, and 7-108. Thus we set

$$\gamma_{\text{TEM}} = \alpha + j\beta = \sqrt{j\omega^2\mu\epsilon'' - \omega^2\mu\epsilon'}$$

Squaring both sides

$$\alpha^2 - \beta^2 + j2\alpha\beta = j\omega^2\mu\epsilon'' - \omega^2\mu\epsilon'$$

Equating real and imaginary components:

$$\alpha^2 - \beta^2 = -\omega^2\mu\epsilon' \tag{8-28}$$

$$2\alpha\beta = \omega^2\mu\epsilon'' \tag{8-29}$$

Solving the second equation for α and substituting into the first yields

$$4\beta^4 - 4\omega^2\mu\epsilon'\beta^2 - \omega^4\mu^2\epsilon''^2 = 0$$

Solving this as a quadratic in β^2 and then taking the square root yields

$$\beta = \sqrt{\tfrac{1}{2}\left(\omega^2\mu\epsilon' \pm \omega^2\mu\sqrt{\epsilon'^2 + \epsilon''^2}\right)}$$

Factoring out $\omega^2\mu\epsilon'$ results in

$$\beta = \omega\sqrt{\mu\epsilon'}\sqrt{\frac{1}{2}\left[1 \pm \sqrt{1 + \left(\frac{\epsilon''}{\epsilon'}\right)^2}\right]}$$

Now since the imaginary part of γ_{TEM} must reduce to real k for the lossless case, we must take the positive sign on the inner radical ($\epsilon'' = 0$ for $\sigma = 0$) to obtain the physically admissible result

$$\beta = \omega\sqrt{\mu\epsilon'}\sqrt{\frac{1}{2}\left[1 + \sqrt{1 + \left(\frac{\epsilon''}{\epsilon'}\right)^2}\right]} \text{ rad/m} \tag{8-30}$$

A similar process using Eqs. 8-28 and 8-29 again, beginning by solving Eq. 8-29 for β first instead of α, yields

$$\alpha = \omega\sqrt{\mu\epsilon'}\sqrt{\frac{1}{2}\left[-1 + \sqrt{1 + \left(\frac{\epsilon''}{\epsilon'}\right)^2}\right]} \text{ nep/m} \tag{8-31}$$

The solution for the positive traveling component then becomes

$$\vec{E} = E_{0x}e^{-(\alpha + j\beta)z}\hat{x} = E_{0x}e^{-\alpha z}e^{-j\beta z}\hat{x} \tag{8-32}$$

For both positive and negative traveling components we would then have

$$\vec{E} = (E_{0x}^{+}e^{-\alpha z}e^{-j\beta z} + E_{0x}^{-}e^{\alpha z}e^{j\beta z})\hat{x} \tag{8-33}$$

This equation is also readily interpreted from transmission line theory as a wave propagating with attenuation. This can be more apparent by converting the positive traveling component into explicit time form as

$$\begin{aligned} \vec{\mathscr{E}}^{+}(z, t) &= \text{Re}\{\vec{E}^{+}e^{j\omega t}\} = \text{Re}\{E_{0x}^{+}e^{-\alpha z}e^{-j\beta z}e^{j\omega t}\hat{x}\} \\ &= |E_{0x}^{+}|e^{-\alpha z}\cos(\omega t + \theta^{+} - \beta z)\hat{x} \end{aligned} \tag{8-34}$$

where θ^{+} is the angle of E_{0x}^{+}. The amplitude decrease is readily seen here.

The wave impedance can be obtained several ways, the most direct being to replace ϵ by $\overline{\epsilon}$ in the lossless form. Doing so yields

$$\overline{\eta} = \sqrt{\frac{\mu}{\overline{\epsilon}}} = \sqrt{\frac{\mu}{\epsilon' - j\epsilon''}} = \sqrt{\frac{\mu}{\epsilon'}}\sqrt{\frac{1}{1 - j\dfrac{\epsilon''}{\epsilon'}}} \tag{8-35}$$

Rationalizing the complex denominator

$$\overline{\eta} = \sqrt{\frac{\mu}{\epsilon'\left[1 + \left(\dfrac{\epsilon''}{\epsilon'}\right)^2\right]}}\sqrt{1 + j\frac{\epsilon''}{\epsilon'}} \tag{8-36}$$

Taking the square root of the complex number using the same technique used to find α and β from γ_{TEM} we obtain

$$\overline{\eta} = \sqrt{\frac{\mu}{\epsilon'}}\frac{1}{\sqrt{2}}\left[\sqrt{\frac{1}{1 + \left(\dfrac{\epsilon''}{\epsilon'}\right)^2} + \frac{1}{\sqrt{1 + \left(\dfrac{\epsilon''}{\epsilon'}\right)^2}}}\right.$$
$$\left. + j\sqrt{-\frac{1}{1 + \left(\dfrac{\epsilon''}{\epsilon'}\right)^2} + \frac{1}{\sqrt{1 + \left(\dfrac{\epsilon''}{\epsilon'}\right)^2}}}\right] \tag{8-37}$$

For lossless media, $\epsilon'' = 0$ and the preceding equation reduces to η as it should.

The expression for $\vec{H}$ can be obtained using either the transmission line analogy, using $\overline{\eta}$ for z_0, or by using Faraday's law from Maxwell's equations on $\vec{E}$, using γ_{TEM} as the exponent coefficient. The results are

$$\vec{H} = \left(\frac{E_{0x}^+}{\overline{\eta}}e^{-\gamma_{\text{TEM}}z} - \frac{E_{0x}^-}{\overline{\eta}}e^{\gamma_{\text{TEM}}z}\right)\hat{y} = \left[\frac{E_{0x}^+}{\left(\dfrac{j\omega\mu}{\gamma}\right)}e^{-\gamma_{\text{TEM}}z} - \frac{E_{0x}^-}{\left(\dfrac{j\omega\mu}{\gamma}\right)}e^{\gamma_{\text{TEM}}z}\right]\hat{y} \tag{8-38}$$

If we compare the time equation form of the electric field intensity (Eq. 8.34) with the lossless case (Eq. 8-5) we see that β takes on the role of k in the cosine function. We then have the more general result for phase velocity

$$V_p = \frac{\omega}{\beta} = \frac{\omega}{\text{Im}\{\gamma\}} \tag{8-39}$$

as suggested by Eq. 8-12.

Similarly, for wavelength Eq. 8-13 would give

$$\lambda = \frac{2\pi}{\beta} = \frac{2\pi}{\text{Im}\{\gamma\}} \tag{8-40}$$

as the general expression. Note we have used λ here rather than λ_0 to indicate the generality of the result, as it applies to any case for which β can be obtained.

Although we have taken the point of view of describing losses as attributable to a lossy dielectric, we could use the same results for any material for which an ϵ' and σ can be obtained. For example, a conductor such as copper has a well-known conductivity. If we use $\epsilon' = \epsilon = \epsilon_0$ for metals, we need not derive separate results. Thus we can use the preceding results for a wide range of materials, as we shall show later in this section.

Up to this point we have considered only a single-frequency sinusoid. To transmit information some type of modulation digital or analog—must be placed on the sine wave (carrier). However, once this is done the composite waveform contains many frequencies. Now to compute the velocity of propagation we must first determine β as required by Eq. 8-40. Since β is frequency dependent, which frequency does one use? We must then conclude that if β is a function of frequency the phase velocity is not able to give us the signal velocity. Thus rather than concerning ourselves with a single frequency we now must consider some ways of describing the propagation velocity of a group of frequencies. There are several different ways of approximating this velocity; see [56]. We will use the more commonly accepted method for describing such a velocity, called the *group velocity*. To obtain this velocity we suppose a wave has two frequencies that differ from some reference frequency ω_0 by an amount $\Delta\omega$. The corresponding β values will also differ from the reference β_0 by an amount $\Delta\beta$. As a function of time such a wave could be expressed as (setting angle of the phasor amplitude to zero)

$$\begin{aligned}\mathscr{E}_x(z, t) &= |E_{0x}| \cos[(\omega_0 + \Delta\omega)t + (\beta_0 + \Delta\beta)z] \\ &\quad + |E_{0x}| \cos[(\omega_0 - \Delta\omega)t + (\beta_0 - \Delta\beta)z]\end{aligned}$$

Applying the trig identity for the sum of cosines yields

$$\begin{aligned}\mathscr{E}_x(z, t) &= 2|E_{0x}| \cos(\Delta\omega t - \Delta\beta z) \cos(\omega_0 t - \beta_0 z) \\ &= 2|E_{0x}| \cos\left[\Delta\omega\left(t - \frac{\Delta\beta}{\Delta\omega} z\right)\right] \cos\left[\omega_0\left(t - \frac{\beta_0}{\omega_0} z\right)\right]\end{aligned} \tag{8-41}$$

Since coefficients of z in the preceding equation must be reciprocal velocity, we define the *group velocity* as

$$V_g = \lim_{\Delta\beta \to 0} \frac{\Delta\omega}{\Delta\beta} = \frac{d\omega}{d\beta} \tag{8-42}$$

Note that in some sense this represents the propagation velocity of the amplitude (i.e., modulation) of the carrier ω_0. Note also that this is the slope of a plot of ω versus β.

The relationship between V_p and V_g is most easily seen from Figure 8-5. The phase velocity is the slope of the line from the *origin* to the frequency of interest, whereas the group velocity is the slope of the curve at the point ω_0. These plots are called the ω-β diagrams, or Brillouin diagrams. The axes may seem to be reversed since ω is usually

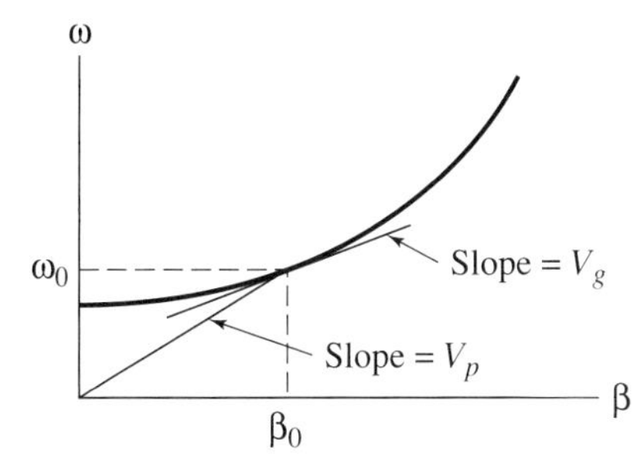

Figure 8-5. Plot illustrating the concepts of phase and group velocities from an ω-β diagram.

considered to be the independent variable; however, such plots show the velocities as slopes in the conventional sense.

The physical effects of a frequency-dependent β can be explained qualitatively by considering the propagation of a pulse as shown in Figure 8-6*a*. At time zero the pulse is assumed to be a perfect rectangular shape. The pulse time waveform will contain many frequency components, and each component will propagate at a different velocity since β is frequency dependent. Thus at a time t_1 greater than zero some frequencies have progressed farther down the line than others. Adding up the frequency components would then produce a pulse like that shown in Figure 8-6*b*. Note that the pulse appears to have spread out or dispersed spatially. This phenomenon is in fact called *dispersion*. Although the pulse contains a wide spectrum of frequencies that violate our assumption of only $\Delta\omega$ deviation, the concept is at least qualitatively correct.

Two types of dispersion are commonly identified, based upon the algebraic sign of the derivative

$$\frac{dV_p}{d\omega}$$

If we expand this derivative using the chain rule and then insert the expression for phase velocity, Eq. 8-40, we have

$$\frac{dV_p}{d\omega} = \frac{dV_p}{d\beta}\frac{d\beta}{d\omega} = \frac{d\left(\dfrac{\omega}{\beta}\right)}{d\beta} \cdot \frac{d\beta}{d\omega} = \left(\frac{\beta\dfrac{d\omega}{d\beta} - \omega}{\beta^2}\right)\frac{d\beta}{d\omega} = \frac{1}{\beta}\left(1 - \frac{V_p}{V_g}\right) \tag{8-43}$$

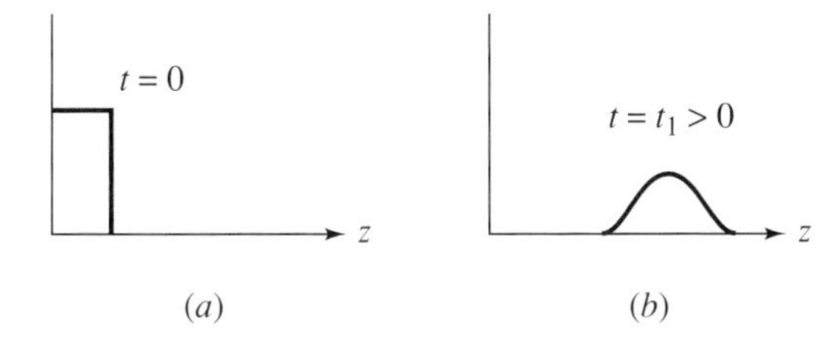

Figure 8-6. Effect of a frequency-dependent β on a programming pulse (dispersion); propagation in a lossy medium.

Solving the first and last parts of the equalities for V_g gives

$$V_g = \frac{V_p}{1 - \beta \dfrac{dV_p}{d\omega}} \tag{8-44}$$

Now when $dV_p/d\omega < 0$, V_g will always have the same sign as V_p. This is called *normal dispersion*. If, however, the derivative is positive, there is the possibility that the denominator could be negative for some frequencies. In this case V_g and V_p can have opposite signs for some frequencies and the same signs for others. This is called *anomalous dispersion*. Those frequencies that cause V_g and V_p to have opposite signs are said to produce *backward waves*.

Although our introduction to normal and anomalous dispersion arose naturally from a consideration of propagation in lossy media, the definitions apply equally well to any system where the sign of the derivative can be generated.

Example 8-5

An example of a transmission system which exhibits the property of anomalous dispersion is shown in Figure 8-7. From transmission-line theory, Eq. 1-22,

$$\gamma = \sqrt{ZY} = \sqrt{\left(\frac{1}{j\omega C}\right)\left(\frac{1}{j\omega L}\right)} = j\frac{1}{\omega\sqrt{LC}}$$

Thus

$$\beta = \text{Im}\{\gamma\} = \frac{1}{\omega\sqrt{LC}}$$

Then

$$V_p = \frac{\omega}{\beta} = \omega^2\sqrt{LC}$$

and

$$V_g = \frac{d\omega}{d\beta} = \frac{1}{\dfrac{d\beta}{d\omega}} = -\omega^2\sqrt{LC}$$

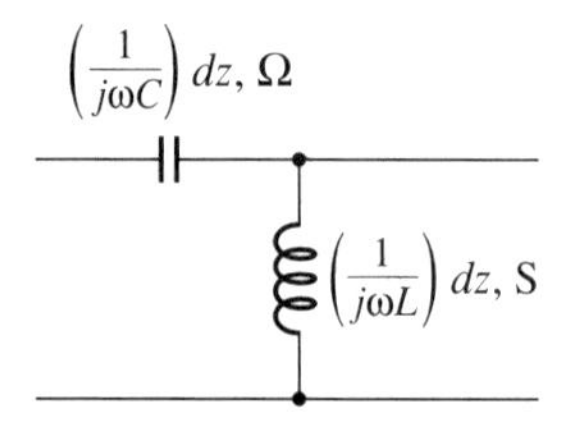

Figure 8-7. A transmission system that exhibits anomalous dispersion and can produce backward waves.

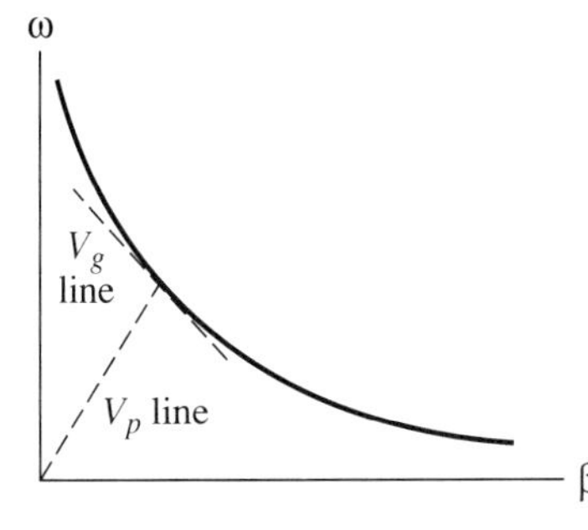

Figure 8-8. The ω-β diagram for the transmission system of Figure 8-7.

Thus the group and phase velocities are of opposite signs and backward waves can exist. This can be shown to be anomalous dispersion by evaluating the derivative of V_p:

$$\frac{dV_p}{d\omega} = \frac{d(\omega^2\sqrt{LC})}{d\omega} = 2\omega\sqrt{LC} > 0$$

The ω-β diagram is given in Figure 8-8.

Example 8-6

Compute the attenuation of a uniform plane wave (in decibels per meter) at frequencies of 3 GHz and 1 MHz and the corresponding wave impedances for distilled water. Use $\mu = \mu_0$.

At 3 GHz Table 8-1 gives the following values for distilled water:

$$\epsilon' = 76.7\epsilon_0 = 6.78 \times 10^{-10} \text{ F/m}$$

$$\frac{\epsilon''}{\epsilon'} = 0.157$$

Equation 8-31 yields

$$\begin{aligned}\alpha &= 2\pi \times 3 \times 10^9\sqrt{4\pi \times 10^{-7} \times 6.78 \times 10^{-10}}\sqrt{\tfrac{1}{2}[-1 + \sqrt{1 + (0.157)^2}]} \\ &= 43 \text{ nep/m}\end{aligned}$$

$$\alpha_{\text{dB/m}} = 8.686\alpha = 374 \text{ dB/m}$$

Note this shows that propagation of microwaves in water is not very successful, so water is a high-loss material. Using Eq. 8-35,

$$\overline{\eta} = \sqrt{\frac{4\pi \times 10^{-7}}{6.78 \times 10^{-10}}}\sqrt{\frac{1}{1 - j0.157}} = 42.6 + j3.32 = 42.7\angle 4.46^\circ \ \Omega$$

For 1 MHz the material parameter values are

$$\epsilon' = 78.2\epsilon_0 = 6.91 \times 10^{-10} \text{ F/m}$$

$$\frac{\epsilon''}{\epsilon'} = 0.04$$

Equation 8-31 gives

$$\alpha = 0.0037 \text{ nep/m} \qquad \text{or} \quad 0.014 \text{ dB/m}$$

Thus lowering the frequency has helped considerably. (As a side note, how would you explain the result at 100 MHz where the loss tangent is only 0.005 compared with 0.04 at 1 MHz?)

Exercises

8-3.0a Carry out the steps suggested in the text and derive Eq. 8-31. Be sure to justify the sign on the inner radical.

8-3.0b The negative traveling wave component in Eq. 8-33 seems to increase in amplitude according to the factor containing α. Explain this apparent anomaly.

8-3.0c Determine the types of dispersion for the systems of Figure 8-9. Also plot the ω-β diagrams.

8-3.0d Teflon is a low-loss dielectric. At a frequency of 3 GHz (see Table 8-1) compute α, β, λ, and $\overline{\eta}$. Now compute the same quantities assuming that the loss tangent is zero. Can you suggest any approximations for low-loss materials?

8-3.0e Suppose we assume that polystyrene has constant parameters $\epsilon_r = 2.56$ and $\sigma = 10^{-16}$ S/m. Plot α, β, and tan δ as functions of frequency from 10 Hz to 10 GHz. Compare with Table 8-1 and give possible reasons for differences.

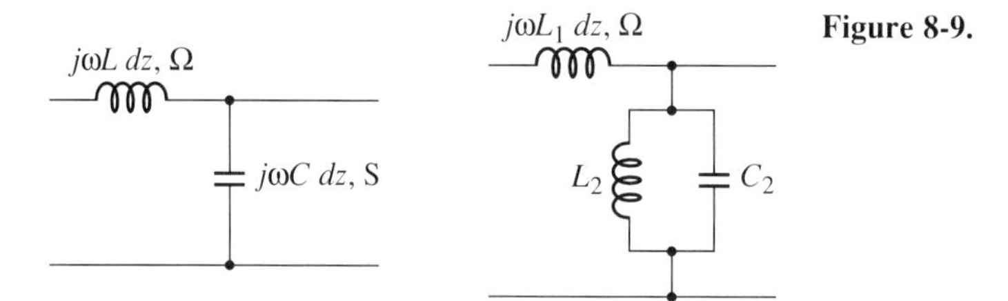

Figure 8-9.

8-4 Propagation of Uniform Plane Waves in Good Conductors

The wide range of materials encountered in engineering practice makes it convenient to identify or classify those materials from a wave propagation point of view. Also, two important classes occur frequently enough in practice to merit separate discussion, namely, good conductors and good (not perfect) dielectrics. In this section we examine the properties of good conductors.

Material classification is made on the basis of the values of the loss tangent, in terms of conductivity, and frequency, using

$$\tan\,\delta = \frac{\epsilon''}{\epsilon'} = \frac{\dfrac{\sigma}{\omega}}{\epsilon'} = \frac{\sigma}{\omega\epsilon'}$$

The usual material classes are:

$\tan\delta = 0 \quad \sigma = 0 \quad (\epsilon'' = 0)$	perfect dielectric
$\tan\delta \ll 0.1 \quad \sigma \ll \omega\epsilon'$	good dielectric
$\tan\delta \approx 1 \quad \sigma \approx \omega\epsilon'$	lossy dielectric
$\sigma \approx 10^{-5}$ *to* 10^6 S/m	semiconductor
$\tan\delta =$ large $\quad \sigma \gg \omega\epsilon' \; (\sigma \approx 10^6$ to 10^8 S/m)	good conductor
$\tan\delta = \infty \quad \sigma = \infty$	perfect conductor ($\vec{\mathcal{B}} = 0, \vec{B} \neq 0$)
$\tan\delta = \infty \quad \sigma = \infty$	superconductor ($\vec{\mathcal{B}} = 0, \vec{B} = 0$)

We first consider a good conductor. With the inequality given in the preceding list we write, using Eq. 8-27,

$$\gamma = \alpha + j\beta = \sqrt{j\omega^2\mu\epsilon'' - \omega^2\mu\epsilon'} = \omega\sqrt{\mu\epsilon'}\sqrt{j\frac{\epsilon''}{\epsilon'} - 1} = \omega\sqrt{\mu\epsilon'}\sqrt{j\frac{\sigma}{\omega\epsilon'} - 1}$$

$$\approx \omega\sqrt{\mu\epsilon'}\sqrt{j\frac{\sigma}{\omega\epsilon'}} = \sqrt{\omega\mu\sigma}\sqrt{j} = \sqrt{\omega\mu\sigma}\sqrt{1\angle 90^\circ} = \sqrt{\omega\mu\sigma}\angle 45^\circ$$

$$\approx \sqrt{\frac{\omega\mu\sigma}{2}} + j\sqrt{\frac{\omega\mu\sigma}{2}} = \sqrt{\pi f\mu\sigma} + j\sqrt{\pi f\mu\sigma} \text{ nep/m}$$

Thus

$$\begin{aligned} \alpha &= \sqrt{\pi f\mu\sigma} \text{ nep/m} \\ \beta &= \sqrt{\pi f\mu\sigma} \text{ rad/m} \end{aligned} \qquad \text{(good conductor } \sigma \gg \omega\epsilon') \tag{8-45}$$

The solution for the positive traveling component is then

$$E_x = E_{0x}e^{-\sqrt{\pi f\mu\sigma}z}e^{-j\sqrt{\pi f\mu\sigma}z} \tag{8-46}$$

where E_{0x} is the field value at $z = 0$. For attenuation in good conductors it is usual to define a quantity called the *skin depth*, which is the distance in the direction of propagation at which the field has decreased by a factor of e^{-1}. From the preceding equation this distance would be $1/\sqrt{\pi f\mu\sigma}$. This is denoted by the symbol δ, not to be confused with the loss angle in $\tan\delta$.

$$\text{skin depth} \equiv \delta \equiv \frac{1}{\sqrt{\pi f\mu\sigma}} \text{ m} \qquad \text{good conductor} \tag{8-47}$$

Note that the wave still exists beyond this distance. The skin depth is also called the *penetration depth*, which yields an attenuation of 8.68 dB. See Table 8-2 for typical materials. (The last column in Table 8-2 is discussed shortly.)

Table 8-2. Material Parameters and Frequency-Dependent Properties of Good Conductors and Composite Materials

Material	Relative Permittivity $\epsilon_r = \frac{\epsilon}{\epsilon_0}$	Conductivity σ (S/m)	Frequency at Which Loss Tangent $(\sigma/\omega\epsilon')$ Is 10 (Hz)	Permeability μ, H/m	Skin Depth δ, m (frequency in Hz)	Surface Resistance R_s (Ω/square)
Aluminum	1	3.82×10^7	6.9×10^{16}	$4\pi \times 10^{-7}$	$0.081/\sqrt{f}$	$43.21 \times 10^{-7}\sqrt{f}$
Beryllium	1	2.19×10^7	3.9×10^{16}	$4\pi \times 10^{-7}$	$0.11/\sqrt{f}$	$4.24 \times 10^{-7}\sqrt{f}$
Brass (66% Cu, 34% Ni)	1	2.56×10^7	4.6×10^{16}	$4\pi \times 10^{-7}$	$0.099/\sqrt{f}$	$3.93 \times 10^{-7}\sqrt{f}$
Constantan (55% Cu, 45% Ni)	1	0.226×10^7	4.1×10^{16}	$4\pi \times 10^{-7}$	$0.33/\sqrt{f}$	$13.22 \times 10^{-7}\sqrt{f}$
Copper	1	5.8×10^7	1×10^{17}	$4\pi \times 10^{-7}$	$0.066/\sqrt{f}$	$2.61 \times 10^{-7}\sqrt{f}$
Gold	1	4.1×10^7	7.4×10^{16}	$4\pi \times 10^{-7}$	$0.079/\sqrt{f}$	$3.1 \times 10^{-7}\sqrt{f}$
Iron (elemental)	1	1.03×10^7	1.8×10^{16}	$\approx 4\pi \times 10^{-2}$	$0.0005/\sqrt{f}$	$1958 \times 10^{-7}\sqrt{f}$
Lead	1	0.457×10^7	8.2×10^{15}	$4\pi \times 10^{-7}$	$0.24/\sqrt{f}$	$9.29 \times 10^{-7}\sqrt{f}$

The frequencies at which the loss tangents $\sigma/\omega\epsilon'$ equal 10 are given so that the good conductor approximation $\sigma \gg \omega\epsilon$ can be assured for frequencies below those values. The table shows that for conducting materials the approximation is valid theoretically up to light frequencies. Red light has wavelengths in the region of 650 nm or a frequency of 4.6×10^{14} Hz in vacuum, so we may use the approximations easily at microwave frequencies for metals.

The equation for the conduction current is, using the skin depth δ in place of the radical,

$$J_x = \sigma E_x = \sigma E_{0x} e^{-z/\delta} e^{-jz/\delta} = J_{0x} e^{-z/\delta} e^{-jz/\delta} \tag{8-48}$$

and

$$\gamma = \alpha + j\beta = \frac{1}{\delta} + j\frac{1}{\delta} \tag{8-49}$$

Figure 8-10 is a plot of the current density as a function of z. Note that the phase of the current also reverses at $z = \pi\delta$. Of course, the phase does not make an abrupt change, but the intermediate phases are difficult to show on a single plot.

For the wave impedance, we begin with the basic Eq. 8-35 and impose the condition $\epsilon'' = \sigma/\omega \gg \epsilon'$:

$$\begin{aligned} \bar{\eta} &\approx \sqrt{\frac{\mu}{\epsilon'}} \sqrt{\frac{1}{-j\dfrac{\sigma}{\omega\epsilon'}}} = \sqrt{\frac{\omega\mu}{\sigma}}\sqrt{j} = \sqrt{\frac{\omega\mu}{\sigma}}\left(\frac{1}{\sqrt{2}} + j\frac{1}{\sqrt{2}}\right) \\ &\approx \sqrt{\frac{\omega\mu}{2\sigma}}(1 + j1) = \sqrt{\frac{\pi f\mu}{\sigma}}(1 + j1) = \sqrt{\frac{\pi f\mu\sigma}{\sigma^2}}(1 + j1) = \frac{1}{\delta\sigma}(1 + j1) \end{aligned} \tag{8-50}$$

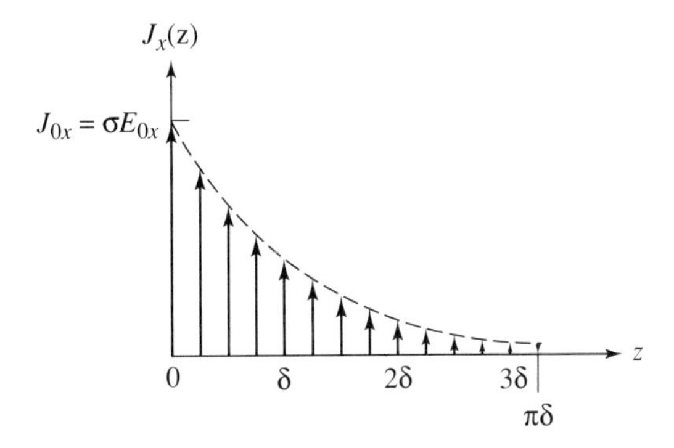

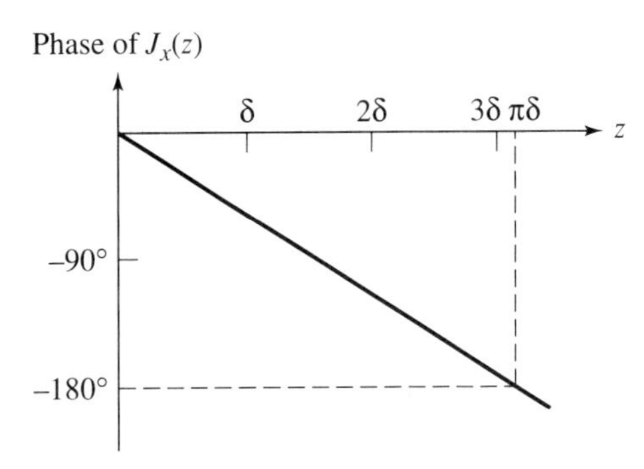

Figure 8-10. Amplitude and phase of current density of a propagating wave in a good conductor.

Since the dimension of the coefficient is ohms we call it *surface resistance* for conductors and set

$$\text{surface resistance} = R_s \equiv \frac{1}{\delta\sigma} = \sqrt{\frac{\pi f \mu}{\sigma}} \ \Omega/\text{square} \tag{8-51}$$

The reason for this name and its units will be explained shortly. Values of R_s for various materials are also given in Table 8-2. The wave impedance may now be written as

$$\overline{\eta} = R_s + jR_s = R_s + jX_s \tag{8-52}$$

from which a conductor *internal* inductance (Section 6-14) can be defined as

$$L_{s_i} = \frac{X_s}{\omega} = \frac{1}{\omega\sigma\delta} = \frac{1}{2}\sqrt{\frac{\mu}{\pi f \sigma}} \ \text{H/square} \tag{8-53}$$

The dimension per square implies that the resistance and inductance of any size square normal to the direction of propagation will have those values. That this is so can be seen from Figure 8-11. Suppose we take a square of material that has resistance R across opposite faces as shown in Figure 8-11*a*. Placing two of these in series will double the resistance as indicated in Figure 8-11*b*. If we then parallel two of the (*b*) units as shown in Figure 8-11*c* the total resistance is again R even though the square is twice the edge

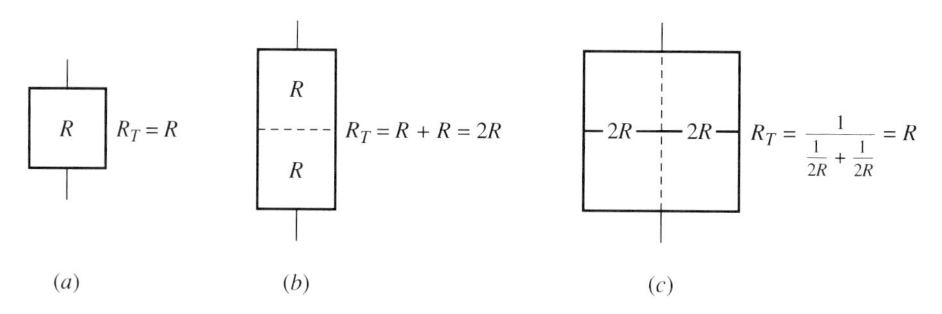

Figure 8-11. Illustration of the concept of ohms per square.

dimension. So, if you are asked "ohms per square what?" simply respond with "square anything!"

We can derive the surface resistance parameter by considering the propagating current directly as shown in Figure 8-12. In the figure we have assumed that the xy plane is the surface of a good conductor and for negative z there is another medium, say air, although that is not critical here. We suppose that the value of the current density at the boundary just inside the metal, $z = 0^+$, is J_{0x}, which we assume to be positive and real for convenience. This current density is constant in a given plane since it is produced by a uniform plane wave. Let us find the total current passing through the surface defined by length 0 to b on the y axis and extending to $z = \infty$. The side x from 0 to a will have no effect since the current lines are parallel to planes that are parallel to the xy plane. The current density will decrease as z increases as shown by a few current density vectors along z. The current is then

$$\bar{I} = \bar{I}_x = \int_0^b \int_0^\infty J_{0x} e^{-\gamma z} dz dy = J_{0x} \int_0^b dy \int_0^\infty e^{-\gamma z} dz = bJ_{0x}\left(-\frac{1}{\gamma} e^{-\gamma z}\right)\Bigg|_0^\infty$$

At the upper limit $e^{-\gamma z}$ is zero, so

$$\bar{I} = \frac{b}{\gamma} J_{0x} \tag{8-54}$$

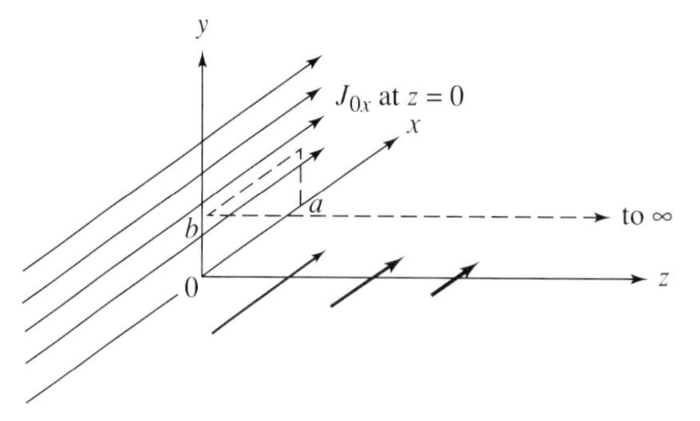

Figure 8-12. Configuration for determining the total current flow through a cross section normal to the current density, $0 - b - \infty$.

The electric field intensity at $z = 0$ is

$$E_{0x} = \frac{J_{0x}}{\sigma} \tag{8-55}$$

and since static field laws hold for fields in the transverse plane for TEM modes the voltage drop from 0 to a will be

$$\overline{V}_a = \alpha E_{0x} = \frac{a}{\sigma} J_{0x}$$

The impedance at the surface ($z = 0$) is then, using Eq. 8-49 for γ,

$$\overline{Z} = \frac{\overline{V}_a}{\overline{I}} = \frac{\dfrac{a}{\sigma} J_{0x}}{\dfrac{b}{\gamma} J_{0x}} = \frac{a}{b}\frac{\gamma}{\sigma} = \frac{a}{b}\frac{1}{\sigma}\left(\frac{1}{\delta} + j\frac{1}{\delta}\right)$$

$$= \frac{a}{b}\left(\frac{1}{\sigma\delta} + j\frac{1}{\sigma\delta}\right) = \frac{a}{b}(R_s + jR_s) = \frac{a}{b}(R_s + jX_s)$$

Now if we let $a = b$ (note it doesn't matter what length unit they are both measured in) we have a square on the surface. This is the surface impedance

$$\overline{Z}_s = \frac{1}{\sigma\delta} + j\frac{1}{\sigma\delta} = R_s + jX_s \ \Omega/\text{square} \tag{8-56}$$

Thus, the wave impedance is the same as the surface impedance (Eq. 8-52).

The Poynting vector provides an additional insight into the power loss in the conductor. Using Eq. 7-56,

$$\vec{S} = \frac{1}{2}\vec{E} \times \vec{H}^* = \frac{1}{2}E_{0x}e^{-z/\delta}e^{-j(z/\delta)}\hat{x} \times \left(\frac{E_{0x}}{\overline{\eta}}e^{-z/\delta}e^{-j(z/\delta)}\hat{y}\right)^*$$

$$= \frac{1}{2}\frac{|E_{0x}|^2}{\overline{\eta}^*}e^{-2z/\delta}\hat{z} \ \text{W/m}^2$$

Substituting for $\overline{\eta}^*$ using Eq. 8-52, and rationalizing the complex denominator

$$\vec{S} = 1/2\,\frac{|E_{0x}|^2}{\dfrac{1}{\sigma\delta}(1 - j1)}e^{-2(z/\delta)}\hat{z} = \frac{\delta\sigma}{4}|E_{0x}|^2(1 + j1)e^{-2(z/\delta)}\hat{z}$$

Substituting for E_{0x}:

$$\vec{S} = \frac{\delta}{4\sigma}|J_{0x}|^2(1 + j1)e^{-2(z/\delta)}\hat{z} \tag{8-57}$$

The sinusoidal power dissipated is the time average, which is the real part, or

$$P_{\text{diss}} = \frac{\delta |J_{0x}|^2}{4\sigma} e^{-2z/\delta} \text{ W/m}^2 \tag{8-58}$$

The dissipated power entering the surface at $z = 0$ is thus

$$P_{\text{diss},s} = \frac{\delta |J_{0x}|^2}{4\sigma} \text{ W/m}^2 \tag{8-59}$$

Now suppose we consider a resistor consisting of 1 m^2 of surface and of depth δ in the z direction as shown in Figure 8-13. Further, suppose that the entire current I given by Eq. 8-54 is uniformly distributed over the 0 to δ dimension and zero outside. The resistance (DC for uniform current distribution) is

$$\begin{aligned} R_{\text{DC}} &= \frac{\text{length}}{\sigma \cdot \text{area through which current flows}} \\ &= \frac{1}{\sigma \times 1 \times \delta} = \frac{1}{\sigma\delta} \end{aligned}$$

The power dissipated is then, for $b = 1$ m in Eq. 8-54,

$$\begin{aligned} P_{\text{diss}} &= \frac{1}{2} |\bar{I}|^2 R_{\text{DC}} = \frac{1}{2} \frac{|J_{0x}|^2}{|\gamma|^2} \frac{1}{\sigma\delta} = \frac{1}{2} \frac{|J_{0x}|^2}{\sigma\delta \left| \frac{1}{\delta} + j\frac{1}{\delta} \right|^2} \\ &= \frac{1}{4} \frac{\delta}{\sigma} |J_{0x}|^2 \text{ W/m}^2 \end{aligned}$$

This is identical to Eq. 8-59, so we conclude that as far as power computations are concerned we may consider all the current flowing in the x direction to be concentrated *uniformly* within one skin depth and zero outside.

One final concept is also useful from an engineering point of view. Suppose we make

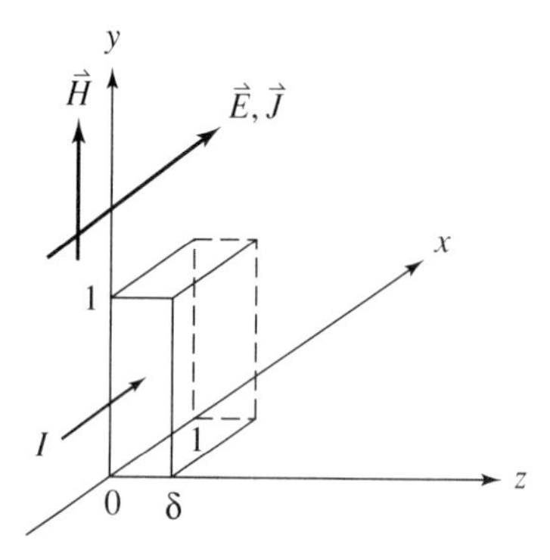

Figure 8-13. Configuration for computing conductor power dissipation per meter square of surface.

the assumption that the conductor is perfect. The boundary condition that applies yields an equivalent surface current $\vec{J}_s$ (not $\vec{J}$!) given by

$$\vec{J}_s = \hat{n} \times \vec{H}$$

If we use the $\vec{H}$ equation at $z = 0$ this gives, considering the metal to be Medium 1 and $z < 0$ to be Medium 2,

$$\vec{J}_s = -\hat{z} \times H_{0y}\hat{y} = H_{0y}\hat{x} \text{ A/m}$$

Now using H_{0y} in the metal (approximately so)

$$\vec{J}_s \approx \frac{E_{0x}}{\eta}\hat{x} = \frac{E_{0x}}{\left(\frac{1}{\sigma\delta} + j\frac{1}{\sigma\delta}\right)}\hat{x} = \frac{\sigma\delta E_{0x}(1 - j1)}{2}\hat{x} \text{ A/m}$$

The current crossing 1 m ($0 \leq y \leq 1$ m) would be, in magnitude,

$$|\bar{I}| = \frac{\sigma\delta}{2}|E_{0x}||1 - j1| = \frac{\sigma\delta}{\sqrt{2}}|E_{0x}| = \frac{\delta}{\sqrt{2}}|J_{0x}| \tag{8-60}$$

The power loss would then be

$$\begin{aligned} P_{\text{diss}} &= \frac{1}{2}|\bar{I}|^2 R = \frac{1}{2}\frac{\sigma^2\delta^2}{2}|E_{0x}|^2\frac{1}{\sigma\delta} = \frac{1}{4\sigma\delta}|E_{0x}|^2 \\ &= \frac{1}{4}\sigma\delta\frac{|J_{0x}|^2}{\sigma^2} = \frac{1}{4}\frac{\delta}{\sigma}|J_{0x}|^2 \text{ W} \end{aligned} \tag{8-61}$$

as before. Thus we can consider good conductors to be perfect conductors as far as currents corresponding to fields are concerned for power computations. This approximation will be found very helpful in the next chapter when we determine attenuation in more complex configurations.

Example 8-7

To quantify the preceding discussions, let's consider copper at frequencies of 1 MHz and 1 GHz. From Table 8-2 we obtain

$$\delta_{\text{Cu}}(1 \text{ MHz}) = \frac{0.066}{\sqrt{10^6}} = 66\ \mu\text{m}$$

$$\delta_{\text{Cu}}(1 \text{ GHz}) = 2.1\ \mu\text{m}$$

Thus it doesn't take much metal to shield from waves. However, at 60 Hz

$$\Delta_{\text{Cu}}(60 \text{ Hz}) = 8.5 \text{ mm}$$

To reduce penetration at these low frequencies we could use iron, whose permeability decreases the skin depth to

$$\delta_{\text{Fe}}(60 \text{ Hz}) = 65\ \mu\text{m}$$

Also from Table 8-2

$$R_s(1 \text{ MHz}) = 2.61 \times 10^{-7}\sqrt{10^6} = 2.61 \times 10^{-4}\ \Omega/\text{square}$$

$$R_s(1 \text{ GHz}) = 8.25 \times 10^{-3}\ \Omega/\text{square}$$

Note the wave sees essentially a short circuit.

Example 8-8

Although we have assumed that the metal surface at $z = 0$ in Figure 8-12 is a plane (the xy plane), since the skin depth is very small we may apply the results even if the surface has curvature. The requirement for making this engineering approximation is that $\delta \ll$ radius of curvature; or, more specifically $\delta < 0.1$ radius of curvature. In Example 8-7 we found that $\delta_{\text{Cu}}(1 \text{ MHz}) = 66\ \mu\text{m}$, so any surface whose curvature is greater than 660 μm, 0.66 mm, is appropriate for plane wave approximation. By way of comparison, number 12 AWG copper wire used in home electrical wiring has a radius of 1.025 mm.

Suppose we have AWG #22 with a radius of 0.322 mm, and that it carries 1 mA of current at 300 MHz (#22 wire is about the center conductor size on the cable TV 75-Ω coax). Determine the DC and AC resistances of the conductor, the cross-sectional current distribution, and the power loss per meter. From Table 8-2,

$$\delta_{\text{Cu}}(300 \text{ MHz}) = 3.8\ \mu\text{m} \qquad \text{or} \qquad 0.0038 \text{ mm} \ll 0.322 \text{ mm}$$

So the plane wave approximation is valid. The field configuration is shown in Figure 8-14. For the current flow as shown, the right-hand rule gives the magnetic field intensity and since the conductor is lossy there is an electric field intensity as shown. Note that the Poynting vector is directed *into* the conductor, so we can imagine the $\vec{E}$ and $\vec{H}$ to be a plane wave at the surface, to the approximations we are using. Note that the z coordinate for the plane wave is the radial or ρ coordinate here. The x coordinate for the plane wave is the z coordinate here. The DC resistance of the conductor uses the usual formula from physics (see Eq. 5-35). Thus for 1 m,

$$R_{\text{DC}} = \frac{l}{\sigma A} = \frac{1}{5.8 \times 10^7 \times \pi \times (0.322 \times 10^{-3})^2} = 0.053\ \Omega/\text{m}$$

The AC resistance is obtained using the concept developed in this section which was that the resistance can be computed from the real part of the wave impedance, Eq. 8-56. Using R_s we then have, using the surface of the conductor and 1 m length, and radius a,

$$R_{\text{AC}} = R_s \times \frac{1 \text{ m}}{2\pi a} = 2.61 \times 10^{-7}\sqrt{300 \times 10^6} \times \frac{1}{2\pi \times 0.322 \times 10^{-3}}$$
$$= 2.23\ \Omega/\text{m}$$

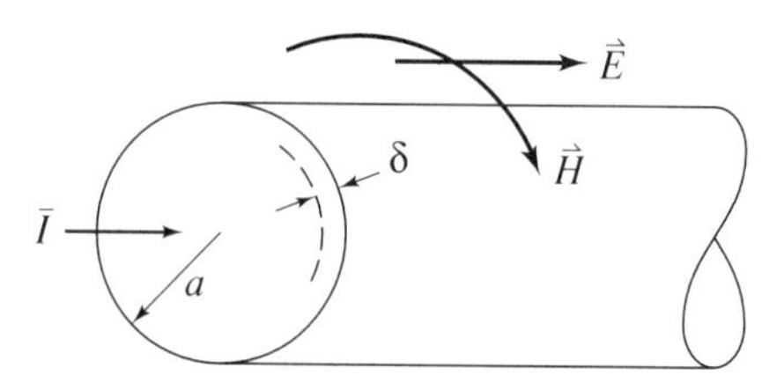

Figure 8-14. Configuration for a solid conductor carrying current $\bar{I}$ for Example 8-8.

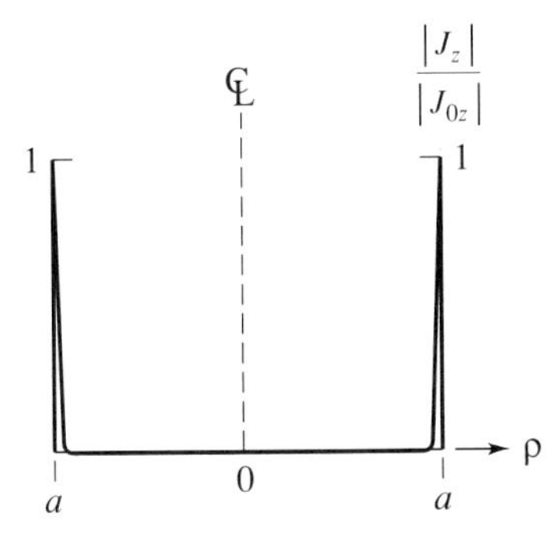

Figure 8-15. Plot of current density as a function of radial coordinate for a solid conductor.

This value is 42 times the DC value, so the skin effect is very significant and would be the resistance parameter for the inner conductor of a coaxial transmission line or of one side of a twin lead line (see Figures 1-1 and 1-2).

The assumption of a plane wave allows us to use Eq. 8-48 to obtain the current density in the cross section, using the z coordinate here as x in the plane wave and ρ coordinate here as the z coordinate in the plane wave as previously explained. Thus, with these coordinate identifications,

$$J_z(\rho) = J_{0z} e^{-\frac{\rho}{3.8\times10^{-6}}} e^{-j\frac{\rho}{3.8\times10^{-6}}} \text{ A/m}^2$$

Taking the magnitude and normalizing by dividing our $|J_{0z}|$ we obtain the plot in Figure 8-15. From this figure it is evident that the current crowds to the outside, or skin, of the conductor with very little penetration. Thus all the current flows in a very narrow annular ring around the periphery resulting in a high resistance.

For the DC resistance case, 1 mA of rf current would have a predicted average power of

$$P_{\text{DC}} = \tfrac{1}{2}(10^{-3})^2(0.053) = 0.027 \ \mu\text{W/m}$$

We compute an AC resistance power for 1 mA flowing in a DC equivalent of one skin depth (see discussion following Eq. 8-59) and approximate the annular ring by a rectangle having dimensions δ by $2\pi a$:

$$\begin{aligned} P_{\text{AC}} &= \tfrac{1}{2}(10^{-3})^2\left(\frac{1}{\sigma 2\pi a\delta}\right) \\ &= 5 \times 10^{-6} \times \frac{1}{5.8 \times 10^7 \times 2\pi \times 0.322 \times 10^{-3} \times 3.8 \times 10^{-6}} \\ &= 1.12 \ \mu\text{W/m} \end{aligned}$$

Or, using the R_{AC} computed earlier in this example:

$$P_{\text{AC}} = \tfrac{1}{2} \times (1 \times 10^{-3})^2 \times 2.23 = 1.12 \ \mu\text{W/m}$$

No wonder cables heat up at higher frequencies, since the power (as expected) as increased 42 times for the same current!

Example 8-9

For the system of Example 8-8 approximate the Poynting vector in the fields at the surface and compute the average power per meter flowing into the wire.

Since all of the current is very near the surface we make the approximation that all the current is on the surface only so that we can compute surface current $\vec{J}_s$. Since the current in the $\hat{z}$ direction flows on the cylindrical contour,

$$\vec{J}_s = J_{sz}\hat{z} = \frac{I_z}{2\pi a}\hat{z} = \frac{10^{-3}}{2\pi \times 0.322 \times 10^{-3}}\hat{z} = 0.494\hat{z} \text{ A/m}$$

From the boundary condition for a perfect (assumed) conductor,

$$\vec{J}_s = \hat{n} \times \vec{H} = \hat{\rho} \times \vec{H} = 0.494\hat{z} \Rightarrow \vec{H} = 0.494\hat{\phi} \text{ A/m}$$

At the surface we may use the $\vec{J}$ obtained by assuming all the current flows uniformly distributed to depth δ, and approximate the annular ring through which the current flows by a rectangle δ by $2\pi a$ as before, so that

$$\vec{J}(\rho = a) = \frac{I_z}{\delta 2\pi a}\hat{z}$$

$$= \frac{10^{-3}}{3.8 \times 10^{-6} \times 2\pi \times 0.322 \times 10^{-3}}\hat{z} = 1.3 \times 10^5\hat{z} \text{ A/m}^2$$

At the surface

$$\vec{E}(\rho = a) = \frac{\vec{J}(\rho = a)}{\sigma} = \frac{1.3 \times 10^5}{5.8 \times 10^7}\hat{z} = 0.00224\hat{z} \text{ V/m}$$

From these,

$$\vec{S} = \tfrac{1}{2}\vec{E} \times \vec{H}^* = \tfrac{1}{2} \times 0.00224\hat{z} \times 494\hat{\phi} = 0.554(-\hat{\rho}) \text{ mW/m}^2$$

The power flowing into the cylindrical surface for a 1-m length is then

$$\bar{S} = 0.554 \times 10^{-3}(-\hat{\rho}) \cdot (1 \times 2\pi \times 0.322 \times 10^{-3})(-\hat{\rho}) = 1.12 \ \mu\text{W/m}$$

This is the same as the power computed on the basis of resistive power loss, so the Poynting vector into the wire is dissipated as heat.

We conclude this section by examining the phase velocity, group velocity, and wavelength of uniform plane waves in good conductors. Using previously given definitions and derived results the phase velocity is

$$V_p = \frac{\omega}{\beta} = \frac{\omega}{\frac{1}{\delta}} = \delta\omega = 2\sqrt{\frac{\pi f}{\mu\sigma}} \text{ m/s} \tag{8-62}$$

In aluminum at a frequency of 100 MHz the phase velocity is 2.56 km/s, while at 100 Hz it is only 2.56 m/s. For a point of reference, a four-minute-mile run requires 6.7 m/s. These velocities are quite different from the velocity in vacuum.

The group velocity is

$$V_g = \frac{d\omega}{d\beta} = \frac{1}{\frac{d\beta}{d\omega}} = \frac{1}{\frac{d(1/\delta)}{d\omega}} = \frac{1}{\frac{d(\sqrt{\pi f \mu \sigma})}{d\omega}} = 4\sqrt{\frac{\pi f}{\mu\sigma}} \text{ m/s} \tag{8-63}$$

which is always twice the phase velocity. For the wavelength we have

$$\lambda = \frac{2\pi}{\beta} = 2\pi\delta = 2\sqrt{\frac{\pi}{f\mu\sigma}} \text{ m} \tag{8-64}$$

Since the skin depth is very small, the wavelengths in good conductors are very short.

Specific applications of waves in conductors will be made when we study guided wave propagation in the next chapter.

Exercises

8-4.0a Determine the required thicknesses of aluminum sheets that will produce 100 dB attenuation of plane waves at 3 MHz, 30 MHz, and 300 MHz.

8-4.0b Perform a unit analysis on the expressions for δ and V_p, Eqs. 8-47 and 8-62, to show that the respective units are m and m/s.

8-4.0c Repeat the computations of Exercise 8-8 for a frequency of 1000 Hz and where the wire is made of iron (see Table 8-2).

8-4.0d Develop the time functions for $\vec{\mathcal{J}}$, $\vec{\mathcal{E}}$, and $\vec{\mathcal{H}}$ for the fields of Example 8-8.

8-4.0e Gold wires are often used to connect semiconductor surfaces to the chip pins. Suppose one uses a 2-mil diameter (0.001 in.) wire of length 2 mm, and that the device is to operate at 950 MHz. What is the impedance of the wire?

8-4.0f Prove the criterion stated for a good conductor by considering the statement: In a good conductor the conduction current is much greater than displacement current.

8-5 Propagation of Uniform Plane Waves in Good Dielectrics

Many materials used in engineering practice have conductivities that are small enough so the loss tangent is very small at frequencies of interest (i.e., $\sigma \approx 0.1\omega\epsilon'$). These are called *good dielectrics*. It is possible to use approximations on the general expressions for α and β given in Eqs. 8-30 and 8-31 so that computations are not so cumbersome and so that some physical insights into the properties of low-loss materials can be developed.

To determine the propagation constant, the attenuation constant, and the phase constant we begin with the general results given by Eqs. 8-30 and 8-31. For the attenuation constant the general equation is

$$\alpha = \omega\sqrt{\mu\epsilon'}\sqrt{\frac{1}{2}\left[-1 + \sqrt{1 + \left(\frac{\epsilon''}{\epsilon'}\right)^2}\right]}$$

where we have assumed all losses are due to the dielectric so μ is real. Now since the loss tangent ϵ''/ϵ' is much less than one we cannot simply set it to zero since we would lose any effect of ϵ''. What we do is use the first terms of the binomial expansion for any power p:

$$(1 + x)^p = 1 + px + \frac{p(p-1)}{2!}x^2 + \frac{p(p-1)(p-2)}{3!}x^3 + \cdots, \qquad |x| < 1$$

For x very much less than one we can use the first two terms and apply the expansion to the inner square root, where $p \Rightarrow \frac{1}{2}$ and $x \Rightarrow (\epsilon''/\epsilon')^2$, to obtain

$$\alpha \approx \omega\sqrt{\mu\epsilon'}\sqrt{\frac{1}{2}\left[-1 + \left(1 + \frac{1}{2}\left(\frac{\epsilon''}{\epsilon'}\right)^2\right)\right]} = \omega\sqrt{\mu\epsilon'}\sqrt{\frac{1}{4}\left(\frac{\epsilon''}{\epsilon'}\right)^2}$$

$$\approx \frac{1}{2}\,\omega\sqrt{\mu\epsilon'}\left(\frac{\epsilon''}{\epsilon'}\right) \qquad \text{good dielectric}$$

For dielectric conductivity σ, $\epsilon'' = \sigma/\omega$ so that an alternate form is

$$\alpha \approx \frac{1}{2}\,\omega\sqrt{\frac{\mu}{\epsilon'}\frac{\sigma}{\omega}} = \frac{1}{2}\,\sigma\sqrt{\frac{\mu}{\epsilon'}}$$

Thus

$$\alpha \approx \frac{1}{2}\,\omega\sqrt{\mu\epsilon'}\left(\frac{\epsilon''}{\epsilon'}\right) = \frac{1}{2}\,\sigma\sqrt{\frac{\mu}{\epsilon'}} \text{ nep/m, good dielectric} \tag{8-65}$$

This shows that the attenuation is very small for good dielectrics that have small conductivity. Although frequency does not occur explicitly here, don't forget that the material parameters are frequency dependent (see Table 8-1 for ϵ_r).

For the propagation (phase) constant we need to use the binomial approximation twice on Eq. 8-30 to get

$$\beta = \omega\sqrt{\mu\epsilon'}\sqrt{\frac{1}{2}\left[1 + \sqrt{1 + \left(\frac{\epsilon''}{\epsilon'}\right)^2}\right]}$$

$$\approx \omega\sqrt{\mu\epsilon'}\sqrt{\frac{1}{2}\left[1 + \left(1 + \frac{1}{2}\left(\frac{\epsilon''}{\epsilon'}\right)^2\right)\right]}$$

$$\approx \omega\sqrt{\mu\epsilon'}\sqrt{\frac{1}{2}\left(2 + \frac{1}{2}\left(\frac{\epsilon''}{\epsilon'}\right)^2\right)} = \omega\sqrt{\mu\epsilon'}\sqrt{1 + \frac{1}{4}\left(\frac{\epsilon''}{\epsilon'}\right)^2}$$

$$\approx \omega\sqrt{\mu\epsilon'}\left[1 + \left(\frac{1}{2}\right)\left(\frac{1}{4}\right)\left(\frac{\epsilon''}{\epsilon'}\right)^2\right] = k\left[1 + \frac{1}{8}\left(\frac{\epsilon''}{\epsilon'}\right)^2\right] \text{ rad/m} \tag{8-66}$$

where k is the free-medium or lossless-medium propagation constant defined earlier (note we are using $\epsilon' = \epsilon$). This result shows that β is essentially unaffected by losses, since for loss tangents on the order of 0.1 we would have

$$\frac{1}{8}\left(\frac{\epsilon''}{\epsilon'}\right)^2 \approx 0.00125$$

Thus for only 0.1% error for loss tangents as large as 0.1 we may use

$$\beta \approx \omega\sqrt{\mu\epsilon'} = k \text{ rad/m, good dielectric} \tag{8-67}$$

which tells us that the lossless-medium phase constant is a very good approximation. In fact, Table 8-1 shows that the loss tangents of most common dielectrics are on the order of 10^{-4}.

For the wave impedance we begin with the general Eq. 8-35

$$\overline{\eta} = \sqrt{\frac{\mu}{\epsilon'}}\sqrt{\frac{1}{1 - j\dfrac{\epsilon''}{\epsilon'}}} = \sqrt{\frac{\mu}{\epsilon'}}\left[1 - j\frac{\epsilon''}{\epsilon'}\right]^{-1/2}$$

In the binomial expansion we identify $p \Rightarrow -1/2$ and $x \Rightarrow j\epsilon''/\epsilon'$. Now since the x is complex we will need to take an additional term in the series so we can obtain approximations to the real and imaginary parts. Thus

$$\begin{aligned}
\overline{\eta} &\approx \sqrt{\frac{\mu}{\epsilon'}}\left[1 + \left(-\frac{1}{2}\right)\left(-j\frac{\epsilon''}{\epsilon'}\right) + \frac{(-\frac{1}{2})(-\frac{3}{2})}{2}\left(-j\frac{\epsilon''}{\epsilon'}\right)^2\right] \\
&= \sqrt{\frac{\mu}{\epsilon'}}\left[1 - \frac{3}{8}\left(\frac{\epsilon''}{\epsilon'}\right)^2 + j\frac{1}{2}\left(\frac{\epsilon''}{\epsilon'}\right)\right] \\
&\approx \sqrt{\frac{\mu}{\epsilon'}}\left[1 - \frac{3}{8}\left(\frac{\epsilon''}{\epsilon'}\right)^2\right] + j\frac{1}{2}\sqrt{\frac{\mu}{\epsilon'}}\left(\frac{\epsilon''}{\epsilon'}\right)
\end{aligned}$$

Again, for usual dielectrics we may neglect the square term and use

$$\overline{\eta} \approx \sqrt{\frac{\mu}{\epsilon'}} + j\frac{1}{2}\sqrt{\frac{\mu}{\epsilon'}}\left(\frac{\epsilon''}{\epsilon'}\right) \quad \text{good dielectric} \tag{8-68}$$

Thus all the small loss does is add a small reactive component to the lossless-medium wave impedance.

The phase velocity, group velocity, and wavelength are easily obtained from the foregoing:

$$\mathrm{V}_p = \frac{\omega}{\beta} \approx \frac{\omega}{k} = \frac{\omega}{\omega\sqrt{\mu\epsilon'}} = \frac{1}{\sqrt{\mu\epsilon'}} \tag{8-69}$$

$$V_g = \frac{d\omega}{d\beta} = \frac{1}{\dfrac{d\beta}{d\omega}} \approx \frac{1}{\dfrac{dk}{d\omega}} = \frac{1}{\sqrt{\mu\epsilon'}} \tag{8-70}$$

$$\lambda = \frac{2\pi}{\beta} \approx \frac{2\pi}{k} = \frac{2\pi}{\omega\sqrt{\mu\epsilon'}} = \frac{1}{f\sqrt{\mu\epsilon'}} = \frac{V_p}{f} \tag{8-71}$$

We see that small losses result in essentially no dispersion and hence there is no significant waveform distortion. The wavelength is essentially that in a lossless medium. The frequency dependences of the relative permittivity ϵ' and the loss tangent for common dielectrics are given in Table 8-1.

In engineering practice whenever "much greater than" or "much less than" is encountered, the usual rule is to use a factor of 10 difference between the quantities involved. It turns out that our approximations are good for larger loss tangents than 0.1, as the next example shows, and still have very small errors. This example also shows how to determine specific limits on "smallness" and "largeness" criteria.

Example 8-10

Determine the maximum loss tangent permissible so that the error in using the approximation for α is 1%.

Using the approximate form for α which has the loss tangent explicitly shown, just before Eq. 8-65, and the exact expression, Eq. 8-31.

$$\%\ \text{error} = \frac{\left|\dfrac{1}{2}\,\omega\sqrt{\mu\epsilon'}\left(\dfrac{\epsilon''}{\epsilon'}\right) - \omega\sqrt{\mu\epsilon'}\sqrt{\dfrac{1}{2}\left[-1 + \sqrt{1 + \left(\dfrac{\epsilon''}{\epsilon'}\right)^2}\right]}\right|}{\omega\sqrt{\mu\epsilon'}\sqrt{\dfrac{1}{2}\left[-1 + \sqrt{1 + \left(\dfrac{\epsilon''}{\epsilon'}\right)^2}\right]}} \times 100\% = 1\%$$

$$\left|\frac{1}{2}\left(\frac{\epsilon''}{\epsilon'}\right) - \sqrt{\frac{1}{2}\left[-1 + \sqrt{1 + \left(\frac{\epsilon''}{\epsilon'}\right)^2}\right]}\right| = 0.01\sqrt{\frac{1}{2}\left[-1 + \sqrt{1 + \left(\frac{\epsilon''}{\epsilon'}\right)^2}\right]}$$

To remove the absolute value signs we need to know which is the larger of the two terms to determine if the difference is positive or negative. The quickest way is to put in some loss tangent value and check. For a loss tangent 0.1, the radical term is 0.499, which is less than 0.5 for the other term. Thus the difference is positive as it is, so we remove the absolute value bars and collect terms to obtain

$$\left(\frac{\epsilon''}{\epsilon'}\right) = 2.02\sqrt{\frac{1}{2}\left[-1 + \sqrt{1 + \left(\frac{\epsilon''}{\epsilon'}\right)^2}\right]}$$

Squaring, collecting terms, squaring again, and solving gives

$$\text{loss tangent} = \frac{\epsilon''}{\epsilon'} = 0.28$$

Example 8-11

Determine the effective conductivity of polyethylene at 100 MHz and also the wave impedance at that frequency. By definition,

$$\text{loss tangent} = \tan \delta = \frac{\epsilon''}{\epsilon'} = \frac{\sigma}{\omega\epsilon'}$$

Solving for conductivity,

$$\sigma = \omega\epsilon' \tan \delta$$

Using values for polyethylene from Table 8-1

$$\sigma = 2\pi \times 10^8 \times \left(2.26 \times \frac{1}{36\pi} \times 10^{-9}\right) \times 2 \times 10^{-4} = 2.51 \times 10^{-6} \text{ S/m}$$

The lossless medium wave impedance is

$$\eta = \sqrt{\frac{4\pi \times 10^{-7}}{2.26 \times \dfrac{1}{36\pi} \times 10^{-9}}} = 251 \ \Omega$$

Using the approximate form for $\overline{\eta}$, Eq. 8-68

$$\overline{\eta} = 251 + j\tfrac{1}{2} \times 251 \times 2 \times 10^{-4} = 251 + j0.0251 = 251\angle 0.006^\circ \ \Omega$$

Example 8-12

Compute the propagation constant for polyethylene at 100 MHz.

Using the approximate forms for α and β:

$$\beta = \omega\sqrt{\mu\epsilon'} = 2\pi \times 10^8 \sqrt{4\pi \times 10^{-7} \times 2.26 \times \frac{1}{36\pi} \times 10^{-9}}$$

$$= 3.15 \text{ rad/m}$$

$$\alpha = \omega\sqrt{\mu\epsilon'}\left(\frac{\epsilon''}{\epsilon'}\right) = 3.15 \times 2 \times 10^{-4} = 6.3 \times 10^{-4} \text{ nep/m}$$

Thus

$$\gamma = 6.3 \times 10^{-4} + j3.15 \text{ nep/m}$$

More specific applications of propagation in dielectrics having losses are given in our study of guided wave propagation in the next chapter.

To this point we have considered only three classes of materials, but those are usually the most important or common cases of engineering interest. A complete matrix of material

Table 8-3. Material Classification and Parameter Equations or Values

	Attenuation Constant	Wave Impedance	Skin Depth	Phase Velocity
	$\omega\sqrt{\mu\epsilon}\sqrt{\frac{1}{2}\left[1+\left(\frac{\sigma}{\omega\epsilon}\right)-1\right]}$, α	$\sqrt{\frac{\mu}{\epsilon}}$, $\overline{\eta}$	$\frac{1}{\alpha}$ δ	$\frac{\omega}{\beta}$,V_p
Perfect dielectric $\sigma = 0$	0	$\sqrt{\frac{\mu}{\epsilon}} = \eta$	∞	$\frac{1}{\sqrt{\mu\epsilon}}$
Good dielectric $\sigma \lll \omega\epsilon$	$\frac{1}{2}\sigma\sqrt{\frac{\mu}{\epsilon'}}$	$\sqrt{\frac{\mu}{\epsilon'}} + j\frac{1}{2}\sqrt{\frac{\mu}{\epsilon'}}\left(\frac{\epsilon}{\epsilon'}\right)$		$\frac{1}{\sqrt{\mu\epsilon}}$
Lossy dielectric $\sigma \approx \omega\epsilon$				
Semiconductor $\sigma \sim$ $10^{-5} - 10^{-6}$ S/m				
Good conductor $\sigma \gg \omega\epsilon$ $\sigma \sim 10^{6} - 10^{8}$ S/m				
Perfect conductor $\sigma = \infty$ $\vec{\mathcal{B}} = 0$, $\vec{B}_{DC} \neq 0$	∞	0	0	
Superconductor $\sigma = \infty$ $\vec{\mathcal{B}} = 0$, $\vec{B}_{DC} = 0$				

classification and important parameters is given in Tables 8-3 and 8-4 along with the parameter equations developed so far. A good exercise would be to complete the table with appropriate expressions or values.

Exercises

8-5.0a Use the binomial series approximation directly on γ, Eq. 8-27, to obtain the approximations for α and β.

8-5.0b Compute exact values of $\overline{\eta}$ and γ for polyethylene at 100 MHz and compare with the results of Examples 8-11 and 8-12.

8-5.0c Compute γ, $\overline{\eta}$, σ, and V_p for a 3-GHz uniform plane wave propagating in aluminum oxide.

8-5.0d A dielectric has $\mu_r = 1$, $\epsilon_r = 3$, and $\sigma = 6 \times 10^{-5}$ s/m when the frequency is 30 MHz. Compute the loss tangent, propagation constant, and wave impedance.

8-5.0e The criterion that defines a good dielectric was given in Section 8-4. Prove that criterion by using the following statement: In a good dielectric the displacement current is much greater than the conduction current.

Table 8-4. Material Classification and Parameter Equations or Values

	Phase or Propagation Constant	Group Velocity	Loss Tangent	Wavelength
	$\omega\sqrt{\mu\epsilon}\sqrt{\frac{1}{2}\left[\sqrt{1-\left(\frac{\sigma}{\omega\epsilon}\right)^2}+1\right]}$, β	$\frac{d\omega}{d\beta}=\frac{1}{\frac{d\beta}{d\omega}}$ V_g	$\frac{\epsilon''}{\epsilon'}$ (tan δ)	$\frac{2\pi}{\beta}$, λ
Perfect dielectric $\sigma = 0$	$\omega\sqrt{\mu\epsilon} = k$	$\frac{1}{\sqrt{\mu\epsilon}}$	0	$\frac{1}{f\sqrt{\mu\epsilon}}$
Good dielectric $\sigma \ll \omega\epsilon$	$\omega\sqrt{\mu\epsilon}$	$\frac{1}{\sqrt{\mu\epsilon}}$		$\frac{1}{f\sqrt{\mu\epsilon'}}$
Lossy dielectric $\sigma \approx \omega\epsilon$				
Semiconductor $\sigma \sim$ $10^{-5} - 10^{-6}$ S/m				
Good conductor $\sigma \gg \omega\epsilon$ $\sigma \sim 10^6 - 10^8$ S/m				
Perfect conductor $\sigma = \infty$ $\vec{\mathcal{B}} = 0$, $\vec{B}_{DC} \neq 0$			∞	
Superconductor $\sigma = \infty$ $\vec{\mathcal{B}} = 0$, $\vec{B}_{DC} = 0$				

8-6 Reflection of Uniform Plane Waves Propagating Normal to a Boundary Between Media

To use radio waves to detect objects or measure material properties, it is usually necessary to monitor a reflected signal. We begin our study with the simplest case of plane waves reflecting from and transmitting into a medium at normal incidence; that is, the wave is propagating normal to the boundary between two media.

The basic configuration is shown in Figure 8-16, where we have kept the coordinate system oriented as in the previous sections with the z axis to the right and we look down along the y axis, which is out of the page. For ease of visualization, we orient the electric field intensities in the $+\hat{x}$ directions and then assign the directions of the magnetic fields so that $\vec{E} \times \vec{H}$ gives the correct directions of propagation denoted by the serpentine arrows. If any of our assumed field orientations are in error, the solutions will have negative signs, much like that for currents and voltages in circuit theory. We use the superscripts + and − to denote the positive traveling (incident and transmitted) and negative (reflected) waves respectively. Usually either $\vec{E}_1^+$ or $\vec{H}_1^+$ is known, or at least assumed to be.

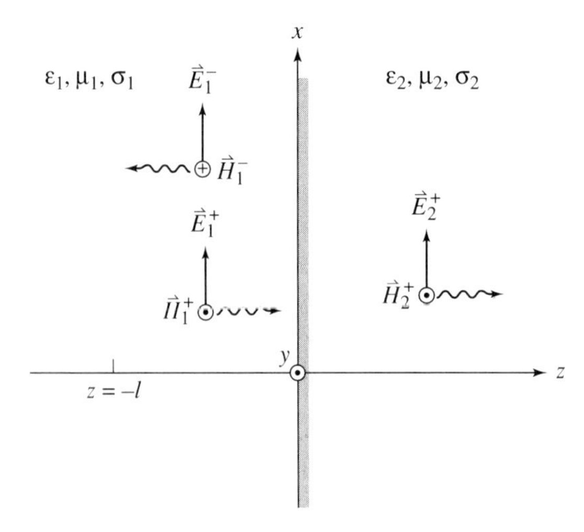

Figure 8-16. System configuration for reflection and transmission computations for normal incidence at a plane boundary.

Using the concepts in earlier sections, we write each wave as a separate equation and so it agrees with the coordinate axes. These are:

$$\vec{E}_1^+ = E_1^+ e^{-\gamma_1 z}\hat{x}, \qquad \vec{H}_1^+ = H_1^+ e^{-\gamma_1 z}\hat{y}, \qquad \frac{E_1^+}{H_1^+} = \overline{\eta}_1 \tag{8-72}$$

$$\vec{E}_1^- = E_1^- e^{\gamma_1 z}\hat{x}, \qquad \vec{H}_1^- = H_1^- e^{\gamma_1 z}\hat{y}, \qquad \frac{E_1^-}{H_1^-} = -\overline{\eta}_1 \tag{8-73}$$

$$\vec{E}_2^+ = E_2^+ e^{-\gamma_2 z}\hat{x}, \qquad \vec{H}_2^+ = H_2^+ e^{-\gamma_2 z}\hat{y}, \qquad \frac{E_2^+}{H_2^+} = \overline{\eta}_1 \tag{8-74}$$

The scalar constants are complex in general, Note that H_1^- is negative by virtue of the negative sign with the wave impedance.

Since the boundary condition on tangential electric field requires the total tangential electric field at the left at $z = 0$ to equal the total tangential electric field on the right $a + z = 0$, the preceding equations give, at $z = 0$,

$$E_1^+ + E_1^- = E_2^+ \tag{8-75}$$

Since there is no *surface* current at the boundary (this could happen only if $\sigma_2 = \infty$, perfect conductor) total tangential magnetic field intensity must be continuous at $z = 0$. From the field equations this requires

$$H_1^+ + H_1^- = H_2^+ \tag{8-76}$$

Using the wave impedances given with the field equations preceding we write this last equation as

$$\frac{E_1^+}{\overline{\eta}_1} - \frac{E_1^-}{\overline{\eta}_1} = \frac{E_2^+}{\overline{\eta}_2} \tag{8-77}$$

These equations are of identical form to the transmission line current and voltage Eqs. 1-37 and 1-56 if we let the transmission medium be the load on Medium 1. Thus all of our transmission line ideas apply here, including the application of the Smith chart. We then are led to look at $\overline{\eta}_1$ as the "characteristic impedance" of the left "line" and $\overline{\eta}_2$ as the "load impedance." You should watch for all these analogies in what follows.

To obtain a reflection coefficient at the boundary we first substitute E_2^+ from Eq. 8-75 into 8-77 and combine terms. The result is

$$\left(\frac{1}{\overline{\eta}_1} - \frac{1}{\overline{\eta}_2}\right)E_1^+ = \left(\frac{1}{\overline{\eta}_1} + \frac{1}{\overline{\eta}_2}\right)E_1^- \tag{8-78}$$

We next define a reflection coefficient at the boundary as in Eq. 1-58 to obtain the complex value

$$\Gamma_L \equiv \frac{E_1^-}{E_1^+} = \frac{\overline{\eta}_2 - \overline{\eta}_1}{\overline{\eta}_2 + \overline{\eta}_1} \equiv \rho_L e^{j\theta_L}, \qquad 0 \leq \rho_L \leq 1 \tag{8-79}$$

which is again like the transmission line formula. We can normalize with respect to $\overline{\eta}_1$ (the "Z_0") and write, by dividing numerator and denominator by $\overline{\eta}_1$,

$$\Gamma_L = \frac{\tilde{\eta}_2 - 1}{\tilde{\eta}_2 + 1} \tag{8-80}$$

where the tilde represents normalized values.

We can obtain the transmission coefficient by first dividing Eq. 8-75 by E_1^+ and then substituting Eq. 8-79

$$1 + \Gamma_L = \frac{E_2^+}{E_1^+} \equiv \tau \tag{8-81}$$

Using Eq. 8-79 again to eliminate Γ_L gives

$$\tau = \frac{2\overline{\eta}_2}{\overline{\eta}_2 + \overline{\eta}_1} = \frac{2\tilde{\eta}_2}{\tilde{\eta}_2 + 1} \tag{8-82}$$

The standing wave ratio is defined as the ratio of the total electric field maximum in Medium 1 to the adjacent total electric field minimum in Medium 1 (Eq. 1-80):

$$S \equiv \frac{|\vec{E}_{1,\,\max}|}{|\vec{E}_{1,\,\min}|} \tag{8-83}$$

The total electric field in Medium 1 is

$$\vec{E}_1 = \vec{E}_1^+ + \vec{E}_1^- = E_1^+ e^{-\gamma_1 z}\hat{y} + E_1^- e^{\gamma_1 z}\hat{y} = (E_1^+ e^{-\gamma_1 z} + \Gamma_L E_1^+ e^{\gamma_1 z})\hat{y} \tag{8-84}$$

In Medium 1 distances z are negative so we evaluate the last equation at $z = -l$ and take the magnitude, which is the equation for the standing wave pattern:

$$|\vec{E}_1(z = -l)| = |E_1^+ e^{\gamma_1 l} + \Gamma_L E_1^+ e^{-\gamma_1 l}| \tag{8-85}$$

We next take E_1^+ as reference so it has 0° angle. Substituting the polar form for Γ_L from Eq. 8-79 and also using γ in terms of α and β,

$$\begin{aligned}|\vec{E}_1(-l)| &= E_1^+|e^{\alpha_1 l}e^{j\beta_1 l} + \rho_L e^{j\theta_L}e^{-\alpha_1 l}e^{-j\beta_1 l}| \\ &= E_1^+|e^{\alpha_1 l}e^{j\beta_1 l}||1 + \rho_L e^{j\theta_L}e^{-2\alpha_1 l}e^{-j2\beta_1 l}| \\ &= E_1^+ e^{\alpha_1 l}|1 + \rho_L e^{-2\alpha_1 l}e^{-j(2\beta_1 l - \theta_L)}|\end{aligned}$$

Using Euler's theorem on the complex exponential, taking the magnitude, and combining terms yields

$$|\vec{E}_1(-l)| = E_1^+ e^{\alpha_1 l}\sqrt{1 + \rho_L^2 e^{-4\alpha_1 l} + 2\rho_L e^{-2\alpha_1 l}\cos(2\beta_1 l - \theta_L)} \tag{8-86}$$

The electric field standing wave pattern in Medium 1 can be plotted using this result. Now since ρ_L is always positive, the field will be maximum when

$$\cos(2\beta_1 l_{\max} - \theta_L) = +1$$

from which

$$2\beta_1 l_{\max} - \theta_L = 2n\pi, \qquad n = 0, 1, \ldots, \qquad l_{\max} > 0$$

thus

$$l_{\max} = \frac{2n\pi + \theta_L}{2\beta_1} \tag{8-87}$$

The field is minimum when

$$\cos(2\beta_1 l_{\min} - \theta_L) = -1$$

from which

$$2\beta_1 l_{\min} - \theta_L = (2m + 1)\pi, \qquad m = 0, 1, 2, \ldots, \qquad l_{\min} > 0$$

Then

$$l_{\min} = \frac{(2m + 1)\pi + \theta_L}{2\beta_1} \tag{8-88}$$

Note there are multiple maxima and minima just as on a transmission line.

With these distances Eq. 8-86 gives

$$|\vec{E}_1(-l_{\max})| = E_1^+ e^{\alpha_1 l_{\max}}(1 + \rho_L e^{-2\alpha_1 l_{\max}}) \tag{8-89}$$

$$|\vec{E}_1(-l_{\min})| = E_1^+ e^{\alpha_1 l_{\min}}(1 - \rho_L e^{-2\alpha_1 l_{\min}}) \tag{8-90}$$

The standing wave ratio is then the ratio of these:

$$S = e^{\alpha_1(l_{\max} - l_{\min})}\frac{1 + \rho_L e^{-2\alpha_1 l_{\max}}}{1 - \rho_L e^{-2\alpha_1 l_{\min}}} \tag{8-91}$$

Engineering approximations can be used to obtain a more useful form. Using Eqs. 8-87 and 8-88 and the same values of m and n to get adjacent maximum and minimum

$$|l_{\max} - l_{\min}| = \frac{\pi}{2\beta_1}$$

Using Eq. 8-40 to eliminate β_1 gives

$$|l_{\max} - l_{\min}| = \frac{\lambda_1}{4} \tag{8-92}$$

so the adjacent maxima and minima are one-quarter wavelength apart as on transmission lines. Adjacent maxima are thus one-half wavelength apart (as are nulls or minima). The coefficient exponential in S is then

$$e^{\alpha_1 \lambda_1/4}$$

Since good dielectrics have attenuation constants on the order of 10^{-4}, this exponent is 1.0005 for a λ_1 as long as 20 m. Thus we can replace that exponent by 1 with very little error to obtain

$$S \approx \frac{1 + \rho_L e^{-2\alpha_1 l_{\max}}}{1 - \rho_L e^{-2\alpha_1 l_{\min}}} \tag{8-93}$$

Note this approaches 1 as we take the distances farther from the load. It is usual practice to drop the subscripts *max* and *min* even though the two distances are really not the same, but this gives us some measure of how S varies along the line. Then we obtain the usual expression:

$$S \approx \frac{1 + \rho_L e^{-2\alpha_1 l}}{1 - \rho_L e^{-2\alpha_1 l}}, \qquad 0 \le \rho_L \le 1 \tag{8-94}$$

which is useful for Smith chart applications as in Chapter 2. Note that for a lossless medium, α_1 is zero and Eq. 8-94 gives the exact expression

$$S = \frac{1 + \rho_L}{1 - \rho_L}, \qquad 0 \le \rho_L \le 1, \qquad \alpha_1 = 0 \tag{8-95}$$

which we obtained for lossless transmission lines.

Plots of the standing wave patterns for the limit values of ρ_L are given in Figure 8-17.

For the normal incident case under consideration it is possible to define a wave impedance in terms of *total* electric field and *total* magnetic field. Using Eqs. 8-72 and 8-73 we define, using $z = -l$,

$$\begin{aligned} \overline{Z}^{w}_{\text{total}}(-l) \equiv \frac{E_{1x}}{H_{1y}} &= \frac{E_1^+ e^{\gamma_1 l} + E_1^- e^{-\gamma_1 l}}{H_1^+ e^{\gamma_1 l} + H_1^- e^{-\gamma_1 l}} = \frac{E_1^+ e^{\gamma_1 l} + \Gamma_L E_1^+ e^{-\gamma_1 l}}{\dfrac{E_1^+}{\overline{\eta}_1} e^{\gamma_1 l} - \dfrac{\Gamma_L E_1^+}{\overline{\eta}_1} e^{-\gamma_1 l}} \\ &= \overline{\eta}_1 \frac{1 + \Gamma_L e^{-2\gamma_1 l}}{1 - \Gamma_L e^{-2\gamma_1 l}} \end{aligned} \tag{8-96}$$

This again in agreement with transmission line theory, where we define the reflection coefficient at any point as

$$\Gamma(z = -l) = \Gamma_L e^{-2\gamma_2 l} \tag{8-97}$$

Applications of these results will next be given by way of examples.

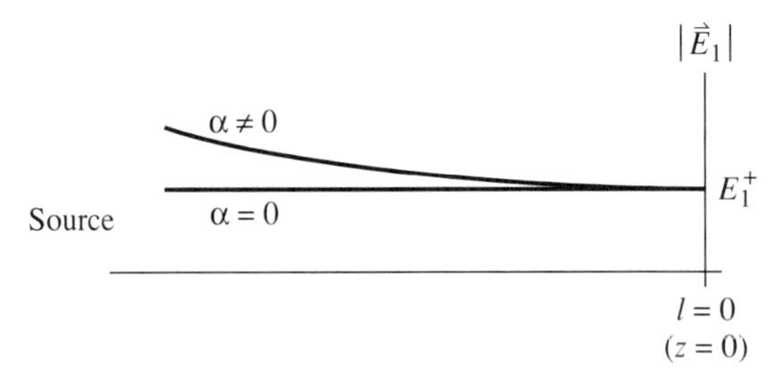

(a) $\rho_L = 0$ ($\overline{\eta}_1 = \overline{\eta}_2$)

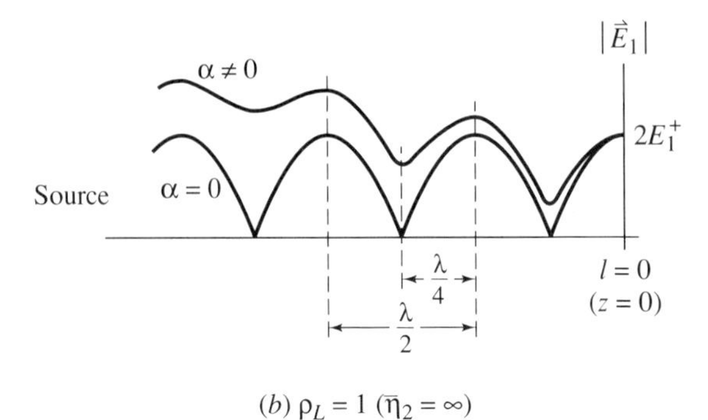

(b) $\rho_L = 1$ ($\overline{\eta}_2 = \infty$)

Figure 8-17. Plots of the electric field standing wave patterns for limiting values of ρ_L and for lossless and lossy media.

Example 8-13

Determine the reflection properties of a plane wave in a perfect dielectric normally incident on a perfect conductor as shown in Figure 8-18.

Using Eq. 8-35,

$$\overline{\eta}_2 = \sqrt{\frac{\mu}{\epsilon - j\epsilon''}} = \sqrt{\frac{\mu_0}{\epsilon' - j\dfrac{\sigma_2}{\omega\epsilon'}}} \Rightarrow 0 \qquad \text{for} \quad \sigma_2 = \infty$$

Similarly,

$$\overline{\eta}_1 = \sqrt{\frac{\mu_0}{\epsilon'}} = \sqrt{\frac{\mu_0}{\epsilon_1}} = \eta_1, \qquad \text{real}$$

The reflection coefficient is then, from Eq. 8-79,

$$\Gamma_L = \frac{0 - \eta_1}{0 + \eta_1} = -1 = 1e^{j\pi}, \qquad \rho_L = 1, \qquad \theta_L = \pi$$

Since the amplitude of the incident is assumed known and of 0° phase,

$$E_1^- = -1 \cdot E_1^+ = -E_1^+$$

The dielectric is lossless so $\alpha_1 = 0$, and Eq. 8-86 gives the standing wave pattern of total electric field

$$|\vec{E}_1(-l)| = E_1^+\sqrt{1 + 1 + 2\cos(2\beta_1 l - \pi)} = E_1^+\sqrt{2}\sqrt{1 - \cos 2\beta_1 l}$$

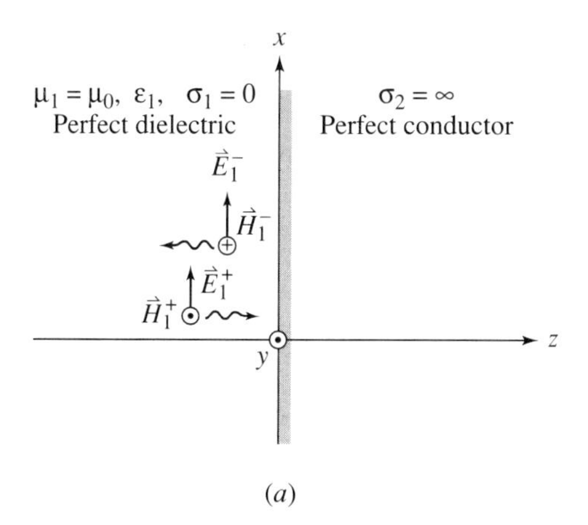

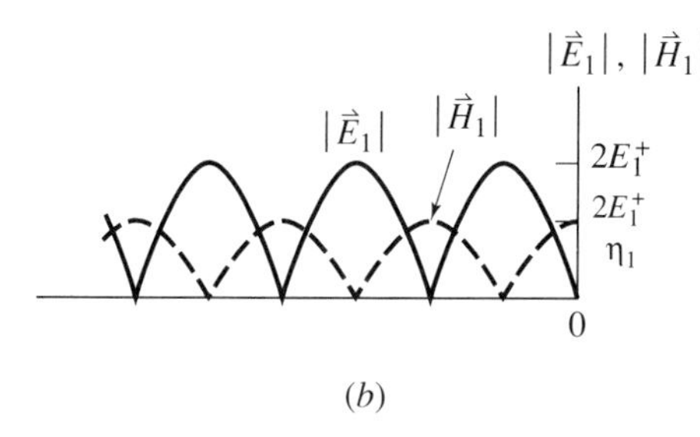

Figure 8-18. System for Example 8-13 and the resulting standing wave pattern. (*a*) Dielectric conductor boundary. (*b*) Standing wave pattern.

Using trig identities, this reduces to

$$|\vec{E}_1| = 2E_1^+|\sin\beta_1 l| \text{ V/m} \tag{8-98}$$

This is plotted in Figure 8-18*b*. Remember that increasing l moves to the left since $z = -l$. This standing wave pattern is identical to that of a shorted ($\overline{Z}_L = 0$) transmission line.

The magnetic field standing wave pattern is obtained by adding the incident and reflect fields and taking the magnitude, using $\alpha_1 = 0$ as before:

$$|\vec{H}_1| = |\vec{H}_1^+ + \vec{H}_1^-| = \left|\frac{E_1^+}{\overline{\eta}_1}e^{+j\beta_1 l}\hat{y} - \frac{E_1^-}{\overline{\eta}_1}e^{-j\beta l}\hat{y}\right| = \left|\frac{E_1^+}{\overline{\eta}_1}e^{j\beta_1 l} - \frac{-E_1^+}{\overline{\eta}_1}e^{-j\beta l}\right|$$

Since E_1^+ and $\overline{\eta}_1$ are both real and positive, and using Euler's theorem for complex exponentials,

$$|\vec{H}_1| = \frac{E_1^+}{\eta_1}|e^{j\beta_1 l} + e^{-j\beta_1 l}| = \frac{2E_1^+}{\eta_1}|\cos\beta_2 l| \text{ A/m} \tag{8-99}$$

This is the dashed curve in Figure 8-18*b*.

Since $\overline{\eta}_2 = 0$, the transmission coefficient is zero and $\vec{E}_2$ and $\vec{H}_2$ are zero. There is then a surface current density

$$\vec{J}s = \hat{n} \times \vec{H}_1|_{z=-l=0} = -\hat{z} \times \frac{2E_1^+}{\eta_1}\cos\beta_1 l\hat{y}\Big|_{l=0} = \frac{2E_1^+}{\eta_1}\hat{x} \text{ A/m} \tag{8-100}$$

If we look at the *total* electric field vector rather than just its magnitude we have, in the time domain,

$$\begin{aligned}\vec{\mathscr{E}}_1(-l, t) &= \text{Re}\{\vec{E}_1 e^{j\omega t}\} = \text{Re}\{2E_1^+ \sin \beta_1 l e^{j\omega t}\hat{x}\} \\ &= 2E_1^+ \sin \beta_1 l \cos \omega t\hat{x}\end{aligned} \tag{8-101}$$

Since there is no combination $(\omega t - \beta_1 z)$ or $(\omega t + \beta_1 l)$ this is *not* a propagating wave; hence the name standing wave.

The total impedance in the z direction is found using Eq. 8-96 with $\alpha_1 = 0$:

$$\begin{aligned}\bar{Z}_{\text{total}}^w &= \eta_1 \frac{1 + (-1)e^{-j2\beta_1 l}}{1 - (-1)e^{-j2\beta_1 l}} = \eta_1 \frac{1 - e^{-j2\beta_1 l}}{1 + e^{-j2\beta_1 l}} \\ &= \eta_1 \frac{e^{j\beta_1 l} - e^{-j\beta_1 l}}{e^{j\beta_1 l} + e^{-j\beta_1 l}} = j\sqrt{\frac{\mu_0}{\epsilon_1}} \tan \beta_1 l\end{aligned} \tag{8-102}$$

This is purely reactive just as it was for the shorted transmission line.

Finally, the standing wave ratio is

$$S = \frac{1 + \rho_L}{1 - \rho_L} = \frac{1 + 1}{1 - 1} = \infty$$

as expected.

Example 8-14

A wave in a perfect dielectric impinges on a good conductor as shown in Figure 8-19. For a little variety, we'll use numerical values this time as shown in the figure for a vacuum-copper boundary at 500 MHz, $E_1^+ = 0.01\angle 0°$ V/m. The conductor is assumed to extend to infinity.

For vacuum:

$$\bar{\eta}_1 = \eta_1 = \sqrt{\frac{4\pi \times 10^{-7}}{\frac{1}{36\pi} \times 10^{-9}}} = 377\ \Omega$$

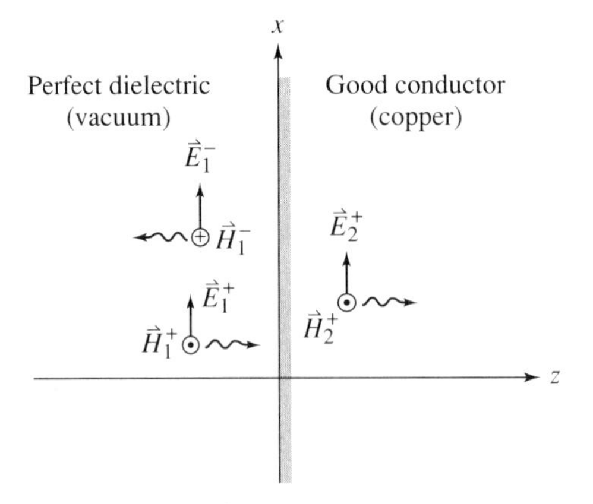

Figure 8-19. Uniform plane wave impinging on a good conductor at normal incidence.

For cooper, Table 8-2 gives the equations:

$$\delta_2 = \frac{0.066}{\sqrt{500 \times 10^6}} = 2.95 \times 10^{-6} \text{ m}$$

$$\sigma_2 = 5.8 \times 10^7 \text{ S/m}$$

$$R_s = 2.61 \times 10^{-7}\sqrt{500 \times 10^6} = 0.00584\ \Omega$$

Then

$$\overline{\eta}_2 = 0.00584 + j0.00584\ \Omega$$

$$\Gamma = \frac{0.00584 + j0.00584 - 377}{0.00584 + j0.00584 + 377} = -0.9997 + j3.098 \times 10^{-5}$$

$$= 0.99997\angle 179.99^\circ$$

This is very nearly -1 as we would expect. Therefore,

$$E_1^- = \Gamma E_1^+ = \Gamma \times 0.01 = 9.99 \times 10^{-3}\angle 179.99^\circ \text{ V/m}$$

$$\tau = \frac{2\overline{\eta}_2}{\overline{\eta}_2 + \eta_1} = \frac{2(0.00584 + j0.00584)}{0.00584 + j0.00584 + 377} = 3.098 \times 10^{-5} + j3.098 \times 10^{-5}$$

$$= 4.381 \times 10^{-5}\angle 45^\circ$$

Therefore,

$$E_2^+ = \tau E_1^+ = 4.381 \times 10^{-7}\angle 45^\circ \text{ V/m}$$

From the electric field components

$$H_1^+ = \frac{E_1^+}{\eta_1} = 2.652 \times 10^{-5} \text{ A/m}$$

$$H_1^- = -\frac{E_1^-}{\eta_1} = -2.65 \times 10^{-5}\angle 179.99^\circ = 2.649 \times 10^{-5}\angle 0^\circ \text{ A/m}$$

$$H_2^+ = \frac{E_2^+}{\overline{\eta}_2} = 5.30 \times 10^{-5}\angle 0^\circ \text{ A/m}$$

(Note the tangential magnetic field intensity is continuous across the boundary at $z = 0$.) In a vacuum at 500 MHz,

$$\alpha_1 = 0$$

$$\beta_1 = k_1 = \omega\sqrt{\mu_1\epsilon_1} = 2\pi \times 5 \times 10^8 \times \sqrt{4\pi \times 10^{-7} \times \frac{1}{36\pi} \times 10^{-9}}$$

$$= 10.47 \text{ rad/m}$$

For the good conductor copper,

$$\alpha_2 = \frac{1}{\delta_2} = 3.39 \times 10^5 \text{ nep/m}$$

$$\beta_2 = \frac{1}{\delta_2} = 3.39 \times 10^5 \text{ rad/m}$$

The total field equations can now be constructed from the parameter values (using $z = -l$):

$$\vec{E}_1 = \vec{E}_1^+ + \vec{E}_1^- = (0.01e^{j10.47l} - 0.00999e^{-j10.47l})\hat{x} \text{ V/m}$$

$$\vec{E}_2 = \vec{E}_2^+ = 4.38 \times 10^{-7}e^{j45°}e^{3.39\times10^5 l}e^{j3.39\times10^5 l}\hat{x} \text{ V/m}$$

Note that l is *negative* to the right of the boundary.

$$\vec{H}_1 = \vec{H}_1^+ + \vec{H}_1^- = (2.652 \times 10^{-5}e^{j10.47l} + 2.649 \times 10^{-5}e^{-j10.47l})\hat{y} \text{ A/m}$$

$$\vec{H}_2 = \vec{H}_2^+ = 5.30 \times 10^{-5}e^{3.39\times10^5 l}e^{j3.39\times10^5 l}\hat{y} \text{ A/m}$$

Note that l is *negative* to the right of the boundary.

As we expect, the wave in Medium 2 is attenuated very rapidly, so our assumption of infinite Medium 2 could be easily met approximately by even a thin sheet of copper ($\delta_2 = 2.95\ \mu\text{m}$).

It is instructive to compute the wavelengths to show the wide variation:

$$\lambda_1 = \frac{2\pi}{\beta_1} = \frac{2\pi}{10.47} = 0.6 \text{ m}$$

$$\lambda_2 = \frac{2\pi}{\beta_2} = 2\pi\delta_2 = 18.5\ \mu\text{m}$$

The results for $\vec{E}_1$ and $\vec{E}_2$ seem to be inconsistent because at $l = 0$ ($z = 0$) the two are not exactly equal. The problem arises from the round-off used to express E_1^-. The exact value using a more precise reflection coefficient Γ would be 0.009999562 V/m.

Example 8-15

We next consider the case of a boundary between two perfect dielectrics as shown in Figure 8-20 and use a frequency of 1 GHz. We assume that Medium 2 extends to infinity so there will be only a positive traveling wave in that dielectric.

$$|\vec{E}_1^+| = E_1^+ = 10 \text{ V/m}$$

For Medium 1:

$$\overline{\eta}_1 = \eta_1 = \sqrt{\frac{4\pi \times 10^{-7}}{2 \times \frac{1}{36\pi} \times 10^{-9}}} = 267\ \Omega$$

$$\beta_1 = k_1 = \omega\sqrt{\mu\epsilon} = 2\pi \times 10^9 \sqrt{4\pi \times 10^{-7} \times 2 \times \frac{1}{36\pi} \times 10^{-9}}$$

$$= 29.62 \text{ rad/m}$$

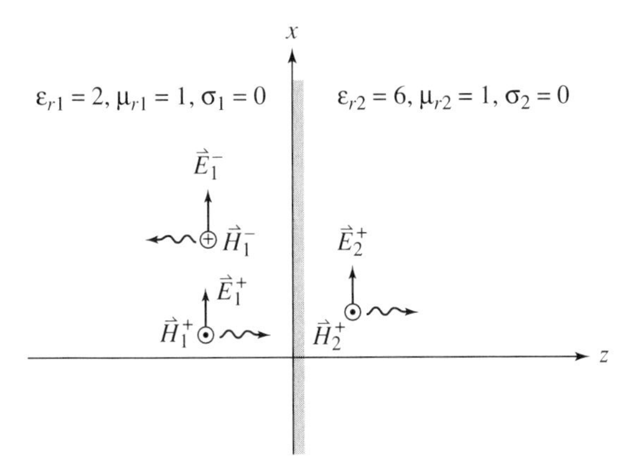

Figure 8-20. Reflection of plane wave at the boundary of two perfect dielectrics.

For Medium 2:

$$\overline{\eta}_2 = \eta_2 = \sqrt{\frac{4\pi \times 10^{-7}}{6 \times \frac{1}{36\pi} \times 10^{-9}}} = 154\ \Omega$$

Then

$$\Gamma = \frac{154 - 267}{154 + 267} = -0.268 = 0.268e^{j180^\circ} = \rho_L e^{j\theta_L}$$

$$\tau = \frac{2 \times 154}{154 + 267} = 0.732$$

$$S = \frac{1 + \rho_L}{1 - \rho_L} = \frac{1 + 0.268}{1 - 0.268} = 1.73$$

Using Γ:

$$E_1^- = \Gamma E_1^+ = 2.68e^{j180^\circ}\ \text{V/m}$$

Using τ:

$$E_2^+ = \tau E_1^+ = 7.32\ \text{V/m}$$

The standing wave pattern is given by Eq. 8-86, with $\alpha_1 = \alpha_2 = 0$ for this case:

$$\begin{aligned}|\vec{E}_1(-l)| = |\vec{E}_1^+ + \vec{E}_1^-| &= 10\sqrt{1 + (0.268)^2 + 2 \times 0.268 \cos(2 \times 29.62l - 180^\circ)} \\ &= 10\sqrt{1.072 - 0.536 \cos 59.24l}\end{aligned}$$

Since there is no reflected wave in Medium 2, ρ_L is zero there so

$$|\vec{E}_2(z > 0)| = |\vec{E}_2^+| = 7.32\ \text{V/m}$$

There results are plotted in Figure 8-21.

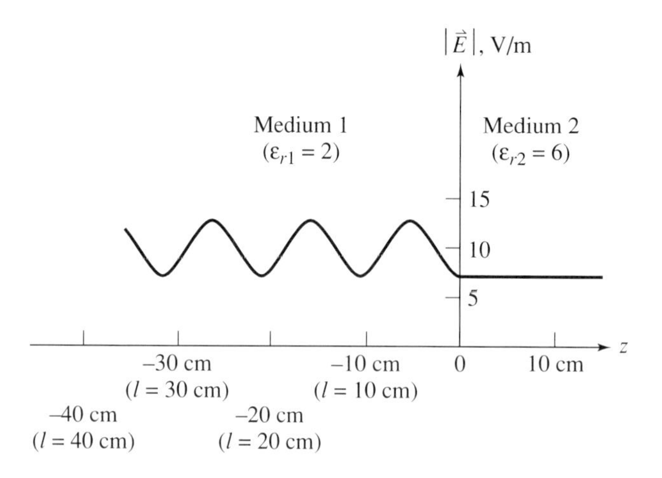

Figure 8-21. Standing wave pattern for the system of Figure 8-20 at 1 GHz. $\lambda_1 = 21.2$ cm.

Example 8-16

To protect an antenna from the environment, it is usual to surround it with a dielectric through which the waves may pass. Such a structure is called a *radome* (contraction of *radar dome*). Although such structures are not planar we will examine a simplified planar model shown in Figure 8-22. The concept can be extended to multiple dielectric interfaces.

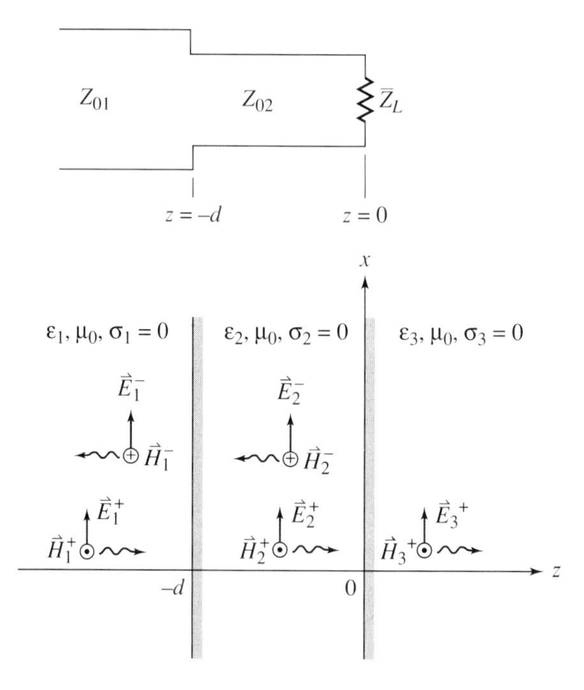

Figure 8-22. Plane wave reflection at multiple dielectric boundary system. The transmission line analogy is also shown.

Although a radome is basically an air-dielectric-air configuration we will generalize that to three different dielectrics as shown. We also assume perfect dielectrics. The analogous transmission-line system is shown above the dielectric system.

For the general case with given dielectric and frequency values it is easiest to use the Smith chart. The basic process would be to normalize $\overline{Z}_{L3}$ to Z_{02}, compute λ_2, and then rotate $\tilde{Z}_{L3}$ toward the generator by d/λ_2. Then determine the actual load at $z = -d$ and normalize it to Z_{01}. To find the impedance at any point in Medium 1 we would compute λ_1 and rotate an additional desired amount toward the generator, taking $z = -d$ to be the new reference location $z' = 0$.

Let's examine this process using equations.

Let η_1, η_2, and η_3 be the wave impedances of the corresponding media. They will be real since we have assumed the lossless case. We use η_3 as the load on Medium 2 and use Eq. 8-96 with the appropriate subscripts to obtain the wave impedance (remember this is the total field ratio) at location $z = -d(d \Rightarrow l)$:

$$\overline{Z}^w_{\text{total}}(-d) = \eta_2 \frac{1 + \dfrac{\eta_3 - \eta_2}{\eta_3 + \eta_2} e^{-j2\beta_2 d}}{1 - \dfrac{\eta_3 - \eta_2}{\eta_3 + \eta_2} e^{-j2\beta_2 d}} = \eta_2 \frac{\eta_3 + \eta_2 \dfrac{1 - e^{-j2\beta_2 d}}{1 + e^{-j2\beta_2 d}}}{\eta_2 + \eta_3 \dfrac{1 - e^{-j2\beta_2 d}}{1 + e^{-j2\beta_2 d}}}$$

$$\overline{Z}^w_{\text{total}}(-d) = \eta_2 \frac{\eta_3 + j\eta_2 \tan \beta_2 d}{\eta_2 + j\eta_3 \tan \beta_2 d} \tag{8-103}$$

Compare with Eq. 1-98 for transmission lines.

Suppose we want to find the thickness d of Medium 2 so that there will be no reflection in Medium 1. This requires the load at the boundary $-d$ to be η_1. The preceding equation gives the constraint for this to be the case:

$$\overline{Z}^w_{\text{total}}(-d) = \eta_1 = \eta_2 \frac{\eta_3 + j\eta_2 \tan \beta_2 d}{\eta_2 + j\eta_3 \tan \beta_2 d}$$

Solving this we find that

$$j \tan \beta_2 d = \frac{\eta_1\eta_2 - \eta_2\eta_3}{\eta_2^2 - \eta_1\eta_3}$$

This requires a pure imaginary number to be a pure real number. The paradox can be resolved two ways.

Case 1: Both Sides Equal to Zero.

Left side:

$$\tan \beta_2 d = 0$$

$$\text{Then } \beta_2 d = n\pi, \text{ or}$$

$$d = \frac{n\pi}{\beta_2} = \frac{n\pi}{\dfrac{2\pi}{\lambda_2}} = n\frac{\lambda_2}{2} \tag{8-104}$$

Right side:

$$\eta_1\eta_2 - \eta_2\eta_3 = 0$$

or

$$\eta_1 = \eta_3 \tag{8-105}$$

This is called the *half-wave dielectric plate*, or window, and requires Media 1 and 3 to be the same. Both requirements must be met. From transmission-line theory we recall that on a line the impedance is periodic every half wavelength.

Case 2: Both Sides Equal to Infinity.

Left side:

$$\tan \beta_2 d = \infty$$

Then $\beta_2 d = \pi/2$, or

$$d = \frac{\pi}{2\beta_2} = \frac{\pi}{2 \times \dfrac{2\pi}{\lambda_2}} = \frac{\lambda_2}{4} \tag{8-106}$$

Right side:

$$\eta_2^2 - \eta_1\eta_3 = 0$$

or

$$\eta_2 = \sqrt{\eta_1\eta_3} \tag{8-107}$$

This is called the *quarter-wave plate*, or window, and requires Medium 2 to be a quarter-wave thick and to have a wave impedance equal to the geometric mean of 1 and 3. This is frequently used in optical lens coating and corresponds to the transmission line quarter-wave matching section.

Example 8-17

Design a quarter-wave matching plate to match materials having $\epsilon_{r1} = 1$ and $\epsilon_{r3} = 3$ at a frequency of 8 GHz. Sketch the standing wave pattern for the system. Let $E_1^+ = 10$ V/m. See Figure 8-23*a*.

$$\eta_1 = \sqrt{\frac{\mu_0}{\epsilon_0}} = 377\ \Omega \qquad \eta_3 = \sqrt{\frac{\mu_0}{3\epsilon_0}} = 218\ \Omega$$

From Case 2 of Example 8-16,

$$\eta_2 = \sqrt{\eta_1\eta_3} = 286\ \Omega$$

The material required for Medium 2 must then satisfy

$$\sqrt{\frac{\mu_0}{\epsilon_{r2}\epsilon_0}} = \frac{377}{\sqrt{\epsilon_{r2}}} = 286$$

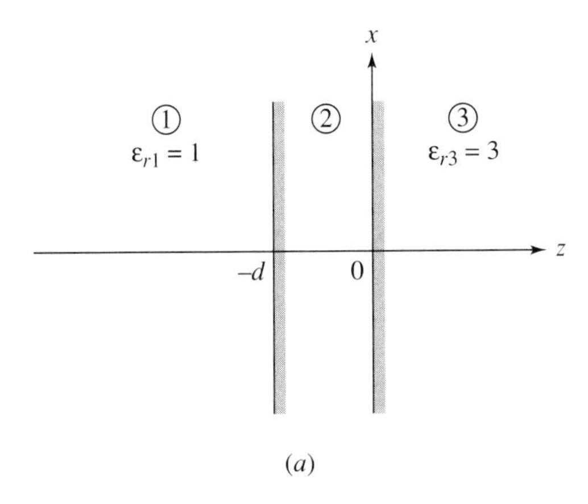

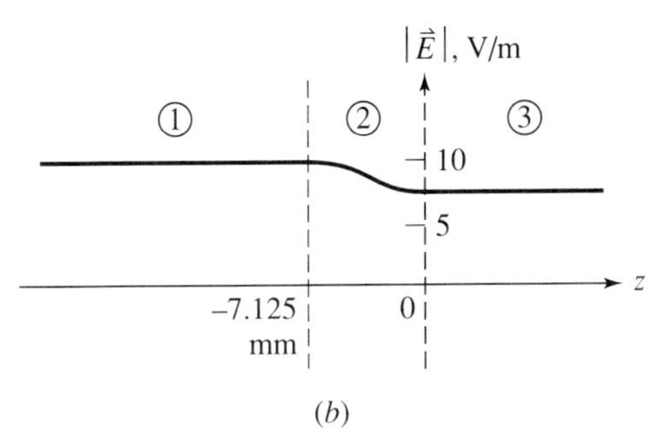

Figure 8-23. Standing wave pattern for the quarter-wave matching section system of Example 8-17. (*a*) Two-boundary system. (*b*) Standing wave pattern.

Thus

$$\epsilon_{r2} = 1.732$$

For Medium 2:

$$\beta_2 = \omega\sqrt{\mu_2\epsilon_2} = 2\pi \times 8 \times 10^9 \sqrt{4\pi \times 10^{-7} \times 1.732 \times \frac{1}{36\pi} \times 10^{-9}}$$

$$= 220.5 \text{ rad/m}$$

$$\lambda_2 = \frac{2\pi}{\beta_2} = 0.0285 \text{ m}$$

The dielectric thickness is then

$$d = \frac{0.0285}{4} = 0.007125 \quad \text{or} \quad 7.125 \text{ mm}$$

The reflection coefficient between Media 2 and 3 is

$$\Gamma_{L23} = \frac{\eta_3 - \eta_2}{\eta_3 + \eta_2} = 0.135e^{j\pi} = \frac{E_2^-}{E_2^+}$$

$$S_2 = \frac{1 + 0.135}{1 - 0.135} = 1.31$$

$$\tau_{23} = \frac{2\eta_3}{\eta_2 + \eta_3} = 0.865$$

Since there is no reflected component in Medium 1, $\gamma_1 = 0$, $S_1 = 1$, and $\tau_{12} = 1$. Thus

$$E_1^+ = 10 \text{ V/m}, \qquad E_1^- = 0, \qquad E_{2\text{total}} - \tau_{12}E_1^+ = 10 \text{ V/m}$$

In Medium 2 there are both incident (positive) and reflected (negative) components. Thus at the boundary between 1 and 2 (which is at $z = -d$) the continuity of tangential electric fields requires

$$\begin{aligned} E_1^+ &= E_2^+ e^{j\beta_2 d} + E_2^- e^{-j\beta_2 d} = E_2^+ e^{j\pi/2} + E_2^- e^{-j\pi/2} \\ &= jE_2^+ - jE_2^- \end{aligned}$$

Using Γ_{L23} calculated earlier

$$E_1^+ = jE_2^+ - j(-0.137)(E_2^+) = j1.137E_2^+$$

Since E_1^+ is known, 10 V/m,

$$\begin{aligned} 10 &= j1.137E_2^+ \\ E_2^+ &= -j8.8 = 8.8e^{-j\pi/2} \text{ V/m} \end{aligned}$$

From this,

$$E_2^- = \Gamma_{L23}E_2^+ = (0.135e^{j\pi})(8.8e^{-j\pi/2}) = 1.19e^{j\pi/2} \text{ V/m}$$

Using the value of τ_{23} computed above,

$$E_3^+ = \tau_{23}E_2^+ = (0.865)(8.8e^{-j\pi/2}) = 7.61e^{-j\pi/2} \text{ V/m}$$

This shows us that the field in Medium 3 is not the same as Medium 1, but delayed by 90° as expected. The standing wave equations are (using Eq. 8-86 for $|\vec{E}_2|$)

$$\begin{aligned} |\vec{E}_1| &= |\vec{E}_1^+| = 10 \text{ V/m} \\ |\vec{E}_2| &= 8.8\sqrt{1 + (0.135)^2 + 2 \times 0.135 \cos(2 \times 220.5l - 180°)} \\ &= 8.8\sqrt{1.018 - 0.27 \cos 441l} \text{ V/m}, \qquad 0 \le l \le \frac{\lambda_1}{4} = 0.007125 \text{ m} \\ |\vec{E}_3| &= |\vec{E}_3^+| = 7.61 \text{ V/m} \end{aligned}$$

These are plotted in Figure 8-23*b*.

Exercises

8-6.0a An incident plane wave has an amplitude $E_1^+ = 10^{-2}$ V/m in Medium 1 ($z < 0$) whose parameters are $\epsilon_{r1} = 1$, $\mu_{r1} = 2$, and $\sigma_1 = 0$. The medium to the right of the plane boundary has $\epsilon_{r2} = 2$, $\mu_{r2} = 3$, and $\sigma_2 = 0$. At a frequency of 100 MHz determine the time waveforms of every incident and reflected component and total fields in both media.

8-6.0b Show that for a lossless-plane dielectric boundary with a normally incident wave

$$\Gamma(-l) = \frac{E_1^-(-l)}{E_1^+(-l)} = \Gamma_L e^{j2\beta_1 l}$$

8-6.0c For a uniform wave normally incident on a perfect conductor (see Example 8-13) compute the three Poynting vectors $\vec{S}_1^+$, $\vec{S}_1^-$, and $\vec{S}_1$. Comment on the results.

8-6.0d Interchange the two media in Example 8-15 and plot the standing wave pattern.

8-6.0e Determine the thickness of a radome that has air on both sides and is made of Teflon. The operating frequency is 2 GHz.

8-6.0f In Figure 8-19 Medium 1 is air and Medium 2 has $\mu_{r2} = 1$, $\epsilon_{r2} = 2$, and $\sigma_2/\omega\epsilon_2 = 0.5$ at 300 MHz. Determine the VSWR in air and the location of the first null on the standing wave pattern.

8-6.0g Use the Smith chart and scales to solve the system of Example 8-15 for: S, τ, return loss, reflected loss, and percent power transmitted. Using these results, compute E_1^- and E_2^+.

8-6.0h Use the Smith chart and scales to solve the system of Example 8-16 for: S_1, S_2, τ_{12}, and τ_{23}. Use the following parameter values: $\mu_1 = \mu_2 = \mu_3 = \mu_0$, $\epsilon_1 = \epsilon_3 = \epsilon_0$, $\epsilon_{2r} = 2.5$, $f = 4$ GHz, and $d = 80$ mm.

8-7 Reflection of Uniform Plane Waves Propagating at Oblique Incidence to a Boundary Between Media: Laws of Reflection and Refraction

We next investigate the reflection and transmission properties of uniform plane waves that impinge obliquely on a boundary between media. This study will lead us very naturally to the laws of reflection and refraction.

One might expect that the approach would be to start with a plane boundary between general media and then apply the results to specific cases. It will be shown that difficulties arise in the general approach that require us to place some restrictions on the classes of materials we may easily solve. To show the restrictions necessary we begin with the general two-media configuration shown in Figure 8-24. There are two basic configurations of interest shown. In Figure 8-24*a* the electric field intensity is directed out of the plane of the page, that is, perpendicular to the figure. This is called *perpendicular polarization*, the term *polarization* referring to the electric field vector orientation. The orientation shown in Figure 8-24*b* has the electric field lying in the plane of the page, that is, parallel to the figure. This is called *parallel polarization.*

These two orientations are really all we need since any other electric field vector can be decomposed into a perpendicular component and a parallel component and the results of the two reflections combined to give the total result. Note that since the plane waves are not propagating in the z direction we have defined new coordinates ξ to indicate the coordinate along which propagation occurs.

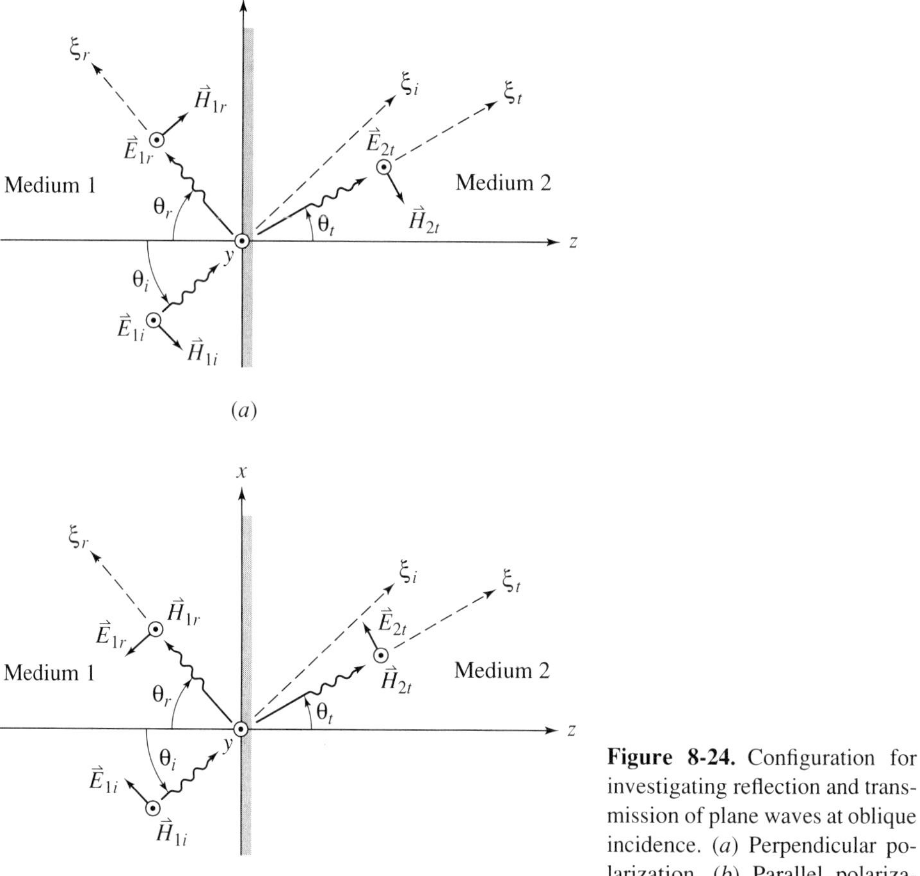

Figure 8-24. Configuration for investigating reflection and transmission of plane waves at oblique incidence. (*a*) Perpendicular polarization. (*b*) Parallel polarization.

8-7.1 Perpendicular Polarization at Oblique Incidence

For the perpendicular polarization case shown in Figure 8-24*a* we arbitrarily assign all electric field components out of the page and then determine each of the magnetic field components so that the Poynting vector cross products give the correct directions of propagation. Thus all of the results we derive must be used in accordance with the figure. The angles of the various wave directions are measured from the normal to the boundary, in this case the z axis, to the lines along the propagation directions (shown as serpentine arrows as before). The serpentine arrows are called *rays* or *ray paths* in optics. The subscripts i, r, and t are used to denote incident, reflected, and transmitted quantities, respectively. It is usual to assume that the incident quantities are known since that will usually come from a transmitter.

Since a plane wave propagating in the plus z direction is associated with the exponential $e^{-\gamma z}$, a wave propagating in the ξ direction will be described by $e^{-\gamma\xi}$. Thus the three electric fields will have the equations

$$\vec{E}_{1i} = E_{1i}e^{-\gamma_1\xi_i}\hat{y} \tag{8-108}$$

$$\vec{E}_{1r} = E_{1r}e^{-\gamma_1\xi_r}\hat{y} \tag{8-109}$$

$$\vec{E}_{2}t = E_{2}te^{-\gamma_2\xi_t}\hat{y} \tag{8-110}$$

To get the ξ coordinates in terms of the rectangular coordinates we use the constructions shown in Figure 8-25. From each of the three figures we obtain (note particularly the negative z part in the reflected wave)

$$\xi_i = x \sin\theta_i + z \cos\theta_i \tag{8-111}$$

$$\xi_t = x \sin\theta_t + z \cos\theta_t \tag{8-112}$$

$$\xi_r = x \sin\theta_r - z \cos\theta_r \tag{8-113}$$

The *total* field expressions in each medium are obtained by substituting these last three equations into the component field equations and combining components in each medium.

Figure 8-25. Construction for obtaining propagation directions in rectangular coordinates. (*a*) Incident wave. (*b*) Transmitted wave. (*c*) Reflected wave.

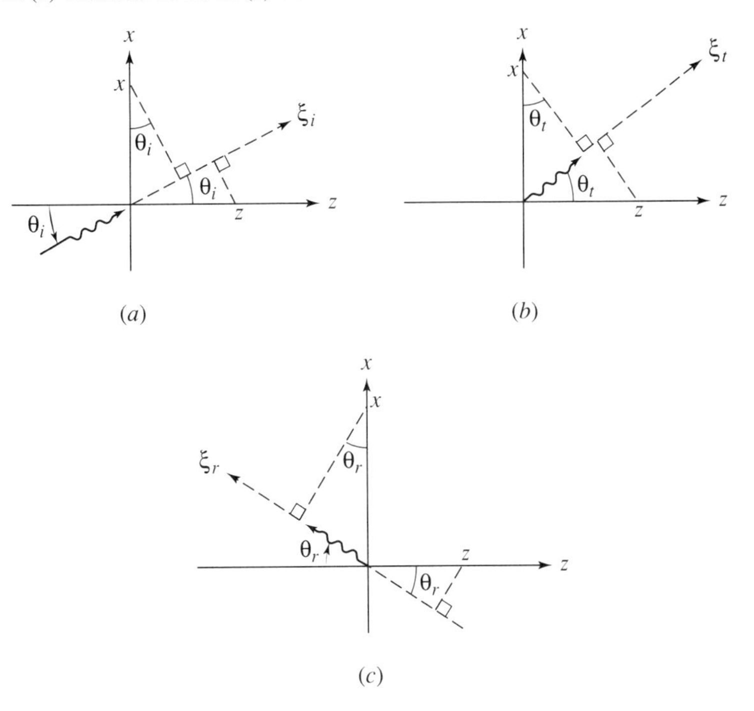

The results are

$$\vec{E}_1 = \vec{E}_{1i} + \vec{E}_{1r} = [E_{1i}e^{-(\gamma_1 x\sin\theta_i + \gamma_1 z\cos\theta_i)} + E_{1r}e^{-(\gamma_1 x\sin\theta_r - \gamma_1 z\cos\theta_r)}]\hat{y} \quad (8\text{-}114)$$

$$\vec{E}_2 = \vec{E}_{2t} = E_{2t}e^{-(\gamma_2 x\sin\theta_t + \gamma_2 z\cos\theta_t)}\hat{y} \quad (8\text{-}115)$$

The corresponding magnetic field intensities are found using Faraday's law from Maxwell's equations:

$$\vec{H}_1 = \vec{H}_{1i} + \vec{H}_{1r} = j\frac{1}{\omega\mu_1}\nabla\times\vec{E}_1$$

$$= \left[\frac{j\gamma_1\cos\theta_i}{\omega\mu_1}E_{1i}e^{-(\gamma_1 x\sin\theta_i + \gamma_1 z\cos\theta_i)} - \frac{j\gamma_1\cos\theta_r}{\omega\mu_1}E_{1r}e^{-(\gamma_1 x\sin\theta_r - \gamma_1 z\cos\theta_r)}\right]\hat{x}$$

$$+ \left[\frac{-j\gamma_1\sin\theta_i}{\omega\mu_1}E_{1i}e^{-(\gamma_1 x\sin\theta_i + \gamma_1 z\cos\theta_i)} - \frac{j\gamma_1\sin\theta_r}{\omega\mu_1}E_{1r}e^{-(\gamma_1 x\sin\theta_r - \gamma_1 z\cos\theta_r)}\right]\hat{z} \quad (8\text{-}116)$$

$$\vec{H}_2 = \vec{H}_{2t} = j\frac{1}{\omega\mu_2}\nabla\times\vec{E}_2$$

$$= \frac{j\gamma_2\cos\theta_t}{\omega\mu_2}E_{2t}e^{-(\gamma_2 x\sin\theta_t + \gamma_2 z\cos\theta_t)}\hat{x} - \frac{j\gamma_2\sin\theta_t}{\omega\mu_2}E_{2t}e^{-(\gamma_2 x\sin\theta_t + \gamma_2 z\cos\theta_t)}\hat{z} \quad (8\text{-}117)$$

At the boundary, $z = 0$, tangential electric field must be continuous. Since the y components are tangential the preceding field equations require

$$E_{1i}e^{-\gamma_1 x\sin\theta_i} + E_{1r}e^{-\gamma_1 x\sin\theta_r} = E_{2t}e^{-\gamma_2 x\sin\theta_t} \quad (8\text{-}118)$$

Now this must be true for any value of x, and since E_{1i}, E_{1r}, and E_{2t} are constants (maybe complex) the exponentials must cancel out in the equation. This process is called *phase matching* at the boundary. We then equate the exponents to obtain

$$-\gamma_1 x\sin\theta_i = -\gamma_1 x\sin\theta_r = -\gamma_2 x\sin\theta_t$$

or

$$\gamma_1\sin\theta_i = \gamma_1\sin\theta_r = \gamma_2\sin\theta_t \quad (8\text{-}119)$$

Since the propagation constants are complex the explicit form is

$$\alpha_1\sin\theta_i + j\beta_1\sin\theta_i = \alpha_1\sin\theta_r + j\beta_1\sin\theta_r = \alpha_2\sin\theta_t + j\beta_2\sin\theta_t \quad (8\text{-}120)$$

The right equality requires, equating real and imaginary parts,

$$\alpha_1\sin\theta_r = \alpha_2\sin\theta_t \quad (8\text{-}121)$$

$$\beta_1\sin\theta_r = \beta_2\sin\theta_t \quad (8\text{-}122)$$

The left equality requires

$$\alpha_1\sin\theta_i = \alpha_1\sin\theta_r$$

$$\beta_1\sin\theta_i = \beta_1\sin\theta_r$$

From either of these last two equations we obtain

$$\sin\theta_i = \sin\theta_r$$

or

$$\theta_r = \theta_i \tag{8-123}$$

This is the familiar *law of reflection* encountered in optics.

The law of reflection gives a dilemma from Eqs. 8-121 and 8-122, however, since for a given θ_i, θ_r is fixed, whereas the equations require

$$\sin\theta_t = \frac{\alpha_1}{\alpha_2}\sin\theta_r$$

and

$$\sin\theta_t = \frac{\beta_1}{\beta_2}\sin\theta_r$$

These give different values for θ_t, except for the degenerate case where the two ratios are equal. We have two ways of avoiding this situation. One is to require α_1 and α_2 to be zero (perfect dielectric media). The other is to require α_2 to be infinite (perfect conductor), which means there would be no transmitted field since there can be no electric field in a perfect conductor. These are the two cases we shall consider from this point on. In passing, however, one may still wonder what will happen if a plane wave does impinge on a lossy boundary. Two points of view may be taken in that case. One is to let the angles be complex and interpret the results. The second is to suppose that pure TEM (plane) waves will not be transmitted and reflected and allow other TE or TM components to be generated. These will not be examined in this text.

With the restrictions developed previously we consider only the dielectric–perfect conductor boundary and the perfect dielectric–perfect dielectric boundary, and $\gamma_1 \to j\beta_1$, $\gamma_2 \to j\beta_2$.

Case 1: Perfect Dielectric–Perfect Conductor Boundary with Perpendicular Polarization. For this case $\gamma_1 = 0$, $\beta_1 = k_1$, and $\sigma_2 = \infty$. We have already shown that (Eq. 8-123)

$$\theta_r = \theta_i \tag{8-124}$$

Since the exponentials cancel in Eq. 8-118 and since $E_{2t} = 0$ for a perfect conductor we obtain

$$E_{1i} + E_{1r} = 0$$

or

$$E_{1r} = -E_{1i} \tag{8-125}$$

Using these results in Eq. 8-114 and $\alpha_1 = 0$ the total electric field is

$$\begin{aligned}\vec{E}_1 &= E_{1i}[e^{-(j\beta_1 x\sin\theta_i + j\beta_1 z\cos\theta_i)} = e^{-(j\beta_1 x\sin\theta_i - j\beta_1 z\cos\theta_i)}]\hat{y}\\ &= E_{1i}e^{-j\beta_1 x\sin\theta_i}(e^{-j\beta_1 z\cos\theta_i} - e^{j\beta_1 z\cos\theta_i})\hat{y}\end{aligned}$$

Using Euler's theorem on the expression in parentheses,

$$\vec{E}_1 = -2jE_{1i}\sin(\beta_1 z\cos\theta_i)e^{-j\beta_1 \sin\theta_i x}\hat{y} \tag{8-126}$$

We could define an x propagation (phase) constant as $\beta_{1x} = \beta_1 \sin\theta_i$. The total magnetic field can be obtained by either using Maxwell's equation (Faraday's law in differential form) or by writing the equations for $\vec{H}_{1i}$ and $\vec{H}_{1r}$ and combining. If we use Faraday's law we have

$$\vec{H}_1 = j\frac{1}{\omega\mu_1}\nabla\times\vec{E}_1 = j\frac{1}{\omega\mu_1}\left[-\frac{\partial E_{1y}}{\partial z}\hat{x} + \frac{\partial E_{1y}}{\partial x}\hat{z}\right]$$

which yields, using $\beta_1 = k_1 = \omega\sqrt{\mu_1\epsilon_1}$ and $\eta_1 = \sqrt{\mu_1/\epsilon_1}$ for lossless dielectrics,

$$\vec{H}_1 = \left[-2\frac{E_{1i}}{\eta_1}\cos\theta_i\cos(k_1 z\cos\theta_i)\hat{x} - j2\frac{E_{1i}}{\eta_1}\sin\theta_i\sin(k_1 z\cos\theta_i)\hat{y}\right]e^{-jk_1\sin\theta_i x} \tag{8-127}$$

Both field equations show that there is a standing wave pattern in the z direction (no $e^{jk_1 z}$ terms) but a propagating wave in the x direction (along the boundary).

It is instructive to look at the system along the boundary using the diagram in Figure 8-26. The plane waves are shown as the edges of the planes (called *wavefronts*) containing the fields. If we want to determine how fast the wavefront moves along the x axis, that is, along the boundary, this would yield a phase velocity in the x direction. From the exponent of the field we then have

$$V_{p1x} = \frac{\omega}{\beta_{1x}} = \frac{\omega}{k_1\sin\theta_i} = \frac{\omega}{\omega\sqrt{\mu_1\epsilon_1}\sin\theta_i} = \frac{V_{p1}}{\sin\theta_i} \tag{8-128}$$

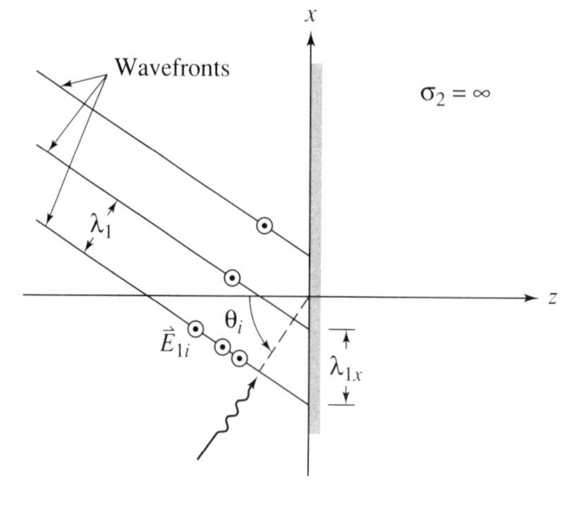

Figure 8-26. Wavefront picture of a plane wave at a perfect conductor boundary.

For $\theta_i = 0$ (normal incidence) the phase velocity along the boundary is infinite as predicted since the wave front will then strike the entire x axis simultaneously everywhere, so it appears to have moved along the boundary infinitely fast.

We can also define a wavelength in the x direction, which would give the distance between points on the boundary that have the same phase, that is, differ by 360°:

$$\lambda_{1x} = \frac{2\pi}{\beta_{1x}} = \frac{2\pi}{\omega\sqrt{\mu_1\epsilon_1}\,\sin\theta_i} = \frac{V_{p1}}{f\sin\theta_i} = \frac{\lambda_1}{\sin\theta_i} \tag{8-129}$$

Considering next the z direction, the distance between nulls (or maxima, or any corresponding values) of the field *magnitude* occurs for the distance d along the standing wave pattern for which

$$k_1 z\cos\theta_i\big|_{z=d} = \pi$$

or

$$d = \frac{\pi}{k_1\cos\theta_i} = \frac{V_{p1}}{2f\cos\theta_i} = \frac{\lambda_1}{2\cos\theta_i} \tag{8-130}$$

The standing wave pattern $|\vec{E}_1|$ is plotted in Figure 8-27. Note that for $\theta_i = 0$ we obtain the shorted transmission line result.

For the wave impedance it is now most important to specify the direction we wish to look, whereas we really had no option in the previous section on normal incidence. To look in the $+z$ direction we use the field components that produce propagation in that direction, namely, the y component of $\vec{E}_1$ and the positive x component of $\vec{H}_1$. From Eqs. 8-126 and 8-127 we obtain

$$Z^w_{z,\perp} = \frac{-2jE_{1i}\sin(k_1 z\cos\theta_i)}{-2\dfrac{E_{1i}}{\eta_1}\cos\theta_i\cos(k_1 z\cos\theta_i)}$$

$$= j\eta_1\sec\theta_i\tan(k_1\cos\theta_i z) \tag{8-131}$$

This is reactive as expected for a short circuit at $z = 0$ and varies with distance z from the short similar to the transmission line. We have used the symbol $\perp$ to denote perpendicular polarization.

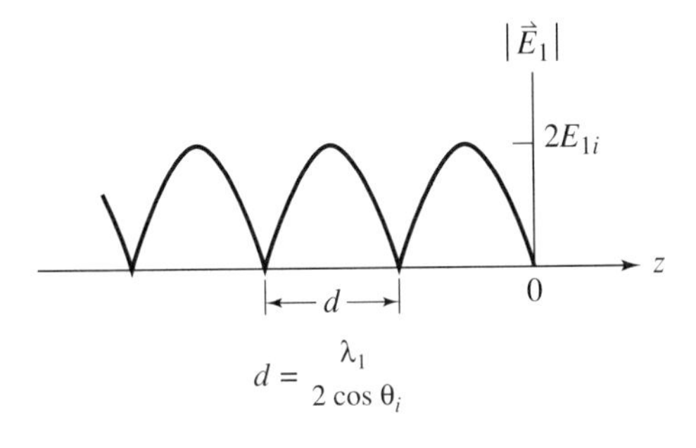

Figure 8-27. Standing wave pattern of a plane wave obliquely incident on a perfect conductor at an angle θ_i.

Similarly, the wave impedance in the $+x$ direction would use the y component of $\vec{E}_1$ and the positive z component of $\vec{H}_1$:

$$Z^w_{x,\perp} = \frac{-2jE_{1i} \sin(k_1 z \cos\theta_i)}{-2 \dfrac{E_{1i}}{\eta_1} \sin\theta_i \sin(k_1 z \cos\theta_i)} = \eta_1 \csc\theta_i \tag{8-132}$$

Case 2: Perfect Dielectric–Perfect Dielectric Boundary with Perpendicular Polarization. For this case $\alpha_1 = \alpha_2 = 0$, $\beta_1 = k_1$, $\beta_2 = k_2$. We have already shown that (Eq. 8-123)

$$\theta_r = \theta_i \tag{8-133}$$

In this case we can have a field in Medium 2 so we must find the transmitted angle θ_t. Using Eq. 8-122 we have from the phase-matching condition

$$\sin\theta_t = \frac{k_1}{k_2} \sin\theta_r = \frac{\omega\sqrt{\mu_1\epsilon_1}}{\omega\sqrt{\mu_2\epsilon_2}} \sin\theta_i = \sqrt{\frac{\mu_1\epsilon_1}{\mu_2\epsilon_2}} \sin\theta_i$$

We can cancel μ_0 and ϵ_0 from the constants to obtain

$$\sin\theta_t = \sqrt{\frac{\mu_{1r}\epsilon_{1r}}{\mu_{2r}\epsilon_{2r}}} \sin\theta_i \tag{8-134}$$

This is Snell's law for refraction, and it is most usually written for the case where the dielectrics are nonmagnetic.

$$\sin\theta_t = \sqrt{\frac{\epsilon_{1r}}{\epsilon_{2r}}} \sin\theta_i \tag{8-135}$$

From this last form of Snell's law, since $\sin\theta_t \leq 1$ we must have

$$\sqrt{\frac{\epsilon_{1r}}{\epsilon_{2r}}} \sin\theta_i \leq 1 \qquad \text{or} \qquad \sin\theta_i \leq \sqrt{\frac{\epsilon_{2r}}{\epsilon_{1r}}}$$

If this condition does not exist, there cannot be a propagating transmitted component. For the limiting case of equality we have

$$\sin\theta_t = 1 \qquad \text{or} \qquad \theta_t = \frac{\pi}{2}$$

We see that the wave moves along the boundary in the $+x$ direction. However, if there is a wave at the boundary, there must be some field component in Medium 2 since we still must require tangential electric field to be continuous. This will be explained in more detail shortly.

There is thus some critical angle at which there is no signal propagation in the z direction in Medium 2. This occurs, as explained earlier, when

$$\sin\theta_{ic} = \sqrt{\frac{\epsilon_2}{\epsilon_1}}$$

or

$$\theta_{ic} = \sin^{-1} \sqrt{\frac{\epsilon_2}{\epsilon_1}} \tag{8-136}$$

This is called the *critical angle* for total reflection. Note this can only happen if $\epsilon_2 < \epsilon_1$, that is, if Medium 1 has a smaller wave impedance η_1 than Medium 2. Use is made of this fact in the design of fiber optic components where one wants to confine a wave to Medium 1. We then conclude that total reflection occurs whenever

$$\theta_i \geq \theta_{ic}$$

To see what is happening in Medium 2 for total reflection we assume an angle $\theta_i > \theta_{ic}$ so that Snell's law yields

$$\sin \theta_t = \sqrt{\frac{\epsilon_{1r}}{\epsilon_{2r}}} \sin(\theta + \theta_{ic}) = K - 1$$

The only way this can be true is for θ_t to be complex so we can get hyperbolic functions, say

$$\theta_t = U + jV$$

Then

$$\begin{aligned} \sin \theta_t &= \sin(U + jV) = \sin V \cos jV + \sin jV \cos U \\ &= \sin U \cosh V + j \sinh V \cos U = K = K + j0 \end{aligned}$$

Equating real and imaginary parts

$$\sinh V \cos U = 0$$

$$\sin U \cosh V = K$$

There are two possibilities for the first equation:

$$V = 0 \qquad \text{or} \quad U = \frac{\pi}{2}$$

For $V = 0$ the second equation would require

$$\sin U \cosh 0 = \sin U = K$$

which is not possible since $K > 1$ and U is real.

For $U = \pi/2$ the second equation of the set gives

$$\cosh V = K$$

from which

$$V = \cosh^{-1} K$$

Thus

$$\theta_t = \frac{\pi}{2} + j \cosh^{-1} K$$

With this, the propagation exponent of the field in Medium 2, Eq. 8-109 with $\gamma_2 = \alpha_2 + j\beta_2 = jk_2$, becomes

$$\begin{aligned} -jk_2(x \sin\theta_t + z \cos\theta_t) &= -jk_2\left[x \sin\left(\frac{\pi}{2} + j \cosh^{-1}K\right) + z \cos\left(\frac{\pi}{2} + j \cosh^{-1} K\right)\right] \\ &= -jk_2[x \cos(j \cosh^{-1} K) - z \sin(j \cosh^{-1} K)] \\ &= -jk_2[x \cosh(\cosh^{-1} K) - jz \sinh(\cosh^{-1} K)] \\ &= -jk_2xK - k^2z \sinh(\cosh^{-1} K) \end{aligned}$$

The exponent of the Medium 2 field is then

$$e^{[-jk_2Kx - k_2\sinh(\cosh^{-1}K)]} = e^{-jk_2Kx}e^{-k_2\sinh(\cosh^{-1}K)}$$

Physically, this tells us that while there is a field in Medium 2 it is attenuating only in the z direction, but propagating in the x direction. This attenuation is not due to dielectric losses since we assumed perfect dielectrics.

The terms used to describe these effects are:

1. Waves that attenuate in the direction of a coordinate are called *evanescent waves*.
2. Waves at a boundary that propagate in one direction but attenuate in the direction normal to the propagation direction at a boundary are called *surface waves*.

Thus for incident angles greater than θ_{ic} surface waves are generated. Another way of looking at this result is that the incident wave attaches itself to the boundary by producing an evanescent wave in the second medium. Thus, even though we call it "total" reflection there is still some field in the second medium.

Returning to our specific problem, since the exponentials cancel the boundary condition Eq. 8-118 becomes

$$E_{1i} + E_{1r} = E_{2t} \tag{8-137}$$

Also, at the boundary there is no surface current $\vec{J}_s$, so the tangential x components of $\vec{H}$ must be continuous. From the x components of Eq.s 8-116 and 8-117 at $z = 0$ and $\gamma = j\beta$ as before, and again canceling the exponents since they are the same as those in the electric field,

$$\frac{-\beta_1}{\omega\mu_1} \cos\theta_i E_{1i} + \frac{\beta_1}{\omega\mu_1} \cos\theta_r E_{1r} = \frac{-\beta_2}{\omega\mu_2} \cos\theta_t E_{2t}$$

This may be simplified using

$$\frac{\beta_1}{\omega\mu_1} = \frac{\omega\sqrt{\mu_1\epsilon_1}}{\omega\mu_1} = \frac{1}{\eta_1} \quad \text{and} \quad \frac{\beta_2}{\omega\mu_2} = \frac{1}{\eta_2}$$

and using the secant in place of the cosine. The result is

$$\frac{E_{1i}}{\eta_1 \sec\theta_i} - \frac{E_{1r}}{\eta_1 \sec\theta_r} = \frac{E_{2t}}{\eta_2 \sec\theta_t} \tag{8-138}$$

The denominators of these are the wave impedances in the $+z$ direction, which we denote by

$$Z^w_{1z,\perp} = \frac{E_{1i}}{-H_{1i,x}} = \eta_1 \sec\theta_i \qquad \text{and} \quad Z^w_{2z,\perp} = \frac{E_{2t}}{-H_{2t,x}} = \eta_2 \sec\theta_t \tag{8-139}$$

The minus signs are required because those components, while yielding positive z-directed cross products with $\vec{E}_1$ are directed in negative x. This assures that the impedances are positive.

With these wave impedances the boundary condition may be written, remembering that $\theta_r = \theta_i$,

$$\frac{E_{1i}}{Z^w_{1z,\perp}} - \frac{E_{1r}}{Z^w_{1z,\perp}} = \frac{E_{2t}}{Z^w_{2z,\perp}} \tag{8-140}$$

Solving this equation for E_{2t}, substituting into the electric field boundary condition (Eq. 8-137), and solving for the ratio E_{1r}/E_{1i} we obtain a reflection coefficient

$$\Gamma_\perp = \frac{E_{1r}}{E_{1i}} = \frac{Z^w_{2z,\perp} - Z^w_{1z,\perp}}{Z^w_{2z,\perp} + Z^w_{1z,\perp}} \tag{8-141}$$

Note this again looks like the transmission line equation for the reflection coefficient where Medium 2 is the load on Medium 1.

We can obtain the transmission coefficient from Eq. 8-137 by first dividing it by E_{1i}, then substituting Eq. 8-141 and solving for the ratio E_{2t}/E_{1i}. The result is

$$\frac{E_{2t}}{E_{1i}} = \tau_\perp = 1 + \Gamma_\perp = \frac{2Z^w_{2z,\perp}}{Z^w_{2z,\perp} + Z^w_{1z,\perp}} \tag{8-142}$$

Example 8-18

A uniform plane wave in air with perpendicular polarization is incident upon a large block of polystyrene (assumed lossless) at an angle of 35°. The frequency is 250 MHz. The incoming amplitude is 10^{-4} V/m. Determine the amplitudes of all field components and also the angle at which total reflection may occur. The configuration is shown in Figure 8-28. From the law of reflection

$$\theta_r = \theta_i = 35°$$

From Snell's law for nonmagnetic dielectrics, Eq. 8-135,

$$\sin\theta_t = \sqrt{\frac{1}{2.56}}\sin 35° = 0.3585 \Rightarrow \theta_t = 21°$$

$$\eta_1 = \sqrt{\frac{\mu_0}{\epsilon_0}} = 377\ \Omega, \qquad \eta_2 = \sqrt{\frac{\mu_0}{2.56\epsilon_0}} = 235.6\ \Omega$$

The z-directed wave impedances are obtained from Eq. 8-139:

$$Z^w_{1z,\perp} = (377)(\sec 35°) = 460.2\ \Omega, \qquad Z^w_{2z,\perp} = (235.6)(\sec 21°) = 252.4\ \Omega$$

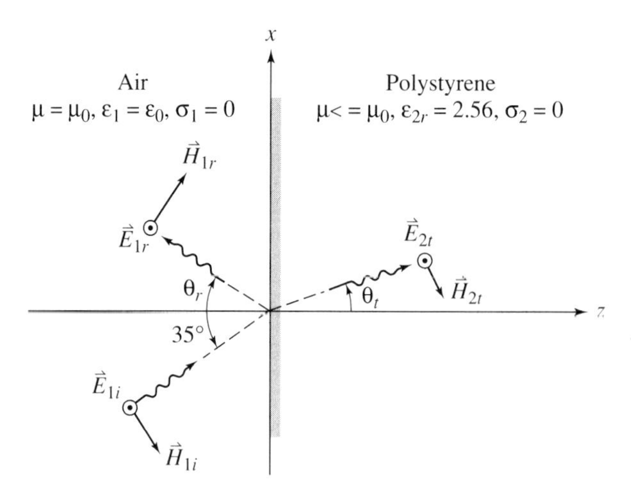

Figure 8-28. Field and material configuration for Example 8-18.

The reflection coefficient is

$$\Gamma_{\perp} = \frac{252.4 - 460.2}{252.4 + 460.2} = -0.2916$$

From this,

$$E_{1r} = \Gamma_{\perp} E_{1i} = (-0.292)(10^{-4}) = -2.92 \times 10^{-5} \text{ V/m}$$

The transmission coefficient is

$$\tau_{\perp} = 1 + \Gamma_{\perp} = 0.7084$$

Then

$$E_{2t} = \tau_{\perp} E_{1i} = (0.7084)(10^{-4}) = 7.08 \times 10^{-5} \text{ V/m}$$

From the three electric field values, the total magnetic fields are

$$H_{1i} = \frac{E_{1i}}{\eta_1} = \frac{10^{-4}}{377} = 2.65 \times 10^{-7} \text{ A/m}$$

$$H_{1r} = \frac{E_{1r}}{\eta_1} = \frac{-2.92 \times 10^{-5}}{377} = -7.74 \times 10^{-8} \text{ A/m}$$

$$H_{2t} = \frac{E_{2t}}{\eta_2} = \frac{7.08 \times 10^{-5}}{235.6} = 3 \times 10^{-7} \text{ A/m}$$

The critical angle for total reflection, Eq. 8-136, is

$$\theta_{ic} = \sin^{-1}\sqrt{\frac{2.56}{1}} \Rightarrow \text{undefined} \qquad (\text{note } \epsilon_2 > \epsilon_1)$$

Thus for this configuration we cannot obtain total reflection. Note that these results are independent of frequency. The frequency enters the problem only if we wish to reconstruct the space-dependent field equations using Eqs. 8-114 through 8-117, where $\gamma \rightarrow j\beta = j\omega\sqrt{\mu\epsilon}$.

8-7.2 Parallel Polarization at Oblique Incidence

We will not repeat the proof that led to restricting our discussions to perfect dielectric–perfect conductor boundaries and perfect dielectric–perfect dielectric boundaries. The same analysis would give the identical result. The basic system to be analyzed is that of Figure 8-24*b*. Note that in this case we assume all magnetic field intensity vectors are out of the page for the three waves and then assign the electric field intensity vectors in the directions that produce Poynting vectors in the appropriate direction for each wave.

The expressions for magnetic fields are derived using the identical processes used for the electric fields. Basically we merely replace all Es by Hs since they are directed in the same direction y. The results are

$$\vec{H}_{1i} = H_{1i}e^{-j\beta_1\xi_i}\hat{y} = H_{1i}e^{-j(\beta_1\sin\theta_i x + \beta_1\cos\theta_i z)}\hat{y} \tag{8-143}$$

$$\vec{H}_{1r} = H_{1r}e^{-j\beta_1\xi_r}\hat{y} = H_{1r}e^{-j(\beta_1\sin\theta_r x - \beta_1\cos\theta_r z)}\hat{y} \tag{8-144}$$

$$\vec{H}_{2t} = H_{2t}e^{-j\beta_2\xi_t}\hat{y} = H_{2t}e^{-j(\beta_2\sin\theta_t x + \beta_2\cos\theta_t z)}\hat{y} \tag{8-145}$$

The total fields in each medium are then

$$\vec{H}_1 = \vec{H}_{1i} + \vec{H}_{1r} = [H_{1i}e^{-j(\beta_1\sin\theta_i x + \beta_1\cos\theta_i z)} + H_{1r}e^{-j(\beta_1\sin\theta_r x - \beta_1\cos\theta_r z)}]\hat{y} \tag{8-146}$$

$$\vec{H}_2 = \vec{H}_{2t} = H_{2t}e^{-j(\beta_2\sin\theta_t x + \beta_2\cos\theta_t z)}\hat{y} \tag{8-147}$$

The corresponding electric field intensities can be obtained by using Ampere's law in differential form from Maxwell's equations:

$$\begin{aligned} \vec{E}_1 &= \vec{E}_{1i} + \vec{E}_{1r} = -j\frac{1}{\omega\epsilon_1}\nabla \times \vec{H}_1 \\ &= [\eta_1 \cos\theta_i H_{1i}e^{-(\beta_1\sin\theta_i x + \beta_1\cos\theta_i z)} - \eta_1 \cos\theta_r H_{1r}e^{-j(\beta_1\sin\theta_r x - \beta_1\cos\theta_r z)}]\hat{x} \\ &\quad + [-\eta_1 \sin\theta_i H_{1i}e^{-j(\beta_1\sin\theta_i x + \beta_1\cos\theta_i z)} - \eta_1 \sin\theta_r H_{1r}e^{-j(\beta_1\sin\theta_r x - \beta_2\cos\theta_r z)}]\hat{z} \end{aligned} \tag{8-148}$$

$$\begin{aligned} \vec{E}_2 &= \vec{E}_{2t} = -j\frac{1}{\omega\epsilon_2}\nabla \times \vec{H}_2 \\ &= \eta_2 \cos\theta_t H_{2t}e^{-j(\beta_2\sin\theta_t x + \beta_2\cos\theta_t z)}\hat{x} - \eta_2 \sin\theta_2 H_{2t}e^{-j(\beta_2\sin\theta_{i\rightarrow t} x + \beta_2\cos\theta_t z)}\hat{y} \end{aligned} \tag{8-149}$$

For perfect dielectrics we will use $\beta_1 = k_1$, $\beta_2 = k_2$.

We next apply these results to the same two cases as we did for perpendicular polarization.

Case 1: Perfect Dielectric–Perfect Conductor Boundary with Parallel Polarization. We have already shown that for any two media (see Eq. 8-123)

$$\theta_r = \theta_i \tag{8-150}$$

In the perfect conductor the electric and magnetic fields are zero, so we set $E_{2t} = 0$ and $H_{2t} = 0$. Also, at the metal boundary the tangential electric field must be zero, which is for the x components. Since $\vec{E}_2$ is zero we use the x component of Eq. 8-148 evaluated at $z = 0$. This requires (using k in place of β for perfect dielectrics)

$$\eta_1 \cos\theta_i H_{1i} e^{-jk_1 \sin\theta_i x} - \eta_1 \cos\theta_r H_{1r} e^{-jk_1 \sin\theta_r x} = 0$$

Due to the phase matching condition (8-150) we cancel terms to obtain

$$H_{1i} - H_{1r} = 0$$

or

$$H_{1r} = H_{1i} \tag{8-151}$$

Using the preceding results in Eq. 8-146, again using k_1 for β_1 for lossless dielectrics, we obtain

$$\begin{aligned} \vec{H}_1 &= H_{1i} e^{-jk_1 \sin\theta_i x} (e^{-jk_1 \cos\theta_i z} + e^{jk_1 \cos\theta_i z}) \hat{y} \\ \vec{H}_1 &= 2H_{1i} \cos(k_1 \cos\theta_i) e^{-jk_1 \sin\theta_i x} \hat{y} \end{aligned} \tag{8-152}$$

The form of this result is similar to that for perpendicular polarization where we have a standing wave pattern in z and a propagating characteristic in x. The corresponding electric field is obtained using Ampere's law and is

$$\begin{aligned} \vec{E}_1 &= -j \frac{1}{\omega \epsilon_1} \nabla \times \vec{H}_1 \\ &= [-j2\eta_1 \cos\theta_i H_{1i} \sin(k_1 \cos\theta_i z) \hat{x} \\ &\quad -2\eta_1 \sin\theta_i H_{1i} \cos(k_1 \cos\theta_i z) \hat{z}] e^{-jk_1 \sin\theta_i x} \end{aligned} \tag{8-153}$$

As for perpendicular polarization we have the phase velocity

$$V_{p_{1x}} = \frac{\omega}{\beta_{1x}} = \frac{\omega}{k_1 \sin\theta_i} = \frac{V_{p1}}{\sin\theta_i} \tag{8-154}$$

and the wavelength

$$\lambda_{1x} = \frac{2\pi}{\beta_{1x}} = \frac{\lambda_1}{\sin\theta_i} \tag{8-155}$$

The distance between nulls in the z direction is also given by

$$d = \frac{\lambda_1}{2 \cos\theta_i} \tag{8-156}$$

The wave impedance in the positive z direction would require us to use E_{1x} and H_{1y}. Taking these components from Eqs. 8-152 and 8-153 and using $\|$ to represent parallel polarization,

$$Z^w_{z,\|} = \frac{E_{1x}}{H_{1y}} = \frac{-j2\eta_1 \cos\theta_i H_{1i} \sin(k_1 \cos\theta_i z)}{2H_{2i} \cos(k_1 \cos\theta_i z)} = -j \cos\theta_i \tan(k_1 \cos\theta_i z) \tag{8-157}$$

This is reactive as expected, but not of identical functional form as before.

For the impedance in the propagating x direction

$$Z^{w}_{x,\|} = \frac{-E_{1z}}{H_{1y}} = 2\eta_1 \sin\theta_i \tag{8-158}$$

The minus sign is used because the x component is in the negative direction.

Case 2: Perfect Dielectric–Perfect Dielectric Boundary with Parallel Polarization. As with the other cases, the incident and reflected angles are equal. Since we now have a wave in Medium 2 we must match the x components of the exponents at $z = 0$ to give the phase match equation

$$-jk_1 \sin\theta_i = -jk_2 \sin\theta_t$$

Thus

$$\sin\theta_t = \frac{k_1}{k_2}\sin\theta_i = \sqrt{\frac{\mu_1\epsilon_1}{\mu_2\epsilon_2}}\sin\theta_i \tag{8-159}$$

This is Snell's law of refraction as obtained before.

The term *refractive index* is used to describe dielectric materials that have $\mu_1 = \mu_2$ or, more usually, $\mu_1 = \mu_2 = \mu_0$. If Medium 1 is vacuum, the refractive index is defined in terms of plane wave velocities in each separate medium:

$$\text{refractive index} \equiv n \equiv \frac{V_{p_{\text{vacuum}}}}{V_{p_{\text{dielectric}}}} = \frac{\sqrt{\mu_0\epsilon_{\text{diel}}}}{\sqrt{\mu_0\epsilon_0}} = \sqrt{\frac{\epsilon_{\text{diel}}}{\epsilon_0}} = \sqrt{\epsilon_{1r}} \tag{8-160}$$

With this definition, for two dielectric Media 1 and 2 in general we would have (still using $\mu_1 = \mu_2 = \mu_0$)

$$\sqrt{\frac{\epsilon_1}{\epsilon_2}} = \sqrt{\frac{\mu_0\epsilon_1}{\mu_0\epsilon_2}} = \sqrt{\frac{\dfrac{\mu_0\epsilon_1}{\mu_0\epsilon_0}}{\dfrac{\mu_0\epsilon_2}{\mu_0\epsilon_0}}} = \frac{\sqrt{\dfrac{\epsilon_1}{\epsilon_0}}}{\sqrt{\dfrac{\epsilon_2}{\epsilon_0}}} = \frac{n_1}{n_2}$$

Thus for nonmagnetic dielectrics Snell's law (Eq. 8-159) takes the form

$$\sin\theta_t = \frac{n_1}{n_2}\sin\theta_i \qquad (\mu_1 = \mu_2 = \mu_0) \tag{8-161}$$

There is also a critical angle for this polarization that is the same as before,

$$\theta_{ic} = \sin^{-1}\sqrt{\frac{\epsilon_2}{\epsilon_1}} \tag{8-162}$$

We next apply the boundary condition that the tangential electric field must be continuous at $z = 0$. Thus using the x components of Eqs. 8-148 and 8-149 at $z = 0$ and canceling the exponentials since the phase terms are equal,

$$\eta_1 \cos\theta_i H_{1i} - \eta_1 \cos\theta_r H_{1r} = \eta_2 \cos\theta_t H_{2t}$$

Since $\theta_r = \theta_i$,

$$\eta_1 \cos\theta_i H_{1i} - \eta_1 \cos\theta_i H_{1r} = \eta_2 \cos\theta_t H_{2t} \tag{8-163}$$

Tangential magnetic field intensity must also be continuous. Using Eqs. 8-146 and 8-147 the y components must satisfy

$$H_{1i} + H_{1r} = H_{2t} \tag{8-164}$$

The wave impedance in the positive z direction would involve (see Figure 8-23) the components E_x and H_y for the positive traveling components in both media. Since all components point in positive coordinate directions we use components from Eqs. 8-146 and 8-148 to obtain

$$Z^w_{1z,\|} = \frac{E_{1i,x}}{H_{1i,y}} = \frac{E_{1i,x}}{H_{1i}} = \eta_1 \cos\theta_i \tag{8-165}$$

For Medium 2:

$$Z^w_{2z,\|} = \frac{E_{2t,x}}{H_{2t,y}} = \frac{E_{2t,x}}{H_{2t}} = \eta_2 \cos\theta_t \tag{8-166}$$

For the reflected field the x component of E_{1r} is directed in the negative x direction so we need to use components from Eqs. 8-146 and 8-148 as follows:

$$Z^w_{1(-z),\|} = \frac{-E_{1r,x}}{H_{1ry}} = \frac{-E_{1r,x}}{H_{1r}} = \eta_1 \cos\theta_r = \eta_1 \cos\theta_i = Z^w_{1z,\|} \tag{8-167}$$

With these impedances the two boundary conditions, Eqs. 8-163 and 8-164, may be written

$$E_{1i,x} + E_{1r,x} = E_{2t,x} \tag{8-168}$$

$$\frac{E_{1i,x}}{Z^w_{1z,\|}} - \frac{E_{1r,x}}{Z^w_{1z,\|}} = \frac{E_{2t,x}}{Z^w_{2z,\|}} \tag{8-169}$$

Substituting the first equation for $E_{2t,x}$ into the second equation and then solving for the ratio $E_{1r,x}/E_{1i,x}$, we obtain a reflection coefficient

$$\Gamma_{\|} = \frac{E_{1r,x}}{E_{1i,x}} = \frac{Z^w_{2z,\|} - Z^w_{1z,\|}}{Z^w_{2z,\|} + Z^w_{2z,\|}} \tag{8-170}$$

This is again a familiar basic form where we consider Medium 2 as the load on Medium 1. The corresponding transmission coefficient is obtained by dividing Eq. 8-168 by $E_{1i,x}$, using the reflection coefficient, and solving for $E_{2t,x}/E_{1i,x}$. The results are

$$\frac{E_{2t,x}}{E_{1i,x}} = \tau_{\|} = 1 + \Gamma_{\|} = \frac{2Z^w_{2z,\|}}{Z^w_{1z,\|} + Z^w_{2z,\|}} \tag{8-171}$$

Example 8-19

A uniform wave in air having amplitude 10 V/m and frequency 200 MHz is parallel polarized and obliquely incident upon a plane polyethylene boundary at 20°. The system is shown in Figure 8-29. Assume lossless media and ϵ_{2r} of 2.26. Determine the amplitudes

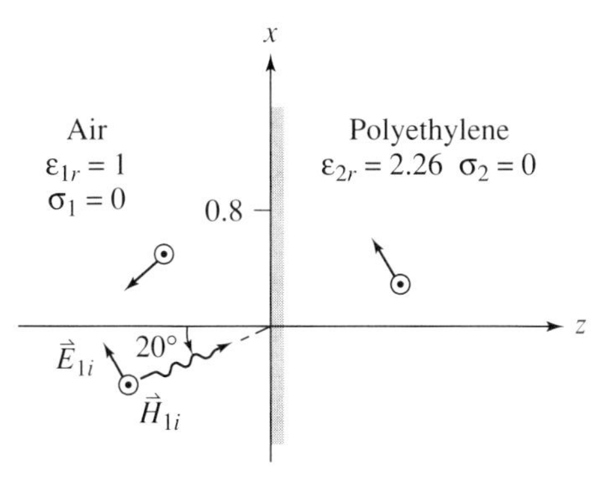

Figure 8-29.

of incident, reflected, and transmitted fields, the real-time expression for $\mathcal{H}_{1r}$, and the value of $E_{1x}(0.8, 0, 0)$.

From the problem statement

$$E_{1i} = 10 \text{ V/m}, \qquad \theta_i = 20°$$

Then

$$E_{1i,x} = (10)(\cos 20°) = 9.4 \text{ V/m} \qquad \text{and} \quad \theta_r = 20°$$

$$\eta_1 = \sqrt{\frac{\mu_0}{\epsilon_0}} = 377 \ \Omega, \qquad \eta_2 = \sqrt{\frac{\mu_0}{2.26\epsilon_0}} = 251 \ \Omega$$

From Snell's law,

$$\sin\theta_t = \sqrt{\frac{\epsilon_0}{2.26\epsilon_0}}\sin 20° = 0.2275 \qquad \text{or} \quad \theta_t = 13.2°$$

The wave impedances are then

$$Z^w_{1z,\|} = 377 \cos 20° = 354 \ \Omega \qquad \text{and} \quad Z^w_{2z,\|} = 251 \cos 13.2° = 244 \ \Omega$$

$$\Gamma_\| = \frac{244 - 354}{244 + 354} = -0.184 = \frac{E_{1r,x}}{E_{1i,x}}$$

Then

$$E_{1r,x} = (-0.184)(9.4) = -1.73 \text{ V/m}$$

Using trigonometry,

$$|E_{1r,x}| = E_{1r}\cos\theta_r = E_{1r}\cos\theta_i$$

Thus

$$E_{1r} = \frac{|-1.73|}{\cos 20°} = 1.84 \text{ V/m}$$

Since the incident wave is by itself a uniform plane wave,

$$\frac{E_{1i}}{H_{1i}} = \eta_1$$

from which

$$H_{1i} = \tfrac{10}{377} = 0.0265 \text{ A/m}$$

Similarly,

$$H_{1r} = \frac{E_{1r}}{\eta_1} = \frac{1.84}{377} = 0.00488 \text{ A/m}$$

For the transmitted wave,

$$\tau_{\|} = 1 + \Gamma_{\|} = 0.816 = \frac{E_{2t,x}}{E_{1i,x}}$$

$$E_{2t,x} = (0.816)(9.4) = 7.67 \text{ V/m}$$

Using the same procedure as for the reflected wave:

$$E_{2t} = \frac{E_{2t,x}}{\cos\theta_t} = \frac{7.67}{\cos 13.3^\circ} = 7.88 \text{ V/m}$$

$$H_{2t} = \frac{E_{2t}}{\eta_2} = \frac{7.88}{251} = 0.0314 \text{ A/m}$$

As a check (good engineering practice) the tangent electric field (x components) at the boundary are continuous:

$$9.4 + (-1.73) = 7.67 = E_{2t,x}$$

Similarly, for the magnetic field y components

$$0.0265 + 0.00488 = 0.0314 = H_{2t,y} = H_{2t}$$

For the lossless dielectric (air)

$$\beta_1 = k_1 = \omega\sqrt{\mu_0\epsilon_0} = \frac{2\pi \times 200 \times 10^6}{3 \times 10^8} = 4.19 \text{ rad/m}$$

Equation 8-144 then gives

$$\begin{aligned}\vec{H}_{1r} &= 0.00488e^{-j(4.19\sin 20^\circ x - 4.19\cos 20^\circ z)}\hat{y} \text{ A/m}\\ &= 4.88e^{-j(1.433x - 3.937z)}\hat{y} \text{ mA/m}\end{aligned}$$

Converting to the time domain, for $\omega = 2\pi \times 200 \times 10^6 = 1.257$ Grad/s

$$\vec{\mathscr{H}}_{1r} = \text{Re}\{\vec{H}_{1r}e^{j\omega t}\} = 4.88 \cos(1.257 \times 10^9 t - 1.433x + 3.937z)\hat{y} \text{ mA/m}$$

To evaluate $E_{1x}(0.8, 0, 0)$ we use the x component of Eq. 8-149 to obtain

$$\begin{aligned}E_{1x}(0.8, 0, 0) &= 377 \cos 20^\circ \times 0.0265e^{-j(1.433)(0.8)} - 377 \cos 20^\circ\\ &\quad \times 0.00488e^{-j(1.433)(0.8)}\\ &= 7.67e^{-j1.1464} = 7.67e^{-j65.7^\circ} \text{ V/m}\end{aligned}$$

Note that the tangential field amplitude stays constant at the boundary, but the phase is delayed as would be expected since it takes the waveform time to propagate to that location. This delay can be obtained by applying the phase velocity in the x direction, Eq. 8-128:

$$V_{p_{1x}} = \frac{V_{p1}}{\sin \theta_i} = \frac{\frac{1}{\sqrt{\mu_0 \epsilon_0}}}{\sin 20^\circ} = 8.77 \times 10^8 \text{ m/s}$$

Time required to move to $x = 0.8$ m is

$$t_{0.8} = \frac{0.8}{8.77 \times 10^8} = 0.912 \text{ ns}$$

For the given frequency of 200 MHz, the period is

$$T = \frac{1}{f} = 5 \text{ ns}$$

Thus the fraction of 1 cycle is a delay of

$$\frac{0.912}{5} \times 360^\circ = 65.7^\circ$$

8-7.3 Total Transmission: Brewster (Polarizing) Angle

For both perpendicular and parallel polarizations we learned that when $\epsilon_1 > \epsilon_2$ there was a critical incident angle θ_{ic} for which (and beyond which) there was total reflection. We next inquire as to whether or not there is some incident angle (or angles) at which the entire incident wave is transmitted. This will be the case if the reflection coefficient is zero. The resulting angle is called the *Brewster*, or *polarizing*, angle.

For parallel polarization the reflection coefficient given in Eq. 8-170 will be zero if the numerator is zero or

$$Z^w_{2z,\|} - Z^w_{1z,\|} = 0$$

Substituting for the impedances from Eqs. 8-165 and 8-166, using θ_{iB} to denote the Brewster angle, and $\|$ for parallel,

$$\eta_2 \cos \theta_{t\|} - \eta_1 \cos \theta_{iB\|} = 0$$

To get $\theta_{iB\|}$ we first express the equation using sine functions and then use Snell's law, Eq. 8-159, to eliminate $\theta_{t\|}$:

$$\eta_2 \sqrt{1 - \sin^2 \theta_{t\|}} - \eta_1 \sqrt{1 - \sin^2 \theta_{iB\|}} = 0$$

$$\eta_2 \sqrt{1 - \frac{\mu_1 \epsilon_1}{\mu_2 \epsilon_2} \sin^2 \theta_{iB\|}} = \eta_1 \sqrt{1 - \sin^2 \theta_{iB\|}}$$

$$\sqrt{\frac{\mu_2}{\epsilon_2}} \sqrt{1 - \frac{\mu_1 \epsilon_1}{\mu_2 \epsilon_2} \sin^2 \theta_{iB\|}} = \sqrt{\frac{\mu_1}{\epsilon_1}} \sqrt{1 - \sin^2 \theta_{iB\|}}$$

Solving:

$$\sin\theta_{iB\parallel} = \sqrt{\frac{\dfrac{\epsilon_2}{\epsilon_1} - \dfrac{\mu_2}{\mu_1}}{\dfrac{\epsilon_2}{\epsilon_1} - \dfrac{\epsilon_1}{\epsilon_2}}} \tag{8-172}$$

Similar steps on the perpendicular polarization reflection coefficient, Eq. 8-141, give

$$\sin\theta_{iB\perp} = \sqrt{\frac{\dfrac{\epsilon_2}{\epsilon_1} - \dfrac{\mu_2}{\mu_2}}{\dfrac{\mu_1}{\mu_2} - \dfrac{\mu_2}{\mu_1}}} \tag{8-173}$$

Most applications of these results are for $\mu_1 = \mu_2$, which are usually both μ_0. For this case, there is *no Brewster angle for the perpendicular polarization*, but for parallel polarization Eq. 8-172 reduces to

$$\sin\theta_{iB\parallel} = \sqrt{\frac{\epsilon_2}{\epsilon_1 + \epsilon_2}} \qquad (\mu_1 = \mu_2) \tag{8-174}$$

Note there is just one unique angle for total transmission, whereas for total reflection there is a range of angles. The application of this phenomenon is to select (or reject, depending upon one's point of view) a particular polarization out of a general orientation plane wave. For a general electric field orientation for a plane wave we can resolve it into perpendicular and parallel components. If the original wave were incident at the Brewster angle, all of the parallel component passes into Medium 2 while only part of the perpendicular component will be transmitted with the remainder reflected. This effect is frequently used in laser design so that all of one polarization can exit the system.

Example 8-20

A uniform plane wave in air whose electric field intensity is 100 V/m is directed out of the *xz* plane and at an angle of 45° with respect to the plane. It is obliquely incident at the Brewster angle onto a slab of glass (SiO_2), as shown in Figure 8.30. The frequency is 300 MHz. Determine the percentage of the parallel polarization power transmitted and reflected and the percentage of the perpendicular power transmitted and reflected.

We first decompose the total field into its parallel and perpendicular components:

$$|\vec{E}_{1\perp}| = 100 \sin 45° = 70.7 \text{ V/m}$$

$$|\vec{E}_{1\parallel}| = 100 \cos 45° = 70.7 \text{ V/m}$$

Using Eq. 8-174 we find the incident angle, θ_{iB},

$$\sin\theta_{iB\parallel} = \sqrt{\frac{5.04\epsilon_0}{\epsilon_0 + 5.04\epsilon_0}} \Rightarrow \theta_{iB\parallel} = 66°$$

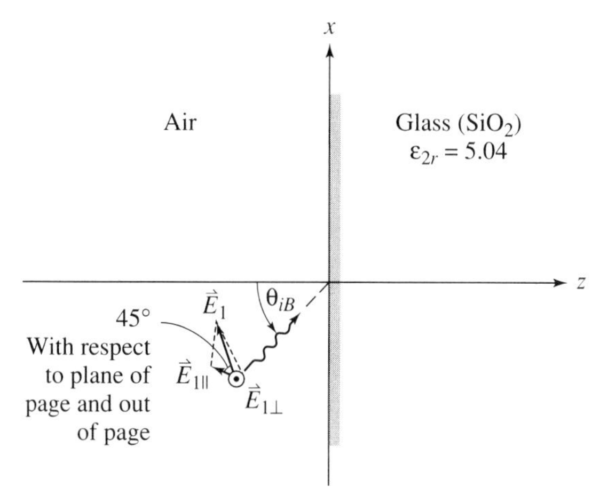

Figure 8-30.

The electric field vectors are

$$\vec{E}_{1\perp} = 70.7\hat{y} \text{ V/m}$$
$$\vec{E}_{1\|} = +70.7 \cos 66°\hat{x} - 70.7 \sin 66°\hat{z} = 28.8\hat{x} - 64.9\hat{z} \text{ V/m}$$

Using the Poynting vector (Eq. 8-16), where the electric field is the total field for the component being evaluated,

$$|\vec{S}_{1\perp}| = \frac{1}{2}\frac{|\vec{E}_{1\perp}|^2}{\eta_1} = \frac{1}{2}\frac{(70.7)^2}{377} = 6.63 \text{ W/m}^2$$
$$|\vec{S}_{1\|}| = \frac{1}{2}\frac{|\vec{E}_{1\|}|^2}{\eta_1} = 6.63 \text{ W/m}^2$$

We next compute the transmission coefficients for each polarization. For perpendicular polarization Eq. 8-139 yields

$$Z^w_{1z,\perp} = \eta_1 \sec\theta_i = 377 \sec 66° = 927 \ \Omega$$

and, using Snell's law (Eq. 8-135),

$$Z^w_{2z,\perp} = \eta_2 \sec\theta_t = \sqrt{\frac{\mu_0}{5.04\epsilon_0}} \sec\left[\sin^{-1}\left(\sqrt{\frac{1}{5.04}} \sin 66°\right)\right]$$
$$= 168 \sec[24°] = 183 \ \Omega \qquad (\eta_2 = 168 \ \Omega, \qquad \theta_t = 24°)$$

Using Eq. 8-142

$$\frac{E_{2t,\perp}}{E_{1i,\perp}} = \tau_\perp = \frac{2 \times 183}{183 + 927} = 0.33$$
$$E_{2t,\perp} = (0.33)(70.7) = 23.3 \text{ V/m}$$

Then

$$|\vec{S}_{2\perp}| = \frac{1}{2}\frac{(23.3)^2}{168} = 1.62 \text{ W/m}^2$$

Fraction of input power transmitted in the perpendicular polarization is given by

$$\text{Transmitted fraction } \perp = \frac{1.62}{6.33} \times 100\% = 24.4\%$$

For parallel polarization Eqs. 8-165 and 8-166 give

$$Z^w_{1z,\|} = \eta_1 \cos\theta_i = 377 \cos 66° = 153 \ \Omega$$

$$Z^w_{2z,\|} = \eta_2 \cos\theta_t = 168 \cos 24° = 153 \ \Omega$$

Using these the transmission coefficient, Eq. 8-171, gives

$$\tau_{\|} = \frac{E_{2t,x}}{E_{1i,x}} = \frac{2 \times 153}{153 + 153} = 1 \qquad \text{(as expected)}$$

Thus

$$E_{2t,x} = (1)(28.8) = 28.8 \text{ V/m}$$

From this, the total parallel component in Medium 2 is

$$|E_{2t,\|}| = \frac{|E_{2t,x}|}{\cos\theta_t} = \frac{28.8}{\cos 24°} = 31.5 \text{ V/m}$$

Then

$$|\vec{S}_{2\|}| = \frac{1}{2}\frac{(31.5)^2}{168} = 2.95 \text{ W/m}^2$$

Thus

$$\text{Transmitted fraction } \| = \frac{2.95}{6.63} \times 100\% = 44\%$$

The reflected component of the perpendicular polarization is, from Eq. 8-141,

$$E_{1r,\perp} = \Gamma_\perp E_{1i,\perp} = \frac{183 - 927}{183 + 927} \times 70.7 = -47.4 \text{ V/m}$$

$$|\vec{S}_{1r,\perp}| = \frac{1}{2}\frac{(47.4)^2}{377} = 2.98 \text{ W/m}^2$$

$$\text{Reflected fraction } \perp = \frac{298}{6.63} \times 100\% = 45\%$$

The reflected component of the parallel polarization is zero ($\Gamma_\|$ at θ_{iB} is 0) so there is no reflected component. Thus

$$\text{Reflected fraction } \| = 0\%$$

This tells us that the reflected wave will be a pure, perpendicularly polarized wave.

Exercises

8-7.3a In Figure 8-26 suppose that the distance between nulls in the z direction is 5 cm and the frequency is known to be 10 GHz. What is the angle of incidence θ_i? Medium 1 is air.

8-7.3b For Example 8-18, show that the tangential field components of $\vec{E}$ and $\vec{H}$ are continuous at the boundary.

8-7.3c For Example 8-18, interchange the two media and solve.

8-7.4d Prove Eq. 8-173.

8-7.3e For Example 8-20 show that normal $\vec{D}$ is continuous at the boundary.

8-7.3f A wave $10^{-3}e^{-j(0.423x+0.906z)}\hat{y}$ in air is incident on a dielectric with $\epsilon_r = 2$. The polarization is perpendicular as in Figure 8-24*a*. Determine the fraction of power transmitted into Medium 2 and the equation for $\vec{E}_{2t}$. What are the Brewster and critical angles for this system? $f = 900$ MHz.

8-7.3g If we define the total wave impedance in the z direction in a given medium as the ratio of total field values by

$$Z_z^w = \frac{E_x}{H_y}$$

determine the equation for Z_{1z}^w for both perpendicular and parallel polarizations for oblique incidence. The results will be functions of l, Γ, and so on.

8-8 Polarization of Uniform Plane Waves

In the preceding derivations and examples each of the incident, reflected, and transmitted components of field had an electric field vector with a unique spatial direction for a given system. As each component propagated away from the boundary the direction of its electric field vector did not change except for sinusoidal time variation. There are important types of plane waves for which the electric field vector direction may vary with time and/or space. The concept of polarization has been developed to describe the various types of electric field variation and orientation.

Suppose we have a plane wave in the xy plane that propagates in the z direction as shown in Figure 8-31. We can express the wave in terms of its two components by

$$\vec{E} = \vec{E}_x + \vec{E}_y = (E_{x0}\hat{x} + E_{y0}e^{j\psi}\hat{y})e^{-j\beta z} \qquad (8\text{-}175)$$

In this expression the fields amplitudes E_{x0} and E_{y0} may be complex with the same angle and the angle ψ is the phase angle by which the y component leads the x component (in sinusoidal steady state). By selecting values of the constant ψ and values E_{x0} and E_{y0} we obtain various polarizations as will be shown next.

Case 1: $\psi = 0$ or π: Linear Polarization. For this case, the time expression is obtained from

$$\vec{E} = (E_{x0}\hat{x} + E_{y0}\hat{y})e^{-j\beta z} \qquad (8\text{-}176)$$

$$\vec{\mathscr{E}}(z, t) = \text{Re}\{\vec{E}e^{j\omega t}\} = E_{x0}\cos(\omega t - \beta z)\hat{x} + E_{y0}\cos(\omega t - \beta z)\hat{y} \qquad (8\text{-}177)$$

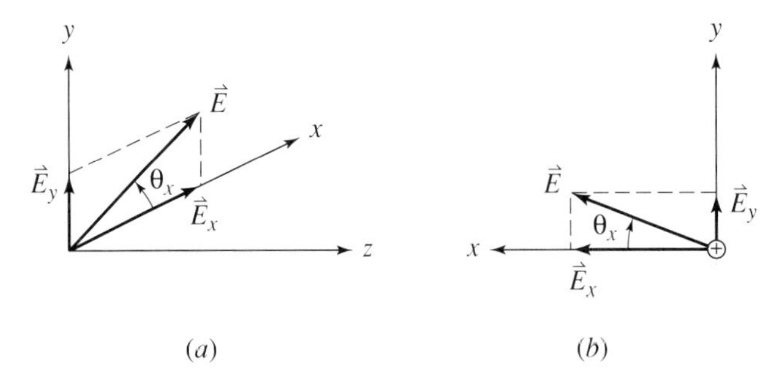

Figure 8-31. (*a*) Convention for describing a general electric field vector of a uniform plane wave propagating in the z direction. (*b*) View of electric field vector behind (to the left of) the xy plane looking in the $+z$ direction.

From this result

$$\theta_x = \tan^{-1}\left(\frac{E_{y0}\cos(\omega t - \beta z)}{E_{x0}\cos(\omega t - \beta z)}\right) = \tan^{-1}\left(\frac{E_{y0}}{E_{x0}}\right) \tag{8-178}$$

$$\begin{aligned} |\mathscr{E}(z, t)| &= \sqrt{E_{x0}^2 \cos^2(\omega t - \beta z) + E_{y0}^2 \cos^2(\omega t - \beta z)} \\ &= \sqrt{E_{0x}^2 + E_{0y}^2}\,|\cos(\omega t - \beta z)| \end{aligned} \tag{8-179}$$

Since the angle that the electric field vector makes with the x axis is independent of z and t, the field vector is always along the line defined by θ_x. Note that at a given location z the amplitude varies sinusoidally along that line. This is shown in Figure 8-32 for $z = 0$, using the components from Eq. 8-177, and for $\theta_x = 20°$. Note that the total electric field intensity always lies along the same straight line.

Two particular cases of linear polarization are often identified. These are named by identifying the position of the electric field vector with respect to the surface of the earth. If we let the xz plane be the surface of the earth, then a wave which has only an E_x component (parallel to the earth) is called a *horizontally polarized linear plane wave*. This is frequently contracted to *horizontal wave*. Similarly, if the wave has only an E_y component (perpendicular to the earth) it is a *vertically polarized linear plane wave*, or simply a *vertical wave*. This is important to know, since an antenna must be properly oriented (parallel) to the field. For example, television stations transmit horizontally polarized waves, so the receiving antenna rods are horizontal for largest signal strength.

Case 2: $\psi = \pm\pi/2$, $E_{x0} = E_{y0} = E_0$: Circular Polarization. The time expression for the electric field is obtained from Eq. 8-175 using the identity $e^{\pm j\pi/2} = \pm j$:

$$\begin{aligned} \vec{\mathscr{E}}(z, t) &= \text{Re}\{(E_0\hat{x} \pm jE_0\hat{y})e^{-j\beta z}e^{j\omega t}\} \\ &= E_0 \cos(\omega t - \beta z)\hat{x} \mp E_0 \sin(\omega t - \beta z)\hat{y} \end{aligned} \tag{8-180}$$

From this result the angle with respect to the x axis is given by

$$\theta_x = \tan^{-1}\left[\frac{\mp E_0 \sin(\omega t - \beta z)}{E_0 \cos(\omega t - \beta z)}\right] = \tan^{-1}[\mp\tan(\omega t - \beta z)] = \mp(\omega t - \beta z) \tag{8-181}$$

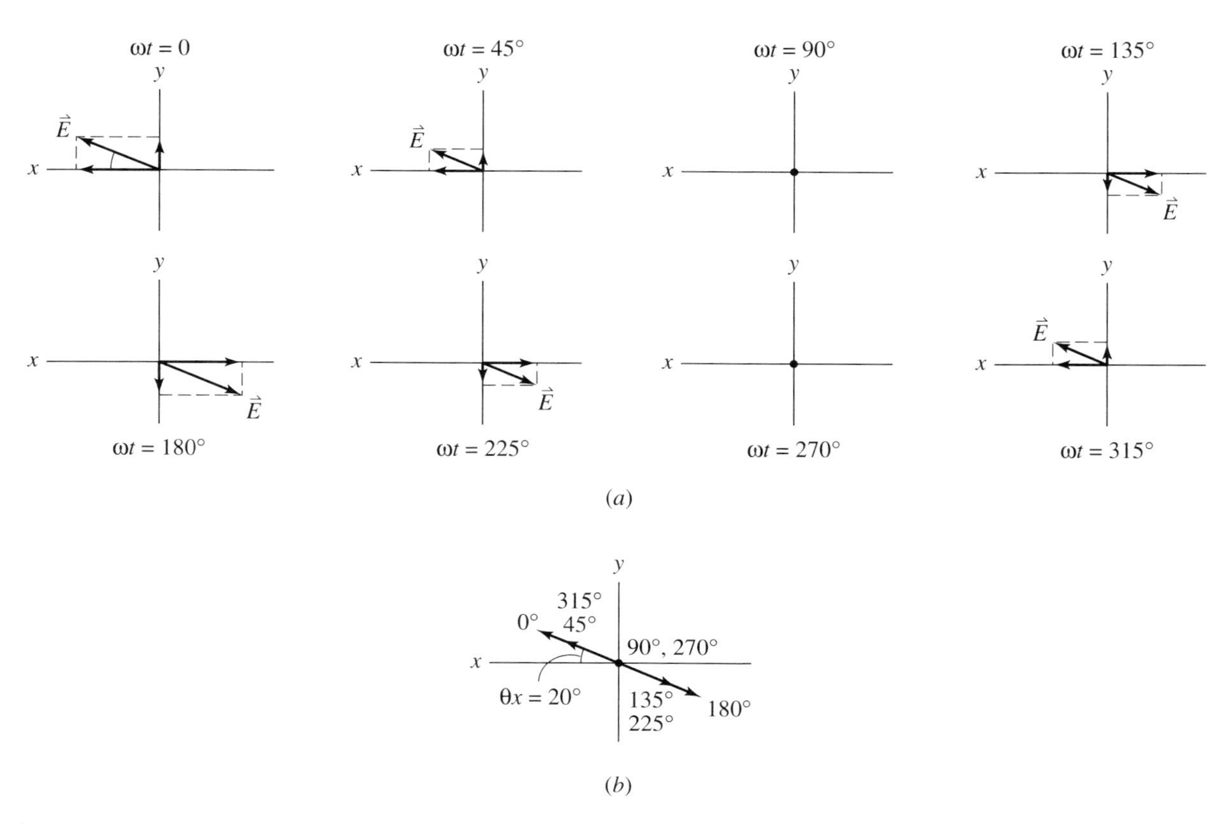

Figure 8-32. Linearly polarized wave. (*a*) Time sequence of the electric field vector and its components. (*b*) Superposition of the field vectors onto a single line.

Note that the minus (top) sign in this expression corresponds to $+\pi/2$. For these values of ψ θ_x is not a constant, being dependent upon both time and location along the z axis. For the field amplitude we have

$$|\vec{\mathscr{E}}(z, t)| = \sqrt{E_0^2 \cos^2(\omega t - \beta z) + E_0^2 \sin^2(\omega t - \beta z)} = E_0 \tag{8-182}$$

Thus the field intensity amplitude is independent of time, but its direction in space changes. Suppose we again stand behind the xy plane (on the $-z$ axis) and look in the $+z$ direction. As before, set $z = 0$ and plot the electric field intensity as a function of time, shown in Figure 8-33*a*, which is for the particular case $\psi = +\pi/2$ with $\theta_x = -\omega t$. This rotational direction is called *counterclockwise* or *left-hand circular polarization*. To be more precise, we say that as the wave propagates *away* from us (recedes) and the field vector rotates in

Figure 8-33. Left-hand circularly polarized wave. (*a*) Electric field vector time sequence ($z = 0$), counterclockwise rotation. $\psi = \pi/2$ ($\theta_x = -\omega t$). (*b*) Electric field vector spatial orientations ($t = 0$). $\psi = \pi/2(\theta_x = \beta z)$.

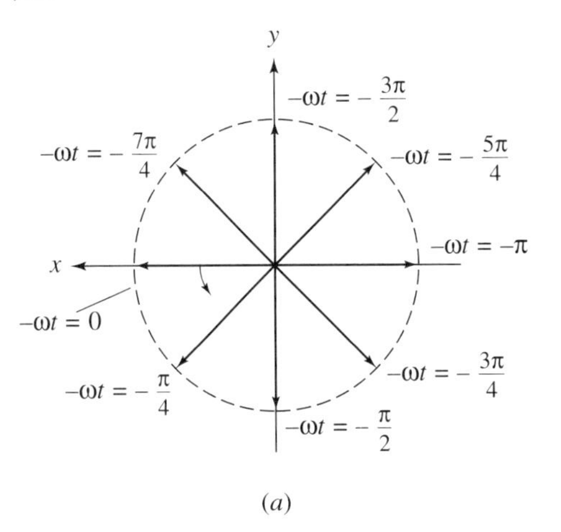

(*a*)

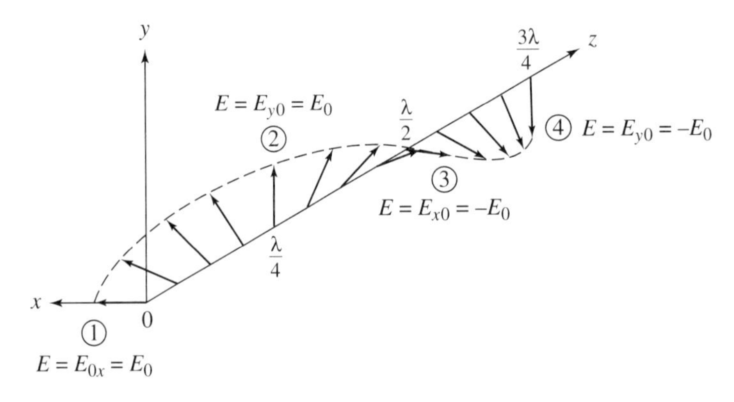

(*b*)

time counterclockwise at a fixed location z we have left-hand circular polarization. The way one views the wave is important, since other fields of science occasionally use other definitions.

We now look at the electric field vector direction as a function of location z for fixed time. Again for $\psi = +\pi/2$, using $t = 0$, the angle of the field is $\theta_x = \beta z$. This is shown in Figure 8-33*b*. The vector arrows are actually all the same length, but the perspective changes them as they are displaced. This is visualized in three dimensions by beginning at point 1 and raising the vector to the positive vertical at point 2. The vector then moves out of the page to point 3 where the field has only an x component directed along the $-x$ axis. From that point the vector moves down to point 4. One must remember that this is *not* a time sequence, for all the vectors shown exist at time $t = 0$ as a snapshot. If a person starts at 1 and walks along the dotted path that connects the field vector arrowheads a spiral (or corkscrew) would be traced out. Note that the person would move clockwise around the z axis. This does not change the definition of "left handedness," which is based on the *time* rotation direction. For the particular case $\psi = -\pi/2$ a similar analysis would lead to a time-rotating field in the clockwise direction. This is called *clockwise* or *right-hand circular polarization.*

As before, this clockwise rotation is observed as the wave propagates away from the viewer. As a point of interest, waves of these types are often received using a helically wound wire antenna. Qualitatively, such a configuration always has a wire component parallel to the electric field. Dimensions and direction of winding are of course important.

Case 3: $\psi \neq 0$ and $E_{x0} \neq E_{y0}$: Elliptical Polarization. For this general case Eq. 8-175 results in the time form

$$\vec{\mathscr{E}}(z, t) = E_{x0} \cos(\omega t - \beta z)\hat{x} + E_{y0} \cos(\omega t - \beta z + \psi)\hat{y} \tag{8-183}$$

$$\theta_x = \tan^{-1}\left[\frac{E_{y0} \cos(\omega t - \beta z + \psi)}{E_{x0} \cos(\omega t - \beta z)}\right] \tag{8-184}$$

$$|\mathscr{E}(z, t)| = \sqrt{E_{x0}^2 \cos^2(\omega t - \beta z) + E_{y0}^2 \cos^2(\omega t - \beta z + \psi)} \tag{8-185}$$

In this case both the angle of the electric field vector and its magnitude vary with time and location. As will be shown the vector length variation traces an ellipse as it rotates in time at a given location z.

For convenience we examine the rotation of the electric field vector looking in the $+z$ direction for z constant, say $z = 0$. For this case using a trigonometric identity in the bracket term of Eq. 8-184 results in

$$\begin{aligned}\theta_x(0, t) &= \tan^{-1}\left[\frac{E_{y0}}{E_{x0}} \frac{\cos \omega t \cos \psi - \sin \omega t \sin \psi}{\cos \omega t}\right] \\ &= \tan^{-1}\left[\frac{E_{y0}}{E_{x0}} \cos \psi - \frac{E_{y0}}{E_{x0}} \sin \psi \tan \omega t\right]\end{aligned}$$

Since E_{y0}, E_{x0}, and ψ are constants this can be written

$$\theta_x(0, t) = \tan^{-1}[A - B \tan \omega t]$$

where we assume $\psi > 0$ so that $B > 0$. For $t = 0$ we see that the field vector makes an angle of value $\tan^{-1}(A)$ with the positive x axis. As time increases the term in brackets gets smaller and then goes negative so that $\theta_x(0, t)$ goes negative. This is the condition similar to that shown in Figure 8-33a, which is a counterclockwise rotation of the field vector. This is called *left-hand elliptical polarization*.

A similar analysis shows that for $\psi < 0$ we would have $\sin \psi$ and hence B negative so that

$$\theta_x(0, t) = \tan^{-1}[A + B \tan \omega t]$$

which yields *right-hand elliptical polarization*.

We still have to show that Eq. 8-185 describes an ellipse. For convenience, we select $z = 0$, for which the components of electric field are

$$\left.\begin{aligned} \mathscr{E}_x &= \mathscr{E}_x(0, t) = E_{x0} \cos \omega t \\ \mathscr{E}_y &= \mathscr{E}_y(0, t) = E_{y0} \cos(\omega t + \psi) = E_{y0} \cos \omega t \cos \psi - E_{y0} \sin \omega t \sin \psi \end{aligned}\right\} \quad (8\text{-}186)$$

These two equations, in fact, describe an ellipse parametrically, but it is convenient to have the explicit relationship between $\mathscr{E}_x$ and $\mathscr{E}_y$ to determine the properties of the ellipse. In fundamental algebraic terms a general ellipse is expressed as

$$Ax^2 + Bxy + Cy^2 + Dx + Ey + F = 0 \qquad \text{with} \quad B^2 - 4AC < 0$$

To eliminate the parameter ωt we solve the $\mathscr{E}_x$ component for $\cos \omega t$ and substitute the result into the $\mathscr{E}_y$ component. This gives, using $\sin^2 \omega t = 1 - \cos^2 \omega t$,

$$\mathscr{E}_y = E_{y0} \frac{\mathscr{E}_x}{E_{x0}} \cos \psi - E_{y0} \sin \psi \sqrt{1 - \left(\frac{\mathscr{E}_x}{E_{x0}}\right)^2}$$

Isolating the square root term:

$$E_{y0} \sin \psi \sqrt{1 - \left(\frac{\mathscr{E}_x}{E_{x0}}\right)^2} = \frac{E_{y0}}{E_{x0}} \mathscr{E}_x \cos \psi - \mathscr{E}_y$$

Squaring and collecting coefficients on $\mathscr{E}_x$ and $\mathscr{E}_y$:

$$\left(\frac{E_{y0}}{E_{x0}}\right)^2 \mathscr{E}_x^2 - 2 \frac{E_{y0}}{E_{x0}} \cos \psi \, \mathscr{E}_x \mathscr{E}_y + \mathscr{E}_y^2 - E_{y0}^2 \sin^2 \psi = 0 \qquad (8\text{-}187)$$

Comparing this with the general algebraic form given earlier and using the variable correspondences $\mathscr{E}_x \approx x$ and $\mathscr{E}_y \approx y$ we then have the constant correspondences

$$A \approx \left(\frac{E_{y0}}{E_{x0}}\right)^2, \qquad B \approx -2 \frac{E_{y0}}{E_{x0}} \cos \psi, \qquad C \approx 1,$$

$$D = E \approx 0, \qquad F \approx -E_{y0}^2 \sin^2 \psi$$

Using these expressions, we obtain the required constraint

$$B^2 - 4AC \approx -4\left(\frac{E_{y0}}{E_{x0}}\right)^2 \sin^2\psi < 0$$

Thus at constant z the tip of the electric field vector is of magnitude

$$|\vec{\mathscr{E}}| = \sqrt{\mathscr{E}_x^2 + \mathscr{E}_y^2}$$

and describes an ellipse.

The lengths of the major and minor axes of the ellipse can be determined by first finding the values of $(\omega t - \beta z)$ that maximize and minimize the magnitude, Eq. 8-185. Performing the differentiation and setting that result to zero we find the (not very attractive) result

$$(\omega t - \beta z)_{\text{extrema}} = \tan^{-1}\left[\left(\cos 2\psi + \frac{E_{0x}^2}{E_{y0}^2 \sin 2\psi}\right) \pm \frac{\sqrt{1 + 2\cos 2\psi \dfrac{E_{x0}^2}{E_{y0}^2} + \dfrac{E_{x0}^4}{E_{y0}^4}}}{|\sin 2\psi|}\right] \tag{8-188}$$

From this result one can use specific values of a particular problem to find these extrema and then compute the corresponding components $\mathscr{E}_x$ and $\mathscr{E}_y$ using Eq. 8-183. Once these values are known, the angle of the major (or minor) axis can be determined. The tilt of the ellipse is usually defined as the angle that the major axis of the ellipse makes with the positive x axis at location $z = 0$. Figure 8-34 gives an example of an elliptically polarized wave. It is often convenient to be able to express a wave of one given polarization in terms of a sum of waves having other types of polarization. For example, if a single straight wire were used to receive a circularly polarized wave, it would be convenient to have the wave expressible as the sum of two linearly polarized waves. It is also much easier to solve reflection problems using the linear decomposition components since those results were presented earlier in this chapter. See the general exercises. The following examples show how this is done.

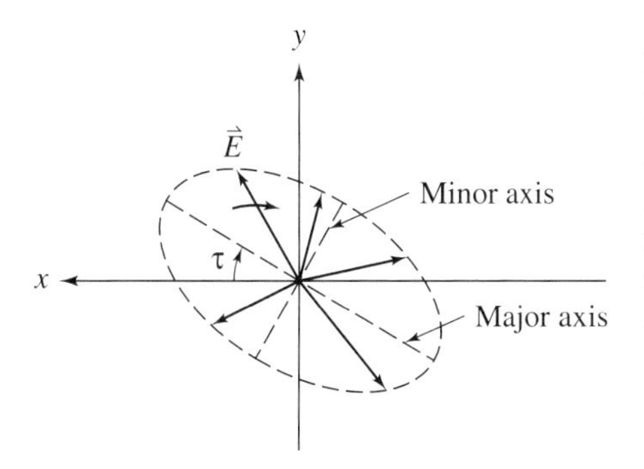

Figure 8-34. An elliptically polarized wave. A few electric field vectors are known. The small arrow on the vector labeled $\vec{E}$ shows the rotation direction for right-hand polarization (looking in the $+z$ direction). τ is the tilt angle at $z = 0$.

Example 8-21

A right-hand circularly polarized wave is given by

$$\vec{E} = E_0(\hat{x} - j\hat{y})e^{-j\beta z}$$

Express the electric field as the sum of two linearly polarized waves.

We simply write the given field in two distinct terms, each with its own propagation factor:

$$\vec{E} = E_0 e^{-j\beta z}\hat{x} + E_0 e^{-j\pi/2} e^{-j\beta z}\hat{y} = E_0 e^{-j\beta z}\hat{x} + E_0 e^{-j(\beta z + \pi/2)}\hat{y}$$

Note that each component is always along a straight line, in this case along the x axis and y axis. The two components are out of phase 90° in time also as one might expect.

This decomposition is not unique, however, since we could rotate both linearly polarized components by an angle ϕ and still generate a circle. This can also be viewed as a rotation of the coordinate axes. To show this, multiply the electric field vector by $e^{j\phi}$ and expand using Euler's theorem. The magnitude is unaffected by this since the exponential has magnitude 1:

$$\begin{aligned}\vec{E} &= E_0(\hat{x} - j\hat{y})e^{-j\beta z}e^{j\phi} = E_0(\hat{x} - j\hat{y})(\cos\phi + j\sin\phi)e^{-j\beta z} \\ &= E_0(\cos\phi\hat{x} - j\cos\phi\hat{y} + \sin\phi\hat{x} + \sin\phi\hat{y})e^{-j\beta z} \\ &= E_0(\cos\phi\hat{x} + \sin\phi\hat{y})e^{-j\beta z} - jE_0(-\sin\phi\hat{x} + \cos\phi\hat{y})e^{-j\beta z}\end{aligned} \tag{8-189}$$

Since ϕ is a constant, the factors in parentheses are linearly polarized waves having angles ϕ with respect to the x axis and y axis, respectively. These results show that a straight wire antenna may have any orientation in the plane of the wave and detect some of the signal. In practice, two wires at right angles are used to detect both linear polarizations. See General Exercise GE8-47.

Example 8-22

Show that a linearly polarized wave may be expressed as the sum of two circularly polarized waves.

The linearly polarized wave may be written in another form by breaking up the components of Eq. 8-176 as follows:

$$\vec{E} = (E_{x0}\hat{x} + E_{y0}\hat{y})e^{-j\beta z} = (\tfrac{1}{2}E_{x0}\hat{x} + \tfrac{1}{2}E_{x0}\hat{x} + \tfrac{1}{2}E_{y0}\hat{y} + \tfrac{1}{2}E_{y0}\hat{y})e^{-j\beta z}$$

Now add and subtract $(1/2)jE_{y0}\hat{x}$ and $(1/2)jE_{x0}\hat{y}$ inside the parentheses (which leaves the expression unchanged):

$$\begin{aligned}\vec{E} = (&\tfrac{1}{2}E_{x0}\hat{x} + \tfrac{1}{2}E_{x0}\hat{x} + \tfrac{1}{2}E_{y0}\hat{y} + \tfrac{1}{2}E_{y0}\hat{y} + \tfrac{1}{2}jE_{y0}\hat{x} - \tfrac{1}{2}jE_{y0}\hat{x} \\ &+ \tfrac{1}{2}jE_{x0}\hat{y} - \tfrac{1}{2}jE_{x0}\hat{y})e^{-j\beta z}\end{aligned}$$

Now combine terms so that circularly polarized components may be identified and factor out $\frac{1}{2}$.

$$\begin{aligned}\vec{E} &= \tfrac{1}{2}[(E_{x0} - jE_{y0})\hat{x} + j(E_{x0} - jE_{y0})\hat{y} + (E_{x0} + jE_{y0})\hat{x} - j(E_{x0} + jE_{y0})\hat{y}]e^{-j\beta z} \\ &= \tfrac{1}{2}[(E_{x0} - jE_{y0})\hat{x} + j(E_{x0} - jE_{y0})\hat{y}]e^{-j\beta z} + \tfrac{1}{2}[(E_{x0} + jE_{y0})\hat{x} - j(E_{x0} + jE_{y0})\hat{y}]e^{-j\beta z}\end{aligned}$$

Since the components within each set of parentheses have the same magnitudes the bracket terms can be identified as two oppositely polarized circular waves.

As a final observation on polarization we refer back to Section 8.7 where the terms *parallel polarization* and *perpendicular polarization* were used. These simply referred to the orientations of the electric field intensity vector of a linearly polarized uniform plane wave with respect to the plane of the page in which oblique incident system was drawn.

Exercises

8-8.0a For circular and elliptical polarization we define the angular velocity of the electric field vector as

$$\omega_x = \frac{d\theta_x}{dt} \text{ rad/s}$$

Determine ω_x expressions for both types of polarization.

8-8.0b For circular polarization develop equations for $\vec{H}$ and $\vec{\mathcal{H}}$. Maxwell's equation works well here. Put the results in terms of intrinsic impedance η.

8-8.0c Develop the equation for the Poynting vector $\vec{S}$ for a circularly polarized wave.

8-8.0d For the linearly polarized wave $\vec{E} = E_0 e^{-j\beta z}\hat{y}$, obtain an expression in terms of two circularly polarized waves.

8-8.0e Show that an elliptically polarized wave can be expressed as the sum of two circularly polarized waves.

8-8.0f Plot a few vectors for each of the following plane waves and state the polarizations represented. Use $z = 0$ and compute τ where appropriate.

(a) $E_{0x} = 1$, $E_{0y} = 0.5$, $\psi = 0$
(b) $E_{0x} = 1$, $E_{0y} = 0.5$, $\psi = \pi$
(c) $E_{0x} = 1$, $E_{0y} = 1$, $\psi = -\pi/4$
(d) $E_{0x} = 1$, $E_{0y} = 0.5$, $\psi = -\pi/4$
(e) $E_{0x} = 1$, $E_{0y} = 0.5$, $\psi = \pi/2$
(f) $E_{0x} = 0.5$, $E_{0y} = 0.5$, $\psi = \pi/2$

8-8.0g The electric field vectors in Figure 8-32 were drawn at fixed location, $z = 0$, with time the variable. Draw a diagram for fixed time, $t = 0$, with z the variable using coordinate axes as in Figure 8-33*b*. Let the range of z be $0 \leq z \leq \lambda$.

8-8.0h The electric field vectors shown in Figure 8-34 were drawn at fixed location, $z = 0$, with time the variable. For $\psi = \pi/4$ and $E_{x0} = \frac{1}{2}E_{y0}$ draw two ellipses corresponding to $z = 0$ and $z = \lambda/2$ identifying $t = 0$ and the directions of rotation in both cases.

Summary, Objectives, and General Exercises

Summary

The focus of this chapter has been on the propagation, reflection, and transmission of uniform plane waves. We have considered the cases of normal and oblique incidence of plane waves at conductor and dielectric boundaries and the use of the impedance concept to develop transmission line analogies. From these studies the laws of reflection and refraction (Snell's law) were obtained. The key concepts used to obtain these results were the wave equation and the boundary conditions for electric and magnetic fields.

Propagation characteristics of uniform plane waves in both lossless (perfect dielectrics) and lossy materials were investigated to determine signal attenuation. The definitions of group and phase velocity were given so that the effect of losses on signal waveforms (dispersion) could be evaluated. Penetration depth (skin depth) in conducting materials was developed.

The final section of this chapter defined and gave the propagation properties of three polarization types: linear, circular, and elliptical. The specific conventions used to determine right- and left-handed circular and elliptical polarizations were given. These were still TEM waves.

Objectives

1. Be able to write the general functional form of a linearly polarized uniform plane wave propagating in a given direction. This includes both $\vec{E}$ and $\vec{H}$.
2. Know the definition of loss tangent and its relationship to the complex dielectric constant.
3. Be able to explain how the complex dielectric constant $\bar{\epsilon}$ can have two formulations for ϵ'', namely, one where the conductivity appears explicitly and another where the conductivity is included as a *part* of ϵ''.
4. Be able to use series approximations to obtain expressions for γ, α, and β for cases of a TEM wave with:
   (a) High and low frequency values
   (b) High and low conductivity values.
5. Be able to describe the difference between conductive losses and polarization losses in a dielectric.
6. Know the physical significance of skin depth and its relationship to the propagation constant for z-directed propagation.
7. Be able to define, compute, and use the concept of intrinsic (characteristic) impedance for a given linearly polarized, uniform-plane wave field configuration.
8. Be able to define wave impedance and to compute it for a given field configuration and in a given direction.
9. Know and be able to use the definition of wavelength and to compute it for a given direction in a linearly polarized uniform plane wave system.
10. Be able to compute the phase velocity and the group velocity of a given wave in a given direction.
11. For low-loss systems be able to explain how one can modify an assumed lossless dielectric solution to account for losses (attenuation).
12. Be able to define surface impedance, surface resistance, and surface reactance and to use them in plane wave field computations at conductor boundaries.
13. Be able to sketch qualitatively and to solve for the results of a linearly polarized uniform plane wave incident at an arbitrary angle at a perfect conducting wall and an imperfectly conducting wall. Consider both polarizations. The imperfectly conducting wall is considered only for normal incidence. Give special cases on the angle of incidence.
14. Be able to sketch qualitatively and to solve the results of a linearly polarized uniform plane wave incident at an arbitrary angle on a lossless dielectric–dielectric boundary. Consider both polarizations and give special cases on the angle of incidence.
15. Be able to solve for the field components in a multiple lossless dielectric boundary system for a TEM wave incident in the first dielectric at an arbitrary angle.
16. Be able to use the Smith chart to solve reflections of uniform plane waves from plane boundaries.
17. Be able to give the equations for and definitions of linear polarization, right- and left-hand

circular polarization and right- and left-hand elliptical polarization.

18. Be able to decompose a linearly polarized wave into two circularly polarized waves and vice versa. Be able to decompose an elliptically polarized wave into two circularly polarized waves.

General Exercises

GE8-1 A uniform plane wave propagates in vacuum in the negative z direction. The magnetic field intensity has a magnitude of 10^{-8} A/m and makes an angle of 60° with respect to the y axis in the first quadrant of the xy plane. The frequency is 30 MHz. Obtain equations for $\vec{E}$, $\vec{H}$, $\vec{\mathscr{E}}$, $\vec{\mathscr{H}}$, $\vec{S}$, and $\vec{\mathscr{S}}$. If an oscilloscope were used to look at the time waveforms at $z = 0$ and $z = 500$ m, what would be the phase lag of the wave at 500 m with respect to that at $z = 0$? Determine the ratios of the orthogonal *components* of $\vec{E}$ and $\vec{H}$ and the ratio of total $|\vec{E}|$ to total $|\vec{H}|$. Comment on the signs of the results.

GE8-2 A uniform plane wave of frequency 10.5 GHz propagates in polystyrene ($\epsilon_r = 2.5$), assumed lossless. The magnetic field intensity is 10 mA/m. Determine the phase velocity, the wavelength, the propagation (phase) constant, the amplitude of $\vec{E}$, and the intrinsic impedance.

GE8-3 In Chapter 7 it was shown that TEM waves have field structures that obey static field laws in the transverse plane. For the electric field intensity of Example 8-3 in Section 8.2, show that each side of Faraday's law in integral form is zero for a square path having one corner at the origin, contained in the xy plane. Also show that $\vec{E}$ is conservative. Evaluate $\nabla_t \times \vec{E}$.

GE8-4 Suppose we have a uniform plane wave in lossless medium that is modulated with a square wave of amplitude A_0 for $nT < t < (n + 1)T$, $n = 0, 1, 2, 3$, and is zero otherwise. The electric field is in the x direction and propagates in the $+z$ direction and propagates with velocity V_p. Make plots of $E_x(z)$ for $t = T/4$ and $T/2$.

GE8-5 Show that material losses in a dielectric can be attributed to a complex χ_e. (Hint: $\epsilon = \epsilon_0\epsilon_r = \epsilon_0(1 + \chi_e)$.)

GE8-6 Derive Eq. 8-37 for $\overline{\eta}$ beginning with Eq. 8-36 and justify the signs on the inner radicals.

GE8-7 Plot $|\hat{\eta}|$, $\mathrm{Re}\{\hat{\eta}\}$, and $\mathrm{Im}\{\hat{\eta}\}$, where $\hat{\eta} = \overline{\eta}/\sqrt{\mu/\epsilon'}$, as a function of the loss tangent, $\tan\delta$. Use Eq. 8-37. Use values from 10^{-4} to 10 on five-cycle log paper.

GE8-8 Using the cross-product expression for the wave impedance obtain Z_z^w of a uniform plane wave in a lossy medium.

GE8-9 Express the lossy medium γ, Eq. 8-27, in polar form. Answer:

$$\omega\sqrt{\mu\epsilon'}\left[1 + \left(\frac{\epsilon''}{\epsilon'}\right)^2\right]^{1/4} e^{j1/2\tan^{-1}\left(\frac{\epsilon''}{-\epsilon'}\right)}$$

Why is it important that the minus sign is in the denominator in the $\tan^{-1}$?

GE8-10 For a uniform plane wave propagating in a lossy medium prove that the Poynting vector attenuation varies as the square of the electric field attenuation.

GE8-11 Suppose that a uniform plane wave propagates in a region of space that contains ρ_v excess charge density. Show that the region may be considered as a "lossy" medium with an equivalent conductivity in the sinusoidal steady state,

$$\sigma_e = -j\,\frac{\rho_v q}{\omega m}$$

where m is the mass of each of the charges in ρ_v and q is the charge of each particle in ρ_v. Use $\vec{\mathscr{J}} = \rho_v\vec{v}$, $\vec{v}$ = velocity of charges, $\vec{\mathscr{F}} = q\vec{\mathscr{E}} = m\vec{a} = m(d\vec{v}/dt)$. What is the expression for the complex dielectric constant? What are the expressions for γ, α, and β? If we define the plasma frequency as the frequency at which propagation ceases, develop the equation for that frequency, and call it ω_p.

GE8-12 A uniform plane wave propagates in loamy soil at 1 MHz. The magnetic field intensity is 30 nA/m at the surface. What is the amplitude at 10 m into the soil? Compute the Poynting vectors at the two locations.

GE8-13 Determine the ratio of displacement current to conduction current in a good conductor. Evaluate this ratio for silver, gold, and solder.

GE8-14 Derive the wave equation for the conduction current $\vec{J}$ in a good conductor. Assume the wave is propagating in the positive z direction. Solve the equation for a uniform plane wave with J_x component only.

GE8-15 Show that in the time domain the electric field intensity in a good conductor satisfies the diffusion equation (first-order time derivative) rather than the wave equation (second-order time derivative).

GE8-16 Determine the internal wire inductance per meter of AWG #22 wire at 10 MHz and 100 MHz. See Example 8-8.

GE8-17 Compute the internal inductance per meter for the conductor system of Example 8-8 and comment on this result with respect to Figure 0-1.

GE8-18 A uniform plane wave has been determined to have the equation

$$\vec{E} = (10\hat{x} + 2\hat{y})e^{-j\beta z} \text{ V/m}$$

The frequency is 1 GHz and the material is Teflon, assumed lossless. Determine V_p, β, λ_0, $\overline{S}$, and $\vec{\mathscr{H}}$.

GD8-19 A uniform plane wave is to be used to determine the properties of an unknown but nonmagnetic material. The magnetic field intensity attenuates 0.3 dB at 200 MHz through a 0.5-m slab of the material. The phase of the field is delayed 300° passing through the material. For the material determine the values of σ, ϵ_r, and $\overline{\eta}$.

GE8-20 A manufacturer produces a ferrite magnetic material that has $\mu_r = 250$, $\sigma = 6.7 \times 10^{-7}$ S/m, and $\epsilon_r = 30$. At a frequency of 3 GHz what are the values of V_p, $\overline{\eta}$, λ_0, α, and β?

GE8-21 Suppose a material has two complex parameters $\overline{\mu}$ and $\overline{\epsilon}$. Derive the general equations for $\overline{\eta} = R + jX$, $\gamma = \alpha + j\beta$, and the group and phase velocities.

GE8-22 For a lossless medium show that

$$\eta = 120\pi\sqrt{\frac{\mu_r}{\epsilon_r}} \qquad V_p = \frac{3 \times 10^8}{\sqrt{\mu_r \epsilon_r}}$$

GE8-23 A uniform plane wave propagates in the z direction and has an electric field intensity vector of magnitude 100 mV/m directed along the unit vector $0.423\hat{x} + 0.906\hat{y}$. The material is Teflon (assumed lossless) and the frequency is 200 MHz. Obtain the real time equations for $\vec{\mathscr{E}}$ and $\vec{\mathscr{H}}$.

GE8-24 Repeat GE8-23 for a wave propagating in the $-\hat{z}$ direction.

GE8-25 A wave in the air (free space, vacuum) has a measured wavelength of 2 cm. The wave is propagating in the positive z direction and has a magnetic flux density of 35 pT/m^2 in the positive x direction. Determine the time expressions for $\vec{\mathscr{E}}$ and $\vec{\mathscr{H}}$.

GE8-26 Determine the amplitude of a uniform plane wave at $z = 20$ m and $t = 10$ ms traveling in ice (lossy) at 100 MHz if

$$\vec{E} = 10^{-3}e^{-\gamma z}\hat{y}$$

Also compute the Poynting vector at $z = 0$ and $z = 20$ m.

GE8-27 Design a material in which the phase velocity is three-tenths of the velocity of light and which attenuates at 0.87 dB/m at 200 MHz. The wave impedance is to have magnitude 150 Ω. (You will need to specify ϵ, μ, and σ.)

GE8-28 It is desired to have a power flow of 0.1 W/m^2 in the z direction in air at 27 MHz. Determine the equations for $\vec{E}$ and $\vec{H}$ if $\vec{E}$ is to be in the $-\hat{y}$ direction.

GE8-29 Determine the largest permissible loss tangent for a good dielectric such that the error in β is 1% from an assumed lossless dielectric. Assume a nonmagnetic material. Compare with that from Example 8-10.

GE8-30 Suppose a wave is propagating in a good conductor. Determine the percentage of power remaining in the wave after one skin depth. Show that for n skin depths the percentage of power remaining in the wave is, for $n \geq 1$,

$$\frac{(\%\ \text{power at}\ \delta)^n}{(100)^{n-1}}\ \%$$

GE8-31 Determine the phase difference between $\vec{E}$ and $\vec{H}$ at $z = 3$ m for a uniform plane wave in a medium that has $\epsilon_r = 3$, $\mu_r = 1.5$, and $\sigma = 4 \times 10^{-4}$ S/m at 30 MHz.

GE8-32 Suppose a single uniform plane wave is split into two equal electric field components and the two waves enter two different dielectrics. The dielectrics are lossless with $\epsilon_{r1} = 2$ and $\epsilon_{r2} = 4$. For a frequency of 500 MHz, what is the phase of the wave in Medium 2 with respect to Medium 1 at a distance of 50 cm from the input face?

GE8-33 Using the cross-product general definition of wave impedance on the total electric and magnetic fields for the configuration of Figure 8-16, show that Eq. 8-86 results.

GE8-34 A uniform plane wave in air impinges normally on a lossless, nonmagnetic dielectric plane boundary. The standing wave ratio is observed to be 4. What is the dielectric constant of the material?

GE8-35 A uniform plane wave in a medium has

$$\vec{\mathscr{E}} = 10^{-3} \cos(6 \times 10^9 t - 30\pi z)\hat{x}\ \text{V/m}$$

and is normally incident on a plane boundary which is a perfect dielectric with $\epsilon_r = 2.5$. All materials are nonmagnetic. Find $\vec{E}$ and $\vec{H}$ in both media.

GE8-36 A uniform plane wave in a lossless nonmagnetic dielectric with $\epsilon_r = 2$ impinges normal to a plane boundary with another lossless nonmagnetic boundary with $\epsilon_r = 4$. Determine the fraction of the incident *power* that is transmitted.

GE8-37 A uniform plane wave exits a dielectric plane boundary into air. The amplitude of the electric field intensity in the air is 10^{-5} V/m. What are the incident and total electric fields in the (lossless, nonmagnetic) dielectric if $\epsilon_r = 3$?

GE8-38 Design a quarter-wave matching section that will match air to Teflon at 300 MHz. Assume uniform plane waves and normal incidence on a plane boundary.

GE8-39 Prove that if a material of surface resistance (lossy) equal to η of the lossless medium in which a plane wave is propagating is placed parallel to a perfectly conducting boundary at a distance $\lambda/4$ from it, no reflections occur if

$$\frac{1}{\sigma d} = \eta$$

where d is the sheet thickness and $d \ll \delta$. (Hint: Think of putting a pure resisatnce on a shorted transmission line at $\lambda/4$ from the short.)

GE8-40 A light wave of wavelength 500 nm strikes a silicon dioxide plate and also a silicon plate. Silicon dioxide has $\epsilon_r = 2.3$ and silicon has $\epsilon_r = 9$. Determine the percentage power reflected for each material.

GE8-41 Determine the thickness and dielectric constant of a material needed to eliminate reflections from silicon (wave is in air) at frequencies of 2 GHz and also at a wavelength (in air) of 2 μm. For the 2-GHz signal plot power reflection coefficient (in %) as $\rho_L^2 \times 100\%$ for the frequency range 1 GHz $\leq f \leq$ 3 GHz, using the thickness as constant as designed for the 2-GHz signal.

GE8-42 For the system of Figure 8-22, derive the equations for $\Gamma(-d)$ and $\tau(-d)$. These will be functions of ω, d, and material parameters.

GE8-43 Suppose we have a plane boundary between a poor conductor ($z < 0$) and a perfect conductor ($z > 0$). The poor conductor has $\sigma/\omega\epsilon_0 = 3$. At a frequency of 500 MHz plot the standing wave pattern for several wavelengths.

GE8-44 A 100-MHz uniform plane wave in air strikes normal to a plane dielectric material (very thick) of Bakelite. The incident electric field is 10 V/m. Determine the amplitude and location of the electric field maximum nearest the dielectric.

GE8-45 For the configuration of Figure 8-22 prove that for $\beta_2 d \ll 1$ (called an *electrically thin medium*) and $\eta_1 = \eta_3$:

$$\overline{Z}_{\text{total}}(-d) \approx \eta_1\left[1 + j\beta_2 d\left(\frac{\eta_2}{\eta_1} - \frac{\eta_1}{\eta_2}\right)\right]$$

$$\text{and} \quad \Gamma_{L12} \approx j\frac{\beta_2 d}{2}\left(\frac{\eta_2}{\eta_1} - \frac{\eta_1}{\eta_2}\right)$$

GE8-46 Determine the equations for the phase velocity in the z direction, $V_{p1,z}$, for both polarizations with oblique incidence (linearly polarized waves in both cases).

GE8-47 Two wires are placed at right angles in a right-hand circularly polarized plane wave as shown in Figure 8-35. Suppose that the two voltages at small gaps in the centers of the wires are directly proportional to the components of electric field parallel to the wires. Plot the magnitude of $V = V_1 + V_2$ as a function of time. Reverse the sign of V_2 and repeat. The wave propagates into the page.

GE8-48 In Figure 8-18a let wave $\vec{E}_1^+$ be a circularly polarized wave, polarized counterclockwise looking in the $+z$ direction. Determine the equation for $\vec{E}_1^-$. (Hint: Break the wave up into two linearly polarized components.) Next determine the equation for the total electric field intensity in Medium 1. Does the total wave have an identifiable polarization? If so, what is it? If not, plot the magnitude of E_1 total as a function of z, for $z \leq 0$. What is the polarization of the reflected wave?

Figure 8-35.

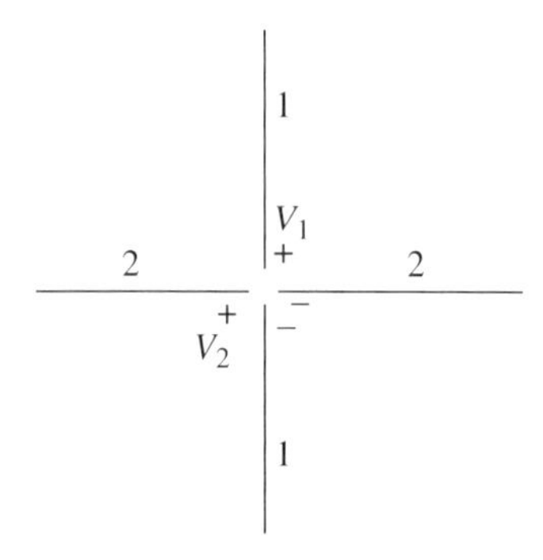

GE8-49 A right-hand circularly polarized plane wave in vacuum is incident at an angle of 30° onto a perfect conductor. Determine the equations of the incident wave and the reflected wave. What is the polarization of the reflected wave?

GE8-50 In Figure 8-20 let wave $\vec{E}_1^+$ be a right-hand circularly polarized wave with magnitude 10 V/m. Determine the average Poynting vector in Medium 1 and in the transmitted wave. What is the percentage power transmitted? Compute the power ratio for Example 8-15 and compare. (Hint: Use linearly polarized components of the incident wave.)

GE8-51 A right-hand circularly polarized wave is incident at an angle θ_i onto a dielectric–dielectric boundary. Let the incident Medium 1 be identified by ϵ_1, μ_0, $\sigma_1 = 0$ and the transmission Medium 2 by ϵ_2, μ_0, $\sigma_2 = 0$. Determine the equations for and polarization types of the reflected and transmitted waves. At what angle or angles of incidence might it be possible to have a linearly polarized wave in Medium 2? What is in the equation for and the polarization type of the reflected wave for such case(s)? (Hint: Use linearly polarized components of the incident wave.)

GE8-52 Complete the missing entries in Table 8-3.

9 Waveguides and Cavities

9-1 Introduction

The uniform plane wave study of Chapter 8 assumed that the waves were of infinite extent in the transverse plane. This description was adequate from the point of view of intercepting a portion of a larger wavefront as shown in Figure 8-1. However, if we want to generate electromagnetic energy and send (guide) all of it in a given direction we need to have some method for confining the region in which the energy flows. Structures used to confine the direction of wave propagation are called *waveguides*. In this chapter we investigate the properties of waveguides in terms of the physical configurations and the fields within. Propagation will be taken in the z direction so that the transmission line analogies can be shown.

We will study only those waveguide structures that have constant cross-sectional configurations (normal to z). These are called *uniform* waveguides and are frequently referred to generically as *cylindrical* waveguides.

In circuit theory the phenomenon of resonance is significant because of applications for energy storage, filtering, and detecting. As the operating frequency of these applications increases, the wavelength of the signal is of the order of the network element dimensions and significant energy is radiated. The radiated energy is a loss component to the network, so the quality factor Q is reduced. Most such networks have Q values on the order of 300 at most. In microwave systems a resonant configuration is obtained by constructing a closed cavity, such as a box in rectangular coordinates. For these cavities the energy is essentially confined within the volume because penetration of waves into metal is governed by skin depth—typically a few microns for good conductors. Additionally, values of Q on the order of 10^4 are not uncommon. A discussion is also given of some techniques used to couple energy into and out of the cavities.

Many of the differential equations that will arise in the analyses cannot be solved by simple direct integration. Also, functions that are solutions of the differential equations are not normally encountered in other basic engineering courses. Some techniques for obtaining solutions to the differential equations are covered in Appendices K and L. One can either study these appendices before continuing or simply digress from the text to the appropriate appendix as the need arises.

9-2 The Wave Equations and Field Component Equations for z-Directed Propagation

In this chapter we consider waveguide systems that have a z coordinate as the direction of propagation. Since we are using right-hand orthogonal coordinate systems, the other two coordinates will be transverse to the z direction. We follow the convention of Chapter 7 and denote those two coordinates by t_1 and t_2. For example, in rectangular coordinate $t_1 \sim x$ and $t_2 \sim y$; in cylindrical, $t_1 \sim \rho$ and $t_2 \sim \phi$.

The results required for this chapter have already been derived in Sections 7-9 and 7-10, so they will be stated here for convenient reference. For waves propagating in the positive z direction we use the propagation factor

$$\epsilon^{-\gamma z} = e^{-(\alpha+j\beta)z} = e^{-\alpha z}e^{-j\beta z} \tag{9-1}$$

where γ must be determined for each waveguide configuration. For propagation to occur β must be a real number.

The wave equations were shown to be expressible in terms of the transverse components of the Laplacian as

$$\nabla_t^2\vec{E}(t_1, t_2) + k_c^2\vec{E}(t_1, t_2) = 0 \tag{9-2}$$

$$\nabla_t^2\vec{H}(t_1, t_2) + k_c^2\vec{H}(t_1, t_2) = 0 \tag{9-3}$$

where

$$k_c^2 \equiv \gamma^2 + \bar{k}^2 = \gamma^2 + (k^2 - j\omega\mu\sigma_d) = \gamma^2 + (\omega^2\mu\epsilon - j\omega\mu\sigma_d) \tag{9-4}$$

with μ, ϵ, and σ_d being properties of the material within the guide, usually a dielectric. ∇_t^2 is the transverse coordinate derivatives part of ∇^2. As we will see, it is from this last equation that we obtain the appropriate values of γ for the waveguides.

In general the field vectors have three components (although they are functions only of the transverse coordinates); namely, $\hat{t}_1$, $\hat{t}_2$, and $\hat{z}$ components. As was shown in Chapter 7 the $\hat{z}$ components satisfy the scalar wave equations:

$$\nabla_t^2 E_z(t_1, t_2) + k_c^2 E_z(t_1, t_2) = 0 \tag{9-5}$$

$$\nabla_t^2 H_z(t_1, t_2) + k_c^2 H_z(t_1, t_2) = 0 \tag{9-6}$$

These last two equations are important because we proved in Section 7-9 that if we find solutions for E_z and H_z, we can determine the solutions of all other transverse components by appropriately combining derivatives of these as follows (see also Appendix J), where h_1 and h_2 are the transverse-coordinate scale factors of the coordinate system (see Section 4-12):

$$E_1(t_1, t_2) = -\frac{\gamma}{h_1 k_c^2}\frac{\partial E_z}{\partial t_1} - j\frac{\omega\mu}{h_2 k_c^2}\frac{\partial H_z}{\partial t_2} \tag{9-7}$$

$$E_2(t_1, t_2) = -\frac{\gamma}{h_2 k_c^2}\frac{\partial E_z}{\partial t_2} + j\frac{\omega\mu}{h_1 k_c^2}\frac{\partial H_z}{\partial t_1} \tag{9-8}$$

$$H_1(t_1, t_2) = j\frac{\omega\epsilon}{h_2k_c^2}\frac{\partial E_z}{\partial t_2} - \frac{\gamma}{h_1k_c^2}\frac{\partial H_z}{\partial t_1} \tag{9-9}$$

$$H_2(t_1, t_2) = -j\frac{\omega\epsilon}{h_1k_c^2}\frac{\partial E_z}{\partial t_1} - \frac{\gamma}{h_2k_c^2}\frac{\partial H_z}{\partial t_2} \tag{9-10}$$

For waves propagating in the negative z direction the exponential factor would be $e^{\gamma z}$, so we can still use the preceding results by simply replacing γ by $-\gamma$.

The solution for a given waveguide configuration would then proceed, in general, according to the following outline:

1. Solve Eqs. 9-5 and 9-6 for $E_z(t_1, t_2)$ and $H_z(t_1,t_2)$.
2. Apply boundary conditions to E_z and H_z to determine as many of the solution constants as possible. This may require using Eqs. 9-7 to 9-10 to obtain field components for which boundary conditions are known. It turns out that during this step we obtain values for k_c from which the propagation constant γ can be obtained from Eq. 9-4.
3. Compute the transverse field components using Eqs. 9-7 to 9-10.
4. Construct the total field equations using

$$\vec{E}(t_1, t_2\ z) = [E_1(t_1, t_2)\hat{t}_1 + E_2(t_1, t_2)\hat{t}_2 + E_z(t_1, t_2)\hat{z}]e^{-\gamma z} \tag{9-11}$$

$$\vec{H}(t_1, t_2, z) = [H_1(t_1, t_2)\hat{t}_1 + H_2(t_1, t_2)\hat{t}_2 + H_z(t_1, t_2)\hat{z}]e^{-\gamma z} \tag{9-12}$$

5. If the time domain expressions are desired they can be obtained in the usual way

$$\vec{\mathscr{E}}(t_1, t_2, z) = \mathrm{Re}\{\vec{E}(t_1, t_2, z)e^{j\omega t}\} \tag{9-13}$$

In Chapter 7 it was pointed out that TEM modes have both E_z and H_z equal to zero so the preceding outline could not be used. An alternative solution was given for that case, which is reviewed in the next section.

Exercises

9-2.0a Express ∇_t^2 in rectangular, cylindrical, and parabolic cylindrical coordinates.

9-2.0b Express Eqs. 9-7 through 9-10 explicitly in rectangular, cylindrical, and parabolic cylindrical coordinates.

9-2.0c For systems we shall use, the boundaries are parallel to the z axis. Write the appropriate boundary conditions for E_z and H_z at such a boundary. Comment on the difficulty of using the boundary condition on H_z (remember, there are unknown constants in H_z).

9-2.0d For systems having boundaries parallel to the z axis, what boundary conditions can be used if equations for E_1, E_2, H_1, and H_2 have been obtained, but unknown constants still remain? (Hint: $\hat{t}_1$ and $\hat{t}_2$ are normal to $\hat{z}$.)

9-3 Modes of Propagation: Physical Descriptions, Basic Equations, and Solution Procedures

In our study of waveguides we will investigate the same modes defined in Section 7-10. A review of these modes is given here for convenient reference. A list of what we want to obtain from the solutions is also given in the last subsection. This list justifies the effort involved in obtaining those solutions.

9-3.1 TEM Modes: $E_z = H_z = 0$.

For transverse electromagnetic (TEM) modes the key results are

$$\gamma_{\text{TEM}} = -\bar{k}^2 = -(\omega^2\mu\epsilon - j\omega\mu\sigma_d) \tag{9-14}$$

and

$$\vec{E} = -\nabla_t\Phi \tag{9-15}$$

where Φ is the scalar electric potential (time varying) in volts, which is the solution of

$$\nabla_t^2\Phi = 0 \tag{9-16}$$

Since we need to solve Laplace's equation for a scalar potential the boundary conditions will be voltages. Because the equation is of the second order we need to have the potentials on at least two distinct (nonconnected) surfaces. Since a perfect conductor is an equipotential surface, this means in practical cases that we must have at least two separate conducting surfaces. This is why TEM modes are referred to as *transmission-line modes*.

The solution process is

1. Solve $\nabla_t^2\Phi = 0$ for Φ and apply potential boundary conditions.
2. Compute the electric field intensity from

$$\vec{E}(t_1, t_2) = -\nabla_t\Phi \tag{9-17}$$

3. Construct the total electric field using Eq. 9-11 with $E_z = 0$ and γ from Eq. 9-14.
4. Develop the equation for the total magnetic field intensity using Faraday's law (one of Maxwell's equations) and the total ∇ operator

$$\vec{H}(t_1, t_2, z) = j\frac{1}{\omega\mu}\nabla \times \vec{E}(t_1, t_2, z) \tag{9-18}$$

9-3.2 TM modes: $H_z = 0$

Since the only z component of field is E_z, transverse magnetic (TM) modes are often called *E-type* modes.

The solution process is:

1. Solve

$$\nabla_t^2 E_z + k_{c,\text{TM}}^2 E_z = 0 \tag{9-19}$$

for $E_z(t_1, t_2)$.

2. Apply boundary conditions to E_z to determine as many unknown constants as possible. As the boundary conditions are applied, values of $k_{c,\text{TM}}$ will be obtained. Use Eq. 9-4 to obtain γ_{TM} values.

3. Use the solution for $E_z(t_1, t_2)$ and with $H_z = 0$ obtain the solutions for the other field components using Eqs. 9-7 through 9-10.
4. Construct the total field solutions using Eqs. 9-11 and 9-12 with $H_z = 0$ and γ_{TM} from Step 2.

9-3.3 TE Modes: $E_z = 0$

Since the only z component of field is H_z, transverse electric (TE) modes are often called *H-type* modes.

The solution process is:

1. Solve

$$\nabla_t^2 H_z + k_{c,\text{TE}}^2 H_z = 0 \tag{9-20}$$

for $H_z(t_1, t_2)$.
2. Apply boundary conditions. For these modes it is most often easier to first compute the other components of electric and magnetic fields using Eqs. 9-7 through 9-10, with $E_z = 0$, and apply known boundary conditions on tangential $\vec{E}$ or normal $\vec{B}$ to determine the unknown constants. As the boundary conditions are applied values of $k_{c,\text{TE}}$ will be obtained. These values are used in Eq. 9-4 to obtain values of the propagation constant γ_{TE}.
3. Use the solution for $H_z(t_1, t_2)$ and $E_z = 0$ and obtain the solutions for the other field components using Eqs. 9-7 through 9-10.
4. Construct the total field solutions using Eqs. 9-11 and 9-12 with $E_z = 0$ and γ_{TE} from Step 2.

9-3.4 Information Obtained from Solutions

When one examines the amount of work required to obtain field solutions, the question "Why are we doing this?" may seem appropriate. Another way of posing the question might be "What can we learn from the solutions?" The following list responds to these queries, the order not being significant:

1. It is important to know the range of frequencies that will allow waves to propagate. We already know that propagation can only occur when γ has an imaginary component so that the exponential factor has a term $e^{-j\beta z}$. If γ is pure real, then the wave is merely attenuated in the z direction. Solutions allow us to predict operating frequencies that one must know for design work.
2. It is important to know the direction of the field vectors within the guide so that we can design a feed system required to establish the propagating waves. Examples of this will be given.
3. The results of waveguide analysis allow us to predict field configurations inside a closed cavity. The field solutions tell us how the resonant frequency depends upon cavity dimensions, a requirement for design. The field solutions also tell us how to couple into and out of a resonant cavity.
4. To use the Smith chart to solve transmission line problems using waveguides, we need

to know the wavelength along the z direction. We call this the *guide wavelength*, denoted by λ_g. Field solutions give us this value.

5. Whenever delay and distortion of signals are critical we need to know the phase and group velocities. These are defined as before:

$$V_p = \frac{\omega}{\beta} = \frac{\omega}{\text{Im}\{\gamma\}}$$

$$V_g = \frac{d\omega}{d\beta}$$

6. If we need to know the exact value of wave impedance at a point in a waveguide system, we can use the Smith chart if we know a wave "characteristic impedance." The characteristic impedance would be the z direction wave impedance seen in a waveguide having only a positive traveling wave (corresponding to zero reflection). To obtain this we need to know the field equations.
7. Power loss along the waveguide can be determined if we know the currents in any metal boundaries or in lossy dielectrics. If we know the equations for the field components we can determine this loss.
8. To determine how much power can be transmitted down the guide we must be able to compute the maximum electric field intensity that occurs in the waveguide and ensure that the medium in the guide does not break down.
9. There are occasions when a given waveguide system may not have a nice, closed-form solution. One can frequently obtain approximate or qualitative results by knowing how the fields are distributed in a closely related system. For example, knowing the exact field structure in a metal circular cylindrical waveguide would suggest approximate field structures in one having elliptical cross section.

The details of developing these various factors are covered in the examples of the following sections.

9-4 Wave Propagation on Two-Conductor Systems

Waveguide systems consisting of two separate conductors are used extensively in practice. Examples are coaxial cable, TV twin lead (parallel wire), twisted wire pair, and printed circuit boards (microstrip and stripline). In this section we study wave propagation on such systems.

9-4.1 TEM Modes: Parallel Planes and Coax

We begin by examining the transmission line modes. These modes are most often used with two-conductor systems.

Parallel Planes

The basic structure is shown in Figure 9-1. The system is defined by $-\infty < x < \infty$ and $z > 0$.

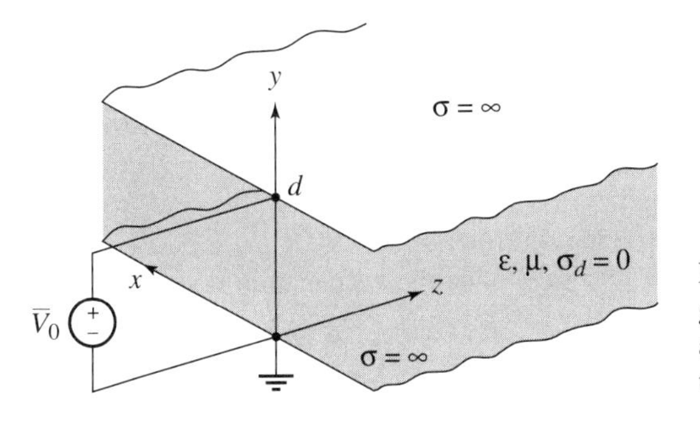

Figure 9-1. Parallel plane waveguide. Material between plates is assumed lossless. Plates extend to $\pm\infty$ in x and 0 to $+\infty$ in z.

We follow the outline given in Section 9-3.1, and begin by solving Laplace's equation in the transverse plane:

$$\nabla_t^2 \Phi = 0$$

First expand the Laplacian in the appropriate coordinate system, in this case rectangular, using only the transverse part:

$$\frac{\partial^2 \Phi}{\partial x^2} + \frac{\partial^2 \Phi}{\partial y^2} = 0$$

The next thing to do is look at the physical system to see if any of the terms in the Laplacian can be removed. We assume that the source is applied all along the edges at $z = 0$, which means the potential will not vary with x. (This assumption eliminates the more difficult case where propagation starts at the source terminals and propagates radially away.) Thus Φ will be a function of y only so the partials with respect to y may be replaced by the total derivative. This leaves

$$\frac{d^2 \Phi}{dy^2} = 0$$

Integrating twice:

$$\Phi(y) = Ay + B$$

If we suppose that the sinusoidal source is $\overline{V}_0 = V_0 \angle 0°$, and use the plate at $y = 0$ as reference we can apply potential boundary conditions. It is usually best to use all the zero values first, both of coordinate and potential. Thus we first use

$$\Phi(y = 0) = 0 = B$$

Then, with B set to zero in the general solution, we use

$$\Phi(y = d) = V_0 = Ad, \qquad \text{so } A = \frac{V_0}{d}$$

The solution reduces to

$$\Phi(y) = \frac{V_0}{d} y \tag{9-21}$$

Using this the electric field intensity is

$$\vec{E}(x, y) = -\nabla_t \Phi(y) = -\frac{\partial \Phi}{\partial y}\hat{y} = -\frac{V_0}{d}\hat{y} = E_y\hat{y} \tag{9-22}$$

Note this is what one would expect physically if the top plate were positive ($V_0 > 0$). Of course, the actual voltage varies sinusoidally.

From Eq. 9-14, with $\sigma_d = 0$, we have the propagation constant

$$\gamma_{\text{TEM}}^2 = -\omega^2\mu\epsilon \qquad \text{or} \qquad \gamma_{\text{TEM}} = j\omega\sqrt{\mu\epsilon} = j\beta = jk \tag{9-23}$$

If ϵ is complex (lossy dielectric) then $\beta = \text{Im}\{\gamma_{\text{TEM}}\}$. The total electric field intensity is constructed using Eq. 9-11 with E_z and $E_x = 0$:

$$\vec{E}(x, y, z) = -\frac{V_0}{d}e^{-j\omega\sqrt{\mu\epsilon}z}\hat{y} = -\frac{V_0}{d}e^{-jkz}\hat{y} \tag{9-24}$$

Using one of Maxwell's equations (Faraday's law),

$$\begin{aligned}\vec{H}(x, y, z) &= j\frac{1}{\omega\mu}\nabla \times \vec{E}(x, y, z) = j\frac{1}{\omega\mu}\left(-\frac{\partial E_y}{\partial z}\hat{x}\right) \\ &= -j\frac{1}{\omega\mu}\left(-\frac{V_0}{d}\right)(-jk)e^{-jkz}\hat{x} = \frac{V_0 k}{\omega\mu d}e^{-jkz}\hat{x} = \frac{V_0}{\eta d}e^{-jkz}\hat{x}\end{aligned} \tag{9-25}$$

The magnetic field intensity is directed along the $+x$ axis. This also agrees with what one might expect since $\vec{E} \times \vec{H}^*$ is in the $+z$ direction. In the coefficient of this last equation the following simplification was made:

$$\frac{V_0 k}{\omega\mu d} = \frac{V_0\omega\sqrt{\mu\epsilon}}{\omega\mu d} = \frac{V_0}{\sqrt{\dfrac{\mu}{\epsilon}}\,d} = \frac{V_0}{\eta d} \tag{9-26}$$

where η is the medium intrinsic (wave) impedance in ohms.

Thus the magnitudes of $\vec{E}$ and $\vec{H}$ are related by the intrinsic impedance of the medium between the planes. This is the same result that we obtained for uniform plane (TEM) waves. Since the fields are related by η, which is analogous to Z_0 for the transmission line where $I^+ = V^+/Z_0$, TEM modes are often called transmission line modes. We will see that other modes do not have this simple relationship.

We are now in a position to examine some of the information items of importance listed in Section 9-3.4. First, we note that there is no frequency constraint to obtain operation in the TEM mode since $\gamma(= jk)$ is always imaginary for any frequency.

Suppose we want to excite TEM propagation by means other than an applied voltage source. One method is to place a small element into the region that has a strong field component along the direction or directions of the field solutions $\vec{E}$ and $\vec{H}$. For example, if we want to couple to the electric field, the element must have a significant $\hat{y}$ electric field component. This could be done by inserting the center conductor of a coax into the region as shown in Figure 9-2*a*. Note that the voltage difference between the coax inner and outer conductors would produce a field (qualitatively) as indicated. There is a strong

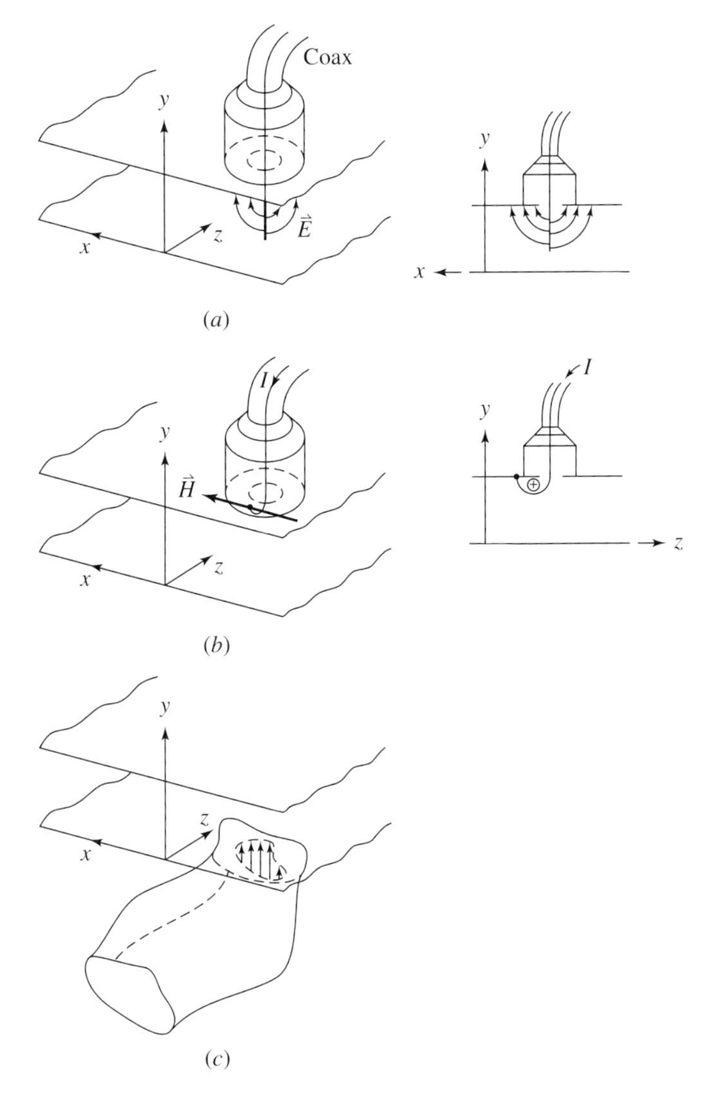

Figure 9-2. Methods for exciting TEM modes between parallel planes. (*a*) Electric field probe coupling. (*b*) Magnetic field loop coupling. (*c*) Aperture coupling.

vertical ($\hat{y}$) component. This element is called a *probe coupler*. The outer conductor of the connector is physically connected to the plate and the center conductor passes through a small hole.

Coupling to the magnetic field is accomplished by causing current to flow such that the current generates a magnetic field component in the direction required by the mode equations. In this case we need an $\hat{x}$ component of $\vec{H}$. This can be done as shown in Figure 9-2*b*. Here, the center conductor is inserted through a small hole, bent into a loop, and physically connected to the top plate on the underside. The right-hand rule shows that within the loop there is a strong $\hat{x}$ magnetic-field component. This element is called a *loop coupler*.

A third way of coupling into the system is by generating either the electric or magnetic field over a hole in a plate or some other device. In a following section we will see that a rectangular cross section guide can generate just such a field structure. This coupling mechanism is shown in Figure 9-2*c*. A hole (aperture) is cut in the generating structure and placed by the plates so that the electric field is in the proper direction. This element is called an *aperture coupler*.

It turns out that electron beams and electrical discharges can also be used to generate the waves, but these will not be described here.

The guide wavelength is defined as the distance in the z direction required to obtain 2π radians phase shift. We denote this by λ_g. This is easily obtained from the exponent of Eq. 9-24 or Eq. 9-25. The definition requires that for $z = \lambda_g$

$$k\lambda_g = 2\pi \tag{9-27}$$

Then

$$\lambda_g = \frac{2\pi}{k} = \frac{2\pi}{\omega\sqrt{\mu\epsilon}} = \frac{1}{f\sqrt{\mu\epsilon}} \text{ m} \tag{9-28}$$

We can obtain the "characteristic impedance" in the z direction using Eq. 7-67, where $\hat{\tau} \sim \hat{z}$, for the wave impedance:

$$Z^w_{z,\text{TEM}} = \frac{(\vec{E} \times \vec{H}^*) \cdot \hat{z}}{(\hat{z} \times \vec{H}) \cdot (\hat{z} \times \vec{H})^*} = \frac{\dfrac{+V_0^2}{\eta d^2}\,\hat{z} \cdot \hat{z}}{\dfrac{V_0}{\eta d}\,e^{-jk_z}\hat{y} \cdot \dfrac{V_0}{\eta d}\,e^{jk_z}\hat{y}} = \eta\ \Omega \tag{9-29}$$

This could also have been obtained by taking the ratio of E_t to H_t. However, to use this approach requires careful attention to the sign. If either component is negative we must negate the ratio to get the correct wave impedance. In this case the electric field is in the $-\hat{y}$ direction so we have

$$Z^w_{z,\text{TEM}} = -\frac{E_y}{H_x} = -\frac{-\dfrac{V_0}{d}}{\dfrac{V_0}{\eta d}} = \eta$$

Now that we have "Z_0" and the wavelength we can use the Smith chart to solve planar waveguide transmission problems.

The propagation velocities are now easily evaluated. For the phase velocity, using $\beta = k$ for TEM:

$$V_{p,\text{TEM}} = \frac{\omega}{\beta} = \frac{\omega}{k} = \frac{\omega}{\omega\sqrt{\mu\epsilon}} = \frac{1}{\sqrt{\mu\epsilon}} \text{ m/s} \tag{9-30}$$

Since the phase velocity is independent of frequency the system is dispersionless.

The group velocity is

$$V_{g,\text{TEM}} = \frac{d\omega}{d\beta} = \frac{d\omega}{dk} = \frac{d\omega}{d(\omega\sqrt{\mu\epsilon})} = \frac{1}{\sqrt{\mu\epsilon}}\frac{d\omega}{d\omega} = \frac{1}{\sqrt{\mu\epsilon}} \text{ m/s} \tag{9-31}$$

The transmitted power can be determined from the Poynting vector using the sinusoidal steady-state forms in Eq. 7-56:

$$\vec{S} = \frac{1}{2}\vec{E} \times \vec{H}^* = \frac{1}{2}\frac{V_0^2}{\eta d^2} \text{ W/m}^2 \tag{9-32}$$

We obtain the surface currents in the planes by applying the boundary condition on tangential magnetic field intensity. From Table 7-5 for the perfect conductor and the sinusoidal steady state (where we replace the script letters by the block phasor representations) we have for the *upper* plate

$$\vec{J}_s = \hat{n} \times \vec{H} = (-\hat{y}) \times \frac{V_0}{\eta d}\hat{x}\bigg|_{y=d} = \frac{V_0}{\eta d}\hat{z} \text{ A/m} \tag{9-33}$$

This is the current density on the *inner surface* of the upper plate. Current flows in the z direction, as expected, since this is a transmission line system of two conductors. To obtain the time domain form, we multiply by the propagation factor e^{-jkz} and write

$$\vec{\mathcal{J}}_s = \text{Re}\{\vec{J}_s e^{-jkz} e^{j\omega t}\} = \frac{V_0}{\eta d}\cos(\omega t - kz)\hat{z} \text{ A/m} \tag{9-34}$$

If we freeze time we can plot the instantaneous current pattern. Set $t = 0$ and use $k = 2\pi/\lambda_g$ from Eq. 9-27:

$$\vec{\mathcal{J}}_s(y = d, z, t = 0) = \frac{V_0}{\eta d}\cos\left(\frac{2\pi}{\lambda_g}z\right)\hat{z} \tag{9-35}$$

Similarly, for the bottom plate we have

$$\vec{\mathcal{J}}_s(y = 0, z, t = 0) = -\frac{V_0}{\eta d}\cos\left(\frac{2\pi}{\lambda_g}z\right)\hat{z} \tag{9-36}$$

These are plotted in Figure 9-3, where the width of the line represents the current amplitude at a location z. Note only one current line has been drawn. There would be parallel sets of these lines all along the x dimension.

Note there seems to be a problem at $z = \lambda_g/4$ because the current is zero there, but two currents flow toward that point. Where did the current go? What about Kirchhoff's current law? To resolve this dilemma we look at the electric field quantities at $t = 0$.

In the time domain the electric field intensity is

$$\vec{\mathcal{E}} = \text{Re}\{\vec{E}e^{j\omega t}\} = -\frac{V_0}{d}\cos(\omega t - kz)\hat{y} \tag{9-37}$$

At $t = 0$:

$$\vec{\mathcal{E}}(y, z, t = 0) = -\frac{V_0}{d}\cos\left(\frac{2\pi}{\lambda_g}z\right)\hat{y} \text{ V/m} \tag{9-38}$$

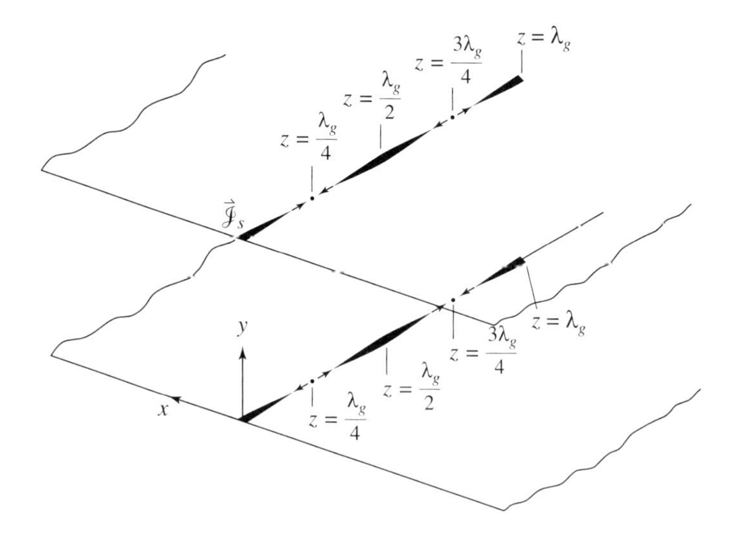

Figure 9-3. Surface current density on parallel planes supporting a TEM mode. (Snapshot at $t = 0$.)

This is plotted in Figure 9-4. The electric field is maximum when the surface current is maximum.

The displacement current is easily obtained from the electric field intensity Eq. 9-37 as

$$\vec{\mathscr{J}}_D = \epsilon \frac{\partial \vec{\mathscr{E}}}{\partial t} = \frac{\epsilon \omega V_0}{d} \sin(\omega t - kz)\hat{y} \text{ A/m}^2 \tag{9-39}$$

At $t = 0$:

$$\vec{\mathscr{J}}_D(y, z, t = 0) = -\frac{\epsilon \omega V_0}{d} \sin\left(\frac{2\pi}{\lambda_g} z\right) \text{ A/m}^2 \tag{9-40}$$

The displacement current is plotted in Figure 9-5. Note that the displacement current is maximum when the surface current is a minimum.

To compare this current with the surface current, we compute the displacement current leaving the top plate from an area 1-m wide and of general length z. This would be an effective displacement current per meter of width at location z. Using Eq. 9-39 with $k = \omega\sqrt{\mu\epsilon}$ and $\eta = \sqrt{\mu/\epsilon}$,

$$\mathscr{J}_D \text{ per meter} = \int_0^1 \int_0^z \vec{\mathscr{J}}_D \cdot dx dz(-\hat{y}) = \frac{V_0}{\eta d} [1 - \cos(\omega t - kz)] \text{ A/m} \tag{9-41}$$

At a given location z the total current per meter is (disregarding the vector nature of the currents) obtained by adding the surface current per meter, Eq. 9-34, and the displacement current per meter, Eq. 9-41. This gives

$$\mathscr{J}_{\text{total}} \text{ per meter} = \mathscr{J}_S + \mathscr{J}_D = \frac{V_0}{\eta d} \text{ A/m} \tag{9-42}$$

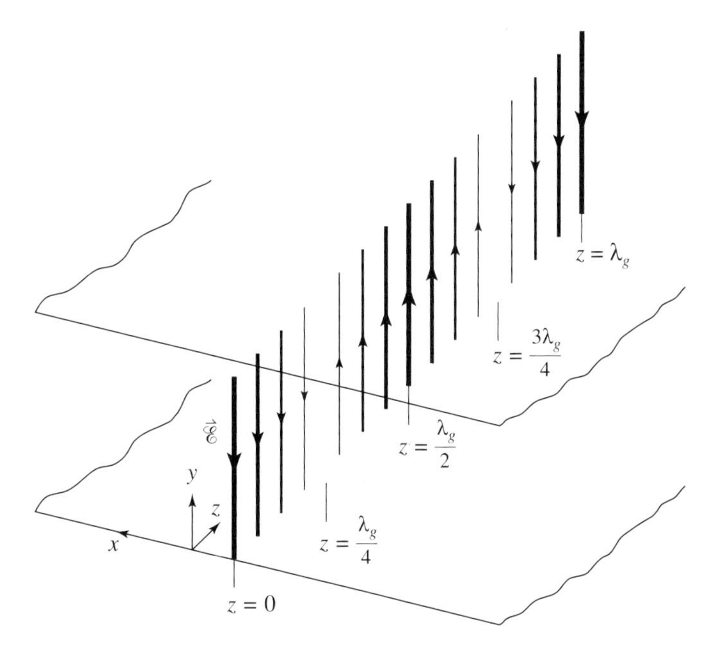

Figure 9-4. Electric field intensity between parallel planes supporting a TEM mode. (Snapshot at $t = 0$.)

Figure 9-5. Displacement current density between parallel planes supporting a TEM mode. (Snapshot at $t = 0$.)

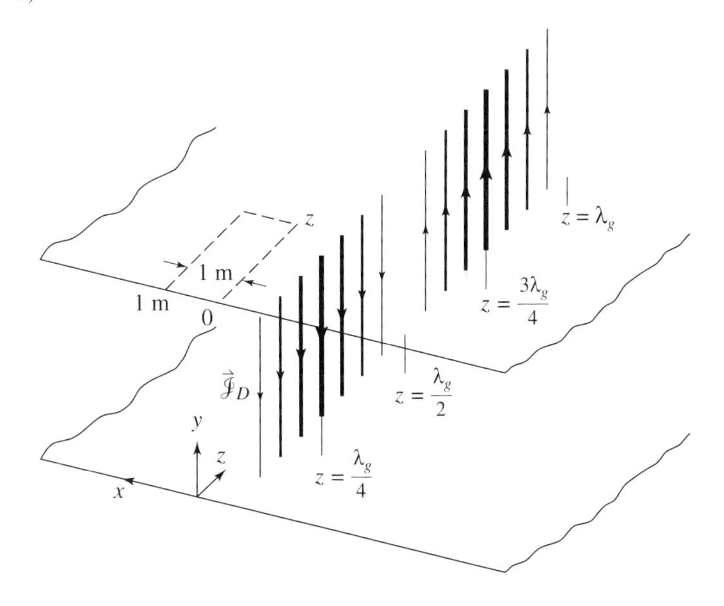

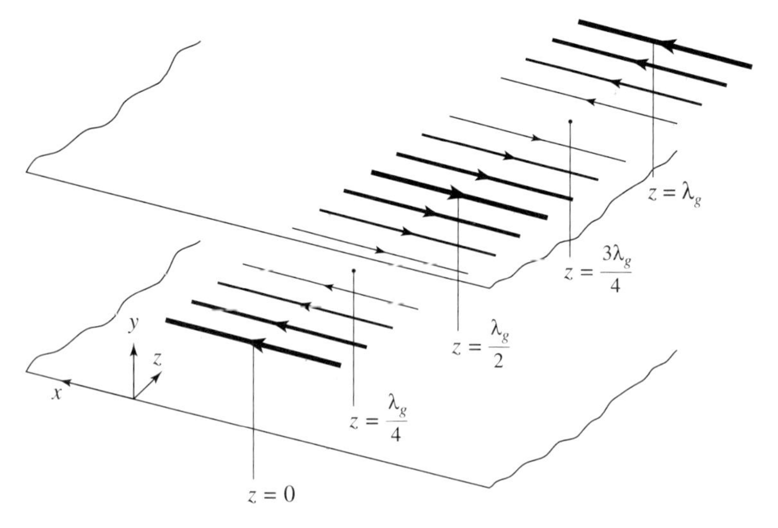

Figure 9-6. Magnetic field intensity between parallel planes supporting a TEM mode. (Snapshot at $t = 0$.)

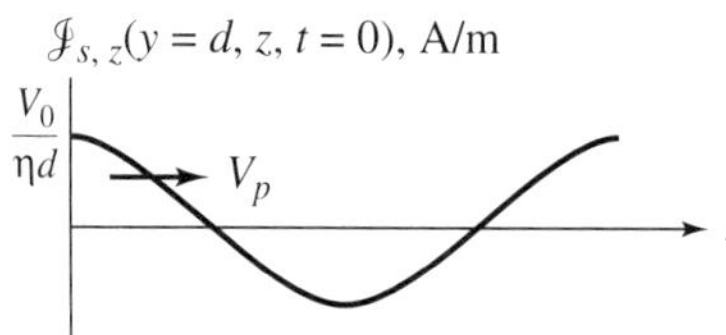

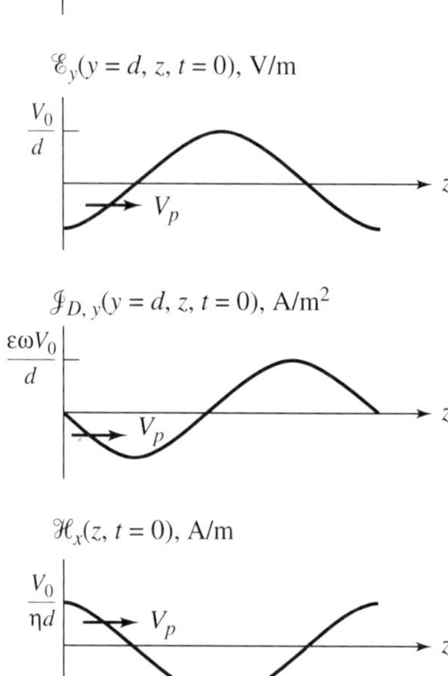

Figure 9-7. Wave plots of selected field components for parallel planes supporting a TEM wave.

Thus the current per meter is constant for any values of z and t! Qualitatively, this means that the total current is divided between surface current and displacement current, so at a given location z Kirchhoff's current law is satisfied at each instant of time. This resolves the visual dilemma of the surface current in Figure 9-3.

To complete the field pictures of the system, the magnetic field intensity is obtained from Eq. 9-25 to be

$$\vec{\mathcal{H}} = \frac{V_0}{\eta d}\cos(\omega t - kz)\hat{x} \tag{9-43}$$

At $t = 0$,

$$\vec{\mathcal{H}}(y, z, t = 0) = \frac{V_0}{\eta d}\cos(kz)\hat{x} = \frac{V_0}{\eta d}\cos\left(\frac{2\pi}{\lambda_g}z\right) \tag{9-44}$$

This is plotted in Figure 9-6.

From transmission line theory we know that the variable $(\omega t - kz)$ represents a wave traveling in the positive z direction. Thus all of the field configurations in the figures propagate along as a unit. The wave nature of the field quantities can be presented more compactly as given in Figure 9-7. All of these waves are also drawn for $t = 0$ using the field equations previously developed. These waves propagate along the z axis at the phase velocity.

Exercises

9-4.1a Using the phasor form of $\vec{E}$, determine the equation for the displacement current $\vec{J}_D$. From this determine the time form $\vec{\mathcal{J}}_{\mathcal{D}}$.

9-4.1b Using the appropriate conductor boundary condition determine the real time equations for the surface charge densities on the top and bottom plates, $\rho_s(x, y, z, t)$. Evaluate these at $t = 0$ and make a plot like those of Figures 9-3 through 9-6, using dots of different sizes to indicate the relative magnitude of the charge density along the z direction. Explain how your plot is consistent physically with those of $\vec{\mathcal{J}}_s(t = 0)$ and $\vec{\mathcal{E}}(t = 0)$.

9-4.1c A TEM wave is launched between parallel plates at a frequency of 30 MHz. The dielectric between the plates is assumed lossless with $\mu = \mu_0$, and $\epsilon_r = 2.6$. The plate spacing is 3 mm. What voltage source is required to obtain a power density of 50 kW/m^2? What is the peak electric field intensity in the dielectric? What is the displacement current density peak value? Determine the guide wavelength and the phase velocity. What are the peak values of the magnetic field intensity and the magnetic flux density?

Coax

The next configuration to be solved is shown in Figure 9-8. The solution process is identical to that of the parallel-plate system. The conductors at $\rho = \alpha$ and $\rho = b$ are assumed perfect and the dielectric is assumed lossless.

Writing the transverse Laplacian in cylindrical coordinates we will begin with

$$\frac{1}{\rho}\frac{\partial}{\partial \rho}\left(\rho\frac{\partial \Phi}{\partial \rho}\right) + \frac{1}{\rho^2}\frac{\partial^2 \Phi}{\partial \phi^2} = 0 \tag{9-45}$$

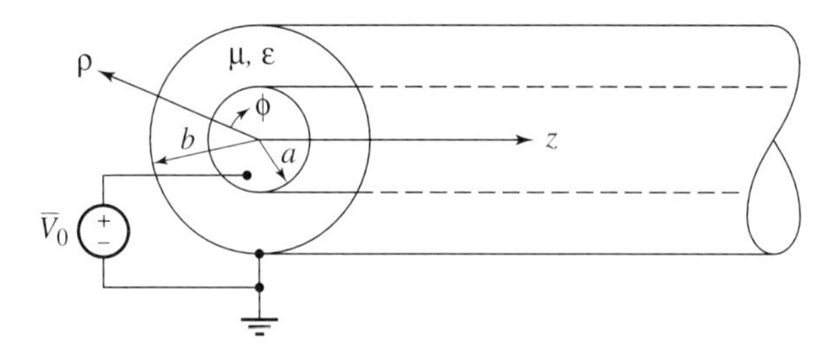

Figure 9-8. Coaxial cable conventions for wave propagation.

Since TEM waves have uniform plane wave characteristics and since the coax is symmetric in the coordinate ϕ we would suppose that the potential would not vary with ϕ. Thus all the ϕ derivatives are zero and Φ is a function of ρ only so that the preceding reduces to

$$\frac{1}{\rho}\frac{d}{d\rho}\left(\rho\frac{d\Phi}{d\rho}\right) = 0 \tag{9-46}$$

Multiply the equation through by ρ:

$$\frac{d}{d\rho}\left(\rho\frac{d\Phi}{d\rho}\right) = 0 \tag{9-47}$$

Since the derivative is zero we must have

$$\rho\frac{d\Phi}{d\rho} = \text{constant} = A$$

or

$$\frac{d\Phi}{d\rho} = \frac{A}{\rho} \tag{9-48}$$

Integrating

$$\Phi(\rho) = A\ \ln\rho + B \tag{9-49}$$

(This equation is exactly what we obtained when solving the static coax problem to determine the capacitance.)

Applying the outer conductor boundary condition on potential:

$$\Phi(\rho = b) = 0 = A\ \ln b + B \Rightarrow B = -A\ \ln b$$

Substituting this result into the preceding equation and using the subtraction property of logarithms we obtain

$$\Phi(\rho) = A\ \ln\left(\frac{\rho}{b}\right) \tag{9-50}$$

On the outer conductor the boundary condition is $\overline{V}_0 = V_0\angle 0°$.

$$\Phi(\rho = a) = V_0 = A\ \ln\left(\frac{a}{b}\right) \Rightarrow A = \frac{V_0}{\ln\left(\frac{a}{b}\right)}.$$

Substituting this into Eq. 9-50 we finally have

$$\Phi(\rho) = \frac{V_0}{\ln\left(\frac{a}{b}\right)} \ln\left(\frac{\rho}{b}\right) \tag{9-51}$$

The electric field is then

$$\vec{E} = -\nabla_t \Phi = -\frac{d\Phi}{d\rho}\hat{\rho} = \frac{-V_0}{\ln\left(\frac{a}{b}\right)} \cdot \frac{1}{\rho/b}\frac{1}{b}\hat{\rho} = -\frac{V_0}{\rho \ln\left(\frac{a}{b}\right)}\hat{\rho}$$

Since $a < b$, the logarithm will be negative, but we can use the reciprocation property of logarithms to write the result as

$$\vec{E} = \frac{V_0}{\rho \ln\left(\frac{b}{a}\right)}\hat{\rho} \tag{9-52}$$

This is in the $+\hat{\rho}$ direction as one would expect if the center conductor were positive.

For the lossless dielectric Eq. 9-14 gives

$$\gamma_{\text{TEM}} = \sqrt{-\omega^2\mu\epsilon} = j\omega\sqrt{\mu\epsilon} = jk = j\beta \tag{9-53}$$

Then

$$\vec{E}(\rho, \phi, z) = \frac{V_0}{\rho \ln\left(\frac{b}{a}\right)} e^{-jkz}\hat{\rho} \text{ V/m} \tag{9-54}$$

Using Faraday's law we obtain

$$\vec{H}(\rho, \phi, z) = j\frac{1}{\omega\mu}\nabla \times \vec{E} = j\frac{1}{\omega\mu}\frac{dE_\rho}{dz}\hat{\phi} = \frac{kV_0}{\rho\omega\mu \ln\left(\frac{b}{a}\right)} e^{-jkz}\hat{\phi} \tag{9-55}$$

The corresponding time expressions are

$$\vec{\mathscr{E}}(\rho,t) = \frac{V_0}{\rho \ln\left(\frac{b}{a}\right)} \cos(\omega t - kz)\hat{\rho} \tag{9-56}$$

$$\vec{\mathscr{H}}(\rho, t) = \frac{V_0}{\rho\eta \ln\left(\frac{b}{a}\right)} \cos(\omega t - kz)\hat{\phi} \tag{9-57}$$

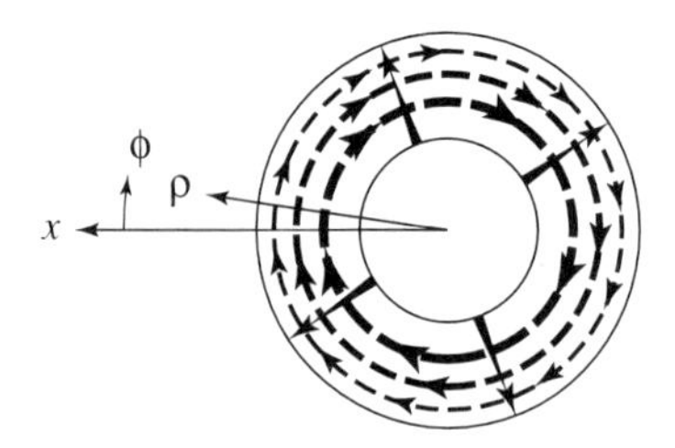

Figure 9-9. Electric (solid) field lines and magnetic (dashed) field lines in a coaxial cable supporting a TEM mode (shown for $t = 0$ and $z = 0$), looking along the z direction. Line widths indicate field magnitudes.

These are plotted in Figure 9-9 for $t = 0$ and $z = 0$ in the transverse plane looking in the $+z$ direction. Both fields are strongest at the inner conductor and become weaker toward the outer conductor.

The determinations of the properties of the waves follow the same procedures as discussed for the parallel-plane guide. We list some of them and reserve some for exercises.

$$\lambda_g = \frac{2\pi}{\beta} = \frac{2\pi}{k} = \frac{1}{f\sqrt{\mu\epsilon}} \tag{9-58}$$

$$Z^W_{z,\text{TEM}} = \frac{E_\rho}{H_\phi} = \eta \tag{9-59}$$

$$\vec{S} = \frac{1}{2}\vec{E} \times \vec{H}^* = \frac{V_0^2}{2\eta\rho^2 \ln^2\left(\dfrac{b}{a}\right)}\hat{z} \tag{9-60}$$

$$\vec{J}_s(\rho = a) = \hat{n} \times \vec{H}(\rho = a) = \hat{\rho} \times \vec{H}(\rho = a) = \frac{V_0}{\eta a \ln\left(\dfrac{b}{a}\right)} e^{-jkz}\hat{z} \tag{9-61}$$

The results allow us to make the connection between the fields and the current and voltage on a transmission line. For example, at $z = 0$ the total current in the $+z$ direction is the source current. Integrating the surface current around the periphery of the inner conductor, for which an increment of length normal to $\vec{J}_s$ is $a\,d\phi$. Then, at $z = 0$,

$$I_0 = \int_0^{2\pi} \frac{V_0}{\eta a \ln\left(\dfrac{b}{a}\right)} a\,d\phi = \frac{2\pi V_0}{\eta \ln\left(\dfrac{b}{a}\right)} \text{ A} \tag{9-62}$$

Since η is real for lossless dielectric, the power being delivered by the source is

$$P_0 \frac{1}{2} V_0 I_0 = \frac{1}{2} V_0 \frac{2\pi V_0}{\eta \ln\left(\dfrac{b}{a}\right)} = \frac{\pi V_0^2}{\eta \ln\left(\dfrac{b}{a}\right)} \text{ W} \tag{9-63}$$

Now let's integrate the Poynting vector over the cross section between the conductors:

$$P_{\text{fields}} = \int\int \vec{S} \cdot d\vec{a} = \int_a^b \int_0^{2\pi} \frac{V_0^2}{2\eta\rho^2 \ln^2\left(\frac{b}{a}\right)} \hat{z} \cdot \rho d\phi d\rho \hat{z}$$

$$= \frac{\pi V_0^2}{\eta \ln\left(\frac{b}{a}\right)} \text{ W}$$

The two powers are identical, and we conclude that the power delivered by the source can be thought of as being transferred to the propagating fields. The voltage and current waves propagate with velocity $1/\sqrt{LC}$ and the electric and magnetic field waves propagate with velocity $1/\sqrt{\mu\epsilon}$. If we use the equations for the capacitance per meter and inductance per meter of the coaxial cable (Exercise 5-14.0a; also Eq. 5-95 and Chapter 6, Example 34) we have

$$\frac{1}{\sqrt{LC}} = \frac{1}{\sqrt{\frac{\mu}{2\pi}\ln\left(\frac{b}{a}\right) \cdot \frac{2\pi\epsilon}{\ln\left(\frac{b}{a}\right)}}} = \frac{1}{\sqrt{\mu\epsilon}} \tag{9-64}$$

Thus the fields and current-voltage waves propagate with the same velocity. The preceding equation is actually true for any TEM system even though it was derived from the coax equations here. Note that we are considering positive traveling waves in each case.

Some care must be used with the analogy. If we wish to use current and voltage as the quantities for Smith chart work the appropriate characteristic impedance would be

$$Z_0 = \frac{V^+}{I^+} = \frac{V_0}{I_0} = \frac{\eta \ln\left(\frac{b}{a}\right)}{2\pi} \quad \Omega \text{ (coax)} \tag{9-65}$$

Exercises

9-4.1d For the coaxial cable show how to excite the TEM mode using probe, loop, and aperture couplings, rather than using a voltage source.

9-4.1e What range of frequencies can be used to propagate the TEM mode in a coax? (Answer: 0 to ∞)

9-4.1f Determine the phase velocity and group velocity for fields in a coax. (Answer: $1/\sqrt{\mu\epsilon}$ for both)

9-4.1g Determine the voltage source amplitude (V_0) required to transmit 2 kW along an RG58C/U cable. See Figure 1-18 of Chapter 1, and assume coax is lossless. Also compute the corresponding maximum electric field intensity. Compare this last result with the dielectric strength (breakdown) of the dielectric.

9-4.1h Derive the expression for the displacement current, $\vec{\mathcal{J}}_D$, and then show that the current per meter at any location z is a constant (also time invariant).

An important class of two-conductor systems which support TEM modes is the printed circuit board configuration. These structures were discussed briefly in Section 1-9, Chapter 1. This is frequently found on hybrid circuits, usually on the silicon directly. Such lines may be intentionally placed in a circuit or may be parasitic elements which arise by packaging of integrated circuits. These systems operate well into the millimeter wave frequency range (> 30 GHz).

The difficulty in analyzing these systems is that there are finite-width dimensions of the planar surfaces and multiple dielectrics. These require other kinds of analysis techniques, such as complex variable theory, and will not be covered here. This is why the results only were given in Chapter 1.

Note: For the remainder of this chapter extensive use is made of the separation of variables technique for solving partial differential equations. Additionally, the various functions that are obtained using the series solution method on ordinary differential equations are introduced. Appendices K, L, and M cover these topics in more detail than the solution examples that follow. A study of those appendices will help significantly if these solution techniques and functions have not been learned elsewhere.

9-4.2 TM Modes: Parallel Planes

The system to be studied is the same as Figure 9-1. We follow the solution procedure of Section 9-3.2, where $t_1 \sim x$ and $t_2 \sim y$.

We expand the wave Eq. 9-19 in rectangular coordinates. Note we only use the transverse components.

$$\frac{\partial^2 E_z}{\partial x^2} + \frac{\partial^2 E_z}{\partial y^2} + k_{c,\mathrm{TM}}^2 E_z = 0$$

Since there is no x variation the partials are zero and E_z is a function of y only. Remember that the z variation is in the exponential $e^{-\gamma z}$, which does not appear explicitly at this point. The preceding equation reduces to

$$\frac{d^2 E_z}{dy^2} + k_{c,\mathrm{TM}}^2 E_z = 0$$

The characteristic equation is

$$m^2 + k_{c,\mathrm{TM}}^2 = 0$$

from which

$$m = \pm jk_{c,\mathrm{TM}}$$

The corresponding trigonometric solution is

$$E_z(y) = A\cos(k_{c,\mathrm{TM}}y) + B\sin(k_{c,\mathrm{TM}}y) \tag{9-66}$$

At $y = 0$ and $y = d$, E_z is a tangential component, which must be zero for a perfect conductor. Using as many zeros as possible first—$y = 0$, $E_z = 0$—Eq. 9-66 requires

$$E_{\tan}(y = 0) = E_z(y = 0) = 0 = A$$

The solution reduces to

$$E_z(y) = B \sin(k_{c,\mathrm{TM}} y) \tag{9-67}$$

At the upper plate

$$E_z(y = d) = 0 = B \sin(k_{c,\mathrm{TM}} d)$$

The constant B cannot be zero because we would not have any solution left. Thus we require

$$\sin(k_{c,\mathrm{TM}} d) = 0$$

This can be satisfied if

$$k_{c,\mathrm{TM}} d = n\pi, \qquad n = 1, 2, 3, \ldots$$

From this,

$$k_{c,\mathrm{TM}} = \frac{n\pi}{d}, \qquad n = 1, 2, 3, \ldots \tag{9-68}$$

These are the eigenvalues of the partial differential equation. We do not use $n = 0$ because the electric field would be identically zero.

The solution for E_z is then

$$E_z(y) = B \sin \frac{n\pi}{d} y, \qquad n = 1, 2, 3, \ldots \tag{9-69}$$

The value of B cannot be determined since it depends upon the generator amplitude. This is not the voltage source directly as it was for the TEM mode solution for Φ. We usually leave this last single constant in the solution.

Now if we assume a lossless dielectric, $\sigma_d = 0$, Eq. 9-4 gives the propagation constant as follows:

$$\gamma_{\mathrm{TM}} = \sqrt{k_{c,\mathrm{TM}}^2 - k^2} = \sqrt{\left(\frac{n\pi}{d}\right)^2 - \omega^2 \mu \epsilon} = k \sqrt{\left(\frac{k_{c,\mathrm{TM}}}{k}\right)^2 - 1}$$

Since propagation can occur only for imaginary γ, it is usual practice to factor out a -1 under the radical and use the form

$$\gamma_{\mathrm{TM}} = jk \sqrt{1 - \left(\frac{k_{c,\mathrm{TM}}}{k}\right)^2} = j\omega\sqrt{\mu\epsilon} \sqrt{1 - \left(\frac{n\pi}{d\omega\sqrt{\mu\epsilon}}\right)^2} = j\beta_{\mathrm{TM}} \tag{9-70}$$

Next use Eq. 9-7 with $t_1 \sim x$, $h_1 = 1$ (rectangular), and $H_z = 0$ (TM mode)

$$E_x(x, y) = -\frac{\gamma_{\mathrm{TM}}}{k_{c,\mathrm{TM}}^2} \frac{dE_z}{dx} = 0$$

From Eq. 9-8

$$E_y(x, y) = -\frac{\gamma_{\text{TM}}}{k_{c,\text{TM}}^2}\frac{dE_z}{dy} = -jB\frac{k\sqrt{1-\left(\frac{k_{c,\text{TM}}}{k}\right)^2}}{\left(\frac{n\pi}{d}\right)}\cos\frac{n\pi}{d}y \tag{9-71}$$

From Eq. 9-9

$$H_x(x, y) = j\frac{\omega\epsilon}{k_{c,\text{TM}}^2}\frac{dE_z}{dy} = jB\frac{\omega\epsilon}{\left(\frac{n\pi}{d}\right)}\cos\frac{n\pi}{d}y \tag{9-72}$$

From Eq. 9-10

$$H_y(x, y) = 0 \tag{9-73}$$

We can now construct the total solutions by multiplying by the exponential factor $e^{-\gamma_{\text{TM}}}$,

$$\begin{aligned}\vec{E}(y, z) = &-j\frac{d}{n\pi}k\sqrt{1-\left(\frac{k_{c,\text{TM}}}{k}\right)^2}\,B\cos\frac{n\pi}{d}\,ye^{-j\sqrt{\omega^2\mu\epsilon-(n\pi/d)^2}z}\hat{y}\\ &+B\sin\frac{n\pi}{d}\,ye^{-j\sqrt{\omega^2\mu\epsilon-(n\pi/d)^2}z}\hat{z}, \qquad n = 1, 2, 3, \ldots\end{aligned} \tag{9-74}$$

$$\vec{H}(y, z) = j\frac{\omega\epsilon d}{n\pi}B\cos\frac{n\pi}{d}\,ye^{-j\sqrt{\omega^2\mu\epsilon-(n\pi/d)^2}z}\hat{x}, \qquad n = 1, 2, 3, \ldots \tag{9-75}$$

Since there are many solutions, one for each n, we need to have a way of identifying them. We use a subscript to indicate the value of n being used and write TM_n, and we refer to this as the *nth order mode* between parallel planes. In practice, the mode that exists in a given system depends upon how the system is excited *and* the frequency of the source as discussed next when we investigate the operating properties of the modes.

TM modes can propagate only for imaginary γ_{TM}. From Eq. 9-70 this requires

$$d\omega\sqrt{\mu\epsilon} > n\pi$$

or

$$\omega > \frac{n\pi}{d\sqrt{\mu\epsilon}} \tag{9-76}$$

The cutoff frequency occurs when $\beta = 0$, for which we define

$$\omega_{c,n} = \frac{n\pi}{d\sqrt{\mu\epsilon}}\ \text{rad/s} \tag{9-77}$$

Thus no frequencies *below* this value can propagate. If we want to operate at a lower frequency than this predicts for a given system we must increase plate spacing d or excite the system to allow smaller n (shown shortly).

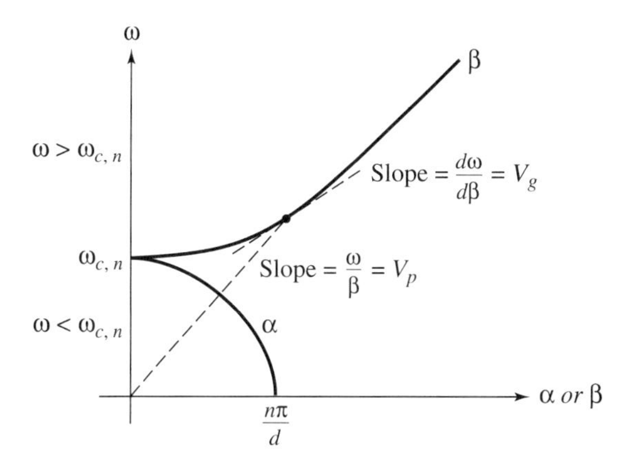

Figure 9-10. Variations of α and β as functions of frequency for TM waves between parallel planes.

For frequencies below cutoff γ_{TM} is a real number and so represents attenuation, $e^{-\alpha z}$. Waves that attenuate in the direction of propagation are called *evanescent waves.*

Plots of α and β as functions of frequency are given in Figure 9-10. Using the independent variable along the vertical axis may appear a little awkward, but the definitions of group and phase velocities make this convenient, as shown in the figure. From Eq. 9-69 the phase velocity is computed as

$$V_p = \frac{\omega}{\beta_{\text{TM}}} = \frac{\omega}{\sqrt{\omega^2 \mu \epsilon - \left(\frac{n\pi}{d}\right)^2}} \tag{9-78}$$

Since the phase velocity is a function of frequency, field signal waveforms that contain several frequencies will be distorted as discussed for dispersion on transmission lines. Suppose we have a pulse modulated on a high-frequency carrier illustrated in Figure 9-11*a*. This waveform consists of several frequencies clustered near the carrier, as can be determined using the Fourier transform. Now as long as ω is greater than cutoff, the equation for phase velocity shows that lower frequencies near cutoff propagate at very high veloc-

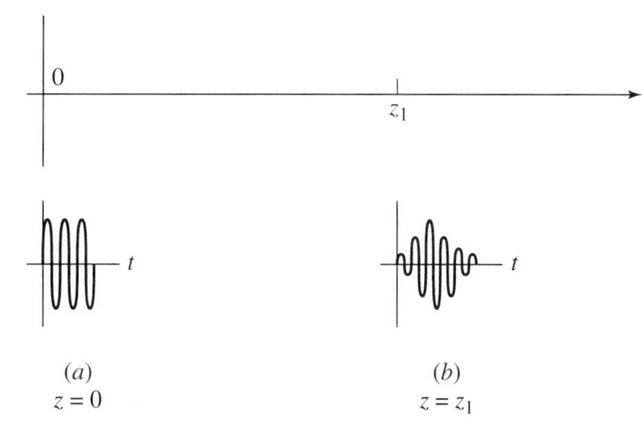

Figure 9-11. Illustration of dispersion in systems that have wave frequency-dependent phase velocity.

ities and higher frequencies propagate at lower velocities. The waveform will then "spread out," or disperse, as the pulse propagates in the $+z$ direction as shown in Figure 9-11*b*. This is called *dispersion*, and it occurs in all systems that have a frequency-dependent phase velocity. Note that the larger z is, the more severe will be the dispersion. If two pulses were transmitted serially in time, they would both broaden out or disperse and eventually merge so that it would appear that only one pulse was transmitted. This illustrates the trade-off between data rate and distance of transmission. The longer the distance, the slower the pulse rate must be to avoid pulse interference. The corresponding group velocity is

$$V_g = \frac{d\omega}{d\beta_{\text{TM}}} = \frac{1}{\dfrac{d\beta_{\text{TM}}}{d\omega}} = \frac{\sqrt{\omega^2\mu\epsilon - \left(\dfrac{n\pi}{d}\right)^2}}{\omega\mu\epsilon} \tag{9-79}$$

This is the approximate velocity of the group of frequencies in a pulse and is frequently used as signal velocity.

Note that for large ω ($\omega \gg \omega_{c,n}$) the group and phase velocities become equal to the velocity of a TEM wave in the unbounded medium (Chapter 8):

$$V_p = V_g = \frac{1}{\sqrt{\mu\epsilon}} \qquad \text{m/s} \qquad (\omega \gg \omega_{c,n}) \tag{9-80}$$

Also, from the equations for V_p and V_g:

$$V_p V_g = \frac{1}{\mu\epsilon} \qquad (\text{any } \omega > \omega_{c,n}) \tag{9-81}$$

We next examine the significance of the mode number n. If we plot the amplitude of E_z, Eq. 9-69, at $z = 0$ and $t = 0$, we obtain the results in Figure 9-12. From the figure we learn that the mode number n determines the number of amplitude variations of the field between the plates (vertically). These diagrams also show us how to couple a particular mode. Figure 9-13 shows possible ways for using probes to excite the TM_1 and TM_2 modes from a coaxial cable. The important thing to remember here is that whichever mode is desired to be excited must also have a frequency greater than the cutoff of that mode.

It is also common in practice to describe the cutoff of a waveguide in terms of a cutoff wavelength. This is defined in terms of the phase velocity of a TEM wave in an unbounded medium of the material between the plates. From Chapter 8 this would be

$$V_{p,\text{TEM}} \equiv v_0 \equiv \frac{1}{\sqrt{\mu\epsilon}} \tag{9-82}$$

If the material were vacuum this would be the velocity of light. Using the general relation velocity = wavelength × frequency we *define* the cutoff wavelength referred to the medium as

$$\lambda_{c,n} \equiv \frac{v_0}{f_{c,n}} = \frac{2\pi v_0}{\omega_{c,n}} \tag{9-83}$$

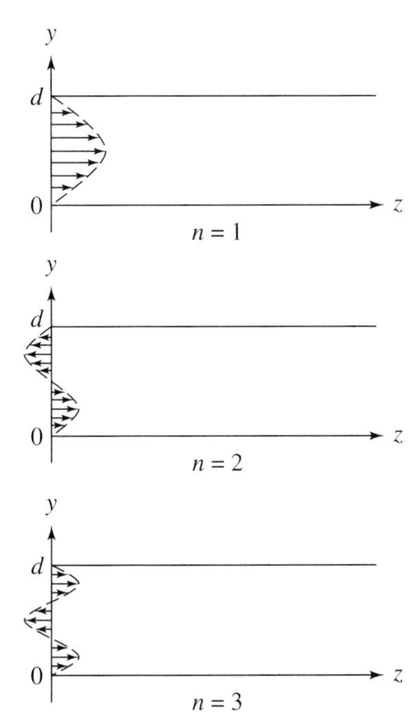

Figure 9-12. Field amplitude variation of E_z for the first three modes for TM_n modes between parallel planes.

Note this is not the guide wavelength. Substituting Eq. 9-77 into this

$$\lambda_{c,n} = \frac{2\pi v_0}{\dfrac{n\pi}{d}\dfrac{1}{\sqrt{\mu\epsilon}}} = \frac{2d}{n} \tag{9-84}$$

This gives us another interesting interpretation if we solve for d:

$$d = n\left(\frac{\lambda_{c,n}}{2}\right) \tag{9-85}$$

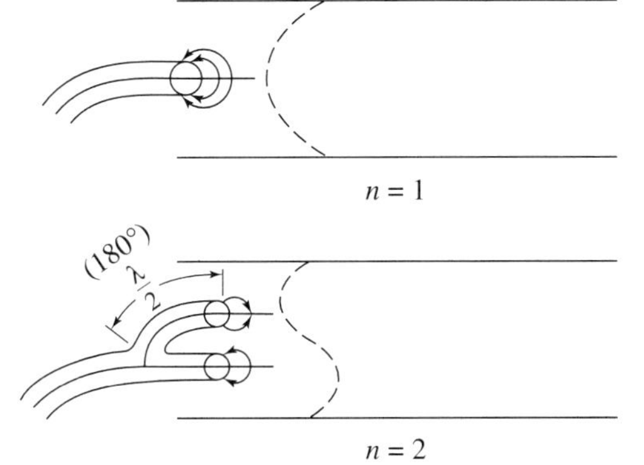

Figure 9-13. Probe coupling for exciting TM_n, $n = 1, 2$, modes between parallel planes.

Since the smallest d value is for $n = 1$, we conclude that propagation can occur only when the plate spacing is greater than half the cutoff wavelength and is an integral multiple of it.

The guide wavelength is obtained in the usual way:

$$\lambda_{g,n} = \frac{2\pi}{\beta_{\mathrm{TM}}} = \frac{2\pi}{\sqrt{\omega^2\mu\epsilon - \left(\frac{n\pi}{d}\right)^2}} = \frac{2\pi}{\omega\sqrt{\mu\epsilon}\sqrt{1 - \left(\frac{\omega_{c,n}}{\omega}\right)^2}}$$
$$= \frac{v_0}{f\sqrt{1 - \left(\frac{\omega_{c,n}}{\omega}\right)^2}} = \frac{\lambda_0}{\sqrt{1 - \left(\frac{\omega_{c,n}}{\omega}\right)^2}} \tag{9-86}$$

Here λ_0 is the wavelength of a TEM wave in the unbounded medium, v_0/f, used in Chapter 8.

The wave impedance in the z direction is obtained using the general impedance formula from Chapter 7:

$$Z^w_{z,\mathrm{TM}} = \frac{(\vec{E} \times \vec{H}^*) \cdot \hat{z}}{(\hat{z} \times \vec{H}) \cdot (\hat{z} \times \vec{H}^*)} = \frac{-E_y}{H_x} = \frac{\sqrt{\omega^2\mu\epsilon - \left(\frac{n\pi}{d}\right)^2}}{\omega\epsilon}$$
$$= \sqrt{\frac{\mu}{\epsilon}}\sqrt{1 - \left(\frac{\omega_{c,n}}{\omega}\right)^2} = \eta\sqrt{1 - \left(\frac{\omega_{c,n}}{\omega}\right)^2} \tag{9-87}$$

Below cutoff the wave impedance is imaginary as expected since no wave (energy) is propagating. This is plotted in Figure 9-14. (Note that the wave impedance has a minus sign along with the ratio of the transverse components. This is because $\vec{E}_y \times \vec{H}^*_x$ is in the $-z$ direction and we want the $+z$ direction.)

Plots of the electric and magnetic field lines are obtained by using the time expressions for $\vec{\mathscr{E}}$ and $\vec{\mathscr{H}}$. It is usual to plot a snapshot at $t = 0$ and plot the field lines as functions of

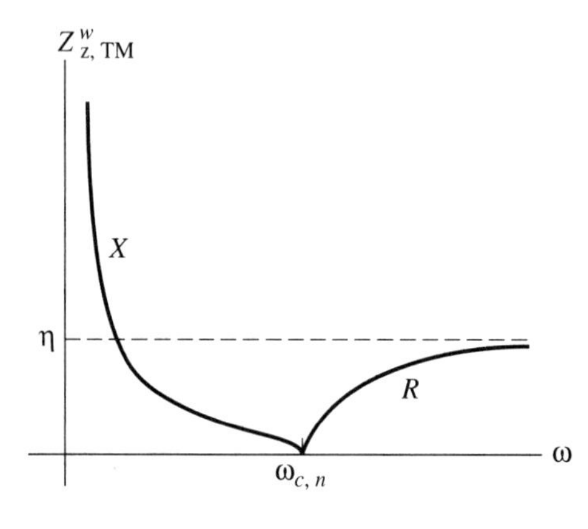

Figure 9-14. Wave impedance for TM waves between parallel planes.

z. Multiplying Eqs. 9-74 and 9-75 by $e^{j\omega t}$ and taking the real part gives the time expressions. At $t = 0$ the fields are, using $\sqrt{(n\pi/d)^2 - \omega^2\mu\epsilon} = j\beta_{\text{TM}}$,

$$\vec{\mathscr{E}}(y, z, t = 0) = -\frac{\beta_{\text{TM}} d}{n\pi} B \cos\frac{n\pi}{d} y \sin \beta_{\text{TM}} z\hat{y} + B \sin\frac{n\pi}{d} y \cos \beta_{\text{TM}} z\hat{z} \quad \text{(9-88)}$$

$$\vec{\mathscr{H}}(y, z, t = 0) = \frac{\omega\epsilon d}{n\pi} B \cos\frac{n\pi}{d} y \sin \beta_{\text{TM}} z\hat{x} \quad \text{(9-89)}$$

The physical structure of the electric field lines can be obtained using the technique presented in Chapter 5, Section 5-6. In this case it turns out that the derivative is integrable so numerical procedures are not required.

$$\frac{dy}{dz} = \frac{E_y}{E_z} = -\frac{\beta_{\text{TM}} d}{n\pi} \cot\frac{n\pi}{d} y \tan \beta_{\text{TM}} z$$

Separating variables:

$$\frac{n\pi}{d} \tan\frac{n\pi}{d} y\, dy = -\beta_{\text{TM}} \tan \beta_{\text{TM}} z\, dz$$

Integrating:

$$-\ln \cos\frac{n\pi}{d} y = \ln \cos \beta_{\text{TM}} z + \ln K_n$$

$$\frac{1}{\cos\frac{n\pi}{d} y} = K_n \cos \beta_{\text{TM}} z$$

The constant K_n is determined by selecting a point through which it is desired to have a field line pass, so

$$K_n = \frac{1}{\cos n\pi\left(\frac{y_0}{d}\right) \cos(\beta_{\text{TM}} z_0)} \quad \text{(9-90)}$$

For convenience we have selected normalized plot variables $(\beta_{\text{TM}} z)$ and (y/d), and selected values on a desired field line. Since we start the plot at $z = 0$, it is convenient to use $z = 0$ for reference, $(\beta_{\text{TM}} z_0) = 0$, and then plot lines passing through various (y/d) values. The electric field lines for the TM_1 mode as shown in Figure 9-15*a*. It is difficult to indicate total field intensity since the electric field has two components. At $\beta_{\text{TM}} z = \pi$ the electric field has only a z component so at that one location different line widths have been used to show the relative values, being greatest in the center, $y/d = 0.5$, and decreasing toward both plates.

The magnetic field intensity is shown in Figure 9-15*b*, with the crosses indicating vectors into the page (tail of the arrows) and dots indicating vectors out of the page. The sizes of the dots and crosses indicate field amplitude. For $t > 0$, these patterns propagate along together in the $+z$ direction.

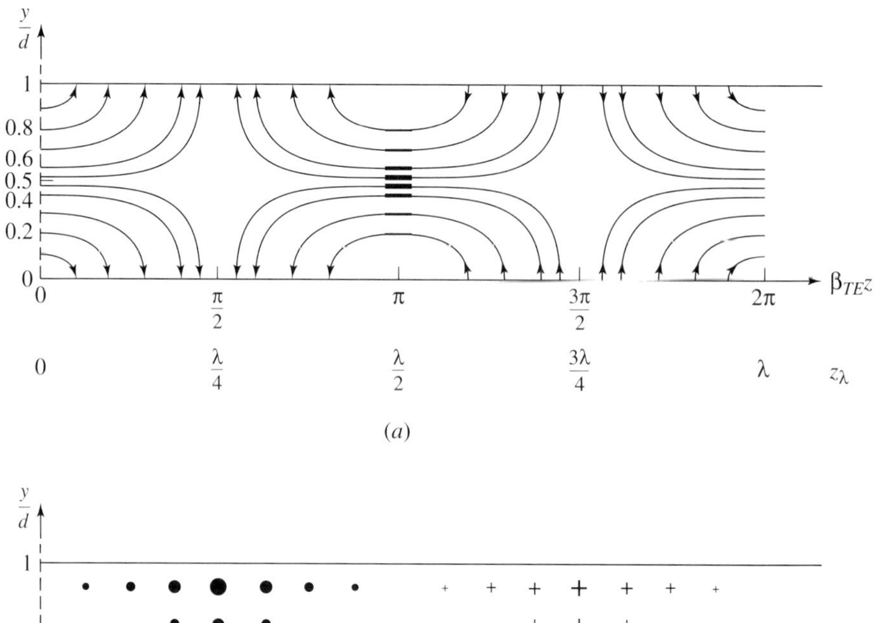

Figure 9-15. (*a*) Electric and (*b*) magnetic field lines of the TM_1 mode between parallel planes, $t = 0$.

An interesting physical interpretation of wave propagation can be obtained from the Poynting vector. Using Eqs. 9-74 and 9-75

$$\vec{S} = \frac{1}{2} \vec{E} \times \vec{H}^*$$

$$= \frac{\omega \epsilon d^2}{2n^2\pi^2} k \sqrt{1 - \left(\frac{k_{c,\text{TM}}}{k}\right)^2} B^2 \cos^2 \frac{n\pi}{d} y\hat{z} - j \frac{\omega \epsilon d^2}{2n\pi} B^2 \cos \frac{n\pi}{d} y \sin \frac{n\pi}{d} y\hat{y} \qquad \text{(9-91)}$$

This shows that the Poynting vector is not directed completely down the guide in the z direction but is at an angle with respect to the z axis. This means that the power flow can be imagined as reflecting back and forth between the two metal plates as it progresses down the guide. Looking at the real and imaginary parts of $\vec{S}$ we find

$$\text{Re}\{\vec{S}\} = \vec{P}(y) = \frac{\omega\epsilon d^2}{2n^2\pi^2} k \sqrt{1 - \left(\frac{k_{c,\text{TM}}}{k}\right)^2} B^2 \cos^2 \frac{n\pi}{d} y\hat{z} \tag{9-92}$$

$$\text{Im}\{\vec{S}\} = \vec{Q}(y) = -\frac{\omega\epsilon d^2}{2n\pi} B^2 \cos \frac{n\pi}{d} y \sin \frac{n\pi}{d} y\hat{y} \tag{9-93}$$

Thus in the z direction we have real power flow and in the y direction a reactive component. This is what one might expect physically since in the y direction no power can be transmitted through the conductors. This is an energy storage or reactive component of power. In fact, at the cutoff frequency $k = \omega_{c,n}\sqrt{\mu\epsilon} = k_{c,\text{TM}}$ so $\vec{P}(y)$ is zero and no power is transmitted in the z direction. At that frequency $\vec{S}$ is purely reactive, as it would also be at frequencies below cutoff. Because of this reactive $\vec{S}$ the evanescent fields are frequently called *reactive field regions*.

The preceding power flow discussion can be interpreted in terms of a plane wave propagating at oblique incidence onto the parallel planes. This is shown in Figure 9-16, as an equivalent direction coordinate ξ with the component of $\vec{E}$ and $\vec{H}$ of the TM wave combined to give the total $\vec{E}$ and total $\vec{H}$ of a TEM wave in a direction denoted by the propagation vector $\vec{k} = \omega\sqrt{\mu\epsilon}\hat{\xi}$. The exponential variation of the TEM component would be, from Chapter 8,

$$\begin{aligned} e^{-jk\xi} &= e^{-jk(z\sin\theta + y\cos\theta)} \\ &= e^{-jk\sin\theta z}e^{-jk\cos\theta y} \end{aligned} \tag{9-94}$$

Since we originally developed the TM modes of propagation using β in the z direction we have

$$\beta = k\sin\theta \tag{9-95}$$

From Eq. 9-70, however,

$$\beta = k\sqrt{1 - \left(\frac{k_{c,\text{TM}}}{k}\right)^2}$$

Since these have to be the same, we require

$$\sin\theta = \sqrt{1 - \left(\frac{k_{c,\text{TM}}}{k}\right)^2} \tag{9-96}$$

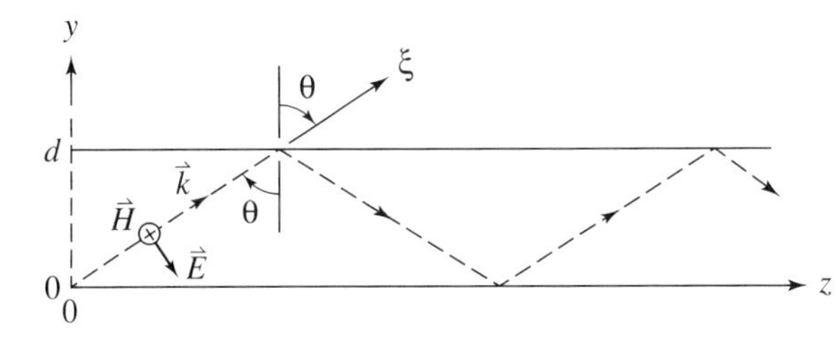

Figure 9-16. TEM wave direction of propagation between parallel conducting planes as a composite of TM wave components shown as $\vec{E}$ and $\vec{H}$ ($E_z \neq 0$ and $H_z = 0$).

In terms of frequency we may write this in another equivalent form using

$$\frac{k_{c,\mathrm{TM}}}{k} = \frac{\dfrac{n\pi}{d}}{\omega\sqrt{\mu\epsilon}} = \frac{\omega_{c,n}}{\omega} \tag{9-97}$$

Equation 9-96 is then

$$\sin\theta = \sqrt{1 - \left(\frac{\omega_{c,n}}{\omega}\right)^2} \tag{9-98}$$

This result is consistent with the Poynting vector analysis, for at the cutoff frequency $\theta = 0°$. This means that the TEM plane wave components bounce back and forth vertically between the two plates.

Example 9-1

A guided TEM wave is to be established between two large parallel planes while ensuring that the TM_1 mode cannot propagate. The plate spacing is 2 cm and the dielectric is polyethylene ($\epsilon_r = 2.6$). What is the highest possible operating frequency?

Using Eq. 9-77 the lowest cutoff frequency for $n = 1$ is

$$\omega_{c,1} = \frac{\pi}{0.02\sqrt{\mu_0 2.26\epsilon_0}} = 3.13 \times 10^{10} \text{ rad/s or } 4.99 \text{ GHz}$$

Since any frequency above this could generate the TM_1 mode this is the highest frequency for which only TEM can propagate.

Example 9-2

Using the system of Example 9-1, what frequency range will allow only TM_1 mode to propagate from all possible TM_n modes?

The cutoff frequency closest to the TM_1 mode is the TM_2 mode for which

$$\omega_{c,2} = \frac{2\pi}{0.02\sqrt{\mu_0 2.26\epsilon_0}} = 6.26 \times 10^{10} \text{ rad/s or } 9.98 \text{ GHz}$$

Thus for TM_1 mode only,

$$4.99 \text{ GHz} < f \leq 9.98 \text{ GHz}$$

Example 9-3

Determine the surface charge density on the top plate for TM_n modes and show that for TM_1 the equation is consistent with the electric field intensity direction at the top plate at $t = 0$.

On the top plate the perfect conductor boundary condition requires (using field subscript n for normal here):

$$\rho_s(y = d) = \mathcal{D}_n(y = d) = \epsilon\mathcal{E}_n(y = d) = \epsilon\vec{\mathcal{E}} \cdot \hat{n}\big|_{y=d} = \epsilon\vec{\mathcal{E}} \cdot (-\hat{y})\big|_{y=d}$$

Using Eq. 9-88 valid for $t = 0$,

$$\rho_s(y = d,\, t = 0) = \frac{\epsilon\beta_{\text{TM}} d}{n\pi} B \cos\frac{n\pi}{d} y \sin\beta_{\text{TM}} z\Big|_{y=d}$$

$$= (-1)^n \frac{\epsilon\beta_{\text{TM}} d}{n\pi} B \sin\beta_{\text{TM}} z$$

For the $n = 1$ mode this gives

$$\rho_s(y = d,\, t = 0) = -\frac{\epsilon\beta_{\text{TM}} d}{\pi} B \sin\beta_{\text{TM}} z$$

On the top plate and for $\beta_{\text{TM}} z < \pi/2$, ρ_s is negative. In Figure 9-15 at that location the field lines are directed *into* the plate as required for negative charge there. Make the positive charge coat test!

Exercises

9-4.2a Determine expressions for $\vec{J}_s$ and $\vec{\mathcal{J}}_s$. Plot $\vec{\mathcal{J}}_s$ on the top plate at $t = 0$ and show that the result is consistent with the magnetic field intensity in Figure 9-15.

9-4.2b Determine expressions for $\vec{J}_D$ and $\vec{\mathcal{J}}_D$. Plot $\vec{\mathcal{J}}_D$ in the yz plane and explain how this result is consistent with the magnetic field intensity of Figure 9-15. A comparison with the plot of Exercise 9-4.2a should also be made.

9-4.2c Show the proper way to couple the TM_1 mode using loop coupling. The loop does not need to be placed at $\beta_{\text{TM}} z = \pi/2$. Why? Remember the plots are at $t = 0$.

9-4.2d For a frequency of 500 MHz, what is the largest allowable plate spacing d to ensure that only a TEM wave will propagate between parallel planes? Assume air dielectric.

Although a coax is a two-conductor system in which TM modes can propagate, a study of these is most easily made in conjunction with circular cross section waveguides. This is done in Section 9-6.3.

9-4.3 TE Modes: Parallel Planes

The configuration of Figure 9-1 is also used for this case. Section 9-3.3 gives the solution procedure for these modes and follows closely that for TM modes. The one major difference is the fact that since boundary conditions on H_z are not always known other field components must be obtained for which boundary conditions are known. The most commonly used condition is the vanishing of tangential electric field intensity at the boundary. Setting normal magnetic flux density (B_n) to zero can also be used.

In the general solution procedure of Section 9-3.3 we use x and y for transverse coordinates t_1 and t_2, respectively.

We begin by expanding the wave Eq. 9-20 in rectangular coordinates with $H_z(x, y)$:

$$\frac{\partial^2 H_z}{\partial x^2} + \frac{\partial^2 H_z}{\partial y^2} + k_{c,\text{TE}}^2 H_z = 0$$

Since the system has infinite extent in the x direction we suppose that there is no x variation. Then H_z is a function of y only (the z variation $e^{-\gamma_{\mathrm{TE}} z}$ to be multiplied later) so that

$$\frac{d^2 H_z}{dy^2} + k_{c,\mathrm{TE}}^2 H_z = 0$$

This has the same form as the reduced wave equation for the TM modes, so we have

$$H_z(y) = A \cos k_{c,\mathrm{TE}} y + B \sin k_{c,\mathrm{TE}} y \tag{9-99}$$

Using Eqs. 9-7 to 9-10:

$$E_x(x, y) = -j \frac{\omega\mu}{k_{c,\mathrm{TE}}^2} \frac{dH_z}{dy} = -j \frac{\omega\mu}{k_{c,\mathrm{TE}}^2} (-Ak_{c,\mathrm{TE}} \sin k_{c,\mathrm{TE}} y + Bk_{c,\mathrm{TE}} \cos k_{c,\mathrm{TE}} y) \tag{9-100}$$

$$E_y(x, y) = j \frac{\omega\mu}{k_{c,\mathrm{TE}}^2} \frac{dH_z}{dx} = 0, \qquad H_x(x, y) = -\frac{\gamma_{\mathrm{TE}}}{k_{c,\mathrm{TE}}^2} \frac{dH_z}{dx} = 0$$

$$H_y = -\frac{\gamma_{\mathrm{TE}}}{k_{c,\mathrm{TE}}^2} \frac{dH_z}{dy} = -\frac{\gamma_{\mathrm{TE}}}{k_{c,\mathrm{TE}}^2} (-Ak_{c,\mathrm{TE}} \sin k_{c,\mathrm{TE}} y + Bk_{c,\mathrm{TE}} \cos k_{c,\mathrm{TE}} y) \tag{9-101}$$

Since E_x is a tangential component at the boundary, we first require

$$E_x(x, y = 0) = 0 = -j \frac{\omega\mu}{k_{c,\mathrm{TE}}^2} (Bk_{c,\mathrm{TE}}) = -j \frac{\omega\mu}{k_{c,\mathrm{TE}}} B$$

This requires $B = 0$, and Eq. 9-100 reduces to

$$E_x(y) = j \frac{\omega\mu}{k_{c,\mathrm{TE}}} A \sin k_{c,\mathrm{TE}} y \tag{9-102}$$

At the top plate

$$E_x(y = d) = 0 = j \frac{\omega\mu}{k_{c,\mathrm{TE}}} A \sin k_{c,\mathrm{TE}} d$$

Now A cannot be zero because we would have no solution at all, so we must have

$$\sin k_{c,\mathrm{TE}} d = 0 \Rightarrow k_{c,\mathrm{TE}} d = n\pi, \qquad n = 1, 2, 3, \ldots$$

$$k_{c,\mathrm{TE}} = \frac{n\pi}{d}, \qquad n = 1, 2, 3, \ldots \tag{9-103}$$

Why is $n = 0$ excluded?

As before, we have many modes denoted by TE_n. Eq. 9-4 gives the propagation constant (with $\sigma_d = 0$ here):

$$\gamma_{\mathrm{TE}} = \sqrt{k_{c,\mathrm{TE}}^2 - \omega^2\mu\epsilon} = \sqrt{\left(\frac{n\pi}{d}\right)^2 - k^2} = j\sqrt{k^2 - \left(\frac{n\pi}{d}\right)^2} = j\beta_{\mathrm{TE}} \tag{9-104}$$

The field components then reduce to the following, with the z variation included, and $k = \omega\sqrt{\mu\epsilon}$ as usual:

$$H_z(y, z) = A \cos\frac{n\pi}{d}y\, e^{-j\sqrt{k^2-\left(\frac{n\pi}{d}\right)^2}z} \tag{9-105}$$

$$E_x(y, z) = j\frac{\omega\mu d}{n\pi} A \sin\frac{n\pi}{d}y\, e^{-j\sqrt{k^2-\left(\frac{n\pi}{d}\right)^2}z} \tag{9-106}$$

$$H_y(y, z) = j\frac{\sqrt{k^2 - \left(\frac{n\pi}{d}\right)^2}\, d}{n\pi} A \sin\frac{n\pi}{d}y\, e^{-j\sqrt{k^2-\left(\frac{n\pi}{d}\right)^2}z} \tag{9-107}$$

Since the expression for γ_{TE} is identical to that for the TM waves we can write the cutoff frequency directly:

$$\omega_{c,n} = \frac{n\pi}{d\sqrt{\mu\epsilon}} \tag{9-108}$$

(Frequencies greater than this propagate.) Similarly,

$$V_p = \frac{\omega}{\beta_{\text{TE}}} = \frac{\omega}{\sqrt{k^2 - \left(\frac{n\pi}{d}\right)^2}} \tag{9-109}$$

$$V_g = \frac{d\omega}{d\beta_{\text{TE}}} = \frac{\sqrt{k^2 - \left(\frac{n\pi}{d}\right)^2}}{\omega\mu\epsilon} \tag{9-110}$$

$$V_p V_g = \frac{1}{\mu\epsilon} \tag{9-111}$$

$$\lambda_{c,n} = \frac{2d}{n} \tag{9-112}$$

$$\lambda_{g,n} = \frac{\lambda_0}{\sqrt{1 - \left(\frac{\omega_{c,n}}{\omega}\right)^2}} \qquad \left(\lambda_0 = \frac{v_0}{f} = \frac{1}{f\sqrt{\mu\epsilon}}\right) \tag{9-113}$$

The wave impedance is

$$Z^w_{z,\text{TE}} = \frac{(\vec{E} \times \vec{H}^*) \cdot \hat{z}}{(\hat{z} \times \hat{H}) \cdot (\hat{z} \times \hat{H}^*)} = \frac{E_x}{H_y} = \frac{\eta}{\sqrt{1 - \left(\frac{\omega_{c,n}}{\omega}\right)^2}}, \qquad \eta = \sqrt{\frac{\mu}{\epsilon}} \tag{9-114}$$

Another way for determining the wave impedance in waveguide systems is given in General Exercise GE9-22 at the end of the chapter. The wave impedances of TE_n and

TM_n have an interesting relationship. If we multiply the wave impedances together we obtain

$$Z^{w}_{z,\mathrm{TM}} Z^{w}_{z,\mathrm{TE}} = \eta^2 = \frac{\mu}{\epsilon} \tag{9-115}$$

Since the results for TE_n and TM_m modes are modes essentially identical (except for the wave impedances) how do we know which mode type will propagate? The answer is in how we excite the system. To obtain TM modes we excited the system so that a strong E_z was produced. Similarly, to obtain TE modes we introduce a strong H_z component. The amplitude patterns for H_z, modes $n = 1, 2, 3$, are shown in Figure 9-17. Loop couplings for $n = 1$ and $n = 2$ are also shown. For a mode to propagate it must be excited in the appropriate way *and* be at a frequency above cutoff. Examples 9-71 and 9-72 apply to TE modes as well.

Figure 9-17. Field amplitude variations of H_z for the first three modes for TE_n modes between parallel planes. Loop couplings for $n = 1$ and $n = 2$ shown.

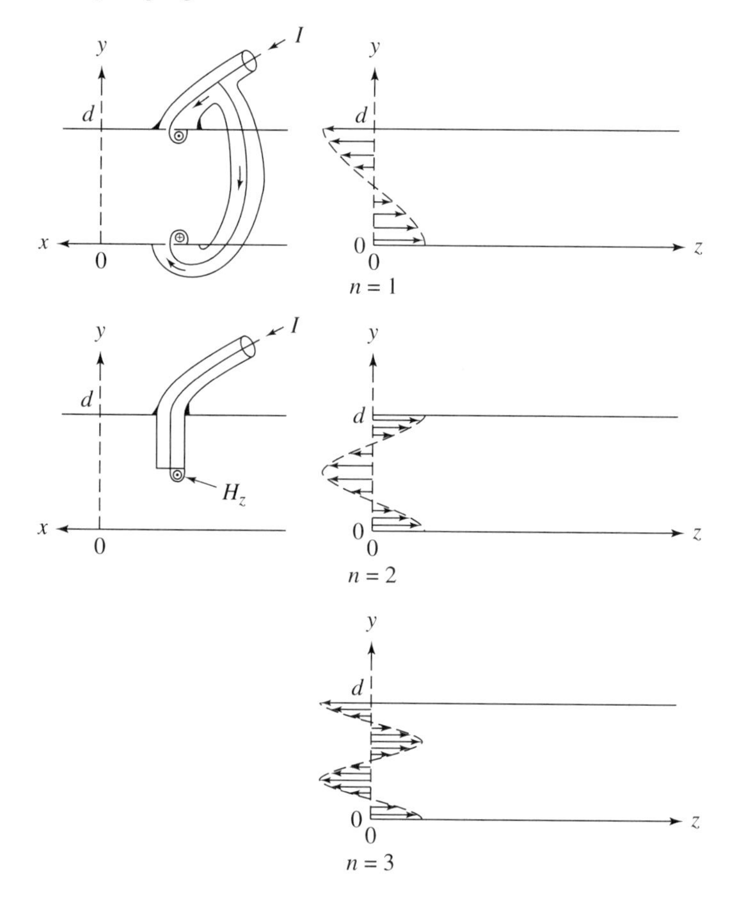

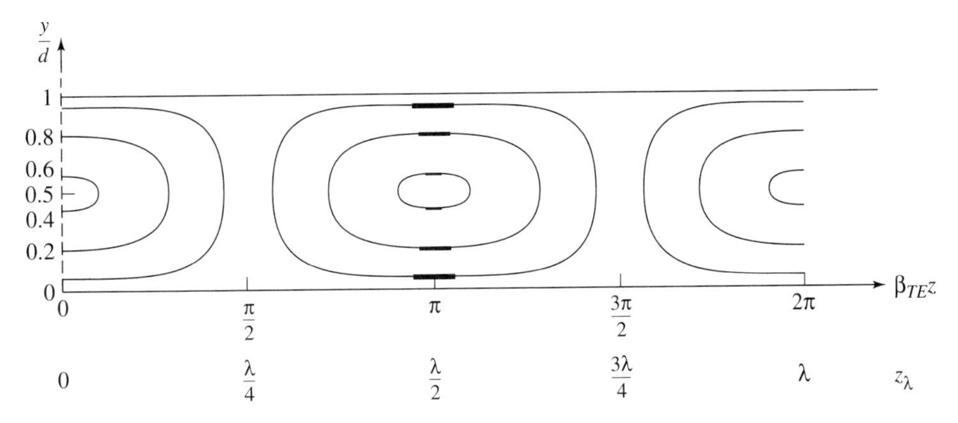

Figure 9-18. Magnetic field lines of the TE_1 mode between parallel planes, $t = 0$. Bar widths suggest field amplitudes.

The magnetic field line configuration for the TE_1 mode shown in Figure 9-18 was obtained using the same procedure given in Section 9-4.2.

Exercises

9-4.3a Using field equations developed in the text, carry out the steps to prove the right two expressions for wave impedance of Eq. 9-114.

9-4.3b Derive the equations for $\vec{J}_s$ and $\vec{\mathcal{J}}_s(t = 0)$ and plot on the top plane (xz plane at $y = d$) using $\beta_{\text{TE}}z$ as the variable. Now prove that on the top plate $\rho_s \equiv 0$. Explain how there can be a current with zero surface charge.

9-4.3c Derive the equation for the Poynting vector $\vec{S}$ and discuss what happens to $\vec{S}$ at $\omega = \omega_{c,n}$. Discuss direction and nature of $\vec{S}$.

9-4.3d You have been asked to design a parallel plane system to support the TE_1 mode over the frequency range 1 to 1.7 GHz. The dielectric is Teflon, which is assumed lossless. Design the structure.

9-4.3e Since $V_p = f(\omega)$ for TM and TE modes there is dispersion even though the system is lossless. Suppose that two signals of frequencies 1 GHz and 5 GHz are simultaneously applied to a parallel plane waveguide having a plate displacement of 2 cm. For both air and Teflon, estimate the time difference between the two signals at a point 1 m down the guide. Use phase velocity.

9-5 Wave Propagation in Rectangular Cross Section Waveguides: One-Conductor System

It was shown in Section 9-3.1 that TEM modes cannot exist within single-conductor systems. In this section we begin directly with the TE mode study for the closed rectangular waveguide shown in Figure 9-19.

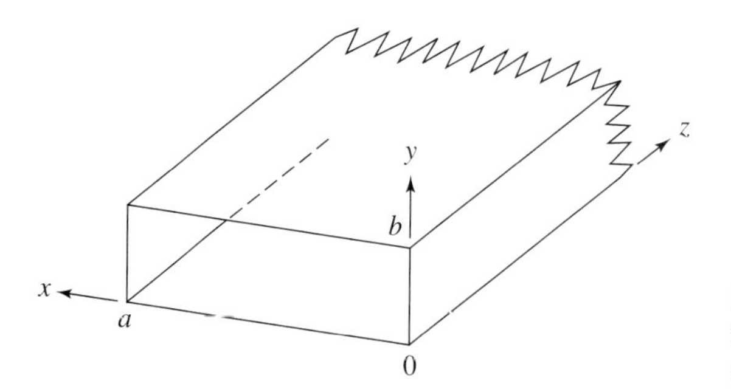

Figure 9-19. Closed metal rectangular waveguide coordinate system and dimensions.

9-5.1 TE Modes: Rectangular Cross Section

Following the solution outline given in Section 9-3.3 we first expand the wave equation in rectangular coordinates for the transverse coordinates:

$$\frac{\partial^2 H_z}{\partial x^2} + \frac{\partial^2 H_z}{\partial y^2} + k_{c,\text{TE}}^2 H_z = 0$$

In this problem the fields will have to have the same values at the left and right walls, so we assume the fields will be periodic in x and thus functions of x. Similarly, there will be y variations, so we must keep both partial derivatives. We must solve for $H_z(x, y)$ from the wave equation.

Using the separation of variables technique we assume the product solution (see Appendix K):

$$H_z(x, y) = X(x)Y(y) \tag{9-116}$$

Substituting this into the wave equation:

$$YX'' + XY'' + k_{c,\text{TE}}^2 = 0 \tag{9-117}$$

Dividing by XY:

$$\frac{X''}{X} + \frac{Y''}{Y} + k_{c,\text{TE}}^2 = 0 \tag{9-118}$$

Setting each ratio to a constant as described in the appendix,

$$\frac{X''}{X} = -k_x^2$$

$$\frac{Y''}{Y} = -k_y^2$$

We have used negative squared constants since we anticipate periodic (trigonometric) solutions in x and y. If our assumption is incorrect, the boundary conditions will correct us—Nature will care for us. Equation 9-118 then reduces to the *constraint relation*:

$$-k_x^2 - k_y^2 + k_{c,\text{TE}}^2 = 0 \tag{9-119}$$

In standard form the two separated differential equations are

$$X'' + k_x^2 X = 0$$

$$Y'' + k_y^2 Y = 0$$

The solutions of these are

$$X(x) = A' \cos k_x x + B' \sin k_x x$$

$$Y(y) = C' \cos k_y y + D' \sin k_y y$$

The solution for $H_z(x, y)$ from Eq. 9-116 becomes

$$H_z(x, y) = (A' \cos k_x x + B' \sin k_x x)(C' \cos k_y y + D' \sin k_y y) \tag{9-120}$$

Since we don't know the values of tangential $\vec{H}$ (H_z here) we must first determine the electric field components using Eqs. 9-7 and 9-8 with $t_1 \sim x$, $t_2 \sim y$, $h_1 = h_2 = 1$, and $E_z = 0$ (we could have used normal $\vec{B} = 0$ at the boundary: H_x and H_y):

$$\begin{aligned} E_x(x, y) &= -j \frac{\omega\mu}{k_{c,\mathrm{TE}}^2} \frac{\partial H_z}{\partial y} \\ &= -j \frac{\omega\mu k_y}{k_{c,\mathrm{TE}}^2} (A' \cos k_x x + B' \sin k_x x)(-C' \sin k_y y + D' \cos k_y y) \end{aligned} \tag{9-121}$$

$$\begin{aligned} E_y(x, y) &= j \frac{\omega\mu}{k_{c,\mathrm{TE}}^2} \frac{\partial H_z}{\partial x} \\ &= j \frac{\omega\mu k_x}{k_{c,\mathrm{TE}}^2} (-A' \sin k_x x + B' \cos k_x x)(C' \cos k_y y + D' \sin k_y y) \end{aligned} \tag{9-122}$$

We apply boundary conditions to these using as many ''zeros'' as we can for coordinate and field values. At $x = 0$,

$$E_y(0, y) = 0 = j \frac{\omega\mu k_x}{k_{c,\mathrm{TE}}^2} \cdot B' \cdot (C' \cos k_y y + D' \sin k_y y), \qquad \text{for all } y$$

Now k_x cannot be zero because all x variation would disappear. Also, C' and D' cannot *both* be zero since the entire solution for fields would be zero.

We then require

$$B' = 0$$

At $y = 0$,

$$E_x(x, 0) = 0 = -j \frac{\omega\mu k_y}{k_{c,\mathrm{TE}}^2} A' \cos k_x x \cdot D' \qquad \text{for all } x$$

Considering all the possibilities only one condition gives an acceptable solution:

$$D' = 0$$

The three field equations (9-120, 9-121, and 9-122) reduce to, letting the product $A'C'$ be a new constant A,

$$H_z(x, y) = A'C' \cos k_x x \cos k_y y = A \cos k_x x \cos k_y y \tag{9-123}$$

$$E_x(x, y) = j \frac{\omega\mu k_y}{k_{c,\mathrm{TE}}^2} A \cos k_x x \sin k_y y \tag{9-124}$$

$$E_y(x, y) = -j \frac{\omega\mu k_x}{k_{c,\mathrm{TE}}^2} A \sin k_x x \cos k_y y \tag{9-125}$$

For the side wall at $x = a$

$$E_y(a, y) = -j \frac{\omega\mu k_x}{k_{c,\mathrm{TE}}^2} A \sin k_x a \cos k_y y = 0$$

The only possibility which gives a nontrivial solution is

$$\sin k_x a = 0 \qquad \text{or} \quad k_x a = m\pi$$

$$k_x = \frac{m\pi}{a}, \qquad m = 0, 1, 2, \ldots \tag{9-126}$$

The value $m = 0$ is allowed since H_z will not vanish. Similarly, for the top surface at $y = b$:

$$E_x(x, b) = j \frac{\omega\mu k_y}{k_{c,\mathrm{TE}}^2} A \cos k_x x \sin k_y b = 0$$

This requires

$$\sin k_y b = 0, \qquad \text{or} \quad k_y b = n\pi$$

$$k_y = \frac{n\pi}{b}, \qquad n = 0, 1, 2, \ldots \tag{9-127}$$

Substituting the values of k_x and k_y in the constraint relation, Eq. 9-119, gives

$$k_{c,\mathrm{TE}}^2 = \left(\frac{m\pi}{a}\right)^2 + \left(\frac{n\pi}{b}\right)^2 \tag{9-128}$$

We can now obtain γ_{TE} using Eq. 9-4, for a lossless dielectric $\sigma_d = 0$, as

$$\gamma_{\mathrm{TE}} = \sqrt{k_{c,\mathrm{TE}}^2 - \omega^2\mu\epsilon} = j\sqrt{\omega^2\mu\epsilon - k_{c,\mathrm{TE}}^2}$$

$$= j\sqrt{k^2 - \left(\frac{m\pi}{a}\right)^2 - \left(\frac{n\pi}{b}\right)^2} = j\beta_{\mathrm{TE}} \tag{9-129}$$

We have assumed ϵ is real. For lossy dielectric $\beta_{\mathrm{TE}} = \mathrm{Im}\{\gamma_{\mathrm{TE}}\}$ see Section 9-8.1. The equations for the three field components are then, with the z variation included,

$$H_z(x, y, z) = A \cos \frac{m\pi}{a} x \cos \frac{n\pi}{b} y e^{-j\beta_{\mathrm{TE}} z} \tag{9-130}$$

$$E_x(x, y, z) = j \frac{\omega\mu n\pi A}{b\left[\left(\frac{m\pi}{a}\right)^2 + \left(\frac{n\pi}{b}\right)^2\right]} \cos\frac{m\pi}{a}x \sin\frac{n\pi}{b}y\, e^{-j\beta_{TE}z} \tag{9-131}$$

$$E_y(x, y, z) = -j \frac{\omega\mu m\pi A}{a\left[\left(\frac{m\pi}{a}\right)^2 + \left(\frac{n\pi}{b}\right)^2\right]} \sin\frac{m\pi}{a}x \cos\frac{n\pi}{b}y\, e^{-j\beta_{TE}z} \tag{9-132}$$

The remaining components of $\vec{H}$ are found using Eqs. 9-9 and 9-10:

$$H_x(x, y, z) = j \frac{\sqrt{k^2 - \left(\frac{m\pi}{a}\right)^2 - \left(\frac{n\pi}{b}\right)^2}\; m\pi A}{a\left[\left(\frac{m\pi}{a}\right)^2 + \left(\frac{n\pi}{b}\right)^2\right]} \sin\frac{m\pi}{a}x \cos\frac{n\pi}{b}y\, e^{-j\beta_{TE}z} \tag{9-133}$$

$$H_y(x, y, z) = j \frac{\sqrt{k^2 - \left(\frac{m\pi}{a}\right)^2 - \left(\frac{n\pi}{b}\right)^2}\; n\pi A}{b\left[\left(\frac{m\pi}{a}\right)^2 + \left(\frac{n\pi}{b}\right)^2\right]} \cos\frac{m\pi}{a}x \sin\frac{n\pi}{b}y\, e^{-j\beta_{TE}z} \tag{9-134}$$

The remaining constant A is determined by the amplitude of the source that drives the waveguide. We will not persue that evaluation.

Since there are many combinations of m and n we identify a particular mode by the mode convention

$$\text{TE}_{mn}, \qquad m, n = 0, 1, 2, \ldots$$

Note that m and n cannot be simultaneously zero since all the fields would be zero or constant. In practice, rectangular guides are most frequently used in the lowest order mode, TE_{10}, where $a > b$. This is called the dominant mode. It has the lowest attenuation of all the modes, as will be shown in Section 9-8.2, Exercises 9-8.2f and g.

The operating frequency range consists of those frequencies for which β_{TE} is a real value. From Eq. 9-129 this requires

$$\omega^2\mu\epsilon > k_{c,TE}^2 = \left(\frac{m\pi}{a}\right)^2 + \left(\frac{n\pi}{b}\right)^2$$

or

$$\omega > \frac{1}{\sqrt{\mu\epsilon}}\sqrt{\left(\frac{m\pi}{a}\right)^2 + \left(\frac{n\pi}{b}\right)^2} \tag{9-135}$$

The cutoff frequency, where β_{TE} is zero, occurs at

$$\omega_{c,TE} = \frac{1}{\sqrt{\mu\epsilon}}\sqrt{\left(\frac{m\pi}{a}\right)^2 + \left(\frac{n\pi}{b}\right)^2} \tag{9-136}$$

A *cutoff wavelength* that corresponds with the cutoff frequency can be defined as

$$\lambda_c \equiv \frac{v_0}{f_c} \tag{9-137}$$

where v_0 is the characteristic velocity of the medium inside the guide and is the velocity of light in that medium. This is the phase velocity of a uniform plane wave in the medium; that is, $1/\sqrt{\mu\epsilon} \cdot f_c$ is the cutoff frequency $\omega_0/2\pi$.

For TE modes the cutoff wavelength is then

$$\lambda_{c,\mathrm{TE}} = \frac{2\pi}{\sqrt{\left(\frac{m\pi}{a}\right)^2 + \left(\frac{n\pi}{b}\right)^2}} \tag{9-138}$$

Note that this wavelength is *not* the guide wavelength. The two should not be confused. For the TE_{10} mode, particularly,

$$\lambda_{c,\mathrm{TE}_{10}} = 2a \tag{9-139}$$

This equation is helpful in waveguide design since wavelengths longer than this (lower frequencies) will not propagate.

Example 9-4

X-band waveguide is nominally rated for 8.2 to 12.4 GHz propagation for the TE_{10} mode. The inside guide dimensions are (x and y) 2.286 cm by 1.016 cm. What are the cutoff wavelength and frequency of an air-filled guide for the TE_{10} mode?

$$\lambda_{c,\mathrm{TE}_{10}} = (2)(2.286) = 4.572 \text{ cm}$$

$$f_{c,\mathrm{TM}_{10}} = \frac{3 \times 10^8 \text{ m/s}}{0.04572 \text{ m}} = 6.56 \text{ GHz}$$

The guide wavelength is obtained easily as follows:

$$\lambda_{g,\mathrm{TE}} = \frac{2\pi}{\beta_{\mathrm{TE}}} = \frac{2\pi}{\sqrt{k^2 - \left(\frac{m\pi}{a}\right)^2 - \left(\frac{n\pi}{b}\right)^2}} = \frac{2\pi}{k\sqrt{1 - \frac{\left(\frac{m\pi}{a}\right)^2 + \left(\frac{n\pi}{b}\right)^2}{k^2}}}$$

$$= \frac{2\pi}{\omega\sqrt{\mu\epsilon}\sqrt{1 - \frac{\left(\frac{m\pi}{a}\right)^2 + \left(\frac{n\pi}{b}\right)^2}{\omega^2\mu\epsilon}}} = \frac{v_0}{f\sqrt{1 - \frac{\frac{v_0^2}{f^2}\left[\left(\frac{m\pi}{a}\right)^2 + \left(\frac{n\pi}{b}\right)^2\right]}{(2\pi)^2}}}$$

$$= \frac{\lambda_0}{\sqrt{1 - \left(\frac{\lambda_0}{\lambda_{c,\mathrm{TE}}}\right)^2}}, \qquad \lambda_0 \equiv \frac{v_0}{f} = \frac{1}{f\sqrt{\mu\epsilon}} \tag{9-140}$$

The radical expression in this last result will be seen to appear in all parameters for waveguide systems carrying TE or TM modes. The radical can also be expressed in terms of frequency as

$$\sqrt{1-\left(\frac{\lambda_0}{\lambda_{c,\mathrm{TE}}}\right)^2}=\sqrt{1-\left(\frac{\omega_{c,\mathrm{TE}}}{\omega}\right)^2} \tag{9-141}$$

Using these radicals the propagation constant can be written in the alternate forms:

$$\beta_{\mathrm{TE}}=k\sqrt{1-\left(\frac{\lambda_0}{\lambda_{c,\mathrm{TE}}}\right)^2}=k\sqrt{1-\left(\frac{\omega_{c,\mathrm{TE}}}{\omega}\right)^2} \tag{9-142}$$

where $k=\omega\sqrt{\mu\epsilon}$.

The field configurations within the guide for the three lowest-order TE modes are shown in Figure 9-20. In Views 1 and 2 are shown electric and magnetic fields within the guide. Views 3 and 4 show the magnetic fields and surface current densities at the surfaces viewed from outside the guide.

At first glance the field views may seem to convey little useful information, but it is from these views that we are able to design methods for coupling into a particular mode and to take samples from the guide. The best way to analyze these patterns is to write the equations using given m and n values and evaluate them at the planes and surfaces indicated. Then examine each reduced equation with the arrows drawn in the parts of the figure.

(The equations for the surface currents are developed from the boundary condition $\vec{J}=\hat{n}\times\vec{H}$, where $\hat{n}$ is the unit vector directed from the guide wall toward the interior of the guide; see Exercise 9-5.1b.) After examining the views one should mentally construct a three-dimensional picture of them.

For the TE_{10} mode pattern shown in Figure 9-20 we can devise a way to excite the mode by examining View 1. Since the electric field is vertical and has its largest value at the center (arrows close together) we could use probe coupling via a coax center conductor as shown in Figure 9-21*a*. This arrangement generates a significant amount of a y component of $\vec{E}$ and can excite the mode *if* the frequency is above cutoff. If we wish to excite wave in only the positive z direction we could close off (short) one end of the guide and place the probe at $x=a/2$ and a quarter wavelength from the shorted end. Note that a vertical electric field would be zero at the shorting end plate and hence be maximum at $z=\lambda/4$. This is shown in Figure 9-21*b*. What has just been described is the qualitative design of a coax to waveguide adaptor.

Loop coupling can be accomplished by placing a current loop inside the guide which generates an appropriate magnetic field. Since the TE_{10} mode has only an H_z component and since that component is maximum at the side walls (Eq. 9-130), one possibility is as shown in Figure 9-21*c*.

Slot or aperture coupling would follow the same concept as was discussed in preceding sections.

The methods for exciting modes can also be used to sample the fields in the guide. However, to be able to sample electric field amplitude along the z direction the probe must be movable in that direction. This gives us the equivalent of a coaxial slotted line from which standing wave measurements can be made, and transmission line techniques learned in earlier chapters can be used. Such a waveguide slotted line is made by cutting a narrow

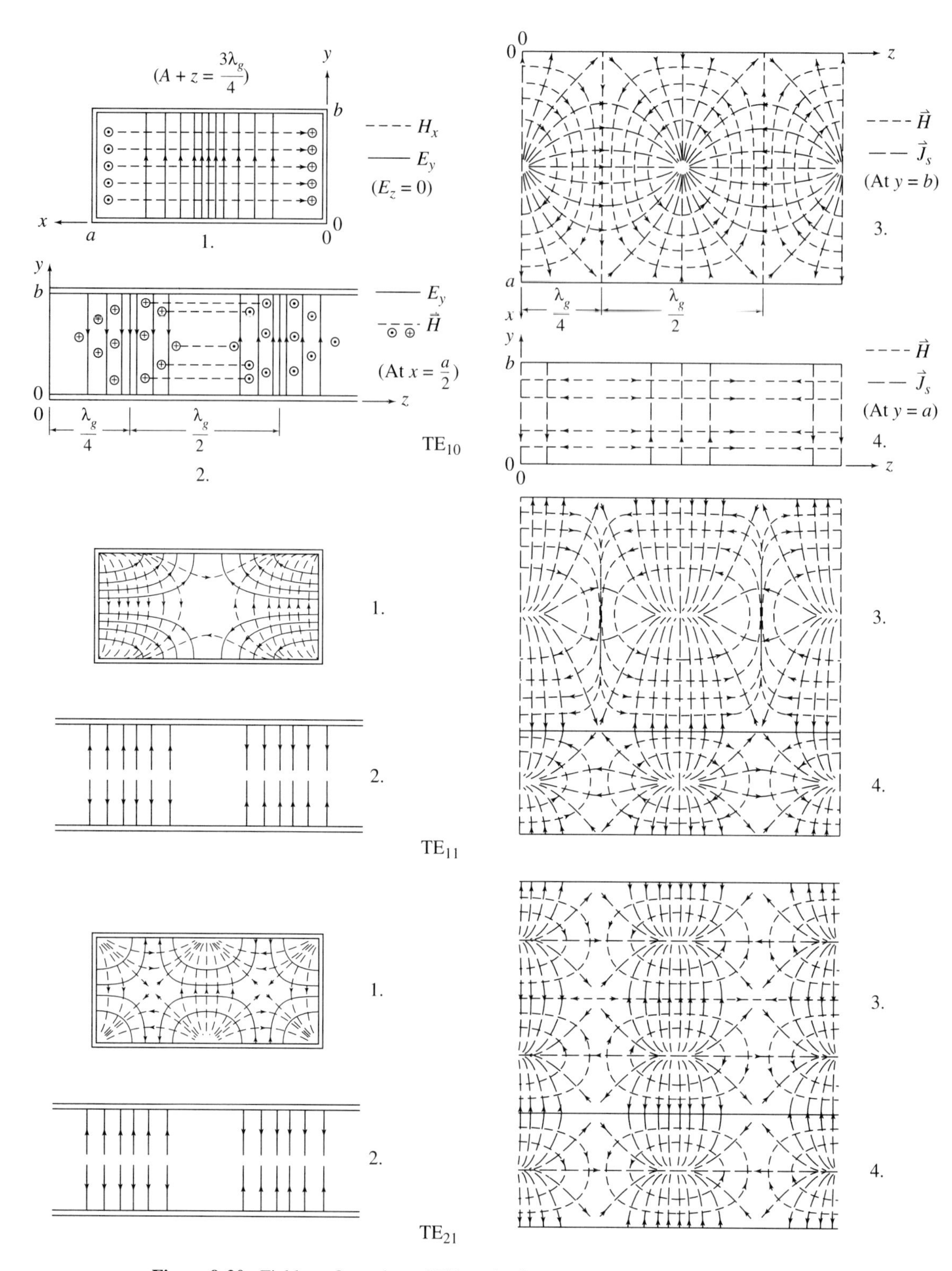

Figure 9-20. Field configuration of TE modes in a metal rectangular waveguide ($t = 0$).

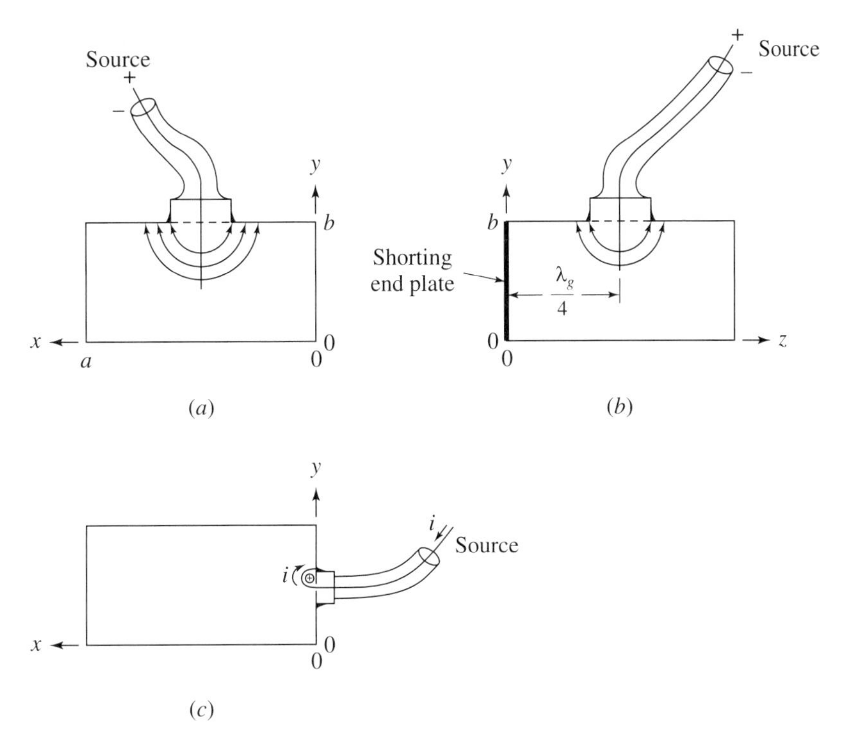

Figure 9-21. Exciting the TE_{10} mode in rectangular waveguides. (*a*) Probe coupling. (*b*) Coax to waveguide adaptor. (*c*) Loop coupling.

slit that is parallel to the *z* axis and then placing the probe on a movable carriage. The output of the coax is then applied to a suitable detector.

Expressions for the group and phase velocities are obtained directly from the definitions. Alternate forms that use the radical expressions developed earlier are possible. The results are

$$V_P^{\text{TE}} = \frac{\omega}{\beta_{\text{TE}}} = \frac{v_0}{\sqrt{1 - \dfrac{\left(\dfrac{m\pi}{a}\right)^2 + \left(\dfrac{n\pi}{b}\right)^2}{\omega^2\mu\epsilon}}} = \frac{v_0}{\sqrt{1 - \left(\dfrac{\lambda_0}{\lambda_{c,\text{TE}}}\right)^2}}$$

$$= \frac{v_0}{\sqrt{1 - \left(\dfrac{\omega_{c,\text{TE}}}{\omega}\right)^2}} \text{ m/s} \tag{9-143}$$

$$V_g^{\text{TE}} = \frac{d\omega}{d\beta_{\text{TE}}} = v_0\sqrt{1 - \dfrac{\left(\dfrac{m\pi}{a}\right)^2 + \left(\dfrac{n\pi}{b}\right)^2}{\omega^2\mu\epsilon}} = v_0\sqrt{1 - \left(\dfrac{\lambda_0}{\lambda_{c,\text{TE}}}\right)^2}$$

$$= v_0\sqrt{1 - \left(\dfrac{\omega_{c,\text{TE}}}{\omega}\right)^2} \text{ m/s} \tag{9-144}$$

where v_0 = velocity of light in waveguide medium $= 1/\sqrt{\mu\epsilon}$ m/s. Note that

$$V_P^{\mathrm{TE}} V_g^{\mathrm{TE}} = v_0^2 = \frac{1}{\mu\epsilon} \tag{9-145}$$

The *characteristic impedance* of a waveguide is defined as the wave impedance in the z direction. Using the wave impedance definition we obtain (assuming ϵ and μ real for lossless medium)

$$Z_0^{\mathrm{TE}} = Z_{z^+}^{\mathrm{TE}} = \frac{(\vec{E} \times \vec{H}^*) \cdot \hat{z}}{(\hat{z} \times \vec{H}) \cdot (\hat{z} \times \vec{H})^*} = \frac{\omega\mu}{\sqrt{k^2 - \left(\frac{m\pi}{a}\right)^2 - \left(\frac{n\pi}{b}\right)^2}}$$

$$= \frac{\omega\mu}{\omega\sqrt{\mu\epsilon}\sqrt{1 - \dfrac{\left(\frac{m\pi}{a}\right)^2 + \left(\frac{n\pi}{b}\right)^2}{\omega^2\mu\epsilon}}}$$

$$Z_0^{\mathrm{TE}} = \frac{\sqrt{\dfrac{\mu}{\epsilon}}}{\sqrt{1 - \dfrac{\left(\frac{m\pi}{a}\right)^2 + \left(\frac{n\pi}{b}\right)^2}{\omega^2\mu\epsilon}}} = \frac{\eta}{\sqrt{1 - \left(\dfrac{\lambda_0}{\lambda_{c,\mathrm{TE}}}\right)^2}} = \frac{\eta}{\sqrt{1 - \left(\dfrac{\omega_{c,\mathrm{TE}}}{\omega}\right)^2}}\ \Omega \tag{9-146}$$

Another method for determining the wave impedance is given in General Exercise GE9-22. The method makes direct use of the field components.

For the power flow we use the Poynting vector expression. The result (again assuming ϵ and μ are real) is

$$\begin{aligned}
\vec{S}_{\mathrm{TE}} &= \tfrac{1}{2}\vec{E} \times \vec{H}^* = \vec{P} + j\vec{Q} \\
&= \frac{\omega\mu\pi^2\sqrt{k^2 - k_{c,\mathrm{TE}}^2}}{2k_{c,\mathrm{TE}}^4} A^2 \left(\frac{m^2}{a^2} \sin^2 \frac{m\pi}{a}x \cos^2 \frac{n\pi}{b}y \right. \\
&\quad \left. + \frac{n^2}{b^2} \cos^2 \frac{m\pi}{a}x \sin^2 \frac{n\pi}{b}y \right) \hat{z} \\
&\quad - j \frac{\omega\mu A^2}{2k_{c,\mathrm{TE}}^2} \left(\frac{m}{a} \sin \frac{m\pi}{a}x \cos \frac{m\pi}{a}x \cos^2 \frac{n\pi}{b}y\, \hat{x} \right. \\
&\quad \left. + \frac{n}{b} \cos^2 \frac{m\pi}{a}x \sin \frac{n\pi}{b}y \cos \frac{n\pi}{b}y\, \hat{y} \right) \ \mathrm{W/m^2}
\end{aligned} \tag{9-147}$$

This power expression contains both real and imaginary components. The real ($\hat{z}$) component represents power flow down the guide and the imaginary ($\hat{x}$ and $\hat{y}$) components represent power that bounces back and forth between the side walls and corresponds to an energy storage or reactive component of power. At the cutoff frequency $k = k_c$ and the z component is zero. Below cutoff the z component also becomes imaginary and no power is transferred in any direction. The TE_{10} mode has an easily visualized interpretation of the ''bouncing'' plane waves propagating between the left and right surfaces similar to that discussed for the parallel plane system. For the TE_{10} mode the preceding equation reduces to

$$\vec{S}_{\text{TE}_{10}} = \frac{\omega\mu A^2 \sqrt{k^2 - \left(\frac{\pi}{a}\right)^2}\, a^2}{2\pi^2} \sin^2 \frac{\pi}{a} x\hat{z} - j\frac{\omega\mu A^2 a}{2\pi} \sin\frac{\pi}{a} x \cos\frac{\pi}{a} x\hat{x} \tag{9-148}$$

This shows components in the x and z directions. The x component is always imaginary and the z component becomes imaginary for frequencies that make

$$k^2 = \omega^2\mu\epsilon \le \left(\frac{\pi}{a}\right)^2 \tag{9-149}$$

From this we again observe that there is a cutoff frequency.

It is instructive to examine the TE_{10} mode Poynting vector from a plane wave point of view; specifically, at the location $x = a/4$ as shown in Figure 9-22. A plane wave would propagate with the velocity of light, or characteristic velocity, in the medium

$$v_0 = \frac{1}{\sqrt{\mu\epsilon}}$$

From the figure and Eq. 9-148 we have

$$\sin\theta = \left.\frac{|\vec{S}_z|}{|\vec{S}_{\text{TE}_{10}}|}\right|_{x=a/4} = \left.\frac{|\vec{S}_z|}{\sqrt{\vec{S}_{\text{TE}_{10}} \cdot \vec{S}^*_{\text{TE}_{10}}}}\right|_{x=a/4} = \frac{\sqrt{k^2 - \left(\frac{\pi}{a}\right)^2}}{k} \tag{9-150}$$

At the cutoff frequency given by $k = \pi/a$ the preceding equation yields $\theta = 0°$. Physically, this means that the wave is bouncing back and forth between the side walls of the guide so no power propagates in the z direction.

The integers m and n have important physical interpretations. The value of m gives the number of amplitude variations as one moves from 0 to a (x direction) across the guide.

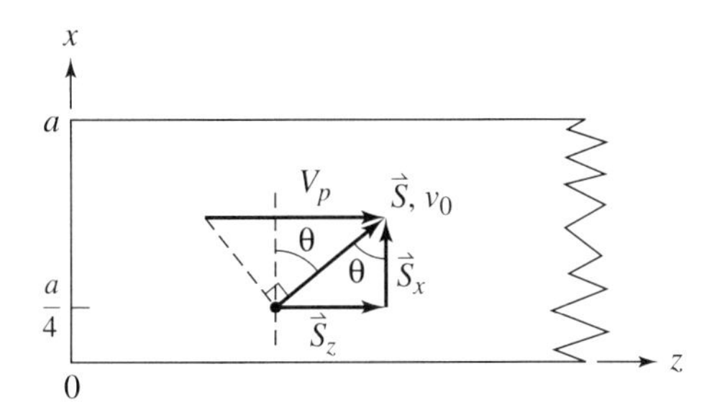

Figure 9-22. Interpretation of TE_{10} mode as a composition of plane waves. Only one component shown.

Similarly, n gives the number of amplitude variations as one moves from 0 to b (y direction). For example, the TE_{10} mode has one amplitude variation along the x direction (zero-maximum-zero) but no variations along the y direction (note field is independent of y for $n = 0$). Examine the other modes in Figure 9-20 to demonstrate this interpretation. Note this is the same as counting the number of half-cycle variations across the guide.

Exercises

9-5.1a Design a rectangular waveguide which has a TE_{10} mode cutoff frequency of 3 GHz and a TE_{11} mode cutoff of 6 GHz. State any assumptions clearly. Can any other TE mode propagate between 3 and 6 GHz?

9-5.1b Plot the surface current density and magnetic field intensity on the bottom surface of the guide ($y = 0$) for the TE_{10} mode in the rectangular guide. This is like View 3 of Figure 9-20, TE_{10}, seen from inside the guide.

9-5.1c Prove the equivalencies given in Eqs. 9-142 and 9-143.

9-5.1d Develop a coupling system for the TE_{11} and TE_{21} modes in the rectangular waveguide.

9-5.1e Plot the field structure at $t = 0$, $\beta z = 3\pi/2$, for the TE_{10} mode in the xy plane at the cutoff frequency.

9-5.1f Plot the displacement current in Views 1. and 2. of Figure 9-20, using $t = 0$.

9-5.1g By checking the units of the terms in the coefficients of the $\hat{x}$, $\hat{y}$, $\hat{z}$ components of $\vec{S}_{\text{TE}}$, show that $\vec{S}_{\text{TE}}$ has units of W/m^2 (Eq. 9-147).

9-5.1h For TE_{mn} modes in a rectangular waveguide determine the location and amplitude of the electric fields maxima. Hint: Must maximize with respect to both x and y.

9-5.1i Evanescent waves are characterized by pure attenuation in the direction of assumed propagation ($e^{-\alpha z}$ in our case). For a waveguide with $a = 2b = 2.286$ cm operating at 5.5 GHz, TE_{10}, determine the waveguide cutoff frequency and the distance in the z direction (at 5 GHz) at which the electric field intensity is decreased to 10% of its value at $z = 0$. Also, compute Z_0^{TE} and explain the result physically.

9-5.1j Explain why the $\hat{x}$ and $\hat{y}$ components of the Poynting vector are imaginary.

9-5.1k For the TE_{11} mode in Figure 9-20, show how section View 1 would be located in Views 2 and 3 and how View 2 would be located in Views 1 and 3. Identify the coordinate locations of the various views; i.e., View 1 is at what value of z, 2 at what value of x, etc.

9-5.1l A microwave slotted line is used to sample the electric field intensity as a function of this z coordinate. This is shown in Figure 9-23 for the TE_{10} mode. The field sample is then applied to a detector, the detector output being proportional to the square of the amplitude of $\vec{E}$ at the probe location. The waveguide inside dimensions are 2.286 cm by 1.016 cm and the guide is air-filled. Probe measurements given two adjacent field minima 1.5 cm apart, with one minimum being located 13 cm from the load. The detector gives a maximum reading of 4 and a minimum reading of 2.28. Determine the load impedance and the operating frequency. Note we have both positive and negative z traveling waves in this problem.

9-5.1m What electric field intensity must exist at $x = a/2$ for the TE_{10} mode in order to deliver 200 mW to a matched (Z_0-terminated) waveguide 2.286 cm by 1.016 cm at a frequency of 10.5 GHz? Using this information determine the constant A in Eq. 9-130.

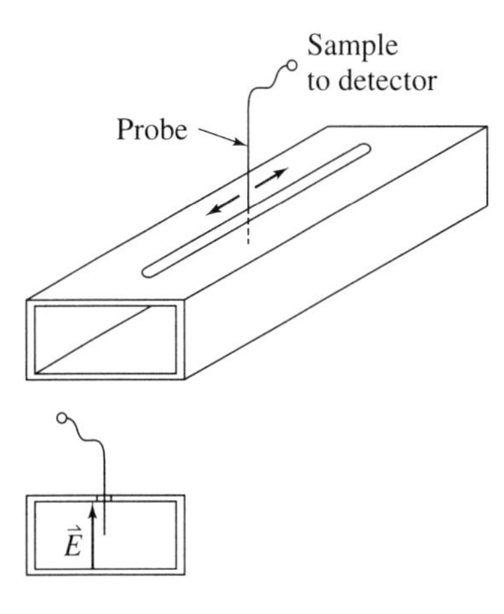

Figure 9-23. Waveguide slotted line.

9-5.2 TM Modes: Rectangular Cross Section

The solution for these modes in rectangular guides follows an outline identical to that for the TE modes. Details of the work are left as Exercise 9-5.2a. These steps should be carried out by the reader for experience in the handling of details in a lengthy solution process. The solution steps are given in Section 9-3.2. From the solution the items of interest (cutoff frequency, λ_g, etc.) can be obtained. Figure 9-19 is again used as the physical configuration with wall conductivity infinite. For reference, the results of the solution are given here as a summary.

Constraint relation (periodic solutions in x and y):

$$-k_x^2 - k_y^2 + k_{c,\text{TM}} = 0 \tag{9-151}$$

$$k_x = \frac{m\pi}{\alpha} \quad k_y = \frac{n\pi}{b} \tag{9-152}$$

$$k_{c,\text{TM}}^2 = \left(\frac{m\pi}{a}\right)^2 + \left(\frac{n\pi}{b}\right)^2 \tag{9-153}$$

The propagation constant is given by

$$\gamma_{\text{TM}} = j\sqrt{k^2 - k_{c,\text{TM}}^2} = j\sqrt{\omega^2\mu\epsilon - \left(\frac{m\pi}{a}\right)^2 - \left(\frac{n\pi}{b}\right)^2} = j\beta_{\text{TM}} \tag{9-154}$$

The field components are

$$E_z(x, y, z) = B \sin\frac{m\pi}{a}x \sin\frac{n\pi}{b}ye^{-j\beta_{\text{TM}}z} \tag{9-155}$$

$$E_x(x, y, z) = -j \frac{\sqrt{k^2 - \left(\frac{m\pi}{a}\right)^2 - \left(\frac{n\pi}{b}\right)^2}\left(\frac{m\pi}{a}\right)}{\left[\left(\frac{m\pi}{a}\right)^2 + \left(\frac{n\pi}{b}\right)^2\right]} B \cos\frac{m\pi}{a}x \sin\frac{n\pi}{b}y\, e^{-j\beta_{\text{TM}}z} \tag{9-156}$$

$$E_y(x, y, z) = -j \frac{\sqrt{k^2 - \left(\frac{m\pi}{a}\right)^2 - \left(\frac{n\pi}{b}\right)^2}\left(\frac{n\pi}{b}\right)}{\left[\left(\frac{m\pi}{a}\right)^2 + \left(\frac{n\pi}{b}\right)^2\right]} B \sin\frac{n\pi}{a}x \cos\frac{n\pi}{b}y\, e^{-j\beta_{\text{TM}}z} \tag{9-157}$$

$$H_x(x, y, z) = j \frac{\omega\epsilon\left(\frac{n\pi}{b}\right)}{\left[\left(\frac{m\pi}{a}\right)^2 + \left(\frac{n\pi}{b}\right)^2\right]} B \sin\frac{m\pi}{\alpha}x \cos\frac{n\pi}{b}y\, e^{-j\beta_{\text{TM}}z} \tag{9-158}$$

$$H_y(x, y, z) = -j \frac{\omega\epsilon\left(\frac{m\pi}{\alpha}\right)}{\left[\left(\frac{m\pi}{\alpha}\right)^2 + \left(\frac{n\pi}{b}\right)^2\right]} B \cos\frac{m\pi}{a}x \sin\frac{n\pi}{b}y\, e^{-j\beta_{\text{TM}}z} \tag{9-159}$$

The modes are identified using the mode convention

$$\text{TM}_{mn}, \qquad m, n = 1, 2, 3, \ldots$$

Why can't m or n be zero?

The operating frequency range, where $\gamma = j\beta_{\text{TM}}$ (pure imaginary) is

$$\omega > \frac{1}{\sqrt{\mu\epsilon}}\sqrt{\left(\frac{m\pi}{a}\right)^2 + \left(\frac{n\pi}{b}\right)^2} \tag{9-160}$$

The cutoff frequency and cutoff wavelength are

$$\omega_{c,\text{TM}} = \frac{1}{\sqrt{\mu\epsilon}}\sqrt{\left(\frac{m\pi}{a}\right)^2 + \left(\frac{n\pi}{b}\right)^2} \tag{9-161}$$

$$\lambda_{c,\text{TM}} = \frac{v_0}{f_c} = \frac{2\pi}{\sqrt{\left(\frac{m\pi}{a}\right)^2 + \left(\frac{n\pi}{b}\right)^2}} \tag{9-162}$$

The guide wavelength is

$$\lambda_{g,\text{TM}} = \frac{2\pi}{\sqrt{k^2 - \left(\frac{m\pi}{a}\right)^2 - \left(\frac{n\pi}{b}\right)^2}} = \frac{\lambda_0}{\sqrt{1 - \left(\frac{\lambda_0}{\lambda_{c,\text{TM}}}\right)^2}} \tag{9-163}$$

where

$$\lambda_0 = \frac{v_0}{f} = \frac{1}{f\sqrt{\mu\epsilon}}$$

The phase and group velocities are

$$V_P^{\mathrm{TM}} = \frac{v_0}{\sqrt{1 - \left(\dfrac{\lambda_0}{\lambda_{c,\mathrm{TM}}}\right)^2}} \qquad \text{and} \qquad V_g^{\mathrm{TM}} = v_0\sqrt{1 - \left(\frac{\lambda_0}{\lambda_{c,\mathrm{TM}}}\right)^2} \tag{9-164}$$

The "characteristic impedance" is

$$\begin{aligned} Z_0^{\mathrm{TM}} = Z_{z^+}^{\mathrm{TM}} &= \sqrt{\frac{\mu}{\epsilon}}\sqrt{1 - \frac{\left(\dfrac{m\pi}{a}\right)^2 + \left(\dfrac{n\pi}{b}\right)^2}{\omega^2\mu\epsilon}} \\ &= \eta\sqrt{1 - \left(\frac{\lambda_0}{\lambda_{c,\mathrm{TM}}}\right)^2} = \eta\sqrt{1 - \left(\frac{\omega_{c,\mathrm{TM}}}{\omega}\right)^2}\ \Omega \end{aligned} \tag{9-165}$$

Finally, the Poynting vector is

$$\begin{aligned} \vec{S}_{\mathrm{TM}} &= \tfrac{1}{2}\vec{E} \times \vec{H}^* = \vec{P} + j\vec{Q} \\ &= \frac{\omega\epsilon\pi^2\sqrt{k^2 - k_{c,\mathrm{TM}}^2}B^2}{k_{c,\mathrm{TM}}^4}\left(\frac{m^2}{a^2}\cos^2\frac{m\pi}{a}x\,\sin^2\frac{n\pi}{b}y\right. \\ &\quad \left. + \frac{n^2}{b^2}\sin^2\frac{m\pi}{a}x\,\cos^2\frac{n\pi}{b}y\right)\hat{z} \\ &\quad + j\frac{\omega\epsilon B^2}{k_{c,\mathrm{TM}}^2}\left(\frac{m}{a}\sin\frac{m\pi}{a}x\,\sin^2\frac{n\pi}{b}y\,\cos\frac{m\pi}{a}x\,\hat{x}\right. \\ &\quad \left. + \frac{n}{b}\sin^2\frac{m\pi}{a}x\,\sin\frac{n\pi}{b}y\,\cos\frac{n\pi}{b}y\,\hat{y}\right) \end{aligned} \tag{9-166}$$

In Figure 9-24 cross-sectional views of the three lowest-order TM modes are shown. Views 1 and 2 show electric and magnetic fields within the guide. Views 3 and 4 show the magnetic fields and current densities at the surface viewed from inside the guide.

One should spend a few moments developing a mental construction of the field patterns and visualizing how the views are oriented with respect to one another, and to what values of x, y, or z each view corresponds in the companion views. See, for example, Exercise 9-5.2e.

Exercises

9-5.2a Derive all of the TM mode quantities given by the equations summary of this section.

9-5.2b Show that the x and y components of the fields satisfy the boundary conditions required for $\vec{E}$ and $\vec{B}$.

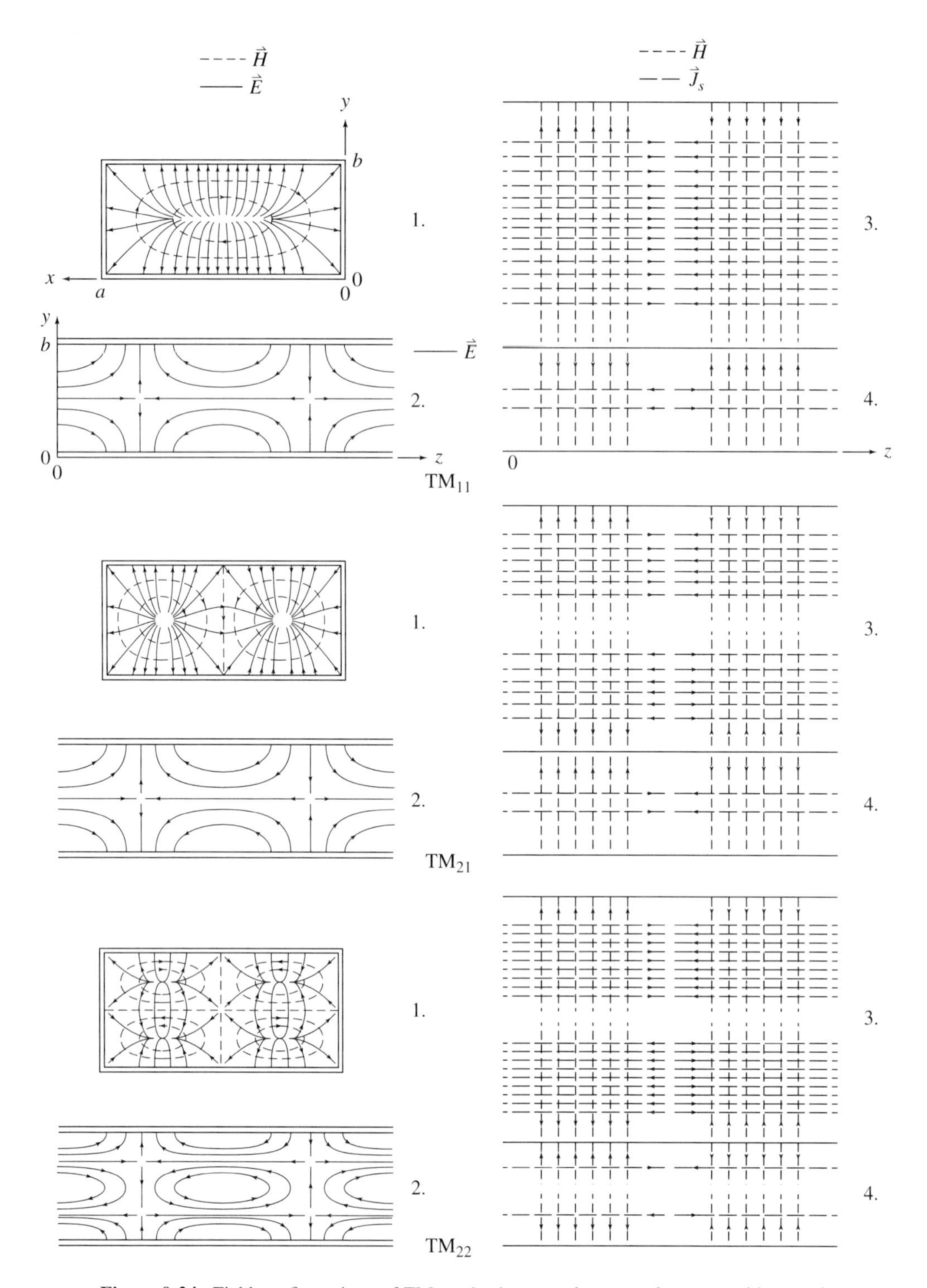

Figure 9-24. Field configurations of TM modes in a metal rectangular waveguide ($t = 0$).

9-5.2c Prove that alternate expressions for β_{TM} are

$$\beta_{\text{TM}} = \sqrt{1 - \left(\frac{\lambda_0}{\lambda_{c,\text{TM}}}\right)^2} = \sqrt{1 - \left(\frac{\omega_{c,\text{TM}}}{\omega_0}\right)^2}$$

9-5.2d Design feed systems for exciting the two lowest-order TM modes in a rectangular waveguide. Make both loop and probe designs.

9-5.2e For Figure 9-24, TM_{11} mode, determine the time and locate distances z in Views 2, 3, and 4 at which the field patterns given are applicable. Your labeling should be similar to that given for the TE_{10} mode of Figure 9-20. It is easiest to begin by converting all field component equations to the time domain. Are the surface currents correct according to the boundary condition on $\vec{H}$?

9-5.2f Design a rectangular waveguide which has a TM_{11} mode cutoff frequency of 10 GHz and a TM_{21} mode cutoff of 15 GHz. Show how you would excite the TM_{11} mode.

9-5.2g Plot the displacement current for the TM_{11} mode in a rectangular waveguide in Views 1 and 2 of Figure 9-24.

9-5.2h For TM_{mn} modes in a rectangular waveguide determine the location of an electric field maximum.

9-5.2i Repeat Exercise 9-5.1k for the TM_{21} mode of Figure 9-24.

9-6 Wave Propagation in Circular Cross Section Waveguides: One- and Two-Conductor Systems

In radar systems, the antenna must rotate to survey all of space. Rectangular waveguides will obviously not allow a transition from a stationary section to a rotating section of waveguide. This is one important application of circular cross section waveguides. In Section 9-4.1 TEM mode propagation in the coaxial system was covered. We now extend that section to include TE and TM modes and show how TE and TM modes can propagate in a single hollow pipe of circular cross section.

As the solutions are constructed we meet a class of functions called *Bessel functions* or *circular-cylindrical* functions. A detailed account of these is given in Appendices L, M, and N. This is an appropriate time to study these appendices, for the results are used in the following subsections.

9-6.1 TE Modes: Single Hollow Cylindrical Tube

The basic configuration for this case along with the definitions of the coordinate directions is given in Figure 9-25. Note these are consistent with the right-hand orthogonal convention. Radius of the tube is a.

For the solution we follow the outline of Section 9-3.3. We begin by expanding the

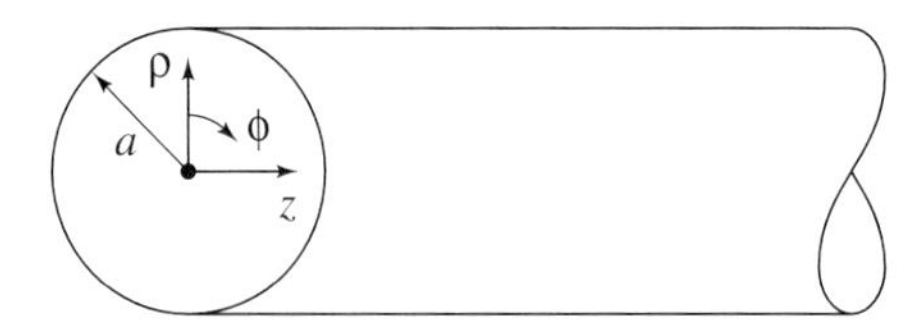

Figure 9-25. Configuration and coordinate convention for circular hollow tube waveguides.

transverse component wave equation (9-20) in polar coordinates, where the transverse coordinates are identified by $t_1 \sim \rho$ and $t_2 \sim \phi$. The h factors are $h_1 = 1$ and $h_2 = \rho$.

$$\frac{1}{\rho}\frac{\partial}{\partial \rho}\left(\rho \frac{\partial H_z}{\partial \rho}\right) + \frac{1}{\rho^2}\frac{\partial^2 H_z}{\partial \phi^2} + k_{c,\mathrm{TE}}^2 H_z = 0$$

where $k_{c,\mathrm{TE}}^2 = \gamma^2 + k^2 = \gamma^2 + \omega^2 \mu\epsilon$ (assume lossless medium inside). Using the separation of variables, we let

$$H_z(\rho, \phi) = R(\rho)\Phi(\phi) \tag{9-167}$$

Substituting this into the wave equation above,

$$\frac{\Phi(\phi)}{\rho}\frac{d\left(\rho \dfrac{dR}{d\rho}\right)}{d\rho} + \frac{R}{\rho^2}\frac{d^2\Phi}{d\phi^2} + k_{c,\mathrm{TE}}^2 R\Phi = 0$$

Multiplying by ρ^2 and dividing by $R\Phi$:

$$\frac{\rho}{R}\frac{d\left(\rho \dfrac{dR}{d\rho}\right)}{d\rho} + \rho^2 k_{c,\mathrm{TE}}^2 + \frac{1}{\Phi}\frac{d^2\Phi}{d\phi^2} = 0 \tag{9-168}$$

As described in the Appendix K the ϕ coordinate term must be a constant. Since ϕ varies from 0 to 2π in this configuration we assume periodicity in ϕ. This will give $H_z(\phi = 0) = H_z(\phi = 2\pi)$. Let the separation constant be $-n^2$ for the Φ part so that

$$\frac{1}{\Phi}\frac{d^2\Phi}{d\phi^2} = -n^2 \qquad \text{or} \qquad \frac{d^2\Phi}{d\phi^2} + n^2\Phi = 0$$

The solution of this ordinary differential equation is

$$\Phi(\phi) = A' \cos n\phi + B' \sin n\phi \tag{9-169}$$

Putting $-n^2$ into Eq. 9-168 for the Φ portion yields a differential equation for R only:

$$\frac{\rho}{R}\frac{d\left(\rho \dfrac{dR}{d\rho}\right)}{d\rho} + \rho^2 k_{c,\mathrm{TE}}^2 - n^2 = 0$$

Using the derivative of a product rule to expand the left term, multiplying by R and dividing by ρ^2 yields

$$\frac{d^2R}{d\rho^2} + \frac{1}{\rho}\frac{dR}{d\rho} + \left(k_{c,\mathrm{TE}}^2 - \frac{n^2}{\rho^2}\right)R = 0 \tag{9-170}$$

This is Bessel's equation, and for n an integer the product solution suggested in Appendix L gives the solution as Eq. L-30:

$$R(\rho) = C'J_n(k_{c,\mathrm{TE}}\rho) + D'N_n(k_{c,\mathrm{TE}}\rho) \tag{9-171}$$

$J_n(x)$ is the Bessel function of the first kind of order n and $N_n(x)$ is the Bessel function of the second kind of order n (or the Neumann function, which explains the use of the terminology N_n). Some authors use $Y_n(x)$ for the second kind. Plots and tables of these functions are found in Appendix L.

Using the solutions for $R(\rho)$ and $\Phi(\phi)$ the product solution is

$$H_z(\rho, \phi) = (A' \cos n\phi + B' \sin n\phi)[C'J_n(k_{c,\mathrm{TE}}\rho) + D'N_n(k_{c,\mathrm{TE}}\rho)]$$

From Appendix Figure L-4 we see that for $\rho = 0$, which is in the region of interest, $N_n(0) = -\infty$ for every n. Since fields must remain finite within we are required to set $D' = 0$. Absorbing the constant C' into the A' and B' to give new constants A and B, the preceding solution reduces to

$$H_z(\rho, \phi) = J_n(k_{c,\mathrm{TE}}\rho)(A \cos n\phi + B \sin n\phi) \tag{9-172}$$

In Figure 9-25 note that the ρ axis need not have been vertical as shown. It could have been taken horizontal in the $z = 0$ plane as well. This essentially means that we can select any location where we want to call $\phi = 0$. We choose to select it so that we have only the cosine variation, which allows us to set $B = 0$ to obtain the final general solution as

$$H_z(\rho, \phi) = AJ_n(k_{c,\mathrm{TE}}\rho) \cos n\phi \tag{9-173}$$

As for the rectangular case we use Eqs. 9-7 through 9-10 to obtain the other field components. To use those equations we will need to take the derivative of the Bessel function. It is customary to express the derivative by using the chain rule as follows:

$$\begin{aligned}\frac{dJ_n(k_{c,\mathrm{TE}}\rho)}{d\rho} = \frac{dJ_n(k_{c,\mathrm{TE}}\rho)}{d(k_{c,\mathrm{TE}}\rho)} \cdot \frac{d(k_{c,\mathrm{TE}}\rho)}{d\rho} &= k_{c,\mathrm{TE}}\frac{dJ_n(k_{c,\mathrm{TE}}\rho)}{d(k_{c,\mathrm{TE}}\rho)} \\ &= k_{c,\mathrm{TE}}J_n'(k_{c,\mathrm{TE}}\rho)\end{aligned} \tag{9-174}$$

It is important to remember that the prime means the derivative with respect to the entire argument:

$$J_n'(x) \equiv \frac{dJ_n(x)}{dx}$$

Since our experience with Bessel functions is not as extensive as that for exponentials, trigonometric functions, and so on, these derivatives must be looked up in a table. Such a table is given in Appendix M.

Using this convention the remaining field components are obtained using Eqs. 9-7 through 9-10 with $h_1 = 1$ and $h_2 = \rho$:

$$E_\rho(\rho, \phi) = j \frac{\omega \mu n}{k_{c,\mathrm{TE}}^2 \rho} A J_n(k_{c,\mathrm{TE}}\rho) \sin n\phi \tag{9-175}$$

$$E_\phi(\rho, \phi) = j \frac{\omega \mu}{k_{c,\mathrm{TE}}} A J_n'(k_{c,\mathrm{TE}}\rho) \cos n\phi \tag{9-176}$$

$$H_\rho(\rho, \phi) = -\frac{\gamma}{k_{c,\mathrm{TE}}} J_n'(k_{c,\mathrm{TE}}\rho) \cos n\phi \tag{9-177}$$

$$H_\phi(\rho, \phi) = \frac{n\gamma}{k_{c,\mathrm{TE}}^2 \rho} A J_n(k_{c,\mathrm{TE}}\rho) \sin n\phi \tag{9-178}$$

We can now obtain the values of $k_{c,\mathrm{TE}}$ and γ by applying either normal $\vec{B} = 0$ ($\mu H_\rho = 0$) at the boundary or tangential $\vec{E} = 0$ ($E_\phi = 0$) at the boundary. Using the tangential electric field requires (Eq. 9-176 for $\rho = a$)

$$j \frac{\omega \mu}{k_{c,\mathrm{TE}}} A J_n'(k_{c,\mathrm{TE}} a) \cos n\phi = 0$$

The constant A cannot be zero because the solution would go to zero. Since this must be zero for all values of ϕ the only constant part that can be zero is the derivative of the Bessel function evaluated at $\rho = a$,

$$J_n'(k_{c,\mathrm{TE}} a) = 0 \tag{9-179}$$

For given values of n and a we can find $k_{c,\mathrm{TE}}$. This was easy for the sine function since our familiarity with trigonometry gave the zeros. We are here required to consult tables to find zeros of the function. Most references use the symbol p_{nl}' for the lth-*nonzero* root of the derivative of the nth-order Bessel function. We then need values (nonzero) of p_{nl}' such that

$$J_n'(p_{nl}') = 0$$

Values of p_{nl}' are given in Table 9-1, and the mode convention corresponding to those values are denoted by

$$\mathrm{TE}_{nl}, \qquad n = 0, 1, 2, \ldots, \quad l = 1, 2, 3, \ldots \tag{9-180}$$

The physical interpretations of n and l will be given later in this section.

Now that we have values for p_{nl}' we obtain the values for $k_{c,\mathrm{TE}}$ by setting

$$k_{c,\mathrm{TE}} a = p_{nl}'$$

from which

$$k_{c,\mathrm{TE}} = \frac{p_{nl}'}{a} \tag{9-181}$$

Table 9-1. lth-Nonzero Roots (p'_{nl}) of the Derivatives of the nth-Order Bessel Functions: $J'_n(p'_{nl}) = 0,\ p'_{nl} \neq 0$

	n							
l	0	1	2	3	4	5	6	7
1	3.832	1.841	3.054	4.201	5.317	6.416	7.501	8.578
2	7.016	5.331	6.706	8.015	9.282	10.520	11.735	12.932
3	10.173	8.536	9.969	11.346	12.682	13.987	15.268	16.529
4	13.324	11.706	13.170	14.586	15.964	17.313	18.637	19.942

Using Eq. 9-4 we obtain the propagation constant as (dielectric loss $\sigma_d = 0$)

$$\gamma_{\mathrm{TE}} = \sqrt{k_{c,\mathrm{TE}}^2 - k^2} = j\sqrt{k^2 - k_{c,\mathrm{TE}}^2} = j\sqrt{\omega^2\mu\epsilon - \left(\frac{p'_{nl}}{a}\right)^2} = j\beta_{\mathrm{TE}} \quad (9\text{-}182)$$

The field components then become

$$H_z(\rho, \phi, z) = AJ_n\left(\frac{p'_{nl}}{a}\rho\right)\cos n\phi \; e^{-j\beta_{\mathrm{TE}}z} \quad (9\text{-}183)$$

$$E_\rho(\rho, \phi, z) = j\frac{\omega\mu n A}{\left(\frac{p'_{nl}}{a}\right)^2 \rho} J_n\left(\frac{p'_{nl}}{a}\rho\right)\sin n\phi \; e^{-j\beta_{\mathrm{TE}}z} \quad (9\text{-}184)$$

$$E_\phi(\rho, \phi, z) = j\frac{\omega\mu A}{\left(\frac{p'_{nl}}{a}\right)} J'_n\left(\frac{p'_{nl}}{a}\rho\right)\cos n\phi \; e^{-j\beta_{\mathrm{TE}}z} \quad (9\text{-}185)$$

$$H_\rho(\rho, \phi, z) = -j\frac{\sqrt{k^2 - \left(\frac{p'_{nl}}{a}\right)^2}}{\left(\frac{p'_{nl}}{a}\right)} AJ'_n\left(\frac{\rho'_{nl}}{a}\rho\right)\cos n\phi \; e^{-j\beta_{\mathrm{TE}}z} \quad (9\text{-}186)$$

$$H_\phi(\rho, \phi, z) = j\frac{n\sqrt{k^2 - \left(\frac{p'_{nl}}{a}\right)^2}}{\left(\frac{p'_{nl}}{a}\right)^2 \rho} AJ_n\left(\frac{p'_{nl}}{a}\rho\right)\sin n\phi \; e^{-j\beta_{\mathrm{TE}}z} \quad (9\text{-}187)$$

The cutoff frequency occurs when β_{TE} is zero. For Eq. 9-182 this will be at

$$\omega_{c,\mathrm{TE}}\sqrt{\mu\epsilon} = \frac{p'_{nl}}{a} \quad \text{or} \quad \omega_{c,\mathrm{TE}} = \frac{p'_{nl}}{a\sqrt{\mu\epsilon}} \quad (9\text{-}188)$$

The cutoff wavelength is computed using the definition Eq. 9-137:

$$\lambda_{c,\text{TE}} = \frac{v_0}{f_{c,\text{TE}}} = \frac{2\pi\left(\dfrac{1}{\sqrt{\mu\epsilon}}\right)}{\omega_{c,\text{TE}}} = \frac{2\pi a}{p'_{nl}} \tag{9-189}$$

(Remember this is the *unbounded medium* wavelength corresponding to the cutoff frequency. The guide wavelength at cutoff is infinite.)

The guide wavelength is

$$\lambda_{g,\text{TE}} = \frac{2\pi}{\beta_{\text{TE}}} = \frac{2\pi}{\sqrt{\omega^2\mu\epsilon - \left(\dfrac{p'_{nl}}{a}\right)^2}} = \frac{2\pi}{\omega\sqrt{\mu\epsilon}\sqrt{1-\left(\dfrac{p'_{nl}}{a\omega\sqrt{\mu\epsilon}}\right)^2}}$$

$$= \frac{1}{f\sqrt{\mu\epsilon}\sqrt{1-\left(\dfrac{1}{\lambda_{c,\text{TE}}f\sqrt{\mu\epsilon}}\right)^2}} = \frac{\lambda_0}{\sqrt{1-\left(\dfrac{\lambda_0}{\lambda_{c,\text{TE}}}\right)^2}}$$

$$= \frac{\lambda_0}{\sqrt{1-\left(\dfrac{\omega_{c,\text{TE}}}{\omega}\right)^2}} \tag{9-190}$$

where

$$\lambda_0 = \frac{v_0}{f} = \frac{1}{f\sqrt{\mu\epsilon}}$$

The expression for β_{TE} can also be written in terms of the radicals as

$$\beta_{\text{TE}} = k\sqrt{1-\left(\frac{\lambda_0}{\lambda_{c,\text{TE}}}\right)^2} = k\sqrt{1-\left(\frac{\omega_{c,\text{TE}}}{\omega}\right)^2} \tag{9-191}$$

For the group and phase velocities we obtain

$$V_p^{\text{TE}} = \frac{\omega}{\beta_{\text{TE}}} = \frac{v_0}{\sqrt{1-\left(\dfrac{p'_{nl}}{a\omega\sqrt{\mu\epsilon}}\right)^2}} = \frac{v_0}{\sqrt{1-\left(\dfrac{\lambda_0}{\lambda_{c,\text{TE}}}\right)^2}} = \frac{v_0}{\sqrt{1-\left(\dfrac{\omega_{c,\text{TE}}}{\omega}\right)^2}} \tag{9-192}$$

$$V_g^{\text{TE}} = \frac{d\omega}{d\beta_{\text{TE}}} = v_0\sqrt{1-\left(\frac{p'_{nl}}{a\omega\sqrt{\mu\epsilon}}\right)^2} = v_0\sqrt{1-\left(\frac{\lambda_0}{\lambda_{c,\text{TE}}}\right)^2}$$
$$= v_0\sqrt{1-\left(\frac{\omega_{c,\text{TE}}}{\omega}\right)^2} \tag{9-193}$$

For the wave or characteristic impedance in the z direction we have, using similar steps as in the rectangular case,

$$Z_0^{\mathrm{TE}} = Z_{z^+}^{\mathrm{TE}} = \frac{(\vec{E} \times \vec{H}^*) \cdot \hat{z}}{(\hat{z} \times \vec{H}) \cdot (\hat{z} \times \vec{H})^*} = \frac{\sqrt{\dfrac{\mu}{\epsilon}}}{\sqrt{1 - \left(\dfrac{p'_{nl}}{a\omega\sqrt{\mu\epsilon}}\right)^2}}$$

$$= \frac{\eta}{\sqrt{1 - \left(\dfrac{\lambda_0}{\lambda_{c,\mathrm{TE}}}\right)^2}} = \frac{\eta}{\sqrt{1 - \left(\dfrac{\omega_{c,\mathrm{TE}}}{\omega}\right)^2}} \ \Omega \tag{9-194}$$

The Poynting vector is obtained in the usual way:

$$\vec{S}_{\mathrm{TE}} = \tfrac{1}{2}\vec{E} \times \vec{H}^* = P + jQ$$

$$= j\frac{\omega\mu A^2}{2\left(\dfrac{p'_{nl}}{a}\right)} J_n\left(\frac{p'_{nl}}{a}\rho\right) J'_n\left(\frac{p'_{nl}}{a}\rho\right)\cos^2 n\phi\hat{\rho}$$

$$+ j\frac{\omega\mu n A^2}{2\left(\dfrac{p'_{nl}}{a}\right)^2 \rho} J_n^2\left(\frac{p'_{nl}}{a}\rho\right)\cos n\phi \, \sin n\phi\hat{\phi} \tag{9-195}$$

$$+ \left(\frac{\omega\mu n^2 A^2 \sqrt{k^2 - \left(\dfrac{p'_{nl}}{a}\right)^2}}{2\left(\dfrac{p'_{nl}}{a}\right)^4 \rho^2} J_n^2\left(\frac{p'_{nl}}{a}\rho\right)\sin^2 n\phi \right.$$

$$\left. + \frac{\omega\mu A^2 \sqrt{k^2 - \left(\dfrac{p'_{nl}}{a}\right)^2}}{2\left(\dfrac{p'_{nl}}{a}\right)^2} \left[J'_n\left(\frac{p'_{nl}}{a}\rho\right)\right]^2 \cos^2 n\phi \right)\hat{z}$$

(Note: To obtain the real values for the z component we have removed $\sqrt{-1}$ from the radicals in H_ρ and H_ϕ as evident from the change in position of k^2 within the radical.)

Let's look at the physical interpretations of the indices n and l. The integer n appears in the trigonometric functions $\cos n\phi$ and $\sin n\phi$. For $n = 1$ the functions go through one complete cycle for $0 \leq \phi \leq 2\pi$; for $n = 2$ they go through two complete cycles, and so on. Thus the integer n represents the number of complete cycles of amplitude variation a field quantity goes through around the guide at constant ρ value.

The value of l determines which root of J'_n one selects. In Appendix L, Figure L-3, one observes that the Bessel functions (and their derivatives, too, since derivatives yield Bessel functions; see Appendix M) have sinusoidal-like variation. If we use J_1 as an example ($n = 1$) we see that for $l = 1$ (first nonzero root at 3.8) H_z would have one amplitude variation zero-maximum-zero as ρ varies from 0 to a. As with the rectangular guide this could be thought of as one half "cycle" variation from 0 to a. Similarly, for $l = 2$ the second root is at 7, so we would obtain two half "cycle" variations. Thus the value of l tells us how many half "cycle" amplitude variations occur for $0 \le \rho \le a$. One could also interpret l as the number of *full* "cycle" variations across the guide *diameter*.

Circular waveguide field distributions for the three lowest-order TE_{nl} modes are given in Figure 9-26. Views 1 and 2 show electric and magnetic fields within the guide. View 2 is a longitudinal view through plane $l - l'$. View 3 shows the magnetic fields and surface current density on the surface viewed from inside the guide at Section $s - s'$.

By way of example we consider the TE_{01} mode to show how the Bessel function derivatives, zeros, and integrals are used.

Example 9-5

For a circular waveguide determine the equations for the field components, the average power transmitted down the guide, and the cutoff frequency, using the TE_{01} mode. The guide is air-filled.

The given mode specifies $n = 0$ and $l = 1$. Equations 9-183 through 9-187 give the following field components ($\mu = \mu_0$ and $\epsilon = \epsilon_0$):

$$H_z(\rho, \phi, z) = AJ_0\left(\frac{p'_{01}}{a}\rho\right)e^{-j\beta_{TE}z}$$

$$E_\phi(\rho, \phi, z) = j\frac{\omega\mu_0 A}{\left(\frac{p'_{01}}{a}\right)}J'_0\left(\frac{p'_{01}}{a}\rho\right)e^{-j\beta_{TE}z}$$

$$H_\rho(\rho, \phi, z) = -j\frac{\sqrt{k^2 - \left(\frac{p'_{01}}{a}\right)^2}A}{\left(\frac{p'_{01}}{a}\right)}J'_0\left(\frac{p'_{01}}{a}\rho\right)e^{-j\beta_{TE}z}$$

with

$$\beta_{TE} = \sqrt{k^2 - \left(\frac{p'_{01}}{a}\right)^2} \quad \text{and} \quad k^2 = \omega^2\mu_0\epsilon_0 = \frac{\omega^2}{9 \times 10^{16}}$$

also

$$\omega_{c,TE} = \frac{p'_{01}}{a\sqrt{\mu_0\epsilon_0}} = \frac{3 \times 10^8 p'_{01}}{a}$$

From Appendix M, Eq. M-2, we have $J'_0(x) = -J_1(x)$. From Table 9-1 we have $p'_{01} = 3.83$. Using these, the preceding equations become

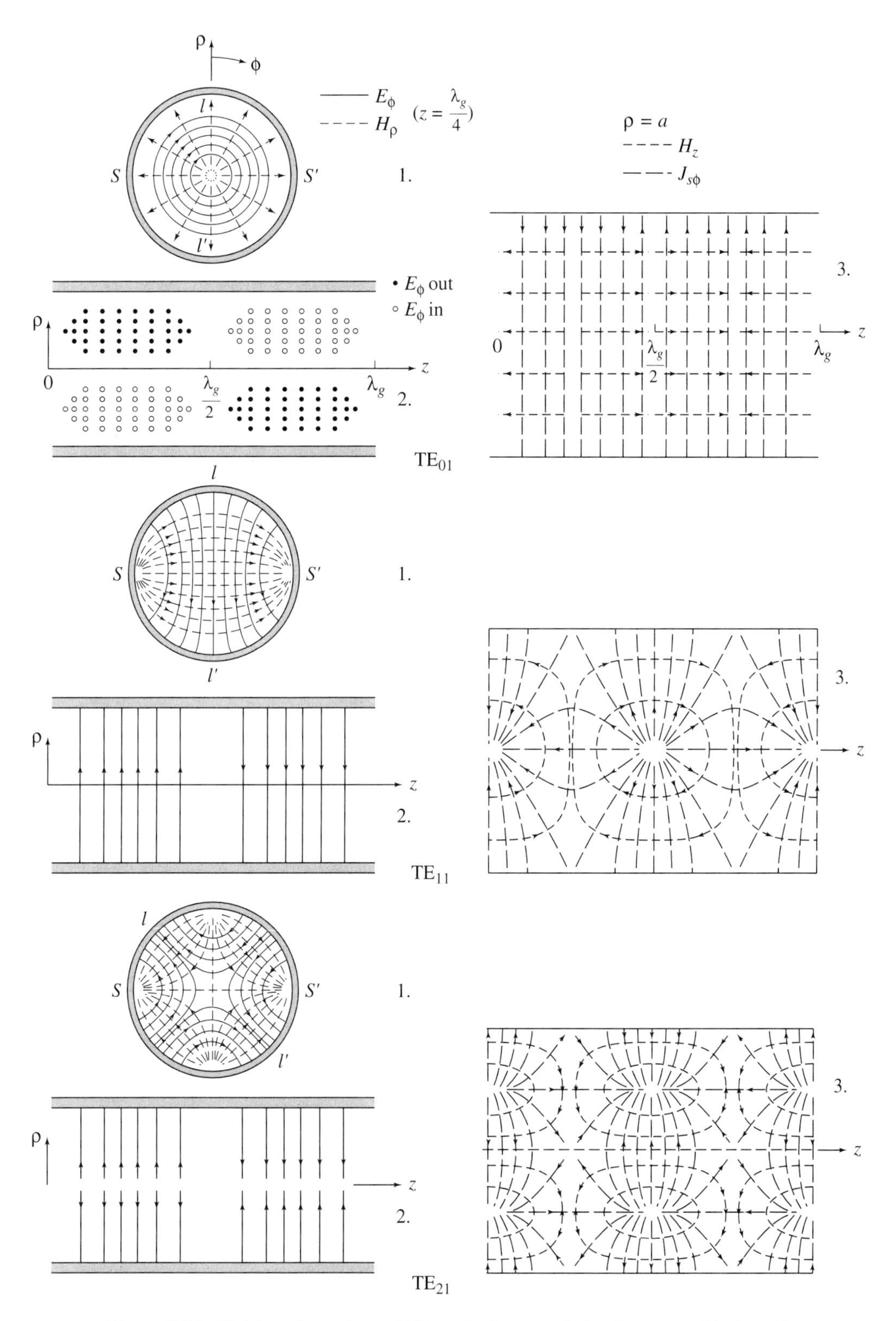

Figure 9-26. Field configurations of TE modes in a metal circular waveguide ($t = 0$).

$$H_z(\rho, \phi, z) = AJ_0\left(\frac{3.83}{a}\rho\right)e^{-j\beta_{\mathrm{TE}}z}$$

$$E_\phi(\rho, \phi, z) = -j3.281 \times 10^{-7}\omega a A J_1\left(\frac{3.83}{a}\rho\right)e^{-j\beta_{\mathrm{TE}}z}$$

$$H_\rho(\rho, \phi, z) = j0.261aA\sqrt{1.11 \times 10^{-17}\omega^2 - \frac{14.67}{a^2}}\,J_1\left(\frac{3.83}{a}\rho\right)e^{-j\beta_{\mathrm{TE}}z}$$

with

$$\beta_{\mathrm{TE}_{01}} = \sqrt{1.11 \times 10^{-7}\omega^2 - \frac{14.67}{a^2}} \qquad \text{and} \qquad \omega_{c,\mathrm{TE}_{01}} = \frac{1.149 \times 10^9}{a}$$

The time domain expressions are obtained in the usual way by multiplying by $e^{j\omega t}$ and taking the real part. Recall we do not multiply by $\sqrt{2}$ since in field theory we deal with peak values.

$$\mathcal{H}_z(\rho, \phi, z, t) = AJ_0\left(\frac{3.83}{a}\rho\right)\cos(\omega t - \beta_{\mathrm{TE}_{01}}z)$$

$$\mathcal{E}_\phi(\rho, \phi, z, t) = 3.281 \times 10^{-7}\omega a A J_1\left(\frac{3.83}{a}\rho\right)\sin(\omega t - \beta_{\mathrm{TE}_{01}}z)$$

$$\mathcal{H}_\rho(\rho, \phi, z, t) = -0.261aA\sqrt{1.11 \times 10^{-17}\omega^2 - \frac{14.67}{a^2}}\,J_1\left(\frac{3.83}{a}\rho\right)\sin(\omega t - \beta_{\mathrm{TE}_{01}}z)$$

Power traveling down the guide is obtained by integrating the Poynting vector over the guide cross section. Using $h_1 = 1$ and $h_2 = \rho$ for cylindrical coordinates:

$$P_z = \int_S \vec{S} \cdot d\vec{a}_z = \int_S \frac{1}{2}\vec{E} \times \vec{H} \cdot (h_1 h_2 du_1 du_2 \hat{z}) = \int_S \frac{1}{2}\vec{E} \times \vec{H}^* \cdot (\rho d\rho d\phi \hat{z})$$

$$= \frac{1}{2}\int_0^a \int_0^{2\pi} \rho[E_\phi\hat{\phi} \times (H_\rho\hat{\rho} + H_z\hat{z})^*] \cdot \hat{z}d\rho d\phi = \frac{1}{2}\int_0^a \int_0^{2\pi} -\rho E_\phi H_\rho^* d\rho d\phi$$

$$= \frac{1}{2}\int_0^a \int_0^{2\pi} \rho\, 8.563 \times 10^{-8}\omega a^2 A^2 \sqrt{1.11 \times 10^{-17}\omega^2 - \frac{14.67}{a^2}}\,J_1^2\left(\frac{3.83}{a}\rho\right)d\rho d\phi$$

$$= \frac{2\pi \times 8.563 \times 10^{-8}\omega a^2 A^2\sqrt{1.11 \times 10^{-17}\omega^2 - \frac{14.67}{a^2}}}{2}\int_0^a \rho J_1^2\left(\frac{3.83}{a}\rho\right)d\rho$$

To integrate the Bessel function we consult the table of integrals in Appendix M, specifically Eq. M-10. The result is

$$P_z = 2.691 \times 10^{-7}\omega a^2 A^2\sqrt{1.11 \times 10^{-17}\omega^2 - \frac{14.67}{a^2}}\left\{\frac{\rho^2}{2}\left[J_1^2\left(\frac{3.83}{a}\rho\right) - J_0\left(\frac{3.83}{a}\rho\right)J_2\left(\frac{3.83}{a}\rho\right)\right]\right\}\Bigg|_0^a$$

Putting in the upper and lower limits as usual:

$$P_z = 1.346 \times 10^{-7}\omega a^4 A^2 \sqrt{1.11 \times 10^{-17}\omega^2 - \frac{14.67}{a^2}}\,[J_1^2(3.83) - J_0(3.83)J_2(3.83)]$$

Using the Bessel function tables in Appendix L to obtain numerical values by interpolation (it turns out that $J_1(3.83) = 0$ and $J_0(3.83)$ has a minimum) we get

$$P_z = 2.183 \times 10^{-8}\omega a^4 A^2 \sqrt{1.11 \times 10^{-17}\omega^2 - \frac{14.67}{a^2}}\ \text{W}$$

To gain some experience at visualizing field patterns, use the time expressions for the TE_{01} mode in Example 9-5 evaluated at $t = 0$ and verify the patterns for that mode given in Figure 9-26.

Exercises

9-6.1a For TE modes in a circular waveguide show that using the normal $\vec{B}$ boundary condition results in the same values of $k_{c,\text{TE}}$.

9-6.1b One of the ''boundary'' conditions used for the TE solution was that the fields remain finite at the center of the guide, $\rho = 0$. The equations for E_ρ and H_ϕ, however, have the coordinate ρ in the denominator. Explain how this is allowed. Hint: Look at the equations as $\rho \to 0$ and also plots of the Bessel functions in Appendix L.

9-6.1c Evaluate the following in terms of Bessel functions that do not have derivatives:

$$(i)\ \frac{dJ_3(b\rho)}{d\rho}; \qquad (ii)\ \frac{d^2J_0(c\rho)}{d\rho^2}$$

9-6.1d Prove that the set of functions $\{J_0[(p_i/a)\rho]\}$, $i = 1, 2, \ldots,$ is an orthogonal function set with respect to the weight function ρ; that is, show that

$$\int_0^a \rho J_0\left(\frac{p_i}{a}\rho\right)J_0\left(\frac{p_j}{a}\rho\right)d\rho = \begin{cases} 0 & i \neq j \\ \text{constant} & i = j \end{cases}$$

where p_i is the ith root of J_0: $J_0(p_i) = 0$. (See Appendix N for details on orthogonal functions.)

9-6.1e Develop the time domain expressions for the field components of the circular waveguide TE_{11} mode, and determine the time and/or coordinate values of the field configurations shown in Figure 9-26.

9-6.1f Evaluate the Bessel function integral in Example 9-5 using Appendix M, Eq. M-12, instead of the one used in the example.

9-6.1g Determine the expression for the wave impedance in the waveguide of Example 9-5.

9-6.1h The dielectric strength of air (breakdown electric field intensity) is 3×10^6 V/m. For the TE_{01} mode determine the location of the electric field maximum and the maximum power that can theoretically be transmitted in a 3-cm radius guide at a frequency 1.5 times cutoff.

9-6.1i Develop probe and loop coupling methods for exciting the TE_{01} mode in a circular guide. Consider coupling to H_z. Use a coax as a feed line.

9-6.1j Compare the cutoff frequencies of TE_{01} modes for an air-filled guide and a polystyrene-filled guide.

9-6.2 TM Modes: Single Hollow Cylindrical Tube

The details of the solution for these modes are left for Exercise 9-6.2a. The solution process is the same as for the TM modes in a rectangular guide, following the steps of Section 9-3.2. This solution uses Figure 9-25 again as the basic configuration.

For the cutoff value of k we obtain

$$k_{c,\mathrm{TM}} = \frac{p_{nl}}{a}, \qquad n = 0, 1, 2, \ldots \tag{9-196}$$

where p_{nl} is the lth nonzero root of the nth order Bessel function, $J_n(p_{nl}) = 0$. Values of p_{nl} are given in Table 9-2.

The corresponding modes are denoted using the mode convention TM_{nl}, where n and l have the same physical interpretations as for the TE_{nl} modes. The propagation constant is

$$\gamma_{\mathrm{TM}} = \sqrt{k_{c,\mathrm{TM}}^2 - k^2} = j\sqrt{k^2 - k_{c,\mathrm{TM}}^2} = j\sqrt{\omega^2\mu\epsilon - \left(\frac{p_{nl}}{a}\right)^2} = j\beta_{\mathrm{TM}} \tag{9-197}$$

The field components are

$$E_z(\rho, \phi, z) = AJ_n\left(\frac{p_{nl}}{a}\rho\right)\cos n\phi e^{-j\beta_{\mathrm{TM}}z} \tag{9-198}$$

$$E_\rho(\rho, \phi, z) = -j\frac{\sqrt{k^2 - \left(\frac{p_{nl}}{a}\right)^2}}{\left(\frac{p_{nl}}{a}\right)}AJ_n'\left(\frac{p_{nl}}{a}\rho\right)\cos n\phi e^{-j\beta_{\mathrm{TM}}z} \tag{9-199}$$

Table 9-2. lth-Nonzero Roots (p_{nl}) of the nth-Order Bessel Function of the First Kind: $J_n(p_{nl}) = 0$, $p_{nl} \neq 0$

	n							
l	0	1	2	3	4	5	6	7
1	2.405	3.832	5.136	6.380	7.588	8.771	9.936	11.086
2	5.520	7.016	8.417	9.761	11.065	12.339	13.589	14.821
3	8.654	10.173	11.620	13.015	14.372	15.700	17.004	18.288
4	11.792	13.323	14.796	16.223	17.616	18.980	20.321	21.642

$$E_\phi(\rho, \phi, z) = j\frac{n\sqrt{k^2 - \left(\frac{p_{nl}}{a}\right)^2}}{\left(\frac{p_{nl}}{a}\right)^2 \rho} AJ_n\left(\frac{p_{nl}}{a}\rho\right) \sin n\phi e^{-j\beta_{\text{TM}}z} \tag{9-200}$$

$$H_\rho(\rho, \phi, z) = -j\frac{\omega\epsilon n}{\left(\frac{p_{nl}}{a}\right)^2 \rho} AJ_n\left(\frac{p_{nl}}{a}\rho\right) \sin n\phi e^{-j\beta_{\text{TM}}z} \tag{9-201}$$

$$H_\phi(\rho, \phi, z) = -j\frac{\omega\epsilon}{\left(\frac{p_{nl}}{a}\right)} AJ_n'\left(\frac{p_{nl}}{a}\rho\right) \cos n\phi e^{-j\beta_{\text{TM}}z} \tag{9-202}$$

As before, the primes mean the derivative with respect to the entire argument of the function

$$J_n'\left(\frac{p_{nl}}{a}\rho\right) = \frac{dJ_n\left(\frac{p_{nl}}{a}\rho\right)}{d\left(\frac{p_{nl}}{a}\rho\right)}$$

The cutoff frequency for which β_{TM} is zero is given by

$$\omega_{c,\text{TM}} = \frac{p_{nl}}{a\sqrt{\mu\epsilon}} \tag{9-203}$$

The cutoff wavelength is

$$\lambda_{c,\text{TM}} = \frac{v_0}{f_{c,\text{TM}}} = \frac{2\pi a}{p_{nl}} \tag{9-204}$$

The guide wavelength is given by

$$\lambda_{g,\text{TM}} = \frac{2\pi}{\beta_{\text{TM}}} = \frac{2\pi}{\sqrt{\omega^2\mu\epsilon - \left(\frac{p_{nl}}{a}\right)^2}} = \frac{\lambda_0}{\sqrt{1 - \left(\frac{\lambda_0}{\lambda_{c,\text{TM}}}\right)^2}} = \frac{\lambda_0}{\sqrt{1 - \left(\frac{\omega_{c,\text{TM}}}{\omega}\right)^2}} \tag{9-205}$$

where

$$\lambda_0 = \frac{v_0}{f} = \frac{1}{f\sqrt{\mu\epsilon}}, \qquad v_0 = \frac{1}{\sqrt{\mu\epsilon}}$$

Other expressions for β_{TM} are

$$\beta_{\text{TM}} = k\sqrt{1 - \left(\frac{\lambda_0}{\lambda_{c,\text{TM}}}\right)^2} = k\sqrt{1 - \left(\frac{\omega_{c,\text{TM}}}{\omega}\right)^2} \tag{9-206}$$

$$V_p^{\mathrm{TM}} = \frac{\omega}{\beta_{\mathrm{TM}}} = \frac{v_0}{\sqrt{1 - \left(\dfrac{p_{nl}}{a\omega\sqrt{\mu\epsilon}}\right)^2}} = \frac{v_0}{\sqrt{1 - \left(\dfrac{\lambda_0}{\lambda_{c,\mathrm{TM}}}\right)^2}} = \frac{v_0}{\sqrt{1 - \left(\dfrac{\omega_{c,\mathrm{TM}}}{\omega}\right)^2}} \tag{9-207}$$

$$\begin{aligned} V_g^{\mathrm{TM}} &= \frac{d\omega}{d\beta_{\mathrm{TM}}} = v_0\sqrt{1 - \left(\frac{p_{nl}}{a\omega\sqrt{\mu\epsilon}}\right)^2} = v_0\sqrt{1 - \left(\frac{\lambda_0}{\lambda_{c,\mathrm{TM}}}\right)^2} \\ &= v_0\sqrt{1 - \left(\frac{\omega_{c,\mathrm{TM}}}{\omega}\right)^2} \end{aligned} \tag{9-208}$$

$$\begin{aligned} Z_0^{\mathrm{TM}} &= Z_{z^+}^{\mathrm{TM}} = \sqrt{\frac{\mu}{\epsilon}}\sqrt{1 - \left(\frac{p_{nl}}{a\omega\sqrt{\mu\epsilon}}\right)^2} = \eta\sqrt{1 - \left(\frac{\lambda_0}{\lambda_{c,\mathrm{TM}}}\right)^2} \\ &= \eta\sqrt{1 - \left(\frac{\omega_{c,\mathrm{TM}}}{\omega}\right)^2} \end{aligned} \tag{9-209}$$

$$\begin{aligned} \vec{S}_{\mathrm{TM}} = {} & j\frac{\omega_\epsilon}{2\left(\dfrac{p_{nl}}{a}\right)} A^2 J_n\left(\frac{p_{nl}}{a}\rho\right) J_n'\left(\frac{p_{nl}}{a}\rho\right)\cos^2 n\phi\hat{\rho} \\ & + j\frac{\omega\epsilon n}{2\left(\dfrac{p_{nl}}{a}\right)^2\rho} A^2 J_n^2\left(\frac{p_{nl}}{a}\rho\right)\cos n\phi\ \sin n\phi\hat{\phi} \\ & + \left(\frac{\omega\epsilon A^2\sqrt{k^2 - \left(\dfrac{p_{nl}}{a}\right)^2}}{2\left(\dfrac{p_{nl}}{a}\right)^2}\left[J_n'\left(\frac{p_{nl}}{a}\rho\right)\right]^2\cos^2 n\phi \right. \\ & \left. + \frac{\omega\epsilon n^2 A^2\sqrt{k^2 - \left(\dfrac{p_{nl}}{a}\right)^2}}{2\left(\dfrac{p_{nl}}{a}\right)^4\rho^2} J_n^2\left(\frac{p_{nl}}{a}\rho\right)\sin^2 n\phi\right)\hat{z} \end{aligned} \tag{9-210}$$

The field configurations for the lowest-order TM modes are shown in Figure 9-27. Views 1 and 2 show electric and magnetic fields within the guide. View 2 is a longitudinal view through plane l–l'. View 3 shows the magnetic fields and surface currents on the surfaces viewed from inside the guide from cross-sectional surface s–s'.

Exercises

9-6.2a Derive all of the TM mode quantities given by the equations summary of this section.

9-6.2b Show that the boundary conditions on normal $\vec{B}$ are satisfied.

9-6.2c Determine the values of z coordinates at key points in field configurations of all views for the TM_{01} mode in Figure 9-27.

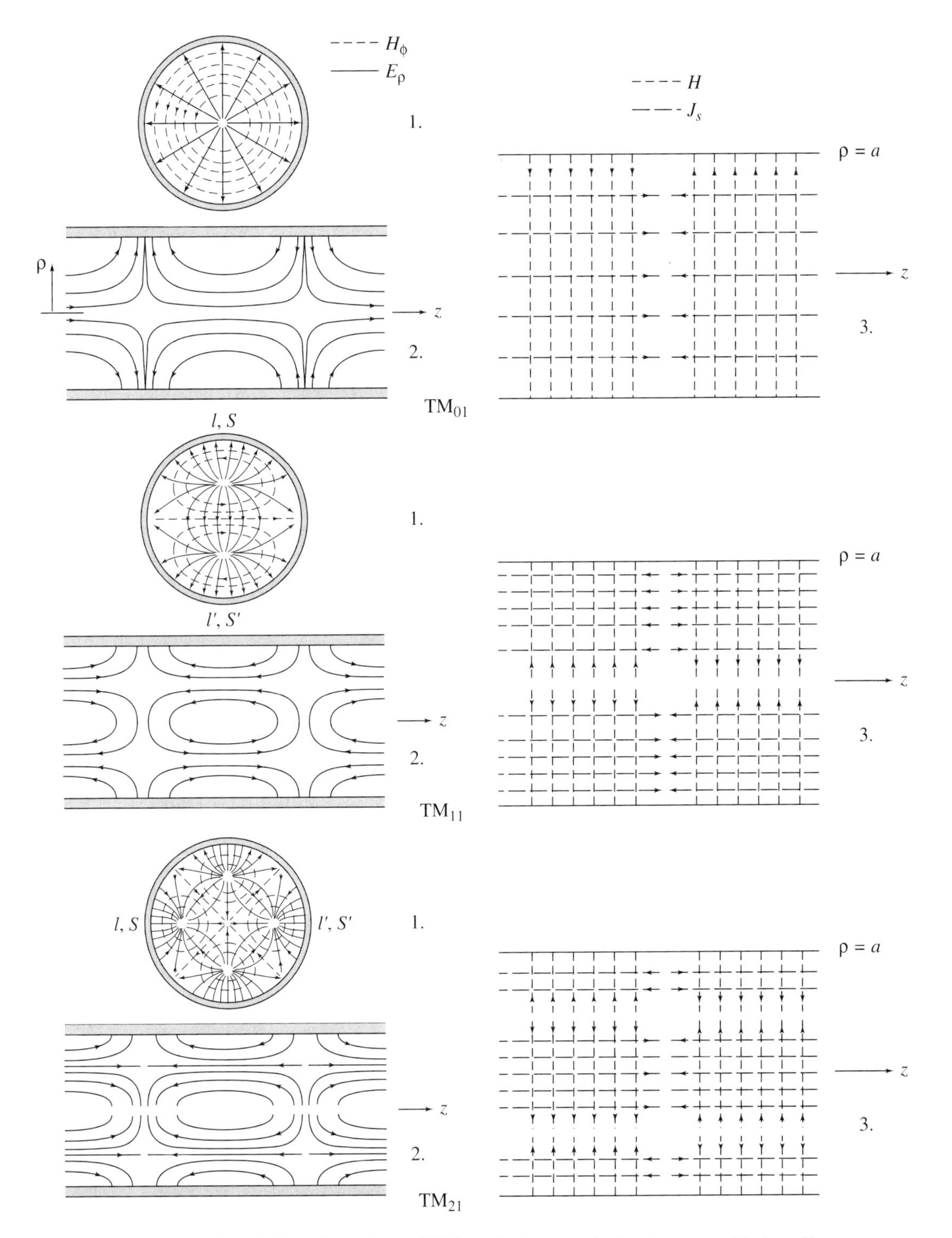

Figure 9-27. Field configurations of TM modes in a metal circular waveguide ($t = 0$).

9-6.2d Repeat Exercise 9-6.2c for the TM_{11} mode.

9-6.2e For the TM_{01} mode develop the equation for the power flow in the z direction, P_z, in watts. See Example 9-5.

9-6.2f Develop probe and loop coupling methods for exciting the TM_{01} and TM_{11} modes in a circular waveguide. Consider coupling to E_z where possible. Use a coax as a feed line.

9-6.2g Derive the equation for the surface current for the TM_{nl} modes. Show that View 3 for the TM_{01} mode in Figure 9-26 results from your general equation. For this last part you will need to use the time domain expression.

9-6.2h Determine the location of the maximum electric field intensity in the TM_{01} mode.

9-6.2i Design a cylindrical waveguide that will operate in the TM_{01} mode at 10 GHz but in which the TE_{01}, TE_{21}, and TM_{11} modes will not propagate. (Hint: Examine ω_c for all modes.) Guide is air filled.

9-6.2j Work Exercise 9-6.2i for a cylindrical waveguide filled with Teflon.

9-6.3 TM Modes: Coaxial Cylinders

A picture of the basic structure to be investigated in this section is given in Figure 9-28. This time we'll do the TM mode and leave the TE mode as an exercise at the end of the next section. We assume that both conductors at $\rho = a$ and $\rho = b$ are perfect. This configuration could represent either rigid coaxial cylinders or the common idealized coaxial cable.

Since we have two conductors this system can support the TEM mode. This has already been discussed in Section 9-4.1. The results of this section and the next one will show that higher-order field structures (modes) can exist in a coaxial cable if the frequency is high enough, that is, above cutoff.

We will use the solution outline given in Section 9-3.2. We thus begin by expanding the transverse component wave equation for E_z in cylindrical coordinates:

$$\frac{1}{\rho}\frac{\partial}{\partial \rho}\left(\rho\frac{\partial E_z}{\partial \rho}\right) + \frac{1}{\rho^2}\frac{\partial^2 E_z}{\partial \phi^2} + k_{c,\text{TM}}^2 E_z = 0$$

Since this is exactly the same form used for the cylindrical waveguide we need not go through the complete separation of variables process again. The result is

$$E_z(\rho, \phi) = (A' \cos n\phi + B' \sin n\phi)[C' J_n(k_{c,\text{TM}}\rho) + D' N_n(k_{c,\text{TE}}\rho)]$$

As before we select the reference for $\phi = 0$ so we have only the cosine variation. Setting

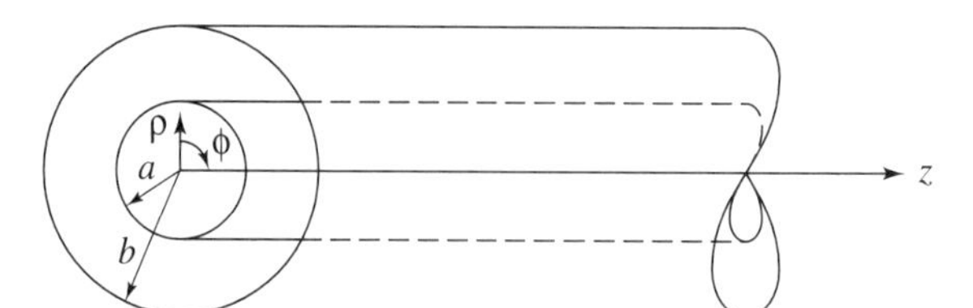

Figure 9-28. Basic configuration and coordinate conventions for coaxial cylinder systems. Conductors at $\rho = a$ and $\rho = b$ assumed to be perfect conductors.

$B' = 0$ and multiplying the constant A' into the C' and D' to get new constants C and D yields

$$E_z(\rho, \phi) = [CJ_n(k_{c,\text{TM}}\rho) + DN_n(k_{c,\text{TM}}\rho)] \cos n\phi \tag{9-211}$$

Since E_z will be a tangential component of electric field at the metal boundaries at $\rho = a$ and $\rho = b$ it must be zero there. This yields the two boundary conditions

$$E_z(a, \phi) = [CJ_n(k_{c,\text{TM}}a) + DN_N(k_{c,\text{TM}}a)] \cos n\phi = 0$$

$$E_z(b, \phi) = [CJ_n(k_{c,\text{TM}}b) + DN_N(k_{c,\text{TM}}b)] \cos n\phi = 0$$

Since these must be true for any ϕ, the $\cos n\phi$ coefficients must be zero:

$$\left.\begin{aligned} CJ_n(k_{c,\text{TM}}a) + DN_n(k_{c,\text{TM}}a) &= 0 \\ CJ_n(k_{c,\text{TM}}b) + DN_n(k_{c,\text{TM}}b) &= 0 \end{aligned}\right\} \tag{9-212}$$

We have used all the independent boundary conditions (using normal $\vec{B}$ from H_ρ would result in the same set of equations; see Exercise 9-6.3a). There are, however, three unknowns, C, D, and $k_{c,\text{TM}}$. One obvious solution is to have $C = D = 0$. This will not do since E_z from Eq. 9-211 would be zero everywhere. Thus C and D are not zero. From linear algebra we know that a set of equations in unknowns C and D that are all equal to zero can have nonzero solutions for C and D if the determinant of the coefficients is zero. Therefore

$$\begin{vmatrix} J_n(k_{c,\text{TM}}a) & N_n(k_{c,\text{TM}}a) \\ J_{n(k_{c,\text{TM}}}b) & N_n(k_{c,\text{TM}}b) \end{vmatrix} = 0$$

$$J_n(k_{c,\text{TM}}a)N_n(k_{c,\text{TM}}b) - J_n(k_{c,\text{TM}}b)N_n(k_{c,\text{TM}}a) = 0 \tag{9-213}$$

This is called the Bessel-Neumann combination boundary condition. For a given integer n (circumferential mode number) and given guide dimensions a and b, only $k_{c,\text{TM}}$ remains to be determined. Since the equation is transcendental it must be solved either numerically or graphically. The graphical solution is obtained by first moving the negative term to the right side of the equation and then naming each side as a function of $k_{c,\text{TM}}$. We then set

$$y_1(k_{c,\text{TM}}) = J_n(k_{c,\text{TM}}a)N_n(k_{c,\text{TM}}b) = J_n(k_{c,\text{TM}}b)N_n(k_{c,\text{TM}}a) = y_2(k_{c,\text{TM}})$$

We next plot y_1 and y_2 versus $k_{c,\text{TM}}$ (for given n, a, and b) on the same set of axes. The intersections of y_1 and y_2 identify the allowed values of $k_{c,\text{TM}}$. Since there will be several intersections for each value of n we identify them with a second subscript l as before. The modes would be denoted by the mode convention

$$\text{TM}_{nl}, \qquad n = 0, 1, 2, \ldots, \quad l = 1, 2, 3, \ldots$$

The lower-order values of $k_{c,\text{TM}}$ are given in Table 9-3. Note that the table does not give $k_{c,\text{TM}}$ directly but depends upon a modification of the Bessel-Neumann combination. Eq. 9-213. This is done by letting

$$x = k_{c,\text{TM}}a$$

from which

$$k_{c,\text{TM}} = \frac{x}{a} \tag{9-214}$$

Table 9-3. Roots of the Bessel-Neumann Boundary Equation (Eq. 9-215): Table Values Are $(b/a - 1)x_{nl}$

$\frac{b}{a}$	nl							
	01	11	21	31	02	12	22	32
1.0	3.142	3.142	3.142	3.142	6.283	6.283	6.283	6.283
1.1	3.141	3.143	3.147	3.154	6.283	6.284	6.286	6.289
1.2	3.140	3.146	3.161	3.187	6.282	6.285	6.293	6.306
1.3	3.139	3.150	3.182	3.236	6.282	6.287	6.304	6.331
1.4	3.137	3.155	3.208	3.294	2.281	2.290	2.317	6.362
1.5	3.135	3.161	3.237	3.36	6.280	6.293	6.332	6.387
1.6	3.133	3.168	3.27	3.43	6.279	6.296	6.349	6.437
1.8	3.128	3.182	3.36	3.6	6.276	6.304	6.387	6.523
2.0	3.123	3.197	3.4	3.7	6.273	6.312	6.43	6.62
2.5	3.110	3.235			6.266	6.335		6.9
3.0	3.097	3.271			6.258	6.357		
3.5	3.085	3.305			6.243	6.403		

$\frac{b}{a}$	nl							
	03	13	23	33	04	14	24	34
1.0	9.425	9.425	9.425	9.425	12.566	12.566	12.566	12.566
1.1	9.425	9.425	9.427	9.429	12.566	12.567	12.568	12.569
1.2	9.424	9.426	9.431	9.440	12.566	12.567	12.571	12.578
1.3	9.424	9.427	9.438	9.457	12.566	12.568	12.577	12.590
1.4	9.423	9.429	9.447	9.478	12.565	12.570	12.583	12.606
1.5	9.423	9.431	9.458	9.502	12.565	12.571	12.591	12.624
1.6	9.422	9.434	9.469	9.528	12.564	12.573	12.600	12.644
1.8	9.420	9.439	9.495	9.587	12.563	12.577	12.619	12.689
2.0	9.418	9.444	9.523	9.652	12.561	12.581	12.640	12.738
2.5	9.413	9.460		9.83	12.558	12.593		12.874
3.0	9.408	9.476		10.0	12.553	12.605		13.02
3.5	9.402	9.493		10.2	12.549	12.619		13.2
4.0	9.396	9.509			12.545	12.631		13.3

Substituting this into Eq. 9-213

$$J_n(x)N_n\left(\frac{b}{a}x\right) - J_n\left(\frac{b}{a}x\right)N_n(x) = 0 \tag{9-215}$$

The values in the table are $(b/a - 1)x_{nl}$, so for given values of a, b, n, and l we read the value from the table and then obtain $k_{c,\mathrm{TM}_{nl}}$ using

$$k_{c,\mathrm{TM}_{nl}} = \frac{x_{nl}}{a} = \frac{\left[\dfrac{\text{table mode value}}{\left(\dfrac{b}{a} - 1\right)}\right]}{a} = \frac{\text{table mode value}}{b - a} \tag{9-216}$$

Once the value of $k_{c,\text{TM}}$ has been determined the value of D (or C, the choice is immaterial) can be obtained using either equation in the boundary condition set of Eq. 9-212. Solving the second of those equations for D we obtain

$$D = -\frac{J_n(k_{c,\text{TM}}b)}{N_n(k_{c,\text{TM}}b)}C \tag{9-217}$$

We finally obtain the solution for $E_z(\rho, \phi)$ using this result in Eq. 9-211:

$$E_z(\rho, \phi) = C\left[J_n(k_{c,\text{TM}}\rho) - \frac{J_n(k_{c,\text{TM}}b)}{N_n(k_{c,\text{TM}}b)}N_n(k_{c,\text{TM}}\rho)\right]\cos n\phi \tag{9-218}$$

As in all previous solutions we have one remaining constant C. Its value depends upon the amplitude of the driving source feeding the guide. See Example 9-7 following.

The propagation constant is, using Eq. 9-4,

$$\gamma_{\text{TM}} = \sqrt{k_{c,\text{TM}}^2 - k^2} = j\sqrt{\omega^2\mu\epsilon - k_{c,\text{TM}}^2} = j\beta_{\text{TM}} \tag{9-219}$$

where we use Eq. 9-216 for $k_{c,\text{TM}}$.

The remaining field components are obtained by using Eqs. 9-7 through 9-10 as usual. The results are

$$E_z(\rho, \phi, z) = C\left[J_n(k_{c,\text{TM}}\rho) - \frac{J_n(k_{c,\text{TM}}b)}{N_n(k_{c,\text{TM}}b)}N_n(k_{c,\text{TM}}\rho)\right]\cos n\phi e^{-j\beta_{\text{TM}}z} \tag{9-220}$$

$$E_\rho(\rho, \phi, z) = -j\frac{\beta_{\text{TM}}C}{k_{c,\text{TM}}}\left[J_n'(k_{c,\text{TM}}\rho) - \frac{J_n(k_{c,\text{TM}}b)}{N_n(k_{c,\text{TM}}b)}N_n'(k_{c,\text{TM}}\rho)\right]\cos n\phi e^{-j\beta_{\text{TM}}z} \tag{9-221}$$

$$E_\phi(\rho, \phi, z) = j\frac{\beta_{\text{TM}}nC}{k_{c,\text{TM}}^2\rho}\left[J_n(k_{c,\text{TM}}\rho) - \frac{J_n(k_{c,\text{TM}}b)}{N_n(k_{c,\text{TM}}b)}N_n(k_{c,\text{TM}}\rho)\right]\sin n\phi e^{-j\beta_{\text{TM}}z} \tag{9-222}$$

$$H_\rho(\rho, \phi, z) = -j\frac{\omega\epsilon nC}{k_{c,\text{TM}}^2\rho}\left[J_n(k_{c,\text{TM}}\rho) - \frac{J_n(k_{c,\text{TM}}b)}{N_n(k_{c,\text{TM}}b)}N_n(k_{c,\text{TM}}\rho)\right]\sin n\phi e^{-j\beta_{\text{TM}}z} \tag{9-223}$$

$$H_\phi(\rho, \phi, z) = -j\frac{\omega\epsilon C}{k_{c,\text{TM}}}\left[J_n'(k_{c,\text{TM}}\rho) - \frac{J_n(k_{c,\text{TM}}b)}{N_n(k_{c,\text{TM}}b)}N_n'(k_{c,\text{TM}}\rho)\right]\cos n\phi e^{-j\beta_{\text{TM}}z} \tag{9-224}$$

For the cutoff frequency at which $\beta_{\text{TM}} = 0$, Eq. 9-219 gives

$$\omega_{c,\text{TM}} = \frac{k_{c,\text{TM}}}{\sqrt{\mu\epsilon}} = \frac{\text{Table 9-3 mode value}}{(b-a)\sqrt{\mu\epsilon}} \tag{9-225}$$

The corresponding cutoff wavelength is

$$\lambda_{c,\text{TM}} = \frac{v_0}{f_{c,\text{TM}}} = \frac{2\pi}{\omega_{c,\text{TM}}\sqrt{\mu\epsilon}} = \frac{2\pi(b-a)}{\text{table mode value}} \tag{9-226}$$

For the guide wavelength

$$\lambda_{g,\text{TM}} = \frac{2\pi}{\beta_{\text{TM}}} = \frac{2\pi}{\sqrt{\omega^2\mu\epsilon - k_{c,\text{TM}}^2}} = \frac{\lambda_0}{\sqrt{1 - \left(\dfrac{\lambda_0}{\lambda_{c,\text{TM}}}\right)^2}} = \frac{\lambda_0}{\sqrt{1 - \left(\dfrac{\omega_{c,\text{TM}}}{\omega}\right)^2}} \tag{9-227}$$

where

$$\lambda_0 = \frac{v_0}{f} = \frac{1}{f\sqrt{\mu\epsilon}}$$

The phase and group velocities are

$$V_p^{\text{TM}} = \frac{\omega}{\beta_{\text{TM}}} = \frac{v_0}{\sqrt{1 - \left(\dfrac{k_{c,\text{TM}}^2}{\omega^2\mu\epsilon}\right)}} = \frac{v_0}{\sqrt{1 - \left(\dfrac{\lambda_0}{\lambda_{c,\text{TM}}}\right)^2}} = \frac{v_0}{\sqrt{1 - \left(\dfrac{\omega_{c,\text{TM}}}{\omega}\right)^2}} \tag{9-228}$$

$$\begin{aligned} V_g^{\text{TM}} &= \frac{d\omega}{d\beta_{\text{TM}}} = v_0\sqrt{1 - \left(\frac{k_{c,\text{TM}}^2}{\omega^2\mu\epsilon}\right)} = v_0\sqrt{1 - \left(\frac{\lambda_0}{\lambda_{c,\text{TM}}}\right)^2} \\ &= v_0\sqrt{1 - \left(\frac{\omega_{c,\text{TM}}}{\omega}\right)^2} \end{aligned} \tag{9-229}$$

For the wave or characteristic impedance in the z direction we have

$$\begin{aligned} Z_0^{\text{TM}} = Z_{z^+}^{\text{TM}} &= \frac{(\vec{E} \times \vec{H}^*) \cdot \hat{z}}{(\hat{z} \times \vec{H}) \cdot (\hat{z} \times \vec{H})^*} = \sqrt{\frac{\mu}{\epsilon}}\sqrt{\omega^2\mu\epsilon - k_{c,\text{TM}}^2} \\ &= \eta\sqrt{1 - \left(\frac{\lambda_0}{\lambda_{c,\text{TM}}}\right)^2} = \eta\sqrt{1 - \left(\frac{\omega_{c,\text{TM}}^2}{\omega}\right)^2} \end{aligned} \tag{9-230}$$

And, last but not least, the Poynting vector is

$$\begin{aligned} \vec{S}_{\text{TM}} = &-j\frac{\omega\epsilon C^2}{2k_{c,\text{TM}}}\left[J_n(k_{c,\text{TM}}\rho) - \frac{J_n(k_{c,\text{TM}}b)}{N_n(k_{c,\text{TM}}b)}N_n(k_{c,\text{TM}}\rho)\right] \\ &\cdot\left[J_n'(k_{c,\text{TM}}\rho) - \frac{J_n(k_{c,\text{TM}}b)}{N_n(k_{c,\text{TM}}b)}N_n'(k_{c,\text{TM}}\rho)\right]\cos^2 n\phi\hat{\rho} \\ &+ j\frac{\omega\epsilon n C^2}{2k_{c,\text{TM}}^2\rho^2}\left[J_n(k_{c,\text{TM}}\rho) - \frac{J_n(k_{c,\text{TM}}b)}{N_n(k_{c,\text{TM}}b)}N_n(k_{c,\text{TM}}\rho)\right]^2\cos n\phi\,\sin n\phi\hat{\phi} \\ &+ \frac{\beta_{\text{TM}}\omega\epsilon C^2}{2k_{c,\text{TM}}^2}\left\{\left[J_n'(k_{c,\text{TM}}\rho) - \frac{J_n(k_{c,\text{TM}}b)}{N_n(k_{c,\text{TM}}b)}N_n'(k_{c,\text{TM}}\rho)\right]^2\cos^2 n\phi\right. \\ &\left.+ \frac{n^2}{k_{c,\text{TM}}^2\rho^2}\left[J_n(k_{c,\text{TM}}\rho) - \frac{J_n(k_{c,\text{TM}}b)}{N_n(k_{c,\text{TM}}b)}N_n(k_{c,\text{TM}}\rho)\right]^2\sin^2 n\phi\right\}\hat{z} \end{aligned} \tag{9-231}$$

The field configurations for the lowest-order TM modes are shown in Figure 9-29. Views 1 and 2 show electric and magnetic fields within the guide. View 2 is a longitudinal view through plane s–s'. Views 2 show the magnetic fields and surface currents on the surfaces viewed from inside the guide from plane s–s'. For these plots $b/a = 3$ and $\lambda_g/b = 4.24$.

One should convert the preceding equations to the time domain and evaluate them at $t = 0$ and check the validity of the patterns given. It is important to determine values of

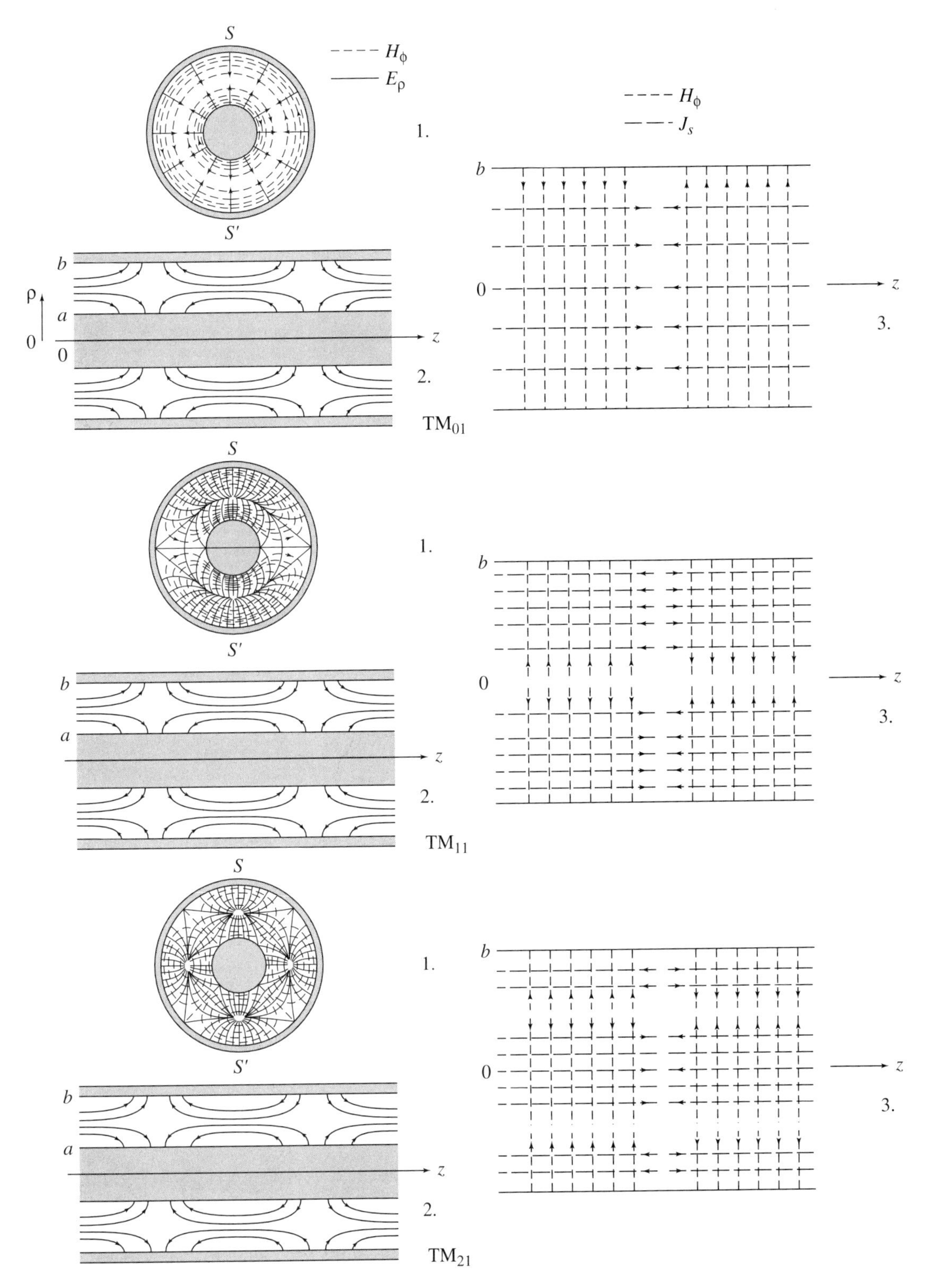

Figure 9-29. Field configurations of TM modes in a coaxial waveguide ($t = 0$).

z and ρ in each view, and to mentally superimpose Views 1 and 2 so the change in field component, that is, from E_ρ to E_z in the TM_{01} mode near $\rho = (a + b)/2$ can be verified or identified on the views.

Example 9-6

A coaxial waveguide has an inner conductor radius of 1.75 cm and an outer conductor radius (inside surface) of 3 cm. The operating frequency is 20 GHz and the guide is air-filled. Determine the cutoff frequency, the cutoff wavelength, the guide wavelength, the phase velocity, and the equation for the z component of the Poynting vector. Use the TM_{01} mode.

$$\frac{b}{a} = \frac{3}{1.75} = 1.714$$

From Table 9-3 we determine the Bessel-Neumann boundary equation root by interpolation as

$$\text{Table mode value} = 3.133 + (3.128 - 3.133)\left(\frac{1.714 - 1.6}{1.8 - 1.6}\right) = 3.130$$

Using Eqs. 9-225 and 9-226:

$$f_{c,\text{TM}_{01}} = \frac{3.130}{2\pi(0.03 - 0.0175)\sqrt{4\pi \times 10^{-7} = \dfrac{1}{36\pi} \times 10^{-4}}} = 11.96 \text{ GHz}$$

$$\lambda_{c,\text{TM}_{01}} = \frac{3 \times 10^8}{11.96 \times 10^9} = 2.51 \text{ cm}$$

At 20 GHz,

$$\lambda_0 = \frac{1}{f\sqrt{\mu_0 \epsilon_0}} = \frac{3 \times 10^8}{20 \times 10^9} = 1.5 \text{ cm}$$

Then

$$\sqrt{1 - \left(\frac{\lambda_0}{\lambda_{c,\text{TM}}}\right)^2} = \sqrt{1 - \left(\frac{1.5}{2.51}\right)^2} = 0.8018$$

Using Eqs. 9-226 and 9-227:

$$\lambda_{g,\text{TM}_{01}} = \frac{1.5}{0.8018} \text{ cm} = 1.87 \text{ cm}$$

$$V_p^{\text{TM}_{01}} = \frac{3 \times 10^8}{0.8018} = 3.74 \times 10^8 \text{ m/s}$$

(Remember the phase velocity can exceed the velocity of light in vacuum since the phase progression along the guide boundary is not parallel to the wave propagation direction. See Figure 9-22, for example.)

Next we use Eqs. 9-216 and 9-219:

$$k_{c,\mathrm{TM}_{01}} = \frac{3.130}{0.03 - 0.0175} = 250.4 \text{ rad/m}$$

$$\beta_{\mathrm{TM}_{01}} = \sqrt{(2\pi \times 20 \times 10^9)^2 \times 4\pi \times 10^{-7} \times \frac{1}{36\pi} \times 10^{-9} - (250.4)^2}$$
$$- 335.8 \text{ rad/m}$$

For this mode $n = 0$ and Eq. 9-231 gives the z component ($\epsilon = \epsilon_0$ here) as

$$S_z = \frac{335.8 \times 2\pi \times 20 \times 10^9 \times \frac{1}{36\pi} \times 10^{-9} C^2}{2 \times (250.4)^2} \times \left[J_0'(250.4\rho) - \frac{J_0(250.4 \times 0.03)}{N_0(250.4 \times 0.03)} N_0'(250.4\rho) \right]^2$$

Using the derivative table in Appendix M, Eq. M-2,

$$S_z = 2.98 \times 10^{-3} C^2 \left[-J_1(250.4\rho) + \frac{J_0(7.512)}{N_0(7.512)} N_1(250.4\rho) \right]^2$$

Obtain values for J_0 and N_0 using interpolation in Appendix L Tables L-1 and L-2:

$$S_z = 2.98 \times 10^{-3} C^2 [2.2 N_1(250.4\rho) - J_1(250.4\rho)]^2 \text{ W/m}^2$$

Example 9-7

For the waveguide of Example 9-6, determine the constant C of the electric field required to transmit 50 mW down the guide.

$$P_z = \int_S \vec{S} \cdot (\rho d\rho d\phi \hat{z}) = \int_0^{2\pi} \int_{0.0175}^{0.03} S_z \rho d\rho d\phi = 2\pi \int_{0.0175}^{0.03} 2.98$$
$$\times 10^{-3} C^2 [2.2 N_1(250.4\rho) - J_1(250.4\rho)]^2 \rho d\rho$$

Change variable to $x = 250.4\rho$. The integral becomes

$$P_z = 6.99 \times 10^{-5} C^2 \int_{4.382}^{7.512} [2.2 N_1(x) - J_1(x)]^2 x dx$$
$$= 6.99 \times 10^{-5} C^2 \int_{4.382}^{7.512} [4.84 x N_1^2(x) - 4.4 x N_1(x) J_1(x) + x J_1^2(x)] dx$$
$$= 6.99 \times 10^{-5} C^2 \left[4.84 \int_{4.382}^{7.512} x N_1^2(x) dx - 4.4 \int_{4.382}^{7.512} x N_1(x) J_1(x) dx + \int_{4.382}^{7.512} x J_1^2(x) dx \right]$$

The first and last integrals can be evaluated using Eq. M-10. However, the middle integral must be evaluated numerically. The results are

$$P_z = 6.99 \times 10^{-5} C^2 [(4.84)(1.011) - (4.4)(0.001) + (1.017)] = 4.13 \times 10^{-4} C^2 \text{ W}$$

To transit 50 mW we must have

$$4.13 \times 10^{-4} C^2 = 50 \times 10^{-3}$$

$$C = 11 \text{ V/m}$$

Exercises

9-6.3a Using Eq. 9-211 develop the equation for H_ρ and show that applying the boundary conditions to this field results in Eq. 9-212.

9-6.3b Determine the equation for $\vec{J}_s$ on the outer conductor. Using this determine the total current in amperes flowing in the $\rho = b$ surface for the TM_{01} mode of Examples 9-6 and 9-7.

9-6.3c Determine ρ_s for the TM_{nl} modes and explain how the equation shows that ρ_s is moving down the guide.

9-6.3d Devise a method for exciting the TM_{01} mode. Design both loop and probe systems.

9-6.3e Design a coaxial waveguide that has a cutoff frequency of 6 GHz for the TM_{01} mode and for which higher-order modes cannot propagate below 7.5 GHz. Can the TEM mode propagate?

9-6.3f Show that for $b/a \leq 1.5$ (approximately) the cutoff wavelengths of the TM_{nl} modes in a coaxial waveguide are essentially the same as the cutoff wavelengths of TM_n modes between parallel planes. What is the ratio $\lambda_{c,\text{TM}}/a$ for this case? Note this means that the coaxial guide appears to be parallel planes. Explain physically how this can be.

9-6.4 TE Modes: Coaxial Cylinders

The details of the solutions for these modes are reserved for Exercise 9-6.4a. The configuration of Figure 9-28 is again used and follows the outline given in Section 9-3.3. Important intermediate results, tables, and final equations are given in this section.

Values of $k_{c,\text{TE}}$ for this system are found from roots of the Bessel-Neumann derivative combination boundary condition

$$J_n'(k_{c,\text{TE}}a)N_n'(k_{c,\text{TE}}b) - J_n'(k_{c,\text{TE}}b)N_n'(k_{c,\text{TE}}a) = 0 \tag{9-232}$$

A form more convenient for tabulating values is obtained by letting

$$x' = k_{c,\text{TE}}a$$

from which $k_{c,\text{TE}} = x'/a$. The boundary condition becomes

$$J_n'(x')N_n'\left(\frac{b}{a}x'\right) - J_n'\left(\frac{b}{a}x'\right)N_n'(x') = 0 \tag{9-233}$$

The boundary equation must be solved graphically, in general, as explained in the preceding section. Table 9-4 supplies the first few values from which the $k_{c,\text{TE}}$ can be determined. The resulting modes are denoted by the usual mode convention:

$$\text{TE}_{nl}, \qquad n = 0, 1, 2, \ldots, \quad l = 1, 2, 3, \ldots$$

Table 9-4. Roots of the Bessel-Neumann Derivative Boundary Equation (Eq. 9-233); Table Values Are $(b/a - 1)x'_{nl}$

$\frac{b}{a}$	*nl* 01	02	03	04	11	12	13	14
1.001	3.14159	6.28318	9.42479	12.5663	3.14159	6.28319	9.42479	12.5663
1.1	3.14268	6.28372	9.42515	12.5666	0.0952739	3.14413	6.28445	9.42564
1.2	4.14555	6.28517	9.4261	12.5673	0.182066	3.15089	6.28782	9.42787
1.3	3.14976	6.28731	9.42753	12.5684	0.261595	3.16092	6.29284	9.43121
1.4	3.15497	6.28996	9.42932	12.5698	0.334829	3.17354	6.29911	9.43539
1.5	3.16094	6.29306	9.43139	12.5714	0.402546	3.18825	6.30643	9.44026
1.6	3.16746	6.29649	9.43372	12.5731	0.465384	3.20468	6.31459	9.44567
1.8	3.18160	6.30407	9.43882	12.5769	0.578451	3.24156	6.33289	9.45785
2.0	3.19658	6.31235	9.44447	12.5812	0.677336	3.28247	6.35321	9.47134
2.5	3.23471	6.33464	9.45987	12.5929	0.877069	3.39546	6.40995	9.50885
3.0	3.27123	6.35768	9.47618	12.6054	1.02724	3.51553	6.47222	9.54988
3.5	3.30494	6.38052	9.49274	12.6183	1.14279	3.63616	6.53804	9.5932
4.0	3.33563	6.40269	9.50922	12.6312	1.23338	3.75334	6.60624	9.63822

$\frac{b}{a}$	*nl* 21	22	23	24	31	32	33	34
1.001	3.14159	6.28319	9.42479	12.5663	3.14159	6.28319	9.42479	12.5663
1.1	0.190547	3.14847	6.28662	9.42708	0.285819	3.1557	6.29023	9.4295
1.2	0.364111	3.16685	6.29579	9.43318	0.546119	3.19328	6.30904	9.44202
1.3	0.523068	3.19418	6.30939	9.44223	0.784295	3.24891	6.33689	9.46058
1.4	0.669236	3.22867	6.32648	9.45361	1.0028	3.31876	6.37183	9.48389
1.5	0.804033	3.26904	6.3464	9.46684	1.20342	3.40004	6.41252	9.51099
1.6	0.928581	3.31431	6.36864	9.4816	1.3875	3.49077	6.45786	9.54119
1.8	1.15044	3.4167	6.41873	9.51474	1.71027	3.69458	6.55984	9.60892
2.0	1.3406	3.53129	6.47471	9.55158	1.97888	3.92005	6.67381	9.68421
2.5	1.70544	3.8496	6.63358	9.65499	2.4649	4.52117	6.99964	9.89616
3.0	1.95499	4.18021	6.81335	9.77013	2.77606	5.08438	7.37439	10.1353
3.5	2.12986	4.49211	7.00958	9.89483	2.98933	5.54626	7.78188	10.4003

For given values of n and l (mode numbers) we take the value from Table 9-4 and compute $k_{c,\mathrm{TE}}$ using

$$k_{c,\mathrm{TE}_{nl}} = \frac{\text{table value}}{\left(\frac{b}{a} - 1\right)a} = \frac{\text{table value}}{b - a} \tag{9-234}$$

The propagation constant is then found using

$$\gamma_{\mathrm{TE}} = \sqrt{k_{c,\mathrm{TE}}^2 - k^2} = j\sqrt{k^2 - k_{c,\mathrm{TE}}^2} = j\sqrt{\omega^2\mu\epsilon - k_{c,\mathrm{TE}}^2} = j\beta_{\mathrm{TE}} \tag{9-235}$$

The field components are then

$$H_z(\rho, \phi, z) = C\left[J_n(k_{c,\text{TE}}\rho) - \frac{j'_n(k_{c,\text{TE}}b)}{N'_n(k_{c,\text{TE}}b)}N_n(k_{c,\text{TE}}\rho)\right]\cos n\phi e^{-j\beta_{\text{TE}}z} \tag{9-236}$$

$$E_\rho(\rho, \phi, z) = j\frac{\omega\mu nC}{\rho k_{c,\text{TE}}^2}\left[J_n(k_{c,\text{TE}}\rho) - \frac{J'_n(k_{c,\text{TE}}b)}{N'_n(k_{c,\text{TE}}b)}N_n(k_{c,\text{TE}}\rho)\right]\sin n\phi e^{-j\beta_{\text{TE}}z} \tag{9-237}$$

$$E_\phi(\rho, \phi, z) = j\frac{\omega\mu C}{k_{c,\text{TE}}}\left[J'_n(k_{c,\text{TE}}\rho) - \frac{J'_n(k_{c,\text{TE}}b)}{N'_n(k_{c,\text{TE}}b)}N'_n(k_{c,\text{TE}}\rho)\right]\cos n\phi e^{-j\beta_{\text{TE}}z} \tag{9-238}$$

$$H_\rho(\rho, \phi, z) = -j\frac{\beta_{\text{TE}}C}{k_{c,\text{TE}}}\left[J'_n(k_{c,\text{TE}}\rho) - \frac{J'_n(k_{c,\text{TE}}b)}{N'_n(k_{c,\text{TE}}b)}N'_n(k_{c,\text{TE}}\rho)\right]\cos n\phi e^{-j\beta_{\text{TE}}z} \tag{9-239}$$

$$H_\phi(\rho, \phi, z) = j\frac{\beta_{\text{TE}}C}{\rho k_{c,\text{TE}}^2}\left[J_n(k_{c,\text{TE}}\rho) - \frac{J'_n(k_{c,\text{TE}}b)}{N'_n(k_{c,\text{TE}}b)}N_n(k_{c,\text{TE}}\rho)\right]\sin n\phi e^{-j\beta_{\text{TE}}z} \tag{9-240}$$

Other system parameters are:

Cutoff frequency:

$$\omega_{c,\text{TE}} = \frac{\text{Table 9-4 mode value}}{(b - a)\sqrt{\mu\epsilon}} \tag{9-241}$$

Cutoff wavelength:

$$\lambda_{c,\text{TE}} = \frac{v_0}{f_{c,\text{TE}}} = \frac{2\pi(b - a)}{\text{Table 9-4 mode value}} \tag{9-242}$$

Guide wavelength:

$$\lambda_{g,\text{TE}} = \frac{2\pi}{\beta_{\text{TE}}} = \frac{2\pi}{\sqrt{k^2 - k_{c,\text{TE}}^2}} = \frac{\lambda_0}{\sqrt{1 - \left(\frac{\lambda_0}{\lambda_{c,\text{TE}}}\right)^2}} = \frac{\lambda_0}{\sqrt{1 - \left(\frac{\omega_{c,\text{TE}}}{\omega}\right)^2}} \tag{9-243}$$

where

$$\lambda_0 = \frac{v_0}{f} = \frac{1}{f\sqrt{\mu\epsilon}}\left(v_0 = \frac{1}{\sqrt{\mu\epsilon}}\right) \tag{9-244}$$

Propagation constant:

$$\beta_{\text{TE}} = k\sqrt{1 - \left(\frac{\lambda_0}{\lambda_{c,\text{TE}}}\right)^2} = k\sqrt{1 - \left(\frac{\omega_{c,\text{TE}}}{\omega}\right)^2} \tag{9-245}$$

Phase velocity:

$$V_p^{\text{TE}} = \frac{\omega}{\beta_{\text{TE}}} = \frac{v_0}{\sqrt{1 - \left(\frac{\lambda_0}{\lambda_{c,\text{TE}}}\right)^2}} = \frac{v_0}{\sqrt{1 - \left(\frac{\omega_{c,\text{TE}}}{\omega}\right)^2}} \tag{9-246}$$

Group velocity:

$$V_g^{\mathrm{TE}} = \frac{d\omega}{d\beta_{\mathrm{TE}}} = v_0\sqrt{1-\left(\frac{\lambda_0}{\lambda_{c,\mathrm{TE}}}\right)^2} = v_0\sqrt{1-\left(\frac{\omega_{c,\mathrm{TE}}}{\omega}\right)^2} \tag{9-247}$$

Wave or characteristic impedance:

$$Z_0^{\mathrm{TE}} = \frac{\sqrt{\dfrac{\mu}{\epsilon}}}{\sqrt{1-\left(\dfrac{\lambda_0}{\lambda_{c,\mathrm{TE}}}\right)^2}} = \frac{\eta}{\sqrt{1-\left(\dfrac{\omega_{c,\mathrm{TE}}}{\omega}\right)^2}} \tag{9-248}$$

Poynting vector:

$$\begin{aligned}\vec{S}_{\mathrm{TE}} = {}& j\,\frac{\omega\mu C^2}{2k_{c,\mathrm{TE}}}\left[J_n(k_{c,\mathrm{TE}}\rho) - \frac{J_n'(k_{c,\mathrm{TE}}b)}{N_n'(k_{c,\mathrm{TE}}b)}N_n(k_{c,\mathrm{TE}}\rho)\right] \\ & \cdot\left[J_n'(k_{c,\mathrm{TE}}\rho) - \frac{J_n'(k_{c,\mathrm{TE}}b)}{N_n'(k_{c,\mathrm{TE}}b)}N_n'(k_{c,\mathrm{TE}}\rho)\right]\cos^2 n\phi\,\hat{\rho} \\ & + j\,\frac{\omega\mu n C^2}{2\rho k_{c,\mathrm{TE}}^2}\left[J_n(k_{c,\mathrm{TE}}\rho) - \frac{J_n'(k_{c,\mathrm{TE}}b)}{N_n'(k_{c,\mathrm{TE}}b)}N_n(k_{c,\mathrm{TE}}\rho)\right]^2\cos n\phi\,\sin n\phi\,\hat{\phi} \\ & + \frac{\omega\mu\beta_{\mathrm{TE}}C^2}{2k_{c,\mathrm{TE}}^2}\left\{\frac{1}{\rho^2k_{c,\mathrm{TE}}^2}\left[J_n(k_{c,\mathrm{TE}}\rho) - \frac{J_n'(k_{c,\mathrm{TE}}b)}{N_n'(k_{c,\mathrm{TE}}b)}N_n(k_{c,\mathrm{TE}}\rho)\right]^2\sin^2 n\phi\right. \\ & \left. + \left[J_n'(k_{c,\mathrm{TE}}\rho) - \frac{J_n'(k_{c,\mathrm{TE}}b)}{N_n'(k_{c,\mathrm{TE}}b)}N_n'(k_{c,\mathrm{TE}}\rho)\right]\cos^2 n\phi\right\}\hat{z}\end{aligned} \tag{9-249}$$

The field configurations for three low-order TE_{nl} modes are given in Figure 9-30. Views 1 and 2 show electric and magnetic fields within the guide. View 2 is a longitudinal view through plane l–l'. Views 3 show the magnetic fields and surface currents on the surfaces viewed from inside the guide from plane s–s'. For these plots $b/a = 3$ and $\lambda_g/b = 4.24$.

Conspicuously absent in the figure is the lowest-order TE_{01} mode. Computations of the cutoff frequencies of several of the lowest-order modes show that the TE_{01} mode has a higher cutoff frequency than those of higher order modes. See Exercise 9-6.4b.

Exercises

9-6.4a Derive all the TE mode quantities given by the equations of this section. Pay particular attention to obtaining the Bessel-Neumann derivative combination boundary condition.

9-6.4b Determine the TE_{nl} modes that have cutoff frequencies less than the TE_{01} mode. Of these modes which one has the lowest cutoff frequency? This is the dominant mode for this configuration.

9-6.4c Repeat Example 9-6 of the preceding section for the TE_{11} mode. Use a frequency of 15 GHz. Now repeat Example 9-7. Note that C now has units of A/m since it appears in H_z rather than E_z.

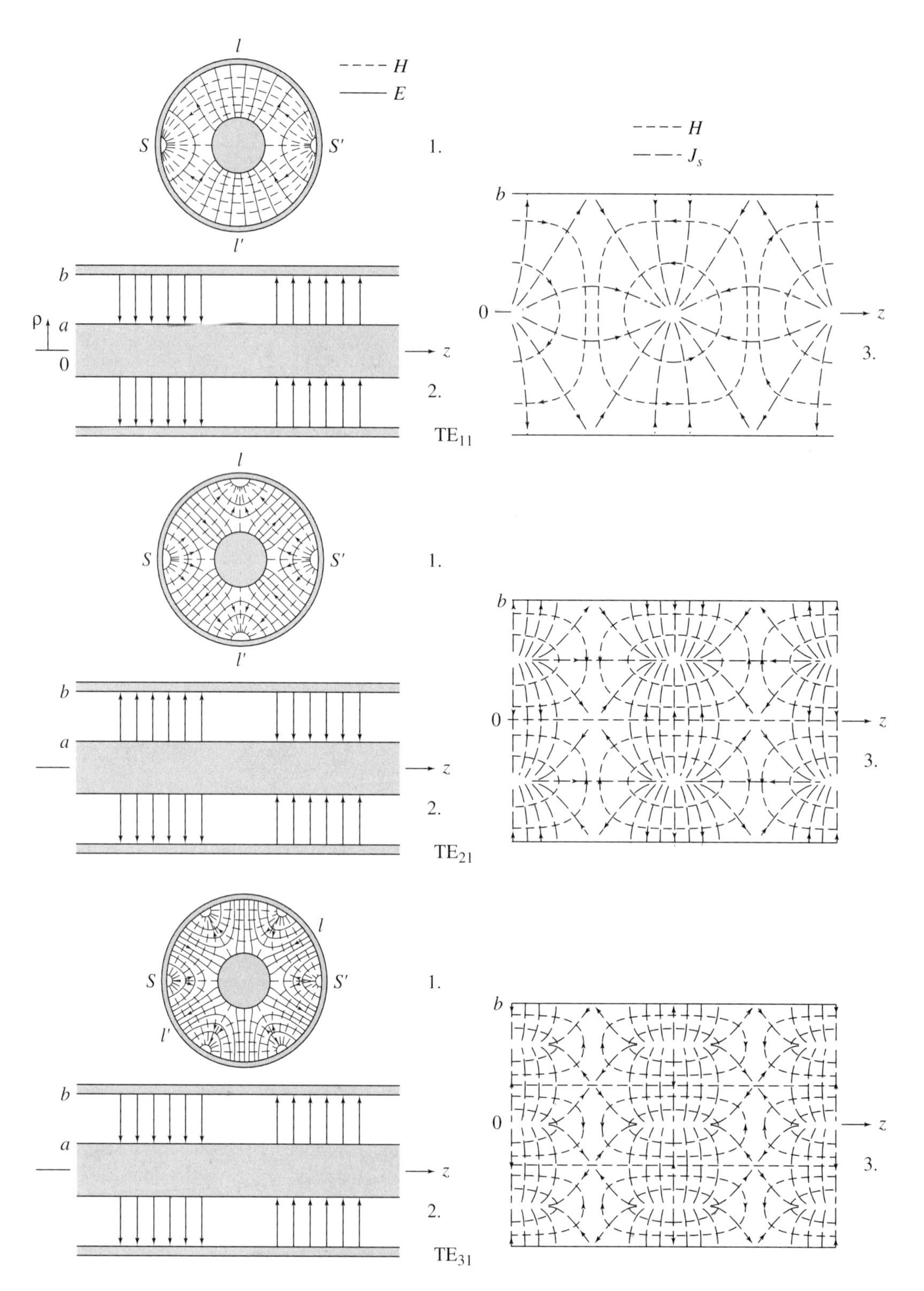

Figure 9-30. Field configurations of TE modes in a coaxial waveguide ($t = 0$).

9-6.4d Using the time domain equations determine the z location in View 2 (TE_{11} mode) that corresponds to View 1 at $t = 0$. See Figure 9-30.

9-6.4e Determine the equation for the surface current at $\rho = b$ for the TE_{11} mode. Using the time domain equation verify view 3 in Figure 9-30. For $t = 0$ where is $z = 0$ in this view?

9-6.4f Design a coaxial waveguide that has a cutoff frequency of 2.8 GHz and for which other low-order modes will not propagate below 5 GHz. Can the TEM mode propagate?

9-6.4g Design loop and probe coupling methods for exciting the TE_{11} mode.

9-7 Aperture (Slot) Coupling

The techniques for loop and probe coupling discussed in the preceding section assumed that waveguide excitation was to be accomplished using a coaxial feed line. There are important cases where it is desirable to couple directly between two waveguide structures. For example, if one wanted to sample the field in a main guide using another waveguide section, it would be advantageous to couple the guides together using a common ''hole,'' or aperture, between them. A waveguide directional coupler is an example of this.

A microwave frequency meter provides another example of waveguide-to-waveguide coupling. The frequency meter is often a section of circular waveguide formed into a resonant cavity (see Section 9-11). A sample of the main waveguide energy is coupled by an aperture into the resonant cavity. When the cavity dimensions are such that resonance occurs energy is absorbed by the cavity. This small energy decrease can be observed at the main guide output.

The aperture coupling concepts will be qualitative only. Quantitative details are to be found in [14].

There are two types of aperture coupling: *electric* and *magnetic*. The type is determined by the orientation of the aperture with respect to the fields present. Of course since the fields are varying with time we always have both $\vec{E}$ and $\vec{H}$, but one component will be dominant in the description of the coupling process. Whichever description is used, the key idea is that the physical orientation of the devices to be coupled must be such that the modes in each device have the coupled field component in common.

Since an aperture extends across a finite extent of the wall any field component in the aperture will have a field component present for several modes. However, if the operating frequency is below cutoff for all modes except one, only one mode will propagate. Thus the operating frequency is also an important consideration.

Electric field coupling has the easiest explanation when the electric field is either normal to the surface of the aperture or parallel to the surface of the aperture. When the slot is cut normal to the electric field lines, there are no surface charges to provide the termination of the line. The electric field lines then pass through the slot and induce terminating charges on external metal surfaces. This is the situation shown in Figure 9-31. In Figure 9-31*b* the field lines move through and then propagate away from the slot. See Exercise 9-7a for the electric field parallel to the aperture.

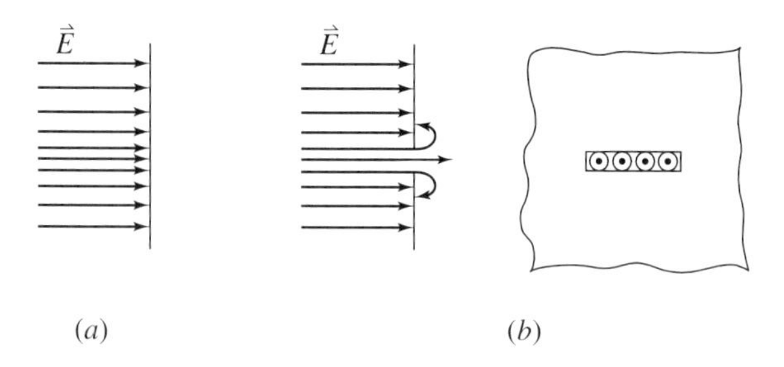

Figure 9-31. Coupling of electric field lines through an aperture. $\vec{E}$ Normal to plane of aperture.

Example 9-8

Develop a method for coupling a TE_{10} mode in one rectangular waveguide to a TE_{10} mode in a second identical waveguide. This device is useful when we want to sample the field in the first guide.

We cut the apertures in the guides so that E_y is normal to the aperture planes. Also, the slots are cut so that they have minimal interference with the surface current $\vec{J}_s$, requiring that the slots run along the z direction. The final configuration is shown in Figure 9-32 slot number 1. Note that there are several ways that the slots could be cut as indicated in the figure. However, Case 1 shown provides minimum current interference.

It turns out that the distance d is important. It should be about $\lambda_g/4$ for optimum coupling since the part of the wave going to the left hits the short and returns back to the aperture to reinforce the coupled field.

Magnetic field coupling is the dominant effect when the aperture is oriented so that its longest sides are normal to the surface current or, equivalently, when they are parallel to the surface magnetic field lines. This situation is illustrated physically in Figure 9-33.

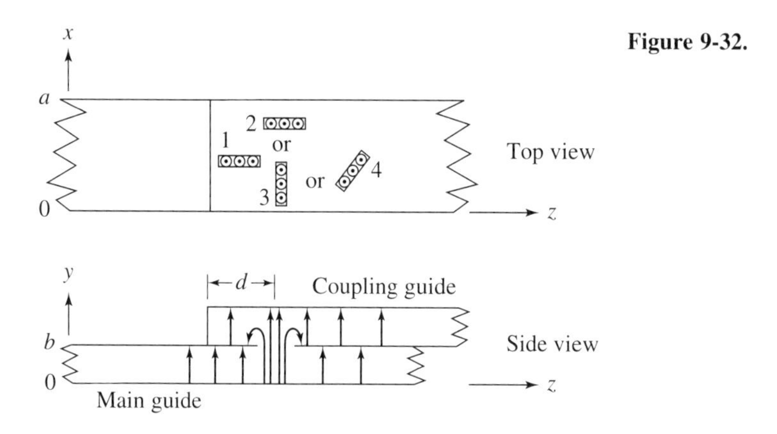

Figure 9-32.

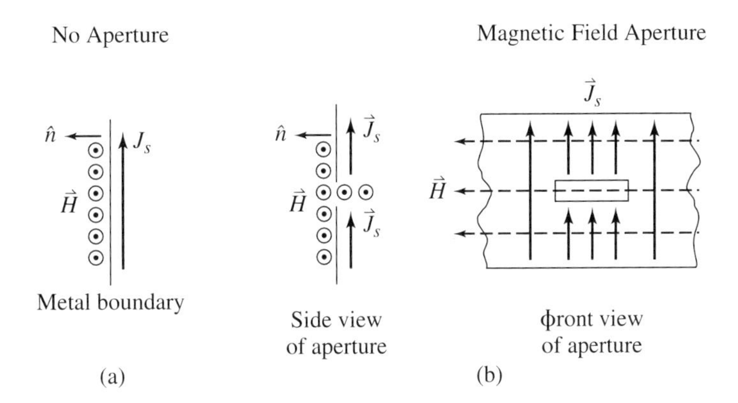

Figure 9-33. Magnetic coupling to aperture. (*a*) Metal surface boundary conditions. (*b*) Aperture orientation for maximum magnetic field coupling.

Example 9-9

Develop a method for coupling the TE_{10} mode in a rectantangular waveguide to the TM_{01} mode in a cylindrical waveguide. Use magnetic field coupling.

The design procedure will be explained using Figure 9-34. Figure 9-34*a* shows the field pattern for the TM_{01} mode in a cylindrical guide. The longitudinal (z axis) view is from the *outside* of the guide and shows the magnetic field and surface current density on the inner wall surface. The surface normal is into the page ($-\rho$ direction). To produce magnetic coupling the aperture must be oriented as shown. Note that one could also use the electric field concept here since the electric field is normal to the plane of the aperture. However, the connection to the rectangular guide will be at an electric field minimum so the process will be predominantly magnetic.

The rectangular TE_{10} mode field pattern is shown in Figure 9-34*b*. On the vertical side wall narrow dimension the electric field would be zero (E_y is tangential $\vec{E}$ there). If we cut the aperture as shown in the longitudinal view we produce magnetic coupling. Now if we line up the two apertures as shown in Figure 9-34*c* the magnetic field components would couple. Note that the electric field lines in the cylindrical guide have the incorrect orientation to produce a y component of electric field in the rectangular guide.

Exercises

9-7.0a Draw a figure like that of Figure 9-30 where $\vec{E}$ is parallel to the aperture. Explain the dominant field direction on the other side of the aperture. Explain why an aperture size on the order of $\lambda_g/8$ is desired for good coupling. Consider $E_{\tan} = 0$.

9-7.0b Develop a method for coupling the rectangular TE_{10} mode into a cylindrical TM_{01} mode using electric field coupling. Give two methods. Note you can couple E_y in two ways.

9-7.0c If two rectangular waveguides are joined along the narrow ($y = b$) sides, what modes could be coupled with an electric coupling aperture? Draw diagrams.

9-7.0d The two systems shown in Figure 9-35 are intended to couple from rectangular TE_{10} to cylindrical TM_{01}. Comment on the relative electric field amplitudes.

9-7.0e A coaxial waveguide which supports the TEM mode is attached to the end of a rectangular guide as shown in Figure 9-36. Show how a slot would be cut in the coax to couple the TM_{11} mode into the rectangular guide. Consider both electric and magnetic couplings.

9-7.0f For Example 9-9 devise another aperture orientation which would produce proper coupling between the modes indicated.

Figure 9-34. Rectangular TE_{10} coupling to cylindrical TM_{01}.

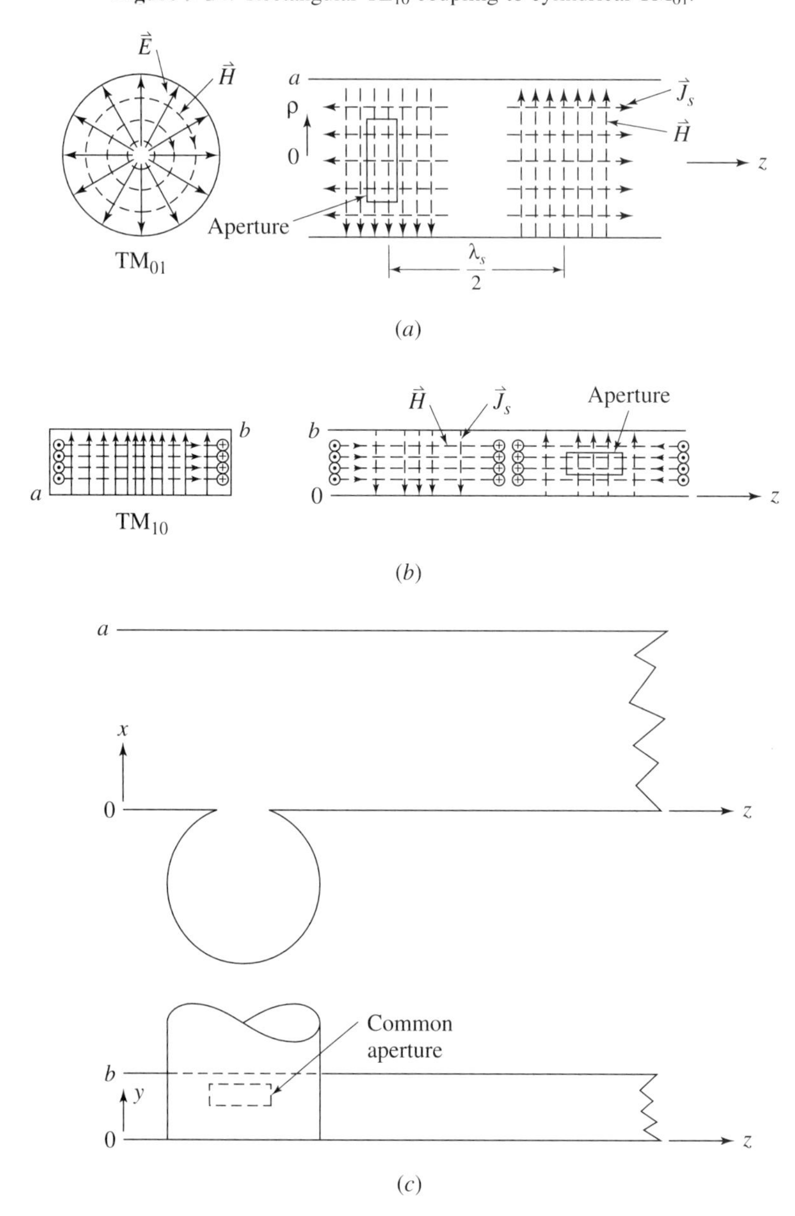

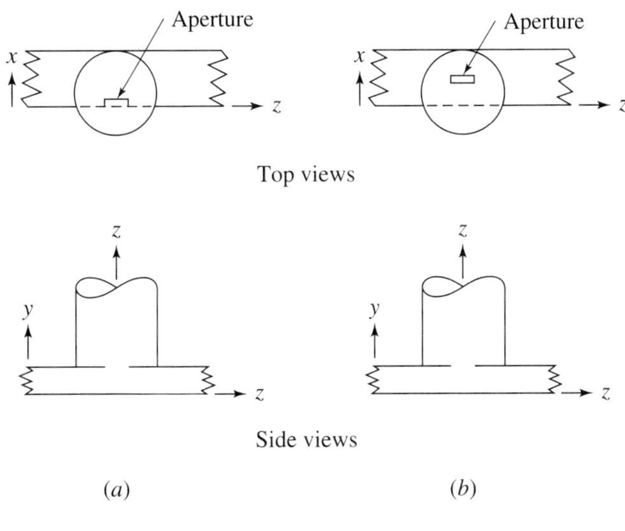

Figure 9-35.

9-7.0g Show how to couple two rectangular waveguides, both supporting the TE_{10} mode, using a magnetic coupling aperture.

9-7.0h In Figure 9-31 explain why d is desired to be $\lambda_g/4$ rather than $\lambda_g/2$ for field reinforcement by the reflected field.

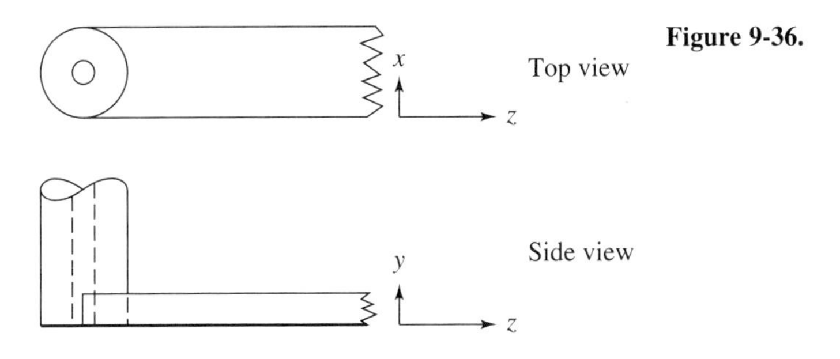

Figure 9-36.

9-8 Attenuation in Waveguides

All the derivations of mode structures in waveguides neglected losses in the system. The dielectrics were assumed lossless (loss tangent equal to zero) and the guide walls were assumed to have infinite conductivity. In actual practice an attenuation must be accounted for.

To include those losses one would theoretically have to resolve all of the waveguide configurations. For finite-wall conductivity there would be $\vec{E}$, $\vec{H}$, and $\vec{J}$ within the walls, which would be described by Maxwell's equations. This would require writing the wave equation solutions in both the conductor and the dielectric (using $\bar{\epsilon} = \epsilon' - j\epsilon'$) and then matching the two solutions at the boundary. When one considers the amount of detail involved with one solution only in the lossless case—where fields exist only within the guide—it is evident that the work required to match two-field solutions is significantly larger.

Fortunately we can avoid having to carry out a complete solution because attenuation is usually very small. For example, the dielectric loss tangent is on the order of 10^{-4}, which is very small compared with 1, and wall conductivities are very high, being on the order of 10^7 S/m. What we will discover is that for small losses the value of β is essentially unchanged as it was for transmission lines. This means that we may approximate β by its lossless waveguide value so that the propagation constant can be written as

$$\gamma = \alpha + j\beta \approx \alpha + j\beta_{\text{lossless}} \tag{9-250}$$

This tells us that for practical systems we can evaluate α independent of β. Consequently we can express the z variation of the fields as

$$e^{\gamma z} = e^{-j(\alpha + j\beta)\mathbf{z}} = e^{-\alpha z}e^{-j\beta z} \approx e^{-\alpha z}e^{-j\beta_{\text{lossless}} z} \tag{9-251}$$

Thus all we need to do is obtain an expression for α and then multiply all the lossless solutions of the preceding sections by $e^{-\alpha z}$. This is obviously much easier than resolving the whole problem.

One may wonder why even bother with the attenuation at all if it is so small. In some microwave systems the signals are very small, such as radar return signals, remote sensing of earth microwave signals, or signals in radio astronomy. Since the waveguide has losses it absorbs energy, and for a system to be in thermal equilibrium it must also radiate energy into the guide. The energy radiated into the guide will contain the spectrum of frequencies that can propagate in the guide. These spectral components corrupt and mask the desired signal as noise. Such small signal noise components must be accounted for in an overall system design.

For small losses we compute the dielectric and wall losses separately and then add the two results to obtain the total attenuation as

$$\alpha = \alpha_{\text{dielectric}} + \alpha_{\text{wall}} \equiv \alpha_d + \alpha_w \tag{9-252}$$

We next develop expressions for these attenuation constants.

9-8.1 Dielectric Attenuation

Dielectric losses are usually accounted for using the complex dielectric constant and loss tangent for conductive losses. By way of review these are obtained as follows, where we use the subscript d for dielectric:

$$\begin{aligned} \nabla \times \vec{H} &= \vec{J}_d + j\omega\epsilon\vec{E} = \sigma_d\vec{E} + j\omega\epsilon\vec{E} = (j\omega\epsilon + \sigma_d)\vec{E} \\ &= j\omega\left(\epsilon - j\frac{\sigma_d}{\omega}\right)\vec{E} \equiv j\omega(\epsilon' - j\epsilon'')\vec{E} \equiv j\omega\bar{\epsilon}\vec{E} \end{aligned} \tag{9-253}$$

$$\text{loss tangent} = \tan\delta \equiv \frac{\epsilon''}{\epsilon'} = \frac{\sigma_d}{\omega\epsilon'} \tag{9-254}$$

For TEM modes or uniform plane waves we begin with the propagation constant, Eq. 9-14, and write, with $\epsilon' = \epsilon$,

$$\gamma_{\text{TEM}} = j\sqrt{\omega^2\mu\epsilon' - j\omega\mu\sigma_d} = j\sqrt{\omega^2\mu\epsilon' - j\omega^2\mu\epsilon''} = j\omega\sqrt{\mu\epsilon' - j\mu\epsilon''} = \alpha_d + j\beta$$

Squaring both sides of the last equality,

$$-\omega^2\mu\epsilon' + j\omega^2\mu\epsilon'' = \alpha_d^2 - \beta^2 + j2\alpha_d\beta$$

Equating real and imaginary components,

$$\alpha_d^2 - \beta^2 = -\omega^2\mu\epsilon' \tag{9-255}$$

$$\omega^2\mu\epsilon'' = 2\alpha_d\beta \tag{9-256}$$

Solving the last equation for β,

$$\beta = \frac{\omega^2\mu\epsilon''}{2\alpha_d} \tag{9-257}$$

Substituting this result into Eq. 9-255 and rearranging,

$$\alpha_d^4 + \omega^2\mu\epsilon'\alpha_d^2 - \frac{\omega^4\mu^2\epsilon''^2}{4} = 0$$

Solving this as a quadratic in α_d^2,

$$\alpha_d^2 = \frac{-\omega^2\mu\epsilon' \pm \sqrt{(\omega^2\mu\epsilon')^2 + \omega^4\mu^2\epsilon''^2}}{2}$$

Since α_d must be a positive real number we use the plus sign on the radical. Then taking the square root,

$$\alpha_d = \sqrt{\frac{-\omega^2\mu\epsilon' + \sqrt{(\omega^2\mu\epsilon')^2 + \omega^4\mu^2\epsilon''^2}}{2}}$$

$$\alpha_d = \omega\sqrt{\mu\epsilon'}\sqrt{\frac{-1 + \sqrt{1 + \left(\frac{\epsilon''}{\epsilon'}\right)^2}}{2}} \text{ nep/m}$$

$$\alpha_d = \omega\sqrt{\frac{-1 + \sqrt{1 + \tan^2\delta}}{2}} \text{ nep/m} \qquad \text{(TEM)} \qquad \delta = \text{loss angle} \tag{9-258}$$

Substituting this result into Eq. 9-255 we obtain β as

$$\beta = \omega\sqrt{\mu\epsilon'}\sqrt{\frac{1 + \sqrt{1 + \tan^2\delta}}{2}} \text{ rad/m} \qquad \text{(TEM)} \tag{9-259}$$

For practical cases we may use the usual ϵ in place of ϵ'. To illustrate the consequence of small losses on β consider the usual case (δ is the loss angle not skin depth):

$$\tan\,\delta = \frac{\epsilon''}{\epsilon'} \sim 10^{-4} \ll 1$$

We can then use the first two terms in the series expansion:

$$(1 + x)^n \approx 1 + nx, \qquad x \ll 1 \tag{9-260}$$

Applying this to the inner radical in Eq. 9-259 for $n = \frac{1}{2}$,

$$\beta \approx \omega\sqrt{\mu\epsilon'}\sqrt{\frac{1 + 1 + \frac{1}{2}\left(\frac{\epsilon''}{\epsilon'}\right)^2}{2}} = \omega\sqrt{\mu\epsilon'}\sqrt{1 + \frac{1}{4}\left(\frac{\epsilon''}{\epsilon'}\right)^2}$$

Now for a loss tangent of 10^{-4}, the second term in the radical is 2.5×10^{-9}, clearly negligible with respect to 1. Therefore

$$\beta_{\text{TEM}} \approx \omega\sqrt{\mu\epsilon'} = \omega\sqrt{\mu\epsilon} \tag{9-261}$$

This is the same as the lossless β given in Eq. 9-23, for example, and justifies our using the lossless β as previously described.

We proceed in a similar way for TE and TM modes. We begin with Eq. 9-4 and then identify α_d and β:

$$\gamma = \sqrt{k_c^2 - \bar{k}^2} = \sqrt{k_c^2 - \omega^2\mu\epsilon' + j\omega\mu\sigma_d} = \sqrt{k_c^2 - \omega^2\mu\epsilon' + j\omega^2\mu\epsilon''} = \alpha_d + j\beta$$

Squaring both sides of the last equality:

$$k_c^2 - \omega^2\mu\epsilon' + j\omega^2\mu\epsilon'' = \alpha_d^2 - \beta^2 + j2\alpha_d\beta$$

Equating real and imaginary parts and solving the resulting two equations as before we obtain

$$\alpha_d = \omega\sqrt{\mu\epsilon'}\sqrt{\frac{\left(\frac{k_c^2}{\omega^2\mu\epsilon'} - 1\right) + \sqrt{\left(1 - \frac{k_c^2}{\omega^2\mu\epsilon'}\right)^2 + \left(\frac{\epsilon''}{\epsilon'}\right)^2}}{2}} \ \text{nep/m} \qquad \omega > \omega_c \quad \text{for TE or TM} \tag{9-262}$$

$$\beta = \omega\sqrt{\mu\epsilon'}\sqrt{\frac{\left(1 - \frac{k_c^2}{\omega^2\mu\epsilon'}\right) + \sqrt{\left(1 - \frac{k_c^2}{\omega^2\mu\epsilon'}\right)^2 + \left(\frac{\epsilon''}{\epsilon'}\right)^2}}{2}} \ \text{rad/m} \qquad \omega > \omega_c \quad \text{for TE or TM} \tag{9-263}$$

9-8.2 Waveguide Wall Attenuation

In this section we first derive the general integral expression for the wall attenuation, α_w, and then apply it to specific cases.

Using the Poynting vector we can compute the power flow per unit area in any guide system as follows:

$$\vec{S} = \tfrac{1}{2}\vec{E} \times \vec{H}^* = \tfrac{1}{2}(\vec{E}_t + \vec{E}_z) \times (\vec{H}_t^* + \vec{H}_z^*)$$
$$= \tfrac{1}{2}\vec{E}_t \times \vec{H}_t^* + \tfrac{1}{2}\vec{E}_t \times \vec{H}_z^* + \tfrac{1}{2}\vec{E}_z \times \vec{H}_t^* \text{ W/m}^2$$

Note that although $\vec{E}_t$ and $\vec{H}_t$ are the total transverse fields they will be orthogonal in the transverse plane, making the angle between them 90°. The first term will then be in the z direction whereas the last two terms will be in the transverse direction. Therefore,

$$S_z = \vec{S} \cdot \hat{z} = \tfrac{1}{2}|\vec{E}_t \times \vec{H}_t^*| = \tfrac{1}{2}|\vec{E}_t||\vec{H}_t^*| \sin\theta_{E_tH_t} = \tfrac{1}{2}|\vec{E}_t||\vec{H}_t^*| = \tfrac{1}{2}|\vec{E}_t||\vec{H}_t| \quad (9\text{-}264)$$

As explained, we assume the lossless solution and multiply by the attenuation factor. Thus for $\vec{E}_t$:

$$\vec{E}_t = E_{t_1}e^{-j\beta z}e^{-\alpha_w z}\hat{t}_1 + E_{t_2}e^{-j\beta z}e^{-\alpha_w z}\hat{t}_2$$

Therefore,

$$|\vec{E}_t| = \sqrt{\vec{E}_t \cdot \vec{E}_t^*} = \sqrt{E_{t_1}^2 e^{-2\alpha_w z} + E_{t_2}^2 e^{-2\alpha_w z}} = E_t e^{-\alpha_w z} \quad (9\text{-}265)$$

Similarly, using Z^w for the wave impedance and the concept of General Exercise GE9-22 we may express $|\vec{H}_t|$ as

$$|\vec{H}_t| = \sqrt{H_{t_1}^2 + H_{t_2}^2}e^{-\alpha_w z} = \sqrt{\frac{E_{t_2}^2}{|Z^w|^2} + \frac{E_{t_1}^2}{|Z^w|^2}}\, e^{-\alpha_w z} = \frac{E_t}{|Z^w|}\, e^{-\alpha_\omega z} \quad (9\text{-}266)$$

Substituting Eqs. 9-265 and 9-266 into Eq. 9-264 there results

$$S_z = \frac{E_t^2}{2|Z^w|}\, e^{-2\alpha_w z} \text{ W/m}^2 \quad (9\text{-}267)$$

The total power propagating in the z direction is then

$$P_z = \int_{\text{cross section}} S_z da_t = \int_{\text{cross section}} \frac{E_t^2}{2|Z^w|}\, e^{-2\alpha_w z} da_t \text{ W}$$

Since the cross section differential area does not involve the coordinate $z (ds_t = h_1 h_2 dt_1 dt_2)$ the exponent may be removed from the integrand, resulting in

$$P_z = e^{-2\alpha_w z}\int_{\text{cross section}} \frac{E_t^2}{2|Z^w|}\, da_t \text{ W}$$

The power loss (decrease) *per meter of waveguide* can be determined as

$$P_L = -\frac{dP_z}{dz} = -(2\alpha_w)e^{-2\alpha_w z}\int_{\text{cross section}} \frac{E_t^2}{2|Z^w|}\, da_t = 2\alpha_w P_z \text{ W/m} \quad (9\text{-}268)$$

Solving for α_w:

$$\alpha_w = \frac{P_L}{2P_z} \tag{9-269}$$

As before, under the assumption of small losses we compute P_z using the lossless solution. Thus we may compute P_z using the z components of the Poynting vectors obtained for the lossless waveguide modes in previous sections. We are left now with the actual computation of the power loss per meter P_L.

The power loss per meter P_L is evaluated by taking the lossless solution wall surface current J_s and then assuming that this is an equivalent current flowing in a lossy conductor to produce an I^2R-type loss. We compute the power loss for 1 m to obtain P_L. The details of this computation follow, and one should note carefully the approximations invoked.

The power absorbed by the wall can be represented by the Poynting vector into the wall, which for waveguides will involve the appropriate transverse components of $\vec{S}$. Remember that the wall boundary is parallel to the z direction. This concept is shown in Figure 9-37. Note that since the conductor is *not* perfect we can have both tangential $\vec{H}$ and tangential $\vec{E}$, as well as a finite current density $\vec{J}_w$ within the wall. The wall thicknesses have been exaggerated to show all the vector quantities. These would be duplicated on the lower wall. The actual penetration depth is only microns, as was discussed in Section 8-4. Thus this same figure could represent a cylindrical guide as well since the radius of curvature of the wall boundary is much larger than the penetration depth.

Figure 9-37. Field quantities in waveguide walls for imperfect conducting walls.

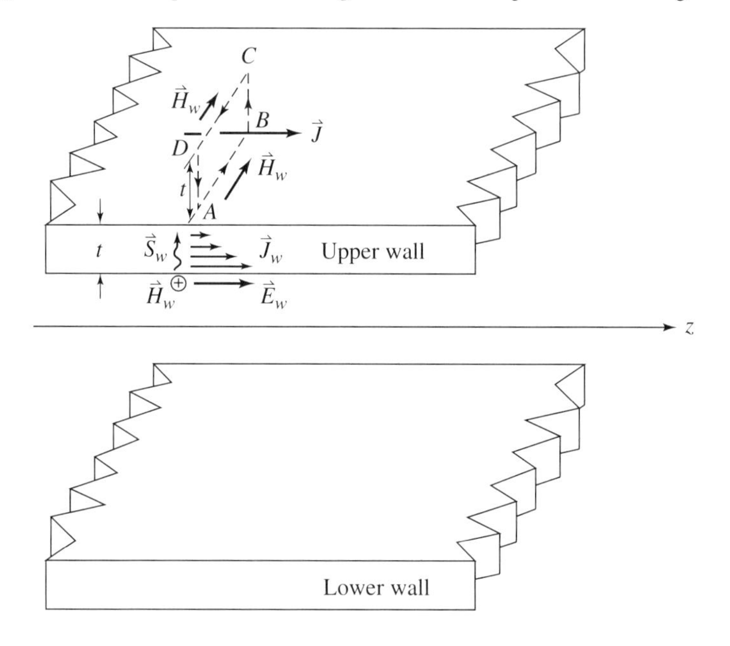

The power loss in the wall would be that due to the real part of $\vec{S}_w$. Since $\vec{E}$ and $\vec{H}$ are orthogonal,

$$\vec{P}_{\text{wall}} = \text{Re}\{\vec{S}_w\} = \text{Re}\{\tfrac{1}{2}\vec{E}_w \times \vec{H}_w^*\} = \text{Re}\{\tfrac{1}{2}E_w H_w^* \hat{t}\} \tag{9-270}$$

Although we have used surface values, $\vec{E}$ and $\vec{H}$ also exist inside the conductor due to the required continuity of tangential fields ($\vec{J}_s = 0$ now). We can then apply Ampere's law inside the wall around contour $ABCD$, with corners located on the guide surfaces so that the entire thickness t is included:

$$\oint \vec{H}_w \cdot d\vec{l} = \int_A^B \vec{H}_w \cdot d\vec{l} + \int_B^C \vec{H}_w \cdot d\vec{l} + \int_C^D \vec{H}_w \cdot d\vec{l} + \int_D^A \vec{H} \cdot d\vec{l} = I_{\text{enclosed}}$$

The integrals from B to C and D to A are zero since $\vec{H}$ is normal to the paths there. Since the fields decay very rapidly the value of $\vec{H}_w$ along path CD is essentially zero also. We are then left with only one integral. Ampere's law reduces to

$$\int_A^B H_w dl_{AB} = I_{\text{enclosed}} \qquad (\vec{H}_w \text{ parallel to } d\vec{l}_{AB})$$

If we take the path AB to be small, H_w will not vary significantly, so it may be removed from the integral. We then have

$$H_w(B - A) = I_{\text{enclosed}}$$

Now since the conductivity is very high the conduction current dominates over the displacement current ($\sigma_w \vec{E}_w \gg j\omega\epsilon \vec{E}_w$). Thus we can approximate the enclosed current by the current crossing path AB on the inner surface, which is

$$I_{\text{enclosed}} \approx (B - A)J_s$$

Ampere's law then yields

$$H_w(B - A) \approx (B - A)J_s$$

or

$$H_w \approx J_s \tag{9-271}$$

This is, of course, what one would expect on a perfect conductor, but here we have shown it to be essentially true for good conductors.

To complete the computation we need an expression for E_w. From Section 8-4 we have the following relationship between $\vec{E}$ and $\vec{H}$ in a good conductor:

$$E_w = Z_s H_w$$

where

$$Z_s = \frac{1 + j}{\delta\sigma_w} = \frac{1}{\delta\sigma_w} + j\frac{1}{\delta\sigma_w} = R_s + jR_s \qquad \text{with} \quad \delta = \frac{1}{\sqrt{\pi f \mu \sigma_w}}$$

The wall electric field can then be expressed as

$$E_w = H_w(R_s + jR_s) \approx J_s(R_s + jR_s) \tag{9-272}$$

Substituting Eqs. 9-271 and 9-272 into Eq. 9-270 we obtain

$$P_w = |\vec{P}_w| = |\text{Re}\{\tfrac{1}{2}J_s(R_s + jR_s)J_s^*\hat{t}\}| = \tfrac{1}{2}|\vec{J}_s|^2 R_s \text{ W/m}^2 \qquad (9\text{-}273)$$

(Note: we have used equal signs even though approximations were used.)

The power loss per meter is then obtained by integrating this Poynting vector component over the transverse contour of the guide (i.e., around the interior guide surface) and for 1 m in the z direction. This gives an area of the guide wall into which the Poynting vector flows that is of unit length:

$$P_L = \int_{C_t}\int_0^1 P_w dl_t dz = \int_{C_t}\int_0^1 \frac{1}{2}|\vec{J}_s|^2 R_s dl_t dz = \int_{C_t}\int_0^1 \frac{1}{2}|\vec{H}_{\tan}|^2 R_s dl_t dz \qquad (9\text{-}274)$$

It is important to note that while the contour C_t is a closed path around the guide walls it is not the usual contour integral. All wall loss contributions will be positive, so if the integral is written as the sum of the integrals around the contour each section must make a positive contribution. Reversing the current direction does not negate loss.

The transmitted power P_z is obtained using Eq. 9-264 where we end up with the z component of $\vec{S}$ integrated over the transverse cross section. Formally we write this as

$$P_z = \int_{\text{cross section}} S_z da_t = \int_{\text{cross section}} \text{Re}\{\tfrac{1}{2}\vec{E} \times \vec{H}^*\}_z da_t \qquad (9\text{-}275)$$

where $\vec{E}$ and $\vec{H}$ are the lossless solutions within the guide. Substituting these last two results into Eq. 9-269 we obtain, canceling $\frac{1}{2}$ factors,

$$\alpha_w = \frac{\int_{C_t}\int_0^1 |\vec{J}_s|^2 R_s dl_t dz}{2\int_{\text{cross section}} \text{Re}\{\vec{E} \times \vec{H}^*\}_z da_t} = \frac{\int_{C_t}\int_0^1 R_s |\vec{H}_{\tan}|^2 dl_t dz}{2\int_{\text{cross section}} \text{Re}\{2S_z\} da_t} \text{ dB/m} \qquad (9\text{-}276)$$

any mode, lossless field solutions

Example 9-10

A rectangular waveguide measures 5 cm by 2 cm and is filled with Teflon. The waveguide is constructed of aluminum and is being operated at 3 GHz. For the TE_{10} mode determine the total attenuation in dB/m, the phase constant β, and the lossless phase constant β_{lossless}.

For Teflon the loss tangent at 3 GHz is 0.00015 and the relative permittivity is 2.1. Aluminum conductivity is 3.82 at 10^8 S/m. For the TE_{10} mode ($m = 1$, $n = 0$) Eq. 9-128 yields

$$k_{c,\text{TE}_{10}} = \frac{\pi}{0.05} = 62.832$$

Using the constants given:

$$\omega^2\mu\epsilon' = (2\pi \times 3 \times 10^9)^2 \times 4\pi \times 10^{-7} \times 2.1 \times \frac{1}{36\pi} \times 10^{-9} = 8290$$

$$\frac{\epsilon''}{\epsilon'} = \tan\delta = \text{loss tangent} = 0.0015$$

Using the preceding values in Eqs. 9-262 and 9-263 gives the dielectric attenuation and phase (propagation) constant:

$$\alpha_d = 91.05\sqrt{\frac{\left(\frac{3948}{8290} - 1\right) + \sqrt{\left(1 - \frac{3948}{8290}\right)^2 + 2.25 \times 10^{-8}}}{2}}$$

$$= 9.44 \times 10^{-3} \text{ nep/m}$$

$$\beta = 91.05\sqrt{\frac{\left(1 - \frac{3948}{8290}\right) + \sqrt{\left(1 - \frac{3948}{8290}\right)^2 + 2.25 \times 10^{-8}}}{2}}$$

$$= 65.8943 \text{ rad/m}$$

The lossless β is obtained using Eq. 9-129:

$$\beta_{\text{lossless}} = \sqrt{8290 - 62.832^2} = 65.3949 \text{ rad/m}$$

These two values of β support the statement made earlier that practical losses have little effect on β. The error is

$$\Delta\beta = \beta_{\text{lossless}} - \beta = 6 \times 10^{-4} \text{ rad/m} \qquad \text{or} \quad 0.03 \text{ degrees phase shift/meter}$$

For the wall attenuation we begin by developing values of constants and field quantities required to use Eq. 9-276:

$$R_s = \frac{1}{\delta\sigma_\omega} = \frac{\sqrt{\pi f \mu \sigma_w}}{\sigma_w} = \sqrt{\frac{\pi f \mu}{\sigma_w}} = \sqrt{\frac{\pi \times 3 \times 10^9 \times 4\pi \times 10^{-7}}{3.82 \times 10^8}}$$

$$= 5.568 \times 10^{-3} \ \Omega/\text{square}$$

For the TE_{10} mode Eqs. 9-130 to 9-134 and Eq. 9-147 reduce to

$$H_z(x, y, z) = A \cos\left(\frac{\pi}{0.05}x\right)e^{-j65.8949z} = A\cos(62.832x)e^{-j65.8949z}$$

$$E_y(x, y, z) = -j\frac{2\pi \times 3 \times 10^9 \times 4\pi \times 10^{-7} \times \pi A}{0.05 \times \left(\frac{\pi}{0.05}\right)^2}\sin\left(\frac{\pi}{0.05}x\right)e^{-j65.8949z}$$

$$= -j377A\sin(62.832x)e^{-j65.895z}$$

$$H_x(x, y, z) = j\frac{\sqrt{8290 - \left(\frac{\pi}{0.05}\right)^2}\,\pi A}{0.05 \times \left(\frac{\pi}{0.05}\right)^2}\sin\left(\frac{\pi}{0.05}x\right)e^{-j65.895z}$$

$$= j1.049A\sin(62.832x)e^{-j65.895z}$$

$$(\vec{E} \times \vec{H}^*)_z = 2S_z = \frac{2\pi \times 3 \times 10^9 \times 4\pi \times 10^{-7}\pi^2\sqrt{8290 - 62.832^2}}{62.832^4 \times (0.05)^2}\sin^2\left(\frac{\pi}{0.05}x\right)$$

$$2S_z = 395.36A^2\sin^2(62.832x)$$

The wall surface currents (squared) are

$$|\vec{J}_s(x = 0, y)|^2 = |H_{\tan}(x = 0, y)|^2 = A^2$$

$$|\vec{J}_s(x, y = b)|^2 = |H_{\tan}(x, y = b)|^2 = |H_z(x, y = b)|^2 + |H_x(x, y = b)|^2$$
$$= A^2 \cos^2(62.832x) + 1.1A^2 \sin^2(62.832x)$$

$$|\vec{J}_s(x = a, y)|^2 = A^2$$

$$|\vec{J}_s(x, y = 0)|^2 = A^2 \cos^2(62.832x) + 1.1A^2 \sin^2(62.832x)$$

Since the top and bottom walls have the same expression and the left and right walls have the same expression we can simply double the integration around half the guide. If this observation is not used careful attention must be paid to the comment following Eq. 9-274.

The surface currents are not functions of z (i.e., $|\vec{H}|^2 = \vec{H} \cdot \vec{H}^*$) so the z integration in the numerator is 1 (1 m). We then have for the numerator of Eq. 9-276:

$$\int_{C_t} \int_0^1 |\vec{J}_s|^2 R_s dl_t dz = 2R_s \left\{ \int_0^{0.02} A^2 dy + \int_0^{0.05} [A^2 \cos^2(62.832x) + 1.1A^2 \sin^2(62.832x)] dx \right\}$$
$$= 8.074 \times 10^{-4} A^2$$

The denominator of Eq. 9-276 is then

$$2 \int_{\text{cross section}} \text{Re}\{\vec{E} \times \vec{H}^*\}_z da_t = 2 \int_0^{0.02} \int_0^{0.05} 395.36A^2 \sin^2(62.832x) dx dy$$
$$= 0.395A^2$$

Equation 9-276 gives α_w as

$$\alpha_w = \frac{8.074 \times 10^{-4} A^2}{0.395A^2} = 2.04 \times 10^{-3} \text{ nep/m}$$

$$\alpha_{\text{total}} = \alpha_d + \alpha_w = 0.01148 \text{ nep/m}$$

$$\alpha_{\text{dB/m}} = 8.686\alpha_{\text{total}} = 0.0997 \text{ dB/m}$$

Exercises

9-8.2a Carry out the steps that are necessary to verify Eqs. 9-262 and 9-263.

9-8.2b Using the approximation given by Eq. 9-260 twice for each of α_d and β for TEM modes show that for small loss tangent (where δ is the loss angle, not skin depth)

$$\alpha_d \approx \frac{k}{2} \tan \delta \qquad \beta \approx k \left[1 + \frac{1}{8} \tan^2 \delta \right] \qquad \text{(TEM)}$$

9-8.2c Using the approximation given by Eq. 9-269 twice for each of α_d and β for TE and TM modes show that for small loss tangent

$$\alpha_d \approx \frac{k \tan \delta}{2\sqrt{1 - \dfrac{k_c^2}{k^2}}}, \qquad \beta \approx k \sqrt{1 - \frac{k_c^2}{k^2}} \left[1 + \frac{1}{8} \frac{\tan^2 \delta}{\left(1 - \dfrac{k_c^2}{k^2}\right)^2} \right] \qquad \text{(TE, TM)}$$

Also show that k_c^2/k^2 may be replaced by ω_c^2/ω^2 or f_c^2/f^2.

9-8.2d To reduce losses in waveguides, interior silver plating is often used. How thick should the silver layer be so that in the silver the penetrating electric field is reduced to 1% of the value at the surface?

9-8.2e For TEM, TE, and TM modes between parallel planes determine expressions for the wall attenuation. Since the planes are infinite in width use 1-m width. This will result in attenuation in nep/m per meter width, or nep/m^2. For cooper plates with lossless dielectric ($\epsilon_r = 2$) and plate separation of 3 mm plot these attenuations for the frequency range $f_c < f \le 5f_c$ for the TE_1, TM_1, and TEM modes. Answer:

$$\alpha_{w,\text{TEM}} = \frac{R_s}{\eta d}; \qquad \alpha_{w,\text{TM}} = \frac{2R_s}{\eta d\sqrt{1-\left(\frac{f_c}{f}\right)^2}}; \qquad \alpha_{w,\text{TE}} = \frac{2R_s\left(\frac{f_c}{f}\right)^2}{\eta d\sqrt{1-\left(\frac{f_c}{f}\right)^2}}$$

9-8.2f For a rectangular waveguide derive the expressions for the wall attenuations for TE_{mn} and TM_{mn} modes. Answer:

$$\alpha_{w,\text{TM}} = \frac{2R_s}{b\eta\sqrt{1-\left(\frac{f_c}{f}\right)^2}}\left[\frac{m^2\left(\frac{b}{a}\right)^3 + n^2}{m^2\left(\frac{b}{a}\right)^2 + n^2}\right]$$

$$\alpha_{w,\text{TE}} = \frac{2R_s}{b\eta\sqrt{1-\left(\frac{f_c}{f}\right)^2}}\left\{\left(1+\frac{b}{a}\right)\left(\frac{f_c}{f}\right)^2 + \left[1-\left(\frac{f_c}{f}\right)^2\right]\left[\frac{m^2\left(\frac{b}{a}\right)^2 + \frac{b}{a}n^2}{m^2\left(\frac{b}{a}\right)^2 + n^2}\right]\right\}$$

9-8.2g For the attenuation constants given in Exercise 9-8.2f plot the attenuation constants for TE_{10} and TM_{11} modes for b/a ratios of 1, $\frac{1}{2}$, $\frac{1}{4}$, for the following waveguide:

Dimension $a = 2.286$ cm, air filled
Aluminum guide
Frequency range: cutoff to 25 GHz

Note that for the TE_{10} mode the dimension b does not appear in the field equations but has significant effect on the attenuation.

9-8.2h Derive the wall attenuation constants for a cylindrical waveguide of radius a, for the TE_{01}, TE_{11}, and TM_{01} modes. Plot these for an air-filled, 2.54-cm waveguide for $f_c < f < 5f_c$, where f_c is the lowest cutoff frequency among the three modes. Partial Answer:

$$\alpha_{c,\text{TE}_{01}} = \frac{R_s\left(\frac{f_c}{f}\right)^2}{a\eta\sqrt{1-\left(\frac{f_c}{f}\right)^2}}$$

9-8.2i Design an air-filled, rectangular copper waveguide that will support the TE_{10} mode and have an attenuation less than 0.025 nep/m at a frequency of 5 GHz.

9-9 Dielectric Waveguides

There are practical cases where the waveguide is not a completely closed metal system. In these cases the wave is confined to (guided by) a dielectric portion of the system. In such waveguides there are two or more dielectrics and the guides are designed so that most of the energy is contained within one of the dielectric sections.

One example of this type of guide is the optical fiber. This is a dielectric rod (very pure silica, SiO_2) within which the light energy is confined. Such dielectric rods are displacing the conventional coaxial cables as cost and performance of the fibers are very competitive with those of cables. In the dielectric surrounding the rod, the fields are evanescent and decay very rapidly normal to and away from the boundary between the dielectrics. In this case the central dielectric rod is called the *core* and the surrounding dielectric the *cladding*.

In this section we investigate the properties of several dielectric waveguides. These are in reality merely additional examples of solving the wave equation using the separation of variables technique and applying boundary conditions. The additional complexity we encounter arises from the fact that at a dielectric-dielectric boundary the tangential and normal field components satisfy continuity conditions rather than have zero values (tangential $\vec{E}$ and normal $\vec{B}(\vec{H})$).

One may wonder why even bother with dielectric waveguides when the closed-metal types work so well. Suppose that one designs a waveguide to operate in the rectangular TE_{10} mode with a cutoff frequency of 80 GHz. The wide dimension a for such a guide would then be given by Eq. 9-139 as

$$a = \frac{\lambda_{c,\mathrm{TE}}}{2} = \frac{3 \times 10^8/80 \times 10^9}{2} = 1.875 \text{ mm}$$

The construction problem is evident, especially since dimension b is typically half of this. The impossibility of light transmission is easy to imagine since those frequencies are on the order of 2.3×10^{14} Hz (230 terahertz) at a wavelength of 1.3 μm. We will find that dielectrics handle these cases nicely.

9-9.1 Dielectric-Coated Conductors

In this section we consider two cases: a dielectric-coated flat metal sheet and a dielectric-coated metal cylinder.

The configuration for the dielectric coated metal sheet is given in Figure 9-38. In most cases the permeability will be the same for Materials 1 and 2, but it is not required for our general solution. We assume that the conductor is perfect. To obtain energy confinement in Dielectric 1 (the coating, or core here) we want to obtain solutions that produce the field amplitude variation shown in Figure 9-38*b*. Applications of this configuration generally have $\epsilon_2 < \epsilon_1$, and in fact frequently $\epsilon_2 = \epsilon_0$, $\mu_2 = \mu_0$. The resulting wave then travels along the surface of the metal and hence the title surface wave for this configuration.

Waves on this structure can also be described by either TM or TE modes. The TM modes will be obtained here, with TE left as an exercise. Since we have two different

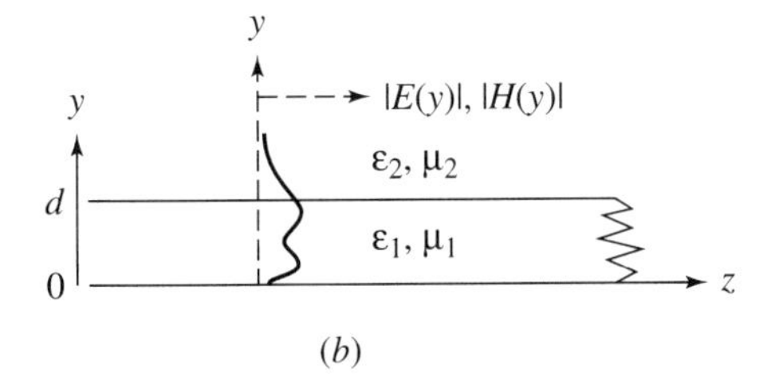

Figure 9-38. Propagation in a dielectric over a perfectly conducting plane. (*a*) Basic configuration. (*b*) Desired field amplitude distribution.

dielectrics we will need to solve the equation in each medium and then match the solutions at the boundary. We will carry out the solution in parallel columns for Regions 1 and 2, using the TM solution process of Section 9-3.2.

Region 1 ($0 \le y \le t$):

$$\nabla_t^2 E_{z1} + k_{c1,\mathrm{TM}}^2 E_{z1} = 0$$

with

$$\gamma_1^2 = k_{c1,\mathrm{TM}}^2 - k_1^2; \qquad k_1^2 = \omega^2 \mu_1 \epsilon_1$$

Region 2 ($d \le y < \infty$):

$$\nabla_t^2 E_{z2} + k_{c2,\mathrm{TM}}^2 E_{z2} = 0$$

with

$$\gamma_1^2 = k_{c2,\mathrm{TM}}^2 - k_2^2; \qquad k_2^2 = \omega^2 \mu_2 \epsilon_2$$

For a system extending from $-\infty$ to $+\infty$ in x we assume no x variations so that $\partial/\partial x = 0$. The transverse Laplacians then reduce to the ordinary derivatives:

Region 1

$$\frac{d^2 E_{z1}}{dy^2} + k_{c1,\mathrm{TM}}^2 E_{z1} = 0$$

Region 2

$$\frac{d^2 E_{z2}}{dy^2} + k_{c2,\mathrm{TM}}^2 E_{z2} = 0$$

To have an exponentially decaying E_z, $k_{c2,\mathrm{TM}}$ must be a negative number, thus

$$k_{c2,\mathrm{TM}}^2 = -\alpha_2^2, \tag{9-277}$$

and then

$$\gamma_2^2 = -\alpha_2^2 - k_2^2 \tag{9-278}$$

Solving these two ordinary differential equations:

Region 1:

$E_{z1} = A \cos k_{c1,\mathrm{TM}} y + B \sin k_{c1,\mathrm{TM}} y$
At $y = 0$, $E_{z1} = 0$ (metal boundary)
$A \cdot 1 + B \cdot 0 = 0 \Rightarrow A = 0$
Then $E_{z1} = B \sin k_{c1,\mathrm{TM}} y$

Region 2:

$E_{z2} = Ce^{\alpha_2 y} + De^{-\alpha_2 y}$
At $y = \infty$, E_{z2} must vanish
Therefore $C = 0$
Then $E_{z2} = De^{-\alpha_2 y}$

At $y = d$ tangential $\vec{E}$ (or normal $\vec{B}$) must be continuous:

$$B \sin k_{c1,\mathrm{TM}} d = De^{-\alpha_2 d}$$

Solve for B (or D just as well):

$$B = \frac{De^{-\alpha d}}{\sin(k_{c1,\mathrm{TM}} d)}$$

Then the solutions become

$$E_{z1} = \frac{De^{-\alpha_2 d}}{\sin(k_{c1,\mathrm{TM}} d)} \sin k_{c1,\mathrm{TM}} y, \qquad E_{z2} = De^{-\alpha_2 y} \tag{9-279}$$

The constant D is determined by the amplitude of the source driving the system. This leaves α_2 and $k_{c1,\mathrm{TM}}$ unknown. Since the fields all propagate as $e^{-\gamma z}$, we need to match the propagation constants at the boundary. This is necessary since the *total* tangential electric field must be continuous. Then we require

$$E_{z1} e^{-\gamma_1 z}\big|_{y=d} = E_{z2} e^{-\gamma_2 z}\big|_{y=d}$$

Substituting the expressions for E_{z1}, E_{z2}, γ_1, and γ_2,

$$\frac{De^{-\alpha_2 d}}{\sin(k_{c1,\mathrm{TM}} d)} \sin(k_{c1,\mathrm{TM}} d) e^{-\sqrt{k_{c1,\mathrm{TM}}^2 - k_1^2}\,d} = De^{-\alpha_2 d} e^{-\sqrt{-\alpha^2 - k_2^2}\,d}$$

Canceling terms where possible and factoring -1 out of each exponent radicand the preceding equation reduces to

$$e^{-j\beta_1 d} = e^{-j\sqrt{k_1^2 - k_{c1,\mathrm{TM}}^2}\,d} = e^{-j\sqrt{\alpha_2^2 + k_2^2}\,d} = e^{-j\beta_2 d}$$

Equating the phase constants (called *phase matching*) we obtain

$$k_1^2 - k_{c1,\mathrm{TM}}^2 = \alpha_2^2 + k_2^2$$

Substituting for k_1 and k_2 and rearranging,

$$\alpha_2^2 + k_{c1,\mathrm{TM}}^2 = \omega^2(\mu_1 \epsilon_1 - \mu_2 \epsilon_2) \tag{9-280}$$

Since the left side is always a positive constant, this constraint condition can be satisfied only if $\mu_2 \epsilon_2 < \mu_1 \epsilon_1$. In the usual case where $\mu_1 = \mu_2$ the restriction would be $\epsilon_2 < \epsilon_1$. This is the useful situation since frequently $\epsilon_2 = \epsilon_0$. From Chapter 8 the condition $\epsilon_2 < \epsilon_1$ is the requirement for total internal reflection of plane waves which would confine energy to Medium 1.

Equation 9-280 is one relationship between α_2 and $k_{c1,\mathrm{TM}}$, and is an equation of a circle if we think of α_2 and $k_{c1,\mathrm{TM}}$ as the variables.

To obtain a second relationship between α_2 and $k_{c1,\mathrm{TM}}$ we use the continuity of tangential $\vec{H}$ at $y = d$. Using Eqs. 9-277 and 9-279 in Eq. 9-9 (for $t_1 \sim x$ and $t_2 \sim y$) to find H_x and equating at $y = d$:

$$j\,\frac{\omega \epsilon_1 D e^{-\alpha_2 d} k_{c1,\mathrm{TM}}}{k_{c1,\mathrm{TM}}^2 \sin(k_{c1,\mathrm{TM}} d)} \cos(k_{c1,\mathrm{TM}} d) = j\,\frac{\omega \epsilon_2 D}{-\alpha_2^2}(-\alpha_2) e^{-\alpha_2 d}$$

Again canceling terms where possible and rearranging:

$$k_{c1,\mathrm{TM}} \tan(k_{c1,\mathrm{TM}}d) = (\alpha_2)\,\frac{\epsilon_1}{\epsilon_2} \tag{9-281}$$

Since the argument of the tangent function involves the dielectric thickness d we can incorporate the d into the variables by multiplying by d on both sides of this equation and by d^2 on both sides of Eq. 9-280. The results are

$$(k_{c1,\mathrm{TM}}d) \tan(k_{c1,\mathrm{TM}}d) = \frac{\epsilon_1}{\epsilon_2}\,(\alpha_2 d) \tag{9-282}$$

$$(\alpha_2 d)^2 + (k_{c1,\mathrm{TM}}d)^2 = (\omega\sqrt{\mu_1\epsilon_1 - \mu_2\epsilon_2}\,d)^2 \tag{9-283}$$

Here we have two equations in two unknowns, which we solve graphically as shown in Figure 9-39. The intersections are common value solutions. The appropriate circle is drawn by evaluating the right side of Eq. 9-283 and using that as the radius. Equation 9-282 is then plotted by assuming a value for $(k_{c1,\mathrm{TM}}d)$ and calculating the corresponding $(\alpha_2 d)$. The intersections 1 through 4 shown in the figure yield α_2 and $k_{c1,\mathrm{TM}}$ values for which a surface wave can be supported. The intersections at negative or zero values of $(\alpha_2 d)$ are not appropriate since they give negative or zero α_2 values, which means no attenuation in Medium 2. Note also that the greater the difference between $\mu_1\epsilon_1$ and $\mu_2\epsilon_2$ the larger α_2 will be, resulting in more rapid decay in Medium 2. However, for large α_2 the circle will have large radius and therefore multiple intersections or modes could develop.

From Figure 9-39 it can also be seen that there is no cutoff frequency since there will always be at least one intersection no matter how small the circle radius is. The larger values of $k_{c1,\mathrm{TM}}$ also mean that there are more amplitude variations of field between 0 and d. Note that there will not be an integral number of variations of amplitude since the boundary conditions at $y = 0$ and $y = d$ are not periodic (equal). Once the values of α_2 and $k_{c1,\mathrm{TM}}$ have been obtained they are inserted into Eq. 9-279 and the remaining field components determined using Eqs. 9-7 through 9-10. The results are

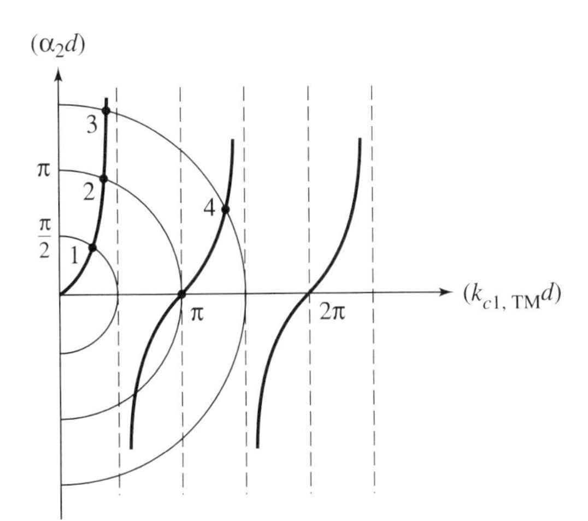

Figure 9-39. Graphical solutions for TM modes on a dielectric-coated sheet. Semicircles are from Eq. 9-283. Modified tangent curve from Eq. 9-282.

Region 1 ($0 \le y \le d$)

$$E_{z1} = \frac{De^{-\alpha_2 d}}{\sin(k_{c1,\text{TM}}d)} \sin(k_{c1,\text{TM}}y)e^{-j\beta_1 z}$$

$$E_{y1} = \frac{-j\sqrt{k_1^2 - k_{c1,\text{TM}}^2}De^{-\alpha_2 d}}{k_{c1,\text{TM}} \sin(k_{c1,\text{TM}}d)} \cos(k_{c1,\text{TM}}y)e^{-j\beta_1 z}$$

$$H_{x1} = j\frac{\omega\epsilon_1 De^{-\alpha_2 d}}{k_{c1,\text{TM}} \sin(k_{c1,\text{TM}}d)} \cos(k_{c1,\text{TM}}y)e^{-j\beta_1 z}$$

Region 2 ($d \le y < \infty$)

$$E_{z2} = De^{-\alpha_2 y}e^{-j\beta_2 z}$$

$$E_{y2} = -j\frac{\sqrt{\alpha_2^2 + k_2^2}}{\alpha_2} De^{-\alpha_2 y}e^{-j\beta_2 z}$$

$$H_{x2} = j\frac{\omega\epsilon_2 D}{\alpha_2} e^{-\alpha_2 y}e^{-j\beta_2 z}$$

where

$$\beta_1 = \beta_2 = \sqrt{k_1^2 - k_{c1,\text{TM}}^2} = \sqrt{\alpha_2^2 + k_2^2}$$

Now that we have the field components we can derive expressions for $\lambda_{c,\text{TM}}$, $\lambda_{g,\text{TM}}$, V_p^{TM}, V_g^{TM}, Z_0^{TM}, and $\vec{S}_{\text{TM}}$ in the same manner as for the waveguides studied in previous sections. Note these are to be determined in both media.

The solution for TE modes in a dielectric-coated plane conductor is left as an exercise.

Guided wave propagation can also be described using a concept called *transverse resonance*. This concept is based on the idea that if no real power is propagating normal to the guide boundary (transverse or normal to z) the impedance normal to z must be reactive. Let's evaluate the impedances at $y = d$, looking in both the $+\hat{y}$ (Region 2) and $-\hat{y}$ direction (Region 1). For Region 2 the general wave impedance equation from Section 7-8 reduces to

$$Z_{y+}^w = \left.\frac{(\vec{E}_2 \times \vec{H}_2^*) \cdot \hat{y}}{(\hat{y} \times \vec{H}_2) \cdot (\hat{y} \times \vec{H}_2)^*}\right|_{y=d} = \frac{E_{z2}}{H_{x2}} = -j\frac{\alpha_2}{\omega\epsilon_2} \tag{9-284}$$

For Region 1:

$$Z_{y-}^w = \left.\frac{(\vec{E}_1 \times \vec{H}_1^*) \cdot (-\hat{y})}{(-\hat{y} \times \vec{H}_1) \cdot (-\hat{y} \times \vec{H}_1)^*}\right|_{y=d} = -\frac{E_{z1}}{H_{x1}} = j\frac{k_{c1,\text{TM}} \tan(k_{c1,\text{TM}}d)}{\omega\epsilon_1} \tag{9-285}$$

As expected the two impedances are pure imaginary quantities since no power is leaving transverse to z. This could have been anticipated by observing that the y component of $\vec{S}_{\text{TM}}$ is also imaginary (reactive). See Exercise 9-9.1a.

The consequences of these two impedances are best explained if we first add the two expressions together:

$$Z_{y+}^w + Z_{y-}^w = j\frac{1}{\omega}\left[-\frac{\alpha_2}{\epsilon_2} + \frac{k_{c1,\text{TM}} \tan(k_{c1,\text{TM}}d)}{\epsilon_1}\right]$$

Next using one of the boundary condition results, Eq. 9-281, the sum becomes

$$Z_{y+}^w + Z_{y-}^w = j\frac{1}{\omega}\left[-\frac{\alpha_2}{\epsilon_2} + \frac{\alpha_2\epsilon_1}{\epsilon_1\epsilon_2}\right] = 0$$

This result tells us that the transverse impedances are complex conjugates of each other. In a resonant system, the reactive parts cancel, leading to the term *transverse resonance*, which is used to describe this concept.

Transverse resonance gives an alternative way of getting the second relationship between the unknown separation constants α_2 and $k_{c1,\text{TM}}$. One would first write expressions for all the field components of $\vec{E}$ and $\vec{H}$ and then develop the transverse impedances. Adding the two impedances and setting the sum to zero results in the desired relationship.

The transverse resonance technique can be applied to any guided wave system. If the system is not lossless, some care must be taken, but this will not be persued here.

Example 9-11

Design a dielectric guide for a 3-GHz signal for a single TM mode. The dielectric is Teflon ($\epsilon_r = 2.1$) and the cover dielectric is air $\epsilon_2 = \epsilon_0$. Also, $\mu_1 = \mu_2 = \mu_0$.

From Figure 9-39 we see there will be only one mode when the circle radius is less than $\pi/2$. Eq. 9-283 gives the radius as

$$\omega\sqrt{\mu_1\epsilon_1 - \mu_2\epsilon_2}d = 2\pi \times 3 \times 10^9 \times \sqrt{\mu_0\epsilon_0}\sqrt{2.1 - 1}d = 65.9d$$

Therefore

$$65.9d < \frac{\pi}{2} \qquad \text{or} \qquad d < 0.024 \text{ m} = 2.4 \text{ cm}$$

Select $d = 0.01$ m. Equations 9-282 and 9-283 reduce to

$$(0.01k_{c1,\text{TM}}) \tan(0.01k_{c1,\text{TM}}) = 2.1(.01\alpha_2)$$

$$(0.01\alpha_2)^2 + (0.01k_{c1,\text{TM}})^2 = 0.4343$$

These are plotted in Figure 9-40. The solutions given at the intersection are $0.01\alpha_2 = 0.21$ ($\alpha_2 = 21$ nep/m) and $0.01\, k_{c1,\text{TM}} = 0.61$ ($k_{c1,\text{TM}} = 61$ rad/m). The accuracy of the graphical

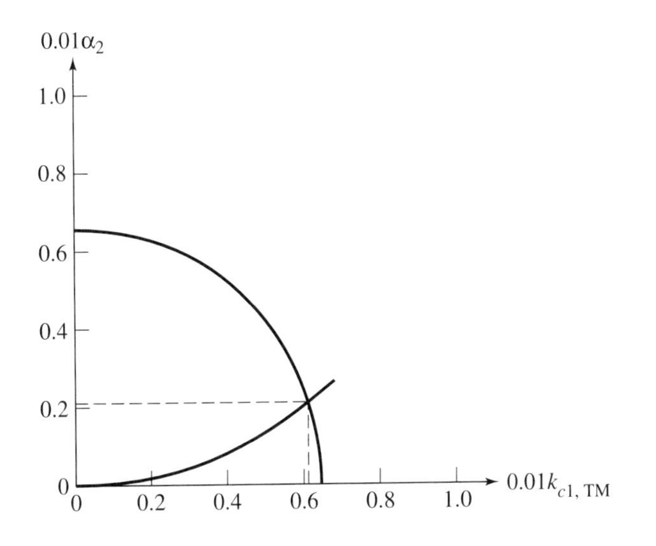

Figure 9-40.

solution can be estimated by computing γ_1 and γ_2 since they should be equal (phase matching):

$$\gamma_1 = \sqrt{k_{c1,\mathrm{TM}}^2 - k_1^2} = j67.598 \text{ rad/m}$$

$$\gamma_2 = \sqrt{-\alpha_2^2 - k_2^2} = j66.248 \text{ rad/m}$$

These values differ by about 2%.

Using these values of d, α_2, $k_{c1,\mathrm{TM}}$ and frequency in the field component expressions developed earlier we obtain

Region 1 ($0 \leq y \leq d$)

$$E_{z1} = 1.415D \sin 61y\, e^{-j67.6z}$$

$$E_{y1} = 1.568D \cos 61y\, e^{-j67.6z}$$

$$H_{x1} = 0.00812D \cos 61y\, e^{-j67.6z}$$

Region 2 ($d \leq y < \infty$)

$$E_{z2} = De^{-21y}e^{-j66.25z}$$

$$E_{y2} = -j3.15De^{-21y}e^{-j66.25z}$$

$$H_{x2} = j0.00794De^{-21y}e^{-j66.25z}$$

Another measure of the accuracy of the graphical solution is to see if tangential $\vec{H}$ is continuous at $y = d$. Evaluating the H_x components we obtain

$$H_{x1}(y = 0.01) = 0.00666 \text{ A/m} \quad H_{x2}(y = 0.01) = 0.00644 \text{ A/m}$$

The difference error is about 3%.

For the next and final application in this section we consider a dielectric-coated cylindrical metal rod as shown in Figure 9-41. We will investigate the TE modes and leave the TM modes for an exercise. The central metal portion is assumed to have infinite conductivity. The dielectrics are assumed lossless. In most practical cases the dielectric Medium 2 (ϵ_2, μ_2) is air, but our derivation will not impose that constraint.

The basic equations are

Region 1 ($a \leq \rho \leq b$)

$$\nabla_t^2 H_{z1} + k_{c1,\mathrm{TE}}^2 H_{z1} = 0$$

with

$$\gamma_1^2 = k_{c1,\mathrm{TE}}^2 - k_1^2 \qquad k_1^2 = \omega^2\mu_1\epsilon_1$$

Region 2 ($b \leq \rho < \infty$)

$$\nabla_t^2 H_{z2} + k_{c2,\mathrm{TE}}^2 H_{z2} = 0$$

with

$$\gamma_2^2 = k_{c2,\mathrm{TE}}^2 - k_2^2 \qquad k_2^2 = \omega^2\mu_2\epsilon_2 \quad (9\text{-}286)$$

Figure 9-41. Dielectric-coated metal rod as a dielectric guide.

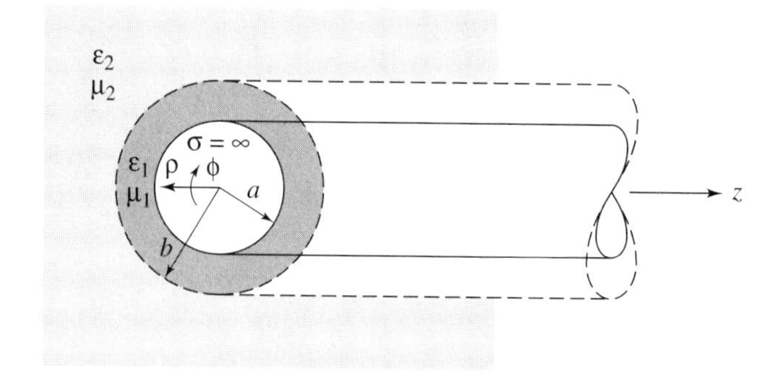

Expanding these in cylindrical coordinates and applying the product solution (separation of variables) technique results in

Region 1 ($a \le \rho \le b$)

$$\frac{1}{\rho}\frac{\partial}{\partial \rho}\left(\rho \frac{\partial H_{z1}}{\partial \rho}\right) + \frac{1}{\rho^2}\frac{\partial^2 H_{z1}}{\partial \phi^2} + k_{c1,\mathrm{TE}}^2 H_{z1} = 0$$

From this:

$$\frac{d^2 R_1}{d\rho^2} + \frac{1}{\rho}\frac{dR_1}{d\rho} + \left(k_{c1,\mathrm{TE}}^2 - \frac{n^2}{\rho^2}\right) R_1 = 0$$

$$\frac{d^2 \Phi_1}{d\phi^2} + n^2 \Phi_1 = 0$$

n is an integer.

Region 2 ($b \le \rho < \infty$)

$$\frac{1}{\rho}\frac{\partial}{\partial \rho}\left(\rho \frac{\partial H_{z2}}{\partial \rho}\right) + \frac{1}{\rho^2}\frac{\partial^2 H_{z2}}{\partial \phi^2} + k_{c2,\mathrm{TE}}^2 H_{z2} = 0$$

From this:

$$\frac{d^2 R_2}{d\rho^2} + \frac{1}{\rho}\frac{dR_2}{d\rho} + \left(k_{c2,\mathrm{TE}}^2 - \frac{m^2}{\rho^2}\right) R_2 = 0$$

$$\frac{d^2 \Phi_2}{d\phi^2} + m^2 \Phi_2 = 0$$

m is an integer.

To have a decaying field in ρ away from the dielectric boundary we must use the modified Bessel function of the second kind, K_m (see Appendix L, Figure L-6). This requires us to set

$$k_{c2,\mathrm{TE}}^2 = -\alpha_2^2 \tag{9-287}$$

so that the equation for R_2 is

$$\frac{d^2 R_2}{d\rho^2} + \frac{1}{\rho}\frac{dR_2}{d\rho} - \left(\alpha_2^2 + \frac{m^2}{\rho^2}\right) R_2 = 0$$

(Note we do not use the I_m functions since they become infinite for $\rho \to \infty$)

Both $\Phi(\phi)$ solutions are trigonometric, having both $\sin \phi$ and $\cos \phi$. We select the reference for $\phi = 0$ so that we have only the cosine term as we did for the cylindrical waveguides. Using this convention and the Bessel solutions from Appendix L we obtain the product solutions (keeping the Neumann functions since $\rho = 0$ is not in the solution region):

Region 1 ($a \le \rho \le b$)

$$H_{z1}(\rho, \phi, z) = \cos n\phi [A_1 J_n(k_{c1,\mathrm{TE}}\rho) + B_1 N_n(k_{c1,\mathrm{TE}}\rho) e^{-j\beta_1 z}]$$

Region 2 ($b \le \rho < \infty$)

$$H_{z2}(\rho, \phi, z) = A_2 \cos m\phi K_m(\alpha_2 \rho) e^{-j\beta_2 z}$$

We will discover shortly that n and m must both be zero for pure TE modes, but we will carry them along so we can show what forces them to zero.

At the dielectric–dielectric boundary H_z must be continuous. Equating these two expressions at $\rho = b$:

$$\cos n\phi [A_1 J_n(k_{c1,\mathrm{TE}} b) + B_1 N_n(k_{c1,\mathrm{TE}} b)] e^{-j\beta_1 z} = A_2 \cos m\phi K_m(\alpha_2 b) e^{-j\beta_2 z} \tag{9-288}$$

Since this result must be independent of z and ϕ the cosines and exponentials must cancel.

This requires

$$\beta_1 = \beta_2 \qquad \text{(phase matching condition)}$$

and

$$n = m$$

From Eqs. 9-283 and 9-284,

$$\gamma_1 = \sqrt{k_{c1,\mathrm{TE}}^2 - k_1^2} = j\sqrt{k_1^2 - k_{c1,\mathrm{TE}}^2} = j\beta_1$$
$$\gamma_2 = \sqrt{-\alpha_2^2 - k_2^2} = j\sqrt{\alpha_2^2 + k_2^2} = j\beta_2$$

Therefore,

$$k_1^2 - k_{c1,\mathrm{TE}}^2 = \alpha_2^2 + k_2^2$$

Rearranging:

$$\alpha_2^2 + k_{c1,\mathrm{TE}}^2 = k_2^2 - k_1^2 = \omega^2(\mu_1\epsilon_1 - \mu_2\epsilon_2) \tag{9-289}$$

The boundary condition (Eq. 9-288) reduces to

$$A_1 J_n(k_{c1,\mathrm{TE}}b) + B_1 N_n(k_{c1,\mathrm{TE}}b) = A_2 K_n(\alpha_2 b)$$

from which

$$A_2 = \frac{A_1 J_n(k_{c1,\mathrm{TE}}b) + B_1 N_n(k_{c1,\mathrm{TE}}b)}{K_n(\alpha_2 b)}$$

The two field components now become (leaving off the z variation)

$$H_{z1} = \cos n\phi\,[A_1 J_n(k_{c1,\mathrm{TE}}\rho) + B_1 N_n(k_{c1,\mathrm{TE}}\rho)]$$
$$H_{z2} = \frac{A_1 J_n(k_{c1,\mathrm{TE}}b) + B_2 N_n(k_{c1,\mathrm{TE}}b)}{K_n(\alpha_2 b)} \cos n\phi\, K_n(\alpha_2\rho)$$

Next, obtain the other field components using Eqs. 9-7 to 9-10. The results are:

Region 1 ($a \leq \rho \leq b$):

$$\left.\begin{aligned}
E_{\rho 1} &= j\frac{\omega\mu_1 n}{\rho k_{c1,\mathrm{TE}}^2} \sin n\phi\,[A_1 J_n(k_{c1,\mathrm{TE}}\rho) + B_1 N_n(k_{c1,\mathrm{TE}}\rho)]\\
E_{\phi 1} &= j\frac{\omega\mu_1}{k_{c1,\mathrm{TE}}} \cos n\phi\,[A_1 J_n'(k_{c1,\mathrm{TE}}\rho) + B_1 N_n'(k_{c1,\mathrm{TE}}\rho)]\\
H_{\rho 1} &= -j\frac{\sqrt{k_1^2 - k_{c1,\mathrm{TE}}^2}}{k_{c1,\mathrm{TE}}} \cos n\phi\,[A_1 J_n'(k_{c1,\mathrm{TE}}\rho) + B_1 N_n'(k_{c1,\mathrm{TE}}\rho)]\\
H_{\phi 1} &= j\frac{\sqrt{k_1^2 - k_{c1,\mathrm{TE}}^2}\,n}{\rho k_{c1,\mathrm{TE}}^2} \sin n\phi\,[A_1 J_n(k_{c1,\mathrm{TE}}\rho) + B_1 N_n(k_{c1,\mathrm{TE}}\rho)]
\end{aligned}\right\} \tag{9-290}$$

Region 2 ($b \le \rho < \infty$):

$$\left.\begin{aligned}
E_{\rho 2} &= -j\frac{\omega\mu_2 n}{\rho\alpha_2^2}\frac{A_1 J_n(k_{c1,\mathrm{TE}}b) + B_1 N_n(k_{c1,\mathrm{TE}}b)}{K_n(\alpha_2 b)}\sin n\phi K_n(\alpha_2\rho) \\
E_{\phi 2} &= -j\frac{\omega\mu_2}{\alpha_2}\frac{A_1 J_n(k_{c1,\mathrm{TE}}b) + B_1 N_n(k_{c1,\mathrm{TE}}b)}{K_n(\alpha_2 b)}\cos n\phi K_n'(\alpha_2\rho) \\
H_{\rho 2} &= j\frac{\sqrt{k_2^2 - k_{c1,\mathrm{TE}}^2}}{\alpha_2}\frac{A_1 J_n(k_{c1,\mathrm{TE}}b) + B_1 N_n(k_{c1,\mathrm{TE}}b)}{K_n(\alpha_2 b)}\cos n\phi K_n'(\alpha_2\rho) \\
H_{\phi 2} &= -j\frac{\sqrt{k_2^2 - k_{c1,\mathrm{TE}}^2}\,n}{\rho\alpha_2^2}\frac{A_1 J_n(k_{c1,\mathrm{TE}}b) + B_1 N_n(k_{c1,\mathrm{TE}}b)}{K_n(\alpha_2 b)}\sin n\phi K_n'(\alpha_2\rho)
\end{aligned}\right\} \quad (9\text{-}291)$$

At $\rho = a$, the perfect conductor requires tangential $\vec{E}$ to be zero. This is the $E_{\phi 1}$ component. From Eq. 9-290 this results in

$$A_1 J_n'(k_{c1,\mathrm{TE}}a) + B_1 N_n'(k_{c1,\mathrm{TE}}a) = 0$$

or

$$B_1 = -\frac{J_n'(k_{c1,\mathrm{TE}}a)}{N_n'(k_{c1,\mathrm{TE}}a)}A_1 \quad (9\text{-}292)$$

This is used to eliminate B_1 in Eqs. 9-290 and 9-291.

As before, the value of A_1 cannot be found explicitly since its value depends upon the amplitude of the driving source. We have now only to obtain values for α_2 and $k_{c1,\mathrm{TE}}$. At $\rho = b$ tangential $\vec{E}$ must be continuous at the dielectric–dielectric boundary: $E_{\phi 1}(\rho = b) = E_{\phi 2}(\rho = b)$. Equating these from Eqs. 9-290 and 9-291, substituting for B_1 using Eq. 9-292, and canceling common factors:

$$\frac{\mu_1}{k_{c1,\mathrm{TE}}}\left[J_n'(k_{c1,\mathrm{TE}}b) - \frac{J_n'(k_{c1,\mathrm{TE}}a)}{N_n'(k_{c1,\mathrm{TE}}a)}N_n'(k_{c1,\mathrm{TE}}b)\right] = -\frac{\mu_2}{\alpha_2}\left[\frac{J_n(k_{c1,\mathrm{TE}}b) - \dfrac{J_n'(k_{c1,\mathrm{TE}}a)}{N_n'(k_{c1,\mathrm{TE}}a)}N_n(k_{c1,\mathrm{TE}}b)}{K_n(\alpha_2 b)}\right]K_n'(\alpha_2 b)$$

Multiplying both sides by $N_n'(k_{c1,\mathrm{TE}}a)$ and rearranging:

$$\frac{\mu_1}{k_{c1,\mathrm{TE}}}\frac{J_n'(k_{c1,\mathrm{TE}}b)N_n'(k_{c1,\mathrm{TE}}a) - J_n'(k_{c1,\mathrm{TE}}a)N_n'(k_{c1,\mathrm{TE}}b)}{J_n(k_{c1,\mathrm{TE}}b)N_n'(k_{c1,\mathrm{TE}}a) - J_n'(k_{c1,\mathrm{TE}}a)N_n(k_{c1,\mathrm{TE}}b)} = -\frac{\mu_2}{\alpha_2}\frac{K_n'(\alpha_2 b)}{K_n(\alpha_2 b)} \quad (9\text{-}293)$$

This result along with Eq. 9-289 provides two equations in the two remaining unknowns α_2 and $k_{c1,\mathrm{TE}}$. There are other boundary conditions, however, that need to be checked. For example, at $\rho = b$ normal $\vec{D}$ must be continuous: $D_{\rho 1}(\rho = b) = D_{\rho 2}(\rho = b)$ or $\epsilon_1 E_{\rho 1}(\rho = b) = \epsilon_2 E_{\rho 2}(\rho = b)$. From Eqs. 9-290 and 9-291 we obtain

$$\frac{\mu_1\epsilon_1 n}{k_{c1,\mathrm{TE}}^2} = -\frac{\mu_2\epsilon_2 n}{\alpha_2^2}$$

This leads to a serious problem since it is a *third* equation relating the two unknowns α_2 and $k_{c1,\mathrm{TE}}$. Also, the left side is always positive and the right side negative. The only way out of this dilemma is to require $n = 0$ in all of the preceding results. Thus, as stated earlier, a pure TE mode in this system can exist only if $n = 0$. To have modes with $n \neq 0$ requires *both* E_z and H_z in order to satisfy all the boundary conditions. Such modes are called hybrid modes and are denoted by EH and HE. A discussion of these modes takes us further than we intend to go so will not be presented here. See [23], [52], and [18] for details.

For the TE mode we set $n = 0$ in the preceding equations and use the identity for the derivatives of J_0, N_0, and K_0. The results are:

Region 1 ($a \le \rho \le b$)

$$\left.\begin{aligned}
H_{z1} &= A_1\left[J_0(k_{c1,\mathrm{TE}}\rho) - \frac{J_1(k_{c1,\mathrm{TE}}a)}{N_1(k_{c1,\mathrm{TE}}a)}N_0(k_{c1,\mathrm{TE}}\rho)\right]e^{-j\sqrt{k_1^2-k_{c1,\mathrm{TE}}^2}z}\\
E_{\phi 1} &= -j\frac{\omega\mu_1 A_1}{k_{c1,\mathrm{TE}}}\left[J_1(k_{c1,\mathrm{TE}}\rho) - \frac{J_1(k_{c1,\mathrm{TE}}a)}{N_1(k_{c1,\mathrm{TE}}a)}N_1(k_{c1,\mathrm{TE}}\rho)\right]e^{-j\sqrt{k_1^2-k_{c1,\mathrm{TE}}^2}z}\\
H_{\rho 1} &= j\frac{\sqrt{k_1^2 - k_{c1,\mathrm{TE}}^2}A_1}{k_{c1,\mathrm{TE}}}\left[J_1(k_{c1,\mathrm{TE}}\rho) - \frac{J_1(k_{c1,\mathrm{TE}}a)}{N_1(k_{c1,\mathrm{TE}}a)}N_1(k_{c1,\mathrm{TE}}\rho)\right]e^{-j\sqrt{k_1^2-k_{c1,\mathrm{TE}}^2}z}
\end{aligned}\right\} \quad (9\text{-}294)$$

Region 2 ($b \le \rho < \infty$)

$$\left.\begin{aligned}
H_{z2} &= A_1\left[\frac{J_0(k_{c1,\mathrm{TE}}b)N_1(k_{c1,\mathrm{TE}}a) - J_1(k_{c1,\mathrm{TE}}a)N_0(k_{c1,\mathrm{TE}}b)}{N_1(k_{c1,\mathrm{TE}}a)K_0(\alpha_2 b)}\right]K_0(\alpha_2\rho)e^{-j\sqrt{\alpha_2^2+k_2^2}z}\\
E_{\phi 2} &= j\frac{\omega\mu_2 A_1}{\alpha_2}\left[\frac{J_0(k_{c1,\mathrm{TE}}b)N_1(k_{c1,\mathrm{TE}}a) - J_1(k_{c1,\mathrm{TE}}a)N_0(k_{c1,\mathrm{TE}}b)}{N_1(k_{c1,\mathrm{TE}}a)K_0(\alpha_2 b)}\right]K_1(\alpha_2\rho)e^{-j\sqrt{\alpha_2^2+k_2^2}z}\\
H_{\rho 2} &= -j\frac{\sqrt{k_2^2 - k_{c1,\mathrm{TE}}^2}A_1}{\alpha_2}\left[\frac{J_0(k_{c1,\mathrm{TE}}b)N_1(k_{c1,\mathrm{TE}}a) - J_1(k_{c1,\mathrm{TE}}a)N_0(k_{c1,\mathrm{TE}}b)}{N_1(k_{c1,\mathrm{TE}}a)K_0(\alpha_2 b)}\right]K_1(\alpha_2\rho)e^{-j\sqrt{\alpha_2^2+k_2^2}z}
\end{aligned}\right\} \quad (9\text{-}295)$$

Where α_2 and $k_{c1,\mathrm{TE}}$ are simultaneous solutions (graphical/numerical) of

$$\frac{\mu_1}{k_{c1,\mathrm{TE}}}\frac{J_1(k_{c1,\mathrm{TE}}b)N_1(k_{c1,\mathrm{TE}}a) - J_1(k_{c1,\mathrm{TE}}a)N_1(k_{c1,\mathrm{TE}}b)}{J_0(k_{c1,\mathrm{TE}}b)N_1(k_{c1,\mathrm{TE}}a) - J_1(k_{c1,\mathrm{TE}}a)N_0(k_{c1,\mathrm{TE}}b)} = -\frac{\mu_2}{\alpha_2}\frac{K_1(\alpha_2 b)}{K_0(\alpha_2 b)} \quad (9\text{-}296)$$

and

$$\alpha_2^2 + k_{c1,\mathrm{TE}}^2 = k_2^2 - k_1^2 = \omega^2(\mu_1\epsilon_1 - \mu_2\epsilon_2) \quad (9\text{-}297)$$

Exercises

9-9.1a For the TM modes prove that $\lambda_{g1} = \lambda_{g2}$, $Z_{01}^{\mathrm{TM}} = Z_{02}^{\mathrm{TM}}$, $V_{p1}^{\mathrm{TM}} = V_{p2}^{\mathrm{TM}}$, and $V_{g1}^{\mathrm{TM}} = V_{g2}^{\mathrm{TM}}$. Also compute $\vec{S}_{\mathrm{TM}} \cdot \hat{z}$ in both dielectrics.

9-9.1b Polyethylene is used to cover a perfect plane conductor. The thickness is 5 mm and a TM mode is to be propagated at 60 GHz. Obtain expressions for the $\vec{E}$ and $\vec{H}$ fields in both materials for the lowest-order mode and then determine the power flow in watts (per meter depth in the x direction) in each dielectric. In Medium 2 you will need to integrate from d to ∞ in y. How would you excite this mode?

9-9.1c Solve for the TE modes in a dielectric coated metal plane. Use Figure 9-38 configuration. Develop the necessary constraint equations and sketch several of these similar to Figure 9-39. Is there a cutoff frequency for these modes? If so, develop experssions for λ_c and f_c. Apply the transverse resonance technique to the field equations and show that a relationship between the separation constants is obtained.

9-9.1d Determine the distance above d at which the power flow (Poynting vector) has decreased to 10% for TM modes in a dielectric coated flat metal plate. What is the percent decrease in $|\vec{E}|$ at that same distance? (Hint: $\vec{S}$ is proportional to the square of $\vec{E}$.) Determine this distance for the results of Example 9-11.

9-9.1e Determine the guide wavelength λ_g and the phase velocity V_p^{TM} for Example 9-11. Compare V_p^{TM} with the values of phase velocities of a uniform plane wave in Medium 1 alone and Medium 2 alone.

9-9.1f For the dielectric coated rod show that the boundary conditions not used to solve the system are satisfied for $n = 0$.

9-9.1g Determine the cutoff frequency and guide wavelength at $1.3\,f_c$ for the TE mode on a dielectric coated rod configured as follows:

Metal rod: #10 copper wire (house wiring) 2.54 mm diameter
Dielectric: 2 mm thick

Air surrounds the system. Assume the conductor is perfect.

9-9.1h Solve for TM only modes on a dielectric coated metal cylinder of Figure 9-41. Are all boundary conditions satisfied?

9-9.1i A 1-cm diameter metal rod is coated with a 1-cm thick layer of Teflon. A 20-GHz source drives the dielectric in the TM mode. Determine how far from the center of the rod one must be to have E_1 decreased to 10% of the air–Teflon boundary value.

9-9.1j Using the transverse resonance technique for TE only mode on a dielectric coated rod, derive one of the relationships between α_2 and $k_{c\,1,\text{TE}}$.

9-9.2 Semi-Infinite Dielectric Slab Waveguide

In this section we investigate one of a class of waveguides that have no metal boundaries. This is the plane dielectric slab waveguide. The section following considers the cylindrical dielectric rod waveguide or optical fiber. Such pure dielectric waveguide systems have primary applications at optical frequencies. Present technology for constructing such waveguides using pure silica produces attenuations on the order of $\frac{1}{100}$ dB/km, considerably less than the attenuation in a metallic waveguide at microwave frequencies by a factor of 1000 or more.

The configuration for the plane dielectric slab waveguide is given in Figure 9-42. Note we have taken $y = 0$ at the center of the slab so that there can be symmetry in the y direction when Mediums 2 and 3 are identical. As one might suspect, the physical concept of having a wave confined essentially to Medium 1 is that of total internal reflection. There will need to be some fields existing in Mediums 2 and 3 in order to satisfy boundary continuity conditions. Basically, confinement can occur when η_2 and η_3 are greater than

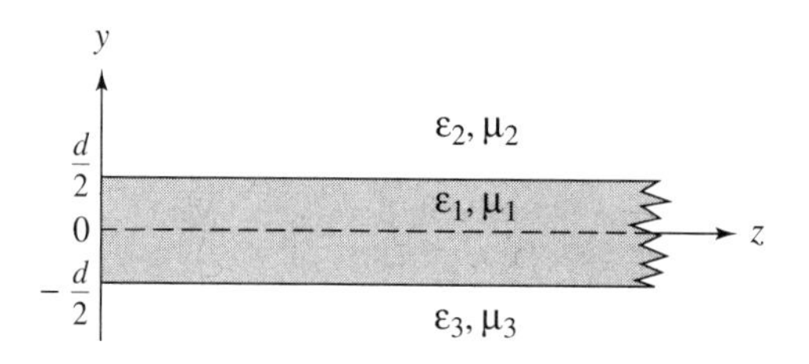

Figure 9-42. Plane dielectric slab waveguide.

η_1 and the incident angle in the material is such that total reflection occurs at both boundaries. This will be demonstrated in detail later. When the product solution method is applied to the wave equation in rectangular coordinates we obtain the following general solutions in the three regions for TM modes having $\partial/\partial x = 0$:

Region 3 ($y < -d/2$)

$$E_{z3} = A_3 e^{\alpha_3 y} e^{-j\beta_3 z}$$

where

$$k_{c3,\text{TM}}^2 = -\alpha_3^2$$

and

$$\gamma_3 = \sqrt{-\alpha_3^2 - k_3^2} = j\sqrt{\alpha_3^2 + k_3^2} = j\beta_3$$

Note we use the positive exponential here since y is negative in this region.

Region 1 ($-d/2 \le y \le d/2$)

$$E_{z1} = (A_1 \cos k_{c1,\text{TM}} y + B_1 \sin k_{c1,\text{TM}} y) e^{-j\beta_1 z}$$

where

$$\gamma_1 = \sqrt{k_{c1,\text{TM}}^2 - k_1^2} = j\sqrt{k_1^2 - k_{c1,\text{TM}}^2} = j\beta_1$$

Region 2 ($d/2 \le y$)

$$E_{z2} = A_2 e^{-\alpha_2 y} e^{-j\beta_2 z}$$

where

$$k_{c2,\text{TM}}^2 = -\alpha_2^2$$

and

$$\gamma_2 = \sqrt{-\alpha_2^2 - k_2^2} = j\sqrt{\alpha_2^2 + k_2^2} = j\beta_2$$

Note we use the negative exponential here since y is positive in this region.

Since Regions 2 and 3 have only one amplitude constant the tangential electric field continuity boundary conditions can be met with either the cosine or sine portion of the Region 1 solution. The choice of the cosine solution yields an even TM variation in y and the sine choice an odd variation in y. These are called the even and odd TM modes, respectively. These will be identified by the superscripts e and o. However, for either even or odd modes the phase constants β must be equal since at the boundaries the exponents must cancel to produce boundary conditions that are independent of z (phase matching). Thus for even and odd modes

$$\beta_3 = \beta_1 \quad \text{and} \quad \beta_2 = \beta_1$$

or

$$\alpha_3^2 + k_3^2 = k_1^2 - k_{c1,\text{TM}}^2 \tag{9-298}$$

and

$$\alpha_2^2 + k_2^2 = k_1^2 - k_{c1,\text{TM}}^2 \tag{9-299}$$

(even or odd modes). Here α_2, α_3 and $k_{c1,\text{TM}}$ are unknown.

Next, using Eqs. 9-8 and 9-9 we obtain expressions for the remaining field components for even and odd modes:

Region 3 ($y < -d/2$)

$$E^e_{z3} = A^e_3 e^{\alpha^e_3 y}$$

$$E^e_{y3} = j\,\frac{\sqrt{(\alpha^e_3)^2 + k_3^2}}{\alpha^e_3}\,A^e_3 e^{\alpha^e_3 y}$$

$$H^e_{x3} = -j\,\frac{\omega\epsilon_3}{\alpha^e_3}\,A^e_3 e^{\alpha^e_3 y}$$

$$E^o_{z3} = A^o_3 e^{\alpha^o_3 y}$$

$$E^o_{y3} = j\,\frac{\sqrt{(\alpha^o_3)^2 + k_3^2}}{\alpha^o_3}\,A^o_3 e^{\alpha^o_3 y}$$

$$H^o_{x3} = -j\,\frac{\omega\epsilon_3}{\alpha^o_3}\,A^o_3 e^{\alpha^o_3 y}$$

Region 1 ($-d/2 \le y \le d/2$)

$$E^e_{z1} = A^e_1 \cos k^e_{c1,\mathrm{TM}} y$$

$$E^e_{y1} = j\,\frac{\sqrt{k_1^2 - (k^e_{c1,\mathrm{TM}})^2}}{k^e_{c1,\mathrm{TM}}}\,A^e_1 \sin k^e_{c1,\mathrm{TM}} y$$

$$H^e_{x1} = -j\,\frac{\omega\epsilon_1}{k^e_{c1,\mathrm{TM}}}\,A^e_1 \sin k^e_{c1,\mathrm{TM}} y$$

$$E^o_{z1} = B^o_1 \sin k^o_{c1,\mathrm{TM}} y$$

$$E^o_{y1} = -j\,\frac{\sqrt{k_1^2 - (k^o_{c1,\mathrm{TM}})^2}}{k^o_{c1,\mathrm{TM}}}\,B^o_1 \cos k^o_{c1,\mathrm{TM}} y$$

$$H^o_{x1} = j\,\frac{\omega\epsilon_1}{k^o_{c1,\mathrm{TM}}}\,B^o_1 \cos k^o_{c1,\mathrm{TM}} y$$

Region 2 ($d/2 \le y$)

$$E^e_{z2} = A^e_2 e^{-\alpha^e_2 y}$$

$$E^3_{y2} = -j\,\frac{\sqrt{(\alpha^e_2)^2 + k_2^2}}{\alpha^e_2}\,A^e_2 e^{-\alpha^e_2 y}$$

$$H^e_{x2} = j\,\frac{\omega\epsilon_2}{\alpha^e_2}\,A^e_2 e^{-\alpha^e_2 y}$$

$$E^o_{z2} = A^o_2 e^{-\alpha^o_2 y}$$

$$E^o_{y2} = -j\,\frac{\sqrt{(\alpha^o_2)^2 + k_2^2}}{\alpha^o_2}\,A^o_2 e^{-\alpha^o_2 y}$$

$$H^o_{x2} = j\,\frac{\omega\epsilon_2}{\alpha^o_2}\,A^o_2 e^{-\alpha^o_2 y}$$

At both boundaries tangential $\vec{E}$ must be continuous. This allows us to determine the relationships among the amplitude constants.

Even Modes

$$A^e_3 e^{-d/2\alpha^e_3} = A^e_1 \cos\left(-k^e_{c1,\mathrm{TM}}\frac{d}{2}\right); \qquad A^e_2 e^{-d/2\alpha^e_2} = A^e_1 \cos\left(k^e_{c1,\mathrm{TM}}\frac{d}{2}\right)$$

Solving for constants in terms of Medium 1 amplitude:

$$A^e_3 = A^e_1 e^{d\alpha^e_3/2} \cos\left(k^e_{c1,\mathrm{TM}}\frac{d}{2}\right); \qquad A^e_2 = A^e_1 e^{d\alpha^e_2/2} \cos\left(k^e_{c1,\mathrm{TM}}\frac{d}{2}\right) \tag{9-300}$$

Odd Modes

$$A^o_3 e^{-d/2\alpha^o_3} = B^o_1 \sin\left(-k_{c1,\mathrm{TM}}\frac{d}{2}\right); \qquad A^o_2 e^{-d/2\alpha^o_2} = B^o_1 \sin\left(k_{c1,\mathrm{TM}}\frac{d}{2}\right)$$

Solving for constants in terms of Medium 1 amplitude:

$$A^o_3 = -B^o_1 e^{d\alpha^o_3/2} \sin\left(k_{c1,\mathrm{TM}}\frac{d}{2}\right); \qquad A^o_2 = B^o_1 e^{d\alpha^o_2/2} \sin\left(k_{c1,\mathrm{TM}}\frac{d}{2}\right) \tag{9-301}$$

We finally need another relationship for α_2, α_3, and $k_{c1,\mathrm{TM}}$ to go with Eqs. 9-298 and 9-299. This can be obtained by requiring normal $\vec{D}$ to be continuous at the boundaries or tangential $\vec{H}$ continuous at the boundaries. The results are the same for both conditions. We use $H_{\tan} \sim H_x$.

Even Modes

$$-j\frac{\omega\epsilon_3 A_3^e}{\alpha_3^e}e^{-d\alpha_3^e/2} = -j\frac{\omega\epsilon_1 A_1^e}{k_{c1,\mathrm{TM}}^e}\sin\left(-k_{c1,\mathrm{TM}}^e\frac{d}{2}\right);$$

$$j\frac{\omega\epsilon_2 A_2^e}{\alpha_2^e}e^{-d\alpha_2^e/2} = -j\frac{\omega\epsilon_1 A_1^e}{k_{c1,\mathrm{TM}}^e}\sin\left(k_{c1,\mathrm{TM}}^e\frac{d}{2}\right)$$

Substituting for the amplitude constants from Equation (9-300) and solving:

$$\alpha_3^e = -\frac{\epsilon_3}{\epsilon_1}k_{c1,\mathrm{TM}}^e\cos\left(k_{c1,\mathrm{TM}}^e\frac{d}{2}\right) \tag{9-302}$$

$$\alpha_2^e = -\frac{\epsilon_2}{\epsilon_1}k_{c1,\mathrm{TM}}^e\cot\left(k_{c1,\mathrm{TM}}^e\frac{d}{2}\right) \tag{9-303}$$

These results can be used in a couple of ways with Eqs. 9-295 and 9-299. One is to substitute Eq. 9-302 into Eq. 9-298 and solve for $k_{c1,\mathrm{TM}}^e$ (graphically or numerically). Then substitute this $k_{c1,\mathrm{TM}}^e$ into Eq. 9-299 and solve for α_2. Otherwise, substitute Eq. 9-303 into Eq. 9-299 first and solve for $k_{c1,\mathrm{TM}}^e$, and so on.

Odd Modes

A similar process yields:

$$\alpha_3^o = \frac{\epsilon_3}{\epsilon_1}k_{c1,\mathrm{TM}}^o\tan\left(k_{c1,\mathrm{TM}}^o\frac{d}{2}\right) \tag{9-304}$$

$$\alpha_2^o = \frac{\epsilon_2}{\epsilon_1}k_{c1,\mathrm{TM}}^o\tan\left(k_{c1,\mathrm{TM}}^o\frac{d}{2}\right) \tag{9-305}$$

From either Eqs. 9-302 and 9-303 or Eqs. 9-304 and 9-305 we obtain, by dividing,

$$\frac{\alpha_3}{\alpha_2} = \frac{\epsilon_3}{\epsilon_2} \qquad \text{(even or odd)} \tag{9-306}$$

Since even and odd modes are defined in terms of the variation of the z component of field, we can illustrate such modes by sketching examples of $E_z(y)$ and $H_z(y)$. Examples of some of the lower-order modes are given in Figure 9-43.

The concept of dielectric guiding from a total internal reflection point of view can be demonstrated by expressing the field inside the dielectric as the superposition of two plane waves as shown in Figure 9-44. Using the technique of Chapter 8 we can express the plane wave in the + arrow direction as

$$\vec{E}_{1+} = (E_{z1}\cos\theta\hat{z} - E_{y1}\sin\theta\hat{y})e^{-j(k_1 y\cos\theta + k_1 z\sin\theta)} \tag{9-307}$$

The z-directed portion of the propagation constant is

$$k_{z1} = k_1\sin\theta$$

from which

$$\sin\theta = \frac{k_{z1}}{k_1} \tag{9-308}$$

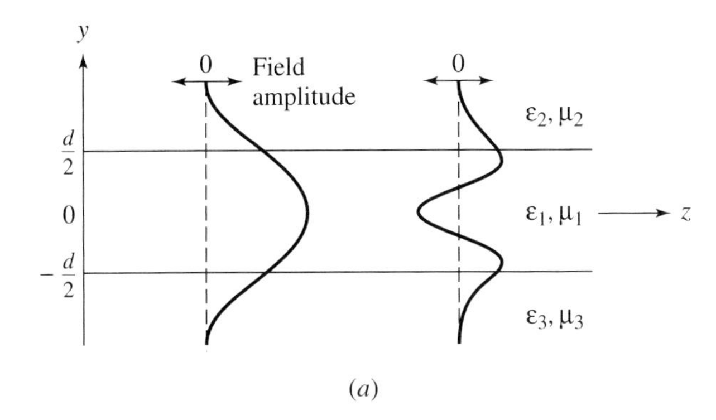

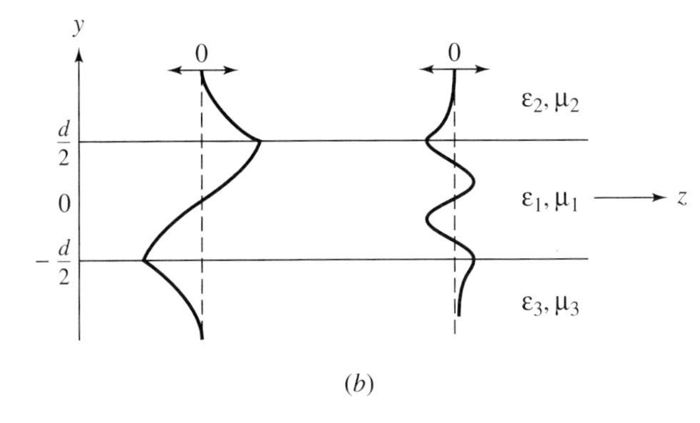

Figure 9-43. Examples of even and odd mode patterns in a dielectric slab waveguide. (*a*) even modes. (*b*) odd modes. E_z for TM, H_z for TE.

From the solutions for the modes, the z-directed propagation constant is $\beta_1 = \sqrt{k_1^2 - k_{c1,\mathrm{TM}}^2} = k_{z1}$. Using this Eq. 9-308 becomes

$$\sin\theta = \frac{\sqrt{k_1^2 - k_{c1,\mathrm{TM}}^2}}{k_1} \tag{9-309}$$

The critical value of θ will be that value for which there is no attenuation in Medium 2 (or Medium 3). Setting α_2 to zero in Eq. 9-299 we obtain

$$k_2^2 = k_1^2 - k_{c1,\mathrm{TM}}^2$$

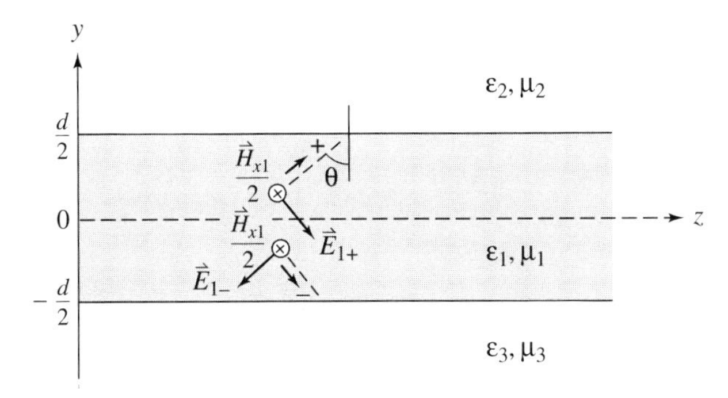

Figure 9-44. Decomposition of total field into two uniform plane wave fields. Dielectric slab guide.

Solving this for $k_{c1,\mathrm{TM}}^2$ and substituting into Eq. 9-309 we obtain the critical angle for total internal reflection at the Medium 1–Medium 2 boundary:

$$\sin\theta_{c,12} = \frac{\sqrt{k_1^2 - (k_1^2 - k_2^2)}}{k_2} = \frac{k_2}{k_1} = \frac{\omega\sqrt{\mu_2\epsilon_2}}{\omega\sqrt{\mu_1\epsilon_1}} = \sqrt{\frac{\mu_2\epsilon_2}{\mu_1\epsilon_1}} \tag{9-310}$$

This is the same result that was obtained for total reflection in Chapter 8.

Similarly, for the Medium 1 Medium 3 boundary:

$$\sin\theta_{c,13} = \sqrt{\frac{\mu_3\epsilon_3}{\mu_1\epsilon_1}} \tag{9-311}$$

Thus to have total internal reflection at both boundaries, we must ensure that the actual angle is equal to or greater than the larger value resulting from Eqs. 9-310 and 9-311. Note these results are independent of the even or odd mode type.

As a final note let's examine what happens when there is no attenuation in Medium 2 and/or 3. When this occurs these fields will also have sinusoidal amplitudes as functions of y so that power extends far away from the boundaries. We now have a dielectric antenna that radiates into regions external to Medium 1. By proper selection of dielectrics it is possible to radiate into only one region and still have attenuation in the other due to the different values of critical angles.

Exercises

9-9.2a For the configuration of Figure 9-42 derive the equations necessary to determine the even and odd TE modes. This includes equations from which α_3, α_2, and $k_{c1,\mathrm{TE}}$ can be obtained. Give equations for the field components.

9-9.2b Derive the equation for the cutoff frequency for the TM modes in a dielectric slab waveguide. Does this result hold for TE modes? Explain.

9-9.2c An alumina ceramic (85% Al_2O_3) is used as a very thick substrate ($\epsilon_r = 8.3$) upon which a 1-mm layer of gallium arsenide (GaAs has $\epsilon_r = 11.1$) is deposited. Air is above that layer. An excitation frequency of 60 GHz is used to excite a TM odd mode. How many modes are possible? Are any of these modes confined to the GaAs? At what frequency will attenuation cease in the substrate? Is it possible to radiate into the air but not the substrate? Assume $\mu_1 = \mu_2 = \mu_0$.

9-9.2d Repeat Exercise 9-9.2c for TE modes.

9-9.2e A parallel plane waveguide (Figure 9-1) carries a TM_1 mode and is flanged at one end, into which a dielectric slab is inserted. The parallel plane guide excites a TM mode into

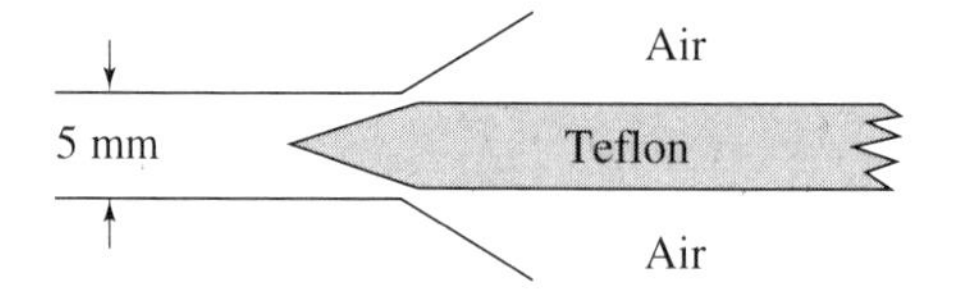

Figure 9-45.

the Teflon dielectric ($\epsilon_r = 2.1$) as shown in Figure 9-45. Will the mode in the Teflon be even or odd?

Determine the frequency at which radiation from the dielectric will begin ($\alpha_2 = \alpha_3 = 0$). How high can the frequency be so only the lowest-order mode can exist in the parallel plane guide?

9-9.3 Circular Dielectric Rod (Optical Fiber)

As the demand for communications channels increases the available frequency bands that can be allocated decreases. This has forced consideration of using higher regions of the spectrum. As a result it is now possible and even desirable to consider optical frequencies. Optical frequencies are less susceptible to radio frequency interference (RFI) and electromagnetic interference (EMI). Fiber optic cables, which are dielectric rod waveguides, can be installed adjacent to power lines without producing noise interference. The advent of low-loss optical fibers has opened a new field of activity called *photonic engineering*.

In this section we shall study the circular cross section dielectric rod guide. The results, of course, are applicable to any part of the spectrum, the diameter of the guide being a key factor of performance. For most microwave applications the diameter is in the millimeter range, whereas for light it is in microns (micrometers).

Optical frequencies are on the order of 10^{14} Hz. These magnitudes are not in the usual realm of discussion, and it is conventional to refer to wavelengths rather than frequencies in the optical region. This wavelength reference is taken with respect to a TEM wave in vacuum so that the equivalency is given by

$$\lambda = \frac{3 \times 10^8}{f} \tag{9-312}$$

Another convention is to refer to dielectric properties in terms of the index of refraction, familiar from physics. The definition is based on the phase velocities of TEM waves in vacuum and in the dielectric as follows:

$$\text{Index of refraction} = n = \frac{c}{V_p} = \frac{\dfrac{1}{\sqrt{\mu_0\epsilon_0}}}{\dfrac{1}{\sqrt{\mu_d\epsilon_d}}} = \sqrt{\frac{\mu_d\epsilon_d}{\mu_0\epsilon_0}} \tag{9-313}$$

In most practical cases $\mu_d = \mu_0$ so that it is customary to use the definition

$$n = \sqrt{\frac{\epsilon_d}{\epsilon_0}} = \sqrt{\epsilon_r} \tag{9-314}$$

Thus expressing material properties in terms of index of refraction is essentially the same as using dielectric constant. One should also remember that ϵ_r can be complex as described earlier in the text.

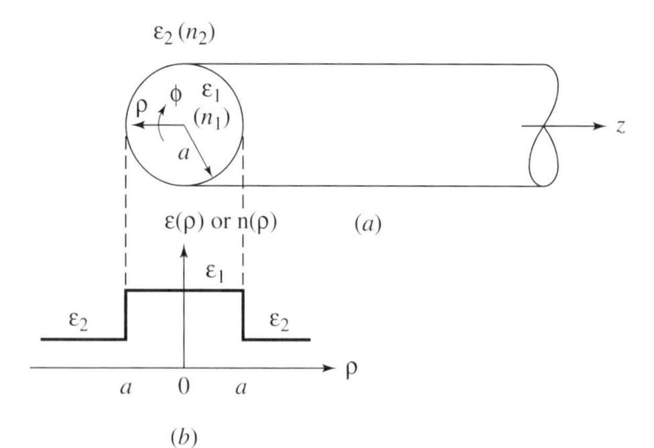

Figure 9-46. Dielectric rod waveguide. (*a*) Basic configuration. (*b*) Step index profile.

The dielectric rod configuration we shall discuss is shown in Figure 9-46. We assume that both dielectrics are uniform as shown in the figure. This results in what is called the *step index profile*. There are many other profiles which will not be discussed here. In optical fiber terminology Medium 1 is called the *core* and Medium 2 the *cladding*.

In Section 9-9.1 we learned that to obtain the most general solution for the cylindrical system we must allow both E_z and H_z, which produce the hybrid modes. Since we have already used the separation of variables for the cylindrical case we will not repeat all steps in detail. We know that we want a decaying field for $\rho > a$ so we use the modified Bessel function of the second kind, K_n. Also, for $\rho < a$ we use only J_n since $\rho = 0$ is in the desired solution region where $N_n \to -\infty$. We will keep both the sine and cosine variations for greatest generality.

To simplify the writing we place one trig function above the other rather than express them as a sum. This means that in all equations to be derived we either take all the top trig functions or the lower trig functions, with the corresponding top or lower sign in the $\pm$ symbol wherever it appears.

We begin as usual with the wave Eqs. 9-19 and 9-21 with the k_cs defined so we obtain the correct Bessel functions in the regions.

Region 1 ($\rho \le a$)

$$\frac{1}{\rho}\frac{\partial}{\partial\rho}\left(\rho\frac{\partial E_{z1}}{\partial\rho}\right) + \frac{1}{\rho^2}\frac{\partial^2 E_{z1}}{\partial\phi^2} + k_{c1}^2 E_{z1} = 0$$

$$\frac{1}{\rho}\frac{\partial}{\partial\rho}\left(\rho\frac{\partial H_{z1}}{\partial\rho}\right) + \frac{1}{\rho^2}\frac{\partial^2 H_{z1}}{\partial\phi^2} + k_{c1}^2 H_{z1} = 0$$

where

$$\gamma_1 = \sqrt{k_{c1}^2 - k_1^2} = j\sqrt{k_1^2 - k_{c1}^2} = j\beta_1$$

Region 2 ($\rho \le a$)

$$\frac{1}{\rho}\frac{\partial}{\partial\rho}\left(\rho\frac{\partial E_{z2}}{\partial\rho}\right) + \frac{1}{\rho^2}\frac{\partial^2 E_{z2}}{\partial\phi^2} - \alpha_2^2 E_{z2} = 0$$

$$\frac{1}{\rho}\frac{\partial}{\partial\rho}\left(\rho\frac{\partial H_{z2}}{\partial\rho}\right) + \frac{1}{\rho^2}\frac{\partial^2 H_{z2}}{\partial\phi^2} - \alpha_2^2 H_{z2} = 0$$

where

$$k_{c2}^2 = -\alpha_2^2$$

with z dependence

$$e^{-\gamma_1 z}$$

and

$$\gamma_2 = \sqrt{k_{c2}^2 - k_2^2} = \sqrt{-\alpha_2^2 - k_2^2}$$
$$= j\sqrt{k_2^2 + \alpha_2^2} = j\beta_2$$

with z dependence

$$e^{-\gamma_2 z}$$

The phase matching condition in z requires that

$$\beta_1 = \beta_2 \tag{9-315}$$

The separation of variables (product solution) produces the following solutions subject to the constraints previously mentioned.

Region 1 ($\rho \leq a$)

$$E_{z1} = A_1 J_n(k_{c1}\rho)\begin{cases}\cos l\phi \\ \sin l\phi\end{cases}$$

$$H_{z1} = B_1 J_n(k_{c1}\rho)\begin{cases}\sin l\phi \\ \cos l\phi\end{cases}$$

Region 2 ($\rho \geq a$)

$$E_{z2} = A_2 K_n(\alpha_2\rho)\begin{cases}\cos l\phi \\ \sin l\phi\end{cases}$$

$$H_{z2} = B_2 K_n(\alpha_2\rho)\begin{cases}\sin l\phi \\ \cos l\phi\end{cases}$$

It is customary to use l in place of n for round fibers to avoid confusing the mode order with the index of refraction when applied to optical cases.

Note that in the solutions for H_z we have placed the sine function on top. This is necessary since as we generate the remaining field components using Eqs. 9-7 through 9-10 (using $t_1 \sim \rho$ and $t_2 \sim \phi$) we differentiate H_z by ϕ and E_z by ρ (or vice versa) so to obtain the single sine or cosine in the ρ and ϕ components H_z and E_z must have opposite functions. Using Eqs. 9-7 through 9-10 the other field components are

Region 1 ($\rho \leq a$)

$$\left.\begin{aligned}
E_{\rho 1} &= j\left[-\frac{\sqrt{k_1^2 - k_{c1}^2}}{k_{c1}} A_1 J_l'(k_{c1}\rho) \mp \frac{\omega\mu_1 l}{\rho k_{c1}^2} B_1 J_l(k_{c1}\rho)\right]\begin{cases}\cos l\phi \\ \sin l\phi\end{cases} \\
E_{\phi 1} &= j\left[\pm\frac{l\sqrt{k_1^2 - k_{c1}^2}}{\rho k_{c1}^2} A_1 J_l(k_{c1}\rho) + \frac{\omega\mu_1}{k_{c1}} B_1 J_l'(k_{c1}\rho)\right]\begin{cases}\sin l\phi \\ \cos l\phi\end{cases} \\
H_{\rho 1} &= j\left[\mp\frac{\omega\epsilon_1 l}{\rho k_{c1}^2} A_1 J_l(k_{c1}\rho) - \frac{\sqrt{k_1^2 - k_{c1}^2}}{k_{c1}} B_1 J_l'(k_{c1}\rho)\right]\begin{cases}\sin l\phi \\ \cos l\phi\end{cases} \\
H_{\phi 1} &= j\left[-\frac{\omega\epsilon_1}{k_{c1}} A_1 J_l'(k_{c1}\rho) \mp \frac{\sqrt{k_1^2 - k_{c1}^2}}{\rho k_{c1}^2} l B_1 J_l(k_{c1}\rho)\right]\begin{cases}\cos l\phi \\ \sin l\rho\end{cases}
\end{aligned}\right\} \tag{9-316}$$

Region 2 ($\rho \geq a$)

$$\left.\begin{aligned}
E_{\rho 2} &= j\left[\frac{\sqrt{\alpha_2^2 + k_2^2}}{\alpha_2} A_2 K_l'(\alpha_2\rho) \pm \frac{\omega\mu_2 l}{\rho\alpha_2^2} B_2 K_l(\alpha_2\rho)\right]\begin{Bmatrix}\cos l\phi \\ \sin l\phi\end{Bmatrix} \\
E_{\phi 2} &= j\left[\mp \frac{\sqrt{\alpha_2^2 + k_2^2}\, l}{\rho\alpha_2^2} A_2 K_l(\alpha_2\rho) - \frac{\omega\mu_2}{\alpha_2} B_2 K_l'(\alpha_2\rho)\right]\begin{Bmatrix}\sin l\phi \\ \cos l\phi\end{Bmatrix} \\
H_{\rho 2} &= j\left[\pm \frac{\omega\epsilon_2 l}{\rho\alpha_2^2} A_2 K_l(\alpha_2\rho) + \frac{\sqrt{\alpha_2^2 + k_2^2}}{\alpha_2} B_2 K_l'(\alpha_2\rho)\right]\begin{Bmatrix}\sin l\phi \\ \cos l\phi\end{Bmatrix} \\
H_{\phi 2} &= j\left[\frac{\omega\epsilon_2}{\alpha_2} A_2 K_l'(\alpha_2\rho) \pm \frac{\sqrt{\alpha_2^2 + k_2^2}}{\rho\alpha_2^2} l B_2 K_l(\alpha_2\rho)\right]\begin{Bmatrix}\cos l\phi \\ \sin l\phi\end{Bmatrix}
\end{aligned}\right\} \quad (9\text{-}317)$$

Next we apply boundary conditions to relate the amplitude constants A_1, B_1, A_2, and B_2. By doing this we can eliminate all but one of them. As usual, the value of this remaining constant depends on the amplitude of the driving source. First, at the boundary $\rho = a$ the tangential fields must be continuous. This requires $E_{z1}(\rho = a) = E_{z2}(\rho = a)$, $H_{z1}(\rho = a) = H_{z2}(\rho = a)$, $E_{\phi 1}(\rho = a) = E_{\phi 2}(\rho = a)$, and $H_{\phi 1}(\rho = a) = H_{\phi 2}(\rho = a)$. It turns out that only one of the ϕ-component boundary conditions can be used since they both give the same result, so there is only one additional independent relationship. Using both the z-component boundary conditions and the H_ϕ boundary condition we obtain (remembering that $\beta_1 = \beta_2$ or $\sqrt{k_1^2 - k_{c1}^2} = \sqrt{\alpha_2^2 + k_2^2}$):

$$A_2 = A_1 \frac{J_l(k_{c1}a)}{K_l(\alpha_2 a)} \qquad B_1 = \mp A_1 \frac{\omega a \alpha_2^2 k_{c1}^2}{l\sqrt{k_1^2 - k_{c1}^2}(\alpha_2^2 + k_{c1}^2)}\left[\frac{\epsilon_1}{k_{c1}}\frac{J_l'(k_{c1}a)}{J_l(k_{c1}a)} + \frac{\epsilon_2}{\alpha_2}\frac{K_l'(\alpha_2 a)}{K_l(\alpha_2 a)}\right]$$

$$B_2 = \mp A_1 \frac{\omega a \alpha_2^2 k_{c1}^2 J_l(k_{c1}a)}{l\sqrt{k_1^2 - k_{c1}^2}(\alpha_2^2 + k_{c1}^2)K_l(\alpha_2 a)}\left[\frac{\epsilon_1}{k_{c1}}\frac{J_l'(k_{c1}a)}{J_l(k_{c1}a)} + \frac{\epsilon_2}{\alpha_2}\frac{K_l'(\alpha_2 a)}{K_l(\alpha_2 a)}\right]$$

Using these in the field component Eqs. 9-316 and 9-317 we obtain the amplitude constant A_1 only in every component. The results are

Region 1 ($\rho \leq a$)

$$E_{z1} = A_1 J_l(k_{c1}\rho)\begin{Bmatrix}\cos l\phi \\ \sin l\phi\end{Bmatrix}$$

$$H_{z1} = \mp A_1 \frac{\omega a \alpha_2^2 k_{c1}^2}{l\sqrt{k_1^2 - k_{c1}^2}(\alpha_2^2 + k_{c1}^2)}\left[\frac{\epsilon_1}{k_{c1}}\frac{J_l'(k_{c1}a)}{J_l(k_{c1}a)} + \frac{\epsilon_2}{\alpha_2}\frac{K_l'(\alpha_2 a)}{K_l(\alpha_2 a)}\right] J_l(k_{c1}\rho)\begin{Bmatrix}\sin l\phi \\ \cos l\phi\end{Bmatrix}$$

$$E_{\rho 1} = jA_1\left[-\frac{\sqrt{k_1^2 - k_{c1}^2}}{k_{c1}} J_l'(k_{c1}\rho) + \frac{\omega^2\mu_1 a\alpha_2^2}{\rho\sqrt{k_1^2 - k_{c1}^2}(\alpha_2^2 + k_{c1}^2)}\left(\frac{\epsilon_1}{k_{c1}}\frac{J_l'(k_{c1}a)}{J_l(k_{c1}a)}\right.\right.$$

$$\left.\left. + \frac{\epsilon_2}{\alpha_2}\frac{K_l'(\alpha_2 a)}{K_l(\alpha_2 a)}\right)J_l(k_{c1}\rho)\right]\begin{Bmatrix}\cos l\phi\\ \sin l\phi\end{Bmatrix}$$

$$E_{\phi 1} = jA_1\left[\pm\frac{l\sqrt{k_1^2 - k_{c1}^2}}{\rho k_{c1}^2} J_l(k_{c1}\rho) \mp \frac{\omega^2\mu_1 a\alpha_2^2 k_{c1}}{l\sqrt{k_1^2 - k_{c1}^2}(\alpha_2^2 + k_{c1}^2)}\left(\frac{\epsilon_1}{k_{c1}}\frac{J_l'(k_{c1}a)}{J_l(k_{c1}a)}\right.\right.$$

$$\left.\left. + \frac{\epsilon_2}{\alpha_2}\frac{K_l'(\alpha_2 a)}{K_l(\alpha_2 a)}\right)J_l'(k_{c1}\rho)\right]\begin{Bmatrix}\sin l\phi\\ \cos l\phi\end{Bmatrix}$$

$$H_{\rho 1} = jA_1\left[\mp\frac{\omega\epsilon l}{\rho k_{c1}^2} J_l(k_{c1}\rho) \pm \frac{\omega a\alpha_2^2 k_{c1}}{l(\alpha_2^2 + k_{c1}^2)}\left(\frac{\epsilon_1}{k_{c1}}\frac{J_l'(k_{c1}a)}{J_l(k_{c1}a)}\right.\right.$$

$$\left.\left. + \frac{\epsilon_2}{\alpha_2}\frac{K_l'(\alpha_2 a)}{K_l(\alpha_2 a)}\right)J_l'(k_{c1}\rho)\right]\begin{Bmatrix}\sin l\phi\\ \cos l\phi\end{Bmatrix}$$

$$H_{\phi 1} = jA_1\left[-\frac{\omega\epsilon_1}{k_{c1}} J_l'(k_{c1}\rho) + \frac{\omega a\alpha_2^2}{\rho(\alpha_2^2 + k_{c1}^2)}\left(\frac{\epsilon_1}{k_{c1}}\frac{J_l'(k_{c1}a)}{J_l(k_{c1}a)}\right.\right.$$

$$\left.\left. + \frac{\epsilon_2}{\alpha_2}\frac{K_l'(\alpha_2 a)}{K_l(\alpha_2 a)}\right)J_l(k_{c1}\rho)\right]\begin{Bmatrix}\cos l\phi\\ \sin l\phi\end{Bmatrix} \tag{9-318}$$

Region 2 ($\rho \geq a$)

$$E_{z2} = A_1\frac{J_l(k_{c1}a)}{K_l(\alpha_2 a)} K_l(\alpha_2\rho)\begin{Bmatrix}\cos l\phi\\ \sin l\phi\end{Bmatrix}$$

$$H_{z2} = \mp A_1\frac{J_l(k_{c1}a)}{K_l(\alpha_2 a)}\frac{\omega a\alpha_2^2 k_{c1}^2}{l\sqrt{k_1^2 - k_{c1}^2}(\alpha_2^2 + k_{c1}^2)}\left[\frac{\epsilon_1}{k_{c1}}\frac{J_l'(k_{c1}a)}{J_l(k_{c1}a)}\right.$$
$$\left. + \frac{\epsilon_2}{\alpha_2}\frac{K_l'(\alpha_2 a)}{K_l(\alpha_2 a)}\right]K_l(\alpha_2\rho)\begin{Bmatrix}\sin l\phi\\ \cos l\phi\end{Bmatrix}$$

$$E_{\rho 2} = jA_1\frac{J_l(k_{c1}a)}{K_l(\alpha_2 a)}\left[\frac{\sqrt{\alpha_2^2 + k_2^2}}{\alpha_2} K_l'(\alpha_2\rho)\right.$$
$$\left. - \frac{\omega^2\mu_2 a k_{c1}^2}{\rho\sqrt{k_1^2 - k_{c1}^2}(\alpha_2^2 + k_{c1}^2)}\left(\frac{\epsilon_1}{k_{c1}}\frac{J_l'(k_{c1}a)}{J_l(k_{c1}a)} + \frac{\epsilon_2}{\alpha_2}\frac{K_l'(\alpha_2 a)}{K_l(\alpha_2 a)}\right)K_l(k_{c1}\rho)\right]\begin{Bmatrix}\cos l\phi\\ \sin l\phi\end{Bmatrix}$$

$$E_{\phi 2} = jA_1\frac{J_l(k_{c1}a)}{K_l(\alpha_2 a)}\left[\mp\frac{\sqrt{\alpha_2^2 + k_2^2}l}{\rho\alpha_2^2} K_l(\alpha_2\rho)\right.$$

$$\pm \frac{\omega^2 \mu_2 a \alpha_2 k_{c1}^2}{l\sqrt{k_1^2 - k_{c1}^2}(\alpha_2^2 + k_{c1}^2)} \left(\frac{\epsilon_1}{k_{c1}} \frac{J_l'(k_{c1}a)}{J_l(k_{c1}a)} + \frac{\epsilon_2}{\alpha_2} \frac{K_l'(\alpha_2 a)}{K_l(\alpha_2 a)} \right) K_l'(\alpha_2 \rho) \Bigg] \begin{Bmatrix} \sin l\phi \\ \cos l\phi \end{Bmatrix}$$

$$H_{\rho 2} = jA_1 \frac{J_l(k_{c1}a)}{K_l(\alpha_2 a)} \Bigg[\pm \frac{\omega \epsilon_2 l}{\rho \alpha_2^2} K_l(\alpha_2 \rho)$$

$$\mp \frac{\omega a \alpha_2 k_{c1}^2}{l(\alpha_2^2 + k_{c1}^2)} \left(\frac{\epsilon_1}{k_{c1}} \frac{J_l'(k_{c1}a)}{J_l(k_{c1}a)} + \frac{\epsilon_2}{\alpha_2} \frac{K_l'(\alpha_2 a)}{K_l(\alpha_2 a)} \right) K_l'(\alpha_2 \rho) \Bigg] \begin{Bmatrix} \sin l\phi \\ \cos l\phi \end{Bmatrix}$$

$$H_{\phi 2} = jA_1 \frac{J_l(k_{c1}a)}{K_l(\alpha_2 a)} \Bigg[\frac{\omega \epsilon_2}{\alpha_2} K_l'(\alpha_2 \rho)$$

$$- \frac{\omega a k_{c1}^2}{\rho(\alpha_2^2 + k_{c1}^2)} \left(\frac{\epsilon_1}{k_{c1}} \frac{J_l'(k_{c1}a)}{J_l(k_{c1}a)} + \frac{\epsilon_2}{\alpha_2} \frac{K_l'(\alpha_2 a)}{K_l(\alpha_2 a)} \right) K_l(\alpha_2 \rho) \Bigg] \begin{Bmatrix} \cos l\phi \\ \sin l\phi \end{Bmatrix} \quad (9\text{-}319)$$

Top signs with top trigonometric function, bottom signs with bottom trigonometric function. Single signs go with either choice of trig function.

The only task remaining is to determine values for α_2 and k_{c1}. We have not yet used the continuity of E_ϕ at the boundary $\rho = a$. Thus setting $E_{\phi 1}(\rho = a) = E_{\phi 2}(\rho = a)$ we obtain, after collecting terms,

$$\left[\mu_2 k_{c1} \frac{K_l'(\alpha_2 a)}{K_l(\alpha_2 a)} + \mu_1 \alpha_2 \frac{J_l'(k_{c1}a)}{J_l(k_{c1}a)} \right] \left[\frac{\epsilon_1}{k_{c1}} \frac{J_l'(k_{c1}a)}{J_l(k_{c1}a)} + \frac{\epsilon_2}{\alpha_2} \frac{K_l'(k_{c1}a)}{K_l(k_{c1}a)} \right] = \frac{l^2(k_1^2 - k_{c1}^2)^2(\alpha_2^2 + k_{c1}^2)^2}{\omega^2 a^2 \alpha_2^3 k_{c1}^3} \quad (9\text{-}320)$$

The second equation relating α_2 and k_{c1} comes from the phase condition $\beta_1 = \beta_2$ or $\beta_1^2 = \beta_2^2$:

$$k_1^2 - k_{c1}^2 = \alpha_2^2 + k_2^2$$

Rearranging,

$$\alpha_2^2 + k_{c1}^2 = k_1^2 - k_2^2 = \omega^2(\mu_1 \epsilon_1 - \mu_2 \epsilon_2) \quad (9\text{-}321)$$

For guided waves we must have $\mu_2 \epsilon_2 < \mu_1 \epsilon_1$.

The process would be to solve this last equation for α_2 or k_{c1} (after evaluating the right-hand side) and substituting into Eq. 9-320. This gives one equation in one unknown. Either a graphical or numerical technique described earlier is required to obtain the unknown value.

For most commonly used fiber materials $\mu_1 = \mu_2 = \mu_0$ and ϵ_1 is almost equal to ϵ_2. For this case the boundary is not very abrupt, which leads to what is called the *weakly guided case*. In this case some approximations can be imposed on Eq. 9-320 to obtain a simpler result but this will not be explored in this text. For details see [23], [35], [52], and [18].

Exercises

9-9.3a For the case $l = 0$ show that the modes in a dielectric rod guide can be expressed as pure TE_{0m} modes. Give the equations for the field components present in the mode. Develop the expressions from which α_2 and k_{c1} can be determined. Show that one of the two relationships between α_2 and k_{c1} can be obtained from Eq. 9-320 for $l = 0$. Partial answers: components present: H_z, E_ϕ, H_ρ:

$$\frac{J_1(k_{c1}a)}{J_0(k_{c1}a)} = -\frac{\epsilon_2 k_{c1}}{\epsilon_1 \alpha_2}\frac{K_1(\alpha_2 a)}{K_0(\alpha_2 a)}; \qquad \alpha_2^2 + k_{c1}^2 = k_1^2 - k_2^2$$

9-9.3b Repeat Exercise 9-9.3a for pure TM_{0m} modes. Partial answers: components present: E_z, E_ρ, H_ϕ:

$$\frac{J_1(k_{c1}a)}{J_0(k_{c1}a)} = -\frac{\epsilon_2 k_{c1}}{\epsilon_1 \alpha_2}\frac{K_1(\alpha_2 a)}{K_0(\alpha_2 a)}; \qquad \alpha_2^2 + k_{c1}^2 = k_1^2 - k_2^2$$

9-9.3c For a step index fiber which has a core diameter of 1 μm and made of glass with $\epsilon_1 = 2.35$ and a cladding having $\epsilon_2 = 2.30$ determine the pure TE_{0m} modes that can propagate at an operating wavelength of 1.3 mm. See Exercise 9-9.2a. For the lowest-order mode how far outside the core surface is H_z equal to 10% of the value at the surface? Assume $\mu_1 = \mu_2 = \mu_0$.

9-9.3d Repeat Exercise 9-9.3c for pure TM_{0m} modes.

9-9.3e Design a dielectric rod guide using Teflon in air that will transmit only the lowest possible order TE_{0m} mode at 40 GHz such that the field intensity H_z is reduced to 10% of the core surface value at a distance of 1 cm from the core surface.

9-9.3f Repeat Exercise 9-9.3e for TM_{0m} mode.

9-9.3g For the case $\mu_1 = \mu_2$ show that the requirement for guided propagation $\mu_1\epsilon_1 > \mu_2\epsilon_2$ is equivalent to requiring $\eta_1 < \eta_n$.

9-10 Mode Dispersion Diagrams for Guided Wave Systems

The design of waveguide systems of any type is greatly facilitated by knowing those frequencies for which propagation can occur, that is, frequencies above cutoff. In earlier sections we discussed the ω-β diagram (ω versus β) with regard to determining the group and phase velocities. Such diagrams also had the cutoff frequency indicated at which $\beta = 0$. However, since the cutoff frequencies are dependent upon the guide dimensions it would be advantageous to have a plot of a normalized frequency and a normalized propagation constant as shown in Figure 9-47, which is the convention for plotting the mode dispersion diagram. In these plots the ω and β axes are switched with respect to those ω-β diagrams

β' Normalized propagation constant

ω' Normalized frequency variable

Figure 9-47. Convention for plotting waveguide mode dispersion diagrams.

of earlier chapters. Thus in mode diagrams the dimensions and constants of the system do not appear explicitly but are parameters that can be varied to compare system designs.

Example 9-12

For the first example we consider wave propagation between parallel planes. For TEM modes the phase constant was developed in Section 9-4.1 as

$$\gamma_{\text{TEM}} = j\omega\sqrt{\mu\epsilon} = j\beta_{\text{TEM}}$$

To obtain the mode dispersion diagram we let

$$\beta' = \beta \qquad \text{and} \quad \omega' = \omega\sqrt{\mu\epsilon} \tag{9-322}$$

and obtain

$$\beta' = \omega'$$

Not a very exciting result, but it illustrates the idea. This is plotted in Figure 9-48*a*.

Figure 9-48. Mode dispersion diagrams for parallel-plane waveguide. (*a*) TEM modes. (*b*) TM_n modes.

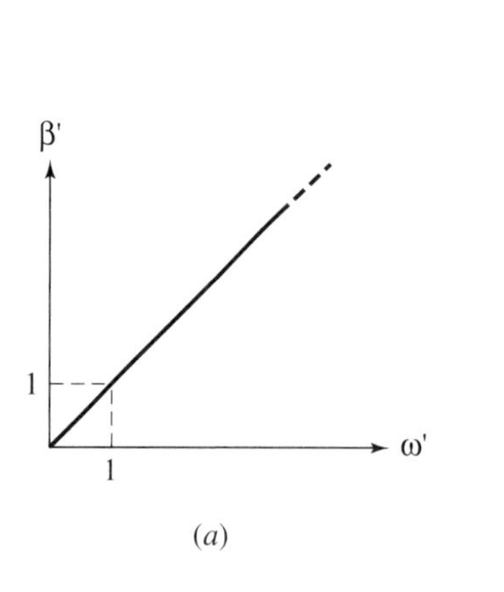

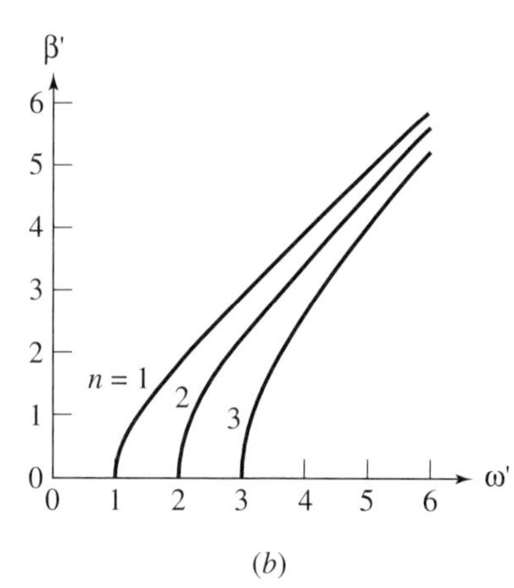

Example 9-13

The TM mode propagation constant for parallel planes is given by Eq. 9-70.

$$\beta_{\text{TM}} = \omega\sqrt{\mu\epsilon}\sqrt{1 - \left(\frac{n\pi}{\omega d\sqrt{\mu\epsilon}}\right)^2} = \sqrt{\omega^2\mu\epsilon - \left(\frac{n\pi}{d}\right)^2}$$

$$\beta_{\text{TM}} = \frac{\pi}{d}\sqrt{\omega^2\frac{\mu\epsilon d^2}{\pi^2} - n^2}$$

$$\frac{\beta_{\text{TM}} d}{\pi} = \sqrt{\left(\frac{\omega d\sqrt{\mu\epsilon}}{\pi}\right)^2 - n^2}$$

Let $\beta' = \beta_{\text{TM}} d/\pi$ and $\omega' = (\omega d\sqrt{\mu\epsilon}/\pi)$. The mode diagram equation is then

$$\beta' = \sqrt{(\omega')^2 - n^2} \tag{9-323}$$

This is plotted in Figure 9-48*b* for modes $n = 1, 2, 3$.

As an example of the application of this diagram, Figure 9-48*b* shows that to propagate TM_2 but not TM_3 we must select $\omega' < 3$. This could have been obtained rather easily from the equation in this basic case, but for more complex configurations the plot is indispensable. The plot shows us some additional features of TM_2 propagation. First, there is the possibility for simultaneous TM_1 propagation. Second, for minimal distortion we would want to select the operating conditions so that ω' is close to 3 so that if any TM_1 mode were excited it would have nearly the same propagation characteristics as the TM_2 mode.

Example 9-14

The next step in complexity is exemplified by the rectangular waveguide. For the TE_{mn} modes Eq. 9-129 gives, for Figure 9-19,

$$\beta_{\text{TE}} = \sqrt{\omega^2\mu\epsilon - \left(\frac{m\pi}{a}\right)^2 - \left(\frac{n\pi}{b}\right)^2}$$

In this case we cannot set up the normalization so that the a and b dimensions are both absorbed into other terms, so we select an a/b ratio. This means that we will have a family of mode dispersion diagrams, one for each a/b ratio. Define $a/b \equiv r$, so we can use

$$a = rb \tag{9-324}$$

The expression for β can be rearranged as follows:

$$\beta_{\text{TE}} = \sqrt{\omega^2\mu\epsilon - \left(\frac{m\pi}{rb}\right)^2 - \left(\frac{n\pi}{b}\right)^2} = \frac{\pi}{b}\sqrt{\left(\frac{\omega\sqrt{\mu\epsilon}b}{\pi}\right)^2 - \left(\frac{m}{r}\right)^2 - n^2}$$

Now let

$$\beta' = \frac{b}{\pi}\beta_{\text{TE}} \qquad \text{and} \qquad \omega' = \frac{\omega b\sqrt{\mu\epsilon}}{\pi} \tag{9-325}$$

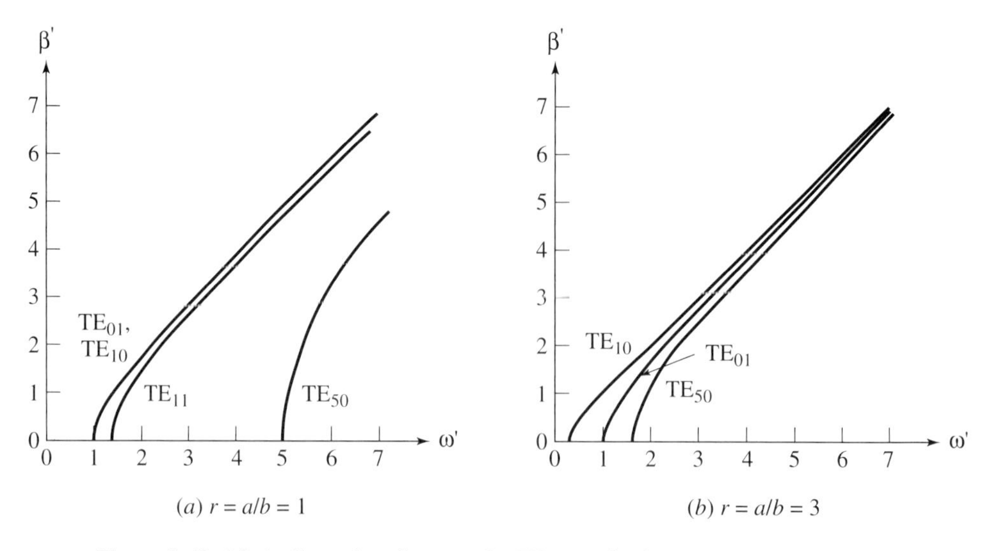

Figure 9-49. Mode dispersion diagrams for TE_{mn} modes in a rectangular waveguide.

This results in

$$\beta' = \sqrt{(\omega')^2 - \left(\frac{m}{r}\right)^2 - n^2} \tag{9-326}$$

Plots of this result for the cases $r = 1$ and $r = 3$ are given in Figures 9-49*a* and *b*. From these plots we can see that if TE_{10} and TE_{01} modes are to be separated the constraint $r > 1$ must be imposed. Also, if the system is to operate over a frequency range rather than a single fixed frequency we would want to use $r = 1$ so that the cutoff for TE_{50}, for example, would be quite high.

Mode dispersion diagrams are very important in the design of optical fiber lines since for optical wavelengths the modes are very close together. References [23], [35], [53], and [18] make extensive use of mode dispersion diagrams.

From these examples it is evident that the normalizations might be done in more than one way. Because of this it is important to define the quantities carefully. When one is reading the literature many different symbols are used for what have been denoted by β' and ω' here and the normalization variables are more complex.

Exercises

9-10.0a Develop the mode dispersion diagram for the TE_n modes for parallel metal plane waveguides.

9-10.0b Prove that for large ω' the mode dispersion diagram for TM_n modes between parallel planes becomes the same as that for the TEM modes.

9-10.0c Plot the mode dispersion diagram for a TE_{mn} rectangular waveguide with $a/b = 2(a = 2b)$. Make plots only for TE_{10}, TE_{01}, and TE_{11}. If the waveguide is filled with air and $a = 2.286$ cm, what frequency range can be used to have only TE_{10} propagate?

9-10.0d Plot the mode dispersion diagram for TM_{01} and TM_{11} modes for a circular metal waveguide of radius a.

9-10.0e Plot the mode dispersion diagram for the three lowest-order TE modes in a coaxial waveguide for which $b/a = 2$ and $b/a = 3$. What do these figures show with respect to mode separation as a function of bandwidth of operation?

9-11 Resonant Cavities

In circuit theory the phenomenon of resonance is employed to design frequency-selective networks (filters) and phase-shifting networks. There it was learned that any losses in the system (resistive components) produced a network that operated over a band of frequencies rather than ideal operation at one frequency. This finite bandwidth was described in terms of a quality factor Q, higher Q being associated with narrower bandwidth (smaller resistive power losses). The values of Q for lumped element passive networks are on the order of 50 to 200 for frequencies up to 500 MHz. For frequencies higher than 500 MHz, we enter the microwave region where losses increase as $\sqrt{f}$ so that frequency selectivity is drastically reduced (low Q). The smaller component sizes restrict the current and voltage levels to values that produce milliwatts of power. Also, energy radiates from the open system (additional losses). For microwave frequencies, systems use resonant cavities as passive frequency-selective devices. Operating Q values in tens of thousands are possible, and power levels in the watt range can be used.

One important application of resonant cavities is the measurement of frequency. When the frequency of the signal applied to the cavity (by loop, probe, or aperture coupling) is at the cavity-resonant frequency, measurable power absorption occurs. With such high Q values, the measurement is very accurate. In this section we develop the equations for the resonant frequency and the Q of such cavities.

The basic approach is to take a section of waveguide and place metal end plates on that section. This will create a reflected field component analogous to that of a shorted transmission line. We then simply take the previously developed field solutions for the waveguide and add a negative traveling component required to make tangential $\vec{E}$ (or normal $\vec{B}$) zero at the end walls. Fortunately, we do not have to re-solve all the waveguide systems for negatively traveling waves. This is avoided by observing that since positive traveling waves are described by $e^{-j\beta z}$ and negative traveling waves by $e^{j\beta z}$ all we need to do is replace β by $-\beta$ in the positive traveling wave solutions to obtain negative traveling wave results.

9-11.1 The Rectangular Resonant Cavity

We begin by examining the rectangular cavity shown in Figure 9-50. All sides are assumed to have infinite conductivity. For the TE resonant modes we can use the H_z component

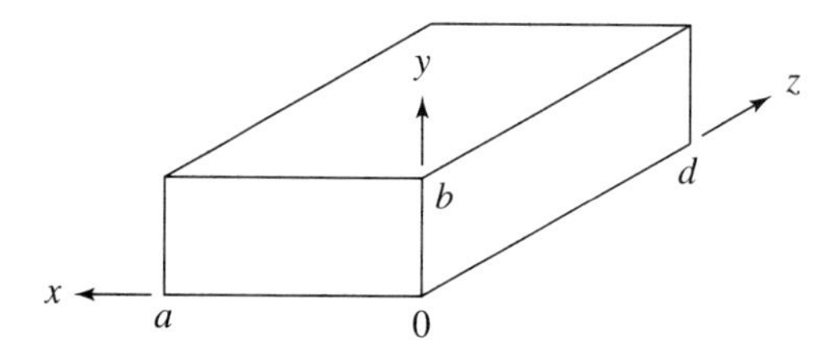

Figure 9-50. Configuration and dimensions for metal rectangular cavity.

and apply normal magnetic field boundary conditions at the end walls $z = 0$ and $z = d$. We will use a superscript $+$ to identify positive z traveling components and a superscript $-$ to identify negative traveling components.

Using the rectangular waveguide equations (9-130 through 9-134) we obtain the following field equations:

Positive traveling components:

$$H_z^+(x, y, z) = A^+ \cos \frac{m\pi}{a} \pi \cos \frac{n\pi}{b} y e^{-j\beta_{\text{TE}} z}$$

$$H_x^+(x, y, z) = j \frac{\beta_{\text{TE}} m\pi A^+}{a\left[\left(\frac{m\pi}{a}\right)^2 + \left(\frac{n\pi}{b}\right)^2\right]} \sin \frac{m\pi}{a} x \cos \frac{n\pi}{b} y e^{-j\beta_{\text{TE}} z}$$

$$H_y^+(x, y, z) = j \frac{\beta_{\text{TE}} n\pi A^+}{b\left[\left(\frac{m\pi}{a}\right)^2 + \left(\frac{n\pi}{b}\right)^2\right]} \cos \frac{m\pi}{a} x \sin \frac{n\pi}{b} y e^{-j\beta_{\text{TE}} z}$$

$$E_x^+(x, y, z) = j \frac{\omega\mu n\pi A^+}{b\left[\left(\frac{m\pi}{a}\right)^2 + \left(\frac{n\pi}{b}\right)^2\right]} \cos \frac{m\pi}{a} x \sin \frac{n\pi}{b} y e^{-j\beta_{\text{TE}} z}$$

$$E_y^+(x, y, z) = -j \frac{\omega\mu m\pi A^+}{a\left[\left(\frac{m\pi}{a}\right)^2 + \left(\frac{n\pi}{b}\right)^2\right]} \sin \frac{m\pi}{a} x \cos \frac{n\pi}{b} y e^{-j\beta_{\text{TE}} z}$$

Negative traveling components:

$$H_z^-(x, y, z) = A^- \cos \frac{m\pi}{a} x \cos \frac{n\pi}{b} y e^{j\beta_{\text{TE}} z}$$

$$H_x^-(x, y, z) = -j \frac{\beta_{\text{TE}} m\pi A^-}{a\left[\left(\frac{m\pi}{a}\right)^2 + \left(\frac{n\pi}{b}\right)^2\right]} \sin \frac{m\pi}{a} x \cos \frac{n\pi}{b} y e^{j\beta_{\text{TE}} z}$$

$$H_y^-(x, y, z) = -j \frac{\beta_{TE} n\pi A^-}{b\left[\left(\frac{m\pi}{a}\right)^2 + \left(\frac{n\pi}{b}\right)^2\right]} \cos \frac{m\pi}{a}x \sin \frac{n\pi}{b}y\ e^{j\beta_{TE}z}$$

$$E_x^-(x, y, z) = j \frac{\omega\mu n\pi A^-}{b\left[\left(\frac{m\pi}{a}\right)^2 + \left(\frac{n\pi}{b}\right)^2\right]} \cos \frac{m\pi}{a}x \sin \frac{n\pi}{b}y\ e^{j\beta_{TE}z}$$

$$E_y^-(x, y, z) = -j \frac{\omega\mu m\pi A^-}{a\left[\left(\frac{m\pi}{a}\right)^2 + \left(\frac{n\pi}{b}\right)^2\right]} \sin \frac{m\pi}{a}x \cos \frac{n\pi}{b}y\ e^{j\beta_{TE}z}$$

Next we apply the boundary condition that normal $\vec{B}$ at a perfect conductor for time-varying fields is zero. This is total $\vec{B}$ that is obtained by adding the positive and negative waves. Also, since normal $\vec{B}$ is zero and $\vec{B} = \mu\vec{H}$, we may set normal $\vec{H}$ equal to zero as well. Using H_z we then have

$$H_z(x, y, 0) = 0 = H_z^+(x, y, 0) + H_z^-(x, y, 0) = A^+ \cos \frac{m\pi}{a} x \cos \frac{n\pi}{b} y$$

$$+ A^- \cos \frac{m\pi}{a} x \cos \frac{n\pi}{b} y$$

Dividing through by the cosine product, the boundary condition reduces to

$$A^+ + A^- = 0$$

from which

$$A^- = -A^+ \tag{9-327}$$

Using this result the total H_z becomes

$$H_z(x, y, z) = H_z^+(x, y, z) + H_z^-(x, y, z) = A^+ \cos \frac{m\pi}{a}x \cos \frac{n\pi}{b}y(e^{-j\beta_{TE}z} - e^{j\beta_{TE}z})$$

From the exponential relation

$$\sin\theta = \frac{e^{j\theta} - e^{-j\theta}}{2j}$$

we have

$$e^{-j\theta} - e^{j\theta} = -j2 \sin\theta$$

The total H_z may then be written

$$H_z(x, y, z) = -j2A^+ \cos \frac{m\pi}{a} x \cos \frac{n\pi}{b} y \sin\beta_{TE}z$$

This must also be zero at $z = d$, so that

$$H_z(x, y, d) = -j2A^+ \cos \frac{m\pi}{a}x \cos \frac{n\pi}{b}y \sin \beta_{\text{TE}}d = 0$$

To be satisfied for all x and y we require

$$\sin \beta_{\text{TE}}d = 0 \qquad \text{or} \qquad \beta_{\text{TE}}d = l\pi, \qquad l = 1, 2, \ldots$$

from which

$$\beta_{\text{TE}} = \frac{l\pi}{d} \qquad l = 1, 2, \ldots \tag{9-328}$$

Note l cannot be zero since H_z would then be zero everywhere. The total z component of $\vec{H}$ finally becomes

$$H_z(x, y, z) = -j2A^+ \cos \frac{m\pi}{a}x \cos \frac{n\pi}{b}y \sin \frac{l\pi}{d}z \tag{9-329}$$

Constructing the other total field components by adding positive and negative traveling components and using Eqs. 9-327 and 9-328 we obtain

$$H_x(x, y, z) = j\frac{2\left(\frac{l\pi}{d}\right)\left(\frac{m\pi}{a}\right)A^+}{\left[\left(\frac{m\pi}{a}\right)^2 + \left(\frac{n\pi}{b}\right)^2\right]} \sin \frac{m\pi}{a}x \cos \frac{n\pi}{b}y \cos \frac{l\pi}{d}z \tag{9-330}$$

$$H_y(x, y, z) = j\frac{2\left(\frac{l\pi}{d}\right)\left(\frac{n\pi}{b}\right)A^+}{\left[\left(\frac{m\pi}{a}\right)^2 + \left(\frac{n\pi}{b}\right)^2\right]} \cos \frac{m\pi}{a}x \sin \frac{n\pi}{b}y \cos \frac{l\pi}{d}z \tag{9-331}$$

$$E_x(x, y, z) = \frac{2\omega\mu n\pi A^+}{b\left[\left(\frac{m\pi}{a}\right)^2 + \left(\frac{n\pi}{b}\right)^2\right]} \cos \frac{m\pi}{a}x \sin \frac{n\pi}{b}y \sin \frac{l\pi}{d}z \tag{9-332}$$

$$E_y(x, y, z) = -\frac{2\omega\mu m\pi A^+}{a\left[\left(\frac{m\pi}{a}\right)^2 + \left(\frac{n\pi}{b}\right)^2\right]} \sin \frac{m\pi}{a}x \cos \frac{n\pi}{b}y \sin \frac{l\pi}{d}z \tag{9-333}$$

In the preceding equations, $m, n = 0, 1, 2, \ldots$ (not both zero simultaneously) and $l = 1, 2, 3, \ldots$

One important thing we need to determine is the frequency at which this resonant condition exists. This is most easily obtained by noting that we have two values of β_{TE} for the same waveguide, namely, Eqs. 9-129 and 9-328. These must be equal so we have, setting $\omega = \omega_{\text{res}}$ in Eq. 9-129 and equating to Eq. 9-328:

$$\sqrt{\omega_{\text{res}}^2\mu\epsilon - \left(\frac{m\pi}{a}\right)^2 - \left(\frac{n\pi}{b}\right)^2} = \frac{l\pi}{d}$$

Solving for ω_{res} and using $\omega_{res} = 2\pi f_{res}$ the resonant frequency is

$$f_{res}^{TE} = \frac{1}{\sqrt{\mu\epsilon}}\sqrt{\left(\frac{m}{2a}\right)^2 + \left(\frac{n}{2b}\right)^2 + \left(\frac{l}{2d}\right)^2}, \quad \begin{array}{r} l = 1, 2, 3, \ldots \\ m, n = 0, 1, 2, \ldots \\ \text{(but not both zero)} \end{array} \tag{9-334}$$

Recognizing $1/\sqrt{\mu\epsilon}$ as the phase velocity, v_0, of a plane TEM wave in the material filling the waveguide (velocity of light in the unbounded medium) we can calculate the unbounded (free space) wavelength corresponding to this frequency as

$$\lambda_0^{res} = \frac{v_0}{f_{res}} = \frac{1}{\sqrt{\mu\epsilon} f_{res}} = \frac{1}{\sqrt{\left(\frac{m}{2a}\right)^2 + \left(\frac{n}{2b}\right)^2 + \left(\frac{l}{2d}\right)^2}} \tag{9-335}$$

We can also compute the guide wavelength as usual to obtain

$$\lambda_g^{res} = \frac{2\pi}{\beta_{TE}} = \frac{2\pi}{\left(\frac{l\pi}{d}\right)} = \frac{2d}{l} \tag{9-336}$$

The preceding equation gives a good physical picture of resonance if we solve for the cavity length at resonance:

$$d = l\left(\frac{\lambda_g^{res}}{2}\right)$$

This tells us that for resonance the cavity must be an integral number of half guide wavelengths long. We might have expected this result from transmission line theory where we learned that standing wave patterns repeat every half line wavelength.

Since there are three integers we identify the modes using the symbol

$$\text{TE}_{mnl}$$

The subscript integers have the same physical interpretations as they had for the rectangular waveguide. The integer m represents the number of half sine wave variations in the x direction, n the number of half sine wave variations in the y direction, and l the number of half sine wave variations in z.

The lowest-order modes would be TE_{011} and TE_{101}. To identify this lowest-order mode with the dominant TE_{10} waveguide mode it is usual to select the TE_{101} resonant cavity mode as the lowest order. The corresponding resonant parameters are

$$\left.\begin{aligned} \lambda_0^{res} &= \frac{2ad}{\sqrt{a^2 + d^2}} \\ \lambda_g^{res} &= 2d \\ f_{res}^{TE} &= \frac{\sqrt{a^2 + d^2}}{2ad\sqrt{\mu\epsilon}} \end{aligned}\right\} \text{TE}_{101} \tag{9-337}$$

Physically one can observe that for the lowest-order mode the cavity length d would be $\lambda_g/2$. This is consistent with the fact that the distance between nulls (in this case the end walls at $z = 0$ and $z = d$) is one-half wavelength.

For the TE_{101} mode the field components are:

$$\left.\begin{aligned} H_z(x, y, z) &= -j2A^+ \cos\frac{\pi}{a}x \sin\frac{\pi}{d}z \\ H_x(x, y, z) &= j\frac{2aA^+}{d} \sin\frac{\pi}{a}x \cos\frac{\pi}{d}z \\ E_y(x, y, z) &= -\frac{2\omega\mu aA^+}{\pi} \sin\frac{\pi}{a}x \sin\frac{\pi}{d}z \end{aligned}\right\} TE_{101} \tag{9-338}$$

Since there is no term of the form $e^{-j\beta z}$ the total field equations represent standing wave patterns as expected. Converting these to the time domain in the usual way we obtain at the resonant frequency

$$\left.\begin{aligned} \mathcal{H}_z(x, y, z) &= 2A^+ \cos\frac{\pi}{a}x \sin\frac{\pi}{d}z \sin\omega_{res}t \\ \mathcal{H}_x(x, y, z) &= -2\frac{a}{d}A^+ \sin\frac{\pi}{a}x \cos\frac{\pi}{d}z \sin\omega_{res}t \\ \mathcal{E}_y(x, y, z) &= -\frac{2\omega_{res}\mu aA^+}{\pi} \sin\frac{\pi}{a}x \sin\frac{\pi}{d}z \cos\omega_{res}t \end{aligned}\right\} TE_{101} \tag{9-339}$$

The plotting of field lines is accomplished using the technique developed in Section 5-6. Since there is no y variation the field lines for $\vec{\mathcal{H}}$ are best shown in the xz plane. We develop the equations for the magnetic field lines (dashed) in Figure 9-51b as follows, using $\omega_{res}t = \pi/4$ in Eq. 9-339 for convenience in both $\vec{\mathcal{E}}$ and $\vec{\mathcal{H}}$:

$$\frac{dx}{dz} = \frac{\mathcal{H}_x}{\mathcal{H}_z} = -\frac{a}{d}\tan\frac{\pi}{a}x \cot\frac{\pi}{d}z$$

To make the plot easy let's select $a = d = \pi$. Then separating the variables we have

$$-\cot x\,dx = \cot z\,dz$$

Integrating and using $\ln k$ as the integration constant:

$$-\ln(\sin x) + \ln k = \ln(\sin z)$$

or

$$\begin{aligned} \ln k &= \ln(\sin z \sin x) \\ k &= |\sin z \sin x| \leq 1 \end{aligned} \tag{9-340}$$

Selecting $k = 0.1, 0.4, 0.8$, and 0.96 four magnetic field lines are obtained as shown in Figure 9-51b, where the smallest area (inner) line corresponds to $K = 0.96$ (large crosses covering it).

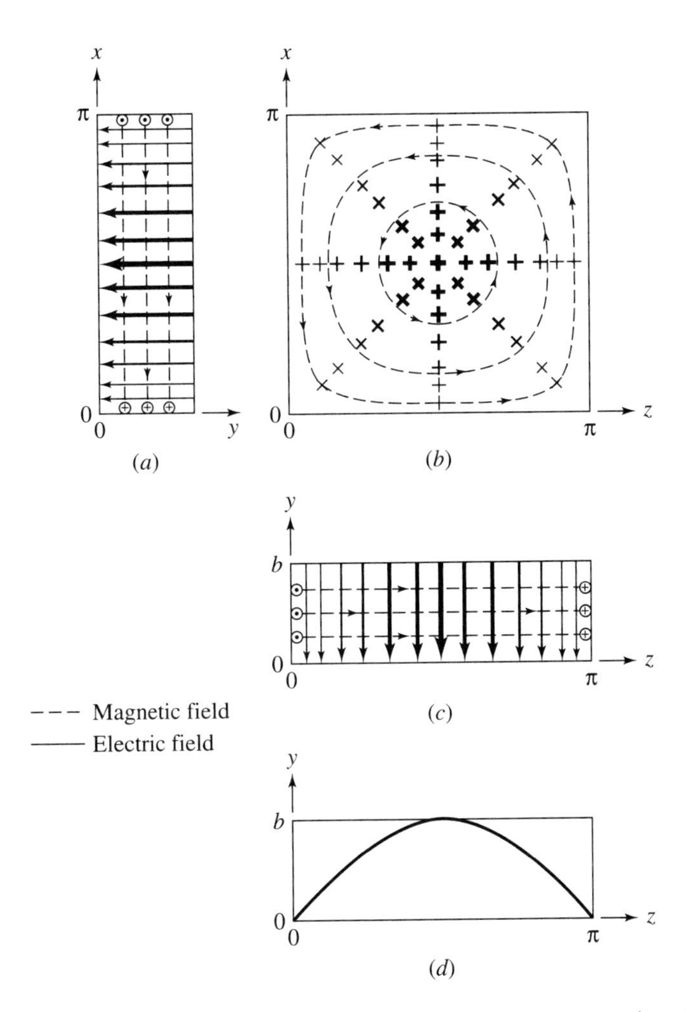

Figure 9-51. Field lines in a TE_{101} square metal cavity. (*a*) Side view. (*b*) Top view. (*c*) Side view. (*d*) Alternate method for indicating electric field amplitude.

The plots of the electric field intensity lines are given in Figures 9-51*a* and 9-51*c* where the widths of the lines indicate the amplitude. At the walls $\mathscr{E}_y$ vanishes as required by the tangential boundary condition. An alternative way of indicating the amplitude of $\mathscr{E}_y$ is shown in Figure 9-51*d*.

Other important quantities for a resonant cavity are the charge distributions on the walls, the surface currents on the inside surfaces of the walls, and the displacement current. These are left as exercises.

One may wonder how the cavity can ever be used since it was assumed that the cavity was a completely closed system. This is the same problem encountered with waveguides.

The solution is to use loop, probe, or slot (aperture) coupling. These were described for the waveguides.

For the TM_{mnl} modes in a rectangular cavity we leave the details for an exercise. The results are, for the configuration shown in Figure 9-50,

$$E_z(x, y, z) = 2B^+ \sin \frac{m\pi}{a} x \sin \frac{n\pi}{b} y \cos \frac{l\pi}{d} z \tag{9-341}$$

$$E_x(x, y, z) = -\frac{2\left(\frac{m\pi}{a}\right)\left(\frac{l\pi}{d}\right)}{\left[\left(\frac{m\pi}{a}\right)^2 + \left(\frac{n\pi}{b}\right)^2\right]} B^+ \cos \frac{m\pi}{a} x \sin \frac{n\pi}{b} y \sin \frac{l\pi}{d} z \tag{9-342}$$

$$E_y(x, y, z) = -\frac{2\left(\frac{n\pi}{b}\right)\left(\frac{l\pi}{d}\right)}{\left[\left(\frac{m\pi}{a}\right)^2 + \left(\frac{n\pi}{b}\right)^2\right]} B^+ \sin \frac{m\pi}{a} x \cos \frac{n\pi}{b} y \sin \frac{l\pi}{d} z \tag{9-343}$$

$$H_x(x, y, z) = j\frac{2\omega\epsilon\left(\frac{n\pi}{b}\right)}{\left[\left(\frac{m\pi}{a}\right)^2 + \left(\frac{n\pi}{b}\right)^2\right]} B^+ \sin \frac{m\pi}{a} x \cos \frac{n\pi}{b} y \cos \frac{l\pi}{d} z \tag{9-344}$$

$$H_y(x, y, z) = -j\frac{2\omega\epsilon\left(\frac{m\pi}{a}\right)}{\left[\left(\frac{m\pi}{a}\right)^2 + \left(\frac{n\pi}{b}\right)^2\right]} B^+ \cos \frac{m\pi}{a} x \sin \frac{n\pi}{b} y \cos \frac{l\pi}{d} z \tag{9-345}$$

$$f_{\text{res}}^{\text{TM}} = \frac{1}{\sqrt{\mu\epsilon}} \sqrt{\left(\frac{m}{2a}\right)^2 + \left(\frac{n}{2b}\right)^2 + \left(\frac{l}{2d}\right)^2} \tag{9-346}$$

$$\lambda_0^{\text{res}} = \frac{v_0}{f_{\text{res}}} = \frac{1}{\sqrt{\mu\epsilon} f_{\text{res}}} = \frac{1}{\sqrt{\left(\frac{m}{2a}\right)^2 + \left(\frac{n}{2b}\right)^2 + \left(\frac{l}{2d}\right)^2}} \tag{9-347}$$

$$\omega = \omega_{\text{res}}^{\text{TM}} = 2\pi f_{\text{res}}^{\text{TM}}$$

In the preceding, $m, n = 1, 2, 3, \ldots$, and $l = 0, 1, 2, 3, \ldots$.

Exercises

9-11.1a For the TE_{101} modes for the square cavity presented in this section derive the equations for and plot on the crossections of Figure 9-51 the following: displacement current, $\vec{\mathcal{J}}_D$ at

$\omega_{\text{res}}t = \pi/4$ (check the right-hand rule for current and $\vec{\mathcal{H}}$); surface charge density, $\rho_s(x, y, z)$ at $\omega_{\text{res}}t = \pi/4$; surface current density, $\vec{\mathcal{J}}_s$ at $\omega_{\text{res}} = \pi/4$.

9-11.1b For the TE_{mnl} modes develop equations for $k_{c\text{TE}}, \gamma_{\text{TE}}$.

9-11.1c Show that the same results for TE_{mnl} modes are obtained if either E_x or E_y is set to zero at $z = 0$ and $z =$ d.

9-11.1d Describe the effects of μ and ϵ on f_{res} and compare with the analogous effects of L and C resonance in circuit theory.

9-11.1e Make plots like those of Figure 9-51 for the TE_{102} mode. Use $a = d = \pi$. Give the equations for λ_0^{res} for TE_{101} and TE_{102} for this cavity. Note there will be two sets of loops in the xz plane view.

9-11.1f Design a square, air-filled cavity that will have a resonant frequency of 10 GHz in the TE_{101} mode.

9-11.1g For the TE_{101} mode in a square cavity show how you would couple energy into the cavity by three methods: loop, probe, and aperture.

9-11.1h For a TE_{101} mode cavity whose length is twice the width and four times the height, determine the required dimensions for a 3-GHz resonant frequency for an air-filled cavity.

9-11.1i Derive Eqs. 9-341 to 9-347 beginning with the waveguide Eqs. 9-154 to 9-159. Use the boundary condition on E_x at $z = 0$ and d. Be sure to identify β_{TM} in the equations and note $m, n \neq 0$.

9-11.1j What is the lowest-order TM_{mnl} mode that can exist in a resonant cavity? Show three ways of exciting this mode. Make a drawing for this mode like that of Figure 9-51.

9-11.2 The Circular Cylindrical Resonant Cavity

To measure an unknown frequency in a microwave system it is desirable to have a cavity whose dimensions can be easily changed. The circular cylindrical cavity provides such a configuration, since the cavity length can be varied by moving the end circular disk along the cavity axis. This basic system, shown in Figure 9-52, is simply a section of a perfectly conducting metal cylindrical waveguide with perfectly conducting end plates.

The procedure for solving for the resonant field configurations and determining the operating properties of resonant frequency, resonant wavelength, and resonant guide wavelength is identical to that used for the rectangular cavity. Thus we begin by adding the positive and negative traveling wave components from the cylindrical waveguide equations. For the TM modes we use Eqs. 9-198 to 9-202 using the expression for β_{TM} from Eq. 9-197. The results are (where p_{nl} is the lth root of the nth-order Bessel function of the

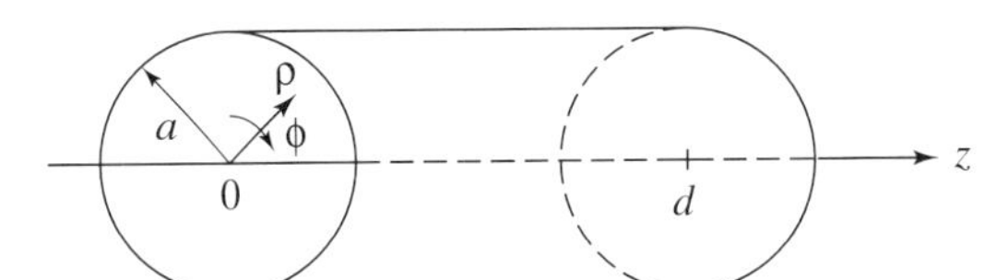

Figure 9-52. Configuration and dimensions for metal circular cylindrical cavity.

first kind and we replace β_{TM} by $-\beta_{\text{TM}}$ in the negative traveling wave solution):

$$E_z(\rho, \phi, z) = J_n\left(\frac{p_{nl}}{a}\rho\right)\cos n\phi\,(A^+e^{-j\beta_{\text{TM}}z} + A^-e^{j\beta_{\text{TM}}z}) \tag{9-348}$$

$$E_\rho(\rho, \phi, z) = j\frac{\beta_{\text{TM}}}{\left(\frac{p_{nl}}{a}\right)}J_n'\left(\frac{p_{nl}}{a}\rho\right)\cos n\phi\,(-A^+e^{-j\beta_{\text{TM}}z} + A^-e^{j\beta_{\text{TM}}z}) \tag{9-349}$$

$$E_\phi(\rho, \phi, z) = j\frac{n\beta_{\text{TM}}}{\left(\frac{p_{nl}}{a}\right)^2\rho}J_n\left(\frac{p_{nl}}{a}\rho\right)\sin n\phi\,(A^+e^{-j\beta_{\text{TM}}z} - A^-e^{j\beta_{\text{TM}}z}) \tag{9-350}$$

$$H_\rho(\rho, \phi, z) = j\frac{\omega n\epsilon}{\left(\frac{p_{nl}}{a}\right)^2\rho}J_n\left(\frac{p_{nl}}{a}\rho\right)\sin n\phi\,(-A^+e^{-j\beta_{\text{TM}}z} - A^-e^{j\beta_{\text{TM}}z}) \tag{9-351}$$

$$H_\phi(\rho, \phi, z) = j\frac{\omega\epsilon}{\left(\frac{p_{nl}}{a}\right)}J_n'\left(\frac{p_{nl}}{a}\rho\right)\cos n\phi\,(-A^+e^{-j\beta_{\text{TM}}z} - A^-e^{j\beta_{\text{TM}}z}) \tag{9-352}$$

In all of the equations $n = 0, 1, 2, \ldots$, and $l = 1, 2, 3, \ldots$. At $z = 0$, E_ρ (and E_ϕ) must be zero since it is tangent at a metal end plate boundary. Equation 9-349 at $z = 0$ then gives $A^- = A^+$. Making this substitution in the preceding five equations, factoring out A^+, and substituting the corresponding sine or cosine function for the exponential combinations in parentheses yields the reduced forms as follows:

$$E_z(\rho, \phi, z) = 2A^+J_n\left(\frac{p_{nl}}{a}\rho\right)\cos n\phi\cos\beta_{\text{TM}}z \tag{9-353}$$

$$E_\rho(\rho, \phi, z) = -\frac{2\beta_{\text{TM}}A^+}{\left(\frac{p_{nl}}{a}\right)}J_n'\left(\frac{p_{nl}}{a}\rho\right)\cos n\phi\sin\beta_{\text{TM}}z \tag{9-354}$$

$$E_\phi(\rho, \phi, z) = \frac{2n\beta_{\text{TM}}A^+}{\left(\frac{p_{nl}}{a}\right)^2\rho}J_n\left(\frac{p_{nl}}{a}\rho\right)\sin n\phi\sin\beta_{\text{TM}}z \tag{9-355}$$

$$H_\rho(\rho, \phi, z) = -j\frac{2\omega n\epsilon A^+}{\left(\frac{p_{nl}}{a}\right)^2\rho}J_n\left(\frac{p_{nl}}{a}\rho\right)\sin n\phi\cos\beta_{\text{TM}}z \tag{9-356}$$

$$H_\phi(\rho, \phi, z) = -j\frac{2\omega\epsilon A^+}{\left(\frac{p_{nl}}{a}\right)}J_n'\left(\frac{p_{nl}}{a}\rho\right)\cos n\phi\cos\beta_{\text{TM}}z \tag{9-357}$$

The final boundary condition is that at $z = dE_\rho$ (and E_ϕ) must again be zero. From Eq. 9-354 this requires

$$\sin \beta_{\mathrm{TM}} d = 0$$

or

$$\beta_{\mathrm{TM}} = \frac{q\pi}{d}, \qquad q = 0, 1, 2, \ldots \tag{9-358}$$

The modes resulting from these TM field configurations are denoted by TM_{nlq}. The lowest-order modes would then be TM_{010}, TM_{011}, and TM_{110}.

For the resonant frequency we use Eq. 9-197 for β_{TM} and the value of β_{TM} above to obtain, by equating them,

$$\sqrt{\omega_{\mathrm{res}}^2 \mu\epsilon - \left(\frac{p_{nl}}{a}\right)^2} = \frac{q\pi}{d}$$

from which

$$f_{\mathrm{res}}^{\mathrm{TM}} = \frac{1}{\sqrt{\mu\epsilon}} \sqrt{\left(\frac{q}{2d}\right)^2 + \left(\frac{p_{nl}}{2\pi a}\right)^2} \tag{9-359}$$

Then

$$\lambda_0^{\mathrm{res}} = \frac{v_0}{f_{\mathrm{res}}} = \frac{1}{\sqrt{\mu\epsilon} f_{\mathrm{res}}} = \frac{1}{\sqrt{\left(\frac{q}{2d}\right)^2 + \left(\frac{p_{nl}}{2\pi a}\right)^2}} \tag{9-360}$$

As usual we compute the guide wavelength as

$$\lambda_g^{\mathrm{res}} = \frac{2\pi}{\beta_{\mathrm{TM}}} = \frac{2d}{q} \tag{9-361}$$

This last result has the same physical interpretation as that for the rectangular guide; that is, for resonance the cavity must be an integral number of wavelengths long. For the cylindrical cavity, however, q can be zero so there is one special case. This is the electrically short cavity, where typically $d < \lambda_g^{\mathrm{res}}/2$, for TM_{010}.

The time-dependent functions corresponding to Eqs. 9-353 to 9-357 are obtained by multiplying by $e^{j\omega t}$ and taking the real part. The results are

$$\mathscr{E}_z(\rho, \phi, z, t) = 2A^+ J_n\left(\frac{p_{nl}}{a}\rho\right) \cos n\phi \cos \beta_{\mathrm{TM}} z \cos \omega t \tag{9-362}$$

$$\mathscr{E}_\rho(\rho, \phi, z, t) = -\frac{2\beta_{\mathrm{TM}} A^+}{\left(\frac{p_{nl}}{a}\right)} J_n'\left(\frac{p_{nl}}{a}\rho\right) \cos n\phi \sin \beta_{\mathrm{TM}} z \cos \omega t \tag{9-363}$$

$$\mathscr{E}_\phi(\rho, \phi, z, t) = \frac{2n\beta_{TM}A^+}{\left(\frac{p_{nl}}{a}\right)^2 \rho} J_n\left(\frac{p_{nl}}{a}\rho\right) \sin n\phi \, \sin\beta_{TM}z \, \cos\omega t \tag{9-364}$$

$$\mathscr{H}_\rho(\rho, \phi, z, t) = \frac{2\omega n\epsilon A^+}{\left(\frac{p_{nl}}{a}\right)^2 \rho} J_n\left(\frac{p_{nl}}{a}\rho\right) \sin n\phi \, \cos\beta_{TM}z \, \sin\omega t \tag{9-365}$$

$$\mathscr{H}_\phi(\rho, \phi, z, t) = \frac{2\omega\epsilon A^+}{\left(\frac{p_{nl}}{a}\right)} J_n'\left(\frac{p_{nl}}{a}\rho\right) \cos n\phi \, \cos\beta_{TM}z \, \sin\omega t \tag{9-366}$$

As a visual example, let's consider the lowest-order mode which is TM_{010}. For this mode $\beta_{TM} = 0$ ($q = 0$). If we take an instant in time, say $\omega_{res}t = \pi/4$, the time equations are (for $\omega = \omega_{res} = (1/\sqrt{\mu\epsilon})(p_{01}/a)$):

$$\mathscr{E}_z^{010}(\rho, \phi, z) = \sqrt{2}A^+J_0\left(\frac{p_{01}}{a}\rho\right) \tag{9-367}$$

$$\mathscr{H}_\phi^{010}(\rho, \phi, z) = \sqrt{2}\sqrt{\frac{\epsilon}{\mu}}A^+J_0'\left(\frac{p_{01}}{a}\rho\right) \tag{9-368}$$

From Table 9-2, $p_{01} \approx 2.405$. If we let $a = 2.405$ we obtain forms that allow for easy plotting:

$$\mathscr{E}_z^{010}(\rho, \phi, z) = \sqrt{2}A^+J_0(\rho) \tag{9-369}$$

$$\mathscr{H}_\phi^{010}(\rho, \phi, z) = \sqrt{2}\sqrt{\frac{\epsilon}{\mu}}A^+J_0'(\rho) = -\sqrt{2}\sqrt{\frac{\epsilon}{\mu}}A^+J_1(\rho) \tag{9-370}$$

These are plotted in Figure 9-53.

The displacement current for this mode and cavity size is obtained from Eq. 9-362 since all other components of $\vec{\mathscr{E}}$ are zero:

$$\begin{aligned}
\vec{\mathscr{J}}_D^{010}(\rho, \phi, z, t) &= \epsilon\frac{\partial\mathscr{E}_z^{010}}{\partial t}\hat{z} = -2\omega_{res}\epsilon A^+J_0(\rho)\sin\omega_{res}t\,\hat{z} \text{ A/m}^2 \\
\vec{\mathscr{J}}_D^{010}\left(\rho, \phi, z, \omega_{res}t = \frac{\pi}{4}\right) &= -\sqrt{2}\frac{1}{\sqrt{\mu\epsilon}}\epsilon A^+J_0(\rho)\hat{z} \\
&= -\sqrt{2}\sqrt{\frac{\epsilon}{\mu}}A^+J_0(\rho)\hat{z} \text{ A/m}^2
\end{aligned} \tag{9-371}$$

These field lines are in the $-\hat{z}$ direction, which is consistent with the right-hand rule where the $\vec{\mathscr{H}}$ lines circulate around $\vec{\mathscr{J}}_D$ lines. Note particularly that the displacement current is strongest at $\rho = 0$ whereas the magnetic field is a minimum.

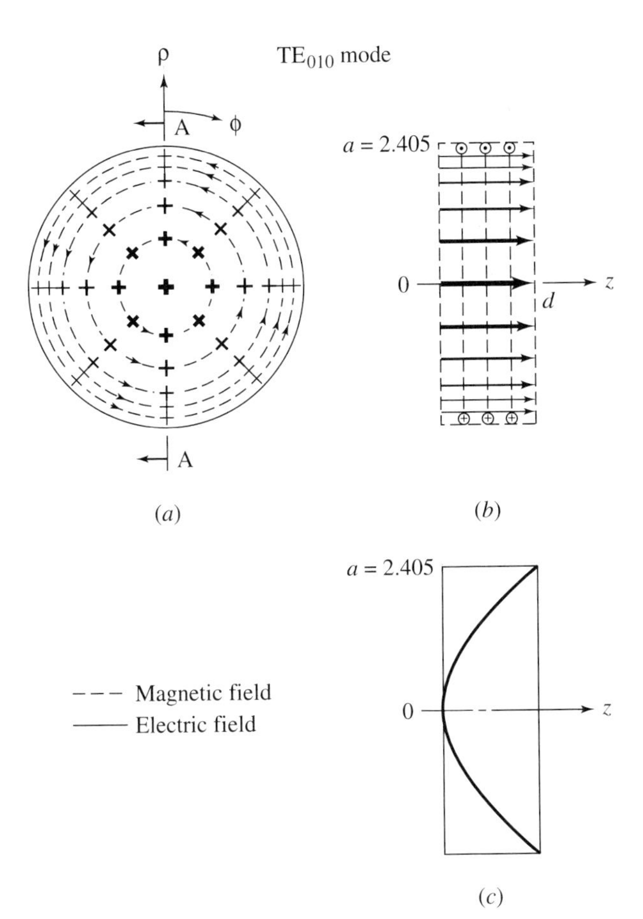

Figure 9-53. Electric and magnetic field lines for lowest-order mode in a circular cylindrical resonant cavity. (*a*) Looking in $+z$ direction. (*b*) Section *A-A*. (*c*) Section *A-A* alternative representation of electric field amplitude.

This particular mode, however, is *not* the best one that could be used for a variable cavity frequency for measurement. To see why this is so, let's determine the surface currents on the *inside* of the cavity walls. On the circumferential walls we have, from the boundary condition on $\vec{\mathcal{H}}$ at a perfect conductor using Eq. 9-366,

$$\begin{aligned}\vec{\mathcal{J}}_s(a, \phi, z, t) &= \hat{n}x\vec{\mathcal{H}}\big|_{\rho=a} = (-\hat{\rho}) \times \mathcal{H}_\phi\hat{\phi}\big|_{\rho=a} \\ &= -\frac{2\omega_{es}\epsilon A^+}{\left(\dfrac{p_{01}}{a}\right)} J_0'\left(\frac{p_{01}}{a}a\right)\sin\omega t\ \hat{z}\end{aligned} \tag{9-372}$$

For the cavity size given, $a = p_{01} = 2.405$,

$$\vec{\mathcal{J}}_s\left(a, \phi, z, \omega_{\text{res}}t = \frac{\pi}{4}\right) = -\sqrt{2}\sqrt{\frac{\epsilon}{\mu}}A^+J_0'(2.405)\hat{z}$$

$$= \sqrt{2}\sqrt{\frac{\epsilon}{\mu}}J_1(2.405)A^+\hat{z}$$

$$= \sqrt{2}\sqrt{\frac{\epsilon}{\mu}}(0.519)A^+\hat{z} = 0.734\sqrt{\frac{\epsilon}{\mu}}A^+\hat{z}\ \text{A/m} \qquad (9\text{-}373)$$

On the end plate at $z = 0$ the current is

$$\vec{\mathcal{J}}_s\left(\rho, \phi, o, \omega_{\text{res}}t = \frac{\pi}{4}\right) = \hat{n} \times \vec{\mathcal{H}} = \hat{z} \times \mathcal{H}_\phi\hat{\phi} = \sqrt{2}\sqrt{\frac{\mu}{\epsilon}}A^+J_1(\rho)\hat{\rho}\ \text{A/m} \qquad (9\text{-}374)$$

On the end plate at $z = d$ the current is

$$\vec{\mathcal{J}}_s\left(\rho, \phi, d, \omega_{\text{res}}t = \frac{\pi}{4}\right) = (-\hat{z}) \times \mathcal{H}_\phi\hat{\phi} = -\sqrt{2}\sqrt{\frac{\mu}{\epsilon}}A^+J_1(\rho)\hat{\rho}\ \text{A/m} \qquad (9\text{-}375)$$

Plots of these surface currents and the displacement current are given in Figure 9-54. Note particularly at $z = d$ and $\rho = 0$ that when the surface currents end displacement currents begin so that Kirchhoff's current law holds at the center of the right end plate. Thus the surface current doesn't simply disappear mysteriously.

Note that if the resonant frequency of the cavity is to be variable the dimension a must be varied, not an easy task. Also, since the surface current flows around the corners of the cavity any break there would have to be a very tight fit corner to avoid interrupting the current flow. It turns out that the best mode to use for frequency meter cavity is the TE_{011} mode for two reasons. First, the resonant frequency can be changed by varying the length

Figure 9-54. Displacement and surface currents for the lowest-order mode in a circular cylindrical cavity corresponding to Figure 9-53.

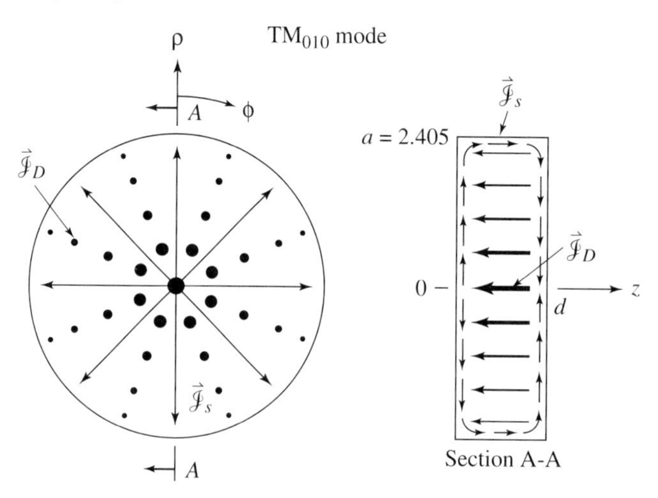

d, which can be accomplished by having an end plate that can be adjusted inside the cavity with a reasonably close fit. Second, there are no surface currents crossing the cavity corners so some small discontinuity there will have little effect on cavity losses.

As expected, the details of the development of the TE_{nlq} modes are left to the exercises. The waveguide equations used are Eqs. 9-181 through 9-187. The solutions are (where p'_{nl} is the lth root (zero) of the derivative of the nth-order Bessel function of the first kind, and the boundary conditions used are $H_z(z = 0) = H_z(z = d) = 0$):

$$H_z(\rho, \phi, z) = -j2A^+ J_n\left(\frac{p'_{nl}}{a}\rho\right) \cos n\phi \sin \beta_{\text{TE}} z \tag{9-376}$$

$$E_\rho(\rho, \phi, z) = 2\frac{\omega\mu n A^+}{\left(\frac{p'_{nl}}{a}\right)^2 \rho} J_n\left(\frac{p'_{nl}}{a}\rho\right) \sin n\phi \sin \beta_{\text{TE}} z \tag{9-377}$$

$$E_\phi(\rho, \phi, z) = 2\frac{\omega\mu A^+}{\left(\frac{p'_{nl}}{a}\right)} J'_n\left(\frac{p'_{nl}}{a}\rho\right) \cos n\phi \sin \beta_{\text{TE}} z \tag{9-378}$$

$$H_\rho(\rho, \phi, z) = -j2\frac{\beta_{\text{TE}} A^+}{\left(\frac{p'_{nl}}{a}\right)} J'_n\left(\frac{p'_{nl}}{a}\rho\right) \cos n\phi \cos \beta_{\text{TE}} z \tag{9-379}$$

$$H_\phi(\rho, \phi, z) = j2\frac{n\beta_{\text{TE}} A^+}{\left(\frac{p'_{nl}}{a}\right)^2 \rho} J_n\left(\frac{p'_{nl}}{a}\rho\right) \sin n\phi \cos \beta_{\text{TE}} z \tag{9-380}$$

In all the preceding equations, $n = 0, 1, 2, 3, \ldots$, and $l = 1, 2, 3, \ldots$.

$$\beta_{\text{TE}} = \frac{q\pi}{d}, \qquad q = 1, 2, 3, \ldots \tag{9-381}$$

$$f_{\text{res}} = \frac{1}{\sqrt{\mu\epsilon}} \sqrt{\left(\frac{q}{2d}\right)^2 + \left(\frac{p'_{nl}}{2\pi a}\right)^2} \tag{9-382}$$

$$\lambda_0^{\text{res}} = \frac{v_0}{f_{\text{res}}} = \frac{1}{\sqrt{\mu\epsilon} f_{\text{res}}} = \frac{1}{\sqrt{\left(\frac{q}{2d}\right)^2 + \left(\frac{p'_{nl}}{2\pi a}\right)^2}} \tag{9-383}$$

$$\lambda_g^{\text{res}} = \frac{2d}{q} \tag{9-384}$$

The corresponding time domain equations are:

$$\mathcal{H}_z(\rho, \phi, z, t) = 2A^+ J_n\left(\frac{p'_{nl}}{a}\rho\right) \cos n\phi \sin \beta_{\text{TE}} z \sin \omega t \tag{9-385}$$

$$\mathscr{E}_\rho(\rho, \phi, z, t) = 2\frac{\omega\mu n A^+}{\left(\frac{p'_{nl}}{a}\right)^2 \rho} J_n\left(\frac{p'_{nl}}{a}\rho\right) \sin n\phi \sin\beta_{\text{TE}}z \cos\omega t \tag{9-386}$$

$$\mathscr{E}_\phi(\rho, \phi, z, t) = 2\frac{\omega\mu A^+}{\left(\frac{p'_{nl}}{a}\right)} J'_n\left(\frac{p'_{nl}}{a}\rho\right) \cos n\phi \sin\beta_{\text{TE}}z \cos\omega t \tag{9-387}$$

$$\mathscr{H}_\rho(\rho, \phi, z, t) = 2\frac{\beta_{\text{TE}}A^+}{\left(\frac{p'_{nl}}{a}\right)} J'_n\left(\frac{p'_{nl}}{a}\rho\right) \cos n\phi \cos\beta_{\text{TE}}z \sin\omega t \tag{9-388}$$

$$\mathscr{H}_\phi(\rho, \phi, z, t) = -2\frac{\beta_{\text{TE}}A^+}{\left(\frac{p'_{nl}}{a}\right)^2 \rho} J_n\left(\frac{p'_{nl}}{a}\rho\right) \sin n\phi \cos\beta_{\text{TE}}z \sin\omega t \tag{9-389}$$

Plots of the lowest-order modes for other TE and TM modes are given in Figure 9-55. The TE_{011} mode is used for microwave frequency meters. Note that the TE_{111} mode is much like that of the rectangular cavity.

Exercises

9-11.2a Explain, on physical grounds, why q cannot be zero for TE_{nlq} modes.

9-11.2b Beginning with the cylindrical waveguide equations derive the field component equations and operating constants for the TE_{nlq} modes.

9-11.2c Plot the field structures for the TE_{011}, TE_{111}, and TM_{011} modes at $\omega_{\text{res}}t = \pi/4$. Plot these on a cavity drawing.

9-11.2d For the TE_{011} mode in a cylindrical cavity of radius $a = 3.832$ (arbitrary units) determine equations for and plot on cavity cross sections the fields $\vec{\mathscr{J}}_s$ and $\vec{\mathscr{J}}_D$.

9-11.2e For the TM_{011} mode cavity of Figure 9-55*c*. Draw the field lines and arrow ends at cross section *C*-*C*.

9-11.2f Draw figures of cavities that show how to use probe, loop, and aperture couplings to excite TM_{010}, TM_{011}, TE_{011}, and TE_{111} modes in a circular cylindrical cavity.

9-11.2g Design a circular cylindrical cavity that operates in the TE_{011} mode at 10 GHz. Compute the cavity tuning sensitivity defined as $S = \partial\omega_{\text{res}}/\partial d$. How far must the end plate be moved to produce a 100-MHz (1%) change in the resonant frequency?

9-11.3 The Circular Cylindrical Coaxial Resonant Cavity

Another important resonant structure is the coaxial circular cylindrical cavity. Since the solution process is identical to that already presented the details are left for general exercises at the end of the chapter.

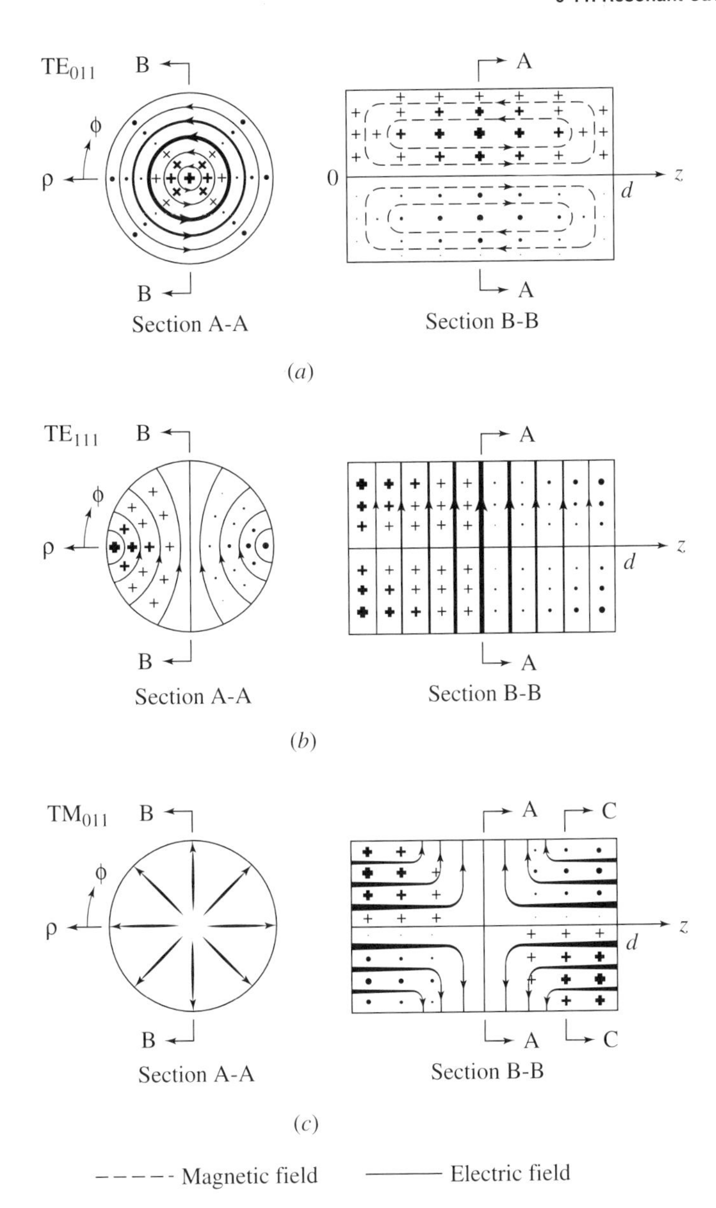

Figure 9-55. Plots of the lowest-order modes in a circular cylindrical cavity. Arrow dots and feathers are magnetic fields.

9-11.4 Energy Storage in a Resonant Cavity

The time forms of the field components in a resonant cavity help us devise a way to compute the energy stored in a resonant cavity. Since we have assumed perfectly conducting walls, the energy in the cavity must be constant at every instant of time. For any of the cases studied one finds that the electric field components all have one time function variation (all $\sin \omega t$ or $\cos \omega t$) and the magnetic field components also all have the same time function but *opposite* that of the electric field (all $\cos \omega t$ or $\sin \omega t$). Thus we can compute the energy stored most conveniently by considering the instant of time when either the electric field is zero or the magnetic field is zero, since all of the energy must be in the other nonzero field.

This result is probably not too surprising since the identical result is obtained for a resonant *LC* circuit studied in network theory. The energy transfers back and forth between the L and C such that $\frac{1}{2}(V_C)^2_{\text{max}} = \frac{1}{2}(I_L)^2_{\text{max}}$.

Exercises

9-11.4a For the TM_{011} mode in a cylindrical cavity compute the energy stored (in joules) in the electric field at the instant of time the electric field is a maximum. The cavity is air-filled. Repeat for the maximum magnetic field and compare. (Hint: Integrate $\frac{1}{2}\epsilon|\vec{\mathscr{E}}_{\text{max}}(t)|^2$ and $\frac{1}{2}\mu|\vec{\mathscr{H}}_{\text{max}}(t)|^2$ over the cavity volume.

9-11.4b Repeat Exercise 9-11.4a for the TE_{101} mode in a rectangular resonant cavity.

9-11.5 Resonant Cavity Q

Since practical conductors have resistance an actual cavity operates over a narrow range of frequencies, not just at the resonant frequency ω_{res}. The dielectric may also have losses. As in circuit theory we define a cavity quality factor Q, which is a measure of how selective the cavity is at indicating the resonant frequency. We also define cavity bandwidth and upper and lower 3-dB frequencies in the usual way. The convention adopted is illustrated in Figure 9-56, where ω_1 and ω_2 are the lower and upper 3-dB points.

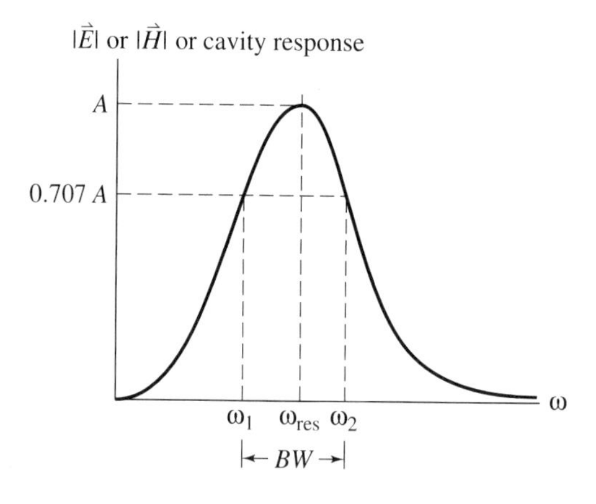

Figure 9-56. Convention for defining the Q of a resonant cavity.

From the definition of Q we obtain the following sequence of expressions:

$$Q = \frac{2\pi \text{ energy stored}}{\text{energy lost per period}} = \frac{2\pi U_{\max}}{\text{average power loss} \times T_{\text{res}}} = \frac{2\pi U_{\max}}{P_{\text{diss}} \times T_{\text{res}}} \tag{9-390}$$

where $U_{\max}$ is the maximum energy in joules at resonance, P_{diss} is the average power loss in watts, and T_{res} is the period of the resonant frequency. Since $f_{\text{res}} = 1/T_{\text{res}}$ we can extend the preceding equation to

$$Q = \frac{2\pi f_{\text{res}} U_{\max}}{P_{\text{diss}}} = \frac{\omega_{\text{res}}}{\left(\dfrac{P_{\text{diss}}}{U_{\max}}\right)} = \frac{\omega_{\text{res}}}{\omega_2 - \omega_1} = \frac{\omega_{\text{res}}}{BW} \tag{9-391}$$

For Q greater than about 5 (essentially always the case),

$$\omega_{\text{res}} \approx \frac{\omega_1 + \omega_2}{2} \qquad \text{and} \qquad \omega_{\text{res}} - \omega_1 \approx \omega_2 - \omega_{\text{res}} = \frac{BW}{2} \tag{9-392}$$

As expected, when $P_{\text{diss}} = 0$ the bandwidth is also zero. This yields a single spike as the response, which occurs at ω_{res}.

To this point all of our resonant cavity solutions were based on the assumption of perfect (lossless) conductor walls. Strictly speaking our first step would be to re-solve all of those cases using lossy walls, not a pleasant prospect. However, as discussed in Section 9-8 practical microwave systems have small losses (wall conductivities on the order of 10^7 S/m (resistivity, 10^{-7} $\Omega \cdot$ m) so that we can use the lossless solutions for the fields to compute losses in the system!

To compute the energy stored we use the ideas of Section 9-11.4 and take the value of time when all of the energy is stored in either $\vec{\mathscr{E}}$ or $\vec{\mathscr{H}}$. Selecting $\vec{\mathscr{E}}$ we would then have

$$U_{\max} = \int_V \frac{1}{2}\, \epsilon |\vec{\mathscr{E}}_{\max}(t)|^2 dv \tag{9-393}$$

where the volume V is the cavity volume. If the dielectric is lossy we would approximate ϵ by its lossless value, $\epsilon' \approx \epsilon_0 \epsilon_r$.

The power dissipated in the system arises from four primary sources:

1. Wall conductor losses due to current flow
2. Dielectric losses due to conduction current flow, $\sigma_d \vec{E}$
3. Small holes in the cavity used to inject or remove field samples
4. Cavity perturbations such as dents or nonuniform dielectric

The wall losses are determined by computing the power flows into the separate walls and adding the results. This was discussed in Section 9-8.2. For waveguide wall losses, for the assumption of small losses, the result given there is

$$P_w = \tfrac{1}{2}|\vec{J}_s|^2 R_s = \tfrac{1}{2}|\vec{H}_{\tan}|^2 R_s \ \text{W/m}^2 \tag{9-394}$$

where

$$R_s = \frac{1}{\delta \sigma_w} = \sqrt{\frac{\pi f \mu}{\sigma_w}}\ \Omega$$

The dielectric losses would be determined using the results of Chapter 7, the loss part of Eq. 7-59:

$$P_d = \tfrac{1}{2}\sigma_d|\vec{E}|^2 \text{ W/m}^2 \tag{9-395}$$

Losses due to holes depend upon the type of coupling used to extract energy from or put energy into the cavity. These losses will not be considered here. From all of the preceding losses we would compute the power dissipation required by Eq. 9-391 as

$$P_{\text{diss}} = \underbrace{\oint_s \tfrac{1}{2}|\vec{H}_{\tan}|^2 R_s da_w}_{\text{cavity walls}} + \underbrace{\int_V \tfrac{1}{2}\sigma|\vec{E}|^2 dv}_{\text{cavity volume}} + P_{\text{holes}} + P_{\text{perturbation}} \text{ W} \tag{9-396}$$

Example 9-15

Determine the expression for the Q_{101} of the TE_{101} mode in a rectangular cavity caused by wall losses only. Evaluate for a square cavity at 11 GHz, for copper walls.

For this mode the field equations show that only H_x, H_z, and E_y are present, and $\beta = \pi/d$. Then from Eq. 9-339 at ω_{res}:

$$\mathscr{E}_y = -\frac{2\omega_{\text{res}}\mu a A^+}{\pi}\sin\frac{\pi}{a}x\,\sin\frac{\pi}{d}z\,\cos\omega_{\text{res}}t$$

$$\mathscr{H}_x = j\frac{2aA^+}{d}\sin\frac{\pi}{a}x\,\cos\frac{\pi}{d}z\,\sin\omega_{\text{res}}t$$

$$\mathscr{H}_z = -j2A^+\cos\frac{\pi}{a}x\,\sin\frac{\pi}{d}z\,\sin\omega_{\text{res}}t$$

Then

$$\begin{aligned}
U_{\max} &= \int_0^d\int_0^b\int_0^a \frac{1}{2}\epsilon|\mathscr{E}_{y,\max}(t)|^2 dx\,dy\,dz \\
&= \frac{\epsilon}{2}\int_0^d\int_0^b\int_0^a \frac{4\omega_{\text{res}}^2\mu^2a^2A^{+2}}{\pi^2}\sin^2\frac{\pi}{a}x\,\sin^2\frac{\pi}{a}z\,dx\,dy\,dz \\
&= \left(\frac{2\epsilon\omega_{\text{res}}^2\mu^2a^2}{\pi^2}A^{+2}\right)(b)\int_0^d \sin^2\frac{\pi}{d}z\,dz\int_0^a \sin^2\frac{\pi}{a}x\,dx \\
&= \frac{2\epsilon\omega_{\text{res}}^2\mu^2a^2b}{\pi^2}A^{+2}\left(\frac{1}{2}z - \frac{1}{4\frac{\pi}{d}}\sin\frac{2\pi}{d}z\right)\Bigg|_o^d\left(\frac{1}{2}x - \frac{1}{4\frac{\pi}{\alpha}}\sin\frac{2\pi}{a}x\right)\Bigg|_0^a \\
&= \frac{2\epsilon\omega_{\text{res}}^2\mu^2a^2b}{\pi^2}A^{+2}\left(\frac{d}{2}\right)\frac{a}{2} = \frac{\epsilon\omega_{\text{res}}^2\mu^2a^3bd}{2\pi^2}A^{+2}\text{ J}
\end{aligned}$$

$$\begin{aligned}
P_{\text{diss}} &= \int_{\substack{\text{top}\\ y=b}} P_w da + \int_{\substack{\text{bottom}\\ y=0}} P_w da + \int_{\substack{\text{front}\\ z=0}} P_w da \\
&\quad + \int_{\substack{\text{back}\\ z=d}} P_w da + \int_{\substack{\text{left}\\ x=z}} P_w da + \int_{\substack{\text{right}\\ x=0}} P_w da \text{ W}
\end{aligned}$$

The fields are symmetric at opposite walls, pairs of integrals are equal, so we have only three integrals to evaluate:

$$P_{\text{diss}} = 2\int_{y=0} P_w da + 2\int_{z=0} P_w da + 2\int_{x=0} P_w da \text{ W}$$

The Poynting vector into the walls is, from the discussion of attenuation,

$$\tfrac{1}{2}R_s|\vec{J}_s|^2 = \tfrac{1}{2}R_s|\vec{H}_{\tan}|^2 \text{ W/m}^2$$

Then

$$P_{\text{diss}} = 2\int_0^d\int_0^a \frac{1}{2}R_s[H_x^2 + H_z^2]\bigg|_{y=0} dxdz + 2\int_0^b\int_0^a \frac{1}{2}R_s H_x^2\bigg|_{z=0} dxdy$$

$$+ 2\int_0^d\int_0^b \frac{1}{2}R_s\left|H_z\right|^2\bigg|_{x=0} dydz$$

$$= R_s A^{+2}\left[\int_0^d\int_0^a\left(\frac{4a^2}{d^2}\sin^2\frac{\pi}{a}x\cos^2\frac{\pi}{d}z + 4\cos^2\frac{\pi}{a}x\sin^2\frac{\pi}{d}z\right)dxdz\right.$$

$$\left. + \int_0^b\int_0^a \frac{4a^2}{d^2}\sin^2\frac{\pi}{a}xdxdy + \int_0^d\int_0^b 4\sin^2\frac{\pi}{d}zdydz\right]$$

$$P_{\text{diss}} = R_s A^{+2}\left[\frac{4a^2}{d^2}\cdot\frac{a}{2}\cdot\frac{d}{2} + 4\cdot\frac{a}{2}\cdot\frac{d}{2} + \frac{4a^2}{d^2}\cdot\frac{a}{2}\cdot b + 4\cdot b\cdot\frac{d}{2}\right]$$

$$= R_s A^{+2}\left(\frac{a^3}{d} + ad + \frac{2a^3b}{d^2} + 2bd\right) \text{ W}$$

Then

$$Q_{101} = \frac{\omega_{\text{res}}\dfrac{\epsilon\omega_{\text{res}}^2\mu^2a^3bd}{2\pi^2}A^{+2}}{R_sA^{+2}\left(\dfrac{a^3}{d} + ad + \dfrac{2a^3b}{d^2} + 2bd\right)} = \frac{\epsilon\omega_{\text{res}}^3\mu^2a^3bd}{2\pi^2R_s\left(\dfrac{a^3}{d} + ad + \dfrac{2a^3b}{d^2} + 2bd\right)}$$

Using ω_{res} from Eq. 9-334 as $2\pi f_{\text{res}}$,

$$\omega_{\text{res}} = \frac{1}{\sqrt{\mu\epsilon}}\sqrt{\left(\frac{\pi}{a}\right)^2 + \left(\frac{\pi}{d}\right)^2}$$

$$Q_{101} = \frac{\epsilon\mu^2a^3bd\dfrac{1}{\mu\epsilon\sqrt{\mu\epsilon}}\pi^2\left(\dfrac{1}{a^2} + \dfrac{1}{d^2}\right)\pi\left(\dfrac{1}{a^2} + \dfrac{1}{d^2}\right)^{1/2}}{2\pi^2R_s\left(\dfrac{a^3}{d} + ad + \dfrac{2a^3b}{d^2} + 2bd\right)}$$

$$= \frac{\pi a^3bd\eta\left(\dfrac{1}{a^2} + \dfrac{1}{d^2}\right)^{3/2}}{2R_s\left(\dfrac{a^3}{d} + ad + \dfrac{2a^3b}{d^2} + 2bd\right)}$$

For a square cavity, $a = b = d$:

$$Q_{101} = \frac{\pi a^5 \eta \left(\frac{2}{a^2}\right)^{3/2}}{2R_s(a^2 + a^2 + 2a^2 + 2a^2)} = \frac{\pi\sqrt{2}\eta}{6R_s}$$

At 11 GHz, for copper of conductivity 6×10^7 S/m, and an air-filled cavity

$$Q_{101} = \frac{\pi\sqrt{2}\sqrt{\dfrac{4\pi \times 10^{-7}}{\dfrac{1}{36\pi} \times 10^{-9}}}}{6\sqrt{\dfrac{\pi \times 11 \times 10^9 \times 4\pi \times 10^{-7}}{6 \times 10^7}}} = 10{,}376$$

This is a conservative (high) value since we have neglected dielectric losses, and so on.

The bandwidth is $(11 \times 10^9)/10{,}376 = 1.06$ MHz, which means that the 3-dB frequencies are at 11.00053 and 10.99947 GHz, indeed a narrow response.

Exercises

9-11.5a Determine the expression for Q_{011} for the TM_{011} mode in a circular cylindrical cavity of radius a and length d. Evaluate the Q for copper walls, at 11 GHz, with $d = 2a$. Compare with the numerical value from Example 9-15. Answer:

$$Q_{011} = \frac{\sqrt{\dfrac{\mu}{\epsilon}}\, p_{01}}{2R_s\left(\dfrac{a}{d} + 1\right)}$$

9-11.5b Repeat Exercise 9-11.5a for the TE_{011} mode.

9-11.5c Derive the expression for the Q or the lowest-order resonant mode in a coaxial cylindrical cavity in the TE mode.

9-11.5d Suppose we couple energy (at resonance) out of a circular cylindrical resonant cavity as shown in Figure 9-57*d*. Assuming the loop area A is small so the field is essentially constant, determine the power removed from the cavity in the TM_{010} mode. Z_0 of the coax is real. Use $v = d\lambda/dt$ volts where λ is the flux linkage. (In the sinusoidal steady state this is $\overline{V} = j\omega\lambda$.) Neglect the hole losses. Explain what would happen to the Q of the cavity.

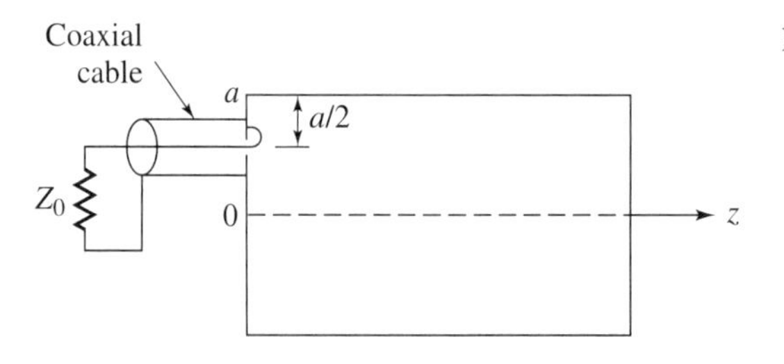

Figure 9-57.

Summary, Objectives, and General Exercises

Summary

In this chapter we learned how to analyze and design waveguides that are used to confine the propagation of energy in a desired direction. This is in contrast to the study of Chapter 8 where waves spread out or extended through all of space. We studied two main categories of waveguides: the metallic wall guides and the dielectric guides.

The introductions of mode structure and mode terminology for transverse magnetic (TM) and transverse electric (TE) types were given. Here we learned that depending upon the operating frequency, guide dimensions, and method of excitation we could select a desired mode. To assist in the selection of operating frequency and guide configuration the mode dispersion diagram was developed, which allowed us to select or eliminate modes of interest.

Qualitative ideas for coupling into and out of waveguides and cavities were given. Loop, probe, and aperture coupling principles were presented.

The chapter concluded with a detailed study of resonant cavities. The procedure for using the waveguide field equations to obtain the resonant frequencies and resonant field structures was given by way of several examples from rectangular and cylindrical cavity modes.

Since real materials used in waveguide construction and those used to fill the guide both have losses methods for computing the resulting attenuations and power losses were studied. This allowed us to predict the attenuation in a waveguide and to compute the power dissipated in a resonant cavity. For the resonant cavity this power loss computation made it possible to determine the quality factor Q of the cavity.

Objectives

1. Be able to use the product solution method on the scalar wave equations for $E_z(t_1, t_2, z)$ and $H_z(t_1, t_2, z)$ to show that these z components may be written as the product of two functions: one that is a function of z only and a second that is a function of transverse coordinates (t_1, t_2) only. Be able to solve the differential equation for the function of z, which is the general solution.
2. Be able to solve for TEM guide wave solutions (field equations and applications) for a given two-conductor system.
3. Be able to identify common configurations of TEM transmission systems and how to excite this mode in the configurations.
4. For a given TEM solution in a two-conductor system be able to compute ρ_s, $\vec{J}_s$, $\vec{P}$, and Z_0.
5. Be able to distinguish between surface current and current density at a conductor boundary.
6. Know how to use the solution procedure for TE and TM modes for z-directed propagation and be able to apply it to a given system. Be able to calculate the wave impedance. This would be for metal waveguides and dielectric waveguides.
7. Be able to calculate the cutoff frequency for a given solution to a TE or TM configuration. Be able to compute the equations for group and phase velocities.
8. Know the conventions for the designations TE_{mn} and TM_{mn} and be able to describe their physical significances in waveguides.
9. Be able to show that $V_{p\mathrm{TEM}} = V_{g\mathrm{TEM}} = 1/\sqrt{\mu\epsilon}$ for a lossless dielectric, beginning with the wave equation for $\vec{E}$ and knowing that $\vec{E} = -\Delta_t\Phi = 0$.
10. Be able to compute the attenuation in a waveguide that includes wall losses and dielectric losses. Be able to explain how and why the lossless solutions can be used to compute losses.
11. Be able to obtain the mode dispersion diagram for a given waveguide system and to use the

diagram to predict the presence (or absence) of a given mode in that system.

12. Know how to obtain the field configurations in a resonant cavity and how to compute the resonant frequency, guide resonant wavelength, and free-space resonant wavelength.
13. Know the sources of power (energy) loss in a resonant cavity and be able to compute wall and dielectric losses.
14. Be able to state the definition of Q and be able to use the extended results of that definition to obtain the expression for the Q of a given cavity.
15. Be able to design physical feed systems for exciting a given mode in a transmission system by three methods (for waveguides and cavities)
    (a) Slot (aperture)
    (b) Loop
    (c) Probe
16. Know the definitions of evanescent waves and surface waves and be able to describe the differences and similarities.
17. Be able to follow the step-by-step solution method to obtain surface wave solutions for TE and TM modes.
18. Be able to give the physical description of phase velocity and wavelength, and to give the relationship among the parameters propagation constant, frequency, and wavelength.

General Exercises

GE9-1 The equality $\nabla^2 = \nabla \cdot \nabla$ is only true in rectangular coordinates. Show that if we use the general form $\nabla^2\Phi = \nabla \cdot (\nabla\Phi)$ any generalized coordinate system that has a z coordinate ($h_3 = 1$) still has component forms

$$\nabla = \nabla_t + \nabla_z \quad \text{and} \quad \nabla^2 = \nabla_t^2 + \nabla_z^2$$

GE9-2 Beginning with $\nabla^2\vec{H} + \bar{k}^2\vec{H} = 0$, show that for rectangular coordinates

$$\nabla^2 H_i + \bar{k}^2 H_i = 0, \qquad i = x, y, z$$

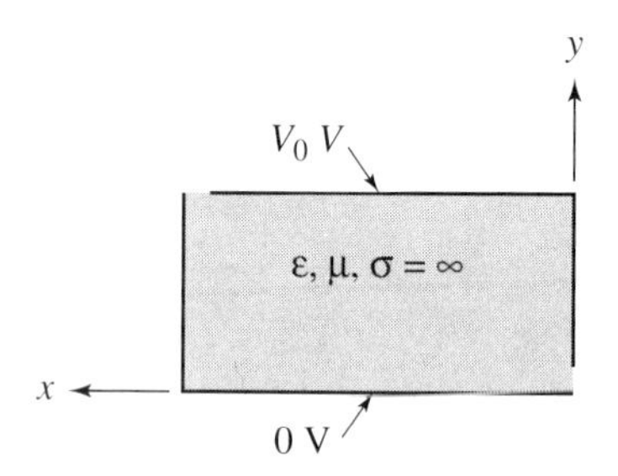

Figure 9-58.

but that for a generalized coordinate system the result holds only for $i = z$. (Hint: How does one handle unit vectors inside the ∇^2 operator?)

GE9-3 Solve for the TEM mode $\vec{E}$ and $\vec{H}$ for the configuration shown in Figure 9-58. Note that there will be field variations in both x and y. Determine expressions for λ_g, $\vec{S}$, and $Z^w_{z,\text{TEM}}$. Sketch $\vec{\mathscr{E}}(x, y, z = 0, t = 0)$. Explain what happens to this pattern for $z > 0$, but $t = 0$. Derive the equation for the surface charge density on the 0-V section.

GE9-4 A two-wire parallel line (twin lead) is shown in Figure 9-59. Using bipolar coordinates solve for the TEM mode structure. Note: $d/2$ is not the location at a in the coordinate system; see Exercise 4-12.1b and the accompanying figure for the coordinate system. You will need to determine values of the coordinate $v(\pm)$ that correspond to the wire surfaces of radii b. Begin by justifying that Φ will be a function of the coordinate u only, and then integrate. Plot one $\vec{E}$ line and one $\vec{H}$ line for $y \neq 0$ and obtain the equation for ρ_s on the right wire. The conductors are at potentials $\pm V_0/2$ so that the potential between the wires is V_0. Assume the medium around the wires extends to infinity in all directions. Determine the wave impedance in the z direction, the guide wavelength, and the group velocity. Determine the transmission line Z_0.

GE9-5 When solving for TEM modes the potential Φ was a function only of transverse coordinates. How does the potential (voltage) vary with z? Obtain the total potential function Φ_T as an ex-

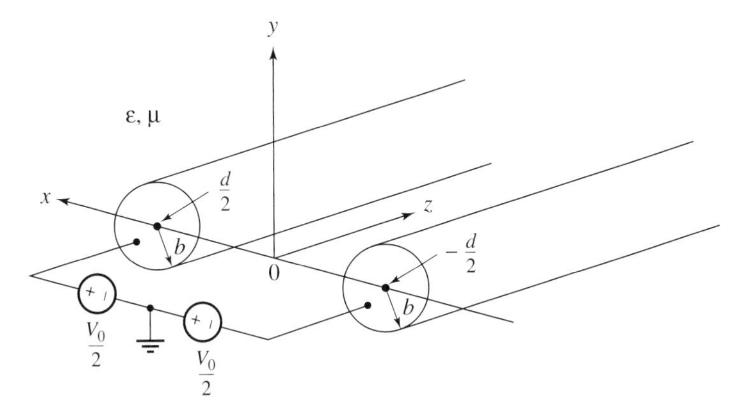

Figure 9-59.

plicit time function and relate the result to transmission line theory.

GE9-6 Using the parameters of the RG58C/U cable given in Figure 1-11 of Chapter 1, and assuming lossless dielectric and conductors, compute the characteristic impedance Z_0 for TEM modes using Eq. 9-63.

GE9-7 For infinite parallel planes, plot the electric and magnetic field lines, similar to those in Figure 9-15, for the TM_2 mode. There will be closed loops in the middle portion and half loops at the two planes for the electric field. Show how you could couple this mode using probes and loops.

GE9-8 A parallel plane waveguide is being used to transmit 6-GHz TM waves. The plate spacing is 4 cm. What dielectric should be used to allow only TM_1 to propagate? (Answer: $\epsilon_r < 1.56$)

GE9-9 The periodic waveform shown in Figure 9-60 is used to excite the TE_1 mode between parallel planes. What is the output waveform at $z = 5$ cm? The waveform consists of a fundamental frequency and the third harmonic. Since the system is linear, superposition applies. Fundamental amplitude is 1, harmonic is 0.5.

GE9-10 Plot $\mathcal{H}_z(y, z, t = 0)$ for the TE_2 mode.

GE9-11 Determine the surface charge density for the TE_2 mode.

GE9-12 For the parallel line of Figure 9-59 carry out, to the extent possible, the solution for the TE_1 mode.

GE9-13 For the elliptic cylindrical coordinate system shown in General Exercise GE5-5 of Chapter 5 determine, to the extent possible, the solution for the TM modes between concentric ellipses $u = 1$ and $u = \frac{3}{2}$.

GE9-14 For the elliptic cylindrical coordinate system shown in General Exercise GE5-5 determine, to the extent possible, the solution for the TE modes between the hyperbolas defined by $v = 2\pi/3, 4\pi/3$ and $v = \pi/3, 5\pi/3$.

Figure 9-60.

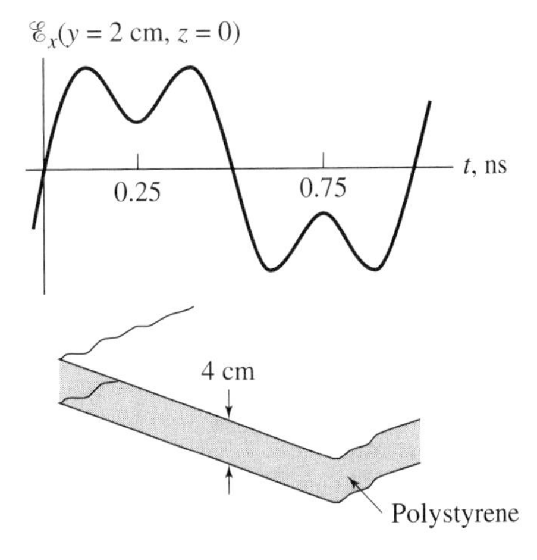

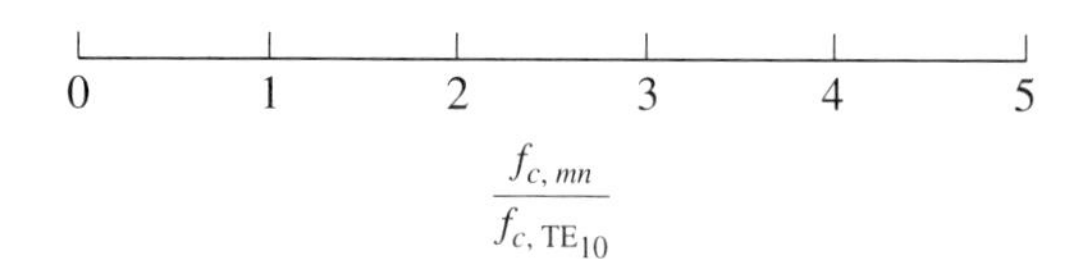

Figure 9-61.

GE9-15 Repeat GE9-13 except solve for the TEM mode. Give equations for all field components.

GE9-16 Repeat GE9-14 except solve for the TEM mode. Give equations for all field components.

GE9-17 The solution for TE modes between parallel planes (Section 9-4.3) used tangential electric field boundary conditions to solve for the constants in the general solution. Show that the same results are obtained by applying the boundary conditions on normal $\vec{B}$.

GE9-18 For the TE_{10} mode in a rectangular waveguide, plot the electric field intensity similar to Views 1 and 2 of Figure 9-20*a* for the cases $\omega t = \pi/2$ and $\omega t = \pi$. Explain how this illustrates wave propagation.

GE9-19 Is it possible to design a rectangular waveguide for which the TM_{11} mode will propagate but the TE_{10} will not? Prove your answer. Show how you would excite that mode if it is possible.

GE9-20 Plot the surface current density and magnetic field intensity on the $y = 0$ surface of a rectangular waveguide carrying the TM_{11} mode. See Figure 9-24. This is to be as viewed from inside the waveguide.

GE9-21 Let $f_{c,mn}$ be the cutoff frequency of TE_{mn} or TM_{mn} modes in a rectangular waveguide. On the line in Figure 9-61 show the locations of the three lowest TE_{mn} modes and the three lowest TM_{mn} modes. Use $a = 2b$. Explain how this is of use in waveguide design.

GE9-22 Another method for determining wave impedance that has been successful makes direct use of the equations for the field components. One selects a transverse component of $\vec{E}$ and the orthogonal transverse component of $\vec{H}$. Thus one would select E_{t_1},and H_{t_2} or E_{t_2} and H_{t_1} for the positive traveling wave. If the selected components yield $\vec{E}_{\text{comp}} \times \vec{H}_{\text{comp}}$ in the $+\hat{z}$ direction the ratio of those components is Z_0^w. If the cross product is in the $-\hat{z}$ direction the ratio of those components is $-Z_0^w$. Apply this concept to TE and TM modes in rectangular, cylindrical, and coaxial cylinders waveguides and verify the "characteristic impedances."

GE9-23 Let $f_{c,nl}$ be the cutoff frequency of TE_{nl} or TM_{nl} modes in a cylindrical (circular) waveguide. On the line of Figure 9-62 show the locations of the four lowest TE_{nl} modes and the four lowest TM_{nl} modes. Explain how this is of use in waveguide design.

GE9-24 An RG218/U coaxial cable has an inner conductor diameter of 4.95 mm and a polyethylene dielectric diameter of 17.27 mm. What is the lowest frequency for which a TE coaxial mode will propagate?

GE9-25 Prove that the wave impedances in the z direction are given by

$$Z_0^{\text{TM}} = \eta \frac{\lambda_0}{\lambda_g} \quad \text{and} \quad Z_0^{\text{TE}} = \eta \frac{\lambda_g}{\lambda_0}$$

for any waveguide system. (Hint: Use Eqs. 9-7 through 9-10 for TE and TM modes, along with Eq. 9-4 for the lossless case.)

GE9-26 Compute values for the cutoff frequencies and cutoff wavelengths and place them in the table of Figure 9-63.

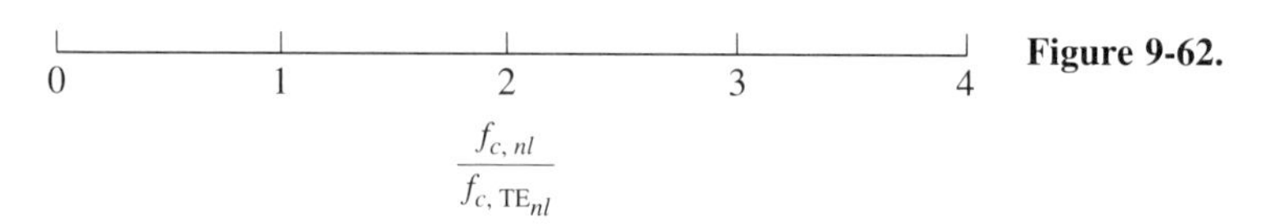

Figure 9-62.

Rectangular Guide $a = 2b$			Circular Guide radius $= a$		
	λ_c	ω_c		λ_c	ω_c
TE_{10}			TE_{11}		
TE_{01}			TM_{01}		
TE_{20}			TE_{21}		
TE_{11}			TE_{01}		
TM_{11}			TM_{11}		

Figure 9-63.

GE9-27 Prove that for any TE or TM mode the guide wavelength is

$$\lambda_g = \frac{\lambda_0}{\sqrt{1 - \left(\frac{\lambda_0}{\lambda_c}\right)^2}}$$

where $\lambda_0 = v_0/f = 1/f\sqrt{\mu\epsilon}$, the unbounded or TEM wavelength in the guide medium. Hint: For any mode

$$\lambda_g = \frac{2\pi}{\beta} = \frac{2\pi}{\text{IM}\{\gamma\}}$$

GE9-28 For waveguide structures operating below cutoff prove that the attenuation is

$$\alpha_{\text{dB/m}} = 54.6\sqrt{\mu\epsilon}f_c\sqrt{1 - \left(\frac{f}{f_c}\right)^2}$$

Neglect dielectric and wall losses. Begin with the general expression for γ and use $\alpha_{\text{dB/m}} = 8.686\alpha$.

GE9-29 Let $J(o)$ be the current density (A/m^2) just inside the surface of a conductor and flowing parallel to the surface. Determine the expression for the depth d into the material one must go so that the average current density from 0 to d times an area 1 m by d normal to the average J produces the same current as a given surface J_s flowing across the 1-m dimension. Your answer will be in terms of skin depth δ, $J(o)$, and J_s. Determine the value of this parameter d for copper, silver, aluminum, and gold at a frequency of 3 GHz. From these results compute the average J and suggest an appropriate value for $J(o)$ as some fraction of J_s that is given.

GE9-30 Suppose two circular plates are to be used as a transmission waveguide with waves propagating in the ρ direction. There are two possibilities for driving the system as shown in Figure 9-64. In one case propagation occures as $e^{-j\beta\rho}$ and in the other as $e_{j\beta\rho}$. Beginning with Maxwell's equations in cylindrical coordinates determine the differential equations the field components must satisfy for TEM ($E_\rho = H_\rho = 0$), TE ($E_\rho = 0$), and TM ($H_\rho = 0$) modes. Carry the solutions as far as possible. For TEM modes assume there is no ϕ variation ($\partial/\partial\phi = 0$), and the source amplitude is V_0. The loads are assumed to match the line so that there are no reflections, and sources and loads are assumed uniformly distributed around the guide rims.

GE9-31 Determine the equations for the field components of TEM waves propagating outward from the two cones shown in Figure 9-65. Prop-

Figure 9-64.

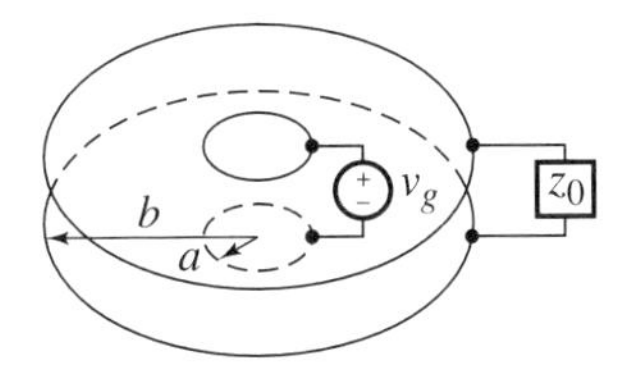

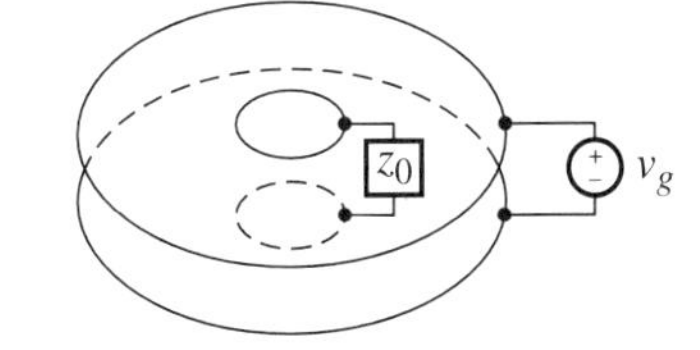

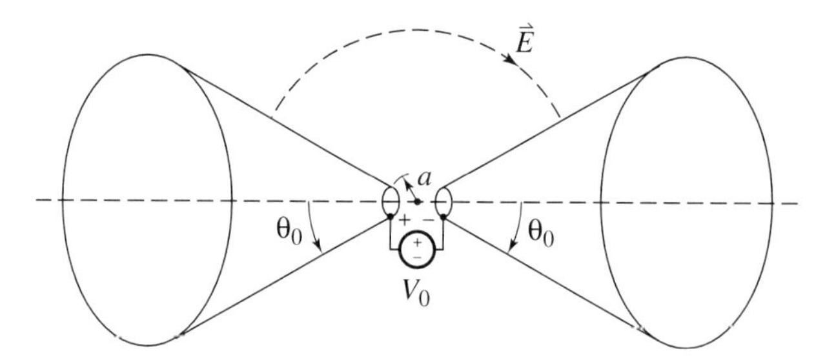

Figure 9-65.

agation occurs as $e^{-j\beta r}$. Begin with Maxwell's equations and assume there is no ϕ variation. Also determine the expression for β.

GE9-32 Determine the expressions for the wall attenuation constants α_w for TEM, TE, and TM modes in a coaxial waveguide. Also obtain explicit expressions for the two lowest-order TE and TM modes (one TE, one TM).

GE9-33 For the dielectric-coated flat metal sheet of Section 9-9.1, show by combining the two separation constant constraint equations (9-280 and 9-281) that for TM modes the quantity $(k_{c1,\text{TM}}d)$ can be determined graphically by plotting both members of the equation

$$\tan(k_{c1,\text{TM}}d) = \frac{\epsilon_1}{\epsilon_2} \frac{\sqrt{(k_1 d)^2 - (k_2 d)^2 - (k_{c1,\text{TM}}d)^2}}{(k_{c1,\text{TM}}d)}$$

and using the intersection of the two plots. Sketch a few curves. This also shows a maximum value for $(k_{c1,\text{TM}}d)$ for guided waves. What is that value?

GE9-34 For TM modes in a dielectric-coated flat sheet shown in Figure 9-38, show that using the normal $\vec{D}$ boundary condition results in the same separation constant constraint equation.

GE9-35 In Section 9-9.1 it was shown that for pure TE or TM modes on a dielectric-coated cylinder there could be no circumferential field variations ($n = 0$). For $n \neq 0$ only hybrid EH or HE modes were possible. Assuming both E_z and H_z are present develop the equations for all field components and the equations from which α_2 and k_{c1} can be determined.

GE9-36 For the two-dielectric system shown in Figure 9-66 prove that TEM modes cannot exist even with the two separate conducting plates. Derive the equations for the TE_{nm} modes. Assume the system extends $-\infty < x < +\infty$ and that there is no x variation. Do this by trying to solve for TEM modes and observing that if $\mu_1 = \mu_2$ the phase condition $\beta_1 = \beta_2$ cannot be met.

Figure 9-66.

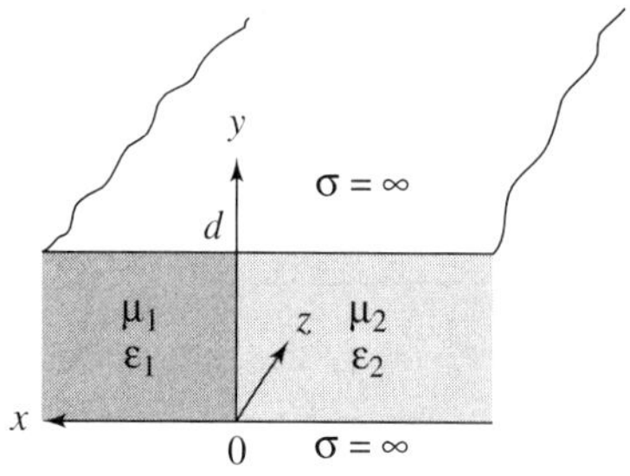

Figure 9-67.

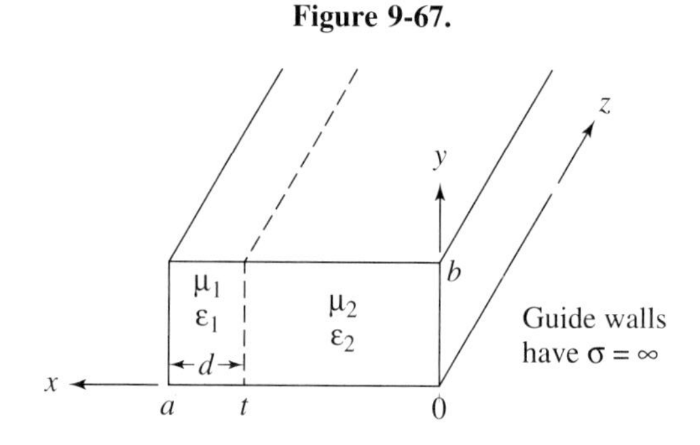

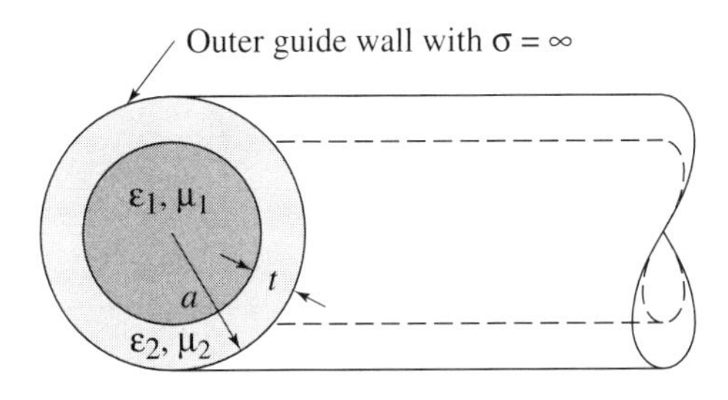

Figure 9-68.

GE9-37 Derive the equations for the TM_{nm} modes for the system of Figure 9-66.

GE9-38 Derive the equations for the TM_{nm} modes in the two-dielectric-filled *metal* rectangular guide shown in Figure 9-67. Partial answer: One of the equations from which the eigenvalues of the separation constants are determined is

$$\frac{-k_x 1 \epsilon_1 \left[\omega_2 \epsilon_2 - \mu_1 \epsilon_1) + k_{x1}^2 + \left(\dfrac{m\pi}{b} \right)^2 \right]}{\epsilon_2 \left[k_{x1}^2 + \left(\dfrac{m\pi}{b} \right)^2 \right] \sqrt{\omega^2(\mu_2\epsilon_2 - \mu_1\epsilon_1) + k_{x1}^2}} = \frac{\cot(\sqrt{\omega^2(\mu_2\epsilon_2 - \mu_1\epsilon_1) + k_{x1}^2}\, t)}{\cot(k_{x1} d)}$$

GE9-39 Repeat Exercise GE9-38 for TE_{nm}.

GE9-40 Use the transverse resonance technique to obtain one of the equations relating the separation constants and the material properties for the configuration of GE9-38, TM_{nm} modes.

GE9-41 Solve the system of Figure 9-67 for the TE_{n0} modes. Next set $\epsilon_2 = \epsilon_0$ and $n = 1$ and obtain the ratio of the phase shift per meter (rad/m) in this configuration to that for a complete air-filled guide. Also compare the cutoff frequencies.

GE9-42 A cylindrical metal waveguide ($\sigma = \infty$) is filled internally with two concentric dielectrics as shown in Figure 9-68. Solve for the field equations for the TM_{0l} modes.

GE9-43 A variable phase shifter is to be made by constructing a waveguide for which the dielectric insert (assumed to be infinite in z for simplicity) can be moved as indicated in Figure 9-69. Obtain the equations for the field components in the TE_{n0} modes, and the equation for the phase shift as a function of w. Determine the value of w for which the phase shift is maximum for the TE_{10} mode and TE_{20} mode.

GE9-44 Beginning with Maxwell's two curl equations, show that for TE modes in systems having a z coordinate

$$\begin{aligned} \nabla^2 H_z(t_1, t_2, z) &= -\omega^2 \mu\epsilon H_z(t_1, t_2, z) \\ &= -k^2 H_z(t_1, t_2, z) \end{aligned}$$

where t_1 and t_2 are the transverse coordinates of the coordinate system, (x, y) or (ρ, ϕ), for example. For a rectangular resonant cavity, solve for the TE_{mnl} modes using the product solution $H_z(x, y, z) = X(x)Y(y)Z(z)$ and the normal magnetic field boundary condition on the preceding wave equation.

GE9-45 Repeat GE9-44 for TM modes.

GE9-46 Show how a rectangular waveguide carrying the TE_{10} mode can be aperture coupled to a TE_{101} rectangular cavity to obtain:
(a) E field coupling
(b) H field coupling

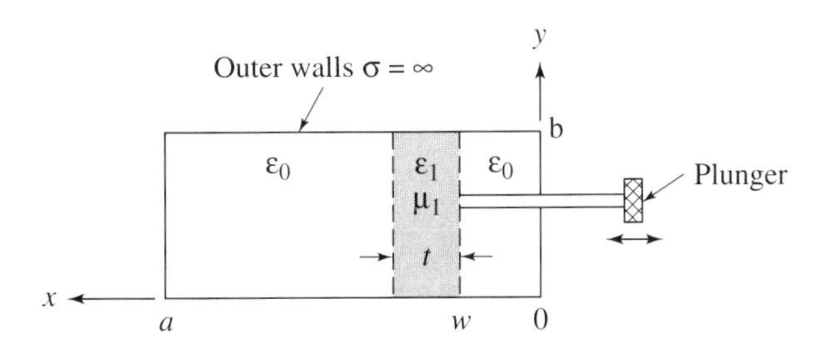

Figure 9-69.

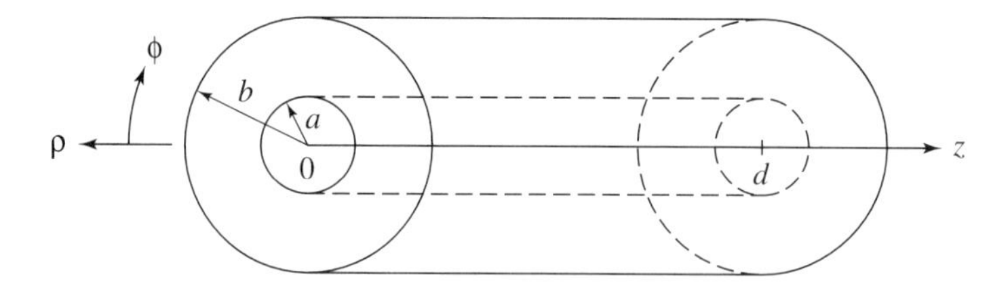

Figure 9-70.

GE9-47 A sphere of radius a is to be used as a resonant cavity. Derive equations for permissible modes TE_{mnl} and TM_{mnl} (transverse with respect to $\hat{r}$, so ''propagation'' has the variation $e^{\pm\gamma r}$ form) for two cases:

(a) $\partial/\partial\phi = 0$ (axial (azimuthal) symmetry)

(b) $\partial/\epsilon\theta = 0$ (polar symmetry; $\theta =$ polar angle)

Since there is no z coordinate you will need to start with Maxwell's equations in sinusoidal steady-state form. Give equations for the resonant frequencies.

GE9-48 For the coaxial circular cylindrical cavity shown in Figure 9-70 derive the equations for the resonant frequency and all field components for the TE_{nlq} modes.

GE9-49 Repeat Exercise GE9-48 for TM_{nlq} modes.

GE9-50 Note: This exercise should be worked with GE9-48. Derive the expression for the Q of the TE_{nlq} coaxial cavity of GE9-48 for the lowest-order mode.

GE9-51 Note: This exercise should be worked with GE9-49. Repeat Exercise GE9-50 for TM_{nlq} lowest-order mode.

GE9-52 A coaxial cable is used to probe couple energy out of a circular cylindrical resonant cavity operating in the TE_{111} mode. Show how you would place the probe in the cavity. For a short probe of length l determine the power extracted from the cavity if the coax is lossless and terminated in Z_0. Assume that the voltage induced in the probe is $|\vec{E}|l$.

GE9-53 Do Exercise GE9-52 for probe coupling from the TM_{010} mode along the axis of the cavity (probe along z axis).

Appendix A. Sinusoidal Steady-State Formulation

This appendix presents a review of the theorems and techniques used in sinusoidal steady-state formulation. The ideas are the same as those used in circuit theory and follow the same pattern of analysis.

The basic idea is to assume that all network or system sources are sinusoidal in form, that all have the same frequency, and that the sources have been applied for a sufficiently long time that any transients caused by prior switchings have disappeared. From the theory of differential equations we know that in the steady state, all system responses will be of the same frequency as the source but will in general have different phases. (Incidentally, if all sources are not of the same frequency we can determine the time response for each different frequency present and, assuming the system is linear, add them to obtain the total time response.)

We first introduce the concept of a real operator, denoted symbolically by $\text{Re}\{\cdot\}$. What this operator does is take a real part of the quantity within the braces.

Example A-1

$$\text{Re}\{6 - j5\} = 6$$

$$\text{Re}\{4\angle 20°\} = \text{Re}\{3.76 + j1.37\} = 3.76$$

$$\text{Re}\{A\angle\theta\} = \text{Re}\{A \cos\theta + jA \sin\theta\} = A \cos\theta$$

Perhaps the most important theorem used in sinusoidal work is Euler's theorem, given by

$$e^{j\theta} = \cos\theta + j \sin\theta \tag{A-1}$$

where $j = \sqrt{-1}$. This theorem is most easily developed by considering the series expansion of the left-hand member and collecting terms. Since $e^x = 1 + x + x^2/2! + x^3/3! + \cdots$ we begin by writing

$$e^{j\theta} = 1 + j\theta + \frac{(j\theta)^2}{2!} + \frac{(j\theta)^2}{3!} + \frac{(j\theta)^4}{4!} + \frac{(j\theta)^5}{5!} + \cdots$$

$$= 1 + j\theta + (j^2)\frac{\theta^2}{2!} + (j^3)\frac{\theta^3}{3!} + (j^4)\frac{\theta^4}{4!} + (j^5)\frac{\theta^5}{5!} + \cdots$$

Using $j = \sqrt{-1}$ and raising to appropriate powers this becomes

$$e^{j\theta} = 1 + j\theta - \frac{\theta^2}{2!} - j\frac{\theta^3}{3!} + \frac{\theta^4}{4!} + j\frac{\theta^5}{5!} + \cdots$$
$$= \left(1 - \frac{\theta^2}{2!} + \frac{\theta^4}{4!} - \cdots\right) + j\left(\theta - \frac{\theta^3}{3!} + \frac{\theta^5}{5!} + \cdots\right)$$

The two expressions in parentheses are next identified as the series expansions for the cosine and sine so we have

$$e^{j\theta} = \cos\theta + j\sin\theta \qquad \textbf{Q.E.D.}$$

Using this theorem we may develop the companion result

$$e^{-j\theta} = e^{j(-\theta)} = \cos(-\theta) + j\sin(-\theta) = \cos\theta - j\sin\theta \tag{A-2}$$

The two results above can be combined into a single statement:

$$e^{\pm j\theta} = \cos\theta \pm j\sin\theta \tag{A-3}$$

The basis for the phasor notation is Euler's theorem. One simply multiplies the Euler theorem by a constant A (which may be positive or negative) to obtain

$$Ae^{j\theta} = A(\cos\theta + j\sin\theta) = A\cos\theta + jA\sin\theta \tag{A-4}$$

In electrical engineering we introduce a shorthand notation for the exponential form by defining

$$\overline{A} \equiv A\angle\theta \equiv Ae^{j\theta} \tag{A-5}$$

which is the polar form of a complex number.

Another useful concept in sinusoidal steady-state work is the conjugate of a complex number. This operation is defined by an asterisk and means that we change the sign of the imaginary part of the complex number (i.e., negate all js).

Example A-2

$$(6 - j5)^* = 6 + j5$$
$$(4\angle 20°)^* = (4\cos 20° + j4\sin 20°)^* = 3.76 - j1.37 = 4\angle -20°$$

A particularly useful result can be obtained by applying the definition to Eq. A-4:

$$(Ae^{j\theta})^* = (A\cos\theta + jA\sin\theta)^* = A\cos\theta - jA\sin\theta$$

Now using Eq. A-2 we obtain

$$(Ae^{j\theta})^* = Ae^{-j\theta} \tag{A-6}$$

In the shorthand notation this would be written

$$(A\angle\theta)^* = A\angle -\theta \tag{A-7}$$

The main idea to be gained here is that *if we want to take the complex conjugate of a complex number or expression we simply need to replace j by −j wherever it appears, or negate the angle of polar form.* This may appear to be simple, but it proves to be of great utility as the following example shows.

Example A-3

$$\left(\frac{6 + j4}{1 - j2}\right)^* = \frac{6 - j4}{1 + j2}$$

Note that we have spared ourselves the agony of having to rationalize the denominator to put the expression in the form $(\alpha + j\beta)^*$.

The next idea is the expression of a real time waveform in complex number form. This is done by using Euler's theorem and the other preceding definitions as follows:

$$\begin{aligned} k(t) = K\cos(\omega t + \phi) &= \text{Re}\{K\cos(\omega t + \phi) + jK\sin(\omega t + \phi)\} \\ &= \text{Re}\{Ke^{j(\omega t+\phi)}\} = \text{Re}\{Ke^{j\phi}e^{j\omega t}\} \\ &= \text{Re}\{\overline{K}e^{j\omega t}\} \end{aligned} \tag{A-8}$$

In this context the complex number $\overline{K}$ is said to be the phasor representation of $k(t)$ in the sinusoidal steady state. An important result of the phasor representation is that *if by any means we can solve for the phasor $\overline{K}$ we can immediately transform back to the real-world time expression by multiplying by $e^{j\omega t}$ and taking the real part of the result.*

Example A-4

Suppose that an electrical network has been analyzed in the sinusoidal steady state and one of the currents has been found to be $\overline{I} = 6\angle -40°$ A. What is the time expression for that current? The frequency of the source is 60 Hz.

Using Eq. A-8 we have

$$\begin{aligned} i(t) &= \text{Re}\{\overline{I}e^{j\omega t}\} = \text{Re}\{6\angle -40° e^{j2\pi ft}\} \\ &= \text{Re}\{6e^{-j40°}e^{j2\pi 60t}\} = \text{Re}\{6e^{j(2\pi 60t - 40°)}\} \\ &= \text{Re}\{6e^{j(377t - 40°)}\} = 6\cos(377t - 40°)\text{ A} \end{aligned}$$

We next state and prove some important theorems involving phasors.

Theorem 1. For a complex number $\overline{A}$, $\text{Re}\{\overline{A}\} = \frac{1}{2}(\overline{A} + \overline{A}^*)$.

Proof

For the left member, $\text{Re}\{\overline{A}\} = A\cos\theta$. For the right member, $\frac{1}{2}(\overline{A} + \overline{A}^*) = \frac{1}{2}(A\cos\theta + jA\sin\theta + A\cos\theta - jA\sin\theta) = A\cos\theta$. Thus

$$\text{Re}\{\overline{A}\} = \tfrac{1}{2}(\overline{A} + \overline{A}^*) \qquad \textbf{Q.E.D.}$$

Theorem 2. For a sinusoidal steady-state function $a(z, t)$ having frequency ω and where z is a space (coordinate) variable and t is time, there is a corresponding phasor $\overline{A}(z)$ such that

$$(1) \qquad \frac{\partial a(z, t)}{\partial t} = \frac{\partial[\text{Re}\{\overline{A}(z)e^{j\omega t}\}]}{\partial t}$$

and $\qquad$ (2) $\quad \dfrac{\partial a(z, t)}{\partial t} = \text{Re}\{j\omega\overline{A}(z)e^{j\omega t}\}$

(Two important things to notice are that $\overline{A}$ is a function of coordinate z and that the effect of taking the time derivative is to multiply by $j\omega$.)

Proof
For Part 1 we simply need to show that there is an appropriate phasor representation for a sinusoid that has z as an additional variable. The usual form for the sinusoid is $A\cos(\omega t + \phi)$. Now if we let ϕ be some function of z, $\phi(z)$, we have

$$a(z, t) = A\cos(\omega t + \phi(z)) = \text{Re}\{Ae^{j(\omega t + \phi(z))}\} = \text{Re}\{Ae^{j\phi(z)}e^{j\omega t}\} = \text{Re}\{\overline{A}(z)e^{j\omega t}\}$$

which verifies Part 1 (Note that A could also be a function of z.)
For Part 2 we evaluate each member as follows.

Left member:

$$\frac{\partial a(z, t)}{\partial t} = \frac{\partial A\cos[\omega t + \phi(z)]}{\partial t} = -\omega A\sin[\omega t + \phi(z)]$$

Right member:

$$\begin{aligned}\text{Re}\{j\omega\overline{A}(z)e^{j\omega t}\} &= \text{Re}\{j\omega Ae^{j\phi(z)}e^{j\omega t}\} = \text{Re}\{j\omega Ae^{j(\omega t + \phi(z))}\} \\ &= \text{Re}\{j\omega A[\cos(\omega t + \phi(z)) + j(j\omega)\sin(\omega t + \phi(z))]\} \\ &= \text{Re}\{j\omega A\cos(\omega t + \phi(z)) - \omega A\sin(\omega t + \phi(z))\} \\ &= -\omega A\sin(\omega t + \phi(z)) \qquad \textbf{Q.E.D.}\end{aligned}$$

Theorem 3. For a sinusoidal steady-state function $a(z, t)$ having frequency ω and where z is a space (coordinate) variable and t is time, there is a corresponding phasor $\overline{A}(z)$ such that

$$\frac{\partial^2 a(z, t)}{\partial t^2} = \text{Re}\{-\omega^2\overline{A}(z)e^{j\omega t}\}$$

(Note that this is really a second application of Theorem 2.)

Proof
The existence of an appropriate $\overline{A}(z)$ has already been shown in Theorem 2, so we proceed to the second derivative proof.

$$\frac{\partial^2 a(z, t)}{\partial t^2} = \frac{\partial\left(\dfrac{\partial a(z, t)}{\partial t}\right)}{\partial t} = \frac{\partial(\text{Re}\{j\omega\overline{A}(z)e^{j\omega t}\})}{\partial t}$$

by Theorem 2. Now $j\omega\overline{A}(z)$ is just another phasor that is not a function of time, so call it $\overline{B}(z)$. Then

$$\frac{\partial^2 a(z, t)}{\partial t^2} = \frac{\partial(\text{Re}\{\overline{B}(z)e^{j\omega t}\})}{\partial t} = \frac{\partial b(z, t)}{\partial t}$$

by Theorem 2 Part 1. Next using Theorem 2 Part 2 on $b(z, t)$ we obtain

$$\frac{\partial^2 a(z,t)}{\partial t^2} = \text{Re}\{j\omega\overline{B}(z)e^{j\omega t}\} = \text{Re}\{(j\omega)(j\omega\overline{A}(z))e^{j\omega t}\}$$
$$= \text{Re}\{-\omega^2\overline{A}(z)e^{j\omega t}\} \qquad \textbf{Q.E.D.}$$

Theorem 4. For phasors $\overline{A}(z)$ and $\overline{B}(z)$ of the same frequency ω the statement

$$\text{Re}\{\overline{A}(z)e^{j\omega t}\} = \text{Re}\{\overline{B}(z)e^{j\omega t}\}$$

also implies that

$$\overline{A}(z)e^{j\omega t} = \overline{B}(z)e^{j\omega t} \qquad \text{or} \quad \overline{A}(z) = \overline{B}(z)$$

(Note that it is not obvious since it is easy to have $\text{Re}\{1 + j2\} = \text{Re}\{1 - j\beta\}$, but $\overline{A} \neq \overline{B}$.)

Basically this theorem says that for phasors of the same frequency the real operator may be removed in an equality.

Proof
Since the equality must hold for all values of time we first suppose $t = 0$ so we obtain from the first equality

$$\text{Re}\{\overline{A}(z)\} = \text{Re}\{\overline{B}(z)\}$$

Writing the complex phasors in terms of real and imaginary parts:

$$\text{Re}\{A_r(z) + jA_i(z)\} = \text{Re}\{B_r + jB_i(z)\}$$

from which

$$A_r(z) = B_r(z)$$

Now for time $t = \pi/2\omega$ we have

$$e^{j\omega t}\big|_{t=\pi/2\omega} = e^{j\pi/2} = \cos\frac{\pi}{2} + j\sin\frac{\pi}{2} = j$$

For this value of time the first equality in the theorem statement becomes

$$\text{Re}\{\overline{A}(z)e^{j\omega t}\} = \text{Re}\{j\overline{A}(z)\} = \text{Re}\{\overline{B}(z)e^{j\omega t}\} = \text{Re}\{j\overline{B}(z)\}$$

Working with this we have, successively,

$$\text{Re}\{j[A_r(z) + jA_i(z)]\} = \text{Re}\{j[B_r(z) + jB_i(z)]\}$$
$$\text{Re}\{jA_r(z) - A_i(z)\} = \text{Re}\{jB_r(z) - B_i(z)\}$$
$$-A_i(z) = -B_i(z)$$

or

$$A_i(z) = B_i(z)$$

Since the two phasors have the same real and imaginary parts then

$$\overline{A}(z) = \overline{B}(z) \qquad \textbf{Q.E.D.}$$

Multiplying by $e^{j\omega t}$ on both sides we then have

$$\overline{A}(z)e^{j\omega t} = \overline{B}(z)e^{j\omega t} \qquad \textbf{Q.E.D.}$$

Theorem 5. For a real quantity K (which may be a function of z)

$$K\ \mathrm{Re}\{\overline{A}(z)e^{j\omega t}\} = \mathrm{Re}\{K\overline{A}(z)e^{j\omega t}\}$$

Proof
We begin by working with the left-hand member:

$$\begin{aligned} K\ \mathrm{Re}\{\overline{A}(z)e^{j\omega t}\} &= K\ \mathrm{Re}\{[A_r(z) + jA_i(z)](\cos\omega t + j\ \sin\omega t)\} \\ &= K[A_r(z)\ \cos\omega t - A_i(Z)\ \sin\omega t] \end{aligned}$$

Now from the right member (recalling that K is real)

$$\begin{aligned} \mathrm{Re}\{K\overline{A}(z)e^{j\omega t}\} &= \mathrm{Re}\{K[A_r(z) + jA_i(z)](\cos\omega t + j\ \sin\omega t)\} \\ &= \mathrm{Re}\{K[A_r(z)\ \cos\omega t - A_i(z)\ \sin\omega t] + jK \\ &\quad [A_r(z)\ \sin\omega t + A_i\ \cos\omega t]\} \\ &= K[A_r(z)\ \cos\omega t - A_i(z)\ \sin\omega t] \end{aligned}$$

This last expression is the same as obtained from the left member. **Q.E.D**

Theorem 6. For phasors $\overline{A}_1(z), \overline{A}_2(z), \overline{A}_3(z), \ldots, \overline{A}_m(z)$,

$$\mathrm{Re}\left\{\sum_{n=1}^{m} \overline{A}_n(z)\right\} = \sum_{n=1}^{m} (\mathrm{Re}\{\overline{A}_n(z)\})$$

In words, the real part of the sum of complex numbers equals the sum of the real parts.

Proof
We begin with the left member:

$$\begin{aligned} \mathrm{Re}\left\{\sum_{n=1}^{m} \overline{A}_n(z)\right\} &= \mathrm{Re}\left\{\sum_{n=1}^{m} [A_{nr}(z) + jA_{nr}(z)]\right\} \\ &= \mathrm{Re}\left\{\sum_{n=1}^{m} A_{nr}(z) + \sum_{n=1}^{m} jA_{nr}(z)\right\} \\ &= \mathrm{Re}\left\{\sum_{n=1}^{m} A_{nr}(z) + j\sum_{n=1}^{m} A_{n=1}A_{ni}(z)\right\} \end{aligned}$$

Since $A_{nr}(z)$ is real part of $\overline{A}_n(z)$ we finally obtain

$$\mathrm{Re}\left\{\sum_{n=1}^{m} \overline{A}_n(z)\right\} = \sum_{n=1}^{m} \mathrm{Re}\{\overline{A}_n(z)\} \qquad \textbf{Q.E.D.}$$

Theorem 7. For phasors $\overline{A}(z)$ and $\overline{B}(z)$

$$\mathrm{Re}\{\overline{A}(z)\}\ \mathrm{Re}\{\overline{B}(z)\} = \tfrac{1}{2}[\mathrm{Re}\{\overline{A}(z)\overline{B}(z)\} + \mathrm{Re}\{\overline{A}(z)\overline{B}^*(z)\}]$$

where the asterisk is the conjugate indicator.

Proof
We work with the right member to obtain the left member. We use $\overline{A}(z) = A_r(z) + jA_i(z)$ and $\overline{B}(z) = B_r(z) + jB_i(z)$.

$$\begin{aligned}\tfrac{1}{2}[\mathrm{Re}\{\overline{A}(z)\overline{B}(z)\} + \mathrm{Re}\{\overline{A}(z)\overline{B}^*(z)\}] &= \tfrac{1}{2}[\mathrm{Re}\{A_r(z)B_r(z) - A_i(z)B_i(z) + jA_r(z)B_i(z) \\ &\quad + jB_r(z)A_i(z)\} + \mathrm{Re}\{A_r(z)B_r(z) + A_i(z)B_i(z) \\ &\quad - jA_r(z)B_i(z) + jB_r(z)A_i(z)\}] \\ &= \tfrac{1}{2}[A_r(z)B_r(z) - A_i(z)B_i(z) + A_r(z)B_r(z) \\ &\quad + A_i(z)B_i(z)] \\ &= A_r(z)B_r(z) = \mathrm{Re}\{\overline{A}(z)\}\ \mathrm{Re}\{\overline{B}(z)\} \qquad \textbf{Q.E.D.}\end{aligned}$$

Theorem 8. For phasors $\overline{A}(z)$ and $\overline{B}(z)$

$$[\mathrm{Re}\{\overline{A}(z)e^{j\omega t}\}\ \mathrm{Re}\{\overline{B}(z)e^{j\omega t}\}]_{\text{average}} = \tfrac{1}{2}\ \mathrm{Re}\{\overline{A}(z)\overline{B}^*(z)\}$$

Where the average denotes the *time* average over one period of the frequency ω, that is, $T = 1/f = 2\pi/\omega$,

$$\frac{1}{T}\int_0^T v(t)dt = V_{\text{average}}$$

Proof
Using the result of Theorem 7 we have

$$\begin{aligned}[\mathrm{Re}\{\overline{A}(z)e^{j\omega t}\}\ \mathrm{Re}\{\overline{B}(z)e^{j\omega t}\}]_{\text{average}} &= \tfrac{1}{2}[\mathrm{Re}\{\overline{A}(z)e^{j\omega t}\overline{B}(z)e^{j\omega t}\} \\ &\quad + \mathrm{Re}\{\overline{A}(z)e^{j\omega t}\overline{B}^*(z)e^{-j\omega t}\}]_{\text{average}} \\ &= \tfrac{1}{2}[\mathrm{Re}\{\overline{A}(z)\overline{B}(z)e^{j2\omega t}\} \\ &\quad + \mathrm{Re}\{\overline{A}(z)\overline{B}^*(z)\}]_{\text{average}}\end{aligned}$$

Since the average of a sine wave over an integral number of periods is zero, the first term of the right member is zero. Since the second term does not involve time t the average is the quantity itself. Thus

$$[\mathrm{Re}\{\overline{A}(z)e^{j\omega t}\}\ \mathrm{Re}\{\overline{B}(z)e^{j\omega t}\}]_{\text{average}} = \tfrac{1}{2}\ \mathrm{Re}\{\overline{A}(z)\overline{B}^*(z)\} \qquad \textbf{Q.E.D.}$$

For some examples we select some circuit theory problems where the phasors are not functions of coordinate z so that the application of the theorems can be easily seen.

Example A-5

For the series circuit shown in Figure A-1 we have the general time equation

$$v(t) = L\frac{di}{dt} + Ri(t)$$

Express it in sinusoidal steady-state form using the theorems and definitions developed earlier.

For sinusoidal applied voltage we use Eq. A-8 to obtain

$$\mathrm{Re}\{\overline{V}e^{j\omega t}\} = L\frac{d(\mathrm{Re}\{\overline{I}e^{j\omega t}\})}{dt} + \mathrm{Re}\{\overline{I}e^{j\omega t}\}R$$

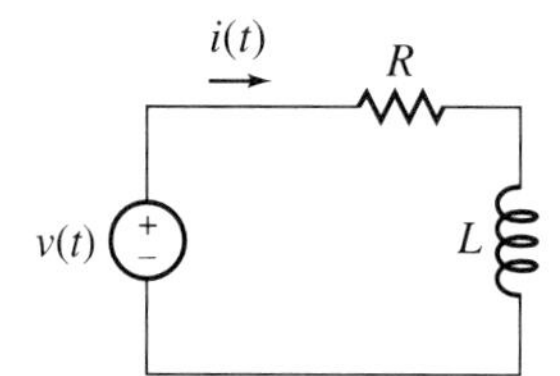

Figure A-1.

Using Theorem 2 on the derivative term:

$$\mathrm{Re}\{\bar{V}e^{j\omega t}\} = L\,\mathrm{Re}\{j\omega\bar{I}e^{j\omega t}\} + \mathrm{Re}\{\bar{I}e^{j\omega t}\}\,R$$

Using Theorem 5 on the two terms on the right-hand side:

$$\mathrm{Re}\{\bar{V}e^{j\omega t}\} = \mathrm{Re}\{j\omega L\bar{I}e^{j\omega t}\} + \mathrm{Re}\{R\bar{I}e^{j\omega t}\}$$

Next use Theorem 6 on the right side:

$$\mathrm{Re}\{\bar{V}e^{j\omega t}\} = \mathrm{Re}\{j\omega\bar{I}e^{j\omega t} + R\bar{I}e^{j\omega t}\}$$

Applying Theorem 4 and then canceling the common factor $e^{j\omega t}$ we have

$$\bar{V} = j\omega L\bar{I} + R\bar{I} = (j\omega L + R)\bar{I}$$

In retrospect we could considerably simplify the conversion to steady state by comparing the preceding result with the original differential equation. We observe that the conversion could be made in two simple steps:

1. Replace each $\partial/\partial t$ by $j\omega$.
2. Replace the script letters for current and voltage by their phasor representations, capital letter complex numbers.

Example A-6

Suppose we wish to obtain the time expression for the sinusoidal steady-state current solution for the preceding network for the particular case $v(t) = V_0 \sin \omega t$. Putting into cosine form we write

$$v(t) = V_0 \cos\left(\omega t - \frac{\pi}{2}\right)$$

Using Eq. A-8 we can identify the phasor representation of this voltage as

$$\bar{V} = V_0 e^{-j(\pi/2)}$$

From the preceding example

$$\bar{I} = \frac{\bar{V}}{j\omega L + R} = \frac{V_0 e^{-j\pi/2}}{R + j\omega L} + \frac{V_0 e^{-j\pi/2}}{\sqrt{R^2 + (\omega L)^2}e^{j\tan^{-1}(\omega L/R)}}$$

$$= \frac{V_0}{\sqrt{R^2 + \omega^2 L^2}} e^{-j[\pi/2 + \tan^{-1}(\omega L/R)]}$$

Using Eq. A-8 directly again

$$i(t) = \operatorname{Re}\{\bar{I}e^{j\omega t}\} = \operatorname{Re}\left\{\frac{V_0}{\sqrt{R^2 + \omega^2 L^2}}\, e^{j[\omega t - \pi/2 - \tan^{-1}(\omega L/R)]}\right\}$$

or

$$i(t) = \frac{V_0}{\sqrt{R^2 + \omega^2 L^2}} \cos\left(\omega t - \frac{\pi}{2} - \tan^{-1}\frac{\omega L}{R}\right)$$

Exercises

A-a Prove that for a sinusoidal steady-state condition

$$\int a(z, t)dt = \int \operatorname{Re}\{\overline{A}(z)e^{j\omega t}\}dt = \operatorname{Re}\left\{\frac{1}{j\omega}\overline{A}(z)e^{j\omega t}\right\}$$

A-b Starting with the general integrodifferential equation, obtain the phasor representation of the voltage $v(t)$ for Figure A-2.

A-c For the circuit of Exercise A-b if $i(t) = I_0 \cos \omega t$ obtain an expression for $v(t)$ using your results of that exercise and applicable theorems.

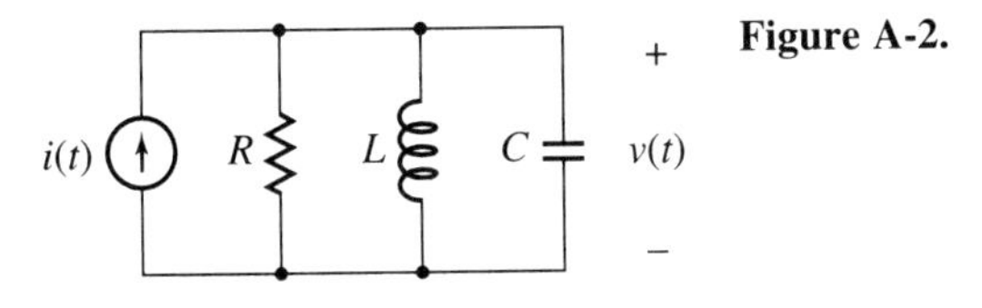

Figure A-2.

Appendix B. Coaxial Cable Data*

ABBREVIATIONS

Dielectric

FEP	Solid fluorinated ethylene propylene
PE	Solid polyethylene
PIB	Polyisobulylene, Type B, per MIL-C-17
PS	Polystyrene
PTFE	Solid polytetrafluoroethylene
Rubber	per MIL-C-17
Sil	Silicone rubber

Conductors and Braid Material

AL	Aluminum
BC	Bare copper
BerC	Berillium-copper alloy
CCA	Copper clad aluminum
CCS	Copper clad steel
CPC	Copper polyester copper laminate
GS	Galvanized steel
HR	High resistance wire
NC	Nickel-covered copper
S	Silver-covered alloy
SC	Silver-covered copper
SCBerC	Silver-covered beryllium-copper alloy
SCCad Br	Silver-covered cadmium bronze
SCCA1	Silver-covered copper-clad aluminum
SCCS	Silver-covered copper-clad steel
SNCSS	Silver-covered nickel-covered copper-clad steel

Jacket Material

E-CTFE-XI	Ethylene chlorotrifluoroethylene copolymer
ETFE-X	Ethylene tetrafluoroethylene copolymer
FEP-IX	Fluorinated ethylene propylene, Type IX, per MIL-C-17

*With permission of Times Microwave Systems, Wallingford, Connecticut.

FG Braid V	Fiberglass, impregnated, Type V, per MIL-C-17
PE-III	Clear polyethylene
PE-IIIA	High molecular weight, black polyethylene, Type IIIA, per MIL-C-17
PFA-XIII	Perfluoroalkoxy, per MIL-C-17
PTFE	Polytetrafluoroethylene, per MIL-C-17
PUR	Polyurethane, black specific compounds
PVC-I	Black polyvinylchloride, contaminating, Type I, per MIL-C-17
PVS-II	Gray polyvinylchloride, non-contaminating, Type II, per MIL-C-17
PVC-IIA	Black polyvinylchloride, non-contaminating, Type IIA, per MIL-C-17
Rubber	Per MIL-C-17
SIL/DAC-VI	Dacron braid over silicon rubber, Type VI, per MIL-C-17
TPE	Thermo-plastic elastomer
XLPE	Crosslinked polyolefin

PROPERTIES OF WIRE AND CABLE-INSULATING DIELECTRIC MATERIALS

Material	Dielectric Constant	Power Factor	Volume Resistivity (Ω-cm)	Normal Operating Temperature Limits (°C)
Cellular FEP	1.50	0.0007	10^{18}	−75 + 200
Cellular polyethylene	1.40–2.10	0.0003	10^{12}	−75 + 80
Cellular TFE	1.40	0.0002	10^{19}	−75 + 250
ECTF	2.5	0.0015	10^{16}	−75 − 150
ETFE	2.6	0.005	10^{16}	−75 + 150
Ethylene propylene	2.24	0.00046	10^{17}	−40 + 105
FEP	2.10	0.0007	10^{18}	−75 + 200
Nylon	4.60–3.50	0.040–0.030	4×10^{14}	−60 + 120
Perforated TFE	1.50	0.0002	10^{19}	−75 + 250
PFA	2.1	0.0010	10^{16}	−75 + 150
Polyethylene	2.3	0.0003	10^{16}	−75 + 80
Polyimide	3.00–3.50	0.002–0.003	10^{13}	−75 + 300
Polyvinylchloride	3.00–8.00	0.0700–0.1600	2×10^{12}	−55 + 105
PTFE	2.1	0.0003	10^{19}	−75 + 250
PVDF	7.6	0.0200	10^{14}	−75 + 125
Silicon rubber	2.08–3.50	0.007–0.016	10^{13}	−70 + 250

FORMULAS COMMON TO ALL COAXIAL CABLE

Formula	Conditions
Capacitance $(C) = \dfrac{7.36\epsilon}{\text{Log}(D/d)}$ pf/ft	α = attenuation in dB/100 ft
Inductance $(L) = 0.140 \log(D/d)\ \mu$H/ft	d = the outside diameter of inner conductor in inches
Impedance $(Z_0) = \sqrt{\dfrac{L}{C}} = \dfrac{130}{\sqrt{\epsilon}} \log(D/d)\ \Omega$	D = the inside diameter of outer conductor in inches
Velocity of propagation: % of speed of light $= \dfrac{100}{\sqrt{\epsilon}}$	S = the maximum voltage gradient of the cable insulation in volts per mil
Time delay $= 1.016\sqrt{\epsilon}$ ns/ft	ϵ = the dielectric constant of the insulation (dielectric) of the cable
Cutoff frequency $(F_{co}) = \dfrac{7.50}{\sqrt{\epsilon}(D + d)}$ GHz	log = logarithm to base 10
Reflection coefficient $= \Gamma = \dfrac{Z_r - Z_0}{Z_r + Z_0} = \dfrac{\text{VSWR} - 1}{\text{VSWR} + 1}$	K = safety factor
$\text{VSWR} = \dfrac{1 + \Gamma}{1 - \Gamma}$	K_1 = strand factor
Peak voltage $= \dfrac{1.15S \times d(\log D/d)}{K}$	K_2 = braid factor
$\alpha = \dfrac{0.435}{DZ_0}\left[\dfrac{D}{d}K_1 + K_2\right]\sqrt{F} + 2.78\sqrt{\epsilon}(\text{P.F.})(\text{F})$	F = frequency in MHz
	P.F. = power factor

SELECTION OF FLEXIBLE AND SEMIFLEXIBLE R.F. COAXIAL CABLE

In choosing the appropriate cable construction for a particular application, the following cable characteristics are to be considered:

A. Characteristic impedance
B. Impedance uniformity (VSWR)
C. Capacitance
D. Capacitance and impedance stability
E. C.W. power rating
F. Maximum operating voltage
G. Attenuation
H. Attenuation uniformity
I. Attenuation stability
J. Velocity of propagation
K. Electrical length stability
L. Pulse response
M. Shielding
N. Cutoff frequency
O. Self-generated cable noise
P. Operating temperature range
Q. Flexibility
R. Environmental resistance
S. Cable strength

RG CABLE DESCRIPTIONS

RG/U Number	Inner Conductor	Dielectric Material	Nominal DOD (in.)	Shielding Number Type	Jacket Material
8	0.0855″ 7/0.0285″ BC	PE	0.285	1: BC	PVC-I
8A	0.0855″ 7/0.0285″ BC	PE	0.285	1: BC	PVC-IIA
9	0.0855″ 7/0.0285″ SC	PE	0.280	2: SC, BC	PVC-II
9A	0.0855″ 7/0.0285″ SC	PE	0.280	2: SC	PVC-II
9B	0.0855″ 7/0.0285″ SC	PE	0.280	2: SC	PVC-IIA
12	0.0477″ 7/0.0159″ TC	PE	0.285	1: BC	PVC-II w. armor
12A	0.0477″ 7/0.0159″ TC	PE	0.285	1: BC	PVC-IIA w. Armor
15	0.0571″ BC	PE	0.370	2: BC	PVC-I
22	2 cond. 0.0456″ 7/0.0152″ BC	PE	0.285	1: TC	PVC-I
22A	2 cond. 0.0456″ 7/0.0152″ BC	PE	0.285	2: TC	PVC-II
22B	2 cond. 0.0456″ 7/0.0152″ BC	PE	0.285	2: TC	PVC-IIA
23	2 cond. 0.0855″ 7/0.0285″ BC	PE 2 Cores	0.380	2: BC	PVC-I
23A	2 cond. 0.0855″ 7/0.0285″ BC	PE 2 cores	0.380	2: BC	PVC-IIA
58	0.0320″ BC	PE	0.116	1: TC	PVC-I
58A	0.0355″ 19/0.0071″ TC	PE	0.116	1: TC	PVC-I
58B	0.0320″ BC	PE	0.116	1: TC	PVC-IIA
58C	0.0355″ 19/0.0071″ TC	PE	0.116	1: TC	PVC-IIA
65	0.0080″ Formex-F 0.1280″-diameter helix	PE	0.285	1: BC	PVC-I
65A	0.0080″ Formex-F 0.1280″-diameter helix	PE	0.285	1: BC	PVC-IIA
72	0.0253″ CCS	Air-space PE	0.460	1: BC	PVC-I Copper

RG/U Number	Inner Conductor	Dielectric Material	Nominal DOD (in.)	Shielding Number Type	Jacket Material
73	0.0650″ BC	PE	0.116	2: BC	braid
86	2 cond. 0.0855″ 7/0.0285″ BC	PE	0.300 × 0.650	None	None
87A	0.0960″ 7/0.0320″ SC	PTFE	0.280	2: SC	FG braid-V
100	0.0735″ 19/0.0147″ BC	PE	0.146	1: BC	PVC-I
108	2 cond. 0.0378″ 7/0.0126″ TC	PE	0.079 each	1: TC	PVC-II
117	0.1880″ BC	PTFE	0.620	1: BC	FG braid-V
117A	0.1880″ BC	PTFE	0.620	1: BC	FG braid-V
125	0.0159″ CCS	Air-space PE	0.460	1: BC	PVC-IIA
176	0.135″ helix over magnetic core	PE	0.285	1: magnet wire	PVC-I
187	0.0120″ 7/0.0040″ SCCS	PTFE	0.060	1: SC	PTFE
211	0.1900″ BC	PTFE	0.620	1: BC	FG braid-V
211A	0.1900″ BC	PTFE	0.620	1: BC	FG braid-V
212	0.0556″ SC	PE	0.185	2: SC	PVC-IIA
213	0.0888″ 7/0.0296″ BC	PE	0.285	1: BC	PVC-IIA
214	0.0888″ 7/0.0296″ SC	PE	0.285	2: SC	PVC-IIA
222	0.0556″ high Resistance wire	PE	0.185	2: SC	PVC-IIA
246	0.1880″ BC	PS helix	0.758	Al. tube	None
247	0.1880″ BC	PS Helix	0.758	0.875″ Al. tube	PE-IIIA
266	0.0113″ cond. over 0.144″ Mag. core	PE	0.285	75 spiral wound wires	PVC-I
294A	2 cond., 0.0808″ 1 BC, 1 TC	PE	0.472	1: SC	PE-IIIA
325	0.1000″ 19/0.0200″ SCC Al.	PE Spline	0.260	2: SC Strip	PUR

RG CABLE DESCRIPTIONS (*continued*)

RG/U Number	Nominal OD (in.)	Nominal Impedance (Ω)	Maximum Attenuation @ 400 MHz (dB/100 ft)	Nominal Capacitance (pf/ft)	Operational Temperature Range (°C)	Maximum Operational Range V, (RMS)
8	0.405	52.0	6.0	29.6	−40 +80	4,000
8A	0.405	52.0	6.0	29.6	−40 +80	5,000
9	0.420	51.0	5.9	30.2	−40 +80	4,000
9A	0.420	51.0	6.1	30.2	−40 +80	4,000
9B	0.420	50.0	6.1	30.8	−40 +80	5,000
12	0.483	76.0	5.7	20.6	−40 +80	4,000
12A	0.463	76.0	5.2	20.6	−40 +80	5,000
15	0.545	76.0	4.5	20.0	−40 +80	5,000
22	0.405	95.0	10.5	16.3	−40 +80	1,000
22A	0.420	95.0	10.5	16.3	−40 +80	1,000
22B	0.420	95.0	10.5	16.3	−40 +80	1,000
23	0.850 × 0.945	125.0	5.2	12.0	−40 +80	3,000
23A	0.650 × 0.945	125.0	5.2	12.0	−40 +80	3,000
58	0.195	53.5	11.7	28.8	−40 +80	1,900
58A	0.195	52.0	13.2	29.6	−40 +80	1,900
58B	0.195	53.5	14.0	28.8	−40 +80	1,900
58C	0.195	50.0	14.0	30.8	−40 +80	1,900
65	0.405	950.0	16.0 @ 5 MHz	44.0	−40 +80	1,000
65A	0.405	950.0	16.0 @ 5 MHz	44.0	−40 +80	1,000
72	0.630	150.0		7.8	−40 + 80	750
73	0.175	25.0	24.0	61.6	−55 +80	1,000
86	0.300 × 0.650	200.0		7.8	−55 +80	10,000
87A	0.425	50.0	5.0	29.4	−55 +250	5,000
100	0.242	35.0	19.0	44.0	−40 +80	2,000
108	0.235	78.0	2.8 @ 10 MHz	19.7	−40 +80	1,000
117	0.730	50.0	2.3	29.4	−55 +250	7,000
117A	0.730	50.0	2.3	29.4	−55 +250	7,000
125	0.600	150.0		7.8	−40 +80	2,000
176	0.405	2,240	11.5 @ 5 MHz	49.0	−40 +80	5,000
187	0.105	75.0	21.0	19.5	−55 +250	1,200
211	0.730	50.0	2.3	29.4	−55 +250	7,000
211A	0.730	50.0	2.3	29.4	−55 +250	7,000
212	0.332	50.0	6.5	29.4	−40 +80	3,000
213	0.405	50.0	5.5	30.8	−40 +80	5,000
214	0.425	50.0	5.5	30.8	−40 +80	5,000
222	0.332	50.0	33.0	30.8	−40 +80	3,000
246	0.875	75.0	1.0	15.2	−55 +80	2,200 peak
247	1.015	75.0	1.0	15.2	−55 +80	2,200 peak
266	0.400	1530.0	7.2 @ 5 MHz	53.0	−40 +80	5,000 DC
294A	0.630	95.0	10.0	16.3	−55 +80	3,000
325	0.350	50.0	3.5	26.3	−55 +80	750

Note: DOD = diameter of dielectric; OD = outer diameter of cable.

CABLE ATTENUATION

Part Number	Maximum Attenuation (dB/100 ft)						Maximum Power	Maximum Frequency
	100 MHz	400 MHz	1 GHz	3 GHz	5 GHz	10 GHz	(W @ 400 MHz)	(GHz)
M17/28 RG 58A, B, C	6.5	17.0	28.0	NA	NA	NA	90	1.0
M17/75 RG9, 9A, B, RG 214	2.6	6.8	12.0	25.0	35.0	56.0	330	11.0
M17/73 RG212	3.0	6.5	12.0	24.0	34.0	54.0	350	11.0
M17/74 RG 8, 8A, 213	2.3	4.8	9.0	NA	NA	NA	320	1.0
RG 325/U	2.2	4.6	7.6	14.4	19.6	30.5	340	12.4
RG 326/U	1.1	2.4	4.1	8.3	11.7	NA	720	7.8
RG 389/U	0.8	1.8	3.2	6.6	9.4	NA	1500	7.6
M17/127 RG 87A	2.4	5.0	8.8	18.0	24.0	37.0	1100	12.4
M17/72 RG 211, 211A, RG 117, 117A	0.85	2.3	4.5	NA	NA	NA	11000	1.0
M17/6 RG 12, 12A	NA	5.2	9.4	NA	NA	NA	290	1.0
RG 15/U	1.8	4.0	6.9	NA	NA	NA	NA	1.0
RG 187 A/U	9.4	19.1	30.8	55.2	NA	NA	NA	3.0

Appendix C. Selected Laplace Transform Pairs

$f(t)$	$\mathcal{L}\{f(t)\} = F(s)$
$\alpha f_1(t) + \beta f_2(t)$ (α and β constants)	$\alpha F_1(s) + \beta F_2(s)$
$\frac{df}{dt} = f'(t)$	$sF(s) - f(0^-)$
$\frac{d^n f}{dt^n} = f^n(t)$	$s^n F(s) - s^{n-1} f(0^-) - s^{n-2} f'(0^-) - \cdots - f^{n-1}(0^-)$
$\int_0^t f(x)dx + c$ (c constant)	$\frac{F(s)}{s} + \frac{c}{s}$
$f(\alpha t)$ $(\alpha > 0)$	$\frac{1}{\alpha} F\left(\frac{s}{\alpha}\right)$, where $F\left(\frac{s}{\alpha}\right) = \mathcal{L}\{f(t)\}\vert_{s \to s/\alpha}$
$f(t - b)$ $(b \geq 0)$	$e^{-bs} F(s)$
$\int_0^\infty f_1(x) f_2(t - x) dx$ (convolution)	$F_1(s) F_2(s)$
$\frac{\partial f(t, x)}{\partial x}$	$\frac{\partial F(s, x)}{\partial x}$
$\int_a^b f(t,x)dx$	$\int_a^b F(s, x)dx$
$e^{-at} f(t)$	$F(s + a) \equiv \mathcal{L}\{f(t)\}\vert_{s \to s+a}$
$U(t - T)$ (unit step at $t = T$) (see Figure C-1*a*)	$\frac{1}{s} e^{-Ts}$

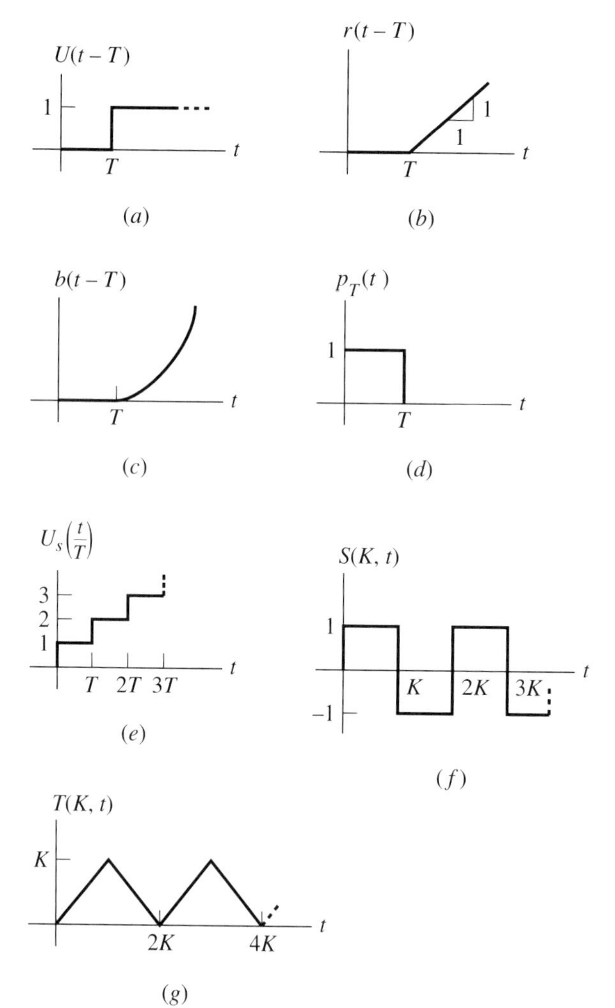

Figure C-1. Plots of important time functions in transmission line transient analysis.

	Important Functions	
	$f(t)$	$F(s)$
1.	$r(t-T) = t - T\ (t \geq T)$ (unit slope ramp at $t = T$) (see Figure C-1b)	$\dfrac{e^{-Ts}}{s^2}$
2.	$b(t-T) = (t-T)^{\alpha-1};\ (\alpha > 0,\ t > T)$ (see Figure C-1c)	$\dfrac{e^{-Ts}}{s^\alpha}\Gamma(\alpha)$, $\Gamma(\alpha)$ = gamma or generalized factorial function
3.	$p_T(t)$ (unit pulse of width T) (see Figure C-1d)	$\dfrac{1-e^{-Ts}}{s}$
4.	$U_s\left(\dfrac{t}{T}\right) = 1 + \left[\dfrac{t}{T}\right] = n$ when $(n-1)T < t < nT$ (unit stairstep) $n = 1, 2, \ldots$ (see Figure C-1e) $\left[\dfrac{t}{T}\right] \equiv$ largest integer $\leq \dfrac{t}{T}$	$\dfrac{1}{s(1-e^{-Ts})} = \dfrac{1+\coth\left(\dfrac{Ts}{2}\right)}{2s}$
5.	$P_T(t) = \begin{cases} 0 & 0 < t < T \\ 1 + a + a^2 + \cdots + a^{n-1} & nT < t < (n+1)T \end{cases}$ (a constant, $n = 1, 2, \ldots$) (unit polynomial function)	$\dfrac{1}{s(e^{Ts}-a)}$
6.	$S(2T, t) = (-1)^{n-1}$ for $2(n-1)T < t < 2nT$, $n = 1, 2, \ldots$ (unit square wave of period 4T) (see Figure c-1f)	$\dfrac{1}{s}\tanh(Ts)$
7.	$\dfrac{1}{2}S(T, t) + \dfrac{1}{2} = \dfrac{1-(-1)^n}{2}$, for $(n-1)T < t < nT$	$\dfrac{1}{s(1+e^{-Ts})}$
8.	$T(2T, t)$ (triangular wave) (see Figure C-1(g))	$\dfrac{1}{s^2}\tanh(Ts)$
9.	$f(t) = 2(n-1)$ for $(2n-3)T < t < (2n-1)T$, $n = 1, 2, 3, \ldots\ (t > 0)$	$\dfrac{1}{s\sinh(Ts)}$
10.	$S(2T, t+3T) + 1 = 1 + (-1)^n$ for $(2n-3)T < t < (2n-1)T$, $n = 1, 2, \ldots\ (t > 0)$	$\dfrac{1}{s\cosh(Ts)}$

	Important Functions	
	$f(t)$	$F(s)$
11.	$f(t) = 2n - 1$ for $2(n-1)T < t < 2nT$	$\frac{1}{s}\coth(Ts)$
12.	$f(t - T)\ (T \geq 0)$	$e^{-Ts}F(s)$
13.	$nt - \frac{n(n-1)}{2}T$ for $(n-1)T < t < nT$, $n = 1, 2, 3, \ldots$	$\frac{1}{s^2(1 + e^{-Ts})}$
14.	$\frac{1-(-1)^n}{2}t + \frac{T}{4}[1 + (-1)^n(2n-1)]$ for $(n-1)T < t < nT$, $n = 1, 2, 3, \ldots$	$\frac{1}{s^2(1 + e^{-Ts})}$
15.	$\frac{e^{-a(t-T)}}{e^{aT} - 1} - \frac{e^{-at}}{e^{aT} - 1}$ for $(n-1)T < t < nT$, $n = 1, 2, 3, \ldots$	$\frac{1}{s(1 - e^{as}e^{-Ts})}$

Appendix D. Expansions of Vector Operators in Rectangular, Spherical, and Cylindrical Coordinates

This appendix is a summary of important expansions in rectangular, cylindrical, and spherical coordinate systems. $\vec{G}$ is a general vector field, and S is a scalar field. Both are functions, in general, of all coordinate variables.

DIVERGENCE

Rectangular:

$$\nabla \cdot \vec{G} = \frac{\partial G_x}{\partial x} + \frac{\partial G_y}{\partial y} + \frac{\partial G_z}{\partial z}$$

Cylindrical:

$$\nabla \cdot \vec{G} = \frac{1}{\rho}\frac{\partial}{\partial \rho}(\rho G_\rho) + \frac{1}{\rho}\frac{\partial G_\phi}{\partial \phi} + \frac{\partial G_z}{\partial z}$$

Spherical:

$$\nabla \cdot \vec{G} = \frac{1}{r^2}\frac{\partial}{\partial r}\left(r^2 G_r\right) + \frac{1}{r\sin\theta}\frac{\partial}{\partial \theta}(\sin\,\theta G_\theta) + \frac{1}{r\,\sin\theta}\frac{\partial G_\phi}{\partial \phi}$$

CURL

Rectangular:

$$\nabla \times \vec{G} = \left(\frac{\partial G_z}{\partial y} - \frac{\partial G_y}{\partial z}\right)\hat{x} + \left(\frac{\partial G_x}{\partial z} - \frac{\partial G_z}{\partial x}\right)\hat{y} + \left(\frac{\partial G_y}{\partial x} - \frac{\partial G_x}{\partial y}\right)\hat{z}$$

Cylindrical:

$$\nabla \times \vec{G} = \left(\frac{1}{\rho}\frac{\partial G_z}{\partial \phi} - \frac{\partial G_\phi}{\partial z}\right)\hat{\rho} + \left(\frac{\partial G_\rho}{\partial z} - \frac{\partial G_z}{\partial \rho}\right)\hat{\phi} + \frac{1}{\rho}\left[\frac{\partial(\rho G_\phi)}{\partial \rho} - \frac{\partial G_\rho}{\partial \phi}\right]\hat{z}$$

Spherical:

$$\nabla \times \vec{G} = \frac{1}{r\sin\theta}\left[\frac{\partial(G_\phi \sin\theta)}{\partial\theta} - \frac{\partial G_\theta}{\partial\phi}\right] + \left[\frac{1}{r}\frac{1}{\sin\theta}\frac{\partial G_r}{\partial\phi} - \frac{1}{r}\frac{\partial(rG_\phi)}{\partial r}\right]\hat{\theta} + \frac{1}{r}\left[\frac{\partial(rG_\theta)}{\partial r} - \frac{\partial G_r}{\partial\theta}\right]\hat{\phi}$$

GRADIENT

Rectangular:

$$\nabla S = \frac{\partial S}{\partial x}\hat{x} + \frac{\partial S}{\partial y}\hat{y} + \frac{\partial S}{\partial z}\hat{z}$$

Cylindrical:

$$\nabla S = \frac{\partial S}{\partial\rho}\hat{\rho} + \frac{1}{\rho}\frac{\partial S}{\partial\phi}\hat{\phi} + \frac{\partial S}{\partial z}\hat{z}$$

Spherical:

$$\nabla S = \frac{\partial S}{\partial r}\hat{r} + \frac{1}{r}\frac{\partial S}{\partial\theta}\hat{\theta} + \frac{1}{r\sin\theta}\frac{\partial S}{\partial\phi}\hat{\phi}$$

LAPLACIAN OPERATOR

Rectangular:

$$\nabla^2 S = \frac{\partial^2 S}{\partial x^2} + \frac{\partial^2 S}{\partial y^2} + \frac{\partial^2 S}{\partial z^2}$$

Cylindrical:

$$\nabla^2 S = \frac{1}{\rho}\frac{\partial}{\partial\rho}\left(\rho\frac{\partial S}{\partial\rho}\right) + \frac{1}{\rho^2}\frac{\partial^2 S}{\partial\phi^2} + \frac{\partial^2 S}{\partial z^2}$$

Spherical:

$$\nabla^2 S = \frac{1}{r^2}\frac{\partial}{\partial r}\left(r^2\frac{\partial S}{\partial r}\right) + \frac{1}{r^2\sin\theta}\frac{\partial}{\partial\theta}\left(\sin\theta\frac{\partial S}{\partial\theta}\right) + \frac{1}{r^2\sin^2\theta}\frac{\partial^2 S}{\partial\phi^2}$$

COORDINATE SYSTEM SCALE FACTORS

Rectangular:

$$h_1 = 1, \qquad h_2 = 1, \qquad h_3 = 1 \qquad (x, z, z)$$

Cylindrical:

$$h_1 = 1, \qquad h_2 = \rho, \qquad h_3 = 1 \qquad (\rho, \phi, z)$$

Spherical:

$$h_1 = 1, \qquad h_2 = r, \qquad h_3 = r \sin\theta \qquad (r, \theta, \phi)$$

Appendix E. Unit Vector Relationships and Partial Derivatives in Rectangular, Spherical, and Cylindrical Coordinates

RECTANGULAR COORDINATES (x, y, z)

Unit Vector Relationships

$$\hat{x} \cdot \hat{x} = \hat{y} \cdot \hat{y} = \hat{z} \cdot \hat{z} = 1$$

All other unit-vector dot products are zero.

$$\hat{x} \times \hat{y} = -(\hat{y} \times \hat{x})$$

$$\hat{y} \times \hat{z} = -(\hat{z} \times \hat{y})$$

$$\hat{z} \times \hat{x} = -(\hat{x} \times \hat{z})$$

$$\hat{x} \times \hat{x} = \hat{y} \times \hat{y} = \hat{z} \times \hat{z} = 0$$

PARTIAL DERIVATIVES

All partials of unit vectors are zero.

$$\frac{\partial \hat{x}}{\partial x} = \frac{\partial \hat{x}}{\partial y} = \frac{\partial \hat{x}}{\partial z} = 0, \ldots$$

CYLINDRICAL COORDINATES (ρ, θ, z)

Unit Vector Relationships

$$\hat{\rho} \cdot \hat{\rho} = \hat{\phi} \cdot \hat{\phi} = \hat{z} \cdot \hat{z} = 1$$

All other unit vector dot products are zero.

$$\hat{\rho} \times \hat{\phi} = \hat{z} = -(\hat{\phi} \times \hat{\rho})$$

$$\hat{\phi} \times \hat{z} = \hat{\rho} = -(\hat{z} \times \hat{\phi})$$

$$\hat{z} \times \hat{\rho} = \hat{\phi} = -(\hat{\rho} \times \hat{z})$$

$$\hat{\rho} \times \hat{\rho} = \hat{\phi} \times \hat{\phi} = \hat{z} \times \hat{z} = 0$$

Partial Derivatives

All partials of unit vectors are zero except:

$$\frac{\partial \hat{\rho}}{\partial \phi} = \hat{\phi} \qquad \text{and} \qquad \frac{\partial \hat{\phi}}{\partial \phi} = -\hat{\rho}$$

SPHERICAL COORDINATES (r, θ, ϕ) θ = polar angle

Unit Vector Relationships

$$\hat{r} \cdot \hat{r} = \hat{\theta} \cdot \hat{\theta} = \hat{\phi} \cdot \hat{\phi} = 1$$

All other unit vector dot products are zero.

$$\hat{r} \times \hat{\theta} = \hat{\phi} = -(\hat{\theta} \times \hat{r})$$

$$\hat{\theta} \times \hat{\phi} = \hat{r} = -(\hat{\phi} \times \hat{\theta})$$

$$\hat{\phi} \times \hat{r} = \hat{\theta} = -(\hat{r} \times \hat{\phi})$$

$$\hat{r} \times \hat{r} = \hat{\theta} \times \hat{\theta} = \hat{\phi} \times \hat{\phi} = 0$$

Partial Derivatives

All partials of unit vectors are zero except:

$$\frac{\partial \hat{r}}{\partial \theta} = \hat{\theta}, \qquad \frac{\partial \hat{\theta}}{\partial \theta} = -\hat{r} \qquad \frac{\partial \hat{r}}{\partial \phi} = \sin\theta \; \hat{\phi}$$

$$\frac{\partial \hat{\theta}}{\partial \phi} = \cos\theta \; \hat{\phi} \qquad \text{and} \qquad \frac{\partial \hat{\phi}}{\partial \phi} = -(\sin\theta \; \hat{r} + \cos\theta \; \hat{\theta})$$

INTERCOORDINATE SYSTEM UNIT VECTOR DOT PRODUCTS

Rectangular-Cylindrical

$$\hat{x} \cdot \hat{\rho} = \cos\phi \qquad \hat{y} \cdot \hat{\rho} = \sin\phi \qquad \hat{z} \cdot \hat{\rho} = 0$$

$$\hat{x} \cdot \hat{\phi} = -\sin\phi \qquad \hat{y} \cdot \hat{\phi} = \cos\phi \qquad \hat{z} \cdot \hat{\phi} = 0$$

$$\hat{x} \cdot \hat{z} = 0 \qquad \hat{y} \cdot \hat{z} = 0 \qquad \hat{z} \cdot \hat{z} = 1$$

Rectangular-Spherical

$$\hat{x} \cdot \hat{r} = \sin\theta \cos\phi \qquad \hat{y} \cdot \hat{r} = \sin\theta \sin\phi \qquad \hat{z} \cdot \hat{r} = \cos\theta$$

$$\hat{x} \cdot \hat{\theta} = \cos\theta \cos\phi \qquad \hat{y} \cdot \hat{\theta} = \cos\theta \sin\phi \qquad \hat{z} \cdot \hat{\theta} = -\sin\theta$$

$$\hat{x} \cdot \hat{\phi} = -\sin\phi \qquad \hat{y} \cdot \hat{\phi} = \cos\phi \qquad \hat{z} \cdot \hat{\phi} = 0$$

Cylindrical-Spherical

$$\hat{\rho} \cdot \hat{r} = \sin\theta \qquad \hat{\phi}_{\text{cyl}} \cdot \hat{r} = 0 \qquad \hat{z} \cdot \hat{r} = \cos\theta$$

$$\hat{\rho} \cdot \hat{\theta} = \cos\theta \qquad \hat{\phi}_{\text{cyl}} \cdot \hat{\theta} = 0 \qquad \hat{z} \cdot \hat{\theta} = -\sin\theta$$

$$\hat{\rho} \cdot \hat{\phi}_{\text{sph}} = 0 \qquad \hat{\phi}_{\text{cyl}} \cdot \hat{\phi}_{\text{sph}} = 1 \qquad \hat{z} \cdot \hat{\phi}_{\text{sph}} = 0$$

F. Vector Helmholtz Theorem

In formal proofs and derivations it is often important to know when a vector field has been completely or uniquely determined. It is also important to know what additional constraints can be imposed on a vector field. The vector Helmholtz theorem gives us such criteria.

STATEMENT OF THE THEOREM

Any continuous vector field $\vec{F}$ is completely specified if its divergence and curl are known.

Alternative Statement

Any continuous vector field $\vec{F}$ can be expressed as the sum of the gradient of a scalar and the curl of a vector. (The scalar is called the *scalar potential* and the vector is called the *vector potential*.)

Before giving the proof of the theorem and its alternate form, a result that is of considerable help in the proof is developed.

THE DELTA FUNCTION AND THE ∇ OPERATOR

We first develop a relationship between the Laplacian ∇^2 and the delta function as applied to a space vector between two points P and P'. The configuration is shown in Figure F-1.

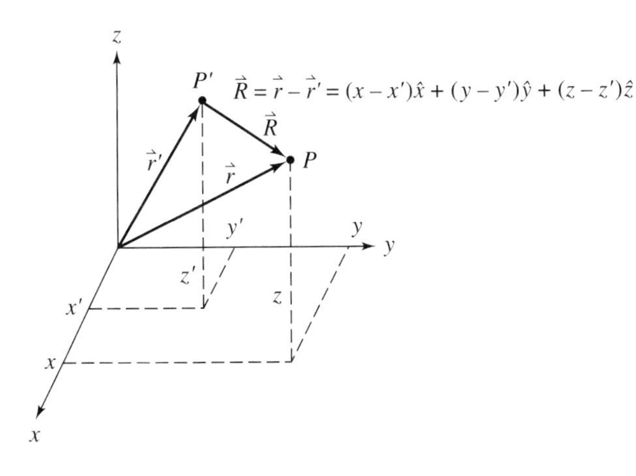

Figure F-1. Distance vector $\vec{R}$ between two points in space.

From the figure we have

$$R = |\vec{R}| = \sqrt{(x - x')^2 + (y - y')^2 + (z - z')^2}$$

Then

$$\nabla\left(\frac{1}{R}\right) = \nabla([(x - x')^2 + (y - y')^2 + (z - z')^2]^{-1/2})$$

$$= \frac{\partial\left(\frac{1}{R}\right)}{\partial x}\hat{x} + \frac{\partial\left(\frac{1}{R}\right)}{\partial y}\hat{y} + \frac{\partial\left(\frac{1}{R}\right)}{\partial z}\hat{z}$$

$$= -\frac{(x - x')}{R^3}\hat{x} - \frac{(y - y')}{R^3}\hat{y} - \frac{(z - z')}{R^3}\hat{z} = -\frac{\vec{R}}{R^3} \tag{F-1}$$

If we let ∇' denote derivatives on the *prime* coordinates, we obtain

$$\nabla'\left(\frac{1}{R}\right) = \frac{(x - x')}{R^3}\hat{x} + \frac{(y - y')}{R^3}\hat{y} + \frac{(z - z')}{R^3}\hat{z} = \frac{\vec{R}}{R^3} \tag{F-2}$$

Thus

$$\nabla\left(\frac{1}{R}\right) = -\nabla'\left(\frac{1}{R}\right) \tag{F-3}$$

Now

$$\nabla^2\left(\frac{1}{R}\right) = \frac{\partial^2\left(\frac{1}{R}\right)}{\partial x^2} + \frac{\partial^2\left(\frac{1}{R}\right)}{\partial y^2} + \frac{\partial^2\left(\frac{1}{R}\right)}{\partial z^2} = \frac{\partial \frac{\partial\left(\frac{1}{R}\right)}{\partial x}}{\partial x} + \frac{\partial \frac{\partial\left(\frac{1}{R}\right)}{\partial y}}{\partial y} + \frac{\partial \frac{\partial\left(\frac{1}{R}\right)}{\partial z}}{\partial z}$$

The partials in parentheses have just been evaluated. Then, using the previous results,

$$\frac{\partial^2\left(\frac{1}{R}\right)}{\partial x^2} = \frac{\partial\left\{-\frac{(x - x')}{[(x - x')^2 + (y - y')^2 + (z - z')^2]^{3/2}}\right\}}{\partial x}$$

$$= \frac{2(x - x')^2 - (y - y')^2 - (z - z')^2}{[(x - x')^2 + (y - y')^2 + (z - z')^2]^{5/2}}$$

Since R is symmetric in x, y, and z, we may then write the remaining Laplacian terms as

$$\frac{\partial^2\left(\frac{1}{R}\right)}{\partial y^2} = \frac{-(x - x')^2 + 2(y - y')^2 - (z - z')^2}{[(x - x')^2 + (y - y')^2 + (z - z')^2]^{5/2}}$$

$$\frac{\partial^2\left(\frac{1}{R}\right)}{\partial z^2} = \frac{-(x - x')^2 - (y - y')^2 + 2(z - z')^2}{[(x - x')^2 + (y - y')^2 + (z - z')^2]^{5/2}}$$

Adding the three second partials, noting that they have the same denominator, the Laplacian expansion becomes

$$\frac{2(x - x')^2 - (y - y')^2 - (z - z')^2 + 2(y - y')^2 - (x - x')^2 - (z - z')^2 - (x - x')^2 - (y - y')^2 + 2(z - z')^2}{[(x - x')^2 + (y - y')^2 + (z - z')^2]^{5/2}}$$

This identical result is obtained if we evaluate the Laplacian for primed coordinates, so that

$$\nabla^2\left(\frac{1}{R}\right) = \nabla'^2\left(\frac{1}{R}\right)$$

If $R \neq 0$ (i.e., if the conditions $x = x'$, $y = y'$, $z = z'$ are not simultaneously satisfied), the right side is 0, since the numerator is zero but the denominator is nonzero. Therefore,

$$\nabla^2\left(\frac{1}{R}\right) = 0, \qquad R \neq 0 \tag{F-4}$$

By similar process, if ∇'^2 is the Laplacian on *prime* coordinates,

$$\nabla'^2\left(\frac{1}{R}\right) = 0, \qquad R \neq 0 \tag{F-5}$$

However, if $R = 0$, the preceding expanded form of $\nabla^2(1/R)$ is the indeterminant form 0/0. We now investigate the value at $R = 0$.

Suppose we have a continuous vector field denoted by $\vec{F}$. Form the integral over a volume V' which includes the point $R = 0$. Divide V' into two regions: one a sphere of radius ϵ, having as its center (x, y, z) called V'_0, and the remaining volume being V'_1 so that $V' = V'_0 + V'_1$. See Figure F-2. Next, integrate the product of the vector field $\vec{F}$ and the unprimed Laplacian over the entire volume V'. This yields

$$\int_{V'} \vec{F}(x', y', z')\, \nabla^2\left(\frac{1}{R}\right) dv' = \int_{V'_0} \vec{F}(x', y', z')\, \nabla^2\left(\frac{1}{R}\right) dv' + \int_{V'_0} \vec{F}(x', y', z')\, \nabla^2\left(\frac{1}{R}\right) dv'$$

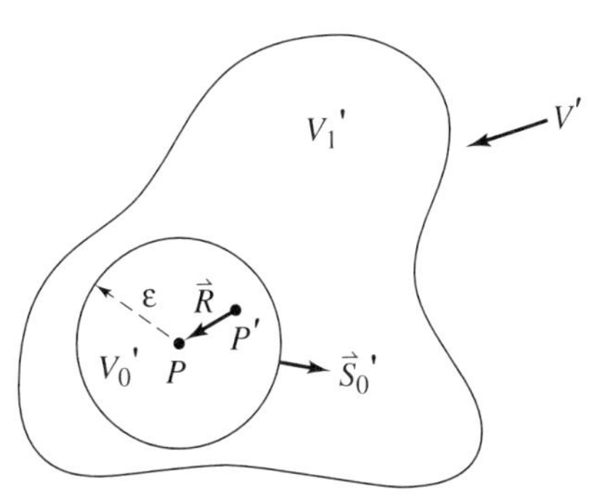

Figure F-2. Division of the re?gion of space around point P into two subregions.

But, the integral on V_1' has $\nabla^2(1/R) = 0$ since the point $R = 0$ is *not* included. (Note: P is considered fixed.) Thus

$$\int_{V'} \vec{F}(x', y', z')\, \nabla^2\left(\frac{1}{R}\right) dv' = \int_{V_0'} \vec{F}(x', y', z')\, \nabla^2\left(\frac{1}{R}\right) dv'$$

The radius ϵ may be taken very small since $\nabla^2(1/R)$ is 0 except at $R = 0$. Also, since $\vec{F}$ is continuous, it has the same value everywhere in the infinitesimal V_0' that it has at the point P(x, y, z) so it may be removed from the integrand. Then

$$\begin{aligned}\int_{V_0'} \vec{F}(x', y', z')\, \nabla^2\left(\frac{1}{R}\right) dv' &= \int_{V_0'} \vec{F}(x, y, z)\, \nabla^2\left(\frac{1}{R}\right) dv' \\ &= \vec{F}(x, y, z) \int_{V_0'} \nabla^2\left(\frac{1}{R}\right) dv'\end{aligned} \tag{F-6}$$

The integral of the Laplacian is evaluated using the fact that $\nabla^2(1/R) = \nabla'^2(1/R)$ as shown earlier and the definition $\nabla^2 \equiv \nabla \cdot \nabla$ along with Eq. F-2.

$$\int_{V_0'} \nabla^2\left(\frac{1}{R}\right) dv' = \int_{V_0'} \nabla'^2\left(\frac{1}{R}\right) dv' = \int_{V_0'} \nabla' \cdot \nabla'\left(\frac{1}{R}\right) dv' = \int_{V_0'} \nabla' \cdot \left(\frac{\vec{R}}{R^3}\right) dv'$$

Using the divergence theorem, we obtain

$$\int_{V_0'} \nabla^2\left(\frac{1}{R}\right) dv' = \int_{V_0'} \frac{\vec{R}}{R^3} \cdot d\vec{s}_0'$$

Since $\vec{R}$ is directed *inward*, $\vec{R} \cdot d\vec{S}_0' = -R dS_0'$ because P' is on the surface, as it must be for evaluating a surface integral. Therefore,

$$\int_{V_0'} \nabla^2\left(\frac{1}{R}\right) dv' = \int_{S_0'} \frac{-R}{R^3}\, ds_0' = -\int_{S_0'} \frac{dS_0'}{R^2} = -\int_{S_0'} d\Omega_0 = -4\pi$$

In the above ds_0'/R^2 is the differential solid angle $d\Omega_0$.

Returning to the integration of $\vec{F}$, we use this last result in Eq. F-6 to obtain

$$\int_{V'} \vec{F} R\, \nabla^2\left(\frac{1}{R}\right) dv' = -4\pi \vec{F}(x, y, z)$$

Thus bringing the constant -4π into the integral,

$$\int_{V'} \vec{F}(x', y', z')\left(-\frac{\nabla^2\left(\frac{1}{R}\right)}{4\pi}\right) dv' = \begin{cases} \vec{F}(x, y, z) & \text{if } v' \text{ includes } R = 0 \\ 0 & \text{if } v' \text{ excludes } R = 0 \end{cases}$$

Recalling the definition of the Dirac delta function, we must have

$$-\frac{\nabla^2\left(\frac{1}{R}\right)}{4\pi} = \delta(R)$$

or

$$\nabla^2\left(\frac{1}{R}\right) = -4\pi\ \delta(R) \tag{F-7}$$

PROOF OF THE VECTOR HELMHOLTZ THEOREM

Let $\vec{F}$ be a continuous vector field function of position (x, y, z). From the properties of the δ function and using Eq. F-7

$$\vec{F}(x, y, z) = \int_{V'} \vec{F}(x', y', z')\ \delta(R)dv' = -\int_{V'} \frac{\vec{F}(x', y', z')}{4\pi}\ \nabla^2\left(\frac{1}{R}\right)dv'$$

Since ∇ operates only on the unprimed coordinates, it can be taken outside the integrand:

$$\vec{F}(x, y, z) = -\nabla^2 \int_{V'} \frac{\vec{F}(x', y', z')}{4\pi}\left(\frac{1}{R}\right)dv'$$

A vector identity states that $-\nabla^2\vec{B} = \nabla \times \nabla \times \vec{B} - \nabla(\nabla \cdot \vec{B})$ (see Section 4-18 item 2) so that the preceding equation expands to

$$F(x, y, z) = \nabla \times \nabla \times \int_{V'} \frac{\vec{F}(x', y', z')}{4\pi R}\ dv' - \nabla\nabla \cdot \int_{V'} \frac{\vec{F}(x', y', z')}{4\pi R}\ dv'$$

$$= \nabla \times \int_{V'} \nabla \times \left(\frac{\vec{F}(x', y', z')}{4\pi R}\right)dv' - \nabla \int_{V'} \nabla \cdot \left(\frac{\vec{F}(x', y', z')}{4\pi R}\right)dv'$$

Let the first integral be denoted by the *vector* $\vec{A}$, since the integrand is a vector quantity. Also, let the second integral be a *scalar* Φ since the integrand is a scalar. We then have

$$\vec{F}(x, y, z) = \nabla \times \vec{A} - \nabla\Phi \tag{F-8}$$

This is the proof of the alternative form of the theorem and also shows us that, in general, a vector field may be broken down into a rotational part (curl) and a gradient part (irrotational). One might argue that this result represents a circular proof since the quantities $\vec{A}$ and Φ are in terms of the vector $\vec{F}$ that is to be determined. However, in practice it is possible to calculate the $\vec{A}$ and Φ from other quantities in the system, such as sources.

Note that the first statement of the theorem has not been proved since the curl and divergence of $\vec{F}$ have not yet been determined. If we take the curl of Eq. F-8 and use the fact that $\nabla \times \nabla\Phi = 0$ for any scalar Φ, we obtain

$$\nabla \times \vec{F} = \nabla \times \nabla \times \vec{A} \tag{F-9}$$

which indicates that if the curl of $\vec{F}$ is known, we may calculate the rotational (curl of $\vec{A}$) component of $\vec{F}$. Similarly, taking the divergence of Eq. F-8 and utilizing the theorem $\nabla \cdot \nabla \times \vec{A} = 0$ for any vector $\vec{A}$, there results

$$\nabla \cdot \vec{F} = -\nabla \cdot \nabla\Phi \tag{F-10}$$

This means that if we know the divergence of $\vec{F}$, we can determine the gradient component of $\vec{F}$.

Therefore, since knowing the curl and divergence of a function also specifies the two components $\nabla \times \vec{A}$ and $\nabla\Phi$ of $\vec{F}$, then the curl and divergence completely determine $\vec{F}$.

Returning momentarily to the integral definitions of $\vec{A}$ and Φ given earlier, it is possible to put these integrals in a form which involves the curl and divergence of $\vec{F}$ directly. The method is covered in Collin [11] pages 35–36. The results are

$$\left.\begin{aligned} \vec{A} &= \int_{V'} \frac{\nabla' \times \vec{F}(x', y', z')}{4\pi R}\, dv' + \oint_{s'} \frac{\vec{F}(x', y', z') \times d\vec{s}\,'}{4\pi R} \\ \Phi &= \int_{V'} \frac{\nabla' \cdot \vec{F}(x', y', z')}{4\pi R}\, dv' - \int_{s'} \frac{\vec{F}(x', y', z') \times d\vec{s}\,'}{4\pi R} \end{aligned}\right\}$$

where s' bounds v' and $\vec{R} = (x - x')\hat{x} + (y - y')\hat{y} + (z - z')\hat{z}$

Appendix G. The Unit Dyad

At the end of Section 4-17, brief mention and definition of the unit dyad were given. Also, in the paragraph just preceding the unit dyad definition, we found it necessary to *define* the Laplacian of a vector as

$$\nabla^2 \vec{A} = \nabla(\nabla \cdot \vec{A}) - \nabla \times \nabla \times \vec{A} \tag{G-1}$$

It may seem unnecessary to have to define $\nabla^2 \vec{A}$ this way when we earlier defined ∇^2 as $\nabla \cdot \nabla$. Thus one might legitimately inquire as to the possibility of simply expanding the equivalent representation

$$\nabla^2 \vec{A} = \nabla \cdot (\nabla \vec{A})$$

in the desired coordinate system. The problem with the evaluation of this expression is that we have no results for the gradient of a *vector*. If we realize that our simplest and *exact* forms involving ∇ occurred in rectangular coordinates, we are led to expand the expression $\nabla \cdot (\nabla \vec{A}) \equiv \nabla^2 \vec{A}$ in *rectangular* form:

$$\nabla^2 \vec{A} \equiv \nabla \cdot (\nabla \vec{A}) = \nabla \cdot \left[\left(\frac{\partial}{\partial x} \hat{x} + \frac{\partial}{\partial y} \hat{y} + \frac{\partial}{\partial z} \hat{z} \right) (A_x \hat{x} + A_y \hat{y} + A_z \hat{z}) \right]$$

Now this next step involves multiplying unit vectors without using either a dot or a cross product. To do this, we must preserve the order of multiplication; that is, commutativity does *not* hold for ordinary multiplication of unit vectors. Then we have

$$\begin{aligned} \nabla^2 \vec{A} = \nabla \cdot \Bigg(& \frac{\partial A_x}{\partial x} \hat{x}\hat{x} + \frac{\partial A_y}{\partial x} \hat{x}\hat{y} + \frac{\partial A_z}{\partial x} \hat{x}\hat{z} + \frac{\partial A_x}{\partial y} \hat{y}\hat{x} + \frac{\partial A_y}{\partial y} \hat{y}\hat{y} \\ & + \frac{\partial A_z}{\partial y} \hat{y}\hat{z} + \frac{\partial A_x}{\partial z} \hat{z}\hat{x} + \frac{\partial A_y}{\partial z} \hat{z}\hat{y} + \frac{\partial A_z}{\partial z} \hat{z}\hat{z} \Bigg) \end{aligned} \tag{G-2}$$

Since we have not defined ordinary multiplication of vectors, we give terms of the form $\hat{x}\hat{y}$, $\hat{z}\hat{x}$, etc., a new name. These are called *unit dyads*. Unfortunately, we are on the verge of going into tensor analysis, which we do not intend to cover here. For our purpose, we need only one rule to accomplish our goal. If a and b are scalars (complex or real) in a generalized coordinate system (u_1, u_2, u_3) we write

$$\begin{gathered} (a\hat{u}_i) \cdot (b\hat{u}_j\hat{u}_k) = ab(\hat{u}_i) \cdot (\hat{u}_j\hat{u}_k) \equiv ab(\hat{u}_i \cdot \hat{u}_j)\hat{u}_k, \\ i = 1, 2, 3, \quad j = 1, 2, 3, \quad k = 1, 2, 3 \end{gathered} \tag{G-3}$$

The important point to notice is that order of the unit vectors *does not* change from left to right. For example,

$$6\hat{x} \cdot 3\hat{x}\hat{z} = 18\hat{x} \cdot \hat{x}\hat{z} = 18(\hat{x} \cdot \hat{x})\hat{z} = 18\hat{z}$$

Using $\nabla = (\partial/\partial x)\hat{x} + (\partial/\partial y)\hat{y} + (\partial/\partial z)\hat{z}$ in the preceding expansion of $\nabla^2\vec{A}$, we then have

$$\nabla^2\vec{A} = \left(\frac{\partial}{\partial x}\hat{x} + \frac{\partial}{\partial y}\hat{y} + \frac{\partial}{\partial z}\hat{z}\right)\cdot\left(\frac{\partial A_x}{\partial x}\hat{x}\hat{x} + \frac{\partial A_y}{\partial x}\hat{x}\hat{y} + \frac{\partial A_z}{\partial x}\hat{x}\hat{z} + \frac{\partial A_x}{\partial y}\hat{y}\hat{x}\right.$$
$$\left. + \frac{\partial A_y}{\partial y}\hat{y}\hat{y} + \frac{\partial A_z}{\partial y}\hat{y}\hat{z} + \frac{\partial A_x}{\partial z}\hat{z}\hat{x} + \frac{\partial A_y}{\partial z}\hat{z}\hat{y} + \frac{\partial A_z}{\partial z}\hat{z}\hat{z}\right)$$

Next, expanding the dot products, using Eq. G-3 (note that there would be 27 terms), and recalling that mixed dot products $\hat{x}\cdot\hat{y}$, $\hat{z}\cdot\hat{y}$, etc., are zero in an orthogonal system, we finally obtain

$$\nabla^2\vec{A} = \frac{\partial^2 A_x}{\partial x^2}\hat{x} + \frac{\partial^2 A_y}{\partial x^2}\hat{y} + \frac{\partial^2 A_z}{\partial x^2}\hat{z} + \frac{\partial^2 A_x}{\partial y^2}\hat{x} + \frac{\partial^2 A_y}{\partial y^2}\hat{y} + \frac{\partial^2 A_z}{\partial y^2}\hat{z}$$
$$+ \frac{\partial^2 A_x}{\partial z^2}\hat{x} + \frac{\partial^2 A_y}{\partial z^2}\hat{y} + \frac{\partial^2 A_z}{\partial z^2}\hat{z} \tag{G-4}$$

Rearranging,

$$\nabla^2\vec{A} = \left(\frac{\partial^2 A_x}{\partial x^2}\hat{x} + \frac{\partial^2 A_y}{\partial y^2}\hat{y} + \frac{\partial^2 A_z}{\partial z^2}\hat{z}\right)$$
$$+ \left(\frac{\partial^2 A_y}{\partial x^2}\hat{y} + \frac{\partial^2 A_z}{\partial x^2}\hat{z} + \frac{\partial^2 A_x}{\partial y^2}\hat{x} + \frac{\partial^2 A_z}{\partial y^2}\hat{z} + \frac{\partial^2 A_x}{\partial z^2}\hat{x} + \frac{\partial^2 A_y}{\partial z^2}\hat{y}\right)$$

$$\nabla^2\vec{A} = \left(\frac{\partial\left(\frac{\partial A_x}{\partial x}\right)}{\partial x}\hat{x} + \frac{\partial\left(\frac{\partial A_y}{\partial y}\right)}{\partial y}\hat{y} + \frac{\partial\left(\frac{\partial A_z}{\partial z}\right)}{\partial z}\hat{z}\right)$$
$$+ \left(\frac{\partial\left(\frac{\partial A_y}{\partial x}\right)}{\partial x}\hat{y} + \frac{\partial\left(\frac{\partial A_z}{\partial x}\right)}{\partial x}\hat{z} + \frac{\partial\left(\frac{\partial A_x}{\partial y}\right)}{\partial y}\hat{x}\right.$$
$$\left. + \frac{\partial\left(\frac{\partial A_z}{\partial y}\right)}{\partial y}\hat{z} + \frac{\partial\left(\frac{\partial A_x}{\partial z}\right)}{\partial z}\hat{x} + \frac{\partial\left(\frac{\partial A_y}{\partial z}\right)}{\partial z}\hat{y}\right)$$

In the first set of parentheses, we see derivatives of terms with respect to one variable and in the unit vector direction of that *same* variable. Then we would hope to set up gradient and dot products in these terms. In the second set of parentheses, we observe that the unit vector directions are different from the variable of differentiation. We would then hope to set up curl-type operations within these terms. To accomplish this, we add and subtract duplicate sets of terms in the first set of parentheses; however, note that what these steps will do is simply return us to Eq. G-1 expanded with no more information. Thus Eq. G-4 is as far as one really needs to go; however, in other coordinate systems one must resort to the definition Eq. G-1 unless unit dyads are used.

Exercises

G-a Making use of unit dyads and using rectangular coordinates, write expressions for

(a) $\nabla\vec{B}$ where $\vec{B}$ is a vector field

(b) $(\vec{A} \cdot \nabla)\vec{B}$ and $\vec{A} \cdot (\nabla\vec{B})$

(c) $\vec{A} \times (\nabla\vec{B})$ and $(\vec{A} \times \nabla)\vec{B}$

G-b Obtain the expansion for $\nabla^2\vec{A}$ in cylindrical coordinates to the point similar to Eq. G-4. Don't forget that in this coordinate system unit vectors cannot always be taken outside derivatives.

G-c Repeat Exercise G-b for spherical coordinates.

Appendix H. Derivation of the Energy in a Magnetic Field

The basic idea used in determining the energy in a magnetic field is to compute the energy extracted from the field as it interacts with a current distribution. We assume that the magnetic field is brought from value zero to the final value so slowly that static conditions may be assumed. The system configuration is shown in Figure H-1. The components will be defined as they are referred to in the derivation. This derivation follows that given by Stratton [56].

We imagine a volume of space, shown by the large ''cylinder'' of cross section S' in the figure, which carries a constant density $\vec{J}$ throughout. Within this region we define an infinitesimal tube of current of cross section $\delta\vec{\sigma}$, which must close on itself since static conditions mean the charge at a given point must remain constant. That is, any current leaving must be compensated for by a current entering. This closed tube defines a contour C. Mathematically, this means the divergence of the current density is zero (see Eq. 5-42) applied to static fields or take the divergence of Eq. 6-24 and apply the theorem $\nabla \cdot \nabla \times \vec{F} \equiv 0$). For simplicity we represent the remainder of the closed path by a conductor which connects the constant current source I_0 to the system. Note that this closed path encloses

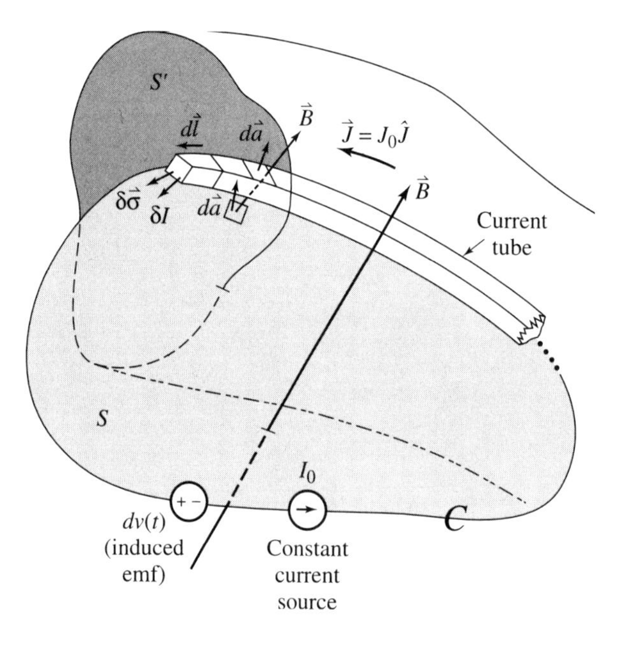

Figure H-1. Surfaces and volumes with associated current I_0 and magnetic flux density $\vec{B}$ used to compute magnetic field energy density.

the area S. A magnetic flux density $\vec{B}$ due to currents external to the current tube is caused to pass through this surface S.

Now as the field $\vec{B}$ increases it will, according to Faraday's law (Chapter 7), induce a voltage into the loop C, which will be in such a direction as to produce a flux opposing that increase (Lenz's law) as shown by $dv(t)$ in the figure. The power delivered to the tube would be that in the source $dv(t)$; that is,

$$\delta p(t) = \delta\left(\frac{dw}{dt}\right) = (\delta I)(dv(t)) = \delta I\,\frac{d\phi}{dt}$$

Multiply by dt:

$$\delta(dw) = \delta I d\phi \tag{H-1}$$

From the figure,

$$\delta I = \vec{J}\cdot\delta\vec{\sigma} \tag{H-2}$$

and

$$d\phi = d\int_S \vec{B}\cdot d\vec{a} = \int_S d\vec{B}\cdot d\vec{a}$$

The last equality is due to the fact that the area is stationary (see Leibniz' rule, Chapter 4). Also note that to obtain the total flux that induces the emf, we have to integrate around the entire tube and all the surface it encloses. Equation H-1 becomes

$$\delta(dw) = (\vec{J}\cdot\delta\vec{\sigma})\left(\int_S d\vec{B}\cdot d\vec{a}\right)$$

From Eq. 6-40 of the text

$$d\vec{B} = \nabla\times d\vec{A} \tag{H-3}$$

so that

$$\delta(dw) = (\vec{J}\cdot\delta\vec{\sigma})\left(\int_S \nabla\times d\vec{A}\cdot d\vec{a}\right)$$

Applying Stokes' theorem to the integrand,

$$\delta(dw) = \vec{J}\cdot\delta\vec{\sigma}\oint_C d\vec{A}\cdot d\vec{l} \tag{H-4}$$

Now the total vector potential at a point within the tube is

$$\vec{A}_{\text{total}} = \vec{A} + \vec{A}_{\delta I}$$

In the limit as $\delta\vec{\sigma}\to 0$, the following limits also occur:

1. $\delta I\to 0$ from Eq. H-2, so $\vec{A}_{\delta I}\to 0$
2. The vectors $\delta\vec{\sigma}$, $\vec{J}$, and $d\vec{l}$ become parallel so we may write

$$\vec{J}\cdot\delta\vec{\sigma}\Rightarrow J_0\delta\sigma$$

$$d\vec{l}\Rightarrow dl\hat{J}$$

With these results Eq. H-4 becomes

$$\delta(dw) = J_0 \delta\sigma \oint_C d\vec{A} \cdot (dl\hat{J})$$

since J_0 is a constant,

$$\delta(dw) = \oint_C J_0 d\vec{A} \cdot (dl\hat{J})\delta\sigma = \oint_C d\vec{A} \cdot (dlJ_0\hat{J})\delta\sigma = \oint_C d\vec{A} \cdot \vec{J} dl\ \delta\sigma$$

Now if we add up all the tubes by integrating over the large cross section of the entire current distribution S',

$$dw = \int_{S'} \oint_C d\vec{A} \cdot \vec{J} dl\ \delta\sigma$$

The combined integrals constitute the entire volume so we write, since $dl\ \delta\sigma = dv$,

$$dw = \int_V d\vec{A} \cdot \vec{J} dv \tag{H-5}$$

Now within the current distribution Eq. 6-25, $\nabla \times \vec{H} = \vec{J}$, applies so this last equation may be written

$$dw = \int_V d\vec{A} \cdot (\nabla \times \vec{H}) dv$$

Using the vector identity $\nabla \cdot (\vec{F} \times \vec{G}) = \vec{G} \cdot (\nabla \times \vec{F}) - \vec{F} \cdot (\nabla \times \vec{G})$ and letting $\vec{H} \sim \vec{F}$ and $\vec{G} \sim d\vec{A}$ the preceding equation becomes

$$dw = \int_V \nabla \cdot (\vec{H} \times d\vec{A}) dv + \int_V \vec{H} \cdot (\nabla \times d\vec{A}) dv$$

Applying the divergence theorem to the left integral and using Eq. H-3 on the right integral, this reduces to

$$dw = \oint_S (\vec{H} \times d\vec{A}) \cdot d\vec{a} + \int_V \vec{H} \cdot d\vec{B} dv$$

For the surface integral, the only requirement is that it enclose V. Thus we let it become large in extent to infinity. For finite-extent currents (no infinite lines, etc.) the fields $\vec{H}$ and $d\vec{A}$ decrease as one moves away from the currents. In fact, the product $\vec{H} \times d\vec{A}$ decreases at least as fast as r^{-3} ($\vec{H}$ decreasing at least r^{-2} and $\vec{A}$ as r^{-1}), but since area increases only as r^2, the left integral will be zero for a surface at infinity. We then finally obtain

$$dw = \int_V \vec{H} \cdot d\vec{B} dv \tag{H-6}$$

Appendix I. Alternative Solution Forms of the Wave Equation

Other forms of the solution to the wave equation may be convenient in some wave propagation problems. These are reviewed here.

The basic equation can be written in general form as

$$\frac{d^2y}{dz^2} + T^2y = 0 \tag{I-1}$$

The characteristic equation is

$$m^2 + T^2 = 0$$

from which

$$m = \pm jT$$

The usual solution is written

$$y(z) = A_1e^{-jTz} + B_1e^{jTz} \tag{I-2}$$

where A_1 and B_1 are in general complex. This form is the most appropriate for propagation studies.

If we apply Euler's theorem to the exponentials we obtain the second form as follows:

$$\begin{aligned} y(z) &= A_1(\cos Tz - j\sin Tz) + B_1(\cos Tz + j\sin Tz) \\ &= (A_1 + B_1)\cos Tz + (jB_1 - jA_1)\sin Tz = A_2\cos Tz + B_2\sin Tz \end{aligned} \tag{I-3}$$

This form is frequently preferred when the system being studied has finite z dimensions and is periodic in z.

When T is pure imaginary, say $j\tau$, the differential equation becomes

$$\frac{d^2y}{dz^2} - \tau^2y = 0 \tag{I-4}$$

for which the solution is

$$y(z) = A_3e^{-\tau z} + B_3e^{\tau z} \tag{I-5}$$

This form is preferred when the field configuration is a purely attenuating one in the z direction *and* where z can become infinite. This last requirement frequently requires B_3 to be zero, which results in a simple solution.

The final form is obtained by using the hyperbolic forms of the exponentials:

$$\begin{aligned} y(z) &= A_3(\cosh \tau z - \sinh \tau z) + B_3(\cosh \tau z + \sinh \tau z) \\ &= (A_3 + B_3)\cosh \tau z + (B_3 - A_3)\sinh \tau z = A_4 \cosh \tau z + B_4 \sinh \tau z \end{aligned} \tag{I-6}$$

This form is usually preferred when the system is not periodic in z *and* when the system is finite in extent z.

Exercise

I-a Showing that substituting $j\tau$ for T in Eqs. I-2 and I-3 results in Eqs. I-5 and I-6

Appendix J. Generalized Coordinate System Expressions for the Transverse Components of the $\vec{E}$ and $\vec{H}$ in Terms of E_z and H_z for Waves Propagating in the z Direction, with Transverse Coordinates t_1 and t_2

We begin with Faraday's law in differential (point) form:

$$\nabla \times \vec{E} = -j\omega\mu\vec{H}$$

Expanding in generalized coordinates, letting $u_1 = t_1$, $u_2 = t_2$, and $u_3 = z$:

$$\frac{1}{h_1 h_2 h_3}\left[h_1\left(\frac{\partial(h_z E_z)}{\partial t_2} - \frac{\partial(h_2 E_2)}{\partial z}\right)\hat{t} + h_2\left(\frac{\partial(h_1 E_1)}{\partial z} - \frac{\partial(h_z E_z)}{\partial t_1}\right)\hat{t}_2 + h_z\left(\frac{\partial(h_2 E_2)}{\partial t_1} - \frac{\partial(h_1 E_1)}{\partial t_2}\right)\right] = -j\omega\mu(H_1\hat{t}_1 + H_2\hat{t}_2 + H_z\hat{z})$$

For systems that have a z coordinate, $h_z = 1$. Also, the scale factors h_1 and h_2 will not be functions of z. With these constraints the preceding equation becomes

$$\frac{1}{h_2}\left(\frac{\partial E_z}{\partial t_2} - h_2\frac{\partial E_2}{\partial z}\right)\hat{t}_1 + \frac{1}{h_1}\left(h_1\frac{\partial E_1}{\partial z} - \frac{\partial E_z}{\partial t_1}\right)\hat{t}_2 + \frac{1}{h_1 h_2}\left(\frac{\partial(h_2 E_2)}{\partial t_1} - \frac{\partial(h_1 E_1)}{\partial t_2}\right)\hat{z} = -j\omega\mu H_1\hat{t}_1 - j\omega\mu H_2\hat{t}_2 - j\omega\mu H_z\hat{z} \tag{J-1}$$

As discussed in Chapter 7 we assume components have the z variations of the form $e^{-\gamma z}$, which is a multiplying factor in each component. This produces components of the general form

$$F(t_1, t_2, z) = F_t(t_1, t_2)e^{-\gamma z} \tag{J-2}$$

We may then replace all $\partial/\partial z$ by $-\gamma$. Since $\partial F(t_1,t_2, z)/\partial z = -\gamma F_t(t_1, t_2)e^{-\gamma z} = -\gamma F(t_1, t_2, z)$, the preceding equation reduces to

$$\left(\frac{1}{h_2}\frac{\partial E_z}{\partial t_2} + \gamma E_2\right)\hat{t}_1 + \left(-\gamma E_1 - \frac{1}{h_1}\frac{\partial E_z}{\partial t_1}\right)\hat{t}_2 + \left(\frac{1}{h_1 h_2}\frac{\partial(h_2 E_2)}{\partial t_1} - \frac{1}{h_1 h_2}\frac{\partial(h_1 E_1)}{\partial t_2}\right)\hat{z} = -j\omega\mu H_1\hat{t}_1 - j\omega\mu H_2\hat{t}_2 - j\omega\mu H_z\hat{z} \tag{J-3}$$

We must now also remember that all field components are functions only of the transverse coordinates t_1 and t_2. We have omitted the t subscripts on the transverse function parts of the components for simplicity, as is usual practice. The exponential must be appended (multiplied) to each field component when they (and γ) have been determined. Equating components:

$$\frac{1}{h_2}\frac{\partial E_z}{\partial t_2} + \gamma E_2 = -j\omega\mu H_1 \tag{J-4}$$

$$\gamma E_1 + \frac{1}{h_1}\frac{\partial E_z}{\partial t_1} = j\omega\mu H_2 \tag{J-5}$$

$$\frac{1}{h_1 h_2}\frac{\partial(h_2 E_2)}{\partial t_1} - \frac{1}{h_1 h_2}\frac{\partial(h_1 E_1)}{\partial t_2} = -j\omega\mu H_2 \tag{J-6}$$

For another of Maxwell's equations $\nabla \times \vec{H} = j\omega\epsilon\vec{E}$ (Ampere's law), we proceed in the identical way to obtain

$$\frac{1}{h_2}\frac{\partial H_z}{\partial t_2} + \gamma H_2 = j\omega\epsilon E_1 \tag{J-7}$$

$$\gamma H_1 + \frac{1}{h_1}\frac{\partial H_z}{\partial t_1} = -j\omega\epsilon E_2 \tag{J-8}$$

$$\frac{1}{h_1 h_2}\frac{\partial(h_2 H_2)}{\partial t_1} - \frac{1}{h_1 h_2}\frac{\partial(h_1 H_1)}{\partial t_2} = j\omega\epsilon E_2 \tag{J-9}$$

We use the preceding set of six equations to obtain equations for the transverse field components in terms of E_z and H_z.

Solve Eq. J-5 for H_2 and substitute the result into Eq. J-7:

$$H_2 = -j\frac{1}{\omega\mu}\left(\gamma E_1 + \frac{1}{h_1}\frac{\partial E_z}{\partial t_1}\right)$$

$$\frac{1}{h_2}\frac{\partial H_z}{\partial t_2} - j\frac{\gamma^2}{\omega\mu}E_1 - j\frac{\gamma}{\omega\mu h_1}\frac{\partial E_z}{\partial t_1} = j\omega\epsilon E_1$$

Solving this for E_1:

$$E_1 = \frac{1}{j\left(\omega\epsilon + \dfrac{\gamma^2}{\omega\mu}\right)}\left(-j\frac{\gamma}{\omega\mu h_1}\frac{\partial E_z}{\partial t_1} + \frac{1}{h_2}\frac{\partial H_z}{\partial t_2}\right)$$

$$= -j\frac{\omega\mu}{\omega^2\mu\epsilon + \gamma^2}\left(-j\frac{\gamma}{\omega\mu h_1}\frac{\partial E_z}{\partial t_1} + \frac{1}{h_2}\frac{H_z}{\partial t_2}\right)$$

Let

$$\omega^2\mu\epsilon + \gamma^2 = k^2 + \gamma^2 = k_c^2 \tag{J-10}$$

Then

$$E_1 = -j\frac{\omega\mu}{k_c^2}\left(-j\frac{\gamma}{\omega\mu h_1}\frac{\partial E_z}{\partial t_1} + \frac{1}{h_2}\frac{\partial H_z}{\partial t_2}\right)$$

$$E_1 = -\frac{\gamma}{k_c^2 h_1}\frac{\partial E_z}{\partial t_1} - j\frac{\omega\mu}{h_2 k_c^2}\frac{\partial H_z}{\partial t_2} \tag{J-11}$$

Next solve Eq. J-4 for H_1 and substitute into Eq. J-8, then solve for E_2. The result is

$$E_2 = -\frac{\gamma}{k_c^2 h_2}\frac{\partial E_z}{\partial t_2} + j\frac{\omega\mu}{h_1 k_c^2}\frac{\partial H_z}{\partial t_1} \tag{J-12}$$

Next solve Eq. J-8 for E_2 and substitute into Eq. J-4, and then solve for H_1. The result is

$$H_1 = j\frac{\omega\epsilon}{k_c^2 h_2}\frac{\partial E_z}{\partial t_2} - \frac{\gamma}{k_c^2 h_1}\frac{\partial H_z}{\partial t_1} \tag{J-13}$$

Finally, solve Eq. J-7 for E_1 and substitute into Eq. J-5 and solve for H_2:

$$H_2 = -j\frac{\omega\epsilon}{k_c^2 h_1}\frac{\partial E_z}{\partial t_1} - \frac{\gamma}{k_c^2 h_2}\frac{\partial H_z}{\partial t_2} \tag{J-14}$$

Exercises

J-a Verify Eq. J-3 by substituting Eq. J-2 (with appropriate letter substitution for F) into Eq. J-1.

J-b Express Eqs. J-11 through J-14 in rectangular and cylindrical form.

J-c Express Eqs. J-11 through J-14 in the elliptic cylindrical coordinate system.

J-d Express Eqs. J-11 through J-14 in the parabolic cylindrical coordinate system.

Appendix K. Solution of Partial Differential Equations by the Method of Separation of Variables

Many complex problems are frequently solved by reducing them to problems for which solutions are known. This is the approach taken with partial differential equations. What is done is to reduce the partial differential equation to ordinary differential equations whose solutions are known or can be obtained using the series solution technique described in detail in Appendix L.

The basic process is as follows:

1. Obtain the partial differential equation in the desired coordinate system.
2. Examine the system physically to see if the solution is constant (invariant) with any of the coordinates. If so, set those partial derivative terms to zero. (If such invariances are not obvious it will not matter if those derivatives are left in. Boundary conditions will eventually produce that invariance if present. Alternatively, the invariance may become evident as the solution proceeds. It simplifies the work if a priori identifications can be made.)
3. Assume that the solution can be written as the product of single-variable functions, one function for each of the remaining derivative (varying) coordinates. This may seem like a restrictive solution. The uniqueness theorem helps here, because if such a product solution can be made to satisfy the boundary conditions and the differential equation it is unique.
4. Substitute the assumed product solution into the partial differential equation and separate the result into single-variable ordinary differential equations.

This process will be carried out on a couple of examples.

Example K-1

Rectangular Coordinates

Suppose we are given

$$\nabla^2\Phi + \frac{\partial^2\Phi}{\partial t^2} = 0$$

Expand in rectangular coordinates

$$\frac{\partial^2\Phi}{\partial x^2} + \frac{\partial^2\Phi}{\partial y^2} + \frac{\partial^2\Phi}{\partial z^2} + \frac{\partial^2\Phi}{\partial t^2} = 0 \tag{K-1}$$

The most general case is variation with all coordinates. Let

$$\Phi(x, y, z, t) \equiv X(x)Y(y)Z(z)T(t) \tag{K-2}$$

When we take partial derivatives all functions but one are constant. Then substituting this solution into the partial differential equation, omitting the parentheses for simplicity,

$$\frac{d^2X}{dx^2} YZT + X \frac{d^2Y}{dy^2} ZT + XY \frac{d^2Z}{dz^2} T + XYZ \frac{d^2T}{dt^2} = 0$$

Now divide by $XYZT$:

$$\frac{1}{X}\frac{d^2X}{dx^2} + \frac{1}{Y}\frac{d^2Y}{dy^2} + \frac{1}{Z}\frac{d^2Z}{dz^2} + \frac{1}{T}\frac{d^2T}{dt^2} = 0 \tag{K-3}$$

Note that we have the sum of four functions, each of which is a function of one variable. For this last equation to be true, each of the terms must be a constant. To show why this is necessary, suppose that we had a set of variables (x_1, y_1, z_1, t_1) for which the sum of the factors were zero. Now if we change only x, say, to value x_2, the sum would no longer be zero if the leftmost term were a different value while the remaining three were unchanged. Thus we set, using the derivative prime notation for ease of writing,

$$\frac{X''}{X} = k_x^2, \qquad \frac{Y''}{Y} = k_y^2, \qquad \frac{Z''}{Z} = k_z^2, \qquad \frac{T''}{T} = k_t^2 \tag{K-4}$$

The constants are called *separation constants*.

We now have four ordinary differential equations to solve, the product of those solutions being the desired solution according to Eq. K-2.

The constants k_x^2, etc., are at this point unknown, but as boundary conditions are applied their values are determined. The reason squared constants are used is because of the fact that the solutions to second-order equations involve the square root of the constant. For example, consider

$$\frac{X''}{X} = k_x^2 \qquad \text{or} \qquad X'' - k_x^2 X = 0 \tag{K-5}$$

Since the function and its derivatives must be of the same functional form to sum to zero we are led to try the solution

$$X(x) = Ae^{mx}$$

in which A and m are constants. Substituting this into the differential equation yields

$$m^2Ae^{mx} - k_x^2Ae^{mx} = 0$$

Since e^{mx} is not zero and A is a constant we may divide them out to obtain what is called the *characteristic equation*

$$m^2 - k_x^2 = 0 \tag{K-6}$$

From this

$$m = \pm k_x$$

for which the solution is written as the sum of all possible solution values m:

$$X(x) = Ae^{k_x x} + Be^{-k_x x} \tag{K-7}$$

The values of m are called *eigenvalues.*

Another observation on the separation constants is that negative values could be used for any or all constants $-k_x^2$, $-k_y^2$, etc. The consequence of this is that the characteristic equation would become

$$m^2 + k_x^2 = 0 \tag{K-8}$$

which gives $m = \pm jk_x$.

The corresponding solution would be

$$X(x) = A_1 e^{jk_x x} + B_1 e^{-jk_x x} \tag{K-9}$$

Using Euler's theorem on the exponentials yields the form

$$X(x) = (A_1 + B_1)\cos k_x x + (jA_1 - jB_1)\sin k_x x = A_2 \cos k_x x + B_2 \sin k_x x \tag{K-10}$$

This solution would be preferred if it could be observed that the solution $\Phi(x, y, z, t)$ in the system would be periodic in x, that is, the same boundary value at two different x values. Thus some separation constants may be positive and some negative.

Substituting Eq. K-4 into the separated variables of Eq. (K-3) we obtain

$$k_x^2 + k_y^2 + k_z^2 + k_t^2 = 0 \tag{K-11}$$

This is called the *separation constant constraint equation.* Thus if we can find any three of the four constants (which come from applying boundary conditions) we can find the fourth from the constraint equation.

Example K-2

Cylindrical Coordinates

Suppose we want to solve the partial differential equation

$$\frac{1}{\rho}\frac{\partial}{\partial \rho}\left(\rho\frac{\partial \Phi}{\partial \rho}\right) + \frac{1}{\rho^2}\frac{\partial^2 \Phi}{\partial \phi^2} + a\frac{\partial \Phi}{\partial t} = 0$$

in which a is constant. Let $\phi(\rho, \phi, t) = R(\rho)F(\phi)T(t)$.

Substituting into the differential equation:

$$\frac{FT}{\rho}\frac{d}{d\rho}\left(\rho\frac{dR}{d\rho}\right) + \frac{RT}{\rho^2}\frac{d^2F}{d\phi^2} + aRF\frac{dT}{dt} = 0$$

Divide by RFT:

$$\frac{1}{\rho R}\frac{d}{d\rho}\left(\rho\frac{dR}{d\rho}\right) + \frac{1}{\rho^2 F}\frac{d^2F}{d\phi^2} + \frac{a}{T}\frac{dT}{dt} = 0 \tag{K-12}$$

Since the last term is a function of t only, it must be a constant.

$$\frac{a}{T}\frac{dT}{dt} = k_t \left(\text{or:} \quad \frac{1}{T}\frac{dT}{dt} = k_t\right)\left(\text{or:}\ \frac{a}{T}\frac{dT}{dt} = -k_t\right) \tag{K-13}$$

Note that the other terms are not necessarily each constant since they both have coordinate ρ in them. One term might increase as the other decreases so the sum could remain constant.

Substitute the preceding equation into Eq. K-12:

$$\frac{1}{\rho R}\frac{d}{d\rho}\left(\rho\frac{dR}{d\rho}\right) + \frac{1}{\rho^2 F}\frac{d^2F}{d\phi^2} + k_t = 0 \qquad \text{(K-14)}$$

Multiply through by ρ^2 and collect terms:

$$\left[\frac{\rho}{R}\frac{d}{d\rho}\left(\rho\frac{dR}{d\rho}\right) + \rho^2 k_t\right] + \frac{1}{F}\frac{d^2F}{d\phi^2} = 0 \qquad \text{(K-15)}$$

This result requires (in the most usual case since solutions are typically periodic in ϕ):

$$\frac{1}{F}\frac{d^2F}{d\phi^2} = -k_\phi^2 \qquad \text{(K-16)}$$

and

$$\frac{\rho}{R}\frac{d}{d\rho}\left(\rho\frac{dR}{d\rho}\right) + \rho^2 k_t = k_\rho^2$$

It is customary not to define the R function equation in terms of a new constant as this last equation suggests, but to use Eq. K-16 in Eq. K-15 to obtain

$$\frac{\rho}{R}\frac{d}{d\rho}\left(\rho\frac{dR}{d\rho}\right) + \rho^2 k_t - k_\phi^2 = 0$$

or

$$\frac{\rho}{R}\frac{d}{d\rho}\left(\rho\frac{dR}{d\rho}\right) + \rho^2 k_t = k_\phi^2 \qquad \text{(K-17)}$$

which is constant as required. This has the advantage of not introducing another unknown.

Note that no constraint relation is obtained here since it was actually already used to eliminate one separation constant. This could have been done in the rectangular case, but the resulting differential equations would not have been of the same single-constant form which is convenient.

Equation K-17 is an ordinary differential equation but with variable coefficients. It must be solved using a series solution. This is discussed in Appendix L.

Exercises

K-a Expand the Laplacian in spherical coordinates and apply the separation of variables technique to obtain three ordinary differential equations. Is there a constraint equation?

K-b Expand the transverse wave equation

$$\nabla_t^2 E_z + k_c^2 E_z = 0$$

in bipolar coordinates and apply the separation of variables technique to obtain two ordinary differential equations.

K-c Determine which of the following partial differential equations can be solved by the method of separation of variables; A, B, and C are constants.

(a) $A \dfrac{\partial^2 \Phi}{\partial x^2} + B \dfrac{\partial^2 \Phi}{\partial x \partial y} + C \dfrac{\partial^2 \Phi}{\partial y^2} = 0$

(b) $A \dfrac{\partial^2 \Phi}{\partial x \partial y} + B\Phi = 0$

(c) $A \dfrac{\partial^2 \Phi}{\partial x^2} + B \dfrac{\partial^2 \Phi}{\partial x \partial y} + C \dfrac{\partial \Phi}{\partial y} = 0$

(d) $x^2 \dfrac{\partial^2 \Phi}{\partial x^2} + y \dfrac{\partial^2 \Phi}{\partial y^2} = 0$

(Answer: (*a*) No; (*b*) Yes; (*c*) No; (*d*) Yes)

K-d Repeat Exercise K-b for parabolic cylindrical coordinates.

K-e Repeat Exercise K-b for elliptic cylindrical coordinates.

Appendix L. Series Solutions of Ordinary Differential Equations and Solution Functions Defined by Them: Trigonometric, Bessel, and Legendre Functions

In field theory we often encounter ordinary differential equations that do not have constant coefficients so that the usual characteristic equation (see Eq. K-6) does not give an easy homogeneous solution. For such equations we obtain solutions using an assumed series and determine the series coefficients. The resulting series can then be defined as a function, with an assigned symbol to identify it. This is called the *method of Frobenius*. The process also works for constant coefficient differential equations, which are special cases.

Since our primary concern is to obtain the solution, we simply state the required definitions and give the applicable theorems without proof. Such proofs are found in advanced mathematics texts.

Definition 1. A singular point of a function $f(r)$ is a value of r for which $f(r)$ is undefined. (Such points are often called poles or singularities.)

Definition 2. A Taylor series expansion of a function $f(r)$ about the point $r = b$ on the interval $a < r < c$ with $b \in (a, c)$ is the convergent series

$$f(r) = \sum_{n=0}^{\infty} \frac{f^{(n)}(b)}{n!} (r - b)^n \tag{L-1}$$

where $f^{(n)}(b)$ is the nth derivative of $f(r)$ evaluated at $r = b$. The *radius of convergence* is the smaller of $|a - b|$ and $|c - b|$.

Definition 3. A function $f(r)$ is said to be analytic at $r = b$ if and only if it has a Taylor series expansion at $r = b$ and which represents $f(r)$ in some neighborhood of b. $r = b$ is called an *ordinary point*.

Theorem 1. A function $f(r)$ is analytic if and only if for a complex variable $r = z = x + iy$ with $U(x, y) = \text{Re}\{f(z)\}$ and $V(x, y) = \text{IM}\{f(z)\}$

$$\frac{\partial U}{\partial x} = \frac{\partial V}{\partial y} \quad \text{and} \quad \frac{\partial U}{\partial y} = -\frac{\partial V}{\partial x} \tag{L-2}$$

These are called the *Cauchy-Riemann Equations*.

If Theorem 1 holds for a function $f(r)$, then it has a convergent Taylor series by Definition 3.

Ezample L-1

Is $f(r) = 6\cos 9r$ analytic? If so, what is the Taylor series expansion about $t = 0$?

$$f(z) = 6\cos 9(x + iy) = 6\cos(9x + i9y)$$

Using trig identities:

$$\begin{aligned} f(z) &= 6\cos 9x \cos i9y - 6\sin 9x \sin i9y \\ &= 6\cos 9x \cosh 9y - 6\sin 9x \sinh 9y \end{aligned}$$

Then

$$U(x, y) = \text{Re}\{f(z)\} = 6\cos 9x \cosh 9y$$

$$V(x, y) = \text{IM}\{f(z)\} = -6\sin 9x \sinh 9y$$

We now test U and V using the Cauchy-Riemann equations (Eq. L-2):

$$\frac{\partial U}{\partial x} = -\sin 9x \cosh 9y$$

$$\frac{\partial V}{\partial y} = -\sin 9x \cosh 9y$$

These are equal:

$$\frac{\partial U}{\partial y} = +54\cos 9x \sinh 9y$$

$$\frac{\partial V}{\partial x} = -54\cos 9x \sinh 9y$$

These are negatives of each other.

Thus $f(r)$ is analytic, so we can evaluate the various derivatives as required by Eq. L-1 with $b = 0$:

$$\begin{aligned} f(0) &= 6 \\ f^{(1)}(0) &= 0 \\ f^{(2)}(0) &= -6 \cdot 9 \cdot 9 = -6 \cdot 9^2 \\ f^{(3)}(0) &= 0 \\ f^{(4)}(0) &= 6 \cdot 9^4 \\ f^{(5)}(0) &= 0 \\ f^{(6)}(0) &= -6 \cdot 9^6 \end{aligned}$$

Then

$$\begin{aligned} f(r) &= \frac{6}{1!} - \frac{6 \cdot 9^2}{2!} r^2 + \frac{6 \cdot 9^4}{4!} r^4 - \frac{6 \cdot 9^6}{6!} r^6 + \cdots \\ &= 6\left(1 - \frac{(9r)^2}{2!} + \frac{(9r)^4}{4!} - \frac{(9r)^6}{6!} + \cdots\right) \end{aligned}$$

The series in parentheses is recognized as the cosine series expansion.

Exercises

L-a Determine whether the following functions are analytic and, if so, develop the series:

$$\cos^2 r,\ 3\sinh(2r),\ 2e^{-r},\ 3e^{-jr},\ |r|^2,\ r^2,\ \frac{1}{r},\ \frac{1}{r+3},\ \frac{r}{r+1},\ \frac{1}{\cos r}\ 4\ln(2r)$$

Hint: For the last one, use $x + iy = \sqrt{x^2 + y^2}e^{j\tan^{-1}(y/x)}$. (Partial answer: yes, yes, yes, yes, no, yes, yes ($r \neq 0$), yes ($r \neq -3$), yes ($r \neq -1$), yes ($-\pi/2 < r < +\pi/2$), yes)

L-b Beginning with the Cauchy-Riemann equations show that $U(x, y)$ satisfies Laplace's equation. Similarly for $V(x, y)$.

Definition 4. For the ordinary differential equation

$$y'' + P(x)y'(x) + Q(x)y = 0$$

1. If $P(x)$ and $Q(x)$ are analytic at $x = a$, then a is an *ordinary point* of the differential equation. Otherwise, a is called a *singular point* of the differential equation.
2. At a singular point $x = a$ of either $P(x)$ or $Q(x)$ if both $(x - a)P(x)$ and $(x - a)^2Q(x)$ are analytic, the singularity is called *regular* or *removable*.
3. If a singularity of either $P(x)$ or $Q(x)$ is not removable, it is called an *irregular* singularity.

Example L-2

Let

$$\frac{d^2y}{dx^2} + \frac{1}{x}\frac{dy}{dx} + (x - 1)y = 0$$

Test this differential equation for singularities and ordinary points. We identify

$$P(x) = \frac{1}{x}; \qquad Q(x) = (x - 1)$$

From these we see:

$x = 3$ is an ordinary point (or any other nonzero value)
$x = 0$ is a singular point

For $x = 0$ test using Definition 4, number 2:

$$(x - 0)\,\frac{1}{x} = 1 \Rightarrow \text{is analytic}$$

$$(x - 0)^2(x - 1) = x^3 - x^2 \Rightarrow \text{analytic (this is the series expansion)}$$

Thus $x = 0$ is a *removable* singularity.

Theorem 2. For the ordinary differential equation

$$y'' + P(x)y'(x) + Q(x)y = 0 \tag{L-3}$$

1. If $x = a$ is a regular point of the differential equation, every solution is expressible as a Taylor series:

$$y(x) = \sum_{i=0}^{\infty} b_i(x - a)^i \tag{L-4}$$

(Radius of convergence is the distance to the nearest (maybe complex) singularity.)

2. If $x = a$ is a regular (removable) singularity, there is at least one solution of the form

$$y(x) = (x - a)^c \sum_{i=0}^{\infty} b_i(x - a)^i \tag{L-5}$$

where c is a constant called the *index* of the differential equation. (The value of c is determined by substituting the series into the differential equation, as shown later.)

3. If $x = a$ is an irregular singularity of the differential equation there are in general no solutions of the form

$$y(x) = (x - a)^c \sum_{i=0}^{\infty} b_i(x - a)^i$$

(There may be one, though.)

4. If any one of the preceding series yields only one value of c (or only one series solution), a second solution can be obtained from the first solution $y_1(x)$ as

$$y_2(x) = \phi(x)y_1(x) \tag{L-6}$$

and substituting this y_2 into the given differential equation and solving for $\phi(x)$. The general solution is then

$$y(x) = Ay_1(x) + B_2(x)$$

This process is valid if the Wronskian $W(y_1, y_2)$ is not zero; that is, if

$$\begin{vmatrix} y_1 & y_2 \\ y_1' & y_2' \end{vmatrix} \neq 0$$

It can be shown, however, that $W(y_1, y_2)$ is never zero for this method.

Before proceeding, we develop a general form for the second solution $y_2(x)$ suggested in Theorem 2, part 4.

Let $y_2(x) = \phi(x)y_1(x)$. Then

$$y_2'(x) = \phi(x)y_1'(x) + \phi'(x)y_1(x) \quad \text{and} \quad y_2''(x) = \phi(x)y_1''(x) + 2\phi'(x)y_1'(x) + \phi''(x)y_1(x)$$

Substituting these results into Eq. L-3 and collecting coefficients on the function $\Phi(x)$ and its derivatives we obtain

$$y_1\phi'' + (2y_1' + P(x)y_1)\phi' + [y_1'' + P(x)y_1' + Q(x)y_1]\phi = 0$$

Since we know that $y_1(x)$ is a solution of the differential equation, the coefficient of ϕ is 0 so the result is

$$y_1\phi'' + (2y_1' + P(x)y_1)\phi' = 0 \tag{L-7}$$

This can be put in an integrable form by letting

$$\phi(x) = \int \psi(x)dx \tag{L-8}$$

From this,

$$\phi'(x) = \psi(x) \qquad \text{and} \qquad \phi''(x) = \psi'(x)$$

Equation L-7 then becomes

$$y_1\psi' + (2y_1' + P(x)y_1)\psi(x) = 0$$

This equation can be separated. Using $\Psi' = d\Psi/dx$ the separated form is

$$\frac{d\psi}{\psi} = -\left(2\frac{y_1'}{y_1} + P(x)\right)dx$$

Integrating

$$\ln \psi = -\int\left(2\frac{y_1'}{y_1} + P(x)\right)dx$$

or

$$\begin{aligned}\psi(x) &= e^{-\int(2\,y_1'/y_1 + P(x))dx} = e^{-\int 2\,y_1'/y_1\,dx}e^{-\int P(x)dx}\\ &= e^{-2\int 1/y_1\,dy_1/dx\,dx}e^{-\int P(x)dx} = e^{-2\ln y_1}e^{-\int P(x)dx}\end{aligned}$$

$$\psi(x) = \frac{1}{y_1^2}\,e^{-\int P(x)dx}$$

Substituting this into Eq. L-8 we obtain $\phi(x)$ as

$$\phi(x) = \int \frac{1}{y_1^2}\,e^{-\int P(x)dx}dx \tag{L-9}$$

The second solution is then

$$y_2(x) = \phi(x)y_1(x) = y_1(x)\int \frac{1}{y_1^2}\,e^{-\int P(x)dx}dx \tag{L-10}$$

Now that the definitions and theorems are available we proceed with some examples and then continue by developing solutions for some differential equations that are important in electromagnetic theory.

Example L-3

Obtain the series solution for the differential equation about the point $x = 0$:

$$\frac{dy}{dx} + ky = 0$$

Although the solution of this equation is well known, we use it as an example that illustrates most of the principles involved. The general principles are identified by a double asterisk and apply to any series solution and shows how solution functions are defined.

**Identify $P(x)$ and $Q(x)$:

$$P(x) = 1 \rightarrow \text{analytic}$$

$$Q(x) = k \rightarrow \text{analytic}$$

**Both of these are analytic, and therefore Theorem 2, part 1 applies so we let, with $a = 0$,

$$y(x) = \sum_{i=0}^{\infty} b_i x^i \tag{L-11}$$

Then, since the derivative of a sum is the sum of the derivatives,

$$y'(x) = \sum_{i=0}^{\infty} i b_i x^{i-1}$$

**Substituting the series and its derivatives into the differential equation

$$\sum_{i=1}^{\infty} i b_i x^{i-1} + k \sum_{i=0}^{\infty} b_i x^i = \sum_{i=0}^{\infty} i b_i x^{i-1} + \sum_{i=0}^{\infty} k b_i x^i = 0$$

**The goal now is to "fix up" the series so they can be combined into one series. This is done by removing (writing explicitly) the lower terms of the series that have powers less than x^i, by removing as many terms as the power is less than i. In this case, this means we remove one term from the left sum, so we remove the term $i = 0$:

$$0 \cdot b_0 x^{-1} + \sum_{i=1}^{\infty} i b_i x^{i-1} + \sum_{i=0}^{\infty} k b_i x^i = 0$$

**In the left sum change the summation index by letting $j = i - 1$:

$$\sum_{j=0}^{\infty} (j + 1) b_{j+1} x^j + \sum_{i=0}^{\infty} k b_i x^i = 0$$

**Since it doesn't really matter what symbol we use as the summation index, we can replace j by i in the left sum:

$$\sum_{i=0}^{\infty} (i + 1) b_{i+1} x^i + \sum_{i=0}^{\infty} k b_i x^i = 0$$

**Now since both sums have the same index ranges *and* the same powers of x, they may be combined:

$$\sum_{i=0}^{\infty} [(i + 1) b_{i+1} + k b_i] x^i = 0$$

**From algebra we know that a power series is zero only if every coefficient of x^i is zero:

$$(i + 1)b_{i+1}kb_i = 0 \qquad \text{for all} \quad i$$

**Next, write out several terms and solve for the b_i and determine a pattern so a general form can be developed:

$$i = 0\text{: } b_1 + kb_0 = 0 \;\Rightarrow b_1 = -kb_0$$

$$i = 1\text{: } 2b_2 + kb_1 = 0 \Rightarrow b_2 = -\tfrac{1}{2}kb_1 = \tfrac{1}{2}k^2b_0$$

$$i = 2\text{: } 3b_3 + kb_2 = 0 \Rightarrow b_3 = -\frac{1}{3}kb_2 = -\frac{1}{2 \cdot 3}k^3b_0$$

$$i = 3\text{: } 4b_4 + kb_3 = 0 \Rightarrow b_4 = -\frac{1}{4}kb_3 = \frac{1}{2 \cdot 3 \cdot 4}k^4b_0$$

The general pattern is then

$$b_i = (-1)^i \frac{k^i}{i!} b_0$$

**Substitute this back into the original assumed series (Eq. L-8):

$$y(x) = \sum_{i=0}^{\infty} (-1)^i \frac{k^i}{i!} b_0x^i = b_0 \sum_{i=0}^{\infty} \frac{(-kx)^i}{i!} \tag{L-12}$$

We can expand the first few terms to see the form:

$$y(x) = b_0\left(1 + \frac{(-kx)}{1!} + \frac{(-kx)^2}{2!} + \frac{(-kx)^3}{3!} + \cdots\right) \tag{L-13}$$

**Since it is awkward to write the series each time, we define a function symbol that represents this series and try to do it in such a way as to obtain a function that is more general than just the solution to this one problem. Since every term contains $(-kx)$ we could generalize to a variable r and give the series the symbolic representation $\exp(r)$ as

$$\exp(r) \equiv 1 + \frac{r}{1!} + \frac{r^2}{2!} + \frac{r^3}{3!} + \cdots \tag{L-14}$$

and call it the exponential function (as expected).

**Now that we have a series for the ''new'' function we can make a table of values and plot it. The tabular values are obtained by carrying out the series to the required precision. These are shown in Figure L-1.

**The solution to the original differential equation may then be written using this series definition in the well-known way:

$$y(x) = b_0 \exp(-kx) \equiv b_0e^{-kx} \qquad (e = \exp(1))$$

**The constant b_0 is determined by applying boundary conditions applicable to the problem.

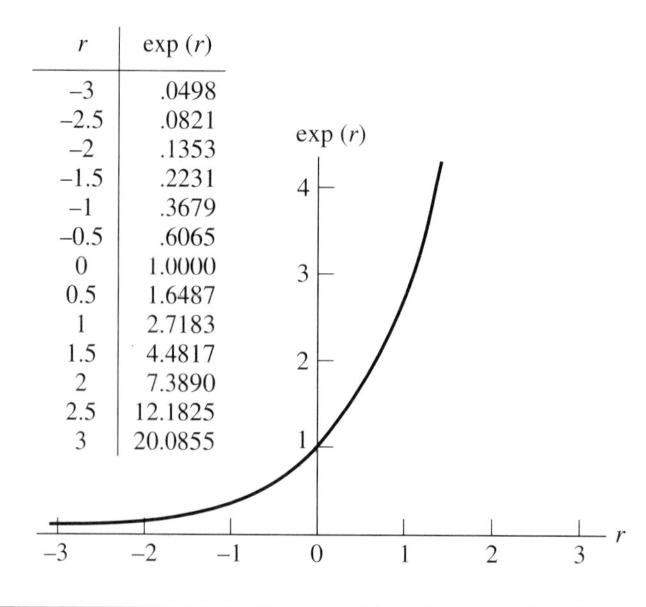

r	exp (r)
−3	.0498
−2.5	.0821
−2	.1353
−1.5	.2231
−1	.3679
−0.5	.6065
0	1.0000
0.5	1.6487
1	2.7183
1.5	4.4817
2	7.3890
2.5	12.1825
3	20.0855

Figure L-1. Tabular values and graph of the solution series defined as exp(r).

Example L-4

Obtain the series solution about the point $x = 0$ for

$$\frac{d^2y}{dx^2} + k^2y = 0, \qquad k^2 \text{ a positive real constant}$$

The same process used in Example L-3 will be followed and the principles again identified by double asterisks

**Identify $P(x)$ and $Q(x)$:

$$P(x) = 0 \rightarrow \text{analytic}$$

$$Q(x) = k^2 \rightarrow \text{analytic}$$

**Both of these are analytic so we set

$$y(x) = \sum_{i=0}^{\infty} b_i x^i$$

**Substituting the series and its second derivative into the differential equation:

$$\sum_{i=0}^{\infty} i(i - 1)b_i x^{i-2} + \sum_{i=0}^{\infty} k^2 b_i x^i = 0$$

**Removing the first two terms explicitly from the left series:

$$(0)(-1)b_0 x^{-2} + (1)(0)b_1 x^{-1} + \sum_{i=2}^{\infty} i(i - 1)b_i x^{i-2} + \sum_{i=0}^{\infty} k^2 b_i x^i = 0$$

**Change the index in the left sum using $j = i - 2$:

$$\sum_{j=0}^{\infty} (j + 2)(j + 1)b_{j+2} x^j + \sum_{i=0}^{\infty} k^2 b_i x^i = 0$$

**Change the index in the left sum using $i = j$:

$$\sum_{i=0}^{\infty} (i + 2)(i + 1)b_{i+2}x^i + \sum_{i=0}^{\infty} k^2 b_i x^i = 0$$

**Since both sums have the same index ranges and powers of x we combine them:

$$\sum_{i=0}^{\infty} [(i + 2)(i + 1)b_{i+2} + k^2 b_i]x^i = 0$$

**All coefficients of x^i must be zero:

$$(i + 2)(i + 1)b_{i+2} + k^2 b_i = 0$$

**Write out a few of these to detect a coefficient pattern:

$$i = 0\text{: } (2)(1)b_2 + k^2 b_0 = 0 \Rightarrow b_2 = -\frac{k^2}{1 \cdot 2} b_0$$

$$i = 1\text{: } (3)(2)b_3 + k^2 b_1 = 0 \Rightarrow b_3 = -\frac{k^2}{2 \cdot 3} b_1$$

$$i = 2\text{: } (4)(3)b_4 + k^2 b_2 = 0 \Rightarrow b_4 = -\frac{k^2}{3 \cdot 4} b_2 = \frac{k^4}{1 \cdot 2 \cdot 3 \cdot 4} b_0$$

$$i = 3\text{: } (5)(4)b_5 + k^2 b_3 = 0 \Rightarrow b_5 = -\frac{k^2}{4 \cdot 5} b_3 = \frac{k^4}{2 \cdot 3 \cdot 4 \cdot 5} b_1$$

$$i = 4\text{: } (6)(5)b_6 + k^2 b_4 = 0 \Rightarrow b_6 = -\frac{k^2}{5 \cdot 6} b_4 = -\frac{k^6}{1 \cdot 2 \cdot 3 \cdot 4 \cdot 5 \cdot 6} b_0$$

$$i = 5\text{: } (7)(6)b_7 + k^2 b_5 = 0 \Rightarrow b_7 = -\frac{k^2}{6 \cdot 7} b_5 = -\frac{k^6}{1 \cdot 2 \cdot 3 \cdot 4 \cdot 5 \cdot 6 \cdot 7} b_1$$

Note all the even bs have b_0 and all the odd ones have b_1.

There are thus two patterns this time

$$b_i = (-1)^{1/2} \frac{k^i}{i!} b_0 \qquad n \text{ even}$$

$$b_i = (-1)^{i-1/2} \frac{k^{i-1}}{i!} b_1 \qquad n \text{ odd}$$

**Our original series can then be written as the sum of and odd and an even series to cover all values of i:

$$\begin{aligned} y(x) &= \sum_{i=0,2,4}^{\infty} (-1)^{i/2} \frac{k^i}{i!} b_0 x^i + \sum_{i=1,3,5}^{\infty} (-1)^{(i-1)/2} \frac{k^{i-1}}{i!} b_1 x^i = 0 \\ &= b_0 \sum_{i=0,2,4}^{\infty} \frac{(-1)^{i/2}}{i!} (kx)^i + b_1 \sum_{i=1,3,5}^{\infty} \frac{(-1)^{(i-1)/2}}{i!} k^{-1}(kx)^i = 0 \end{aligned}$$

Since k^{-1} is a constant it may be taken outside the second sum and combined with the b_1 to get a new constant b_1'. We finally obtain

$$y(x) = b_0 \sum_{i=0,2,4}^{\infty} \frac{(-1)^{i/2}}{i!}(kx)^i + b_1' \sum_{i=1,3,5}^{\infty} \frac{(-1)^{i-1/2}}{i!} k^{-1}(kx)^i = 0$$

Expanding a few terms of each:

$$y(x) = b_0\left[1 - \frac{(kx)^2}{2!} + \frac{(kx)^4}{4!} - \frac{(kx)^6}{6!} + \cdots\right] + b_1'\left[(kx) - \frac{(kx)^3}{3!} + \frac{(kx)^5}{5!} - \frac{(kx)^7}{7!} + \cdots\right]$$

**These series have been given functional names defined by (as expected):

$$\cos(r) \equiv 1 - \frac{r^2}{2!} + \frac{r^4}{4!} - \frac{r^6}{6!} + \cdots \tag{L-15}$$

$$\sin(r) \equiv r - \frac{r^3}{3!} + \frac{r^5}{5!} - \frac{r^7}{7!} + \cdots \tag{L-16}$$

**These ''new'' functions can be put in tabular form and plotted. Note that r is just a number. These are the standard sine and cosine functions using radian measure. Tables and graphs of these are available.

**The solution to the original differential equation may then be written, using the series function definitions, in the well-known form

$$y(x) = b_0 \cos(kx) + b_1' \sin(kx) \tag{L-17}$$

Note we have two arbitrary constants as required by the second-order differential equations. These are found using boundary conditions.

Example L-5

This example will consider a differential equation whose solution is not commonly encountered. Solve the differential equation

$$x\frac{d^2y}{dx^2} + x\frac{dy}{dx} + y = 0 \qquad \text{about} \quad x = 0$$

We first put the differential equation in standard form of Eq. L-3 so the second derivative term has unity coefficient:

$$\frac{d^2y}{dx^2} + \frac{dy}{dx} + \frac{1}{x}y = 0$$

then

$$P(x) = 0 \rightarrow \text{analytic}$$

$$Q(x) = \frac{k}{x} \rightarrow \text{not analytic at } x = 0 \text{ where solution is desired}$$

We next test to see if the singularity is removable. Using Definition 4, part 2:

$$xP(x) = x \rightarrow \text{analytic}$$

$$x^2Q(x) = kx \rightarrow \text{analytic}$$

Since the singularity is removable, we use Theorem 2, part 2 with constant $a = 0$, and write the series

$$y(x) = x^c \sum_{i=0}^{\infty} b_i x^i = \sum_{i=0}^{\infty} b_i x^{i+c} \qquad (c \text{ constant}) \tag{L-18}$$

Substitute this into the standard-form differential equation:

$$\sum_{i=0}^{\infty} (i + c)(i + c - 1)b_i x^{i+c-2} + \sum_{i=0}^{\infty} (i + c)b_i x^{i+c-1} + \frac{1}{x}\sum_{i=0}^{\infty} b_i x^{i+c} = 0$$

In the right sum take the $1/x$ inside:

$$\sum_{i=0}^{\infty} (i + c)(i + c - 1)b_i x^{i+c-2} + \sum_{i=0}^{\infty} (i + c)b_i x^{i+c-1} + \sum_{i=0}^{\infty} b_i x^{i+c-1} = 0$$

Since the center and right sums have the same index ranges and powers of x they can be combined:

$$\sum_{i=0}^{\infty} (i + c)(i + c - 1)b_i x^{i+c-2} + \sum_{i=0}^{\infty} [(i + c) + 1]b_i x^{i+c-1} = 0$$

Take the $i = 0$ and $i = 1$ terms out explicitly from the left sum and the $i = 0$ term from the right sum:

$$b_0c(c - 1)x^{c-2} + b_1c(c + 1)x^{c-1} + \sum_{i=2}^{\infty} b_i(i + c)(i + c - 1)x^{i+c-2}$$
$$+ b_0(c + 1)x^{c-1} + \sum_{i=1}^{\infty} b_i(i + c + 1)x^{i+c-1} = 0$$

Combine coefficients on $x^c - 1$ and let $j = i - 2$ in the middle sum and $j = i - 1$ in the right sum:

$$b_0c(c - 1)x^{c-2} + [b_0(c + 1) + b_1(c + 1)c]x^{c-1}$$
$$+ \sum_{j=0}^{\infty} b_{j+2}(j + c + 2)(j + c + 1)x^{j+c} + \sum_{j=0}^{\infty} b_{j+1}(j + c + 2)x^{j+c} = 0$$

We can now combine the two sums and let $j = i$:

$$b_0c(c - 1)x^{c-2} + [b_0(c + 1) + b_1(c + 1)c]x^{c-1}$$
$$+ \sum_{i=0}^{\infty} [b_{i+2}(i + c + 2)(i + c + 1) + b_{i+1}(i + c + 2)]x^{i+c} = 0$$

All coefficients of powers of x must be zero, so we begin with x^{c-2}:

$$b_0 c(c - 1) = 0 \tag{L-19}$$

This is called the *indicial equation* because values of the index c come from it.

**The constant b_0 cannot be zero, for if it were we could simply factor out an x from the remaining terms in our original series Eq. L-18 and obtain another series that starts with constant b_1. This merely adds 1 to the constant c when and x is factored out. We than have two possibilities:

$$c = 0 \qquad \text{and/or} \quad c = 1$$

The coefficient of x^{c-1} requires

$$b_0(c + 1) + b_1(c + 1)c = (c + 1)(b_0 + b_1 c) = 0$$

**We test the previous values of c in this result:

For $c = 0$ the preceding equation requires to $b_0 = 0$, but this is not allowed.

For $c = 1$ the equation becomes

$$2(b_0 + b_1) = 0$$

Thus

$$b_1 = -b_0 \neq 0$$

For the remaining coefficients in the sum with $c = 1$ we have

$$b_{i+2}(i + 3)(i + 2) + b_{i+1}(i + 3) = 0$$

Since $i > 0$ we may divide out the $(i + 3)$ factor and then solve for b_{i+2}:

$$b_{i+2} = -\frac{b_{i+1}}{i + 2}$$

for all values of i. Writing out the first few terms using b_1,

$$i = 0\colon b_2 = -\frac{b_1}{2} = \frac{b_o}{2}$$

$$i = 1\colon b_3 = -\frac{b_2}{3} = \frac{b_0}{2 \cdot 3}$$

$$i = 2\colon b_4 = -\frac{b_3}{4} = \frac{b_0}{2 \cdot 3 \cdot 4}$$

From this pattern we have the general result:

$$b_i = \frac{(-1)^i}{i!} b_0$$

(Note for $i = 1$ this yields $b_1 = -b_0$ as required.)

The assumed series Eq. L-18 is then

$$y(x) = x \sum_{i=0}^{\infty} \frac{(-1)^i}{i!} b_0 x^i = b_0 x \sum_{i=0}^{\infty} \frac{(-1)^i}{i!} x^i = b_0 x \sum_{i=0}^{\infty} \frac{(-x)^i}{i!}$$

The general function represented by this solution could be called

$$M(r) = r \sum_{i=0}^{\infty} \frac{(-r)^i}{i!} \tag{L-20}$$

The summation is identified as Eq. L-14 for negative r so the solution may be written

$$M(r) = r \exp(-r) = re^{-r}$$

As before we can make a table of values of the function and plot it. This is shown in Figure L-2. Note that we only have one solution where we require two for the second-order differential equation. Theorem 2, part 2 guarantees only one solution, and that's all we obtained. Example 2-4 resulted in both solutions, sine and cosine.

A second solution can be obtained using Eq. L-10, where $M(x)$ is our first solution. Call the second solution $G(x)$:

$$G(x) = M(x) \int \frac{1}{M^2(x)} e^{-\int P(x)dx} dx = xe^{-x} \int \frac{1}{x^2 e^{-2x}} e^{-\int 1 dx} dx$$

$$= -1 + xe^{-x}\left(\ln x + \frac{x}{1!} + \frac{x^2}{2 \cdot 2!} + \frac{x^3}{3 \cdot 3!} + \cdots \right)$$

$$G(x) = -1 + xe^{-x} \ln x + \frac{x^2}{1!} e^{-x} + \frac{x^3}{2 \cdot 2!} e^{-x} + \frac{x^4}{3 \cdot 3!} e^{-x} + \cdots \tag{L-21}$$

This is not defined for negative x. The function $G(x)$ could also be plotted and tabulated.

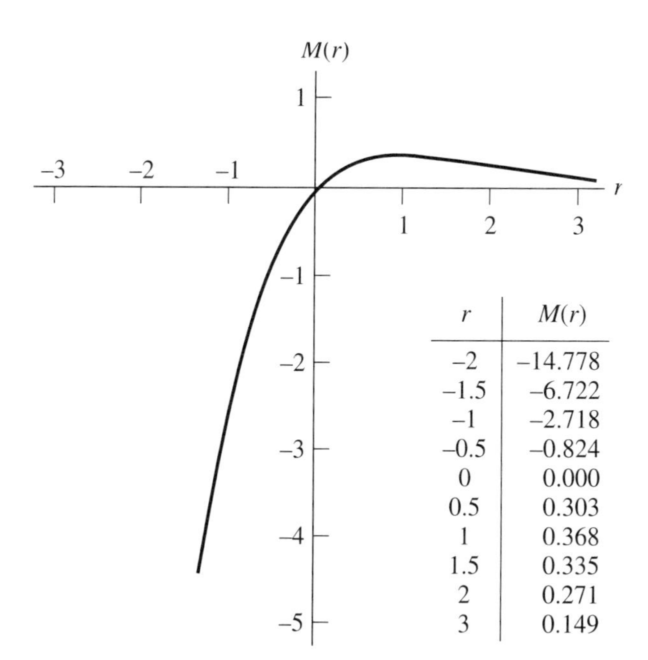

r	$M(r)$
–2	–14.778
–1.5	–6.722
–1	–2.718
–0.5	–0.824
0	0.000
0.5	0.303
1	0.368
1.5	0.335
2	0.271
3	0.149

Figure L-2. Tabular values and graph of the solution series defined by $M(r)$.

The general solution to the differential equation would then be written

$$y(x) = AM(x) + BG(x) \tag{L-22}$$

If the differential equation is to be solved for systems that have negative x, the constant B would be set to zero.

Exercises

L-c Obtain the series solution about $x = 0$ for the differential equation

$$y'' - k^2 y = 0$$

Identify the series, if possible by consulting a handbook having series expansions.

L-d Repeat Exercise L-c for $y' + (1/x)y = 0$

L-e Repeat L-c for $4x^2 y'' + (4x + 1)y = 0$. Be sure to use the standard form of the differential equation.

L-f Repeat L-c for $y'' + (2/x)y' - [m(m + 1)/x^2]y = 0$, where m is a constant. (Answer: $y(x) = Ax^m + Bx^{-(m+1)}$)

With the preceding background we next investigate the solutions of several ordinary differential equations that often arise in electromagnetic theory. We will obtain several new functions which may not be as familiar as the exponential or trigonometric ones. Since the functions are less familiar we need to rely on tables and plots of the functions for obtaining physical and qualitative insights of their properties.

BESSEL'S EQUATION, CYLINDRICAL COORDINATES

When separating the wave equation in cylindrical coordinates the differential equation for the ρ component usually obtained is

$$\rho \frac{d}{d\rho}\left(\rho \frac{dR}{d\rho}\right) + (k^2\rho^2 - n^2)R = 0 \tag{L-23}$$

where k and n are constants, and R is a function of ρ only. Other forms of this same equation are

$$\rho^2 \frac{d^2R}{d\rho^2} + \rho \frac{dR}{d\rho} + (k^2\rho^2 - n^2)R = 0 \tag{L-24}$$

$$\frac{d^2R}{d\rho^2} + \frac{1}{\rho}\frac{dR}{d\rho} + \left(k^2 - \frac{n^2}{\rho^2}\right) = 0 \tag{L-25}$$

This last equation is in the series solution standard form of Eq. L-3 where we identify

$$P(\rho) = \frac{1}{\rho} \quad \text{and} \quad Q(\rho) = \left(k^2 - \frac{n^2}{\rho^2}\right) = 0$$

Both functions have singularities at $\rho = 0$. We use Definitions 4, part 2 to determine the types of singularities:

$$\rho P(\rho) = 1 \rightarrow \text{analytic}$$
$$\rho^2 Q(\rho) = k^2\rho^2 - n^2 \rightarrow \text{analytic}$$

Thus the singularity is removable, so the appropriate series solution from Theorem 2, part 2 is, for $a = 0$,

$$R(\rho) = \rho^c \sum_{i=0}^{\infty} b_i \rho^i \tag{L-26}$$

We will not carry out the details of the substitution of this assumed series as it is algebraically more lengthy. The details can be found in [63]. Two cases arise from this process, and the results are stated.

Case 1: n an Integer. We obtain only one solution directly from the series:

$$J_n(k\rho) \equiv \left(\frac{k\rho}{2}\right)^n \sum_{i=0}^{\infty} \frac{(-i)^i \left(\frac{k\rho}{2}\right)^{2i}}{i!(n+i)!} \tag{L-27}$$

This first solution is called the *Bessel function of the first kind of order* n.

As a general function form it is customary to write this solution using a variable x (not rectangular):

$$J_n(x) = \left(\frac{x}{2}\right)^n \sum_{i=0}^{\infty} \frac{(-i)^i \left(\frac{x}{2}\right)^{2i}}{i!(n+i)!} \tag{L-28}$$

The second solution could be obtained using Eq. L-10 as was done in Example L-5. For $P(\rho) = 1/\rho$ the second solution would be

$$R_2(x) = J_n(x) \int \frac{dx}{x J_n^2(x)}$$

Since $J_n(x)$ is a series this is not an easy evaluation. Numerical methods are obviously called for. This is *not* the second solution in common use, however.

Since a solution can also be constructed from functions of solutions, the most common second solution is defined by

$$N_n(x) = \lim_{\nu \rightarrow n} \frac{J_\nu(x) \cos \nu\pi - J_{-\nu}(x)}{\sin \nu\pi} \tag{L-29}$$

where ν is not an integer but approaches integer value. (In this expression the factorial is replaced by the gamma function; see Case 2.)

This is called the *Bessel function of the second kind of order* n. (Note the terminology for this second solution is not uniform. Some call this the *Neumann function* or *Weber function* and use the symbol $Y_n(x)$. These are the same as our $N_n(x)$ and we use N from the name Neumann.) The series expression for $N_n(x)$ is an extensive equation and will not be given. See [1]. Tables of values and plots of these functions follow, for n an integer only, in Tables L-1 and L-2 and Figures L-3 and L-4.

Table L-1. Bessel Functions of the First Kind of Order n: $J_n(x)$

x	$J_0(x)$	$J_1(x)$	$J_2(x)$	$J_3(x)$	$J_4(x)$	$J_5(x)$
0.0	1.000000	.000000	.000000	.000000	.000000	.000000
0.1	.997502	.049938	.001249	.000021	.000000	.000000
0.2	.990025	.099501	.004983	.000166	.000004	.000000
0.3	.977626	.148319	.011166	.000559	.000021	.000001
0.4	.960398	.196027	.019735	.001320	.000066	.000003
0.5	.938470	.242268	.030604	.002564	.000161	.000008
0.6	.912005	.286701	.043665	.004400	.000331	.000020
0.7	.881201	.328996	.058787	.006930	.000610	.000043
0.8	.846287	.368842	.075818	.010247	.001033	.000083
0.9	.807524	.405950	.094586	.014434	.001641	.000149
1.0	.765198	.440051	.114903	.019563	.002477	.000250
1.1	.719622	.470902	.136564	.025695	.003588	.000399
1.2	.671133	.498289	.159349	.032874	.005023	.000610
1.3	.620086	.522023	.183027	.041136	.006831	.000901
1.4	.566855	.541948	.207356	.050498	.009063	.001290
1.5	.511828	.557937	.232088	.060964	.011768	.001799
1.6	.455402	.569896	.256968	.072523	.014995	.002452
1.7	.397985	.577765	.281739	.085150	.018790	.003275
1.8	.339986	.581517	.306144	.098802	.023197	.004294
1.9	.281819	.581157	.329926	.113423	.028253	.005538
2.0	.223891	.576725	.352834	.128943	.033996	.007040
2.1	.166607	.568292	.374624	.145277	.040453	.008828
2.2	.110362	.555963	.395059	.162325	.047647	.010937
2.3	.055540	.539873	.413915	.179979	.055596	.013397
2.4	.002508	.520185	.430980	.198115	.064307	.016242
2.5	−.048384	.497094	.446059	.216600	.073782	.019502
2.6	−.096805	.470818	.458973	.235294	.084013	.023207
2.7	−.142449	.441601	.469562	.254045	.094984	.027388
2.8	−.185036	.409709	.477685	.272699	.106669	.032069
2.9	−.224312	.375427	.483227	.291093	.119033	.037276
3.0	−.260052	.339059	.486091	.309063	.132034	.043028
3.1	−.292064	.300921	.486207	.326443	.145618	.049345
3.2	−.320188	.261343	.483528	.343066	.159722	.056238
3.3	−.344296	.220663	.478032	.358769	.174275	.063717
3.4	−.364296	.179226	.469723	.373389	.189199	.071785
3.5	−.380128	.137378	.458629	.386770	.204405	.080442
3.6	−.391769	.095466	.444805	.398763	.219799	.089680
3.7	−.399230	.053834	.428330	.409225	.235279	.099485
3.8	−.402556	.012821	.409304	.418026	.250736	.109840

x	$J_0(x)$	$J_1(x)$	$J_2(x)$	$J_3(x)$	$J_4(x)$	$J_5(x)$
3.9	−.401826	−.027244	.387855	.425044	.266059	.120718
4.0	−.397150	−.066043	.364128	.430171	.281129	.132087
4.1	−.388670	−.103273	.338292	.433315	.295827	.143908
4.2	−.376557	−.138647	.310535	.434394	.310029	.156136
4.3	−.361011	−.171897	.281059	.433347	.323611	.168720
4.4	−.342257	−.202776	.250086	.430127	.336450	.181601
4.5	−.320543	−.231060	.217849	.424704	.348423	.194715
4.6	−.296138	−.256553	.184593	.417069	.359409	.207991
4.7	−.269331	−.279081	.150573	.407228	.369292	.221355
4.8	−.240425	−.298500	.116050	.395209	.377960	.234725
4.9	−.209738	−.314695	.081292	.381055	.385307	.248017
5.0	−.177597	−.327579	.046565	.364831	.391232	.261141
5.1	−.144335	−.337097	.012140	.346619	.395647	.274004
5.2	−.110290	−.343223	−.021718	.326517	.398468	.286512
5.3	−.075803	−.345961	−.054748	.304641	.399625	.298567
5.4	−.041210	−.345345	−.086695	.281126	.399058	.310070
5.5	−.006844	−.341438	−.117315	.256118	.396717	.320925
5.6	.026971	−.334333	−.146375	.229779	.392567	.331031
5.7	.059920	−.324148	−.173656	.202284	.386586	.340294
5.8	.091703	−.311028	−.198954	.173818	.378766	.348617
5.9	.122033	−.295142	−.222082	.144579	.369111	.355911
6.0	.150645	−.276684	−.242873	.114768	.357642	.362087
6.1	.177291	−.255865	−.261182	.084598	.344393	.367065
6.2	.201747	−.232917	−.276882	.054283	.329414	.370767
6.3	.223812	−.208087	−.289871	.024042	.312768	.373124
6.4	.243311	−.181638	−.300072	−.005908	.294534	.374075
6.5	.260095	−.153841	−.307430	−.035347	.274803	.373565
6.6	.274043	−.124980	−.311916	−.064060	.253680	.371551
6.7	.285065	−.095342	−.313525	−.091837	.231283	.367996
6.8	.293096	−.065219	−.312278	−.118474	.207742	.362876
6.9	.298102	−.034902	−.308219	−.143775	.183197	.356177
7.0	.300079	−.004683	−.301417	−.167556	.157798	.347896
7.1	.299051	.025153	−.291966	−.189641	.131706	.338042
7.2	.295071	.054327	−.279980	−.209872	.105087	.326635
7.3	.288217	.082570	−.265595	−.228102	.078114	.313706
7.4	.278596	.109625	−.248968	−.244202	.050966	.299301
7.5	.266340	.135248	−.230273	−.258061	.023825	.283474
7.6	.251602	.159214	−.209703	−.269584	−.003126	.266293
7.7	.234559	.181313	−.187465	−.278697	−.029702	.247838
7.8	.215408	.201357	−.163778	−.285346	−.055719	.228198
7.9	.194362	.219179	−.138873	−.289495	−.080996	.207474
8.0	.171651	.234636	−.112992	−.291132	−.105357	.185775
8.1	.147517	.247608	−.086380	−.290264	−.128631	.163222
8.2	.122215	.257999	−.059289	−.286920	−.150653	.139942
8.3	.096006	.265739	−.031973	−.281148	−.171267	.116071
8.4	.069157	.270786	−.004684	−.273017	−.190328	.091752

Table L-1. Bessel Functions of the First Kind of Order n: $J_n(x)$ (*continued*)

x	$J_0(x)$	$J_1(x)$	$J_2(x)$	$J_3(x)$	$J_4(x)$	$J_5(x)$
8.5	.041939	.273122	.022325	−.262616	−.207701	.067133
8.6	.014623	.272755	.048808	−.250053	−.223264	.042366
8.7	−.012523	.269719	.074527	−.235454	−.236909	.017606
8.8	−.039234	.264074	.099251	−.218960	−.248541	−.006987
8.9	−.065253	.255902	.122759	−.200730	−.258083	−.031255
9.0	−.090334	.245312	.144847	−.180935	−.265471	−.055039
9.1	−.114239	.232431	.165323	−.159761	−.270660	−.078182
9.2	−.136748	.217409	.184011	−.137404	−.273622	−.100529
9.3	−.157655	.200414	.200755	−.114068	−.274347	−.121930
9.4	−.176772	.181632	.215417	−.089966	−.272842	−.142240
9.5	−.193929	.161264	.227879	−.065315	−.269131	−.161321
9.6	−.208979	.139525	.238046	−.040339	−.263258	−.179043
9.7	−.221795	.116639	.245845	−.015259	−.255283	−.195284
9.8	−.232276	.092840	.251223	.009700	−.245284	−.209932
9.9	−.240341	.068370	.254153	.034318	−.233354	−.222887

Table L-2. Bessel Functions of the Second Kind of Order n: $N_n(x)$ (or $Y_n(x)$)

x	$N_0(x)$	$N_1(x)$	$N_2(x)$	$N_3(x)$	$N_4(x)$	$N_5(x)$
0.0						
0.1	−1.53424	−6.45895	−127.645	−5099.33	−305832.	−24E+06
0.2	−1.08111	−3.32382	−32.1571	−0639.82	−19162.4	−765857.
0.3	−.807274	−2.29311	−14.4801	−190.775	−3801.02	−101170.
0.4	−.606025	−1.78087	−8.29834	−81.2025	−1209.74	−24113.6
0.5	−.444519	−1.47147	−5.44137	−42.0595	−499.273	−7946.30
0.6	−.308510	−1.26039	−3.89279	−24.6916	−243.023	−3215.61
0.7	−.190665	−1.10325	−2.96148	−15.8195	−132.634	−1500.00
0.8	−.086802	−.978144	−2.35856	−10.8146	−78.7513	−776.698
0.9	.005628	−.873127	−1.94591	−7.77536	−49.8898	−435.690
1.0	.088257	−.781213	−1.65068	−5.82152	−33.2784	−260.406
1.1	.162163	−.698120	−1.43147	−4.50723	−23.1534	−163.881
1.2	.228084	−.621136	−1.26331	−3.58990	−16.6862	−107.651
1.3	.286535	−.548520	−1.13041	−2.92967	−12.3911	−73.3235
1.4	.337895	−.479147	−1.02239	−2.44197	−9.44319	−51.5191
1.5	.382449	−.412309	−.932194	−2.07354	−7.36197	−37.1903
1.6	.420427	−.347578	−.854899	−1.78967	−5.85636	−27.4922
1.7	.452027	−.284726	−.786999	−1.56704	−4.74372	−20.7563
1.8	.477432	−.223665	−.725948	−1.38955	−3.90590	−15.9700
1.9	.496820	−.164406	−.669879	−1.24587	−3.26443	−12.4991
2.0	.510376	−.107032	−.617408	−1.12778	−2.76594	−9.93599
2.1	.518294	−.051679	−.567511	−1.02930	−2.37333	−8.01197
2.2	.520784	.001488	−.519432	−.945909	−2.06032	−6.54617
2.3	.518075	.052277	−.472617	−.874220	−1.80796	−5.41432
2.4	.510415	.100489	−.426674	−.811612	−1.60236	−4.52958

x	$N_0(x)$	$N_1(x)$	$N_2(x)$	$N_3(x)$	$N_4(x)$	$N_5(x)$
2.5	.498070	.145918	−.381336	−.756055	−1.43320	−3.83018
2.6	.481331	.188364	−.336436	−.705957	−1.29270	−3.27157
2.7	.460504	.227632	−.291887	−.660058	−1.17491	−2.82115
2.8	.435916	.263545	−.247669	−.617359	−1.07524	−2.45476
2.9	.407912	.295940	−.203815	−.577064	−.990111	−2.15428
3.0	.376850	.324674	−.160400	−.538542	−.916683	−1.90595
3.1	.343103	.349629	−.117535	−.501288	−.852700	−1.69923
3.2	.307053	.370711	−.075359	−.464910	−.796347	−1.52596
3.3	.269092	.387853	−.034030	−.429101	−.746154	−1.37976
3.4	.229615	.401015	.006276	−.393632	−.700920	−1.25559
3.5	.189022	.410188	.045371	−.358335	−.659661	−1.14946
3.6	.147710	.415392	.083063	−.323099	−.621562	−1.05815
3.7	.106074	.416674	.119155	−.287858	−.585952	−.979065
3.8	.064503	.414115	.153452	−.252586	−.552273	−.910093
3.9	.023376	.407820	.185763	−.217294	−.520062	−.849499
4.0	−.016941	.397926	.215904	−.182022	−.488937	−.795851
4.1	−.056095	.384594	.243701	−.146837	−.458584	−.747962
4.2	−.093751	.368013	.268995	−.111827	−.428748	−.704836
4.3	−.129596	.348394	.291640	−.077101	−.399223	−.665638
4.4	−.163336	.325971	.311505	−.042784	−.369847	−.629665
4.5	−.194705	.300997	.328482	−.009014	−.340500	−.596319
4.6	−.223460	.273745	.342480	.024063	−.311093	−.565094
4.7	−.249388	.244501	.353431	.056291	−.281570	−.535559
4.8	−.272304	.213565	.361289	.087509	−.251903	−.507347
4.9	−.292055	.181247	.366033	.117556	−.222087	−.480147
5.0	−.308518	.147863	.367663	.146267	−.192142	−.453695
5.1	−.321602	.113736	.366205	.173483	−.162107	−.427769
5.2	−.331251	.079190	.361709	.199047	−.132039	−.402184
5.3	−.337437	.044548	.354248	.222809	−.102011	−.376788
5.4	−.340168	.010127	.343919	.244627	−.072111	−.351458
5.5	−.339481	−.023758	.330841	.264370	−.042438	−.326097
5.6	−.335444	−.056806	.315156	.281917	−.013102	−.300635
5.7	−.328157	−.088723	.297026	.297163	.015777	−.275020
5.8	−.317746	−.119234	.276631	.310014	.044073	−.249224
5.9	−.304366	−.148077	.254170	.320396	.071656	−.223235
6.0	−.288195	−.175010	.229858	.328249	.098391	−.197061
6.1	−.269435	−.199812	.203923	.333532	.124142	−.170723
6.2	−.248310	−.222284	.176606	.336223	.148771	−.144260
6.3	−.225062	−.242249	.148157	.336318	.172145	−.117720
6.4	−.199949	−.259560	.118836	.333832	.194132	−.091168
6.5	−.173242	−.274091	.088907	.328803	.214604	−.064675
6.6	−.145226	−.285747	.058636	.321284	.233441	−.038326
6.7	−.116191	−.294459	.028293	.311351	.250529	−.012212
6.8	−.086434	−.300187	−.001856	.299095	.265764	.013568
6.9	−.056254	−.302918	−.031549	.284629	.279052	.038910
7.0	−.025950	−.302667	−.060527	.268081	.290310	.063702

Table L-2. Bessel Functions of the Second Kind of Order n: $N_n(x)$ (or $Y_n(x)$) (*continued*)

x	$N_0(x)$	$N_1(x)$	$N_2(x)$	$N_3(x)$	$N_4(x)$	$N_5(x)$
7.1	.004182	−.299479	−.088542	.249596	.299468	.087833
7.2	.033850	−.293423	−.115357	.229336	.306470	.111186
7.3	.062774	−.284594	−.140745	.207474	.311271	.133645
7.4	.090681	−.273115	−.164496	.184198	.313846	.155094
7.5	.117313	−.259129	−.186414	.159708	.314180	.175418
7.6	.142429	−.242801	−.206324	.134210	.312279	.194505
7.7	.165802	−.224318	−.224066	.107920	.308160	.212246
7.8	.187227	−.203885	−.239505	.081062	.301861	.228539
7.9	.206521	−.181721	−.252526	.053860	.293432	.243287
8.0	.223521	−.158060	−.263037	.026542	.282943	.256401
8.1	.238091	−.133149	−.270968	−.000662	.270477	.267800
8.2	.250118	−.107241	−.276274	−.027527	.256132	.277413
8.3	.259515	−.080598	−.278936	−.053830	.240023	.285177
8.4	.266222	−.053485	−.278956	−.079352	.222276	.291044
8.5	.270205	−.026169	−.276362	−.103884	.203032	.294974
8.6	.271458	.001084	−.271206	−.127226	.182443	.296941
8.7	.269999	.028011	−.263560	−.149188	.160672	.296932
8.8	.265875	.054356	−.253521	−.169593	.137890	.294947
8.9	.259156	.079869	−.241208	−.188277	.114279	.291000
9.0	.249937	.104315	−.226756	−.205095	.090026	.285118
9.1	.238336	.127466	−.210322	−.219915	.065323	.277341
9.2	.224494	.149113	−.192078	−.232625	.040366	.267726
9.3	.208570	.169061	−.172213	−.243131	.015354	.256339
9.4	.190744	.187136	−.150928	−.251360	−.009515	.243262
9.5	.171211	.203180	−.128436	−.257258	−.034043	.228590
9.6	.150180	.217059	−.104960	−.260792	−.058036	.212429
9.7	.127875	.228660	−.080728	−.261950	−.081303	.194896
9.8	.104527	.237893	−.055977	−.260741	−.103660	.176121
9.9	.080377	.244692	−.030944	−.257195	−.124931	.156241

So, for this case (n an integer) the general solution of Bessel's equation is

$$R(\rho) = AJ_n(k\rho) + BN_n(k\rho) \tag{L-30}$$

Case 2: n not an Integer, Denoted by ν. In this case we obtain both solutions from the series substitution. The results are denoted by

$$J_\nu(k\rho) \quad \text{and} \quad J_{-\nu}(k\rho)$$

where

$$J_\nu(x) = \left(\frac{x}{2}\right)^\nu \sum_{i=0}^{\infty} \frac{(-1)^i\left(\frac{x}{2}\right)^{2i}}{i!\Gamma(\nu + i + 1)} \tag{L-31}$$

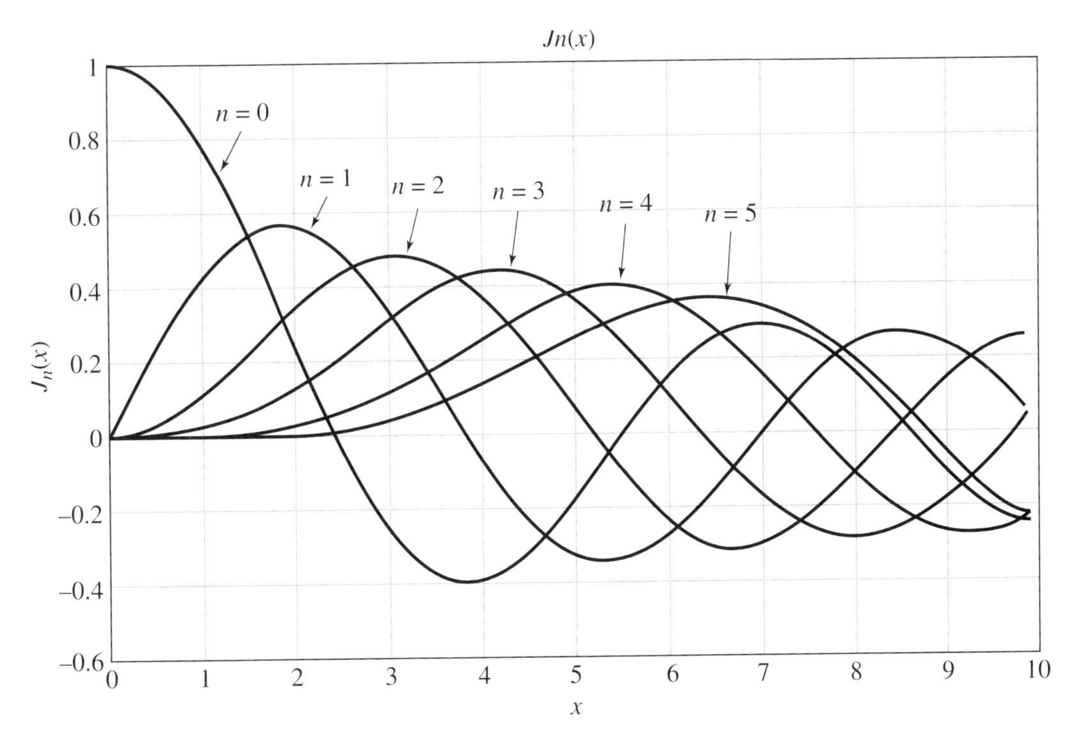

Figure L-3. Bessel functions of the first kind of order n.

Figure L-4. Bessel functions of the second kind of order n.

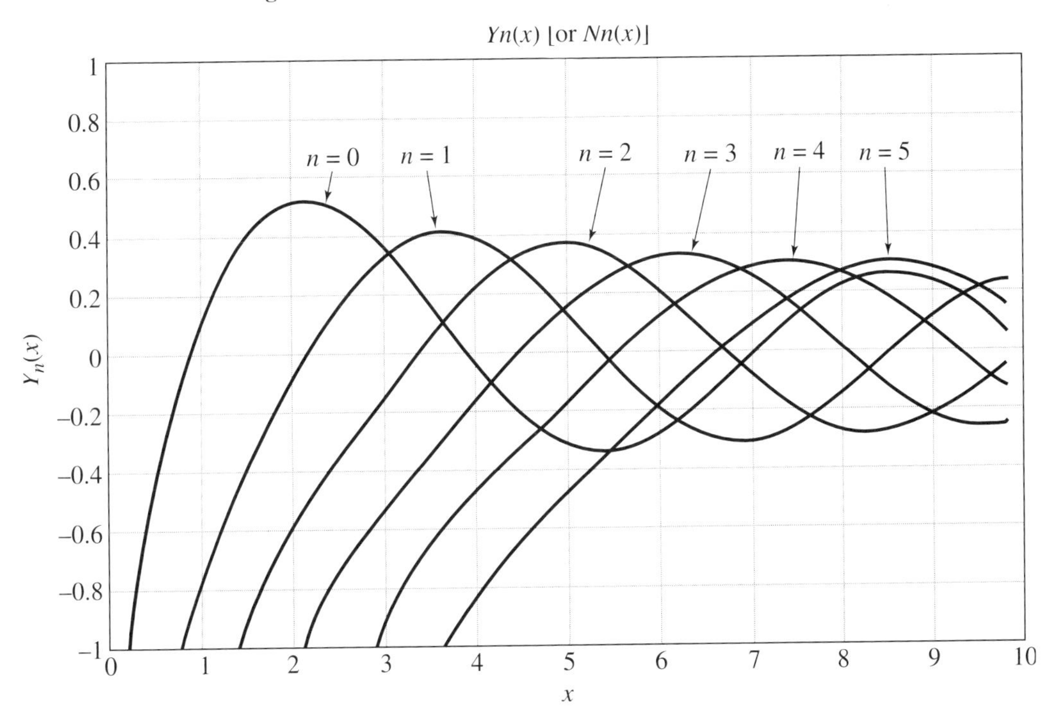

is the Bessel function of the first kind of order ν, and

$$J_{-\nu}(x) = \left(\frac{x}{2}\right)^{-\nu} \sum_{i=0}^{\infty} \frac{(-i)^i \left(\frac{x}{2}\right)^{2i}}{i!\Gamma(-\nu + i + 1)} \tag{L-32}$$

is the Bessel Function of the second kind of order $-\nu$.

In these series

$$\Gamma(m) \equiv \int_0^{\infty} t^{m-1} e^{-t} dt, \ m > 0$$

$$\Gamma(m + 1) = m\Gamma(m)$$

For m an integer, say n,

$$\Gamma(n) = (n - 1)!$$

For $m < 0$,

$$\Gamma(m) \equiv \frac{\Gamma(m + 1)}{m}$$

These two solutions will not work for ν and integer because the two solutions are not linearly independent as two solutions must be. In fact, from Eq. L-32

$$J_{-n}(x) = (-1)^n J_n(x) \qquad \text{for integer } n$$

Since the solutions for ν not an integer are independent, this allows us to construct the solution for n an integer to obtain $N_n(x)$ as in Eq. L-29.

For this case the appropriate solution is

$$R(\rho) = AJ_\nu(k\rho) + BJ_{-\nu}(k\rho) \tag{L-33}$$

As a side note, since the two Bessel functions are linearly independent, we could leave off the limit operation in $N_n(x)$ given by Eq. L-29, and use it as the second solution, since any linear combination of solutions is also a solution. This would then give

$$R(\rho) = AJ_\nu(k\rho) + BN_\nu(k\rho) \tag{L-34}$$

Case 3: n an Integer. Since linear combinations of solutions are also solutions, other functions that turn out to be useful in practice are defined. The two defined solutions are themselves linearly independent:

$$H_n^{(1)}(x) = J_n(x) + jN_n(x) \tag{L-35}$$

is the Hankel function of the first kind of order n, and

$$H_n^{(2)}(x) = J_n(x) - jN_n(x) \tag{L-36}$$

is the Hankel function of the second kind of order n. The solution is then

$$R(\rho) = AH_n^{(1)}(k\rho) + BH_n^{(2)}(k\rho) \tag{L-37}$$

These are given in Tables L-3 and L-4 and Figure L-5. (Since $N_n(0) = -\infty$, this solution cannot be used whenever $\rho = 0$ is in the region of interest.)

Table L-3. Hankel Functions of the First Kind of Order n: $Hn^{(1)}(x)$

	Magnitude							Phase (Radians)					
x	$H_0(x)$	$H_1(x)$	$H_2(x)$	$H_3(x)$	$H_4(x)$	$H_5(x)$	x	$H_0(x)$	$H_1(x)$	$H_2(x)$	$H_3(x)$	$H_4(x)$	$H_5(x)$
0.0							0.0	−1.50302	−1.57080	−1.57080	−1.57080	−1.57080	−1.57080
0.1	1.829999	6.459144	127.6448	5099.332	305832.3	24.E+06	0.1	−0.99431	−1.56306	−1.57079	−1.57080	−1.57080	−1.57080
0.2	1.465926	3.325314	32.15714	639.8191	19162.4	7.7E+05	0.2	−0.82935	−1.54087	−1.57064	−1.57080	−1.57080	−1.57080
0.3	1.267850	2.297897	14.48010	190.7748	3801.016	1.0E+05	0.3	−0.69025	−1.50621	−1.57003	−1.57079	−1.57080	−1.57080
0.4	1.135619	1.791628	8.298359	81.20248	1209.739	24113.6	0.4	−0.56291	−1.46116	−1.56842	−1.57078	−1.57080	−1.57080
0.5	1.038423	1.491283	5.441457	42.05949	499.2726	7946.301	0.5	−0.44236	−1.40762	−1.56517	−1.57074	−1.57080	−1.57080
0.6	.962773	1.292588	3.893040	24.69157	243.0229	3215.614	0.6	−0.32619	−1.34713	−1.55958	−1.57062	−1.57079	−1.57080
0.7	.901592	1.151260	2.962061	15.81948	132.6341	1499.998	0.7	−0.21308	−1.28099	−1.55095	−1.57036	−1.57079	−1.57080
0.8	.850727	1.045376	2.359776	10.81465	78.75129	776.6983	0.8	−0.10221	−1.21020	−1.53866	−1.56985	−1.57078	−1.57080
0.9	.807543	.962884	1.948207	7.775374	49.88983	435.6898	0.9	0.00697	−1.13559	−1.52223	−1.56894	−1.57076	−1.57080
1.0	.770271	.896626	1.654677	5.821550	33.27842	260.4059	1.0	0.11483	−1.05781	−1.50130	−1.56744	−1.57072	−1.57080
1.1	.737667	.842093	1.437971	4.507305	23.15343	163.8813	1.1	0.22164	−0.97737	−1.47568	−1.56510	−1.57064	−1.57079
1.2	.708831	.796305	1.273321	3.590050	16.68619	107.6513	1.2	0.32760	−0.89470	−1.44532	−1.56164	−1.57050	−1.57079
1.3	.683088	.757220	1.145133	2.929959	12.39115	73.32353	1.3	0.43286	−0.81014	−1.41028	−1.55676	−1.57025	−1.57078
1.4	.659923	.723387	1.043206	2.442492	9.443198	51.51913	1.4	0.53754	−0.72397	−1.37070	−1.55012	−1.56984	−1.57077
1.5	.638933	.693752	.960651	2.074437	7.361981	37.19031	1.5	0.64172	−0.63642	−1.32679	−1.54140	−1.56920	−1.57075
1.6	.619798	.667527	.892684	1.791139	5.856384	27.49215	1.6	0.74549	−0.54767	−1.27881	−1.53030	−1.56824	−1.57071
1.7	.602263	.644113	.835909	1.569348	4.743754	20.75634	1.7	0.84889	−0.45788	−1.22702	−1.51651	−1.56684	−1.57064
1.8	.586116	.623047	.787861	1.393062	3.905965	15.96999	1.8	0.95199	−0.36718	−1.17171	−1.49981	−1.56486	−1.57053
1.9	.571185	.603964	.746718	1.251018	3.264555	12.49911	1.9	1.05481	−0.27569	−1.11315	−1.48001	−1.56214	−1.57035
2.0	.557324	.586573	.711115	1.135131	2.766152	9.935992	2.0	1.15740	−0.18350	−1.05161	−1.45696	−1.55851	−1.57009
2.1	.544414	.570637	.680009	1.039497	2.373678	8.011978	2.1	1.25978	−0.09069	−0.98734	−1.43058	−1.55375	−1.56969
2.2	.532350	.555965	.652595	.959736	2.060871	6.546174	2.2	1.36197	0.00268	−0.92057	−1.40084	−1.54767	−1.56913
2.3	.521044	.542398	.628245	.892554	1.808811	5.414340	2.3	1.46400	0.09653	−0.85152	−1.36776	−1.54006	−1.56832
2.4	.510421	.529803	.606461	.835442	1.603646	4.529605	2.4	1.56588	0.19083	−0.78038	−1.33138	−1.53069	−1.56721
2.5	.500415	.518068	.586844	.786470	1.435095	3.830226	2.5	1.66763	0.28552	−0.70733	−1.29178	−1.51936	−1.56570
2.6	.490969	.507100	.569074	.744136	1.295422	3.271650	2.6	1.76927	0.38057	−0.63254	−1.24908	−1.50590	−1.56370
2.7	.482033	.496818	.552889	.707259	1.178741	2.821283	2.7	1.87079	0.47595	−0.55616	−1.20339	−1.49013	−1.56109
2.8	.473562	.487153	.538074	.674905	1.080520	2.454971	2.8	1.97222	0.57162	−0.47832	−1.15485	−1.47192	−1.55773
2.9	.465519	.478044	.524451	.646327	.997241	2.154599	2.9	2.07356	0.66755	−0.39914	−1.10361	−1.45115	−1.55349
3.0	.457868	.469441	.511872	.620924	.926143	1.906432	3.0	2.17483	0.76373	−0.31873	−1.04980	−1.42775	−1.54822
3.1	.450579	.461296	.500212	.598210	.865044	1.699943	3.1	2.27601	0.86013	−0.23719	−0.99357	−1.40166	−1.54176
3.2	.443624	.453572	.489365	.577785	.812207	1.526994	3.2	2.37713	0.95674	−0.15461	−0.93507	−1.37285	−1.53396
3.3	.436979	.446231	.479241	.559323	.766236	1.381227	3.3	2.47819	1.05353	−0.07107	−0.87443	−1.34134	−1.52465
3.4	.430621	.439244	.469764	.542554	.726007	1.257643	3.4	2.57919	1.15050	0.01336	−0.81178	−1.30715	−1.51369
3.5	.424531	.432582	.460868	.527253	.690604	1.152272	3.5	2.68014	1.24762	0.09861	−0.74725	−1.27031	−1.50093
3.6	.418690	.426221	.452495	.513230	.659281	1.061943	3.6	2.78104	1.34490	0.18461	−0.68096	−1.23090	−1.48625
3.7	.413082	.420138	.444594	.500327	.631424	.984107	3.7	2.88190	1.44231	0.27133	−0.61302	−1.18897	−1.46953
3.8	.407691	.414313	.437124	.488411	.606526	.916697	3.8	2.98271	1.53985	0.35869	−0.54353	−1.14461	−1.45069
3.9	.402505	.408729	.430045	.477367	.584167	.858033	3.9	3.08348	1.63750	0.44667	−0.47259	−1.09792	−1.42964
4.0	.397511	.403369	.423325	.467097	.563997	.806738	4.0	−3.09896	1.73527	0.53521	−0.40029	−1.04898	−1.40633
4.1	.392697	.398218	.416932	.457518	.545722	.761680	4.1	−2.99826	1.83313	0.62428	−0.32672	−0.99788	−1.38072
4.2	.388052	.393264	.410841	.448557	.529096	.721923	4.2	−2.89758	1.93110	0.71384	−0.25196	−0.94473	−1.35280
4.3	.383568	.388493	.405028	.440152	.513909	.686688	4.3	−2.79694	2.02915	0.80387	−0.17608	−0.88962	−1.32255
4.4	.379234	.383894	.399473	.432249	.499986	.655330	4.4	−2.69632	2.12728	0.89433	−0.09914	−0.83265	−1.29001
4.5	.375043	.379458	.394155	.424800	.487174	.627304	4.5	−2.59573	2.22550	0.98520	−0.02122	−0.77390	−1.25518
4.6	.370988	.375174	.389059	.417762	.475346	.602156	4.6	−2.49517	2.32379	1.07645	0.05763	−0.71346	−1.21812
4.7	.367060	.371035	.384169	.411100	.464391	.579501	4.7	−2.39462	2.42214	1.16805	0.13736	−0.65143	−1.17886
4.8	.363254	.367032	.379470	.404781	.454212	.559014	4.8	−2.29410	2.52057	1.25999	0.21791	−0.58787	−1.13747
4.9	.359564	.363157	.374951	.398776	.444729	.540420	4.9	−2.19360	2.61905	1.35226	0.29924	−0.52288	−1.09400

Table L-4. Hankel Functions of the Second Kind of Order n: $Hn^{(2)}(x)$

	Magnitude							Phase (radians)					
x	$H_0(x)$	$H_1(x)$	$H_2(x)$	$H_3(x)$	$H_4(x)$	$H_5(x)$	x	$H_0(x)$	$H_1(x)$	$H_2(x)$	$H_3(x)$	$H_4(x)$	$H_5(x)$
0.0							0.0	1.50302	1.57080	1.57080	1.57080	1.57080	1.57080
0.1	1.829999	6.459144	127.6448	5099.332	305832.3	24.E+06	0.1	0.99431	1.56306	1.57079	1.57080	1.57080	1.57080
0.2	1.465926	3.325314	32.15714	639.8191	19162.4	7.7E+05	0.2	0.82935	1.54087	1.57064	1.57080	1.57080	1.57080
0.3	1.267850	2.297897	14.48010	190.7748	3801.016	1.0E+05	0.3	0.69025	1.50621	1.57003	1.57079	1.57080	1.57080
0.4	1 135619	1.791628	8.298359	81.20248	1209.739	24113.6	0.4	0.56291	1.46116	1.56842	1 57078	1.57080	1.57080
0.5	1.038423	1.491283	5.441457	42.05949	499.2726	7946.301	0.5	0.44236	1.40762	1.56517	1.57074	1.57080	1.57080
0.6	.962773	1.292588	3.893040	24.69157	243.0229	3215.614	0.6	0.32619	1.34713	1.55958	1.57062	1.57079	1.57080
0.7	.901592	1.151260	2.962061	15.81948	132.6341	1499.998	0.7	0.21308	1.28099	1.55095	1.57036	1.57079	1.57080
0.8	.850727	1.045376	2.359776	10.81465	78.75129	776.6983	0.8	0.10221	1.21020	1.53866	1.56985	1.57078	1.57080
0.9	.807543	962884	1.948207	7.775374	49.88983	435.6898	0.9	−0.00697	1.13559	1.52223	1.56894	1.57076	1.57080
1.0	.770271	.896626	1.654677	5.821550	33.27842	260.4059	1.0	−0.11483	1.05781	1.50130	1.56744	1.57072	1.57080
1.1	.737667	.842093	1.437971	4.507305	23.15343	163.8813	1.1	−0.22164	0.97737	1.47568	1.56510	1.57064	1.57079
1.2	.708831	.796305	1.273321	3.590050	16.68619	107.6513	1.2	−0.32760	0.89470	1.44532	1.56164	1.57050	1.57079
1.3	.683088	.757220	1.145133	2.929959	12.39115	73.32353	1.3	−0.43286	0.81014	1.41028	1.55676	1.57025	1.57078
1.4	.659923	.723387	1.043206	2.442492	9.443198	51.51913	1.4	−0.53754	0.72397	1.37070	1.55012	1.56984	1.57077
1.5	.638933	.693752	.960651	2.074437	7.361981	37.19031	1.5	−0.64172	0.63642	1.32679	1.54140	1.56920	1.57075
1.6	.619798	.667527	.892684	1.791139	5.856384	27.49215	1.6	−0.74549	0.54767	1.27881	1.53030	1.56824	1.57071
1.7	.602263	.644113	.835909	1.569348	4.743754	20.75634	1.7	−0.84889	0.45788	1.22702	1.51651	1.56684	1.57064
1.8	.586116	.623047	.787861	1.393062	3.905965	15.96999	1.8	−0.95199	0.36718	1.17171	1.49981	1.56486	1.57053
1.9	.571185	.603964	.746718	1.251018	3.264555	12.49911	1.9	−1.05481	0.27569	1.11315	1.48001	1.56214	1.57035
2.0	.557324	.586573	.711115	1.135131	2.766152	9.935992	2.0	−1.15740	0.18350	1.05161	1.45696	1.55851	1.57009
2.1	.544414	.570637	.680009	1.039497	2.373678	8.011978	2.1	−1.25978	0.09069	0.98734	1.43058	1.55375	1.56969
2.2	.532350	.555965	.652595	.959736	2.060871	6.546174	2.2	−1.36197	−0.00268	0.92057	1.40084	1.54767	1.56913
2.3	.521044	.542398	.628245	.892554	1.808811	5.414340	2.3	−1.46400	−0.09653	0.85152	1.36776	1.54006	1.56832
2.4	.510421	.529803	.606461	.835442	1.603646	4.529605	2.4	−1.56588	−0.19083	0.78038	1.33138	1.53069	1.56721
2.5	.500415	.518068	.586844	.786470	1.435095	3.830226	2.5	−1.66763	−0.28552	0.70733	1.29178	1.51936	1.56570
2.6	.490969	.507100	.569074	.744136	1.295422	3.271650	2.6	−1.76927	−0.38057	0.63254	1.24908	1.50590	1.56370
2.7	.482033	.496818	.552889	.707259	1.178741	2.821283	2.7	−1.87079	−0.47595	0.55616	1.20339	1.49013	1.56109
2.8	.473562	.487153	.538074	.674905	1.080520	2.454971	2.8	−1.97222	−0.57162	0.47832	1.15485	1.47192	1.55773
2.9	.465519	.478044	.524451	.646327	.997241	2.154599	2.9	−2.07356	−0.66755	0.39914	1.10361	1.45115	1.55349
3.0	.457868	.469441	.511872	.620924	.926143	1.906432	3.0	−2.17483	−0.76373	0.31873	1.04980	1.42775	1.54822
3.1	.450579	.461296	.500212	.598210	.865044	1.699943	3.1	−2.27601	−0.86013	0.23719	0.99357	1.40166	1.54176
3.2	.443624	.453572	.489365	.577785	.812207	1.526994	3.2	−2.37713	−0.95674	0.15461	0.93507	1.37285	1.53396
3.3	.436979	.446231	.479241	.559323	.766236	1.381227	3.3	−2.47819	−1.05353	0.07107	0.87443	1.34134	1.52465
3.4	.430621	.439244	.469764	.542554	.726007	1.257643	3.4	−2.57919	−1.15050	−0.01336	0.81178	1.30715	1.51369
3.5	.424531	.432582	.460868	.527253	.690604	1.152272	3.5	−2.68014	−1.24762	−0.09861	0.74725	1.27031	1.50093
3.6	.418690	.426221	.452495	.513230	.659281	1.061943	3.6	−2.78104	−1.34490	−0.18461	0.68096	1.23090	1.48625
3.7	.413082	.420138	.444594	.500327	.631424	.984107	3.7	−2.88190	−1.44231	−0.27133	0.61302	1.18897	1.46953
3.8	.407691	.414313	.437124	.488411	.606526	.916697	3.8	−2.98271	−1.53985	−0.35869	0.54353	1.14461	1.45069
3.9	.402505	.408729	.430045	.477367	.584167	.858033	3.9	−3.08348	−1.63750	−0.44667	0.47259	1.09792	1.42964
4.0	.397511	.403369	.423325	.467097	.563997	.806738	4.0	3.09896	−1.73527	−0.53521	0.40029	1.04898	1.40633
4.1	.392697	.398218	.416932	.457518	.545722	.761680	4.1	2.99826	−1.83313	−0.62428	0.32672	0.99788	1.38072
4.2	.388052	.393264	.410841	.448557	.529096	.721923	4.2	2.89758	−1.93110	−0.71384	0.25196	0.94473	1.35280
4.3	.383568	.388493	.405028	.440152	.513909	.686688	4.3	2.79694	−2.02915	−0.80387	0.17608	0.88962	1.32255
4.4	.379234	.383894	.399473	.432249	.499986	.655330	4.4	2.69632	−2.12728	−0.89433	0.09914	0.83265	1.29001
4.5	.375043	.379458	.394155	.424800	.487174	.627304	4.5	2.59573	−2.22550	−0.98520	0.02122	0.77390	1.25518
4.6	.370988	.375174	.389059	.417762	.475346	.602156	4.6	2.49517	−2.32379	−1.07645	−0.05763	0.71346	1.21812
4.7	.367060	.371035	.384169	.411100	.464391	.579501	4.7	2.39462	−2.42214	−1.16805	0.13736	0.65143	1.17886
4.8	.363254	.367032	.379470	.404781	.454212	.559014	4.8	2.29410	−2.52057	−1.25999	−0.21791	0.58787	1.13747
4.9	.359564	.363157	.374951	.398776	.444729	.540420	4.9	2.19360	−2.61905	−1.35226	−0.29924	0.52288	1.09400

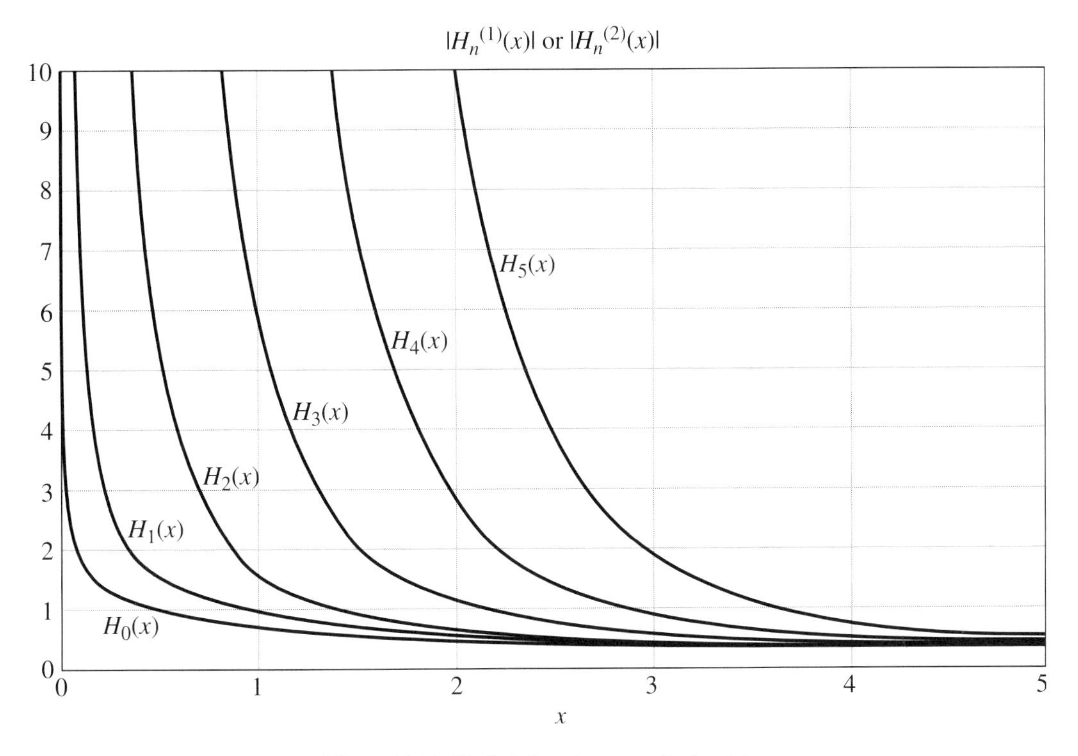

Figure L-5. Hankel function magnitude plot.

This process is much like combining the sine and cosine solutions to obtain other solutions as

$$e^{jx} = \cos x + j\sin x \qquad \text{and} \qquad e^{-jx} = \cos x - j\sin x$$

Case 4: Constant k is Imaginary, say $j\kappa$; n an Integer. If the constant k in Bessel's equation is imaginary the resulting series will have powers of j in them. The results are used to define new solution functions that have real values. The first of these is given by the definition:

$$I_n(\kappa\rho) \equiv j^{-n}J_n(j\kappa\rho) \qquad \text{or} \qquad I_n(x) = j^{-n}J_n(jx) \tag{L-38}$$

This is called the *modified Bessel function of the first kind of order* n.

The second solution is defined by

$$K_n(\kappa\rho) = \frac{\pi}{2}j^{n+1}H_n^{(1)}(\kappa\rho) \qquad \text{or} \qquad K_n(x) = \frac{\pi}{2}j^{n+1}H_n^{(1)}(\kappa\rho) \tag{L-39}$$

This is called the *modified Bessel function of the second kind of order* n. These functions are tabulated and plotted in Tables L-5 and L-6, and Figure L-6.

As with e^x, $\cos x$, $\sin x$, and so on, the Bessel functions have several identities that relate them, and they can be differentiated and integrated. A collection of these formulas is given in Appendix M.

Trigonometric integrals yielding Bessel functions are given in Appendix O.

Table L-5. Modified Bessel Functions of the First Kind; In(x)

x	$H_0(x)$	$H_1(x)$	$H_2(x)$	$H_3(x)$	$H_4(x)$	$H_5(x)$
0.0	1	5E−11	1.25E−21	2.08E−32	2.6E−43	2.6E−54
0.1	1.002502	0.050063	0.001251	2.08E−05	2.61E−07	2.61E−09
0.2	1.010025	0.100501	0.005017	0.000167	4.18E−06	8.35E−08
0.3	1.022627	0.151694	0.011335	0.000566	2.12E−05	6.35E−07
0.4	1.040402	0.204027	0.020268	0.001347	6.72E−05	2.68E−06
0.5	1.063483	0.257894	0.031906	0.002645	0.000165	8.22E−06
0.6	1.092045	0.313704	0.046365	0.004602	0.000344	2.06E−05
0.7	1.126303	0.37188	0.06379	0.007367	0.000641	4.47E−05
0.8	1.166515	0.432865	0.084353	0.0111	0.001101	8.76E−05
0.9	1.212985	0.497126	0.10826	0.015972	0.001779	0.000159
1.0	1.266066	0.565159	0.135748	0.022168	0.002737	0.000271
1.1	1.32616	0.637489	0.167089	0.029891	0.004049	0.000441
1.2	1.393726	0.714678	0.202596	0.039359	0.005801	0.000688
1.3	1.469278	1.797329	0.242617	0.050815	0.008089	0.001037
1.4	1.553395	0.886092	0.287549	0.064522	0.011026	0.001519
1.5	1.646723	0.981666	0.337835	0.080774	0.014738	0.002171
1.6	1.749981	1.084811	0.393967	0.099892	0.019371	0.003036
1.7	1.863965	1.196347	0.456498	0.122233	0.025089	0.004166
1.8	1.989559	1.317167	0.52604	0.148189	0.032077	0.005625
1.9	2.12774	1.448244	0.603272	0.178197	0.040545	0.007483
2.0	2.279585	1.590637	0.688948	0.21274	0.050729	0.009826
2.1	2.446283	1.7455	0.783902	0.252352	0.062895	0.012751
2.2	2.629143	1.914095	0.889057	0.297628	0.077345	0.016374
2.3	2.829606	2.0978	1.005432	0.349223	0.094415	0.020825
2.4	3.049257	2.298124	1.134153	0.407868	0.114483	0.026257
2.5	3.289839	2.516716	1.276466	0.47437	0.137977	0.032843
2.6	3.553269	2.755384	1.433742	0.549627	0.165373	0.040786
2.7	3.841651	3.016108	1.607497	0.63463	0.197207	0.050313
2.8	4.157298	3.301056	1.799401	0.730483	0.234079	0.061686
2.9	4.502749	3.612607	2.011295	0.838407	0.276661	0.075204
3.0	4.880793	3.95337	2.245212	0.959754	0.325705	0.091206
3.1	5.294491	4.326206	2.503391	1.096024	0.382053	0.11008
3.2	5.747207	4.734254	2.788299	1.248881	0.446647	0.132263
3.3	6.24263	5.180959	3.102655	1.420164	0.520538	0.158254
3.4	6.784813	5.670102	3.449459	1.611915	0.604903	0.188615
3.5	7.378203	6.205835	3.832012	1.826393	0.701053	0.223985
3.6	8.027685	6.792715	4.253954	2.066099	0.810456	0.265085
3.7	8.738618	7.435746	4.719295	2.333805	0.934747	0.31273
3.8	9.516888	8.140425	5.232454	2.632578	1.075752	0.367838
3.9	10.36896	8.912787	5.798298	2.965815	1.235505	0.431447
4.0	11.30192	9.759465	6.422189	3.337276	1.416276	0.504724
4.1	12.32357	10.68774	7.110038	3.75112	1.620594	0.588986
4.2	13.44246	11.70562	7.868351	4.211952	1.851277	0.685711
4.3	14.66797	12.82189	8.704302	4.724868	2.111463	0.796564
4.4	16.01044	14.04622	9.625789	5.295504	2.404648	0.923416
4.5	17.48117	15.38922	10.64152	5.930096	2.734722	1.068368
4.6	19.09262	16.86256	11.76107	6.635544	3.106016	1.233778
4.7	20.85846	18.47907	12.99502	7.419478	3.523347	1.422292
4.8	22.79368	20.25283	14.355	8.290337	3.992075	1.636878
4.9	24.91478	22.19935	15.85382	9.257454	4.518162	1.880863

Table L-6. Modified Bessel Functions of the Second Kind: $K_n(x)$

x	$H_0(x)$	$H_1(x)$	$H_2(x)$	$H_3(x)$	$H_4(x)$	$H_5(x)$
0.0	23.14178	1E+10	2E+20	8E+30	4.8E+41	3.84E+52
0.1	2.427069	9.853845	199.504	7990.012	479600.2	38376010
0.2	1.752704	4.775973	49.51243	995.0246	29900.25	1197005
0.3	1.37246	3.055992	21.74574	292.9992	5881.73	157139.1
0.4	1.114529	2.184354	12.0363	122.5474	1850.247	37127.48
0.5	0.924419	1.656441	7.550184	62.05791	752.2451	12097.98
0.6	0.777522	1.302835	5.120305	35.4382	359.5023	4828.803
0.7	0.66052	1.050284	3.66133	21.97217	191.9942	2216.192
0.8	0.565347	0.861782	2.719801	14.46079	111.1757	1126.218
0.9	0.48673	0.716534	2.079027	9.956654	68.45672	618.4609
1.0	0.421024	0.601907	1.624839	7.101263	44.23242	360.9606
1.1	0.365602	0.50976	1.292439	5.209537	29.7081	221.2684
1.2	0.318508	0.434592	1.042829	3.910689	20.59627	141.2192
1.3	0.278248	0.372547	0.851398	2.992233	14.6617	93.21809
1.4	0.243655	0.320836	0.701992	2.326528	10.67282	63.31409
1.5	0.213806	0.277388	0.583656	1.833804	7.918871	44.06778
1.6	0.187955	0.240634	0.488747	1.462502	5.973129	31.32815
1.7	0.165496	0.209362	0.411805	1.178316	4.570567	22.68686
1.8	0.145931	0.182623	0.348846	0.957836	3.541634	16.69843
1.9	0.128846	0.15966	0.296909	0.784732	2.775011	12.46899
2.0	0.113894	0.139866	0.25376	0.647385	2.195916	9.431049
2.1	0.100784	0.122746	0.217685	0.537385	1.75307	7.215746
2.2	0.089269	0.107897	0.187357	0.448546	1.410664	5.578234
2.3	0.07914	0.094982	0.161733	0.376258	1.143276	4.352869
2.4	0.070217	0.083725	0.139988	0.317038	0.932584	3.42565
2.5	0.062348	0.073891	0.12146	0.268227	0.765205	2.716884
2.6	0.055398	0.065284	0.105617	0.227771	0.631243	2.170058
2.7	0.049255	0.057738	0.092025	0.194071	0.523294	1.744571
2.8	0.04382	0.051113	0.080329	0.165868	0.435761	1.410901
2.9	0.039006	0.045286	0.070238	0.142167	0.364376	1.147343
3.0	0.03474	0.040156	0.06151	0.12217	0.305851	0.937774
3.1	0.030955	0.035634	0.053944	0.10524	0.257634	0.770102
3.2	0.027595	0.031643	0.047372	0.090858	0.21773	0.635182
3.3	0.024611	0.028117	0.041651	0.078603	0.184566	0.526036
3.4	0.021958	0.024999	0.036663	0.068132	0.156897	0.437301
3.5	0.019599	0.022239	0.032307	0.059162	0.133727	0.364824
3.6	0.0175	0.019795	0.028497	0.051458	0.11426	0.30537
3.7	0.015631	0.017628	0.025159	0.044827	0.097852	0.2564
3.8	0.013966	0.015706	0.022232	0.039108	0.083981	0.215911
3.9	0.012482	0.013999	0.019661	0.034165	0.072223	0.182314
4.0	0.01116	0.012483	0.017401	0.029885	0.062229	0.154343
4.1	0.00998	0.011136	0.015412	0.026173	0.053714	0.13098
4.2	0.008928	0.009938	0.01366	0.022948	0.046442	0.111409
4.3	0.007988	0.008872	0.012115	0.020142	0.040219	0.094968
4.4	0.007149	0.007923	0.010751	0.017696	0.034882	0.081119
4.5	0.0064	0.007078	0.009546	0.015563	0.030297	0.069424
4.6	0.005731	0.006325	0.00848	0.013699	0.026349	0.059524
4.7	0.005132	0.005654	0.007538	0.012069	0.022945	0.051125
4.8	0.004597	0.005055	0.006704	0.010641	0.020005	0.043984
4.9	0.004119	0.004521	0.005964	0.00939	0.017462	0.0379

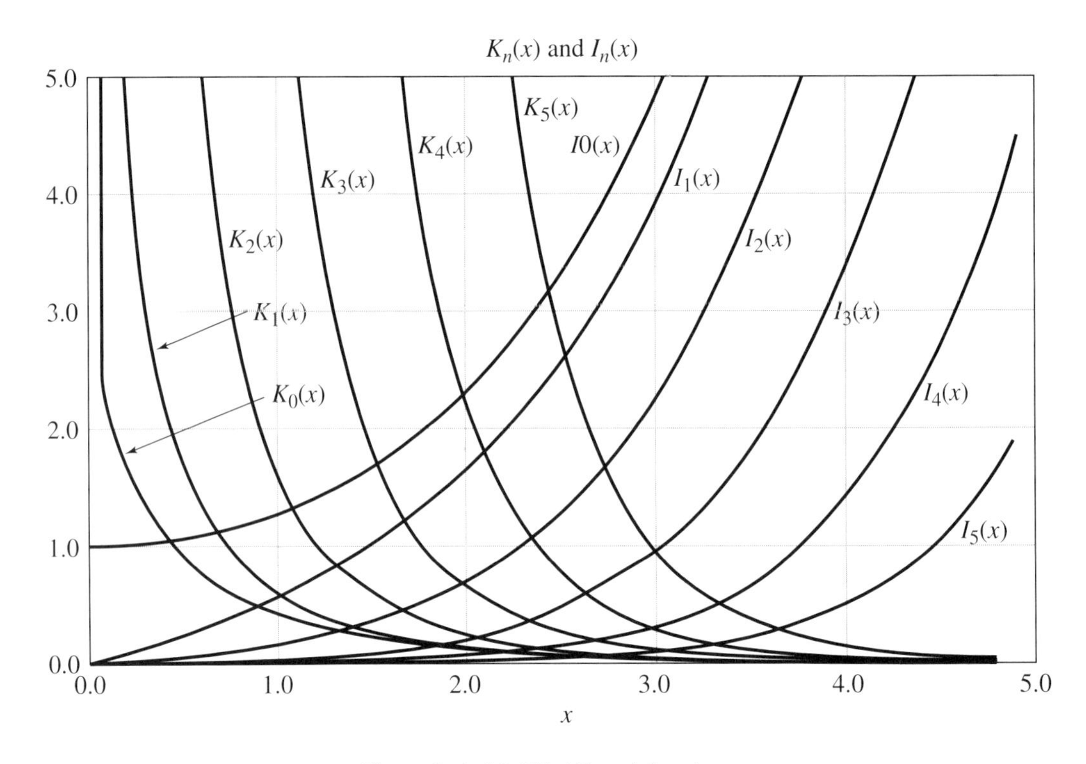

Figure L-6. Modified Bessel functions.

LEGENDRE'S EQUATION, SPHERICAL COORDINATES

When Maxwell's equations are used in spherical coordinates a wavelike equation can be obtained, even though there is no z direction for propagation. In this case propagation frequently occurs in the radial (r) direction as a wave propagates away from or toward the origin. When the separation of variables is used, one of the equations that frequently arises is a differential equation for the θ variation. Letting $\Theta(\theta)$ denote this function we usually obtain an ordinary differential equation of the form

$$\frac{d^2\Theta}{d\theta^2} + \frac{\cos\theta}{\sin\theta}\frac{d\Theta}{d\theta} + \left[n(n + 1) - \frac{m^2}{\sin^2\theta}\right]\Theta = 0 \tag{L-40}$$

where m and n are the separation constants. This is called the *associated Legendre equation*, which is one of the possible forms.

This equation has a singularity at $\theta = 0$, but since the form is not in terms of $(x - a)$ we cannot use the test for removability as the equation stands. To obtain a more tractable form, we change the variable by setting

$$u = \cos\theta \tag{L-41}$$

For which $du = -\sin\theta d\theta = -\sqrt{1 - \cos^2\theta}\, d\theta$. With this substitution the differential equation becomes

$$\frac{d^2\Theta}{du^2} - \frac{2u}{1-u^2}\frac{d\Theta}{du} + \left[\frac{n(n+1)}{1-u^2} - \frac{m^2}{(1-u^2)^2}\right]\Theta = 0 \tag{L-42}$$

This is the *algebraic form of Legendre's equation.*

Here we have two singularities; one at $u = +1$ and one at $u = -1$. These two points define the range of the variable u that is the cosine functions. Testing each of these singularities shows they are removable at the respective points of singularity. The appropriate series solution is then, about the point $u = 0$,

$$\Theta(u) = u^c \sum_{i=0}^{\infty} b_i u^i$$

This series solution results directly in one solution that will not be written here. See [1]. This solution is denoted by

$$P_n^m(u) \tag{L-43}$$

and is called the *Legendre function of the first kind of order* n *and degree* m.

The second solution is written

$$Q_n^m(u) \tag{L-44}$$

and is called the *Legendre function of the second kind of order* n *and degree* m.

As an important special case we set $m = 0$ and for n an integer use the abbreviated notations

$$P_n^0(u) \equiv P_n(u) \tag{L-45}$$

$$Q_n^0(u) \equiv Q_n(u) \tag{L-46}$$

For these we can generate the functions that can be tabulated and plotted using

$$P_n(u) = \frac{(-1)^n}{2^n n!}\frac{d^n(1-u^2)^n}{du^n} \tag{L-47}$$

Note that these will be finite series since for a given n the process stops after n derivatives. Once we have the $P_n(u)$ we can generate all the functions for m and n both integers using the following:

$$P_n^m(u) = (1-u^2)^{m/2}\frac{d^m P_n(u)}{du^m} \tag{L-48}$$

$$Q_n(u) = \frac{1}{2}P_n(u)\ln\left(\frac{1+u}{1-u}\right) - \sum_{i=1}^{n}\frac{1}{i}P_{i-1}(u)P_{n-1}(u) \tag{L-49}$$

$$Q_n^m(u) = (1-u^2)^{m/2}\frac{d^m Q_n(u)}{du^m} \tag{L-50}$$

These functions have been tabulated and plotted [1], Chapter 8. Plots of a few of P_n and Q_n are given in Figures L-7 and L-8. Table L-7 lists specific expressions for some of the Legendre functions. For $m = 0$:

$$P_n(u) = \frac{1}{2^n}\sum_{i=0}^{n/2}(-i)^i\binom{n}{i}\binom{2n-2i}{n}u^{n-2i} \tag{L-51}$$

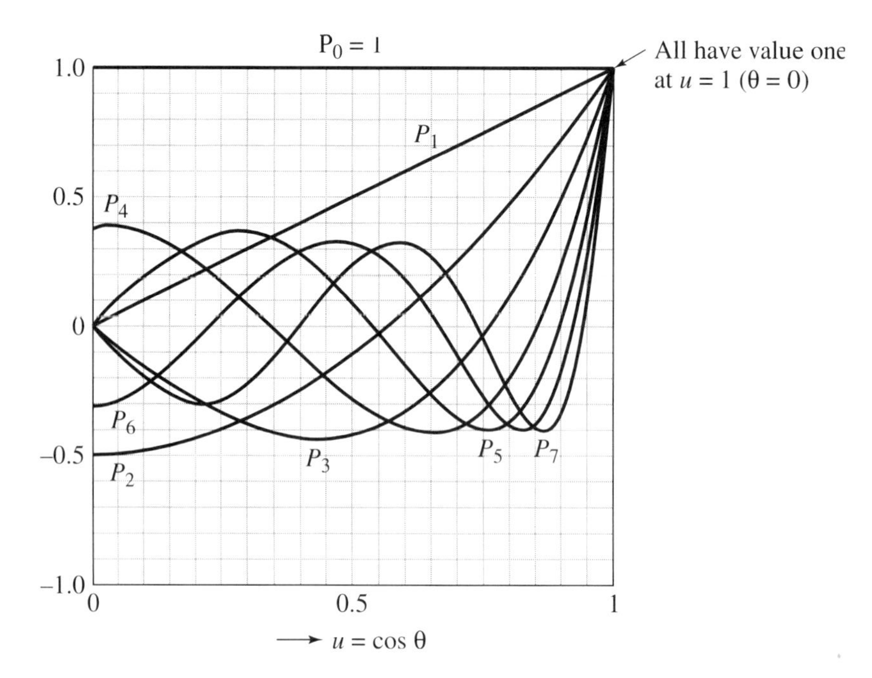

Figure L-7. Legendre polynomials of the first kind, $P_n(u)$.

Figure L-8. Legendre polynomials of the second kind, $Q_n(u)$.

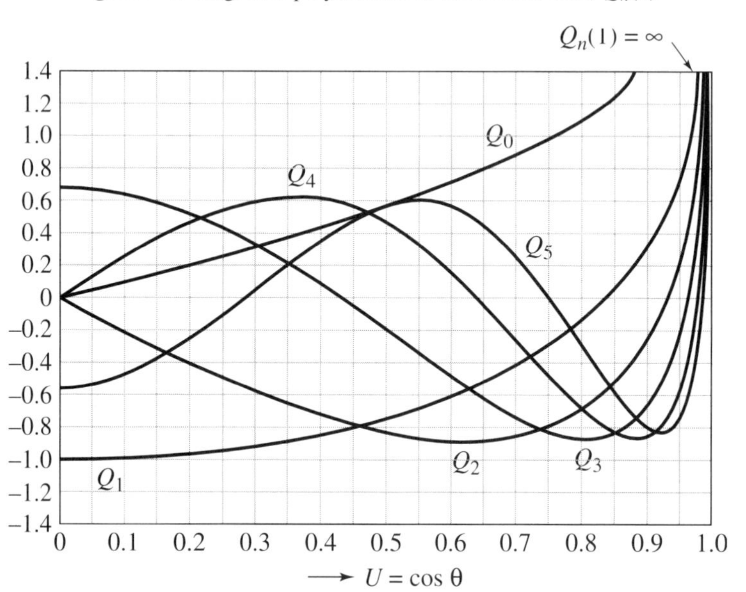

Table L-7. Table of Selected Legendre Functions

$$P_0(u) = 1$$

$$P_1(u) = u = \cos\theta$$

$$P_2(u) = \tfrac{1}{2}(3u^2 - 1) = \tfrac{1}{2}(3\cos^2\theta - 1)$$

$$P_3(u) = \tfrac{1}{2}(5u^3 - 3u)$$

$$P_4(u) = \tfrac{1}{8}(35u^4 - 30u^2 + 3)$$

$$P_5(u) = \tfrac{1}{8}(63u^5 - 70u^3 + 15u)$$

$$P_1^1(u) = \sqrt{1 - u^2} = \sin\theta$$

$$P_2^1(u) = 3u\sqrt{1 - u^2}$$

$$P_3^1(u) = \tfrac{3}{2}\sqrt{1 - u^2}(5u^2 - 1)$$

$$P_4^1(u) = \tfrac{5}{2}\sqrt{1 - u^2}(7u^3 - 3u)$$

$$P_2^2(u) = 3(1 - u^2)$$

$$P_3^2(u) = 15u(1 - u^2)$$

$$P_4^2(u) = \tfrac{15}{2}(7u^2 - 1)(1 - u^2)$$

$$P_3^3(u) = 15(1 - u^2)^{3/2}$$

$$P_4^3(u) = 105u(1 - u^2)^{3/2}$$

$$Q_0(u) = \frac{1}{2}\ln\left(\frac{1 + u}{1 - u}\right)$$

$$Q_1(u) = P_1(u)Q_0(u) - 1$$

$$Q_2(u) = P_2(u)Q_0(u) - \tfrac{3}{2}u$$

$$Q_3(u) = P_3(u)Q_0(u) - \tfrac{5}{2}u^2 + \tfrac{2}{3}$$

$$Q_1^1(u) = \left[Q_0(u) + \frac{u}{1 - u^2}\right](1 - u^2)^{1/2}$$

$$Q_2^1(u) = \left[3uQ_0(u) + \frac{3u^2 - 2}{1 - u^2}\right](1 - u^2)^{1/2}$$

$$Q_2^2(u) = \left[3Q_0(u) - \frac{3u^2 - 5u}{(1 - u^2)^2}\right](1 - u^2)$$

ADDITIONAL FUNCTIONS FROM SPHERICAL COORDINATES

For the spherical coordinate system separation of variables only solutions for the θ variation, Θ, have been obtained. The radial or r variation differential equation is of the form

$$\frac{d^2R}{dr^2} + \left[k^2 - \frac{n(n+1)}{r^2}\right]R = 0 \qquad \text{(L-52)}$$

The constant n in this equation is the same as that in the Θ differential equation (Eq. L-40). This equation is almost of the form of Bessel's equation (Eq. L-25). Equation L-52 can in fact be put into Bessel form by using the new function substitution

$$R(r) = \sqrt{r}R_1(r) \qquad \text{(L-53)}$$

Differentiating twice using the product rule for derivatives and substituting into Eq. L-52 there results

$$\frac{d^2R_1}{dr^2} + \frac{1}{r}\frac{dR_1}{dr} + \left[k^2 - \frac{n(n + \frac{1}{2})^2}{r^2}\right]R_1 = 0 \qquad \text{(L-54)}$$

This is now in Bessel equation form with $(n + \frac{1}{2})^2$ replacing n in Bessel's equation. The solution is then

$$R_1(r) = AJ_{(n+1/2)}(kr) + BN_{(n+1/2)}(kr) \qquad \text{(L-55)}$$

The solution for $R(r)$ is obtained using this in Eq. L-53:

$$R(r) = A\sqrt{r}J_{(n+1/2)}(kr) + B\sqrt{r}N_{(n+1/2)}(kr) \qquad \text{(L-56)}$$

The functions $J_{(n+1/2)}(x)$ and $N_{(n+1/2)}(x)$ are called *Bessel functions of fractional order of the first and second kinds*, respectively. These are tabulated and plotted in [1].

In some applications other functional forms involving the fractional-order Bessel functions are obtained. These are denoted by

$$j_n(kr) = \sqrt{\frac{\pi}{2kr}}J_{(n+1/2)}(kr) \qquad \text{(L-57)}$$

and

$$y_n(kr) = \sqrt{\frac{\pi}{2kr}}N_{(n+1/2)}(kr) \qquad \text{(L-58)}$$

The general function forms

$$j_n(x) = \sqrt{\frac{\pi}{2x}}J_{(n+1/2)}(x) \qquad \text{(L-59)}$$

$$y_n(x) = \sqrt{\frac{\pi}{2x}}N_{(n+1/2)}(x) \qquad \text{(L-60)}$$

are called *spherical Bessel functions of the first and second kinds*, respectively. These are also available in [1] and to a lesser extent in [15].

To complete these function solutions, we note that there are corresponding Hankel function forms:

$$h_n^{(1)}(x) = j_n(x) + jy_n(x) = \sqrt{\frac{\pi}{2x}}H_{(n+1/2)}^{(1)}(x) \qquad \text{(L-61)}$$

$$h_n^{(2)}(x) = j_n(x) - jy_n(x) = \sqrt{\frac{\pi}{2x}}H_{(n+1/2)}^{(2)}(x) \qquad \text{(L-62)}$$

These are *spherical Hankel functions of the first and second kinds*, but they are also sometimes called *spherical Bessel functions of the third kind.*

Extensive use is made of the functions developed in this appendix. Chapter 9 has examples of waveguide and cavity structures for which these functions describe the coordinate variations of the field components.

Appendix M. Identities, Recursion Formulas, Differential and Integral Formulas for Bessel and Legendre Functions

For a more extensive collection of relationships among the various functions consult [15], [54], [1], and [42].

BESSEL FUNCTIONS

Let $Z_\nu(x)$ represent any of the functions $J_\nu(x)$, $\mathrm{N}_\nu(x)$, $H_\nu^{(1)}(x)$, and $H_\nu^{(2)}(x)$. The prime denotes dZ/dx:

$$Z_\nu(x) = \frac{x}{2\nu}\left[Z_{\nu+1}(x) + Z_{\nu-1}(x)\right] \tag{M-1}$$

$$Z_0'(x) = -Z_1(x) \tag{M-2}$$

$$Z_\nu'(x) = \frac{\nu}{x} Z_\nu(x) - Z_{\nu+1}(x) \tag{M-3}$$

$$Z_\nu'(x) = -\frac{\nu}{x} Z_\nu(x) + Z_{\nu-l}(x) \tag{M-4}$$

$$Z_\nu'(x) = \tfrac{1}{2}\left[Z_{\nu-1}(x) + Z_{\nu+1}(x)\right] \tag{M-5}$$

$$\int x^\nu Z_{\nu-1}(x)dx = x^{-\nu} Z_\nu(x) \tag{M-6}$$

$$\int x^{-\nu} Z_{\nu+1}(x)dx = -x^{-\nu} Z_\nu(x) \tag{M-7}$$

$$\int Z_\nu(x)dx = 2\sum_{i=0}^{\infty} Z_{\nu+2i+1}(x) \tag{M-8}$$

$$\int x Z_\nu(ax) Z_\nu(bx)dx = \frac{bxZ_\nu(ax)Z_{\nu-1}(bx) - axZ_{\nu-1}(ax)Z_\nu(bx)}{a^2 - b^2} \qquad a^2 \neq b^2 \tag{M-9}$$

(Note: The two Z_ν do *not* need to be the same function; they can be any two.)

$$\int xZ_\nu^2(ax)dx = \frac{x^2}{2}\{[Z_\nu(ax)]^2 - Z_{\nu-1}(ax)Z_{\nu+1}(ax)\} \tag{M-10}$$

$$\int \frac{1}{x} Z_\nu(ax)Z_\tau(ax)dx = ax\frac{Z_{\nu-1}(ax)Z_\tau(ax) - Z_\nu(ax)Z_{\tau-1}(ax)}{\nu^2 - \tau^2} - \frac{Z_\nu(ax)Z_\tau(ax)}{\nu + \tau} \tag{M-11}$$

(Note: The Z_ν and Z_τ do *not* need to be the same function, can be any two.)

$$\int xZ_\nu^2(ax)dx = \frac{x^2}{2}\left\{[Z'_\nu(ax)]^2 + \left(1 - \frac{\nu^2}{a^2x^2}\right)Z_\nu^2(ax)\right\} \tag{M-12}$$

MODIFIED BESSEL FUNCTIONS

$$I_\nu(x) = \frac{x}{2\nu} I_{\nu-1}(x) - \frac{x}{2\nu} I_{\nu+1}(x) \tag{M-13}$$

$$K_\nu(x) = \frac{x}{2\nu} K_{\nu+1}(x) - \frac{x}{2\nu} K_{\nu-1}(x) \tag{M-14}$$

$$I'_\nu(x) = \tfrac{1}{2}[I_{\nu-1}(x) + I_{\nu+1}(x)] \tag{M-15}$$

$$I'_\nu(x) = I_{\nu-1}(x) - \frac{\nu}{x} I_\nu(x) \tag{M-16}$$

$$I'_\nu(x) = I_{\nu+1}(x) + \frac{\nu}{x} I_\nu(x) \tag{M-17}$$

$$I'_0(x) = I_1(x) \tag{M-18}$$

$$K'_\nu(x) = -\tfrac{1}{2}[K_{\nu-1}(x) + K_{\nu+1}(x)] \tag{M-19}$$

$$K'_\nu(x) = -K_{\nu-1}(x) - \frac{\nu}{x} K_\nu(x) \tag{M-20}$$

$$K'_\nu(x) = -K_{\nu+1}(x) + \frac{\nu}{x} K_\nu(x) \tag{M-21}$$

$$K'_0(x) = -K_1(x) \tag{M-22}$$

LEGENDRE POLYNOMIALS

Both P_n and Q_n satisfy the following recursion relations:

$$Z_{n+1}(x) = \frac{(2n + 1)x}{n + 1} Z_n(x) - \frac{n}{n + 1} Z_{n-1}(x) \tag{M-23}$$

$$Z'_n(x) = \frac{-nx}{1 - x^2} Z_n(x) + \frac{n}{1 - x^2} Z_{n-1}(x) \tag{M-24}$$

$$Z'_n(x) = \frac{n + 1}{1 - x^2} Z_{n+1}(x) - \frac{x}{1 - x^2} Z_n(x) \tag{M-25}$$

Integral formulas:

$$\int_{\cos\theta}^{1} xP_n(x)dx = \frac{-\sin\theta}{(n - 1)(n + 2)} [\sin\theta\, P_n(\cos\theta) + \cos\theta\, P'_n(\cos\ \theta)] \tag{M-26}$$

$$\int_0^1 \frac{[P_n^m(x)]^2}{1 - x^2} dx = \frac{(n + m)!}{2m(n - m)!} \qquad (0 < m \leq n) \tag{M-27}$$

$$\int_{-1}^{1} \frac{P_n^m(x)P_n^k(x)}{1 - x^2} dx = 0 \qquad (0 \leq m \leq n, \quad 0 \leq k \leq n, \quad m \neq k) \tag{M-28}$$

$$\int_{-1}^{1} \frac{P_n(x)}{\sqrt{1 - x^2}} dx = \frac{2^{3/2}}{2n + 1} \tag{M-29}$$

$$\int_0^{\pi} P_{2n}(\cos\theta)d\theta = \frac{\pi}{16^n} \binom{2n}{n}^2 \tag{M-30}$$

$$\int_0^{\pi} P_{2n+1}(\cos\theta)\ \cos\theta\ d\theta = \frac{\pi}{4^{2n+1}} \binom{2n}{n}\binom{2n + 2}{n + 1} \tag{M-31}$$

$$\int_{-1}^{1} P_m(x)P_n(x)dx = \begin{cases} 0 & m \neq n \\ \dfrac{2}{2n + 1} & m = n \end{cases} \tag{M-32}$$

$$\int_0^{\pi} P_m(\cos\theta)P_n(\cos\theta)\ \sin\theta\ d\theta = \begin{cases} 0 & m \neq n \\ \dfrac{2}{2n + 1} & m = n \end{cases} \tag{M-33}$$

$$\int_x^1 P_n(x)dx = \frac{P_{n-1}(x) - P_{n+1}(x)}{2n + 1} \tag{M-34}$$

Appendix N. Orthogonal Functions and Orthogonal Function Series

The expansion of an arbitrary periodic function into a series of sine and cosines, a Fourier series, is one of the most commonly used techniques in engineering. If we know how a system responds to sinusoidal signals, which are usually expressed as a frequency-response function, we can in principle determine the system response to an arbitrary periodic function. Under the usual assumption of linearity we could determine the system response to each frequency component of the signal and then sum the results.

In electromagnetic theory there are cases where a boundary condition cannot be met by a single function. For example, if there is a metallic wall at potential V_0 and we set the product solution equal to that constant, the functional form can frequently be obtained only as a series. The Fourier Series is one such series that can be used.

This appendix is devoted to presenting the general theory of series expansions of arbitrary functions defined over an interval, and the particular functions used in the series are called *orthogonal functions*.

Definition 1. Functions $f(kx)$ and $g(kx)$, k a constant, are said to be orthogonal with respect to the weight function $w(x)$ over interval $[a, b]$ if

$$\int_a^b w(x)f(kx)g(kx)dx \equiv 0 \tag{N-1}$$

Definition 2. A given set of functions which has a parameter n (usually an integer) is denoted by $\{\Phi_n(kx)\}$. This set of functions is called an *orthogonal set* of functions with respect to the weight function $w(x)$ over the interval $[a, b]$ if for all possible pairs of functions in the set

$$\int_a^b w(x)\Phi_n(kx)\Phi_m(kx)dx = K_n \, \delta_{mn} \tag{N-2}$$

where K_n is a nonzero constant and δ_{mn} is the Kronecker delta defined by

$$\delta_{mn} = \begin{cases} 0 & m \neq n \\ 1 & m = n \end{cases} \tag{N-3}$$

Note that the weight function could actually be removed from the definition by defining a new set of functions by

$$\Theta_n(kx) = \sqrt{w(x)}\Phi_n(kx) \tag{N-4}$$

Definition 3. An orthogonal set of functions for which $K_n = 1$ is called an *orthonormal set.* Note that an orthogonal set of functions can be made orthonormal by redefining the set as

$$\Theta_n(kx) = \frac{\Phi_n(kx)}{\sqrt{K_n}} \tag{N-5}$$

Example N-1

Suppose $\Phi_n(kx) = \{\cos nkx\}$, $n = 0, 1, 2, \ldots$. Determine the value of k that will make this an orthogonal set with respect to the weight function $w(x) = 1$ over the interval $[0,l]$.

Using Definition 2 we evaluate the left side of Eq. N-2 using integral tables:

$$\int_0^l \cos nkx \cos mkx\, dx = \begin{cases} \left.\dfrac{\sin(m-n)kx}{2k(m-n)} + \dfrac{\sin(m+n)kx}{2k(m+n)}\right|_0^l & m \neq n \\ \left.\dfrac{\pi}{2} + \dfrac{1}{4nk}\sin 2nkx\right|_0^l & m = n \end{cases}$$

Substituting the limits:

$$\int_0^l \cos nkx \cos mkx\, dx = \begin{cases} \dfrac{\sin(m-n)kl}{2k(m-n)} + \dfrac{\sin(m+n)kl}{2k(m+n)} & m \neq n \\ \dfrac{l}{2} + \dfrac{1}{4nk}\sin 2nkl & m = n \end{cases}$$

Now the definition requires this to be zero for $m \neq n$. Since m and n are integers, their sum and difference are integers. The case $m \neq n$ can then be expressed as

$$\frac{\sin Mkl}{2kM} + \frac{\sin Nkl}{2kN}$$

This can be made zero if

$$Mkl = M2\pi \Rightarrow k = \frac{2\pi}{l} \quad \text{and} \quad Nkl = N2\pi \Rightarrow k = \frac{2\pi}{l}$$

We then have the required value

$$k = \frac{2\pi}{l}$$

With this the integral becomes

$$\int_0^l \cos \frac{2\pi n}{l} x \cos \frac{2\pi m}{l} x dx = \frac{l}{2}\, \delta_{mn} \tag{N-6}$$

The orthogonal set would then be $\{\cos(2\pi n/l)x\}$ for $w(x) = 1$ over the interval $[0, l]$. This can be written as an orthonormal set

$$\left\{\sqrt{\frac{2}{l}} \cos \frac{2\pi n}{l}\right\} \tag{N-7}$$

Exercises

N-a Find the value of k for which $\{\Phi_n(kx)\} = \{\sin nkx\}$ is an orthogonal function set with respect to the weight function $w(x) = 1$ over the interval $[0, l]$.

N-b Prove that Eq. N-7 is an orthonormal set.

N-c Show that the set of functions $\{\cos nx\}$ is an orthogonal function set with respect to the weight function $w(x) = 1$ over the interval $[0, \pi]$. What is the corresponding orthonormal set?

N-d Given the set of functions $\{a_0, a_1 + a_2x, a_3 + a_4x + a_5x^2\}$, determine the values of the constants a_i, $a_0 \neq 0$, so that the set is orthonormal with respect to the weight function $w(x) = 1$ over the interval $[0, 1]$.

N-e Repeat Exercise N-d. Using $w(x) = x$.

N-f Show that the set of functions $\{1, 1 - x, 2 - 4x + x^2\}$ is an orthogonal set with respect to the weight function e^{-x} over the interval $[0, \infty]$.

N-g Show that the set $\{\cos(n \cos^{-1} x)\}$ is an orthogonal set with respect to the weight function $(1 - x^2)^{-1/2}$ over $[-1, 1]$. (Try a change of variable $x = \cos \theta$.)

Definition 4. An orthogonal function set is said to be a *complete* set if the relation

$$\int_a^b w(x)f(x)\Phi_n(x)dx = 0 \qquad n = 0, 1, 2, \ldots \tag{N-8}$$

is satisfied *only* for $f(x) = 0$ (i.e., $\int_a^b f^2(x)dx = 0$, null function).

Example N-2

Is $\{\cos(2n\pi/l)x\}$ a *complete* orthogonal set for $w(x) = 1$ and interval $[0, l]$?

Using Definition 4 as the check, we use $f(x) = \sin(2\pi/l)x$:

$$\int_0^l \sin\frac{2\pi}{l}x \cos\frac{2n\pi}{l}x\, dx = \begin{cases} -\dfrac{l}{2}\cos\dfrac{2\pi}{l}x\Bigg|_0^l = 0 & n = 0 \\[2ex] \left.\dfrac{\cos\left[(1-n)\dfrac{2\pi}{l}\right]}{2(1-n)\dfrac{2\pi}{l}} - \dfrac{\cos\left[(1+n)\dfrac{2\pi}{l}\right]}{2(1+n)\dfrac{2\pi}{l}}\right|_0^l = 0 & n \neq 0 \end{cases}$$

Since we have nonzero function that yields zero value of the integral for every n the set is *not* complete.

The question of set completeness is difficult to answer using the definition, because it is obviously impossible to try every function to demonstrate that only $f(x) = 0$ makes the integral zero for every n. Before pursuing the question further it is important to understand why completeness is even an issue. The following theorem tells why.

Theorem 1. If a set of orthogonal functions $\Phi_n(kx)$ is *complete* over the interval $[a, b]$, then a function $f(x)$ defined over the interval $[a, b]$, can be expressed as a series

$$f(x) = \sum_{n=0}^{\infty} a_n\Phi_n(k) \tag{N-9}$$

where

$$a_n = \frac{1}{K_n} \int_a^b w(x) f(x) \Phi_n(x) dx \tag{N-10}$$

and $w(x)$ is the weight function of the function set.

To show the result given by Eq. N-10 we multiply both sides of Eq. N-9 by $w(x)\Phi_m(x)$ and integrate over $[a, b]$:

$$\int_a^b w(x) f(x) \Phi_m(x) dx = \int_a^b w(x)\ \Phi_m(x) \sum_{n=0}^{\infty} a_n \Phi_n(x) dx$$

Since $w(x)$ and $\Phi_m(x)$ are not functions of n they may be taken inside the summation. Then since the integral of a sum is the sum of the integrals the preceding equation can be written

$$\int_a^b w(x) f(x) \Phi_m(x) dx = \sum_{n=0}^{\infty} a_n \int_a^b w(x) \Phi_m(x) \Phi_n(x) dx$$

Since the function set is orthogonal the right integral is zero except for $m = n$ for which the integral has value K_n.

Thus

$$\int_a^b w(x) f(x) \Phi_n(x) dx = a_n K_n$$

Then

$$a_n = \frac{1}{K_n} \int_a^b w(x) f(x) \Phi_n(x) dx \tag{N-11}$$

The reason completeness is important is that if the set were not complete there is a chance that all a_n might be zero even if $f(x)$ is not zero. This would be covered by Definition 4.

Since we are guaranteed a correct series expansion only if the set is complete we need to examine some aspects of completeness. To simplify what follows we assume that the set has been normalized and modified so that the effective weight function is unity using Eqs. N-4 and N-5; that is,

$$\Theta_n(kx) = \sqrt{\frac{w(x)}{K_n}}\ \Phi_n(kx) \tag{N-12}$$

With this modification we expand a function $f(x)$ as

$$f(x) = \sum_{n=0}^{\infty} a_n \Theta_n(kx) \tag{N-13}$$

Next, square both sides and integrate over $[a, b]$, noting we can use different summation indices:

$$\int_0^l [f(x)]^2 dx = \int_0^l \left[\sum_{i=1}^{\infty} a_i \Theta_i\right]\left[\sum_{n=0}^{\infty} a_n \Theta_n\right] dx = \int_0^l \sum_{i=1}^{\infty} \sum_{n=0}^{\infty} a_i a_n \Theta_i \Theta_n dx$$

$$= \sum_{i=1}^{\infty} \sum_{n=0}^{\infty} a_i a_n \int_0^l \Theta_i \Theta_n dx$$

Since the functions are orthonormal with respect to weight function 1, the right integral is zero except for $i = n$, where it has value 1. Thus

$$\int_0^l [f(x)]^2 dx = \sum_{n=0}^{\infty} a_n^2 \tag{N-14}$$

This is called *Parseval's theorem.*

Suppose we have a function for which all the a_n are zero. Then from Eq. N-14

$$\int_0^l [f(x)]^2 dx = 0$$

But $[f(x)]^2 \geq 0$, so we are forced to set $f(x) = 0$ if all a_n are zero. Thus the set would not be complete if $f(x) \neq 0$.

If we have a function $f(x)$ defined on $[a, b]$ and it has a Taylor series expansion (see Appendix L) we could provide one check for completeness by requiring that the integral

$$\int_a^b \Theta_n(kx) x^i dx \qquad i = 0, 1, 2, \ldots, \infty$$

not be identically zero for every i and each n. This effectively means that there would be some $\Theta_n(kx)$ that would give a nonzero coefficient a_n in the orthogonal series expansion.

Another way of looking at the power series expansion is to divide the series into three subseries, namely, a constant 1, an even sequence $x^2, x^4, x^6, \ldots$, and an odd sequence x, $x^3, x^5, \ldots$.

This suggests that if we had a function $f(x)$ defined on $[a, b]$ that contained all powers of x we could decompose the function into

$$f_{\text{constant}} = \frac{1}{b - a} \int_a^b f(x) dx$$

$$f_{\text{even}} = \frac{f(x) + f(-x)}{2}$$

$$f_{\text{odd}} = \frac{f(x) - f(-x)}{2}$$

Would the set of orthogonal functions be complete if each $\Phi_n(kx)$ gave a nonzero a_n for at least one of these three components? If so, this would suggest using the test function e^{-x}, which contains all three decomposed components, over $[a, b]$.

We return now to the problem of obtaining orthogonal series function expansion of $f(x)$ using Theorem 4.

Example N-3

Given $f(x)$ defined over the interval $[0, l]$, obtain equations for the a_n coefficients for the complete orthogonal function set $\{\cos(2n\pi/l)x, \sin(2n\pi/l)x\}$ with weight function 1 over

[0, l]. The technique is to use Eq. N-5 of Theorem 1 and Eq. N-2 to obtain the required K_n values. For the cosine part of the set:

$$K_n = \int_0^l \cos^2 \frac{2n\pi}{l} x dx = \begin{cases} l & n = 0 \\ \frac{l}{2} & n \neq 0 \end{cases}$$

Therefore,

$$a_n = \frac{1}{K_n} \int_0^l f(x) \cos \frac{2n\pi}{l} x dx = \begin{cases} \frac{1}{l} \int_0^l f(x) dx & n = 0 \\ \frac{2}{l} \int_0^l f(x) \cos \frac{2n\pi}{l} x dx & n \neq 0 \end{cases} \qquad \text{(N-15)}$$

For the sine part of the set:

$$K_n = \int_0^l \sin^2 \frac{2n\pi}{l} x dx = \begin{cases} 0 & n = 0 \\ \frac{l}{2} & n \neq 0 \end{cases}$$

(same K_n for $n \neq 0$ for both sine and cosine parts.)

To avoid confusing the series coefficients for the a_n coefficients for the cosine terms it is common to denote the sine coefficients by b_n. Then

$$b_n = \frac{1}{K_n} \int_0^l f(x) \sin \frac{2n\pi}{l} x dx = \begin{cases} 0 & n = 0 \\ \frac{2}{l} \int_0^l f(x) \sin \frac{2n\pi}{l} x dx & n \neq 0 \end{cases} \qquad \text{(N-16)}$$

Thus the series expansion for the complete set would be

$$f(x) = \sum_{n=0}^{\infty} \left(a_n \cos \frac{2n\pi}{l} + b_n \sin \frac{2n\pi}{l} \right) \qquad \text{(N-17)}$$

This is called the *Fourier series* expansion of $f(x)$ over the interval [0, l].

Example N-4

This next example illustrates a couple of important ideas with respect to series expansions. Suppose we have the function $f(x) = 100$ over [0, l] as shown in Figure N-1. Develop the Fourier series for the function.

Using Eq. N-15 the cosine coefficients are

$$a_n = \frac{1}{K_n} \int_0^l 100 \cos \frac{2n\pi}{l} dx = \begin{cases} \frac{100l}{K_0} = 100 & n = 0 \\ \frac{100l}{K_n 2n\pi} \sin \frac{2n\pi}{l} x \Big|_0^l = 0 & n \neq 0 \end{cases}$$

The sine coefficients are

$$b_n = \frac{1}{K_n} \int_0^l 100 \sin \frac{2n\pi}{l} dx = \frac{-100l}{2n\pi K_n} \cos \frac{2n\pi}{l} \Big|_0^l = 0 \qquad n \neq 0$$

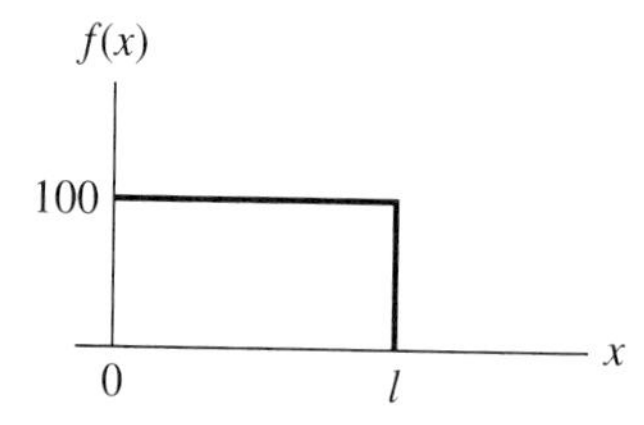

Figure N-1. Function plot for Example N-4.

There is only one nonzero coefficent for the cosine at $n = 0$ so Eq. N-17 gives

$$f(x) = 100$$

This is as one might have expected since one of the orthogonal functions $(\cos(2n\pi/l)x$ for $n = 0)$ gives an exact representation of the given function over $[0, l]$. The consequences of completeness are evident here. Suppose that only the sine functions had been used. All the b_n would be zero so there would be a nonzero function with all zero coefficients. See Definition 4.

Example N-5

Expand the function shown is Figure N-2 into a Fourier series over $[0,l]$.

$$a_n = \frac{1}{K_n}\int_0^l \left(\frac{10}{l}x\right)\cos\frac{2n\pi}{l}\,dx = \begin{cases} \dfrac{1}{l}\displaystyle\int_0^l \frac{10}{l}\,x dx = 5 & n = 0 \\ \left(\dfrac{2}{l}\right)\left(\dfrac{10}{l}\right)\displaystyle\int_0^l x\cos\frac{2n\pi}{l}\,dx = 0 & n \neq 0 \end{cases}$$

$$b_n = \frac{1}{K_n}\int_0^l \left(\frac{10}{l}x\right)\sin\frac{2n\pi}{l}\,dx$$

$$= \frac{20}{l^2}\left[\frac{l^2}{4n^2\pi^2}\sin\frac{2n\pi}{l} - \frac{xl}{2n\pi}\cos\frac{2n\pi}{l}\right]\Bigg|_0^l = -\frac{10}{n\pi} \qquad n \neq 0$$

The series is then

$$f(x) = 5 + \sum_{n=1}^{\infty}\left(-\frac{10}{n\pi}\right)\sin\frac{2n\pi}{l}x$$

It is instructive to plot a few of these terms to see how the function builds up. Take the first four steps:

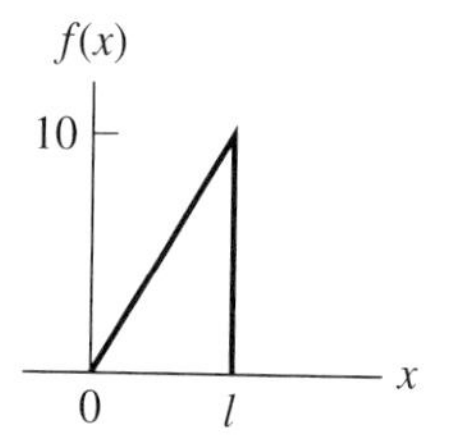

Figure N-2.

$$f_0(x) = 5$$

$$f_1(x) = 5 - \frac{10}{\pi} \sin \frac{2\pi}{l} x$$

$$f_2(x) = 5 - \frac{10}{\pi} \sin \frac{2\pi}{l} x - \frac{10}{2\pi} \sin \frac{4\pi}{l} x$$

$$f_3(x) = 5 - \frac{10}{\pi} \sin \frac{2\pi}{l} x - \frac{10}{2\pi} \sin \frac{4\pi}{l} x - \frac{10}{3\pi} \sin \frac{6\pi}{l} x$$

These are plotted in Figure N-3.

Other functions that form complete orthogonal sets are the Bessel functions and Legendre polynomials developed in Appendix L. The general expression for the K_n of Bessel functions is too unwieldy to be of direct use. The particular form for a given problem is best developed as the problem is solved. The weight function for the Bessel functions $J_n(kx)$ is x [63].

The Legendre polynomials are orthogonal over the interval $[-1, 1]$ for x as the variable, and $[0, \pi]$ for θ as the variable. Specifically,

$$\int_0^{\pi} \sin\theta \, P_n^2(\cos\,\theta)d\theta = \int_{-1}^{1} P_n^2(x)dx = \frac{2}{2n + 1} = K_n$$

For the variable x the weight function is 1 and for the θ variable the weight function is $\sin\,\theta$.

Figure N-3.

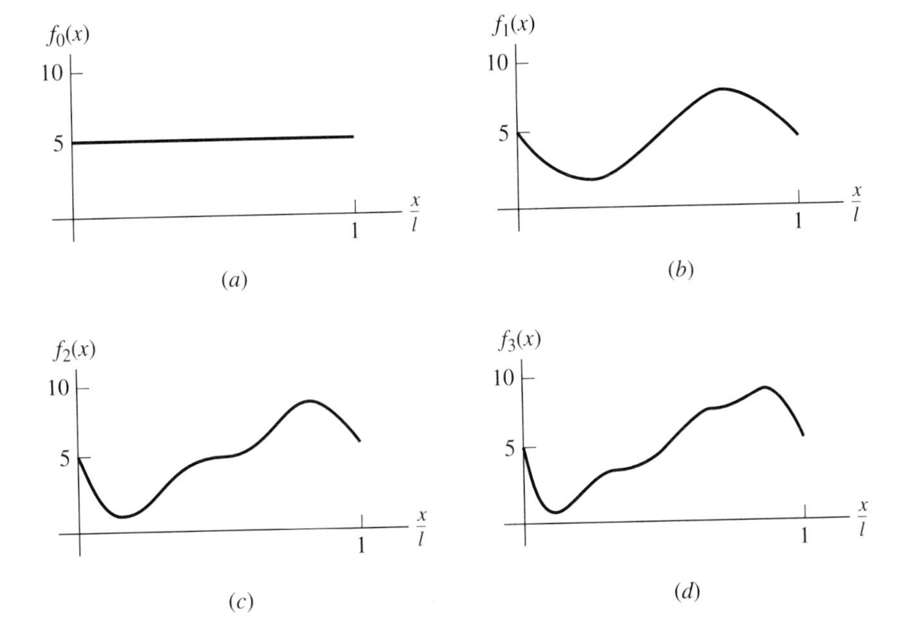

Exercises

N-h For the orthogonal set $\{1, -x, 2 - 4x + x^2\}$, with weight function e^{-x} over $[0, \infty]$, find a non-zero function $f(x)$ for which

$$\int_0^\infty w(x)f(x)\Phi_n(x)dx = 0 \qquad \text{for} \quad n = 0, 1, 2$$

Comment on completeness.

N-i For Figure N-4 obtain the Fourier series and plot the first four series approximations using a normalized x/l scale as the independent variable. Use the interval $[0, l]$.

N-j Repeat Exercise N-i for the function shown in Figure N-5.

N-k Repeat Exercise N-i for the function shown in Figure N-6

N-l Repeat Exercise N-i for the function shown in Figure N-7. Use $l = 4$ and use x as the independent variable.

N-m Suppose that for the Bessel functions of order 0 we let p_i be root of $J_0(x) = 0$; that is, $J_0(p_i) = 0$.

Using the integral table in Appendix M show that the function set $\{J_0[(p_i/a)\rho]\}$ is an orthogonal function set with respect to the weight function $\omega(\rho) = \rho$ over $[0, a]$.

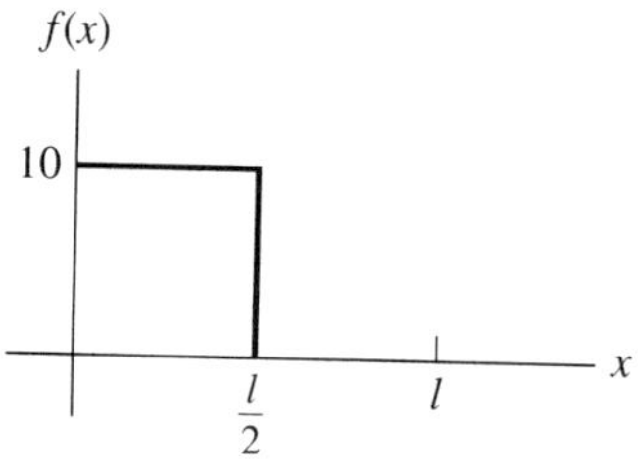

Figure N-4.

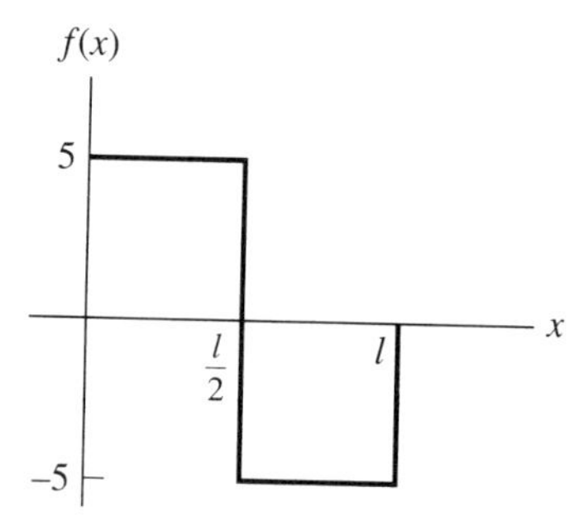

Figure N-5.

Figure N-6.

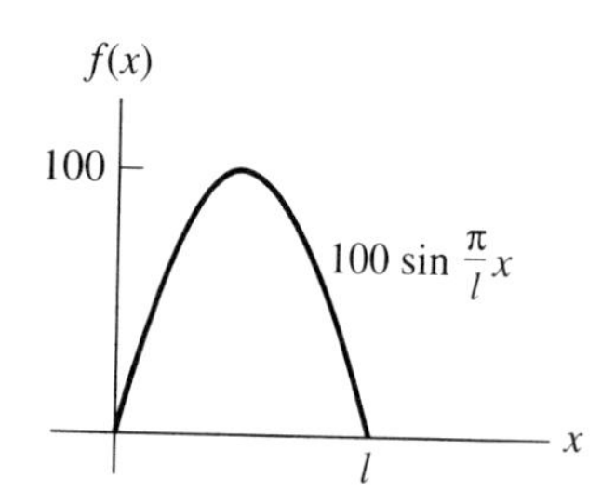

Figure N-7.

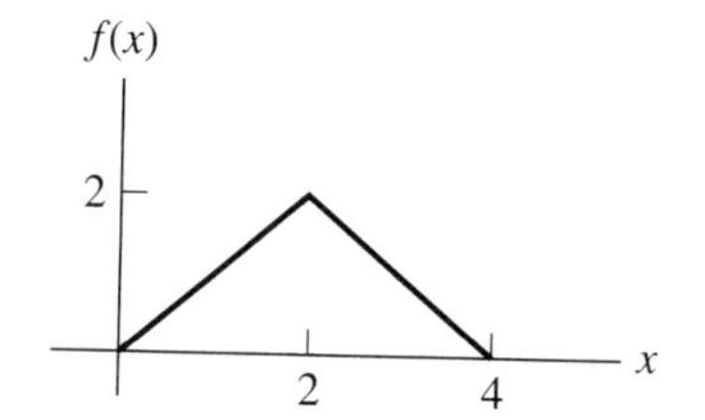

Appendix O. Definite Integrals That Yield Bessel Functions

BASIC RESULTS INVOLVING TRIGONOMETRIC FUNCTIONS

$$\int_0^{\pi} \cos n\phi \cos(x \sin\phi)d\phi = \pi J_n(x), \qquad n \text{ even only} \tag{O-1}$$

Since the integrand is an even function of ϕ:

$$\begin{aligned}\int_0^{2\pi} \cos n\phi \cos(x \sin\phi)d\phi &= 2\int_0^{\pi} \cos n\phi \cos(x \sin\phi)d\phi \\ &= 2\pi J_n(x), \qquad n \text{ even}\end{aligned} \tag{O-2}$$

$$\int_0^{\pi} \sin n\phi \sin(x \sin\phi)d\phi = \pi J_n(x), \qquad n \text{ odd only} \tag{O-3}$$

Since the integrand is an even function of ϕ (odd $\times$ odd = even):

$$\int_0^{2\pi} \sin n\phi \sin(x \sin\phi)d\phi = 2\int_0^{\pi} \sin n\phi \sin(x \sin\phi)d\phi = 2\pi J_n(x), \qquad n \text{ odd} \tag{O-4}$$

PARTICULAR RESULTS FOR BESSEL FUNCTIONS OF ORDER ZERO

$$\int_0^{2\pi} \cos(x \sin\phi)d\phi = 2\pi J_0(x) \tag{O-5}$$

$$\int_0^{\pi} \cos(x \cos\phi)d\phi = \pi J_0(x) \tag{O-6}$$

$$\int_0^{2\pi} \cos(x \cos\phi)d\phi = 2\pi J_0(x) \tag{O-7}$$

$$\int_0^{2\pi} e^{\pm jx\sin\phi}d\phi = 2\pi J_0(x) \tag{O-8}$$

EXPONENTIAL FORMS RESULTING IN BESSEL FUNCTIONS

$$\int_0^{2\pi} e^{jn\phi} e^{jx\cos\phi} d\phi = 2\pi j^n J_n(x) \overset{n=0}{\Rightarrow} \int_0^{2\pi} e^{jx\cos\phi} d\phi = 2\pi J_0(x) \quad \text{(O-9)}$$

$$\int_0^{\pi} \sin^{2n}\phi \, e^{\pm jx\cos\phi} d\phi = \frac{\sqrt{\pi}\,\Gamma(n + \frac{1}{2})}{\left(\frac{x}{2}\right)^n} J_n(x) \quad \text{(O-10)}$$

$$\int_0^{2\pi} \cos n\phi \, e^{jx\cos\phi} d\phi = 2\pi j^n J_n(x) \quad \text{(O-11)}$$

for $n = 1$,

$$\int_0^{2\pi} \cos\phi \, e^{jx\cos\phi} d\phi = j2\pi J_1(x) \quad \text{(O-12)}$$

INTEGRALS CONTAINING BESSEL FUNCTIONS

$$\int_0^{\pi/2} J_\nu(x \sin\phi) \sin^{\nu+1}\phi \cos^{2\mu+1}\phi \, d\phi = \frac{2^\nu \Gamma(\mu + 1)}{x^{\mu+1}} J_{\mu+\nu+1}(x) \quad \text{(O-13)}$$

$$\int_0^{\pi/2} J_\nu^2(x \sin\phi) \sin\phi \, d\phi = \frac{1}{2x} \int_0^{2x} J_{2\nu}(\mu) d\mu$$

$$\int_0^{\pi} J_1^2(x \sin\phi) \sin\phi \, d\phi = \frac{1}{x} \int_0^{2x} J_2(y) dy \quad \text{(O-14)}$$

$$\int_0^{x} J_\nu(x) dx = 2 \sum_{r=0}^{\infty} J_{\nu+2r+1}(x) \quad \text{(O-15)}$$

$$\int_0^{\infty} J_\nu(x) dx = 1 \qquad \text{Re}\{\nu\} > -1 \quad \text{(O-16)}$$

Bibliography

[1] M. Abramowitz and I. Stegun, *Handbook of Mathematical Functions with Formulas, Graphs, and Mathematical Tables,* 5th ed. New York: Dover, 1969.

[2] L. C. Andrews, *Special Functions of Mathematics for Engineers.* 2nd ed. New York: McGraw-Hill, 1992.

[3] I. Bahl and D. Trivedi, ''A Designer's Guide to Microstrip.'' *Microwaves,* May 1977, pp. 174–182.

[4] C. A. Balanis, *Advanced Engineering Electromagnetics.* New York: Wiley, 1989.

[5] W. R. Beam, *Electronics of Solids.* New York: McGraw-Hill, 1965. Chapters 8 and 9 present magnetics.

[6] H. G. Booker, ''Is the Teaching of Electricity and Magnetism in Need of Change?'' *IEEE Transactions on Education,* vol. E-20, 1977, pp. 126–130.

[7] L. Brillouin, *Wave Propagation and Group Velocity.* New York: Academic Press, 1960.

[8] *Microwave Handbook and Buyer's Guide.* 1964. Horizon House Publishers.

[9] R. Chipman, *Transmission Lines.* New York: McGraw-Hill, 1968.

[10] S. B. Cohn, ''Problems in Strip Transmission Lines,'' *Transactions of the IRE, Professional Group on Microwave Theory and Techiques,* vol. MTT-3, 1955, pp. 119–126.

[11] R. E. Collin, *Principles and Applications of Electromagnetic Fields.* Englewood Cliffs, N.J.: Prentice-Hall, 1961.

[12] R. E. Collin and F. J. Zucker, *Antenna Theory.* New York: McGraw-Hill, 1969.

[13] R. E. Collin, *Field Theory of Guided Waves,* 2nd ed. New York: IEEE Press, 1991.

[14] R. E. Collin, *Foundations for Microwave Engineering,* 2nd ed. New York: McGraw-Hill, 1992.

[15] E. William H. Beyer, *CRC Standard Mathematical Tables,* 27th ed. Boca Raton,; Fla.: CRC Press, 1984.

[16] C. Davidson, *Transmission Lines for Communications.* New York: Wiley, 1978.

[17] A. Dekker, *Solid State Physics.* Englewood Cliffs, N.J.: Prentice-Hall, 1963. Presents magnetics; uses CGS units.

[18] P. Diament, *Wave Transmission of Fiber Optics.* New York: Macmillan, 1990.

[19] P. Dirac, *Quantum Mechanics,* 4th ed. London: Oxford University Press, 1958.

[20] L. Dworsky, *Modern Transmission Line Theory and Applications.* New York: Wiley-Interscience, 1979.

[21] J. A. Edminister, *Electromagnetics.* New York: McGraw-Hill, 1979.

[22] T. Edwards, *Foundations for Microstrip Circuit Design.* New York: Wiley-Interscience, 1981.

[23] D. Gloge, ''Weakly Guided Fibers.'' *Applied Optics,* vol. 10, 1971, p. 2252.

[24] I. Gradshteyn and I. Ryshik, *Table of Integrals, Series, and Products,* corrected and enlarged edition. New York: Academic Press, 1980.

[25] E. Hammerstad and O. Jensen, ''Accurate Models for Microstrip Computer-Aided Design,'' *IEEE MTT-S Digest Int. Microwave Symp. USA,* 1980, pp. 407–409.

[26] R. W. Hornbeck, *Numerical Methods.* New York: Quantum Publishers, 1975.

[27] Howe, Harlan, *Stripline Circuit Design.* Norwood, Mass.: Artech House, 1974.

[28] T. Itoh, ed., *Planar Transmission Line Structures.* New York: IEEE Press, 1987.

[29] J. Jackson, *Classical Electrodynamics,* 2nd ed. New York: Wiley, 1975. Uses esu units.

[30] E. Jahnke and F. Emde, *Tables of Functions with Formulae and Curves.* New York: Dover, 1945.

[31] H. Jasik, *Antenna Engineering.* New York: McGraw-Hill, 1961.

[32] E. C. Jordan and K. G. Balmain, *Electromagnetic Waves and Radiating Systems,* 2nd ed. Englewood Cliffs, Prentice-Hall, 1968.

[33] J. H. Kindle, *Analytic Geometry.* New York: McGraw-Hill, 1978.

[34] C. Kittel, *Introduction to Solid State Physics,* 5th ed. New York: Wiley, 1976.

[35] H. Kogelnik and V. Ramaswamy, "Scaling Rules for Thin Film Optical Waveguides." *Applied Optics,* vol. 13, 1974, p. 1857.

[36] J. D. Kraus, *Antennas.* New York: McGraw-Hill, 1950.

[37] L. Landau and E. Lifshitz, *Electrodynamics of Continuous Media.* Reading, Mass.: Pergamon Press and Addison-Wesley, 1960. Course of Theoretical Physics, vol. 8.

[38] N. Marcuvitz, *Waveguide Handbook,* vol. 10. Lexington, Mass.: Boston Technical Publishers, 1964.

[39] J. C. Maxwell, *A Treatise on Electricity and Magnetism,* vols. One and Two. New York: Dover, 1954.

[40] N. McLachlan, *Laplace Transforms and Their Applications to Differential Equations.* New York: Dover, 1962. See particularly Chapter 4.

[41] J. Meixner, "The Behavior of Electromagnetic Fields at Edges." *Inst. Math. Sci. Res. Rept.,* vol. EM-72, December 1954.

[42] P. M. Morse and H. Feshbach, *Methods of Theoretical Physics,* part I and part II. New York: McGraw-Hill, 1953. Shows three-dimensional stereoscopic pictures of coordinate systems.

[43] W. Panofsky and M. Phillips, *Classical Electricity and Magnetism,* 2nd ed. Reading, Mass.: Addison-Wesley, 1962.

[44] S. Ramo, J. Whinnery, and T. VanDuzer, *Fields and Waves in Communication Electronics,* 3rd ed. New York: Wiley, 1994.

[45] Mac E. Van Valkenburg, *Reference Data for Radio Engineers,* 8th ed. Carmel, Ind.: Sams–Prentice Hall Computer Publishing, 1993.

[46] S. Rosenstark, *Transmission Lines in Computer Engineering.* New York: McGraw-Hill, 1994.

[47] S. A. Schelkunoff, *Bell System Technical Journal,* vol. 17, 1938.

[48] M. Schneider, "Microstrip Line for Microwave Integrated Circuits." *Bell System Technical Journal,* vol. 48, May/June 1969, pp. 1421–1444.

[49] S. Silver, *Microwave Antenna Theory and Design.* New York: McGraw-Hill, 1965.

[50] P. Smith, *Electronic Applications of the Smith Chart.* New York: McGraw-Hill, 1969.

[51] W. R. Smythe, *Static and Dynamic Electricity,* 3rd ed. New York: Hemisphere, 1989. A more advanced treatment of the solution of Laplace's equation.

[52] A. W. Snyder and J. D. Love, *Optical Waveguide Theory.* London: Chapman and Hall, 1983.

[53] A Sommerfeld, *Partial Differential Equations in Physics.* New York: Academic Press, 1949, p. 128, para. 28.

[54] J. Spanier and K. Oldham, *An Atlas of Functions.* New York: Hemisphere, 1987.

[55] M. R. Spiegel, *Theory and Problems of Vector Analysis and an Introduction to Tensor Analysis.* New York: McGraw-Hill, 1968.

[56] J. A. Stratton, *Electromagnetic Theory.* New York: McGraw-Hill, 1941.

[57] J. R. Wait, *Electromagnetic Wave Theory.* New York: Harper and Row, 1985.

[58] G. N. Watson, *A Treatise on the Theory of Bessel Functions.* Cambridge: Cambridge University Press, 1966.

[59] W. Weeks, *Antenna Engineering.* New York: McGraw-Hill, 1968.

[60] W. Weeks, *Transmission and Distribution of Electrical Energy.* New York: Harper and Row, 1981.

[61] H. Wheeler, "Transmission Line Properties of Parallel Wide Strips by a Conformal Mapping Approximation." *IEEE Transactions on Microwave Theory and Techniques,* vol. MTT-12, 1964, pp. 280–290.

[62] H. Wheeler, "Transmission-Line Properties of a Strip on a Dielectric Sheet on a Plane." *IEEE Transactions on Microwave Theory and Techniques,* vol. MTT-25, 1977, p. 631.

[63] R. C. Wylie and L. Barrett, *Advanced Engineering Mathematics,* 5th ed. Boca Raton, Fla.: CRC Press, 1984.

Index